Reactions at Supports and Connections for a Two-Dimensional Structure

Support or Connection	Reaction	Number of Unknowns
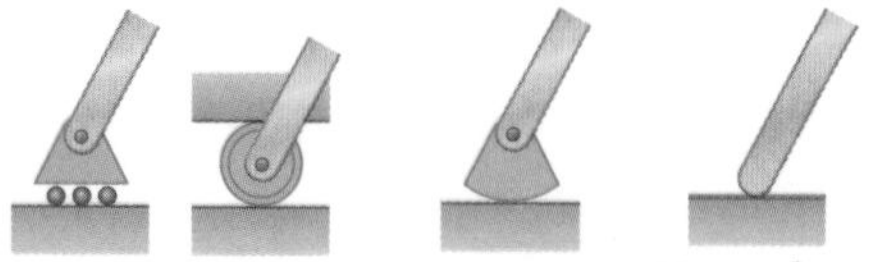 Rollers; Rocker; Frictionless surface	 Force with known line of action	1
 Short cable; Short link	 Force with known line of action	1
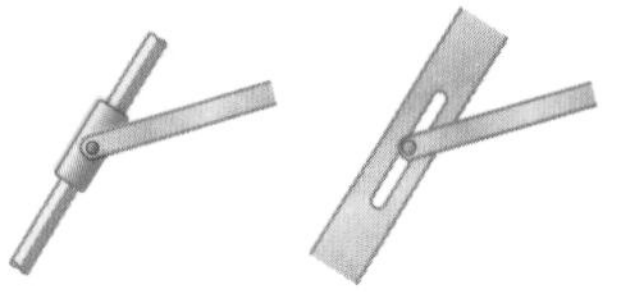 Collar on frictionless rod; Frictionless pin in slot	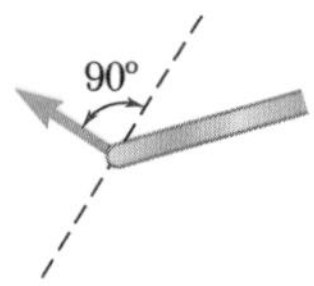 90° Force with known line of action	1
Frictionless pin or hinge; Rough surface	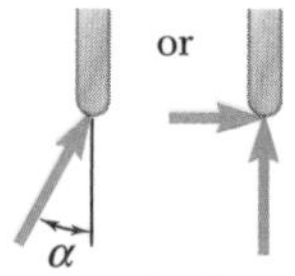 α or Force of unknown direction	2
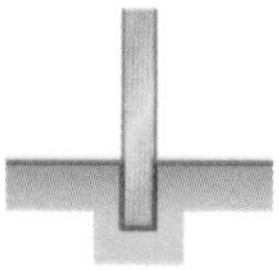 Fixed support	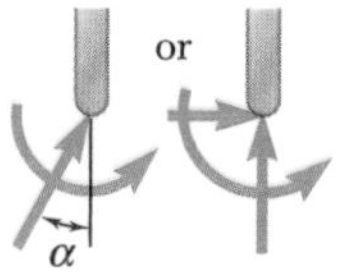 α or Force and couple	3

The first step in the solution of any problem concerning the equilibrium of a rigid body is to construct an appropriate free-body diagram of the body. As part of that process, it is necessary to show on the diagram the reactions through which the ground and other bodies oppose a possible motion of the body. The figures on this and the facing page summarize the possible reactions exerted on two- and three-dimensional bodies.

Reactions at Supports and Connections for a Three-Dimensional Structure

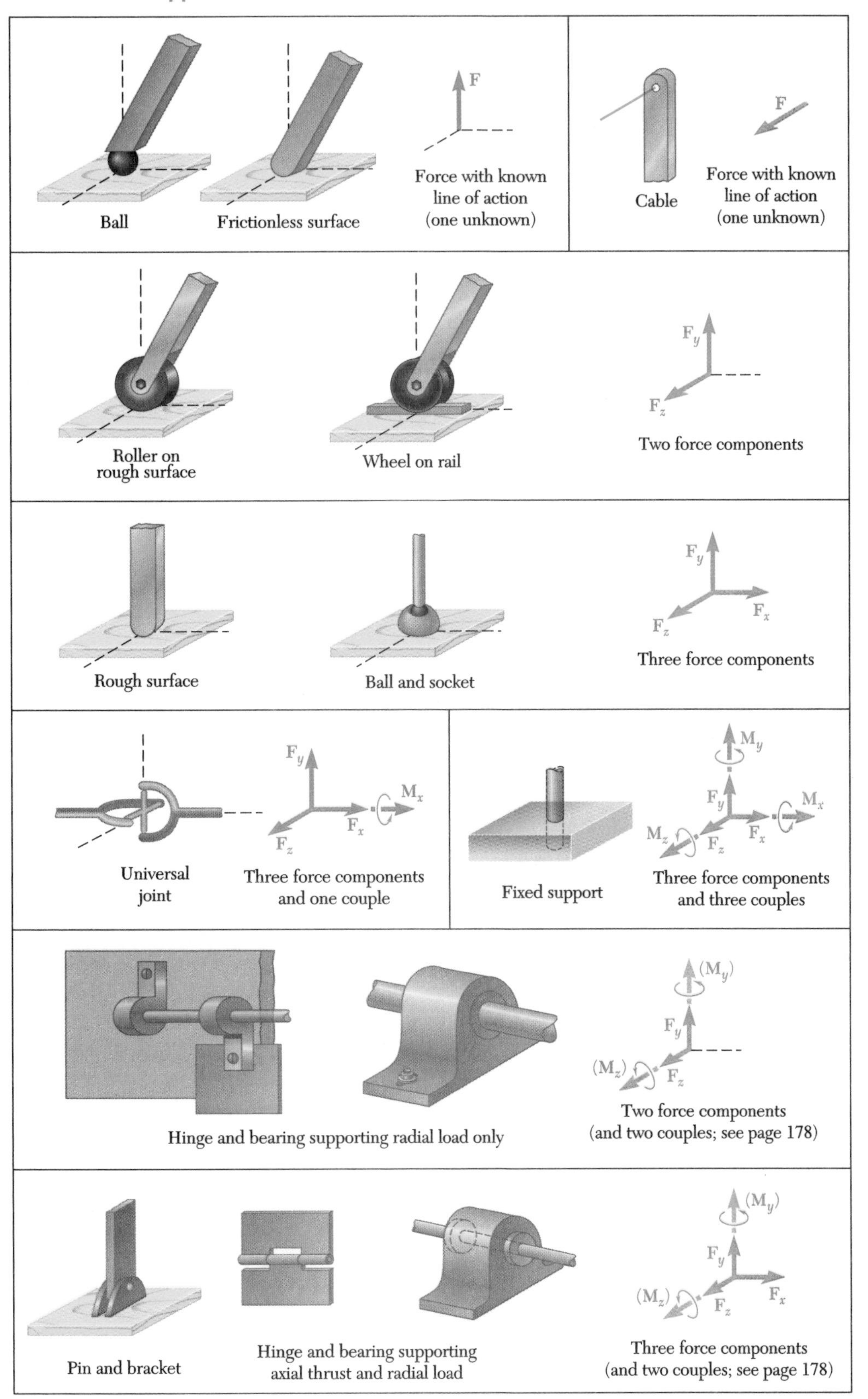

Second Edition

Statics and Mechanics of Materials

Ferdinand P. Beer
Late of Lehigh University

E. Russell Johnston, Jr.
Late of University of Connecticut

John T. DeWolf
University of Connecticut

David F. Mazurek
U.S. Coast Guard Academy

STATICS AND MECHANICS OF MATERIALS, SECOND EDITION

Published by McGraw-Hill Education, 2 Penn Plaza, New York, NY 10121.

This book is printed on acid-free paper.

1 2 3 4 5 6 7 8 9 0 DOW/DOW 1 0 9 8 7 6

ISBN 978-0-07-339816-7
MHID 0-07-339816-0

Senior Vice President, Products & Markets: *Kurt L. Strand*
Vice President, General Manager, Products & Markets: *Marty Lange*
Vice President, Content Design & Delivery: *Kimberly Meriwether David*
Managing Director: *Thomas Timp*
Brand Manager: *Thomas M. Scaife, Ph.D.*
Director, Product Development: *Rose Koos*
Product Developer: *Robin Reed/Jolynn Kilburg*
Marketing Manager: *Nick McFadden*
Digital Product Developer: *Joan Weber*
Director, Content Design & Delivery: *Linda Avenarius*
Program Manager: *Faye M. Herrig*
Content Project Managers: *Melissa M. Leick, Tammy Juran, Sandra Schnee*
Buyer: *Jennifer Pickel*
Design: *Studio Montage, Inc.*
Content Licensing Specialists: *Beth Thole*
Cover Image: *Steve Speller*
Compositor: *SPi Global*
Printer: *R. R. Donnelley*

All credits appearing on page or at the end of the book are considered to be an extension of the copyright page.

Library of Congress Cataloging-in-Publication Data

Names: Beer, Ferdinand P. (Ferdinand Pierre), 1915–2003.
Title: Statics and mechanics of materials / Ferdinand P. Beer, late of Lehigh University [and three others].
Description: Second edition. | New York, NY : McGraw-Hill Education, [2017]
Identifiers: LCCN 2015047086 | ISBN 9780073398167 (alk. paper)
Subjects: LCSH: Statics. | Materials. | Mechanics, Applied.
Classification: LCC TA351 .S68 2017 | DDC 620.1—dc23
LC record available at http://lccn.loc.gov/2015047086

About the Authors

John T. DeWolf, Emeritus Professor of Civil Engineering at the University of Connecticut, joined the Beer and Johnston team as an author on the second edition of *Mechanics of Materials*. John holds a B.S. degree in civil engineering from the University of Hawaii and M.E. and Ph.D. degrees in structural engineering from Cornell University. He is a Fellow of the American Society of Civil Engineers and a member of the Connecticut Academy of Science and Engineering. He is a registered Professional Engineer and a member of the Connecticut Board of Professional Engineers. He was selected as a University of Connecticut Teaching Fellow in 2006. Professional interests include elastic stability, bridge monitoring, and structural analysis and design.

David F. Mazurek, Professor of Civil Engineering at the United States Coast Guard Academy, joined the Beer and Johnston team on the eighth edition of *Statics* and the fifth edition of *Mechanics of Materials*. David holds a B.S. degree in ocean engineering and an M.S. degree in civil engineering from the Florida Institute of Technology and a Ph.D. degree in civil engineering from the University of Connecticut. He is a Fellow of the American Society of Civil Engineers and a member of the Connecticut Academy of Science and Engineering. He is a registered Professional Engineer and has served on the American Railway Engineering & Maintenance-of-Way Association's Committee 15—Steel Structures since 1991. Among his numerous awards, he was recognized by the National Society of Professional Engineers as the Coast Guard Engineer of the Year for 2015. Professional interests include bridge engineering, structural forensics, and blast-resistant design.

Brief Contents

Contents

8 Concept of Stress 337

9 Stress and Strain–Axial Loading 383

10 Torsion 451

11 Pure Bending 491

12 Analysis and Design of Beams for Bending 551

13 Shearing Stresses in Beams and Thin-Walled Members 591

14 Transformations of Stress 625

*Advanced or specialty topics

Preface

Objectives

The main objective of a basic mechanics course should be to develop in the engineering student the ability to analyze a given problem in a simple and logical manner and to apply to its solution a few fundamental and well-understood principles. This text is designed for a course that combines statics and mechanics of materials—or strength of materials—offered to engineering students in the sophomore year.

General Approach

In this text the study of statics and mechanics of materials is based on the understanding of a few basic concepts and on the use of simplified models. This approach makes it possible to develop all the necessary formulas in a rational and logical manner, and to clearly indicate the conditions under which they can be safely applied to the analysis and design of actual engineering structures and machine components.

Practical Applications Are Introduced Early. One of the characteristics of the approach used in this text is that mechanics of *particles* is clearly separated from the mechanics of *rigid bodies.* This approach makes it possible to consider simple practical applications at an early stage and to postpone the introduction of the more difficult concepts. As an example, statics of particles is treated first (Chap. 2); after the rules of addition and subtraction of vectors are introduced, the principle of equilibrium of a particle is immediately applied to practical situations involving only concurrent forces. The statics of rigid bodies is considered in Chaps. 3 and 4. In Chap. 3, the vector and scalar products of two vectors are introduced and used to define the moment of a force about a point and about an axis. The presentation of these new concepts is followed by a thorough and rigorous discussion of equivalent systems of forces, leading, in Chap. 4, to many practical applications involving the equilibrium of rigid bodies under general force systems.

New Concepts Are Introduced in Simple Terms. Because this text is designed for the first course in mechanics, new concepts are presented in simple terms and every step is explained in detail. On the other hand, by discussing the broader aspects of the problems considered and by stressing methods of general applicability, a definite maturity of approach is achieved. For example, the concepts of partial constraints and statical indeterminacy are introduced early and are used throughout.

Fundamental Principles Are Placed in the Context of Simple Applications. The fact that mechanics is essentially a *deductive* science based on a few fundamental principles is stressed. Derivations have been presented in their logical sequence and with all the rigor warranted at this level. However, the learning process being largely *inductive,* simple applications are considered first.

As an example, the statics of particles precedes the statics of rigid bodies, and problems involving internal forces are postponed until Chap. 6. In Chap. 4, equilibrium problems involving only coplanar forces are considered first and solved by ordinary algebra, while problems involving three-dimensional forces and requiring the full use of vector algebra are discussed in the second part of the chapter.

The first four chapters treating mechanics of materials (Chaps. 8, 9, 10, and 11) are devoted to the analysis of the stresses and of the corresponding deformations in various structural members, considering successively axial loading, torsion, and pure bending. Each analysis is based on a few basic concepts, namely, the conditions of equilibrium of the forces exerted on the member, the relations existing between stress and strain in the material, and the conditions imposed by the supports and loading of the member. The study of each type of loading is complemented by a large number of examples, sample problems, and problems to be assigned, all designed to strengthen the students' understanding of the subject.

Free-body Diagrams Are Used Extensively. Throughout the text, free-body diagrams are used to determine external or internal forces. The use of "picture equations" will also help the students understand the superposition of loadings and the resulting stresses and deformations.

Design Concepts Are Discussed Throughout the Text Whenever Appropriate. A discussion of the application of the factor of safety to design can be found in Chap. 8, where the concept of allowable stress design is presented.

The SMART Problem-Solving Methodology Is Employed. New to this edition of the text, students are introduced to the SMART approach for solving engineering problems, whose acronym reflects the solution steps of ***S***trategy, ***M***odeling, ***A***nalysis, and ***R***eflect & ***T***hink. This methodology is used in all Sample Problems, and it is intended that students will apply this in the solution of all assigned problems.

A Careful Balance Between SI and U.S. Customary Units Is Consistently Maintained. Because it is essential that students be able to handle effectively both SI metric units and U.S. customary units, half the examples, sample problems, and problems to be assigned have been stated in SI units and half in U.S. customary units. Since a large number of problems are available, instructors can assign problems using each system of units in whatever proportion they find most desirable for their class.

It also should be recognized that using both SI and U.S. customary units entails more than the use of conversion factors. Because the SI system of units is an absolute system based on the units of time, length, and mass, whereas the U.S. customary system is a gravitational system based on the units of time, length, and force, different approaches are required for the solution of many problems. For example, when SI units are used, a body is generally specified by its mass expressed in kilograms; in most problems of statics it will be necessary to determine the weight of the body in newtons, and an additional calculation will be required for this purpose. On the other hand, when U.S. customary units are used, a body is specified by its weight in pounds and, in dynamics problems (such as would be encountered in a follow-on course in dynamics), an additional calculation will be required to determine its mass in slugs (or lb·s^2/ft). The authors, therefore, believe that problem assignments should include both systems of units.

Optional Sections Offer Advanced or Specialty Topics. A number of optional sections have been included. These sections are indicated by asterisks and thus are easily distinguished from those that form the core of the basic first mechanics course. They may be omitted without prejudice to the understanding of the rest of the text.

The material presented in the text and most of the problems require no previous mathematical knowledge beyond algebra, trigonometry, and elementary calculus; all the elements of vector algebra necessary to the understanding of mechanics are carefully presented in Chaps. 2 and 3. In general, a greater emphasis is placed on the correct understanding of the basic mathematical concepts involved than on the nimble manipulation of mathematical formulas. In this connection, it should be mentioned that the determination of the centroids of composite areas precedes the calculation of centroids by integration, thus making it possible to establish the concept of the moment of an area firmly before introducing the use of integration.

Chapter Organization and Pedagogical Features

Each chapter begins with an introductory section setting the purpose and goals of the chapter and describing in simple terms the material to be covered and its application to the solution of engineering problems.

Chapter Lessons. The body of the text has been divided into units, each consisting of one or several theory sections followed by sample problems and a large number of problems to be assigned. Each unit corresponds to a well-defined topic and generally can be covered in one lesson.

Concept Applications and Sample Problems. Many theory sections include concept applications designed to illustrate the material being presented and facilitate its understanding. The sample problems provided after all lessons are intended to show some of the applications of the theory to the solution of engineering problems. Because they have been set up in much the same form that students will use in solving the assigned problems, the sample problems serve the double purpose of amplifying the text and demonstrating the type of neat and orderly work that students should cultivate in their own solutions.

Homework Problem Sets. Most of the problems are of a practical nature and should appeal to engineering students. They are primarily designed, however, to illustrate the material presented in the text and help the students understand the basic principles used in engineering mechanics. The problems have been grouped according to the portions of material they illustrate and have been arranged in order of increasing difficulty. Answers to problems are given at the end of the book, except for those with a number set in *red italics*.

Chapter Review and Summary. Each chapter ends with a review and summary of the material covered in the chapter. Notes in the margin have been included to help the students organize their review work, and cross references are provided to help them find the portions of material requiring their special attention.

Review Problems. A set of review problems is included at the end of each chapter. These problems provide students further opportunity to apply the most important concepts introduced in the chapter.

New to the Second Edition

We've made some significant changes from the first edition of this text. The updates include:

- **Complete Rewrite.** The text has undergone a complete edit of the language to make the book easier to read and more student-friendly.
- **New Photographs Throughout.** We have updated many of the photos appearing in the second edition.
- **The SMART Problem-Solving Methodology is Employed.** Students are introduced to the SMART approach for solving engineering problems, which is used in all Sample Problems and is intended for use in the solution of all assigned problems.
- **Revised or New Problems.** Over 55% of the problems are revised or new to this edition.
- **Connect with SmartBook.** The second edition is now equipped with Connect, our one-of-a-kind teaching and learning platform that boosts student learning through our adaptive SmartBook and includes access to ALL of the more than 1200 homework problems in the text. Instructors will appreciate having access through Connect to a complete instructor's solutions manual, lecture PowerPoint slides to facilitate classroom discussion of the concepts in the text, and textbook images for repurposing in personalized classroom materials.

Acknowledgments

The authors thank the many companies that provided photographs for this edition. We also wish to recognize the efforts of the team at McGraw-Hill Education, including Thomas Scaife, Brand Manager; Nick McFadden, Marketing Manager; Robin Reed, Product Developer; and Melissa Leick, Content Project Manager. Our special thanks go to Amy Mazurek (B.S. degree in civil engineering from the Florida Institute of Technology, and a M.S. degree in civil engineering from the University of Connecticut) for her work in the checking and preparation of the solutions and answers of all the problems in this edition.

We also gratefully acknowledge the help, comments, and suggestions offered by the many users of previous editions of books in the Beer & Johnston Engineering Mechanics series.

John T. DeWolf
David F. Mazurek

Required=Results

McGraw-Hill Connect®
Learn Without Limits

Connect is a teaching and learning platform that is proven to deliver better results for students and instructors.

Connect empowers students by continually adapting to deliver precisely what they need, when they need it and how they need it, so your class time is more engaging and effective.

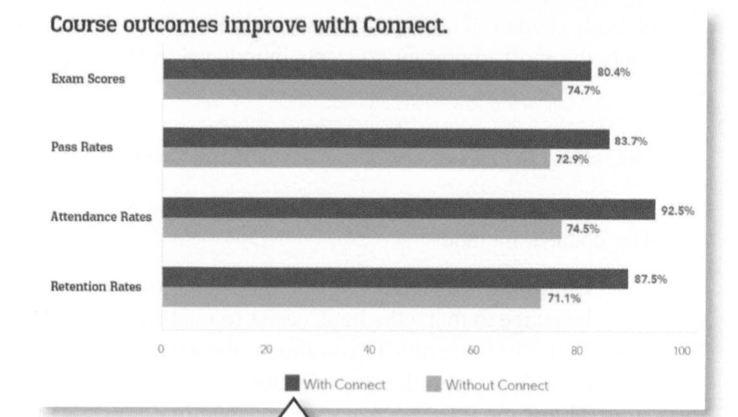

Using **Connect** improves passing rates by **10.8%** and retention by **16.4%**.

88% of instructors who use **Connect** require it; instructor satisfaction **increases** by 38% when **Connect** is required.

Analytics

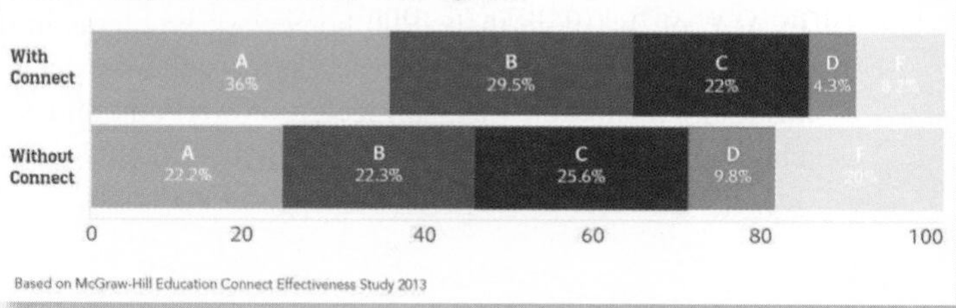

Connect Insight®

Connect Insight is Connect's new one-of-a-kind visual analytics dashboard—now available for both instructors and students—that provides at-a-glance information regarding student performance, which is immediately actionable. By presenting assignment, assessment, and topical performance results together with a time metric that is easily visible for aggregate or individual results, Connect Insight gives the user the ability to take a just-in-time approach to teaching and learning, which was never before available. Connect Insight presents data that empowers students and helps instructors improve class performance in a way that is efficient and effective.

Students can view their results for any **Connect** course.

Mobile

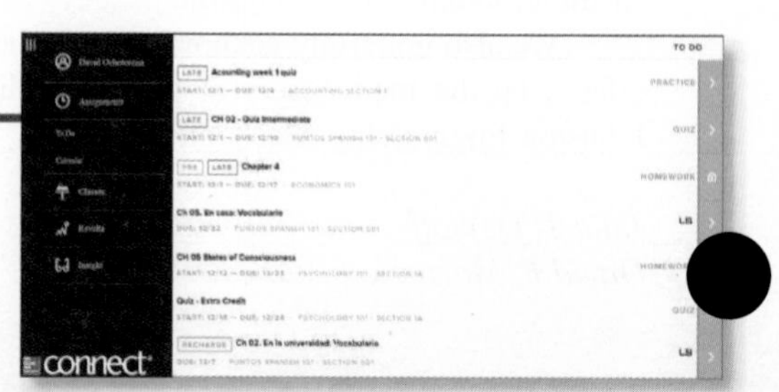

Connect's new, intuitive mobile interface gives students and instructors flexible and convenient, anytime–anywhere access to all components of the Connect platform.

Adaptive

THE FIRST AND ONLY **ADAPTIVE READING EXPERIENCE** DESIGNED TO TRANSFORM THE WAY STUDENTS READ

More students earn **A's** and **B's** when they use McGraw-Hill Education **Adaptive** products.

SmartBook®

Proven to help students improve grades and study more efficiently, SmartBook contains the same content within the print book, but actively tailors that content to the needs of the individual. SmartBook's adaptive technology provides precise, personalized instruction on what the student should do next, guiding the student to master and remember key concepts, targeting gaps in knowledge and offering customized feedback, and driving the student toward comprehension and retention of the subject matter. Available on smartphones and tablets, SmartBook puts learning at the student's fingertips—anywhere, anytime.

Over **4 billion questions** have been answered, making McGraw-Hill Education products more intelligent, reliable, and precise.

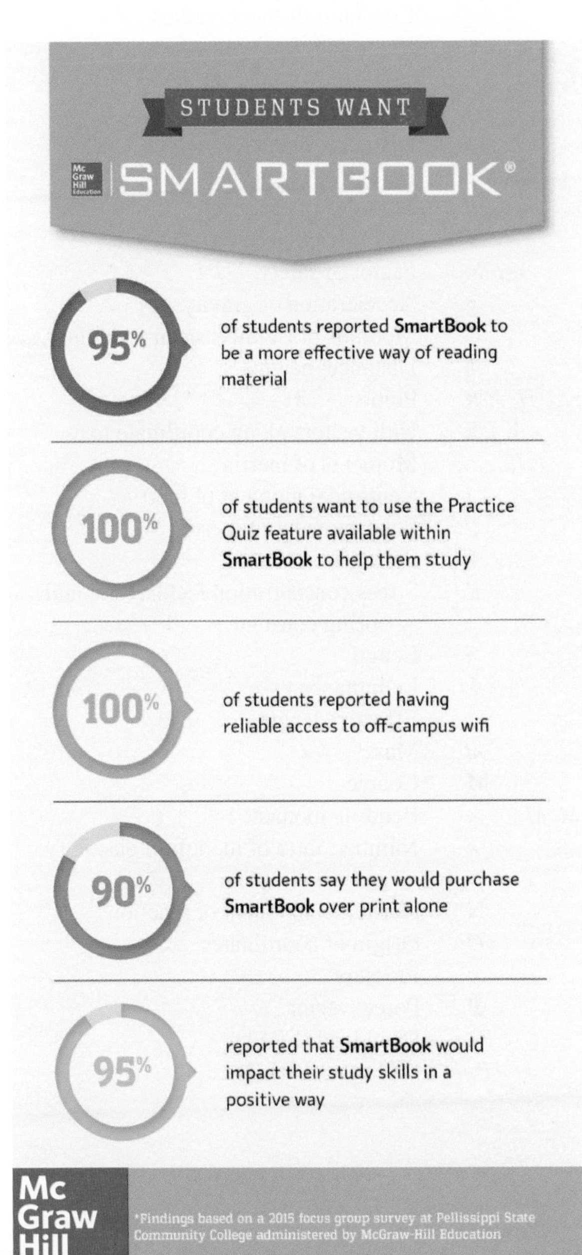

List of Symbols

a	Constant; radius; distance
$\mathbf{A}, \mathbf{B}, \mathbf{C}, \ldots$	Forces; reactions at supports and connections
$A, B, C, \ldots$	Points
A	Area
b	Width; distance
c	Constant; distance; radius
C	Centroid
$C_1, C_2, \ldots$	Constants of integration
C_P	Column stability factor
d	Distance; diameter; depth
e	Distance; eccentricity
E	Modulus of elasticity
$\mathbf{F}$	Force; friction force
$F.S.$	Factor of safety
g	Acceleration of gravity
G	Modulus of rigidity; shear modulus
h	Distance; height
H, J, K	Points
$\mathbf{i}, \mathbf{j}, \mathbf{k}$	Unit vectors along coordinate axes
$I, I_x, \ldots$	Moments of inertia
$\bar{I}$	Centroidal moment of inertia
J	Polar moment of inertia
k	Spring constant
K	Stress concentration factor; torsional spring constant
l	Length
L	Length; span
L_e	Effective length
m	Mass
$\mathbf{M}$	Couple
$M, M_x, \ldots$	Bending moment
n	Number; ratio of moduli of elasticity; normal direction
$\mathbf{N}$	Normal component of reaction
O	Origin of coordinates
p	Pressure
$\mathbf{P}$	Force; vector
P_D	Dead load (LRFD)
P_L	Live load (LRFD)
P_U	Ultimate load (LRFD)
q	Shearing force per unit length; shear flow
$\mathbf{Q}$	Force; vector
Q	First moment of area
$\bar{r}$	Centroidal radius of gyration
$\mathbf{r}$	Position vector
r_x, r_y, r_O	Radii of gyration
r	Radius; distance; polar coordinate
$\mathbf{R}$	Resultant force; resultant vector; reaction
R	Radius of earth
s	Length
$\mathbf{S}$	Force; vector
S	Elastic section modulus
t	Thickness
$\mathbf{T}$	Force; torque
T	Tension; temperature
u, v	Rectangular coordinates
$\mathbf{V}$	Vector product; shearing force
V	Volume; shear
w	Width; distance; load per unit length
$\mathbf{W}, W$	Weight; load
x, y, z	Rectangular coordinates; distances; displacements; deflections
$\bar{x}, \bar{y}, \bar{z}$	Coordinates of centroid
α, β, γ	Angles
α	Coefficient of thermal expansion; influence coefficient
γ	Shearing strain; specific weight
γ_D	load factor, dead load (LRFD)
γ_L	load factor, live load (LRFD)
δ	Deformation; displacement; elongation
ε	Normal strain
θ	Angle; slope
λ	Unit vector along a line
μ	Coefficient of friction
ν	Poisson's ratio
ρ	Radius of cuvature; distance; density
σ	Normal stress
τ	Shearing stress
ϕ	Angle; angle of twist; resistance factor

© Renato Bordoni/Alamy

1 Introduction

The tallest skyscraper in the Western Hemisphere, One World Trade Center is a prominent feature of the New York City skyline. From its foundation to its structural components and mechanical systems, the design and operation of the tower is based on the fundamentals of engineering mechanics.

Introduction

Objectives

- **Define** the science of mechanics and examine its fundamental principles.
- **Discuss** and compare the International System of Units and U.S. Customary Units.
- **Discuss** how to approach the solution of mechanics problems, and introduce the SMART problem-solving methodology.
- **Examine** factors that govern numerical accuracy in the solution of a mechanics problem.

1.1 WHAT IS MECHANICS?

Mechanics is defined as the science that describes and predicts the conditions of rest or motion of bodies under the action of forces. It consists of the mechanics of *rigid bodies,* mechanics of *deformable bodies,* and mechanics of *fluids*.

The mechanics of rigid bodies is subdivided into **statics** and **dynamics**. Statics deals with bodies at rest; dynamics deals with bodies in motion. In this text, we assume bodies are perfectly rigid. In fact, actual structures and machines are never absolutely rigid; they deform under the loads to which they are subjected. However, because these deformations are usually small, they do not appreciably affect the conditions of equilibrium or the motion of the structure under consideration. They are important, though, as far as the resistance of the structure to failure is concerned. Deformations are studied in a course in mechanics of materials, which is part of the mechanics of deformable bodies. The third division of mechanics, the mechanics of fluids, is subdivided into the study of *incompressible fluids* and of *compressible fluids*. An important subdivision of the study of incompressible fluids is *hydraulics,* which deals with applications involving water.

Mechanics is a physical science, since it deals with the study of physical phenomena. However, some teachers associate mechanics with mathematics, whereas many others consider it as an engineering subject. Both these views are justified in part. Mechanics is the foundation of most engineering sciences and is an indispensable prerequisite to their study. However, it does not have the *empiricism* found in some engineering sciences, i.e., it does not rely on experience or observation alone. The rigor of mechanics and the emphasis it places on deductive reasoning makes it resemble mathematics. However, mechanics is not an *abstract* or even a *pure* science; it is an *applied* science.

The purpose of mechanics is to explain and predict physical phenomena and thus to lay the foundations for engineering applications. You need to know statics to determine how much force will be exerted on a point in a bridge design and whether the structure can withstand that force. Determining the force a dam needs to withstand from the water in a river requires statics. You need statics to calculate how much weight a crane can lift, how much force a locomotive needs to pull a freight train, or how much force a circuit board in a computer can withstand. The concepts of dynamics enable you to analyze the flight characteristics of a jet, design a building to resist earthquakes, and mitigate shock and vibration to passengers inside a vehicle. The concepts of dynamics enable you to calculate how much force you need to send a satellite into orbit, accelerate a 200,000-ton cruise ship, or design a toy truck that doesn't break. You will not learn how to do these things in this course, but the ideas and methods you learn here will be the underlying basis for the engineering applications you will learn in your work.

1.2 FUNDAMENTAL CONCEPTS AND PRINCIPLES

1.2A Mechanics of Rigid Bodies

Although the study of mechanics goes back to the time of Aristotle (384–322 B.C.) and Archimedes (287–212 B.C.), not until Newton (1642–1727) did anyone develop a satisfactory formulation of its fundamental principles. These principles were later modified by d'Alembert, Lagrange, and Hamilton. Their validity remained unchallenged until Einstein formulated his **theory of relativity** (1905). Although its limitations have now been recognized, **newtonian mechanics** still remains the basis of today's engineering sciences.

The basic concepts used in mechanics are *space, time, mass,* and *force*. These concepts cannot be truly defined; they should be accepted on the basis of our intuition and experience and used as a mental frame of reference for our study of mechanics.

The concept of **space** is associated with the position of a point *P*. We can define the position of *P* by providing three lengths measured from a certain reference point, or *origin,* in three given directions. These lengths are known as the *coordinates* of *P*.

To define an event, it is not sufficient to indicate its position in space. We also need to specify the **time** of the event.

We use the concept of **mass** to characterize and compare bodies on the basis of certain fundamental mechanical experiments. Two bodies of the same mass, for example, are attracted by the earth in the same manner; they also offer the same resistance to a change in translational motion.

A **force** represents the action of one body on another. A force can be exerted by actual contact, like a push or a pull, or at a distance, as in the case of gravitational or magnetic forces. A force is characterized by its *point of application,* its *magnitude,* and its *direction;* a force is represented by a *vector* (Sec. 2.1B).

In newtonian mechanics, space, time, and mass are absolute concepts that are independent of each other. (This is not true in **relativistic mechanics**, where the duration of an event depends upon its position and the mass of a body varies with its velocity.) On the other hand, the concept of force is not independent of the other three. Indeed, one of the fundamental principles of newtonian mechanics listed below is that the resultant force acting on a body is related to the mass of the body and to the manner in which its velocity varies with time.

In this text, you will study the conditions of rest or motion of particles and rigid bodies in terms of the four basic concepts we have introduced. By **particle**, we mean a very small amount of matter, which we assume occupies a single point in space. A **rigid body** consists of a large number of particles occupying fixed positions with respect to one another.

The study of the mechanics of particles is clearly a prerequisite to that of rigid bodies. Besides, we can use the results obtained for a particle directly in a large number of problems dealing with the conditions of rest or motion of actual bodies.

The study of elementary mechanics rests on six fundamental principles, based on experimental evidence.

- **The Parallelogram Law for the Addition of Forces.** Two forces acting on a particle may be replaced by a single force, called their *resultant,* obtained by drawing the diagonal of the parallelogram with sides equal to the given forces (Sec. 2.1A).
- **The Principle of Transmissibility.** The conditions of equilibrium or of motion of a rigid body remain unchanged if a force acting at a given point of the rigid body is replaced by a force of the same magnitude and same direction, but acting at a different point, provided that the two forces have the same line of action (Sec. 3.1B).
- **Newton's Three Laws of Motion.** Formulated by Sir Isaac Newton in the late seventeenth century, these laws can be stated as follows:

  FIRST LAW. If the resultant force acting on a particle is zero, the particle remains at rest (if originally at rest) or moves with constant speed in a straight line (if originally in motion) (Sec. 2.3B).

  SECOND LAW. If the resultant force acting on a particle is not zero, the particle has an acceleration proportional to the magnitude of the resultant and in the direction of this resultant force.

  This law can be stated as

$$\mathbf{F} = m\mathbf{a} \tag{1.1}$$

  where $\mathbf{F}$, m, and $\mathbf{a}$ represent, respectively, the resultant force acting on the particle, the mass of the particle, and the acceleration of the particle expressed in a consistent system of units.

  THIRD LAW. The forces of action and reaction between bodies in contact have the same magnitude, same line of action, and opposite sense (Chap. 6, Introduction).

- **Newton's Law of Gravitation.** Two particles of mass M and m are mutually attracted with equal and opposite forces $\mathbf{F}$ and $-\mathbf{F}$ of magnitude F (Fig. 1.1), given by the formula

$$F = G\frac{Mm}{r^2} \tag{1.2}$$

  where r = the distance between the two particles and G = a universal constant called the *constant of gravitation.* Newton's law of gravitation introduces the idea of an action exerted at a distance and extends the range of application of Newton's third law: the action $\mathbf{F}$ and the reaction $-\mathbf{F}$ in Fig. 1.1 are equal and opposite, and they have the same line of action.

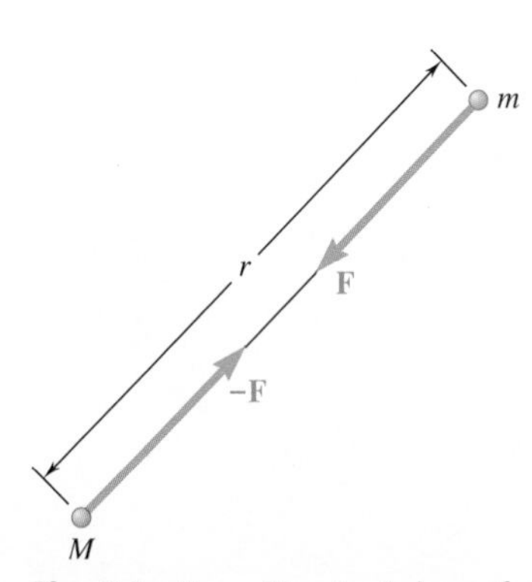

Fig. 1.1 From Newton's law of gravitation, two particles of masses M and m exert forces upon each other of equal magnitude, opposite direction, and the same line of action. This also illustrates Newton's third law of motion.

A particular case of great importance is that of the attraction of the earth on a particle located on its surface. The force $\mathbf{F}$ exerted by the earth on the particle is defined as the **weight W** of the particle. Suppose we set M equal to the mass of the earth, m equal to the mass of the particle, and r equal to the earth's radius R. Then introducing the constant

$$g = \frac{GM}{R^2} \quad (1.3)$$

we can express the magnitude W of the weight of a particle of mass m as[†]

$$W = mg \quad (1.4)$$

The value of R in formula (1.3) depends upon the elevation of the point considered; it also depends upon its latitude, since the earth is not truly spherical. The value of g therefore varies with the position of the point considered. However, as long as the point actually remains on the earth's surface, it is sufficiently accurate in most engineering computations to assume that g equals 9.81 m/s^2 or 32.2 ft/s^2.

The principles we have just listed will be introduced in the course of our study of mechanics as they are needed. The statics of particles carried out in Chap. 2 will be based on the parallelogram law of addition and on Newton's first law alone. We introduce the principle of transmissibility in Chap. 3 as we begin the study of the statics of rigid bodies, and we bring in Newton's third law in Chap. 6 as we analyze the forces exerted on each other by the various members forming a structure.

As noted earlier, the six fundamental principles listed previously are based on experimental evidence. Except for Newton's first law and the principle of transmissibility, they are independent principles that cannot be derived mathematically from each other or from any other elementary physical principle. On these principles rests most of the intricate structure of newtonian mechanics. For more than two centuries, engineers have solved a tremendous number of problems dealing with the conditions of rest and motion of rigid bodies, deformable bodies, and fluids by applying these fundamental principles. Many of the solutions obtained could be checked experimentally, thus providing a further verification of the principles from which they were derived. Only in the twentieth century has Newton's mechanics found to be at fault, in the study of the motion of atoms and the motion of the planets, where it must be supplemented by the theory of relativity. On the human or engineering scale, however, where velocities are small compared with the speed of light, Newton's mechanics have yet to be disproved.

Photo 1.1 When in orbit of the earth, people and objects are said to be *weightless* even though the gravitational force acting is approximately 90% of that experienced on the surface of the earth. This apparent contradiction can be resolved in a course on dynamics when Newton's second law is applied to the motion of particles.

1.2B Mechanics of Deformable Bodies

The concepts needed for mechanics of deformable bodies, also referred to as *mechanics of materials,* are necessary for analyzing and designing various machines and load-bearing structures. These concepts involve the determination of *stresses* and *deformations.*

In Chaps. 8 through 16, the analysis of stresses and the corresponding deformations will be developed for structural members subject to axial loading, torsion, and bending. This requires the use of basic concepts involving the conditions of equilibrium of forces exerted on the member, the relations existing between stress and deformation in the material, and the conditions imposed by the supports and loading of the member. Later chapters expand on these subjects, providing a basis for designing both structures that are statically determinant and those that are indeterminant, i.e., structures in which the internal forces cannot be determined from statics alone.

[†]A more accurate definition of the weight **W** should take into account the earth's rotation.

1.3 SYSTEMS OF UNITS

Associated with the four fundamental concepts just discussed are the so-called *kinetic units,* i.e., the units of *length, time, mass,* and *force*. These units cannot be chosen independently if Eq. (1.1) is to be satisfied. Three of the units may be defined arbitrarily; we refer to them as **basic units**. The fourth unit, however, must be chosen in accordance with Eq. (1.1) and is referred to as a **derived unit**. Kinetic units selected in this way are said to form a **consistent system of units**.

International System of Units (SI Units).[†] In this system, which will be in universal use after the United States has completed its conversion to SI units, the base units are the units of length, mass, and time, and they are called, respectively, the **meter** (m), the **kilogram** (kg), and the **second** (s). All three are arbitrarily defined. The second was originally chosen to represent 1/86 400 of the mean solar day, but it is now defined as the duration of 9 192 631 770 cycles of the radiation corresponding to the transition between two levels of the fundamental state of the cesium-133 atom. The meter, originally defined as one ten-millionth of the distance from the equator to either pole, is now defined as 1 650 763.73 wavelengths of the orange-red light corresponding to a certain transition in an atom of krypton-86. (The newer definitions are much more precise and with today's modern instrumentation, are easier to verify as a standard.) The kilogram, which is approximately equal to the mass of 0.001 m^3 of water, is defined as the mass of a platinum-iridium standard kept at the International Bureau of Weights and Measures at Sèvres, near Paris, France. The unit of force is a derived unit. It is called the **newton** (N) and is defined as the force that gives an acceleration of 1 m/s^2 to a body of mass 1 kg (Fig. 1.2). From Eq. (1.1), we have

$$1 \text{ N} = (1 \text{ kg})(1 \text{ m/s}^2) = 1 \text{ kg·m/s}^2 \qquad \textbf{(1.5)}$$

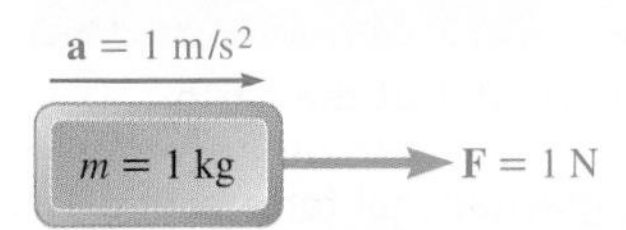

Fig. 1.2 A force of 1 newton applied to a body of mass 1 kg provides an acceleration of 1 m/s^2.

The SI units are said to form an *absolute* system of units. This means that the three base units chosen are independent of the location where measurements are made. The meter, the kilogram, and the second may be used anywhere on the earth; they may even be used on another planet and still have the same significance.

The *weight* of a body, or the *force of gravity* exerted on that body, like any other force, should be expressed in newtons. From Eq. (1.4), it follows that the weight of a body of mass 1 kg (Fig. 1.3) is

$$\begin{aligned} W &= mg \\ &= (1 \text{ kg})(9.81 \text{ m/s}^2) \\ &= 9.81 \text{ N} \end{aligned}$$

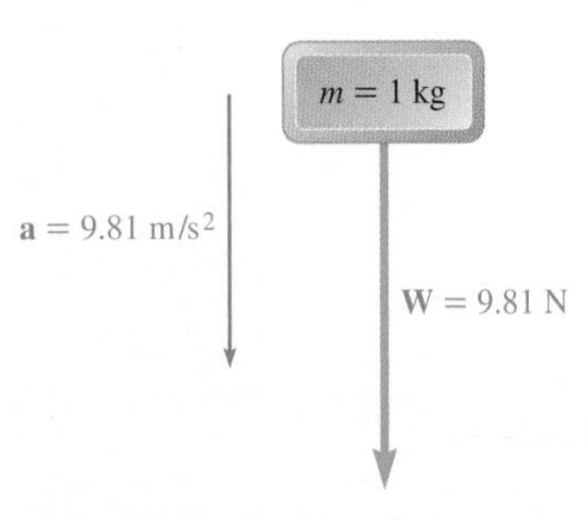

Fig. 1.3 A body of mass 1 kg experiencing an acceleration due to gravity of 9.81 m/s^2 has a weight of 9.81 N.

Multiples and submultiples of the fundamental SI units are denoted through the use of the prefixes defined in Table 1.1. The multiples and submultiples of the units of length, mass, and force most frequently used in engineering are, respectively, the *kilometer* (km) and the *millimeter* (mm); the *megagram*[‡] (Mg) and the *gram* (g); and the *kilonewton* (kN). According to Table 1.1, we have

$$1 \text{ km} = 1000 \text{ m} \quad 1 \text{ mm} = 0.001 \text{ m}$$
$$1 \text{ Mg} = 1000 \text{ kg} \quad 1 \text{ g} = 0.001 \text{ kg}$$
$$1 \text{ kN} = 1000 \text{ N}$$

[†]SI stands for *Système International d'Unités* (French).

[‡]Also known as a *metric ton*.

Table 1.1 SI Prefixes

Multiplication Factor	Prefix†	Symbol
1 000 000 000 000 = 10^{12}	tera	T
1 000 000 000 = 10^{9}	giga	G
1 000 000 = 10^{6}	mega	M
1 000 = 10^{3}	kilo	k
100 = 10^{2}	hecto‡	h
10 = 10^{1}	deka‡	da
0.1 = 10^{-1}	deci‡	d
0.01 = 10^{-2}	centi‡	c
0.001 = 10^{-3}	milli	m
0.000 001 = 10^{-6}	micro	μ
0.000 000 001 = 10^{-9}	nano	n
0.000 000 000 001 = 10^{-12}	pico	p
0.000 000 000 000 001 = 10^{-15}	femto	f
0.000 000 000 000 000 001 = 10^{-18}	atto	a

†The first syllable of every prefix is accented, so that the prefix retains its identity. Thus, the preferred pronunciation of kilometer places the accent on the first syllable, not the second.

‡The use of these prefixes should be avoided, except for the measurement of areas and volumes and for the nontechnical use of centimeter, as for body and clothing measurements.

The conversion of these units into meters, kilograms, and newtons, respectively, can be effected by simply moving the decimal point three places to the right or to the left. For example, to convert 3.82 km into meters, move the decimal point three places to the right:

$$3.82 \text{ km} = 3820 \text{ m}$$

Similarly, to convert 47.2 mm into meters, move the decimal point three places to the left:

$$47.2 \text{ mm} = 0.0472 \text{ m}$$

Using engineering notation, you can also write

$$3.82 \text{ km} = 3.82 \times 10^{3} \text{ m}$$
$$47.2 \text{ mm} = 47.2 \times 10^{-3} \text{ m}$$

The multiples of the unit of time are the *minute* (min) and the *hour* (h). Since 1 min = 60 s and 1 h = 60 min = 3600 s, these multiples cannot be converted as readily as the others.

By using the appropriate multiple or submultiple of a given unit, you can avoid writing very large or very small numbers. For example, it is usually simpler to write 427.2 km rather than 427 200 m and 2.16 mm rather than 0.002 16 m.†

Units of Area and Volume. The unit of area is the *square meter* (m^2), which represents the area of a square of side 1 m; the unit of volume is the *cubic meter* (m^3), which is equal to the volume of a cube of side 1 m. In order to avoid exceedingly small or large numerical values when computing areas and volumes, we use systems of subunits obtained by respectively squaring and cubing not only the millimeter, but also two intermediate

†Note that when more than four digits appear on either side of the decimal point to express a quantity in SI units—as in 427 000 m or 0.002 16 m—use spaces, never commas, to separate the digits into groups of three. This practice avoids confusion with the comma used in place of a decimal point, which is the convention in many countries.

submultiples of the meter: the *decimeter* (dm) and the *centimeter* (cm). By definition,

$$1\text{ dm} = 0.1\text{ m} = 10^{-1}\text{ m}$$
$$1\text{ cm} = 0.01\text{ m} = 10^{-2}\text{ m}$$
$$1\text{ mm} = 0.001\text{ m} = 10^{-3}\text{ m}$$

Therefore, the submultiples of the unit of area are

$$1\text{ dm}^2 = (1\text{ dm})^2 = (10^{-1}\text{ m})^2 = 10^{-2}\text{ m}^2$$
$$1\text{ cm}^2 = (1\text{ cm})^2 = (10^{-2}\text{ m})^2 = 10^{-4}\text{ m}^2$$
$$1\text{ mm}^2 = (1\text{ mm})^2 = (10^{-3}\text{ m})^2 = 10^{-6}\text{ m}^2$$

Similarly, the submultiples of the unit of volume are

$$1\text{ dm}^3 = (1\text{ dm})^3 = (10^{-1}\text{ m})^3 = 10^{-3}\text{ m}^3$$
$$1\text{ cm}^3 = (1\text{ cm})^3 = (10^{-2}\text{ m})^3 = 10^{-6}\text{ m}^3$$
$$1\text{ mm}^3 = (1\text{ mm})^3 = (10^{-3}\text{ m})^3 = 10^{-9}\text{ m}^3$$

Note that when measuring the volume of a liquid, the cubic decimeter (dm^3) is usually referred to as a *liter* (L).

Table 1.2 shows other derived SI units used to measure the moment of a force, the work of a force, etc. Although we will introduce these units in later chapters as they are needed, we should note an important rule at

Table 1.2 Principal SI Units Used in Mechanics

Quantity	Unit	Symbol	Formula
Acceleration	Meter per second squared	. . .	m/s^2
Angle	Radian	rad	†
Angular acceleration	Radian per second squared	. . .	rad/s^2
Angular velocity	Radian per second	. . .	rad/s
Area	Square meter	. . .	m^2
Density	Kilogram per cubic meter	. . .	kg/m^3
Energy	Joule	J	N·m
Force	Newton	N	kg·m/s^2
Frequency	Hertz	Hz	s^{-1}
Impulse	Newton-second	. . .	kg·m/s
Length	Meter	m	‡
Mass	Kilogram	kg	‡
Moment of a force	Newton-meter	. . .	N·m
Power	Watt	W	J/s
Pressure	Pascal	Pa	N/m^2
Stress	Pascal	Pa	N/m^2
Time	Second	s	‡
Velocity	Meter per second	. . .	m/s
Volume			
Solids	Cubic meter	. . .	m^3
Liquids	Liter	L	10^{-3} m^3
Work	Joule	J	N·m

†Supplementary unit (1 revolution = 2π rad = 360°).
‡Base unit.

this time: When a derived unit is obtained by dividing a base unit by another base unit, you may use a prefix in the numerator of the derived unit, but not in its denominator. For example, the constant k of a spring that stretches 20 mm under a load of 100 N is expressed as

$$k = \frac{100 \text{ N}}{20 \text{ mm}} = \frac{100 \text{ N}}{0.020 \text{ m}} = 5000 \text{ N/m} \quad \text{or} \quad k = 5 \text{ kN/m}$$

but never as $k = 5$ N/mm.

U.S. Customary Units. Most practicing American engineers still commonly use a system in which the base units are those of length, force, and time. These units are, respectively, the *foot* (ft), the *pound* (lb), and the *second* (s). The second is the same as the corresponding SI unit. The foot is defined as 0.3048 m. The pound is defined as the *weight* of a platinum standard, called the *standard pound,* which is kept at the National Institute of Standards and Technology outside Washington D.C., the mass of which is 0.453 592 43 kg. Since the weight of a body depends upon the earth's gravitational attraction, which varies with location, the standard pound should be placed at sea level and at a latitude of 45° to properly define a force of 1 lb. Clearly the U.S. customary units do not form an absolute system of units. Because they depend upon the gravitational attraction of the earth, they form a *gravitational* system of units.

Although the standard pound also serves as the unit of mass in commercial transactions in the United States, it cannot be used that way in engineering computations, because such a unit would not be consistent with the base units defined in the preceding paragraph. Indeed, when acted upon by a force of 1 lb—that is, when subjected to the force of gravity—the standard pound has the acceleration due to gravity, $g = 32.2$ ft/s^2 (Fig. 1.4), not the unit acceleration required by Eq. (1.1). The unit of mass consistent with the foot, the pound, and the second is the mass that receives an acceleration of 1 ft/s^2 when a force of 1 lb is applied to it (Fig. 1.5). This unit, sometimes called a *slug,* can be derived from the equation $F = ma$ after substituting 1 lb for F and 1 ft/s^2 for a. We have

$$F = ma \quad 1 \text{ lb} = (1 \text{ slug})(1 \text{ ft/s}^2)$$

This gives us

$$1 \text{ slug} = \frac{1 \text{ lb}}{1 \text{ ft/s}^2} = 1 \text{ lb·s}^2\text{/ft} \tag{1.6}$$

Comparing Figs. 1.4 and 1.5, we conclude that the slug is a mass 32.2 times larger than the mass of the standard pound.

The fact that, in the U.S. customary system of units, bodies are characterized by their weight in pounds rather than by their mass in slugs is convenient in the study of statics, where we constantly deal with weights and other forces and only seldom deal directly with masses. However, in the study of dynamics, where forces, masses, and accelerations are involved, the mass m of a body is expressed in slugs when its weight W is given in pounds. Recalling Eq. (1.4), we write

$$m = \frac{W}{g} \tag{1.7}$$

where g is the acceleration due to gravity ($g = 32.2$ ft/s^2).

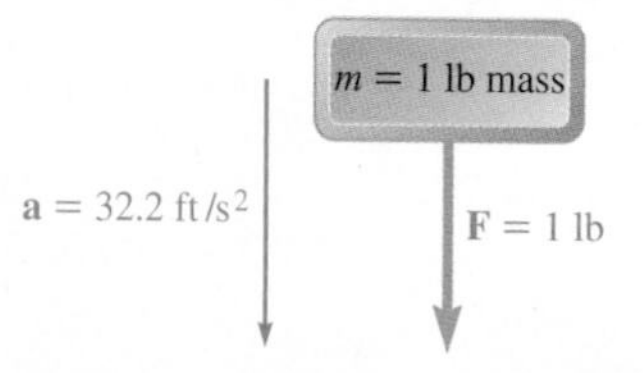

Fig. 1.4 A body of 1 pound mass acted upon by a force of 1 pound has an acceleration of 32.2 ft/s^2.

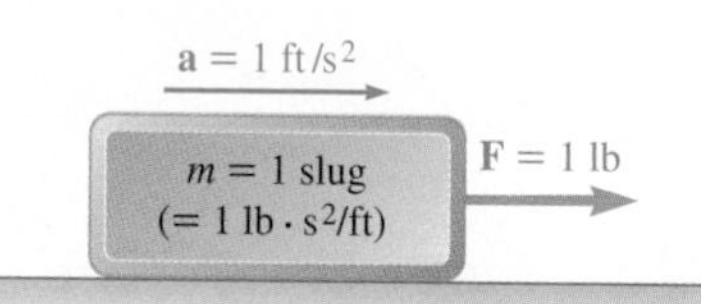

Fig. 1.5 A force of 1 pound applied to a body of mass 1 slug produces an acceleration of 1 ft/s^2.

Other U.S. customary units frequently encountered in engineering problems are the *mile* (mi), equal to 5280 ft; the *inch* (in.), equal to (1/12) ft; and the *kilopound* (kip), equal to 1000 lb. The *ton* is often used to represent a mass of 2000 lb but, like the pound, must be converted into slugs in engineering computations.

The conversion into feet, pounds, and seconds of quantities expressed in other U.S. customary units is generally more involved and requires greater attention than the corresponding operation in SI units. For example, suppose we are given the magnitude of a velocity $v = 30$ mi/h and want to convert it to ft/s. First we write

$$v = 30\frac{\text{mi}}{\text{h}}$$

Since we want to get rid of the unit miles and introduce instead the unit feet, we should multiply the right-hand member of the equation by an expression containing miles in the denominator and feet in the numerator. However, since we do not want to change the value of the right-hand side of the equation, the expression used should have a value equal to unity. The quotient (5280 ft)/(1 mi) is such an expression. Operating in a similar way to transform the unit hour into seconds, we have

$$v = \left(30\frac{\text{mi}}{\text{h}}\right)\left(\frac{5280\text{ ft}}{1\text{ mi}}\right)\left(\frac{1\text{ h}}{3600\text{ s}}\right)$$

Carrying out the numerical computations and canceling out units that appear in both the numerator and the denominator, we obtain

$$v = 44\frac{\text{ft}}{\text{s}} = 44\text{ ft/s}$$

1.4 CONVERTING BETWEEN TWO SYSTEMS OF UNITS

In many situations, an engineer might need to convert into SI units a numerical result obtained in U.S. customary units or vice versa. Because the unit of time is the same in both systems, only two kinetic base units need be converted. Thus, since all other kinetic units can be derived from these base units, only two conversion factors need be remembered.

Units of Length. By definition, the U.S. customary unit of length is

$$1\text{ ft} = 0.3048\text{ m} \tag{1.8}$$

It follows that

$$1\text{ mi} = 5280\text{ ft} = 5280(0.3048\text{ m}) = 1609\text{ m}$$

or

$$1\text{ mi} = 1.609\text{ km} \tag{1.9}$$

Also,

$$1\text{ in.} = \frac{1}{12}\text{ ft} = \frac{1}{12}(0.3048\text{ m}) = 0.0254\text{ m}$$

or

$$1 \text{ in.} = 25.4 \text{ mm} \tag{1.10}$$

Units of Force. Recall that the U.S. customary unit of force (pound) is defined as the weight of the standard pound (of mass 0.4536 kg) at sea level and at a latitude of 45° (where $g = 9.807 \text{ m/s}^2$). Then, using Eq. (1.4), we write

$$W = mg$$
$$1 \text{ lb} = (0.4536 \text{ kg})(9.807 \text{ m/s}^2) = 4.448 \text{ kg}\cdot\text{m/s}^2$$

From Eq. (1.5), this reduces to

$$1 \text{ lb} = 4.448 \text{ N} \tag{1.11}$$

Units of Mass. The U.S. customary unit of mass (slug) is a derived unit. Thus, using Eqs. (1.6), (1.8), and (1.11), we have

$$1 \text{ slug} = 1 \text{ lb}\cdot\text{s}^2/\text{ft} = \frac{1 \text{ lb}}{1 \text{ ft/s}^2} = \frac{4.448 \text{ N}}{0.3048 \text{ m/s}^2} = 14.59 \text{ N}\cdot\text{s}^2/\text{m}$$

Again, from Eq. (1.5),

$$1 \text{ slug} = 1 \text{ lb}\cdot\text{s}^2/\text{ft} = 14.59 \text{ kg} \tag{1.12}$$

Although it cannot be used as a consistent unit of mass, recall that the mass of the standard pound is, by definition,

$$1 \text{ pound mass} = 0.4536 \text{ kg} \tag{1.13}$$

We can use this constant to determine the *mass* in SI units (kilograms) of a body that has been characterized by its *weight* in U.S. customary units (pounds).

To convert a derived U.S. customary unit into SI units, simply multiply or divide by the appropriate conversion factors. For example, to convert the moment of a force that is measured as $M = 47 \text{ lb}\cdot\text{in.}$ into SI units, use formulas (1.10) and (1.11) and write

$$M = 47 \text{ lb}\cdot\text{in.} = 47(4.448 \text{ N})(25.4 \text{ mm})$$
$$= 5310 \text{ N}\cdot\text{mm} = 5.31 \text{ N}\cdot\text{m}$$

You can also use conversion factors to convert a numerical result obtained in SI units into U.S. customary units. For example, if the moment of a force is measured as $M = 40 \text{ N}\cdot\text{m}$, follow the procedure at the end of Sec. 1.3 to write

$$M = 40 \text{ N}\cdot\text{m} = (40 \text{ N}\cdot\text{m})\left(\frac{1 \text{ lb}}{4.448 \text{ N}}\right)\left(\frac{1 \text{ ft}}{0.3048 \text{ m}}\right)$$

Carrying out the numerical computations and canceling out units that appear in both the numerator and the denominator, you obtain

$$M = 29.5 \text{ lb}\cdot\text{ft}$$

The U.S. customary units most frequently used in mechanics are listed in Table 1.3 with their SI equivalents.

Photo 1.2 In 1999, The *Mars Climate Orbiter* entered orbit around Mars at too low an altitude and disintegrated. Investigation showed that the software on board the probe interpreted force instructions in newtons, but the software at mission control on the earth was generating those instructions in terms of pounds.

Table 1.3 U.S. Customary Units and Their SI Equivalents

Quantity	U.S. Customary Unit	SI Equivalent
Acceleration	ft/s^2	$0.3048\ m/s^2$
	$in./s^2$	$0.0254\ m/s^2$
Area	ft^2	$0.0929\ m^2$
	in^2	$645.2\ mm^2$
Energy	ft·lb	1.356 J
Force	kip	4.448 kN
	lb	4.448 N
	oz	0.2780 N
Impulse	lb·s	4.448 N·s
Length	ft	0.3048 m
	in.	25.40 mm
	mi	1.609 km
Mass	oz mass	28.35 g
	lb mass	0.4536 kg
	slug	14.59 kg
	ton	907.2 kg
Moment of a force	lb·ft	1.356 N·m
	lb·in.	0.1130 N·m
Moment of inertia		
Of an area	in^4	$0.4162 \times 10^6\ mm^4$
Of a mass	$lb \cdot ft \cdot s^2$	$1.356\ kg \cdot m^2$
Momentum	lb·s	4.448 kg·m/s
Power	ft·lb/s	1.356 W
	hp	745.7 W
Pressure or stress	lb/ft^2	47.88 Pa
	lb/in^2 (psi)	6.895 kPa
Velocity	ft/s	0.3048 m/s
	in./s	0.0254 m/s
	mi/h (mph)	0.4470 m/s
	mi/h (mph)	1.609 km/h
Volume	ft^3	$0.02832\ m^3$
	in^3	$16.39\ cm^3$
Liquids	gal	3.785 L
	qt	0.9464 L
Work	ft·lb	1.356 J

1.5 METHOD OF SOLVING PROBLEMS

You should approach a problem in mechanics as you would approach an actual engineering situation. By drawing on your own experience and intuition about physical behavior, you will find it easier to understand and formulate the problem. Once you have clearly stated and understood the problem, however, there is no place in its solution for arbitrary methodologies.

The solution must be based on the six fundamental principles stated in Sec. 1.2A or on theorems derived from them.

Every step you take in the solution must be justified on this basis. Strict rules must be followed, which lead to the solution in an almost automatic fashion, leaving no room for your intuition or "feeling." After you have obtained an answer, you should check it. Here again, you may call upon your common sense and personal experience. If you are not completely satisfied with the result, you should carefully check your formulation of the problem, the validity of the methods used for its solution, and the accuracy of your computations.

In general, you can usually solve problems in several different ways; there is no one approach that works best for everybody. However, we have found that students often find it helpful to have a general set of guidelines to use for framing problems and planning solutions. In the Sample Problems throughout this text, we use a four-step method for approaching problems, which we refer to as the SMART methodology: **S**trategy, **M**odeling, **A**nalysis, and **R**eflect and **T**hink.

1. **Strategy.** The statement of a problem should be clear and precise, and it should contain the given data and indicate what information is required. The first step in solving the problem is to decide what concepts you have learned that apply to the given situation and to connect the data to the required information. It is often useful to work backward from the information you are trying to find: Ask yourself what quantities you need to know to obtain the answer, and if some of these quantities are unknown, how can you find them from the given data.
2. **Modeling.** The first step in modeling is to define the system; that is, clearly define what you are setting aside for analysis. After you have selected a system, draw a neat sketch showing all quantities involved with a separate diagram for each body in the problem. For equilibrium problems, indicate clearly the forces acting on each body along with any relevant geometrical data, such as lengths and angles. (These diagrams are known as **free-body diagrams** and are described in detail in Sec. 2.3C and the beginning of Chap. 4.)
3. **Analysis.** After you have drawn the appropriate diagrams, use the fundamental principles of mechanics listed in Sec. 1.2 to write equations expressing the conditions of rest or motion of the bodies considered. Each equation should be clearly related to one of the free-body diagrams and should be numbered. If you do not have enough equations to solve for the unknowns, try selecting another system, or reexamine your strategy to see if you can apply other principles to the problem. Once you have obtained enough equations, you can find a numerical solution by following the usual rules of algebra, neatly recording each step and the intermediate results. Alternatively, you can solve the resulting equations with your calculator or a computer. (For multipart problems, it is sometimes convenient to present the Modeling and Analysis steps together, but they are both essential parts of the overall process.)
4. **Reflect and Think.** After you have obtained the answer, check it carefully. Does it make sense in the context of the original problem? For instance, the problem may ask for the force at a given point of a structure. If your answer is negative, what does that mean for the force at the point?

You can often detect mistakes in *reasoning* by checking the units. For example, to determine the moment of a force of 50 N about a point 0.60 m from its line of action, we write (Sec. 3.3A)

$$M = Fd = (30\ \text{N})(0.60\ \text{m}) = 30\ \text{N}\cdot\text{m}$$

The unit N·m obtained by multiplying newtons by meters is the correct unit for the moment of a force; if you had obtained another unit, you would know that some mistake had been made.

You can often detect errors in *computation* by substituting the numerical answer into an equation that was not used in the solution and verifying that the equation is satisfied. The importance of correct computations in engineering cannot be overemphasized.

1.6 NUMERICAL ACCURACY

The accuracy of the solution to a problem depends upon two items: (1) the accuracy of the given data and (2) the accuracy of the computations performed. The solution cannot be more accurate than the less accurate of these two items.

For example, suppose the loading of a bridge is known to be 75 000 lb with a possible error of 100 lb either way. The relative error that measures the degree of accuracy of the data is

$$\frac{100 \text{ lb}}{75\,000 \text{ lb}} = 0.0013 = 0.13\%$$

In computing the reaction at one of the bridge supports, it would be meaningless to record it as 14 322 lb. The accuracy of the solution cannot be greater than 0.13%, no matter how precise the computations are, and the possible error in the answer may be as large as (0.13/100)(14 322 lb) $\approx$ 20 lb. The answer should be properly recorded as 14 320 $\pm$ 20 lb.

In engineering problems, the data are seldom known with an accuracy greater than 0.2%. It is therefore seldom justified to write answers with an accuracy greater than 0.2%. A practical rule is to use four figures to record numbers beginning with a "1" and three figures in all other cases. Unless otherwise indicated, you should assume the data given in a problem are known with a comparable degree of accuracy. A force of 40 lb, for example, should be read as 40.0 lb, and a force of 15 lb should be read as 15.00 lb.

Electronic calculators are widely used by practicing engineers and engineering students. The speed and accuracy of these calculators facilitate the numerical computations in the solution of many problems. However, you should not record more significant figures than can be justified merely because you can obtain them easily. As noted previously, an accuracy greater than 0.2% is seldom necessary or meaningful in the solution of practical engineering problems.

© Getty Images RF

2

Statics of Particles

Many engineering problems can be solved by considering the equilibrium of a "particle." In the case of this beam that is being hoisted into position, a relation between the tensions in the various cables involved can be obtained by considering the equilibrium of the hook to which the cables are attached.

Objectives

- **Describe** force as a vector quantity.
- **Examine** vector operations useful for the analysis of forces.
- **Determine** the resultant of multiple forces acting on a particle.
- **Resolve** forces into components.
- **Add** forces that have been resolved into rectangular components.
- **Introduce** the concept of the free-body diagram.
- **Use** free-body diagrams to assist in the analysis of planar and spatial particle equilibrium problems.

Introduction

In this chapter, you will study the effect of forces acting on particles. By the word "particle" we do not mean only tiny bits of matter, like an atom or an electron. Instead, we mean that the sizes and shapes of the bodies under consideration do not significantly affect the solutions of the problems. Another way of saying this is that we assume all forces acting on a given body act at the same point. This does not mean the object must be tiny—if you were modeling the mechanics of the Milky Way galaxy, for example, you could treat the Sun and the entire Solar System as just a particle.

Our first step is to explain how to replace two or more forces acting on a given particle by a single force having the same effect as the original forces. This single equivalent force is called the *resultant* of the original forces. After this step, we will derive the relations among the various forces acting on a particle in a state of *equilibrium*. We will use these relations to determine some of the forces acting on the particle.

The first part of this chapter deals with forces contained in a single plane. Because two lines determine a plane, this situation arises any time we can reduce the problem to one of a particle subjected to two forces that support a third force, such as a crate suspended from two chains or a traffic light held in place by two cables. In the second part of this chapter, we examine the more general case of forces in three-dimensional space.

2.1 ADDITION OF PLANAR FORCES

Many important practical situations in engineering involve forces in the same plane. These include forces acting on a pulley, projectile motion, and an object in equilibrium on a flat surface. We will examine this situation first before looking at the added complications of forces acting in three-dimensional space.

2.1A Force on a Particle: Resultant of Two Forces

A force represents the action of one body on another. It is generally characterized by its **point of application**, its **magnitude**, and its **direction**. Forces acting on a given particle, however, have the same point of application. Thus, each force considered in this chapter is completely defined by its magnitude and direction.

The magnitude of a force is characterized by a certain number of units. As indicated in Chap. 1, the SI units used by engineers to measure the magnitude of a force are the newton (N) and its multiple the kilonewton (kN), which is equal to 1000 N. The U.S. customary units used for the same purpose are the pound (lb) and its multiple the kilopound (kip), which is equal to 1000 lb. We saw in Chapter 1 that a force of 445 N is equivalent to a force of 100 lb or that a force of 100 N equals a force of about 22.5 lb.

We define the direction of a force by its **line of action** and the **sense** of the force. The line of action is the infinite straight line along which the force acts; it is characterized by the angle it forms with some fixed axis (Fig. 2.1). The force itself is represented by a segment of that line; through the use of an appropriate scale, we can choose the length of this segment to represent the magnitude of the force. We indicate the sense of the force by an arrowhead. It is important in defining a force to indicate its sense. Two forces having the same magnitude and the same line of action but a different sense, such as the forces shown in Fig. 2.1*a* and *b*, have directly opposite effects on a particle.

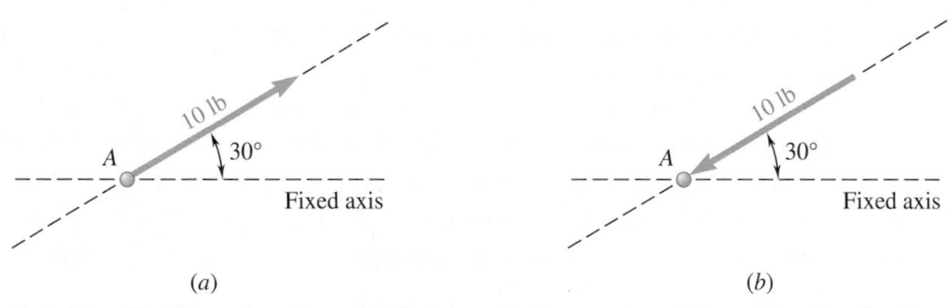

Fig. 2.1 The line of action of a force makes an angle with a given fixed axis. (*a*) The sense of the 10-lb force is away from particle *A*; (*b*) the sense of the 10-lb force is toward particle *A*.

Experimental evidence shows that two forces **P** and **Q** acting on a particle *A* (Fig. 2.2*a*) can be replaced by a single force **R** that has the same effect on the particle (Fig. 2.2*c*). This force is called the **resultant** of the forces **P** and **Q**. We can obtain **R**, as shown in Fig. 2.2*b*, by constructing a parallelogram, using **P** and **Q** as two adjacent sides. **The diagonal that passes through *A* represents the resultant**. This method for finding the resultant is known as the **parallelogram law** for the addition of two forces. This law is based on experimental evidence; it cannot be proved or derived mathematically.

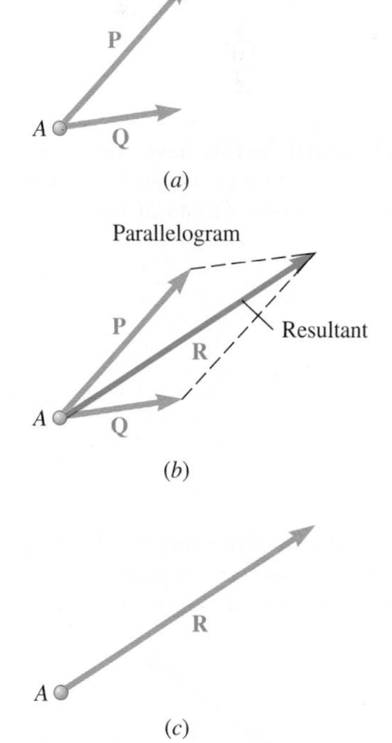

Fig. 2.2 (*a*) Two forces **P** and **Q** act on particle *A*. (*b*) Draw a parallelogram with **P** and **Q** as the adjacent sides and label the diagonal that passes through *A* as **R**. (*c*) **R** is the resultant of the two forces **P** and **Q** and is equivalent to their sum.

2.1B Vectors

We have just seen that forces do not obey the rules of addition defined in ordinary arithmetic or algebra. For example, two forces acting at a right angle to each other, one of 4 lb and the other of 3 lb, add up to a force of

Photo 2.1 In its purest form, a tug-of-war pits two opposite and almost-equal forces against each other. Whichever team can generate the larger force, wins. As you can see, a competitive tug-of-war can be quite intense.

© DGB/Alamy

5 lb acting at an angle between them, *not* to a force of 7 lb. Forces are not the only quantities that follow the parallelogram law of addition. As you will see later, *displacements, velocities, accelerations,* and *momenta* are other physical quantities possessing magnitude and direction that add according to the parallelogram law. All of these quantities can be represented mathematically by **vectors**. Those physical quantities that have magnitude but not direction, such as *volume, mass,* or *energy*, are represented by plain numbers often called **scalars** to distinguish them from vectors.

Vectors are defined as **mathematical expressions possessing magnitude and direction, which add according to the parallelogram law**. Vectors are represented by arrows in diagrams and are distinguished from scalar quantities in this text through the use of boldface type (**P**). In longhand writing, a vector may be denoted by drawing a short arrow above the letter used to represent it ($\vec{P}$). The magnitude of a vector defines the length of the arrow used to represent it. In this text, we use italic type to denote the magnitude of a vector. Thus, the magnitude of the vector **P** is denoted by *P*.

A vector used to represent a force acting on a given particle has a well-defined point of application—namely, the particle itself. Such a vector is said to be a *fixed*, or *bound*, vector and cannot be moved without modifying the conditions of the problem. Other physical quantities, however, such as couples (see Chap. 3), are represented by vectors that may be freely moved in space; these vectors are called *free* vectors. Still other physical quantities, such as forces acting on a rigid body (see Chap. 3), are represented by vectors that can be moved along their lines of action; they are known as *sliding* vectors.

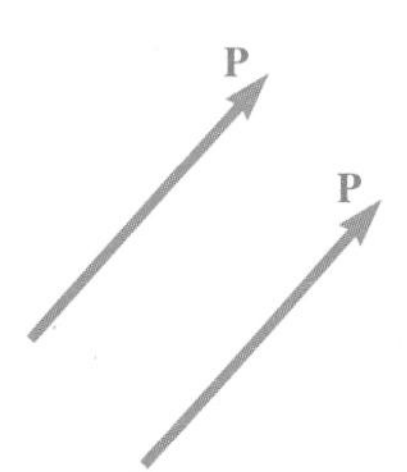

Fig. 2.3 Equal vectors have the same magnitude and the same direction, even if they have different points of application.

Two vectors that have the same magnitude and the same direction are said to be *equal*, whether or not they also have the same point of application (Fig. 2.3); equal vectors may be denoted by the same letter.

The *negative vector* of a given vector **P** is defined as a vector having the same magnitude as **P** and a direction opposite to that of **P** (Fig. 2.4); the negative of the vector **P** is denoted by −**P**. The vectors **P** and −**P** are commonly referred to as **equal and opposite** vectors. Clearly, we have

$$\mathbf{P} + (-\mathbf{P}) = 0$$

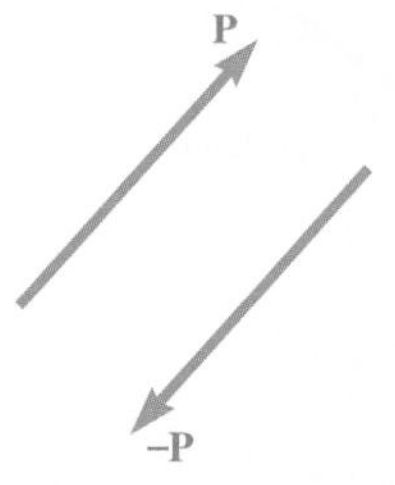

Fig. 2.4 The negative vector of a given vector has the same magnitude but the opposite direction of the given vector.

2.1C Addition of Vectors

By definition, vectors add according to the parallelogram law. Thus, we obtain the sum of two vectors **P** and **Q** by attaching the two vectors to the same point *A* and constructing a parallelogram, using **P** and **Q** as two adjacent sides (Fig. 2.5). The diagonal that passes through *A* represents the sum of the vectors **P** and **Q**, denoted by **P** + **Q**. The fact that the sign + is used for both vector and scalar addition should not cause any confusion if vector and scalar quantities are always carefully distinguished. Note that the magnitude of the vector **P** + **Q** is *not*, in general, equal to the sum $P + Q$ of the magnitudes of the vectors **P** and **Q**.

Since the parallelogram constructed on the vectors **P** and **Q** does not depend upon the order in which **P** and **Q** are selected, we conclude that the addition of two vectors is *commutative*, and we write

$$\mathbf{P} + \mathbf{Q} = \mathbf{Q} + \mathbf{P} \tag{2.1}$$

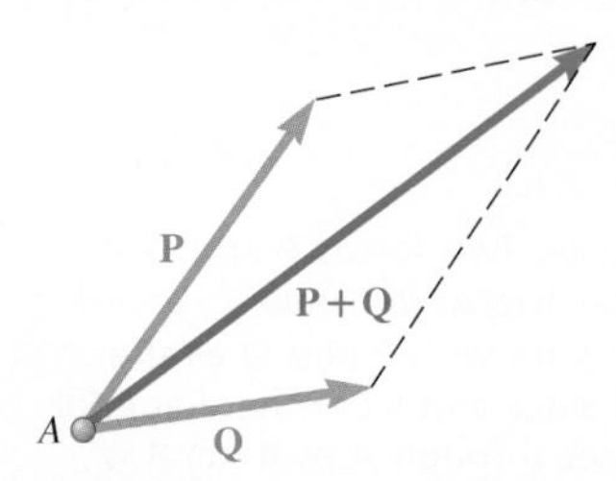

Fig. 2.5 Using the parallelogram law to add two vectors.

From the parallelogram law, we can derive an alternative method for determining the sum of two vectors, known as the **triangle rule**. Consider Fig. 2.5, where the sum of the vectors **P** and **Q** has been determined by the parallelogram law. Since the side of the parallelogram opposite **Q** is equal to **Q** in magnitude and direction, we could draw only half of the parallelogram (Fig. 2.6*a*). The sum of the two vectors thus can be found by **arranging P and Q in tip-to-tail fashion and then connecting the tail of P with the tip of Q**. If we draw the other half of the parallelogram, as in Fig. 2.6*b*, we obtain the same result, confirming that vector addition is commutative.

We define *subtraction* of a vector as the addition of the corresponding negative vector. Thus, we determine the vector **P** − **Q**, representing the difference between the vectors **P** and **Q**, by adding to **P** the negative vector −**Q** (Fig. 2.7). We write

$$\mathbf{P} - \mathbf{Q} = \mathbf{P} + (-\mathbf{Q}) \tag{2.2}$$

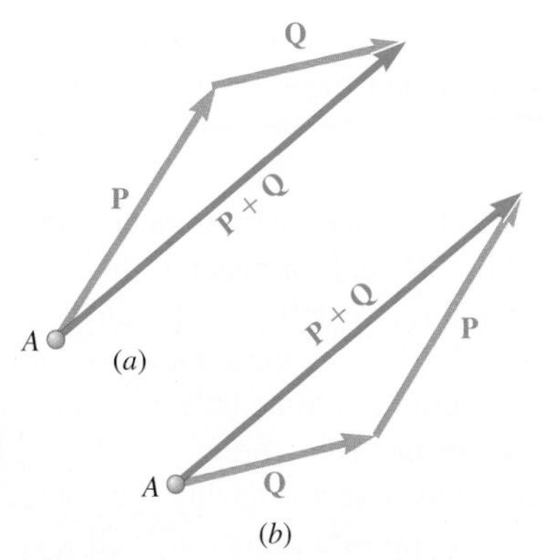

Fig. 2.6 The triangle rule of vector addition. (*a*) Adding vector **Q** to vector **P** equals (*b*) adding vector **P** to vector **Q**.

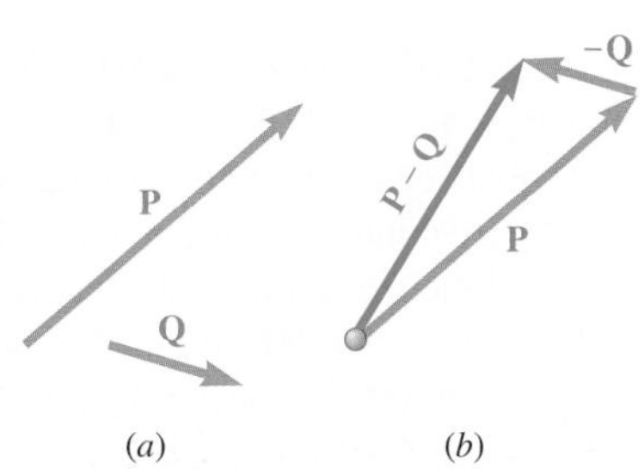

Fig. 2.7 Vector subtraction: Subtracting vector **Q** from vector **P** is the same as adding vector −**Q** to vector **P**.

Here again we should observe that, although we use the same sign to denote both vector and scalar subtraction, we avoid confusion by taking care to distinguish between vector and scalar quantities.

We now consider the *sum of three or more vectors*. The sum of three vectors **P**, **Q**, and **S** is, *by definition*, obtained by first adding the vectors **P** and **Q** and then adding the vector **S** to the vector **P** + **Q**. We write

$$\mathbf{P} + \mathbf{Q} + \mathbf{S} = (\mathbf{P} + \mathbf{Q}) + \mathbf{S} \tag{2.3}$$

Similarly, we obtain the sum of four vectors by adding the fourth vector to the sum of the first three. It follows that we can obtain the sum of any number of vectors by applying the parallelogram law repeatedly to successive pairs of vectors until all of the given vectors are replaced by a single vector.

If the given vectors are *coplanar*, i.e., if they are contained in the same plane, we can obtain their sum graphically. For this case, repeated application of the triangle rule is simpler than applying the parallelogram law. In Fig. 2.8*a*, we find the sum of three vectors **P**, **Q**, and **S** in this manner. The triangle rule is first applied to obtain the sum **P** + **Q** of the

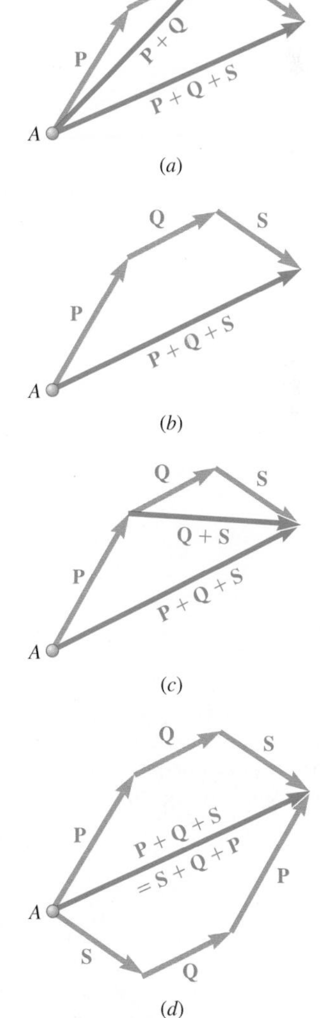

Fig. 2.8 Graphical addition of vectors. (*a*) Applying the triangle rule twice to add three vectors; (*b*) the vectors can be added in one step by the polygon rule; (*c*) vector addition is associative; (*d*) the order of addition is immaterial.

vectors **P** and **Q**; we apply it again to obtain the sum of the vectors **P** + **Q** and **S**. However, we could have omitted determining the vector **P** + **Q** and obtain the sum of the three vectors directly, as shown in Fig. 2.8*b*, by **arranging the given vectors in tip-to-tail fashion and connecting the tail of the first vector with the tip of the last one**. This is known as the **polygon rule** for the addition of vectors.

The result would be unchanged if, as shown in Fig. 2.8*c*, we had replaced the vectors **Q** and **S** by their sum **Q** + **S**. We may thus write

$$\mathbf{P} + \mathbf{Q} + \mathbf{S} = (\mathbf{P} + \mathbf{Q}) + \mathbf{S} = \mathbf{P} + (\mathbf{Q} + \mathbf{S}) \tag{2.4}$$

which expresses the fact that vector addition is *associative*. Recalling that vector addition also has been shown to be commutative in the case of two vectors, we can write

$$\begin{aligned} \mathbf{P} + \mathbf{Q} + \mathbf{S} &= (\mathbf{P} + \mathbf{Q}) + \mathbf{S} = \mathbf{S} + (\mathbf{P} + \mathbf{Q}) \\ &= \mathbf{S} + (\mathbf{Q} + \mathbf{P}) = \mathbf{S} + \mathbf{Q} + \mathbf{P} \end{aligned} \tag{2.5}$$

This expression, as well as others we can obtain in the same way, shows that the order in which several vectors are added together is immaterial (Fig. 2.8*d*).

Product of a Scalar and a Vector. It is convenient to denote the sum **P** + **P** by 2**P**, the sum **P** + **P** + **P** by 3**P**, and, in general, the sum of *n* equal vectors **P** by the product *n***P**. Therefore, we define the product *n***P** of a positive integer *n* and a vector **P** as a vector having the same direction as **P** and the magnitude *nP*. Extending this definition to include all scalars and recalling the definition of a negative vector given earlier, we define the product *k***P** of a scalar *k* and a vector **P** as a vector having the same direction as **P** (if *k* is positive) or a direction opposite to that of **P** (if *k* is negative) and a magnitude equal to the product of *P* and the absolute value of *k* (Fig. 2.9).

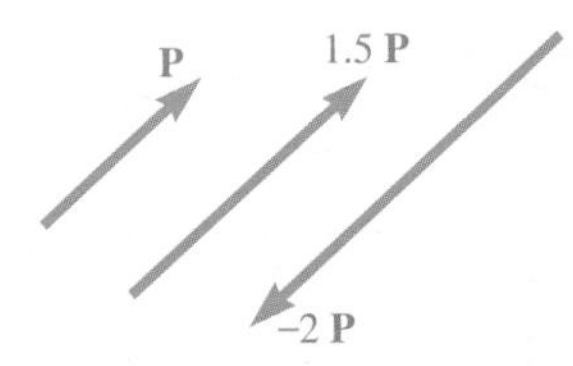

Fig. 2.9 Multiplying a vector by a scalar changes the vector's magnitude, but not its direction (unless the scalar is negative, in which case the direction is reversed).

2.1D Resultant of Several Concurrent Forces

Consider a particle *A* acted upon by several coplanar forces, i.e., by several forces contained in the same plane (Fig. 2.10*a*). Since the forces all pass through *A*, they are also said to be *concurrent*. We can add the vectors representing the forces acting on *A* by the polygon rule (Fig. 2.10*b*). Since the use of the polygon rule is equivalent to the repeated application of the parallelogram law, the vector **R** obtained in this way represents the resultant of the given concurrent forces. That is, the single force **R** has the same effect on the particle *A* as the given forces. As before, the order in which we add the vectors **P**, **Q**, and **S** representing the given forces is immaterial.

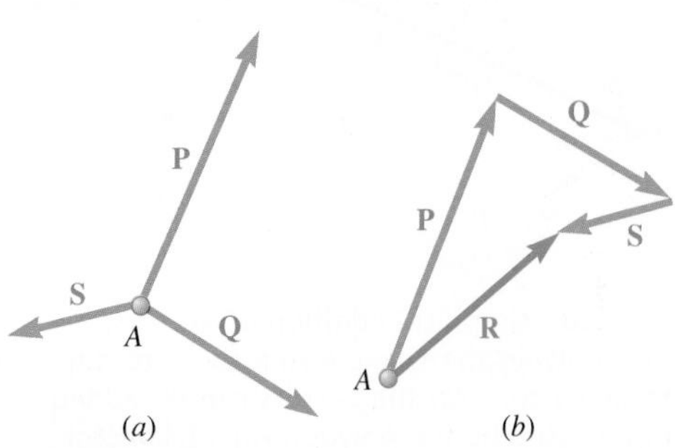

Fig. 2.10 Concurrent forces can be added by the polygon rule.

2.1E Resolution of a Force into Components

We have seen that two or more forces acting on a particle may be replaced by a single force that has the same effect on the particle. Conversely, a single force **F** acting on a particle may be replaced by two or more forces that, together, have the same effect on the particle. These forces are called **components** of the original force **F**, and the process of substituting them for **F** is called **resolving the force F into components**.

Clearly, each force **F** can be resolved into an infinite number of possible sets of components. Sets of *two components* **P** *and* **Q** are the most important as far as practical applications are concerned. However, even then, the number of ways in which a given force **F** may be resolved into two components is unlimited (Fig. 2.11).

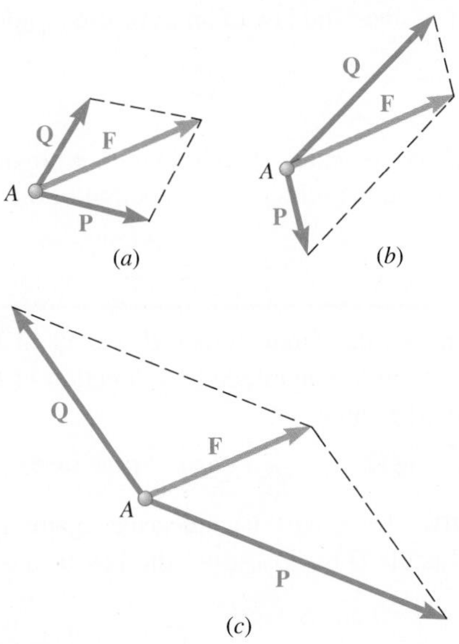

Fig. 2.11 Three possible sets of components for a given force vector **F**.

In many practical problems, we start with a given vector **F** and want to determine a useful set of components. Two cases are of particular interest:

1. **One of the Two Components, P, Is Known.** We obtain the second component, **Q**, by applying the triangle rule and joining the tip of **P** to the tip of **F** (Fig. 2.12). We can determine the magnitude and direction of **Q** graphically or by trigonometry. Once we have determined **Q**, both components **P** and **Q** should be applied at *A*.
2. **The Line of Action of Each Component Is Known.** We obtain the magnitude and sense of the components by applying the parallelogram law and drawing lines through the tip of **F** that are parallel to the given lines of action (Fig. 2.13). This process leads to two well-defined components, **P** and **Q**, which can be determined graphically or computed trigonometrically by applying the law of sines.

You will encounter many similar cases; for example, you might know the direction of one component while the magnitude of the other component is to be as small as possible (see Sample Prob. 2.2). In all cases, you need to draw the appropriate triangle or parallelogram that satisfies the given conditions.

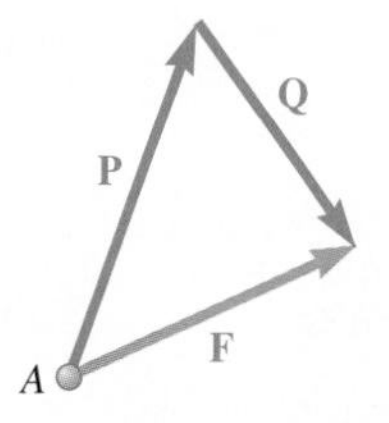

Fig. 2.12 When component **P** is known, use the triangle rule to find component **Q**.

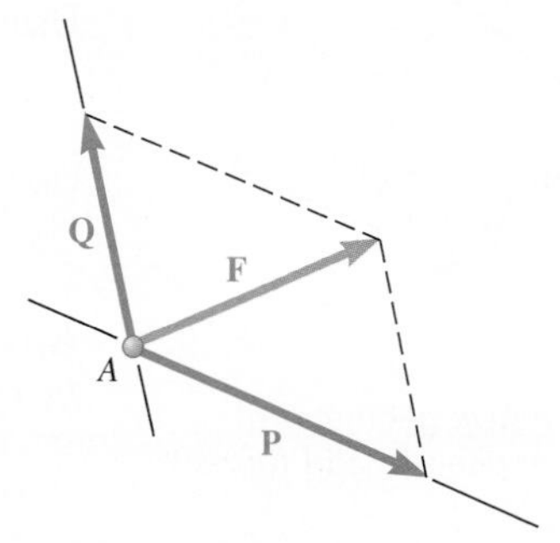

Fig. 2.13 When the lines of action are known, use the parallelogram rule to determine components **P** and **Q**.

Sample Problem 2.1

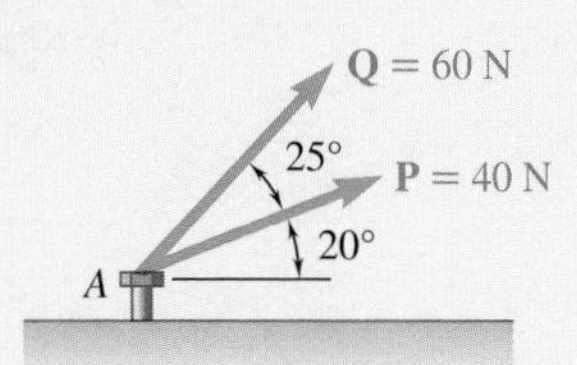

Two forces **P** and **Q** act on a bolt A. Determine their resultant.

STRATEGY: Two lines determine a plane, so this is a problem of two coplanar forces. You can solve the problem graphically or by trigonometry.

MODELING: For a graphical solution, you can use the parallelogram rule or the triangle rule for addition of vectors. For a trigonometric solution, you can use the law of cosines and law of sines or use a right-triangle approach.

ANALYSIS:

Graphical Solution. Draw to scale a parallelogram with sides equal to **P** and **Q** (Fig. 1). Measure the magnitude and direction of the resultant. They are

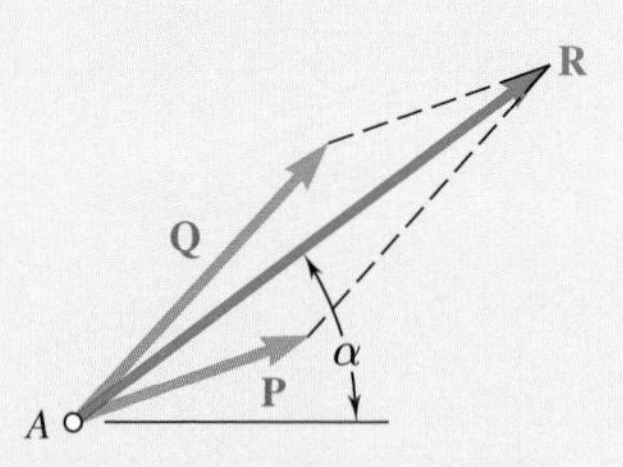

Fig. 1 Parallelogram law applied to add forces **P** and **Q**.

$$R = 98 \text{ N} \qquad \alpha = 35° \qquad \mathbf{R} = 98 \text{ N} \measuredangle 35° \quad ◀$$

You can also use the triangle rule. Draw forces **P** and **Q** in tip-to-tail fashion (Fig. 2). Again measure the magnitude and direction of the resultant. The answers should be the same.

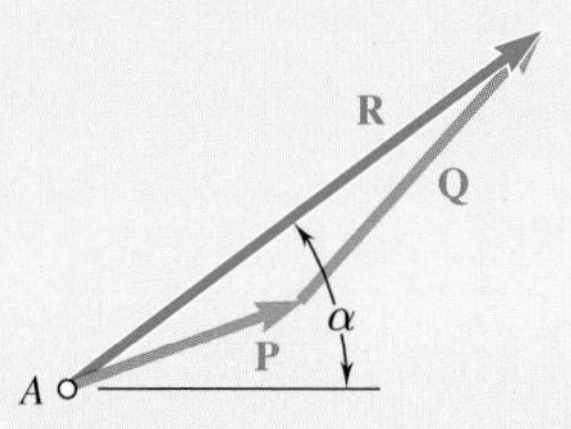

Fig. 2 Triangle rule applied to add forces **P** and **Q**.

$$R = 98 \text{ N} \qquad \alpha = 35° \qquad \mathbf{R} = 98 \text{ N} \measuredangle 35° \quad ◀$$

Trigonometric Solution. Using the triangle rule again, you know two sides and the included angle (Fig. 3). Apply the law of cosines.

$$R^2 = P^2 + Q^2 - 2PQ \cos B$$
$$R^2 = (40 \text{ N})^2 + (60 \text{ N})^2 - 2(40 \text{ N})(60 \text{ N}) \cos 155°$$
$$R = 97.73 \text{ N}$$

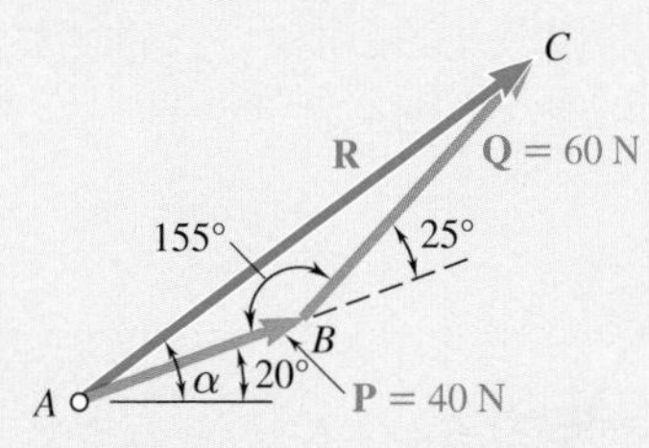

Fig. 3 Geometry of triangle rule applied to add forces **P** and **Q**.

Now apply the law of sines:

$$\frac{\sin A}{Q} = \frac{\sin B}{R} \qquad \frac{\sin A}{60 \text{ N}} = \frac{\sin 155°}{97.73 \text{ N}} \tag{1}$$

Solving Eq. (1) for sin A, you obtain

$$\sin A = \frac{(60 \text{ N}) \sin 155°}{97.73 \text{ N}}$$

Using a calculator, compute this quotient, and then obtain its arc sine:

$$A = 15.04° \qquad \alpha = 20° + A = 35.04°$$

Use three significant figures to record the answer (cf. Sec. 1.6):

$$\mathbf{R} = 97.7 \text{ N} \measuredangle 35.0° \quad ◀$$

Alternative Trigonometric Solution. Construct the right triangle BCD (Fig. 4) and compute

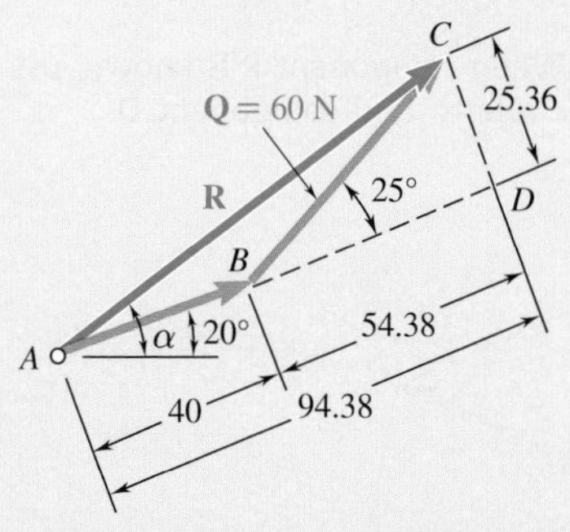

Fig. 4 Alternative geometry of triangle rule applied to add forces **P** and **Q**.

$$CD = (60 \text{ N}) \sin 25° = 25.36 \text{ N}$$
$$BD = (60 \text{ N}) \cos 25° = 54.38 \text{ N}$$

(continued)

Then, using triangle ACD, you have

$$\tan A = \frac{25.36\text{ N}}{94.38\text{ N}} \qquad A = 15.04^\circ$$

$$R = \frac{25.36}{\sin A} \qquad R = 97.73\text{ N}$$

Again,

$$\alpha = 20^\circ + A = 35.04^\circ \quad \mathbf{R} = 97.7\text{ N} \measuredangle 35.0^\circ \blacktriangleleft$$

REFLECT and THINK: An analytical solution using trigonometry provides for greater accuracy. However, it is helpful to use a graphical solution as a check.

Sample Problem 2.2

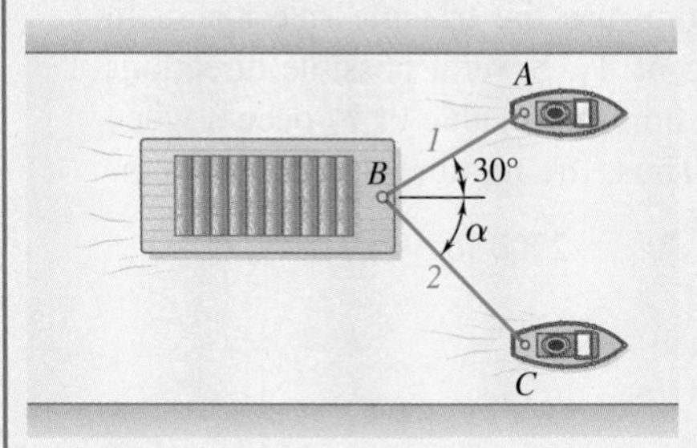

Two tugboats are pulling a barge. If the resultant of the forces exerted by the tugboats is a 5000-lb force directed along the axis of the barge, determine (*a*) the tension in each of the ropes, given that $\alpha = 45^\circ$, (*b*) the value of α for which the tension in rope *2* is minimum.

STRATEGY: This is a problem of two coplanar forces. You can solve the first part either graphically or analytically. In the second part, a graphical approach readily shows the necessary direction for rope 2, and you can use an analytical approach to complete the solution.

MODELING: You can use the parallelogram law or the triangle rule to solve part (a). For part (b), use a variation of the triangle rule.

ANALYSIS: a. Tension for $\alpha = 45^\circ$.

Graphical Solution. Use the parallelogram law. The resultant (the diagonal of the parallelogram) is equal to 5000 lb and is directed to the right. Draw the sides parallel to the ropes (Fig. 1). If the drawing is done to scale, you should measure

$$T_1 = 3700\text{ lb} \qquad T_2 = 2600\text{ lb} \blacktriangleleft$$

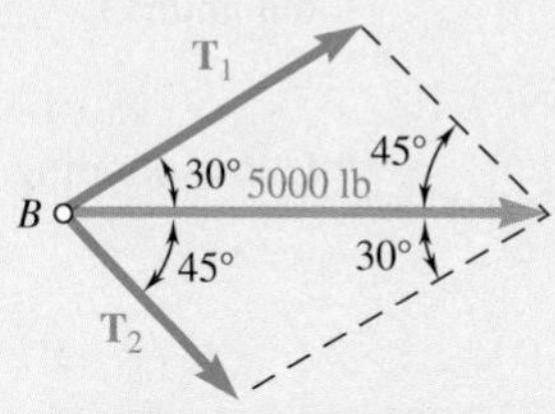

Fig. 1 Parallelogram law applied to add forces $\mathbf{T}_1$ and $\mathbf{T}_2$.

Trigonometric Solution. Use the triangle rule. Note that the triangle in Fig. 2 represents half of the parallelogram shown in Fig. 1. Using the law of sines,

$$\frac{T_1}{\sin 45°} = \frac{T_2}{\sin 30°} = \frac{5000 \text{ lb}}{\sin 105°}$$

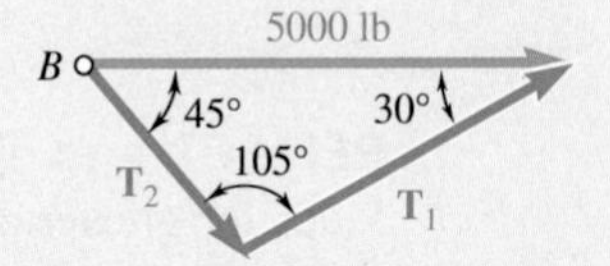

Fig. 2 Triangle rule applied to add forces $\mathbf{T}_1$ and $\mathbf{T}_2$.

With a calculator, compute and store the value of the last quotient. Multiply this value successively by sin 45° and sin 30°, obtaining

$$T_1 = 3660 \text{ lb} \qquad T_2 = 2590 \text{ lb} \blacktriangleleft$$

b. Value of α for Minimum T_2. To determine the value of α for which the tension in rope *2* is minimum, use the triangle rule again. In Fig. 3, line *1-1'* is the known direction of $\mathbf{T}_1$. Several possible directions of $\mathbf{T}_2$ are shown by the lines *2-2'*. The minimum value of T_2 occurs when $\mathbf{T}_1$ and $\mathbf{T}_2$ are perpendicular (Fig. 4). Thus, the minimum value of T_2 is

$$T_2 = (5000 \text{ lb}) \sin 30° = 2500 \text{ lb}$$

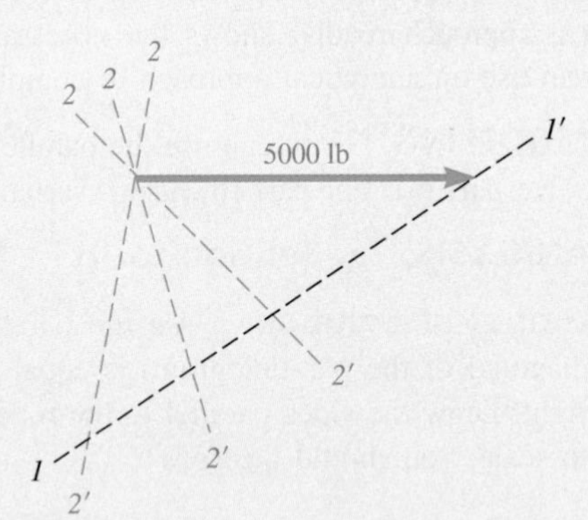

Fig. 3 Determination of direction of minimum $\mathbf{T}_2$.

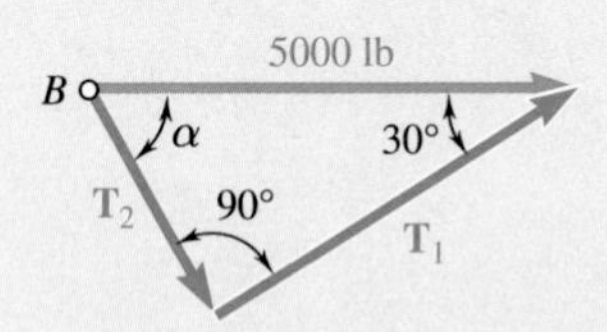

Fig. 4 Triangle rule applied for minimum $\mathbf{T}_2$.

Corresponding values of T_1 and α are

$$T_1 = (5000 \text{ lb}) \cos 30° = 4330 \text{ lb}$$

$$\alpha = 90° - 30° \qquad \alpha = 60° \blacktriangleleft$$

REFLECT and THINK: Part (a) is a straightforward application of resolving a vector into components. The key to part (b) is recognizing that the minimum value of T_2 occurs when $\mathbf{T}_1$ and $\mathbf{T}_2$ are perpendicular.

Problems

2.1 and 2.2 Determine graphically the magnitude and direction of the resultant of the two forces shown using (*a*) the parallelogram law, (*b*) the triangle rule.

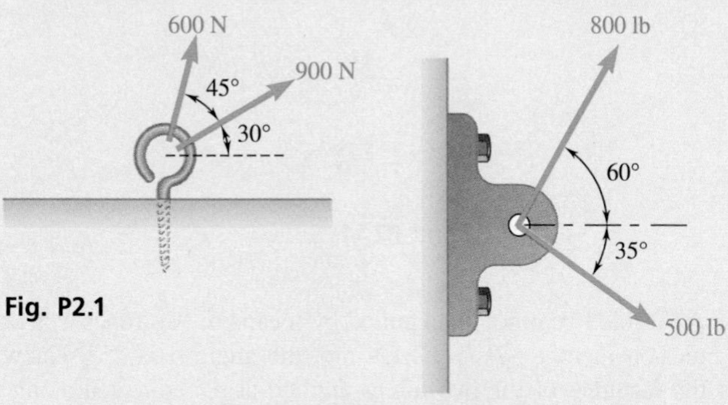

Fig. P2.1

Fig. P2.2

2.3 Two structural members *B* and *C* are bolted to bracket *A*. Knowing that both members are in tension and that $P = 10$ kN and $Q = 15$ kN, determine graphically the magnitude and direction of the resultant force exerted on the bracket using (*a*) the parallelogram law, (*b*) the triangle rule.

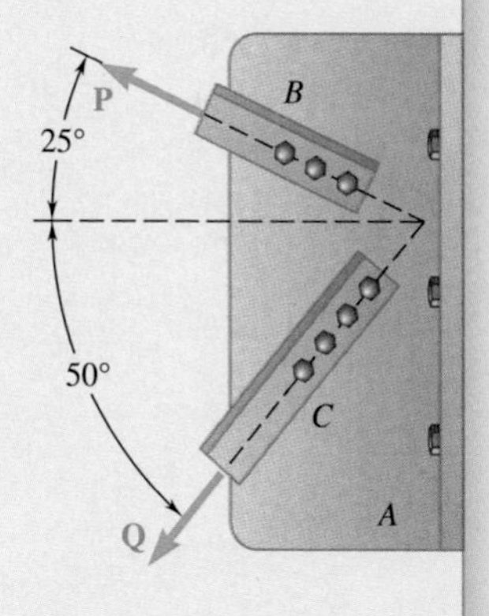

Fig. P2.3 and P2.4

2.4 Two structural members *B* and *C* are bolted to bracket *A*. Knowing that both members are in tension and that $P = 6$ kips and $Q = 4$ kips, determine graphically the magnitude and direction of the resultant force exerted on the bracket using (*a*) the parallelogram law, (*b*) the triangle rule.

2.5 The 300-lb force is to be resolved into components along lines *a–a′* and *b–b′*. (a) Determine the angle α by trigonometry, knowing that the component along line *a–a′* is to be 240 lb. (*b*) What is the corresponding value of the component along *b–b′*?

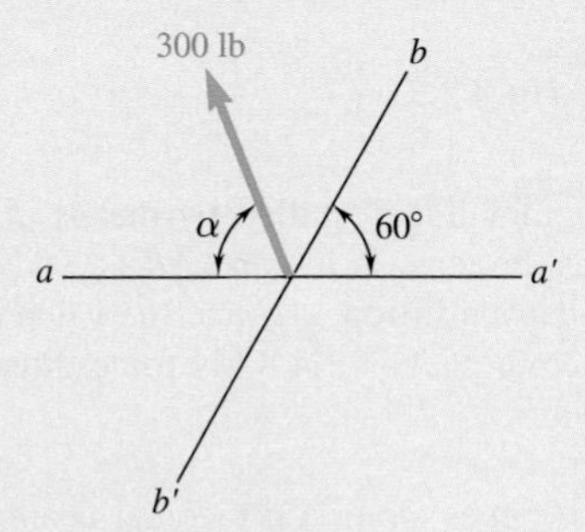

Fig. P2.5 and *P2.6*

2.6 The 300-lb force is to be resolved into components along lines *a–a′* and *b–b′*. (a) Determine the angle α by trigonometry knowing that the component along line *b–b′* is to be 120 lb. (*b*) What is the corresponding value of the component along *a–a′*?

2.7 A trolley that moves along a horizontal beam is acted upon by two forces as shown. (*a*) Knowing that $\alpha = 25°$, determine by trigonometry the magnitude of the force **P** so that the resultant force exerted on the trolley is vertical. (*b*) What is the corresponding magnitude of the resultant?

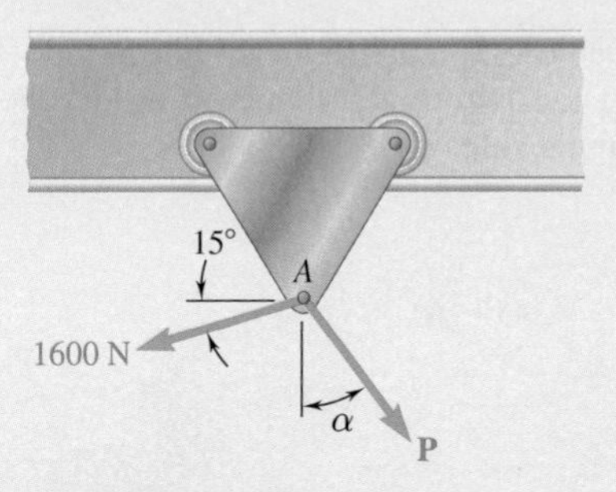

Fig. P2.7 and P2.11

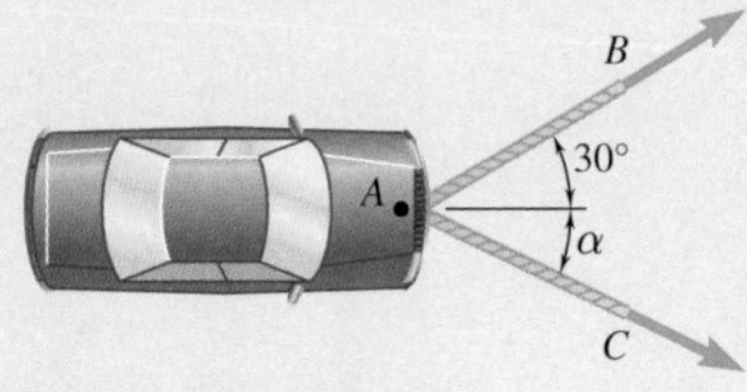

Fig. *P2.8* and P2.10

2.8 A disabled automobile is pulled by means of two ropes as shown. The tension in rope *AB* is 2.2 kN and the angle α is 25°. Knowing that the resultant of the two forces applied at *A* is directed along the axis of the automobile, determine by trigonometry (*a*) the tension in rope *AC*, (*b*) the magnitude of the resultant of the two forces applied at *A*.

2.9 Two forces are applied as shown to a hook support. Knowing that the magnitude of **P** is 35 N, determine by trigonometry (*a*) the required angle α if the resultant **R** of the two forces applied to the support is to be horizontal, (*b*) the corresponding magnitude of **R**.

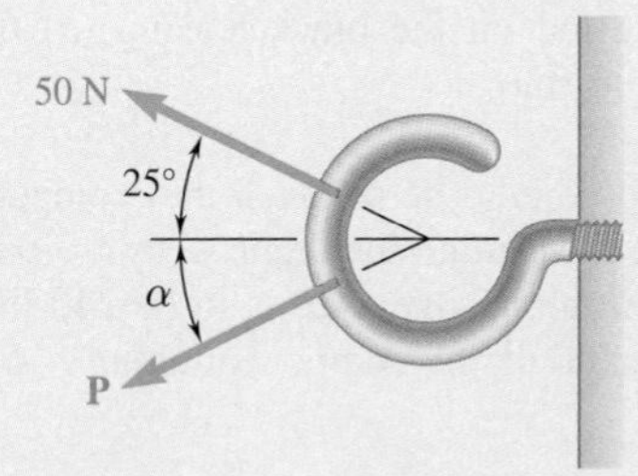

Fig. P2.9

2.10 A disabled automobile is pulled by means of two ropes as shown. Knowing that the tension in rope *AB* is 3 kN, determine by trigonometry the tension in rope *AC* and the value of α so that the resultant force exerted at *A* is a 4.8-kN force directed along the axis of the automobile.

2.11 A trolley that moves along a horizontal beam is acted upon by two forces as shown. Determine by trigonometry the magnitude and direction of the force **P** so that the resultant is a vertical force of 2500 N.

2.12 For the hook support shown, determine by trigonometry the magnitude and direction of the resultant of the two forces applied to the hook.

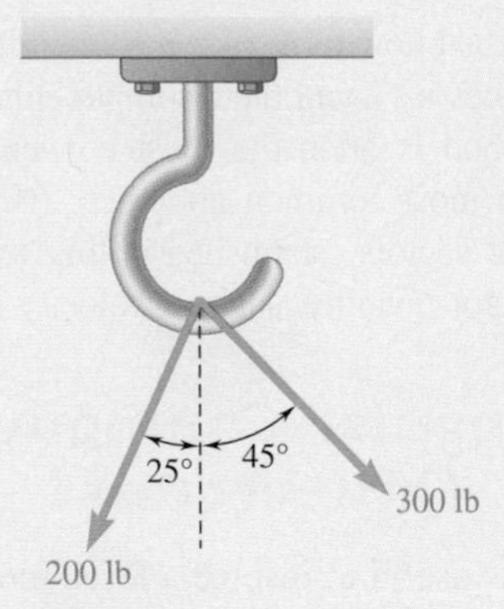

Fig. P2.12

2.13 The cable stays *AB* and *AD* help support pole *AC*. Knowing that the tension is 120 lb in *AB* and 40 lb in *AD*, determine by trigonometry the magnitude and direction of the resultant of the forces exerted by the stays at *A*.

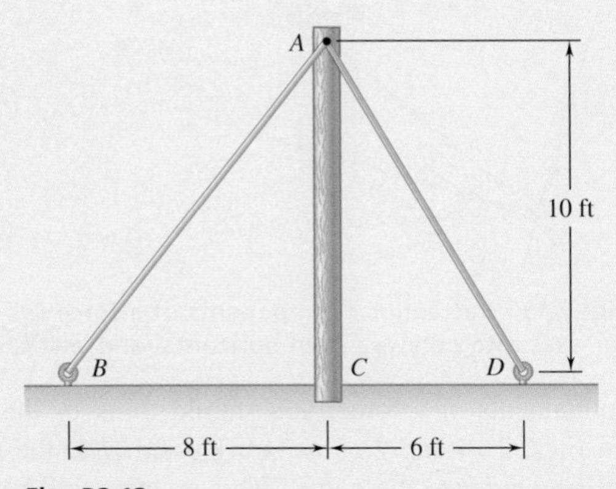

Fig. P2.13

2.14 Solve Prob. 2.4 by trigonometry.

2.15 For the hook support of Prob. 2.9, determine by trigonometry (*a*) the magnitude and direction of the smallest force **P** for which the resultant **R** of the two forces applied to the support is horizontal, (*b*) the corresponding magnitude of **R**.

2.2 ADDING FORCES BY COMPONENTS

In Sec. 2.1E, we described how to resolve a force into components. Here we discuss how to add forces by using their components, especially rectangular components. This method is often the most convenient way to add forces and, in practice, is the most common approach. (Note that we can readily extend the properties of vectors established in this section to the rectangular components of any vector quantity, such as velocity or momentum.)

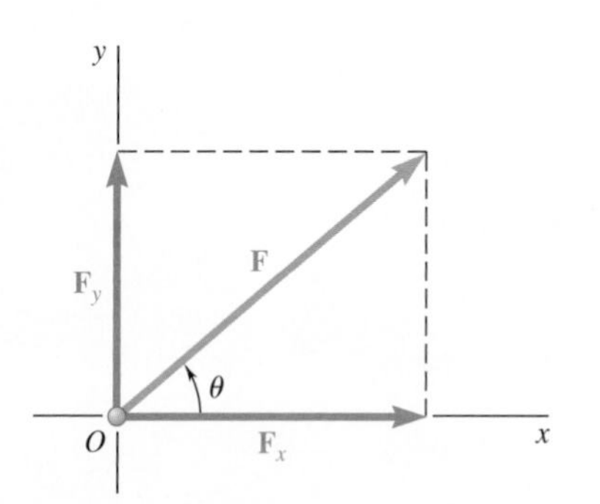

Fig. 2.14 Rectangular components of a force **F**.

2.2A Rectangular Components of a Force: Unit Vectors

In many problems, it is useful to resolve a force into two components that are perpendicular to each other. Figure 2.14 shows a force **F** resolved into a component $\mathbf{F}_x$ along the x axis and a component $\mathbf{F}_y$ along the y axis. The parallelogram drawn to obtain the two components is a rectangle, and $\mathbf{F}_x$ and $\mathbf{F}_y$ are called **rectangular components**.

The x and y axes are usually chosen to be horizontal and vertical, respectively, as in Fig. 2.14; they may, however, be chosen in any two perpendicular directions, as shown in Fig. 2.15. In determining the rectangular components of a force, you should think of the construction lines shown in Figs. 2.14 and 2.15 as being *parallel* to the x and y axes, rather than *perpendicular* to these axes. This practice will help avoid mistakes in determining *oblique* components, as in Sec. 2.1E.

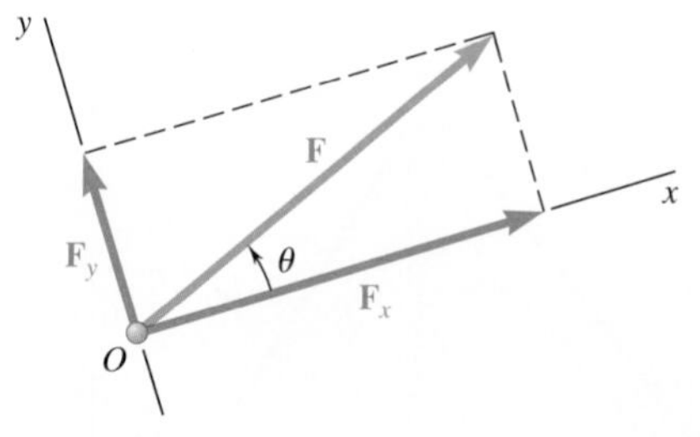

Fig. 2.15 Rectangular components of a force **F** for axes rotated away from horizontal and vertical.

Force in Terms of Unit Vectors. To simplify working with rectangular components, we introduce two vectors of unit magnitude, directed respectively along the positive x and y axes. These vectors are called **unit vectors** and are denoted by **i** and **j**, respectively (Fig. 2.16). Recalling the

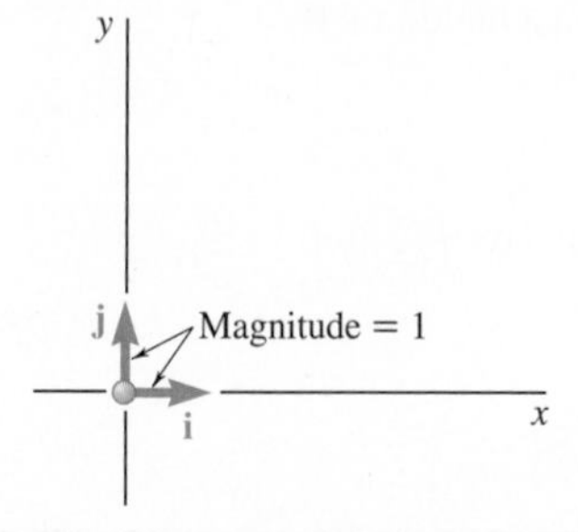

Fig. 2.16 Unit vectors along the x and y axes.

definition of the product of a scalar and a vector given in Sec. 2.1C, note that we can obtain the rectangular components $\mathbf{F}_x$ and $\mathbf{F}_y$ of a force $\mathbf{F}$ by multiplying respectively the unit vectors $\mathbf{i}$ and $\mathbf{j}$ by appropriate scalars (Fig. 2.17). We have

$$\mathbf{F}_x = F_x\mathbf{i} \qquad \mathbf{F}_y = F_y\mathbf{j} \tag{2.6}$$

and

$$\mathbf{F} = F_x\mathbf{i} + F_y\mathbf{j} \tag{2.7}$$

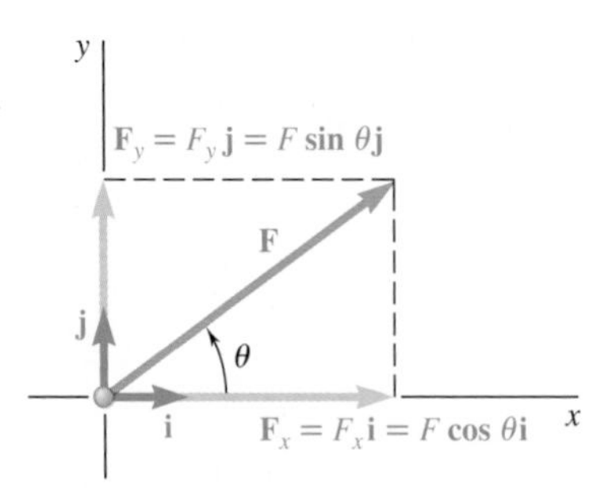

Fig. 2.17 Expressing the components of **F** in terms of unit vectors with scalar multipliers.

The scalars F_x and F_y may be positive or negative, depending upon the sense of $\mathbf{F}_x$ and of $\mathbf{F}_y$, but their absolute values are equal to the magnitudes of the component forces $\mathbf{F}_x$ and $\mathbf{F}_y$, respectively. The scalars F_x and F_y are called the **scalar components** of the force $\mathbf{F}$, whereas the actual component forces $\mathbf{F}_x$ and $\mathbf{F}_y$ should be referred to as the **vector components** of $\mathbf{F}$. However, when there exists no possibility of confusion, we may refer to the vector as well as the scalar components of $\mathbf{F}$ as simply the **components** of $\mathbf{F}$. Note that the scalar component F_x is positive when the vector component $\mathbf{F}_x$ has the same sense as the unit vector $\mathbf{i}$ (i.e., the same sense as the positive x axis) and is negative when $\mathbf{F}_x$ has the opposite sense. A similar conclusion holds for the sign of the scalar component F_y.

Scalar Components. Denoting by F the magnitude of the force $\mathbf{F}$ and by θ the angle between $\mathbf{F}$ and the x axis, which is measured counterclockwise from the positive x axis (Fig. 2.17), we may express the scalar components of $\mathbf{F}$ as

$$F_x = F\cos\theta \qquad F_y = F\sin\theta \tag{2.8}$$

These relations hold for any value of the angle θ from 0° to 360°, and they define the signs as well as the absolute values of the scalar components F_x and F_y.

Concept Application 2.1

A force of 800 N is exerted on a bolt A as shown in Fig. 2.18a. Determine the horizontal and vertical components of the force.

Solution

In order to obtain the correct sign for the scalar components F_x and F_y, we could substitute the value $180° - 35° = 145°$ for θ in Eqs. (2.8). However, it is often more practical to determine by inspection the signs of F_x and F_y (Fig. 2.18b) and then use the trigonometric functions of the angle $\alpha = 35°$. Therefore,

$$F_x = -F\cos\alpha = -(800\text{ N})\cos 35° = -655\text{ N}$$
$$F_y = +F\sin\alpha = +(800\text{ N})\sin 35° = +459\text{ N}$$

The vector components of $\mathbf{F}$ are thus

$$\mathbf{F}_x = -(655\text{ N})\mathbf{i} \qquad \mathbf{F}_y = +(459\text{ N})\mathbf{j}$$

and we may write $\mathbf{F}$ in the form

$$\mathbf{F} = -(655\text{ N})\mathbf{i} + (459\text{ N})\mathbf{j} \blacktriangleleft$$

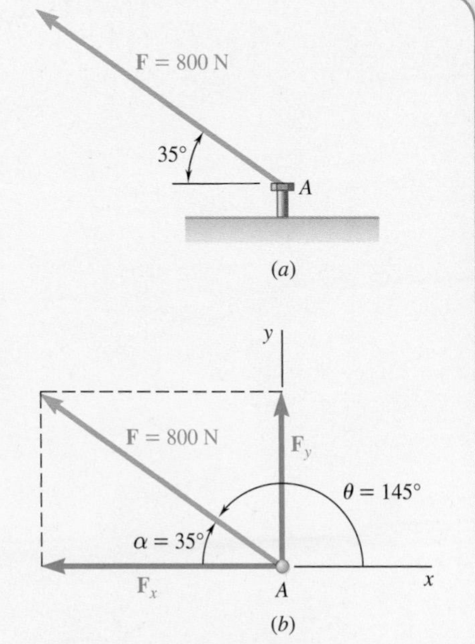

Fig. 2.18 (a) Force **F** exerted on a bolt; (b) rectangular components of **F**.

Concept Application 2.2

A man pulls with a force of 300 N on a rope attached to the top of a building, as shown in Fig. 2.19*a*. What are the horizontal and vertical components of the force exerted by the rope at point *A*?

Solution

You can see from Fig. 2.19*b* that

$$F_x = +(300 \text{ N}) \cos \alpha \qquad F_y = -(300 \text{ N}) \sin \alpha$$

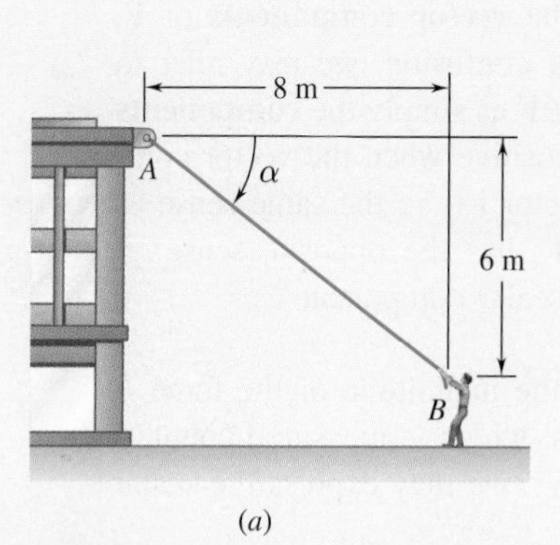

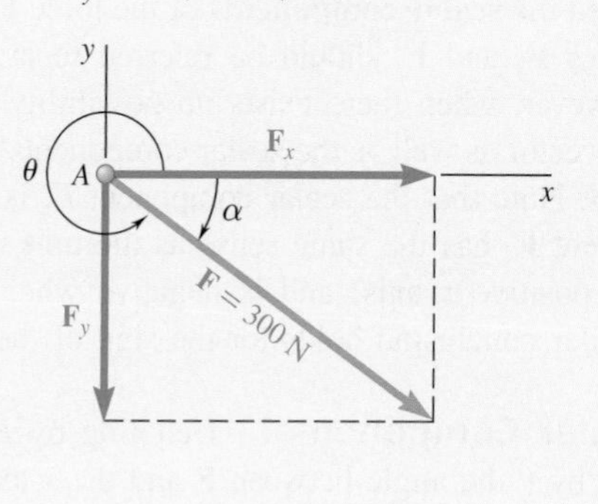

Fig. 2.19 (*a*) A man pulls on a rope attached to a building; (*b*) components of the rope's force **F**.

Observing that $AB = 10$ m, we find from Fig. 2.19*a*

$$\cos \alpha = \frac{8 \text{ m}}{AB} = \frac{8 \text{ m}}{10 \text{ m}} = \frac{4}{5} \qquad \sin \alpha = \frac{6 \text{ m}}{AB} = \frac{6 \text{ m}}{10 \text{ m}} = \frac{3}{5}$$

We thus obtain

$$F_x = +(300 \text{ N})\frac{4}{5} = +240 \text{ N} \qquad F_y = -(300 \text{ N})\frac{3}{5} = -180 \text{ N}$$

This gives us a total force of

$$\mathbf{F} = (240 \text{ N})\mathbf{i} - (180 \text{ N})\mathbf{j}$$

Direction of a Force. When a force **F** is defined by its rectangular components F_x and F_y (see Fig. 2.17), we can find the angle θ defining its direction from

$$\tan\theta = \frac{F_y}{F_x} \tag{2.9}$$

We can obtain the magnitude F of the force by applying the Pythagorean theorem,

$$F = \sqrt{F_x^2 + F_y^2} \tag{2.10}$$

or by solving for F from one of the Eqs. (2.8).

Concept Application 2.3

A force $\mathbf{F} = (700 \text{ lb})\mathbf{i} + (1500 \text{ lb})\mathbf{j}$ is applied to a bolt A. Determine the magnitude of the force and the angle θ it forms with the horizontal.

Solution

First draw a diagram showing the two rectangular components of the force and the angle θ (Fig. 2.20). From Eq. (2.9), you obtain

$$\tan\theta = \frac{F_y}{F_x} = \frac{1500 \text{ lb}}{700 \text{ lb}}$$

Using a calculator, enter 1500 lb and divide by 700 lb; computing the arc tangent of the quotient gives you $\theta = 65.0°$. Solve the second of Eqs. (2.8) for F to get

$$F = \frac{F_y}{\sin\theta} = \frac{1500 \text{ lb}}{\sin 65.0°} = 1655 \text{ lb}$$

The last calculation is easier if you store the value of F_y when originally entered; you may then recall it and divide it by $\sin\theta$.

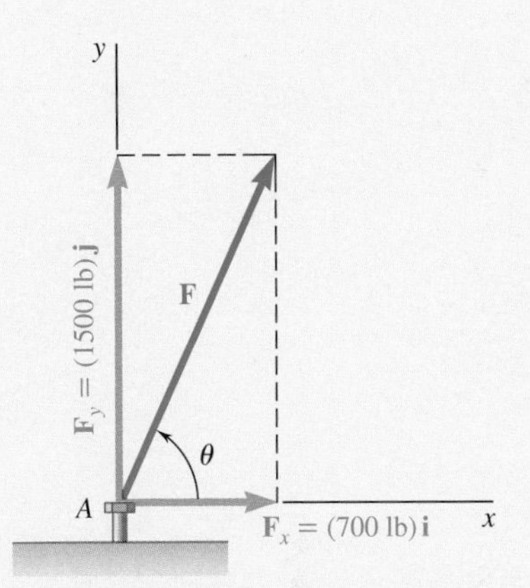

Fig. 2.20 Components of a force **F** exerted on a bolt.

2.2B Addition of Forces by Summing X and Y Components

We described in Sec. 2.1A how to add forces according to the parallelogram law. From this law, we derived two other methods that are more readily applicable to the graphical solution of problems: the triangle rule for the addition of two forces and the polygon rule for the addition of three or more forces. We also explained that the force triangle used to define the resultant of two forces could be used to obtain a trigonometric solution.

However, when we need to add three or more forces, we cannot obtain any practical trigonometric solution from the force polygon that defines the resultant of the forces. In this case, the best approach is to obtain an analytic solution of the problem by resolving each force into two rectangular components.

Consider, for instance, three forces **P**, **Q**, and **S** acting on a particle A (Fig. 2.21a). Their resultant **R** is defined by the relation

$$\mathbf{R} = \mathbf{P} + \mathbf{Q} + \mathbf{S} \qquad (2.11)$$

Resolving each force into its rectangular components, we have

$$\begin{aligned} R_x\mathbf{i} + R_y\mathbf{j} &= P_x\mathbf{i} + P_y\mathbf{j} + Q_x\mathbf{i} + Q_y\mathbf{j} + S_x\mathbf{i} + S_y\mathbf{j} \\ &= (P_x + Q_x + S_x)\mathbf{i} + (P_y + Q_y + S_y)\mathbf{j} \end{aligned}$$

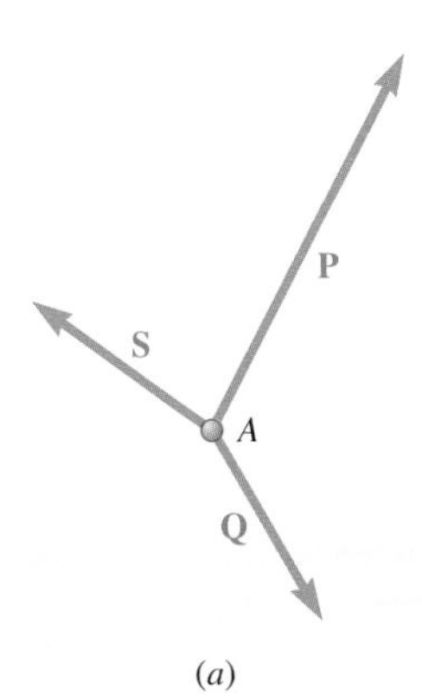

Fig. 2.21 (a) Three forces acting on a particle.

From this equation, we can see that

$$R_x = P_x + Q_x + S_x \qquad R_y = P_y + Q_y + S_y \tag{2.12}$$

or for short,

$$R_x = \Sigma F_x \qquad R_y = \Sigma F_y \tag{2.13}$$

We thus conclude that **when several forces are acting on a particle, we obtain the scalar components R_x and R_y of the resultant R by adding algebraically the corresponding scalar components of the given forces.** *(Clearly, this result also applies to the addition of other vector quantities, such as velocities, accelerations, or momenta.)*

In practice, determining the resultant **R** is carried out in three steps, as illustrated in Fig. 2.21.

1. Resolve the given forces (Fig. 2.21*a*) into their *x* and *y* components (Fig. 2.21*b*).

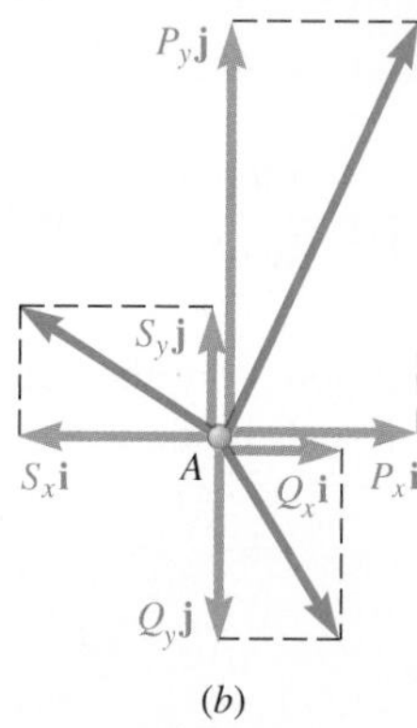

Fig. 2.21 (*b*) Rectangular components of each force.

2. Add these components to obtain the *x* and *y* components of **R** (Fig. 2.21*c*).

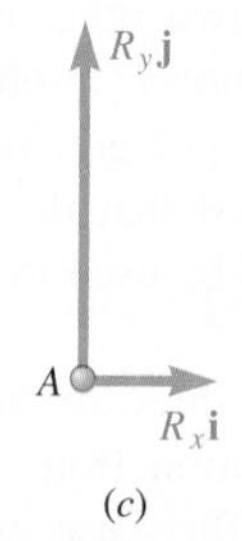

Fig. 2.21 (*c*) Summation of the components.

3. Apply the parallelogram law to determine the resultant $\mathbf{R} = R_x\mathbf{i} + R_y\mathbf{j}$ (Fig. 2.21*d*).

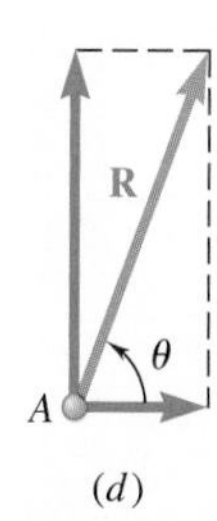

Fig. 2.21 (*d*) Determining the resultant from its components.

The procedure just described is most efficiently carried out if you arrange the computations in a table (see Sample Problem 2.3). Although this is the only practical analytic method for adding three or more forces, it is also often preferred to the trigonometric solution in the case of adding two forces.

Sample Problem 2.3

Four forces act on bolt A as shown. Determine the resultant of the forces on the bolt.

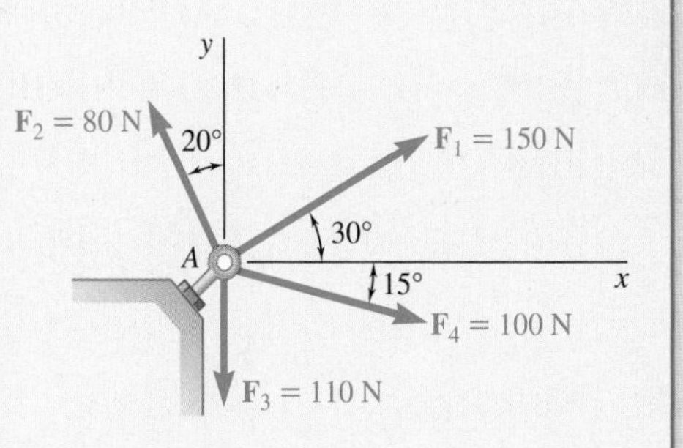

STRATEGY: The simplest way to approach a problem of adding four forces is to resolve the forces into components.

MODELING: As we mentioned, solving this kind of problem is usually easier if you arrange the components of each force in a table. In the table below, we entered the x and y components of each force as determined by trigonometry (Fig. 1). According to the convention adopted in this section, the scalar number representing a force component is positive if the force component has the same sense as the corresponding coordinate axis. Thus, x components acting to the right and y components acting upward are represented by positive numbers.

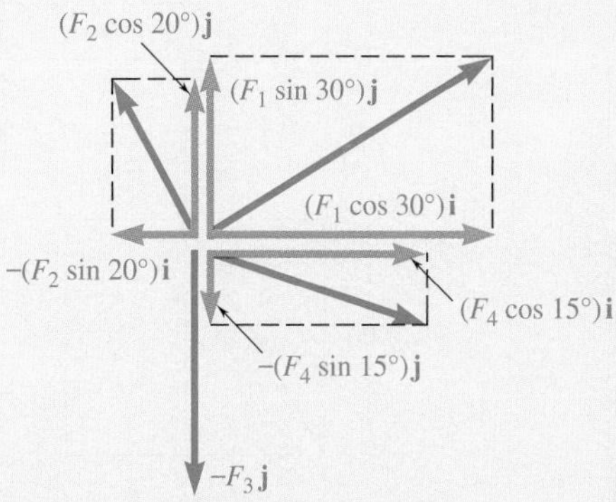

Fig. 1 Rectangular components of each force.

ANALYSIS:

Force	Magnitude, N	x Component, N	y Component, N
$\mathbf{F}_1$	150	+129.9	+75.0
$\mathbf{F}_2$	80	−27.4	+75.2
$\mathbf{F}_3$	110	0	−110.0
$\mathbf{F}_4$	100	+96.6	−25.9
		$R_x = +199.1$	$R_y = +14.3$

Thus, the resultant $\mathbf{R}$ of the four forces is

$$\mathbf{R} = R_x\mathbf{i} + R_y\mathbf{j} \qquad \mathbf{R} = (199.1 \text{ N})\mathbf{i} + (14.3 \text{ N})\mathbf{j} \quad ◀$$

You can now determine the magnitude and direction of the resultant. From the triangle shown in Fig. 2, you have

$$\tan\alpha = \frac{R_y}{R_x} = \frac{14.3 \text{ N}}{199.1 \text{ N}} \qquad \alpha = 4.1°$$

$$R = \frac{14.3 \text{ N}}{\sin\alpha} = 199.6 \text{ N} \qquad \mathbf{R} = 199.6 \text{ N} \measuredangle 4.1° \quad ◀$$

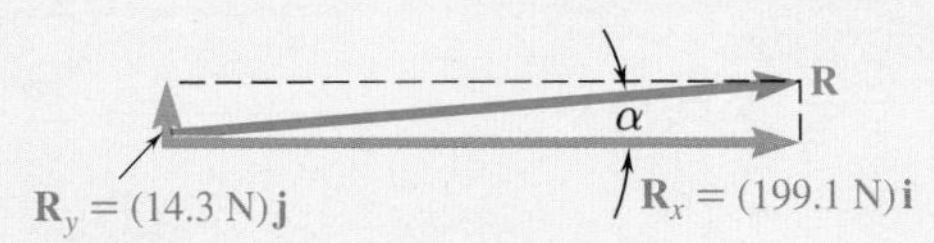

Fig. 2 Resultant of the given force system.

REFLECT and THINK: Arranging data in a table not only helps you keep track of the calculations, but also makes things simpler for using a calculator on similar computations.

Problems

2.16 and 2.17 Determine the x and y components of each of the forces shown.

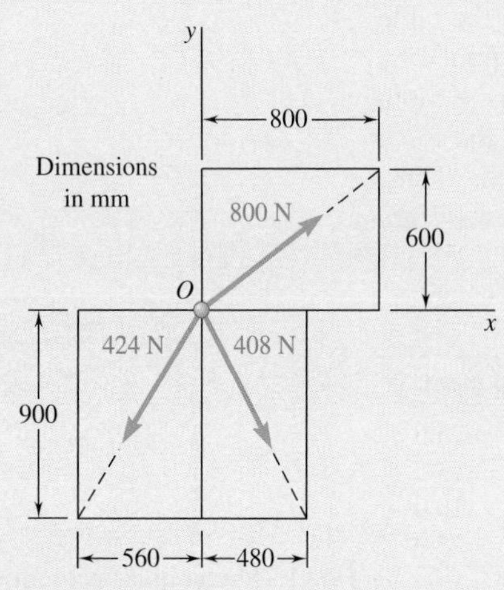

Fig. P2.16

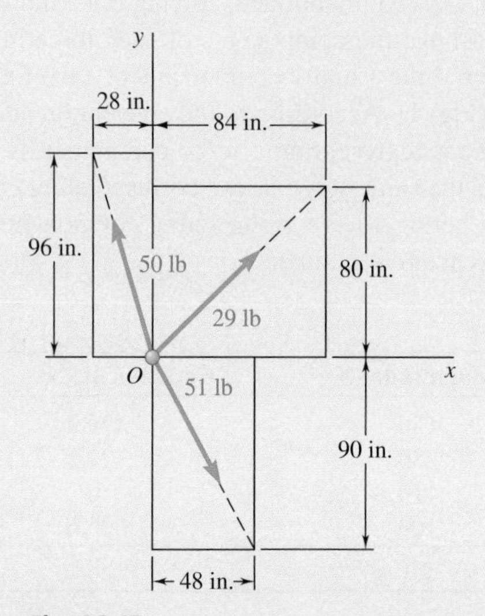

Fig. P2.17

2.18 and 2.19 Determine the x and y components of each of the forces shown.

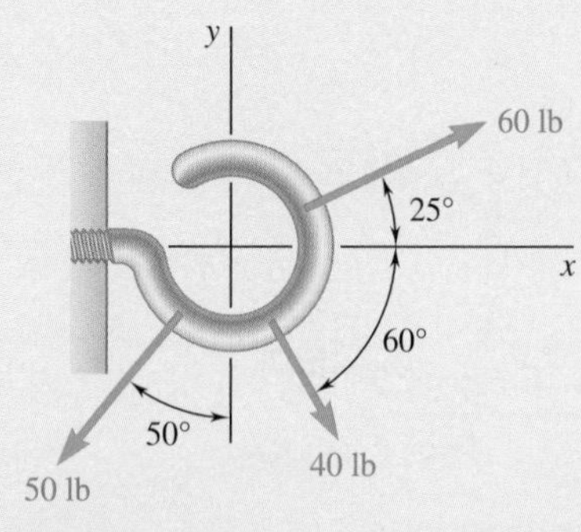

Fig. P2.18

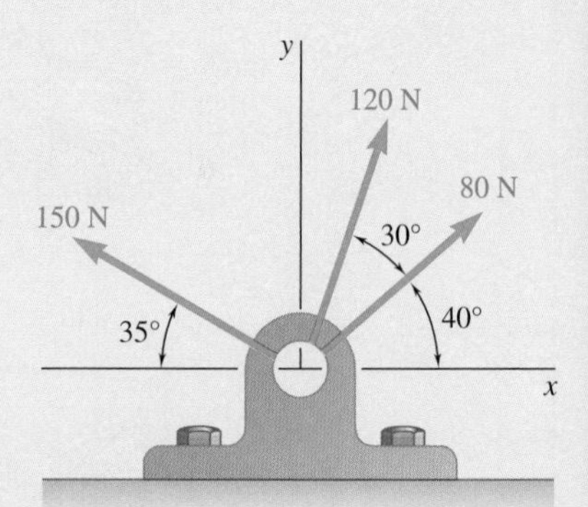

Fig. P2.19

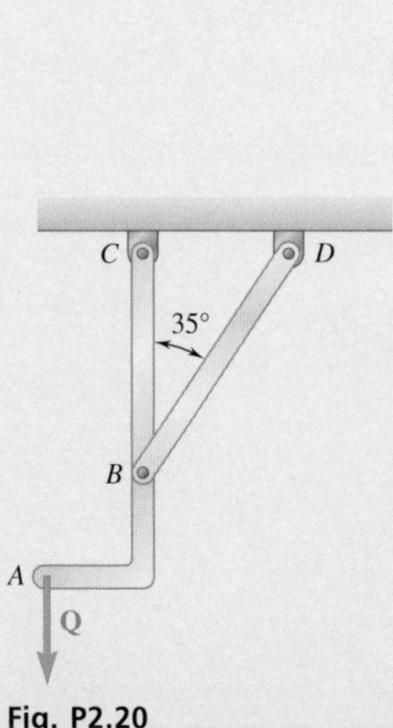

Fig. P2.20

2.20 Member BD exerts on member ABC a force **P** directed along line BD. Knowing that **P** must have a 300-lb horizontal component, determine (a) the magnitude of the force **P**, (b) its vertical component.

2.21 Member BC exerts on member AC a force **P** directed along line BC. Knowing that **P** must have a 325-N horizontal component, determine (*a*) the magnitude of the force **P**, (*b*) its vertical component.

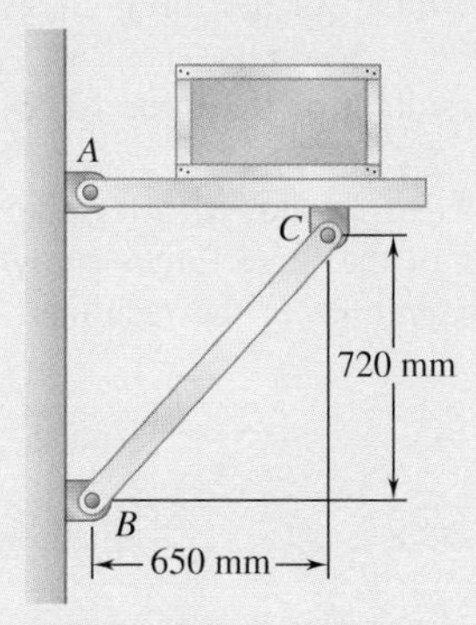

Fig. *P2.21*

2.22 Cable AC exerts on beam AB a force **P** directed along line AC. Knowing that **P** must have a 350-lb vertical component, determine (*a*) the magnitude of the force **P**, (*b*) its horizontal component.

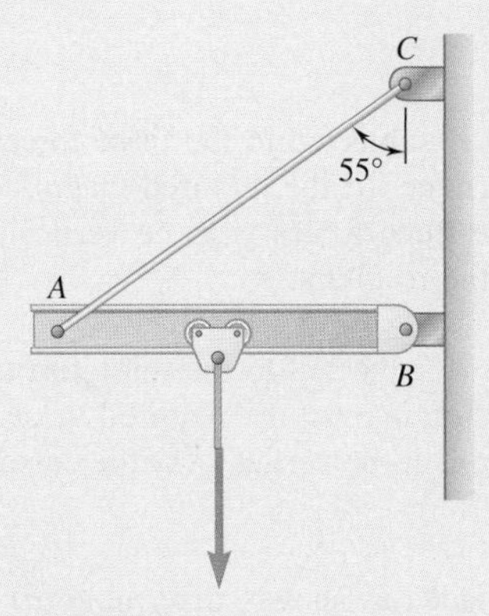

Fig. *P2.22*

2.23 The hydraulic cylinder BD exerts on member ABC a force **P** directed along line BD. Knowing that **P** must have a 750-N component perpendicular to member ABC, determine (*a*) the magnitude of the force **P**, (*b*) its component parallel to ABC.

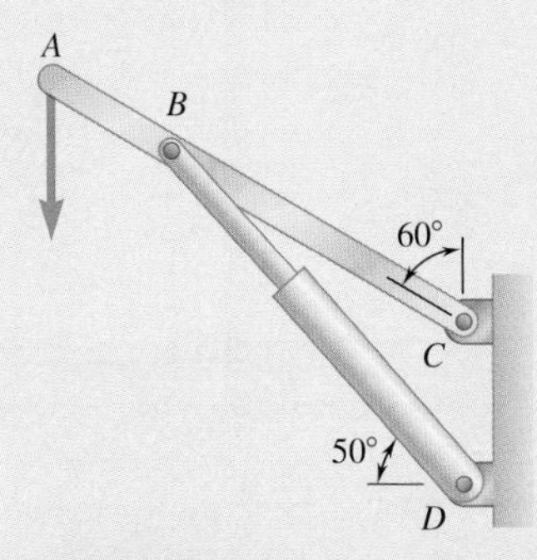

Fig. P2.23

2.24 Determine the resultant of the three forces of Prob. 2.16.

2.25 Determine the resultant of the three forces of Prob. 2.17.

2.26 Determine the resultant of the three forces of Prob. 2.18.

2.27 Determine the resultant of the three forces of Prob. 2.19.

2.28 For the collar loaded as shown, determine (*a*) the required value of α if the resultant of the three forces shown is to be vertical, (*b*) the corresponding magnitude of the resultant.

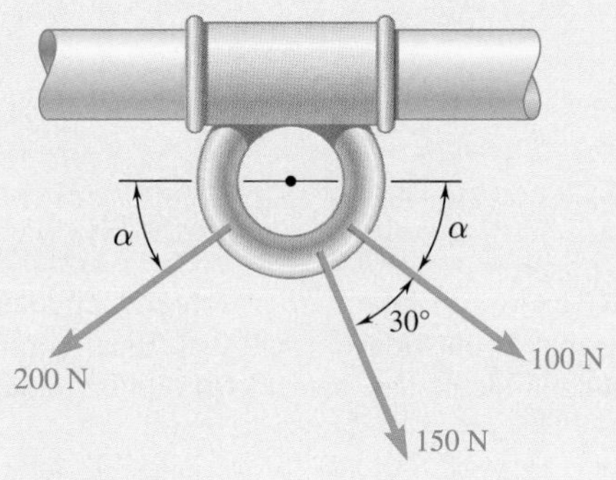

Fig. ***P2.28***

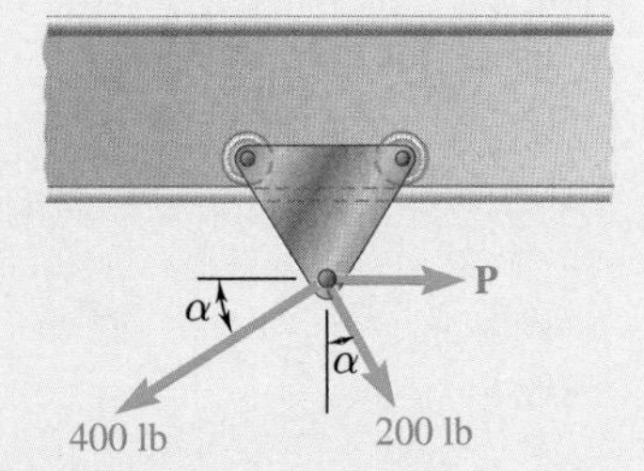

Fig. P2.29 and ***P2.30***

2.29 A hoist trolley is subjected to the three forces shown. Knowing that $\alpha = 40°$, determine (*a*) the required magnitude of the force **P** if the resultant of the three forces is to be vertical, (*b*) the corresponding magnitude of the resultant.

2.30 A hoist trolley is subjected to the three forces shown. Knowing that $P = 250$ lb, determine (*a*) the required value of α if the resultant of the three forces is to be vertical, (*b*) the corresponding magnitude of the resultant.

2.31 For the post loaded as shown, determine (*a*) the required tension in rope *AC* if the resultant of the three forces exerted at point *C* is to be horizontal, (*b*) the corresponding magnitude of the resultant.

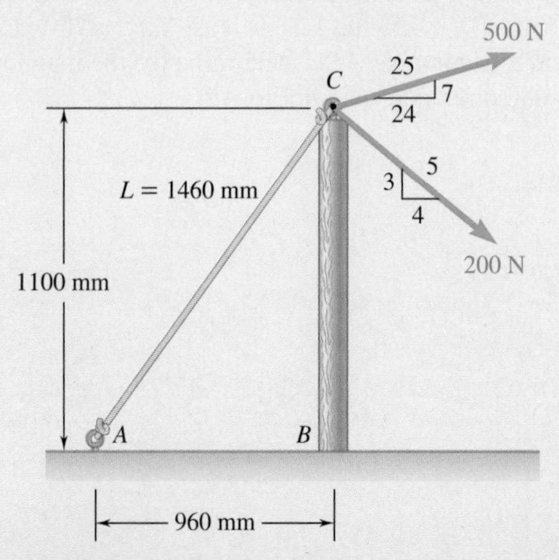

Fig. P2.31

2.3 FORCES AND EQUILIBRIUM IN A PLANE

Now that we have seen how to add forces, we can proceed to one of the key concepts in this course: the equilibrium of a particle. The connection between equilibrium and the sum of forces is very direct: a particle can be in equilibrium only when the sum of the forces acting on it is zero.

2.3A Equilibrium of a Particle

In the preceding sections, we discussed methods for determining the resultant of several forces acting on a particle. Although it has not occurred in any of the problems considered so far, it is quite possible for the resultant to be zero. In such a case, the net effect of the given forces is zero, and the particle is said to be in **equilibrium**. We thus have the definition:

When the resultant of all the forces acting on a particle is zero, the particle is in equilibrium.

A particle acted upon by two forces is in equilibrium if the two forces have the same magnitude and the same line of action but opposite sense. The resultant of the two forces is then zero, as shown in Fig. 2.22.

Another case of equilibrium of a particle is represented in Fig. 2.23*a*, where four forces are shown acting on particle *A*. In Fig. 2.23*b*, we use the polygon rule to determine the resultant of the given forces. Starting from point *O* with $\mathbf{F}_1$ and arranging the forces in tip-to-tail fashion, we find that the tip of $\mathbf{F}_4$ coincides with the starting point *O*. Thus, the resultant **R** of the given system of forces is zero, and the particle is in equilibrium.

Photo 2.2 Forces acting on the carabiner include the weight of the girl and her harness, and the force exerted by the pulley attachment. Treating the carabiner as a particle, it is in equilibrium because the resultant of all forces acting on it is zero.

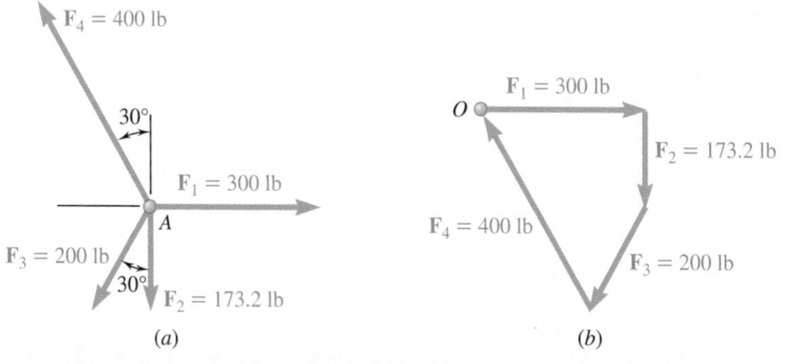

Fig. 2.23 (*a*) Four forces acting on particle *A*; (*b*) using the polygon law to find the resultant of the forces in (*a*), which is zero because the particle is in equilibrium.

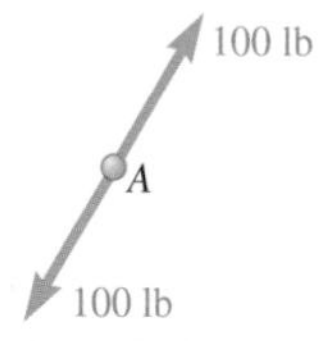

Fig. 2.22 When a particle is in equilibrium, the resultant of all forces acting on the particle is zero.

The closed polygon drawn in Fig. 2.23*b* provides a *graphical* expression of the equilibrium of *A*. To express *algebraically* the conditions for the equilibrium of a particle, we write

Equilibrium of a particle $$\mathbf{R} = \Sigma\mathbf{F} = 0 \qquad (2.14)$$

Resolving each force **F** into rectangular components, we have

$$\Sigma (F_x\mathbf{i} + F_y\mathbf{j}) = 0 \qquad \text{or} \qquad (\Sigma F_x)\mathbf{i} + (\Sigma F_y)\mathbf{j} = 0$$

We conclude that the necessary and sufficient conditions for the equilibrium of a particle are

Equilibrium of a particle (scalar equations)

$$\Sigma F_x = 0 \qquad \Sigma F_y = 0 \tag{2.15}$$

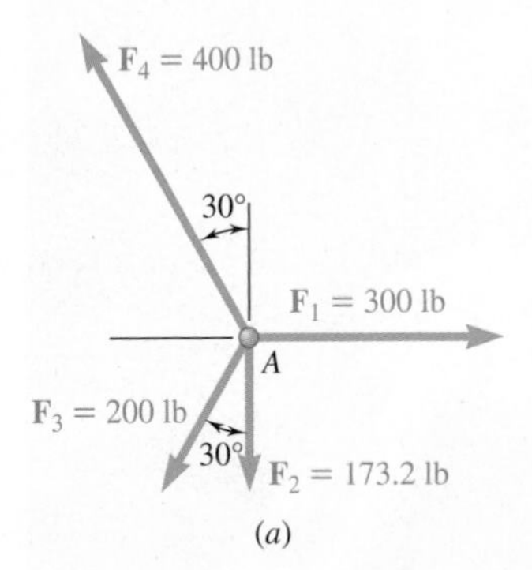

Fig. 2.23(a) (repeated)

Returning to the particle shown in Fig. 2.23, we can check that the equilibrium conditions are satisfied. We have

$$\begin{aligned}\Sigma F_x &= 300 \text{ lb} - (200 \text{ lb}) \sin 30° - (400 \text{ lb}) \sin 30° \\ &= 300 \text{ lb} - 100 \text{ lb} - 200 \text{ lb} = 0 \\ \Sigma F_y &= -173.2 \text{ lb} - (200 \text{ lb}) \cos 30° + (400 \text{ lb}) \cos 30° \\ &= -173.2 \text{ lb} - 173.2 \text{ lb} + 346.4 \text{ lb} = 0\end{aligned}$$

2.3B Newton's First Law of Motion

As we discussed in Section 1.2, Sir Isaac Newton formulated three fundamental laws upon which the science of mechanics is based. The first of these laws can be stated as:

If the resultant force acting on a particle is zero, the particle will remain at rest (if originally at rest) or will move with constant speed in a straight line (if originally in motion).

From this law and from the definition of equilibrium just presented, we can see that a particle in equilibrium is either at rest or moving in a straight line with constant speed. If a particle does not behave in either of these ways, it is not in equilibrium, and the resultant force on it is not zero. In the following section, we consider various problems concerning the equilibrium of a particle.

Note that most of statics involves using Newton's first law to analyze an equilibrium situation. In practice, this means designing a bridge or a building that remains stable and does not fall over. It also means understanding the forces that might act to disturb equilibrium, such as a strong wind or a flood of water. The basic idea is pretty simple, but the applications can be quite complicated.

2.3C Free-Body Diagrams and Problem Solving

In practice, a problem in engineering mechanics is derived from an actual physical situation. A sketch showing the physical conditions of the problem is known as a **space diagram**.

The methods of analysis discussed in the preceding sections apply to a system of forces acting on a particle. A large number of problems involving actual structures, however, can be reduced to problems concerning the equilibrium of a particle. The method is to choose a significant particle and draw a separate diagram showing this particle and all the forces acting on it. Such a diagram is called a **free-body diagram**. (The name derives from the fact that when drawing the chosen body, or particle, it is "free" from all other bodies in the actual situation.)

As an example, consider the 75-kg crate shown in the space diagram of Fig. 2.24*a*. This crate was lying between two buildings, and is now being lifted onto a truck, which will remove it. The crate is supported by a vertical cable that is joined at *A* to two ropes, which pass over pulleys attached to the buildings at *B* and *C*. We want to determine the tension in each of the ropes *AB* and *AC*.

In order to solve this problem, we first draw a free-body diagram showing a particle in equilibrium. Since we are interested in the rope tensions, the free-body diagram should include at least one of these tensions or, if possible, both tensions. You can see that point *A* is a good free body for this problem. The free-body diagram of point *A* is shown in Fig. 2.24*b*. It shows point *A* and the forces exerted on *A* by the vertical cable and the two ropes. The force exerted by the cable is directed downward, and its magnitude is equal to the weight *W* of the crate. Recalling Eq. (1.4), we write

$$W = mg = (75 \text{ kg})(9.81 \text{ m/s}^2) = 736 \text{ N}$$

and indicate this value in the free-body diagram. The forces exerted by the two ropes are not known. Since they are respectively equal in magnitude to the tensions in rope *AB* and rope *AC*, we denote them by $\mathbf{T}_{AB}$ and $\mathbf{T}_{AC}$ and draw them away from *A* in the directions shown in the space diagram. No other detail is included in the free-body diagram.

Since point *A* is in equilibrium, the three forces acting on it must form a closed triangle when drawn in tip-to-tail fashion. We have drawn this **force triangle** in Fig. 2.24*c*. The values T_{AB} and T_{AC} of the tensions in the ropes may be found graphically if the triangle is drawn to scale, or they may be found by trigonometry. If we choose trigonometry, we use the law of sines:

$$\frac{T_{AB}}{\sin 60°} = \frac{T_{AC}}{\sin 40°} = \frac{736 \text{ N}}{\sin 80°}$$

$$T_{AB} = 647 \text{ N} \qquad T_{AC} = 480 \text{ N}$$

When a particle is in equilibrium under three forces, you can solve the problem by drawing a force triangle. When a particle is in equilibrium under more than three forces, you can solve the problem graphically by drawing a force polygon. If you need an analytic solution, you should solve the **equations of equilibrium** given in Sec. 2.3A:

$$\Sigma F_x = 0 \qquad \Sigma F_y = 0 \tag{2.15}$$

These equations can be solved for no more than *two unknowns*. Similarly, the force triangle used in the case of equilibrium under three forces can be solved for only two unknowns.

The most common types of problems are those in which the two unknowns represent (1) the two components (or the magnitude and direction) of a single force or (2) the magnitudes of two forces, each of known direction. Problems involving the determination of the maximum or minimum value of the magnitude of a force are also encountered (see Probs. 2.43 through 2.47).

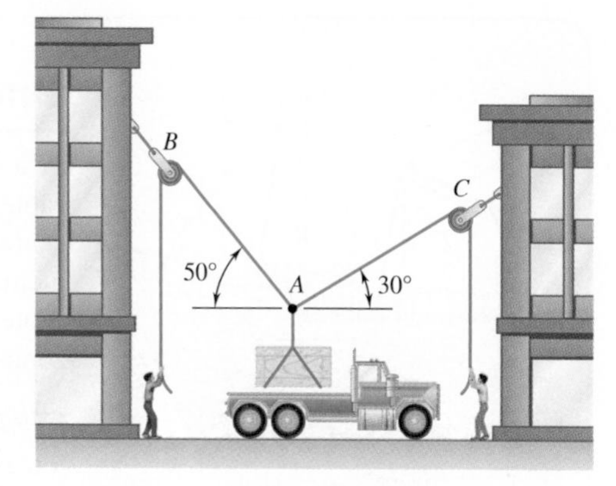

(*a*) Space diagram

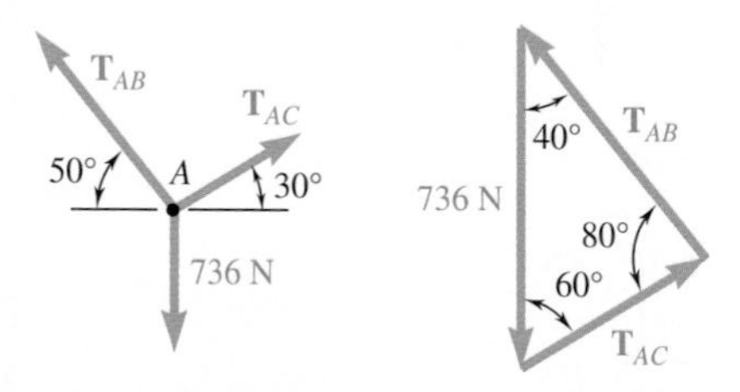

(*b*) Free-body diagram (*c*) Force triangle

Fig. 2.24 (*a*) The space diagram shows the physical situation of the problem; (*b*) the free-body diagram shows one central particle and the forces acting on it; (*c*) the force triangle can be solved with the law of sines. Note that the forces form a closed triangle because the particle is in equilibrium and the resultant force is zero.

Photo 2.3 As illustrated in Fig. 2.24, it is possible to determine the tensions in the cables supporting the shaft shown by treating the hook as a particle and then applying the equations of equilibrium to the forces acting on the hook.

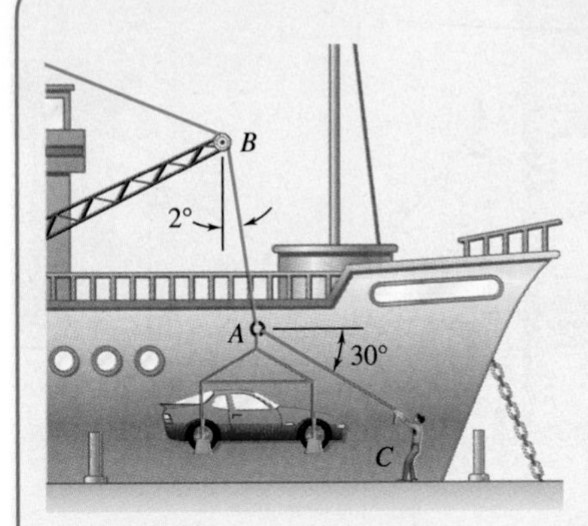

Sample Problem 2.4

In a ship-unloading operation, a 3500-lb automobile is supported by a cable. A worker ties a rope to the cable at A and pulls on it in order to center the automobile over its intended position on the dock. At the moment illustrated, the automobile is stationary, the angle between the cable and the vertical is 2°, and the angle between the rope and the horizontal is 30°. What are the tensions in the rope and cable?

STRATEGY: This is a problem of equilibrium under three coplanar forces. You can treat point A as a particle and solve the problem using a force triangle.

MODELING and ANALYSIS:

Free-Body Diagram. Choose point A as the particle and draw the complete free-body diagram (Fig. 1). T_{AB} is the tension in the cable AB, and T_{AC} is the tension in the rope.

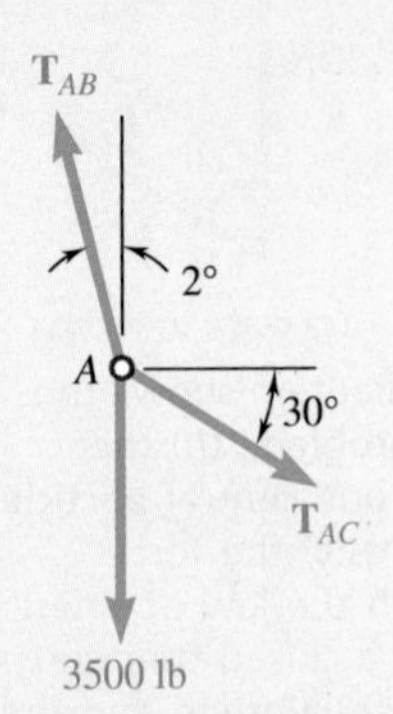

Fig. 1 Free-body diagram of particle A.

Equilibrium Condition. Since only three forces act on point A, draw a force triangle to express that it is in equilibrium (Fig. 2). Using the law of sines,

$$\frac{T_{AB}}{\sin 120°} = \frac{T_{AC}}{\sin 2°} = \frac{3500 \text{ lb}}{\sin 58°}$$

With a calculator, compute and store the value of the last quotient. Multiplying this value successively by sin 120° and sin 2°, you obtain

$$T_{AB} = 3570 \text{ lb} \qquad T_{AC} = 144 \text{ lb} \blacktriangleleft$$

REFLECT and THINK: This is a common problem of knowing one force in a three-force equilibrium problem and calculating the other forces from the given geometry. This basic type of problem will occur often as part of more complicated situations in this text.

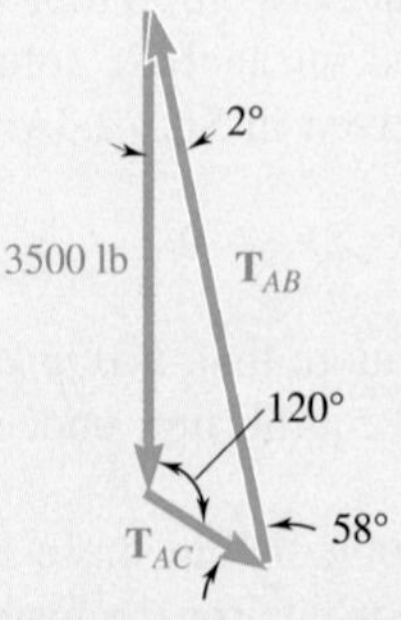

Fig. 2 Force triangle of the forces acting on particle A.

Sample Problem 2.5

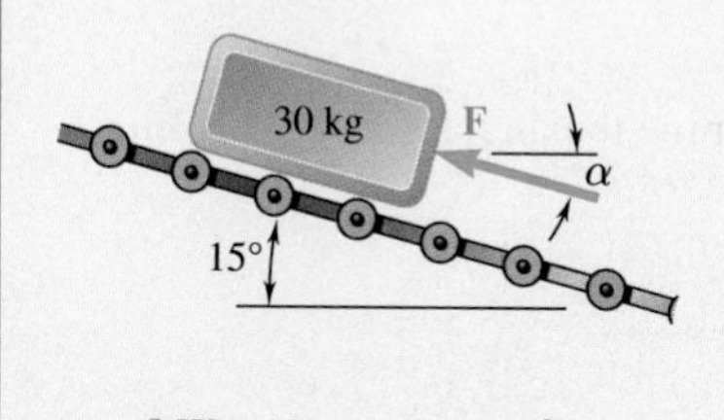

Determine the magnitude and direction of the smallest force **F** that maintains the 30-kg package shown in equilibrium. Note that the force exerted by the rollers on the package is perpendicular to the incline.

STRATEGY: This is an equilibrium problem with three coplanar forces that you can solve with a force triangle. The new wrinkle is to determine a minimum force. You can approach this part of the solution in a way similar to Sample Problem 2.2.

MODELING and ANALYSIS:

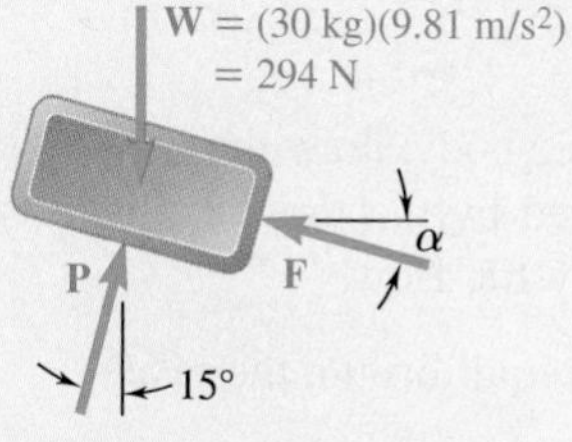

Fig. 1 Free-body diagram of package, treated as a particle.

Free-Body Diagram. Choose the package as a free body, assuming that it can be treated as a particle. Then draw the corresponding free-body diagram (Fig. 1).

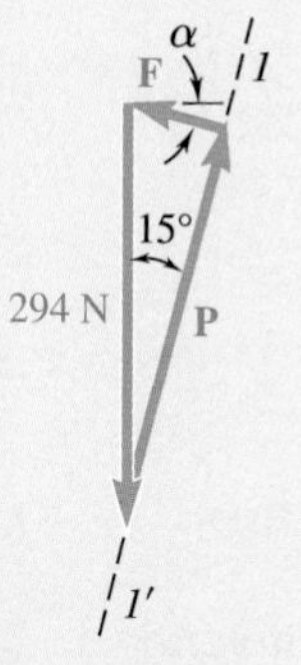

Fig. 2 Force triangle of the forces acting on package.

Equilibrium Condition. Since only three forces act on the free body, draw a force triangle to express that it is in equilibrium (Fig. 2). Line *1-1′* represents the known direction of **P**. In order to obtain the minimum value of the force **F**, choose the direction of **F** to be perpendicular to that of **P**. From the geometry of this triangle,

$$F = (294 \text{ N}) \sin 15° = 76.1 \text{ N} \qquad \alpha = 15°$$

$$\mathbf{F} = 76.1 \text{ N} \measuredangle 15° \quad \blacktriangleleft$$

REFLECT and THINK: Determining maximum and minimum forces to maintain equilibrium is a common practical problem. Here the force needed is about 25% of the weight of the package, which seems reasonable for an incline of 15°.

Sample Problem 2.6

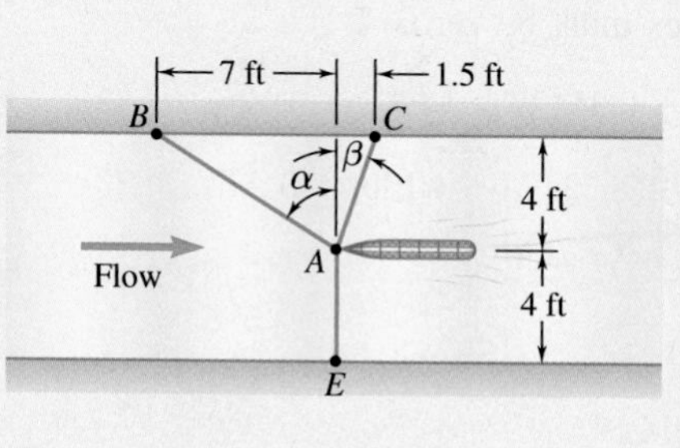

For a new sailboat, a designer wants to determine the drag force that may be expected at a given speed. To do so, she places a model of the proposed hull in a test channel and uses three cables to keep its bow on the centerline of the channel. Dynamometer readings indicate that for a given speed, the tension is 40 lb in cable *AB* and 60 lb in cable *AE*. Determine the drag force exerted on the hull and the tension in cable *AC*.

STRATEGY: The cables all connect at point *A*, so you can treat that as a particle in equilibrium. Because four forces act at *A* (tensions in three cables and the drag force), you should use the equilibrium conditions and sum forces by components to solve for the unknown forces.

(continued)

MODELING and ANALYSIS:

Determining the Angles. First, determine the angles α and β defining the direction of cables AB and AC:

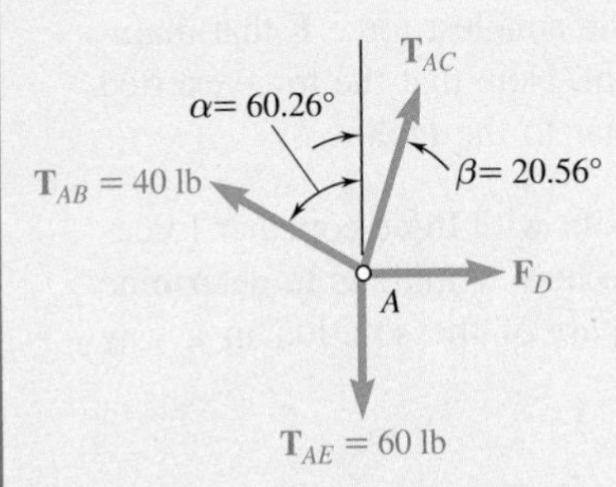

Fig. 1 Free-body diagram of particle A.

$$\tan\alpha = \frac{7\text{ ft}}{4\text{ ft}} = 1.75 \qquad \tan\beta = \frac{1.5\text{ ft}}{4\text{ ft}} = 0.375$$

$$\alpha = 60.26° \qquad \beta = 20.56°$$

Free-Body Diagram. Choosing point A as a free body, draw the free-body diagram (Fig. 1). It includes the forces exerted by the three cables on the hull, as well as the drag force $\mathbf{F}_D$ exerted by the flow.

Equilibrium Condition. Because point A is in equilibrium, the resultant of all forces is zero:

$$\mathbf{R} = \mathbf{T}_{AB} + \mathbf{T}_{AC} + \mathbf{T}_{AE} + \mathbf{F}_D = 0 \tag{1}$$

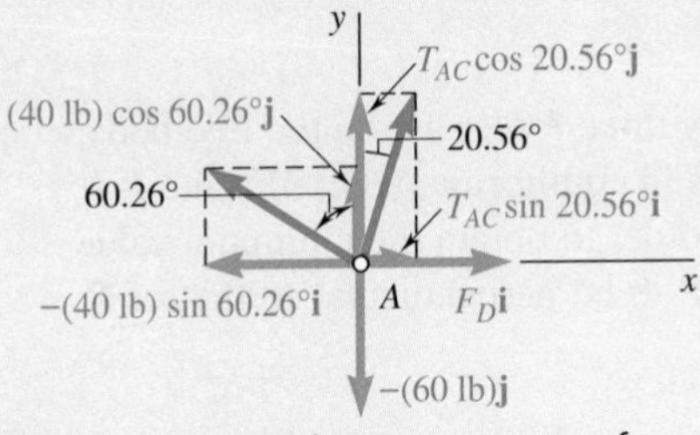

Fig. 2 Rectangular components of forces acting on particle A.

Because more than three forces are involved, resolve the forces into x and y components (Fig. 2):

$$\mathbf{T}_{AB} = -(40\text{ lb})\sin 60.26°\mathbf{i} + (40\text{ lb})\cos 60.26°\mathbf{j}$$
$$= -(34.73\text{ lb})\mathbf{i} + (19.84\text{ lb})\mathbf{j}$$
$$\mathbf{T}_{AC} = T_{AC}\sin 20.56°\mathbf{i} + T_{AC}\cos 20.56°\mathbf{j}$$
$$= 0.3512T_{AC}\mathbf{i} + 0.9363T_{AC}\mathbf{j}$$
$$\mathbf{T}_{AE} = -(60\text{ lb})\mathbf{j}$$
$$\mathbf{F}_D = F_D\mathbf{i}$$

Substituting these expressions into Eq. (1) and factoring the unit vectors **i** and **j**, you have

$$(-34.73\text{ lb} + 0.3512T_{AC} + F_D)\mathbf{i} + (19.84\text{ lb} + 0.9363T_{AC} - 60\text{ lb})\mathbf{j} = 0$$

This equation is satisfied if, and only if, the coefficients of **i** and **j** are each equal to zero. You obtain the following two equilibrium equations, which express, respectively, that the sum of the x components and the sum of the y components of the given forces must be zero.

$$(\Sigma F_x = 0:) \qquad -34.73\text{ lb} + 0.3512T_{AC} + F_D = 0 \tag{2}$$

$$(\Sigma F_y = 0:) \qquad 19.84\text{ lb} + 0.9363T_{AC} - 60\text{ lb} = 0 \tag{3}$$

From Eq. (3), you find

$$T_{AC} = +42.9\text{ lb} \blacktriangleleft$$

Substituting this value into Eq. (2) yields

$$F_D = +19.66\text{ lb} \blacktriangleleft$$

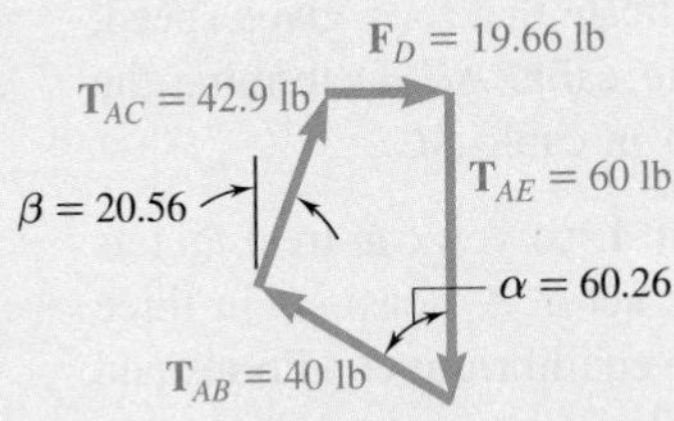

Fig. 3 Force polygon of forces acting on particle A.

REFLECT and THINK: In drawing the free-body diagram, you assumed a sense for each unknown force. A positive sign in the answer indicates that the assumed sense is correct. You can draw the complete force polygon (Fig. 3) to check the results.

Problems

2.32 Two cables are tied together at C and are loaded as shown. Knowing that $\alpha = 30°$, determine the tension (*a*) in cable AC, (*b*) in cable BC.

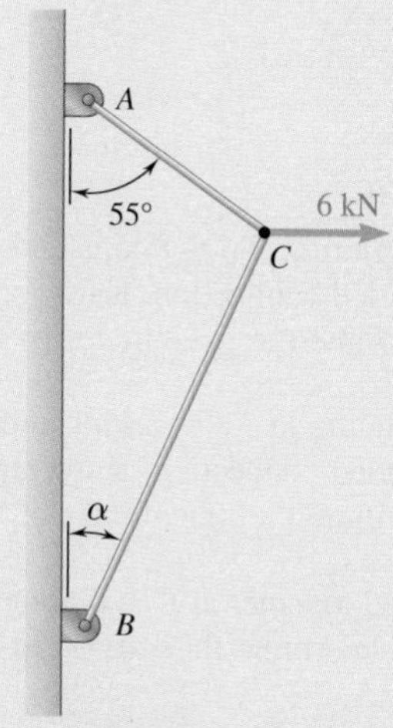

Fig. P2.32

2.33 and 2.34 Two cables are tied together at C and are loaded as shown. Determine the tension (*a*) in cable AC, (*b*) in cable BC.

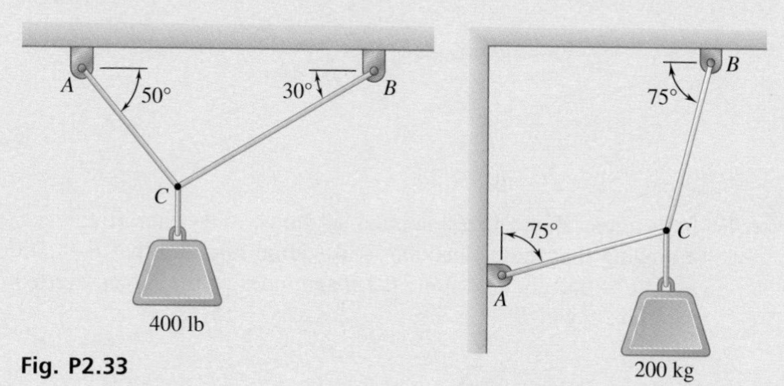

Fig. P2.33

Fig. P2.34

2.35 Two cables are tied together at C and loaded as shown. Determine the tension (*a*) in cable AC, (*b*) in cable BC.

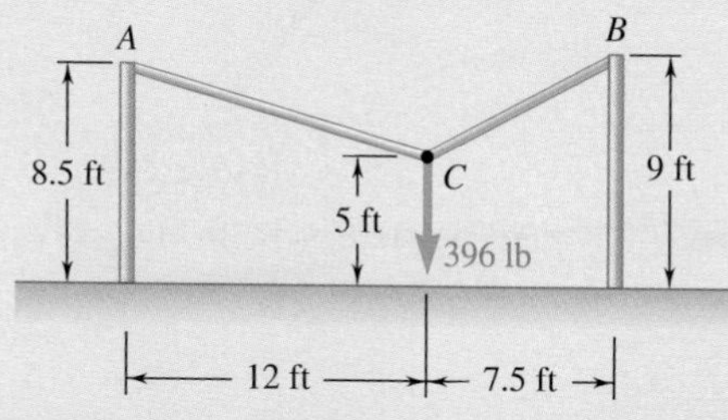

Fig. P2.35

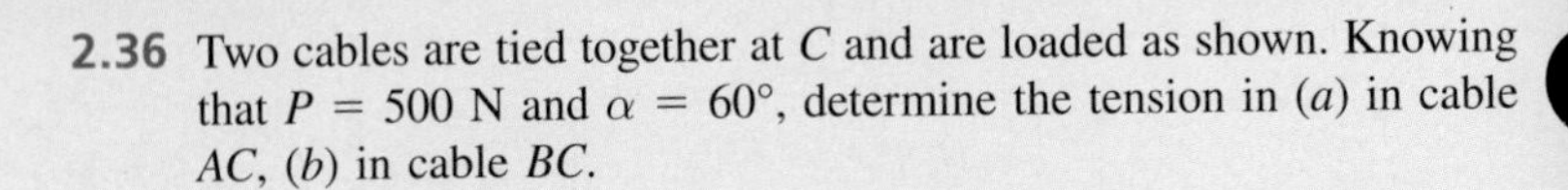
2.36 Two cables are tied together at C and are loaded as shown. Knowing that $P = 500$ N and $\alpha = 60°$, determine the tension in (*a*) in cable AC, (*b*) in cable BC.

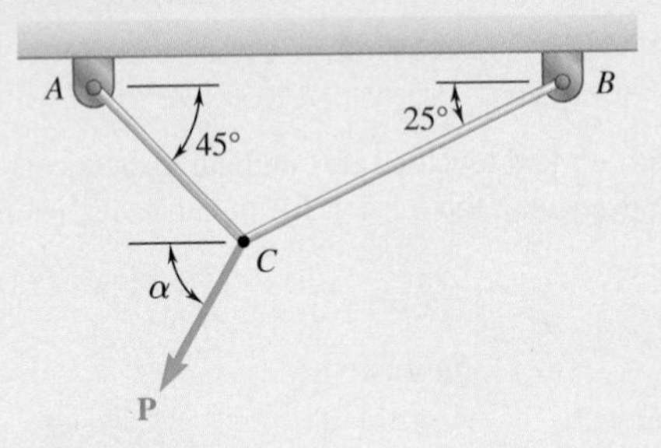

Fig. P2.36

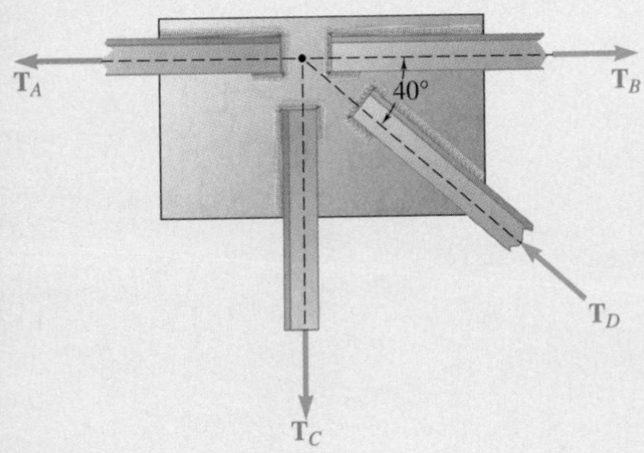

Fig. *P2.37* and P2.38

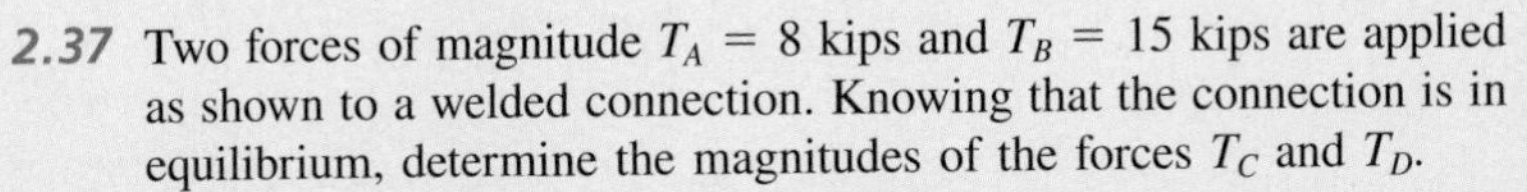
2.37 Two forces of magnitude $T_A = 8$ kips and $T_B = 15$ kips are applied as shown to a welded connection. Knowing that the connection is in equilibrium, determine the magnitudes of the forces T_C and T_D.

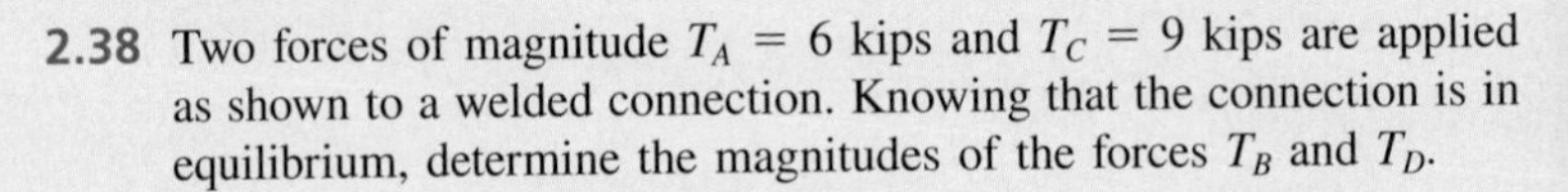
2.38 Two forces of magnitude $T_A = 6$ kips and $T_C = 9$ kips are applied as shown to a welded connection. Knowing that the connection is in equilibrium, determine the magnitudes of the forces T_B and T_D.

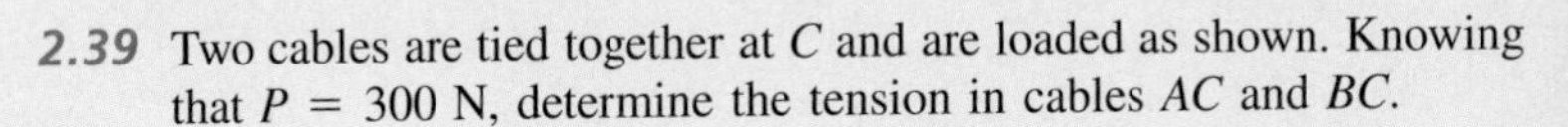
2.39 Two cables are tied together at C and are loaded as shown. Knowing that $P = 300$ N, determine the tension in cables AC and BC.

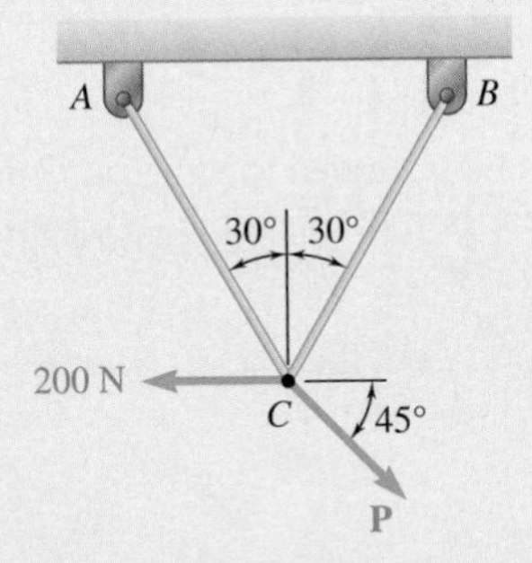

Fig. *P2.39*

2.40 Two forces **P** and **Q** are applied as shown to an aircraft connection. Knowing that the connection is in equilibrium and that $P = 500$ lb and $Q = 650$ lb, determine the magnitudes of the forces exerted on the rods A and B.

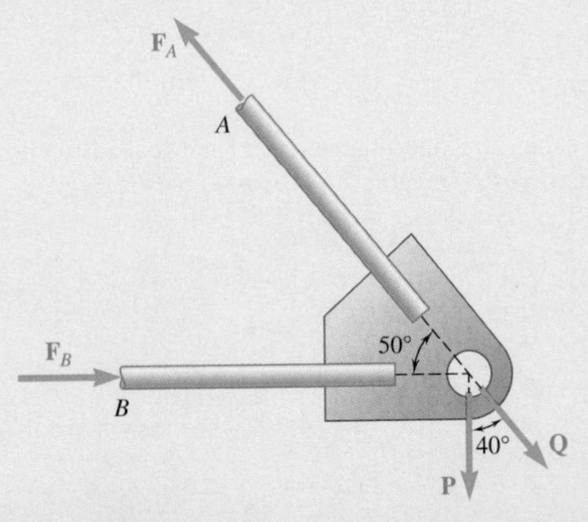

Fig. P2.40

2.41 A sailor is being rescued using a boatswain's chair that is suspended from a pulley that can roll freely on the support cable ACB and is pulled at a constant speed by cable CD. Knowing that $\alpha = 30°$ and $\beta = 10°$ and that the combined weight of the boatswain's chair and the sailor is 200 lb, determine the tension (*a*) in the support cable ACB, (*b*) in the traction cable CD.

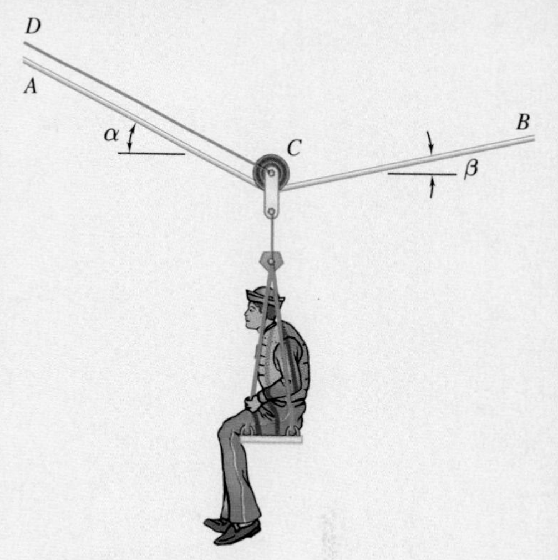

Fig. P2.41 and *P2.42*

2.42 A sailor is being rescued using a boatswain's chair that is suspended from a pulley that can roll freely on the support cable ACB and is pulled at a constant speed by cable CD. Knowing that $\alpha = 25°$ and $\beta = 15°$ and that the tension in cable CD is 20 lb, determine (*a*) the combined weight of the boatswain's chair and the sailor, (*b*) the tension in the support cable ACB.

2.43 For the cables of Prob. 2.32, find the value of α for which the tension is as small as possible (*a*) in cable BC, (*b*) in both cables simultaneously. In each case determine the tension in each cable.

2.44 For the cables of Prob. 2.36, it is known that the maximum allowable tension is 600 N in cable AC and 750 N in cable BC. Determine (*a*) the maximum force **P** that can be applied at C, (*b*) the corresponding value of α.

2.45 Two cables tied together at C are loaded as shown. Knowing that the maximum allowable tension in each cable is 800 N, determine (*a*) the magnitude of the largest force **P** that can be applied at C, (*b*) the corresponding value of α.

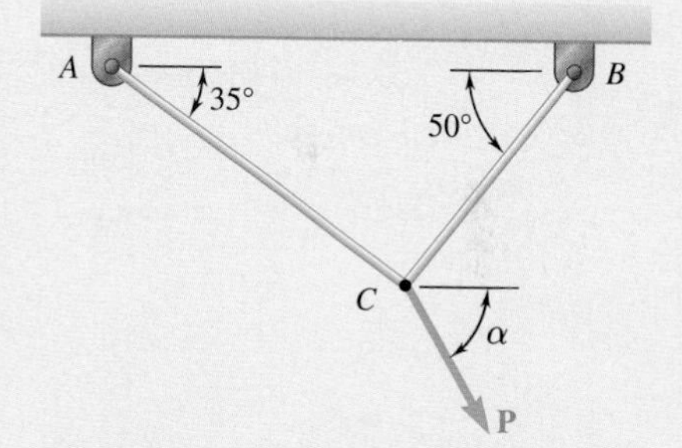

Fig. P2.45 and *P2.46*

2.46 Two cables tied together at C are loaded as shown. Knowing that the maximum allowable tension is 1200 N in cable AC and 600 N in cable BC, determine (*a*) the magnitude of the largest force **P** that can be applied at C, (*b*) the corresponding value of α.

2.47 Two cables tied together at C are loaded as shown. Determine the range of values of Q for which the tension will not exceed 60 lb in either cable.

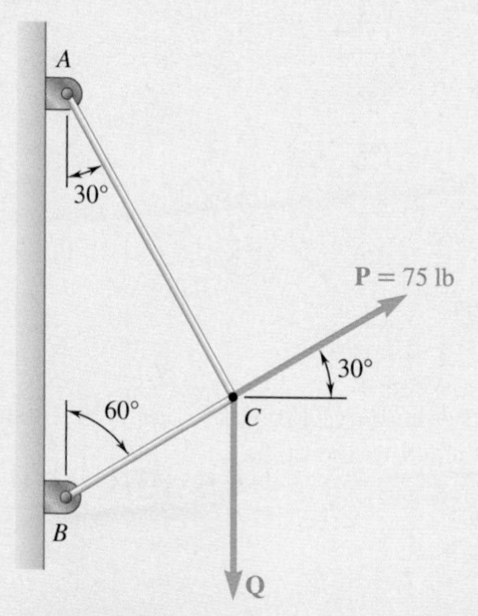

Fig. P2.47

2.48 Collar *A* is connected as shown to a 50-lb load and can slide on a frictionless horizontal rod. Determine the magnitude of the force **P** required to maintain the equilibrium of the collar when (*a*) $x = 4.5$ in., (*b*) $x = 15$ in.

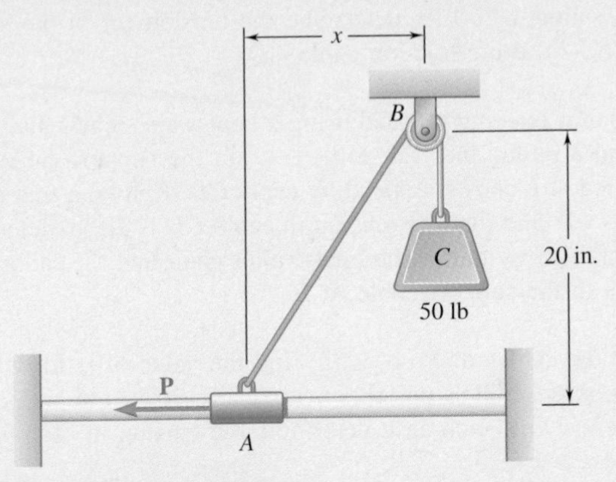

Fig. P2.48 and P2.49

2.49 Collar *A* is connected as shown to a 50-lb load and can slide on a frictionless horizontal rod. Determine the distance *x* for which the collar is in equilibrium when $P = 48$ lb.

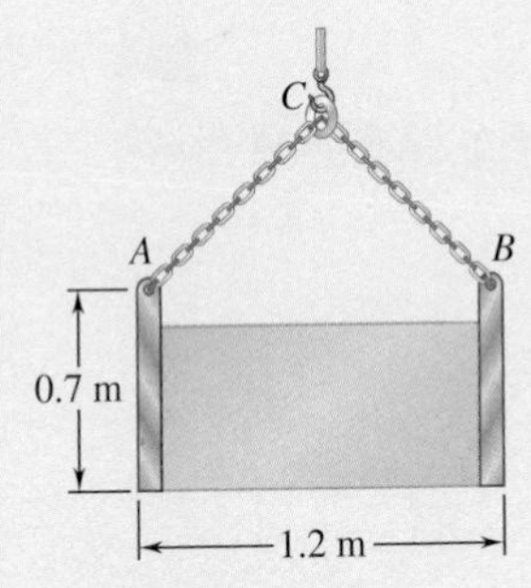

Fig. P2.50

2.50 A movable bin and its contents have a combined weight of 2.8 kN. Determine the shortest chain sling *ACB* that can be used to lift the loaded bin if the tension in the chain is not to exceed 5 kN.

2.51 A 600-lb crate is supported by several rope-and-pulley arrangements as shown. Determine for each arrangement the tension in the rope. (*Hint:* The tension in the rope is the same on each side of a simple pulley. This can be proved by the methods of Chap. 4.)

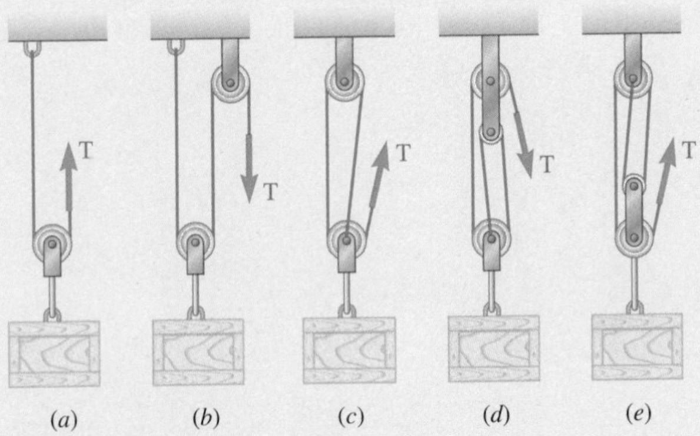

Fig. P2.51

2.52 Solve parts *b* and *d* of Prob. 2.51, assuming that the free end of the rope is attached to the crate.

2.53 A 200-kg crate is to be supported by the rope-and-pulley arrangement shown. Determine the magnitude and direction of the force **P** that must be exerted on the free end of the rope to maintain equilibrium. (See the hint for Prob. 2.51.)

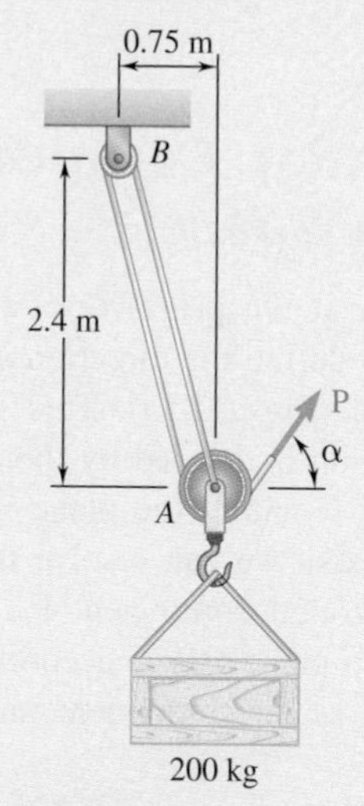

Fig. *P2.53*

2.54 A load **Q** is applied to pulley *C*, which can roll on cable *ACB*. The pulley is held in the position shown by a second cable *CAD*, which passes over pulley *A* and supports a load **P**. Knowing that $P = 750$ N, determine (*a*) the tension in cable *ACB*, (*b*) the magnitude of load **Q**.

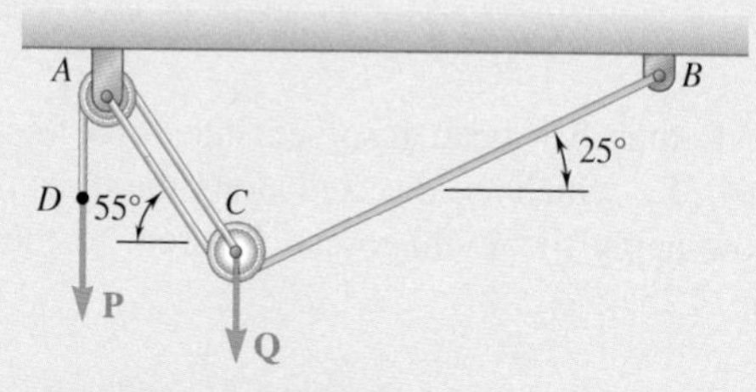

Fig. P2.54 and P2.55

2.55 An 1800-N load **Q** is applied to pulley *C*, which can roll on cable *ACB*. The pulley is held in the position shown by a second cable *CAD*, which passes over pulley *A* and supports a load **P**. Determine (*a*) the tension in cable *ACB*, (*b*) the magnitude of load **P**.

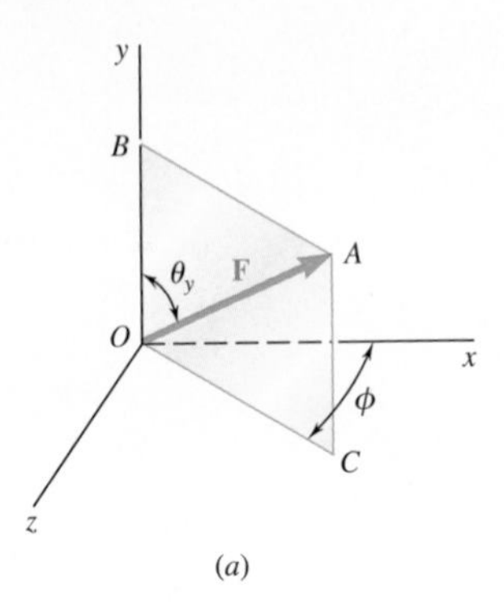

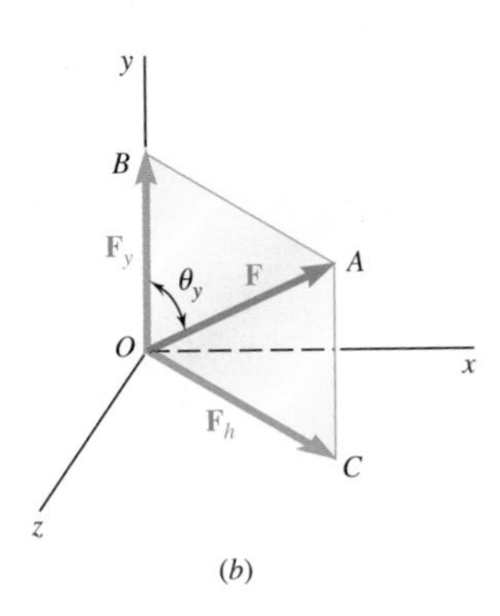

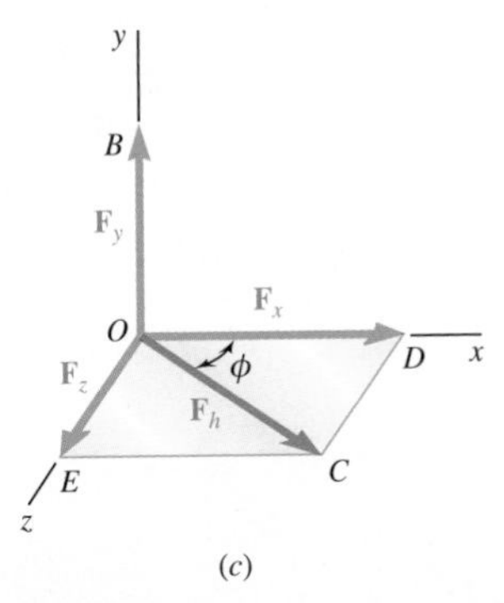

Fig. 2.25 (a) A force **F** in an *xyz* coordinate system; (b) components of **F** along the *y* axis and in the *xz* plane; (c) components of **F** along the three rectangular axes.

2.4 ADDING FORCES IN SPACE

The problems considered in the first part of this chapter involved only two dimensions; they were formulated and solved in a single plane. In the last part of this chapter, we discuss problems involving the three dimensions of space.

2.4A Rectangular Components of a Force in Space

Consider a force **F** acting at the origin *O* of the system of rectangular coordinates *x*, *y*, and *z*. To define the direction of **F**, we draw the vertical plane *OBAC* containing **F** (Fig. 2.25*a*). This plane passes through the vertical *y* axis; its orientation is defined by the angle ϕ it forms with the *xy* plane. The direction of **F** within the plane is defined by the angle θ_y that **F** forms with the *y* axis. We can resolve the force **F** into a vertical component $\mathbf{F}_y$ and a horizontal component $\mathbf{F}_h$; this operation, shown in Fig. 2.25*b*, is carried out in plane *OBAC* according to the rules developed earlier. The corresponding scalar components are

$$F_y = F\cos\theta_y \qquad F_h = F\sin\theta_y \tag{2.16}$$

However, we can also resolve $\mathbf{F}_h$ into two rectangular components $\mathbf{F}_x$ and $\mathbf{F}_z$ along the *x* and *z* axes, respectively. This operation, shown in Fig. 2.25*c*, is carried out in the *xz* plane. We obtain the following expressions for the corresponding scalar components:

$$\begin{aligned} F_x &= F_h\cos\phi = F\sin\theta_y\cos\phi \\ F_z &= F_h\sin\phi = F\sin\theta_y\sin\phi \end{aligned} \tag{2.17}$$

The given force **F** thus has been resolved into three rectangular vector components $\mathbf{F}_x$, $\mathbf{F}_y$, $\mathbf{F}_z$, which are directed along the three coordinate axes.

We can now apply the Pythagorean theorem to the triangles *OAB* and *OCD* of Fig. 2.25:

$$\begin{aligned} F^2 &= (OA)^2 = (OB)^2 + (BA)^2 = F_y^2 + F_h^2 \\ F_h^2 &= (OC)^2 = (OD)^2 + (DC)^2 = F_x^2 + F_z^2 \end{aligned}$$

Eliminating F_h^2 from these two equations and solving for F, we obtain the following relation between the magnitude of **F** and its rectangular scalar components:

Magnitude of a force in space

$$F = \sqrt{F_x^2 + F_y^2 + F_z^2} \tag{2.18}$$

The relationship between the force **F** and its three components $\mathbf{F}_x$, $\mathbf{F}_y$, and $\mathbf{F}_z$ is more easily visualized if we draw a "box" having $\mathbf{F}_x$, $\mathbf{F}_y$, and $\mathbf{F}_z$ for edges, as shown in Fig. 2.26. The force **F** is then represented by the main diagonal *OA* of this box. Figure 2.26*b* shows the right triangle

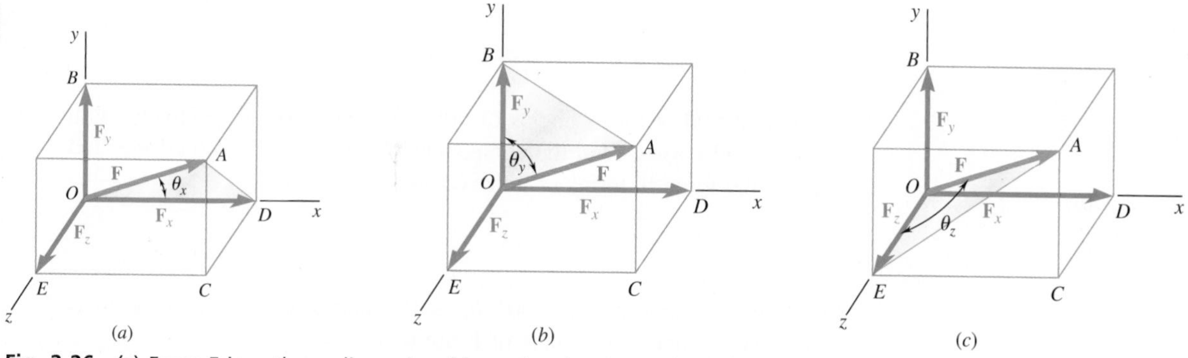

Fig. 2.26 (*a*) Force **F** in a three-dimensional box, showing its angle with the *x* axis; (*b*) force **F** and its angle with the *y* axis; (*c*) force **F** and its angle with the *z* axis.

OAB used to derive the first of the formulas (2.16): $F_y = F \cos \theta_y$. In Fig. 2.26*a* and *c*, two other right triangles have also been drawn: *OAD* and *OAE*. These triangles occupy positions in the box comparable with that of triangle *OAB*. Denoting by θ_x and θ_z, respectively, the angles that **F** forms with the *x* and *z* axes, we can derive two formulas similar to $F_y = F \cos \theta_y$. We thus write

Scalar components of a force F

$$F_x = F \cos \theta_x \qquad F_y = F \cos \theta_y \qquad F_z = F \cos \theta_z \tag{2.19}$$

The three angles θ_x, θ_y, and θ_z define the direction of the force **F**; they are more commonly used for this purpose than the angles θ_y and ϕ introduced at the beginning of this section. The cosines of θ_x, θ_y, and θ_z are known as the **direction cosines** of the force **F**.

Introducing the unit vectors **i**, **j**, and **k**, which are directed respectively along the *x*, *y*, and *z* axes (Fig. 2.27), we can express **F** in the form

Vector expression of a force F

$$\mathbf{F} = F_x\mathbf{i} + F_y\mathbf{j} + F_z\mathbf{k} \tag{2.20}$$

where the scalar components F_x, F_y, and F_z are defined by the relations in Eq. (2.19).

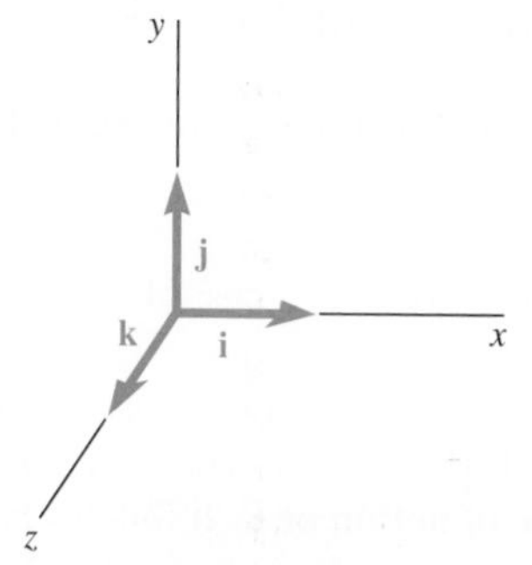

Fig. 2.27 The three unit vectors **i**, **j**, **k** lie along the three coordinate axes *x*, *y*, *z*, respectively.

Concept Application 2.4

A force of 500 N forms angles of 60°, 45°, and 120°, respectively, with the x, y, and z axes. Find the components F_x, F_y, and F_z of the force and express the force in terms of unit vectors.

Solution

Substitute $F = 500$ N, $\theta_x = 60°$, $\theta_y = 45°$, and $\theta_z = 120°$ into formulas (2.19). The scalar components of **F** are then

$$F_x = (500 \text{ N}) \cos 60° = +250 \text{ N}$$
$$F_y = (500 \text{ N}) \cos 45° = +354 \text{ N}$$
$$F_z = (500 \text{ N}) \cos 120° = -250 \text{ N}$$

Carrying these values into Eq. (2.20), you have

$$\mathbf{F} = (250 \text{ N})\mathbf{i} + (354 \text{ N})\mathbf{j} - (250 \text{ N})\mathbf{k}$$

As in the case of two-dimensional problems, a plus sign indicates that the component has the same sense as the corresponding axis, and a minus sign indicates that it has the opposite sense.

The angle a force **F** forms with an axis should be measured from the positive side of the axis and is always between 0 and 180°. An angle θ_x smaller than 90° (acute) indicates that **F** (assumed attached to O) is on the same side of the yz plane as the positive x axis; $\cos \theta_x$ and F_x are then positive. An angle θ_x larger than 90° (obtuse) indicates that **F** is on the other side of the yz plane; $\cos \theta_x$ and F_x are then negative. In Concept Application 2.4, the angles θ_x and θ_y are acute and θ_z is obtuse; consequently, F_x and F_y are positive and F_z is negative.

Substituting into Eq. (2.20) the expressions obtained for F_x, F_y, and F_z in Eq. (2.19), we have

$$\mathbf{F} = F(\cos \theta_x \mathbf{i} + \cos \theta_y \mathbf{j} + \cos \theta_z \mathbf{k}) \qquad \textbf{(2.21)}$$

This equation shows that the force **F** can be expressed as the product of the scalar F and the vector

$$\boldsymbol{\lambda} = \cos \theta_x \mathbf{i} + \cos \theta_y \mathbf{j} + \cos \theta_z \mathbf{k} \qquad \textbf{(2.22)}$$

Clearly, the vector $\boldsymbol{\lambda}$ is a vector whose magnitude is equal to 1 and whose direction is the same as that of **F** (Fig. 2.28). The vector $\boldsymbol{\lambda}$ is referred to as the **unit vector along the line of action** of **F**. It follows from Eq. (2.22) that the components of the unit vector $\boldsymbol{\lambda}$ are respectively equal to the direction cosines of the line of action of **F**:

$$\lambda_x = \cos \theta_x \qquad \lambda_y = \cos \theta_y \qquad \lambda_z = \cos \theta_z \qquad \textbf{(2.23)}$$

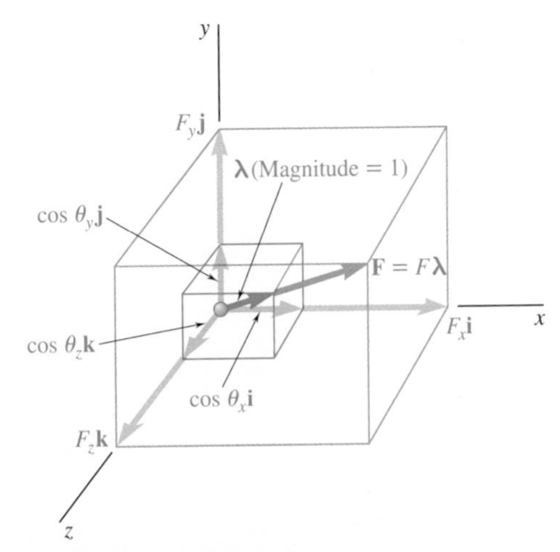

Fig. 2.28 Force **F** can be expressed as the product of its magnitude F and a unit vector $\boldsymbol{\lambda}$ in the direction of **F**. Also shown are the components of **F** and its unit vector.

Note that the values of the three angles θ_x, θ_y, and θ_z are not independent. Recalling that the sum of the squares of the components of a vector is equal to the square of its magnitude, we can write

$$\lambda_x^2 + \lambda_y^2 + \lambda_z^2 = 1$$

Substituting for λ_x, λ_y, and λ_z from Eq. (2.23), we obtain

Relationship among direction cosines

$$\cos^2\theta_x + \cos^2\theta_y + \cos^2\theta_z = 1 \tag{2.24}$$

In Concept Application 2.4, for instance, once the values $\theta_x = 60°$ and $\theta_y = 45°$ have been selected, the value of θ_z *must* be equal to 60° or 120° in order to satisfy the identity in Eq. (2.24).

When the components F_x, F_y, and F_z of a force **F** are given, we can obtain the magnitude F of the force from Eq. (2.18). We can then solve relations in Eq. (2.19) for the direction cosines as

$$\cos\theta_x = \frac{F_x}{F} \qquad \cos\theta_y = \frac{F_y}{F} \qquad \cos\theta_z = \frac{F_z}{F} \tag{2.25}$$

From the direction cosines, we can find the angles θ_x, θ_y, and θ_z characterizing the direction of **F**.

Concept Application 2.5

A force **F** has the components $F_x = 20$ lb, $F_y = -30$ lb, and $F_z = 60$ lb. Determine its magnitude F and the angles θ_x, θ_y, and θ_z it forms with the coordinate axes.

Solution

You can obtain the magnitude of **F** from formula (2.18):

$$\begin{aligned} F &= \sqrt{F_x^2 + F_y^2 + F_z^2} \\ &= \sqrt{(20\text{ lb})^2 + (-30\text{ lb})^2 + (60\text{ lb})^2} \\ &= \sqrt{4900\text{ lb}} = 70\text{ lb} \end{aligned}$$

Substituting the values of the components and magnitude of **F** into Eqs. (2.25), the direction cosines are

$$\cos\theta_x = \frac{F_x}{F} = \frac{20\text{ lb}}{70\text{ lb}} \qquad \cos\theta_y = \frac{F_y}{F} = \frac{-30\text{ lb}}{70\text{ lb}} \qquad \cos\theta_z = \frac{F_z}{F} = \frac{60\text{ lb}}{70\text{ lb}}$$

Calculating each quotient and its arc cosine gives you

$$\theta_x = 73.4° \qquad \theta_y = 115.4° \qquad \theta_z = 31.0°$$

These computations can be carried out easily with a calculator.

2.4B Force Defined by Its Magnitude and Two Points on Its Line of Action

In many applications, the direction of a force **F** is defined by the coordinates of two points, $M(x_1, y_1, z_1)$ and $N(x_2, y_2, z_2)$, located on its line of action (Fig. 2.29). Consider the vector $\overrightarrow{MN}$ joining M and N and of the same

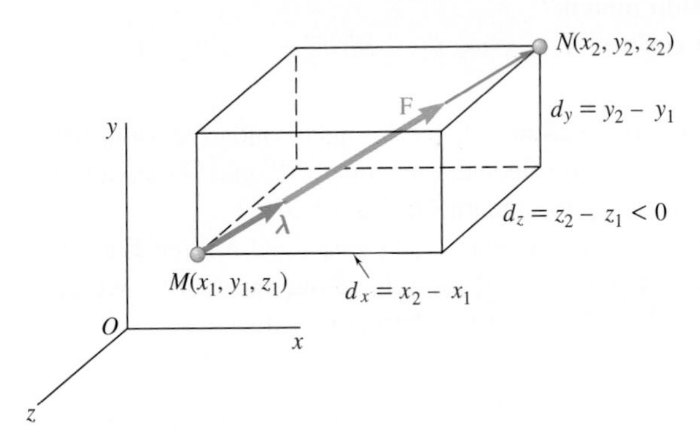

Fig. 2.29 A case where the line of action of force **F** is determined by the two points *M* and *N*. We can calculate the components of **F** and its direction cosines from the vector $\overrightarrow{MN}$.

sense as a force **F**. Denoting its scalar components by d_x, d_y, and d_z, respectively, we write

$$\overrightarrow{MN} = d_x\mathbf{i} + d_y\mathbf{j} + d_z\mathbf{k} \tag{2.26}$$

We can obtain a unit vector $\boldsymbol{\lambda}$ along the line of action of **F** (i.e., along the line MN) by dividing the vector $\overrightarrow{MN}$ by its magnitude MN. Substituting for $\overrightarrow{MN}$ from Eq. (2.26) and observing that MN is equal to the distance d from M to N, we have

$$\boldsymbol{\lambda} = \frac{\overrightarrow{MN}}{MN} = \frac{1}{d}(d_x\mathbf{i} + d_y\mathbf{j} + d_z\mathbf{k}) \tag{2.27}$$

Recalling that **F** is equal to the product of F and $\boldsymbol{\lambda}$, we have

$$\mathbf{F} = F\boldsymbol{\lambda} = \frac{F}{d}(d_x\mathbf{i} + d_y\mathbf{j} + d_z\mathbf{k}) \tag{2.28}$$

It follows that the scalar components of **F** are, respectively,

Scalar components of force F

$$F_x = \frac{Fd_x}{d} \qquad F_y = \frac{Fd_y}{d} \qquad F_z = \frac{Fd_z}{d} \tag{2.29}$$

The relations in Eq. (2.29) considerably simplify the determination of the components of a force **F** of given magnitude F when the line of action of **F** is defined by two points M and N. The calculation consists of first subtracting the coordinates of M from those of N, then determining the components of the vector $\overrightarrow{MN}$ and the distance d from M to N. Thus,

$$d_x = x_2 - x_1 \qquad d_y = y_2 - y_1 \qquad d_z = z_2 - z_1$$

$$d = \sqrt{d_x^2 + d_y^2 + d_z^2}$$

Substituting for F and for d_x, d_y, d_z, and d into the relations in Eq. (2.29), we obtain the components F_x, F_y, and F_z of the force.

We can then obtain the angles θ_x, θ_y, and θ_z that **F** forms with the coordinate axes from Eqs. (2.25). Comparing Eqs. (2.22) and (2.27), we can write

Direction cosines of force F

$$\cos\theta_x = \frac{d_x}{d} \qquad \cos\theta_y = \frac{d_y}{d} \qquad \cos\theta_z = \frac{d_z}{d} \qquad \textbf{(2.30)}$$

In other words, we can determine the angles θ_x, θ_y, and θ_z directly from the components and the magnitude of the vector $\overrightarrow{MN}$.

2.4C Addition of Concurrent Forces in Space

We can determine the resultant **R** of two or more forces in space by summing their rectangular components. Graphical or trigonometric methods are generally not practical in the case of forces in space.

The method followed here is similar to that used in Sec. 2.2B with coplanar forces. Setting

$$\mathbf{R} = \Sigma\mathbf{F}$$

we resolve each force into its rectangular components:

$$R_x\mathbf{i} + R_y\mathbf{j} + R_z\mathbf{k} = \Sigma\,(F_x\mathbf{i} + F_y\mathbf{j} + F_z\mathbf{k})$$
$$= (\Sigma F_x)\mathbf{i} + (\Sigma F_y)\mathbf{j} + (\Sigma F_z)\mathbf{k}$$

From this equation, it follows that

Rectangular components of the resultant

$$R_x = \Sigma F_x \qquad R_y = \Sigma F_y \qquad R_z = \Sigma F_z \qquad \textbf{(2.31)}$$

The magnitude of the resultant and the angles θ_x, θ_y, and θ_z that the resultant forms with the coordinate axes are obtained using the method discussed earlier in this section. We end up with

Resultant of concurrent forces in space

$$R = \sqrt{R_x^2 + R_y^2 + R_z^2} \qquad \textbf{(2.32)}$$

$$\cos\theta_x = \frac{R_x}{R} \qquad \cos\theta_y = \frac{R_y}{R} \qquad \cos\theta_z = \frac{R_z}{R} \qquad \textbf{(2.33)}$$

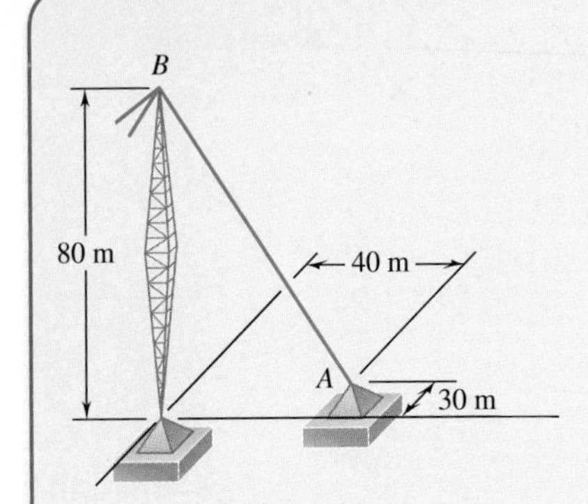

Sample Problem 2.7

A tower guy wire is anchored by means of a bolt at A. The tension in the wire is 2500 N. Determine (*a*) the components F_x, F_y, and F_z of the force acting on the bolt and (*b*) the angles θ_x, θ_y, and θ_z defining the direction of the force.

STRATEGY: From the given distances, we can determine the length of the wire and the direction of a unit vector along it. From that, we can find the components of the tension and the angles defining its direction.

MODELING and ANALYSIS:

a. Components of the Force. The line of action of the force acting on the bolt passes through points A and B, and the force is directed from A to B. The components of the vector $\overrightarrow{AB}$, which has the same direction as the force, are

$$d_x = -40 \text{ m} \qquad d_y = +80 \text{ m} \qquad d_z = +30 \text{ m}$$

The total distance from A to B is

$$AB = d = \sqrt{d_x^2 + d_y^2 + d_z^2} = 94.3 \text{ m}$$

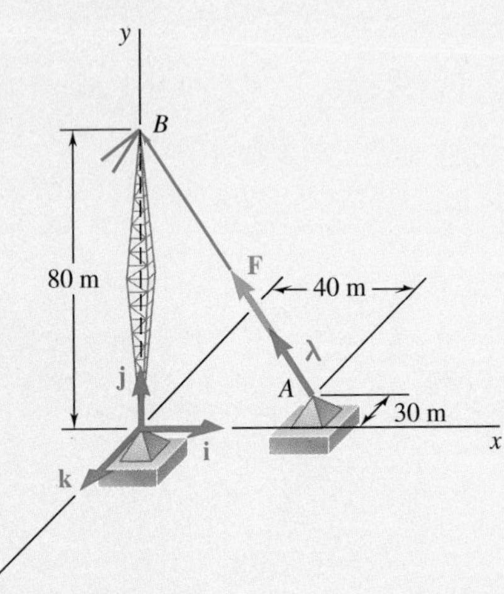

Fig. 1 Cable force acting on bolt at A, and its unit vector.

Denoting the unit vectors along the coordinate axes by **i**, **j**, and **k**, you have

$$\overrightarrow{AB} = -(40 \text{ m})\mathbf{i} + (80 \text{ m})\mathbf{j} + (30 \text{ m})\mathbf{k}$$

Introducing the unit vector $\boldsymbol{\lambda} = \overrightarrow{AB}/AB$ (Fig. 1), you can express **F** in terms of $\overrightarrow{AB}$ as

$$\mathbf{F} = F\boldsymbol{\lambda} = F\frac{\overrightarrow{AB}}{AB} = \frac{2500 \text{ N}}{94.3 \text{ m}}\overrightarrow{AB}$$

Substituting the expression for $\overrightarrow{AB}$ gives you

$$\mathbf{F} = \frac{2500 \text{ N}}{94.3 \text{ m}}\left[-(40 \text{ m})\mathbf{i} + (80 \text{ m})\mathbf{j} + (30 \text{ m})\mathbf{k}\right]$$
$$= -(1060 \text{ N})\mathbf{i} + (2120 \text{ N})\mathbf{j} + (795 \text{ N})\mathbf{k}$$

The components of **F**, therefore, are

$$F_x = -1060 \text{ N} \qquad F_y = +2120 \text{ N} \qquad F_z = +795 \text{ N} \quad \blacktriangleleft$$

b. Direction of the Force. Using Eqs. (2.25), you can write the direction cosines directly (Fig. 2):

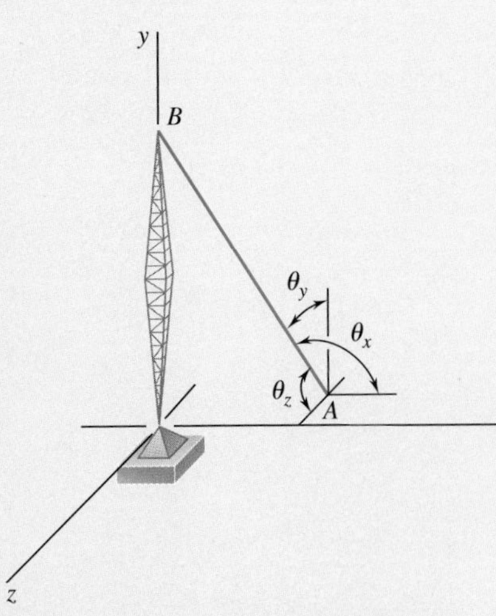

Fig. 2 Direction angles for cable AB.

$$\cos\theta_x = \frac{F_x}{F} = \frac{-1060 \text{ N}}{2500 \text{ N}} \qquad \cos\theta_y = \frac{F_y}{F} = \frac{+2120 \text{ N}}{2500 \text{ N}}$$

$$\cos\theta_z = \frac{F_z}{F} = \frac{+795 \text{ N}}{2500 \text{ N}}$$

Calculating each quotient and its arc cosine, you obtain

$$\theta_x = 115.1^\circ \qquad \theta_y = 32.0^\circ \qquad \theta_z = 71.5^\circ \blacktriangleleft$$

(*Note*. You could have obtained this same result by using the components and magnitude of the vector $\overrightarrow{AB}$ rather than those of the force **F**.)

REFLECT and THINK: It makes sense that, for a given geometry, only a certain set of components and angles characterize a given resultant force. The methods in this section allow you to translate back and forth between forces and geometry.

Sample Problem 2.8

A wall section of precast concrete is temporarily held in place by the cables shown. If the tension is 840 lb in cable *AB* and 1200 lb in cable *AC*, determine the magnitude and direction of the resultant of the forces exerted by cables *AB* and *AC* on stake *A*.

STRATEGY: This is a problem in adding concurrent forces in space. The simplest approach is to first resolve the forces into components and to then sum the components and find the resultant.

MODELING and ANALYSIS:

Components of the Forces. First resolve the force exerted by each cable on stake *A* into *x*, *y*, and *z* components. To do this, determine the components and magnitude of the vectors $\overrightarrow{AB}$ and $\overrightarrow{AC}$, measuring them from *A* toward the wall section (Fig. 1). Denoting the unit vectors along the coordinate axes by **i**, **j**, **k**, these vectors are

$$\overrightarrow{AB} = -(16\text{ ft})\mathbf{i} + (8\text{ ft})\mathbf{j} + (11\text{ ft})\mathbf{k} \qquad AB = 21\text{ ft}$$
$$\overrightarrow{AC} = -(16\text{ ft})\mathbf{i} + (8\text{ ft})\mathbf{j} - (16\text{ ft})\mathbf{k} \qquad AC = 24\text{ ft}$$

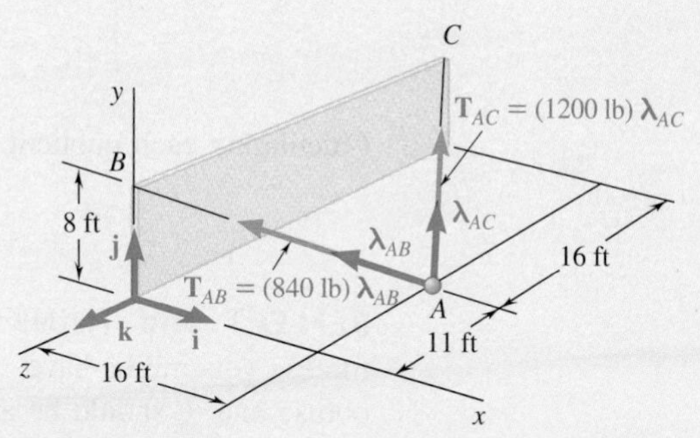

Fig. 1 Cable forces acting on stake at *A*, and their unit vectors.

Denoting by $\boldsymbol{\lambda}_{AB}$ the unit vector along AB, the tension in AB is

$$\mathbf{T}_{AB} = T_{AB}\boldsymbol{\lambda}_{AB} = T_{AB}\frac{\overrightarrow{AB}}{AB} = \frac{840\text{ lb}}{21\text{ ft}}\overrightarrow{AB}$$

Substituting the expression found for $\overrightarrow{AB}$, the tension becomes

$$\mathbf{T}_{AB} = \frac{840\text{ lb}}{21\text{ ft}}[-(16\text{ ft})\mathbf{i} + (8\text{ ft})\mathbf{j} + (11\text{ ft})\mathbf{k}]$$

$$\mathbf{T}_{AB} = -(640\text{ lb})\mathbf{i} + (320\text{ lb})\mathbf{j} + (440\text{ lb})\mathbf{k}$$

Similarly, denoting by $\boldsymbol{\lambda}_{AC}$ the unit vector along AC, the tension in AC is

$$\mathbf{T}_{AC} = T_{AC}\boldsymbol{\lambda}_{AC} = T_{AC}\frac{\overrightarrow{AC}}{AC} = \frac{1200\text{ lb}}{24\text{ ft}}\overrightarrow{AC}$$

$$\mathbf{T}_{AC} = -(800\text{ lb})\mathbf{i} + (400\text{ lb})\mathbf{j} - (800\text{ lb})\mathbf{k}$$

Resultant of the Forces. The resultant $\mathbf{R}$ of the forces exerted by the two cables is

$$\mathbf{R} = \mathbf{T}_{AB} + \mathbf{T}_{AC} = -(1440\text{ lb})\mathbf{i} + (720\text{ lb})\mathbf{j} - (360\text{ lb})\mathbf{k}$$

You can now determine the magnitude and direction of the resultant as

$$R = \sqrt{R_x^2 + R_y^2 + R_z^2} = \sqrt{(-1440)^2 + (720)^2 + (-300)^2}$$

$$R = 1650\text{ lb} \quad \blacktriangleleft$$

The direction cosines come from Eqs. (2.33):

$$\cos\theta_x = \frac{R_x}{R} = \frac{-1440\text{ lb}}{1650\text{ lb}} \qquad \cos\theta_y = \frac{R_y}{R} = \frac{+720\text{ lb}}{1650\text{ lb}}$$

$$\cos\theta_z = \frac{R_z}{R} = \frac{-360\text{ lb}}{1650\text{ lb}}$$

Calculating each quotient and its arc cosine, the angles are

$$\theta_x = 150.8^\circ \qquad \theta_y = 64.1^\circ \qquad \theta_z = 102.6^\circ \quad \blacktriangleleft$$

REFLECT and THINK: Based on visual examination of the cable forces, you might have anticipated that θ_x for the resultant should be obtuse and θ_y should be acute. The outcome of θ_z was not as apparent.

Problems

2.56 Determine (*a*) the *x*, *y*, and *z* components of the 900-N force, (*b*) the angles θ_x, θ_y, and θ_z that the force forms with the coordinate axes.

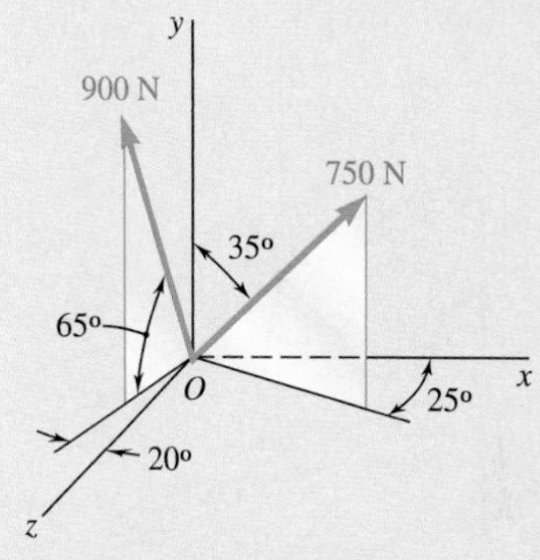

Fig. P2.56 and P2.57

2.57 Determine (*a*) the *x*, *y*, and *z* components of the 750-N force, (*b*) the angles θ_x, θ_y, and θ_z that the force forms with the coordinate axes.

2.58 The end of the coaxial cable *AE* is attached to the pole *AB*, which is strengthened by the guy wires *AC* and *AD*. Knowing that the tension in wire *AC* is 120 lb, determine (*a*) the components of the force exerted by this wire on the pole, (*b*) the angles θ_x, θ_y, and θ_z that the force forms with the coordinate axes.

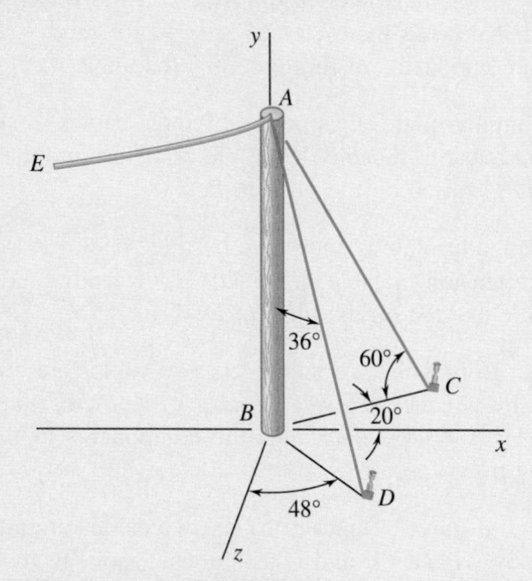

Fig. P2.58 and P2.59

2.59 The end of the coaxial cable *AE* is attached to the pole *AB*, which is strengthened by the guy wires *AC* and *AD*. Knowing that the tension in wire *AD* is 85 lb, determine (*a*) the components of the force exerted by this wire on the pole, (*b*) the angles θ_x, θ_y, and θ_z that the force forms with the coordinate axes.

2.60 A gun is aimed at a point *A* located 35° east of north. Knowing that the barrel of the gun forms an angle of 40° with the horizontal and that the maximum recoil force is 400 N, determine (*a*) the *x*, *y*, and *z* components of that force, (*b*) the values of the angles θ_x, θ_y, and θ_z defining the direction of the recoil force. (Assume that the *x*, *y*, and *z* axes are directed, respectively, east, up, and south.)

2.61 Solve Prob. 2.60, assuming that point *A* is located 15° north of west and that the barrel of the gun forms an angle of 25° with the horizontal.

2.62 Determine the magnitude and direction of the force $\mathbf{F} = (690 \text{ lb})\mathbf{i} + (300 \text{ lb})\mathbf{j} - (580 \text{ lb})\mathbf{k}$.

2.63 Determine the magnitude and direction of the force **F** = (650 N)**i** – (320 N)**j** + (760 N)**k**.

2.64 A force acts at the origin of a coordinate system in a direction defined by the angles $\theta_x = 69.3°$ and $\theta_z = 57.9°$. Knowing that the y component of the force is –174.0 lb, determine (*a*) the angle θ_y, (*b*) the other components and the magnitude of the force.

2.65 A force acts at the origin of a coordinate system in a direction defined by the angles $\theta_x = 70.9°$ and $\theta_y = 144.9°$. Knowing that the z component of the force is –52.0 lb, determine (*a*) the angle θ_z, (*b*) the other components and the magnitude of the force.

2.66 A force acts at the origin of a coordinate system in a direction defined by the angles $\theta_y = 55°$ and $\theta_z = 45°$. Knowing that the x component of the force is –500 lb, determine (*a*) the angle θ_x, (*b*) the other components and the magnitude of the force.

2.67 A force **F** of magnitude 1200 N acts at the origin of a coordinate system. Knowing that $\theta_x = 65°$, $\theta_y = 40°$, and $F_z > 0$, determine (*a*) the components of the force, (*b*) the angle θ_z.

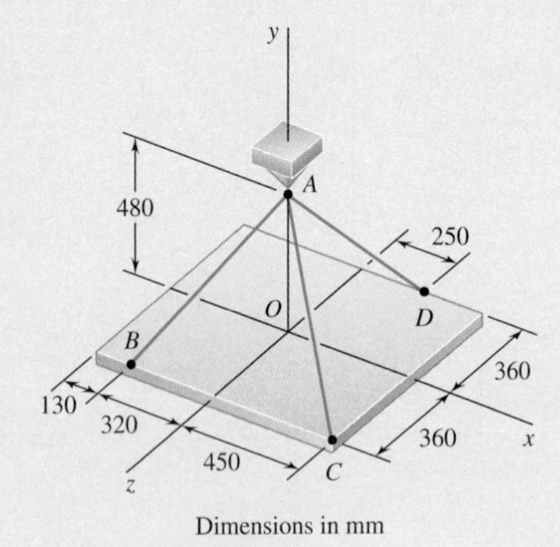

Fig. *P2.68* and P2.69

2.68 A rectangular plate is supported by three cables as shown. Knowing that the tension in cable AB is 408 N, determine the components of the force exerted on the plate at B.

2.69 A rectangular plate is supported by three cables as shown. Knowing that the tension in cable AD is 429 N, determine the components of the force exerted on the plate at D.

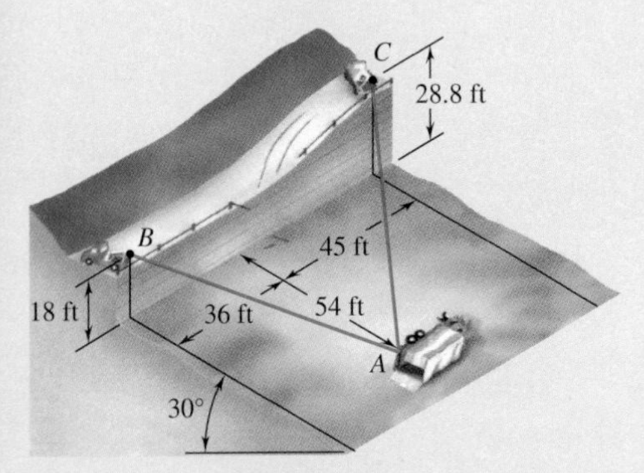

Fig. *P2.70* and P2.71

2.70 In order to move a wrecked truck, two cables are attached at A and pulled by winches B and C as shown. Knowing that the tension in cable AB is 2 kips, determine the components of the force exerted at A by the cable.

2.71 In order to move a wrecked truck, two cables are attached at A and pulled by winches B and C as shown. Knowing that the tension in cable AC is 1.5 kips, determine the components of the force exerted at A by the cable.

2.72 Find the magnitude and direction of the resultant of the two forces shown knowing that P = 300 N and Q = 400 N.

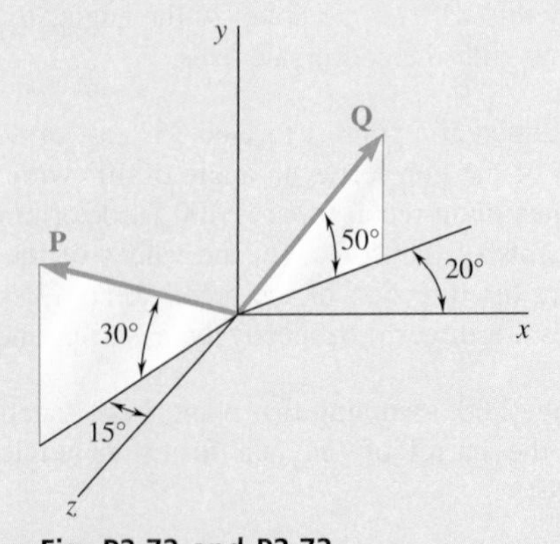

Fig. P2.72 and P2.73

2.73 Find the magnitude and direction of the resultant of the two forces shown knowing that P = 400 N and Q = 300 N.

2.74 Knowing that the tension is 425 lb in cable AB and 510 lb in cable AC, determine the magnitude and direction of the resultant of the forces exerted at A by the two cables.

2.75 Knowing that the tension is 510 lb in cable AB and 425 lb in cable AC, determine the magnitude and direction of the resultant of the forces exerted at A by the two cables.

2.76 A frame ABC is supported in part by cable DBE that passes through a frictionless ring at B. Knowing that the tension in the cable is 385 N, determine the magnitude and direction of the resultant of the forces exerted by the cable at B.

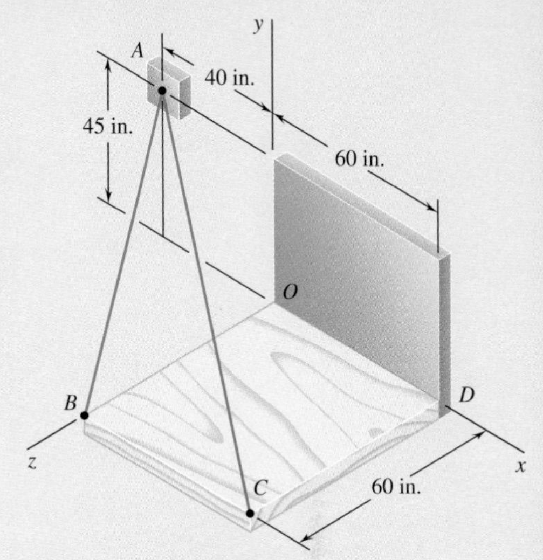

Fig. P2.74 and P2.75

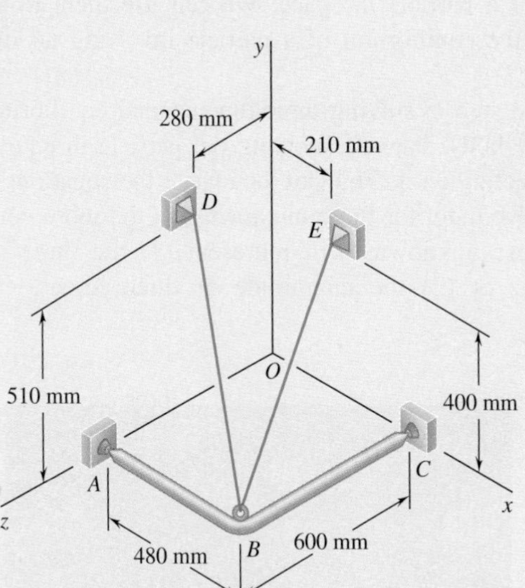
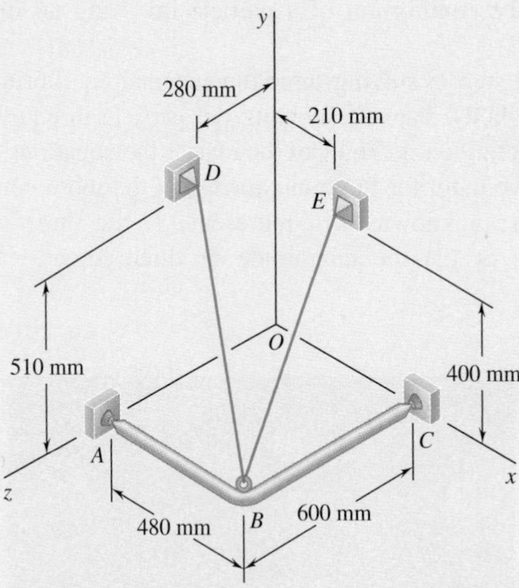

Fig. P2.76

2.77 For the plate of Prob. 2.68, determine the tensions in cables AB and AD knowing that the tension in cable AC is 54 N and that the resultant of the forces exerted by the three cables at A must be vertical.

2.78 The boom OA carries a load **P** and is supported by two cables as shown. Knowing that the tension in cable AB is 183 lb and that the resultant of the load **P** and of the forces exerted at A by the two cables must be directed along OA, determine the tension in cable AC.

2.79 For the boom and loading of Prob. 2.78, determine the magnitude of the load **P**.

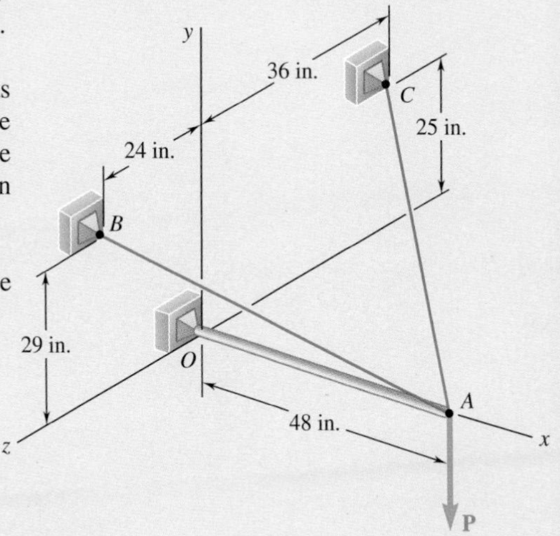

Fig. P2.78

2.5 FORCES AND EQUILIBRIUM IN SPACE

According to the definition given in Sec. 2.3, a particle A is in equilibrium if the resultant of all the forces acting on A is zero. The components R_x, R_y, and R_z of the resultant of forces in space are given by equations (2.31); when the components of the resultant are zero, we have

$$\Sigma F_x = 0 \qquad \Sigma F_y = 0 \qquad \Sigma F_z = 0 \tag{2.34}$$

Equations (2.34) represent the necessary and sufficient conditions for the equilibrium of a particle in space. We can use them to solve problems dealing with the equilibrium of a particle involving no more than three unknowns.

The first step in solving three-dimensional equilibrium problems is to draw a free-body diagram showing the particle in equilibrium and *all* of the forces acting on it. You can then write the equations of equilibrium (2.34) and solve them for three unknowns. In the more common types of problems, these unknowns will represent (1) the three components of a single force or (2) the magnitude of three forces, each of known direction.

Photo 2.4 Although we cannot determine the tension in the four cables supporting the car by using the three equations (2.34), we can obtain a relation among the tensions by analyzing the equilibrium of the hook.

Sample Problem 2.9

A 200-kg cylinder is hung by means of two cables *AB* and *AC* that are attached to the top of a vertical wall. A horizontal force **P** perpendicular to the wall holds the cylinder in the position shown. Determine the magnitude of **P** and the tension in each cable.

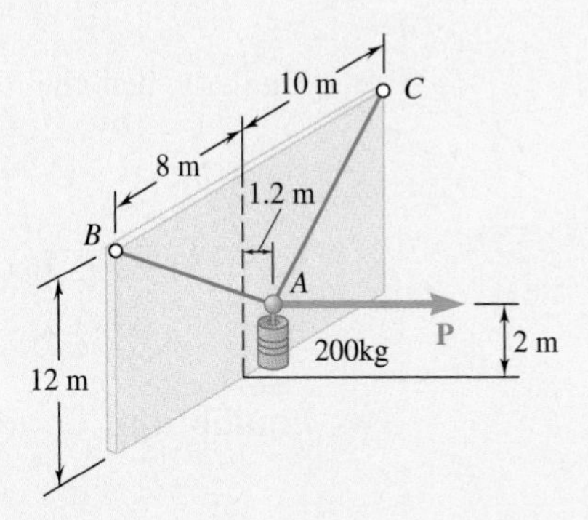

STRATEGY: Connection point *A* is acted upon by four forces, including the weight of the cylinder. You can use the given geometry to express the force components of the cables and then apply equilibrium conditions to calculate the tensions.

MODELING and ANALYSIS:

Free-Body Diagram. Choose point *A* as a free body; this point is subjected to four forces, three of which are of unknown magnitude. Introducing the unit vectors **i**, **j**, and **k**, resolve each force into rectangular components (Fig. 1):

$$\mathbf{P} = P\mathbf{i}$$

$$\mathbf{W} = -mg\mathbf{j} = -(200 \text{ kg})(9.81 \text{ m/s}^2)\mathbf{j} = -(1962 \text{ N})\mathbf{j} \quad (1)$$

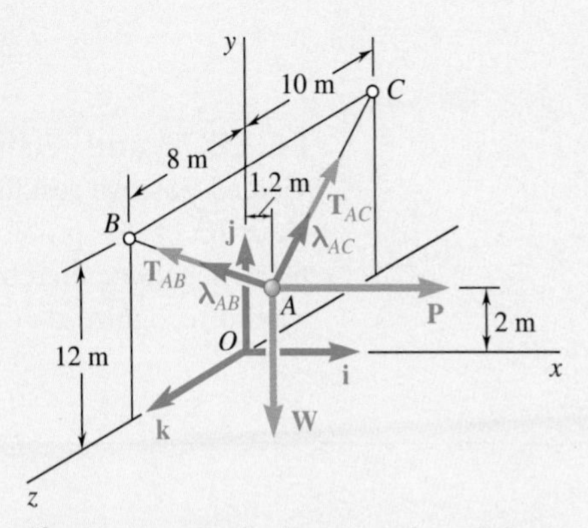

Fig. 1 Free-body diagram of particle *A*.

For $\mathbf{T}_{AB}$ and $\mathbf{T}_{AC}$, it is first necessary to determine the components and magnitudes of the vectors $\overrightarrow{AB}$ and $\overrightarrow{AC}$. Denoting the unit vector along AB by $\boldsymbol{\lambda}_{AB}$, you can write $\mathbf{T}_{AB}$ as

$$\overrightarrow{AB} = -(1.2\text{ m})\mathbf{i} + (10\text{ m})\mathbf{j} + (8\text{ m})\mathbf{k} \qquad AB = 12.862\text{ m}$$

$$\boldsymbol{\lambda}_{AB} = \frac{\overrightarrow{AB}}{12.862\text{ m}} = -0.09330\mathbf{i} + 0.7775\mathbf{j} + 0.6220\mathbf{k}$$

$$\mathbf{T}_{AB} = T_{AB}\boldsymbol{\lambda}_{AB} = -0.09330T_{AB}\mathbf{i} + 0.7775T_{AB}\mathbf{j} + 0.6220T_{AB} \qquad (2)$$

Similarly, denoting the unit vector along AC by $\boldsymbol{\lambda}_{AC}$, you have for $\mathbf{T}_{AC}$

$$\overrightarrow{AC} = -(1.2\text{ m})\mathbf{i} + (10\text{ m})\mathbf{j} - (10\text{ m})\mathbf{k} \qquad AC = 14.193\text{ m}$$

$$\boldsymbol{\lambda}_{AC} = \frac{\overrightarrow{AC}}{14.193\text{ m}} = -0.08455\mathbf{i} + 0.7046\mathbf{j} - 0.7046\mathbf{k}$$

$$\mathbf{T}_{AC} = T_{AC}\boldsymbol{\lambda}_{AC} = -0.08455T_{AC}\mathbf{i} + 0.7046T_{AC}\mathbf{j} - 0.7046T_{AC}\mathbf{k} \qquad (3)$$

Equilibrium Condition. Since A is in equilibrium, you must have

$$\Sigma\mathbf{F} = 0: \qquad \mathbf{T}_{AB} + \mathbf{T}_{AC} + \mathbf{P} + \mathbf{W} = 0$$

or substituting from Eqs. (1), (2), and (3) for the forces and factoring **i**, **j**, and **k**, you have

$$\begin{aligned}(-0.09330T_{AB} - 0.08455T_{AC} + P)\mathbf{i}\\ + (0.7775T_{AB} + 0.7046T_{AC} - 1962\text{ N})\mathbf{j}\\ + (0.6220T_{AB} - 0.7046T_{AC})\mathbf{k} = 0\end{aligned}$$

Setting the coefficients of **i**, **j**, and **k** equal to zero, you can write three scalar equations, which express that the sums of the x, y, and z components of the forces are respectively equal to zero.

$$\begin{aligned}(\Sigma F_x = 0:) &\qquad -0.09330T_{AB} - 0.08455T_{AC} + P = 0\\ (\Sigma F_y = 0:) &\qquad +0.7775T_{AB} + 0.7046T_{AC} - 1962\text{ N} = 0\\ (\Sigma F_z = 0:) &\qquad +0.6220T_{AB} - 0.7046T_{AC} = 0\end{aligned}$$

Solving these equations, you obtain

$$P = 235\text{ N} \qquad T_{AB} = 1402\text{ N} \qquad T_{AC} = 1238\text{ N} \blacktriangleleft$$

REFLECT and THINK: The solution of the three unknown forces yielded positive results, which is completely consistent with the physical situation of this problem. Conversely, if one of the cable force results had been negative, thereby reflecting compression instead of tension, you should recognize that the solution is in error.

Problems

2.80 A container is supported by three cables that are attached to a ceiling as shown. Determine the weight W of the container, knowing that the tension in cable AB is 6 kN.

2.81 A container is supported by three cables that are attached to a ceiling as shown. Determine the weight W of the container, knowing that the tension in cable AD is 4.3 kN.

2.82 Three cables are used to tether a balloon as shown. Knowing that the balloon exerts an 800-N vertical force at A, determine the tension in each cable.

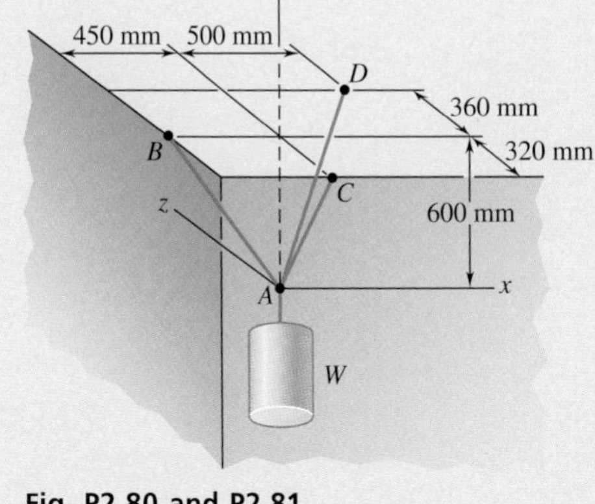

Fig. P2.80 and P2.81

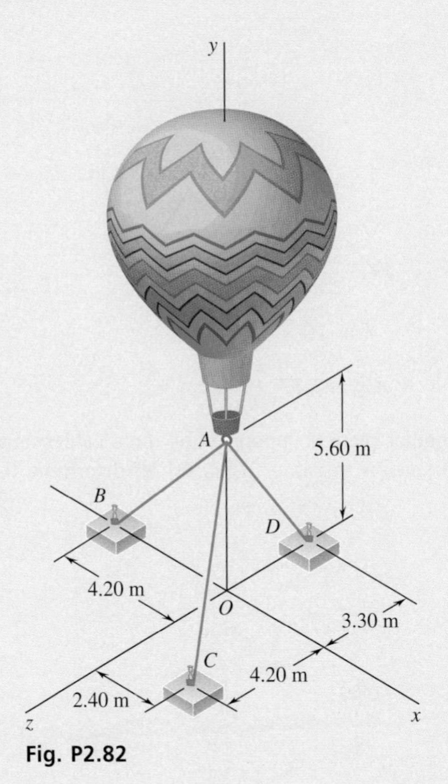

Fig. P2.82

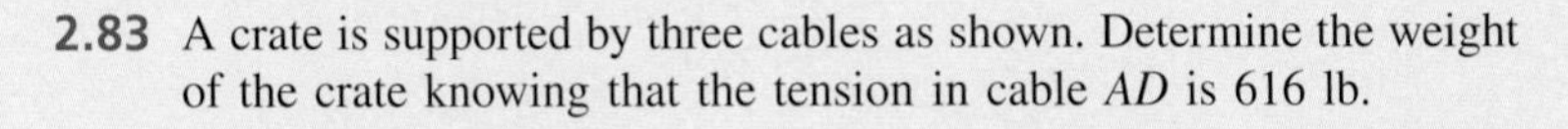
2.83 A crate is supported by three cables as shown. Determine the weight of the crate knowing that the tension in cable AD is 616 lb.

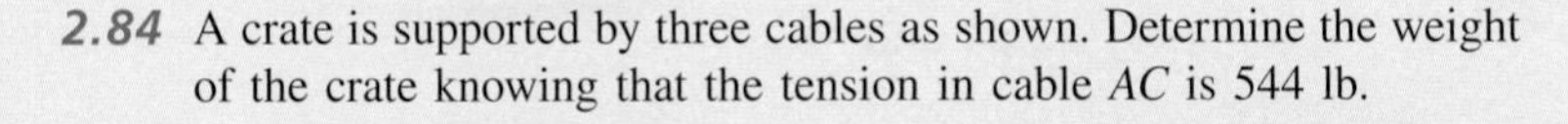
2.84 A crate is supported by three cables as shown. Determine the weight of the crate knowing that the tension in cable AC is 544 lb.

2.85 A 1600-lb crate is supported by three cables as shown. Determine the tension in each cable.

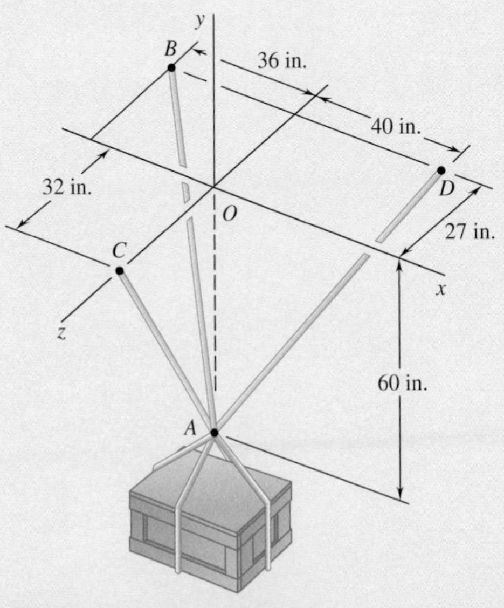

Fig. P2.83, *P2.84* and P2.85

2.86 Three wires are connected at point D, which is located 18 in. below the T-shaped pipe support ABC. Determine the tension in each wire when a 180-lb cylinder is suspended from point D as shown.

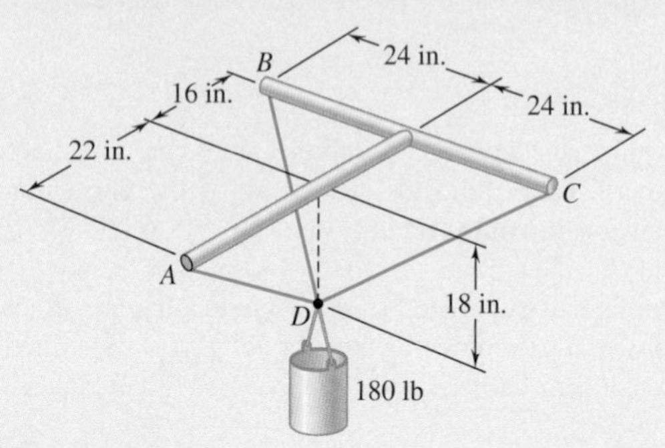

Fig. P2.86

2.87 A 36-lb triangular plate is supported by three wires as shown. Determine the tension in each wire, knowing that $a = 6$ in.

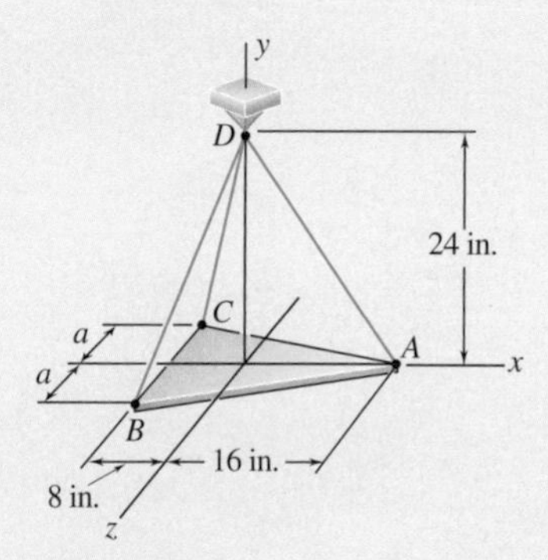

Fig. P2.87

2.88 A rectangular plate is supported by three cables as shown. Knowing that the tension in cable AC is 60 N, determine the weight of the plate.

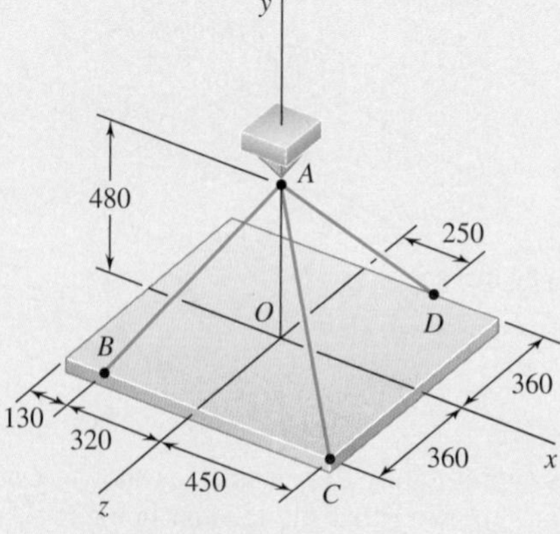

Fig. *P2.88* and P2.89

2.89 A rectangular plate is supported by three cables as shown. Knowing that the tension in cable AD is 520 N, determine the weight of the plate.

2.90 In trying to move across a slippery icy surface, a 175-lb man uses two ropes AB and AC. Knowing that the force exerted on the man by the icy surface is perpendicular to that surface, determine the tension in each rope.

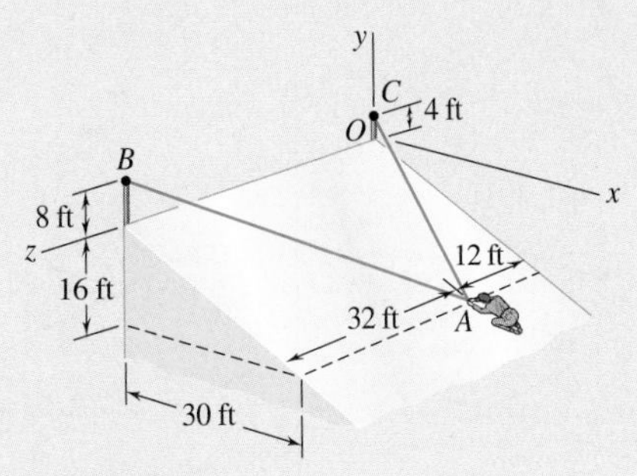

Fig. P2.90

2.91 Solve Prob. 2.90, assuming that a friend is helping the man at A by pulling on him with a force $\mathbf{P} = -(45 \text{ lb})\mathbf{k}$.

2.92 Three cables are connected at A, where the forces $\mathbf{P}$ and $\mathbf{Q}$ are applied as shown. Knowing that $Q = 0$, find the value of P for which the tension in cable AD is 305 N.

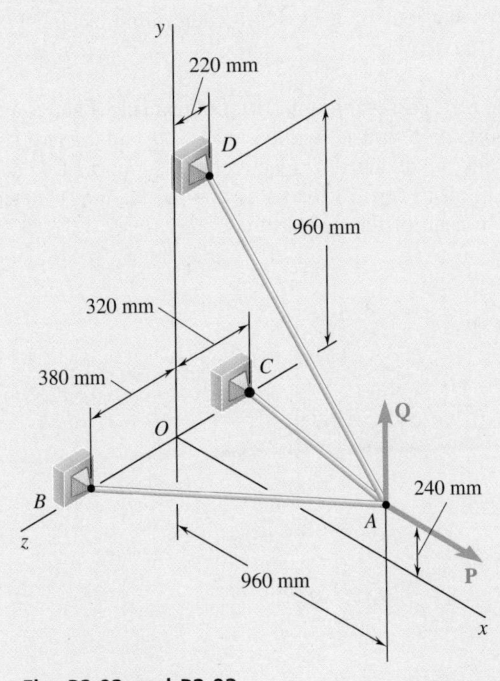

Fig. P2.92 and P2.93

2.93 Three cables are connected at A, where the forces $\mathbf{P}$ and $\mathbf{Q}$ are applied as shown. Knowing that $P = 1200$ N, determine the values of Q for which cable AD is taut.

2.94 A container of weight W is suspended from ring A. Cable BAC passes through the ring and is attached to fixed supports at B and C. Two forces $\mathbf{P} = P\mathbf{i}$ and $\mathbf{Q} = Q\mathbf{k}$ are applied to the ring to maintain the container in the position shown. Knowing that $W = 376$ N, determine P and Q. (*Hint:* The tension is the same in both portions of cable BAC.)

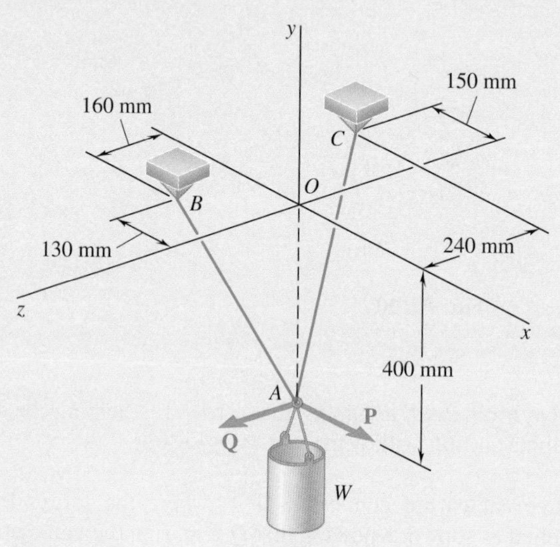

Fig. P2.94

2.95 For the system of Prob. 2.94, determine W and Q knowing that $P = 164$ N.

2.96 Cable BAC passes through a frictionless ring A and is attached to fixed supports at B and C, while cables AD and AE are both tied to the ring and are attached, respectively, to supports at D and E. Knowing that a 200-lb vertical load $\mathbf{P}$ is applied to ring A, determine the tension in each of the three cables.

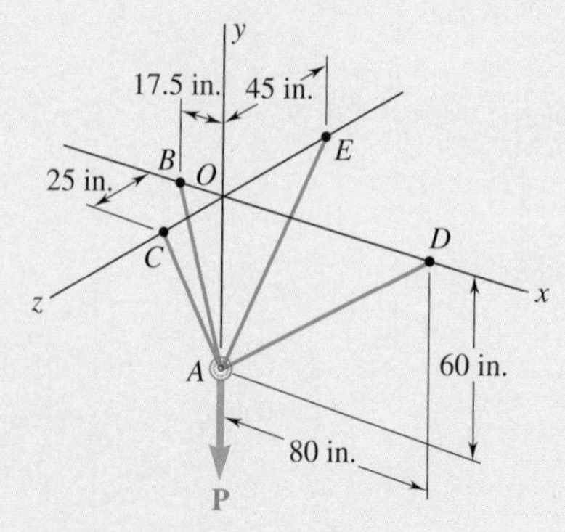

Fig. P2.96

2.97 Knowing that the tension in cable AE of Prob. 2.96 is 75 lb, determine (*a*) the magnitude of the load $\mathbf{P}$, (*b*) the tension in cables BAC and AD.

2.98 A container of weight W is suspended from ring A, to which cables AC and AE are attached. A force **P** is applied to the end F of a third cable that passes over a pulley at B and through ring A and that is attached to a support at D. Knowing that $W = 1000$ N, determine the magnitude of **P**. (*Hint:* The tension is the same in all portions of cable $FBAD$.)

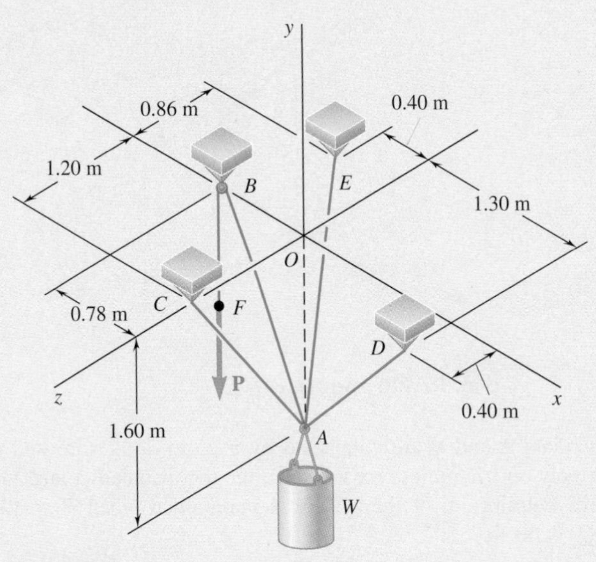

Fig. P2.98

2.99 Using two ropes and a roller chute, two workers are unloading a 200-lb cast-iron counterweight from a truck. Knowing that at the instant shown the counterweight is kept from moving and that the positions of points A, B, and C are, respectively, $A(0, -20 \text{ in.}, 40 \text{ in.})$, $B(-40 \text{ in.}, 50 \text{ in.}, 0)$, and $C(45 \text{ in.}, 40 \text{ in.}, 0)$, and assuming that no friction exists between the counterweight and the chute, determine the tension in each rope. (*Hint:* Because there is no friction, the force exerted by the chute on the counterweight must be perpendicular to the chute.)

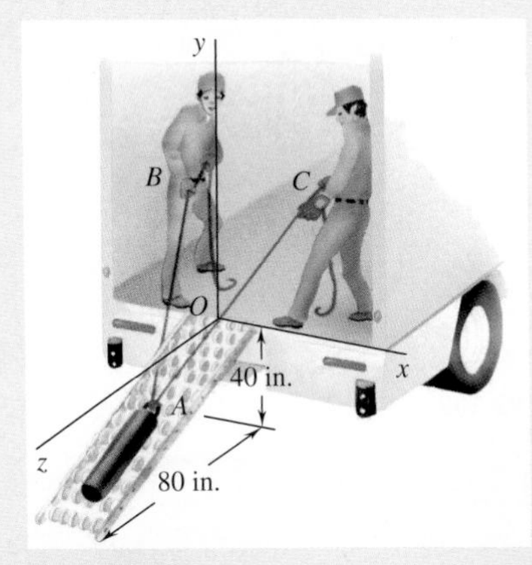

Fig. P2.99

2.100 Collars A and B are connected by a 25-in.-long wire and can slide freely on frictionless rods. If a 60-lb force **Q** is applied to collar B as shown, determine (*a*) the tension in the wire when $x = 9$ in., (*b*) the corresponding magnitude of the force **P** required to maintain the equilibrium of the system.

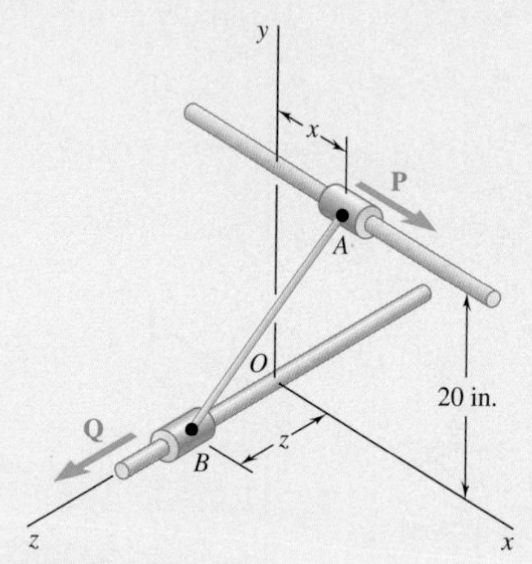

Fig. P2.100 and *P2.101*

2.101 Collars A and B are connected by a 25-in.-long wire and can slide freely on frictionless rods. Determine the distances x and z for which the equilibrium of the system is maintained when $P = 120$ lb and $Q = 60$ lb.

2.102 Collars A and B are connected by a 525-mm-long wire and can slide freely on frictionless rods. If a force $\mathbf{P} = (341\text{ N})\mathbf{j}$ is applied to collar A, determine (*a*) the tension in the wire when $y = 155$ mm, (*b*) the magnitude of the force **Q** required to maintain the equilibrium of the system.

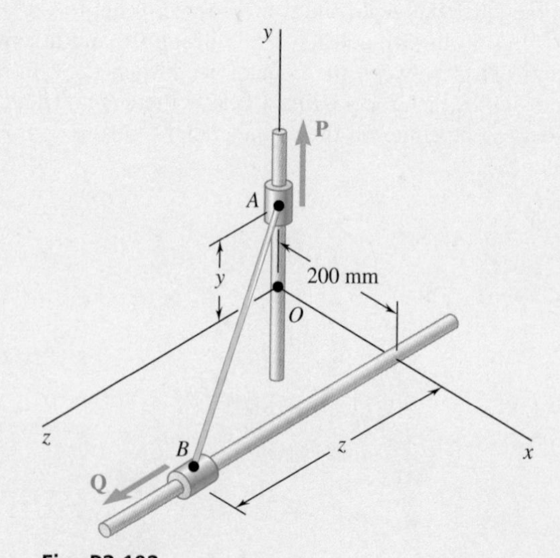

Fig. P2.102

2.103 Solve Prob. 2.102 assuming that $y = 275$ mm.

Review and Summary

In this chapter, we have studied the effect of forces on particles, i.e., on bodies of such shape and size that we may assume all forces acting on them apply at the same point.

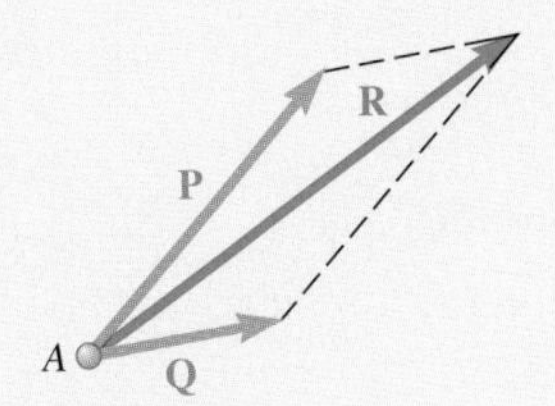

Fig. 2.30

Resultant of Two Forces

Forces are *vector quantities;* they are characterized by a point of application, a magnitude, and a direction, and they add according to the parallelogram law (Fig. 2.30). We can determine the magnitude and direction of the resultant **R** of two forces **P** and **Q** either graphically or by trigonometry using the law of cosines and the law of sines [Sample Prob. 2.1].

Components of a Force

Any given force acting on a particle can be resolved into two or more components, i.e., it can be replaced by two or more forces that have the same effect on the particle. A force **F** can be resolved into two components **P** and **Q** by drawing a parallelogram with **F** for its diagonal; the components **P** and **Q** are then represented by the two adjacent sides of the parallelogram (Fig. 2.31). Again, we can determine the components either graphically or by trigonometry [Sec. 2.1E].

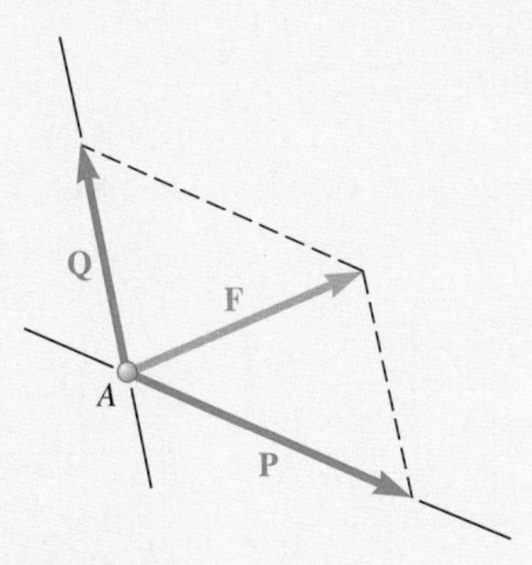

Fig. 2.31

Rectangular Components; Unit Vectors

A force **F** is resolved into two rectangular components if its components $\mathbf{F}_x$ and $\mathbf{F}_y$ are perpendicular to each other and are directed along the coordinate axes (Fig. 2.32). Introducing the unit vectors **i** and **j** along the x and y axes, respectively, we can write the components and the vector as [Sec. 2.2A]

$$\mathbf{F}_x = F_x\mathbf{i} \qquad \mathbf{F}_y = F_y\mathbf{j} \tag{2.6}$$

and

$$\mathbf{F} = F_x\mathbf{i} + F_y\mathbf{j} \tag{2.7}$$

where F_x and F_y are the *scalar components* of **F**. These components, which can be positive or negative, are defined by the relations

$$F_x = F\cos\theta \qquad F_y = F\sin\theta \tag{2.8}$$

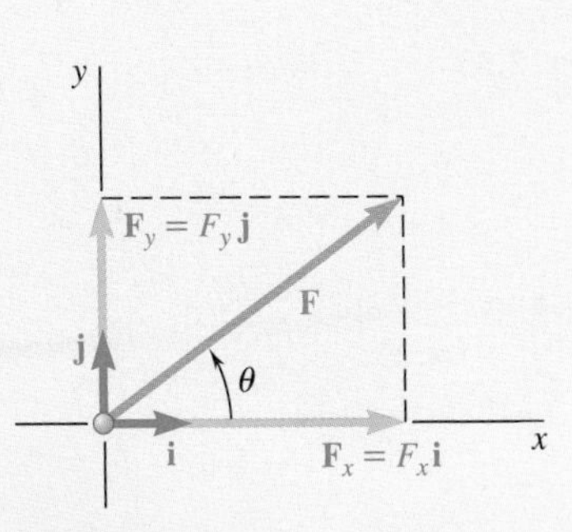

Fig. 2.32

When the rectangular components F_x and F_y of a force **F** are given, we can obtain the angle θ defining the direction of the force from

$$\tan \theta = \frac{F_y}{F_x} \quad \textbf{(2.9)}$$

We can obtain the magnitude F of the force by solving one of the equations (2.8) for F or by applying the Pythagorean theorem:

$$F = \sqrt{F_x^2 + F_y^2} \quad \textbf{(2.10)}$$

Resultant of Several Coplanar Forces

When three or more coplanar forces act on a particle, we can obtain the rectangular components of their resultant **R** by adding the corresponding components of the given forces algebraically [Sec. 2.2B]:

$$R_x = \Sigma F_x \qquad R_y = \Sigma F_y \quad \textbf{(2.13)}$$

The magnitude and direction of **R** then can be determined from relations similar to Eqs. (2.9) and (2.10) [Sample Prob. 2.3].

Forces in Space

A force **F** in three-dimensional space can be resolved into rectangular components $\mathbf{F}_x$, $\mathbf{F}_y$, and $\mathbf{F}_z$ [Sec. 2.4A]. Denoting by θ_x, θ_y, and θ_z, respectively, the angles that **F** forms with the x, y, and z axes (Fig. 2.33), we have

$$F_x = F \cos \theta_x \qquad F_y = F \cos \theta_y \qquad F_z = F \cos \theta_z \quad \textbf{(2.19)}$$

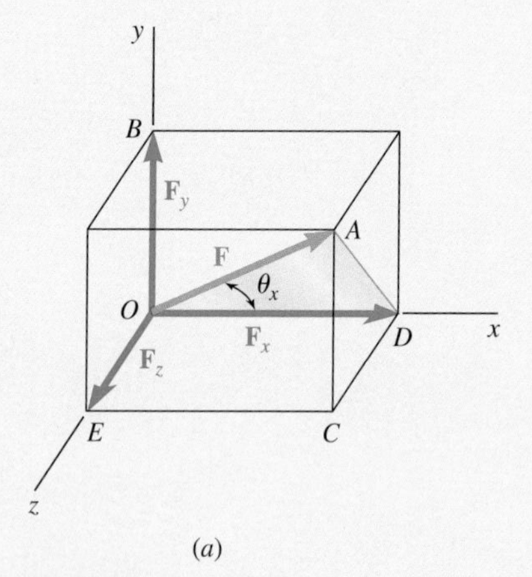

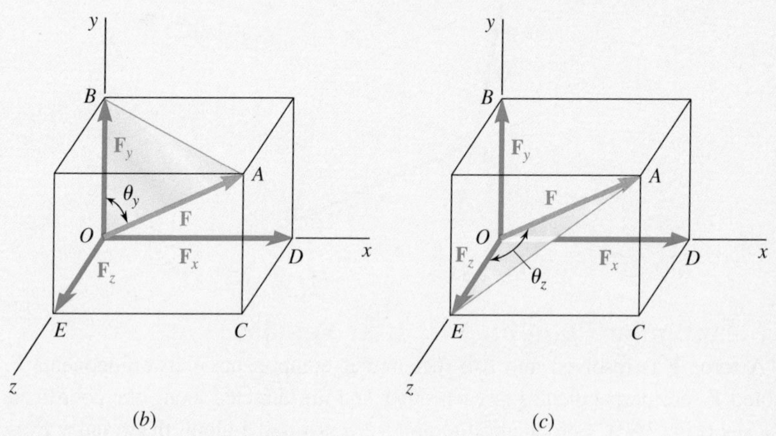

Fig. 2.33

Direction Cosines

The cosines of θ_x, θ_y, and θ_z are known as the *direction cosines* of the force **F**. Introducing the unit vectors **i**, **j**, and **k** along the coordinate axes, we can write **F** as

$$\mathbf{F} = F_x\mathbf{i} + F_y\mathbf{j} + F_z\mathbf{k} \quad \textbf{(2.20)}$$

or

$$\mathbf{F} = F(\cos \theta_x\mathbf{i} + \cos \theta_y\mathbf{j} + \cos \theta_z\mathbf{k}) \quad \textbf{(2.21)}$$

This last equation shows (Fig. 2.34) that **F** is the product of its magnitude F and the unit vector expressed by

$$\boldsymbol{\lambda} = \cos\theta_x\mathbf{i} + \cos\theta_y\mathbf{j} + \cos\theta_z\mathbf{k}$$

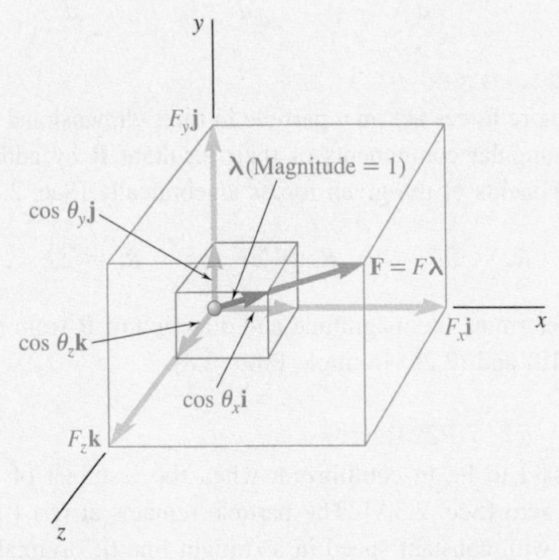

Fig. 2.34

Since the magnitude of $\boldsymbol{\lambda}$ is equal to unity, we must have

$$\cos^2\theta_x + \cos^2\theta_y + \cos^2\theta_z = 1 \quad \textbf{(2.24)}$$

When we are given the rectangular components F_x, F_y, and F_z of a force **F**, we can find the magnitude **F** of the force by

$$F = \sqrt{F_x^2 + F_y^2 + F_z^2} \quad \textbf{(2.18)}$$

and the direction cosines of **F** are obtained from Eqs. (2.19). We have

$$\cos\theta_x = \frac{F_x}{F} \qquad \cos\theta_y = \frac{F_y}{F} \qquad \cos\theta_z = \frac{F_z}{F} \quad \textbf{(2.25)}$$

When a force **F** is defined in three-dimensional space by its magnitude F and two points M and N on its line of action [Sec. 2.4B], we can obtain its rectangular components by first expressing the vector $\overrightarrow{MN}$ joining points M and N in terms of its components d_x, d_y, and d_z (Fig. 2.35):

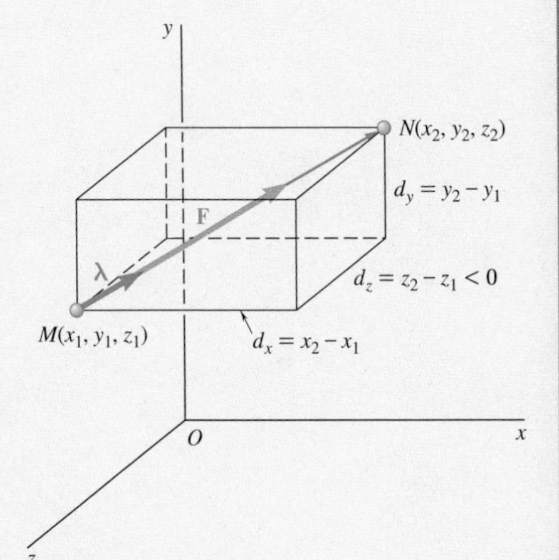

Fig. 2.35

$$\overrightarrow{MN} = d_x\mathbf{i} + d_y\mathbf{j} + d_z\mathbf{k} \quad \textbf{(2.26)}$$

We next determine the unit vector $\boldsymbol{\lambda}$ along the line of action of **F** by dividing $\overrightarrow{MN}$ by its magnitude $MN = d$:

$$\boldsymbol{\lambda} = \frac{\overrightarrow{MN}}{MN} = \frac{1}{d}(d_x\mathbf{i} + d_y\mathbf{j} + d_z\mathbf{k}) \quad \textbf{(2.27)}$$

Recalling that **F** is equal to the product of F and $\boldsymbol{\lambda}$, we have

$$\mathbf{F} = F\boldsymbol{\lambda} = \frac{F}{d}(d_x\mathbf{i} + d_y\mathbf{j} + d_z\mathbf{k}) \quad \textbf{(2.28)}$$

From this equation it follows [Sample Probs. 2.7 and 2.8] that the scalar components of **F** are, respectively,

$$F_x = \frac{Fd_x}{d} \qquad F_y = \frac{Fd_y}{d} \qquad F_z = \frac{Fd_z}{d} \tag{2.29}$$

Resultant of Forces in Space

When two or more forces act on a particle in three-dimensional space, we can obtain the rectangular components of their resultant **R** by adding the corresponding components of the given forces algebraically [Sec. 2.4C]. We have

$$R_x = \Sigma F_x \qquad R_y = \Sigma F_y \qquad R_z = \Sigma F_z \tag{2.31}$$

We can then determine the magnitude and direction of **R** from relations similar to Eqs. (2.18) and (2.25) [Sample Prob. 2.8].

Equilibrium of a Particle

A particle is said to be in equilibrium when the resultant of all the forces acting on it is zero [Sec. 2.3A]. The particle remains at rest (if originally at rest) or moves with constant speed in a straight line (if originally in motion) [Sec. 2.3B].

Free-Body Diagram

To solve a problem involving a particle in equilibrium, first draw a free-body diagram of the particle showing all of the forces acting on it [Sec. 2.3C]. If only three coplanar forces act on the particle, you can draw a force triangle to express that the particle is in equilibrium. Using graphical methods of trigonometry, you can solve this triangle for no more than two unknowns [Sample Prob. 2.4]. If more than three coplanar forces are involved, you should use the equations of equilibrium:

$$\Sigma F_x = 0 \qquad \Sigma F_y = 0 \tag{2.15}$$

These equations can be solved for no more than two unknowns [Sample Prob. 2.6].

Equilibrium in Space

When a particle is in equilibrium in three-dimensional space [Sec. 2.5], use the three equations of equilibrium:

$$\Sigma F_x = 0 \qquad \Sigma F_y = 0 \qquad \Sigma F_z = 0 \tag{2.34}$$

These equations can be solved for no more than three unknowns [Sample Prob. 2.9].

Review Problems

2.104 Two structural members A and B are bolted to a bracket as shown. Knowing that both members are in compression and that the force is 15 kN in member A and 10 kN in member B, determine by trigonometry the magnitude and direction of the resultant of the forces applied to the bracket by members A and B.

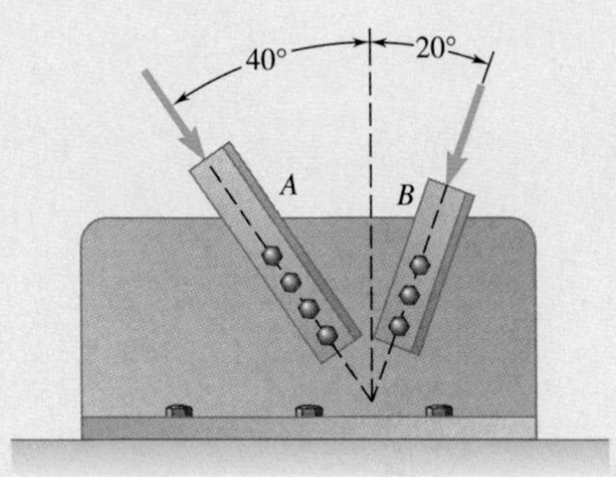

Fig. P2.104

2.105 Determine the x and y components of each of the forces shown.

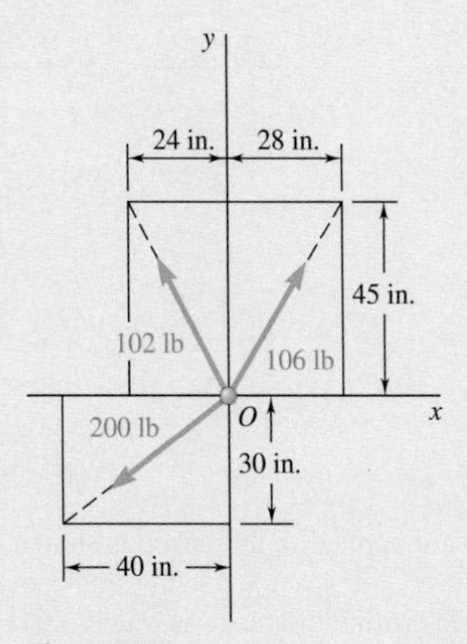

Fig. P2.105

2.106 The hydraulic cylinder BC exerts on member AB a force $\mathbf{P}$ directed along line BC. This force develops due to the moment $\mathbf{M}$ applied at A as shown; the concept of moments will be introduced in Chapter 3. Knowing that $\mathbf{P}$ must have a 600-N component perpendicular to member AB, determine (a) the magnitude of the force $\mathbf{P}$, (b) its component along line AB.

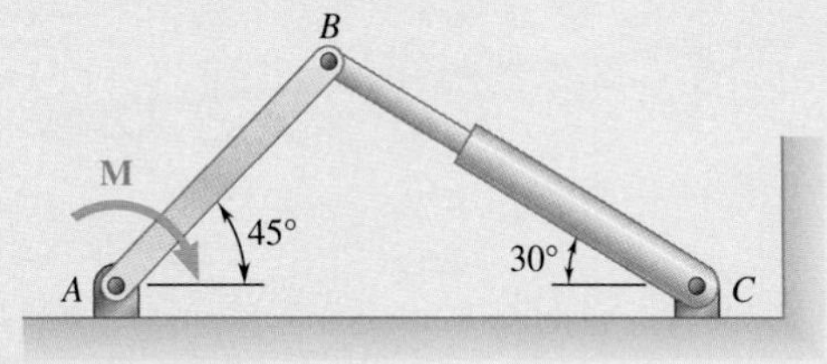

Fig. ***P2.106***

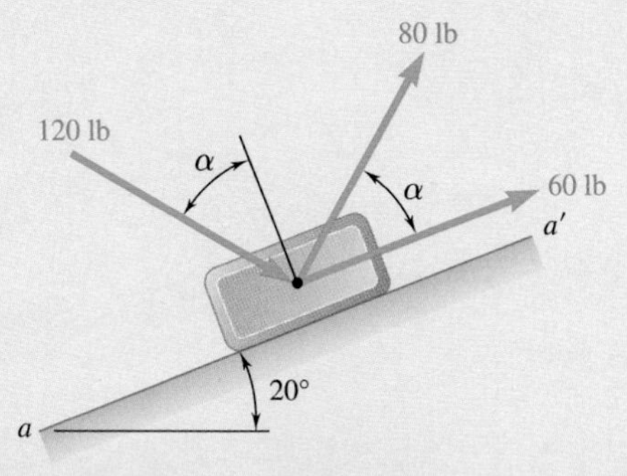

Fig. P2.107

2.107 Knowing that $\alpha = 40°$, determine the resultant of the three forces shown.

2.108 Knowing that $\alpha = 20°$, determine the tension (*a*) in cable *AC*, (*b*) in rope *BC*.

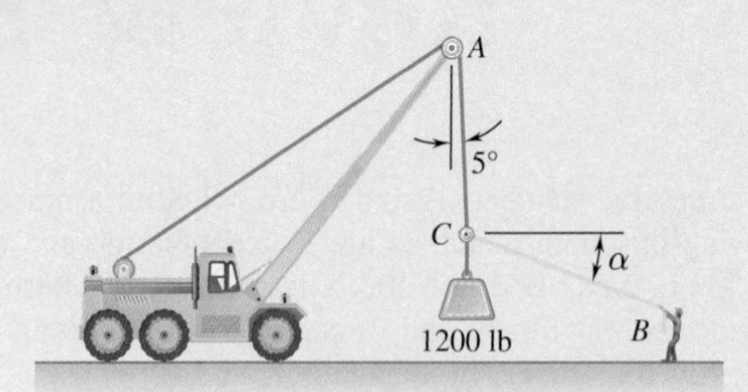

Fig. P2.108

2.109 For $W = 800$ N, $P = 200$ N, and $d = 600$ mm, determine the value of *h* consistent with equilibrium.

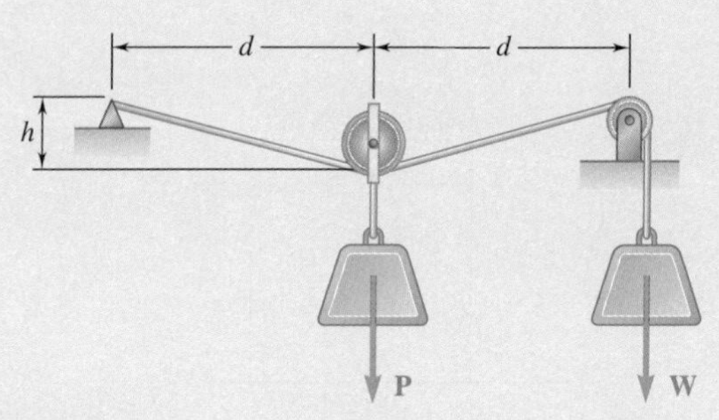

Fig. ***P2.109***

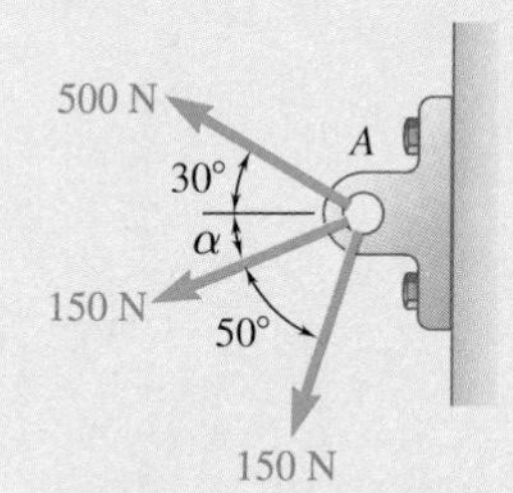

Fig. P2.110

2.110 Three forces are applied to a bracket as shown. The directions of the two 150-N forces may vary, but the angle between these forces is always 50°. Determine the range of values of α for which the magnitude of the resultant of the forces acting at *A* is less than 600 N.

2.111 Cable *AB* is 65 ft long, and the tension in that cable is 3900 lb. Determine (*a*) the *x*, *y*, and *z* components of the force exerted by the cable on the anchor *B*, (*b*) the angles θ_x, θ_y, and θ_z defining the direction of that force.

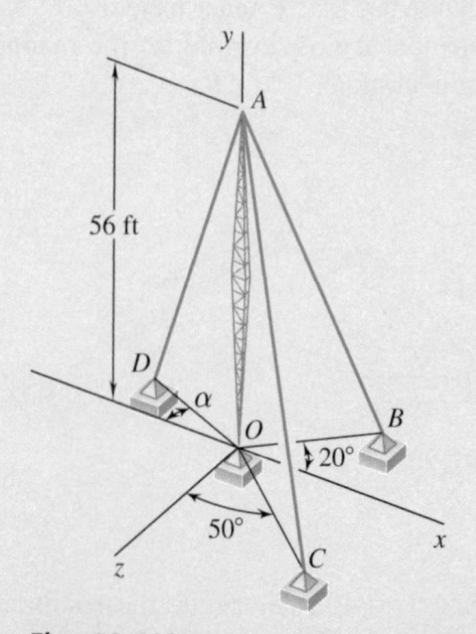

Fig. ***P2.111***

2.112 Three cables are used to tether a balloon as shown. Determine the vertical force **P** exerted by the balloon at *A* knowing that the tension in cable *AB* is 259 N.

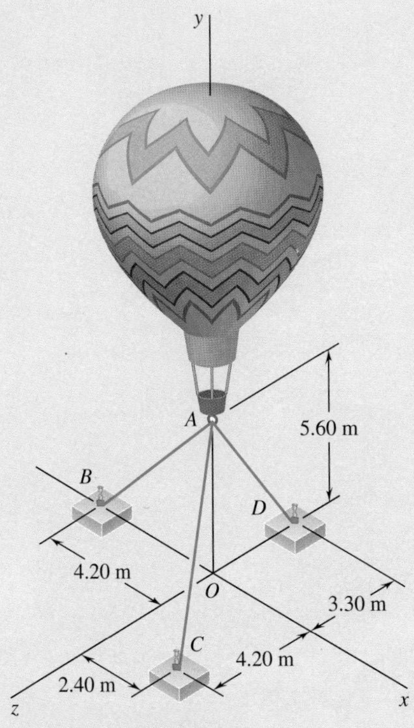

Fig. P2.112 and P2.113

2.113 Three cables are used to tether a balloon as shown. Determine the vertical force **P** exerted by the balloon at *A* knowing that the tension in cable *AC* is 444 N.

2.114 A transmission tower is held by three guy wires attached to a pin at *A* and anchored by bolts at *B*, *C*, and *D*. If the tension in wire *AB* is 630 lb, determine the vertical force **P** exerted by the tower on the pin at *A*.

2.115 A transmission tower is held by three guy wires attached to a pin at *A* and anchored by bolts at *B*, *C*, and *D*. If the tension in wire *AC* is 920 lb, determine the vertical force **P** exerted by the tower on the pin at *A*.

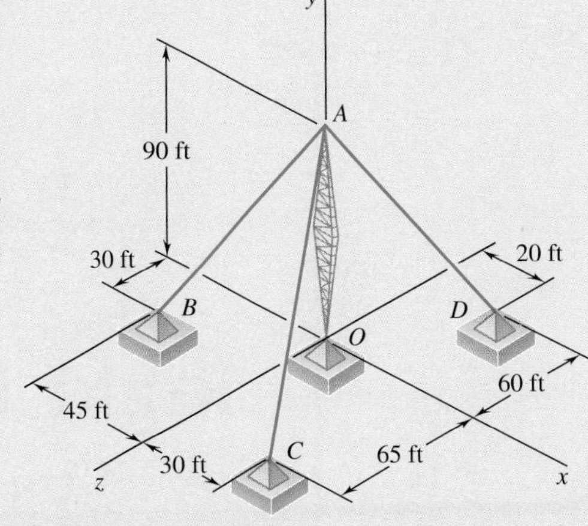

Fig. *P2.114* and P2.115

3

Rigid Bodies: Equivalent Systems of Forces

Four tugboats work together to free the oil tanker Coastal Eagle Point that ran aground while attempting to navigate a channel in Tampa Bay. It will be shown in this chapter that the forces exerted on the ship by the tugboats could be replaced by an equivalent force exerted by a single, more powerful, tugboat.

Introduction

Objectives

- **Discuss** the principle of transmissibility that enables a force to be treated as a sliding vector.
- **Define** the moment of a force about a point.
- **Examine** vector and scalar products, useful in analysis involving moments.
- **Apply** Varignon's Theorem to simplify certain moment analyses.
- **Define** the mixed triple product and use it to determine the moment of a force about an axis.
- **Define** the moment of a couple, and consider the particular properties of couples.
- **Resolve** a given force into an equivalent force-couple system at another point.
- **Reduce** a system of forces into an equivalent force-couple system.
- **Examine** circumstances where a system of forces can be reduced to a single force.

Introduction

In Chapter 2, we assumed that each of the bodies considered could be treated as a single particle. Such a view, however, is not always possible. In general, a body should be treated as a combination of a large number of particles. In this case, we need to consider the size of the body as well as the fact that forces act on different parts of the body and thus have different points of application.

Most of the bodies considered in elementary mechanics are assumed to be rigid. We define a **rigid body** as one that does not deform. Actual structures and machines are never absolutely rigid and deform under the loads to which they are subjected. However, these deformations are usually small and do not appreciably affect the conditions of equilibrium or the motion of the structure under consideration. They are important, though, as far as the resistance of the structure to failure is concerned and are considered in the study of mechanics of materials.

In this chapter, you will study the effect of forces exerted on a rigid body, and you will learn how to replace a given system of forces by a simpler equivalent system. This analysis rests on the fundamental assumption that the effect of a given force on a rigid body remains unchanged if that force is moved along its line of action (*principle of transmissibility*). It follows that forces acting on a rigid body can be represented by *sliding vectors,* as indicated earlier in Sec. 2.1B.

Two important concepts associated with the effect of a force on a rigid body are the *moment of a force about a point* (Sec. 3.1E) and the *moment of a force about an axis* (Sec. 3.2C). The determination of these quantities involves computing vector products and scalar products of two vectors, so in this chapter, we introduce the fundamentals of vector algebra and apply them to the solution of problems involving forces acting on rigid bodies.

Another concept introduced in this chapter is that of a *couple,* i.e., the combination of two forces that have the same magnitude, parallel lines of action, and opposite sense (Sec. 3.3A). As you will see, we can replace any system of forces acting on a rigid body by an equivalent system consisting of one force acting at a given point and one couple. This basic combination is called a *force-couple system.* In the case of concurrent, coplanar, or parallel forces, we can further reduce the equivalent force-couple system to a single force, called the *resultant* of the system, or to a single couple, called the *resultant couple* of the system.

3.1 FORCES AND MOMENTS

The basic definition of a force does not change if the force acts on a point or on a rigid body. However, the effects of the force can be very different, depending on factors such as the point of application or line of action of that force. As a result, calculations involving forces acting on a rigid body are generally more complicated than situations involving forces acting on a point. We begin by examining some general classifications of forces acting on rigid bodies.

3.1A External and Internal Forces

Forces acting on rigid bodies can be separated into two groups: (1) *external forces* and (2) *internal forces*.

1. **External forces** are exerted by other bodies on the rigid body under consideration. They are entirely responsible for the external behavior of the rigid body, either causing it to move or ensuring that it remains at rest. We shall be concerned only with external forces in this chapter and in Chaps. 4 and 5.
2. **Internal forces** hold together the particles forming the rigid body. If the rigid body is structurally composed of several parts, the forces holding the component parts together are also defined as internal forces. We will consider internal forces in Chaps. 6 and 7.

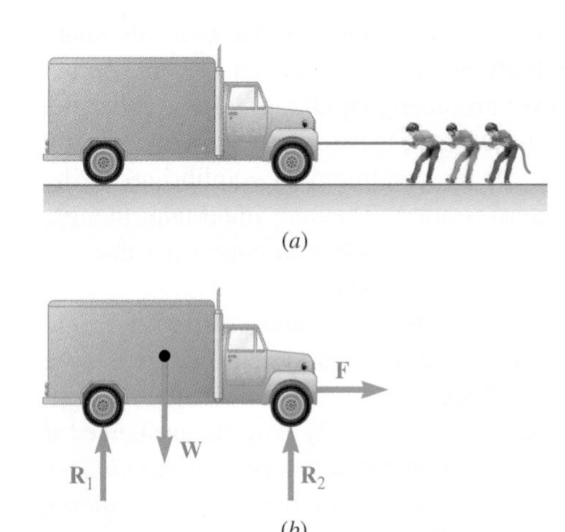

Fig. 3.1 (*a*) Three people pulling on a truck with a rope; (*b*) free-body diagram of the truck, shown as a rigid body instead of a particle.

As an example of external forces, consider the forces acting on a disabled truck that three people are pulling forward by means of a rope attached to the front bumper (Fig. 3.1*a*). The external forces acting on the truck are shown in a **free-body diagram** (Fig. 3.1*b*). Note that this free-body diagram shows the entire object, not just a particle representing the object. Let us first consider the **weight** of the truck. Although it embodies the effect of the earth's pull on each of the particles forming the truck, the weight can be represented by the single force **W**. The **point of application** of this force—that is, the point at which the force acts—is defined as the

center of gravity of the truck. (In Chap. 5, we will show how to determine the location of centers of gravity.) The weight **W** tends to make the truck move vertically downward. In fact, it would actually cause the truck to move downward, i.e., to fall, if it were not for the presence of the ground. The ground opposes the downward motion of the truck by means of the reactions $\mathbf{R}_1$ and $\mathbf{R}_2$. These forces are exerted *by* the ground *on* the truck and must therefore be included among the external forces acting on the truck.

The people pulling on the rope exert the force **F**. The point of application of **F** is on the front bumper. The force **F** tends to make the truck move forward in a straight line and does actually make it move, since no external force opposes this motion. (We are ignoring rolling resistance here for simplicity.) This forward motion of the truck, during which each straight line keeps its original orientation (the floor of the truck remains horizontal, and the walls remain vertical), is known as a **translation**. Other forces might cause the truck to move differently. For example, the force exerted by a jack placed under the front axle would cause the truck to pivot about its rear axle. Such a motion is a **rotation**. We conclude, therefore, that each *external force* acting on a *rigid body* can, if unopposed, impart to the rigid body a motion of translation or rotation, or both.

3.1B Principle of Transmissibility: Equivalent Forces

The **principle of transmissibility** states that the conditions of equilibrium or motion of a rigid body remain unchanged if a force **F** acting at a given point of the rigid body is replaced by a force **F′** of the same magnitude and same direction, but acting at a different point, *provided that the two forces have the same line of action* (Fig. 3.2). The two forces **F** and **F′** have the same effect on the rigid body and are said to be **equivalent forces**. This principle, which states that the action of a force may be *transmitted* along its line of action, is based on experimental evidence. It *cannot* be derived from the properties established so far in this text and therefore must be accepted as an experimental law. Therefore, our study of the statics of rigid bodies is based on the three principles introduced so far: the parallelogram law of vector addition, Newton's first law, and the principle of transmissibility.

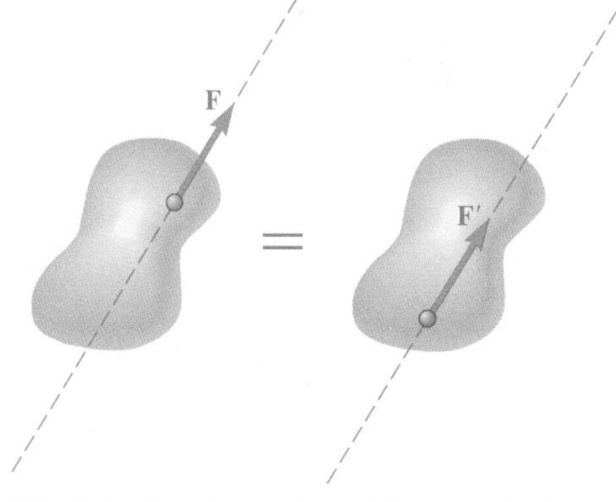

Fig. 3.2 Two forces **F** and **F′** are equivalent if they have the same magnitude and direction and the same line of action, even if they act at different points.

We indicated in Chap. 2 that we could represent the forces acting on a particle by vectors. These vectors had a well-defined point of application—namely, the particle itself—and were therefore fixed, or bound, vectors. In the case of forces acting on a rigid body, however, the point of application of the force does not matter, as long as the line of action remains unchanged. Thus, forces acting on a rigid body must be represented by a different kind of vector, known as a **sliding vector**, since forces are allowed to slide along their lines of action. Note that all of the properties we derive in the following sections for the forces acting on a rigid body are valid more generally for any system of sliding vectors. In order to keep our presentation more intuitive, however, we will carry it out in terms of physical forces rather than in terms of mathematical sliding vectors.

Returning to the example of the truck, we first observe that the line of action of the force **F** is a horizontal line passing through both the front

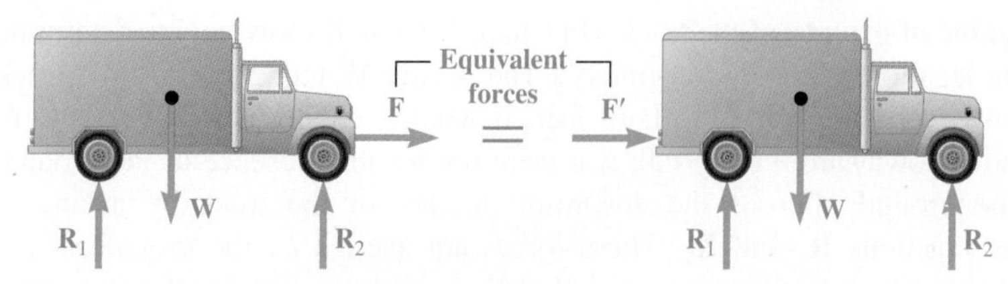

Fig. 3.3 Force **F′** is equivalent to force **F**, so the motion of the truck is the same whether you pull it or push it.

and rear bumpers of the truck (Fig. 3.3). Using the principle of transmissibility, we can therefore replace **F** by an *equivalent force* **F′** acting on the rear bumper. In other words, the conditions of motion are unaffected, and all of the other external forces acting on the truck (**W**, $\mathbf{R}_1$, $\mathbf{R}_2$) remain unchanged if the people push on the rear bumper instead of pulling on the front bumper.

The principle of transmissibility and the concept of equivalent forces have limitations. Consider, for example, a short bar *AB* acted upon by equal and opposite axial forces $\mathbf{P}_1$ and $\mathbf{P}_2$, as shown in Fig. 3.4*a*. According

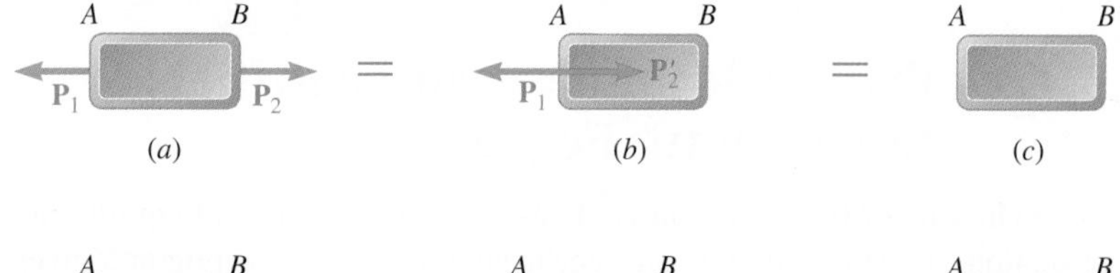

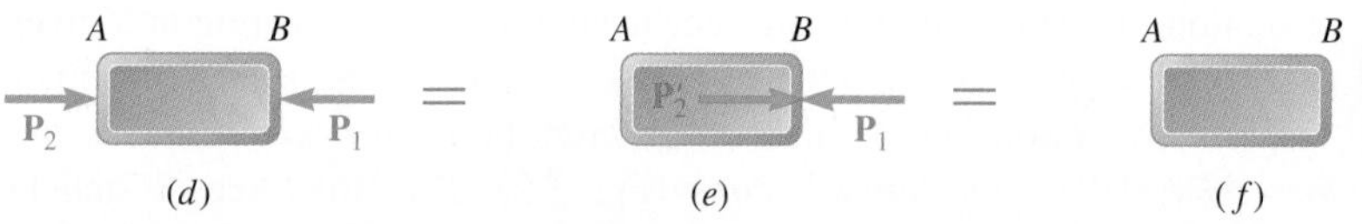

Fig. 3.4 (*a–c*) A set of equivalent forces acting on bar *AB*; (*d–f*) another set of equivalent forces acting on bar *AB*. Both sets produce the same external effect (equilibrium in this case) but different internal forces and deformations.

to the principle of transmissibility, we can replace force $\mathbf{P}_2$ by a force $\mathbf{P}_2'$ having the same magnitude, the same direction, and the same line of action but acting at *A* instead of *B* (Fig. 3.4*b*). The forces $\mathbf{P}_1$ and $\mathbf{P}_2'$ acting on the same particle can be added according to the rules of Chap. 2, and since these forces are equal and opposite, their sum is equal to zero. Thus, in terms of the external behavior of the bar, the original system of forces shown in Fig. 3.4*a* is equivalent to no force at all (Fig. 3.4*c*).

Consider now the two equal and opposite forces $\mathbf{P}_1$ and $\mathbf{P}_2$ acting on the bar *AB* as shown in Fig. 3.4*d*. We can replace the force $\mathbf{P}_2$ by a force $\mathbf{P}_2'$ having the same magnitude, the same direction, and the same line of action but acting at *B* instead of at *A* (Fig. 3.4*e*). We can add forces $\mathbf{P}_1$ and $\mathbf{P}_2'$, and their sum is again zero (Fig. 3.4*f*). From the point of view of the mechanics of rigid bodies, the systems shown in Fig. 3.4*a* and *d* are thus equivalent. However, the *internal forces* and *deformations* produced by the two systems are clearly different. The bar of Fig. 3.4*a* is in *tension* and, if not absolutely rigid, increases in length slightly; the bar of Fig. 3.4*d* is in *compression* and, if not absolutely rigid, decreases in length slightly. Thus, although we can use the principle of transmissibility to determine the conditions of motion or equilibrium of rigid bodies and to compute the external forces acting on these bodies, it should be avoided, or at least used with care, in determining internal forces and deformations.

3.1C Vector Products

In order to gain a better understanding of the effect of a force on a rigid body, we need to introduce a new concept, the *moment of a force about a point*. However, this concept is more clearly understood and is applied more effectively if we first add to the mathematical tools at our disposal the vector product of two vectors.

The **vector product** of two vectors **P** and **Q** is defined as the vector **V** that satisfies the following conditions.

1. The line of action of **V** is perpendicular to the plane containing **P** and **Q** (Fig. 3.5*a*).
2. The magnitude of **V** is the product of the magnitudes of **P** and **Q** and of the sine of the angle θ formed by **P** and **Q** (the measure of which is always 180° or less). We thus have

Magnitude of a vector product

$$V = PQ \sin \theta \tag{3.1}$$

3. The direction of **V** is obtained from the **right-hand rule**. Close your right hand and hold it so that your fingers are curled in the same sense as the rotation through θ that brings the vector **P** in line with the vector **Q**. Your thumb then indicates the direction of the vector **V** (Fig. 3.5*b*). Note that if **P** and **Q** do not have a common point of application, you should first redraw them from the same point. The three vectors **P**, **Q**, and **V**—taken in that order—are said to form a *right-handed triad*.†

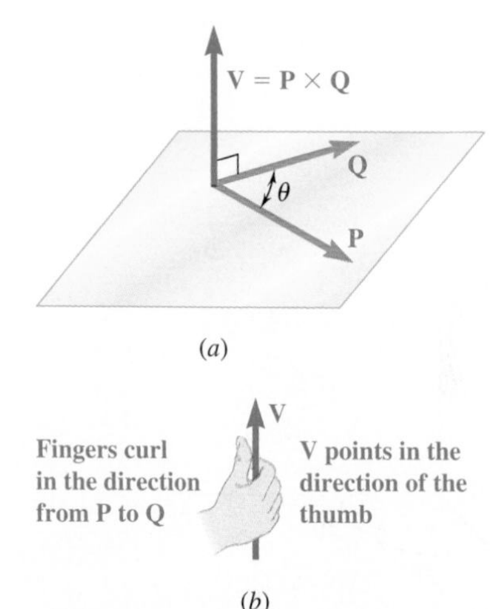

Fig. 3.5 (*a*) The vector product **V** has the magnitude $PQ \sin \theta$ and is perpendicular to the plane of **P** and **Q**; (*b*) you can determine the direction of **V** by using the right-hand rule.

As stated previously, the vector **V** satisfying these three conditions (which define it uniquely) is referred to as the *vector product* of **P** and **Q**. It is represented by the mathematical expression

Vector product

$$\mathbf{V} = \mathbf{P} \times \mathbf{Q} \tag{3.2}$$

Because of this notation, the vector product of two vectors **P** and **Q** is also referred to as the *cross product* of **P** and **Q**.

It follows from Eq. (3.1) that if the vectors **P** and **Q** have either the same direction or opposite directions, their vector product is zero. In the general case when the angle θ formed by the two vectors is neither 0° nor 180°, Eq. (3.1) has a simple geometric interpretation: The magnitude V of the vector product of **P** and **Q** is equal to the area of the parallelogram that has **P** and **Q** for sides (Fig. 3.6). The vector product **P** × **Q** is

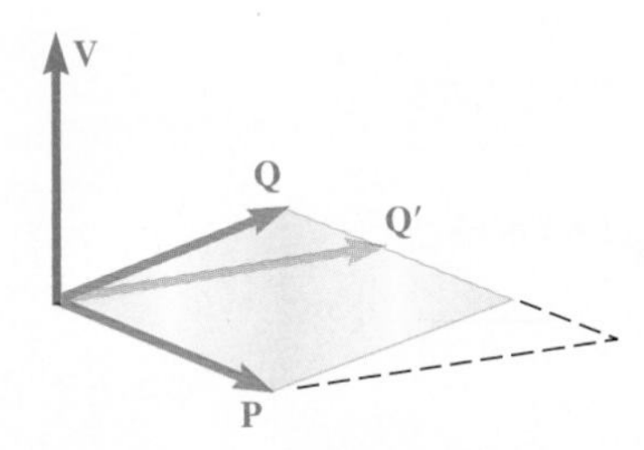

Fig. 3.6 The magnitude of the vector product **V** equals the area of the parallelogram formed by **P** and **Q**. If you change **Q** to **Q′** in such a way that the parallelogram changes shape but **P** and the area are still the same, then the magnitude of **V** remains the same.

†Note that the x, y, and z axes used in Chap. 2 form a right-handed system of orthogonal axes and that the unit vectors **i**, **j**, and **k** defined in Sec. 2.4A form a right-handed orthogonal triad.

therefore unchanged if we replace **Q** by a vector **Q′** that is coplanar with **P** and **Q** such that the line joining the tips of **Q** and **Q′** is parallel to **P**:

$$\mathbf{V} = \mathbf{P} \times \mathbf{Q} = \mathbf{P} \times \mathbf{Q}' \tag{3.3}$$

From the third condition used to define the vector product **V** of **P** and **Q**—namely, that **P**, **Q**, and **V** must form a right-handed triad—it follows that vector products *are not commutative;* i.e., $\mathbf{Q} \times \mathbf{P}$ is not equal to $\mathbf{P} \times \mathbf{Q}$. Indeed, we can easily check that $\mathbf{Q} \times \mathbf{P}$ is represented by the vector $-\mathbf{V}$, which is equal and opposite to **V**:

$$\mathbf{Q} \times \mathbf{P} = -(\mathbf{P} \times \mathbf{Q}) \tag{3.4}$$

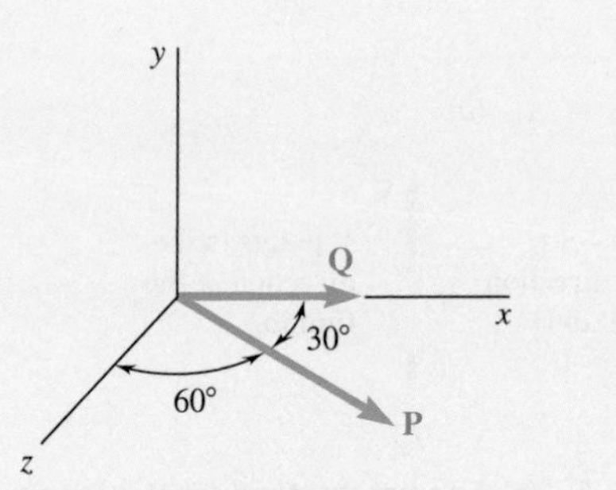

Fig. 3.7 Two vectors **P** and **Q** with angle between them.

Concept Application 3.1

Let us compute the vector product $\mathbf{V} = \mathbf{P} \times \mathbf{Q}$, where the vector **P** is of magnitude 6 and lies in the *zx* plane at an angle of 30° with the *x* axis, and where the vector **Q** is of magnitude 4 and lies along the *x* axis (Fig. 3.7).

Solution

It follows immediately from the definition of the vector product that the vector **V** must lie along the *y* axis, directed upward, with the magnitude

$$V = PQ \sin \theta = (6)(4) \sin 30° = 12 \quad \blacksquare$$

We saw that the commutative property does not apply to vector products. However, it can be demonstrated that the *distributive* property

$$\mathbf{P} \times (\mathbf{Q}_1 + \mathbf{Q}_2) = \mathbf{P} \times \mathbf{Q}_1 + \mathbf{P} \times \mathbf{Q}_2 \tag{3.5}$$

does hold.

A third property, the associative property, does not apply to vector products; we have in general

$$(\mathbf{P} \times \mathbf{Q}) \times \mathbf{S} \neq \mathbf{P} \times (\mathbf{Q} \times \mathbf{S}) \tag{3.6}$$

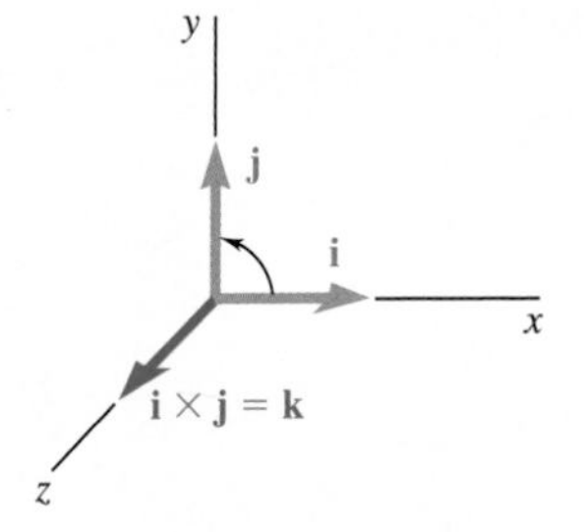

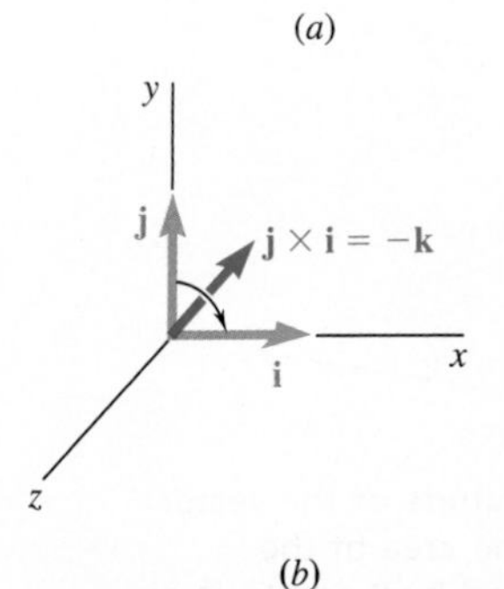

Fig. 3.8 (*a*) The vector product of the **i** and **j** unit vectors is the **k** unit vector; (*b*) the vector product of the **j** and **i** unit vectors is the −**k** unit vector.

3.1D Rectangular Components of Vector Products

Before we turn back to forces acting on rigid bodes, let's look at a more convenient way to express vector products using rectangular components. To do this, we use the unit vectors **i**, **j**, and **k** that were defined in Chap. 2.

Consider first the vector product $\mathbf{i} \times \mathbf{j}$ (Fig. 3.8*a*). Since both vectors have a magnitude equal to 1 and since they are at a right angle to each other, their vector product is also a unit vector. This unit vector must be **k**, since the vectors **i**, **j**, and **k** are mutually perpendicular and form a right-handed triad. Similarly, it follows from the right-hand rule given in Sec. 3.1C that the product $\mathbf{j} \times \mathbf{i}$ is equal to $-\mathbf{k}$ (Fig. 3.8*b*). Finally, note that the vector product of a unit vector with itself, such as $\mathbf{i} \times \mathbf{i}$, is equal to zero, since both vectors have the same direction. Thus, we can list the vector products of all the various possible pairs of unit vectors:

$$\begin{array}{lll} \mathbf{i} \times \mathbf{i} = \mathbf{0} & \mathbf{j} \times \mathbf{i} = -\mathbf{k} & \mathbf{k} \times \mathbf{i} = \mathbf{j} \\ \mathbf{i} \times \mathbf{j} = \mathbf{k} & \mathbf{j} \times \mathbf{j} = \mathbf{0} & \mathbf{k} \times \mathbf{j} = -\mathbf{i} \\ \mathbf{i} \times \mathbf{k} = -\mathbf{j} & \mathbf{j} \times \mathbf{k} = \mathbf{i} & \mathbf{k} \times \mathbf{k} = \mathbf{0} \end{array} \tag{3.7}$$

We can determine the sign of the vector product of two unit vectors simply by arranging them in a circle and reading them in the order of the multiplication (Fig. 3.9). The product is positive if they follow each other in counterclockwise order and is negative if they follow each other in clockwise order.

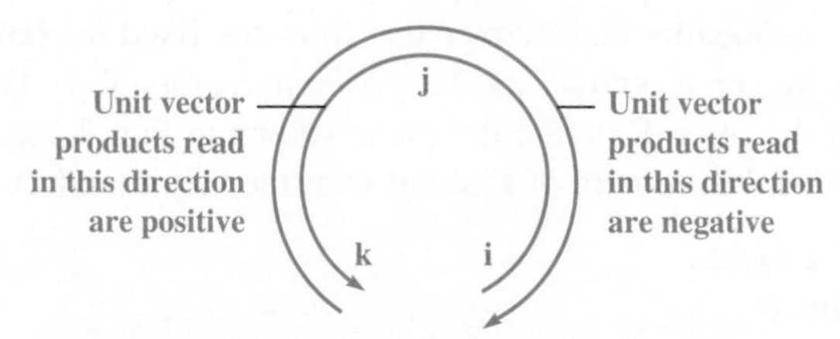

Fig. 3.9 Arrange the three letters **i**, **j**, **k** in a counterclockwise circle. You can use the order of letters for the three unit vectors in a vector product to determine its sign.

We can now easily express the vector product **V** of two given vectors **P** and **Q** in terms of the rectangular components of these vectors. Resolving **P** and **Q** into components, we first write

$$\mathbf{V} = \mathbf{P} \times \mathbf{Q} = (P_x\mathbf{i} + P_y\mathbf{j} + P_z\mathbf{k}) \times (Q_x\mathbf{i} + Q_y\mathbf{j} + Q_z\mathbf{k})$$

Making use of the distributive property, we express **V** as the sum of vector products, such as $P_x\mathbf{i} \times Q_y\mathbf{j}$. We find that each of the expressions obtained is equal to the vector product of two unit vectors, such as $\mathbf{i} \times \mathbf{j}$, multiplied by the product of two scalars, such as P_xQ_y. Recalling the identities of Eq. (3.7) and factoring out **i**, **j**, and **k**, we obtain

$$\mathbf{V} = (P_yQ_z - P_zQ_y)\mathbf{i} + (P_zQ_x - P_xQ_z)\mathbf{j} + (P_xQ_y - P_yQ_x)\mathbf{k} \tag{3.8}$$

Thus, the rectangular components of the vector product **V** are

Rectangular components of a vector product

$$\begin{aligned} V_x &= P_yQ_z - P_zQ_y \\ V_y &= P_zQ_x - P_xQ_z \\ V_z &= P_xQ_y - P_yQ_x \end{aligned} \tag{3.9}$$

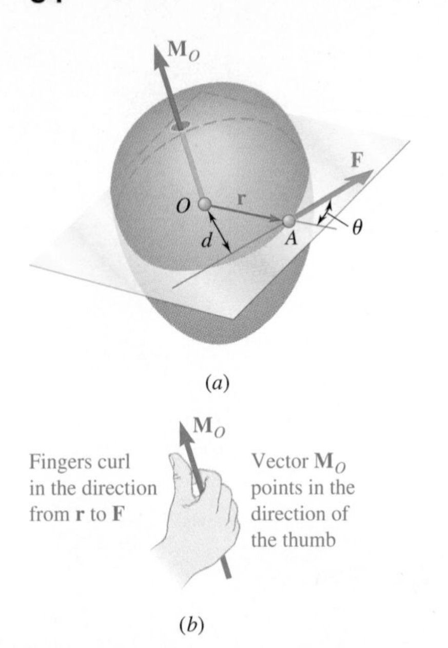

Fig. 3.10 Moment of a force about a point. (*a*) The moment $\mathbf{M}_O$ is the vector product of the position vector **r** and the force **F**; (*b*) a right-hand rule indicates the sense of $\mathbf{M}_O$.

Returning to Eq. (3.8), notice that the right-hand side represents the expansion of a determinant. Thus, we can express the vector product **V** in the following form, which is more easily memorized:†

Rectangular components of a vector product (determinant form)

$$\mathbf{V} = \begin{vmatrix} \mathbf{i} & \mathbf{j} & \mathbf{k} \\ P_x & P_y & P_z \\ Q_x & Q_y & Q_z \end{vmatrix} \tag{3.10}$$

3.1E Moment of a Force about a Point

We are now ready to consider a force **F** acting on a rigid body (Fig. 3.10*a*). As we know, the force **F** is represented by a vector that defines its magnitude and direction. However, the effect of the force on the rigid body depends also upon its point of application A. The position of A can be conveniently defined by the vector **r** that joins the fixed reference point O with A; this vector is known as the *position vector* of A. The position vector **r** and the force **F** define the plane shown in Fig. 3.10*a*.

We define the **moment of F about *O*** as the vector product of **r** and **F**:

Moment of a force about a point *O*

$$\mathbf{M}_O = \mathbf{r} \times \mathbf{F} \tag{3.11}$$

According to the definition of the vector product given in Sec. 3.1C, the moment $\mathbf{M}_O$ must be perpendicular to the plane containing O and force **F**. The sense of $\mathbf{M}_O$ is defined by the sense of the rotation that will bring vector **r** in line with vector **F**; this rotation is observed as *counterclockwise* by an observer located at the tip of $\mathbf{M}_O$. Another way of defining the sense of $\mathbf{M}_O$ is furnished by a variation of the right-hand rule: Close your right hand and hold it so that your fingers curl in the sense of the rotation that **F** would impart to the rigid body about a fixed axis directed along the line of action of $\mathbf{M}_O$. Then your thumb indicates the sense of the moment $\mathbf{M}_O$ (Fig. 3.10*b*).

Finally, denoting by θ the angle between the lines of action of the position vector **r** and the force **F**, we find that the magnitude of the moment of **F** about O is

Magnitude of the moment of a force

$$M_O = rF \sin\theta = Fd \tag{3.12}$$

†Any determinant consisting of three rows and three columns can be evaluated by repeating the first and second columns and forming products along each diagonal line. The sum of the products obtained along the red lines is then subtracted from the sum of the products obtained along the black lines.

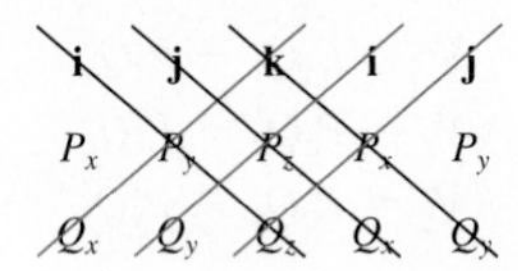

where d represents the perpendicular distance from O to the line of action of $\mathbf{F}$ (see Fig. 3.10). Experimentally, the tendency of a force $\mathbf{F}$ to make a rigid body rotate about a fixed axis perpendicular to the force depends upon the distance of $\mathbf{F}$ from that axis, as well as upon the magnitude of $\mathbf{F}$. For example, a child's breath can exert enough force to make a toy propeller spin (Fig. 3.11*a*), but a wind turbine requires the force of a substantial wind to rotate the blades and generate electrical power (Fig. 3.11*b*). However, the perpendicular distance between the rotation point and the line of action of the force (often called the *moment arm*) is just as important. If you want to apply a small moment to turn a nut on a pipe without breaking it, you might use a small pipe wrench that gives you a small moment arm

(*a*) Small force
© Gavela Montes Productions/Getty Images RF

(*b*) Large force
© Image Source/Getty Images RF

(*c*) Small moment arm
© Valery Voennyy/Alamy RF

(*d*) Large moment arm
© Monty Rakusen/Getty Images RF

Fig. 3.11 (*a*, *b*) The moment of a force depends on the magnitude of the force; (*c*, *d*) it also depends on the length of the moment arm.

(Fig. 3.11*c*). But if you need a larger moment, you could use a large wrench with a long moment arm (Fig. 3.11*d*). Therefore,

> ***The magnitude of* $\mathbf{M}_O$ *measures the tendency of the force* F *to make the rigid body rotate about a fixed axis directed along* $\mathbf{M}_O$.**

In the SI system of units, where a force is expressed in newtons (N) and a distance in meters (m), the moment of a force is expressed in newton-meters (N·m). In the U.S. customary system of units, where a force is expressed in pounds and a distance in feet or inches, the moment of a force is expressed in lb·ft or lb·in.

Note that although the moment $\mathbf{M}_O$ of a force about a point depends upon the magnitude, the line of action, and the sense of the force, it does *not* depend upon the actual position of the point of application of the force along its line of action. Conversely, the moment $\mathbf{M}_O$ of a force **F** does not characterize the position of the point of application of **F**.

However, as we will see shortly, the moment $\mathbf{M}_O$ of a force **F** of a given magnitude and direction *completely defines the line of action of* **F**. Indeed, the line of action of **F** must lie in a plane through O perpendicular to the moment $\mathbf{M}_O$; its distance d from O must be equal to the quotient M_O/F of the magnitudes of $\mathbf{M}_O$ and **F**; and the sense of $\mathbf{M}_O$ determines whether the line of action of **F** occurs on one side or the other of the point O.

Recall from Sec. 3.1B that the principle of transmissibility states that two forces **F** and **F**′ are equivalent (i.e., have the same effect on a rigid body) if they have the same magnitude, same direction, and same line of action. We can now restate this principle:

> ***Two forces* F *and* F′ *are equivalent if, and only if, they are equal* (i.e., have the same magnitude and same direction) *and have equal moments about a given point O*.**

The necessary and sufficient conditions for two forces **F** and **F**′ to be equivalent are thus

$$\mathbf{F} = \mathbf{F}' \quad \text{and} \quad \mathbf{M}_O = \mathbf{M}'_O \tag{3.13}$$

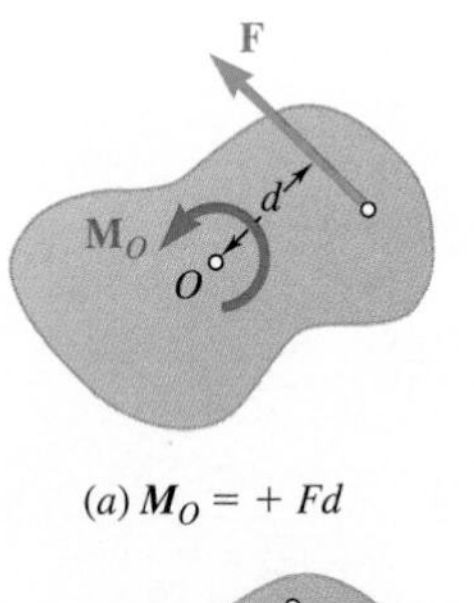

(*a*) $\mathbf{M}_O = +Fd$

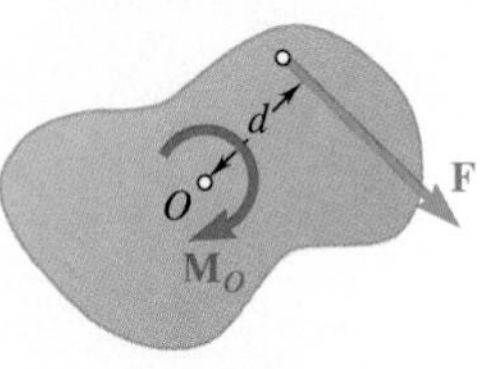

(*b*) $\mathbf{M}_O = -Fd$

Fig. 3.12 (*a*) A moment that tends to produce a counterclockwise rotation is positive; (*b*) a moment that tends to produce a clockwise rotation is negative.

We should observe that if the relations of Eqs. (3.13) hold for a given point O, they hold for any other point.

Two-Dimensional Problems. Many applications in statics deal with two-dimensional structures. Such structures have length and breadth but only negligible depth. Often, they are subjected to forces contained in the plane of the structure. We can easily represent two-dimensional structures and the forces acting on them on a sheet of paper or on a blackboard. Their analysis is therefore considerably simpler than that of three-dimensional structures and forces.

Consider, for example, a rigid slab acted upon by a force **F** in the plane of the slab (Fig. 3.12). The moment of **F** about a point O, which is chosen in the plane of the figure, is represented by a vector $\mathbf{M}_O$ perpendicular to that plane and of magnitude Fd. In the case of Fig. 3.12*a*, the vector $\mathbf{M}_O$ points *out of* the page, whereas in the case of Fig. 3.12*b*, it points *into* the page. As we look at the figure, we observe in the first case that **F** tends to rotate the slab counterclockwise and in the second case that it tends to rotate the slab clockwise. Therefore, it is natural to refer to the sense of the moment of **F** about *O* in Fig. 3.12*a* as counterclockwise ↺, and in Fig. 3.12*b* as clockwise ↻.

Since the moment of a force **F** acting in the plane of the figure must be perpendicular to that plane, we need only specify the *magnitude* and the *sense* of the moment of **F** about *O*. We do this by assigning to the magnitude M_O of the moment a positive or negative sign according to whether the vector $\mathbf{M}_O$ points out of or into the page.

3.1F Rectangular Components of the Moment of a Force

We can use the distributive property of vector products to determine the moment of the resultant of several *concurrent forces*. If several forces $\mathbf{F}_1$, $\mathbf{F}_2$, . . . are applied at the same point *A* (Fig. 3.13) and if we denote by **r** the position vector of *A*, it follows immediately from Eq. (3.5) that

$$\mathbf{r} \times (\mathbf{F}_1 + \mathbf{F}_2 + \cdots) = \mathbf{r} \times \mathbf{F}_1 + \mathbf{r} \times \mathbf{F}_2 + \cdots \quad (3.14)$$

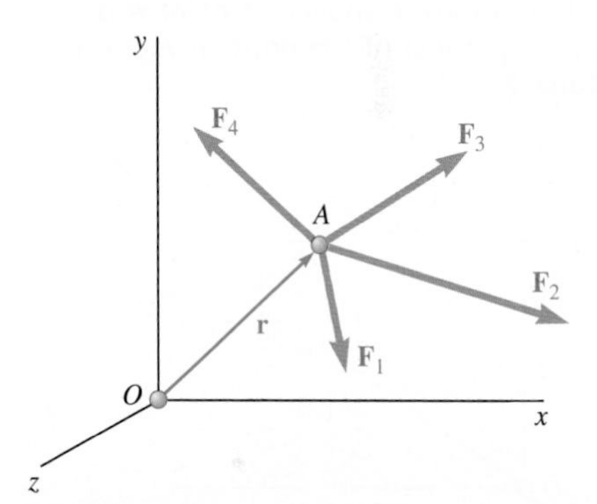

Fig. 3.13 Varignon's theorem says that the moment about point *O* of the resultant of these four forces equals the sum of the moments about point *O* of the individual forces.

In words,

The moment about a given point O of the resultant of several concurrent forces is equal to the sum of the moments of the various forces about the same point O.

This property, which was originally established by the French mathematician Pierre Varignon (1654–1722) long before the introduction of vector algebra, is known as **Varignon's theorem**.

The relation in Eq. (3.14) makes it possible to replace the direct determination of the moment of a force **F** by determining the moments of two or more component forces. As you will see shortly, **F** is generally resolved into components parallel to the coordinate axes. However, it may be more expeditious in some instances to resolve **F** into components that are not parallel to the coordinate axes (see Sample Prob. 3.3).

In general, determining the moment of a force in space is considerably simplified if the force and the position vector of its point of application are resolved into rectangular *x*, *y*, and *z* components. Consider, for example, the moment $\mathbf{M}_O$ about *O* of a force **F** whose components are F_x, F_y, and F_z and that is applied at a point *A* with coordinates *x*, *y*, and *z* (Fig. 3.14). Since the components of the position vector **r** are respectively equal to the coordinates *x*, *y*, and *z* of the point *A*, we can write **r** and **F** as

$$\mathbf{r} = x\mathbf{i} + y\mathbf{j} + z\mathbf{k} \quad (3.15)$$
$$\mathbf{F} = F_x\mathbf{i} + F_y\mathbf{j} + F_z\mathbf{k} \quad (3.16)$$

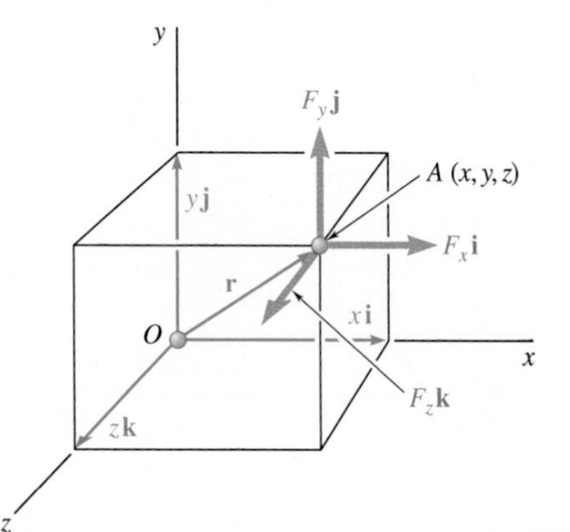

Fig. 3.14 The moment $\mathbf{M}_O$ about point *O* of a force **F** applied at point *A* is the vector product of the position vector **r** and the force **F**, which can both be expressed in rectangular components.

Substituting for **r** and **F** from Eqs. (3.15) and (3.16) into

$$\mathbf{M}_O = \mathbf{r} \times \mathbf{F} \quad (3.11)$$

and recalling Eqs. (3.8) and (3.9), we can write the moment $\mathbf{M}_O$ of **F** about *O* in the form

$$\mathbf{M}_O = M_x\mathbf{i} + M_y\mathbf{j} + M_z\mathbf{k} \quad (3.17)$$

where the components M_x, M_y, and M_z are defined by the relations

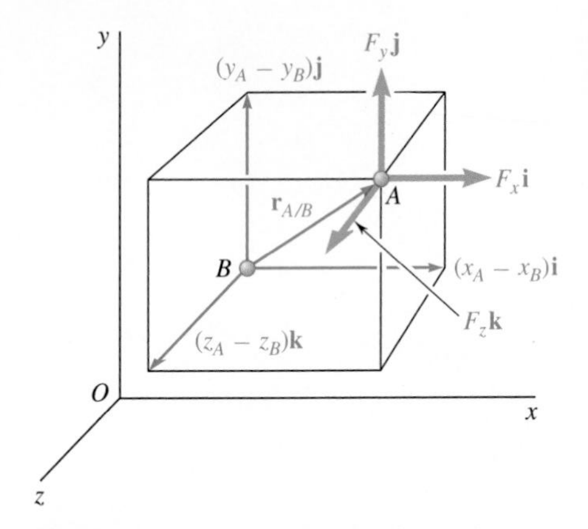

Fig. 3.15 The moment $\mathbf{M}_B$ about the point B of a force **F** applied at point A is the vector product of the position vector $\mathbf{r}_{A/B}$ and force **F**.

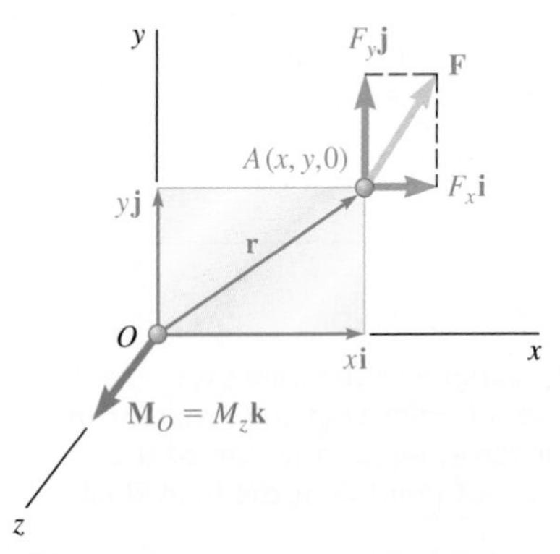

Fig. 3.16 In a two-dimensional problem, the moment $\mathbf{M}_O$ of a force **F** applied at A in the xy plane reduces to the z component of the vector product of **r** with **F**.

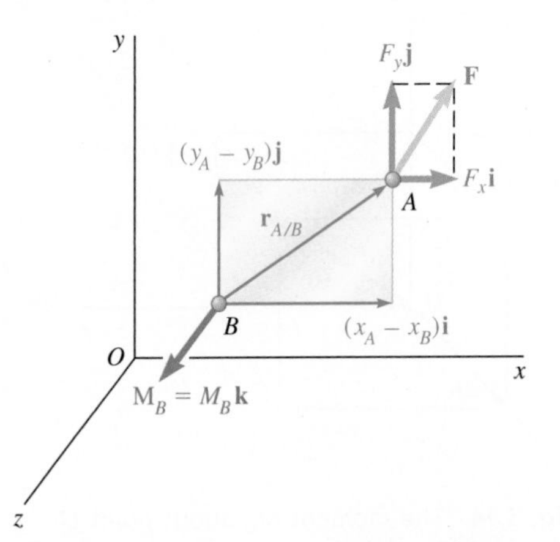

Fig. 3.17 In a two-dimensional problem, the moment $\mathbf{M}_B$ about a point B of a force **F** applied at A in the xy plane reduces to the z component of the vector product of $\mathbf{r}_{A/B}$ with **F**.

Rectangular components of a moment

$$\begin{aligned} M_x &= yF_z - zF_y \\ M_y &= zF_x - xF_z \\ M_z &= xF_y - yF_x \end{aligned} \tag{3.18}$$

As you will see in Sec. 3.2C, the scalar components M_x, M_y, and M_z of the moment $\mathbf{M}_O$ measure the tendency of the force **F** to impart to a rigid body a rotation about the x, y, and z axes, respectively. Substituting from Eq. (3.18) into Eq. (3.17), we can also write $\mathbf{M}_O$ in the form of the determinant, as

$$\mathbf{M}_O = \begin{vmatrix} \mathbf{i} & \mathbf{j} & \mathbf{k} \\ x & y & z \\ F_x & F_y & F_z \end{vmatrix} \tag{3.19}$$

To compute the moment $\mathbf{M}_B$ about an arbitrary point B of a force **F** applied at A (Fig. 3.15), we must replace the position vector **r** in Eq. (3.11) by a vector drawn from B to A. This vector is the *position vector of A relative to B*, denoted by $\mathbf{r}_{A/B}$. Observing that $\mathbf{r}_{A/B}$ can be obtained by subtracting $\mathbf{r}_B$ from $\mathbf{r}_A$, we write

$$\mathbf{M}_B = \mathbf{r}_{A/B} \times \mathbf{F} = (\mathbf{r}_A - \mathbf{r}_B) \times \mathbf{F} \tag{3.20}$$

or using the determinant form,

$$\mathbf{M}_B = \begin{vmatrix} \mathbf{i} & \mathbf{j} & \mathbf{k} \\ x_{A/B} & y_{A/B} & z_{A/B} \\ F_x & F_y & F_z \end{vmatrix} \tag{3.21}$$

where $x_{A/B}$, $y_{A/B}$, and $z_{A/B}$ denote the components of the vector $\mathbf{r}_{A/B}$:

$$x_{A/B} = x_A - x_B \qquad y_{A/B} = y_A - y_B \qquad z_{A/B} = z_A - z_B$$

In the case of two-dimensional problems, we can assume without loss of generality that the force **F** lies in the xy plane (Fig. 3.16). Setting $z = 0$ and $F_z = 0$ in Eq. (3.19), we obtain

$$\mathbf{M}_O = (xF_y - yF_x)\mathbf{k}$$

We can verify that the moment of **F** about O is perpendicular to the plane of the figure and that it is completely defined by the scalar

$$M_O = M_z = xF_y - yF_x \tag{3.22}$$

As noted earlier, a positive value for M_O indicates that the vector $\mathbf{M}_O$ points out of the paper (the force **F** tends to rotate the body counterclockwise about O), and a negative value indicates that the vector $\mathbf{M}_O$ points into the paper (the force **F** tends to rotate the body clockwise about O).

To compute the moment about $B(x_B, y_B)$ of a force lying in the xy plane and applied at $A(x_A, y_A)$ (Fig. 3.17), we set $z_{A/B} = 0$ and $F_z = 0$ in Eq. (3.21) and note that the vector $\mathbf{M}_B$ is perpendicular to the xy plane and is defined in magnitude and sense by the scalar

$$M_B = (x_A - x_B)F_y - (y_A - y_B)F_x \tag{3.23}$$

Sample Problem 3.1

A 100-lb vertical force is applied to the end of a lever, which is attached to a shaft at O. Determine (a) the moment of the 100-lb force about O; (b) the horizontal force applied at A that creates the same moment about O; (c) the smallest force applied at A that creates the same moment about O; (d) how far from the shaft a 240-lb vertical force must act to create the same moment about O; (e) whether any one of the forces obtained in parts b, c, or d is equivalent to the original force.

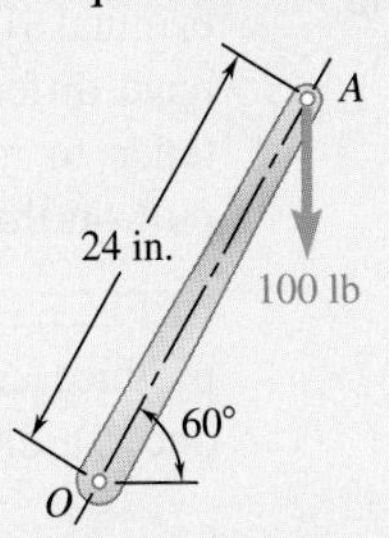

STRATEGY: The calculations asked for all involve variations on the basic defining equation of a moment, $M_O = Fd$.

MODELING and ANALYSIS:

a. Moment about O. The perpendicular distance from O to the line of action of the 100-lb force (Fig. 1) is

$$d = (24 \text{ in.}) \cos 60° = 12 \text{ in.}$$

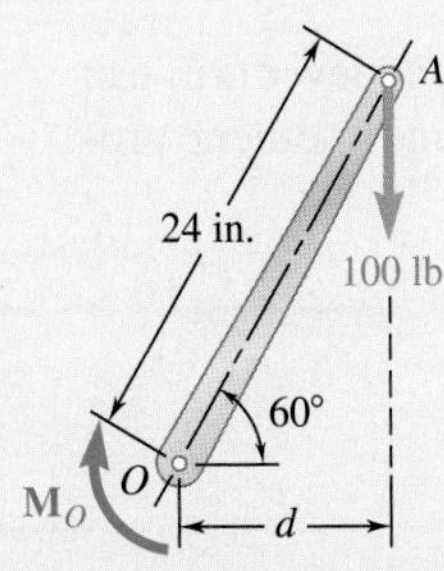

Fig. 1 Determination of the moment of the 100-lb force about O using perpendicular distance d.

The magnitude of the moment about O of the 100-lb force is

$$M_O = Fd = (100 \text{ lb})(12 \text{ in.}) = 1200 \text{ lb·in.}$$

Since the force tends to rotate the lever clockwise about O, represent the moment by a vector $\mathbf{M}_O$ perpendicular to the plane of the figure and pointing *into* the paper. You can express this fact with the notation

$$\mathbf{M}_O = 1200 \text{ lb·in. } \circlearrowright \blacktriangleleft$$

b. Horizontal Force. In this case, you have (Fig. 2)

$$d = (24 \text{ in.}) \sin 60° = 20.8 \text{ in.}$$

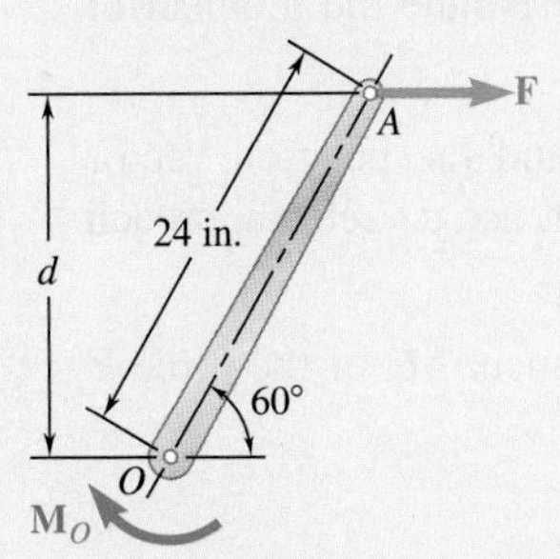

Fig. 2 Determination of horizontal force at A that creates same moment about O.

Since the moment about O must be 1200 lb·in., you obtain

$$M_O = Fd$$
$$1200 \text{ lb·in.} = F(20.8 \text{ in.})$$
$$F = 57.7 \text{ lb} \qquad \mathbf{F} = 57.7 \text{ lb} \rightarrow \blacktriangleleft$$

c. Smallest Force. Since $M_O = Fd$, the smallest value of F occurs when d is maximum. Choose the force perpendicular to OA and note that $d = 24$ in. (Fig. 3); thus

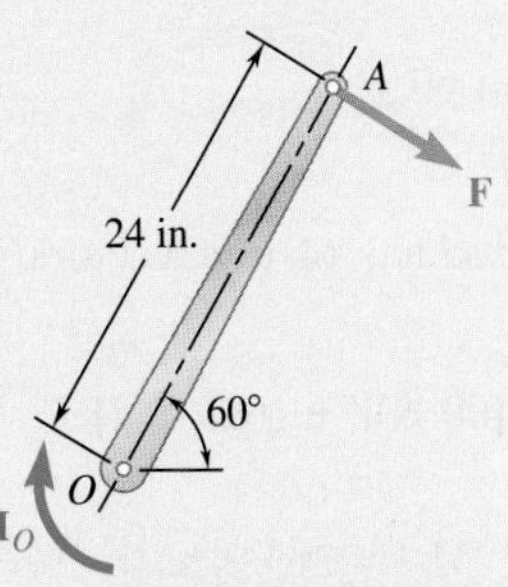

Fig. 3 Determination of smallest force at A that creates same moment about O.

$$M_O = Fd$$
$$1200 \text{ lb·in.} = F(24 \text{ in.})$$
$$F = 50 \text{ lb} \qquad \mathbf{F} = 50 \text{ lb} \measuredangle 30° \blacktriangleleft$$

(continued)

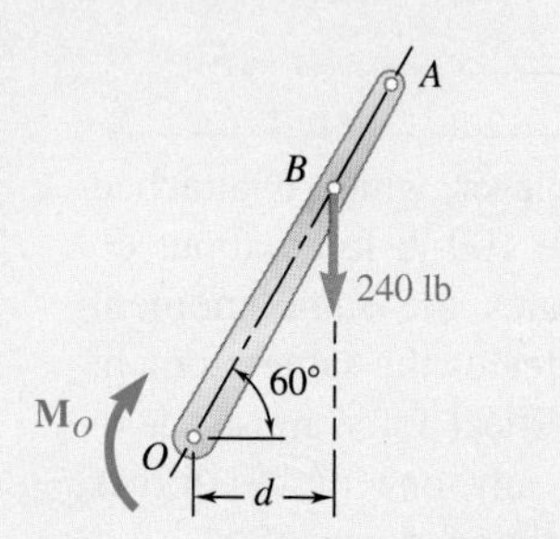

Fig. 4 Position of vertical 240-lb force that creates same moment about O.

d. 240-lb Vertical Force. In this case (Fig. 4), $M_O = Fd$ yields

$$1200 \text{ lb·in.} = (240 \text{ lb})d \qquad d = 5 \text{ in.}$$

but

$$OB \cos 60° = d$$

so

$$OB = 10 \text{ in.}$$ ◀

e. None of the forces considered in parts *b*, *c*, or *d* is equivalent to the original 100-lb force. Although they have the same moment about *O*, they have different *x* and *y* components. In other words, although each force tends to rotate the shaft in the same direction, each causes the lever to pull on the shaft in a different way.

REFLECT and THINK: Various combinations of force and lever arm can produce equivalent moments, but the system of force and moment produces a different overall effect in each case.

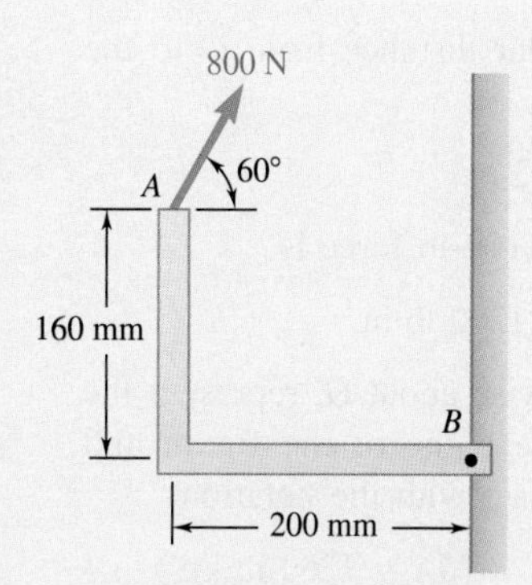

Sample Problem 3.2

A force of 800 N acts on a bracket as shown. Determine the moment of the force about *B*.

STRATEGY: You can resolve both the force and the position vector from *B* to *A* into rectangular components and then use a vector approach to complete the solution.

MODELING and ANALYSIS: Obtain the moment $\mathbf{M}_B$ of the force $\mathbf{F}$ about *B* by forming the vector product

$$\mathbf{M}_B = \mathbf{r}_{A/B} \times \mathbf{F}$$

where $\mathbf{r}_{A/B}$ is the vector drawn from *B* to *A* (Fig. 1). Resolving $\mathbf{r}_{A/B}$ and $\mathbf{F}$ into rectangular components, you have

$$\mathbf{r}_{A/B} = -(0.2 \text{ m})\mathbf{i} + (0.16 \text{ m})\mathbf{j}$$
$$\mathbf{F} = (800 \text{ N}) \cos 60°\mathbf{i} + (800 \text{ N}) \sin 60°\mathbf{j}$$
$$= (400 \text{ N})\mathbf{i} + (693 \text{ N})\mathbf{j}$$

Recalling the relations in Eq. (3.7) for the cross products of unit vectors (Sec. 3.5), you obtain

$$\mathbf{M}_B = \mathbf{r}_{A/B} \times \mathbf{F} = [-(0.2 \text{ m})\mathbf{i} + (0.16 \text{ m})\mathbf{j}] \times [(400 \text{ N})\mathbf{i} + (693 \text{ N})\mathbf{j}]$$
$$= -(138.6 \text{ N·m})\mathbf{k} - (64.0 \text{ N·m})\mathbf{k}$$
$$= -(202.6 \text{ N·m})\mathbf{k} \qquad M_B = 203 \text{ N·m} \curvearrowright$$ ◀

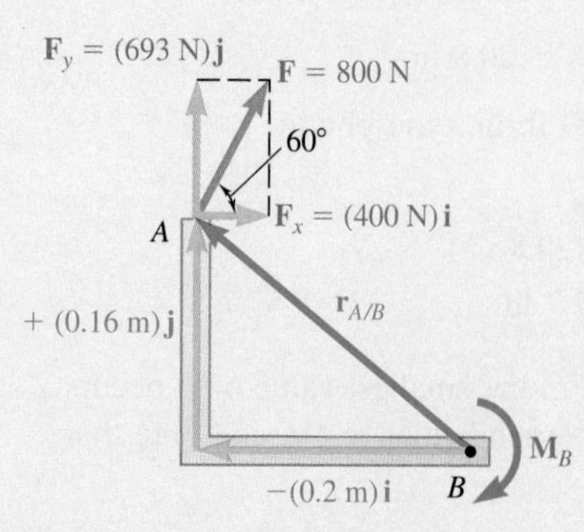

Fig. 1 The moment $\mathbf{M}_B$ is determined from the vector product of position vector $\mathbf{r}_{A/B}$ and force vector **F**.

The moment $\mathbf{M}_B$ is a vector perpendicular to the plane of the figure and pointing *into* the page.

(continued)

REFLECT and THINK: We can also use a scalar approach to solve this problem using the components for the force **F** and the position vector $\mathbf{r}_{A/B}$. Following the right-hand rule for assigning signs, we have

$$+\circlearrowleft M_B = \Sigma M_B = \Sigma Fd = -(400\ \text{N})(0.16\ \text{m}) - (693\ \text{N})(0.2\ \text{m}) = -202.6\ \text{N·m}$$

$$\mathbf{M}_B = 203\ \text{N·m}\ \circlearrowright \blacktriangleleft$$

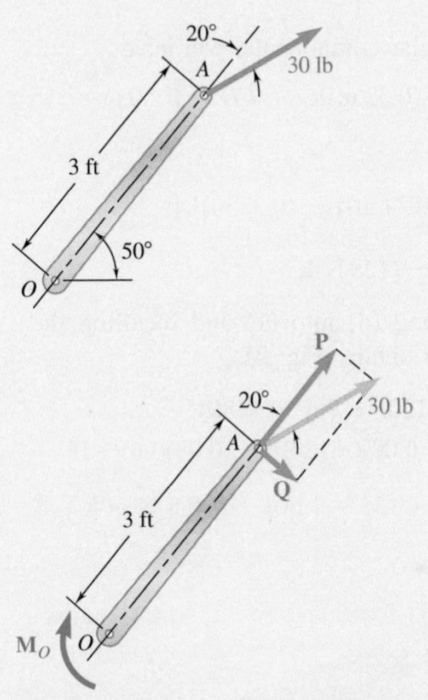

Fig. 1 30-lb force at *A* resolved into components **P** and **Q** to simplify the determination of the moment $\mathbf{M}_O$.

Sample Problem 3.3

A 30-lb force acts on the end of the 3-ft lever as shown. Determine the moment of the force about *O*.

STRATEGY: Resolving the force into components that are perpendicular and parallel to the axis of the lever greatly simplifies the moment calculation.

MODELING and ANALYSIS: Replace the force by two components: one component **P** in the direction of *OA* and one component **Q** perpendicular to *OA* (Fig. 1). Since *O* is on the line of action of **P**, the moment of **P** about *O* is zero. Thus, the moment of the 30-lb force reduces to the moment of **Q**, which is clockwise and can be represented by a negative scalar.

$$Q = (30\ \text{lb})\sin 20° = 10.26\ \text{lb}$$
$$M_O = -Q(3\ \text{ft}) = -(10.26\ \text{lb})(3\ \text{ft}) = -30.8\ \text{lb·ft}$$

Since the value obtained for the scalar M_O is negative, the moment $\mathbf{M}_O$ points *into* the page. You can write it as

$$\mathbf{M}_O = 30.8\ \text{lb·ft}\ \circlearrowright \blacktriangleleft$$

REFLECT and THINK: Always be alert for simplifications that can reduce the amount of computation.

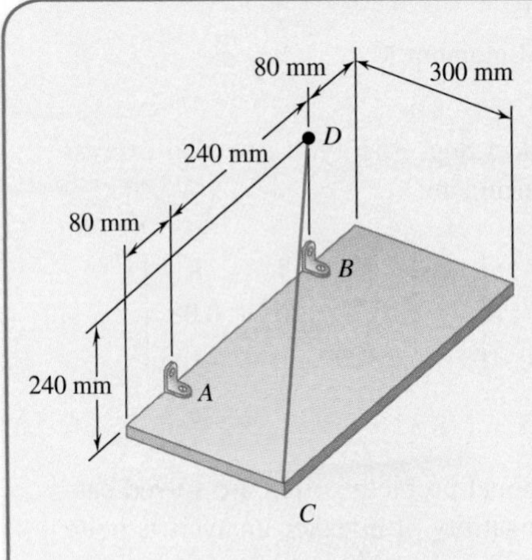

Sample Problem 3.4

A rectangular plate is supported by brackets at *A* and *B* and by a wire *CD*. If the tension in the wire is 200 N, determine the moment about *A* of the force exerted by the wire on point *C*.

STRATEGY: The solution requires resolving the tension in the wire and the position vector from *A* to *C* into rectangular components. You will need a unit vector approach to determine the force components.

MODELING and ANALYSIS: Obtain the moment $\mathbf{M}_A$ about *A* of the force **F** exerted by the wire on point *C* by forming the vector product

$$\mathbf{M}_A = \mathbf{r}_{C/A} \times \mathbf{F} \qquad (1)$$

(continued)

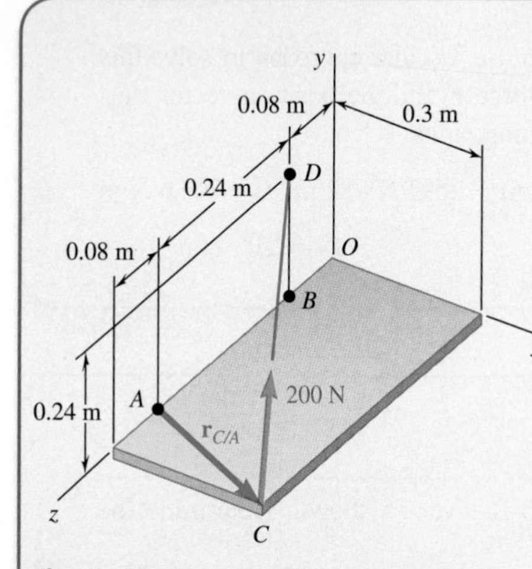

Fig. 1 The moment $\mathbf{M}_A$ is determined from position vector $\mathbf{r}_{C/A}$ and force vector **F**.

where $\mathbf{r}_{C/A}$ is the vector from A to C

$$\mathbf{r}_{C/A} = \overrightarrow{AC} = (0.3\text{ m})\mathbf{i} + (0.08\text{ m})\mathbf{k} \tag{2}$$

and **F** is the 200-N force directed along CD (Fig. 1). Introducing the unit vector

$$\boldsymbol{\lambda} = \overrightarrow{CD}/CD,$$

you can express **F** as

$$\mathbf{F} = F\boldsymbol{\lambda} = (200\text{N})\frac{\overrightarrow{CD}}{CD} \tag{3}$$

Resolving the vector $\overrightarrow{CD}$ into rectangular components, you have

$$\overrightarrow{CD} = -(0.3\text{ m})\mathbf{i} + (0.24\text{ m})\mathbf{j} - (0.32\text{ m})\mathbf{k} \qquad CD = 0.50\text{ m}$$

Substituting into (3) gives you

$$\mathbf{F} = \frac{200\text{N}}{0.50\text{m}}[-(0.3\text{ m})\mathbf{i} + (0.24\text{ m})\mathbf{j} - (0.32\text{ m})\mathbf{k}]$$
$$= -(120\text{ N})\mathbf{i} + (96\text{ N})\mathbf{j} - (128\text{ N})\mathbf{k} \tag{4}$$

Substituting for $\mathbf{r}_{C/A}$ and **F** from (2) and (4) into (1) and recalling the relations in Eq. (3.7) of Sec. 3.1D, you obtain (Fig. 2)

$$\mathbf{M}_A = \mathbf{r}_{C/A} \times \mathbf{F} = (0.3\mathbf{i} + 0.08\mathbf{k}) \times (-120\mathbf{i} + 96\mathbf{j} - 128\mathbf{k})$$
$$= (0.3)(96)\mathbf{k} + (0.3)(-128)(-\mathbf{j}) + (0.08)(-120)\mathbf{j} + (0.08)(96)(-\mathbf{i})$$

$$\mathbf{M}_A = -(7.68\text{ N·m})\mathbf{i} + (28.8\text{ N·m})\mathbf{j} + (28.8\text{ N·m})\mathbf{k} \quad ◀$$

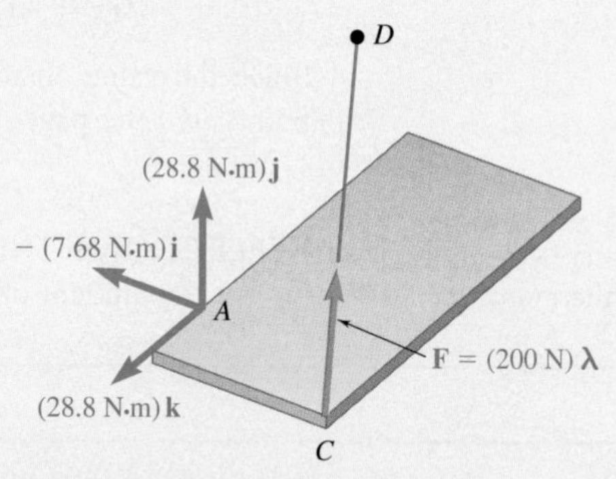

Fig. 2 Components of moment $\mathbf{M}_A$ applied at A.

Alternative Solution. As indicated in Sec. 3.1F, you can also express the moment $\mathbf{M}_A$ in the form of a determinant:

$$\mathbf{M}_A = \begin{vmatrix} \mathbf{i} & \mathbf{j} & \mathbf{k} \\ x_C - x_A & y_C - y_A & z_C - z_A \\ F_x & F_y & F_z \end{vmatrix} = \begin{vmatrix} \mathbf{i} & \mathbf{j} & \mathbf{k} \\ 0.3 & 0 & 0.08 \\ -120 & 96 & -128 \end{vmatrix}$$

$$\mathbf{M}_A = -(7.68\text{N·m})\mathbf{i} + (28.8\text{ N·m})\mathbf{j} + (28.8\text{ N·m})\mathbf{k} \quad ◀$$

REFLECT and THINK: Two-dimensional problems often are solved easily using a scalar approach, but the versatility of a vector analysis is quite apparent in a three-dimensional problem such as this.

Problems

3.1 A 20-lb force is applied to the control rod AB as shown. Knowing that the length of the rod is 9 in. and that $\alpha = 25°$, determine the moment of the force about point B by resolving the force into horizontal and vertical components.

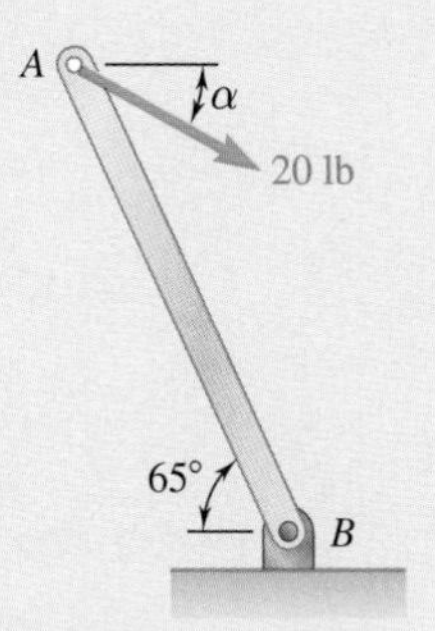

Fig. P3.1 and P3.2

3.2 A 20-lb force is applied to the control rod AB as shown. Knowing that the length of the rod is 9 in. and that the moment of the force about B is 120 lb·in. clockwise, determine the value of α.

3.3 A 300-N force **P** is applied at point A of the bell crank shown. (a) Compute the moment of the force **P** about O by resolving it into horizontal and vertical components. (b) Using the result of part a, determine the perpendicular distance from O to the line of action of **P**.

3.4 A 400-N force **P** is applied at point A of the bell crank shown. (a) Compute the moment of the force **P** about O by resolving it into components along line OA and in a direction perpendicular to that line. (b) Determine the magnitude and direction of the smallest force **Q** applied at B that has the same moment as **P** about O.

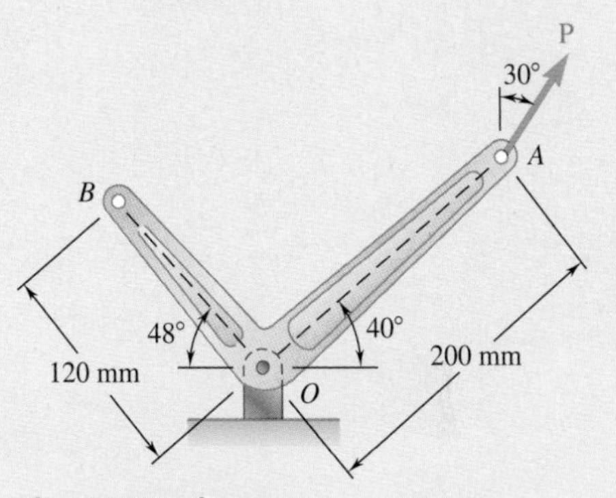

Fig. P3.3 and *P3.4*

3.5 A 300-N force is applied at A as shown. Determine (a) the moment of the 300-N force about D, (b) the smallest force applied at B that creates the same moment about D.

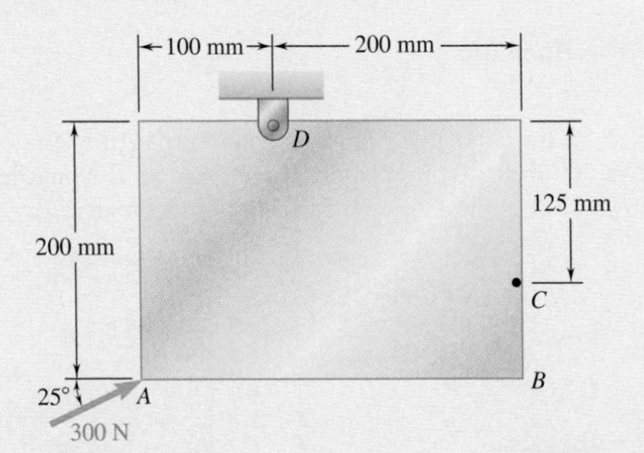

Fig. P3.5 and P3.6

3.6 A 300-N force is applied at A as shown. Determine (a) the moment of the 300-N force about D, (b) the magnitude and sense of the horizontal force applied at C that creates the same moment about D, (c) the smallest force applied at C that creates the same moment about D.

3.7 and *3.8* The tailgate of a car is supported by the hydraulic lift *BC*. If the lift exerts a 125-lb force directed along its centerline on the ball and socket at *B*, determine the moment of the force about *A*.

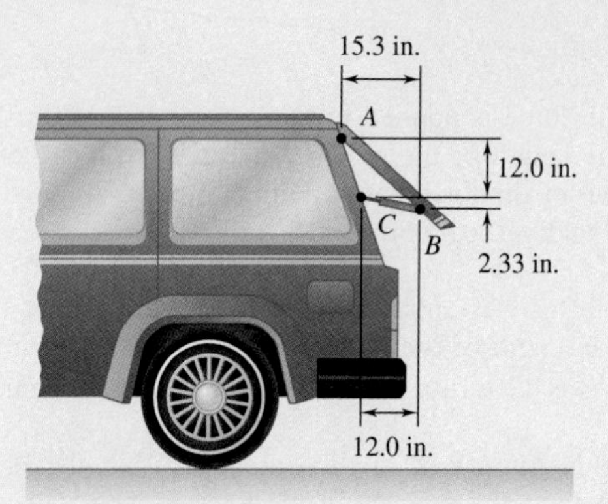

Fig. P3.7

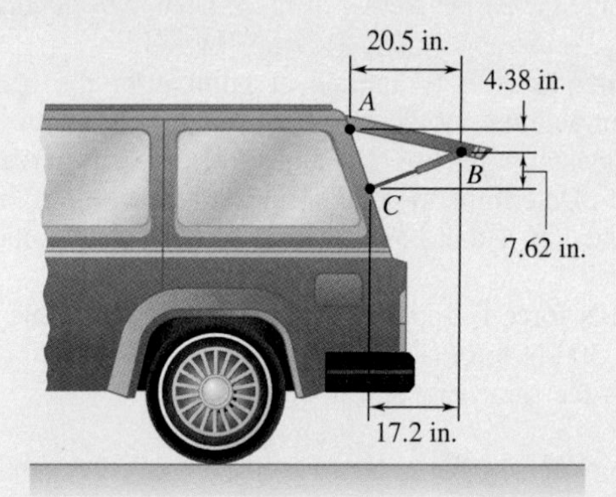

Fig. *P3.8*

3.9 and *3.10* It is known that the connecting rod *AB* exerts on the crank *BC* a 500-lb force directed down and to the left along the centerline of *AB*. Determine the moment of the force about *C*.

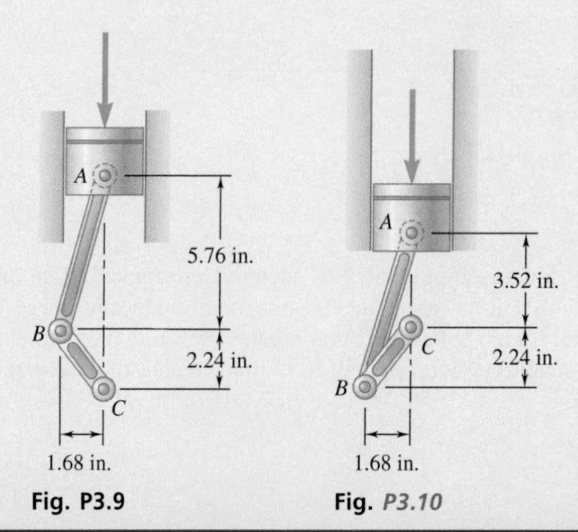

Fig. P3.9 **Fig. *P3.10***

3.11 Rod AB is held in place by the cord AC. Knowing that the tension in the cord is 1350 N and that c = 360 mm, determine the moment about B of the force exerted by the cord at point A by resolving that force into horizontal and vertical components applied (a) at point A, (b) at point C.

3.12 Rod AB is held in place by the cord AC. Knowing that c = 840 mm and that the moment about B of the force exerted by the cord at point A is 756 N·m, determine the tension in the cord.

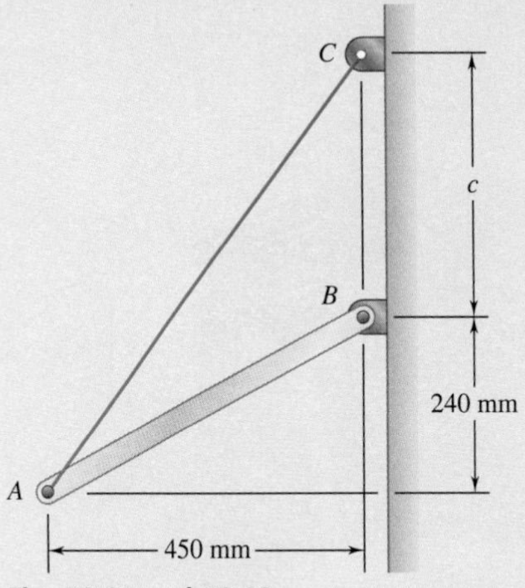

Fig. P3.11 and P3.12

3.13 Determine the moment about the origin O of the force $\mathbf{F} = 4\mathbf{i} - 3\mathbf{j} + 5\mathbf{k}$ that acts at a point A. Assume that the position vector of A is (a) $\mathbf{r} = 2\mathbf{i} + 3\mathbf{j} - 4\mathbf{k}$, ($b$) $\mathbf{r} = -8\mathbf{i} + 6\mathbf{j} - 10\mathbf{k}$, (c) $\mathbf{r} = 8\mathbf{i} - 6\mathbf{j} + 5\mathbf{k}$.

3.14 Determine the moment about the origin O of the force $\mathbf{F} = 2\mathbf{i} + 3\mathbf{j} - 4\mathbf{k}$ that acts at a point A. Assume that the position vector of A is (a) $\mathbf{r} = 3\mathbf{i} - 6\mathbf{j} + 5\mathbf{k}$, ($b$) $\mathbf{r} = \mathbf{i} - 4\mathbf{j} - 2\mathbf{k}$, (c) $\mathbf{r} = 4\mathbf{i} + 6\mathbf{j} - 8\mathbf{k}$.

3.15 A 6-ft-long fishing rod AB is securely anchored in the sand of a beach. After a fish takes the bait, the resulting force in the line is 6 lb. Determine the moment about A of the force exerted by the line at B.

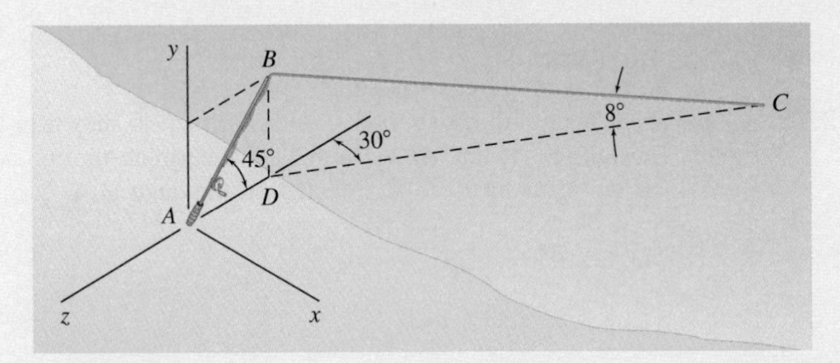

Fig. P3.15

3.16 The wire AE is stretched between the corners A and E of a bent plate. Knowing that the tension in the wire is 435 N, determine the moment about O of the force exerted by the wire (a) on corner A, (b) on corner E.

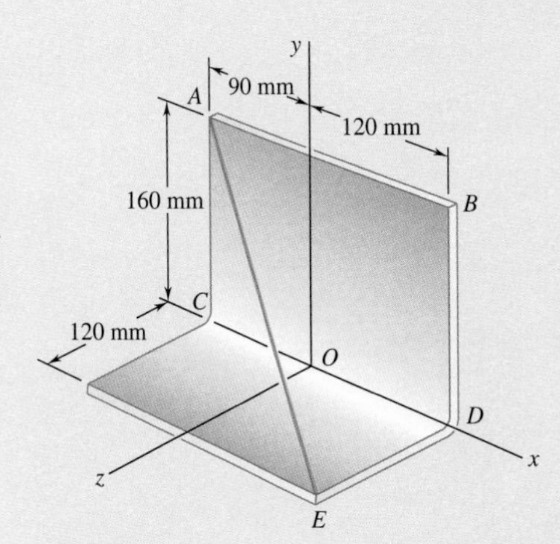

Fig. P3.16

3.17 The 12-ft boom AB has a fixed end A. A steel cable is stretched from the free end B of the boom to a point C located on the vertical wall. If the tension in the cable is 380 lb, determine the moment about A of the force exerted by the cable at B.

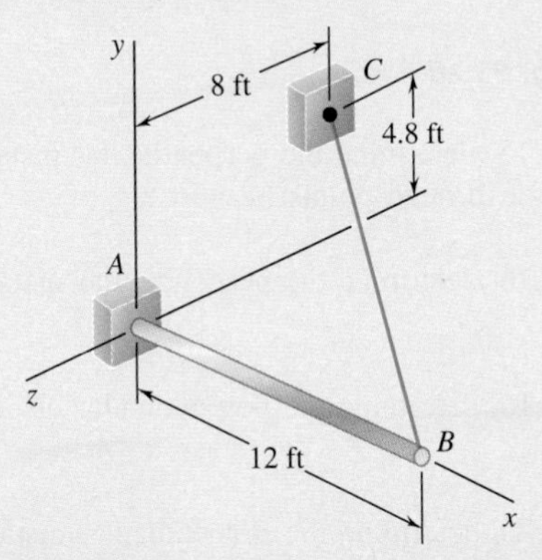

Fig. P3.17

3.18 A wooden board AB, which is used as a temporary prop to support a small roof, exerts at point A of the roof a 57-lb force directed along BA. Determine the moment about C of that force.

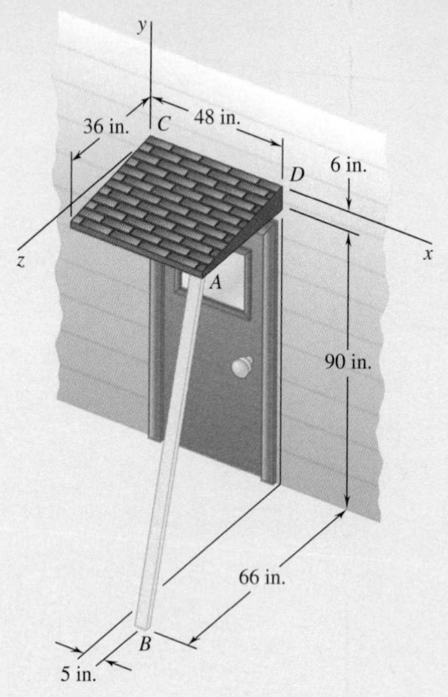

Fig. P3.18

3.19 A 200-N force is applied as shown to the bracket ABC. Determine the moment of the force about A.

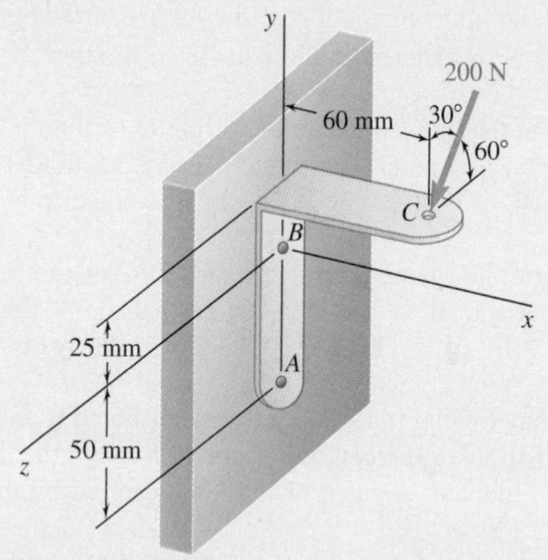

Fig. P3.19

3.20 A small boat hangs from two davits, one of which is shown in the figure. The tension in line $ABAD$ is 82 lb. Determine the moment about C of the resultant force $\mathbf{R}_A$ exerted on the davit at A.

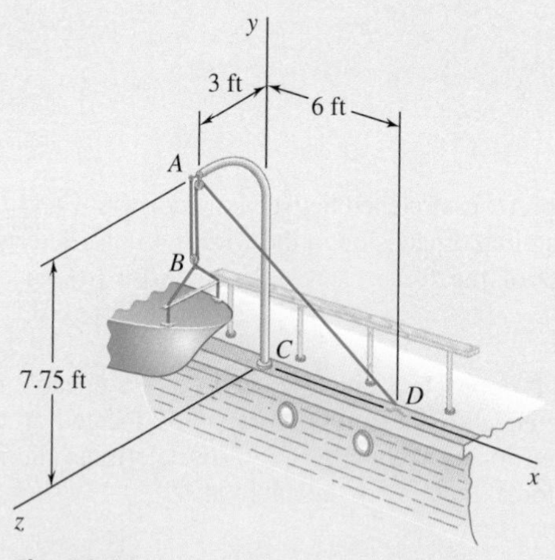

Fig. P3.20

3.21 In Prob. 3.15, determine the perpendicular distance from point A to a line drawn through points B and C.

3.22 In Prob. 3.16, determine the perpendicular distance from point O to wire AE.

3.23 In Prob. 3.16, determine the perpendicular distance from point B to wire AE.

3.24 In Prob. 3.20, determine the perpendicular distance from point C to portion AD of the line $ABAD$.

3.2 MOMENT OF A FORCE ABOUT AN AXIS

We want to extend the idea of the moment about a point to the often useful concept of the moment about an axis. However, first we need to introduce another tool of vector mathematics. We have seen that the vector product multiplies two vectors together and produces a new vector. Here we examine the scalar product, which multiplies two vectors together and produces a scalar quantity.

3.2A Scalar Products

The **scalar product** of two vectors **P** and **Q** is defined as the product of the magnitudes of **P** and **Q** and of the cosine of the angle θ formed between them (Fig. 3.18). The scalar product of **P** and **Q** is denoted by $\mathbf{P} \cdot \mathbf{Q}$.

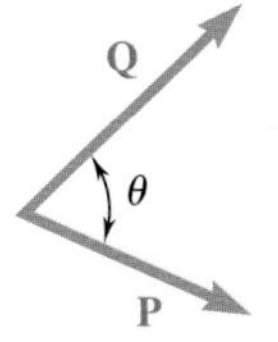

Fig. 3.18 Two vectors **P** and **Q** and the angle θ between them.

Scalar product

$$\mathbf{P} \cdot \mathbf{Q} = PQ\cos\theta \tag{3.24}$$

Note that this expression is not a vector but a *scalar,* which explains the name *scalar product*. Because of the notation used, $\mathbf{P} \cdot \mathbf{Q}$ is also referred to as the *dot product* of the vectors **P** and **Q**.

It follows from its very definition that the scalar product of two vectors is *commutative,* i.e., that

$$\mathbf{P} \cdot \mathbf{Q} = \mathbf{Q} \cdot \mathbf{P} \tag{3.25}$$

It can also be proven that the scalar product is *distributive,* as shown by

$$\mathbf{P} \cdot (\mathbf{Q}_1 + \mathbf{Q}_2) = \mathbf{P} \cdot \mathbf{Q}_1 + \mathbf{P} \cdot \mathbf{Q}_2 \tag{3.26}$$

As far as the associative property is concerned, this property cannot apply to scalar products. Indeed, $(\mathbf{P} \cdot \mathbf{Q}) \cdot \mathbf{S}$ has no meaning, because $\mathbf{P} \cdot \mathbf{Q}$ is not a vector but a scalar.

We can also express the scalar product of two vectors **P** and **Q** in terms of their rectangular components. Resolving **P** and **Q** into components, we first write

$$\mathbf{P} \cdot \mathbf{Q} = (P_x\mathbf{i} + P_y\mathbf{j} + P_z\mathbf{k}) \cdot (Q_x\mathbf{i} + Q_y\mathbf{j} + Q_z\mathbf{k})$$

Making use of the distributive property, we express $\mathbf{P} \cdot \mathbf{Q}$ as the sum of scalar products, such as $P_x\mathbf{i} \cdot Q_x\mathbf{i}$ and $P_x\mathbf{i} \cdot Q_y\mathbf{j}$. However, from the definition of the scalar product, it follows that the scalar products of the unit vectors are either zero or one.

$$\begin{aligned} \mathbf{i} \cdot \mathbf{i} &= 1 \qquad \mathbf{j} \cdot \mathbf{j} = 1 \qquad \mathbf{k} \cdot \mathbf{k} = 1 \\ \mathbf{i} \cdot \mathbf{j} &= 0 \qquad \mathbf{j} \cdot \mathbf{k} = 0 \qquad \mathbf{k} \cdot \mathbf{i} = 0 \end{aligned} \tag{3.27}$$

Thus, the expression for $\mathbf{P} \cdot \mathbf{Q}$ reduces to

Scalar product

$$\mathbf{P} \cdot \mathbf{Q} = P_xQ_x + P_yQ_y + P_zQ_z \tag{3.28}$$

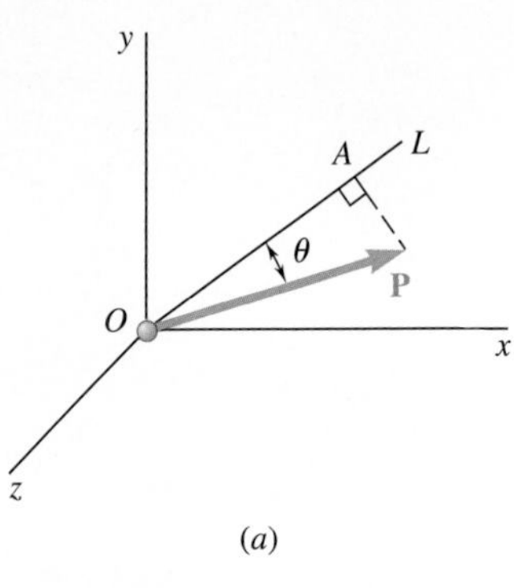

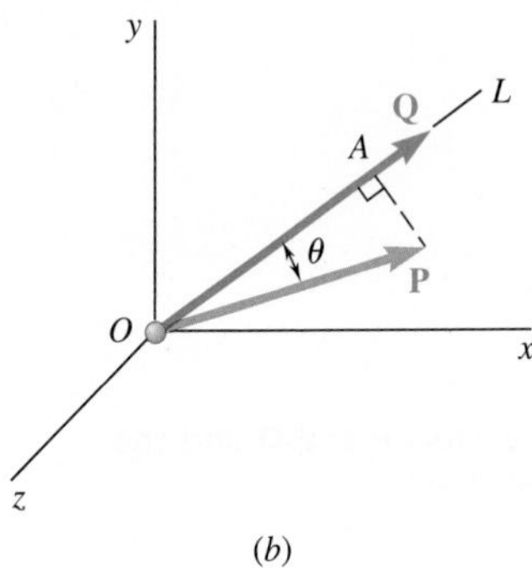

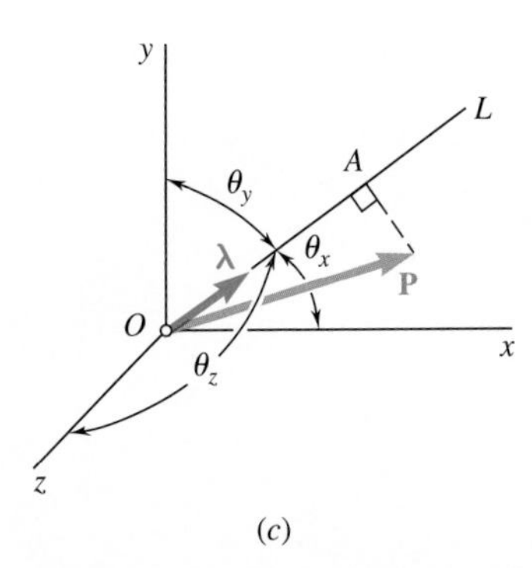

Fig. 3.19 (*a*) The projection of vector **P** at an angle θ to a line *OL*; (*b*) the projection of **P** and a vector **Q** along *OL*; (*c*) the projection of **P**, a unit vector $\boldsymbol{\lambda}$ along *OL*, and the angles of *OL* with the coordinate axes.

In the particular case when **P** and **Q** are equal, we note that

$$\mathbf{P} \cdot \mathbf{P} = P_x^2 + P_y^2 + P_z^2 = P^2 \tag{3.29}$$

Applications of the Scalar Product

1. **Angle formed by two given vectors.** Let two vectors be given in terms of their components:

$$\mathbf{P} = P_x\mathbf{i} + P_y\mathbf{j} + P_z\mathbf{k}$$
$$\mathbf{Q} = Q_x\mathbf{i} + Q_y\mathbf{j} + Q_z\mathbf{k}$$

To determine the angle formed by the two vectors, we equate the expressions obtained in Eqs. (3.24) and (3.28) for their scalar product,

$$PQ \cos \theta = P_xQ_x + P_yQ_y + P_zQ_z$$

Solving for $\cos \theta$, we have

$$\cos \theta = \frac{P_xQ_x + P_yQ_y + P_zQ_z}{PQ} \tag{3.30}$$

2. **Projection of a vector on a given axis.** Consider a vector **P** forming an angle θ with an axis, or directed line, *OL* (Fig. 3.19*a*). We define the *projection of* **P** *on the axis OL* as the scalar

$$P_{OL} = P \cos \theta \tag{3.31}$$

The projection P_{OL} is equal in absolute value to the length of the segment *OA*. It is positive if *OA* has the same sense as the axis *OL*—that is, if θ is acute—and negative otherwise. If **P** and *OL* are at a right angle, the projection of **P** on *OL* is zero.

Now consider a vector **Q** directed along *OL* and of the same sense as *OL* (Fig. 3.19*b*). We can express the scalar product of **P** and **Q** as

$$\mathbf{P} \cdot \mathbf{Q} = PQ \cos \theta = P_{OL}Q \tag{3.32}$$

from which it follows that

$$P_{OL} = \frac{\mathbf{P} \cdot \mathbf{Q}}{Q} = \frac{P_xQ_x + P_yQ_y + P_zQ_z}{Q} \tag{3.33}$$

In the particular case when the vector selected along *OL* is the unit vector $\boldsymbol{\lambda}$ (Fig. 3.19*c*), we have

$$P_{OL} = \mathbf{P} \cdot \boldsymbol{\lambda} \tag{3.34}$$

Recall from Sec. 2.4A that the components of $\boldsymbol{\lambda}$ along the coordinate axes are respectively equal to the direction cosines of *OL*. Resolving **P** and $\boldsymbol{\lambda}$ into rectangular components, we can express the projection of **P** on *OL* as

$$P_{OL} = P_x \cos \theta_x + P_y \cos \theta_y + P_z \cos \theta_z \tag{3.35}$$

where θ_x, θ_y, and θ_z denote the angles that the axis *OL* forms with the coordinate axes.

3.2B Mixed Triple Products

We have now seen both forms of multiplying two vectors together: the vector product and the scalar product. Here we define the **mixed triple product** of the three vectors **S**, **P**, and **Q** as the scalar expression

Mixed triple product

$$\mathbf{S}\cdot(\mathbf{P}\times\mathbf{Q}) \tag{3.36}$$

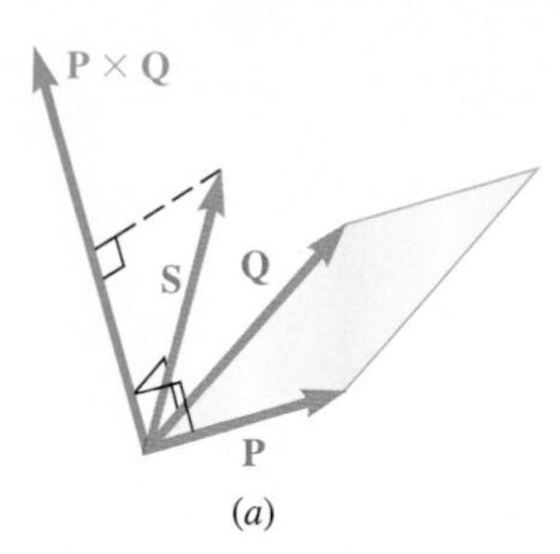

(*a*)

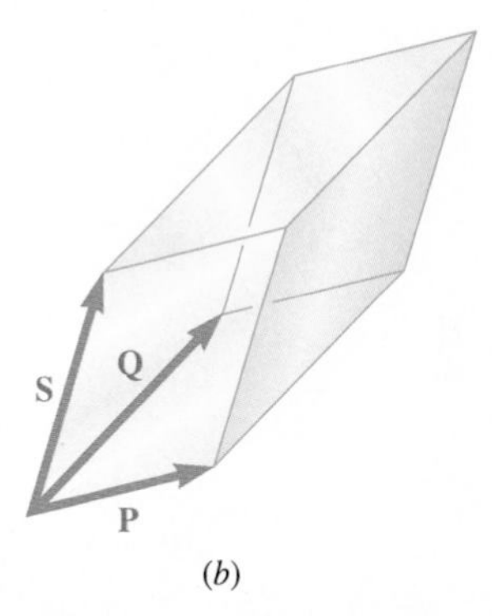

(*b*)

Fig. 3.20 (*a*) The mixed triple product is equal to the magnitude of the cross product of two vectors multiplied by the projection of the third vector onto that cross product; (*b*) the result equals the volume of the parallelepiped formed by the three vectors.

This is obtained by forming the scalar product of **S** with the vector product of **P** and **Q**. [In Chapter 15, we will introduce another kind of triple product, called the vector triple product, **S** × (**P** × **Q**).]

The mixed triple product of **S**, **P**, and **Q** has a simple geometrical interpretation (Fig. 3.20*a*). Recall from Sec. 3.4 that the vector **P** × **Q** is perpendicular to the plane containing **P** and **Q** and that its magnitude is equal to the area of the parallelogram that has **P** and **Q** for sides. Also, Eq. (3.32) indicates that we can obtain the scalar product of **S** and **P** × **Q** by multiplying the magnitude of **P** × **Q** (i.e., the area of the parallelogram defined by **P** and **Q**) by the projection of **S** on the vector **P** × **Q** (i.e., by the projection of **S** on the normal to the plane containing the parallelogram). The mixed triple product is thus equal, in absolute value, to the volume of the parallelepiped having the vectors **S**, **P**, and **Q** for sides (Fig. 3.20*b*). The sign of the mixed triple product is positive if **S**, **P**, and **Q** form a right-handed triad and negative if they form a left-handed triad. [That is, **S**·(**P** × **Q**) is negative if the rotation that brings **P** into line with **Q** is observed as clockwise from the tip of **S**.] The mixed triple product is zero if **S**, **P**, and **Q** are coplanar.

Since the parallelepiped defined in this way is independent of the order in which the three vectors are taken, the six mixed triple products that can be formed with **S**, **P**, and **Q** all have the same absolute value, although not the same sign. It is easily shown that

$$\begin{aligned}\mathbf{S}\cdot(\mathbf{P}\times\mathbf{Q}) &= \mathbf{P}\cdot(\mathbf{Q}\times\mathbf{S}) = \mathbf{Q}\cdot(\mathbf{S}\times\mathbf{P})\\ &= -\mathbf{S}\cdot(\mathbf{Q}\times\mathbf{P}) = -\mathbf{P}\cdot(\mathbf{S}\times\mathbf{Q}) = -\mathbf{Q}\cdot(\mathbf{P}\times\mathbf{S})\end{aligned} \tag{3.37}$$

Arranging the letters representing the three vectors counterclockwise in a circle (Fig. 3.21), we observe that the sign of the mixed triple product remains unchanged if the vectors are permuted in such a way that they still read in counterclockwise order. Such a permutation is said to be a *circular permutation*. It also follows from Eq. (3.37) and from the commutative property of scalar products that the mixed triple product of **S**, **P**, and **Q** can be defined equally well as **S**·(**P** × **Q**) or (**S** × **P**)·**Q**.

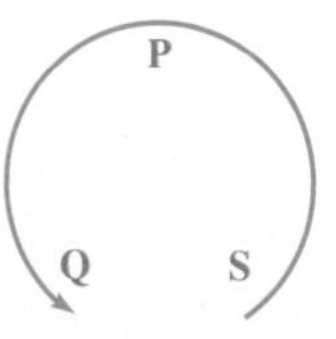

Fig. 3.21 Counterclockwise arrangement for determining the sign of the mixed triple product of three vectors **P**, **Q**, and **S**.

We can also express the mixed triple product of the vectors **S**, **P**, and **Q** in terms of the rectangular components of these vectors. Denoting **P** × **Q** by **V** and using formula (3.28) to express the scalar product of **S** and **V**, we have

$$\mathbf{S}\cdot(\mathbf{P}\times\mathbf{Q}) = \mathbf{S}\cdot\mathbf{V} = S_xV_x + S_yV_y + S_zV_z$$

Substituting from the relations in Eq. (3.9) for the components of **V**, we obtain

$$\begin{aligned}\mathbf{S}\cdot(\mathbf{P}\times\mathbf{Q}) = S_x(P_yQ_z - P_zQ_y) + S_y(P_zQ_x - P_xQ_z)\\ + S_z(P_xQ_y - P_yQ_x)\end{aligned} \tag{3.38}$$

We can write this expression in a more compact form if we observe that it represents the expansion of a determinant:

Mixed triple product, determinant form

$$\mathbf{S}\cdot(\mathbf{P}\times\mathbf{Q}) = \begin{vmatrix} S_x & S_y & S_z \\ P_x & P_y & P_z \\ Q_x & Q_y & Q_z \end{vmatrix} \tag{3.39}$$

By applying the rules governing the permutation of rows in a determinant, we could easily verify the relations in Eq. (3.37), which we derived earlier from geometrical considerations.

3.2C Moment of a Force about a Given Axis

Now that we have the necessary mathematical tools, we can introduce the concept of moment of a force about an axis. Consider again a force **F** acting on a rigid body and the moment $\mathbf{M}_O$ of that force about O (Fig. 3.22). Let OL be an axis through O.

We define the moment M_{OL} of F about OL as the projection OC of the moment M_O onto the axis OL.

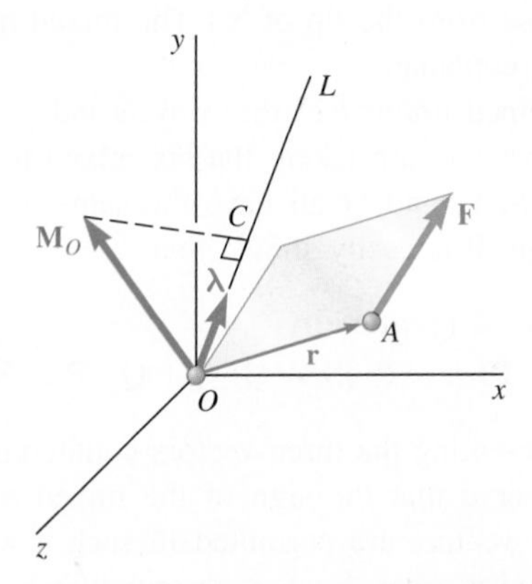

Fig. 3.22 The moment $\mathbf{M}_{OL}$ of a force **F** about the axis OL is the projection on OL of the moment $\mathbf{M}_O$. The calculation involves the unit vector λ along OL and the position vector **r** from O to A, the point upon which the force **F** acts.

Suppose we denote the unit vector along OL by $\boldsymbol{\lambda}$ and recall the expressions (3.34) and (3.11) for the projection of a vector on a given axis and for the moment $\mathbf{M}_O$ of a force **F**. Then we can express M_{OL} as

Moment about an axis through the origin

$$M_{OL} = \boldsymbol{\lambda}\cdot\mathbf{M}_O = \boldsymbol{\lambda}\cdot(\mathbf{r}\times\mathbf{F}) \tag{3.40}$$

This shows that the moment M_{OL} of $\mathbf{F}$ about the axis OL is the scalar obtained by forming the mixed triple product of $\boldsymbol{\lambda}$, $\mathbf{r}$, and $\mathbf{F}$. We can also express M_{OL} in the form of a determinant,

$$M_{OL} = \begin{vmatrix} \lambda_x & \lambda_y & \lambda_z \\ x & y & z \\ F_x & F_y & F_z \end{vmatrix} \tag{3.41}$$

where $\lambda_x, \lambda_y, \lambda_z$ = direction cosines of axis OL
x, y, z = coordinates of point of application of $\mathbf{F}$
F_x, F_y, F_z = components of force $\mathbf{F}$

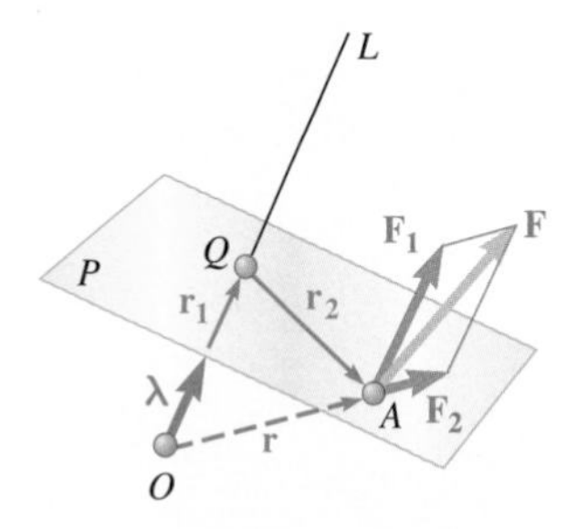

Fig. 3.23 By resolving the force **F** into components parallel to the axis *OL* and in a plane perpendicular to the axis, we can show that the moment $\mathbf{M}_{OL}$ of **F** about *OL* measures the tendency of **F** to rotate the rigid body about the axis.

The physical significance of the moment M_{OL} of a force $\mathbf{F}$ about a fixed axis OL becomes more apparent if we resolve $\mathbf{F}$ into two rectangular components $\mathbf{F}_1$ and $\mathbf{F}_2$, with $\mathbf{F}_1$ parallel to OL and $\mathbf{F}_2$ lying in a plane P perpendicular to OL (Fig. 3.23). Resolving $\mathbf{r}$ similarly into two components $\mathbf{r}_1$ and $\mathbf{r}_2$ and substituting for $\mathbf{F}$ and $\mathbf{r}$ into Eq. (3.40), we get

$$\begin{aligned} M_{OL} &= \boldsymbol{\lambda} \cdot [(\mathbf{r}_1 + \mathbf{r}_2) \times (\mathbf{F}_1 + \mathbf{F}_2)] \\ &= \boldsymbol{\lambda} \cdot (\mathbf{r}_1 \times \mathbf{F}_1) + \boldsymbol{\lambda} \cdot (\mathbf{r}_1 \times \mathbf{F}_2) + \boldsymbol{\lambda} \cdot (\mathbf{r}_2 \times \mathbf{F}_1) + \boldsymbol{\lambda} \cdot (\mathbf{r}_2 \times \mathbf{F}_2) \end{aligned}$$

Note that all of the mixed triple products except the last one are equal to zero because they involve vectors that are coplanar when drawn from a common origin (Sec. 3.2B). Therefore, this expression reduces to

$$M_{OL} = \boldsymbol{\lambda} \cdot (\mathbf{r}_2 \times \mathbf{F}_2) \tag{3.42}$$

The vector product $\mathbf{r}_2 \times \mathbf{F}_2$ is perpendicular to the plane P and represents the moment of the component $\mathbf{F}_2$ of $\mathbf{F}$ about the point Q where OL intersects P. Therefore, the scalar M_{OL}, which is positive if $\mathbf{r}_2 \times \mathbf{F}_2$ and OL have the same sense and is negative otherwise, measures the tendency of $\mathbf{F}_2$ to make the rigid body rotate about the fixed axis OL. The other component $\mathbf{F}_1$ of $\mathbf{F}$ does not tend to make the body rotate about OL, because $\mathbf{F}_1$ and OL are parallel. Therefore, we conclude that

The moment M_{OL} of* F *about OL measures the tendency of the force* F *to impart to the rigid body a rotation about the fixed axis OL.

From the definition of the moment of a force about an axis, it follows that the moment of $\mathbf{F}$ about a coordinate axis is equal to the component of $\mathbf{M}_O$ along that axis. If we substitute each of the unit vectors $\mathbf{i}$, $\mathbf{j}$, and $\mathbf{k}$ for $\boldsymbol{\lambda}$ in Eq. (3.40), we obtain expressions for the *moments of* **F** *about the coordinate axes*. These expressions are respectively equal to those obtained earlier for the components of the moment $\mathbf{M}_O$ of $\mathbf{F}$ about O:

$$\begin{aligned} M_x &= yF_z - zF_y \\ M_y &= zF_x - xF_z \\ M_z &= xF_y - yF_x \end{aligned} \tag{3.18}$$

Just as the components F_x, F_y, and F_z of a force $\mathbf{F}$ acting on a rigid body measure, respectively, the tendency of $\mathbf{F}$ to move the rigid body in the x, y, and z directions, the moments M_x, M_y, and M_z of $\mathbf{F}$ about the coordinate axes measure the tendency of $\mathbf{F}$ to impart to the rigid body a rotation about the x, y, and z axes, respectively.

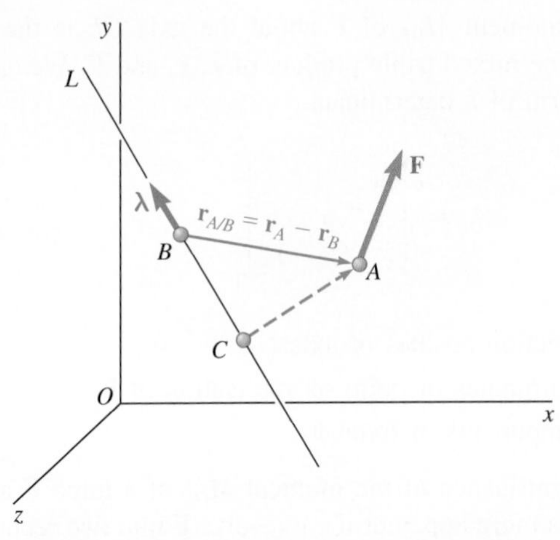

Fig. 3.24 The moment of a force about an axis or line L can be found by evaluating the mixed triple product at a point B on the line. The choice of B is arbitrary, since using any other point on the line, such as C, yields the same result.

More generally, we can obtain the moment of a force $\mathbf{F}$ applied at A about an axis that does not pass through the origin by choosing an arbitrary point B on the axis (Fig. 3.24) and determining the projection on the axis BL of the moment $\mathbf{M}_B$ of $\mathbf{F}$ about B. The equation for this projection is given here.

Moment about an arbitrary axis

$$M_{BL} = \boldsymbol{\lambda} \cdot \mathbf{M}_B = \boldsymbol{\lambda} \cdot (\mathbf{r}_{A/B} \times \mathbf{F}) \tag{3.43}$$

where $\mathbf{r}_{A/B} = \mathbf{r}_A - \mathbf{r}_B$ represents the vector drawn from B to A. Expressing M_{BL} in the form of a determinant, we have

$$M_{BL} = \begin{vmatrix} \lambda_x & \lambda_y & \lambda_z \\ x_{A/B} & y_{A/B} & z_{A/B} \\ F_x & F_y & F_z \end{vmatrix} \tag{3.44}$$

where $\lambda_x, \lambda_y, \lambda_z$ = direction cosines of axis BL

$x_{A/B} = x_A - x_B \qquad y_{A/B} = y_A - y_B \qquad z_{A/B} = z_A - z_B$

F_x, F_y, F_z = components of force $\mathbf{F}$

Note that this result is independent of the choice of the point B on the given axis. Indeed, denoting by M_{CL} the moment obtained with a different point C, we have

$$\begin{aligned} M_{CL} &= \boldsymbol{\lambda} \cdot [(\mathbf{r}_A - \mathbf{r}_C) \times \mathbf{F}] \\ &= \boldsymbol{\lambda} \cdot [(\mathbf{r}_A - \mathbf{r}_B) \times \mathbf{F}] + \boldsymbol{\lambda} \cdot [(\mathbf{r}_B - \mathbf{r}_C) \times \mathbf{F}] \end{aligned}$$

However, since the vectors $\boldsymbol{\lambda}$ and $\mathbf{r}_B - \mathbf{r}_C$ lie along the same line, the volume of the parallelepiped having the vectors $\boldsymbol{\lambda}$, $\mathbf{r}_B - \mathbf{r}_C$, and $\mathbf{F}$ for sides is zero, as is the mixed triple product of these three vectors (Sec. 3.2B). The expression obtained for M_{CL} thus reduces to its first term, which is the expression used earlier to define M_{BL}. In addition, it follows from Sec. 3.1E that, when computing the moment of $\mathbf{F}$ about the given axis, A can be any point on the line of action of $\mathbf{F}$.

Sample Problem 3.5

A cube of side a is acted upon by a force **P** along the diagonal of a face, as shown. Determine the moment of **P** (*a*) about A, (*b*) about the edge AB, (*c*) about the diagonal AG of the cube. (*d*) Using the result of part *c*, determine the perpendicular distance between AG and FC.

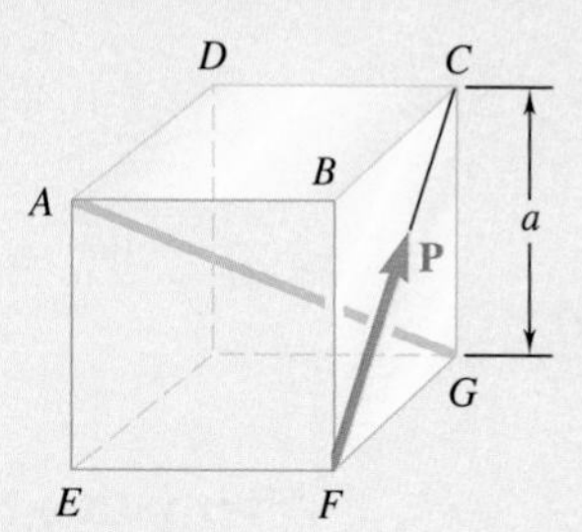

STRATEGY: Use the equations presented in this section to compute the moments asked for. You can find the distance between AG and FC from the expression for the moment M_{AG}.

MODELING and ANALYSIS:

a. Moment about *A*. Choosing x, y, and z axes as shown (Fig. 1), resolve into rectangular components the force **P** and the vector $\mathbf{r}_{F/A} = \overrightarrow{AF}$ drawn from A to the point of application F of **P**.

$$\mathbf{r}_{F/A} = a\mathbf{i} - a\mathbf{j} = a(\mathbf{i} - \mathbf{j})$$
$$\mathbf{P} = (P/\sqrt{2})\mathbf{j} - (P/\sqrt{2})\mathbf{k} = (P/\sqrt{2})(\mathbf{j} - \mathbf{k})$$

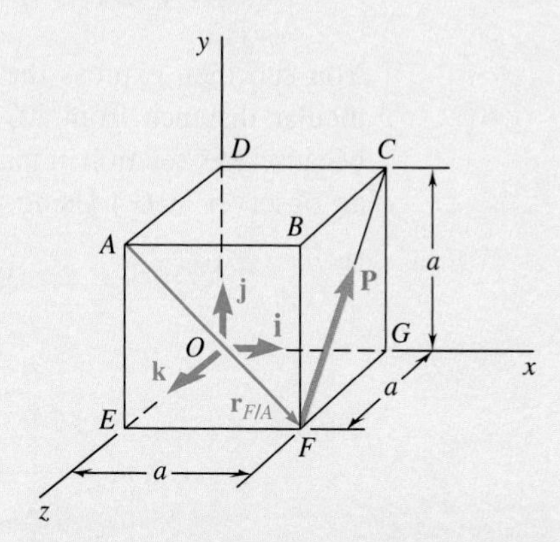

Fig. 1 Position vector $\mathbf{r}_{F/A}$ and force vector **P** relative to chosen coordinate system.

The moment of **P** about A is the vector product of these two vectors:

$$\mathbf{M}_A = \mathbf{r}_{F/A} \times \mathbf{P} = a(\mathbf{i} - \mathbf{j}) \times (P/\sqrt{2})(\mathbf{j} - \mathbf{k})$$
$$\mathbf{M}_A = (aP/\sqrt{2})(\mathbf{i} + \mathbf{j} + \mathbf{k}) \quad ◀$$

b. Moment about *AB*. You want the projection of $\mathbf{M}_A$ on AB:

$$M_{AB} = \mathbf{i} \cdot \mathbf{M}_A = \mathbf{i} \cdot (aP/\sqrt{2})(\mathbf{i} + \mathbf{j} + \mathbf{k})$$
$$M_{AB} = aP/\sqrt{2} \quad ◀$$

(continued)

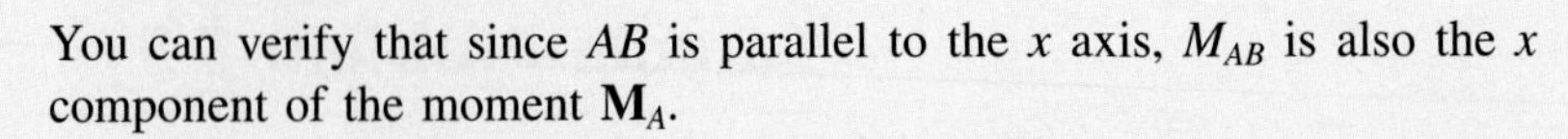

You can verify that since AB is parallel to the x axis, M_{AB} is also the x component of the moment $\mathbf{M}_A$.

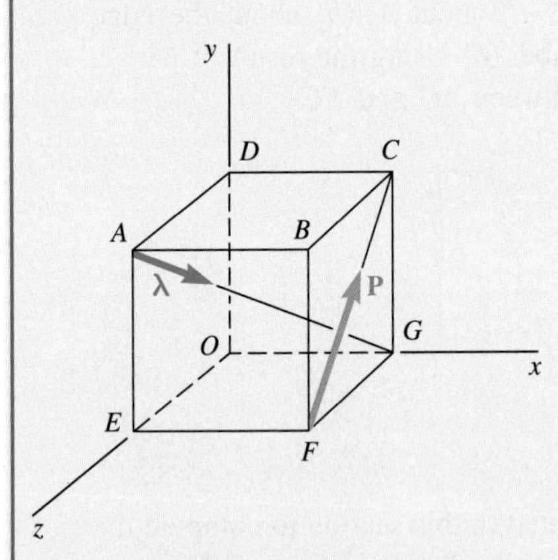

Fig. 2 Unit vector λ used to determine moment of **P** about *AG*.

c. Moment about diagonal *AG*. You obtain the moment of **P** about AG by projecting $\mathbf{M}_A$ on AG. If you denote the unit vector along AG by $\boldsymbol{\lambda}$ (Fig. 2), the calculation looks like this:

$$\boldsymbol{\lambda} = \frac{\overrightarrow{AG}}{AG} = \frac{a\mathbf{i} - a\mathbf{j} - a\mathbf{k}}{a\sqrt{3}} = (1/\sqrt{3})(\mathbf{i} - \mathbf{j} - \mathbf{k})$$

$$M_{AG} = \boldsymbol{\lambda}\cdot\mathbf{M}_A = (1/\sqrt{3})(\mathbf{i} - \mathbf{j} - \mathbf{k})\cdot(aP/\sqrt{2})(\mathbf{i} + \mathbf{j} + \mathbf{k})$$

$$M_{AG} = (aP/\sqrt{6})(1 - 1 - 1) \quad M_{AG} = -aP/\sqrt{6}$$

Alternative Method. You can also calculate the moment of **P** about AG from the determinant form:

$$M_{AG} = \begin{vmatrix} \lambda_x & \lambda_y & \lambda_z \\ x_{F/A} & y_{F/A} & z_{F/A} \\ F_x & F_y & F_z \end{vmatrix} = \begin{vmatrix} 1/\sqrt{3} & -1/\sqrt{3} & -1/\sqrt{3} \\ a & -a & 0 \\ 0 & P/\sqrt{2} & -P/\sqrt{2} \end{vmatrix} = -aP/\sqrt{6}$$

d. Perpendicular Distance between *AG* and *FC*. First note that **P** is perpendicular to the diagonal AG. You can check this by forming the scalar product $\mathbf{P}\cdot\boldsymbol{\lambda}$ and verifying that it is zero:

$$\mathbf{P}\cdot\boldsymbol{\lambda} = (P/\sqrt{2})(\mathbf{j} - \mathbf{k})\cdot(1/\sqrt{3})(\mathbf{i} - \mathbf{j} - \mathbf{k}) = (P\sqrt{6})(0 - 1 + 1) = 0$$

You can then express the moment M_{AG} as $-Pd$, where d is the perpendicular distance from AG to FC (Fig. 3). (The negative sign is needed because the rotation imparted to the cube by **P** appears as clockwise to an observer at G.) Using the value found for M_{AG} in part c,

$$M_{AG} = -Pd = -aP/\sqrt{6} \qquad d = a/\sqrt{6}$$

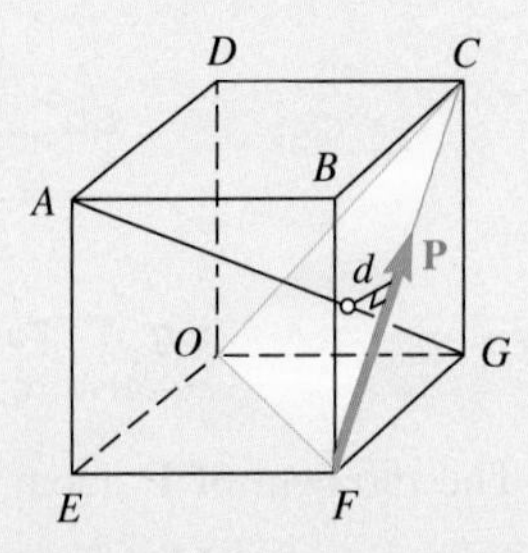

Fig. 3 Perpendicular distance *d* from *AG* to *FC*.

REFLECT and THINK: In a problem like this, it is important to visualize the forces and moments in three dimensions so you can choose the appropriate equations for finding them and also recognize the geometric relationships between them.

Problems

3.25 Given the vectors $\mathbf{P} = 3\mathbf{i} - \mathbf{j} + 2\mathbf{k}$, $\mathbf{Q} = 4\mathbf{i} + 5\mathbf{j} - 3\mathbf{k}$, and $\mathbf{S} = -2\mathbf{i} + 3\mathbf{j} - \mathbf{k}$, compute the scalar products $\mathbf{P} \cdot \mathbf{Q}$, $\mathbf{P} \cdot \mathbf{S}$, and $\mathbf{Q} \cdot \mathbf{S}$.

3.26 Form the scalar product $\mathbf{B} \cdot \mathbf{C}$ and use the result obtained to prove the identity

$$\cos(\alpha - \beta) = \cos\alpha \cos\beta + \sin\alpha \sin\beta$$

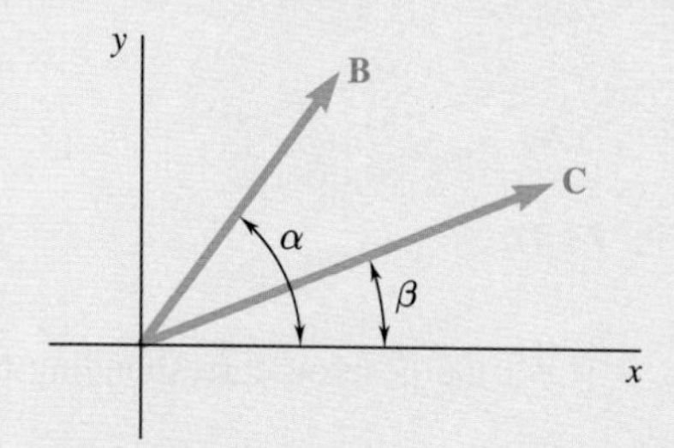

Fig. P3.26

3.27 Knowing that the tension in cable AC is 1260 N, determine (*a*) the angle between cable AC and the boom AB, (*b*) the projection on AB of the force exerted by cable AC at point A.

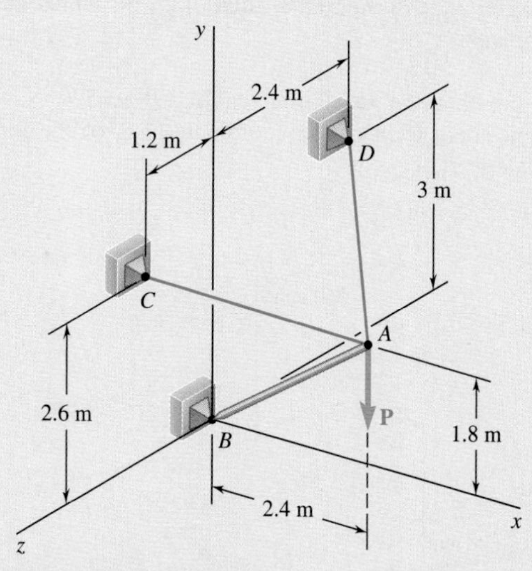

Fig. P3.27 and P3.28

3.28 Knowing that the tension in cable AD is 405 N, determine (*a*) the angle between cable AD and the boom AB, (*b*) the projection on AB of the force exerted by cable AD at point A.

3.29 Three cables are used to support a container as shown. Determine the angle formed by cables AB and AD.

3.30 Three cables are used to support a container as shown. Determine the angle formed by cables AC and AD.

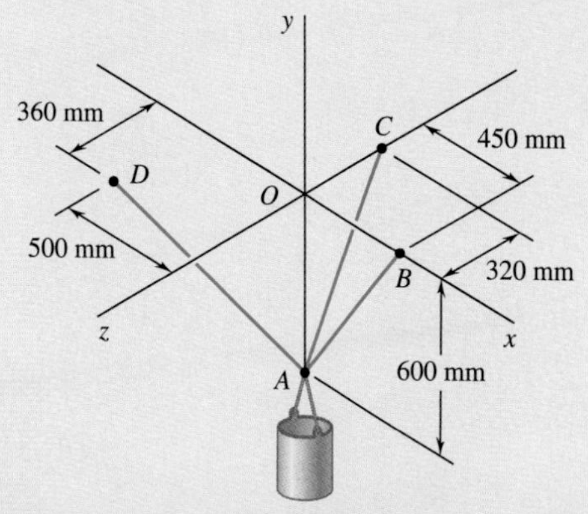

Fig. P3.29 and *P3.30*

3.31 The 20-in. tube AB can slide along a horizontal rod. The ends A and B of the tube are connected by elastic cords to the fixed point C. For the position corresponding to $x = 11$ in., determine the angle formed by the two cords, (*a*) using Eq. (3.30), (*b*) applying the law of cosines to triangle ABC.

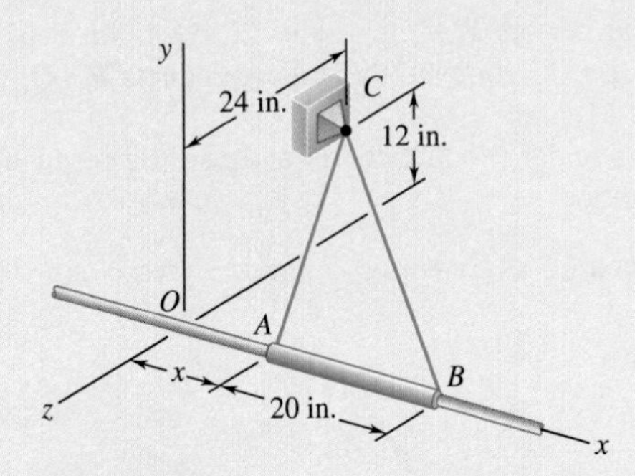

Fig. P3.31

3.32 Solve Prob. 3.31 for the position corresponding to $x = 4$ in.

3.33 Determine the volume of the parallelepiped of Fig. 3.20*b* when (*a*) $\mathbf{P} = 4\mathbf{i} - 3\mathbf{j} + 2\mathbf{k}$, $\mathbf{Q} = -2\mathbf{i} - 5\mathbf{j} + \mathbf{k}$, and $\mathbf{S} = 7\mathbf{i} + \mathbf{j} - \mathbf{k}$, (*b*) $\mathbf{P} = 5\mathbf{i} - \mathbf{j} + 6\mathbf{k}$, $\mathbf{Q} = 2\mathbf{i} + 3\mathbf{j} + \mathbf{k}$, and $\mathbf{S} = -3\mathbf{i} - 2\mathbf{j} + 4\mathbf{k}$.

3.34 Given the vectors $\mathbf{P} = 3\mathbf{i} - \mathbf{j} + \mathbf{k}$, $\mathbf{Q} = 4\mathbf{i} + Q_y\mathbf{j} - 2\mathbf{k}$, and $\mathbf{S} = 2\mathbf{i} - 2\mathbf{j} + 2\mathbf{k}$, determine the value of Q_y for which the three vectors are coplanar.

3.35 Knowing that the tension in cable AB is 570 N, determine the moment about each of the coordinate axes of the force exerted on the plate at B.

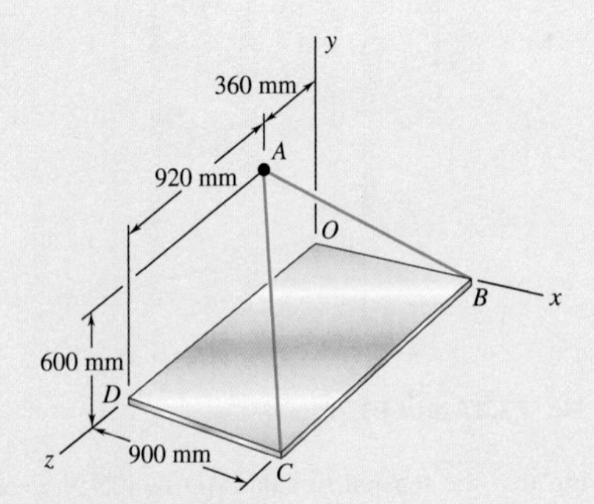

Fig. P3.35 and *P3.36*

3.36 Knowing that the tension in cable AC is 1065 N, determine the moment about each of the coordinate axes of the force exerted on the plate at C.

3.37 A farmer uses cables and winch pullers B and E to plumb one side of a small barn. If it is known that the sum of the moments about the x axis of the forces exerted by the cables on the barn at points A and D is equal to 4728 lb·ft, determine the magnitude of $\mathbf{T}_{DE}$ when $T_{AB} = 255$ lb.

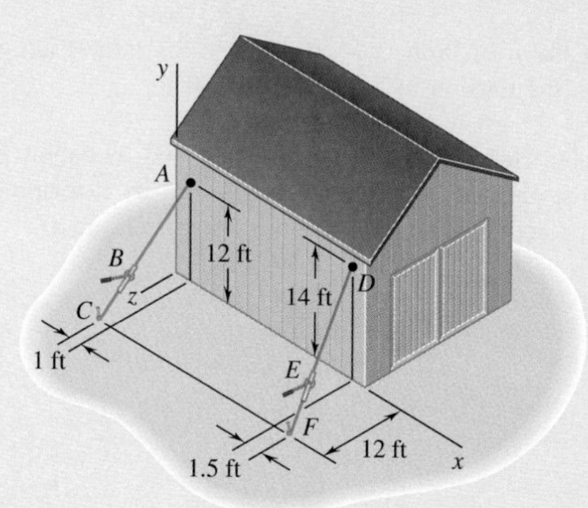

Fig. P3.37

3.38 Solve Prob. 3.37 when the tension in cable AB is 306 lb.

3.39 To lift a heavy crate, a man uses a block and tackle attached to the bottom of an I-beam at hook B. Knowing that the moments about the y and the z axes of the force exerted at B by portion AB of the rope are, respectively, 120 N·m and –460 N·m, determine the distance a.

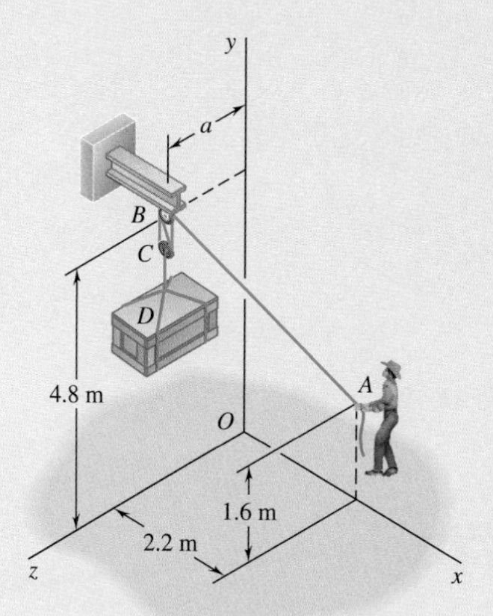

Fig. P3.39 and P3.40

3.40 To lift a heavy crate, a man uses a block and tackle attached to the bottom of an I-beam at hook B. Knowing that the man applies a 195-N force to end A of the rope and that the moment of that force about the y axis is 132 N·m, determine the distance a.

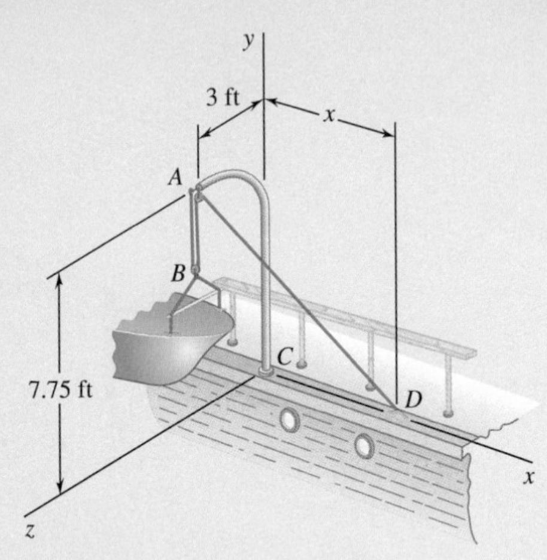

Fig. P3.41

3.41 A small boat hangs from two davits, one of which is shown in the figure. It is known that the moment about the z axis of the resultant force $\mathbf{R}_A$ exerted on the davit at A must not exceed 279 lb·ft in absolute value. Determine the largest allowable tension in line $ABAD$ when $x = 6$ ft.

3.42 For the davit of Prob. 3.41, determine the largest allowable distance x when the tension in line $ABAD$ is 60 lb.

3.43 A sign erected on uneven ground is guyed by cables EF and EG. If the force exerted by cable EF at E is 46 lb, determine the moment of that force about the line joining points A and D.

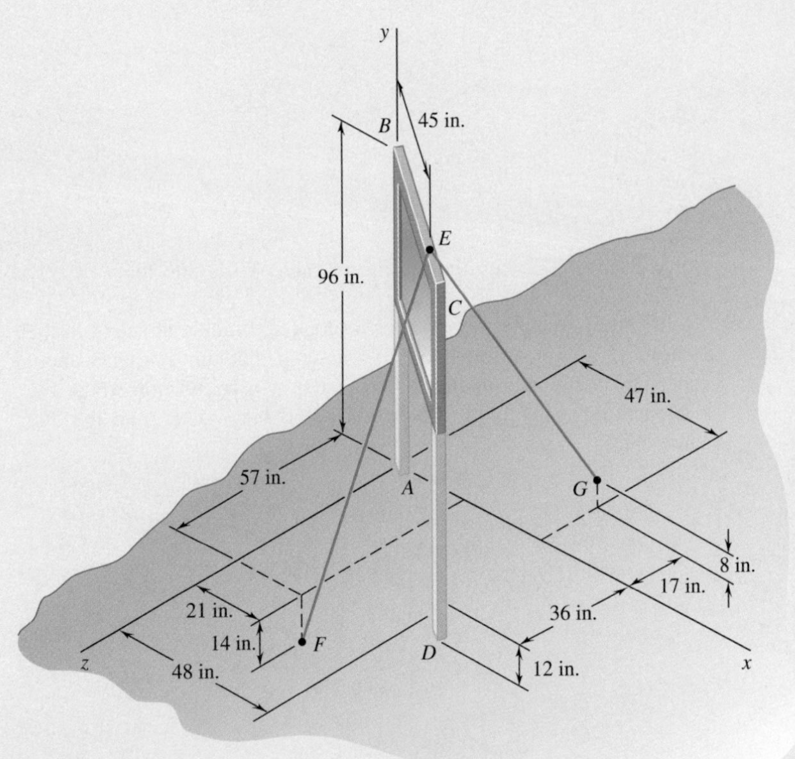

Fig. P3.43 and P3.44

3.44 A sign erected on uneven ground is guyed by cables EF and EG. If the force exerted by cable EG at E is 54 lb, determine the moment of that force about the line joining points A and D.

3.45 The frame ACD is hinged at A and D and is supported by a cable that passes through a ring at B and is attached to hooks at G and H. Knowing that the tension in the cable is 450 N, determine the moment about the diagonal AD of the force exerted on the frame by portion BH of the cable.

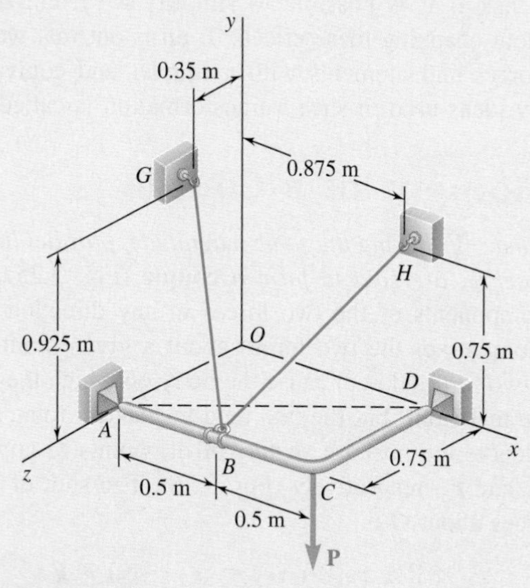

Fig. *P3.45*

3.46 In Prob. 3.45, determine the moment about the diagonal AD of the force exerted on the frame by portion BG of the cable.

3.47 The 23-in. vertical rod CD is welded to the midpoint C of the 50-in. rod AB. Determine the moment about AB of the 235-lb force **P**.

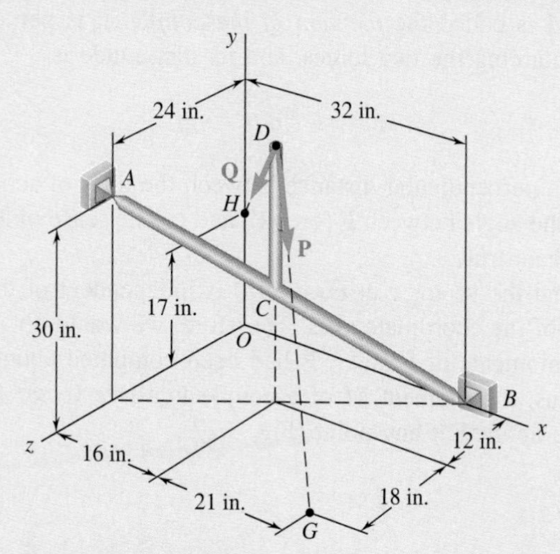

Fig. *P3.47* **and P3.48**

3.48 The 23-in. vertical rod CD is welded to the midpoint C of the 50-in. rod AB. Determine the moment about AB of the 174-lb force **Q**.

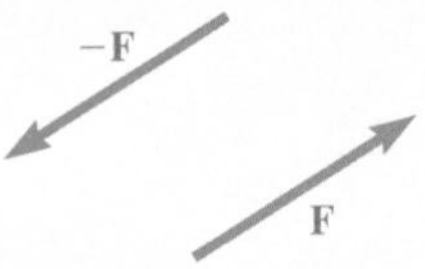

3.3 COUPLES AND FORCE-COUPLE SYSTEMS

Now that we have studied the effects of forces and moments on a rigid body, we can ask if it is possible to simplify a system of forces and moments without changing these effects. It turns out that we *can* replace a system of forces and moments with a simpler and equivalent system. One of the key ideas used in such a transformation is called a couple.

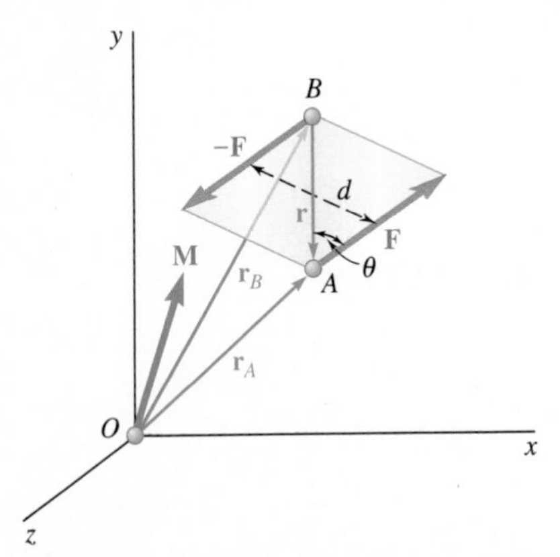

Fig. 3.25 A couple consists of two forces with equal magnitude, parallel lines of action, and opposite sense.

3.3A Moment of a Couple

Two forces **F** *and* −**F**, *having the same magnitude, parallel lines of action, and opposite sense, are said to form a* **couple** (Fig. 3.25). Clearly, the sum of the components of the two forces in any direction is zero. The sum of the moments of the two forces about a given point, however, is not zero. The two forces do not cause the body on which they act to move along a line (translation), but they do tend to make it rotate.

Let us denote the position vectors of the points of application of **F** and −**F** by $\mathbf{r}_A$ and $\mathbf{r}_B$, respectively (Fig. 3.26). The sum of the moments of the two forces about O is

$$\mathbf{r}_A \times \mathbf{F} + \mathbf{r}_B \times (-\mathbf{F}) = (\mathbf{r}_A - \mathbf{r}_B) \times \mathbf{F}$$

Fig. 3.26 The moment **M** of the couple about *O* is the sum of the moments of **F** and of −**F** about *O*.

Setting $\mathbf{r}_A - \mathbf{r}_B = \mathbf{r}$, where **r** is the vector joining the points of application of the two forces, we conclude that the sum of the moments of **F** and −**F** about O is represented by the vector

$$\mathbf{M} = \mathbf{r} \times \mathbf{F} \tag{3.45}$$

Photo 3.1 The parallel upward and downward forces of equal magnitude exerted on the arms of the lug nut wrench are an example of a couple.

The vector **M** is called the *moment of the couple.* It is perpendicular to the plane containing the two forces, and its magnitude is

$$M = rF \sin \theta = Fd \tag{3.46}$$

where d is the perpendicular distance between the lines of action of **F** and −**F** and θ is the angle between **F** (or −**F**) and **r**. The sense of **M** is defined by the right-hand rule.

Note that the vector **r** in Eq. (3.45) is independent of the choice of the origin O of the coordinate axes. Therefore, we would obtain the same result if the moments of **F** and −**F** had been computed about a different point O'. Thus, the moment **M** of a couple is a *free vector* (Sec. 2.1B), which can be applied at any point (Fig. 3.27).

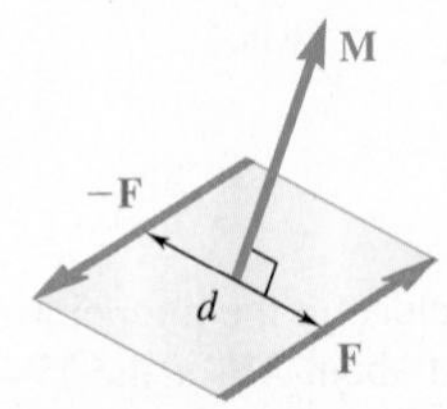

Fig. 3.27 The moment **M** of a couple equals the product of *F* and *d*, is perpendicular to the plane of the couple, and may be applied at any point of that plane.

From the definition of the moment of a couple, it also follows that two couples—one consisting of the forces $\mathbf{F}_1$ and $-\mathbf{F}_1$, the other of the forces $\mathbf{F}_2$ and $-\mathbf{F}_2$ (Fig. 3.28)—have equal moments if

$$F_1d_1 = F_2d_2 \tag{3.47}$$

provided that the two couples lie in parallel planes (or in the same plane) and have the same sense (i.e., clockwise or counterclockwise).

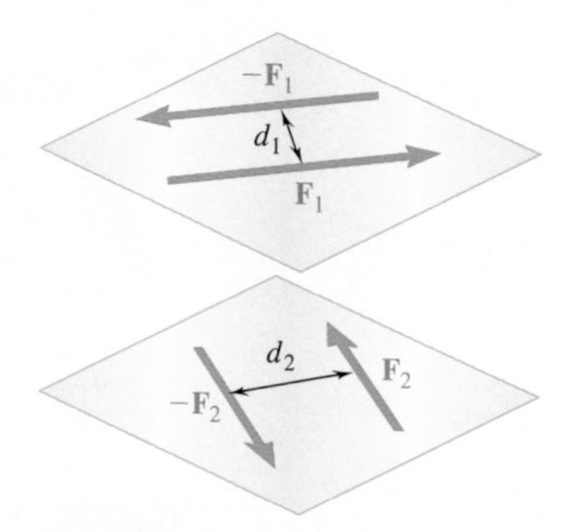

Fig. 3.28 Two couples have the same moment if they lie in parallel planes, have the same sense, and if $F_1d_1 = F_2d_2$.

3.3B Equivalent Couples

Imagine that three couples act successively on the same rectangular box (Fig. 3.29). As we have just seen, the only motion a couple can impart to a rigid body is a rotation. Since each of the three couples shown has the same moment **M** (same direction and same magnitude $M = 120$ lb·in.), we can expect each couple to have the same effect on the box.

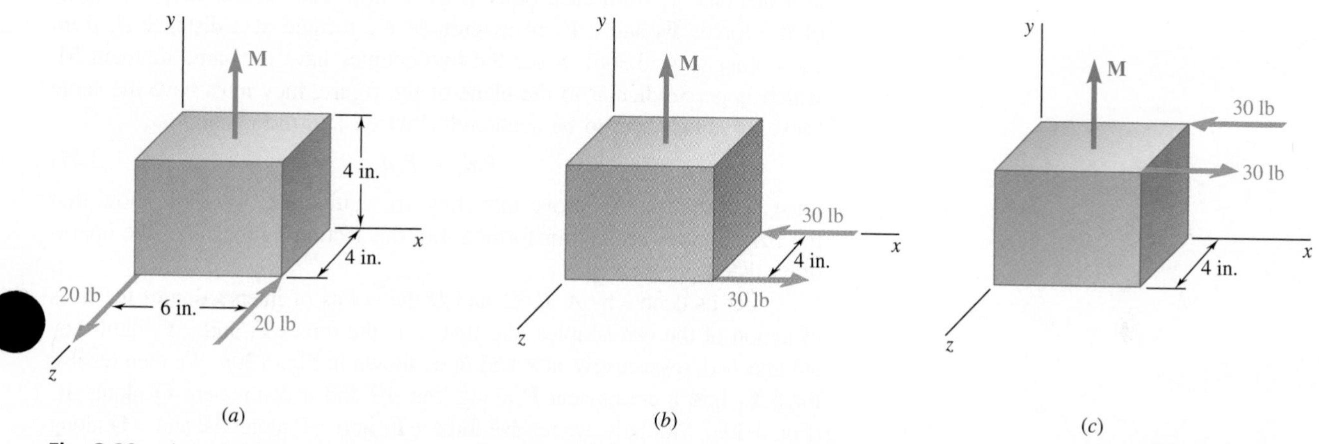

Fig. 3.29 Three equivalent couples. (*a*) A couple acting on the bottom of the box, acting counterclockwise viewed from above; (*b*) a couple in the same plane and with the same sense but larger forces than in (*a*); (*c*) a couple acting in a different plane but same sense.

As reasonable as this conclusion appears, we should not accept it hastily. Although intuition is of great help in the study of mechanics, it should not be accepted as a substitute for logical reasoning. Before stating that two systems (or groups) of forces have the same effect on a rigid body, we should prove that fact on the basis of the experimental evidence introduced so far. This evidence consists of the parallelogram law for the addition of two forces (Sec. 2.1A) and the principle of transmissibility (Sec. 3.1B). Therefore, we state that **two systems of forces are equivalent** (i.e., they have the same effect on a rigid body) **if we can transform one of them into the other by means of one or several of the following operations**: (1) replacing two forces acting on the same particle by their resultant; (2) resolving a force into two components; (3) canceling two equal and opposite forces acting on the same particle; (4) attaching to the same particle two equal and opposite forces; and (5) moving a force along its line of action. Each of these operations is easily justified on the basis of the parallelogram law or the principle of transmissibility.

Let us now prove that **two couples having the same moment M are equivalent**. First consider two couples contained in the same plane, and

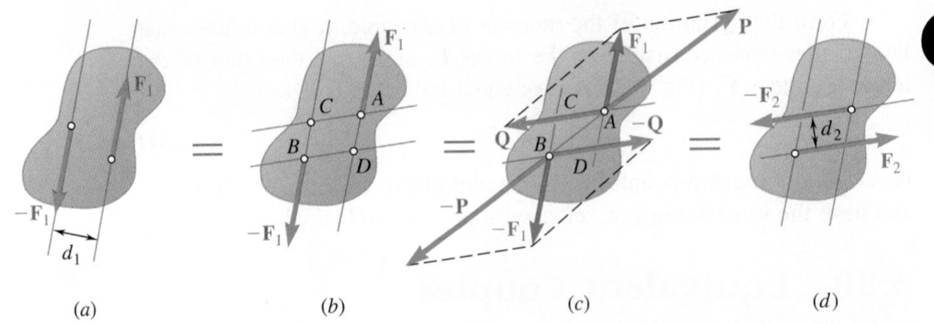

Fig. 3.30 Four steps in transforming one couple to another couple in the same plane by using simple operations. (*a*) Starting couple; (*b*) label points of intersection of lines of action of the two couples; (*c*) resolve forces from first couple into components; (*d*) final couple.

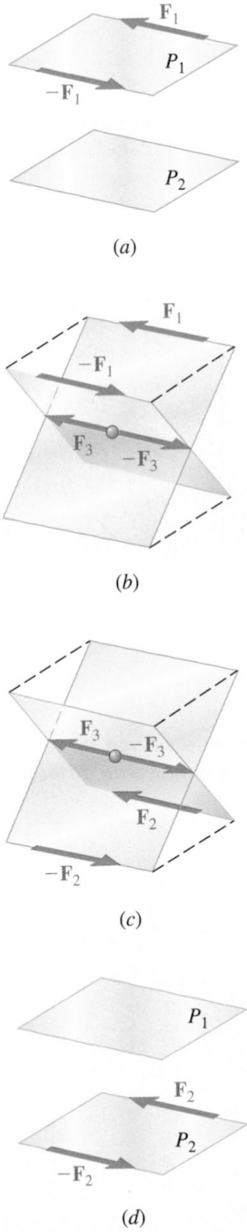

assume that this plane coincides with the plane of the figure (Fig. 3.30). The first couple consists of the forces $\mathbf{F}_1$ and $-\mathbf{F}_1$ of magnitude F_1, located at a distance d_1 from each other (Fig. 3.30*a*). The second couple consists of the forces $\mathbf{F}_2$ and $-\mathbf{F}_2$ of magnitude F_2, located at a distance d_2 from each other (Fig. 3.30*d*). Since the two couples have the same moment **M**, which is perpendicular to the plane of the figure, they must have the same sense (assumed here to be counterclockwise), and the relation

$$F_1 d_1 = F_2 d_2 \qquad (3.47)$$

must be satisfied. To prove that they are equivalent, we shall show that the first couple can be transformed into the second by means of the operations listed previously.

Let us denote by *A*, *B*, *C*, and *D* the points of intersection of the lines of action of the two couples. We first slide the forces $\mathbf{F}_1$ and $-\mathbf{F}_1$ until they are attached, respectively, at *A* and *B*, as shown in Fig. 3.30*b*. We then resolve force $\mathbf{F}_1$ into a component **P** along line *AB* and a component **Q** along *AC* (Fig. 3.30*c*). Similarly, we resolve force $-\mathbf{F}_1$ into $-\mathbf{P}$ along *AB* and $-\mathbf{Q}$ along *BD*. The forces **P** and $-\mathbf{P}$ have the same magnitude, the same line of action, and opposite sense; we can move them along their common line of action until they are applied at the same point and may then be canceled. Thus, the couple formed by $\mathbf{F}_1$ and $-\mathbf{F}_1$ reduces to a couple consisting of **Q** and $-\mathbf{Q}$.

We now show that the forces **Q** and $-\mathbf{Q}$ are respectively equal to the forces $-\mathbf{F}_2$ and $\mathbf{F}_2$. We obtain the moment of the couple formed by **Q** and $-\mathbf{Q}$ by computing the moment of **Q** about *B*. Similarly, the moment of the couple formed by $\mathbf{F}_1$ and $-\mathbf{F}_1$ is the moment of $\mathbf{F}_1$ about *B*. However, by Varignon's theorem, the moment of $\mathbf{F}_1$ is equal to the sum of the moments of its components **P** and **Q**. Since the moment of **P** about *B* is zero, the moment of the couple formed by **Q** and $-\mathbf{Q}$ must be equal to the moment of the couple formed by $\mathbf{F}_1$ and $-\mathbf{F}_1$. Recalling Eq. (3.47), we have

$$Qd_2 = F_1 d_1 = F_2 d_2 \qquad \text{and} \qquad Q = F_2$$

Thus, the forces **Q** and $-\mathbf{Q}$ are respectively equal to the forces $-\mathbf{F}_2$ and $\mathbf{F}_2$, and the couple of Fig. 3.30*a* is equivalent to the couple of Fig. 3.30*d*.

Now consider two couples contained in parallel planes P_1 and P_2. We prove that they are equivalent if they have the same moment. In view of the preceding discussion, we can assume that the couples consist of forces of the same magnitude *F* acting along parallel lines (Fig. 3.31*a* and *d*). We propose to show that the couple contained in plane P_1 can be transformed into the couple contained in plane P_2 by means of the standard operations listed previously.

Fig. 3.31 Four steps in transforming one couple to another couple in a parallel plane by using simple operations. (*a*) Initial couple; (*b*) add a force pair along the line of intersection of two diagonal planes; (*c*) replace two couples with equivalent couples in the same planes; (*d*) final couple.

Let us consider the two diagonal planes defined respectively by the lines of action of $\mathbf{F}_1$ and $-\mathbf{F}_2$ and by those of $-\mathbf{F}_1$ and $\mathbf{F}_2$ (Fig. 3.31*b*). At a point on their line of intersection, we attach two forces $\mathbf{F}_3$ and $-\mathbf{F}_3$, which are respectively equal to $\mathbf{F}_1$ and $-\mathbf{F}_1$. The couple formed by $\mathbf{F}_1$ and $-\mathbf{F}_3$ can be replaced by a couple consisting of $\mathbf{F}_3$ and $-\mathbf{F}_2$ (Fig. 3.31*c*), because both couples clearly have the same moment and are contained in the same diagonal plane. Similarly, the couple formed by $-\mathbf{F}_1$ and $\mathbf{F}_3$ can be replaced by a couple consisting of $-\mathbf{F}_3$ and $\mathbf{F}_2$. Canceling the two equal and opposite forces $\mathbf{F}_3$ and $-\mathbf{F}_3$, we obtain the desired couple in plane P_2 (Fig. 3.31*d*). Thus, we conclude that two couples having the same moment $\mathbf{M}$ are equivalent, whether they are contained in the same plane or in parallel planes.

The property we have just established is very important for the correct understanding of the mechanics of rigid bodies. It indicates that when a couple acts on a rigid body, it does not matter where the two forces forming the couple act or what magnitude and direction they have. The only thing that counts is the *moment* of the couple (magnitude and direction). Couples with the same moment have the same effect on the rigid body.

3.3C Addition of Couples

Consider two intersecting planes P_1 and P_2 and two couples acting respectively in P_1 and P_2. Recall that each couple is a free vector in its respective plane and can be represented within this plane by any combination of equal, opposite, and parallel forces and of perpendicular distance of separation that provides the same sense and magnitude for this couple. Thus, we can assume, without any loss of generality, that the couple in P_1 consists of two forces $\mathbf{F}_1$ and $-\mathbf{F}_1$ perpendicular to the line of intersection of the two planes and acting respectively at A and B (Fig. 3.32*a*). Similarly, we can assume that the couple in P_2 consists of two forces $\mathbf{F}_2$ and $-\mathbf{F}_2$ perpendicular to AB and acting respectively at A and B. It is clear that the resultant $\mathbf{R}$ of $\mathbf{F}_1$ and $\mathbf{F}_2$ and the resultant $-\mathbf{R}$ of $-\mathbf{F}_1$ and $-\mathbf{F}_2$ form a couple. Denoting the vector joining B to A by $\mathbf{r}$ and recalling the definition of the moment of a couple (Sec. 3.3A), we express the moment $\mathbf{M}$ of the resulting couple as

$$\mathbf{M} = \mathbf{r} \times \mathbf{R} = \mathbf{r} \times (\mathbf{F}_1 + \mathbf{F}_2)$$

By Varignon's theorem, we can expand this expression as

$$\mathbf{M} = \mathbf{r} \times \mathbf{F}_1 + \mathbf{r} \times \mathbf{F}_2$$

The first term in this expression represents the moment $\mathbf{M}_1$ of the couple in P_1, and the second term represents the moment $\mathbf{M}_2$ of the couple in P_2. Therefore, we have

$$\mathbf{M} = \mathbf{M}_1 + \mathbf{M}_2 \tag{3.48}$$

We conclude that the sum of two couples of moments $\mathbf{M}_1$ and $\mathbf{M}_2$ is a couple of moment $\mathbf{M}$ equal to the vector sum of $\mathbf{M}_1$ and $\mathbf{M}_2$ (Fig. 3.32*b*). We can extend this conclusion to state that any number of couples can be added to produce one resultant couple, as

$$\mathbf{M} = \Sigma\mathbf{M} = \Sigma(\mathbf{r} \times \mathbf{F})$$

3.3D Couple Vectors

We have seen that couples with the same moment, whether they act in the same plane or in parallel planes, are equivalent. Therefore, we have no need to draw the actual forces forming a given couple in order to define its effect

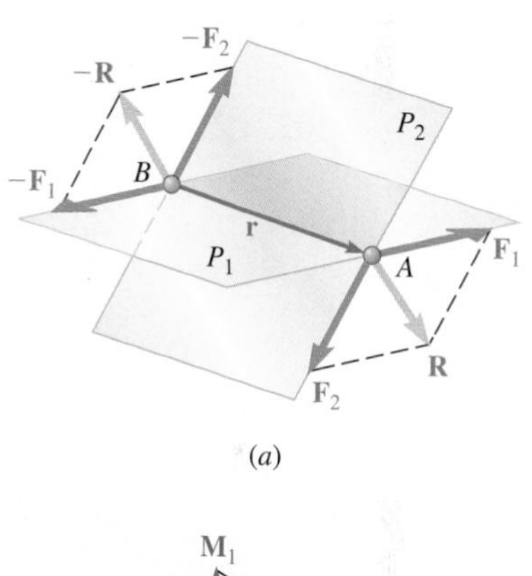

(*a*)

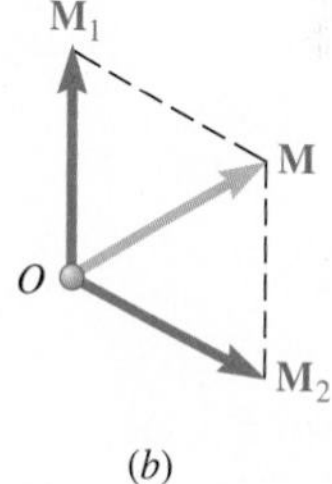

(*b*)

Fig. 3.32 (*a*) We can add two couples, each acting in one of two intersecting planes, to form a new couple. (*b*) The moment of the resultant couple is the vector sum of the moments of the component couples.

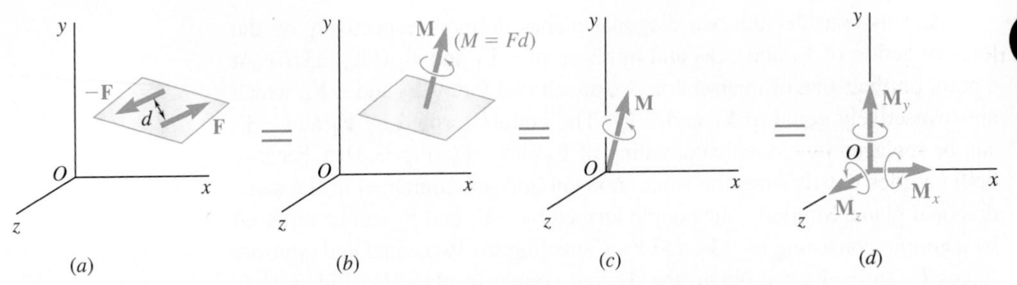

Fig. 3.33 (*a*) A couple formed by two forces can be represented by (*b*) a couple vector, oriented perpendicular to the plane of the couple. (*c*) The couple vector is a free vector and can be moved to other points of application, such as the origin. (*d*) A couple vector can be resolved into components along the coordinate axes.

on a rigid body (Fig. 3.33*a*). It is sufficient to draw an arrow equal in magnitude and direction to the moment **M** of the couple (Fig. 3.33*b*). We have also seen that the sum of two couples is itself a couple and that we can obtain the moment **M** of the resultant couple by forming the vector sum of the moments $\mathbf{M}_1$ and $\mathbf{M}_2$ of the given couples. Thus, couples obey the law of addition of vectors, so the arrow used in Fig. 3.33*b* to represent the couple defined in Fig. 3.33*a* truly can be considered a vector.

The vector representing a couple is called a **couple vector**. Note that, in Fig. 3.33, we use a red arrow to distinguish the couple vector, *which represents the couple itself,* from the *moment* of the couple, which was represented by a green arrow in earlier figures. Also note that we added the symbol ↰ to this red arrow to avoid any confusion with vectors representing forces. A couple vector, like the moment of a couple, is a free vector. Therefore, we can choose its point of application at the origin of the system of coordinates, if so desired (Fig. 3.33*c*). Furthermore, we can resolve the couple vector **M** into component vectors $\mathbf{M}_x$, $\mathbf{M}_y$, and $\mathbf{M}_z$ that are directed along the coordinate axes (Fig. 3.33*d*). These component vectors represent couples acting, respectively, in the *yz, zx,* and *xy* planes.

3.3E Resolution of a Given Force into a Force at *O* and a Couple

Consider a force **F** acting on a rigid body at a point *A* defined by the position vector **r** (Fig. 3.34*a*). Suppose that for some reason it would simplify the analysis to have the force act at point *O* instead. Although we can move **F** along its line of action (principle of transmissibility), we cannot move it to a point *O* that does not lie on the original line of action without modifying the action of **F** on the rigid body.

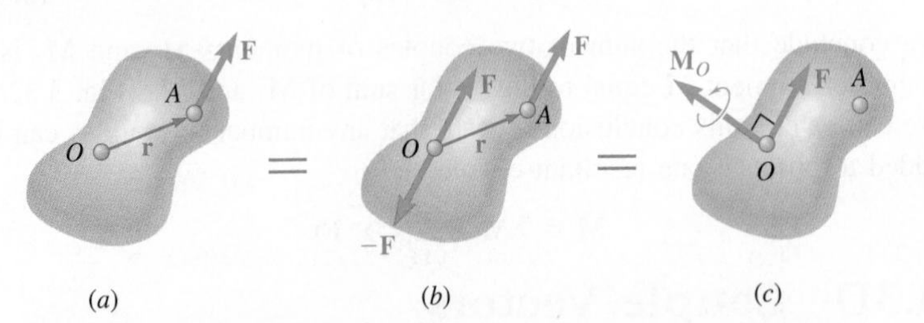

Fig. 3.34 Replacing a force with a force and a couple. (*a*) Initial force **F** acting at point *A*; (*b*) attaching equal and opposite forces at *O*; (*c*) force **F** acting at point *O* and a couple.

We can, however, attach two forces at point O, one equal to $\mathbf{F}$ and the other equal to $-\mathbf{F}$, without modifying the action of the original force on the rigid body (Fig. 3.34b). As a result of this transformation, we now have a force $\mathbf{F}$ applied at O; the other two forces form a couple of moment $\mathbf{M}_O = \mathbf{r} \times \mathbf{F}$. Thus,

Any force F acting on a rigid body can be moved to an arbitrary point *O* provided that we add a couple whose moment is equal to the moment of F about *O*.

The couple tends to impart to the rigid body the same rotational motion about O that force $\mathbf{F}$ tended to produce before it was transferred to O. We represent the couple by a couple vector $\mathbf{M}_O$ that is perpendicular to the plane containing $\mathbf{r}$ and $\mathbf{F}$. Since $\mathbf{M}_O$ is a free vector, it may be applied anywhere; for convenience, however, the couple vector is usually attached at O together with $\mathbf{F}$. This combination is referred to as a **force-couple system** (Fig. 3.34c).

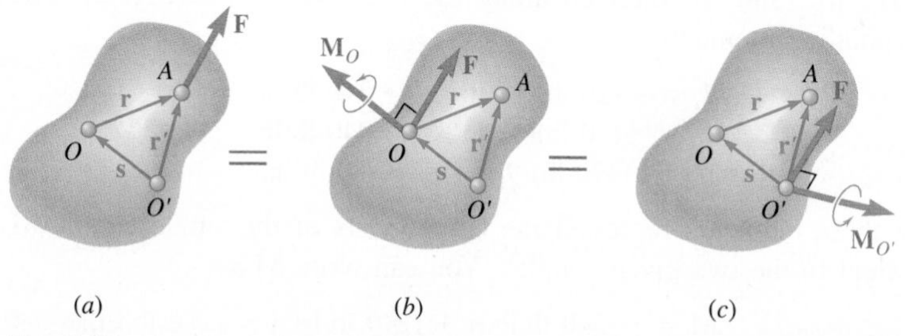

Fig. 3.35 Moving a force to different points. (a) Initial force **F** acting at A; (b) force **F** acting at O and a couple; (c) force **F** acting at O' and a different couple.

If we move force $\mathbf{F}$ from A to a different point O' (Fig. 3.35a and c), we have to compute the moment $\mathbf{M}_{O'} = \mathbf{r}' \times \mathbf{F}$ of $\mathbf{F}$ about O' and add a new force-couple system consisting of $\mathbf{F}$ and the couple vector $\mathbf{M}_{O'}$ at O'. We can obtain the relation between the moments of $\mathbf{F}$ about O and O' as

$$\mathbf{M}_{O'} = \mathbf{r}' \times \mathbf{F} = (\mathbf{r} + \mathbf{s}) \times \mathbf{F} = \mathbf{r} \times \mathbf{F} + \mathbf{s} \times \mathbf{F}$$

$$\mathbf{M}_{O'} = \mathbf{M}_O + \mathbf{s} \times \mathbf{F} \qquad (3.49)$$

where $\mathbf{s}$ is the vector joining O' to O. Thus, we obtain the moment $\mathbf{M}_{O'}$ of $\mathbf{F}$ about O' by adding to the moment $\mathbf{M}_O$ of $\mathbf{F}$ about O the vector product $\mathbf{s} \times \mathbf{F}$, representing the moment about O' of the force $\mathbf{F}$ applied at O.

We also could have established this result by observing that, in order to transfer to O' the force-couple system attached at O (Fig. 3.35b and c), we could freely move the couple vector $\mathbf{M}_O$ to O'. However, to move force $\mathbf{F}$ from O to O', we need to add to $\mathbf{F}$ a couple vector whose moment is equal to the moment about O' of force $\mathbf{F}$ applied at O. Thus, the couple vector $\mathbf{M}_{O'}$ must be the sum of $\mathbf{M}_O$ and the vector $\mathbf{s} \times \mathbf{F}$.

As noted here, the force-couple system obtained by transferring a force $\mathbf{F}$ from a point A to a point O consists of $\mathbf{F}$ and a couple vector $\mathbf{M}_O$ perpendicular to $\mathbf{F}$. Conversely, any force-couple system consisting of a force $\mathbf{F}$ and a couple vector $\mathbf{M}_O$ that are *mutually perpendicular* can be replaced by a single equivalent force. This is done by moving force $\mathbf{F}$ in the plane perpendicular to $\mathbf{M}_O$ until its moment about O is equal to the moment of the couple being replaced.

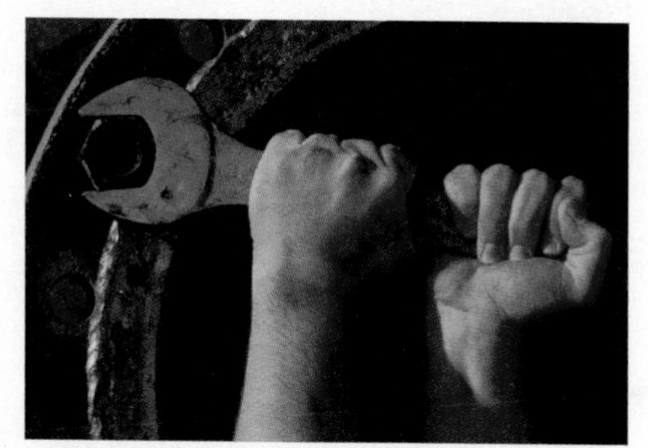

Photo 3.2 The force exerted by each hand on the wrench could be replaced with an equivalent force-couple system acting on the nut.

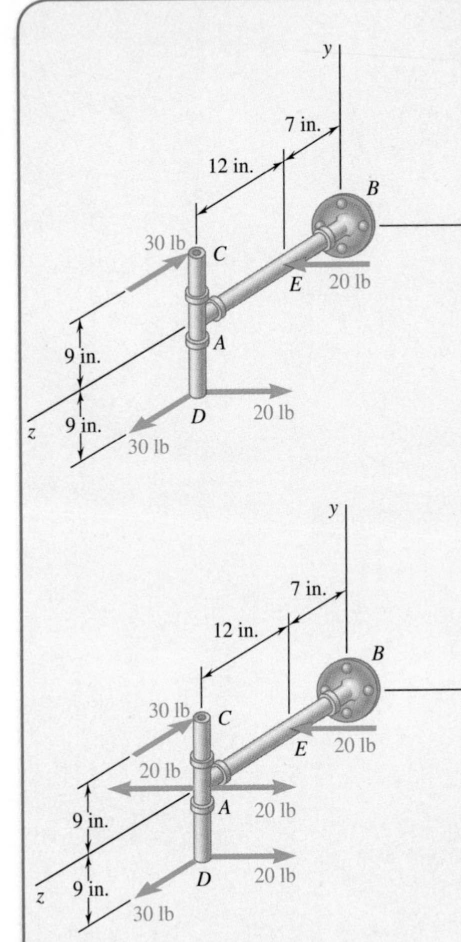

Fig. 1 Placing two equal and opposite 20-lb forces at *A* to simplify calculations.

Sample Problem 3.6

Determine the components of the single couple equivalent to the two couples shown.

STRATEGY: Look for ways to add equal and opposite forces to the diagram that, along with already known perpendicular distances, will produce new couples with moments along the coordinate axes. These can be combined into a single equivalent couple.

MODELING: You can simplify the computations by attaching two equal and opposite 20-lb forces at *A* (Fig. 1). This enables you to replace the original 20-lb-force couple by two new 20-lb-force couples: one lying in the *zx* plane and the other in a plane parallel to the *xy* plane.

ANALYSIS: You can represent these three couples by three couple vectors $\mathbf{M}_x$, $\mathbf{M}_y$, and $\mathbf{M}_z$ directed along the coordinate axes (Fig. 2). The corresponding moments are

$$M_x = -(30 \text{ lb})(18 \text{ in.}) = -540 \text{ lb·in.}$$
$$M_y = +(20 \text{ lb})(12 \text{ in.}) = +240 \text{ lb·in.}$$
$$M_z = +(20 \text{ lb})(9 \text{ in.}) = +180 \text{ lb·in.}$$

These three moments represent the components of the single couple **M** equivalent to the two given couples. You can write **M** as

$$\mathbf{M} = -(540 \text{ lb·in.})\mathbf{i} + (240 \text{ lb·in.})\mathbf{j} + (180 \text{ lb·in.})\mathbf{k}$$ ◀

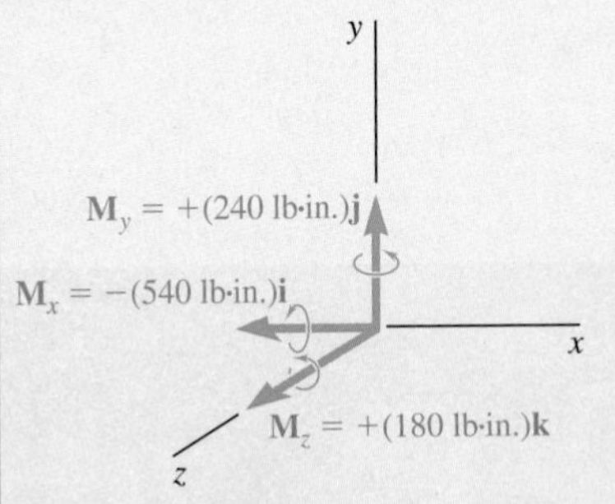

Fig. 2 The three couples represented as couple vectors.

REFLECT and THINK: You can also obtain the components of the equivalent single couple **M** by computing the sum of the moments of the four given forces about an arbitrary point. Selecting point *D*, the moment is (Fig. 3)

$$\mathbf{M} = \mathbf{M}_D = (18 \text{ in.})\mathbf{j} \times (-30 \text{ lb})\mathbf{k} + [(9 \text{ in.})\mathbf{j} - (12 \text{ in.})\mathbf{k}] \times (-20 \text{ lb})\mathbf{i}$$

After computing the various cross products, you get the same result, as

$$\mathbf{M} = -(540 \text{ lb·in.})\mathbf{i} + (240 \text{ lb·in.})\mathbf{j} + (180 \text{ lb·in.})\mathbf{k}$$ ◀

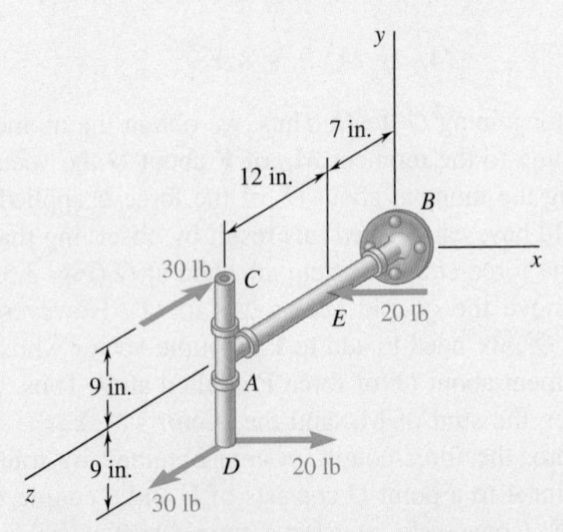

Fig. 3 Using the given force system, the equivalent single couple can also be determined from the sum of moments of the forces about any point, such as point *D*.

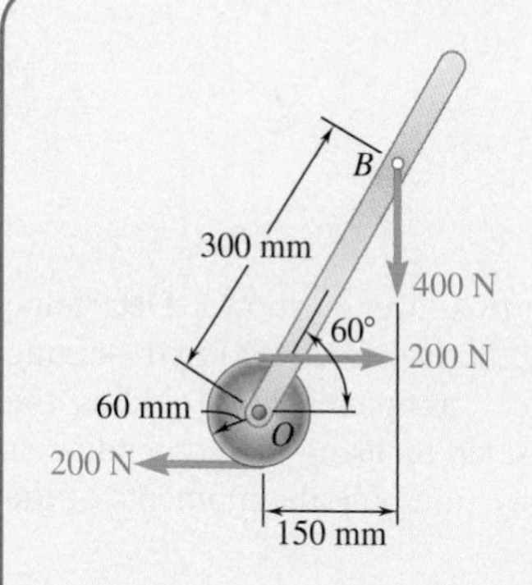

Sample Problem 3.7

Replace the couple and force shown by an equivalent single force applied to the lever. Determine the distance from the shaft to the point of application of this equivalent force.

STRATEGY: First replace the given force and couple by an equivalent force-couple system at O. By moving the force of this force-couple system a distance that creates the same moment as the couple, you can then replace the system with one equivalent force.

MODELING and ANALYSIS: To replace the given force and couple, move the force $\mathbf{F} = -(400\text{ N})\mathbf{j}$ to O, and at the same time, add a couple of moment $\mathbf{M}_O$ that is equal to the moment about O of the force in its original position (Fig. 1). Thus,

$$\mathbf{M}_O = \overrightarrow{OB} \times \mathbf{F} = [(0.150\text{ m})\mathbf{i} + (0.260\text{ m})\mathbf{j}] \times (-400\text{ N})\mathbf{j}$$
$$= -(60\text{ N}\cdot\text{m})\mathbf{k}$$

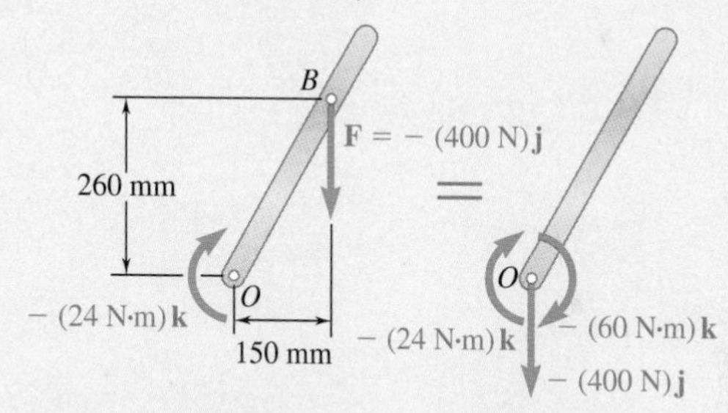

Fig. 1 Replacing given force and couple with an equivalent force-couple at O.

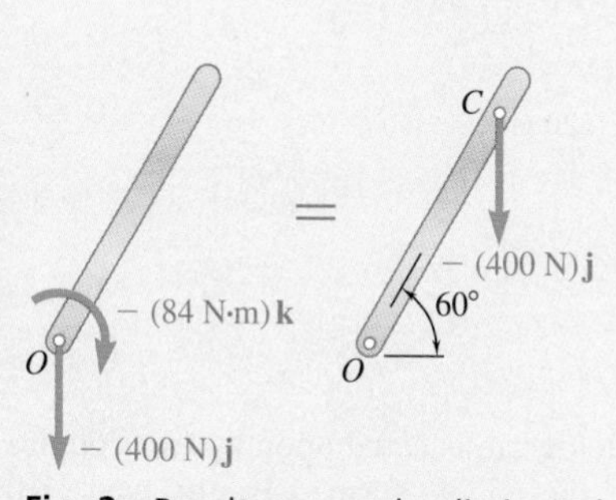

Fig. 2 Resultant couple eliminated by moving force **F**.

When you add this new couple to the couple of moment $-(24\text{ N}\cdot\text{m})\mathbf{k}$ formed by the two 200-N forces, you obtain a couple of moment $-(84\text{ N}\cdot\text{m})\mathbf{k}$ (Fig. 2). You can replace this last couple by applying $\mathbf{F}$ at a point C chosen in such a way that

$$-(84\text{ N}\cdot\text{m})\mathbf{k} = \overrightarrow{OC} \times \mathbf{F}$$
$$= [(OC)\cos 60°\mathbf{i} + (OC)\sin 60°\mathbf{j}] \times (-400\text{ N})\mathbf{j}$$
$$= -(OC)\cos 60°(400\text{ N})\mathbf{k}$$

The result is

$$(OC)\cos 60° = 0.210\text{ m} = 210\text{ mm} \qquad OC = 420\text{ mm} \blacktriangleleft$$

REFLECT and THINK: Since the effect of a couple does not depend on its location, you can move the couple of moment $-(24\text{ N}\cdot\text{m})\mathbf{k}$ to B, obtaining a force-couple system at B (Fig. 3). Now you can eliminate this couple by applying $\mathbf{F}$ at a point C chosen in such a way that

$$-(24\text{ N}\cdot\text{m})\mathbf{k} = \overrightarrow{BC} \times \mathbf{F}$$
$$= -(BC)\cos 60°(400\text{ N})\mathbf{k}$$

The conclusion is

$$(BC)\cos 60° = 0.060\text{ m} = 60\text{ mm} \qquad BC = 120\text{ mm}$$
$$OC = OB + BC = 300\text{ mm} + 120\text{ mm} \qquad OC = 420\text{ mm} \blacktriangleleft$$

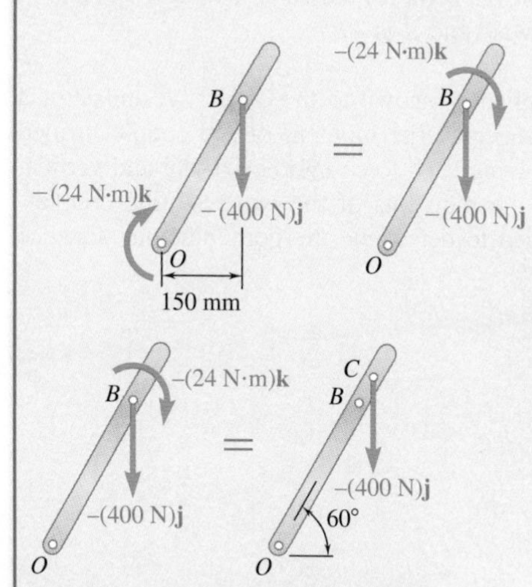

Fig. 3 Couple can be moved to B with no change in effect. This couple can then be eliminated by moving force **F**.

Problems

3.49 Two parallel 60-N forces are applied to a lever as shown. Determine the moment of the couple formed by the two forces (*a*) by resolving each force into horizontal and vertical components and adding the moments of the two resulting couples, (*b*) by using the perpendicular distance between the two forces, (*c*) by summing the moments of the two forces about point *A*.

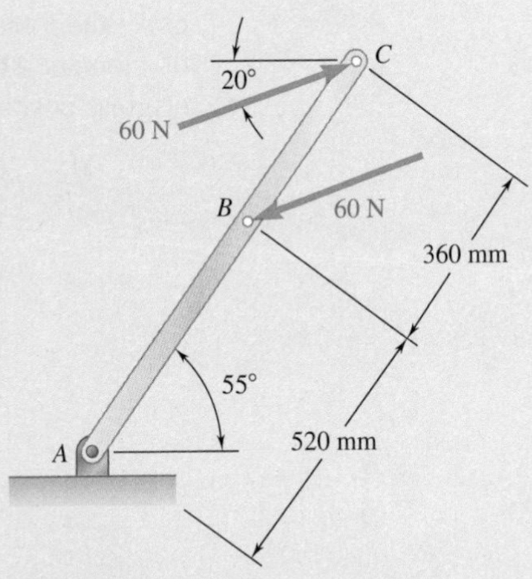

Fig. P3.49

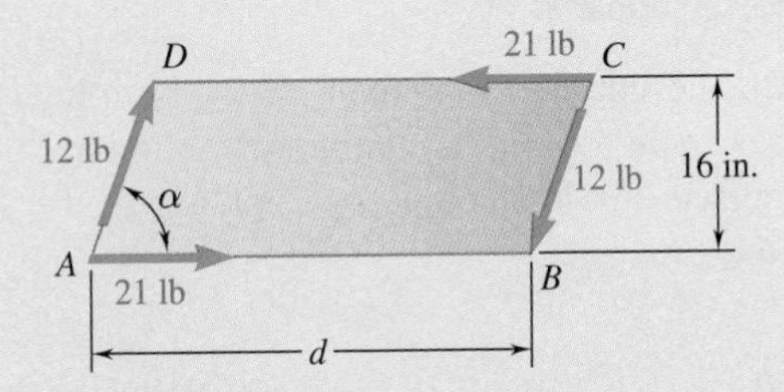

Fig. P3.50

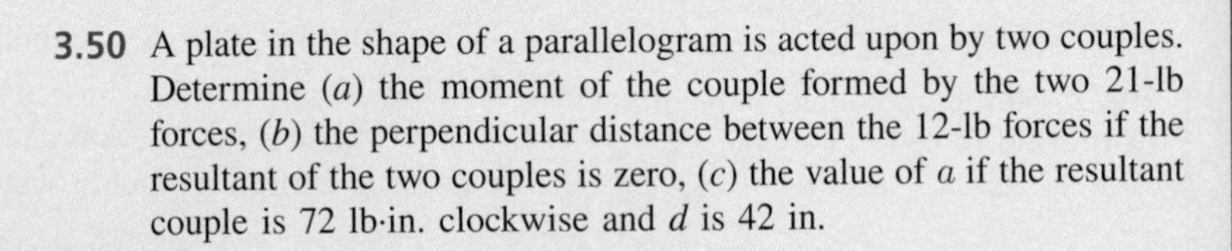

3.50 A plate in the shape of a parallelogram is acted upon by two couples. Determine (*a*) the moment of the couple formed by the two 21-lb forces, (*b*) the perpendicular distance between the 12-lb forces if the resultant of the two couples is zero, (*c*) the value of α if the resultant couple is 72 lb·in. clockwise and *d* is 42 in.

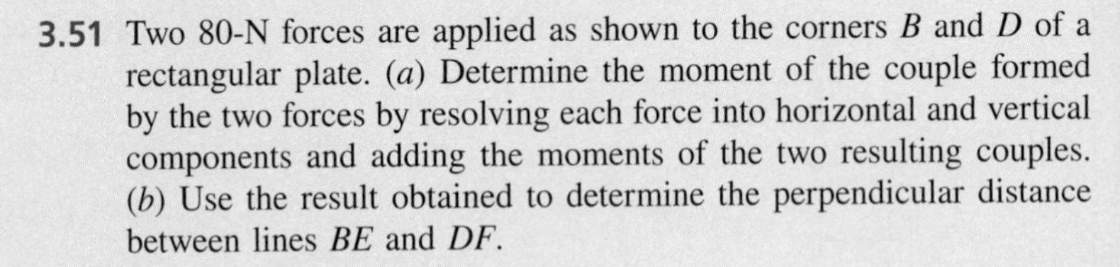

3.51 Two 80-N forces are applied as shown to the corners *B* and *D* of a rectangular plate. (*a*) Determine the moment of the couple formed by the two forces by resolving each force into horizontal and vertical components and adding the moments of the two resulting couples. (*b*) Use the result obtained to determine the perpendicular distance between lines *BE* and *DF*.

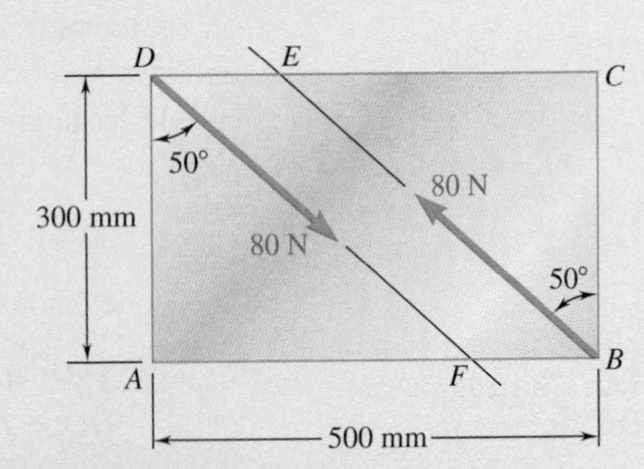

Fig. P3.51

3.52 A piece of plywood in which several holes are being drilled successively has been secured to a workbench by means of two nails. Knowing that the drill exerts a 12-N·m couple on the piece of plywood, determine the magnitude of the resulting forces applied to the nails if they are located (*a*) at *A* and *B*, (*b*) at *B* and *C*, (*c*) at *A* and *C*.

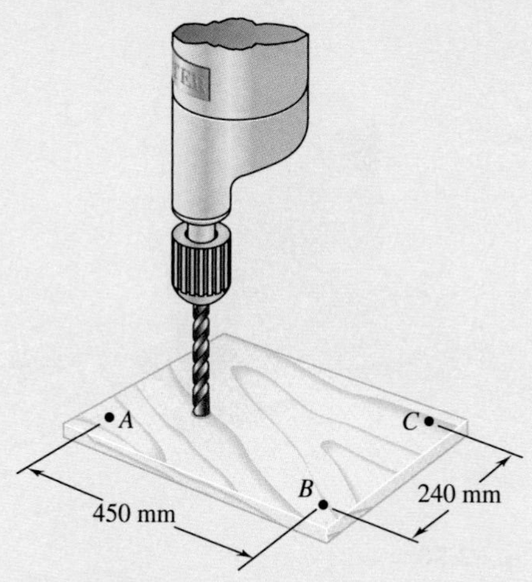

Fig. P3.52

3.53 Four $1\frac{1}{2}$-in.-diameter pegs are attached to a board as shown. Two strings are passed around the pegs and pulled with the forces indicated. (*a*) Determine the resultant couple acting on the board. (*b*) If only one string is used, around which pegs should it pass and in what directions should it be pulled to create the same couple with the minimum tension in the string? (*c*) What is the value of that minimum tension?

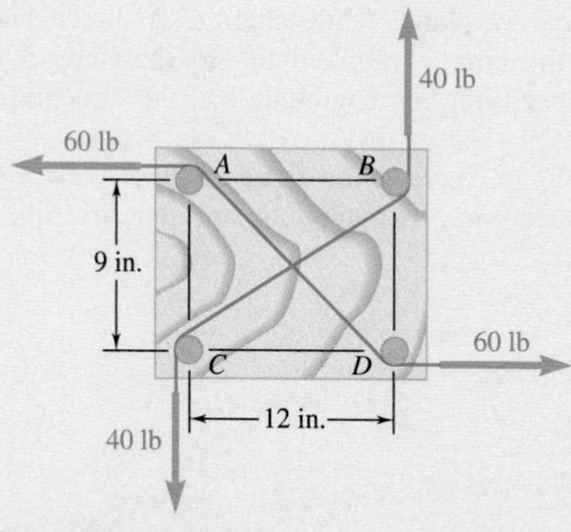

Fig. P3.53 and P3.54

3.54 Four pegs of the same diameter are attached to a board as shown. Two strings are passed around the pegs and pulled with the forces indicated. Determine the diameter of the pegs knowing that the resultant couple applied to the board is 1132.5 lb·in. counterclockwise.

3.55 In a manufacturing operation, three holes are drilled simultaneously in a workpiece. If the holes are perpendicular to the surfaces of the workpiece, replace the couples applied to the drills with a single equivalent couple, specifying its magnitude and the direction of its axis.

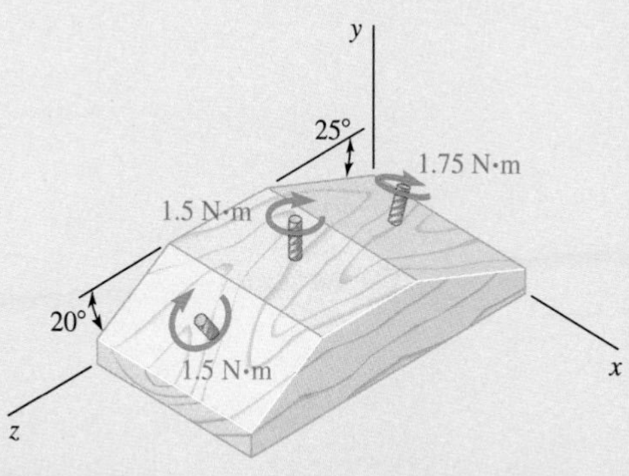

Fig. P3.55

3.56 The two shafts of a speed-reducer unit are subjected to couples of magnitude $M_1 = 15$ lb·ft and $M_2 = 3$ lb·ft, respectively. Replace the two couples with a single equivalent couple, specifying its magnitude and the direction of its axis.

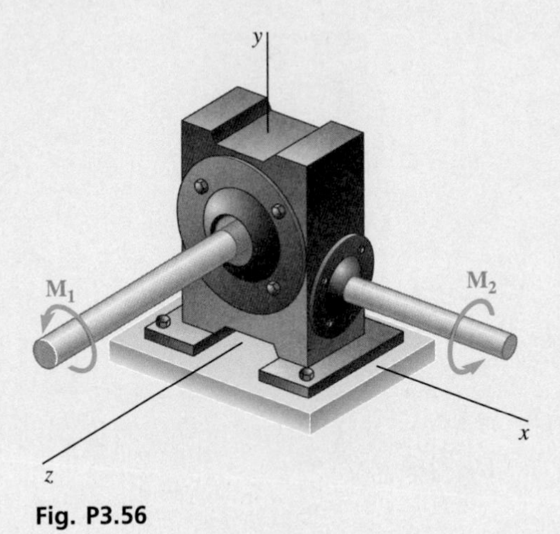

Fig. P3.56

Fig. P3.57

3.57 Replace the two couples shown with a single equivalent couple, specifying its magnitude and the direction of its axis.

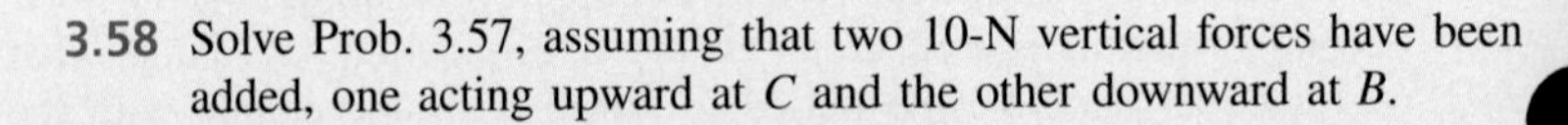

3.58 Solve Prob. 3.57, assuming that two 10-N vertical forces have been added, one acting upward at *C* and the other downward at *B*.

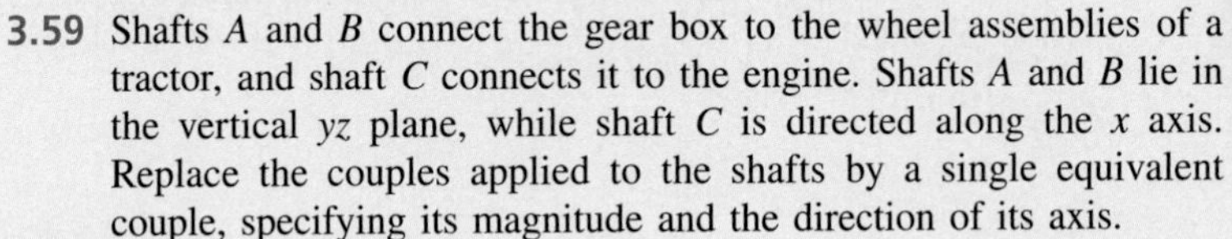

3.59 Shafts *A* and *B* connect the gear box to the wheel assemblies of a tractor, and shaft *C* connects it to the engine. Shafts *A* and *B* lie in the vertical *yz* plane, while shaft *C* is directed along the *x* axis. Replace the couples applied to the shafts by a single equivalent couple, specifying its magnitude and the direction of its axis.

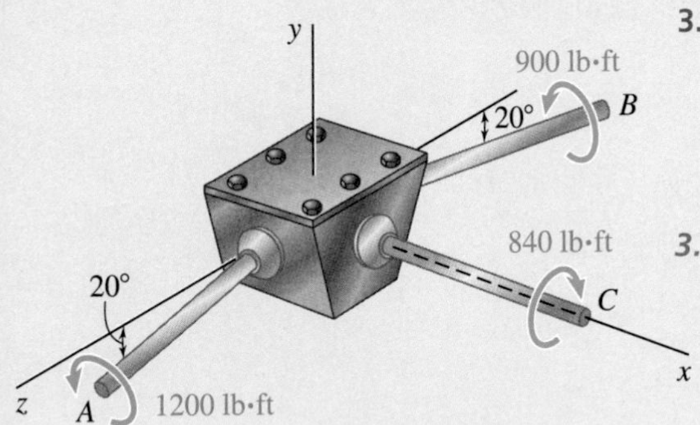

Fig. P3.59

3.60 If $P = 20$ lb, replace the three couples with a single equivalent couple, specifying its magnitude and the direction of its axis.

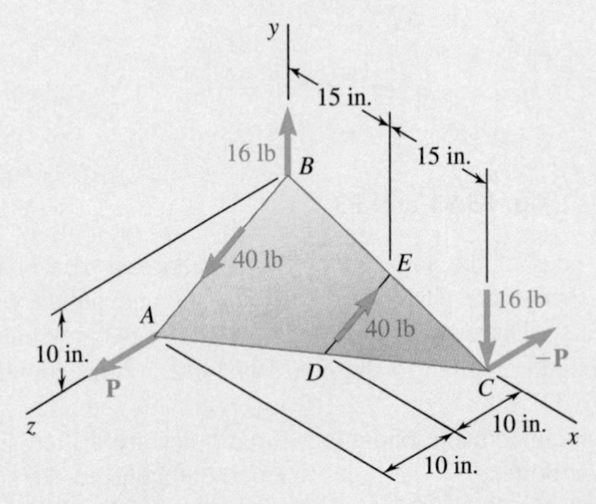

Fig. *P3.60*

3.61 A 30-lb vertical force **P** is applied at A to the bracket shown, which is held by screws at B and C. (a) Replace **P** with an equivalent force-couple system at B. (b) Find the two horizontal forces at B and C that are equivalent to the couple obtained in part a.

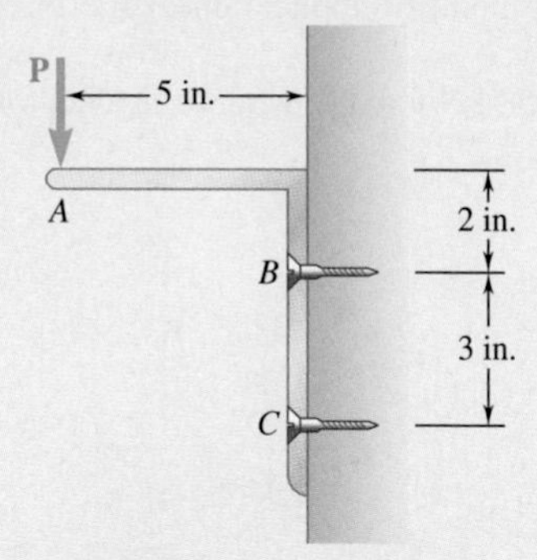

Fig. P3.61

3.62 The force **P** has a magnitude of 250 N and is applied at the end C of a 500-mm rod AC attached to a bracket at A and B. Assuming $\alpha = 30°$ and $\beta = 60°$, replace **P** with (a) an equivalent force-couple system at B, (b) an equivalent system formed by two parallel forces applied at A and B.

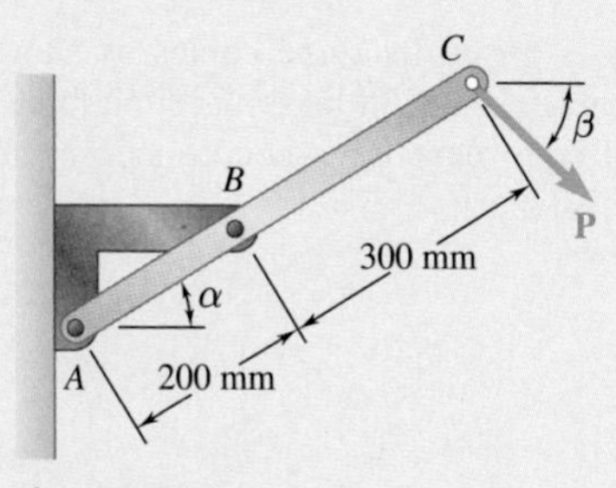

Fig. ***P3.62***

3.63 Solve Prob. 3.62, assuming $\alpha = \beta = 25°$.

3.64 A 260-lb force is applied at A to the rolled-steel section shown. Replace that force with an equivalent force-couple system at the center C of the section.

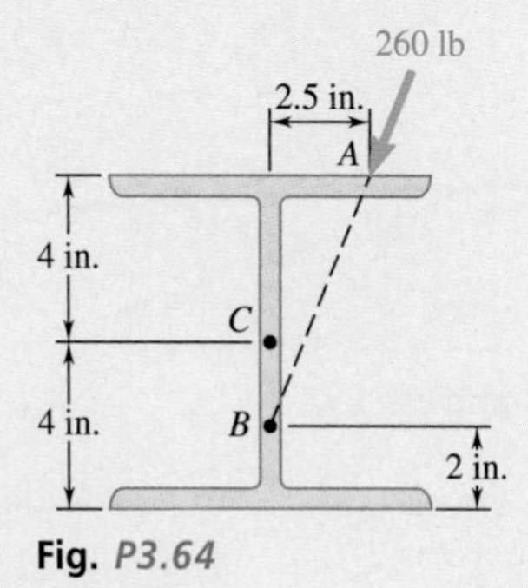

Fig. ***P3.64***

3.65 A dirigible is tethered by a cable attached to its cabin at B. If the tension in the cable is 1040 N, replace the force exerted by the cable at B with an equivalent system formed by two parallel forces applied at A and C.

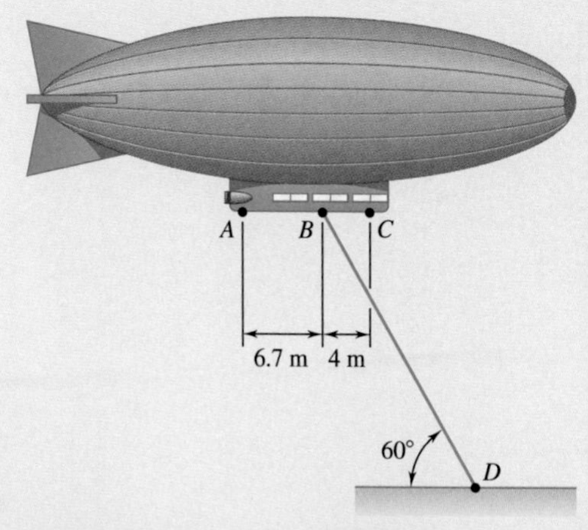

Fig. P3.65

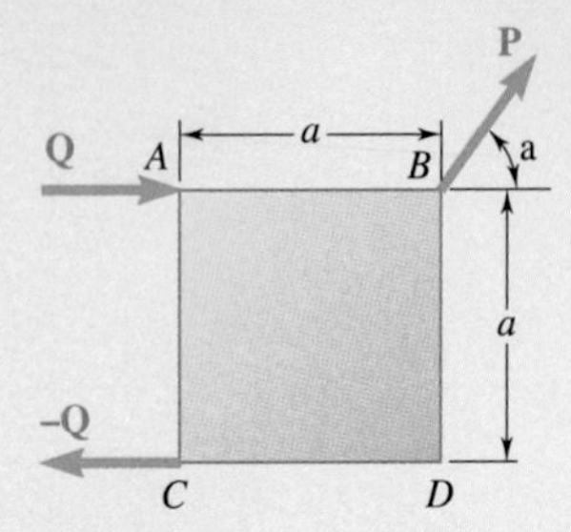

Fig. P3.66

3.66 A force and couple act as shown on a square plate of side $a = 25$ in. Knowing that $P = 60$ lb, $Q = 40$ lb, and $\alpha = 50°$, replace the given force and couple by a single force applied at a point located (*a*) on line *AB*, (*b*) on line *AC*. In each case determine the distance from *A* to the point of application of the force.

3.67 Replace the 250-kN force **P** by an equivalent force-couple system at *G*.

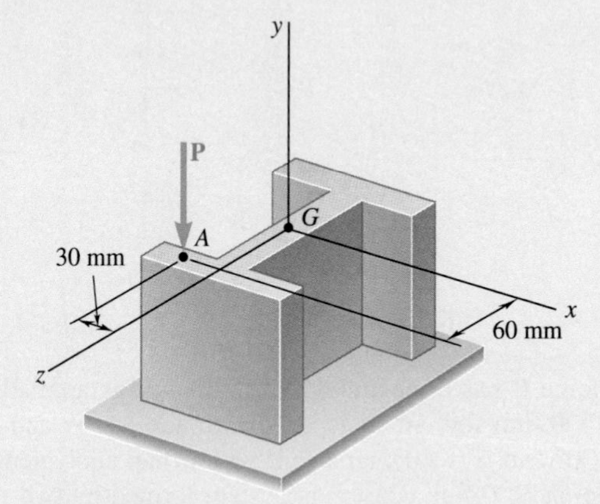

Fig. P3.67

3.68 An antenna is guyed by three cables as shown. Knowing that the tension in cable *AB* is 288 lb, replace the force exerted at *A* by cable *AB* with an equivalent force-couple system at the center *O* of the base of the antenna.

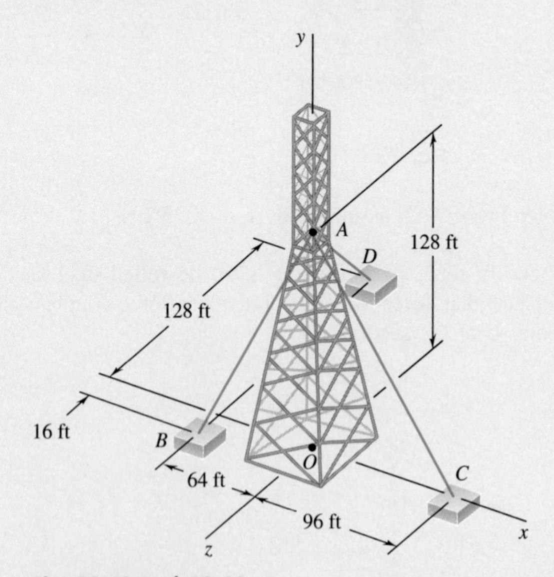

Fig. P3.68 and *P3.69*

3.69 An antenna is guyed by three cables as shown. Knowing that the tension in cable *AD* is 270 lb, replace the force exerted at *A* by cable *AD* with an equivalent force-couple system at the center *O* of the base of the antenna.

3.70 Replace the 150-N force by an equivalent force-couple system at A.

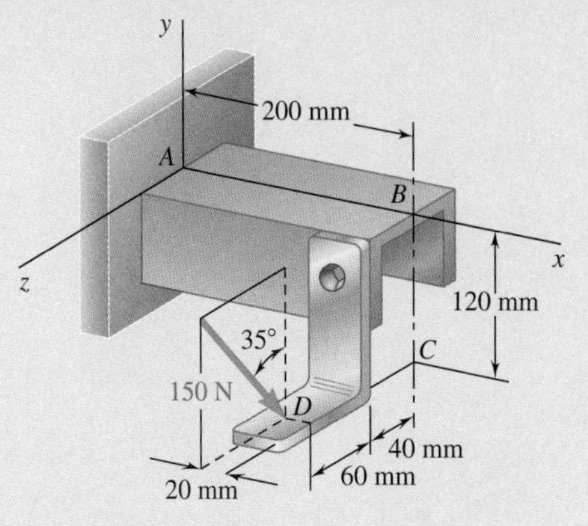

Fig. ***P3.70***

3.71 A 2.6-kip force is applied at point D of the cast-iron post shown. Replace that force with an equivalent force-couple system at the center A of the base section.

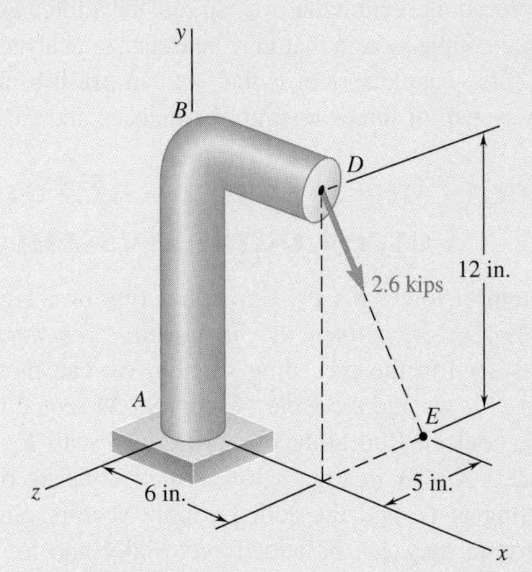

Fig. P3.71

3.72 A 110-N force acting in a vertical plane parallel to the yz plane is applied to the 220-mm-long horizontal handle AB of a socket wrench. Replace the force with an equivalent force-couple system at the origin O of the coordinate system.

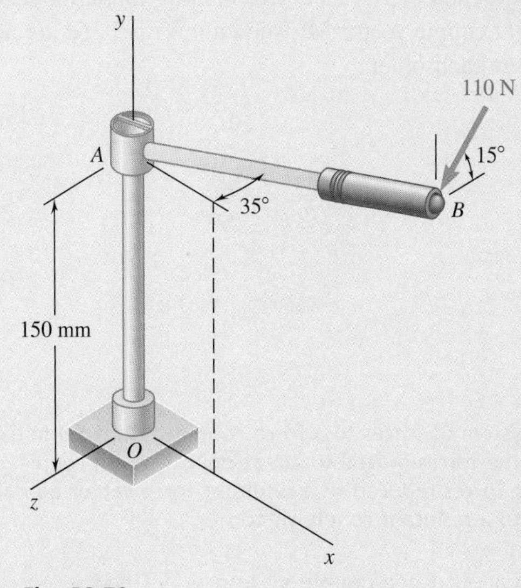

Fig. P3.72

3.4 SIMPLIFYING SYSTEMS OF FORCES

We saw in the preceding section that we can replace a force acting on a rigid body with a force-couple system that may be easier to analyze. However, the true value of a force-couple system is that we can use it to replace not just one force but a system of forces to simplify analysis and calculations.

3.4A Reducing a System of Forces to a Force-Couple System

Consider a system of forces $\mathbf{F}_1$, $\mathbf{F}_2$, $\mathbf{F}_3$, . . . , acting on a rigid body at the points A_1, A_2, A_3, . . . , *defined by the position vectors* $\mathbf{r}_1$, $\mathbf{r}_2$, $\mathbf{r}_3$, *etc.* (Fig. 3.36*a*). As seen in the preceding section, we can move $\mathbf{F}_1$ from A_1 to a given point O if we add a couple of moment $\mathbf{M}_1$ equal to the moment $\mathbf{r}_1 \times \mathbf{F}_1$ of $\mathbf{F}_1$ about O. Repeating this procedure with $\mathbf{F}_2$, $\mathbf{F}_3$, . . . , we obtain the system shown in Fig. 3.36*b*, which consists of the original forces, now acting at O, and the added couple vectors. Since the forces are now concurrent, they can be added vectorially and replaced by their resultant $\mathbf{R}$. Similarly, the couple vectors $\mathbf{M}_1$, $\mathbf{M}_2$, $\mathbf{M}_3$, . . . , can be added vectorially and replaced by a single couple vector $\mathbf{M}_O^R$. Thus,

We can reduce any system of forces, however complex, to an equivalent force-couple system acting at a given point *O*.

Note that, although each of the couple vectors $\mathbf{M}_1$, $\mathbf{M}_2$, $\mathbf{M}_3$, . . . in Fig. 3.36*b* is perpendicular to its corresponding force, the resultant force $\mathbf{R}$ and the resultant couple vector $\mathbf{M}_O^R$ shown in Fig. 3.36*c* are not, in general, perpendicular to each other.

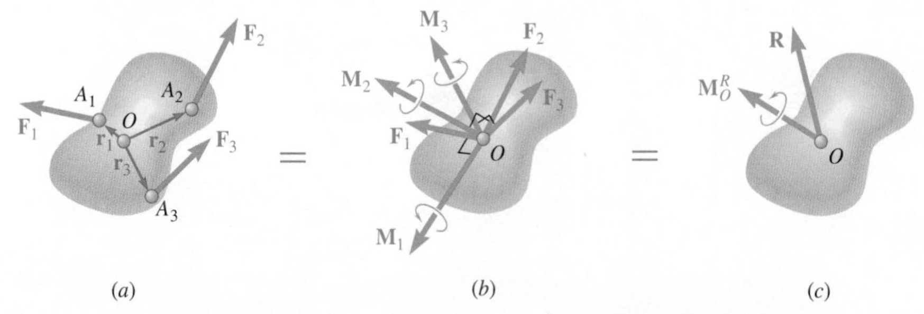

Fig. 3.36 Reducing a system of forces to a force-couple system. (*a*) Initial system of forces; (*b*) all the forces moved to act at point *O*, with couple vectors added; (*c*) all the forces reduced to a resultant force vector and all the couple vectors reduced to a resultant couple vector.

The equivalent force-couple system is defined by

Force-couple system

$$\mathbf{R} = \Sigma\mathbf{F} \qquad \mathbf{M}_O^R = \Sigma\mathbf{M}_O = \Sigma(\mathbf{r} \times \mathbf{F}) \tag{3.50}$$

These equations state that we obtain force $\mathbf{R}$ by adding all of the forces of the system, whereas we obtain the moment of the resultant couple vector $\mathbf{M}_O^R$, called the **moment resultant** of the system, by adding the moments about O of all the forces of the system.

Once we have reduced a given system of forces to a force and a couple at a point O, we can replace it with a force and a couple at another point O'. The resultant force **R** will remain unchanged, whereas the new moment resultant $\mathbf{M}^R_{O'}$ will be equal to the sum of $\mathbf{M}^R_O$ and the moment about O' of force **R** attached at O (Fig. 3.37). We have

$$\mathbf{M}^R_{O'} = \mathbf{M}^R_O + \mathbf{s} \times \mathbf{R} \tag{3.51}$$

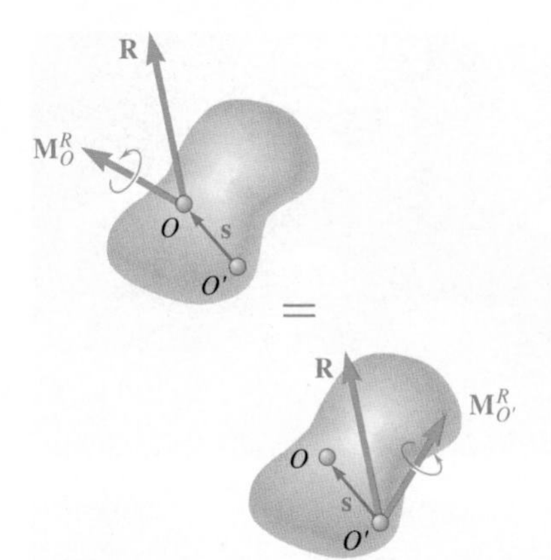

Fig. 3.37 Once a system of forces has been reduced to a force-couple system at one point, we can replace it with an equivalent force-couple system at another point. The force resultant stays the same, but we have to add the moment of the resultant force about the new point to the resultant couple vector.

In practice, the reduction of a given system of forces to a single force **R** at O and a couple vector $\mathbf{M}^R_O$ is carried out in terms of components. Resolving each position vector **r** and each force **F** of the system into rectangular components, we have

$$\mathbf{r} = x\mathbf{i} + y\mathbf{j} + z\mathbf{k} \tag{3.52}$$

$$\mathbf{F} = F_x\mathbf{i} + F_y\mathbf{j} + F_z\mathbf{k} \tag{3.53}$$

Substituting for **r** and **F** in Eq. (3.50) and factoring out the unit vectors **i**, **j**, and **k**, we obtain **R** and $\mathbf{M}^R_O$ in the form

$$\mathbf{R} = R_x\mathbf{i} + R_y\mathbf{j} + R_z\mathbf{k} \qquad \mathbf{M}^R_O = M^R_x\mathbf{i} + M^R_y\mathbf{j} + M^R_z\mathbf{k} \tag{3.54}$$

The components R_x, R_y, and R_z represent, respectively, the sums of the x, y, and z components of the given forces and measure the tendency of the system to impart to the rigid body a translation in the x, y, or z direction. Similarly, the components M^R_x, M^R_y, and M^R_z represent, respectively, the sum of the moments of the given forces about the x, y, and z axes and measure the tendency of the system to impart to the rigid body a rotation about the x, y, or z axis.

If we need to know the magnitude and direction of force **R**, we can obtain them from the components R_x, R_y, and R_z by means of the relations in Eqs. (2.18) and (2.19) of Sec. 2.4A. Similar computations yield the magnitude and direction of the couple vector $\mathbf{M}^R_O$.

3.4B Equivalent and Equipollent Systems of Forces

We have just seen that any system of forces acting on a rigid body can be reduced to a force-couple system at a given point O. This equivalent force-couple system characterizes completely the effect of the given force system on the rigid body.

Two systems of forces are equivalent if they can be reduced to the same force-couple system at a given point *O*.

Recall that the force-couple system at O is defined by the relations in Eq. (3.50). Therefore, we can state that

Two systems of forces, F_1, F_2, F_3, . . . , and F'_1, F'_2, F'_3, . . . , that act on the same rigid body are equivalent if, and only if, the sums of the forces and the sums of the moments about a given point *O* of the forces of the two systems are, respectively, equal.

Mathematically, the necessary and sufficient conditions for the two systems of forces to be equivalent are

Conditions for equivalent systems of forces

$$\Sigma\mathbf{F} = \Sigma\mathbf{F}' \qquad \text{and} \qquad \Sigma\mathbf{M}_O = \Sigma\mathbf{M}'_O \tag{3.55}$$

Photo 3.3 The forces exerted by the children upon the wagon can be replaced with an equivalent force-couple system when analyzing the motion of the wagon.

© Jose Luis Pelaez/Getty Images

Note that to prove that two systems of forces are equivalent, we must establish the second of the relations in Eq. (3.55) with respect to *only one point O*. It will hold, however, with respect to *any point* if the two systems are equivalent.

Resolving the forces and moments in Eqs. (3.55) into their rectangular components, we can express the necessary and sufficient conditions for the equivalence of two systems of forces acting on a rigid body as

$$\begin{aligned} \Sigma F_x &= \Sigma F'_x \qquad & \Sigma F_y &= \Sigma F'_y \qquad & \Sigma F_z &= \Sigma F'_z \\ \Sigma M_x &= \Sigma M'_x \qquad & \Sigma M_y &= \Sigma M'_y \qquad & \Sigma M_z &= \Sigma M'_z \end{aligned} \tag{3.56}$$

These equations have a simple physical significance. They express that

Two systems of forces are equivalent if they tend to impart to the rigid body (1) the same translation in the *x*, *y*, and *z* directions, respectively, and (2) the same rotation about the *x*, *y*, and *z* axes, respectively.

In general, when two systems of vectors satisfy Eqs. (3.55) or (3.56), i.e., when their resultants and their moment resultants about an arbitrary point O are respectively equal, the two systems are said to be **equipollent**. The result just established can thus be restated as

If two systems of forces acting on a rigid body are equipollent, they are also equivalent.

It is important to note that this statement does not apply to *any* system of vectors. Consider, for example, a system of forces acting on a set of independent particles that do *not* form a rigid body. A different system of forces acting on the same particles may happen to be equipollent to the first one; i.e., it may have the same resultant and the same moment resultant. Yet, since different forces now act on the various particles, their effects on these particles are different; the two systems of forces, while equipollent, are *not equivalent*.

3.4C Further Reduction of a System of Forces

We have now seen that any given system of forces acting on a rigid body can be reduced to an equivalent force-couple system at O, consisting of a force $\mathbf{R}$ equal to the sum of the forces of the system, and a couple vector $\mathbf{M}_O^R$ of moment equal to the moment resultant of the system.

When $\mathbf{R} = 0$, the force-couple system reduces to the couple vector $\mathbf{M}_O^R$. The given system of forces then can be reduced to a single couple called the **resultant couple** of the system.

What are the conditions under which a given system of forces can be reduced to a single force? It follows from the preceding section that we can replace the force-couple system at O by a single force $\mathbf{R}$ acting along a new line of action if $\mathbf{R}$ and $\mathbf{M}_O^R$ are mutually perpendicular. The systems of forces that can be reduced to a single force, or *resultant,* are therefore the systems for which force $\mathbf{R}$ and the couple vector $\mathbf{M}_O^R$ are mutually perpendicular. This condition *is generally not satisfied* by systems of forces in space, but it *is satisfied* by systems consisting of (1) concurrent forces, (2) coplanar forces, or (3) parallel forces. Let's look at each case separately.

1. **Concurrent forces** act at the same point; therefore, we can add them directly to obtain their resultant $\mathbf{R}$. Thus, they always reduce to a single force. Concurrent forces were discussed in detail in Chap. 2.

2. **Coplanar forces** act in the same plane, which we assume to be the plane of the figure (Fig. 3.38*a*). The sum **R** of the forces of the system also lies in the plane of the figure, whereas the moment of each force about O and thus the moment resultant $\mathbf{M}_O^R$ are perpendicular to that plane. The force-couple system at O consists, therefore, of a force **R** and a couple vector $\mathbf{M}_O^R$ that are mutually perpendicular (Fig. 3.38*b*).† We can reduce them to a single force **R** by moving **R** in the plane of the figure until its moment about O becomes equal to $\mathbf{M}_O^R$. The distance from O to the line of action of **R** is $d = M_O^R/R$ (Fig. 3.38*c*).

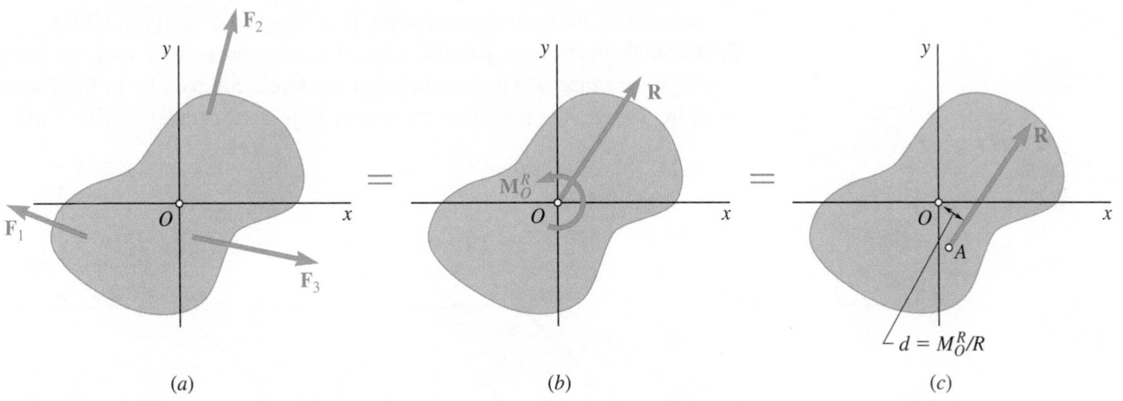

Fig. 3.38 Reducing a system of coplanar forces. (*a*) Initial system of forces; (*b*) equivalent force-couple system at *O*; (*c*) moving the resultant force to a point *A* such that the moment of **R** about *O* equals the couple vector.

As noted earlier, the reduction of a system of forces is considerably simplified if we resolve the forces into rectangular components. The force-couple system at O is then characterized by the components (Fig. 3.39*a*)

$$R_x = \Sigma F_x \qquad R_y = \Sigma F_y \qquad M_z^R = M_O^R = \Sigma M_O \qquad \textbf{(3.57)}$$

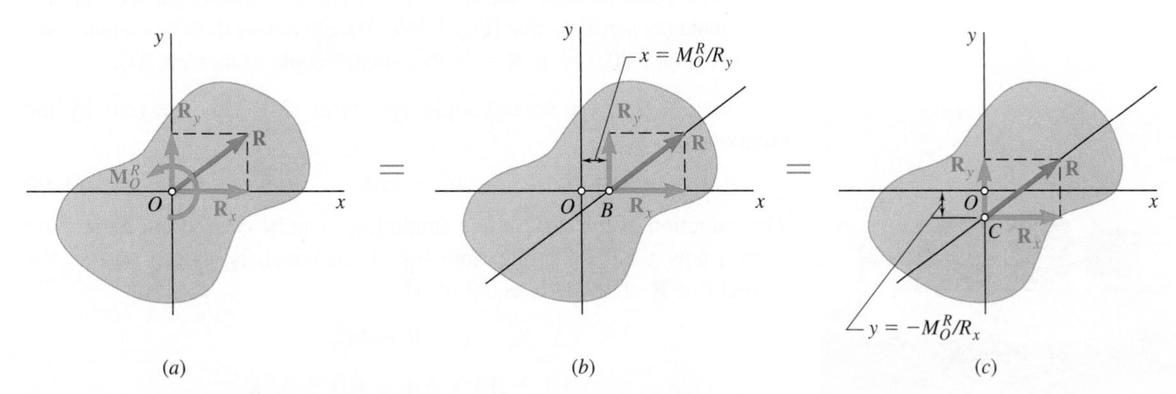

Fig. 3.39 Reducing a system of coplanar forces by using rectangular components. (*a*) From Fig. 3.38(*b*), resolve the resultant into components along the *x* and *y* axes; (*b*) determining the *x* intercept of the final line of action of the resultant; (*c*) determining the *y* intercept of the final line of action of the resultant.

†Because the couple vector $\mathbf{M}_O^R$ is perpendicular to the plane of the figure, we represent it by the symbol ↶. A counterclockwise couple ↶ represents a vector pointing out of the page and a clockwise couple ↻ represents a vector pointing into the page.

To reduce the system to a single force **R**, the moment of **R** about O must be equal to $\mathbf{M}_O^R$. If we denote the coordinates of the point of application of the resultant by x and y and apply equation (3.22) of Sec. 3.1F, we have

$$xR_y - yR_x = M_O^R$$

This represents the equation of the line of action of **R**. We can also determine the x and y intercepts of the line of action of the resultant directly by noting that $\mathbf{M}_O^R$ must be equal to the moment about O of the y component of **R** when **R** is attached at B (Fig. 3.39*b*) and to the moment of its x component when **R** is attached at C (Fig. 3.39*c*).

3. **Parallel forces** have parallel lines of action and may or may not have the same sense. Assuming here that the forces are parallel to the y axis (Fig. 3.40*a*), we note that their sum **R** is also parallel to the y axis.

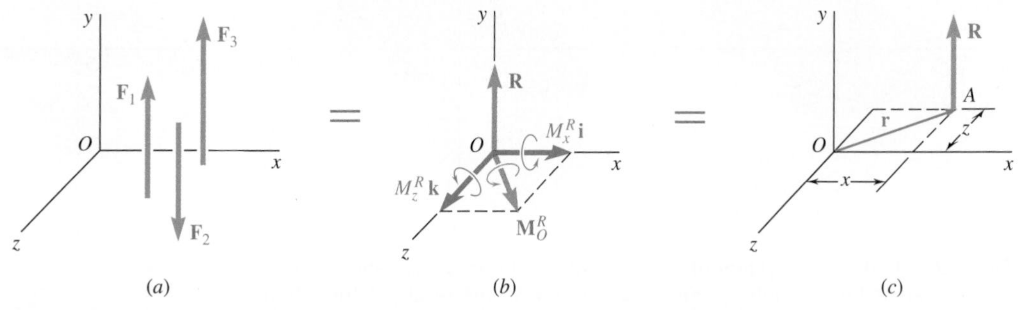

Fig. 3.40 Reducing a system of parallel forces. (*a*) Initial system of forces; (*b*) equivalent force-couple system at *O*, resolved into components; (*c*) moving **R** to point *A*, chosen so that the moment of **R** about *O* equals the resultant moment about *O*.

On the other hand, since the moment of a given force must be perpendicular to that force, the moment about O of each force of the system and thus the moment resultant $\mathbf{M}_O^R$ lie in the zx plane. The force-couple system at O consists, therefore, of a force **R** and a couple vector $\mathbf{M}_O^R$ that are mutually perpendicular (Fig. 3.40*b*). We can reduce them to a single force **R** (Fig. 3.40*c*) or, if $\mathbf{R} = 0$, to a single couple of moment $\mathbf{M}_O^R$.

Photo 3.4 The parallel wind forces acting on the highway signs can be reduced to a single equivalent force. Determining this force can simplify the calculation of the forces acting on the supports of the frame to which the signs are attached.

© Images-USA/Alamy RF

In practice, the force-couple system at O is characterized by the components

$$R_y = \Sigma F_y \qquad M_x^R = \Sigma M_x \qquad M_z^R = \Sigma M_z \tag{3.58}$$

The reduction of the system to a single force can be carried out by moving **R** to a new point of application $A(x, 0, z)$, which is chosen so that the moment of **R** about O is equal to $\mathbf{M}_O^R$.

$$\mathbf{r} \times \mathbf{R} = \mathbf{M}_O^R$$

$$(x\mathbf{i} + z\mathbf{k}) \times R_y\mathbf{j} = M_x^R\mathbf{i} + M_z^R\mathbf{k}$$

By computing the vector products and equating the coefficients of the corresponding unit vectors in both sides of the equation, we obtain two scalar equations that define the coordinates of A:

$$-zR_y = M_x^R \quad \text{and} \quad xR_y = M_z^R$$

These equations express the fact that the moments of **R** about the x and z axes must be equal, respectively, to M_x^R and M_z^R.

Sample Problem 3.8

A 4.80-m-long beam is subjected to the forces shown. Reduce the given system of forces to (*a*) an equivalent force-couple system at *A*, (*b*) an equivalent force-couple system at *B*, (*c*) a single force or resultant. *Note:* Since the reactions at the supports are not included in the given system of forces, the given system will not maintain the beam in equilibrium.

STRATEGY: The *force* part of an equivalent force-couple system is simply the sum of the forces involved. The *couple* part is the sum of the moments caused by each force relative to the point of interest. Once you find the equivalent force-couple at one point, you can transfer it to any other point by a moment calculation.

MODELING and ANALYSIS:

a. Force-Couple System at *A*. The force-couple system at *A* equivalent to the given system of forces consists of a force **R** and a couple $\mathbf{M}_A^R$ defined as (Fig. 1):

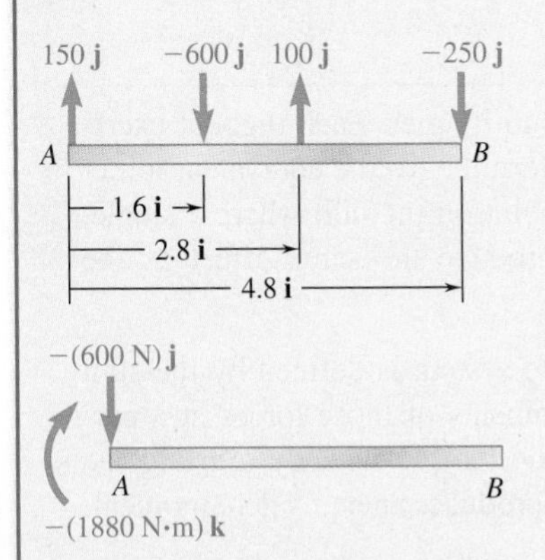

Fig. 1 Force-couple system at *A* that is equivalent to given system of forces.

$$\mathbf{R} = \Sigma\mathbf{F}$$
$$= (150\text{ N})\mathbf{j} - (600\text{ N})\mathbf{j} + (100\text{ N})\mathbf{j} - (250\text{ N})\mathbf{j} = -(600\text{ N})\mathbf{j}$$
$$\mathbf{M}_A^R = \Sigma(\mathbf{r} \times \mathbf{F})$$
$$= (1.6\mathbf{i}) \times (-600\mathbf{j}) + (2.8\mathbf{i}) \times (100\mathbf{j}) + (4.8\mathbf{i}) \times (-250\mathbf{j})$$
$$= -(1880\text{ N·m})\mathbf{k}$$

The equivalent force-couple system at *A* is thus

$$\mathbf{R} = 600\text{ N}\downarrow \qquad \mathbf{M}_A^R = 1880\text{ N·m} \circlearrowright \blacktriangleleft$$

b. Force-Couple System at *B*. You want to find a force-couple system at *B* equivalent to the force-couple system at *A* determined in part *a*. The force **R** is unchanged, but you must determine a new couple $\mathbf{M}_B^R$, the moment of which is equal to the moment about *B* of the force-couple system determined in part *a* (Fig. 2). You have

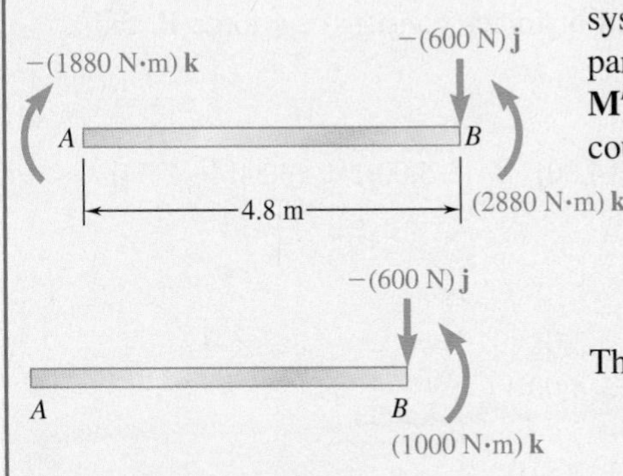

Fig. 2 Finding force-couple system at *B* equivalent to that determined in part *a*.

$$\mathbf{M}_B^R = \mathbf{M}_A^R + \overrightarrow{BA} \times \mathbf{R}$$
$$= -(1880\text{ N·m})\mathbf{k} + (-4.8\text{ m})\mathbf{i} \times (-600\text{ N})\mathbf{j}$$
$$= -(1880\text{ N·m})\mathbf{k} + (2880\text{ N·m})\mathbf{k} = +(1000\text{ N·m})\mathbf{k}$$

The equivalent force-couple system at *B* is thus

$$\mathbf{R} = 600\text{ N}\downarrow \qquad \mathbf{M}_B^R = 1000\text{ N·m} \circlearrowleft \blacktriangleleft$$

c. Single Force or Resultant. The resultant of the given system of forces is equal to **R**, and its point of application must be such that the moment of **R** about *A* is equal to $\mathbf{M}_A^R$ (Fig. 3). This equality of moments leads to

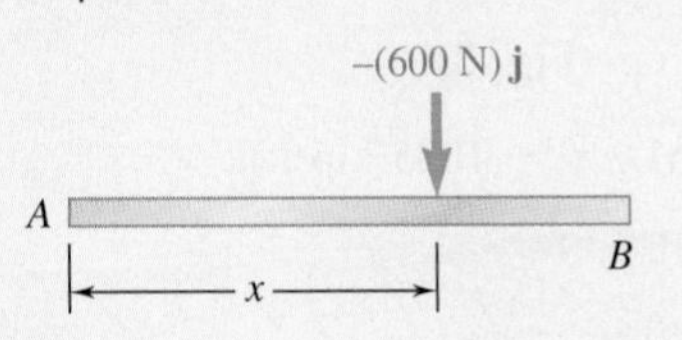

Fig. 3 Single force that is equivalent to given system of forces.

$$\mathbf{r} \times \mathbf{R} = \mathbf{M}_A^R$$
$$x\mathbf{i} \times (-600\text{ N})\mathbf{j} = -(1880\text{ N·m})\mathbf{k}$$
$$-x(600\text{ N})\mathbf{k} = -(1880\text{ N·m})\mathbf{k}$$

(*continued*)

Solving for x, you get $x = 3.13$ m. Thus, the single force equivalent to the given system is defined as

$$\mathbf{R} = 600 \text{ N} \downarrow \qquad x = 3.13 \text{ m}$$

REFLECT and THINK: This reduction of a given system of forces to a single equivalent force uses the same principles that you will use later for finding centers of gravity and centers of mass, which are important parameters in engineering mechanics.

Sample Problem 3.9

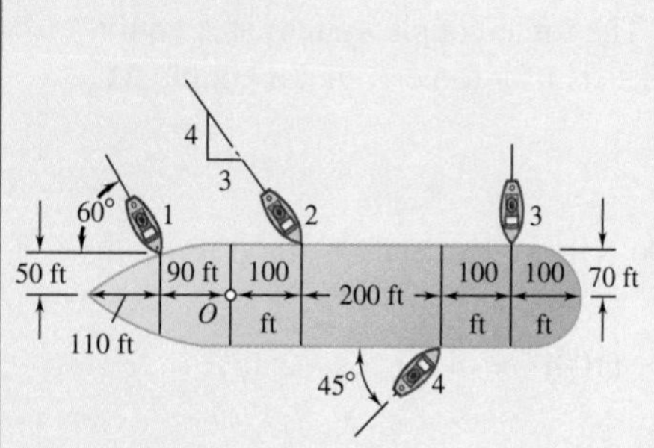

Four tugboats are bringing an ocean liner to its pier. Each tugboat exerts a 5000-lb force in the direction shown. Determine (*a*) the equivalent force-couple system at the foremast *O*, (*b*) the point on the hull where a single, more powerful tugboat should push to produce the same effect as the original four tugboats.

STRATEGY: The equivalent force-couple system is defined by the sum of the given forces and the sum of the moments of those forces at a particular point. A single tugboat could produce this system by exerting the resultant force at a point of application that produces an equivalent moment.

MODELING and ANALYSIS:

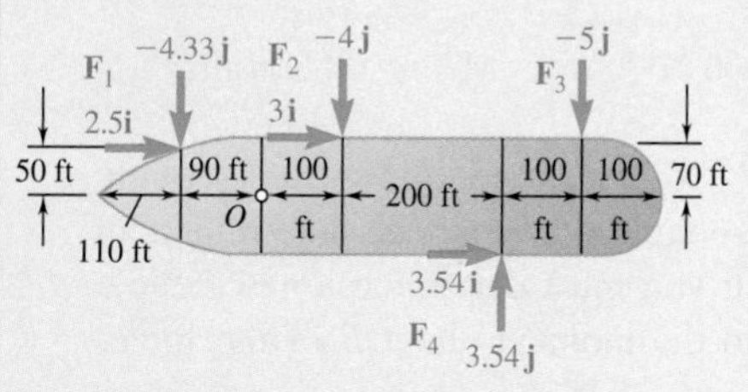

Fig. 1 Given forces resolved into components.

a. Force-Couple System at *O*. Resolve each of the given forces into components, as in Fig. 1 (kip units are used). The force-couple system at O equivalent to the given system of forces consists of a force $\mathbf{R}$ and a couple $\mathbf{M}_O^R$ defined as

$$\begin{aligned}\mathbf{R} &= \Sigma\mathbf{F}\\ &= (2.50\mathbf{i} - 4.33\mathbf{j}) + (3.00\mathbf{i} - 4.00\mathbf{j}) + (-5.00\mathbf{j}) + (3.54\mathbf{i} + 3.54\mathbf{j})\\ &= 9.04\mathbf{i} - 9.79\mathbf{j}\end{aligned}$$

$$\begin{aligned}\mathbf{M}_O^R &= \Sigma(\mathbf{r} \times \mathbf{F})\\ &= (-90\mathbf{i} + 50\mathbf{j}) \times (2.50\mathbf{i} - 4.33\mathbf{j})\\ &\quad + (100\mathbf{i} + 70\mathbf{j}) \times (3.00\mathbf{i} - 4.00\mathbf{j})\\ &\quad + (400\mathbf{i} + 70\mathbf{j}) \times (-5.00\mathbf{j})\\ &\quad + (300\mathbf{i} - 70\mathbf{j}) \times (3.54\mathbf{i} + 3.54\mathbf{j})\\ &= (390 - 125 - 400 - 210 - 2000 + 1062 + 248)\mathbf{k}\\ &= -1035\mathbf{k}\end{aligned}$$

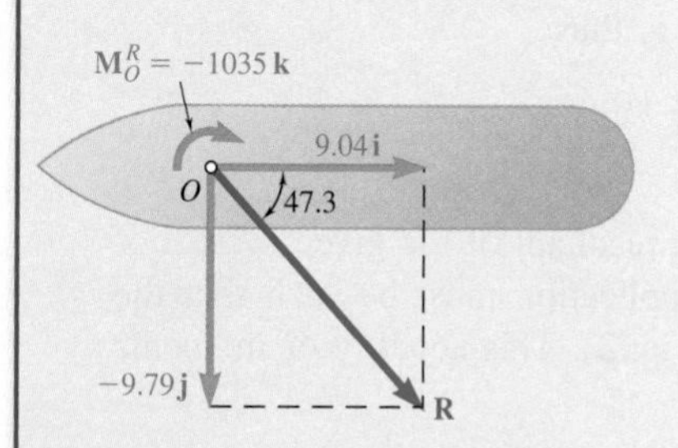

Fig. 2 Equivalent force-couple system at *O*.

The equivalent force-couple system at O is thus (Fig. 2)

$$\mathbf{R} = (9.04 \text{ kips})\mathbf{i} - (9.79 \text{ kips})\mathbf{j} \qquad \mathbf{M}_O^R = -(1035 \text{ kip·ft})\mathbf{k}$$

or

$$\mathbf{R} = 13.33 \text{ kips} \measuredangle 47.3° \qquad \mathbf{M}_O^R = 1035 \text{ kip·ft} \circlearrowright$$

(continued)

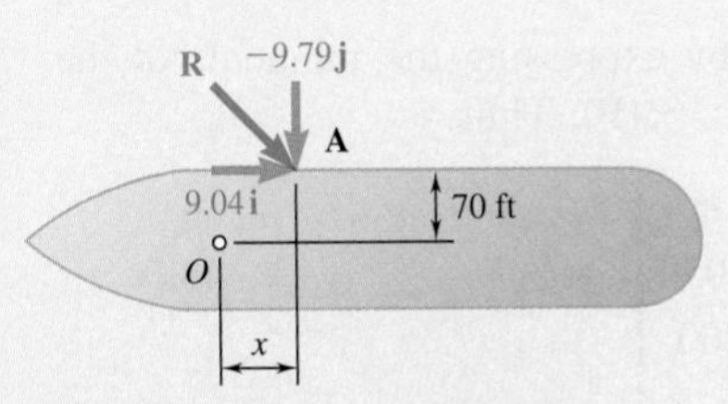

Fig. 3 Point of application of single tugboat to create same effect as given force system.

Remark: Since all the forces are contained in the plane of the figure, you would expect the sum of their moments to be perpendicular to that plane. Note that you could obtain the moment of each force component directly from the diagram by first forming the product of its magnitude and perpendicular distance to O and then assigning to this product a positive or a negative sign, depending upon the sense of the moment.

b. Single Tugboat. The force exerted by a single tugboat must be equal to $\mathbf{R}$, and its point of application A must be such that the moment of $\mathbf{R}$ about O is equal to $\mathbf{M}_O^R$ (Fig. 3). Observing that the position vector of A is

$$\mathbf{r} = x\mathbf{i} + 70\mathbf{j}$$

you have

$$\mathbf{r} \times \mathbf{R} = \mathbf{M}_O^R$$
$$(x\mathbf{i} + 70\mathbf{j}) \times (9.04\mathbf{i} - 9.79\mathbf{j}) = -1035\mathbf{k}$$
$$-x(9.79)\mathbf{k} - 633\mathbf{k} = -1035\mathbf{k} \qquad x = 41.1 \text{ ft} \blacktriangleleft$$

REFLECT and THINK: Reducing the given situation to that of a single force makes it easier to visualize the overall effect of the tugboats in maneuvering the ocean liner. But in practical terms, having four boats applying force allows for greater control in slowing and turning a large ship in a crowded harbor.

Sample Problem 3.10

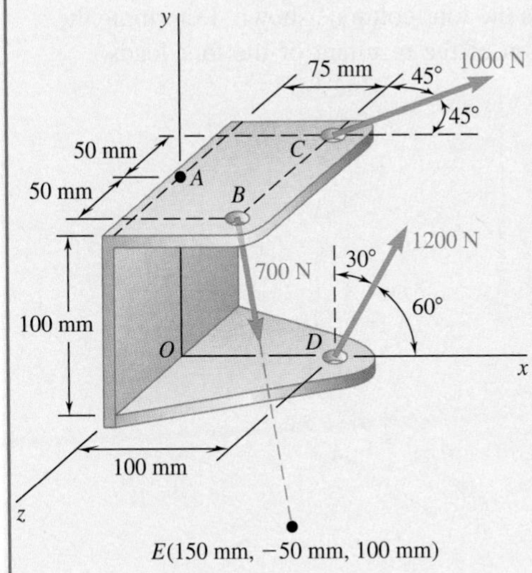

Three cables are attached to a bracket as shown. Replace the forces exerted by the cables with an equivalent force-couple system at A.

STRATEGY: First determine the relative position vectors drawn from point A to the points of application of the various forces and resolve the forces into rectangular components. Then sum the forces and moments.

MODELING and ANALYSIS: Note that $\mathbf{F}_B = (700 \text{ N})\boldsymbol{\lambda}_{BE}$ where

$$\boldsymbol{\lambda}_{BE} = \frac{\overrightarrow{BE}}{BE} = \frac{75\mathbf{i} - 150\mathbf{j} + 50\mathbf{k}}{175}$$

Using meters and newtons, the position and force vectors are

$$\mathbf{r}_{B/A} = \overrightarrow{AB} = 0.075\mathbf{i} + 0.050\mathbf{k} \qquad \mathbf{F}_B = 300\mathbf{i} - 600\mathbf{j} + 200\mathbf{k}$$
$$\mathbf{r}_{C/A} = \overrightarrow{AC} = 0.075\mathbf{i} - 0.050\mathbf{k} \qquad \mathbf{F}_C = 707\mathbf{i} \qquad - 707\mathbf{k}$$
$$\mathbf{r}_{D/A} = \overrightarrow{AD} = 0.100\mathbf{i} - 0.100\mathbf{j} \qquad \mathbf{F}_D = 600\mathbf{i} + 1039\mathbf{j}$$

The force-couple system at A equivalent to the given forces consists of a force $\mathbf{R} = \Sigma\mathbf{F}$ and a couple $\mathbf{M}_A^R = \Sigma(\mathbf{r} \times \mathbf{F})$. Obtain the force $\mathbf{R}$ by adding respectively the x, y, and z components of the forces:

$$\mathbf{R} = \Sigma\mathbf{F} = (1607 \text{ N})\mathbf{i} + (439 \text{ N})\mathbf{j} - (507 \text{ N})\mathbf{k} \blacktriangleleft$$

(continued)

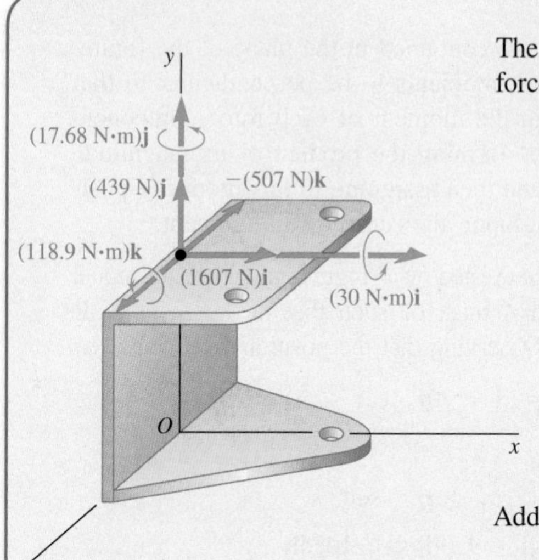

Fig. 1 Rectangular components of equivalent force-couple system at A.

The computation of $\mathbf{M}_A^R$ is facilitated by expressing the moments of the forces in the form of determinants (Sec. 3.1F). Thus,

$$\mathbf{r}_{B/A} \times \mathbf{F}_B = \begin{vmatrix} \mathbf{i} & \mathbf{j} & \mathbf{k} \\ 0.075 & 0 & 0.050 \\ 300 & -600 & 200 \end{vmatrix} = 30\mathbf{i} \qquad -45\mathbf{k}$$

$$\mathbf{r}_{C/A} \times \mathbf{F}_C = \begin{vmatrix} \mathbf{i} & \mathbf{j} & \mathbf{k} \\ 0.075 & 0 & -0.050 \\ 707 & 0 & -707 \end{vmatrix} = \qquad 17.68\mathbf{j}$$

$$\mathbf{r}_{D/A} \times \mathbf{F}_D = \begin{vmatrix} \mathbf{i} & \mathbf{j} & \mathbf{k} \\ 0.100 & -0.100 & 0 \\ 600 & 1039 & 0 \end{vmatrix} = \qquad 163.9\mathbf{k}$$

Adding these expressions, you have

$$\mathbf{M}_A^R = \Sigma(\mathbf{r} \times \mathbf{F}) = (30 \text{ N·m})\mathbf{i} + (17.68 \text{ N·m})\mathbf{j} + (118.9 \text{ N·m})\mathbf{k} \quad ◀$$

Figure 1 shows the rectangular components of the force $\mathbf{R}$ and the couple $\mathbf{M}_A^R$.

REFLECT and THINK: The determinant approach to calculating moments shows its advantages in a general three-dimensional problem such as this.

Sample Problem 3.11

A square foundation mat supports the four columns shown. Determine the magnitude and point of application of the resultant of the four loads.

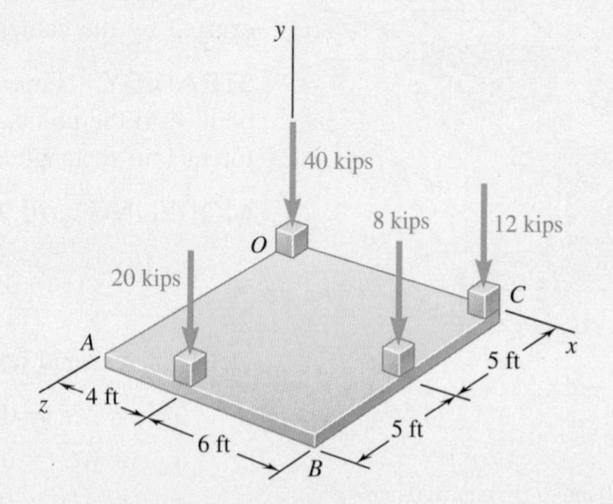

STRATEGY: Start by reducing the given system of forces to a force-couple system at the origin O of the coordinate system. Then reduce the system further to a single force applied at a point with coordinates x, z.

(continued)

MODELING: The force-couple system consists of a force $\mathbf{R}$ and a couple vector $\mathbf{M}_O^R$ defined as

$$\mathbf{R} = \Sigma\mathbf{F} \qquad \mathbf{M}_O^R = \Sigma(\mathbf{r} \times \mathbf{F})$$

ANALYSIS: After determining the position vectors of the points of application of the various forces, you may find it convenient to arrange the computations in tabular form. The results are shown in Fig. 1.

$\mathbf{r}$, ft	$\mathbf{F}$, kips	$\mathbf{r} \times \mathbf{F}$, kip·ft
0	$-40\mathbf{j}$	0
$10\mathbf{i}$	$-12\mathbf{j}$	$-120\mathbf{k}$
$10\mathbf{i} + 5\mathbf{k}$	$-8\mathbf{j}$	$40\mathbf{i} - 80\mathbf{k}$
$4\mathbf{i} + 10\mathbf{k}$	$-20\mathbf{j}$	$200\mathbf{i} - 80\mathbf{k}$
	$\mathbf{R} = -80\mathbf{j}$	$\mathbf{M}_O^R = 240\mathbf{i} - 280\mathbf{k}$

Fig. 1 Force-couple system at O that is equivalent to given force system.

The force $\mathbf{R}$ and the couple vector $\mathbf{M}_O^R$ are mutually perpendicular, so you can reduce the force-couple system further to a single force $\mathbf{R}$. Select the new point of application of $\mathbf{R}$ in the plane of the mat and in such a way that the moment of $\mathbf{R}$ about O is equal to $\mathbf{M}_O^R$. Denote the position vector of the desired point of application by $\mathbf{r}$ and its coordinates by x and z (Fig. 2). Then

$$\mathbf{r} \times \mathbf{R} = \mathbf{M}_O^R$$

$$(x\mathbf{i} + z\mathbf{k}) \times (-80\mathbf{j}) = 240\mathbf{i} - 280\mathbf{k}$$

$$-80x\mathbf{k} + 80z\mathbf{i} = 240\mathbf{i} - 280\mathbf{k}$$

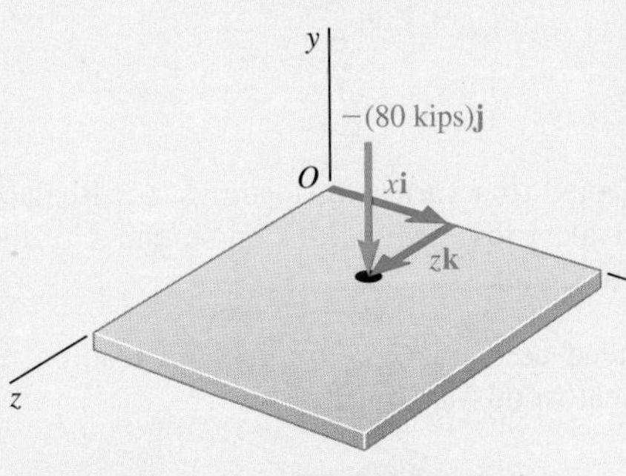

Fig. 2 Single force that is equivalent to given force system.

It follows that

$$-80x = -280 \qquad 80z = 240$$

$$x = 3.50 \text{ ft} \qquad z = 3.00 \text{ ft}$$

The resultant of the given system of forces is

$$\mathbf{R} = 80 \text{ kips} \downarrow \qquad \text{at } x = 3.50 \text{ ft}, z = 3.00 \text{ ft}$$ ◀

REFLECT and THINK: The fact that the given forces are all parallel simplifies the calculations, so the final step becomes just a two-dimensional analysis.

Problems

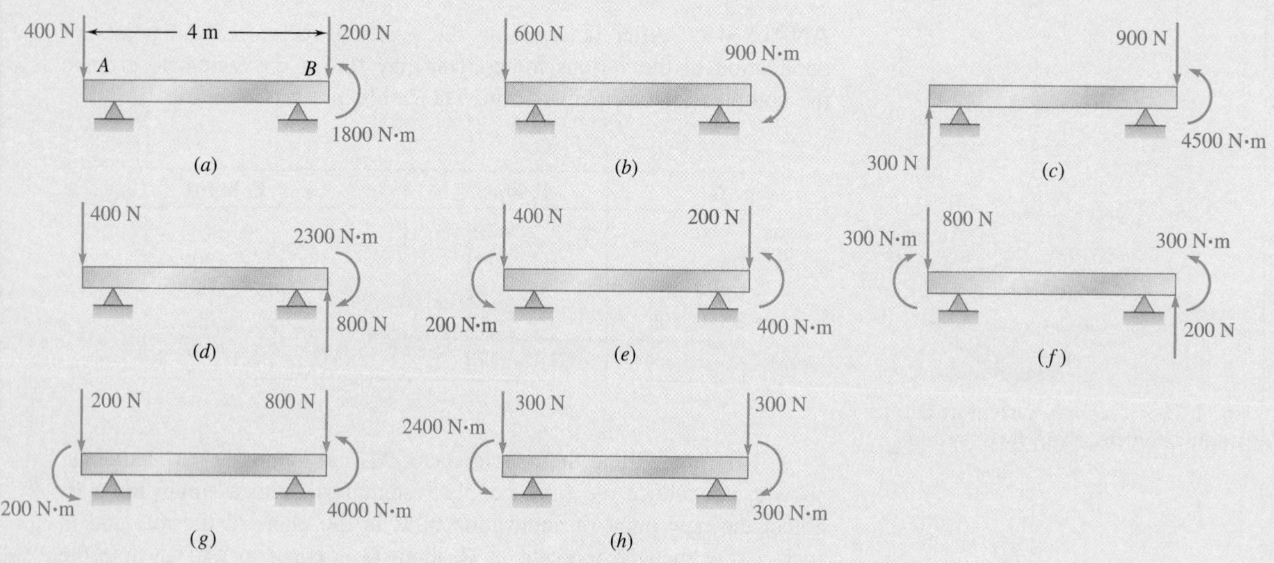

Fig. P3.73

3.73 A 4-m-long beam is subjected to a variety of loadings. (*a*) Replace each loading with an equivalent force-couple system at end *A* of the beam. (*b*) Which of the loadings are equivalent?

3.74 A 4-m-long beam is loaded as shown. Determine the loading of Prob. 3.73 that is equivalent to this loading.

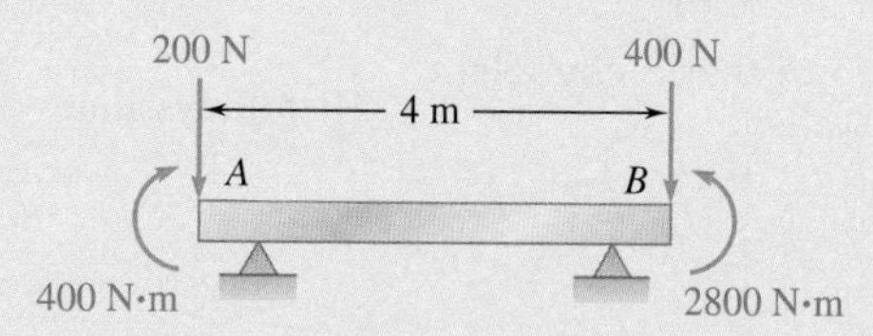

Fig. P3.74

3.75 Determine the single equivalent force and the distance from point *A* to its line of action for the beam and loading of (*a*) Prob. 3.73*b*, (*b*) Prob. 3.73*d*, (*c*) Prob. 3.73*e*.

3.76 The weights of two children sitting at ends *A* and *B* of a seesaw are 84 lb and 64 lb, respectively. Where should a third child sit so that the resultant of the weights of the three children will pass through *C* if she weighs (*a*) 60 lb, (*b*) 52 lb?

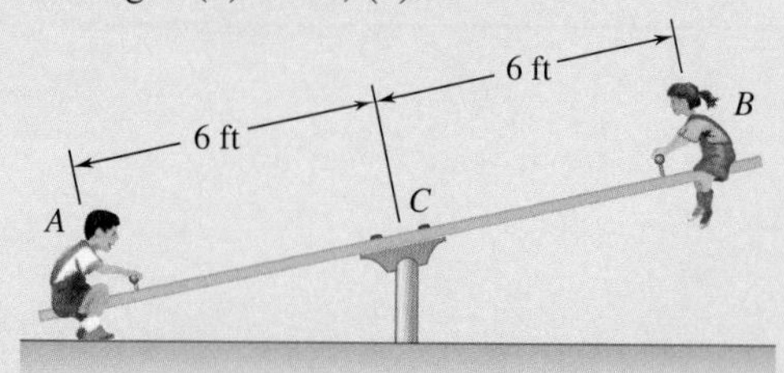

Fig. P3.76

3.77 Three stage lights are mounted on a pipe as shown. The lights at A and B each weigh 4.1 lb, while the one at C weighs 3.5 lb. (*a*) If $d = 25$ in., determine the distance from D to the line of action of the resultant of the weights of the three lights. (*b*) Determine the value of d so that the resultant of the weights passes through the midpoint of the pipe.

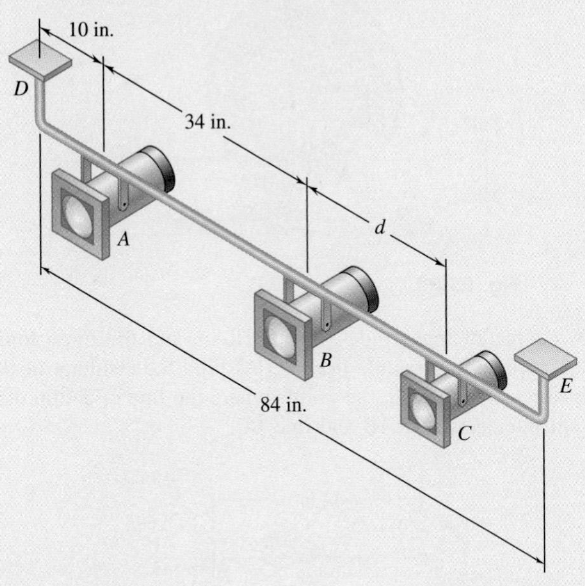

Fig. *P3.77*

3.78 Five separate force-couple systems act at the corners of a piece of sheet metal, which has been bent into the shape shown. Determine which of these systems is equivalent to a force $\mathbf{F} = (10 \text{ lb})\mathbf{i}$ and a couple of moment $\mathbf{M} = (15 \text{ lb·ft})\mathbf{j} + (15 \text{ lb·ft})\mathbf{k}$ located at the origin.

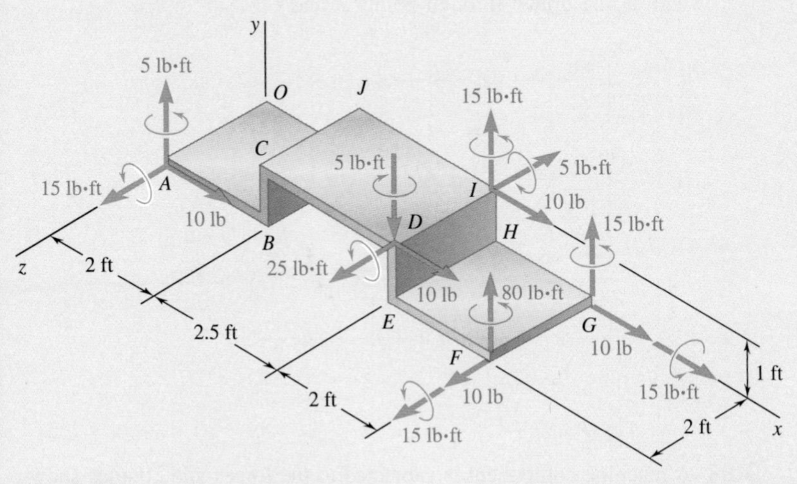

Fig. P3.78

3.79 Four forces act on a 700 × 375-mm plate as shown. (*a*) Find the resultant of these forces. (*b*) Locate the two points where the line of action of the resultant intersects the edge of the plate.

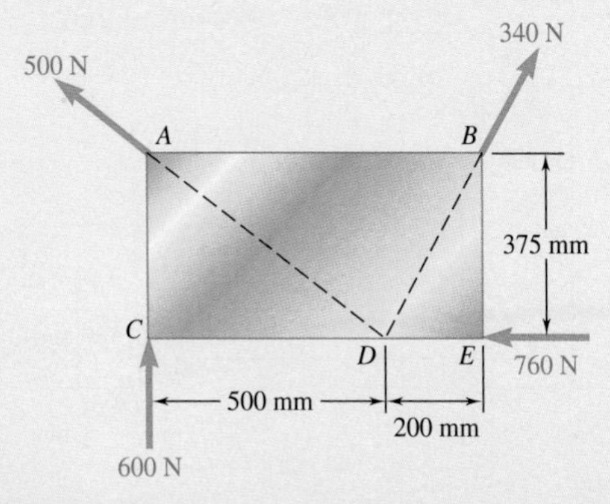

Fig. *P3.79*

3.80 A 32-lb motor is mounted on the floor. Find the resultant of the weight and the forces exerted on the belt, and determine where the line of action of the resultant intersects the floor.

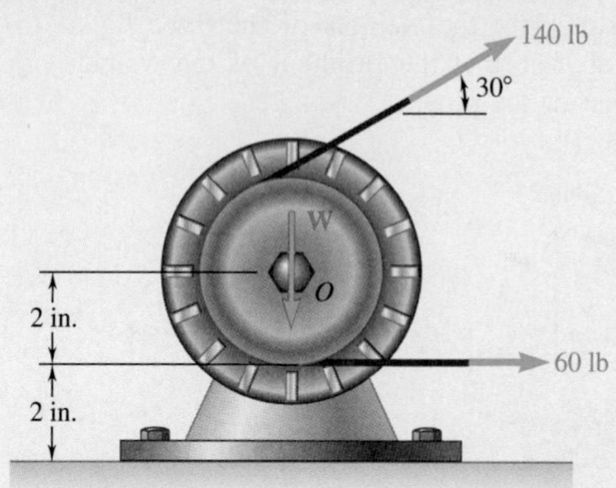

Fig. P3.80

3.81 A couple of magnitude $M = 54$ lb·in. and the three forces shown are applied to an angle bracket. (*a*) Find the resultant of this system of forces. (*b*) Locate the points where the line of action of the resultant intersects line *AB* and line *BC*.

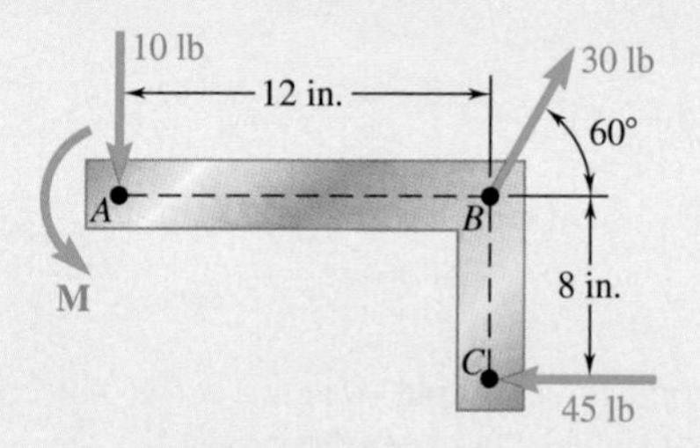

Fig. P3.81

3.82 A truss supports the loading shown. Determine the equivalent force acting on the truss and the point of intersection of its line of action with a line drawn through points *A* and *G*.

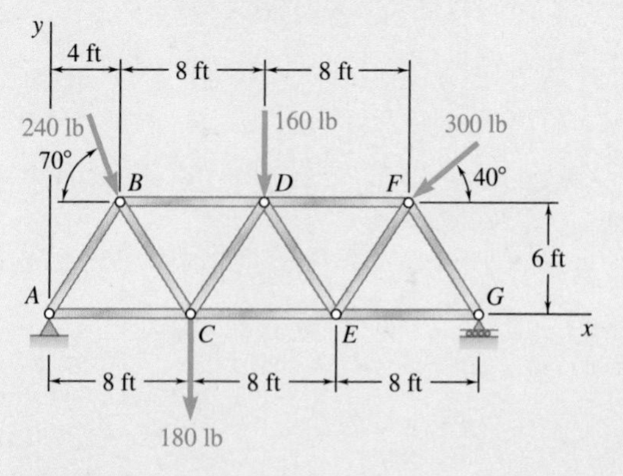

Fig. P3.82

3.83 A machine component is subjected to the forces and couples shown. The component is to be held in place by a single rivet that can resist a force but not a couple. For $P = 0$, determine the location of the rivet hole if it is to be located (*a*) on line *FG*, (*b*) on line *GH*.

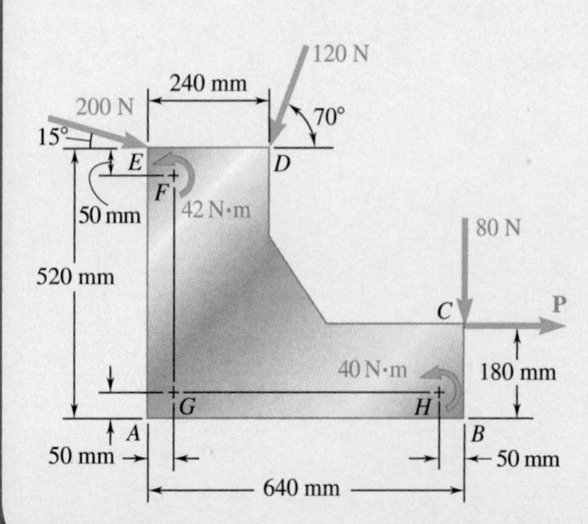

Fig. P3.83

3.84 Solve Prob. 3.83, assuming that $P = 60$ N.

3.85 As an adjustable brace BC is used to bring a wall into plumb, the force-couple system shown is exerted on the wall. Replace this force-couple system with an equivalent force-couple system at A if R = 21.2 lb and M = 13.25 lb·ft.

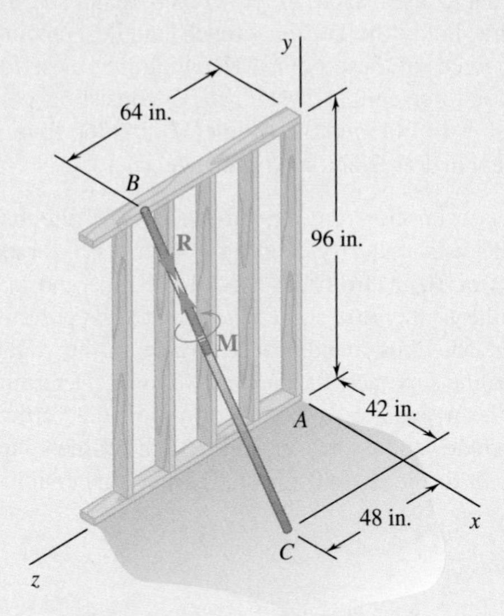

Fig. P3.85

3.86 As plastic bushings are inserted into a 60-mm-diameter cylindrical sheet metal enclosure, the insertion tools exert the forces shown on the enclosure. Each of the forces is parallel to one of the coordinate axes. Replace these forces with an equivalent force-couple system at C.

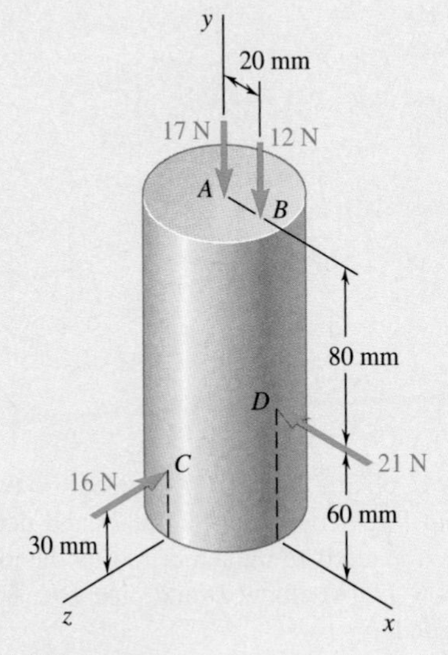

Fig. ***P3.86***

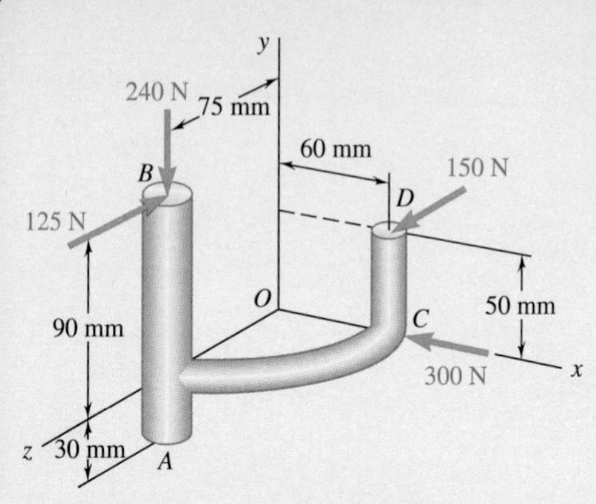

Fig. P3.87

3.87 A machine component is subjected to the forces shown, each of which is parallel to one of the coordinate axes. Replace these forces with an equivalent force-couple system at A.

3.88 A mechanic uses a crowfoot wrench to loosen a bolt at C. The mechanic holds the socket wrench handle at points A and B and applies forces at these points. Knowing that these forces are equivalent to a force-couple system at C consisting of the force $\mathbf{C} = -(8 \text{ lb})\mathbf{i} + (4 \text{ lb})\mathbf{k}$ and the couple $\mathbf{M}_C = (360 \text{ lb·in.})\mathbf{i}$, determine the forces applied at A and at B when $A_z = 2$ lb.

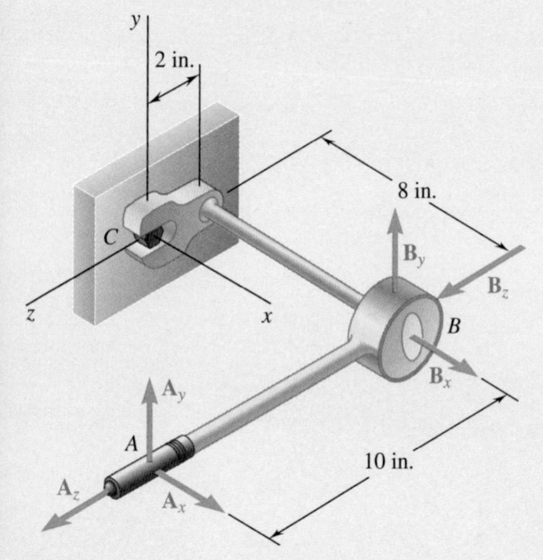

Fig. *P3.88*

3.89 In order to unscrew the tapped faucet A, a plumber uses two pipe wrenches as shown. By exerting a 40-lb force on each wrench, at a distance of 10 in. from the axis of the pipe and in a direction perpendicular to the pipe and to the wrench, he prevents the pipe from rotating, and thus avoids loosening or further tightening the joint between the pipe and the tapped elbow C. Determine (*a*) the angle θ that the wrench at A should form with the vertical if elbow C is not to rotate about the vertical, (*b*) the force-couple system at C equivalent to the two 40-lb forces when this condition is satisfied.

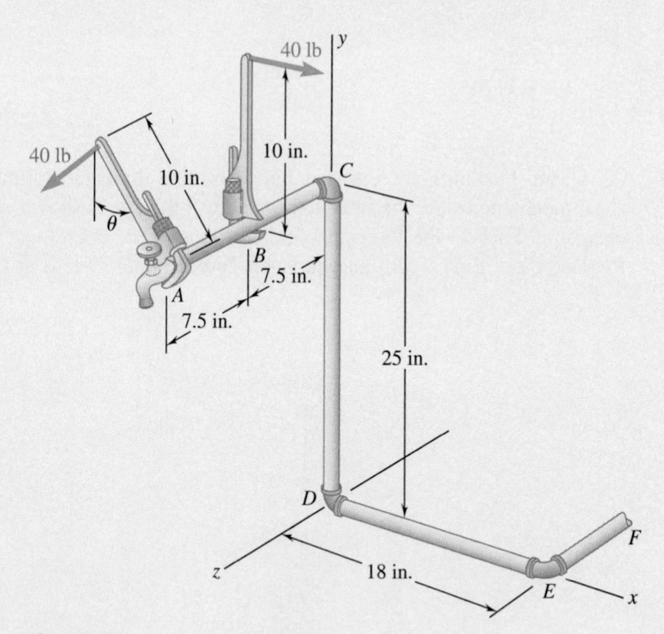

Fig. P3.89

3.90 Assuming $\theta = 60°$ in Prob. 3.89, replace the two 40-lb forces with an equivalent force-couple system at D and determine whether the plumber's action tends to tighten or loosen the joint between (*a*) pipe CD and elbow D, (*b*) elbow D and pipe DE. Assume all threads to be right-handed.

3.91 A blade held in a brace is used to tighten a screw at A. (a) Determine the forces exerted at B and C, knowing that these forces are equivalent to a force-couple system at A consisting of $\mathbf{R} = -(30 \text{ N})\mathbf{i} + R_y\mathbf{j} + R_z\mathbf{k}$ and $\mathbf{M} = -(12 \text{ N} \cdot \text{m})\mathbf{i}$. ($b$) Find the corresponding values of R_y and R_z. (c) What is the orientation of the slot in the head of the screw for which the blade is least likely to slip when the brace is in the position shown?

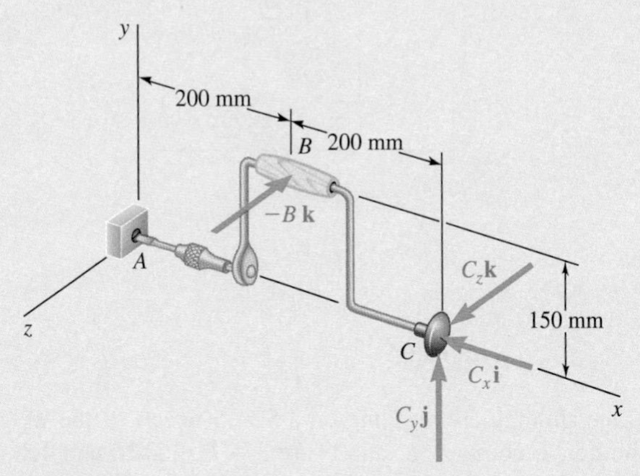

Fig. P3.91

3.92 Four signs are mounted on a frame spanning a highway, and the magnitudes of the horizontal wind forces acting on the signs are as shown. Determine the magnitude and the point of application of the resultant of the four wind forces when $a = 1$ ft and $b = 12$ ft.

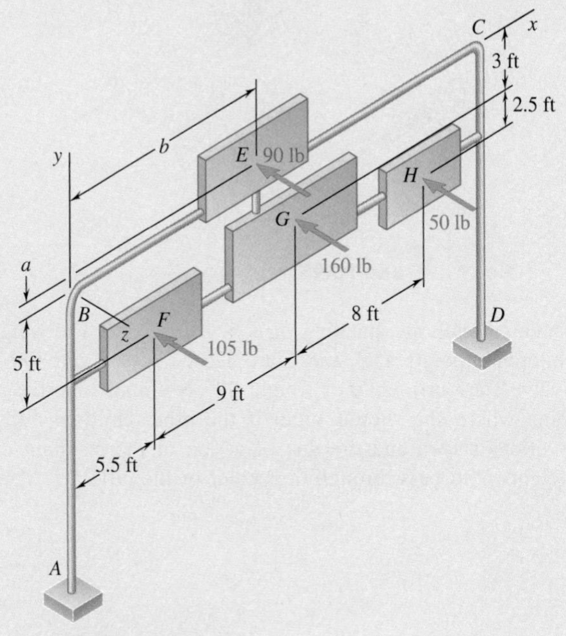

Fig. P3.92 and ***P3.93***

3.93 Four signs are mounted on a frame spanning a highway, and the magnitudes of the horizontal wind forces acting on the signs are as shown. Determine a and b so that the point of application of the resultant of the four forces is at G.

3.94 A concrete foundation mat of 5-m radius supports four equally spaced columns, each of which is located 4 m from the center of the mat. Determine the magnitude and the point of application of the resultant of the four loads.

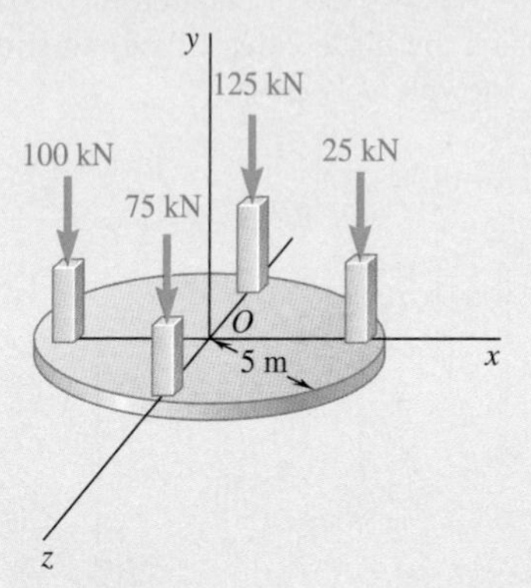

Fig. P3.94

3.95 Three children are standing on a 5 × 5-m raft. If the weights of the children at points *A*, *B*, and *C* are 375 N, 260 N, and 400 N, respectively, determine the magnitude and the point of application of the resultant of the three weights.

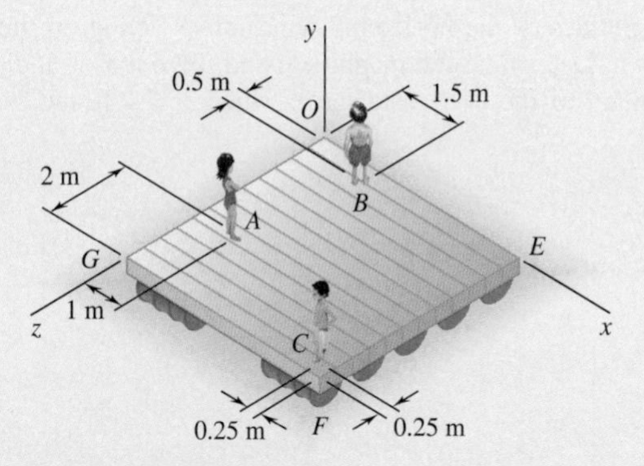

Fig. P3.95 and P3.96

3.96 Three children are standing on a 5 × 5-m raft. The weights of the children at points *A*, *B*, and *C* are 375 N, 260 N, and 400 N, respectively. If a fourth child of weight 425 N climbs onto the raft, determine where she should stand if the other children remain in the positions shown and the line of action of the resultant of the four weights is to pass through the center of the raft.

Review and Summary

Principle of Transmissibility

In this chapter, we presented the effects of forces exerted on a rigid body. We began by distinguishing between **external** and **internal** forces [Sec. 3.1A]. We then explained that, according to the **principle of transmissibility**, the effect of an external force on a rigid body remains unchanged if we move that force along its line of action [Sec. 3.1B]. In other words, two forces **F** and **F′** acting on a rigid body at two different points have the same effect on that body if they have the same magnitude, same direction, and same line of action (Fig. 3.41). Two such forces are said to be **equivalent**.

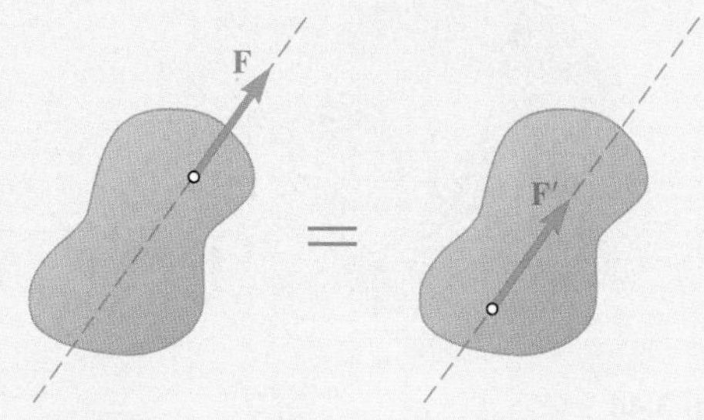

Fig. 3.41

Vector Product

Before proceeding with the discussion of **equivalent systems of forces**, we introduced the concept of the **vector product of two vectors** [Sec. 3.1C]. We defined the vector product

$$\mathbf{V} = \mathbf{P} \times \mathbf{Q}$$

of the vectors **P** and **Q** as a vector perpendicular to the plane containing **P** and **Q** (Fig. 3.42) with a magnitude of

$$V = PQ \sin \theta \qquad (3.1)$$

and directed in such a way that a person located at the tip of **V** will observe the rotation to be counterclockwise through θ, bringing the vector **P** in line with the vector **Q**. The three vectors **P**, **Q**, and **V**—taken in that order—are said to form a *right-handed triad*. It follows that the vector products **Q** × **P** and **P** × **Q** are represented by equal and opposite vectors:

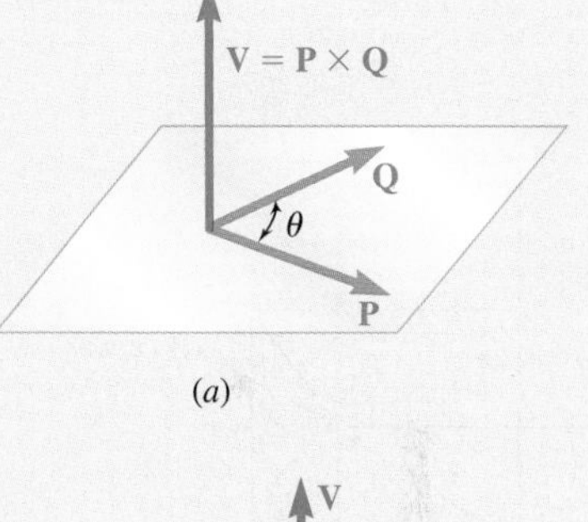

(*a*)

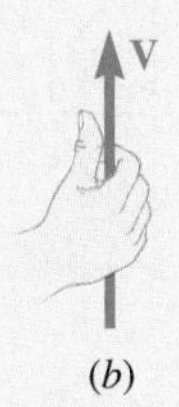

(*b*)

Fig. 3.42

$$\mathbf{Q} \times \mathbf{P} = -(\mathbf{P} \times \mathbf{Q}) \qquad (3.4)$$

It also follows from the definition of the vector product of two vectors that the vector products of the unit vectors **i**, **j**, and **k** are

$$\mathbf{i} \times \mathbf{i} = 0 \qquad \mathbf{i} \times \mathbf{j} = \mathbf{k} \qquad \mathbf{j} \times \mathbf{i} = -\mathbf{k}$$

and so on. You can determine the sign of the vector product of two unit vectors by arranging in a circle and in counterclockwise order the three letters representing the unit vectors (Fig. 3.43): The vector product of two unit vectors is positive if they follow each other in counterclockwise order and negative if they follow each other in clockwise order.

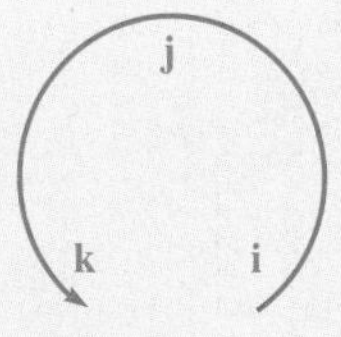

Fig. 3.43

Rectangular Components of Vector Product

The **rectangular components of the vector product V** of two vectors **P** and **Q** are expressed [Sec. 3.1D] as

$$\begin{aligned} V_x &= P_yQ_z - P_zQ_y \\ V_y &= P_zQ_x - P_xQ_z \qquad (3.9) \\ V_z &= P_xQ_y - P_yQ_x \end{aligned}$$

We can also express the components of a vector product as a determinant:

$$\mathbf{V} = \begin{vmatrix} \mathbf{i} & \mathbf{j} & \mathbf{k} \\ P_x & P_y & P_z \\ Q_x & Q_y & Q_z \end{vmatrix} \tag{3.10}$$

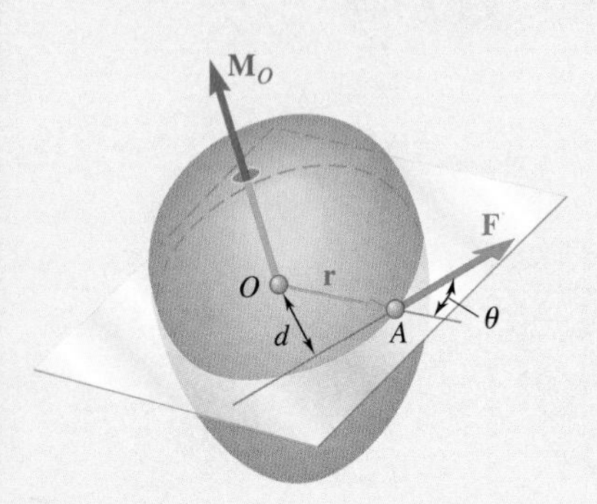

Fig. 3.44

Moment of a Force about a Point

We defined the **moment of a force F about a point** O [Sec. 3.1E] as the vector product

$$\mathbf{M}_O = \mathbf{r} \times \mathbf{F} \tag{3.11}$$

where **r** is the *position vector* drawn from O to the point of application A of the force **F** (Fig. 3.44). Denoting the angle between the lines of action of **r** and **F** as θ, we found that the magnitude of the moment of **F** about O is

$$M_O = rF \sin \theta = Fd \tag{3.12}$$

where d represents the perpendicular distance from O to the line of action of **F**.

Rectangular Components of Moment

The **rectangular components of the moment $\mathbf{M}_O$ of a force F** [Sec. 3.1F] are

$$\begin{aligned} M_x &= yF_z - zF_y \\ M_y &= zF_x - xF_z \\ M_z &= xF_y - yF_x \end{aligned} \tag{3.18}$$

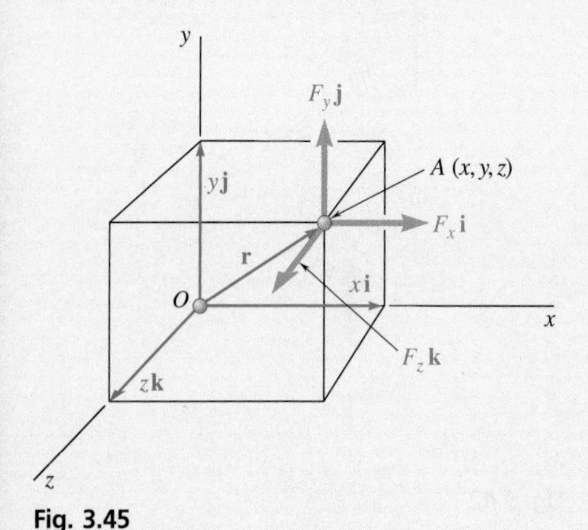

Fig. 3.45

where x, y, and z are the components of the position vector **r** (Fig. 3.45). Using a determinant form, we also wrote

$$\mathbf{M}_O = \begin{vmatrix} \mathbf{i} & \mathbf{j} & \mathbf{k} \\ x & y & z \\ F_x & F_y & F_z \end{vmatrix} \tag{3.19}$$

In the more general case of the moment about an arbitrary point B of a force **F** applied at A, we had

$$\mathbf{M}_B = \begin{vmatrix} \mathbf{i} & \mathbf{j} & \mathbf{k} \\ x_{A/B} & y_{A/B} & z_{A/B} \\ F_x & F_y & F_z \end{vmatrix} \tag{3.21}$$

where $x_{A/B}$, $y_{A/B}$, and $z_{A/B}$ denote the components of the vector $\mathbf{r}_{A/B}$:

$$x_{A/B} = x_A - x_B \qquad y_{A/B} = y_A - y_B \qquad z_{A/B} = z_A - z_B$$

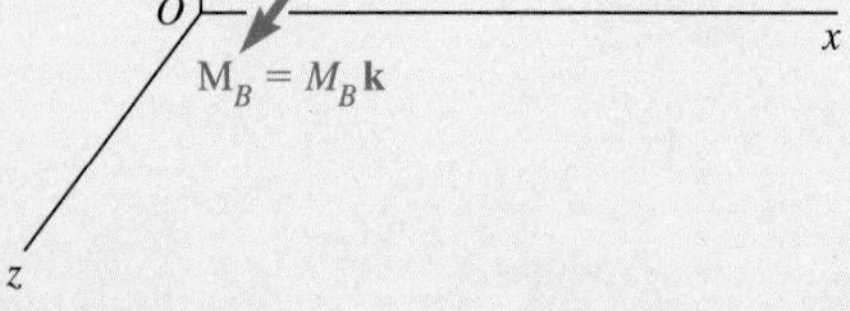

In the case of *problems involving only two dimensions,* we can assume the force **F** lies in the xy plane. Its moment $\mathbf{M}_B$ about a point B in the same plane is perpendicular to that plane (Fig. 3.46) and is completely defined by the scalar

$$M_B = (x_A - x_B)F_y - (y_A - y_B)F_x \tag{3.23}$$

Fig. 3.46

Various methods for computing the moment of a force about a point were illustrated in Sample Probs. 3.1 through 3.4.

Scalar Product of Two Vectors

The **scalar product** of two vectors **P** and **Q** [Sec. 3.2A], denoted by $\mathbf{P} \cdot \mathbf{Q}$, is defined as the scalar quantity

$$\mathbf{P} \cdot \mathbf{Q} = PQ \cos \theta \tag{3.24}$$

where θ is the angle between **P** and **Q** (Fig. 3.47). By expressing the scalar product of **P** and **Q** in terms of the rectangular components of the two vectors, we determined that

$$\mathbf{P} \cdot \mathbf{Q} = P_x Q_x + P_y Q_y + P_z Q_z \qquad (3.28)$$

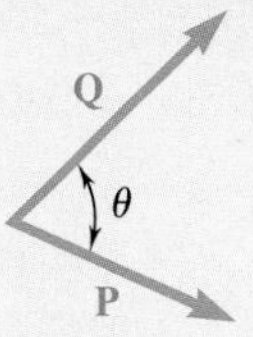

Fig. 3.47

Projection of a Vector on an Axis

We obtain the **projection of a vector P on an axis** *OL* (Fig. 3.48) by forming the scalar product of **P** and the unit vector **λ** along *OL*. We have

$$P_{OL} = \mathbf{P} \cdot \boldsymbol{\lambda} \qquad (3.34)$$

Using rectangular components, this becomes

$$P_{OL} = P_x \cos \theta_x + P_y \cos \theta_y + P_z \cos \theta_z \qquad (3.35)$$

where θ_x, θ_y, and θ_z denote the angles that the axis *OL* forms with the coordinate axes.

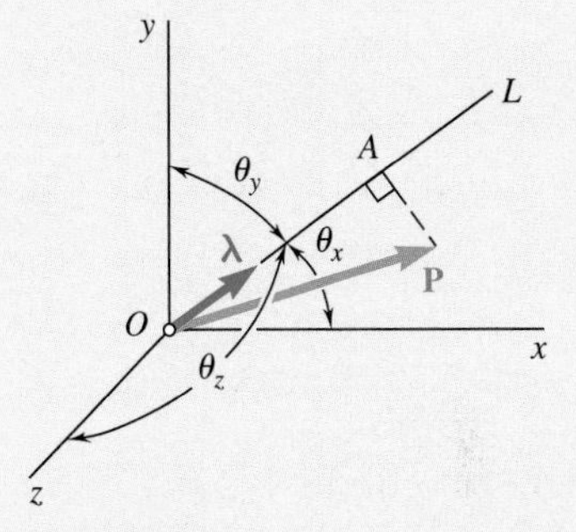

Fig. 3.48

Mixed Triple Product of Three Vectors

We defined the **mixed triple product** of the three vectors **S**, **P**, and **Q** as the scalar expression

$$\mathbf{S} \cdot (\mathbf{P} \times \mathbf{Q}) \qquad (3.36)$$

obtained by forming the scalar product of **S** with the vector product of **P** and **Q** [Sec. 3.2B]. We showed that

$$\mathbf{S} \cdot (\mathbf{P} \times \mathbf{Q}) = \begin{vmatrix} S_x & S_y & S_z \\ P_x & P_y & P_z \\ Q_x & Q_y & Q_z \end{vmatrix} \qquad (3.39)$$

where the elements of the determinant are the rectangular components of the three vectors.

Moment of a Force about an Axis

We defined the **moment of a force F about an axis** *OL* [Sec. 3.2C] as the projection *OC* on *OL* of the moment $\mathbf{M}_O$ of the force **F** (Fig. 3.49), i.e., as the mixed triple product of the unit vector **λ**, the position vector **r**, and the force **F**:

$$M_{OL} = \boldsymbol{\lambda} \cdot \mathbf{M}_O = \boldsymbol{\lambda} \cdot (\mathbf{r} \times \mathbf{F}) \qquad (3.40)$$

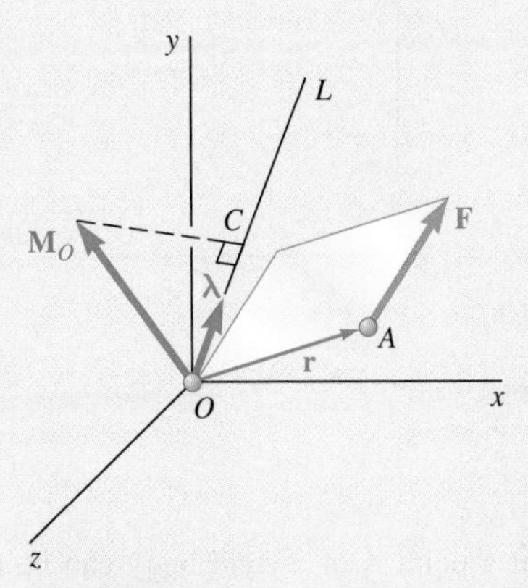

Fig. 3.49

The determinant form for the mixed triple product is

$$M_{OL} = \begin{vmatrix} \lambda_x & \lambda_y & \lambda_z \\ x & y & z \\ F_x & F_y & F_z \end{vmatrix} \tag{3.41}$$

where

$\lambda_x, \lambda_y, \lambda_z =$ direction cosines of axis OL
$x, y, z =$ components of $\mathbf{r}$
$F_x, F_y, F_z =$ components of $\mathbf{F}$

An example of determining the moment of a force about a skew axis appears in Sample Prob. 3.5.

Couples

Two forces **F** *and* **−F** *having the same magnitude, parallel lines of action, and opposite sense are said to form a* **couple** [Sec. 3.3A]. The moment of a couple is independent of the point about which it is computed; it is a vector **M** perpendicular to the plane of the couple and equal in magnitude to the product of the common magnitude F of the forces and the perpendicular distance d between their lines of action (Fig. 3.50).

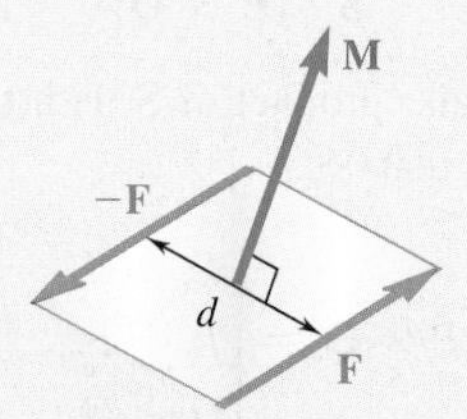

Fig. 3.50

Two couples having the same moment **M** are *equivalent,* i.e., they have the same effect on a given rigid body [Sec. 3.3B]. The sum of two couples is itself a couple [Sec. 3.3C], and we can obtain the moment **M** of the resultant couple by adding vectorially the moments $\mathbf{M}_1$ and $\mathbf{M}_2$ of the original couples [Sample Prob. 3.6]. It follows that we can represent a couple by a vector, called a **couple vector**, equal in magnitude and direction to the moment **M** of the couple [Sec. 3.3D]. A couple vector is a *free vector* that can be attached to the origin O if so desired and resolved into components (Fig. 3.51).

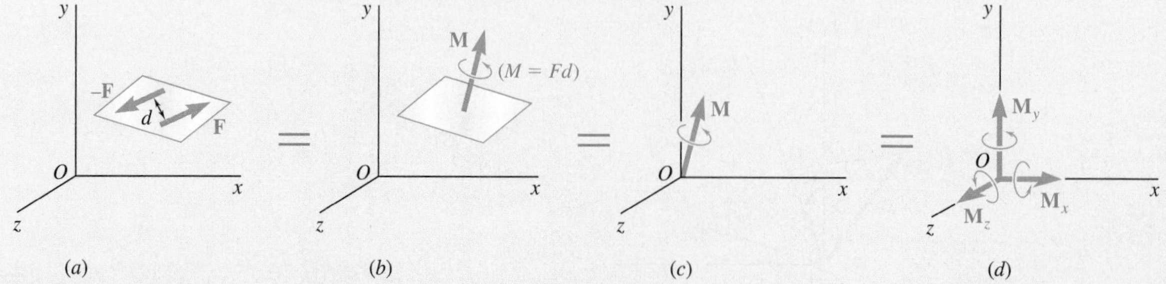

(a) (b) (c) (d)

Fig. 3.51

Force-Couple System

Any force **F** acting at a point A of a rigid body can be replaced by a **force-couple system** at an arbitrary point O consisting of the force **F** applied at O

and a couple of moment $\mathbf{M}_O$, which is equal to the moment about O of the force $\mathbf{F}$ in its original position [Sec. 3.3E]. Note that the force $\mathbf{F}$ and the couple vector $\mathbf{M}_O$ are always perpendicular to each other (Fig. 3.52).

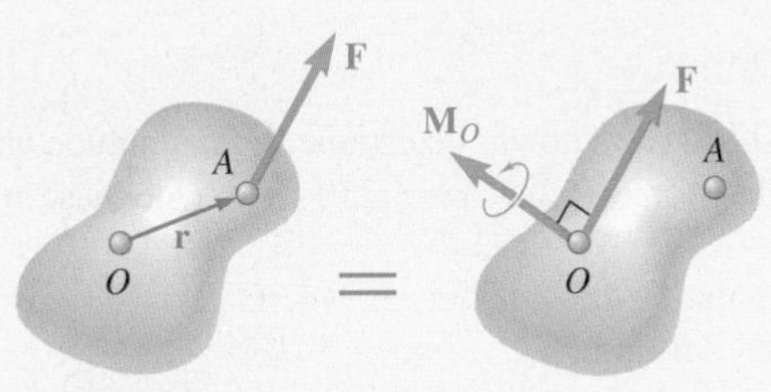

Fig. 3.52

Reduction of a System of Forces to a Force-Couple System

It follows [Sec. 3.4A] that *any system of forces can be reduced to a force-couple system at a given point O* by first replacing each of the forces of the system by an equivalent force-couple system at O (Fig. 3.53) and then adding all of the forces and all of the couples to obtain a resultant force $\mathbf{R}$ and a resultant couple vector $\mathbf{M}_O^R$ [Sample Probs. 3.8 through 3.11]. In general, the resultant $\mathbf{R}$ and the couple vector $\mathbf{M}_O^R$ will not be perpendicular to each other.

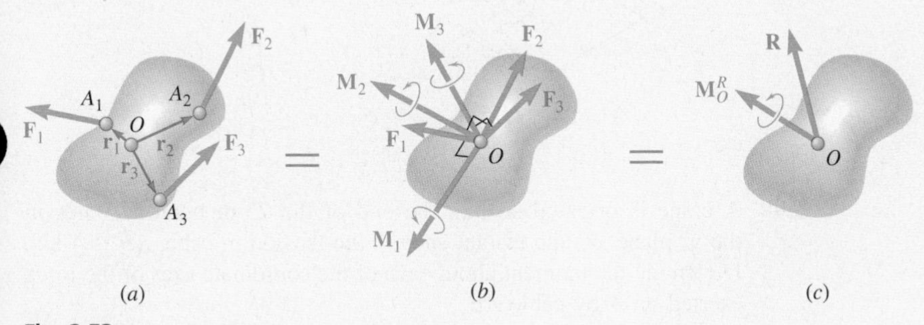

Fig. 3.53

Equivalent Systems of Forces

We concluded [Sec. 3.4B] that, as far as rigid bodies are concerned, *two systems of forces,* $\mathbf{F}_1$, $\mathbf{F}_2$, $\mathbf{F}_3$, . . . *and* $\mathbf{F}'_1$, $\mathbf{F}'_2$, $\mathbf{F}'_3$, . . . , *are equivalent if, and only if,*

$$\Sigma\mathbf{F} = \Sigma\mathbf{F}' \qquad \text{and} \qquad \Sigma\mathbf{M}_O = \Sigma\mathbf{M}'_O \tag{3.55}$$

Further Reduction of a System of Forces

If the resultant force $\mathbf{R}$ and the resultant couple vector $\mathbf{M}_O^R$ are perpendicular to each other, we can further reduce the force-couple system at O to a single resultant force [Sec. 3.4C]. This is the case for systems consisting of (*a*) concurrent forces (cf. Chap. 2), (*b*) coplanar forces [Sample Probs. 3.8 and 3.9], or (*c*) parallel forces [Sample Prob. 3.11]. If the resultant $\mathbf{R}$ and the couple vector $\mathbf{M}_O^R$ are *not* perpendicular to each other, the system *cannot* be reduced to a single force.

Review Problems

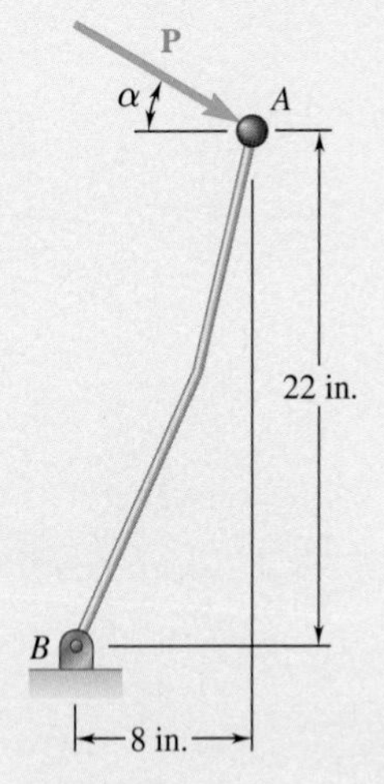

Fig. P3.97

3.97 For the shift lever shown, determine the magnitude and the direction of the smallest force **P** that has a 210-lb·in. clockwise moment about B.

3.98 Consider the volleyball net shown. Determine the angle formed by guy wires AB and AC.

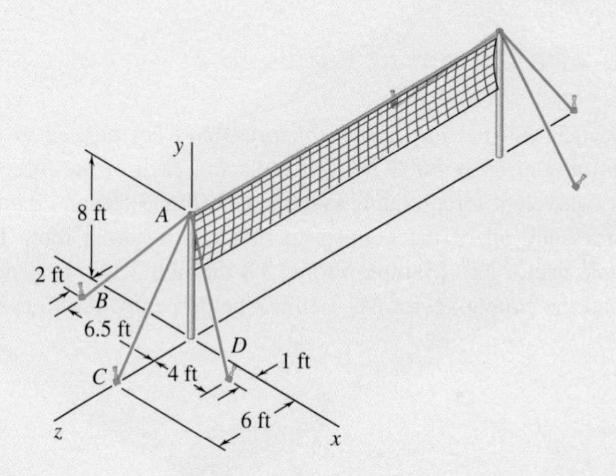

Fig. *P3.98*

3.99 A crane is oriented so that the end of the 25-m boom AO lies in the yz plane. At the instant shown, the tension in cable AB is 4 kN. Determine the moment about each of the coordinate axes of the force exerted on A by cable AB.

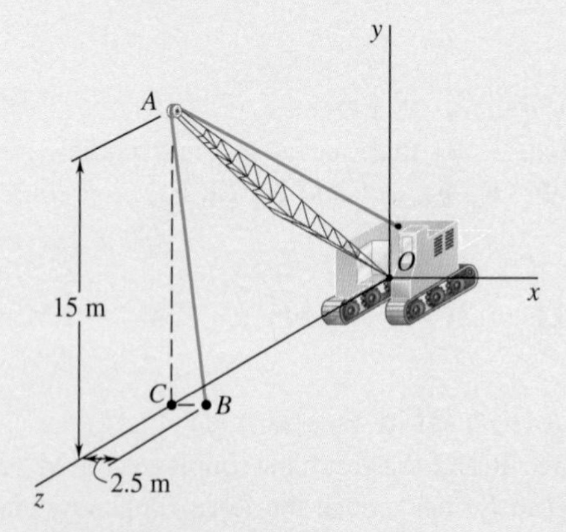

Fig. P3.99 and *P3.100*

3.100 The 25-m crane boom AO lies in the yz plane. Determine the maximum permissible tension in cable AB if the absolute value of moments about the coordinate axes of the force exerted on A by cable AB must be as follows:

$$|M_x| \leq 60 \text{ kN}\cdot\text{m},\ |M_y| \leq 12 \text{ kN}\cdot\text{m},\ |M_z| \leq 8 \text{ kN}\cdot\text{m}.$$

3.101 A single force **P** acts at C in a direction perpendicular to the handle BC of the crank shown. Determine the moment M_x of **P** about the x axis when $\theta = 65°$, knowing that $M_y = -15$ N·m and $M_z = -36$ N·m.

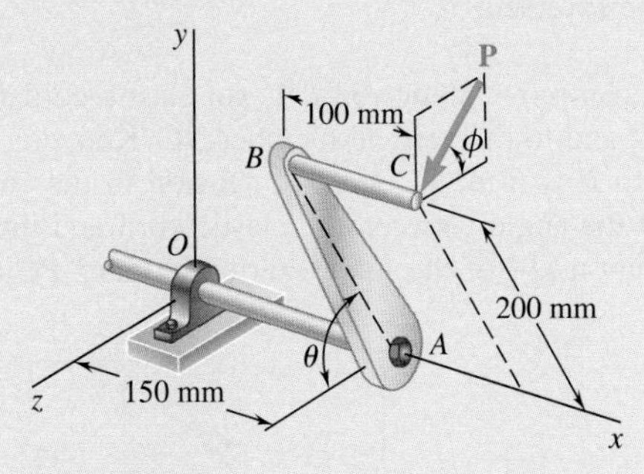

Fig. P3.101

3.102 While tapping a hole, a machinist applies the horizontal forces shown to the handle of the tap wrench. Show that these forces are equivalent to a single force, and specify, if possible, the point of application of the single force on the handle.

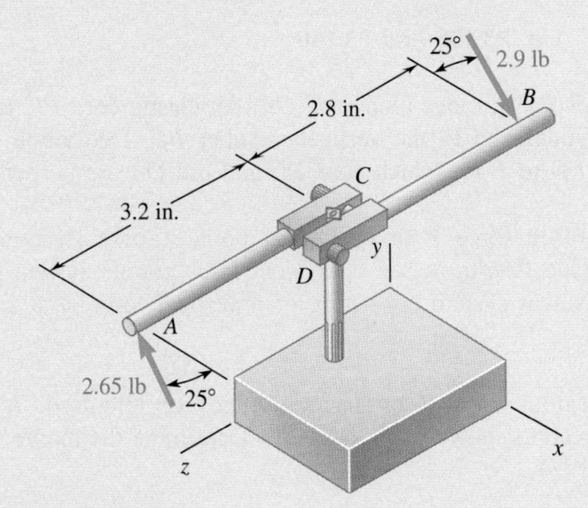

Fig. P3.102

3.103 A 500-N force is applied to a bent plate as shown. Determine (*a*) an equivalent force-couple system at B, (*b*) an equivalent system formed by a vertical force at A and a force at B.

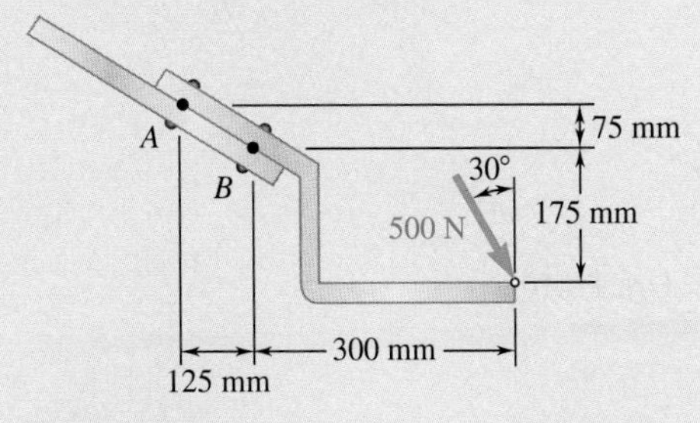

Fig. P3.103

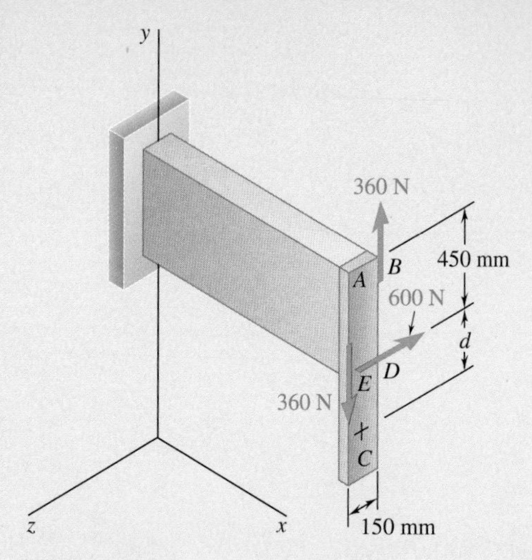

Fig. P3.104

3.104 A force and a couple are applied as shown to the end of a cantilever beam. (*a*) Replace this system with a single force **F** applied at point *C*, and determine the distance *d* from *C* to a line drawn through points *D* and *E*. (*b*) Solve part *a* if the directions of the two 360-N forces are reversed.

3.105 Slider *P* can move along rod *OA*. An elastic cord *PC* is attached to the slider and to the vertical member *BC*. Knowing that the distance from *O* to *P* is 6 in. and that the tension in the cord is 3 lb, determine (*a*) the angle between the elastic cord and the rod *OA*, (*b*) the projection on *OA* of the force exerted by cord *PC* at point *P*.

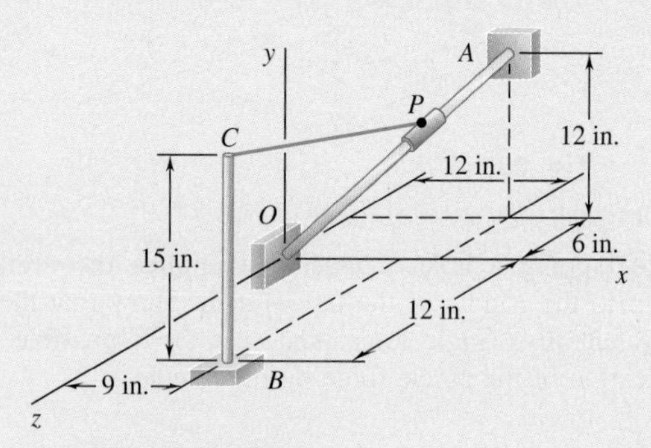

Fig. P3.105 and P3.106

3.106 Slider *P* can move along rod *OA*. An elastic cord *PC* is attached to the slider and to the vertical member *BC*. Determine the distance from *O* to *P* for which cord *PC* and rod *OA* are perpendicular.

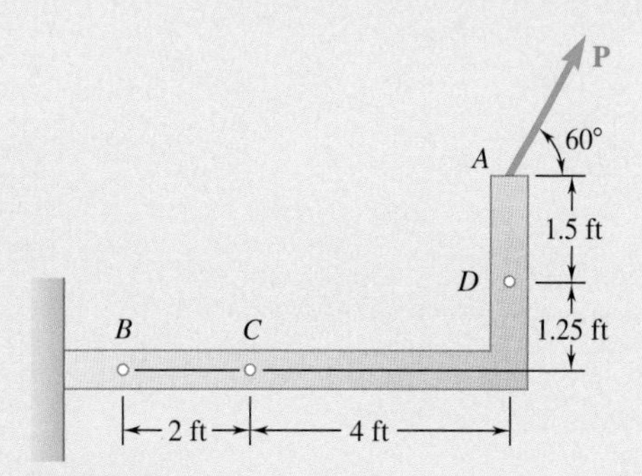

Fig. P3.107

3.107 A 160-lb force **P** is applied at point *A* of a structural member. Replace **P** with (*a*) an equivalent force-couple system at *C*, (*b*) an equivalent system consisting of a vertical force at *B* and a second force at *D*.

3.108 A regular tetrahedron has six edges of length *a*. A force **P** is directed as shown along edge *BC*. Determine the moment of **P** about edge *OA*.

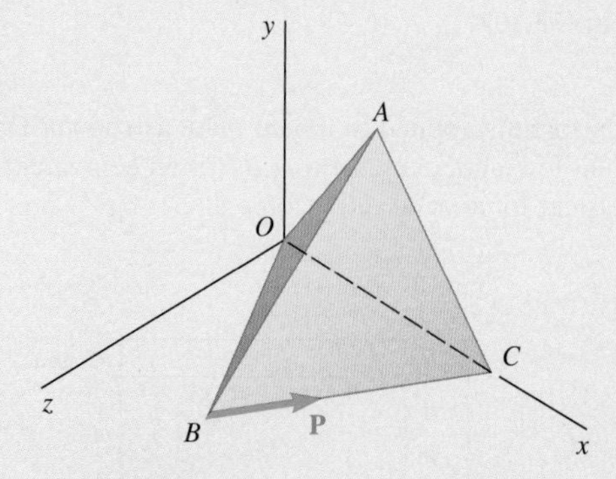

Fig. P3.108

© View Stock/Getty Images RF

4

Equilibrium of Rigid Bodies

The Tianjin Eye is a Ferris wheel that straddles a bridge over the Hai River in China. The structure is designed so that the support reactions at the wheel bearings as well as those at the base of the frame maintain equilibrium under the effects of vertical gravity and horizontal wind forces.

Objectives

- **Analyze** the static equilibrium of rigid bodies in two and three dimensions.
- **Consider** the attributes of a properly drawn free-body diagram, an essential tool for the equilibrium analysis of rigid bodies.
- **Examine** rigid bodies supported by statically indeterminate reactions and partial constraints.
- **Study** two cases of particular interest: the equilibrium of two-force and three-force bodies.
- **Examine** the laws of dry friction and use these to consider the equilibrium of rigid bodies where friction exists at contact surfaces.

Introduction

We saw in Chapter 3 how to reduce the external forces acting on a rigid body to a force-couple system at some arbitrary point O. When the force and the couple are both equal to zero, the external forces form a system equivalent to zero, and the rigid body is said to be in **equilibrium**.

We can obtain the necessary and sufficient conditions for the equilibrium of a rigid body by setting $\mathbf{R}$ and $\mathbf{M}_O^R$ equal to zero in the relations of Eq. (3.50) of Sec. 3.4A:

$$\Sigma \mathbf{F} = 0 \qquad \Sigma \mathbf{M}_O = \Sigma (\mathbf{r} \times \mathbf{F}) = 0 \tag{4.1}$$

Resolving each force and each moment into its rectangular components, we can replace these vector equations for the equilibrium of a rigid body with the following six scalar equations:

$$\Sigma F_x = 0 \qquad \Sigma F_y = 0 \qquad \Sigma F_z = 0 \tag{4.2}$$
$$\Sigma M_x = 0 \qquad \Sigma M_y = 0 \qquad \Sigma M_z = 0 \tag{4.3}$$

We can use these equations to determine unknown forces applied to the rigid body or unknown reactions exerted on it by its supports. Note that Eqs. (4.2) express the fact that the components of the external forces in the x, y, and z directions are balanced; Eqs. (4.3) express the fact that the moments of the external forces about the x, y, and z axes are balanced. Therefore, for a rigid body in equilibrium, the system of external forces imparts no translational or rotational motion to the body.

In order to write the equations of equilibrium for a rigid body, we must first identify all of the forces acting on that body and then draw the corresponding **free-body diagram**. In this chapter, we first consider the equilibrium of *two-dimensional structures* subjected to forces contained in their planes and study how to draw their free-body diagrams. In addition to the forces *applied* to a structure, we must also consider the *reactions* exerted on the structure by its supports. A specific reaction is associated with each type of support. You will see how to determine whether the structure is properly supported, so that you can know in advance whether you can solve the equations of equilibrium for the unknown forces and reactions.

Later in this chapter, we consider the equilibrium of three-dimensional structures, and we provide the same kind of analysis to these structures and their supports. This will be followed by a discussion of equilibrium of rigid bodies supported on surfaces in which friction acts to restrain motion of one surface with respect to the other.

Free-Body Diagrams

In solving a problem concerning a rigid body in equilibrium, it is essential to consider *all* of the forces acting on the body. It is equally important to exclude any force that is *not* directly applied to the body. Omitting a force or adding an extraneous one would destroy the conditions of equilibrium. Therefore, the first step in solving the problem is to draw a **free-body diagram** of the rigid body under consideration.

We have already used free-body diagrams on many occasions in Chap. 2. However, in view of their importance to the solution of equilibrium problems, we summarize here the steps you must follow in drawing a correct free-body diagram.

1. Start with a clear decision regarding the choice of the free body to be analyzed. Mentally, you need to detach this body from the ground and separate it from all other bodies. Then you can sketch the contour of this isolated body.

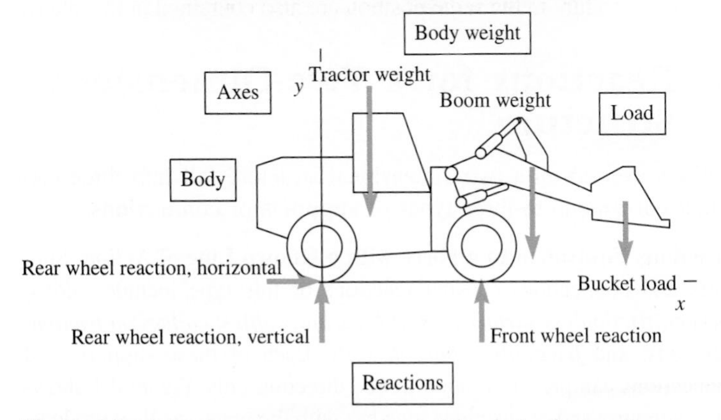

Photo 4.1 A tractor supporting a bucket load. As shown, its free-body diagram should include all external forces acting on the tractor.

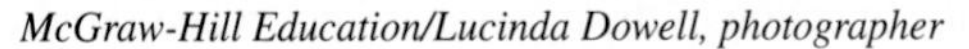
McGraw-Hill Education/Lucinda Dowell, photographer

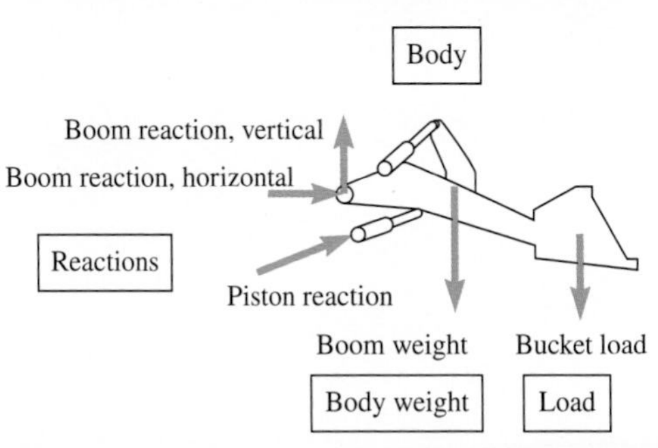

Photo 4.2 Tractor bucket and boom. In Chap. 6, we will see how to determine the internal forces associated with interconnected members such as these using free-body diagrams like the one shown.

© McGraw-Hill Education/Lucinda Dowell, photographer

2. Indicate all external forces on the free-body diagram. These forces represent the actions exerted *on* the free body *by* the ground and *by* the bodies that have been detached. In the diagram, apply these forces at the various points where the free body was supported by the ground or was connected to the other bodies. Generally, you should include the *weight* of the free body among the external forces, since it represents the attraction exerted by the earth on the various particles forming the free body. You will see in Chapter 5 that you should draw the weight so it acts at the center of gravity of the body. If the free body is made of several parts, do *not* include the forces the various parts exert on each other among the external forces. These forces are internal forces as far as the free body is concerned.
3. Clearly mark the magnitudes and directions of the *known external forces* on the free-body diagram. Recall that when indicating the directions of these forces, the forces are those exerted *on,* and not *by,* the free body. Known external forces generally include the *weight* of the free body and *forces applied* for a given purpose.
4. *Unknown external forces* usually consist of the **reactions** through which the ground and other bodies oppose a possible motion of the free body. The reactions constrain the free body to remain in the same position; for that reason, they are sometimes called *constraining forces*. Reactions are exerted at the points where the free body is *supported by* or *connected to* other bodies; you should clearly indicate these points. Reactions are discussed in detail in Secs. 4.1 and 4.3.
5. The free-body diagram should also include dimensions, since these may be needed for computing moments of forces. Any other detail, however, should be omitted.

4.1 EQUILIBRIUM IN TWO DIMENSIONS

In the first part of this chapter, we consider the equilibrium of two-dimensional structures; i.e., we assume that the structure being analyzed and the forces applied to it are contained in the same plane. Clearly, the reactions needed to maintain the structure in the same position are also contained in this plane.

4.1A Reactions for a Two-Dimensional Structure

The reactions exerted on a two-dimensional structure fall into three categories that correspond to three types of **supports** or **connections**.

1. **Reactions Equivalent to a Force with a Known Line of Action.** Supports and connections causing reactions of this type include *rollers, rockers, frictionless surfaces, short links and cables, collars on frictionless rods,* and *frictionless pins in slots*. Each of these supports and connections can prevent motion in one direction only. Figure 4.1 shows these supports and connections together with the reactions they produce. Each reaction involves *one unknown*—specifically, the magnitude of the reaction. In problem solving, you should denote this magnitude by an appropriate letter. The line of action of the reaction is known and should be indicated clearly in the free-body diagram.

Support or Connection	Reaction	Number of Unknowns
Rollers; Rocker; Frictionless surface	Force with known line of action perpendicular to surface	1
Short cable; Short link	Force with known line of action along cable or link	1
Collar on frictionless rod; Frictionless pin in slot	90°; Force with known line of action perpendicular to rod or slot	1
Frictionless pin or hinge; Rough surface	or; α; Force of unknown direction	2
Fixed support	or; α; Force and couple	3

Fig. 4.1 Reactions of supports and connections in two dimensions.

This rocker bearing supports the weight of a bridge. The convex surface of the rocker allows the bridge to move slightly horizontally.

Links are often used to support suspended spans of highway bridges.

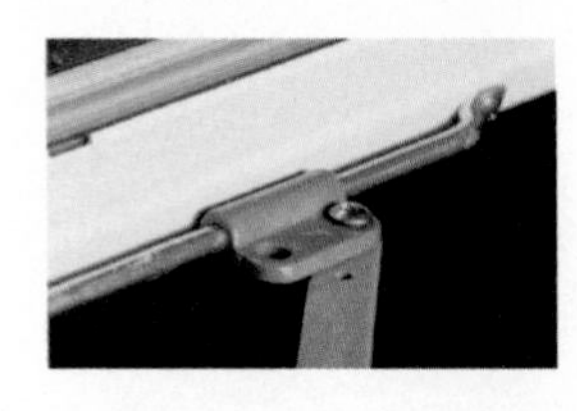

Force applied to the slider exerts a normal force on the rod, causing the window to open.

Pin supports are common on bridges and overpasses.

This cantilever support is fixed at one end and extends out into space at the other end.

The sense of the reaction must be as shown in Fig. 4.1 for cases of a frictionless surface (toward the free body) or a cable (away from the free body). The reaction can be directed either way in the cases of double-track rollers, links, collars on rods, or pins in slots. Generally, we assume that single-track rollers and rockers are reversible, so the corresponding reactions can be directed either way.

2. **Reactions Equivalent to a Force of Unknown Direction and Magnitude.** Supports and connections causing reactions of this type include

frictionless pins in fitted holes, hinges, and *rough surfaces*. They can prevent translation of the free body in all directions, but they cannot prevent the body from rotating about the connection. Reactions of this group involve *two unknowns* and are usually represented by their x and y components. In the case of a rough surface, the component normal to the surface must be directed away from the surface.

3. **Reactions Equivalent to a Force and a Couple.** These reactions are caused by *fixed supports* that oppose any motion of the free body and thus constrain it completely. Fixed supports actually produce forces over the entire surface of contact; these forces, however, form a system that can be reduced to a force and a couple. Reactions of this group involve *three unknowns* usually consisting of the two components of the force and the moment of the couple.

When the sense of an unknown force or couple is not readily apparent, do not attempt to determine it. Instead, arbitrarily assume the sense of the force or couple; the sign of the answer will indicate whether the assumption is correct or not. (A positive answer means the assumption is correct, while a negative answer means the assumption is incorrect.)

4.1B Rigid-Body Equilibrium in Two Dimensions

The conditions stated in Sec. 4.1A for the equilibrium of a rigid body become considerably simpler for the case of a two-dimensional structure. Choosing the x and y axes to be in the plane of the structure, we have

$$F_z = 0 \qquad M_x = M_y = 0 \qquad M_z = M_O$$

for each of the forces applied to the structure. Thus, the six equations of equilibrium stated in Sec. 4.1 reduce to three equations:

$$\Sigma F_x = 0 \qquad \Sigma F_y = 0 \qquad \Sigma M_O = 0 \qquad \textbf{(4.4)}$$

Since $\Sigma M_O = 0$ must be satisfied regardless of the choice of the origin O, we can write the equations of equilibrium for a two-dimensional structure in the more general form

Equations of equilibrium in two dimensions

$$\Sigma F_x = 0 \qquad \Sigma F_y = 0 \qquad \Sigma M_A = 0 \qquad \textbf{(4.5)}$$

where A is any point in the plane of the structure. These three equations can be solved for no more than *three unknowns*.

You have just seen that unknown forces include reactions and that the number of unknowns corresponding to a given reaction depends upon the type of support or connection causing that reaction. Referring to Fig. 4.1, note that you can use the equilibrium equations (4.5) to determine the reactions associated with two rollers and one cable, or one fixed support, or one roller and one pin in a fitted hole, etc.

For example, consider Fig. 4.2*a*, in which the truss shown is in equilibrium and is subjected to the given forces **P**, **Q**, and **S**. The truss is held in place by a pin at A and a roller at B. The pin prevents point A from moving by exerting a force on the truss that can be resolved into the components $\mathbf{A}_x$ and $\mathbf{A}_y$. The roller keeps the truss from rotating about A by exerting the vertical force **B**. The free-body diagram of the truss is shown in Fig. 4.2*b*; it includes the reactions $\mathbf{A}_x$, $\mathbf{A}_y$, and **B** as well as the applied forces **P**, **Q**, and **S** (in x and y component form) and the weight **W** of the truss.

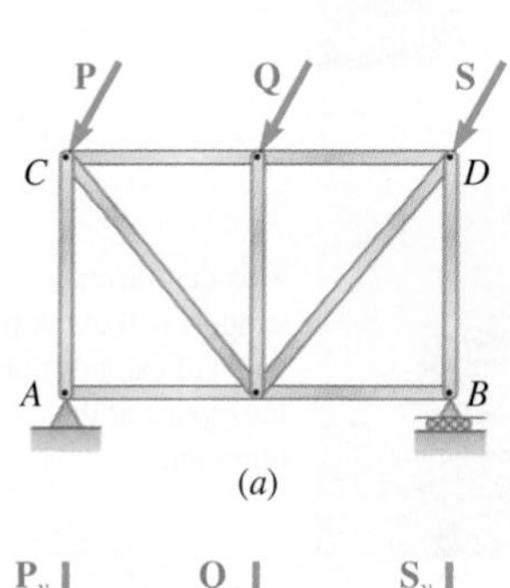

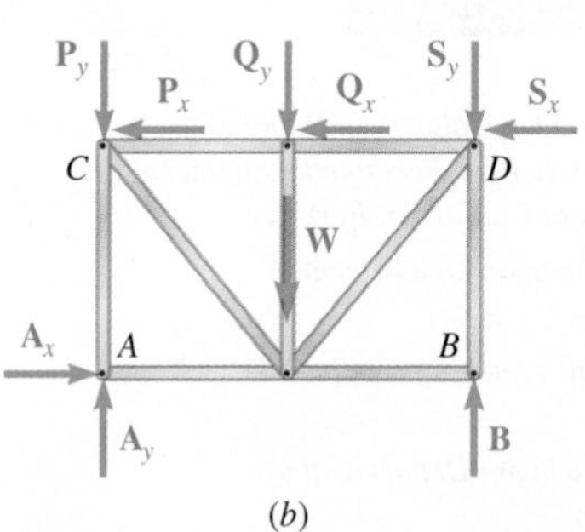

Fig. 4.2 (*a*) A truss supported by a pin and a roller; (*b*) free-body diagram of the truss.

Since the truss is in equilibrium, the sum of the moments about A of all of the forces shown in Fig. 4.2*b* is zero, or $\Sigma M_A = 0$. We can use this equation to determine the magnitude B because the equation does not contain A_x or A_y. Then, since the sum of the x components and the sum of the y components of the forces are zero, we write the equations $\Sigma F_x = 0$ and $\Sigma F_y = 0$. From these equations, we can obtain the components A_x and A_y, respectively.

We could obtain an additional equation by noting that the sum of the moments of the external forces about a point other than A is zero. We could write, for instance, $\Sigma M_B = 0$. This equation, however, does not contain any new information, because we have already established that the system of forces shown in Fig. 4.2*b* is equivalent to zero. The additional equation *is not independent* and cannot be used to determine a fourth unknown. It can be useful, however, for checking the solution obtained from the original three equations of equilibrium.

Although the three equations of equilibrium cannot be *augmented* by additional equations, any of them can be *replaced* by another equation. Properly chosen, the new system of equations still describes the equilibrium conditions but may be easier to work with. For example, an alternative system of equations for equilibrium is

$$\Sigma F_x = 0 \qquad \Sigma M_A = 0 \qquad \Sigma M_B = 0 \tag{4.6}$$

Here the second point about which the moments are summed (in this case, point B) cannot lie on the line parallel to the y axis that passes through point A (Fig. 4.2*b*). These equations are sufficient conditions for the equilibrium of the truss. The first two equations indicate that the external forces must reduce to a single vertical force at A. Since the third equation requires that the moment of this force be zero about a point B that is not on its line of action, the force must be zero, and the rigid body is in equilibrium.

A third possible set of equilibrium equations is

$$\Sigma M_A = 0 \qquad \Sigma M_B = 0 \qquad \Sigma M_C = 0 \tag{4.7}$$

where the points A, B, and C do not lie in a straight line (Fig. 4.2*b*). The first equation requires that the external forces reduce to a single force at A; the second equation requires that this force pass through B; and the third equation requires that it pass through C. Since the points A, B, C do not lie in a straight line, the force must be zero, and the rigid body is in equilibrium.

Notice that the equation $\Sigma M_A = 0$, stating that the sum of the moments of the forces about pin A is zero, possesses a more definite physical meaning than either of the other two equations (4.7). These two equations express a similar idea of balance but with respect to points about which the rigid body is not actually hinged. They are, however, as useful as the first equation. The choice of equilibrium equations should not be unduly influenced by their physical meaning. Indeed, in practice, it is desirable to choose equations of equilibrium containing only one unknown, since this eliminates the necessity of solving simultaneous equations. You can obtain equations containing only one unknown by summing moments about the point of intersection of the lines of action of two unknown forces or, if these forces are parallel, by summing force components in a direction perpendicular to their common direction.

For example, in Fig. 4.3, in which the truss shown is held by rollers at A and B and a short link at D, we can eliminate the reactions at A and B

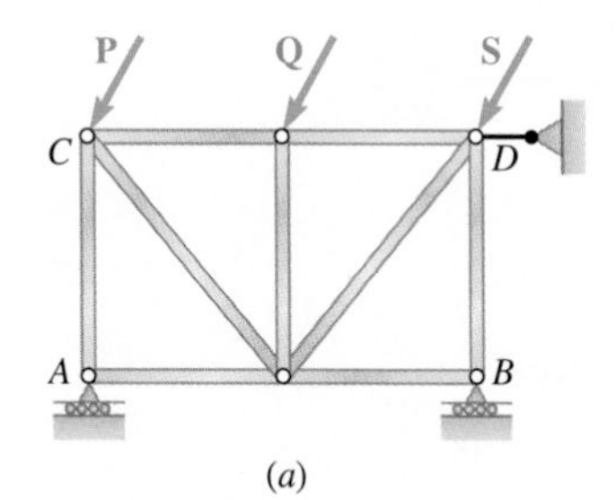

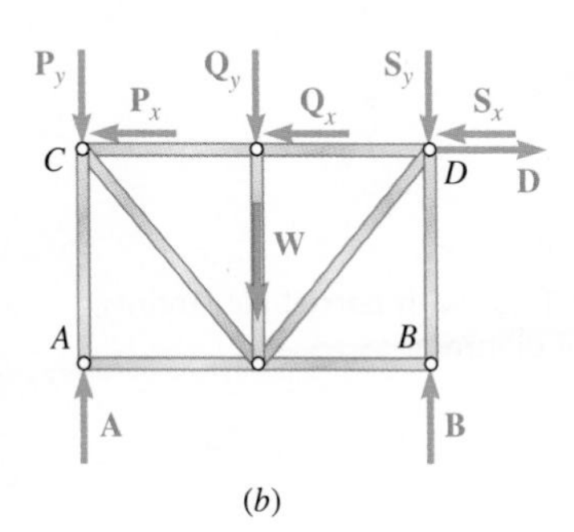

Fig. 4.3 (*a*) A truss supported by two rollers and a short link; (*b*) free-body diagram of the truss.

by summing x components. We can eliminate the reactions at A and D by summing moments about C and the reactions at B and D by summing moments about D. The resulting equations are

$$\Sigma F_x = 0 \qquad \Sigma M_C = 0 \qquad \Sigma M_D = 0$$

Each of these equations contains only one unknown.

4.1C Statically Indeterminate Reactions and Partial Constraints

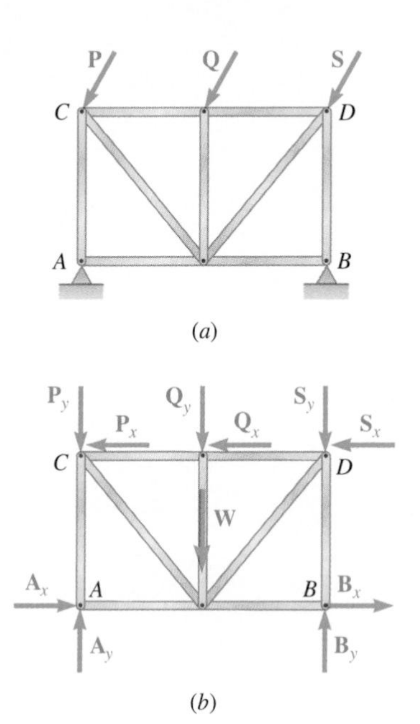

Fig. 4.4 (*a*) Truss with statically indeterminate reactions; (*b*) free-body diagram.

In the two examples considered in Figs. 4.2 and 4.3, the types of supports used were such that the rigid body could not possibly move under the given loads or under any other loading conditions. In such cases, the rigid body is said to be **completely constrained**. Recall that the reactions corresponding to these supports involved *three unknowns* and could be determined by solving the three equations of equilibrium. When such a situation exists, the reactions are said to be **statically determinate**.

Consider Fig. 4.4*a*, in which the truss shown is held by pins at A and B. These supports provide more constraints than are necessary to keep the truss from moving under the given loads or under any other loading conditions. Note from the free-body diagram of Fig. 4.4*b* that the corresponding reactions involve *four unknowns*. We pointed out in Sec. 4.1D that only three independent equilibrium equations are available; therefore, in this case, we have *more unknowns than equations*. As a result, we cannot determine all of the unknowns. The equations $\Sigma M_A = 0$ and $\Sigma M_B = 0$ yield the vertical components B_y and A_y, respectively, but the equation $\Sigma F_x = 0$ gives only the sum $A_x + B_x$ of the horizontal components of the reactions at A and B. The components A_x and B_x are **statically indeterminate**. We could determine their magnitudes by considering the deformations produced in the truss by the given loading, but this method is beyond the scope of statics and belongs to the study of mechanics of materials.

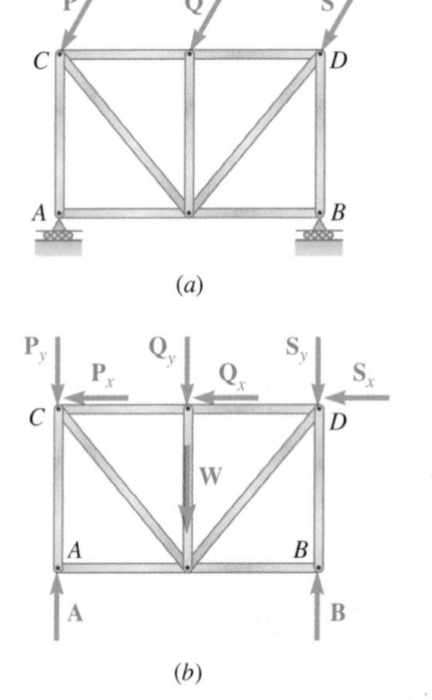

Fig. 4.5 (*a*) Truss with partial constraints; (*b*) free-body diagram.

Let's consider the opposite situation. The supports holding the truss shown in Fig. 4.5*a* consist of rollers at A and B. Clearly, the constraints provided by these supports are not sufficient to keep the truss from moving. Although they prevent any vertical motion, the truss is free to move horizontally. The truss is said to be **partially constrained.**† From the free-body diagram in Fig. 4.5*b*, note that the reactions at A and B involve only *two unknowns*. Since three equations of equilibrium must still be satisfied, we have *fewer unknowns than equations*. In such a case, one of the equilibrium equations will not be satisfied in general. The equations $\Sigma M_A = 0$ and $\Sigma M_B = 0$ can be satisfied by a proper choice of reactions at A and B, but the equation $\Sigma F_x = 0$ is not satisfied unless the sum of the horizontal components of the applied forces happens to be zero. We thus observe that the equilibrium of the truss of Fig. 4.5 cannot be maintained under general loading conditions.

From these examples, it would appear that, if a rigid body is to be completely constrained and if the reactions at its supports are to be statically determinate, **there must be as many unknowns as there are equations of equilibrium**. When this condition is *not* satisfied, we can be certain that either the rigid body is not completely constrained or that the reactions at its supports

†Partially constrained bodies are often referred to as *unstable*. However, to avoid confusion between this type of instability, due to insufficient constraints, and the type of instability considered in Chap. 16, which relates to the behavior of columns, we shall restrict the use of the words *stable* and *unstable* to the latter case.

are not statically determinate. It is also possible that the rigid body is not completely constrained *and* that the reactions are statically indeterminate.

You should note, however, that, although this condition is *necessary,* it is *not sufficient.* In other words, the fact that the number of unknowns is equal to the number of equations is no guarantee that a body is completely constrained or that the reactions at its supports are statically determinate. Consider Fig. 4.6*a*, which shows a truss held by rollers at *A, B,* and *E.* We have three unknown reactions of **A**, **B**, and **E** (Fig. 4.6*b*), but the equation $\Sigma F_x = 0$ is not satisfied unless the sum of the horizontal components of the applied forces happens to be zero. Although there are a sufficient number of constraints, these constraints are not properly arranged, so the truss is free to move horizontally. We say that the truss is **improperly constrained**. Since only two equilibrium equations are left for determining three unknowns, the reactions are statically indeterminate. Thus, improper constraints also produce static indeterminacy.

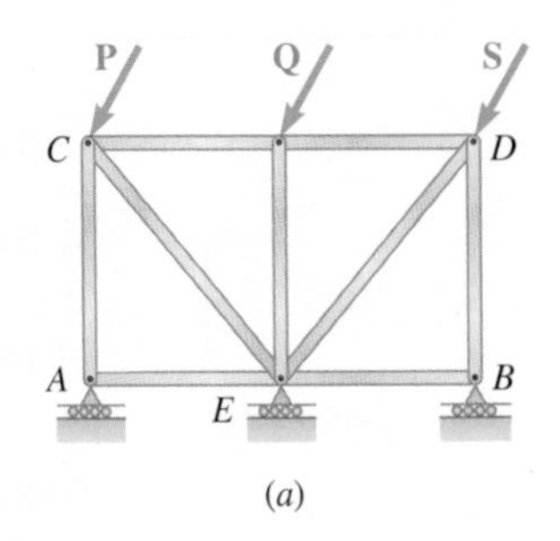

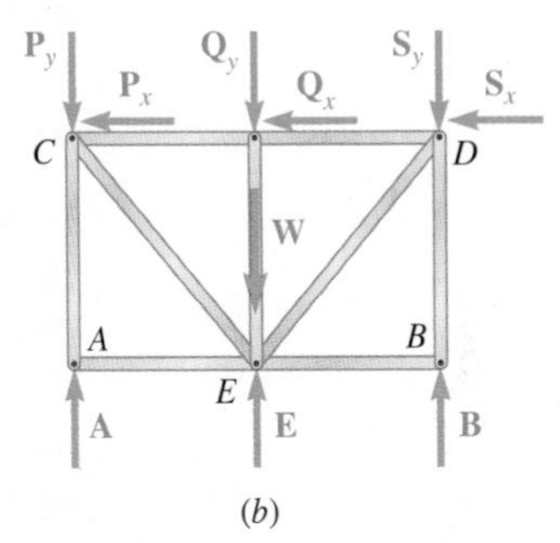

Fig. 4.6 *(a)* Truss with improper constraints; *(b)* free-body diagram.

The truss shown in Fig. 4.7 is another example of improper constraints—and of static indeterminacy. This truss is held by a pin at *A* and by rollers at *B* and *C*, which altogether involve four unknowns. Since only three independent equilibrium equations are available, the reactions at the supports are statically indeterminate. On the other hand, we note that the equation $\Sigma M_A = 0$ cannot be satisfied under general loading conditions, since the lines of action of the reactions **B** and **C** pass through *A*. We conclude that the truss can rotate about *A* and that it is improperly constrained.†

The examples of Figs. 4.6 and 4.7 lead us to conclude that

A rigid body is improperly constrained whenever the supports (even though they may provide a sufficient number of reactions) are arranged in such a way that the reactions must be either concurrent or parallel.‡

In summary, to be sure that a two-dimensional rigid body is completely constrained and that the reactions at its supports are statically determinate, you should verify that the reactions involve three—and only three—unknowns and that the supports are arranged in such a way that they do not require the reactions to be either concurrent or parallel.

Supports involving statically indeterminate reactions should be used with care in the design of structures and only with a full knowledge of the problems they may cause. On the other hand, the analysis of structures possessing statically indeterminate reactions often can be partially carried out by the methods of statics. In the case of the truss of Fig. 4.4, for example, we can determine the vertical components of the reactions at *A* and *B* from the equilibrium equations.

For obvious reasons, supports producing partial or improper constraints should be avoided in the design of stationary structures. However, a partially or improperly constrained structure will not necessarily collapse; under particular loading conditions, equilibrium can be maintained. For example, the trusses of Figs. 4.5 and 4.6 will be in equilibrium if the applied forces **P**, **Q**, and **S** are vertical. Besides, structures designed to move *should* be only partially constrained. A railroad car, for instance, would be of little use if it were completely constrained by having its brakes applied permanently.

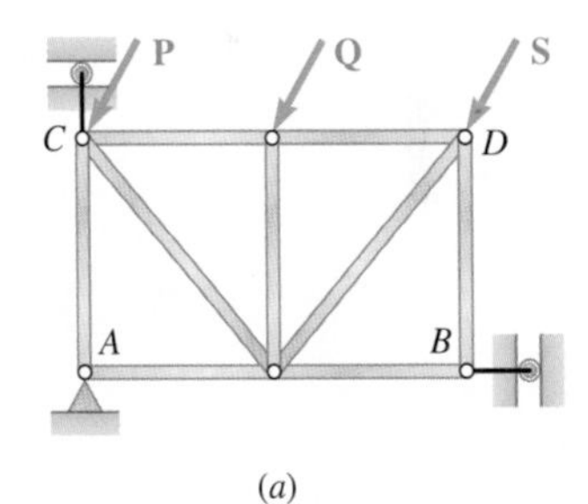

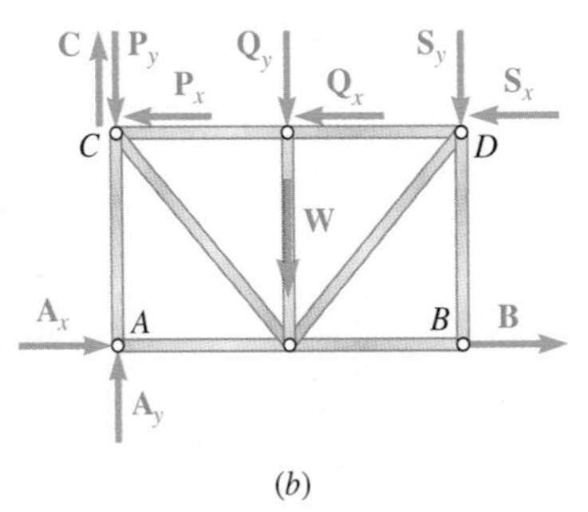

Fig. 4.7 *(a)* Truss with improper constraints; *(b)* free-body diagram.

†Rotation of the truss about *A* requires some "play" in the supports at *B* and *C*. In practice such play will always exist. In addition, we note that if the play is kept small, the displacements of the rollers *B* and *C* and, thus, the distances from *A* to the lines of action of the reactions **B** and **C** will also be small. The equation $\Sigma M_A = 0$ then requires that **B** and **C** be very large, a situation which can result in the failure of the supports at *B* and *C*.

‡Because this situation arises from an inadequate arrangement or *geometry* of the supports, it is often referred to as *geometric instability.*

Sample Problem 4.1

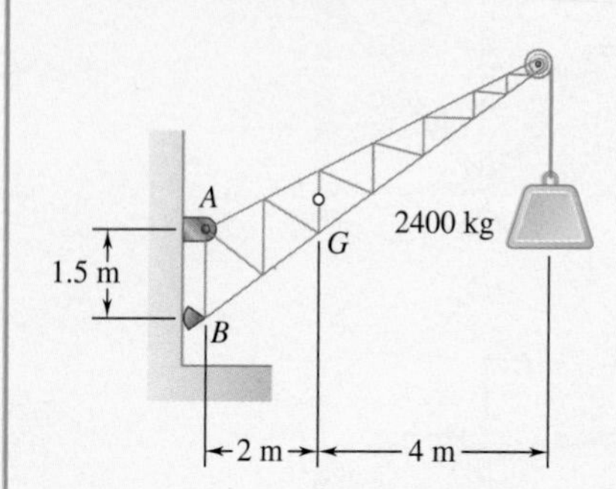

A fixed crane has a mass of 1000 kg and is used to lift a 2400-kg crate. It is held in place by a pin at A and a rocker at B. The center of gravity of the crane is located at G. Determine the components of the reactions at A and B.

STRATEGY: Draw a free-body diagram to show all of the forces acting on the crane, then use the equilibrium equations to calculate the values of the unknown forces.

MODELING:

Free-Body Diagram. By multiplying the masses of the crane and of the crate by $g = 9.81\ \text{m/s}^2$, you obtain the corresponding weights—that is, 9810 N or 9.81 kN, and 23 500 N or 23.5 kN (Fig. 1). The reaction at pin A is a force of unknown direction; you can represent it by components $\mathbf{A}_x$ and $\mathbf{A}_y$. The reaction at the rocker B is perpendicular to the rocker surface; thus, it is horizontal. Assume that $\mathbf{A}_x$, $\mathbf{A}_y$, and $\mathbf{B}$ act in the directions shown.

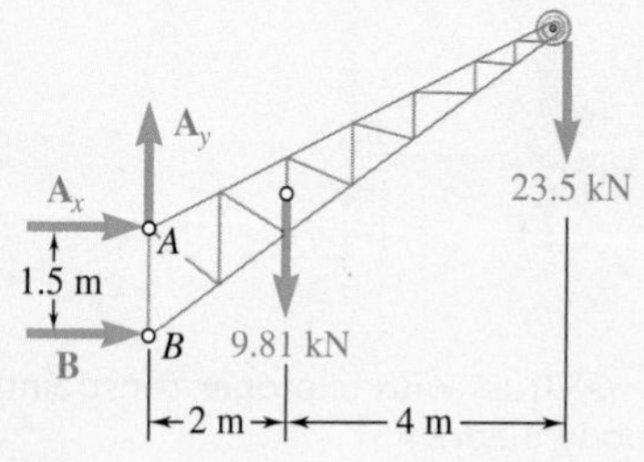

Fig. 1 Free-body diagram of crane.

ANALYSIS:

Determination of *B*. The sum of the moments of all external forces about point A is zero. The equation for this sum contains neither A_x nor A_y, since the moments of $\mathbf{A}_x$ and $\mathbf{A}_y$ about A are zero. Multiplying the magnitude of each force by its perpendicular distance from A, you have

$$+\circlearrowleft\Sigma M_A = 0: \quad +B(1.5\ \text{m}) - (9.81\ \text{kN})(2\ \text{m}) - (23.5\ \text{kN})(6\ \text{m}) = 0$$

$$B = +107.1\ \text{kN} \qquad \mathbf{B} = 107.1\ \text{kN} \rightarrow$$

Since the result is positive, the reaction is directed as assumed.

Determination of $\mathbf{A}_x$. Determine the magnitude of $\mathbf{A}_x$ by setting the sum of the horizontal components of all external forces to zero.

$$\overset{+}{\rightarrow}\Sigma F_x = 0: \quad A_x + B = 0$$

$$A_x + 107.1\ \text{kN} = 0$$

$$A_x = -107.1\ \text{kN} \qquad \mathbf{A}_x = 107.1\ \text{kN} \leftarrow$$

Since the result is negative, the sense of $\mathbf{A}_x$ is opposite to that assumed originally.

Determination of $\mathbf{A}_y$. The sum of the vertical components must also equal zero. Therefore,

$$+\uparrow\Sigma F_y = 0: \quad A_y - 9.81\ \text{kN} - 23.5\ \text{kN} = 0$$

$$A_y = +33.3\ \text{kN} \qquad \mathbf{A}_y = 33.3\ \text{kN} \uparrow$$

Adding the components $\mathbf{A}_x$ and $\mathbf{A}_y$ vectorially, you can find that the reaction at A is 112.2 kN $\measuredangle$17.3°.

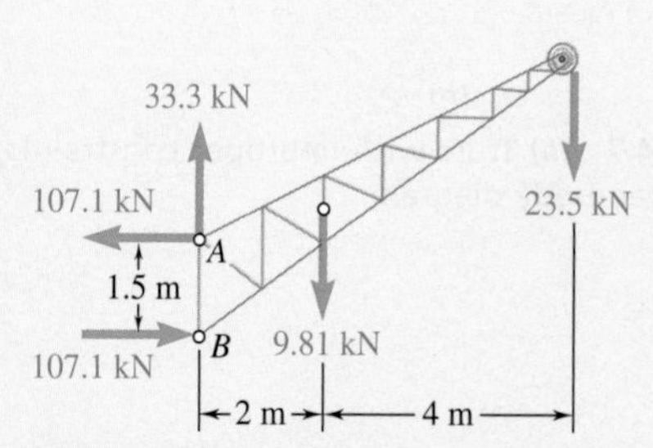

Fig. 2 Free-body diagram of crane with solved reactions.

REFLECT and THINK: You can check the values obtained for the reactions by recalling that the sum of the moments of all the external forces about any point must be zero. For example, considering point B (Fig. 2), you can show

$$+\circlearrowleft\Sigma M_B = -(9.81\ \text{kN})(2\ \text{m}) - (23.5\ \text{kN})(6\ \text{m}) + (107.1\ \text{kN})(1.5\ \text{m}) = 0$$

Sample Problem 4.2

Three loads are applied to a beam as shown. The beam is supported by a roller at A and by a pin at B. Neglecting the weight of the beam, determine the reactions at A and B when $P = 15$ kips.

P 6 kips 6 kips A B 3 ft 6 ft 2 ft 2 ft

STRATEGY: Draw a free-body diagram of the beam, then write the equilibrium equations, first summing forces in the x direction and then summing moments at A and at B.

MODELING:

Free-Body Diagram. The reaction at A is vertical and is denoted by **A** (Fig. 1). Represent the reaction at B by components $\mathbf{B}_x$ and $\mathbf{B}_y$. Assume that each component acts in the direction shown.

15 kips 6 kips 6 kips A A B B_x B_y 3 ft 6 ft 2 ft 2 ft

Fig.1 Free-body diagram of beam.

ANALYSIS:

Equilibrium Equations. Write the three equilibrium equations and solve for the reactions indicated:

$\xrightarrow{+} \Sigma F_x = 0$: $\quad B_x = 0 \qquad \mathbf{B}_x = 0$ ◀

$+\circlearrowleft \Sigma M_A = 0$:

$$-(15 \text{ kips})(3 \text{ ft}) + B_y(9 \text{ ft}) - (6 \text{ kips})(11 \text{ ft}) - (6 \text{ kips})(13 \text{ ft}) = 0$$

$$B_y = +21.0 \text{ kips} \qquad \mathbf{B}_y = 21.0 \text{ kips} \uparrow \blacktriangleleft$$

$+\circlearrowleft \Sigma M_B = 0$:

$$-A(9 \text{ ft}) + (15 \text{ kips})(6 \text{ ft}) - (6 \text{ kips})(2 \text{ ft}) - (6 \text{ kips})(4 \text{ ft}) = 0$$

$$A = +6.00 \text{ kips} \qquad \mathbf{A} = 6.00 \text{ kips} \uparrow \blacktriangleleft$$

REFLECT and THINK: Check the results by adding the vertical components of all of the external forces:

$$+\uparrow \Sigma F_y = +6.00 \text{ kips} - 15 \text{ kips} + 21.0 \text{ kips} - 6 \text{ kips} - 6 \text{ kips} = 0$$

Remark. In this problem, the reactions at both A and B are vertical; however, these reactions are vertical for different reasons. At A, the beam is supported by a roller; hence, the reaction cannot have any horizontal component. At B, the horizontal component of the reaction is zero because it must satisfy the equilibrium equation $\Sigma F_x = 0$ and none of the other forces acting on the beam has a horizontal component.

You might have noticed at first glance that the reaction at B was vertical and dispensed with the horizontal component $\mathbf{B}_x$. This, however, is bad practice. In following it, you run the risk of forgetting the component $\mathbf{B}_x$ when the loading conditions require such a component (i.e., when a horizontal load is included). Also, you found the component $\mathbf{B}_x$ to be zero by using and solving an equilibrium equation, $\Sigma F_x = 0$. By setting $\mathbf{B}_x$ equal to zero immediately, you might not realize that you actually made use of this equation. Thus, you might lose track of the number of equations available for solving the problem.

Sample Problem 4.3

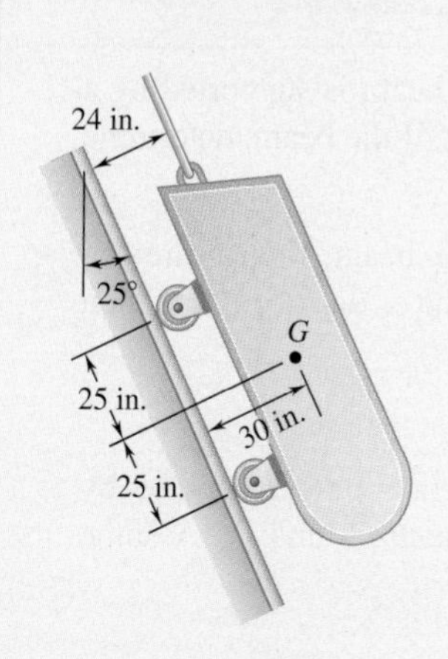

A loading car is at rest on a track forming an angle of 25° with the vertical. The gross weight of the car and its load is 5500 lb, and it acts at a point 30 in. from the track, halfway between the two axles. The car is held by a cable attached 24 in. from the track. Determine the tension in the cable and the reaction at each pair of wheels.

STRATEGY: Draw a free-body diagram of the car to determine the unknown forces, and write equilibrium equations to find their values, summing moments at A and B and then summing forces.

MODELING:

Free-Body Diagram. The reaction at each wheel is perpendicular to the track, and the tension force **T** is parallel to the track. Therefore, for convenience, choose the x axis parallel to the track and the y axis perpendicular to the track (Fig. 1). Then resolve the 5500-lb weight into x and y components.

$$W_x = +(5500 \text{ lb}) \cos 25° = +4980 \text{ lb}$$
$$W_y = -(5500 \text{ lb}) \sin 25° = -2320 \text{ lb}$$

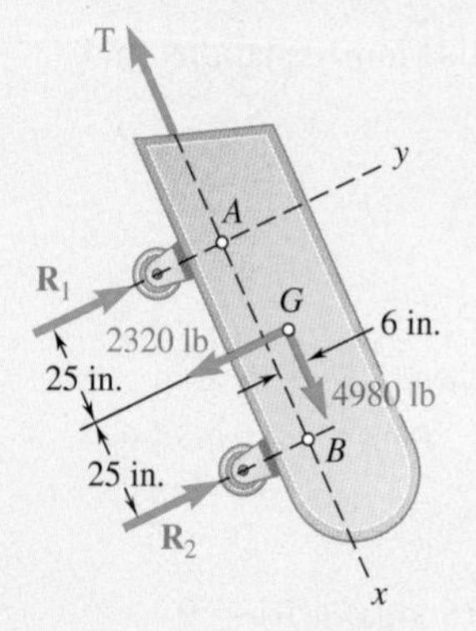

Fig. 1 Free-body diagram of car.

ANALYSIS:

Equilibrium Equations. Take moments about A to eliminate **T** and $\mathbf{R}_1$ from the computation.

$$+\circlearrowleft \Sigma M_A = 0: \quad -(2320 \text{ lb})(25 \text{ in.}) - (4980 \text{ lb})(6 \text{ in.}) + R_2(50 \text{ in.}) = 0$$
$$R_2 = +1758 \text{ lb} \qquad \mathbf{R}_2 = 1758 \text{ lb} \nearrow \blacktriangleleft$$

Then take moments about B to eliminate **T** and $\mathbf{R}_2$ from the computation.

$$+\circlearrowleft \Sigma M_B = 0: \quad (2320 \text{ lb})(25 \text{ in.}) - (4980 \text{ lb})(6 \text{ in.}) - R_1(50 \text{ in.}) = 0$$
$$R_1 = +562 \text{ lb} \qquad \mathbf{R}_1 = 562 \text{ lb} \nearrow \blacktriangleleft$$

Determine the value of T by summing forces in the x direction.

$$\searrow + \Sigma F_x = 0: \quad +4980 \text{ lb} - T = 0$$
$$T = +4980 \text{ lb} \qquad \mathbf{T} = 4980 \text{ lb} \nwarrow \blacktriangleleft$$

Figure 2 shows the computed values of the reactions.

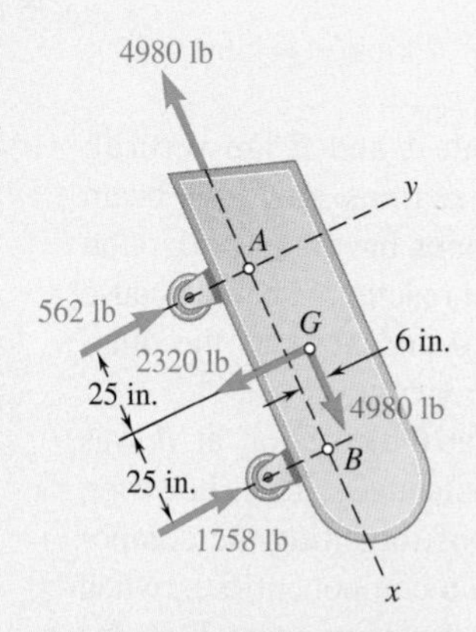

Fig. 2 Free-body diagram of car with solved reactions.

REFLECT and THINK: You can verify the computations by summing forces in the y direction.

$$\nearrow + \Sigma F_y = +562 \text{ lb} + 1758 \text{ lb} - 2320 \text{ lb} = 0$$

You could also check the solution by computing moments about any point other than A or B.

Sample Problem 4.4

The frame shown supports part of the roof of a small building. Knowing that the tension in the cable is 150 kN, determine the reaction at the fixed end E.

STRATEGY: Draw a free-body diagram of the frame and of the cable BDF. The support at E is fixed, so the reactions here include a moment; to determine its value, sum moments about point E.

MODELING:

Free-Body Diagram. Represent the reaction at the fixed end E by the force components $\mathbf{E}_x$ and $\mathbf{E}_y$ and the couple $\mathbf{M}_E$ (Fig. 1). The other forces acting on the free body are the four 20-kN loads and the 150-kN force exerted at end F of the cable.

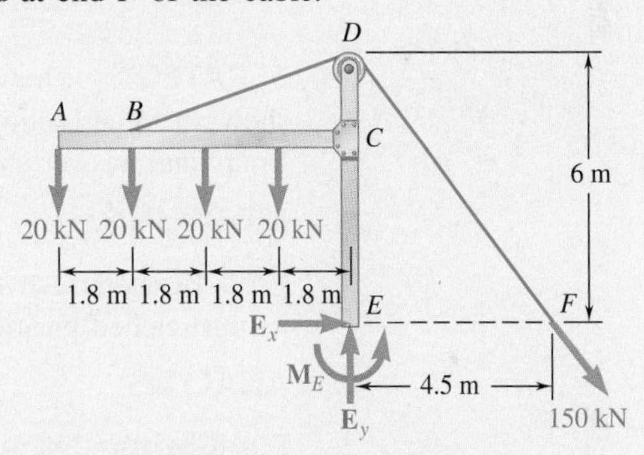

Fig. 1 Free-body diagram of frame.

ANALYSIS:

Equilibrium Equations. First note that

$$DF = \sqrt{(4.5\text{ m})^2 + (6\text{ m})^2} = 7.5\text{ m}$$

Then you can write the three equilibrium equations and solve for the reactions at E.

$$\overset{+}{\rightarrow}\Sigma F_x = 0: \qquad E_x + \frac{4.5}{7.5}(150\text{ kN}) = 0$$

$$E_x = -90.0\text{ kN} \qquad \mathbf{E}_x = 90.0\text{ kN} \leftarrow$$

$$+\uparrow\Sigma F_y = 0: \qquad E_y - 4(20\text{ kN}) - \frac{6}{7.5}(150\text{ kN}) = 0$$

$$E_y = +200\text{ kN} \qquad \mathbf{E}_y = 200\text{ kN} \uparrow$$

$$+\circlearrowleft\Sigma M_E = 0: \qquad (20\text{ kN})(7.2\text{ m}) + (20\text{ kN})(5.4\text{ m}) + (20\text{ kN})(3.6\text{ m})$$

$$+(20\text{ kN})(1.8\text{ m}) - \frac{6}{7.5}(150\text{ kN})(4.5\text{ m}) + M_E = 0$$

$$M_E = +180.0\text{ kN·m} \qquad \mathbf{M}_E = 180.0\text{ kN·m} \circlearrowleft$$

REFLECT and THINK: The cable provides a fourth constraint, making this situation statically indeterminate. This problem therefore gave us the value of the cable tension, which would have been determined by means other than statics. We could then use the three available independent static equilibrium equations to solve for the remaining three reactions.

Sample Problem 4.5

A 400-lb weight is attached at A to the lever shown. The constant of the spring BC is $k = 250$ lb/in., and the spring is unstretched when $\theta = 0$. Determine the position of equilibrium.

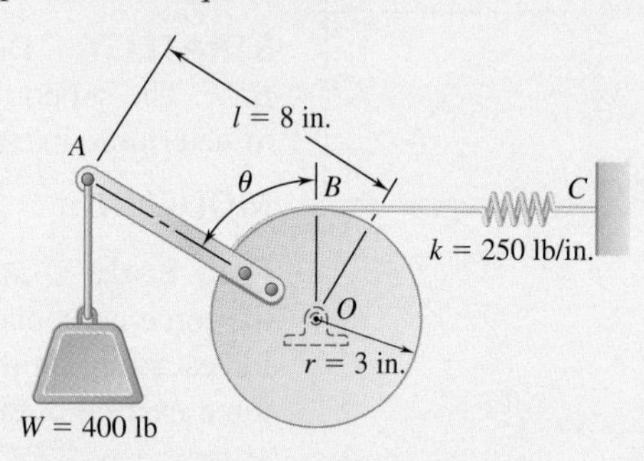

STRATEGY: Draw a free-body diagram of the lever and cylinder to show all forces acting on the body (Fig. 1), then sum moments about O. Your final answer should be the angle θ.

MODELING:

Free-Body Diagram. Denote by s the deflection of the spring from its unstretched position and note that $s = r\theta$. Then $F = ks = kr\theta$.

ANALYSIS:

Equilibrium Equation. Sum the moments of **W** and **F** about O to eliminate the reactions supporting the cylinder. The result is

$$+\circlearrowleft \Sigma M_O = 0: \qquad Wl \sin\theta - r(kr\theta) = 0 \qquad \sin\theta = \frac{kr^2}{Wl}\theta$$

Substituting the given data yields

$$\sin\theta = \frac{(250 \text{ lb/in.})(3 \text{ in.})^2}{(400 \text{ lb})(8 \text{ in.})}\theta \quad \sin\theta = 0.703\,\theta$$

Solving by trial and error, the angle is $\theta = 0$ $\qquad \theta = 80.3°$ ◀

REFLECT and THINK: The weight could represent any vertical force acting on the lever. The key to the problem is to express the spring force as a function of the angle θ.

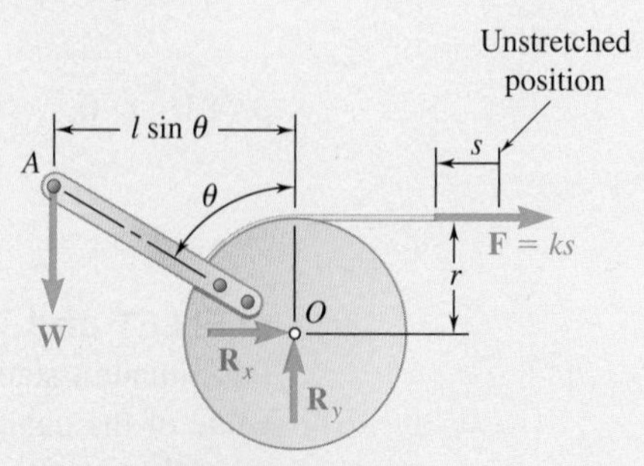

Fig. 1 Free-body diagram of the lever and cylinder.

Problems

4.1 For the beam and loading shown, determine (*a*) the reaction at *A*, (*b*) the tension in cable *BC*.

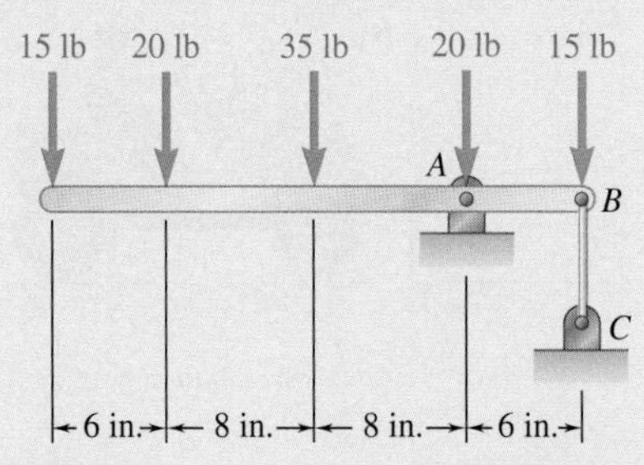

Fig. P4.1

4.2 A 3200-lb forklift truck is used to lift a 1700-lb crate. Determine the reaction at each of the two (*a*) front wheels *A*, (*b*) rear wheels *B*.

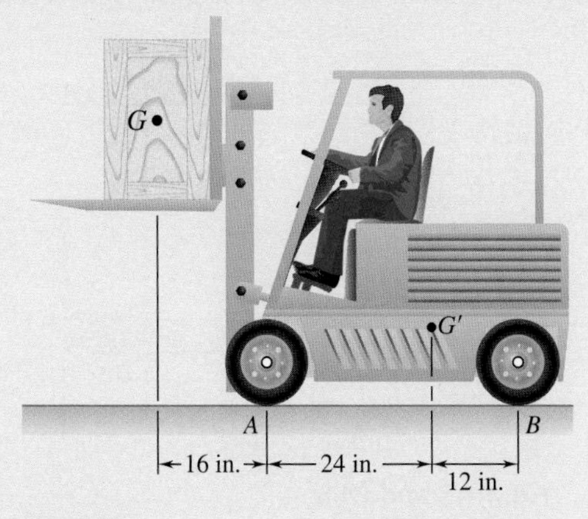

Fig. P4.2

4.3 A gardener uses a 60-N wheelbarrow to transport a 250-N bag of fertilizer. What force must she exert on each handle?

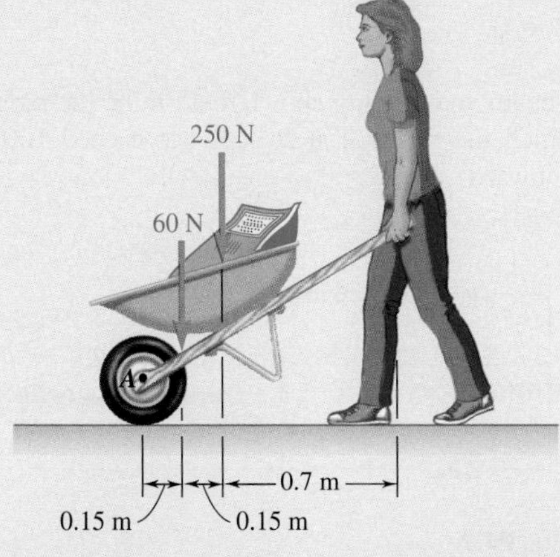

Fig. P4.3

4.4 A load of lumber of weight $W = 25$ kN is being raised by a mobile crane. Knowing that the tension is 25 kN in all portions of cable AEF and that the weight of boom ABC is 3 kN, determine (*a*) the tension in rod CD, (*b*) the reaction at pin B.

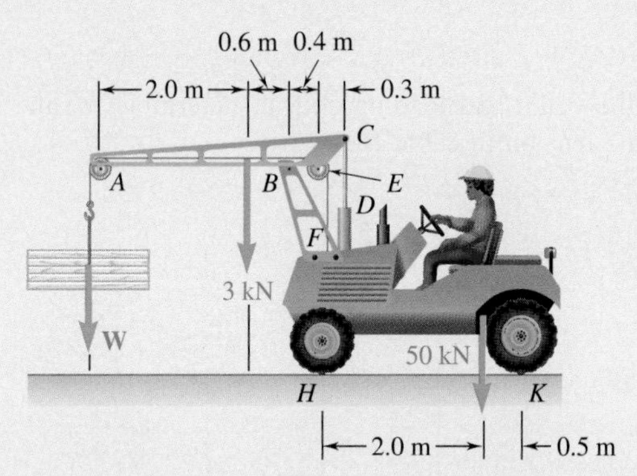

Fig. *P4.4*

4.5 Three loads are applied as shown to a light beam supported by cables attached at B and D. Neglecting the weight of the beam, determine the range of values of Q for which neither cable becomes slack when $P = 0$.

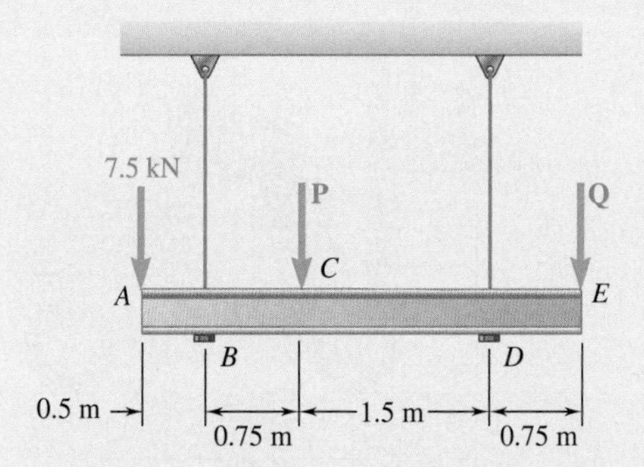

Fig. P4.5 and P4.6

4.6 Three loads are applied as shown to a light beam supported by cables attached at B and D. Knowing that the maximum allowable tension in each cable is 12 kN and neglecting the weight of the beam, determine the range of values of Q for which the loading is safe when $P = 0$.

4.7 For the beam and loading shown, determine the range of the distance a for which the reaction at B does not exceed 100 lb downward or 200 lb upward.

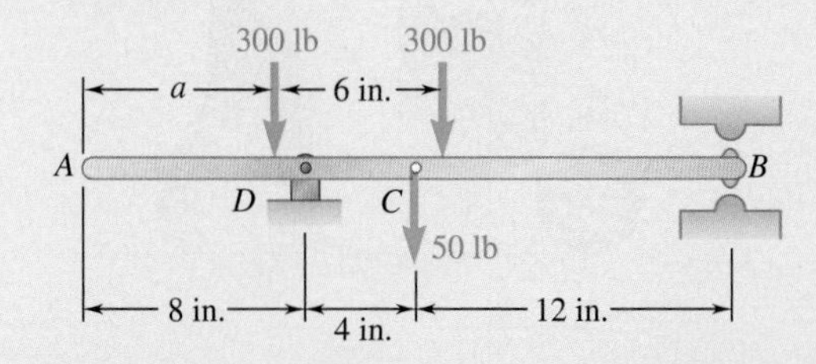

Fig. P4.7

4.8 For the beam of Sample Prob. 4.2, determine the range of values of P for which the beam will be safe, knowing that the maximum allowable value of each of the reactions is 25 kips and that the reaction at A must be directed upward.

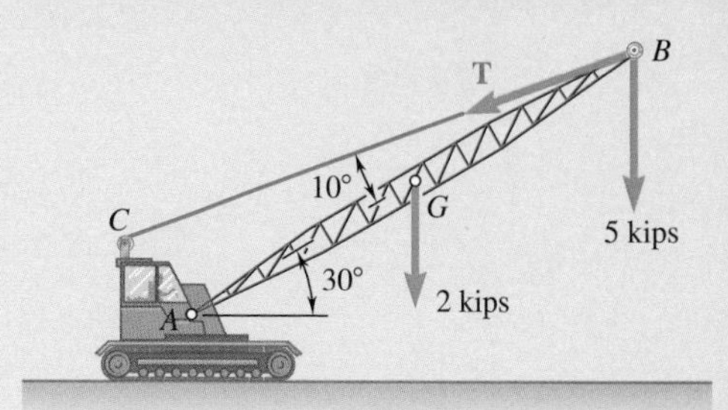

Fig. P4.9

4.9 The 40-ft boom AB weighs 2 kips; the distance from axle A to the center of gravity G of the boom is 20 ft. For the position shown, determine (*a*) the tension T in the cable, (*b*) the reaction at A.

4.10 The lever BCD is hinged at C and attached to a control rod at B. If $P = 100$ lb, determine (*a*) the tension in rod AB, (*b*) the reaction at C.

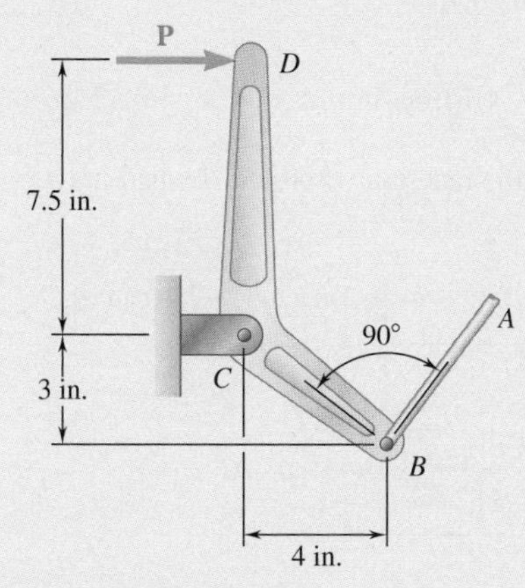

Fig. P4.10 and *P4.11*

4.11 The lever BCD is hinged at C and attached to a control rod at B. Determine the maximum force **P** that can be safely applied at D if the maximum allowable value of the reaction at C is 250 lb.

4.12 A lever AB is hinged at C and attached to a control cable at A. If the lever is subjected to a 500-N horizontal force at B, determine (*a*) the tension in the cable, (*b*) the reaction at C.

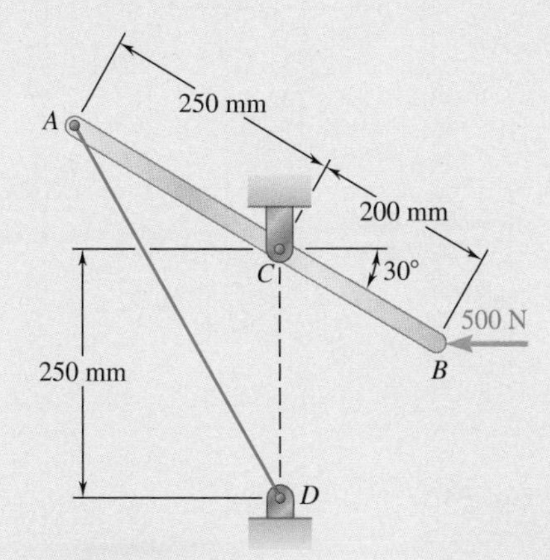

Fig. P4.12

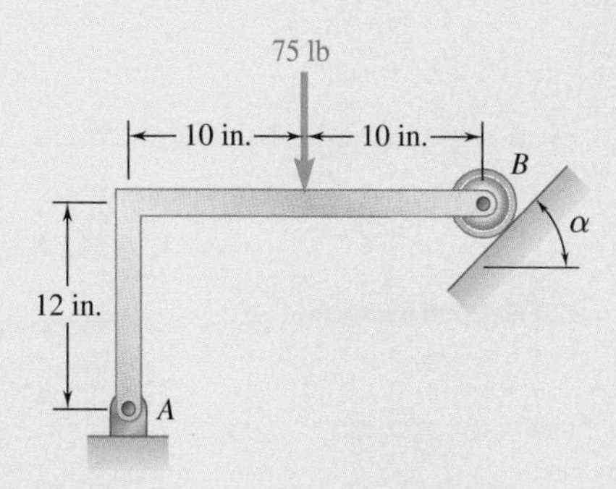

Fig. P4.13

4.13 Determine the reactions at A and B when (*a*) $\alpha = 0$, (*b*) $\alpha = 90°$, (*c*) $\alpha = 30°$.

4.14 The bracket BCD is hinged at C and attached to a control cable at B. For the loading shown, determine (*a*) the tension in the cable, (*b*) the reaction at C.

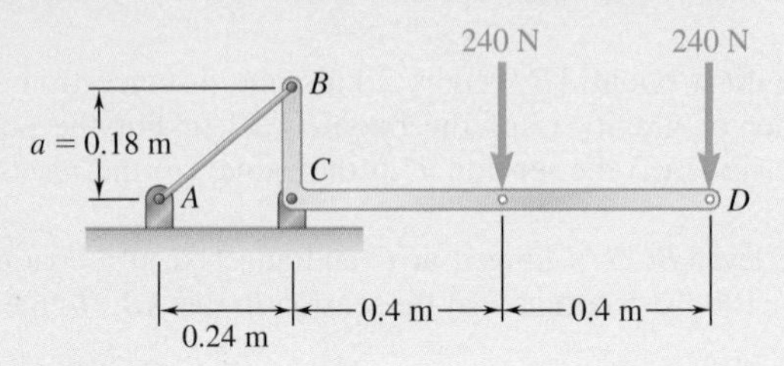

Fig. P4.14

4.15 Solve Prob. 4.14, assuming that $a = 0.32$ m.

4.16 Determine the reactions at A and B when (*a*) $h = 0$, (*b*) $h = 200$ mm.

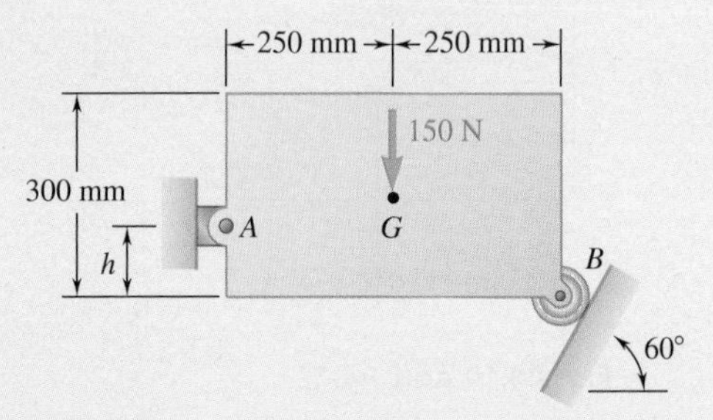

Fig. P4.16

4.17 A light bar AD is suspended from a cable BE and supports a 50-lb block at C. The ends A and D of the bar are in contact with frictionless vertical walls. Determine the tension in cable BE and the reactions at A and D.

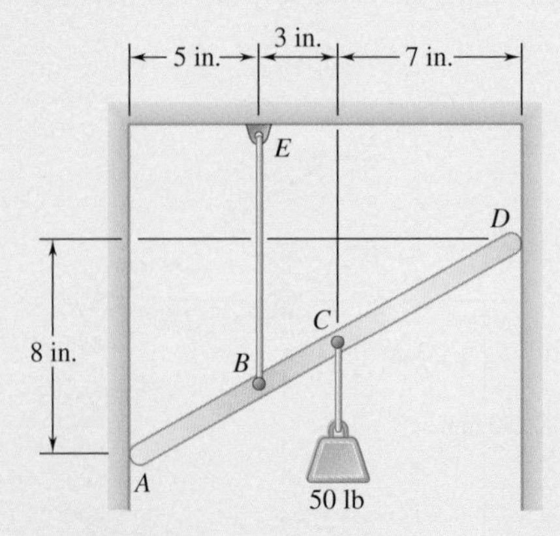

Fig. P4.17

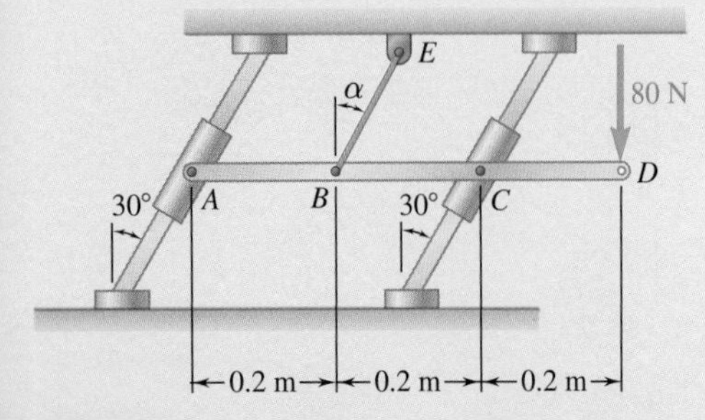

Fig. P4.18

4.18 Bar AD is attached at A and C to collars that can move freely on the rods shown. If the cord BE is vertical ($\alpha = 0$), determine the tension in the cord and the reactions at A and C.

4.19 Solve Prob. 4.18 if the cord BE is parallel to the rods ($\alpha = 30°$).

4.20 Two slots have been cut in plate DEF, and the plate has been placed so that the slots fit two fixed, frictionless pins A and B. Knowing that $P = 15$ lb, determine (*a*) the force each pin exerts on the plate, (*b*) the reaction at F.

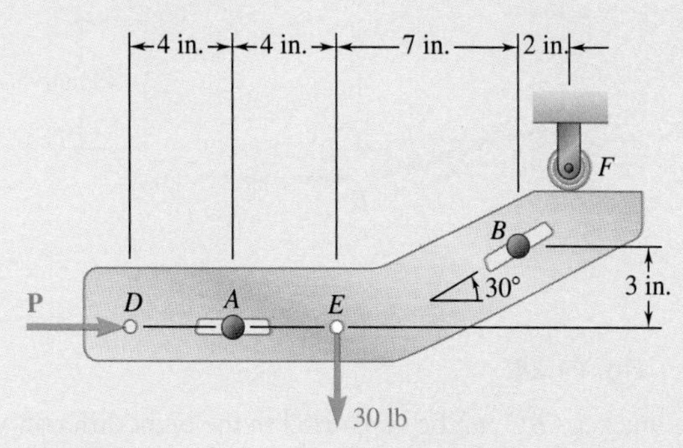

Fig. P4.20

4.21 A 6-m telephone pole weighing 1600 N is used to support the ends of two wires. The wires form the angles shown with the horizontal, and the tensions in the wires are, respectively, $T_1 = 600$ N and $T_2 = 375$ N. Determine the reaction at the fixed end A.

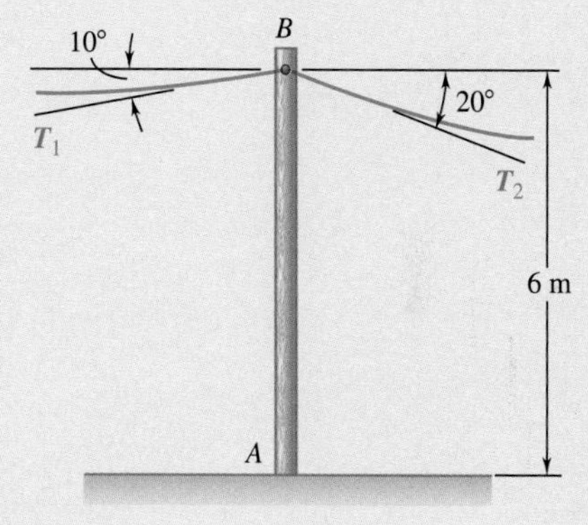

Fig. P4.21

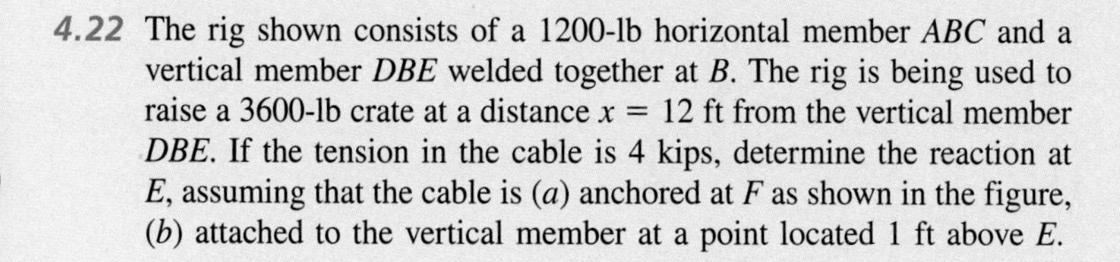

4.22 The rig shown consists of a 1200-lb horizontal member ABC and a vertical member DBE welded together at B. The rig is being used to raise a 3600-lb crate at a distance $x = 12$ ft from the vertical member DBE. If the tension in the cable is 4 kips, determine the reaction at E, assuming that the cable is (*a*) anchored at F as shown in the figure, (*b*) attached to the vertical member at a point located 1 ft above E.

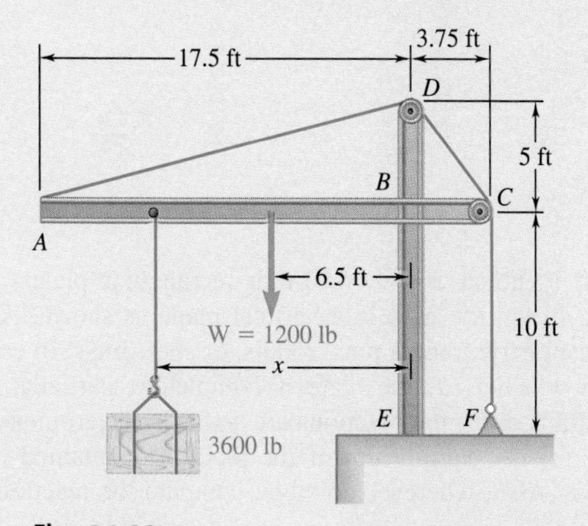

Fig. P4.22

4.23 For the rig and crate of Prob. 4.22 and assuming that the cable is anchored at F as shown, determine (*a*) the required tension in cable $ADCF$ if the maximum value of the couple at E as x varies from 1.5 to 17.5 ft is to be as small as possible, (*b*) the corresponding maximum value of the couple.

4.24 A tension of 20 N is maintained in a tape as it passes through the support system shown. Knowing that the radius of each pulley is 10 mm, determine the reaction at C.

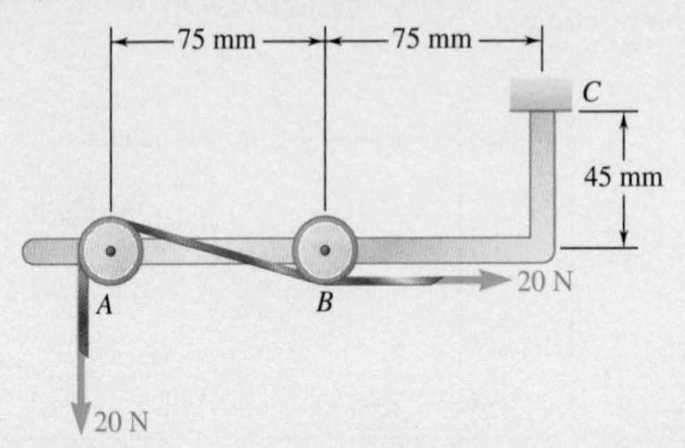

Fig. P4.24

4.25 The bracket ABC can be supported in the eight different ways shown. All connections consist of smooth pins, rollers, or short links. In each case, determine whether (*a*) the plate is completely, partially, or improperly constrained, (*b*) the reactions are statically determinate or indeterminate, (*c*) the equilibrium of the plate is maintained in the position shown. Also, wherever possible, compute the reactions assuming that the magnitude of the force **P** is 100 lb.

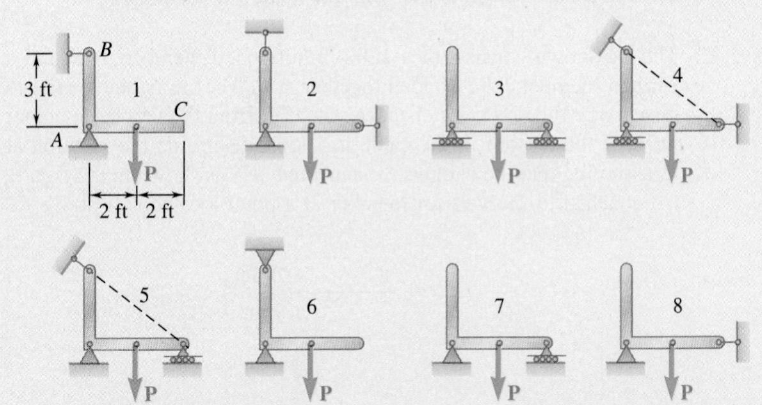

Fig. P4.25

4.26 Eight identical 500 × 750-mm rectangular plates, each of mass $m = 40$ kg, are held in a vertical plane as shown. All connections consist of frictionless pins, rollers, or short links. In each case, determine whether (*a*) the plate is completely, partially, or improperly constrained, (*b*) the reactions are statically determinate or indeterminate, (*c*) the equilibrium of the plate is maintained in the position shown. Also, wherever possible, compute the reactions.

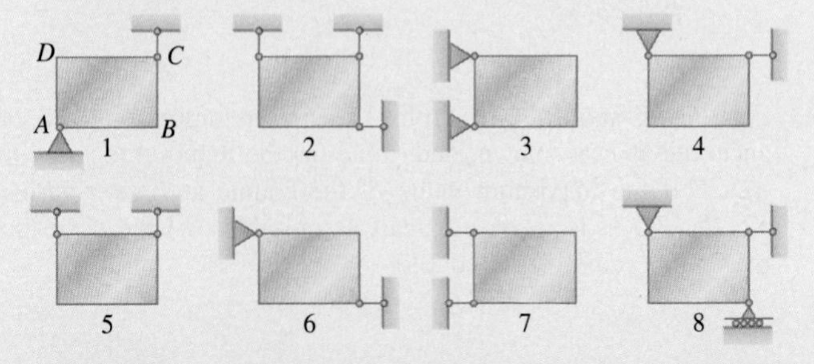

Fig. P4.26

4.2 TWO SPECIAL CASES

In practice, some simple cases of equilibrium occur quite often, either as part of a more complicated analysis or as the complete models of a situation. By understanding the characteristics of these cases, you can often simplify the overall analysis.

4.2A Equilibrium of a Two-Force Body

A particular case of equilibrium of considerable interest in practical applications is that of a rigid body subjected to two forces. Such a body is commonly called a **two-force body**. We show here that, **if a two-force body is in equilibrium, the two forces must have the same magnitude, the same line of action, and opposite sense**.

Consider a corner plate subjected to two forces $\mathbf{F}_1$ and $\mathbf{F}_2$ acting at A and B, respectively (Fig. 4.8*a*). If the plate is in equilibrium, the sum of the moments of $\mathbf{F}_1$ and $\mathbf{F}_2$ about any axis must be zero. First, we sum moments about A. Since the moment of $\mathbf{F}_1$ is obviously zero, the moment of $\mathbf{F}_2$ also must be zero and the line of action of $\mathbf{F}_2$ must pass through A (Fig. 4.8*b*). Similarly, summing moments about B, we can show that the line of action of $\mathbf{F}_1$ must pass through B (Fig. 4.8*c*). Therefore, both forces have the same line of action (line AB). You can see from either of the equations $\Sigma F_x = 0$ and $\Sigma F_y = 0$ that they must also have the same magnitude but opposite sense.

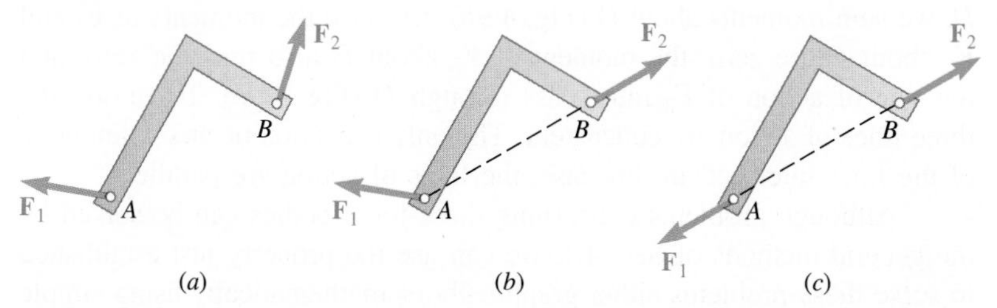

Fig. 4.8 A two-force body in equilibrium. *(a)* Forces act at two points of the body; *(b)* summing moments about point *A* shows that the line of action of $\mathbf{F}_2$ must pass through *A*; *(c)* summing moments about point *B* shows that the line of action of $\mathbf{F}_1$ must pass through *B*.

If several forces act at two points A and B, the forces acting at A can be replaced by their resultant $\mathbf{F}_1$, and those acting at B can be replaced by their resultant $\mathbf{F}_2$. Thus, a two-force body can be more generally defined as **a rigid body subjected to forces acting at only two points**. The resultants $\mathbf{F}_1$ and $\mathbf{F}_2$ then must have the same line of action, the same magnitude, and opposite sense (Fig. 4.8).

Later, in the study of structures, frames, and machines, you will see how the recognition of two-force bodies simplifies the solution of certain problems.

4.2B Equilibrium of a Three-Force Body

Another case of equilibrium that is of great practical interest is that of a **three-force body**, i.e., a rigid body subjected to three forces or, more generally, **a rigid body subjected to forces acting at only three points**. Consider a rigid body subjected to a system of forces that can be reduced to three forces $\mathbf{F}_1$, $\mathbf{F}_2$, and $\mathbf{F}_3$ acting at *A*, *B*, and *C*, respectively (Fig. 4.9*a*). We show that if the body is in equilibrium, **the lines of action of the three forces must be either concurrent or parallel.**

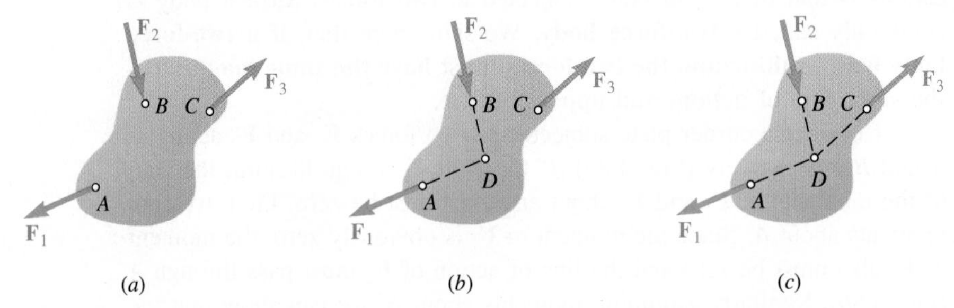

Fig. 4.9 A three-force body in equilibrium. *(a–c)* Demonstration that the lines of action of the three forces must be either concurrent or parallel.

Since the rigid body is in equilibrium, the sum of the moments of $\mathbf{F}_1$, $\mathbf{F}_2$, and $\mathbf{F}_3$ about any axis must be zero. Assuming that the lines of action of $\mathbf{F}_1$ and $\mathbf{F}_2$ intersect and denoting their point of intersection by *D*, we sum moments about *D* (Fig. 4.9*b*). Because the moments of $\mathbf{F}_1$ and $\mathbf{F}_2$ about *D* are zero, the moment of $\mathbf{F}_3$ about *D* also must be zero, and the line of action of $\mathbf{F}_3$ must pass through *D* (Fig. 4.9*c*). Therefore, the three lines of action are concurrent. The only exception occurs when none of the lines intersect; in this case, the lines of action are parallel.

Although problems concerning three-force bodies can be solved by the general methods of Sec. 4.1, we can use the property just established to solve these problems either graphically or mathematically using simple trigonometric or geometric relations (see Sample Problem 4.6).

Sample Problem 4.6

A man raises a 10-kg joist with a length of 4 m by pulling on a rope. Find the tension T in the rope and the reaction at A.

STRATEGY: The joist is acted upon by three forces: its weight **W**, the force **T** exerted by the rope, and the reaction **R** of the ground at A. Therefore, it is a three-force body, and you can compute the forces by using a force triangle.

MODELING: First note that

$$W = mg = (10 \text{ kg})(9.81 \text{ m/s}^2) = 98.1 \text{ N}$$

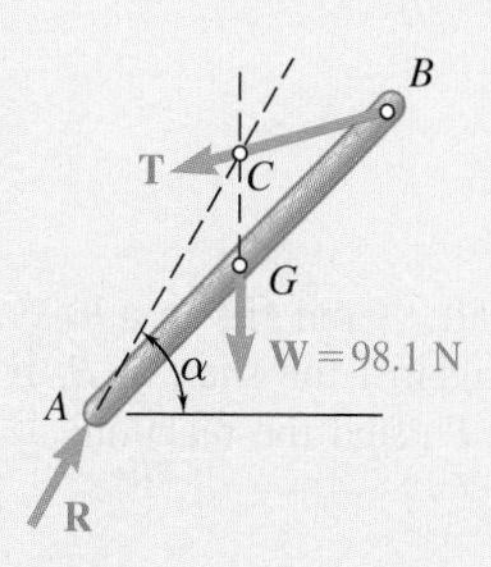

Fig. 1 Free-body diagram of joist.

Since the joist is a three-force body, the forces acting on it must be concurrent. The reaction **R** therefore must pass through the point of intersection C of the lines of action of the weight **W** and the tension force **T**, as shown in the free-body diagram (Fig. 1). You can use this fact to determine the angle α that **R** forms with the horizontal.

ANALYSIS: Draw the vertical line BF through B and the horizontal line CD through C (Fig. 2). Then

$$AF = BF = (AB)\cos 45° = (4 \text{ m})\cos 45° = 2.828 \text{ m}$$
$$CD = EF = AE = \tfrac{1}{2}(AF) = 1.414 \text{ m}$$
$$BD = (CD)\cot(45° + 25°) = (1.414 \text{ m})\tan 20° = 0.515 \text{ m}$$
$$CE = DF = BF - BD = 2.828 \text{ m} - 0.515 \text{ m} = 2.313 \text{ m}$$

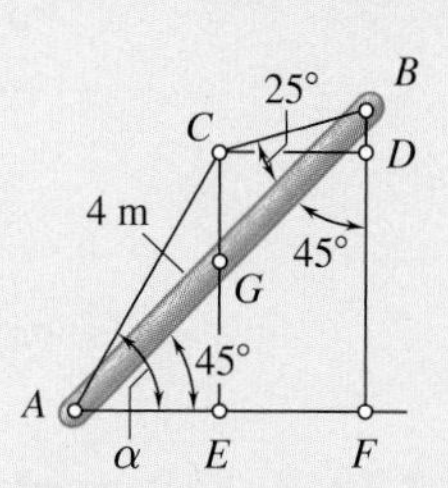

Fig. 2 Geometry analysis of the lines of action for the three forces acting on joist, concurrent at point C.

From these calculations, you can determine the angle α as

$$\tan\alpha = \frac{CE}{AE} = \frac{2.313 \text{ m}}{1.414 \text{ m}} = 1.636$$

$$\alpha = 58.6° \quad ◀$$

You now know the directions of all the forces acting on the joist.

Force Triangle. Draw a force triangle as shown (Fig. 3) with its interior angles computed from the known directions of the forces. You can then use the law of sines to find the unknown forces.

$$\frac{T}{\sin 31.4°} = \frac{R}{\sin 110°} = \frac{98.1 \text{ N}}{\sin 38.6°}$$

$$T = 81.9 \text{ N} \quad ◀$$

$$R = 147.8 \text{ N} \measuredangle 58.6° \quad ◀$$

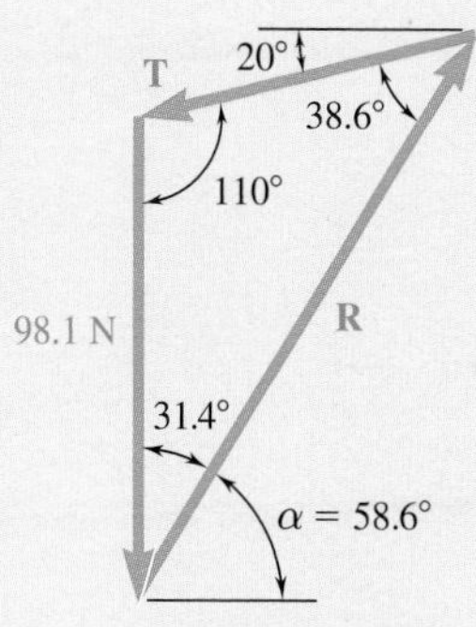

Fig. 3 Force triangle.

REFLECT and THINK: In practice, three-force members occur often, so learning this method of analysis is useful in many situations.

Problems

4.27 Determine the reactions at B and C when $a = 30$ mm.

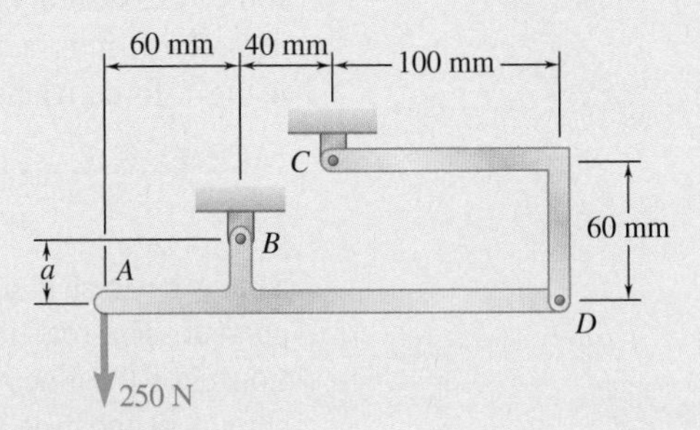

Fig. P4.27

4.28 The spanner shown is used to rotate a shaft. A pin fits in a hole at A, while a flat, frictionless surface rests against the shaft at B. If a 60-lb force **P** is exerted on the spanner at D, find the reactions at A and B.

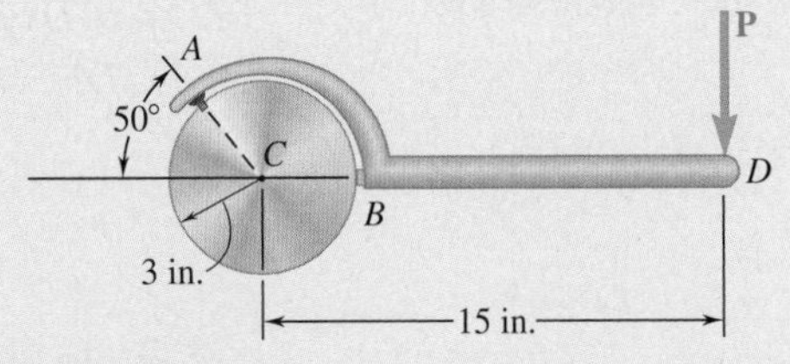

Fig. P4.28

4.29 A 12-ft wooden beam weighing 80 lb is supported by a pin and bracket at A and by cable BC. Find the reaction at A and the tension in the cable.

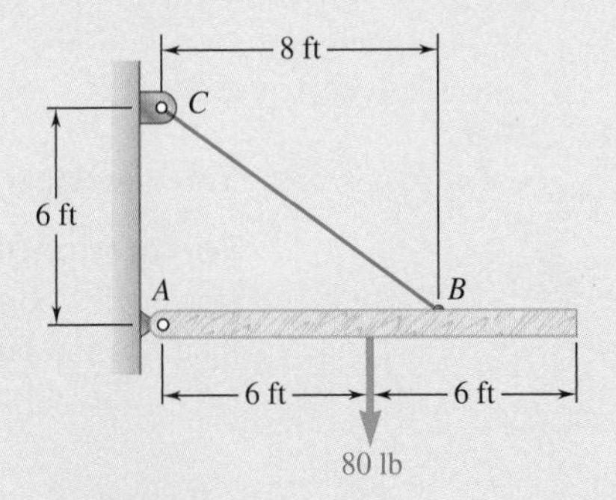

Fig. P4.29

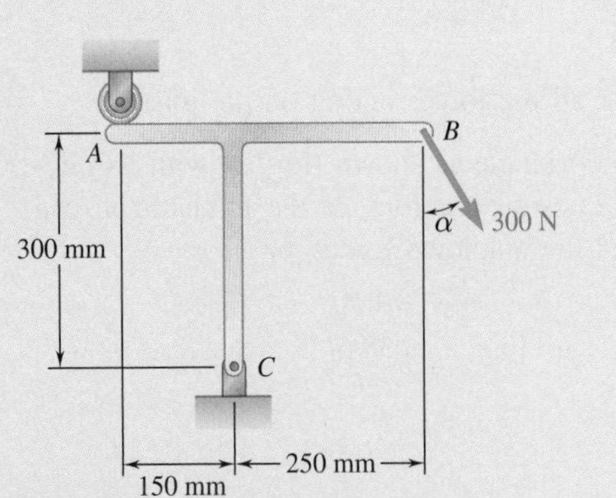

Fig. P4.30

4.30 A T-shaped bracket supports a 300-N load as shown. Determine the reactions at A and C when $\alpha = 45°$.

4.31 One end of rod AB rests in the corner A and the other is attached to cord BD. If the rod supports a 150-N load at its midpoint C, find the reaction at A and the tension in the cord.

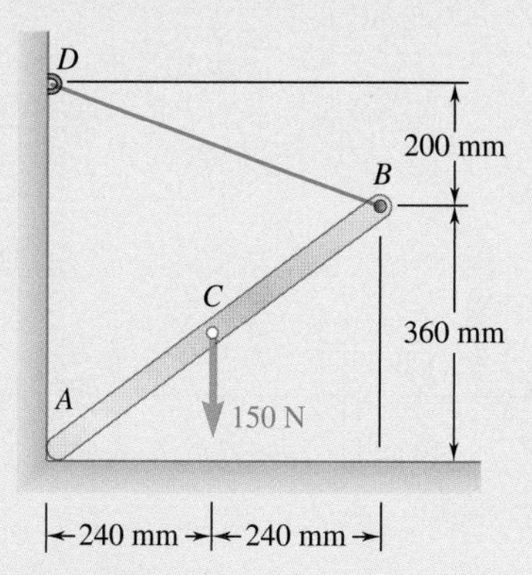

Fig. P4.31

4.32 Using the method of Sec. 4.2B, solve Prob. 4.12.

4.33 Using the method of Sec. 4.2B, solve Prob. 4.16.

4.34 A 40-lb roller of 8-in. diameter, which is to be used on a tile floor, is resting directly on the subflooring as shown. Knowing that the thickness of each tile is 0.3 in., determine the force **P** required to move the roller onto the tiles if the roller is (*a*) pushed to the left, (*b*) pulled to the right.

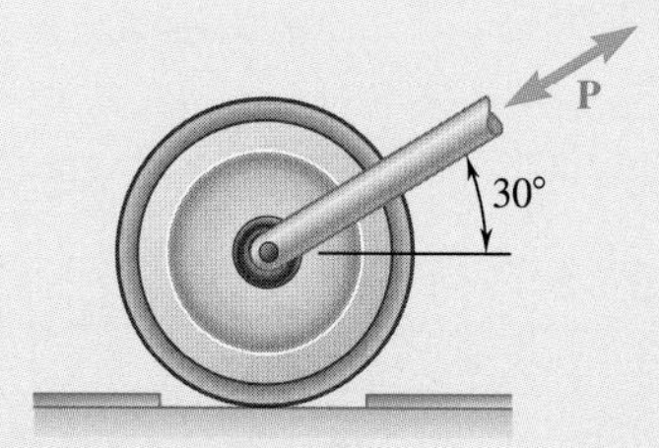

Fig. P4.34

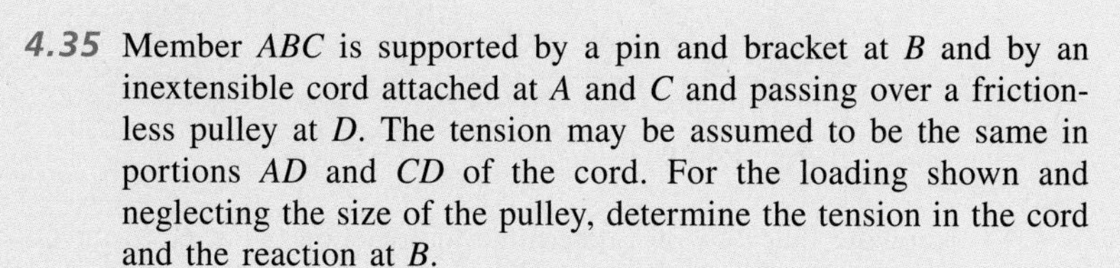

4.35 Member ABC is supported by a pin and bracket at B and by an inextensible cord attached at A and C and passing over a frictionless pulley at D. The tension may be assumed to be the same in portions AD and CD of the cord. For the loading shown and neglecting the size of the pulley, determine the tension in the cord and the reaction at B.

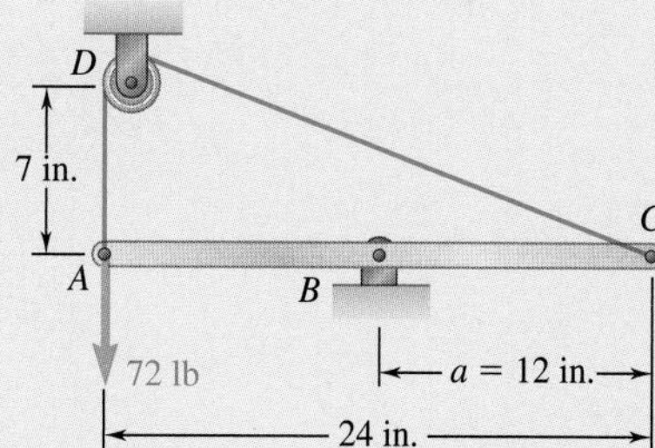

Fig. P4.35

4.36 Determine the reactions at A and B when $a = 150$ mm.

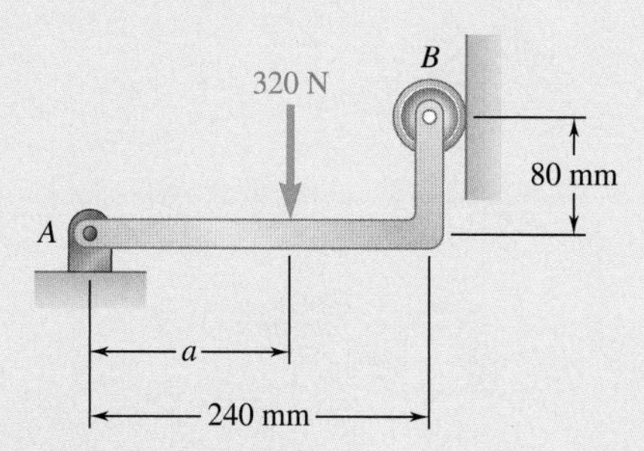

Fig. P4.36 and P4.37

4.37 Determine the value of a for which the magnitude of the reaction at B is equal to 800 N.

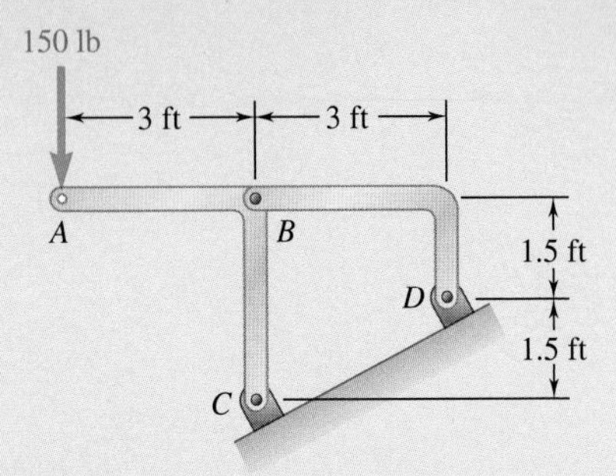

Fig. P4.38

4.38 For the frame and loading shown, determine the reactions at C and D.

4.39 For the boom and loading shown, determine (*a*) the tension in cord BD, (*b*) the reaction at C.

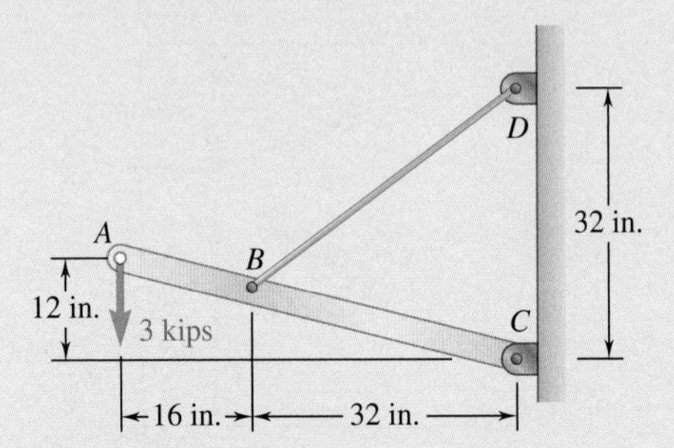

Fig. P4.39

4.40 A slender rod BC of length L and weight W is held by two cables as shown. Knowing that cable AB is horizontal and that the rod forms an angle of 40° with the horizontal, determine (*a*) the angle θ that cable CD forms with the horizontal, (*b*) the tension in each cable.

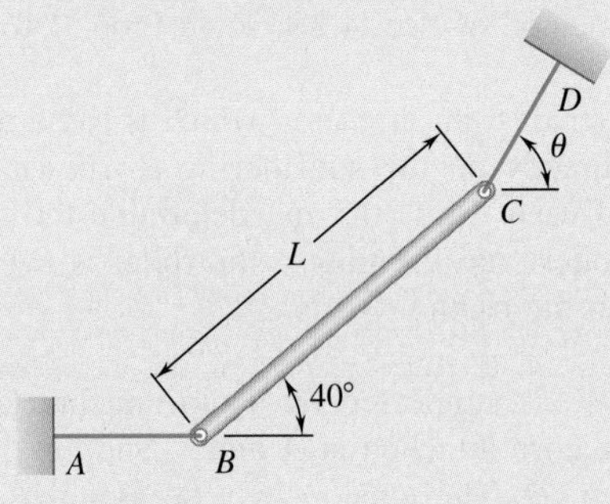

Fig. P4.40

4.41 Knowing that $\theta = 30°$, determine the reaction (*a*) at B, (*b*) at C.

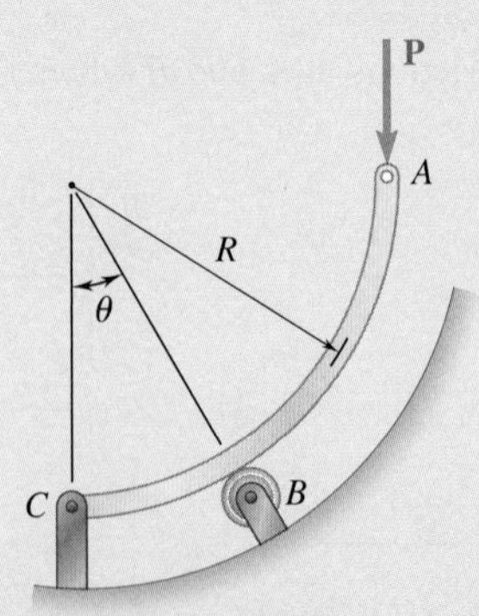

Fig. P4.41

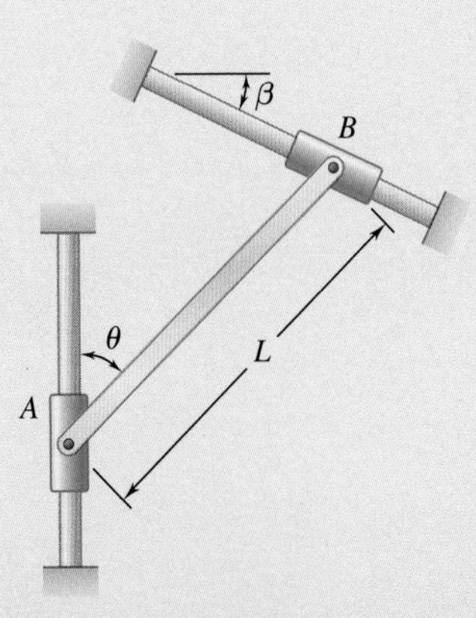

Fig. P4.42 and P4.43

4.42 A slender rod of length L is attached to collars that can slide freely along the guides shown. Knowing that the rod is in equilibrium, derive an expression for the angle θ in terms of the angle β.

4.43 An 8-kg slender rod of length L is attached to collars that can slide freely along the guides shown. Knowing that the rod is in equilibrium and that $\beta = 30°$, determine (*a*) the angle θ that the rod forms with the vertical, (*b*) the reactions at A and B.

4.44 Rod AB is supported by a pin and bracket at A and rests against a frictionless peg at C. Determine the reactions at A and C when a 170-N vertical force is applied at B.

4.45 Solve Prob. 4.44, assuming that the 170-N force applied at B is horizontal and directed to the left.

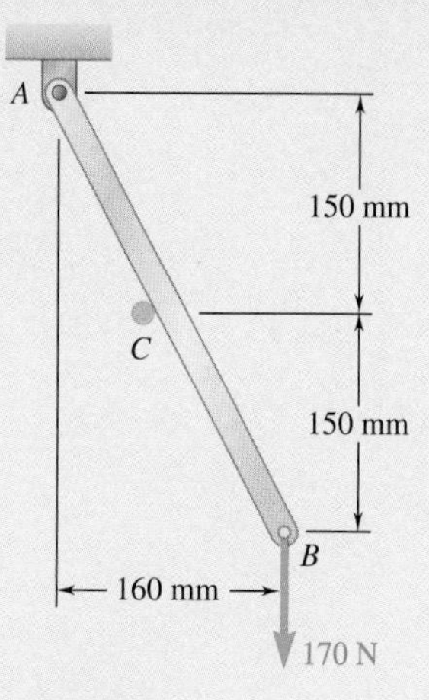

Fig. P4.44

4.46 Determine the reactions at A and B when $\beta = 50°$.

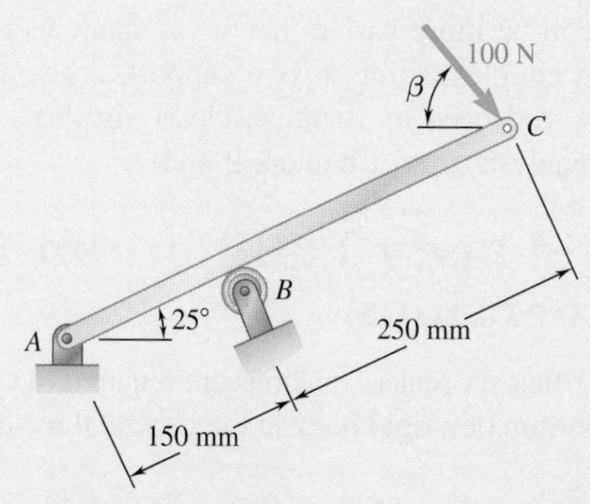

Fig. P4.46 and P4.47

4.47 Determine the reactions at A and B when $\beta = 80°$.

4.48 A slender rod of length L and weight W is attached to a collar at A and is fitted with a small wheel at B. Knowing that the wheel rolls freely along a cylindrical surface of radius R, and neglecting friction, derive an equation in θ, L, and R that must be satisfied when the rod is in equilibrium.

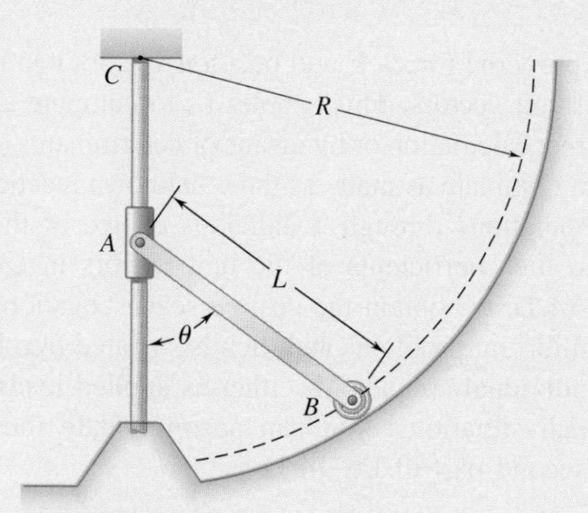

Fig. P4.48

4.49 Knowing that for the rod of Prob. 4.48, $L = 15$ in., $R = 20$ in., and $W = 10$ lb, determine (a) the angle θ corresponding to equilibrium, (b) the reactions at A and B.

4.50 A uniform rod AB of length $2R$ rests inside a hemispherical bowl of radius R as shown. Neglecting friction, determine the angle θ corresponding to equilibrium.

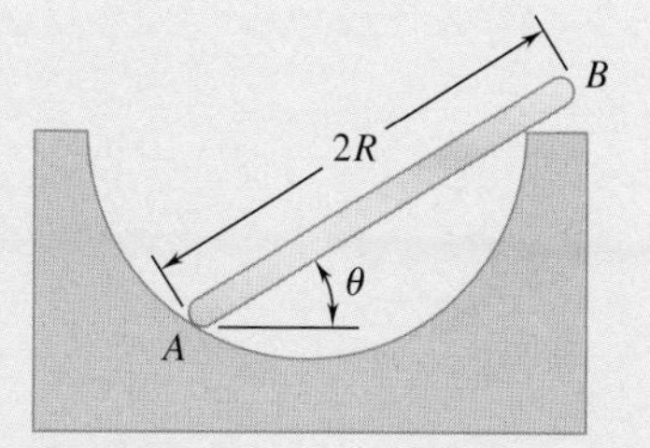

Fig. P4.50

4.3 EQUILIBRIUM IN THREE DIMENSIONS

The most general situation of rigid-body equilibrium occurs in three dimensions. The approach to modeling and analyzing these situations is the same as in two dimensions: Draw a free-body diagram and then write and solve the equilibrium equations. However, you now have more equations and more variables to deal with. In addition, reactions at supports and connections can be more varied, having as many as three force components and three couples acting at one support. As you will see in the Sample Problems, you need to visualize clearly in three dimensions and recall the vector analysis from Chapters 2 and 3.

4.3A Rigid-Body Equilibrium in Three Dimensions

We saw in Sec. 4.1 that six scalar equations are required to express the conditions for the equilibrium of a rigid body in the general three-dimensional case:

$$\Sigma F_x = 0 \qquad \Sigma F_y = 0 \qquad \Sigma F_z = 0 \tag{4.2}$$
$$\Sigma M_x = 0 \qquad \Sigma M_y = 0 \qquad \Sigma M_z = 0 \tag{4.3}$$

We can solve these equations for no more than *six unknowns,* which generally represent reactions at supports or connections.

In most problems, we can obtain the scalar equations (4.2) and (4.3) more conveniently if we first write the conditions for the equilibrium of the rigid body considered in vector form:

$$\Sigma \mathbf{F} = 0 \qquad \Sigma \mathbf{M}_O = \Sigma(\mathbf{r} \times \mathbf{F}) = 0 \tag{4.1}$$

Then we can express the forces **F** and position vectors **r** in terms of scalar components and unit vectors. This enables us to compute all vector products either by direct calculation or by means of determinants (see Sec. 3.1F). Note that we can eliminate as many as three unknown reaction components from these computations through a judicious choice of the point O. By equating to zero the coefficients of the unit vectors in each of the two relations in Eq. (4.1), we obtain the desired scalar equations.[†]

Some equilibrium problems and their associated free-body diagrams might involve individual couples $\mathbf{M}_i$ either as applied loads or as support reactions. In such situations, you can accommodate these couples by expressing the second part of Eq. (4.1) as

$$\Sigma \mathbf{M}_O = \Sigma(\mathbf{r} \times \mathbf{F}) + \Sigma \mathbf{M}_i = 0 \tag{4.1'}$$

4.3B Reactions for a Three-Dimensional Structure

The reactions on a three-dimensional structure range from a single force of known direction exerted by a frictionless surface to a force-couple system

[†]In some problems, it may be convenient to eliminate from the solution the reactions at two points A and B by writing the equilibrium equation $\Sigma M_{AB} = 0$. This involves determining the moments of the forces about the axis AB joining points A and B (see Sample Prob. 4.10).

exerted by a fixed support. Consequently, in problems involving the equilibrium of a three-dimensional structure, between one and six unknowns may be associated with the reaction at each support or connection.

Figure 4.10 shows various types of supports and connections with their corresponding reactions. A simple way of determining the type of reaction corresponding to a given support or connection and the number of unknowns involved is to find which of the six fundamental motions (translation in the x, y, and z directions and rotation about the x, y, and z axes) are allowed and which motions are prevented. The number of motions prevented equals the number of reactions.

Ball supports, frictionless surfaces, and cables, for example, prevent translation in one direction only and thus exert a single force whose line of action is known. Therefore, each of these supports involves one unknown—namely, the magnitude of the reaction. Rollers on rough surfaces and wheels on rails prevent translation in two directions; the corresponding reactions consist of two unknown force components. Rough surfaces in direct contact and ball-and-socket supports prevent translation in three directions while still allowing rotation; these supports involve three unknown force components.

Some supports and connections can prevent rotation as well as translation; the corresponding reactions include couples as well as forces. For example, the reaction at a fixed support, which prevents any motion (rotation as well as translation) consists of three unknown forces and three unknown couples. A universal joint, which is designed to allow rotation about two axes, exerts a reaction consisting of three unknown force components and one unknown couple.

Other supports and connections are primarily intended to prevent translation; their design, however, is such that they also prevent some rotations. The corresponding reactions consist essentially of force components but *may* also include couples. One group of supports of this type includes hinges and bearings designed to support radial loads only (for example, journal bearings or roller bearings). The corresponding reactions consist of two force components but may also include two couples. Another group includes pin-and-bracket supports, hinges, and bearings designed to support an axial thrust as well as a radial load (for example, ball bearings). The corresponding reactions consist of three force components but may include two couples. However, these supports do not exert any appreciable couples under normal conditions of use. Therefore, *only* force components should be included in their analysis *unless* it is clear that couples are necessary to maintain the equilibrium of the rigid body or unless the support is known to have been specifically designed to exert a couple (see Probs. 4.119 through 4.122).

If the reactions involve more than six unknowns, you have more unknowns than equations, and some of the reactions are **statically indeterminate**. If the reactions involve fewer than six unknowns, you have more equations than unknowns, and some of the equations of equilibrium cannot be satisfied under general loading conditions. In this case, the rigid body is only **partially constrained**. Under the particular loading conditions corresponding to a given problem, however, the extra equations often reduce to trivial identities, such as $0 = 0$, and can be disregarded; although only partially constrained, the rigid body remains in equilibrium (see Sample Probs. 4.7 and 4.8). Even with six or more unknowns, it is possible that some equations of equilibrium are not satisfied. This can occur when the reactions associated with the given supports either are parallel or intersect the same line; the rigid body is then **improperly constrained**.

Photo 4.3 Universal joints, seen on the drive shafts of rear-wheel-drive cars and trucks, allow rotational motion to be transferred between two noncollinear shafts.

Photo 4.4 This pillow block bearing supports the shaft of a fan used in an industrial facility.

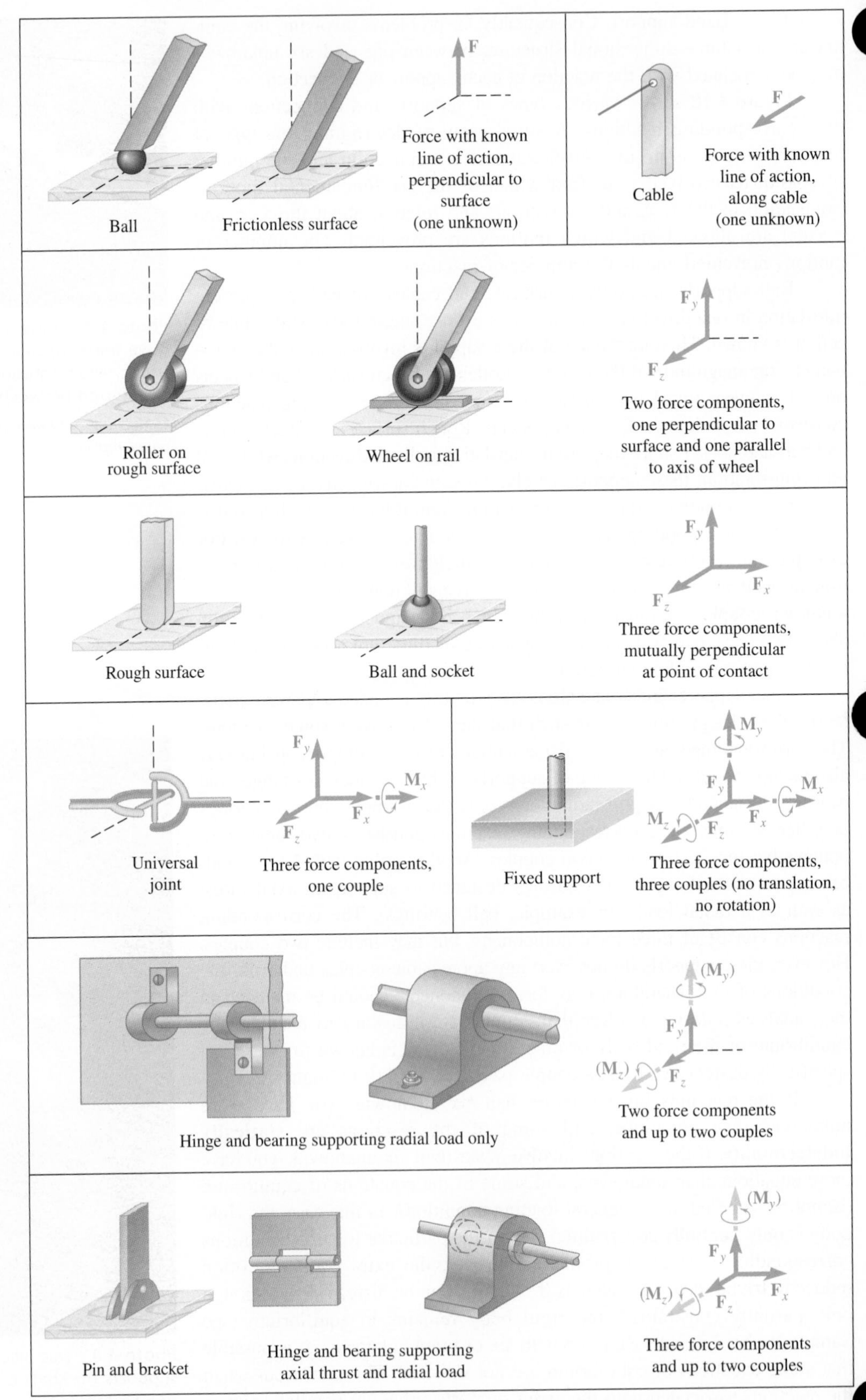

Fig. 4.10 Reactions at supports and connections in three dimensions.

Sample Problem 4.7

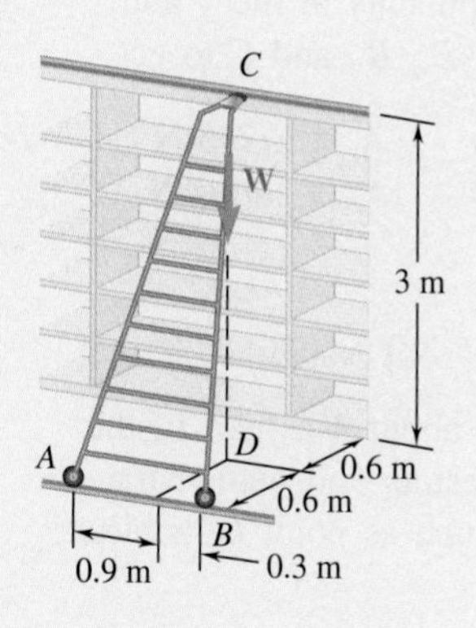

A 20-kg ladder used to reach high shelves in a storeroom is supported by two flanged wheels A and B mounted on a rail and by a flangeless wheel C resting against a rail fixed to the wall. An 80-kg man stands on the ladder and leans to the right. The line of action of the combined weight **W** of the man and ladder intersects the floor at point D. Determine the reactions at A, B, and C.

STRATEGY: Draw a free-body diagram of the ladder, then write and solve the equilibrium equations in three dimensions.

MODELING:

Free-Body Diagram. The combined weight of the man and ladder is

$$\mathbf{W} = -mg\mathbf{j} = -(80\text{ kg} + 20\text{ kg})(9.81\text{ m/s}^2)\mathbf{j} = -(981\text{ N})\mathbf{j}$$

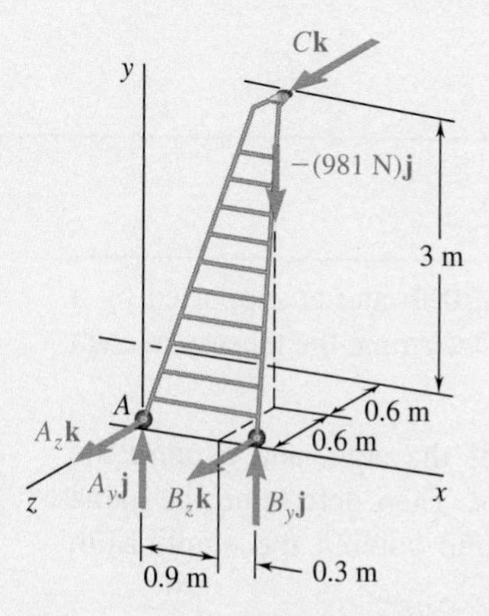

Fig. 1 Free-body diagram of ladder.

You have five unknown reaction components: two at each flanged wheel and one at the flangeless wheel (Fig. 1). The ladder is thus only partially constrained; it is free to roll along the rails. It is, however, in equilibrium under the given load because the equation $\Sigma F_x = 0$ is satisfied.

ANALYSIS:

Equilibrium Equations. The forces acting on the ladder form a system equivalent to zero:

$$\Sigma\mathbf{F} = 0: \quad A_y\mathbf{j} + A_z\mathbf{k} + B_y\mathbf{j} + B_z\mathbf{k} - (981\text{ N})\mathbf{j} + C\mathbf{k} = 0$$

$$(A_y + B_y - 981\text{ N})\mathbf{j} + (A_z + B_z + C)\mathbf{k} = 0 \quad (1)$$

$$\Sigma\mathbf{M}_A = \Sigma(\mathbf{r} \times \mathbf{F}) = 0: \quad 1.2\mathbf{i} \times (B_y\mathbf{j} + B_z\mathbf{k}) + (0.9\mathbf{i} - 0.6\mathbf{k}) \times (-981\mathbf{j}) + (0.6\mathbf{i} + 3\mathbf{j} - 1.2\mathbf{k}) \times C\mathbf{k} = 0$$

Computing the vector products gives you[†]

$$1.2B_y\mathbf{k} - 1.2B_z\mathbf{j} - 882.9\mathbf{k} - 588.6\mathbf{i} - 0.6C\mathbf{j} + 3C\mathbf{i} = 0$$

$$(3C - 588.6)\mathbf{i} - (1.2B_z + 0.6C)\mathbf{j} + (1.2B_y - 882.9)\mathbf{k} = 0 \quad (2)$$

Setting the coefficients of **i**, **j**, **and k** equal to zero in Eq. (2) produces the following three scalar equations, which state that the sum of the moments about each coordinate axis must be zero:

$$3C - 588.6 = 0 \qquad C = +196.2\text{ N}$$

$$1.2B_z + 0.6C = 0 \qquad B_z = -98.1\text{ N}$$

$$1.2B_y - 882.9 = 0 \qquad B_y = +736\text{ N}$$

The reactions at B and C are therefore

$$\mathbf{B} = +(736\text{ N})\mathbf{j} - (98.1\text{ N})\mathbf{k} \qquad \mathbf{C} = +(196.2\text{ N})\mathbf{k} \quad ◀$$

[†]The moments in this sample problem, as well as in Sample Probs. 4.8 and 4.9, also can be expressed as determinants (see Sample Prob. 3.10).

Setting the coefficients of **j** and **k** equal to zero in Eq. (1), you obtain two scalar equations stating that the sums of the components in the y and z directions are zero. Substitute the values above for B_y, B_z, and C to get

$$A_y + B_y - 981 = 0 \qquad A_y + 736 - 981 = 0 \qquad A_y = +245 \text{ N}$$
$$A_z + B_z + C = 0 \qquad A_z - 98.1 + 196.2 = 0 \qquad A_z = -98.1 \text{ N}$$

Therefore, the reaction at A is

$$\mathbf{A} = +(245 \text{ N})\mathbf{j} - (98.1 \text{ N})\mathbf{k} \quad \blacktriangleleft$$

REFLECT and THINK: You summed moments about A as part of the analysis. As a check, you could now use these results and demonstrate that the sum of moments about any other point, such as point B, is also zero.

Sample Problem 4.8

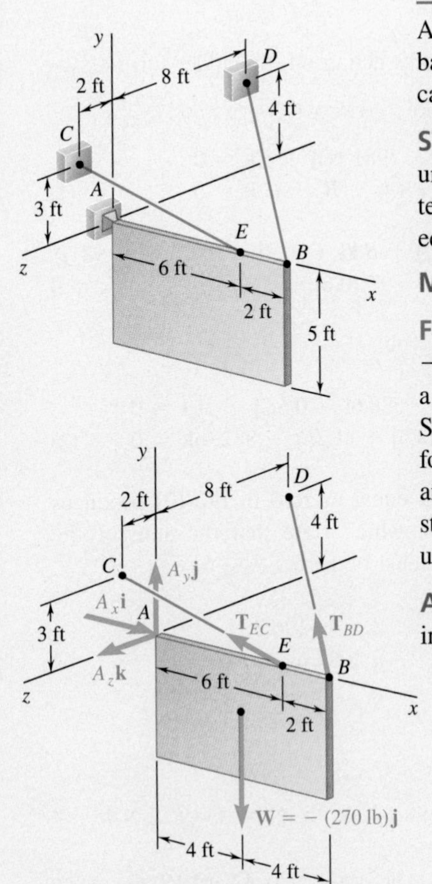

Fig. 1 Free-body diagram of sign.

A 5 × 8-ft sign of uniform density weighs 270 lb and is supported by a ball-and-socket joint at A and by two cables. Determine the tension in each cable and the reaction at A.

STRATEGY: Draw a free-body diagram of the sign, and express the unknown cable tensions as Cartesian vectors. Then determine the cable tensions and the reaction at A by writing and solving the equilibrium equations.

MODELING:

Free-Body Diagram. The forces acting on the sign are its weight $\mathbf{W} = -(270 \text{ lb})\mathbf{j}$ and the reactions at A, B, and E (Fig. 1). The reaction at A is a force of unknown direction represented by three unknown components. Since the directions of the forces exerted by the cables are known, these forces involve only one unknown each: specifically, the magnitudes T_{BD} and T_{EC}. The total of five unknowns means that the sign is partially constrained. It can rotate freely about the x axis; it is, however, in equilibrium under the given loading, since the equation $\Sigma M_x = 0$ is satisfied.

ANALYSIS: You can express the components of the forces $\mathbf{T}_{BD}$ and $\mathbf{T}_{EC}$ in terms of the unknown magnitudes T_{BD} and T_{EC} as follows:

$$\overrightarrow{BD} = -(8 \text{ ft})\mathbf{i} + (4 \text{ ft})\mathbf{j} - (8 \text{ ft})\mathbf{k} \qquad BD = 12 \text{ ft}$$
$$\overrightarrow{EC} = -(6 \text{ ft})\mathbf{i} + (3 \text{ ft})\mathbf{j} + (2 \text{ ft})\mathbf{k} \qquad EC = 7 \text{ ft}$$
$$\mathbf{T}_{BD} = T_{BD}\left(\frac{\overrightarrow{BD}}{BD}\right) = T_{BD}(-\tfrac{2}{3}\mathbf{i} + \tfrac{1}{3}\mathbf{j} - \tfrac{2}{3}\mathbf{k})$$
$$\mathbf{T}_{EC} = T_{EC}\left(\frac{\overrightarrow{EC}}{EC}\right) = T_{EC}(-\tfrac{6}{7}\mathbf{i} + \tfrac{3}{7}\mathbf{j} - \tfrac{2}{7}\mathbf{k})$$

Equilibrium Equations. The forces acting on the sign form a system equivalent to zero:

$$\Sigma \mathbf{F} = 0: \quad A_x\mathbf{i} + A_y\mathbf{j} + A_z\mathbf{k} + \mathbf{T}_{BD} + \mathbf{T}_{EC} - (270 \text{ lb})\mathbf{j} = 0$$

$$(A_x - \tfrac{2}{3}T_{BD} - \tfrac{6}{7}T_{EC})\mathbf{i} + (A_y + \tfrac{1}{3}T_{BD} + \tfrac{3}{7}T_{EC} - 270 \text{ lb})\mathbf{j} + (A_z - \tfrac{2}{3}T_{BD} + \tfrac{2}{7}T_{EC})\mathbf{k} = 0 \quad (1)$$

$$\Sigma \mathbf{M}_A = \Sigma(\mathbf{r} \times \mathbf{F}) = 0:$$

$$(8 \text{ ft})\mathbf{i} \times T_{BD}(-\tfrac{2}{3}\mathbf{i} + \tfrac{1}{3}\mathbf{j} - \tfrac{2}{3}\mathbf{k}) + (6 \text{ ft})\mathbf{i} \times T_{EC}(-\tfrac{6}{7}\mathbf{i} + \tfrac{3}{7}\mathbf{j} + \tfrac{2}{7}\mathbf{k}) + (4 \text{ ft})\mathbf{i} \times (-270 \text{ lb})\mathbf{j} = 0$$

$$(2.667T_{BD} + 2.571T_{EC} - 1080 \text{ lb})\mathbf{k} + (5.333T_{BD} - 1.714T_{EC})\mathbf{j} = 0 \quad (2)$$

Setting the coefficients of **j** and **k** equal to zero in Eq. (2) yields two scalar equations that can be solved for T_{BD} and T_{EC}:

$$T_{BD} = 101.3 \text{ lb} \qquad T_{EC} = 315 \text{ lb} \blacktriangleleft$$

Setting the coefficients of **i**, **j**, and **k** equal to zero in Eq. (1) produces three more equations, which yield the components of **A**.

$$\mathbf{A} = +(338 \text{ lb})\mathbf{i} + (101.2 \text{ lb})\mathbf{j} - (22.5 \text{ lb})\mathbf{k} \blacktriangleleft$$

REFLECT and THINK: Cables can only act in tension, and the free-body diagram and Cartesian vector expressions for the cables were consistent with this. The solution yielded positive results for the cable forces, which confirms that they are in tension and validates the analysis.

Sample Problem 4.9

A uniform pipe cover of radius r = 240 mm and mass 30 kg is held in a horizontal position by the cable CD. Assuming that the bearing at B does not exert any axial thrust, determine the tension in the cable and the reactions at A and B.

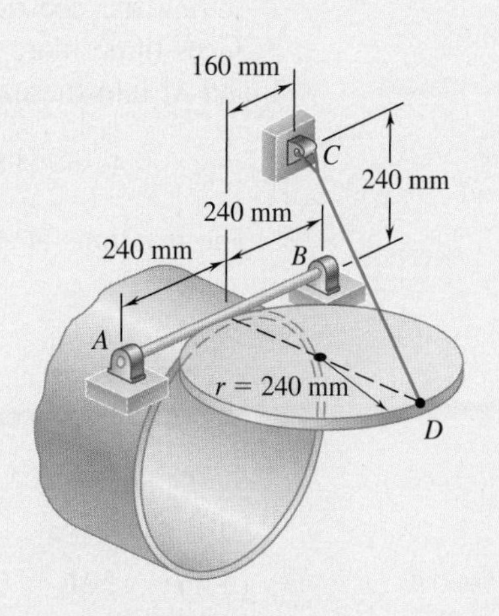

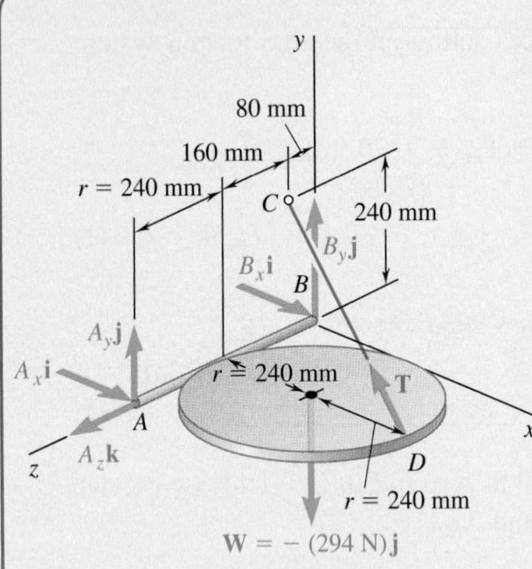

Fig. 1 Free-body diagram of pipe cover.

STRATEGY: Draw a free-body diagram with the coordinate axes shown (Fig. 1) and express the unknown cable tension as a Cartesian vector. Then apply the equilibrium equations to determine this tension and the support reactions.

MODELING:

Free-Body Diagram. The forces acting on the free body include its weight, which is

$$\mathbf{W} = -mg\mathbf{j} = -(30 \text{ kg})(9.81 \text{ m/s}^2)\mathbf{j} = -(294 \text{ N})\mathbf{j}$$

The reactions involve six unknowns: the magnitude of the force **T** exerted by the cable, three force components at hinge A, and two at hinge B. Express the components of **T** in terms of the unknown magnitude T by resolving the vector $\overrightarrow{DC}$ into rectangular components:

$$\overrightarrow{DC} = -(480 \text{ mm})\mathbf{i} + (240 \text{ mm})\mathbf{j} - (160 \text{ mm})\mathbf{k} \qquad DC = 560 \text{ mm}$$

$$\mathbf{T} = T\frac{\overrightarrow{DC}}{DC} = -\tfrac{6}{7}T\mathbf{i} + \tfrac{3}{7}T\mathbf{j} - \tfrac{2}{7}T\mathbf{k}$$

ANALYSIS:

Equilibrium Equations. The forces acting on the pipe cover form a system equivalent to zero. Thus,

$$\Sigma\mathbf{F} = 0: \qquad A_x\mathbf{i} + A_y\mathbf{j} + A_z\mathbf{k} + B_x\mathbf{i} + B_y\mathbf{j} + \mathbf{T} - (294 \text{ N})\mathbf{j} = 0$$

$$(A_x + B_x - \tfrac{6}{7}T)\mathbf{i} + (A_y + B_y + \tfrac{3}{7}T - 294 \text{ N})\mathbf{j} + (A_z - \tfrac{2}{7}T)\mathbf{k} = 0 \quad (1)$$

$$\Sigma\mathbf{M}_B = \Sigma(\mathbf{r} \times \mathbf{F}) = 0:$$

$$2r\mathbf{k} \times (A_x\mathbf{i} + A_y\mathbf{j} + A_z\mathbf{k})$$
$$+ (2r\mathbf{i} + r\mathbf{k}) \times (-\tfrac{6}{7}T\mathbf{i} + \tfrac{3}{7}T\mathbf{j} - \tfrac{2}{7}T\mathbf{k})$$
$$+ (r\mathbf{i} + r\mathbf{k}) \times (-294 \text{ N})\mathbf{j} = 0$$

$$(-2A_y - \tfrac{3}{7}T + 294 \text{ N})r\mathbf{i} + (2A_x - \tfrac{2}{7}T)r\mathbf{j} + (\tfrac{6}{7}T - 294 \text{ N})r\mathbf{k} = 0 \quad (2)$$

Setting the coefficients of the unit vectors equal to zero in Eq. (2) gives three scalar equations, which yield

$$A_x = +49.0 \text{ N} \qquad A_y = +73.5 \text{ N} \qquad T = 343 \text{ N} \blacktriangleleft$$

Setting the coefficients of the unit vectors equal to zero in Eq. (1) produces three more scalar equations. After substituting the values of T, A_x, and A_y into these equations, you obtain

$$A_z = +98.0 \text{ N} \qquad B_x = +245 \text{ N} \qquad B_y = +73.5 \text{ N}$$

The reactions at A and B are therefore

$$\mathbf{A} = +(49.0 \text{ N})\mathbf{i} + (73.5 \text{ N})\mathbf{j} + (98.0 \text{ N})\mathbf{k} \blacktriangleleft$$

$$\mathbf{B} = +(245 \text{ N})\mathbf{i} + (73.5 \text{ N})\mathbf{j} \blacktriangleleft$$

REFLECT and THINK: As a check, you can determine the tension in the cable using a scalar analysis. Assigning signs by the right-hand rule (rhr), we have

$$(+\text{rhr}) \quad \Sigma M_z = 0: \quad \tfrac{3}{7}T(0.48 \text{ m}) - (294 \text{ N})(0.24 \text{ m}) = 0 \quad T = 343 \text{ N} \blacktriangleleft$$

Sample Problem 4.10

A 450-lb load hangs from the corner C of a rigid piece of pipe $ABCD$ that has been bent as shown. The pipe is supported by ball-and-socket joints A and D, which are fastened, respectively, to the floor and to a vertical wall, and by a cable attached at the midpoint E of the portion BC of the pipe and at a point G on the wall. Determine (*a*) where G should be located if the tension in the cable is to be minimum, (*b*) the corresponding minimum value of the tension.

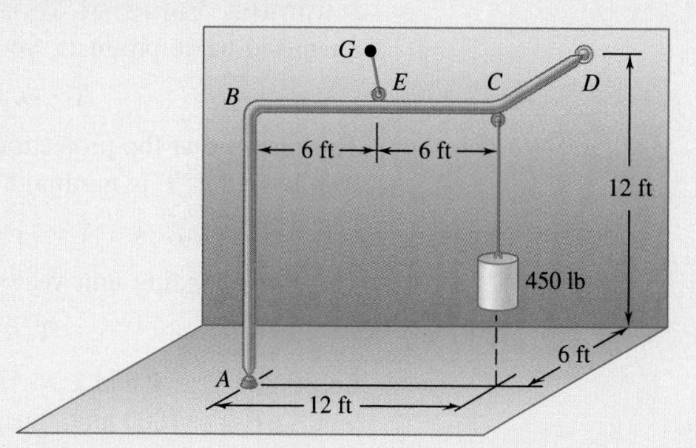

STRATEGY: Draw the free-body diagram of the pipe showing the reactions at A and D. Isolate the unknown tension $\mathbf{T}$ and the known weight $\mathbf{W}$ by summing moments about the diagonal line AD, and compute values from the equilibrium equations.

MODELING and ANALYSIS:

Free-Body Diagram. The free-body diagram of the pipe includes the load $\mathbf{W} = (-450 \text{ lb})\mathbf{j}$, the reactions at A and D, and the force $\mathbf{T}$ exerted by the cable (Fig. 1). To eliminate the reactions at A and D from the computations, take the sum of the moments of the forces about the line AD and set it equal to zero. Denote the unit vector along AD by $\boldsymbol{\lambda}$, which enables you to write

$$\Sigma M_{AD} = 0: \quad \boldsymbol{\lambda} \cdot (\overrightarrow{AE} \times \mathbf{T}) + \boldsymbol{\lambda} \cdot (\overrightarrow{AC} \times \mathbf{W}) = 0 \tag{1}$$

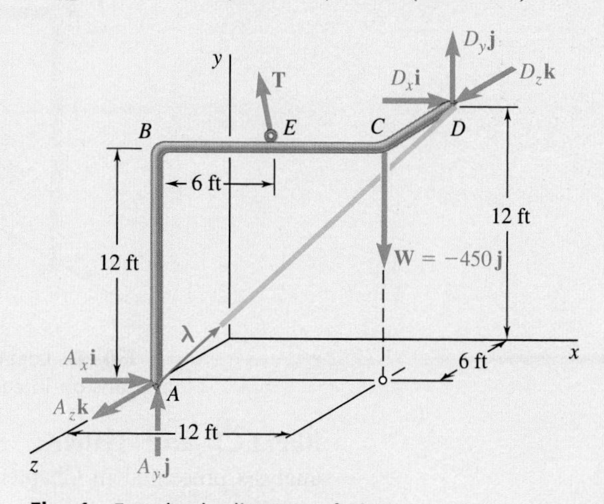

Fig. 1 Free-body diagram of pipe.

You can compute the second term in Eq. (1) as follows:

$$\overrightarrow{AC} \times \mathbf{W} = (12\mathbf{i} + 12\mathbf{j}) \times (-450\mathbf{j}) = -5400\mathbf{k}$$

$$\boldsymbol{\lambda} = \frac{\overrightarrow{AD}}{AD} = \frac{12\mathbf{i} + 12\mathbf{j} - 6\mathbf{k}}{18} = \tfrac{2}{3}\mathbf{i} + \tfrac{2}{3}\mathbf{j} - \tfrac{1}{3}\mathbf{k}$$

$$\boldsymbol{\lambda} \cdot (\overrightarrow{AC} \times \mathbf{W}) = (\tfrac{2}{3}\mathbf{i} + \tfrac{2}{3}\mathbf{j} - \tfrac{1}{3}\mathbf{k}) \cdot (-5400\mathbf{k}) = +1800$$

Substituting this value into Eq. (1) gives

$$\boldsymbol{\lambda} \cdot (\overrightarrow{AE} \times \mathbf{T}) = -1800 \text{ lb·ft} \tag{2}$$

Minimum Value of Tension. Recalling the commutative property for mixed triple products, you can rewrite Eq. (2) in the form

$$\mathbf{T} \cdot (\boldsymbol{\lambda} \times \overrightarrow{AE}) = -1800 \text{ lb·ft} \tag{3}$$

This shows that the projection of **T** on the vector $\boldsymbol{\lambda} \times \overrightarrow{AE}$ is a constant. It follows that **T** is minimum when it is parallel to the vector

$$\boldsymbol{\lambda} \times \overrightarrow{AE} = (\tfrac{2}{3}\mathbf{i} + \tfrac{2}{3}\mathbf{j} - \tfrac{1}{3}\mathbf{k}) \times (6\mathbf{i} + 12\mathbf{j}) = 4\mathbf{i} - 2\mathbf{j} + 4\mathbf{k}$$

The corresponding unit vector is $\tfrac{2}{3}\mathbf{i} - \tfrac{1}{3}\mathbf{j} + \tfrac{2}{3}\mathbf{k}$, which gives

$$\mathbf{T}_{\min} = T(\tfrac{2}{3}\mathbf{i} - \tfrac{1}{3}\mathbf{j} + \tfrac{2}{3}\mathbf{k}) \tag{4}$$

Substituting for **T** and $\boldsymbol{\lambda} \times \overrightarrow{AE}$ in Eq. (3) and computing the dot products yields $6T = -1800$ and, thus, $T = -300$. Carrying this value into Eq. (4) gives you

$$\mathbf{T}_{\min} = -200\mathbf{i} + 100\mathbf{j} - 200\mathbf{k} \qquad T_{\min} = 300 \text{ lb}$$ ◀

Location of G. Since the vector $\overrightarrow{EG}$ and the force $\mathbf{T}_{\min}$ have the same direction, their components must be proportional. Denoting the coordinates of G by x, y, and 0 (Fig. 2), you get

$$\frac{x-6}{-200} = \frac{y-12}{+100} = \frac{0-6}{-200} \qquad x = 0 \qquad y = 15 \text{ ft}$$ ◀

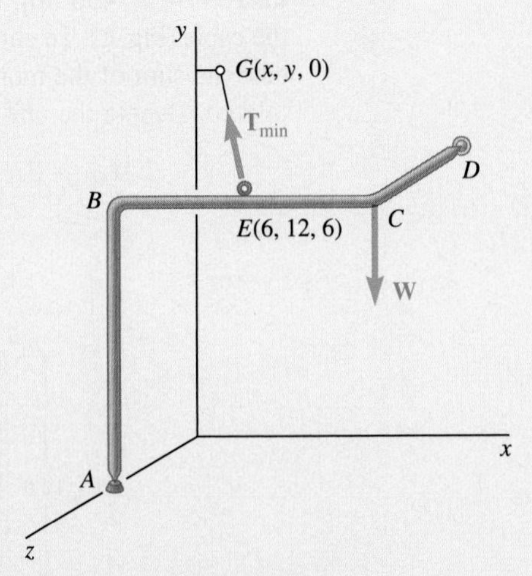

Fig. 2 Location of point G for minimum tension in cable.

REFLECT and THINK: Sometimes you have to rely on the vector analysis presented in Chapters 2 and 3 as much as on the conditions for equilibrium described in this chapter.

Problems

4.51 Two transmission belts pass over a double-sheaved pulley that is attached to an axle supported by bearings at A and D. The radius of the inner sheave is 125 mm and the radius of the outer sheave is 250 mm. Knowing that when the system is at rest, the tension is 90 N in both portions of belt B and 150 N in both portions of belt C, determine the reactions at A and D. Assume that the bearing at D does not exert any axial thrust.

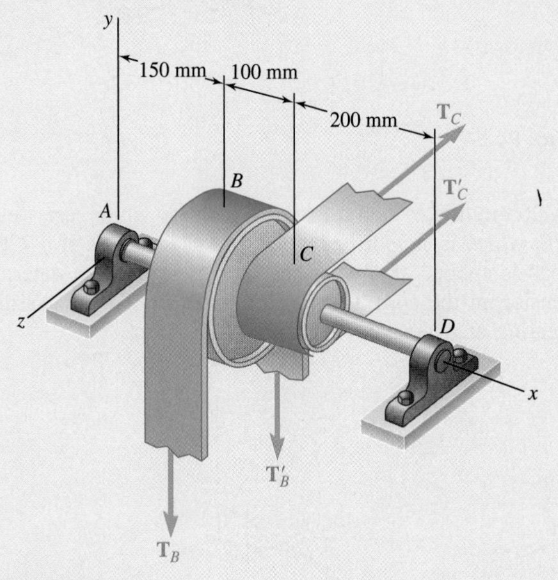

Fig. P4.51

4.52 Solve Prob. 4.51, assuming that the pulley rotates at a constant rate and that $T_B = 104$ N, $T'_B = 84$ N, $T_C = 175$ N.

4.53 A 4 × 8-ft sheet of plywood weighing 40 lb has been temporarily propped against column CD. It rests at A and B on small wooden blocks and against protruding nails. Neglecting friction at all surfaces of contact, determine the reactions at A, B, and C.

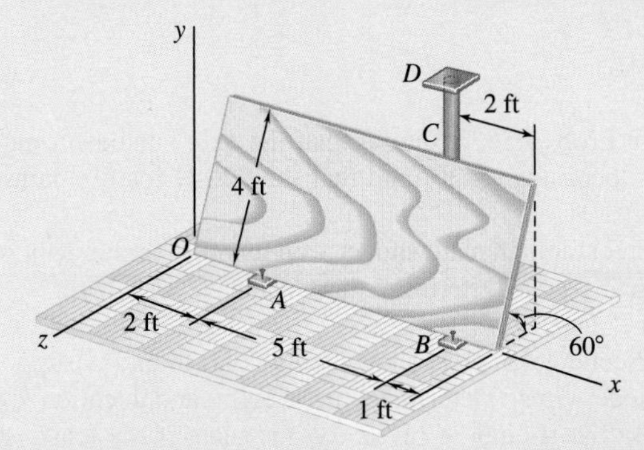

Fig. P4.53

4.54 A small winch is used to raise a 120-lb load. Find (*a*) the magnitude of the vertical force **P** that should be applied at *C* to maintain equilibrium in the position shown, (*b*) the reactions at *A* and *B*, assuming that the bearing at *B* does not exert any axial thrust.

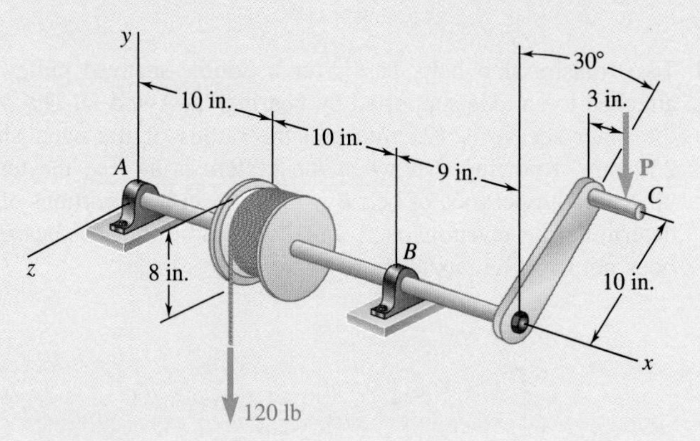

Fig. P4.54

4.55 A 200-mm lever and a 240-mm-diameter pulley are welded to axle *BE*, which is supported by bearings at *C* and *D*. If a 720-N vertical load is applied at *A* when the lever is horizontal, determine (*a*) the tension in the cord, (*b*) the reactions at *C* and *D*. Assume that the bearing at *D* does not exert any axial thrust.

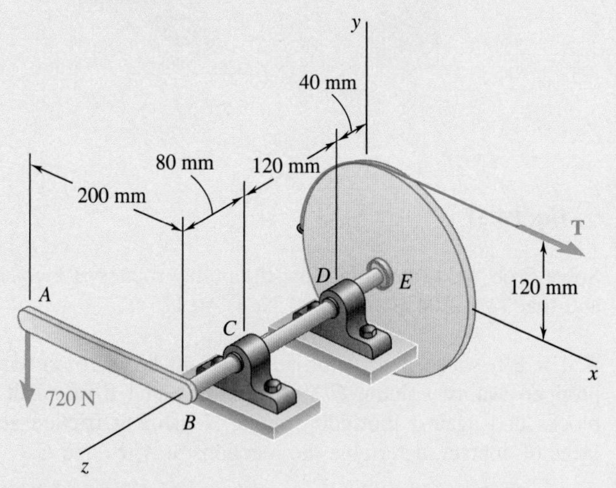

Fig. P4.55

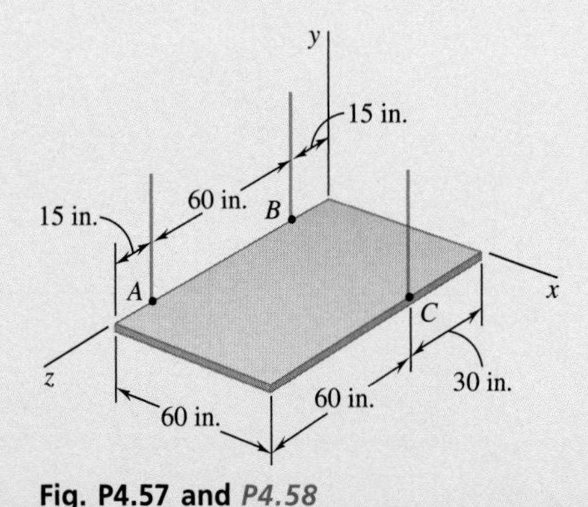

Fig. P4.57 and P4.58

4.56 Solve Prob. 4.55, assuming that the axle has been rotated clockwise in its bearings by 30° and that the 720-N load remains vertical.

4.57 The rectangular plate shown weighs 80 lb and is supported by three vertical wires. Determine the tension in each wire.

4.58 The rectangular plate shown weighs 80 lb and is supported by three vertical wires. Determine the weight and location of the lightest block that should be placed on the plate if the tensions in the three wires are to be equal.

4.59 An opening in a floor is covered by a 1 × 1.2-m sheet of plywood of mass 18 kg. The sheet is hinged at *A* and *B* and is maintained in a position slightly above the floor by a small block *C*. Determine the vertical component of the reaction (*a*) at *A*, (*b*) at *B*, (*c*) at *C*.

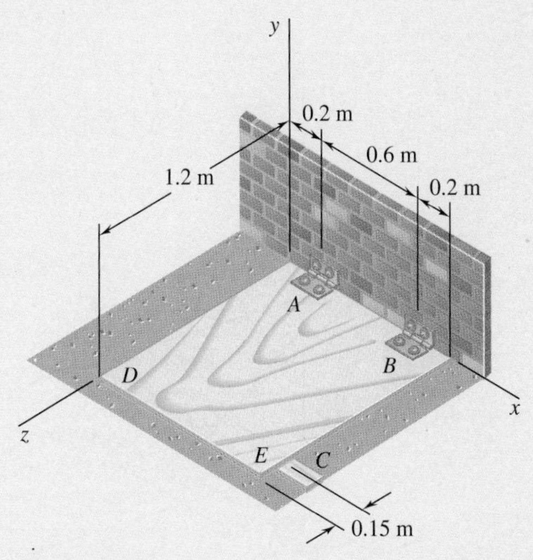

Fig. P4.59

4.60 Solve Prob. 4.59, assuming that the small block *C* is moved and placed under edge *DE* at a point 0.15 m from corner *E*.

4.61 A 48-in. boom is held by a ball-and-socket joint at *C* and by two cables *BF* and *DAE*; cable *DAE* passes around a frictionless pulley at *A*. For the loading shown, determine the tension in each cable and the reaction at *C*.

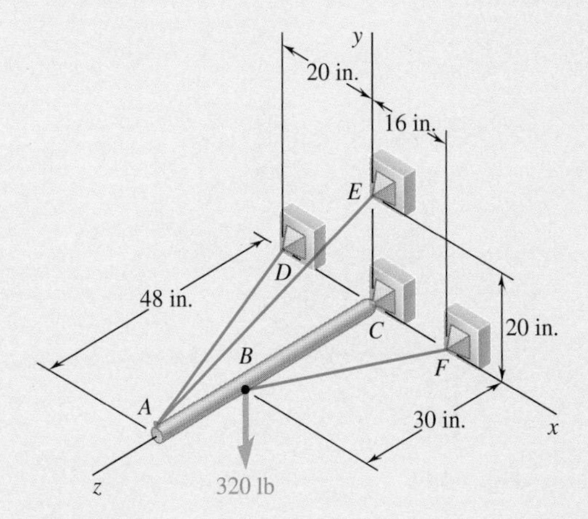

Fig. P4.61

4.62 Solve Prob. 4.61, assuming that the 320-lb load is applied at *A*.

4.63 The 6-m pole *ABC* is acted upon by a 455-N force as shown. The pole is held by a ball-and-socket joint at *A* and by two cables *BD* and *BE*. For *a* = 3 m, determine the tension in each cable and the reaction at *A*.

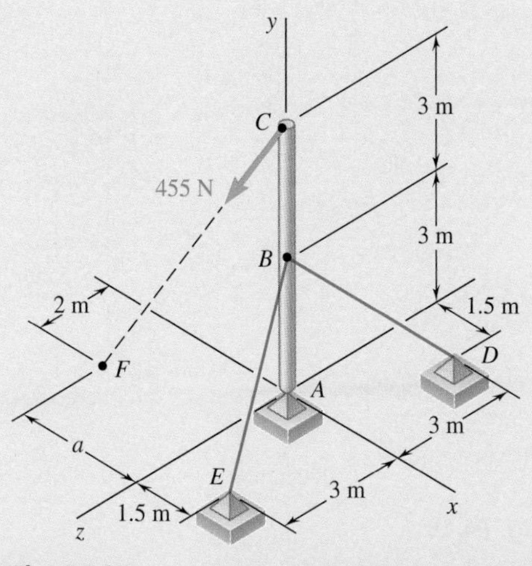

Fig. P4.63

4.64 A 600-lb crate hangs from a cable that passes over a pulley B and is attached to a support at H. The 200-lb boom AB is supported by a ball-and-socket joint at A and by two cables DE and DF. The center of gravity of the boom is located at G. Determine (a) the tension in cables DE and DF, (b) the reaction at A.

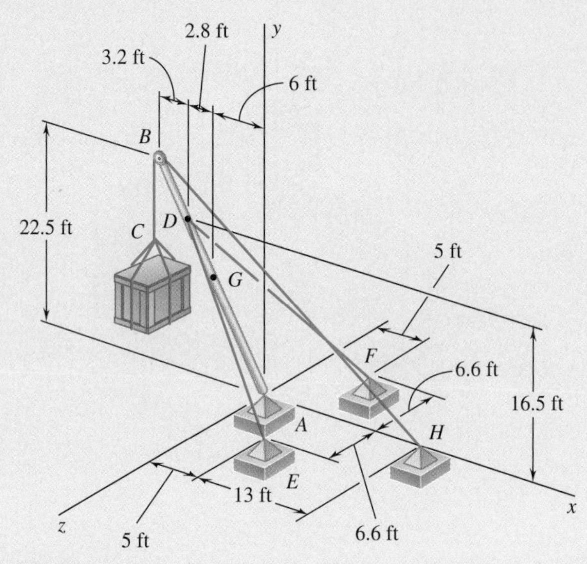

Fig. ***P4.64***

4.65 The horizontal platform $ABCD$ weighs 60 lb and supports a 240-lb load at its center. The platform is normally held in position by hinges at A and B and by braces CE and DE. If brace DE is removed, determine the reactions at the hinges and the force exerted by the remaining brace CE. The hinge at A does not exert any axial thrust.

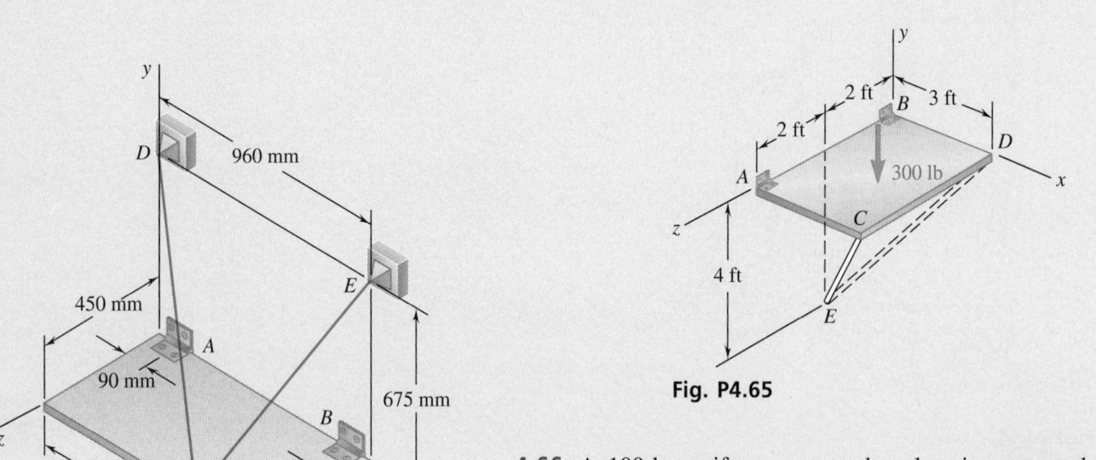

Fig. P4.65

4.66 A 100-kg uniform rectangular plate is supported in the position shown by hinges A and B and by cable DCE that passes over a frictionless hook at C. Assuming that the tension is the same in both parts of the cable, determine (a) the tension in the cable, (b) the reactions at A and B. Assume that the hinge at B does not exert any axial thrust.

Fig. P4.66

4.67 The rectangular plate shown weighs 75 lb and is held in the position shown by hinges at A and B and by cable EF. Assuming that the hinge at B does not exert any axial thrust, determine (*a*) the tension in the cable, (*b*) the reactions at A and B.

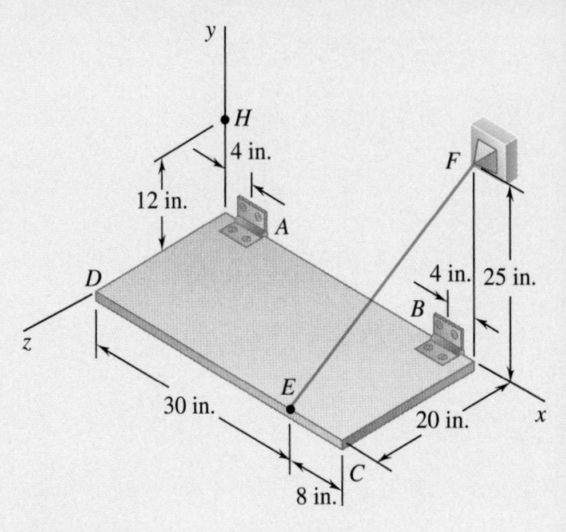

Fig. P4.67

4.68 The lid of a roof scuttle weighs 75 lb. It is hinged at corners A and B and maintained in the desired position by a rod CD pivoted at C; a pin at end D of the rod fits into one of several holes drilled in the edge of the lid. For $\alpha = 50°$, determine (*a*) the magnitude of the force exerted by rod CD, (*b*) the reactions at the hinges. Assume that the hinge at B does not exert any axial thrust.

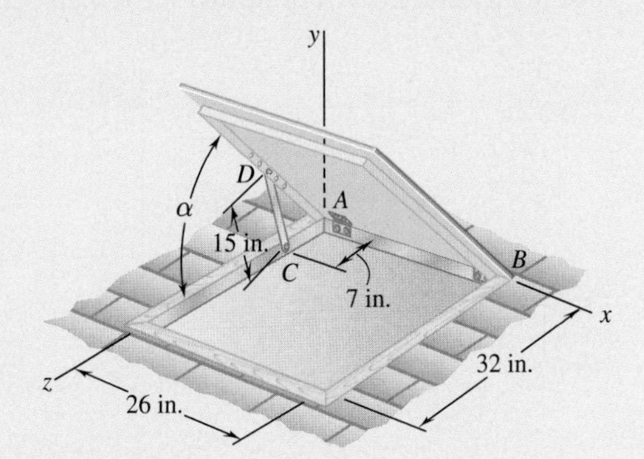

Fig. *P4.68*

4.69 A 10-kg storm window measuring 900 × 1500 mm is held by hinges at A and B. In the position shown, it is held away from the side of the house by a 600-mm stick CD. Assuming that the hinge at A does not exert any axial thrust, determine the magnitude of the force exerted by the stick and the components of the reactions at A and B.

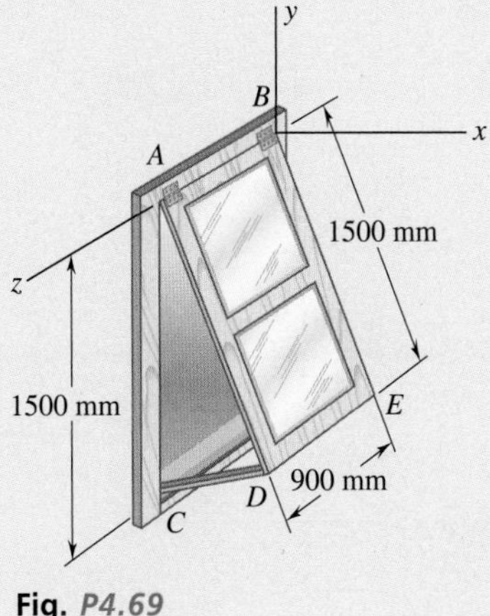

Fig. *P4.69*

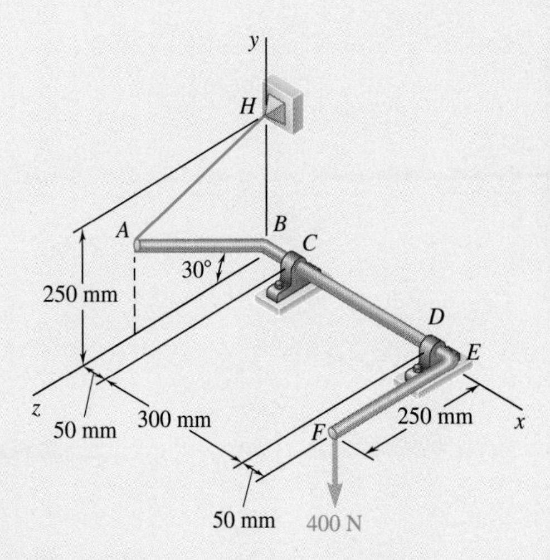

Fig. P4.70

4.70 The bent rod $ABEF$ is supported by bearings at C and D and by wire AH. Knowing that portion AB of the rod is 250 mm long, determine (*a*) the tension in wire AH, (*b*) the reactions at C and D. Assume that the bearing at D does not exert any axial thrust.

4.71 Solve Prob. 4.65, assuming that the hinge at B has been removed and that the hinge at A can exert an axial thrust, as well as couples about axes parallel to the x and y axes.

4.72 Solve Prob. 4.69, assuming that the hinge at A has been removed and that the hinge at B can exert couples about axes parallel to the x and y axes.

4.73 The assembly shown is welded to collar A that fits on the vertical pin shown. The pin can exert couples about the x and z axes but does not prevent motion about or along the y axis. For the loading shown, determine the tension in each cable and the reaction at A.

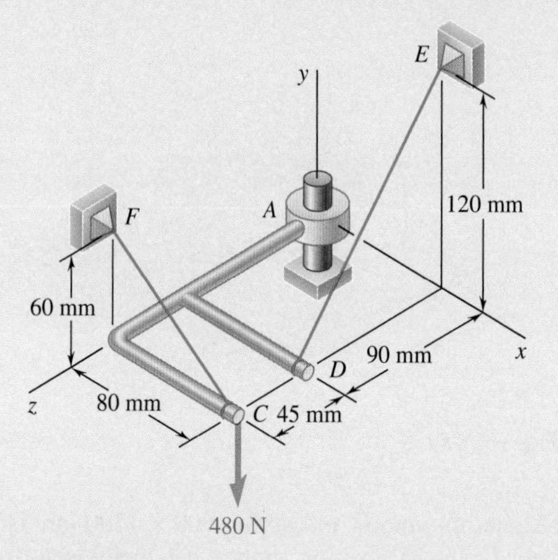

Fig. P4.73

4.74 Three rods are welded together to form a "corner" that is supported by three eyebolts. Neglecting friction, determine the reactions at A, B, and C when $P = 240$ lb, $a = 12$ in., $b = 8$ in., and $c = 10$ in.

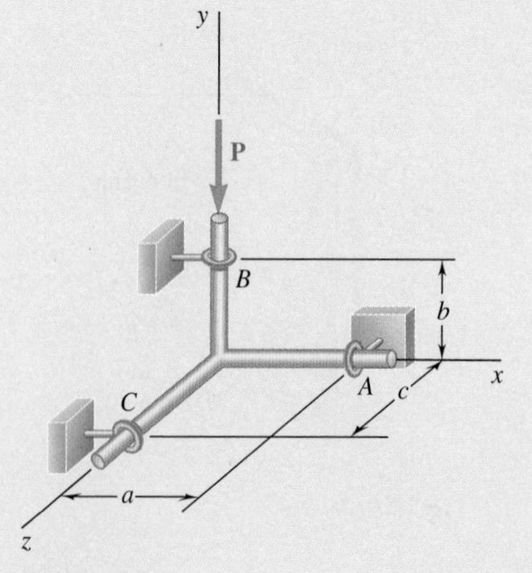

Fig. P4.74

4.4 FRICTION FORCES

In the previous sections, we assumed that surfaces in contact are either *frictionless* or *rough*. If they are frictionless, the force each surface exerts on the other is normal to the surfaces, and the two surfaces can move freely with respect to each other. If they are rough, tangential forces can develop that prevent the motion of one surface with respect to the other.

This view is a simplified one. Actually, no perfectly frictionless surface exists. When two surfaces are in contact, tangential forces, called **friction forces**, always develop if you attempt to move one surface with respect to the other. However, these friction forces are limited in magnitude and do not prevent motion if you apply sufficiently large forces. Thus, the distinction between frictionless and rough surfaces is a matter of degree. You will see this more clearly in this chapter, which is devoted to the study of friction and its applications to common engineering situations.

There are two types of friction: **dry friction**, sometimes called *Coulomb friction*, and **fluid friction** or *viscosity*. Fluid friction develops between layers of fluid moving at different velocities. This is of great importance in analyzing problems involving the flow of fluids through pipes and orifices or dealing with bodies immersed in moving fluids. It is also basic for the analysis of the motion of *lubricated mechanisms*. Such problems are considered in texts on fluid mechanics. The present study is limited to dry friction, i.e., to situations involving rigid bodies that are in contact along *unlubricated* surfaces.

4.4A The Laws of Dry Friction

We can illustrate the laws of dry friction by the following experiment. Place a block of weight **W** on a horizontal plane surface (Fig. 4.11*a*). The forces acting on the block are its weight **W** and the reaction of the surface. Since

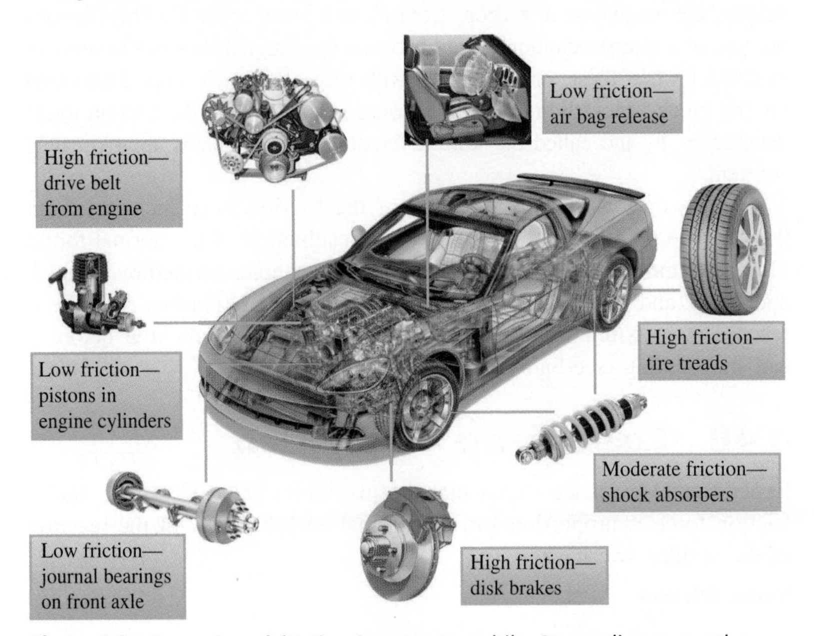

Photo 4.5 Examples of friction in an automobile. Depending upon the application, the degree of friction is controlled by design engineers.

the weight has no horizontal component, the reaction of the surface also has no horizontal component; the reaction is therefore *normal* to the surface and is represented by **N** in Fig. 4.11*a*. Now suppose that you apply a horizontal force **P** to the block (Fig. 4.11*b*). If **P** is small, the block does not move; some other horizontal force must therefore exist, which balances **P**. This other force is the **static-friction force F**, which is actually the resultant of a great number of forces acting over the entire surface of contact between the block and the plane. The nature of these forces is not known exactly, but we generally assume that these forces are due to the irregularities of the surfaces in contact and, to a certain extent, to molecular attraction.

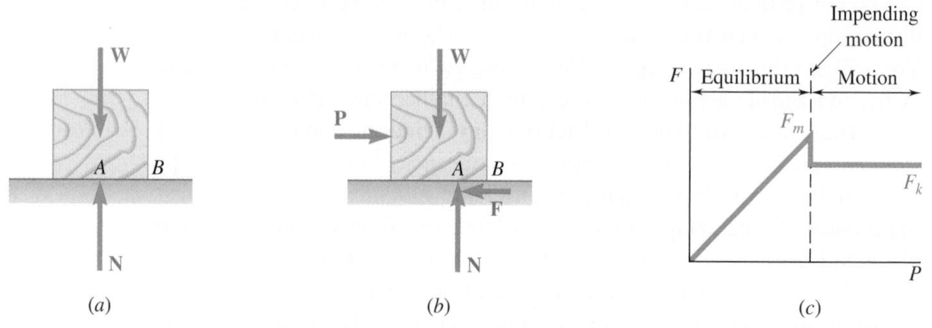

Fig. 4.11 (*a*) Block on a horizontal plane, friction force is zero; (*b*) a horizontally applied force **P** produces an opposing friction force **F**; (*c*) graph of **F** with increasing **P**.

If you increase the force **P**, the friction force **F** also increases, continuing to oppose **P**, until its magnitude reaches a certain *maximum value* F_m (Fig. 4.11*c*). If **P** is further increased, the friction force cannot balance it anymore, and the block starts sliding. As soon as the block has started in motion, the magnitude of **F** drops from F_m to a lower value F_k. This happens because less interpenetration occurs between the irregularities of the surfaces in contact when these surfaces move with respect to each other. From then on, the block keeps sliding with increasing velocity while the friction force, denoted by $\mathbf{F}_k$ and called the **kinetic-friction force**, remains approximately constant.

Note that, as the magnitude F of the friction force increases from 0 to F_m, the point of application A of the resultant **N** of the normal forces of contact moves to the right. In this way, the couples formed by **P** and **F** and by **W** and **N**, respectively, remain balanced. If **N** reaches B before F reaches its maximum value F_m, the block starts to tip about B before it can start sliding (see Sample Prob. 4.14).

4.4B Coefficients of Friction

Experimental evidence shows that the maximum value F_m of the static-friction force is proportional to the normal component N of the reaction of the surface. We have

Static friction

$$F_m = \mu_s N \tag{4.8}$$

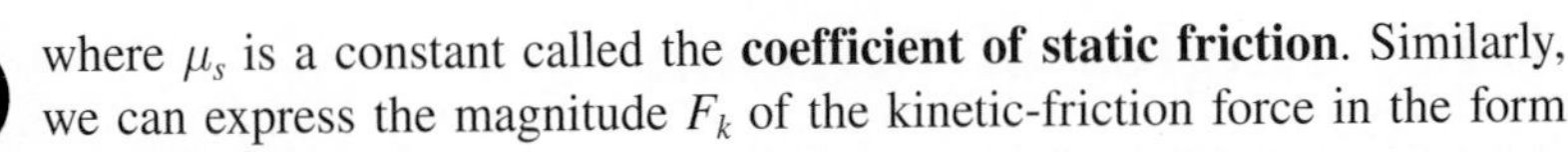

where μ_s is a constant called the **coefficient of static friction**. Similarly, we can express the magnitude F_k of the kinetic-friction force in the form

Kinetic friction

$$F_k = \mu_k N \tag{4.9}$$

where μ_k is a constant called the **coefficient of kinetic friction**. The coefficients of friction μ_s and μ_k do not depend upon the area of the surfaces in contact. Both coefficients, however, depend strongly on the *nature* of the surfaces in contact. Since they also depend upon the exact condition of the surfaces, their value is seldom known with an accuracy greater than 5%. Approximate values of coefficients of static friction for various combinations of dry surfaces are given in Table 4.1. The corresponding values of the coefficient of kinetic friction are about 25% smaller. Since coefficients of friction are dimensionless quantities, the values given in Table 4.1 can be used with both SI units and U.S. customary units.

Table 4.1 Approximate Values of Coefficient of Static Friction for Dry Surfaces

Metal on metal	0.15–0.60
Metal on wood	0.20–0.60
Metal on stone	0.30–0.70
Metal on leather	0.30–0.60
Wood on wood	0.25–0.50
Wood on leather	0.25–0.50
Stone on stone	0.40–0.70
Earth on earth	0.20–1.00
Rubber on concrete	0.60–0.90

From this discussion, it appears that four different situations can occur when a rigid body is in contact with a horizontal surface:

1. The forces applied to the body do not tend to move it along the surface of contact; there is no friction force (Fig. 4.12*a*).
2. The applied forces tend to move the body along the surface of contact but are not large enough to set it in motion. We can find the static-friction force **F** that has developed by solving the equations of equilibrium for the body. Since there is no evidence that **F** has reached its maximum value, the equation $F_m = \mu_s N$ *cannot be used* to determine the friction force (Fig. 4.12*b*).
3. The applied forces are such that the body is just about to slide. We say that *motion is impending*. The friction force **F** has reached its maximum value F_m and, together with the normal force **N**, balances the applied forces. Both the equations of equilibrium and the equation $F_m = \mu_s N$ *can be used*. Note that the friction force has a sense opposite to the sense of impending motion (Fig. 4.12*c*).
4. The body is sliding under the action of the applied forces, and the equations of equilibrium no longer apply. However, **F** is now equal to $\mathbf{F}_k$, and we can use the equation $F_k = \mu_k N$. The sense of $\mathbf{F}_k$ is opposite to the sense of motion (Fig. 4.12*d*).

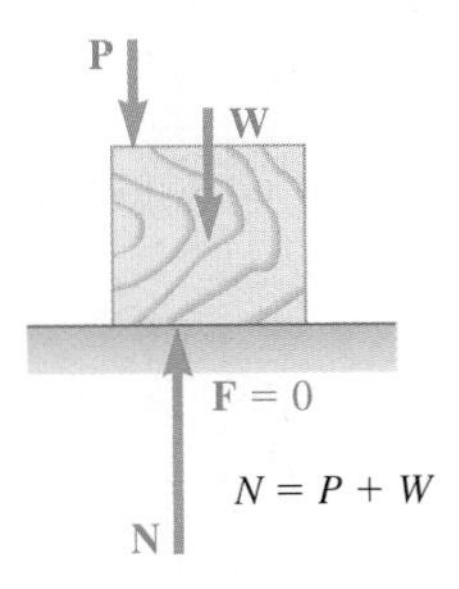

(*a*) No friction ($P_x = 0$)

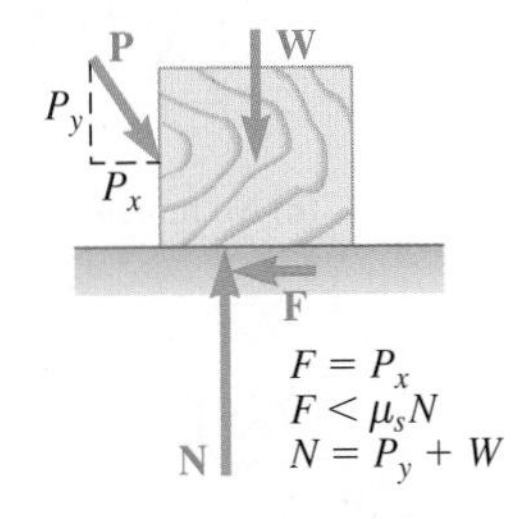

(*b*) No motion ($P_x < F_m$)

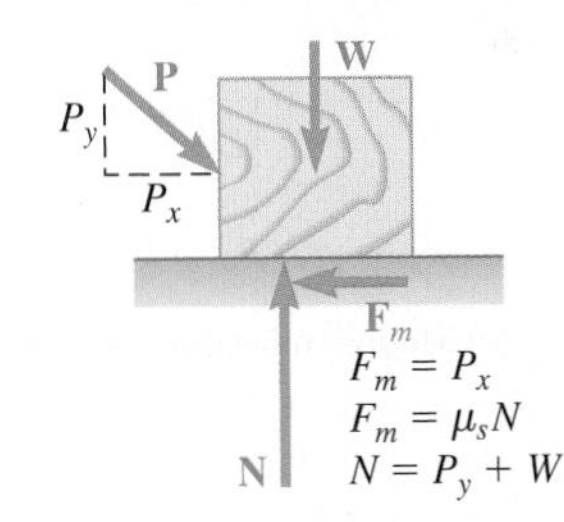

(*c*) Motion impending ⟶ ($P_x = F_m$)

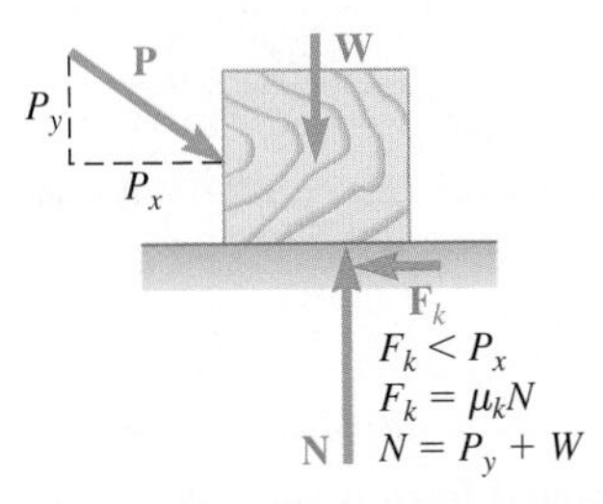

(*d*) Motion ⟶ ($P_x > F_k$)

Fig. 4.12 (*a*) Applied force is vertical, friction force is zero; (*b*) horizontal component of applied force is less than F_m, no motion occurs; (*c*) horizontal component of applied force equals F_m, motion is impending; (*d*) horizontal component of applied force is greater than F_k, forces are unbalanced and motion continues.

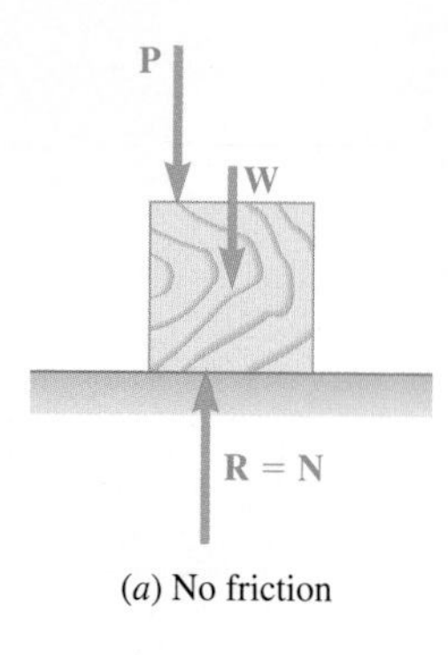

(*a*) No friction

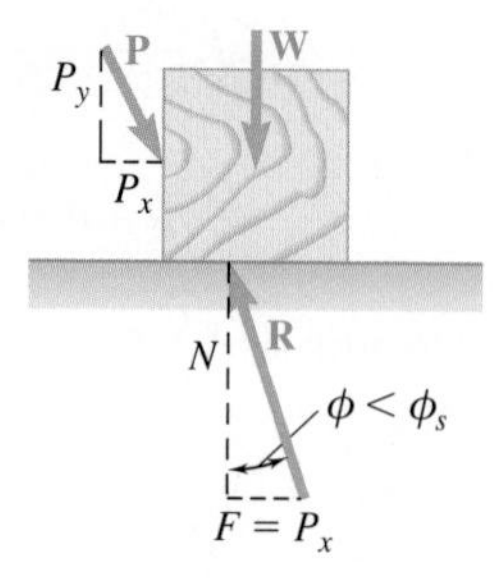

(*b*) No motion

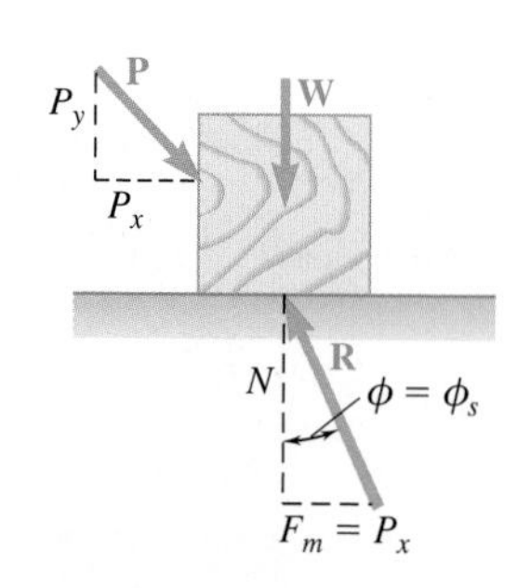

(*c*) Motion impending

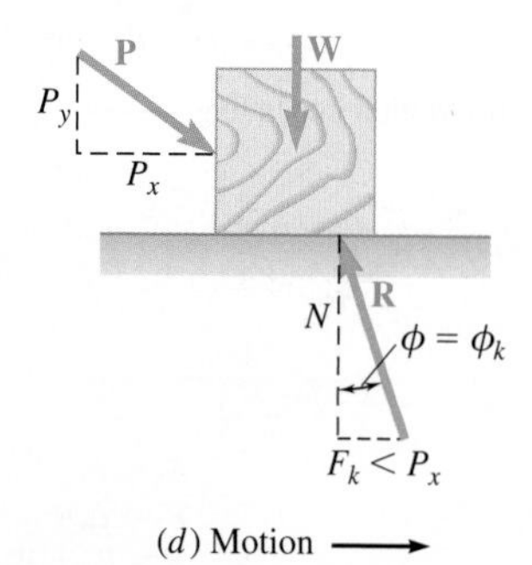

(*d*) Motion

Fig. 4.13 (*a*) Applied force is vertical, friction force is zero; (*b*) applied force is at an angle, its horizontal component balanced by the horizontal component of the surface resultant; (*c*) impending motion, the horizontal component of the applied force equals the maximum horizontal component of the resultant; (*d*) motion, the horizontal component of the resultant is less than the horizontal component of the applied force.

4.4C Angles of Friction

It is sometimes convenient to replace the normal force **N** and the friction force **F** by their resultant **R**. Let's see what happens when we do that.

Consider again a block of weight **W** resting on a horizontal plane surface. If no horizontal force is applied to the block, the resultant **R** reduces to the normal force **N** (Fig. 4.13*a*). However, if the applied force **P** has a horizontal component $\mathbf{P}_x$ that tends to move the block, force **R** has a horizontal component **F** and, thus, forms an angle ϕ with the normal to the surface (Fig. 4.13*b*). If you increase $\mathbf{P}_x$ until motion becomes impending, the angle between **R** and the vertical grows and reaches a maximum value (Fig. 4.13*c*). This value is called the **angle of static friction** and is denoted by ϕ_s. From the geometry of Fig. 4.13*c*, we note that

Angle of static friction

$$\tan \phi_s = \frac{F_m}{N} = \frac{\mu_s N}{N}$$

$$\tan \phi_s = \mu_s \tag{4.10}$$

If motion actually takes place, the magnitude of the friction force drops to F_k; similarly, the angle between **R** and **N** drops to a lower value ϕ_k, which is called the **angle of kinetic friction** (Fig. 4.13*d*). From the geometry of Fig. 4.13*d*, we have

Angle of kinetic friction

$$\tan \phi_k = \frac{F_k}{N} = \frac{\mu_k N}{N}$$

$$\tan \phi_k = \mu_k \tag{4.11}$$

Another example shows how the angle of friction can be used to advantage in the analysis of certain types of problems. Consider a block resting on a board and subjected to no other force than its weight **W** and the reaction **R** of the board. The board can be given any desired inclination. If the board is horizontal, the force **R** exerted by the board on the block is perpendicular to the board and balances the weight **W** (Fig. 4.14*a*). If the board is given a small angle of inclination θ, force **R** deviates from the perpendicular to the board by angle θ and continues to balance **W** (Fig. 4.14*b*). The reaction **R** now has a normal component **N** with a magnitude of $N = W \cos \theta$ and a tangential component **F** with a magnitude of $F = W \sin \theta$.

If we keep increasing the angle of inclination, motion soon becomes impending. At that time, the angle between **R** and the normal reaches its maximum value $\theta = \phi_s$ (Fig. 4.14*c*). The value of the angle of inclination corresponding to impending motion is called the **angle of repose**. Clearly, the angle of repose is equal to the angle of static friction ϕ_s. If we further increase the angle of inclination θ, motion starts and the angle between **R** and the normal drops to the lower value ϕ_k (Fig. 4.14*d*). The reaction **R** is not vertical anymore, and the forces acting on the block are unbalanced.

4.4D Problems Involving Dry Friction

Many engineering applications involve dry friction. Some are simple situations, such as variations on the block sliding on a plane just described.

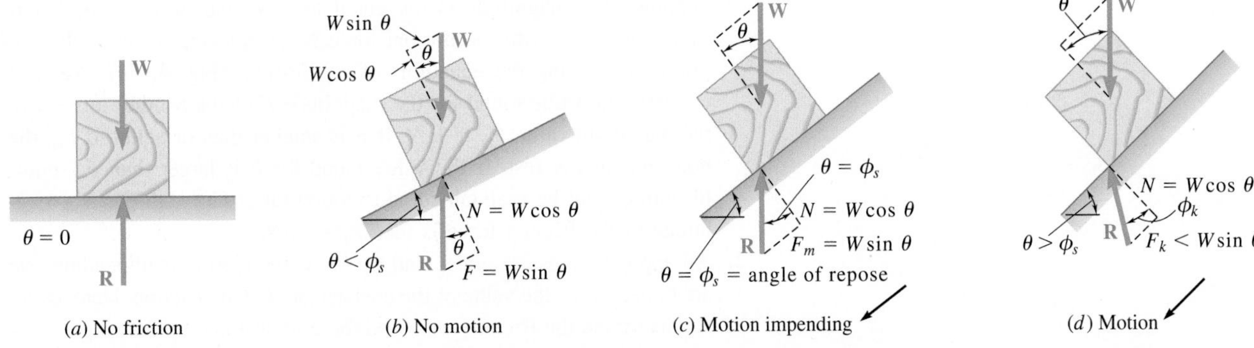

Fig. 4.14 (*a*) Block on horizontal board, friction force is zero; (*b*) board's angle of inclination is less than angle of static friction, no motion; (*c*) board's angle of inclination equals angle of friction, motion is impending; (*d*) angle of inclination is greater than angle of friction, forces are unbalanced and motion occurs.

Others involve more complicated situations, as in Sample Prob. 4.13. Many problems deal with the stability of rigid bodies in accelerated motion and will be studied in dynamics. Also, several common machines and mechanisms can be analyzed by applying the laws of dry friction, including wedges, screws, journal and thrust bearings, and belt transmissions. We will study these applications in the following sections.

The methods used to solve problems involving dry friction are the same that we used in the preceding chapters. If a problem involves only a motion of translation with no possible rotation, we can usually treat the body under consideration as a particle and use the methods of Chap. 2. If the problem involves a possible rotation, we must treat the body as a rigid body and use the methods presented in this chapter.

If the body being considered is acted upon by more than three forces (including the reactions at the surfaces of contact), the reaction at each surface is represented by its components **N** and **F**, and we solve the problem using the equations of equilibrium. If only three forces act on the body under consideration, it may be more convenient to represent each reaction by the single force **R** and solve the problem by using a force triangle.

Most problems involving friction fall into one of the following three groups.

1. All applied forces are given, and we know the coefficients of friction; we are to determine whether the body being considered remains at rest or slides. The friction force **F** *required to maintain equilibrium* is

Photo 4.6 The coefficient of static friction between a crate and the inclined conveyer belt must be sufficiently large to enable the crate to be transported without slipping.

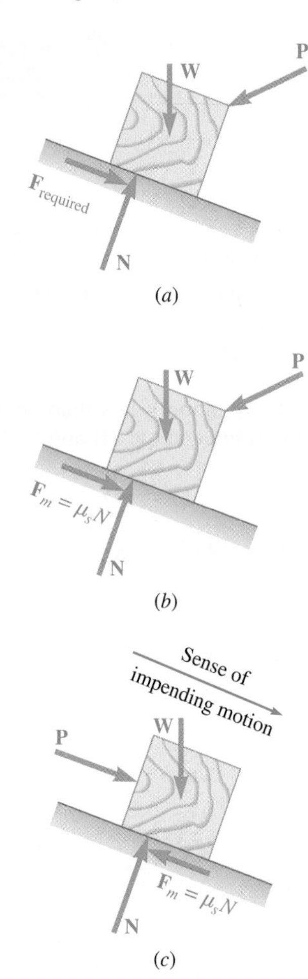

Fig. 4.15 Three types of friction problems: (*a*) given the forces and coefficient of friction, will the block slide or stay? (*b*) given the forces and that motion is pending, determine the coefficient of friction; (*c*) given the coefficient of friction and that motion is impending, determine the applied force.

unknown (its magnitude is *not* equal to $\mu_s N$) and needs to be determined, together with the normal force **N**, by drawing a free-body diagram and solving the equations of equilibrium (Fig. 4.15*a*). We then compare the value found for the magnitude F of the friction force with the maximum value $F_m = \mu_s N$. If F is smaller than or equal to F_m, the body remains at rest. If the value found for F is larger than F_m, equilibrium cannot be maintained and motion takes place; the actual magnitude of the friction force is then $F_k = \mu_k N$.

2. All applied forces are given, and we know the motion is impending; we are to determine the value of the coefficient of static friction. Here again, we determine the friction force and the normal force by drawing a free-body diagram and solving the equations of equilibrium (Fig. 4.15*b*). Since we know that the value found for F is the maximum value F_m, we determine the coefficient of friction by solving the equation $F_m = \mu_s N$.
3. The coefficient of static friction is given, and we know that the motion is impending in a given direction; we are to determine the magnitude or the direction of one of the applied forces. The friction force should be shown in the free-body diagram with a *sense opposite to that of the impending motion* and with a magnitude $F_m = \mu_s N$ (Fig. 4.15*c*). We can then write the equations of equilibrium and determine the desired force.

As noted previously, when only three forces are involved, it may be more convenient to represent the reaction of the surface by a single force **R** and to solve the problem by drawing a force triangle. Such a solution is used in Sample Prob. 4.12.

When two bodies A and B are in contact (Fig. 4.16*a*), the forces of friction exerted, respectively, by A on B and by B on A are equal and opposite (Newton's third law). In drawing the free-body diagram of one of these bodies, it is important to include the appropriate friction force with its correct sense. Observe the following rule: *The sense of the friction force acting on A is opposite to that of the motion (or impending motion) of A as observed from B* (Fig. 4.16*b*). (It is therefore the same as the motion of B as observed from A.) The sense of the friction force acting on B is determined in a similar way (Fig. 4.16*c*). Note that the motion of A as observed from B is a *relative motion*. For example, if body A is fixed and body B moves, body A has a relative motion with respect to B. Also, if both B and A are moving down but B is moving faster than A, then body A is observed, from B, to be moving up.

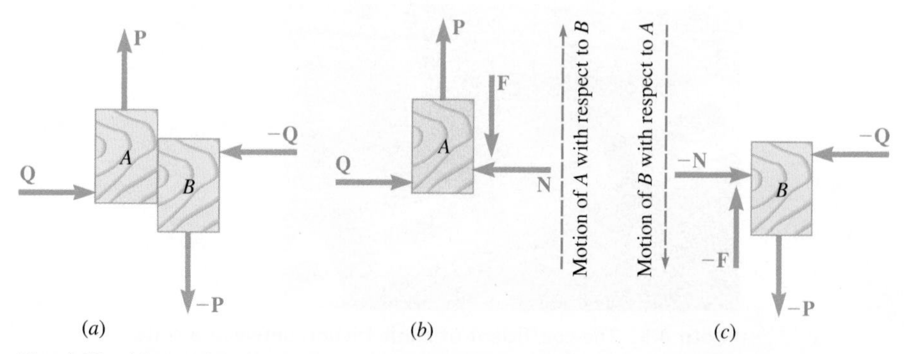

Fig. 4.16 (*a*) Two blocks held in contact by forces; (*b*) free-body diagram for block *A*, including direction of friction force; (*c*) free-body diagram for block *B*, including direction of friction force.

Sample Problem 4.11

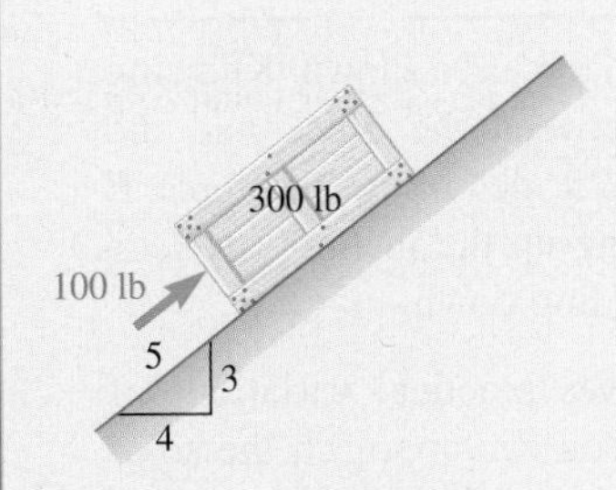

A 100-lb force acts as shown on a 300-lb crate placed on an inclined plane. The coefficients of friction between the crate and the plane are $\mu_s = 0.25$ and $\mu_k = 0.20$. Determine whether the crate is in equilibrium, and find the value of the friction force.

STRATEGY: This is a friction problem of the first type: You know the forces and the friction coefficients and want to determine if the crate moves. You also want to find the friction force.

MODELING and ANALYSIS

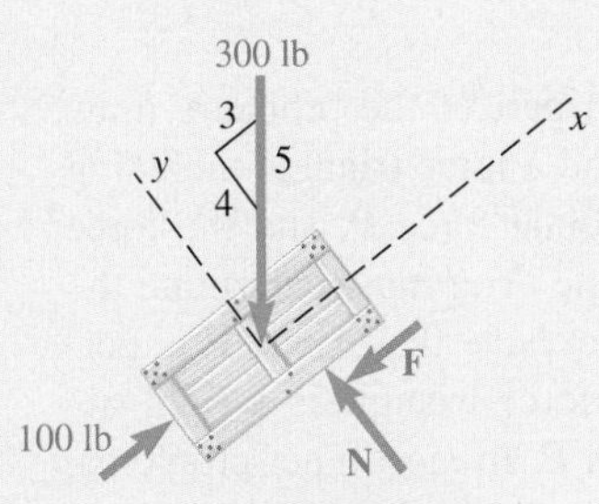

Fig. 1 Free-body diagram of crate showing assumed direction of friction force.

Force Required for Equilibrium. First determine the value of the friction force *required to maintain equilibrium.* Assuming that **F** is directed down and to the left, draw the free-body diagram of the crate (Fig. 1) and solve the equilibrium equations:

$$+\nearrow \Sigma F_x = 0: \qquad 100 \text{ lb} - \tfrac{3}{5}(300 \text{ lb}) - F = 0$$
$$F = -80 \text{ lb} \qquad \mathbf{F} = 80 \text{ lb} \nearrow$$

$$+\nwarrow \Sigma F_y = 0: \qquad N - \tfrac{4}{5}(300 \text{ lb}) = 0$$
$$N = +240 \text{ lb} \qquad \mathbf{N} = 240 \text{ lb} \nwarrow$$

The force **F** required to maintain equilibrium is an 80-lb force directed up and to the right; the tendency of the crate is thus to move down the plane.

Maximum Friction Force. The magnitude of the maximum friction force that may be developed between the crate and the plane is

$$F_m = \mu_s N \qquad F_m = 0.25(240 \text{ lb}) = 60 \text{ lb}$$

Since the value of the force required to maintain equilibrium (80 lb) is larger than the maximum value that may be obtained (60 lb), equilibrium is not maintained and *the crate will slide down the plane.*

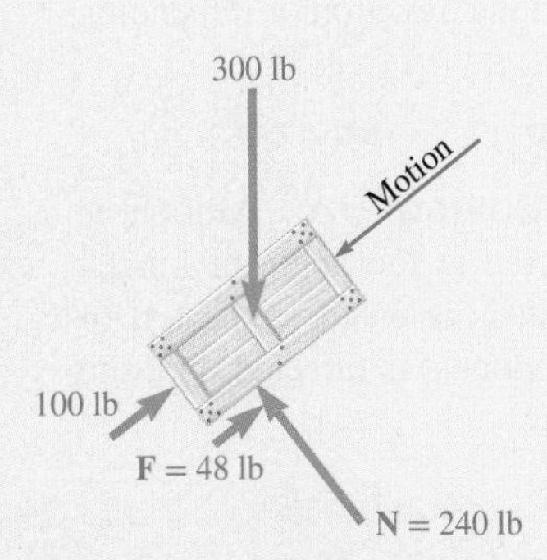

Fig. 2 Free-body diagram of crate showing actual friction force.

Actual Value of Friction Force. The magnitude of the actual friction force is

$$F_{\text{actual}} = F_k = \mu_k N = 0.20(240 \text{ lb}) = 48 \text{ lb}$$

The sense of this force is opposite to the sense of motion; the force is thus directed up and to the right (Fig. 2):

$$\mathbf{F}_{\text{actual}} = 48 \text{ lb} \nearrow \blacktriangleleft$$

Note that the forces acting on the crate are not balanced. Their resultant is

$$\tfrac{3}{5}(300 \text{ lb}) - 100 \text{ lb} - 48 \text{ lb} = 32 \text{ lb} \swarrow$$

REFLECT and THINK: This is a typical friction problem of the first type. Note that you used the coefficient of static friction to determine if the crate moves, but once you found that it does move, you needed the coefficient of kinetic friction to determine the friction force.

Sample Problem 4.12

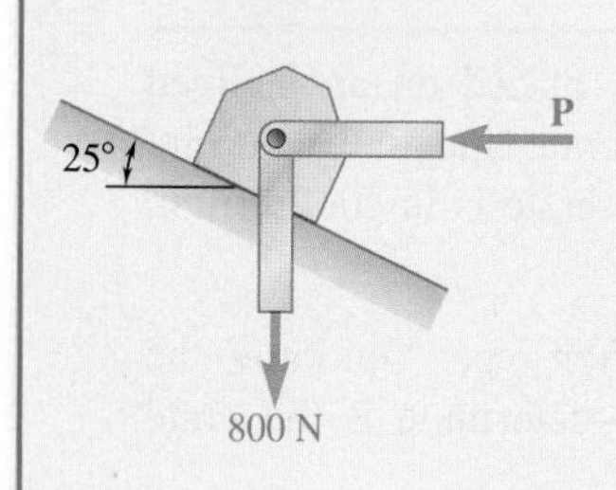

A support block is acted upon by two forces as shown. Knowing that the coefficients of friction between the block and the incline are $\mu_s = 0.35$ and $\mu_k = 0.25$, determine the force **P** required to (*a*) start the block moving up the incline; (*b*) keep it moving up; (*c*) prevent it from sliding down.

STRATEGY: This problem involves practical variations of the third type of friction problem. You can approach the solutions through the concept of the angles of friction.

MODELING:

Free-Body Diagram. For each part of the problem, draw a free-body diagram of the block and a force triangle including the 800-N vertical force, the horizontal force **P**, and the force **R** exerted on the block by the incline. You must determine the direction of **R** in each separate case. Note that, since **P** is perpendicular to the 800-N force, the force triangle is a right triangle, which easily can be solved for **P**. In most other problems, however, the force triangle will be an oblique triangle and should be solved by applying the law of sines.

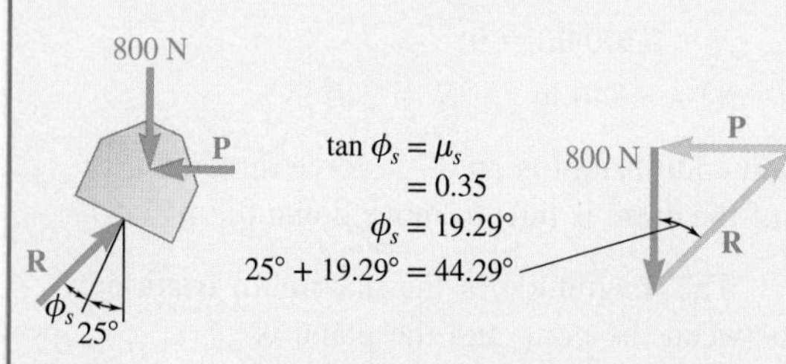

Fig. 1 Free-body diagram of block and its force triangle—motion impending up incline.

ANALYSIS:

a. Force *P* to Start Block Moving Up. In this case, motion is impending up the incline, so the resultant is directed at the angle of static friction (Fig. 1). Note that the resultant is oriented to the left of the normal such that its friction component (not shown) is directed opposite the direction of impending motion.

$$P = (800\text{ N}) \tan 44.29° \qquad \mathbf{P} = 780\text{ N}\leftarrow \blacktriangleleft$$

Fig. 2 Free-body diagram of block and its force triangle—motion continuing up incline.

b. Force *P* to Keep Block Moving Up. Motion is continuing, so the resultant is directed at the angle of kinetic friction (Fig. 2). Again, the resultant is oriented to the left of the normal such that its friction component is directed opposite the direction of motion.

$$P = (800\text{ N}) \tan 39.04° \qquad \mathbf{P} = 649\text{ N}\leftarrow \blacktriangleleft$$

c. Force *P* to Prevent Block from Sliding Down. Here, motion is impending down the incline, so the resultant is directed at the angle of static friction (Fig. 3). Note that the resultant is oriented to the right of the normal such that its friction component is directed opposite the direction of impending motion.

$$P = (800\text{ N}) \tan 5.71° \qquad \mathbf{P} = 80.0\text{ N}\leftarrow \blacktriangleleft$$

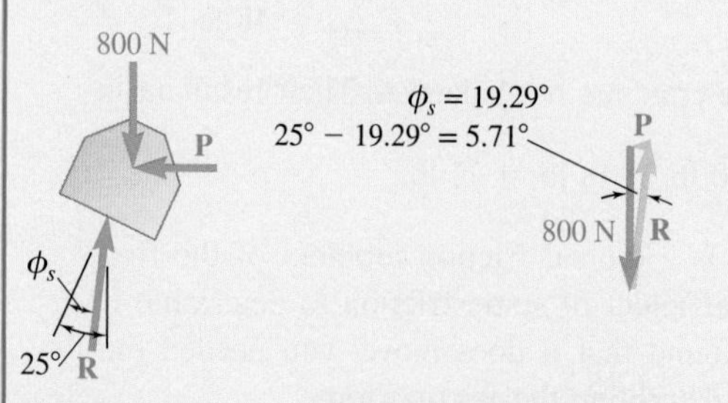

Fig. 3 Free-body diagram of block and its force triangle—motion prevented down the slope.

REFLECT and THINK: As expected, considerably more force is required to begin moving the block up the slope than is necessary to restrain it from sliding down the slope.

Sample Problem 4.13

The movable bracket shown may be placed at any height on the 3-in.-diameter pipe. If the coefficient of static friction between the pipe and bracket is 0.25, determine the minimum distance x at which the load **W** can be supported. Neglect the weight of the bracket.

STRATEGY: In this variation of the third type of friction problem, you know the coefficient of static friction and that motion is impending. Since the problem involves consideration of resistance to rotation, you should apply both moment equilibrium and force equilibrium.

MODELING:

Free-Body Diagram. Draw the free-body diagram of the bracket (Fig. 1). When **W** is placed at the minimum distance x from the axis of the pipe, the bracket is just about to slip, and the forces of friction at A and B have reached their maximum values:

$$F_A = \mu_s N_A = 0.25\ N_A$$
$$F_B = \mu_s N_B = 0.25\ N_B$$

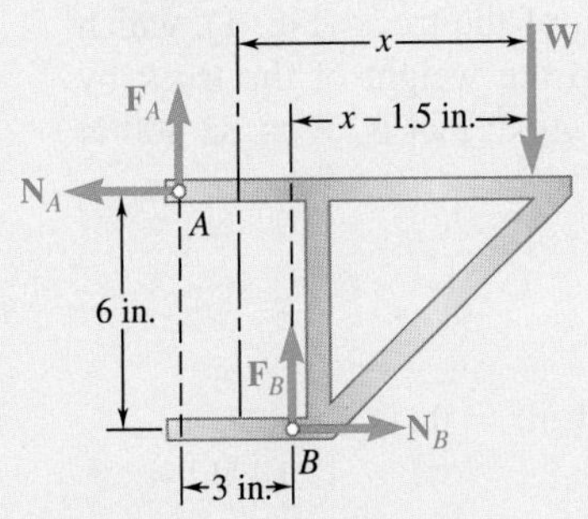

Fig. 1 Free-body diagram of bracket.

ANALYSIS:

Equilibrium Equations.

$$\xrightarrow{+}\Sigma F_x = 0: \qquad N_B - N_A = 0$$
$$N_B = N_A$$

$$+\uparrow\Sigma F_y = 0: \qquad F_A + F_B - W = 0$$
$$0.25N_A + 0.25N_B = W$$

Since N_B is equal to N_A,

$$0.50N_A = W$$
$$N_A = 2W$$

$$+\circlearrowleft\Sigma M_B = 0: \qquad N_A(6\text{ in.}) - F_A(3\text{ in.}) - W(x - 1.5\text{ in.}) = 0$$
$$6N_A - 3(0.25N_A) - Wx + 1.5W = 0$$
$$6(2W) - 0.75(2W) - Wx + 1.5W = 0$$

Dividing through by W and solving for x, you have

$$x = 12\text{ in.} \quad ◀$$

REFLECT and THINK: In a problem like this, you may not figure out how to approach the solution until you draw the free-body diagram and examine what information you are given and what you need to find. In this case, since you are asked to find a distance, the need to evaluate moment equilibrium should be clear.

Sample Problem 4.14

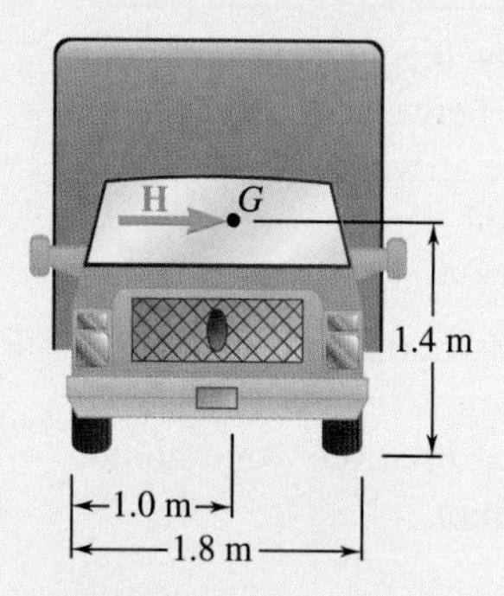

An 8400-kg truck is traveling on a level horizontal curve, resulting in an effective lateral force **H** (applied at the center of gravity G of the truck). Treating the truck as a rigid system with the center of gravity shown, and knowing that the distance between the outer edges of the tires is 1.8 m, determine (*a*) the maximum force **H** before tipping of the truck occurs, (*b*) the minimum coefficient of static friction between the tires and roadway such that slipping does not occur before tipping.

STRATEGY: For the direction of **H** shown, the truck would tip about the outer edge of the right tire. At the verge of tip, the normal force and friction force are zero at the left tire, and the normal force at the right tire is at the outer edge. You can apply equilibrium to determine the value of **H** necessary for tip and the required friction force such that slipping does not occur.

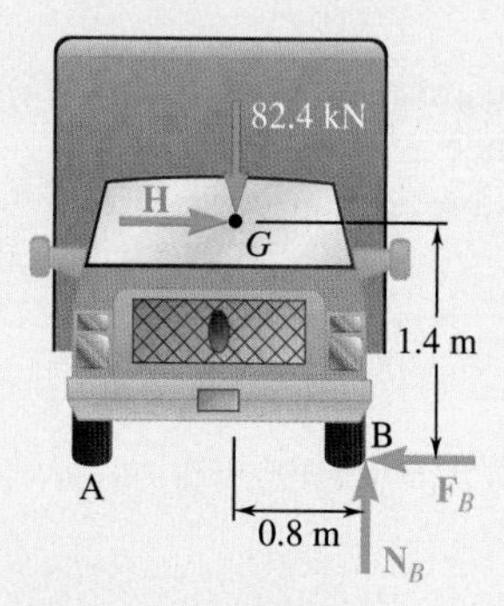

Fig. 1 Free-body diagram of truck.

MODELING: Draw the free-body diagram of the truck (Fig. 1), which reflects impending tip about point B. Obtain the weight of the truck by multiplying its mass of 8400 kg by $g = 9.81 \text{ m/s}^2$; that is, $W = 82\,400$ N or 82.4 kN.

ANALYSIS:

Free Body: Truck (Fig. 1).

$+\circlearrowleft \Sigma M_B = 0$: $\quad (82.4 \text{ kN})(0.8 \text{ m}) - H(1.4 \text{ m}) = 0$

$H = +47.1 \text{ kN} \qquad\qquad H = 47.1 \text{ kN} \rightarrow$ ◀

$\overset{+}{\rightarrow} \Sigma F_x = 0$: $\quad 47.1 \text{ kN} - F_B = 0$

$F_B = +47.1 \text{ kN}$

$+\uparrow \Sigma F_y = 0$: $\quad N_B - 82.4 \text{ kN} = 0$

$N_B = +82.4 \text{ kN}$

Minimum Coefficient of Static Friction. The magnitude of the maximum friction force that can be developed is

$$F_m = \mu_s N_B = \mu_s \,(82.4 \text{ kN})$$

Setting this equal to the friction force required, $F_B = 47.1$ kN, gives

$$\mu_s \,(82.4 \text{ kN}) = 47.1 \text{ kN} \qquad \mu_s = 0.572 \blacktriangleleft$$

REFLECT and THINK: Recall from physics that **H** represents the force due to the centripetal acceleration of the truck (of mass m), and its magnitude is

$$H = m(v^2/\rho)$$

where

v = velocity of the truck
ρ = radius of curvature

In this problem, if the truck was traveling around a curve of 100-m radius (measured to G), the velocity at which it would begin to tip would be 23.7 m/s (or 85.2 km/h). You will learn more about this aspect in the study of dynamics.

Problems

4.75 Determine whether the block shown is in equilibrium and find the magnitude and direction of the friction force when $\theta = 25°$ and $P = 750$ N.

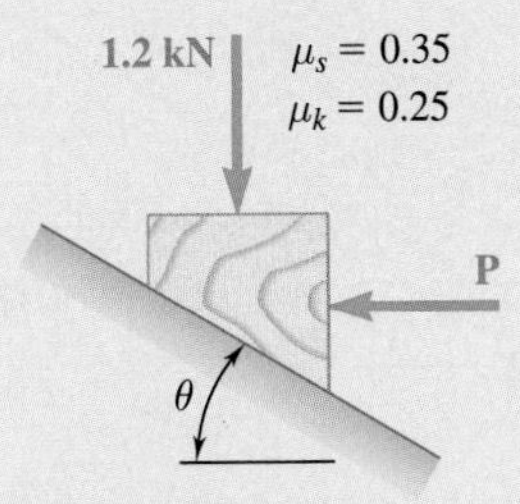

Fig. P4.75

4.76 Solve Prob. 4.75 when $\theta = 30°$ and $P = 150$ N.

4.77 Determine whether the block shown is in equilibrium and find the magnitude and direction of the friction force when $P = 120$ lb.

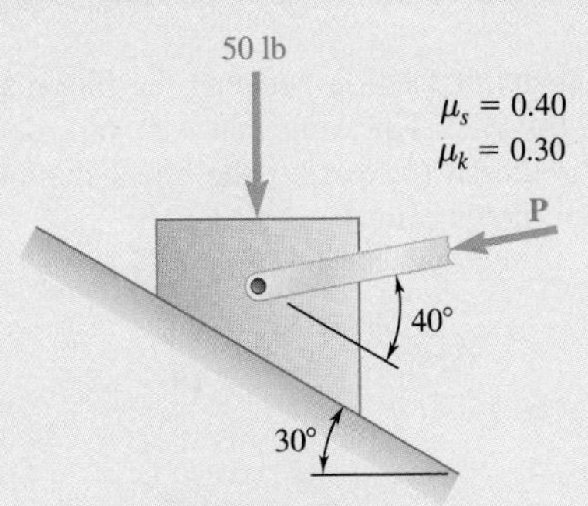

Fig. P4.77, P4.78, and *P4.79*

4.78 Determine whether the block shown is in equilibrium and find the magnitude and direction of the friction force when $P = 80$ lb.

4.79 Determine the smallest value of P required to (*a*) start the block up the incline, (*b*) keep it moving up.

4.80 The 80-lb block is attached to link AB and rests on a moving belt. Knowing that $\mu_s = 0.25$ and $\mu_k = 0.20$, determine the magnitude of the horizontal force **P** that should be applied to the belt to maintain its motion (*a*) to the right, (*b*) to the left.

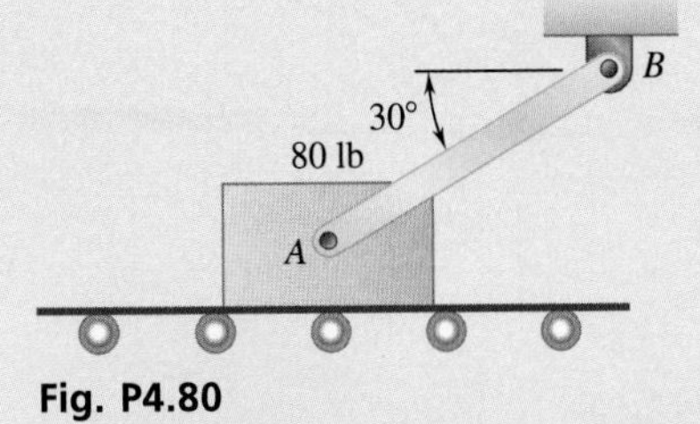

Fig. P4.80

4.81 The 50-lb block A and the 25-lb block B are supported by an incline that is held in the position shown. Knowing that the coefficient of static friction is 0.15 between the two blocks and zero between block B and the incline, determine the value of θ for which motion is impending.

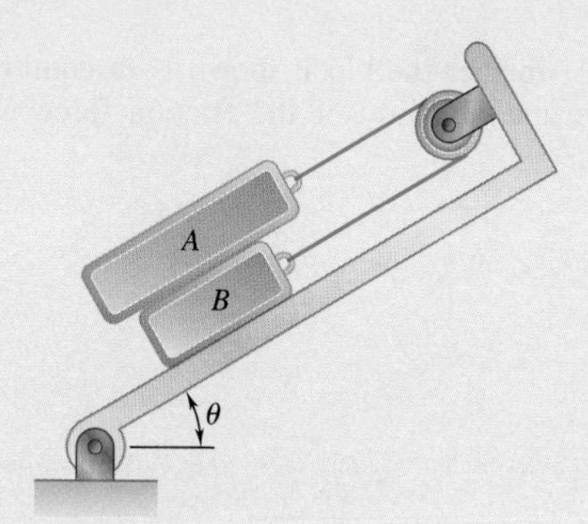

Fig. P4.81 and P4.82

4.82 The 50-lb block A and the 25-lb block B are supported by an incline that is held in the position shown. Knowing that the coefficient of static friction is 0.15 between all surfaces of contact, determine the value of θ for which motion is impending.

4.83 The coefficients of friction between the block and the rail are $\mu_s = 0.30$ and $\mu_k = 0.25$. Knowing that $\theta = 65°$, determine the smallest value of P required (*a*) to start the block moving up the rail, (*b*) to keep it from moving down.

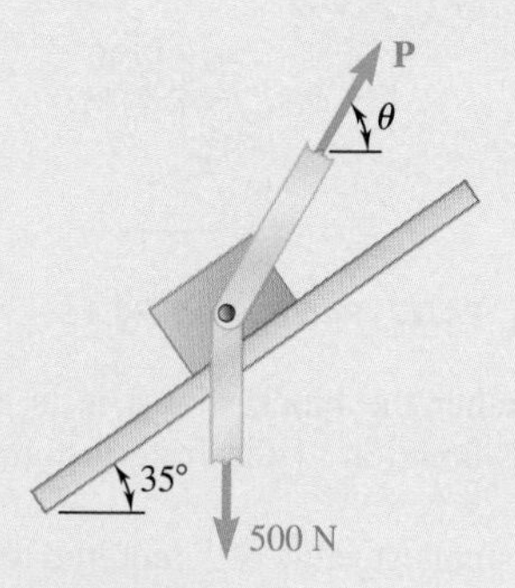

Fig. P4.83

4.84 Knowing that $P = 100$ N, determine the range of values of θ for which equilibrium of the 7.5-kg block is maintained.

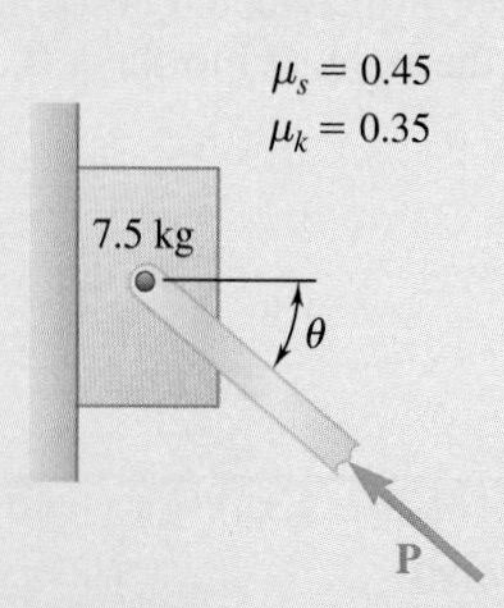

Fig. P4.84

4.85 A 120-lb cabinet is mounted on casters that can be locked to prevent their rotation. The coefficient of static friction between the floor and each caster is 0.30. If h = 32 in., determine the magnitude of the force **P** required to move the cabinet to the right (*a*) if all casters are locked, (*b*) if the casters at *B* are locked and the casters at *A* are free to rotate, (*c*) if the casters at *A* are locked and the casters at *B* are free to rotate.

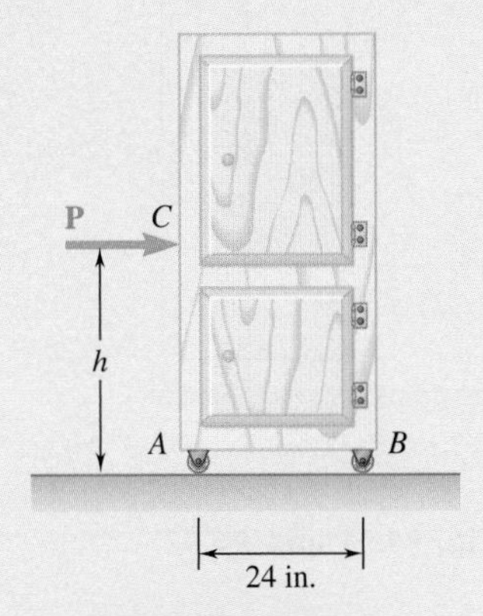

Fig. P4.85 and ***P4.86***

4.86 A 120-lb cabinet is mounted on casters that can be locked to prevent their rotation. The coefficient of static friction between the floor and each caster is 0.30. Assuming that the casters at both *A* and *B* are locked, determine (*a*) the force **P** required to move the cabinet to the right, (*b*) the largest allowable value of h if the cabinet is not to tip over.

4.87 A 40-kg packing crate must be moved to the left along the floor without tipping. Knowing that the coefficient of static friction between the crate and the floor is 0.35, determine (*a*) the largest allowable value of α, (*b*) the corresponding magnitude of the force **P**.

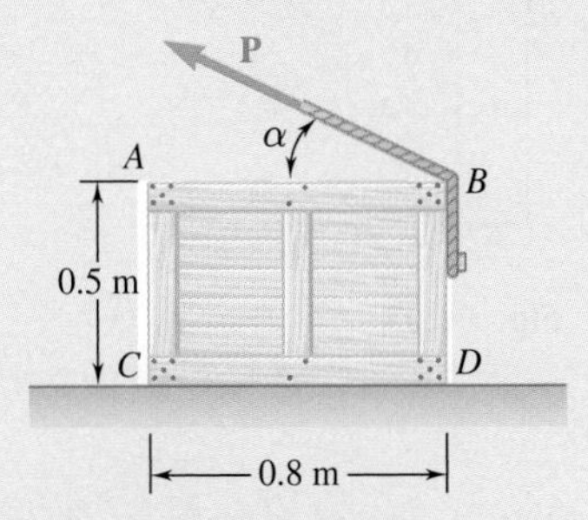

Fig. P4.87 and P4.88

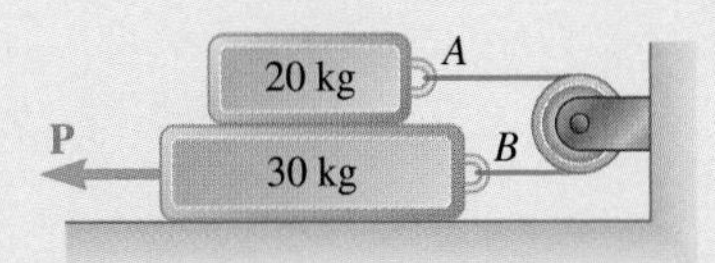

Fig. ***P4.89***

4.88 A 40-kg packing crate is pulled by a rope as shown. The coefficient of static friction between the crate and the floor is 0.35. If $\alpha = 40°$, determine (*a*) the magnitude of the force **P** required to move the crate, (*b*) whether the crate will slide or tip.

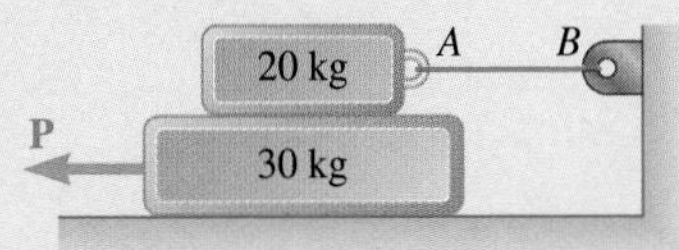

Fig. P4.90

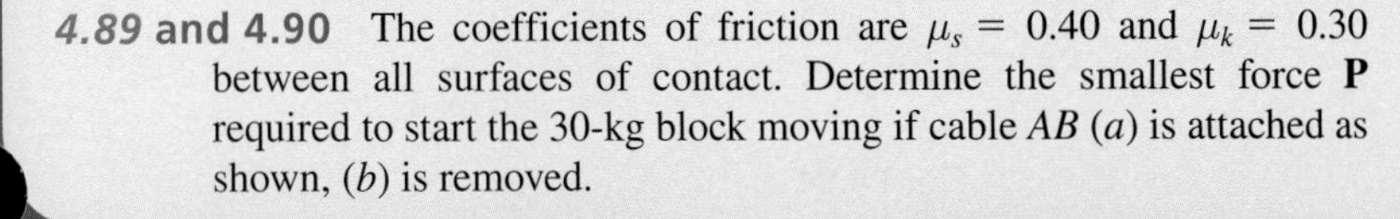

4.89 **and 4.90** The coefficients of friction are $\mu_s = 0.40$ and $\mu_k = 0.30$ between all surfaces of contact. Determine the smallest force **P** required to start the 30-kg block moving if cable *AB* (*a*) is attached as shown, (*b*) is removed.

4.91 A 6.5-m ladder *AB* leans against a wall as shown. Assuming that the coefficient of static friction μ_s is zero at *B*, determine the smallest value of μ_s at *A* for which equilibrium is maintained.

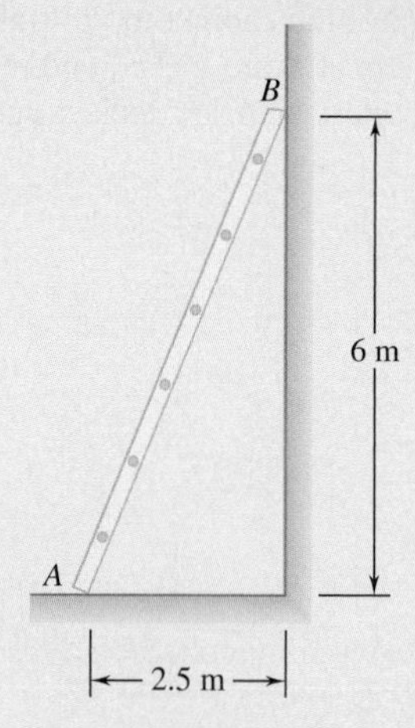

Fig. P4.91 and *P4.92*

4.92 A 6.5-m ladder *AB* leans against a wall as shown. Assuming that the coefficient of static friction μ_s is the same at *A* and *B*, determine the smallest value of μ_s for which equilibrium is maintained.

4.93 and 4.94 End *A* of a slender, uniform rod of length *L* and weight *W* bears on a surface as shown, while end *B* is supported by a cord *BC*. Knowing that the coefficients of friction are $\mu_s = 0.40$ and $\mu_k = 0.30$, determine (*a*) the largest value of θ for which motion is impending, (*b*) the corresponding value of the tension in the cord.

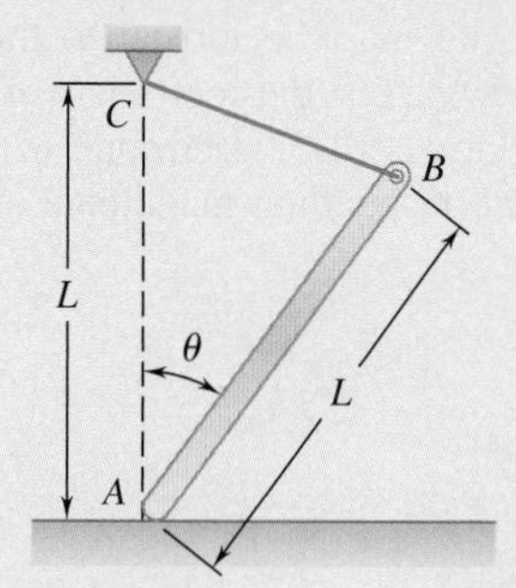

Fig. P4.93

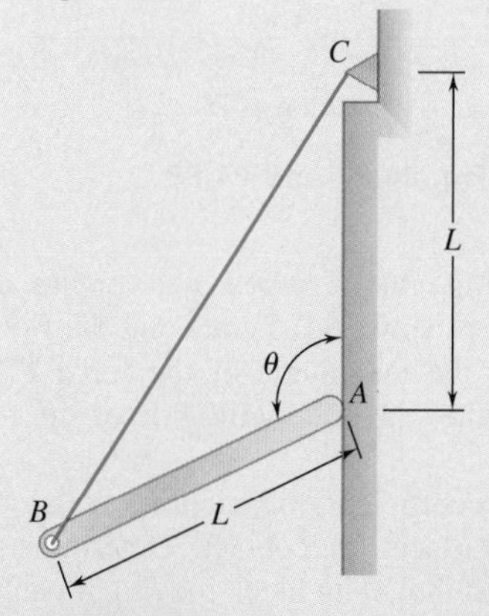

Fig. P4.94

4.95 Two slender rods of negligible weight are pin-connected at C and attached to blocks A and B, each of weight W. Knowing that $\theta = 80°$ and that the coefficient of static friction between the blocks and the horizontal surface is 0.30, determine the largest value of P for which equilibrium is maintained.

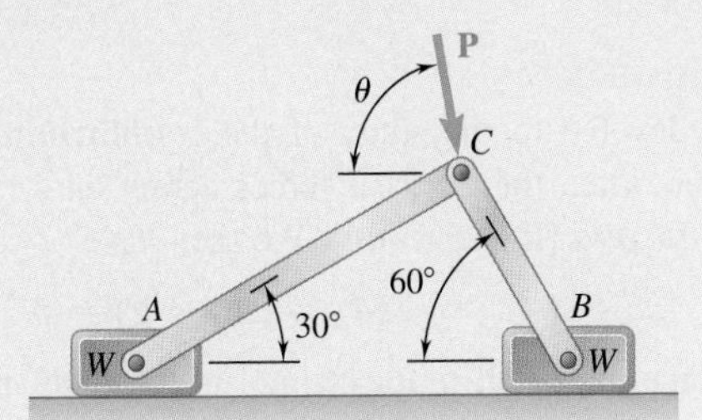

Fig. P4.95 and *P4.96*

4.96 Two slender rods of negligible weight are pin-connected at C and attached to blocks A and B, each of weight W. Knowing that $P = 1.260W$ and that the coefficient of static friction between the blocks and the horizontal surface is 0.30, determine the range of values of θ, between 0 and 180°, for which equilibrium is maintained.

4.97 The cylinder shown is of weight W and radius r, and the coefficient of static friction μ_s is the same at A and B. Determine the magnitude of the largest couple **M** that can be applied to the cylinder if it is not to rotate.

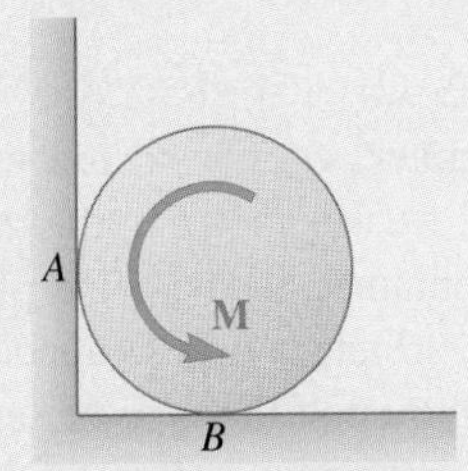

Fig. P4.97 and P4.98

4.98 The cylinder shown is of weight W and radius r. Express in terms W and r the magnitude of the largest couple **M** that can be applied to the cylinder if it is not to rotate, assuming the coefficient of static friction to be (a) zero at A and 0.30 at B, (b) 0.25 at A and 0.30 at B.

Review and Summary

Equilibrium Equations

This chapter was devoted to the study of the **equilibrium of rigid bodies**, i.e., to the situation when the external forces acting on a rigid body *form a system equivalent to zero* [Introduction]. We then have

$$\Sigma\mathbf{F} = 0 \qquad \Sigma\mathbf{M}_O = \Sigma(\mathbf{r} \times \mathbf{F}) = 0 \tag{4.1}$$

Resolving each force and each moment into its rectangular components, we can express the necessary and sufficient conditions for the equilibrium of a rigid body with the following six scalar equations:

$$\Sigma F_x = 0 \qquad \Sigma F_y = 0 \qquad \Sigma F_z = 0 \tag{4.2}$$

$$\Sigma M_x = 0 \qquad \Sigma M_y = 0 \qquad \Sigma M_z = 0 \tag{4.3}$$

We can use these equations to determine unknown forces applied to the rigid body or unknown reactions exerted by its supports.

Free-Body Diagram

When solving a problem involving the equilibrium of a rigid body, it is essential to consider *all* of the forces acting on the body. Therefore, the first step in the solution of the problem should be to draw a **free-body diagram** showing the body under consideration and all of the unknown as well as known forces acting on it.

Equilibrium of a Two-Dimensional Structure

In the first part of this chapter, we considered the **equilibrium of a two-dimensional structure**; i.e., we assumed that the structure considered and the forces applied to it were contained in the same plane. We saw that each of the reactions exerted on the structure by its supports could involve one, two, or three unknowns, depending upon the type of support [Sec. 4.1A].

In the case of a two-dimensional structure, the equations given previously reduce to *three equilibrium equations*:

$$\Sigma F_x = 0 \qquad \Sigma F_y = 0 \qquad \Sigma M_A = 0 \tag{4.5}$$

where A is an arbitrary point in the plane of the structure [Sec. 4.1B]. We can use these equations to solve for three unknowns. Although the three equilibrium equations (4.5) cannot be *augmented* with additional equations, any of them can be *replaced* by another equation. Therefore, we can write alternative sets of equilibrium equations, such as

$$\Sigma F_x = 0 \qquad \Sigma M_A = 0 \qquad \Sigma M_B = 0 \tag{4.6}$$

where point B is chosen in such a way that the line AB is not parallel to the y axis, or

$$\Sigma M_A = 0 \qquad \Sigma M_B = 0 \qquad \Sigma M_C = 0 \tag{4.7}$$

where the points A, B, and C do not lie in a straight line.

Static Indeterminacy, Partial Constraints, Improper Constraints

Since any set of equilibrium equations can be solved for only three unknowns, the reactions at the supports of a rigid two-dimensional structure cannot be completely determined if they involve *more than three unknowns;* they are said to be *statically indeterminate* [Sec. 4.1C]. On the other hand, if the reactions involve *fewer than three unknowns,* equilibrium is not maintained under general loading conditions; the structure is said to be *partially constrained.* The fact that the reactions involve exactly three unknowns is no guarantee that you can solve the equilibrium equations for all three unknowns. If the supports are arranged in such a way that the reactions are *either concurrent or parallel,* the reactions are statically indeterminate, and the structure is said to be *improperly constrained.*

Two-Force Body, Three-Force Body

We gave special attention in Sec. 4.2 to two particular cases of equilibrium of a rigid body. We defined a **two-force body** as a rigid body subjected to forces at only two points, and we showed that the resultants $\mathbf{F}_1$ and $\mathbf{F}_2$ of these forces must have the *same magnitude, the same line of action, and opposite sense* (Fig. 4.17), which is a property that simplifies the solution of certain problems in later chapters. We defined a **three-force body** as a rigid body subjected to forces at only three points, and we demonstrated that the resultants $\mathbf{F}_1$, $\mathbf{F}_2$, and $\mathbf{F}_3$ of these forces must be *either concurrent* (Fig. 4.18) *or parallel.* This property provides us with an alternative approach to the solution of problems involving a three-force body [Sample Prob. 4.6].

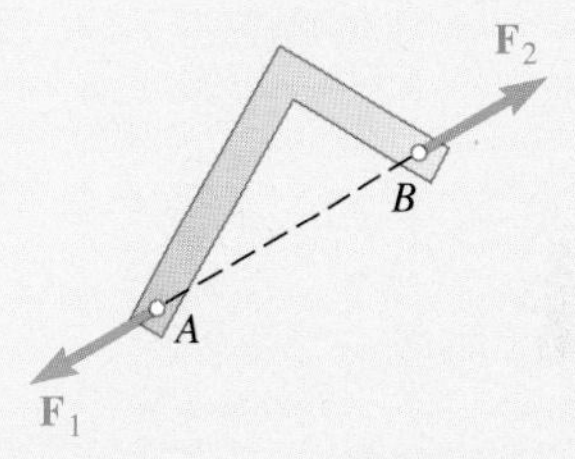

Fig. 4.17

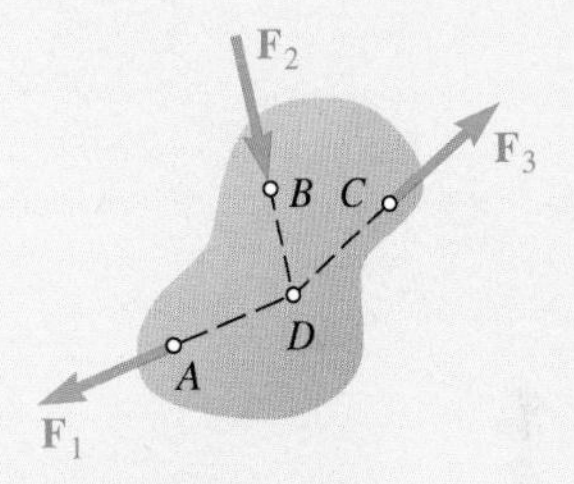

Fig. 4.18

Equilibrium of a Three-Dimensional Body

In the second part of this chapter, we considered the *equilibrium of a three-dimensional body.* We saw that each of the reactions exerted on the body by its supports could involve between one and six unknowns, depending upon the type of support [Sec. 4.3A].

In the general case of the equilibrium of a three-dimensional body, all six of the scalar equilibrium equations (4.2) and (4.3) should be used and solved for *six unknowns* [Sec. 4.3B]. In most problems, however, we can obtain these equations more conveniently if we start from

$$\Sigma\mathbf{F} = 0 \qquad \Sigma\mathbf{M}_O = \Sigma\,(\mathbf{r} \times \mathbf{F}) = 0 \tag{4.1}$$

and then express the forces $\mathbf{F}$ and position vectors $\mathbf{r}$ in terms of scalar components and unit vectors. We can compute the vector products either directly or by means of determinants, and obtain the desired scalar equations by equating to zero the coefficients of the unit vectors [Sample Probs. 4.7 through 4.9].

We noted that we may eliminate as many as three unknown reaction components from the computation of $\Sigma\mathbf{M}_O$ in the second of the relations (4.1) through a judicious choice of point O. Also, we can eliminate the reactions at two points A and B from the solution of some problems by writing the equation $\Sigma M_{AB} = 0$, which involves the computation of the moments of the forces about an axis AB joining points A and B [Sample Prob. 4.10].

We observed that when a body is subjected to individual couples $\mathbf{M}_i$, either as applied loads or as support reactions, we can include these couples by expressing the second part of Eq. (4.1) as

$$\Sigma\mathbf{M}_O = \Sigma(\mathbf{r} \times \mathbf{F}) + \Sigma\mathbf{M}_i = 0 \tag{4.1'}$$

If the reactions involve more than six unknowns, some of the reactions are *statically indeterminate;* if they involve fewer than six unknowns, the rigid body is only *partially constrained.* Even with six or more unknowns, the rigid body is *improperly constrained* if the reactions associated with the given supports are either parallel or intersect the same line.

Static and Kinetic Friction

The final part of this chapter was devoted to the study of **dry friction**, i.e., to problems involving rigid bodies in contact along unlubricated surfaces. If we apply a horizontal force **P** to a block resting on a horizontal surface [Sec. 4.4A], we note that at first the block does not move. This shows that a **friction force F** must have developed to balance **P** (Fig. 4.19). As the magnitude of **P** increases, the magnitude of **F** also increases until it reaches a maximum value F_m. If **P** is further increased, the block starts sliding, and the magnitude of **F** drops from F_m to a lower value F_k. Experimental evidence shows that F_m and F_k are proportional to the normal component N of the reaction of the surface. We have

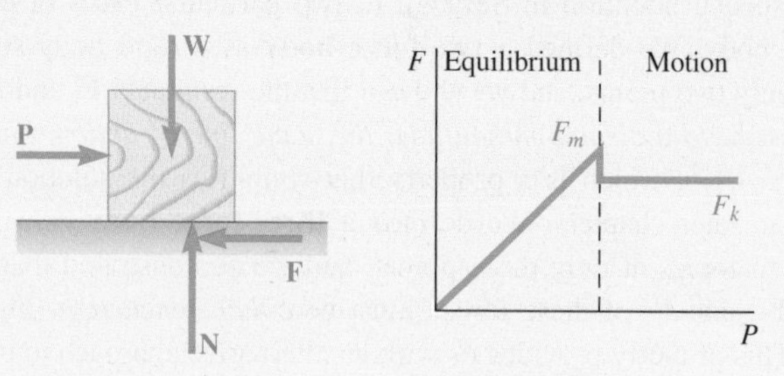

Fig. 4.19

$$F_m = \mu_s N \qquad F_k = \mu_k N \qquad \textbf{(4.8, 4.9)}$$

where μ_s and μ_k are called, respectively, the **coefficient of static friction** and the **coefficient of kinetic friction**. These coefficients depend on the nature and the condition of the surfaces in contact. Approximate values of the coefficients of static friction are given in Table 4.1.

Angles of Friction

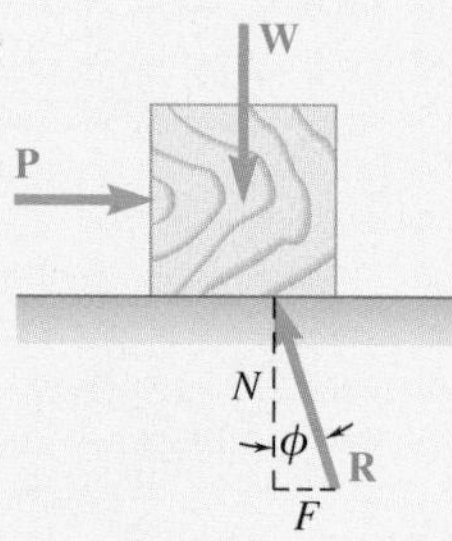

Fig. 4.20

It is sometimes convenient to replace the normal force **N** and the friction force **F** by their resultant **R** (Fig. 4.20). As the friction force increases and reaches its maximum value $F_m = \mu_s N$, the angle ϕ that **R** forms with the normal to the surface increases and reaches a maximum value ϕ_s, which is called the **angle of static friction**. If motion actually takes place, the magnitude of **F** drops to F_k; similarly, the angle ϕ drops to a lower value ϕ_k, which is called the **angle of kinetic friction**. As shown in Sec. 4.4C, we have

$$\tan \phi_s = \mu_s \qquad \tan \phi_k = \mu_k \qquad \textbf{(4.10, 4.11)}$$

Problems Involving Friction

When solving equilibrium problems involving friction, you should keep in mind that the magnitude F of the friction force is equal to $F_m = \mu_s N$ *only if the body is about to slide* [Sec. 4.4D]. *If motion is not impending,* you should treat F and N as independent unknowns to be determined from the equilibrium equations (Fig. 4.21*a*). You should also check that the value of F required to maintain equilibrium is not larger than F_m; if it were, the body would move, and the magnitude of the friction force would be $F_k = \mu_k N$ [Sample Prob. 4.11]. On the other hand, *if motion is known to be impending,* F has reached its maximum value $F_m = \mu_s N$ (Fig. 4.21*b*), and you should substitute this expression for F in the equilibrium equations [Sample Prob. 4.13]. When only three forces are involved in a free-body diagram, including the reaction **R** of the surface in contact with the body, it is usually more convenient to solve the problem by drawing a force triangle [Sample Prob. 4.12]. In some problems, impending motion can be due to tipping instead of slipping; the assessment of this condition requires a moment equilibrium analysis of the body [Sample Prob. 4.14].

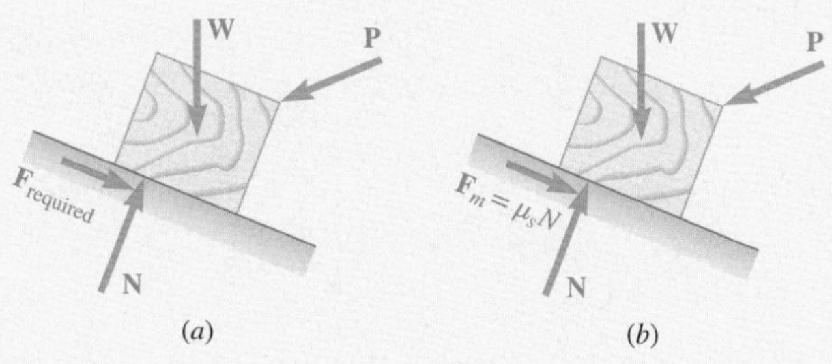

Fig. 4.21

When a problem involves the analysis of the forces exerted on each other by *two bodies A and B*, it is important to show the friction forces with their correct sense. The correct sense for the friction force exerted by B on A, for instance, is opposite to that of the *relative motion* (or impending motion) of A with respect to B [Fig. 4.16].

Review Problems

4.99 A T-shaped bracket supports the four loads shown. Determine the reactions at A and B (*a*) if $a = 10$ in., (*b*) if $a = 7$ in.

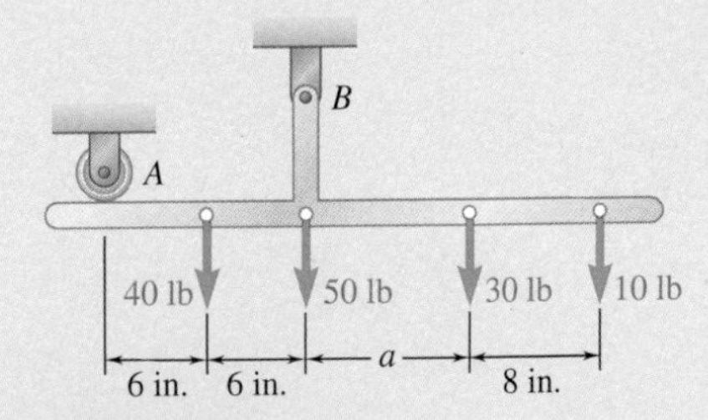

Fig. P4.99

4.100 Neglecting friction and the radius of the pulley, determine (*a*) the tension in cable ADB, (*b*) the reaction at C.

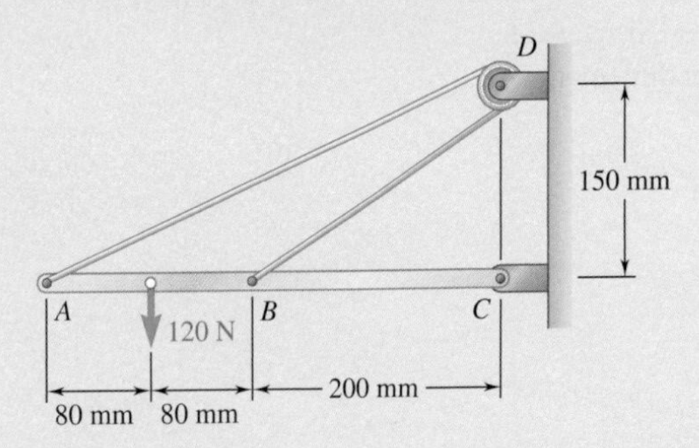

Fig. P4.100

4.101 Member ABC is supported by a pin and bracket at B and by an inextensible cord attached at A and C and passing over a frictionless pulley at D. The tension may be assumed to be the same in portions AD and CD of the cord. For the loading shown and neglecting the size of the pulley, determine the tension in the cord and the reaction at B.

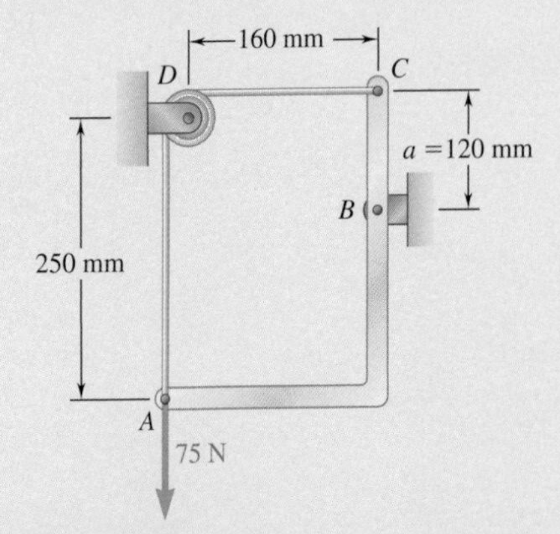

Fig. P4.101

4.102 A movable bracket is held at rest by a cable attached at C and by frictionless rollers at A and B. For the loading shown, determine (a) the tension in the cable, (b) the reactions at A and B.

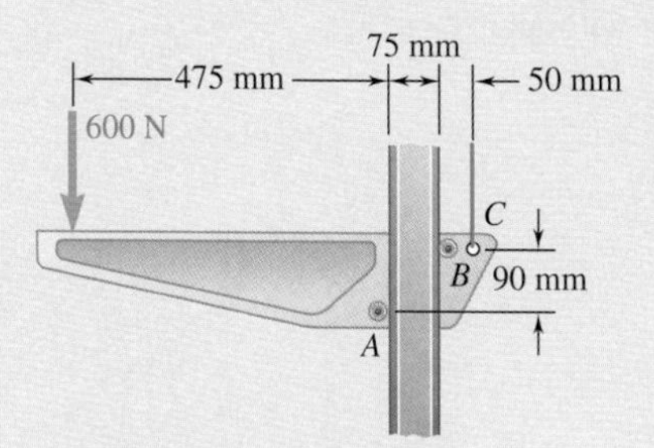

Fig. P4.102

4.103 Rod AB is bent into the shape of an arc of circle and is lodged between two pegs D and E. It supports a load **P** at end B. Neglecting friction and the weight of the rod, determine the distance c corresponding to equilibrium when $a = 20$ mm and $R = 100$ mm.

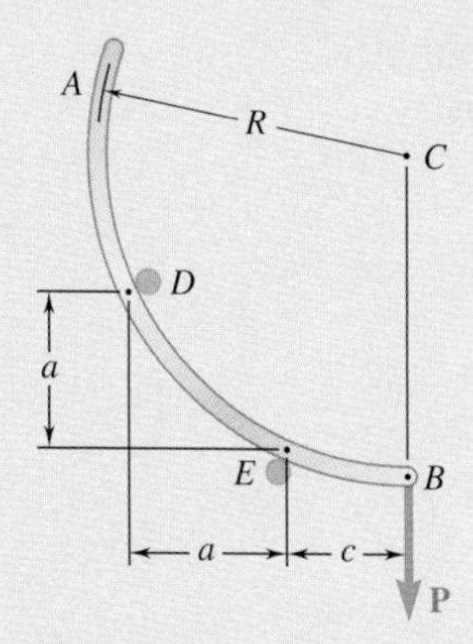

Fig. ***P4.103***

4.104 A thin ring of mass 2 kg and radius $r = 140$ mm is held against a frictionless wall by a 125-mm string AB. Determine (a) the distance d, (b) the tension in the string, (c) the reaction at C.

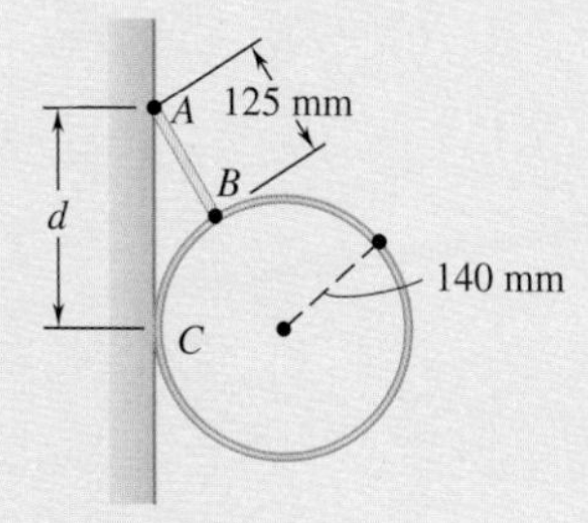

Fig. P4.104

4.105 The table shown weighs 30 lb and has a diameter of 4 ft. It is supported by three legs equally spaced around the edge. A vertical load **P** of magnitude 100 lb is applied to the top of the table at *D*. Determine the maximum value of *a* if the table is not to tip over. Show, on a sketch, the area of the table over which **P** can act without tipping the table.

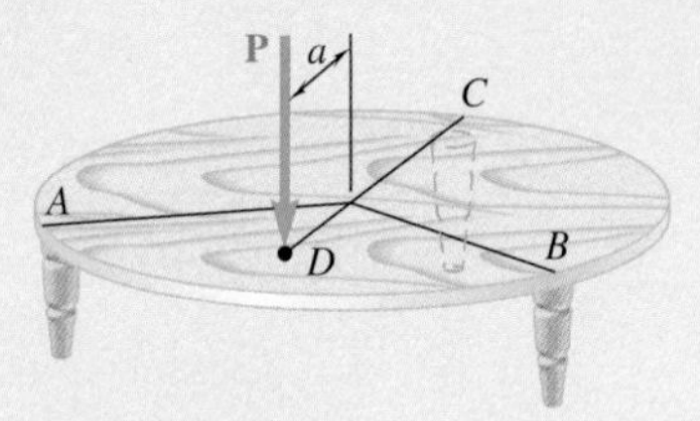

Fig. P4.105

4.106 A vertical load **P** is applied at end *B* of rod *BC*. The constant of the spring is *k*, and the spring is unstretched when $\theta = 60°$. (*a*) Neglecting the weight of the rod, express the angle θ corresponding to the equilibrium position in terms of *P*, *k*, and *l*. (*b*) Determine the value of θ corresponding to equilibrium if $P = \frac{1}{4}kl$.

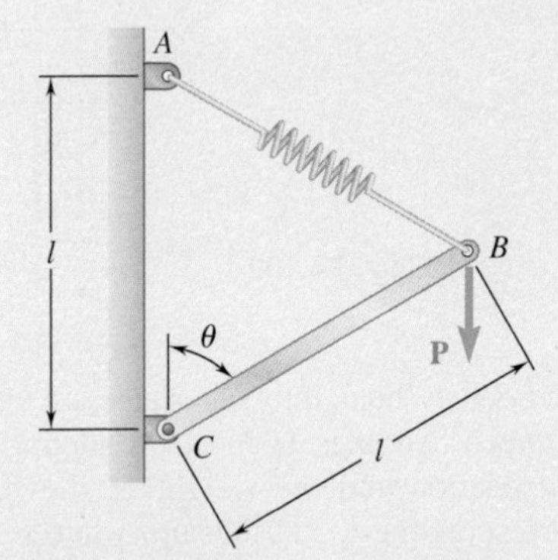

Fig. *P4.106*

4.107 A force **P** is applied to a bent rod *ABC*, which can be supported in four different ways as shown. In each case, if possible, determine the reactions at the supports.

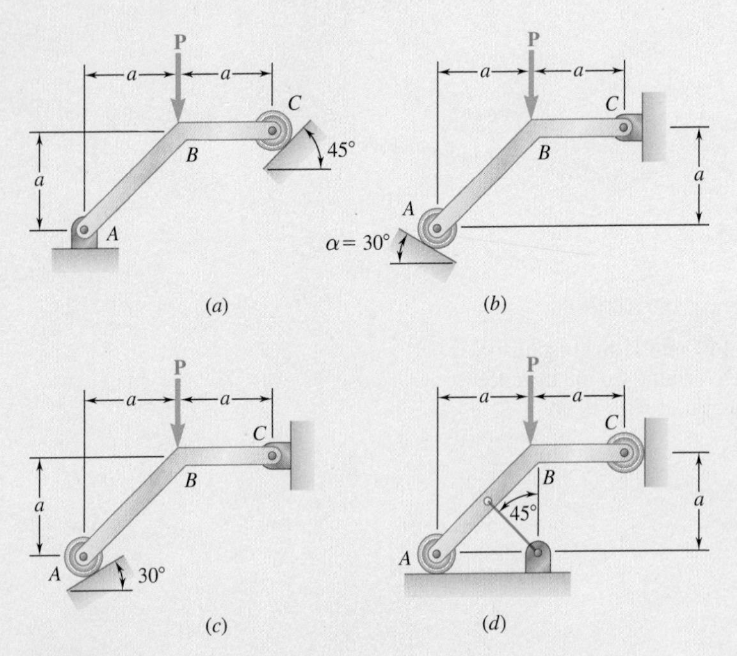

Fig. P4.107

4.108 The rigid L-shaped member ABF is supported by a ball-and-socket joint at A and by three cables. For the loading shown, determine the tension in each cable and the reaction at A.

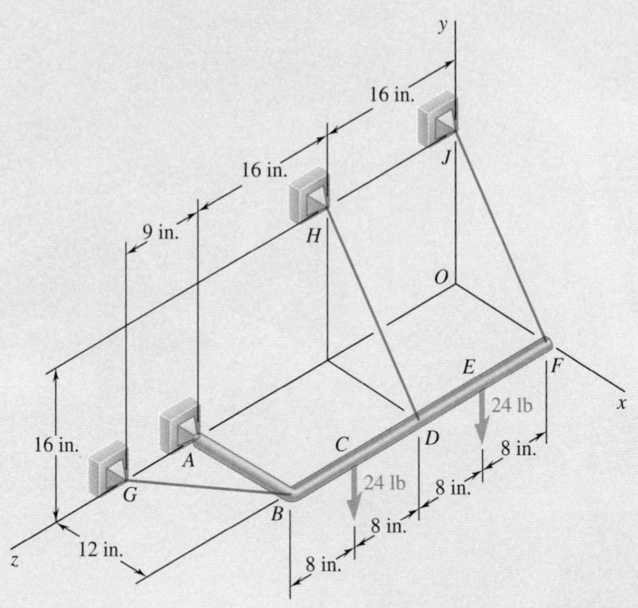

Fig. *P4.108*

4.109 A 1.2-m plank of mass 3 kg rests on two joists. Knowing that the coefficient of static friction between the plank and the joists is 0.30, determine the magnitude of the horizontal force required to move the plank when (*a*) $a = 750$ mm, (*b*) $a = 900$ mm.

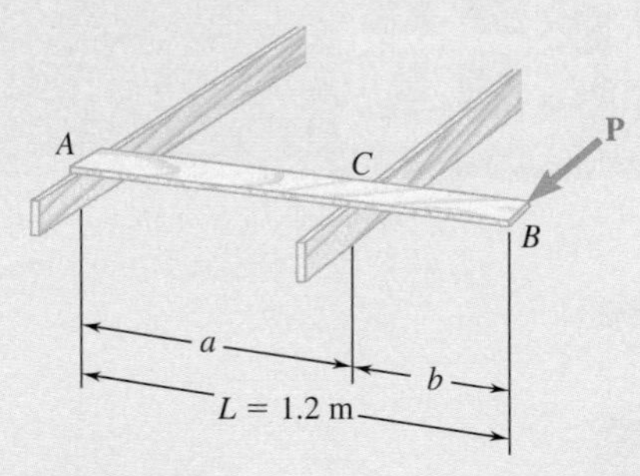

Fig. P4.109

4.110 Two 10-lb blocks A and B are connected by a slender rod of negligible weight. The coefficient of static friction is 0.30 between all surfaces of contact, and the rod forms an angle $\theta = 30°$ with the vertical. (*a*) Show that the system is in equilibrium when $P = 0$. (*b*) Determine the largest value of P for which equilibrium is maintained.

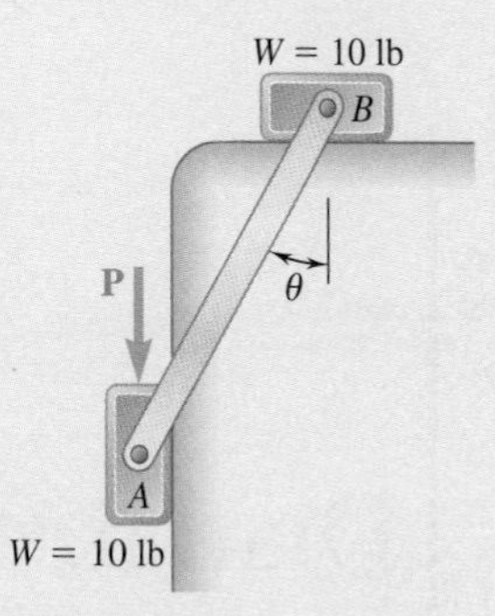

Fig. P4.110

Photo courtesy of Massachussets Turnpike Authority

5

Distributed Forces: Centroids and Centers of Gravity

A precast section of roadway for a new interchange on Interstate 93 is shown being lowered from a gantry crane. In this chapter we will introduce the concept of the centroid of an area; later chapters will establish the relation between the location of the centroid and the behavior of the roadway under loading.

Introduction

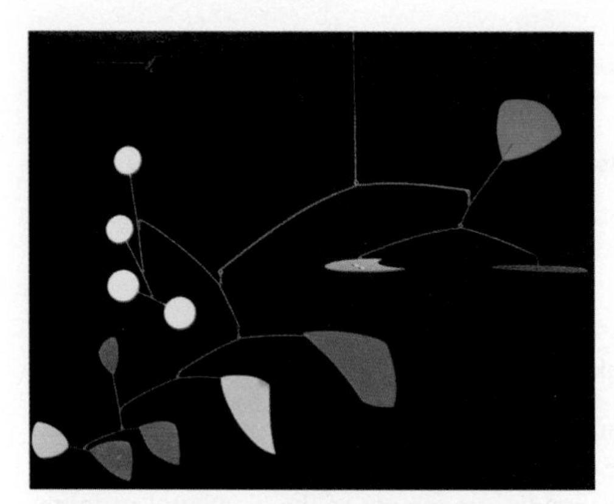

Photo 5.1 The precise balancing of the components of a mobile requires an understanding of centers of gravity and centroids, the main topics of this chapter.

Objectives

- **Describe** the centers of gravity of two and three-dimensional bodies.
- **Define** the centroids of lines, areas, and volumes.
- **Consider** the first moments of lines and areas, and examine their properties.
- **Determine** centroids of composite lines, areas, and volumes by summation methods.
- **Determine** centroids of composite areas by integration.
- **Apply** the theorems of Pappus-Guldinus to analyze surfaces and bodies of revolution.
- **Analyze** distributed loads on beams.

Introduction

We have assumed so far that we could represent the attraction exerted by the earth on a rigid body by a single force **W**. This force, called the force due to gravity or the weight of the body, is applied at the **center of gravity** of the body (Sec. 3.1A). Actually, the earth exerts a force on each of the particles forming the body, so we should represent the attraction of the earth on a rigid body by a large number of small forces distributed over the entire body. You will see in this chapter, however, that all of these small forces can be replaced by a single equivalent force **W**. You will also see how to determine the center of gravity—i.e., the point of application of the resultant **W**—for bodies of various shapes.

In the first part of this chapter, we study two-dimensional bodies, such as flat plates and wires contained in a given plane. We introduce two concepts closely associated with determining the center of gravity of a plate or a wire: the **centroid** of an area or a line and the **first moment** of an area or a line with respect to a given axis. Computing the area of a surface of revolution or the volume of a body of revolution is directly related to determining the centroid of the line or area used to generate that surface or body of revolution (theorems of Pappus-Guldinus). Also, as we show in Sec. 5.3, the determination of the centroid of an area simplifies the analysis of beams subjected to distributed loads.

In the last part of this chapter, you will see how to determine the center of gravity of a three-dimensional body as well as how to calculate the centroid of a volume and the first moments of that volume with respect to the coordinate planes.

5.1 PLANAR CENTERS OF GRAVITY AND CENTROIDS

In Chapter 4, we showed how the locations of the lines of action of forces affect the replacement of a system of forces with an equivalent system of forces and couples. In this section, we extend this idea to show how a distributed system of forces (in particular, the elements of an object's weight) can be replaced by a single resultant force acting at a specific point on an object. The specific point is called the object's center of gravity.

5.1A Center of Gravity of a Two-Dimensional Body

Let us first consider a flat horizontal plate (Fig. 5.1). We can divide the plate into n small elements. We denote the coordinates of the first element by x_1 and y_1, those of the second element by x_2 and y_2, etc. The forces exerted by the earth on the elements of the plate are denoted, respectively, by $\Delta\mathbf{W}_1, \Delta\mathbf{W}_2, \ldots, \Delta\mathbf{W}_n$. These forces or weights are directed toward the center of the earth; however, for all practical purposes, we can assume them to be parallel. Their resultant is therefore a single force in the same direction. The magnitude W of this force is obtained by adding the magnitudes of the elemental weights.

$$\Sigma F_z: \qquad W = \Delta W_1 + \Delta W_2 + \cdots + \Delta W_n$$

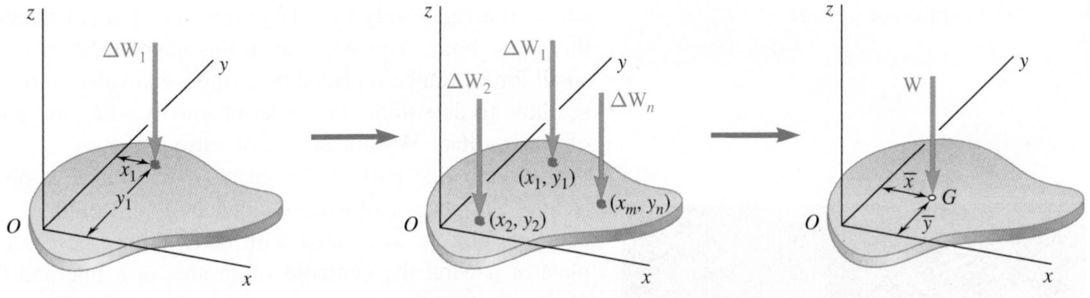

(a) Single element of the plate (b) Multiple elements of the plate (c) Center of gravity

$$\bar{x} = \frac{\int x\, dW}{W} \qquad \bar{y} = \frac{\int y\, dW}{W}$$

Fig. 5.1 The center of gravity of a plate is the point where the resultant weight of the plate acts. It is the weighted average of all the elements of weight that make up the plate.

To obtain the coordinates $\bar{x}$ and $\bar{y}$ of point G where the resultant $\mathbf{W}$ should be applied, we note that the moments of $\mathbf{W}$ about the y and x axes are equal to the sum of the corresponding moments of the elemental weights:

$$\begin{aligned} \Sigma M_y: \quad & \bar{x}W = x_1\Delta W_1 + x_2\Delta W_2 + \cdots + x_n\Delta W_n \\ \Sigma M_x: \quad & \bar{y}W = y_1\Delta W_1 + y_2\Delta W_2 + \cdots + y_n\Delta W_n \end{aligned} \tag{5.1}$$

Solving these equations for $\bar{x}$ and $\bar{y}$ gives us

$$\bar{x} = \frac{x_1\Delta W_1 + x_2\Delta W_2 + \cdots + x_n\Delta W_n}{W}$$

$$\bar{y} = \frac{y_1\Delta W_1 + y_2\Delta W_2 + \cdots + y_n\Delta W_n}{W}$$

We could use these equations in this form to find the center of gravity of a collection of n objects, each with a weight of W_i.

If we now increase the number of elements into which we divide the plate and simultaneously decrease the size of each element, in the limit of infinitely many elements of infinitesimal size, we obtain the expressions

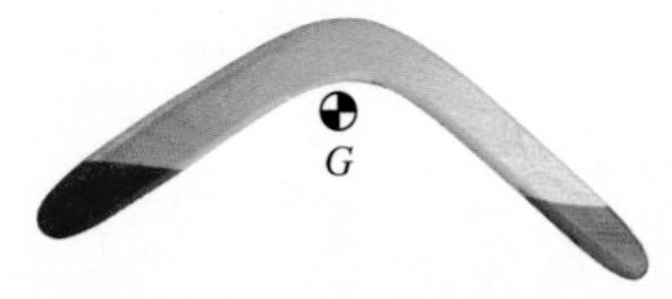

Photo 5.2 The center of gravity of a boomerang is not located on the object itself.

Weight, center of gravity of a flat plate

$$W = \int dW \qquad \bar{x}W = \int x\,dW \qquad \bar{y}W = \int y\,dW \tag{5.2}$$

Or, solving for $\bar{x}$ and $\bar{y}$, we have

$$W = \int dW \qquad \bar{x} = \frac{\int x\,dW}{W} \qquad \bar{y} = \frac{\int y\,dW}{W} \tag{5.2'}$$

These equations define the weight **W** and the coordinates $\bar{x}$ and $\bar{y}$ of the **center of gravity** G of a flat plate. The same equations can be derived for a wire lying in the xy plane (Fig. 5.2). Note that the center of gravity G of a wire is usually not located on the wire.

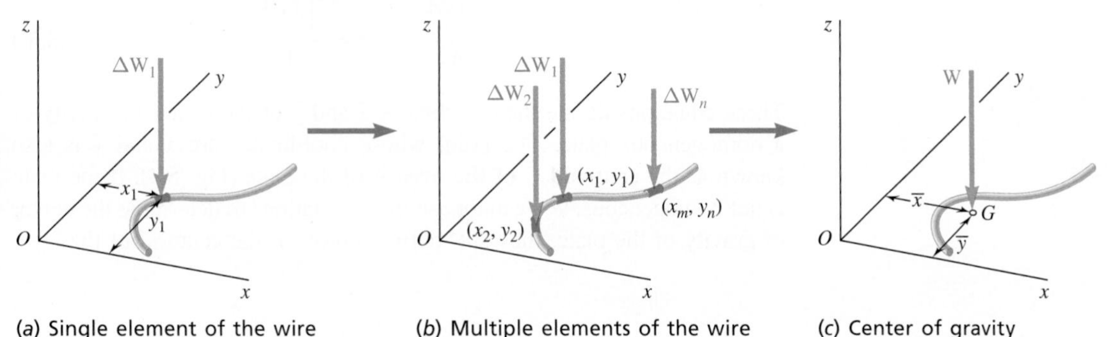

(*a*) Single element of the wire (*b*) Multiple elements of the wire (*c*) Center of gravity

$$\bar{x} = \frac{\int x\,dW}{W} \qquad \bar{y} = \frac{\int y\,dW}{W}$$

Fig. 5.2 The center of gravity of a wire is the point where the resultant weight of the wire acts. The center of gravity may not actually be located on the wire.

5.1B Centroids of Areas and Lines

In the case of a flat homogeneous plate of uniform thickness, we can express the magnitude ΔW of the weight of an element of the plate as

$$\Delta W = \gamma t\,\Delta A$$

where γ = specific weight (weight per unit volume) of the material
t = thickness of the plate
ΔA = area of the element

Similarly, we can express the magnitude W of the weight of the entire plate as

$$W = \gamma t A$$

where A is the total area of the plate.

If U.S. customary units are used, the specific weight γ should be expressed in lb/ft^3, the thickness t in feet, and the areas ΔA and A in square feet. Then ΔW and W are expressed in pounds. If SI units are used, γ should be expressed in N/m^3, t in meters, and the areas ΔA and A in square meters; the weights ΔW and W are then expressed in newtons.†

Substituting for ΔW and W in the moment equations (5.1) and dividing throughout by γt, we obtain

$$\Sigma M_y: \quad \bar{x}A = x_1\,\Delta A_1 + x_2\,\Delta A_2 + \cdots + x_n\,\Delta A_n$$
$$\Sigma M_x: \quad \bar{y}A = y_1\,\Delta A_1 + y_2\,\Delta A_2 + \cdots + y_n\,\Delta A_n$$

If we increase the number of elements into which the area A is divided and simultaneously decrease the size of each element, in the limit we obtain

Centroid of an area A

$$\bar{x}A = \int x\,dA \qquad \bar{y}A = \int y\,dA \tag{5.3}$$

Or, solving for $\bar{x}$ and $\bar{y}$, we obtain

$$\bar{x} = \frac{\int x\,dA}{A} \qquad \bar{y} = \frac{\int y\,dA}{A} \tag{5.3'}$$

These equations define the coordinates $\bar{x}$ and $\bar{y}$ of the center of gravity of a homogeneous plate. The point whose coordinates are $\bar{x}$ and $\bar{y}$ is also known as the **centroid C of the area** A of the plate (Fig. 5.3). If the plate is not homogeneous, you cannot use these equations to determine the center of gravity of the plate; they still define, however, the centroid of the area.

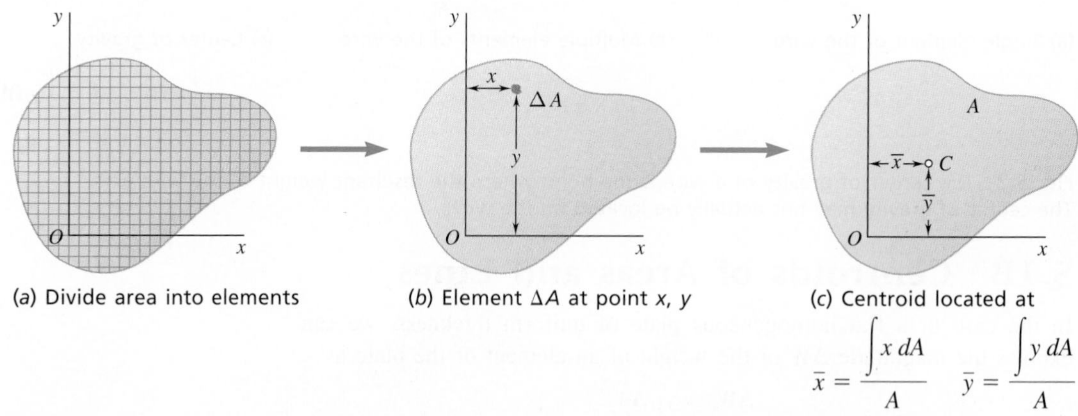

(a) Divide area into elements (b) Element ΔA at point x, y (c) Centroid located at $\bar{x} = \dfrac{\int x\,dA}{A}$ $\bar{y} = \dfrac{\int y\,dA}{A}$

Fig. 5.3 The centroid of an area is the point where a homogeneous plate of uniform thickness would balance.

†We should note that in the SI system of units, a given material is generally characterized by its density ρ (mass per unit volume) rather than by its specific weight γ. You can obtain the specific weight of the material from the relation

$$\gamma = \rho g$$

where $g = 9.81$ m/s^2. Note that since ρ is expressed in kg/m^3, the units of γ are (kg/m^3)(m/s^2), or N/m^3.

In the case of a homogeneous wire of uniform cross section, we can express the magnitude ΔW of the weight of an element of wire as

$$\Delta W = \gamma a \, \Delta L$$

where γ = specific weight of the material
a = cross-sectional area of the wire
ΔL = length of the element

The center of gravity of the wire then coincides with the **centroid C of the line L** defining the shape of the wire (Fig. 5.4). We can obtain the coordinates $\bar{x}$ and $\bar{y}$ of the centroid of line L from the equations

Centroid of a line L

$$\bar{x}L = \int x \, dL \qquad \bar{y}L = \int y \, dL \tag{5.4}$$

Solving for $\bar{x}$ and $\bar{y}$ gives us

$$\bar{x} = \frac{\int x \, dL}{L} \qquad \bar{y} = \frac{\int y \, dL}{L} \tag{5.4'}$$

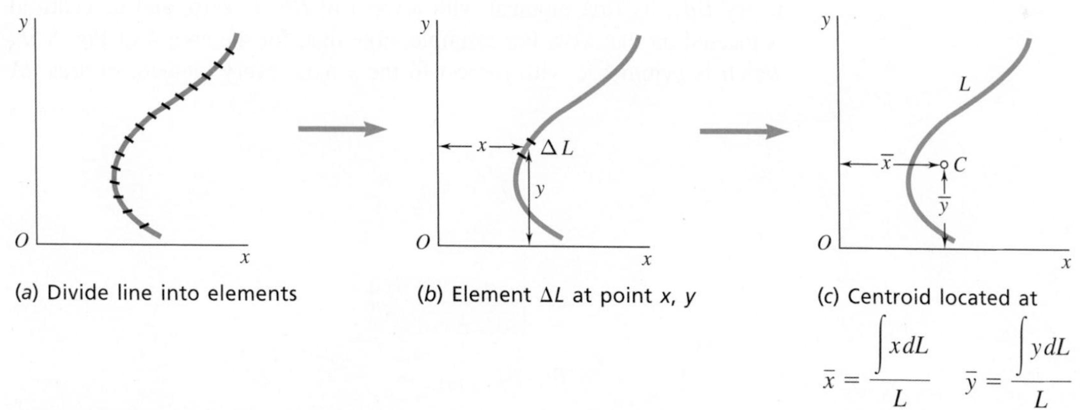

(*a*) Divide line into elements (*b*) Element ΔL at point x, y (*c*) Centroid located at $\bar{x} = \frac{\int x dL}{L}$ $\bar{y} = \frac{\int y dL}{L}$

Fig. 5.4 The centroid of a line is the point where a homogeneous wire of uniform cross section would balance.

5.1C First Moments of Areas and Lines

The integral $\int x \, dA$ in Eqs. (5.3) is known as the **first moment of the area A with respect to the y axis** and is denoted by Q_y. Similarly, the integral $\int y \, dA$ defines the **first moment of A with respect to the x axis** and is denoted by Q_x. That is,

First moments of area A

$$Q_y = \int x dA \qquad Q_x = \int y dA \tag{5.5}$$

Comparing Eqs. (5.3) with Eqs. (5.5), we note that we can express the first moments of the area A as the products of the area and the coordinates of its centroid:

$$Q_y = \bar{x}A \qquad Q_x = \bar{y}A \tag{5.6}$$

It follows from Eqs. (5.6) that we can obtain the coordinates of the centroid of an area by dividing the first moments of that area by the area itself. The first moments of the area are also useful in mechanics of materials for determining the shearing stresses in beams under transverse loadings. Finally, we observe from Eqs. (5.6) that, if the centroid of an area is located on a coordinate axis, the first moment of the area with respect to that axis is zero. Conversely, if the first moment of an area with respect to a coordinate axis is zero, the centroid of the area is located on that axis.

We can use equations similar to Eqs. (5.5) and (5.6) to define the first moments of a line with respect to the coordinate axes and to express these moments as the products of the length L of the line and the coordinates $\bar{x}$ and $\bar{y}$ of its centroid.

An area A is said to be **symmetric with respect to an axis** BB' if for every point P of the area there exists a point P' of the same area such that the line PP' is perpendicular to BB' and is divided into two equal parts by that axis (Fig. 5.5*a*). The axis BB' is called an **axis of symmetry**. A line L is said to be symmetric with respect to an axis BB' if it satisfies similar conditions. When an area A or a line L possesses an axis of symmetry BB', its first moment with respect to BB' is zero, and its centroid is located on that axis. For example, note that, for the area A of Fig. 5.5*b*, which is symmetric with respect to the y axis, every element of area dA

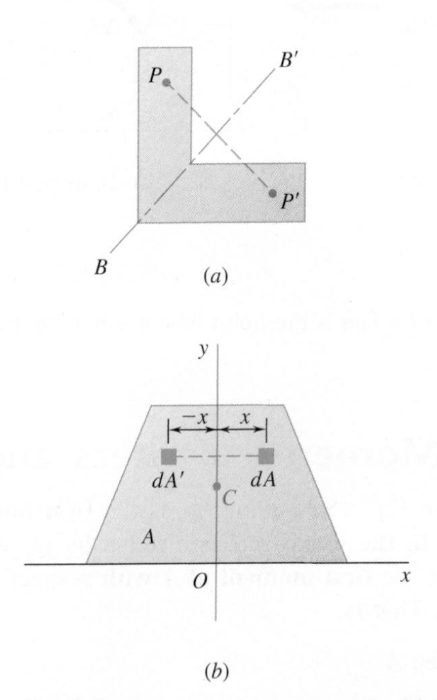

Fig. 5.5 Symmetry about an axis. (*a*) The area is symmetric about the axis *BB'*. (*b*) The centroid of the area is located on the axis of symmetry.

with abscissa x corresponds to an element dA' of equal area and with abscissa $-x$. It follows that the integral in the first of Eqs. (5.5) is zero and, thus, that $Q_y = 0$. It also follows from the first of the relations in Eq. (5.3) that $\bar{x} = 0$. Thus, if an area A or a line L possesses an axis of symmetry, its centroid C is located on that axis.

We further note that if an area or line possesses two axes of symmetry, its centroid C must be located at the intersection of the two axes (Fig. 5.6). This property enables us to determine immediately the centroids of areas such as circles, ellipses, squares, rectangles, equilateral triangles, or other symmetric figures, as well as the centroids of lines in the shape of the circumference of a circle, the perimeter of a square, etc.

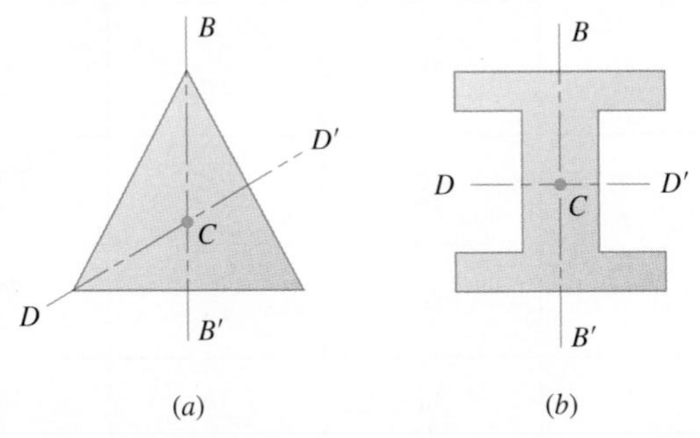

Fig. 5.6 If an area has two axes of symmetry, the centroid is located at their intersection. (*a*) An area with two axes of symmetry but no center of symmetry; (*b*) an area with two axes of symmetry and a center of symmetry.

We say that an area A is **symmetric with respect to a center** O if, for every element of area dA of coordinates x and y, there exists an element dA' of equal area with coordinates $-x$ and $-y$ (Fig. 5.7). It then follows that the integrals in Eqs. (5.5) are both zero and that $Q_x = Q_y = 0$. It also follows from Eqs. (5.3) that $\bar{x} = \bar{y} = 0$; that is, that the centroid of the area coincides with its center of symmetry O. Similarly, if a line possesses a center of symmetry O, the centroid of the line coincides with the center O.

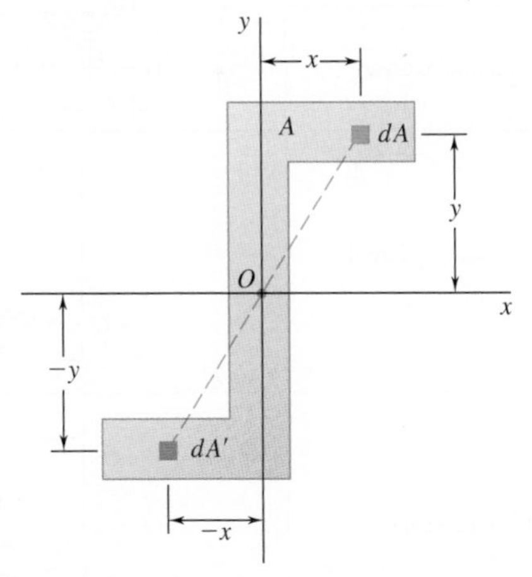

Fig. 5.7 An area may have a center of symmetry but no axis of symmetry.

Note that a figure possessing a center of symmetry does not necessarily possess an axis of symmetry (Fig. 5.7), whereas a figure possessing two axes of symmetry does not necessarily possess a center of symmetry (Fig. 5.6*a*). However, if a figure possesses two axes of symmetry at right angles to each other, the point of intersection of these axes is a center of symmetry (Fig. 5.6*b*).

Determining the centroids of unsymmetrical areas and lines and of areas and lines possessing only one axis of symmetry will be discussed in the next section. Centroids of common shapes of areas and lines are shown in Fig. 5.8A and B.

5.1D Composite Plates and Wires

In many instances, we can divide a flat plate into rectangles, triangles, or the other common shapes shown in Fig. 5.8A. We can determine the abscissa $\bar{X}$ of the plate's center of gravity G from the abscissas $\bar{x}_1, \bar{x}_2, \ldots, \bar{x}_n$ of the centers of gravity of the various parts. To do this, we equate the moment of the weight of the whole plate about the y axis to the sum of

Shape	$\bar{x}$	$\bar{y}$	Area
Triangular area	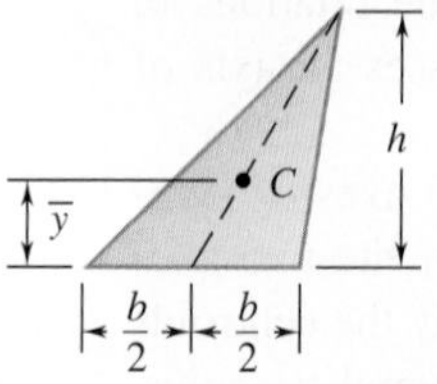	$\frac{h}{3}$	$\frac{bh}{2}$
Quarter-circular area	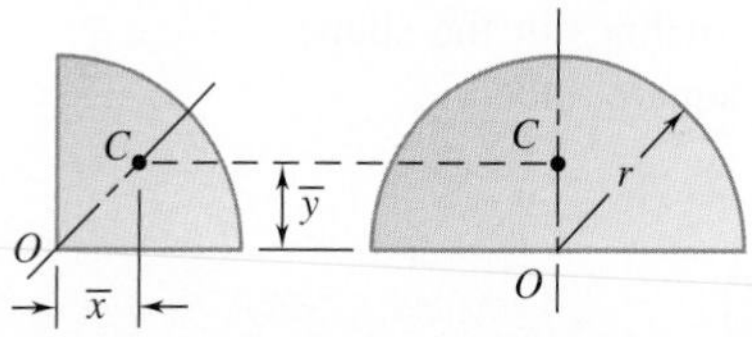 $\frac{4r}{3\pi}$	$\frac{4r}{3\pi}$	$\frac{\pi r^2}{4}$
Semicircular area	0	$\frac{4r}{3\pi}$	$\frac{\pi r^2}{2}$
Quarter-elliptical area	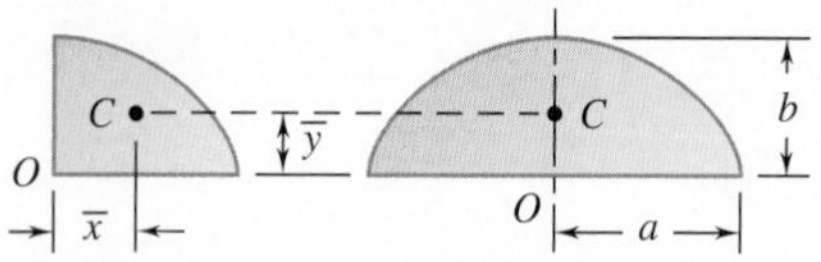 $\frac{4a}{3\pi}$	$\frac{4b}{3\pi}$	$\frac{\pi ab}{4}$
Semielliptical area	0	$\frac{4b}{3\pi}$	$\frac{\pi ab}{2}$
Semiparabolic area	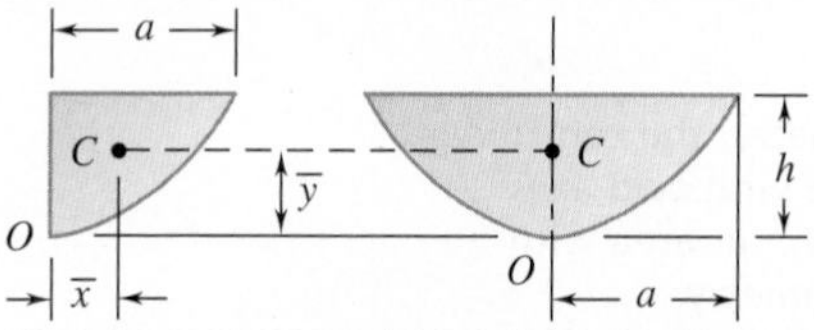 $\frac{3a}{8}$	$\frac{3h}{5}$	$\frac{2ah}{3}$
Parabolic area	0	$\frac{3h}{5}$	$\frac{4ah}{3}$
Parabolic spandrel	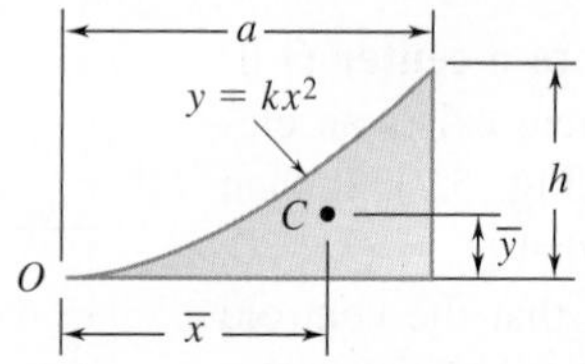 $\frac{3a}{4}$	$\frac{3h}{10}$	$\frac{ah}{3}$
General spandrel	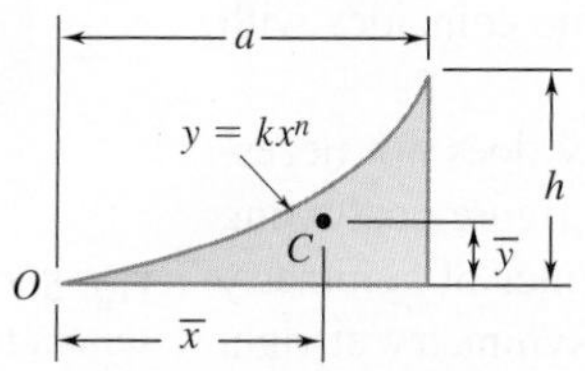 $\frac{n+1}{n+2}a$	$\frac{n+1}{4n+2}h$	$\frac{ah}{n+1}$
Circular sector	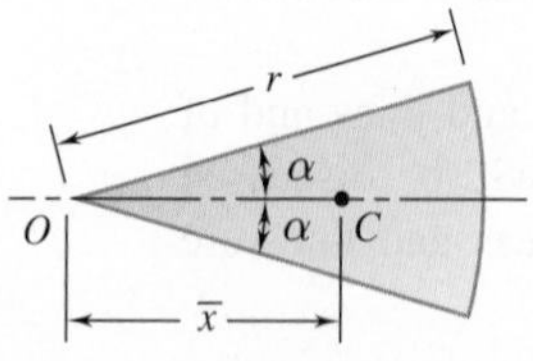 $\frac{2r \sin \alpha}{3\alpha}$	0	αr^2

Fig. 5.8A Centroids of common shapes of areas.

Shape		$\overline{x}$	$\overline{y}$	Length
Quarter-circular arc	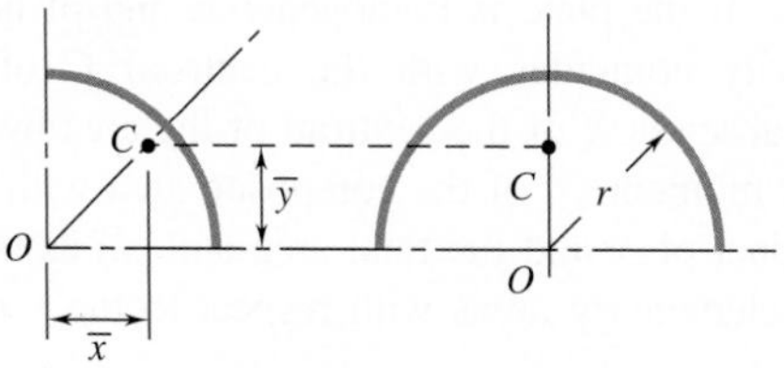	$\frac{2r}{\pi}$	$\frac{2r}{\pi}$	$\frac{\pi r}{2}$
Semicircular arc		0	$\frac{2r}{\pi}$	πr
Arc of circle	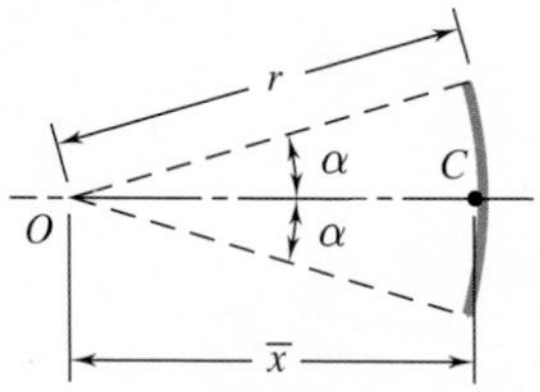	$\frac{r \sin \alpha}{\alpha}$	0	$2\alpha r$

Fig. 5.8B Centroids of common shapes of lines.

the moments of the weights of the various parts about the same axis (Fig. 5.9). We can obtain the ordinate $\overline{Y}$ of the center of gravity of the plate in a similar way by equating moments about the x axis. Mathematically, we have

$$\Sigma M_y:\quad \overline{X}(W_1 + W_2 + \cdots + W_n) = \overline{x}_1 W_1 + \overline{x}_2 W_2 + \cdots + \overline{x}_n W_n$$
$$\Sigma M_x:\quad \overline{Y}(W_1 + W_2 + \cdots + W_n) = \overline{y}_1 W_1 + \overline{y}_2 W_2 + \cdots + \overline{y}_n W_n$$

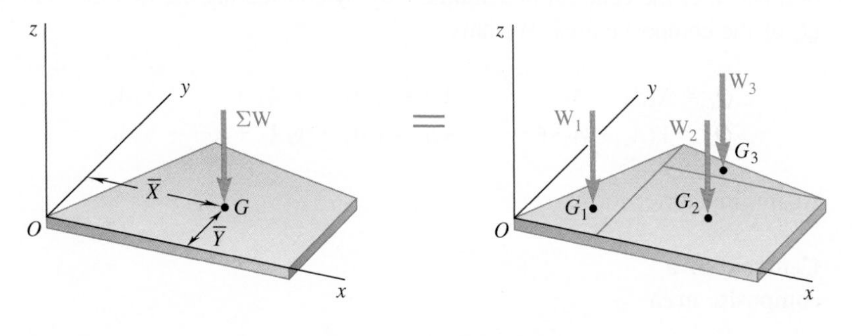

$$\Sigma M_y:\quad \overline{X}\Sigma W = \Sigma \overline{x} W$$
$$\Sigma M_x:\quad \overline{Y}\Sigma W = \Sigma \overline{y} W$$

Fig. 5.9 We can determine the location of the center of gravity G of a composite plate from the centers of gravity G_1, G_2, ... of the component plates.

In more condensed notation, this is

Center of gravity of a composite plate

$$\overline{X} = \frac{\Sigma \overline{x} W}{W} \qquad \overline{Y} = \frac{\Sigma \overline{y} W}{W} \tag{5.7}$$

We can use these equations to find the coordinates $\overline{X}$ and $\overline{Y}$ of the center of gravity of the plate from the centers of gravity of its component parts.

If the plate is homogeneous and of uniform thickness, the center of gravity coincides with the centroid C of its area. We can determine the abscissa $\overline{X}$ of the centroid of the area by noting that we can express the first moment Q_y of the composite area with respect to the y axis as (1) the product of $\overline{X}$ and the total area and (2) as the sum of the first moments of the elementary areas with respect to the y axis (Fig. 5.10). We obtain the

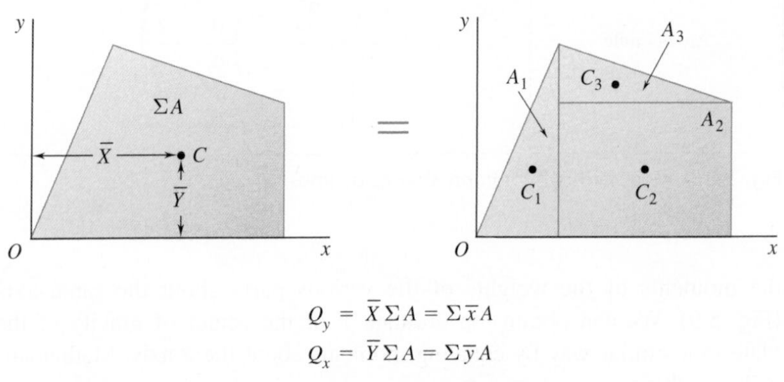

$$Q_y = \overline{X}\,\Sigma A = \Sigma \bar{x} A$$
$$Q_x = \overline{Y}\,\Sigma A = \Sigma \bar{y} A$$

Fig. 5.10 We can find the location of the centroid of a composite area from the centroids of the component areas.

ordinate $\overline{Y}$ of the centroid in a similar way by considering the first moment Q_x of the composite area. We have

$$Q_y = \overline{X}(A_1 + A_2 + \cdots + A_n) = \bar{x}_1 A_1 + \bar{x}_2 A_2 + \cdots + \bar{x}_n A_n$$
$$Q_x = \overline{Y}(A_1 + A_2 + \cdots + A_n) = \bar{y}_1 A_1 + \bar{y}_2 A_2 + \cdots + \bar{y}_n A_n$$

Again, in shorter form,

Centroid of a composite area

$$Q_y = \overline{X}\Sigma A = \Sigma \bar{x} A \qquad Q_x = \overline{Y}\Sigma A = \Sigma \bar{y} A \tag{5.8}$$

These equations yield the first moments of the composite area, or we can use them to obtain the coordinates $\overline{X}$ and $\overline{Y}$ of its centroid.

First moments of areas, like moments of forces, can be positive or negative. Thus, you need to take care to assign the appropriate sign to the moment of each area. For example, an area whose centroid is located to the left of the y axis has a negative first moment with respect to that axis. Also, the area of a hole should be assigned a negative sign (Fig. 5.11).

Similarly, it is possible in many cases to determine the center of gravity of a composite wire or the centroid of a composite line by dividing the wire or line into simpler elements (see Sample Prob. 5.2).

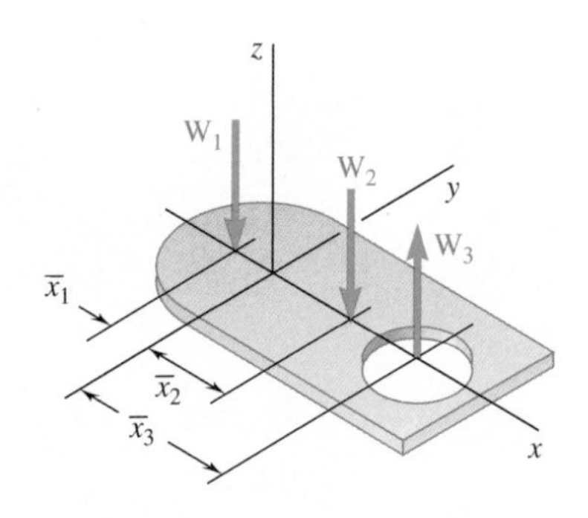

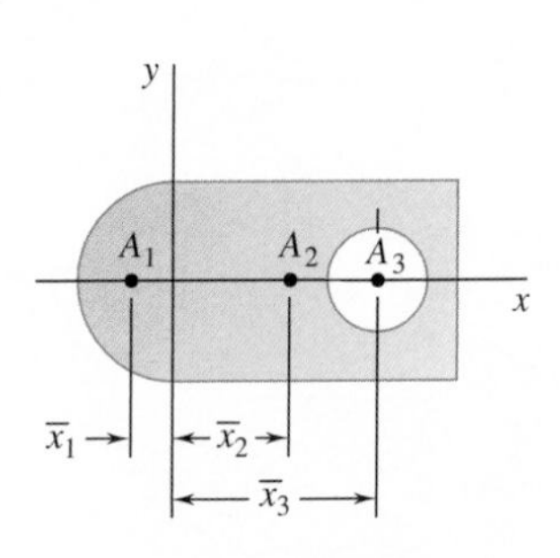

	$\bar{x}$	A	$\bar{x}A$
A_1 Semicircle	–	+	–
A_2 Full rectangle	+	+	+
A_3 Circular hole	+	–	–

Fig. 5.11 When calculating the centroid of a composite area, note that if the centroid of a component area has a negative coordinate distance relative to the origin, or if the area represents a hole, then the first moment is negative.

Sample Problem 5.1

For the plane area shown, determine (*a*) the first moments with respect to the x and y axes; (*b*) the location of the centroid.

STRATEGY: Break up the given area into simple components, find the centroid of each component, and then find the overall first moments and centroid.

MODELING: As shown in Fig. 1, you obtain the given area by adding a rectangle, a triangle, and a semicircle and then subtracting a circle. Using the coordinate axes shown, find the area and the coordinates of the centroid of each of the component areas. To keep track of the data, enter them in a table. The area of the circle is indicated as negative because it is subtracted from the other areas. The coordinate $\bar{y}$ of the centroid of the triangle is negative for the axes shown. Compute the first moments of the component areas with respect to the coordinate axes and enter them in your table.

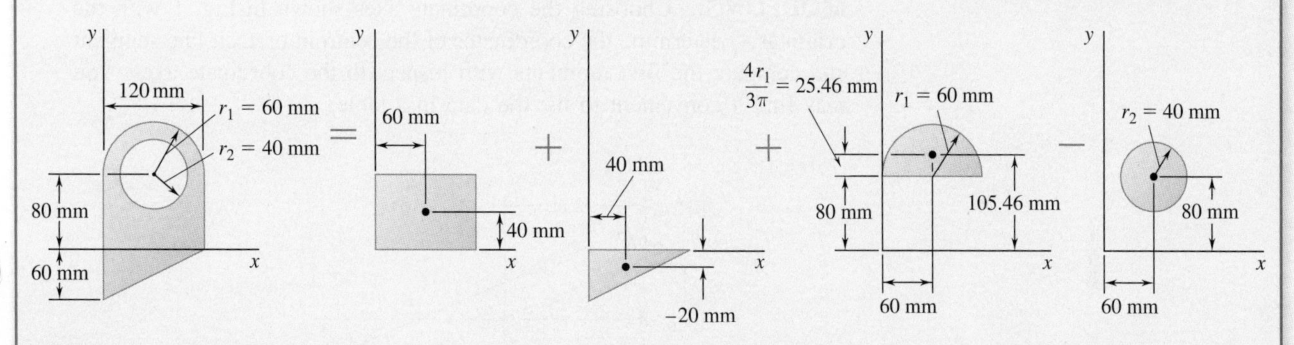

Component	A, mm²	$\bar{x}$, mm	$\bar{y}$, mm	$\bar{x}A$, mm³	$\bar{y}A$, mm³
Rectangle	$(120)(80) = 9.6 \times 10^3$	60	40	$+576 \times 10^3$	$+384 \times 10^3$
Triangle	$\frac{1}{2}(120)(60) = 3.6 \times 10^3$	40	−20	$+144 \times 10^3$	-72×10^3
Semicircle	$\frac{1}{2}\pi(60)^2 = 5.655 \times 10^3$	60	105.46	$+339.3 \times 10^3$	$+596.4 \times 10^3$
Circle	$-\pi(40)^2 = -5.027 \times 10^3$	60	80	-301.6×10^3	-402.2×10^3
	$\Sigma A = 13.828 \times 10^3$			$\Sigma\bar{x}A = +757.7 \times 10^3$	$\Sigma\bar{y}A = +506.2 \times 10^3$

Fig. 1 Given area modeled as the combination of simple geometric shapes.

ANALYSIS:

a. First Moments of the Area. Using Eqs. (5.8), you obtain

$$Q_x = \Sigma\bar{y}A = 506.2 \times 10^3 \text{ mm}^3 \qquad Q_x = 506 \times 10^3 \text{ mm}^3 \blacktriangleleft$$

$$Q_y = \Sigma\bar{x}A = 757.7 \times 10^3 \text{ mm}^3 \qquad Q_y = 758 \times 10^3 \text{ mm}^3 \blacktriangleleft$$

b. Location of Centroid. Substituting the values given in the table into the equations defining the centroid of a composite area yields (Fig. 2)

$$\bar{X}\Sigma A = \Sigma\bar{x}A: \quad \bar{X}(13.828 \times 10^3 \text{ mm}^2) = 757.7 \times 10^3 \text{ mm}^3$$

$$\bar{X} = 54.8 \text{ mm} \blacktriangleleft$$

$$\bar{Y}\Sigma A = \Sigma\bar{y}A: \quad \bar{Y}(13.828 \times 10^3 \text{ mm}^2) = 506.2 \times 10^3 \text{ mm}^3$$

$$\bar{Y} = 36.6 \text{ mm} \blacktriangleleft$$

Fig. 2 Centroid of composite area.

REFLECT and THINK: Given that the lower portion of the shape has more area to the left and that the upper portion has a hole, the location of the centroid seems reasonable upon visual inspection.

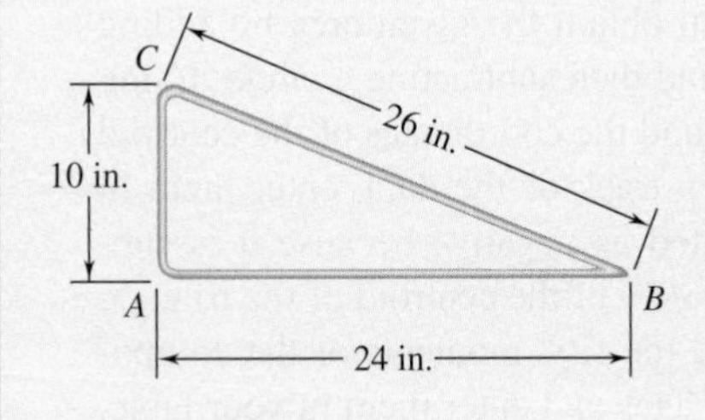

Sample Problem 5.2

The figure shown is made from a piece of thin, homogeneous wire. Determine the location of its center of gravity.

STRATEGY: Since the figure is formed of homogeneous wire, its center of gravity coincides with the centroid of the corresponding line. Therefore, you can simply determine that centroid.

MODELING: Choosing the coordinate axes shown in Fig. 1 with the origin at *A*, determine the coordinates of the centroid of each line segment and compute the first moments with respect to the coordinate axes. You may find it convenient to list the data in a table.

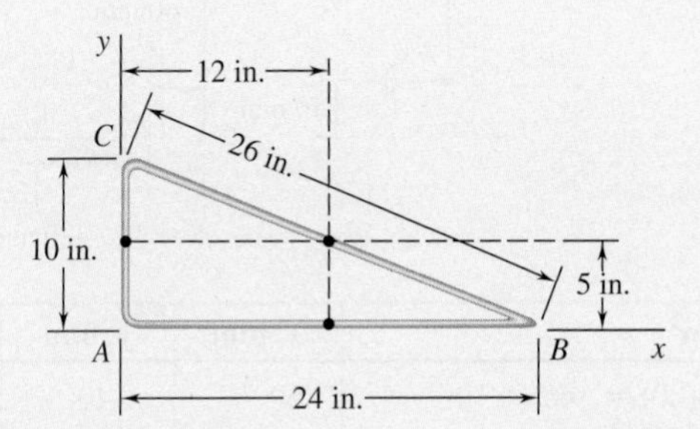

Fig. 1 Location of each line segment's centroid.

Segment	L, in.	$\bar{x}$, in.	$\bar{y}$, in.	$\bar{x}L$, in^2	$\bar{y}L$, in^2
AB	24	12	0	288	0
BC	26	12	5	312	130
CA	10	0	5	0	50
	$\Sigma L = 60$			$\Sigma \bar{x}L = 600$	$\Sigma \bar{y}L = 180$

ANALYSIS: Substituting the values obtained from the table into the equations defining the centroid of a composite line gives

$\bar{X}\Sigma L = \Sigma \bar{x}L$: $\quad \bar{X}(60 \text{ in.}) = 600 \text{ in}^2 \qquad \bar{X} = 10 \text{ in.}$ ◀

$\bar{Y}\Sigma L = \Sigma \bar{y}L$: $\quad \bar{Y}(60 \text{ in.}) = 180 \text{ in}^2 \qquad \bar{Y} = 3 \text{ in.}$ ◀

REFLECT and THINK: The centroid is not on the wire itself, but it is within the area enclosed by the wire.

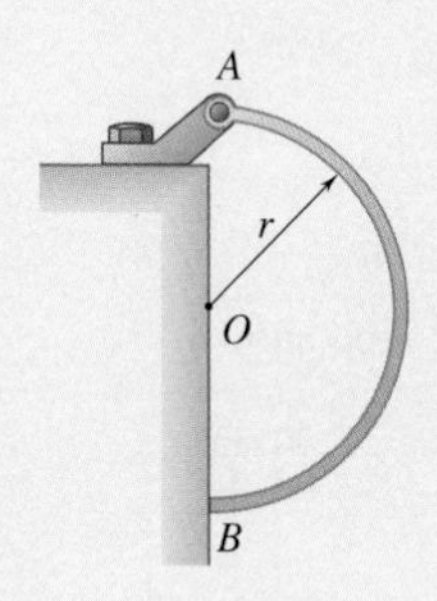

Sample Problem 5.3

A uniform semicircular rod of weight W and radius r is attached to a pin at A and rests against a frictionless surface at B. Determine the reactions at A and B.

STRATEGY: The key to solving the problem is finding where the weight W of the rod acts. Since the rod is a simple geometrical shape, you can look in Fig. 5.8 for the location of the wire's centroid.

MODELING: Draw a free-body diagram of the rod (Fig. 1). The forces acting on the rod are its weight $\mathbf{W}$, which is applied at the center of gravity G (whose position is obtained from Fig. 5.8B); a reaction at A, represented by its components $\mathbf{A}_x$ and $\mathbf{A}_y$; and a horizontal reaction at B.

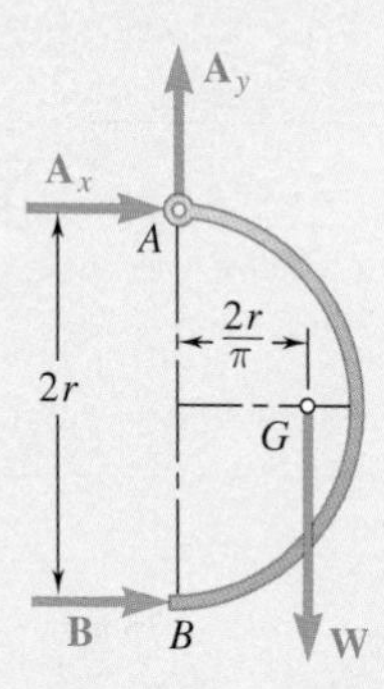

Fig. 1 Free-body diagram of rod.

ANALYSIS:

$$+\circlearrowleft \Sigma M_A = 0: \quad B(2r) - W\left(\frac{2r}{\pi}\right) = 0$$

$$B = +\frac{W}{\pi} \qquad \mathbf{B} = \frac{W}{\pi} \rightarrow \blacktriangleleft$$

$$\xrightarrow{+} \Sigma F_x = 0: \quad A_x + B = 0$$

$$A_x = -B = -\frac{W}{\pi} \qquad \mathbf{A}_x = \frac{W}{\pi} \leftarrow$$

$$+\uparrow \Sigma F_y = 0: \quad A_y - W = 0 \qquad \mathbf{A}_y = W\uparrow$$

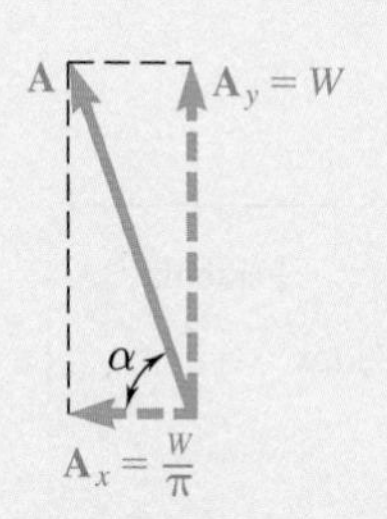

Fig. 2 Reaction at A.

Adding the two components of the reaction at A (Fig. 2), we have

$$A = \left[W^2 + \left(\frac{W}{\pi}\right)^2\right]^{1/2} \qquad A = W\left(1 + \frac{1}{\pi^2}\right)^{1/2} \blacktriangleleft$$

$$\tan \alpha = \frac{W}{W/\pi} = \pi \qquad \alpha = \tan^{-1}\pi \blacktriangleleft$$

The answers can also be expressed as

$$\mathbf{A} = 1.049W \measuredangle 72.3° \qquad \mathbf{B} = 0.318W \rightarrow \blacktriangleleft$$

REFLECT and THINK: Once you know the location of the rod's center of gravity, the problem is a straightforward application of the concepts in Chapter 4.

Problems

5.1 through *5.8* Locate the centroid of the plane area shown.

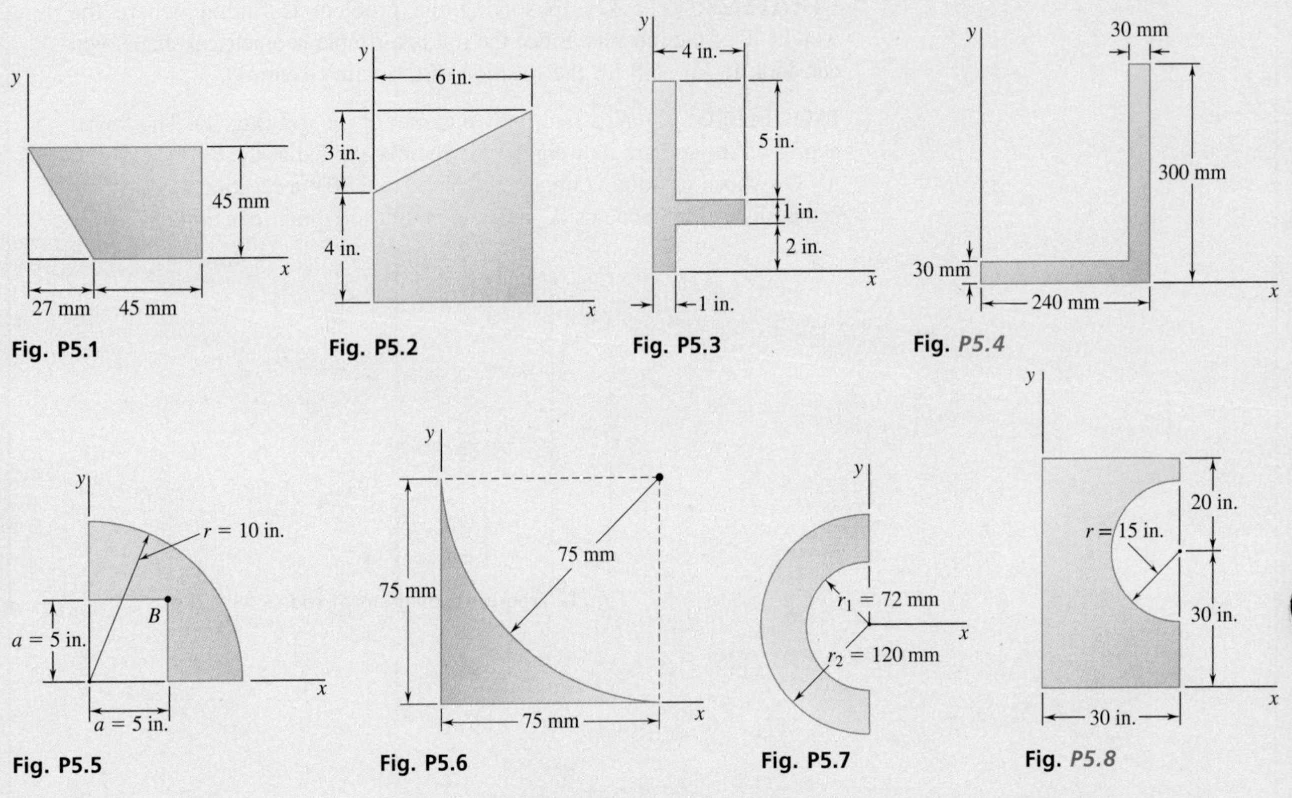

Fig. P5.1

Fig. P5.2

Fig. P5.3

Fig. *P5.4*

Fig. P5.5

Fig. P5.6

Fig. P5.7

Fig. *P5.8*

5.9 through 5.12 Locate the centroid of the plane area shown.

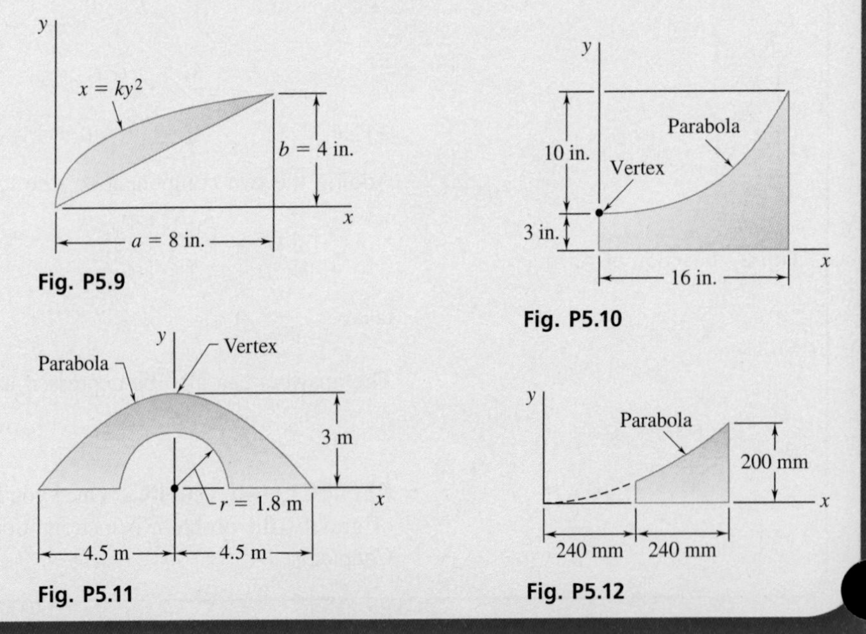

Fig. P5.9

Fig. P5.10

Fig. P5.11

Fig. P5.12

5.13 and 5.14 The horizontal x axis is drawn through the centroid C of the area shown and divides it into two component areas A_1 and A_2. Determine the first moment of each component area with respect to the x axis, and explain the results obtained.

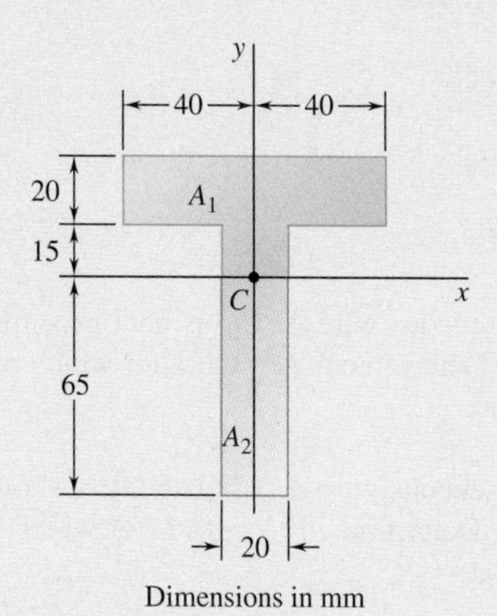

Fig. P5.13

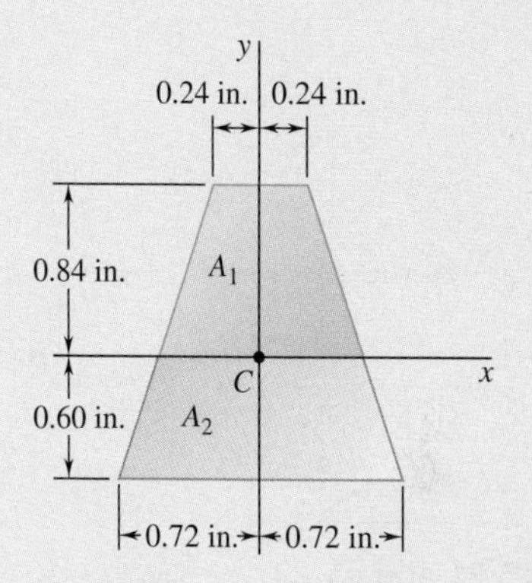

Fig. P5.14

5.15 The first moment of the shaded area with respect to the x axis is denoted by Q_x. (a) Express Q_x in terms of b, c, and the distance y from the base of the shaded area to the x axis. (b) For what value of y is Q_x maximum, and what is that maximum value?

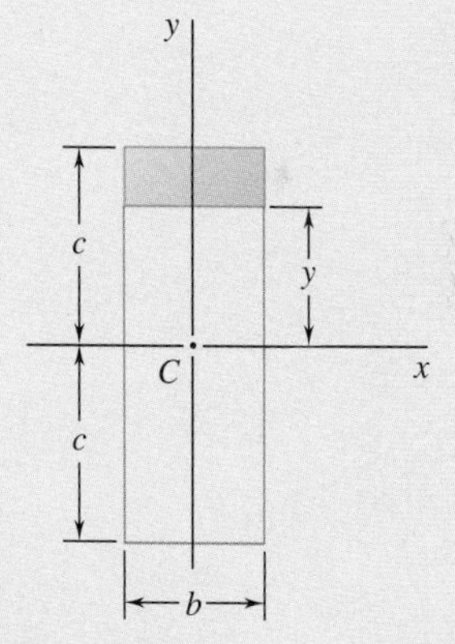

Fig. ***P5.15***

5.16 A composite beam is constructed by bolting four plates to four $60 \times 60 \times 12$-mm angles as shown. The bolts are equally spaced along the beam, and the beam supports a vertical load. As proved in mechanics of materials, the shearing forces exerted on the bolts at A and B are proportional to the first moments with respect to the centroidal x axis of the red shaded areas shown, respectively, in parts a and b of the figure. Knowing that the force exerted on the bolt at A is 280 N, determine the force exerted on the bolt at B.

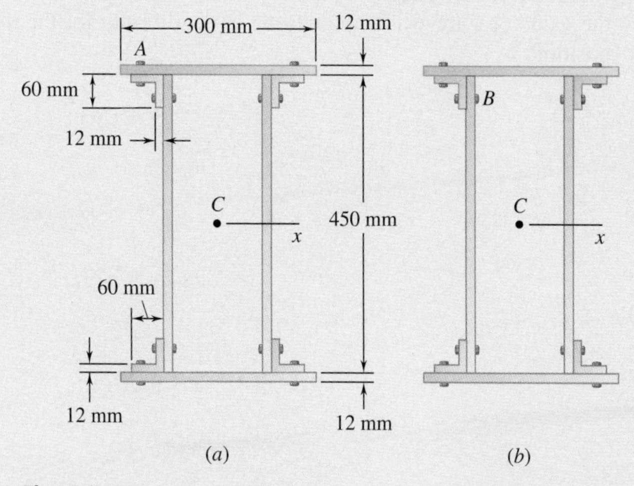

Fig. ***P5.16***

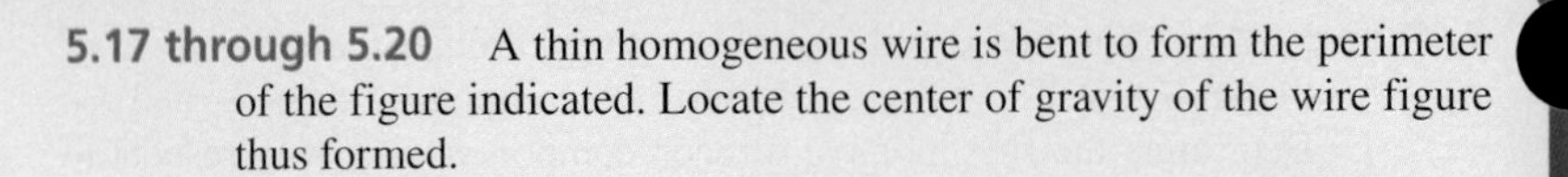
5.17 through 5.20 A thin homogeneous wire is bent to form the perimeter of the figure indicated. Locate the center of gravity of the wire figure thus formed.

5.17 Fig. P5.1.

5.18 Fig. P5.2.

5.19 Fig. P5.4.

5.20 Fig. P5.5.

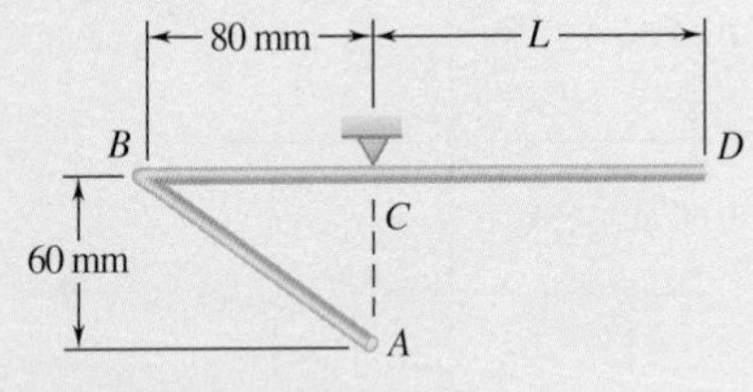

Fig. P5.21 and ***P5.22***

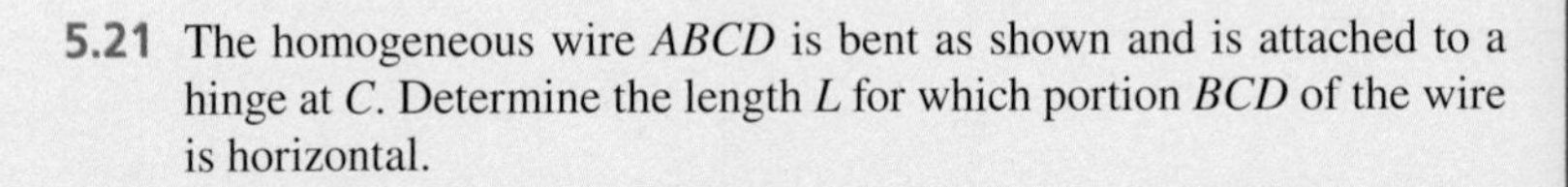
5.21 The homogeneous wire *ABCD* is bent as shown and is attached to a hinge at *C*. Determine the length *L* for which portion *BCD* of the wire is horizontal.

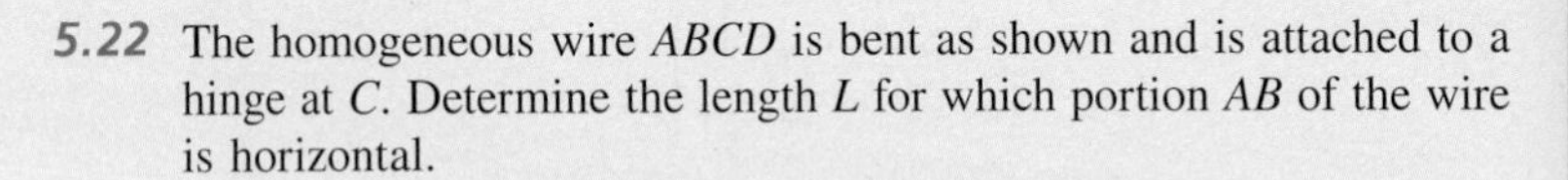
5.22 The homogeneous wire *ABCD* is bent as shown and is attached to a hinge at *C*. Determine the length *L* for which portion *AB* of the wire is horizontal.

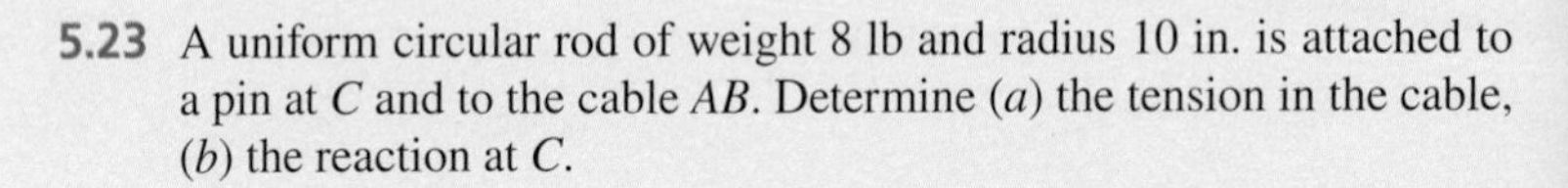
5.23 A uniform circular rod of weight 8 lb and radius 10 in. is attached to a pin at *C* and to the cable *AB*. Determine (*a*) the tension in the cable, (*b*) the reaction at *C*.

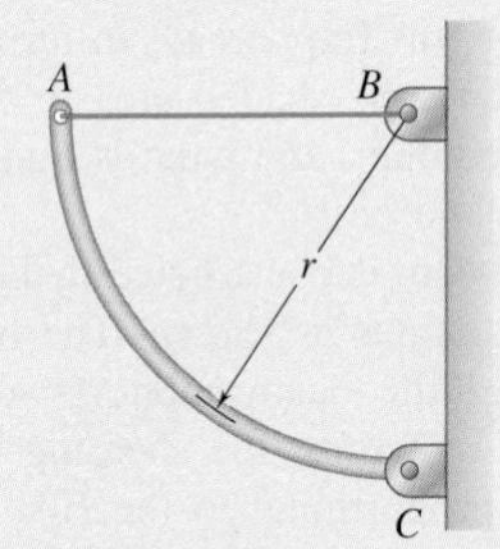

Fig. P5.23

5.24 The homogeneous wire *ABC* is bent into a semicircular arc and a straight section as shown and is attached to a hinge at *A*. Determine the value of θ for which the wire is in equilibrium for the indicated position.

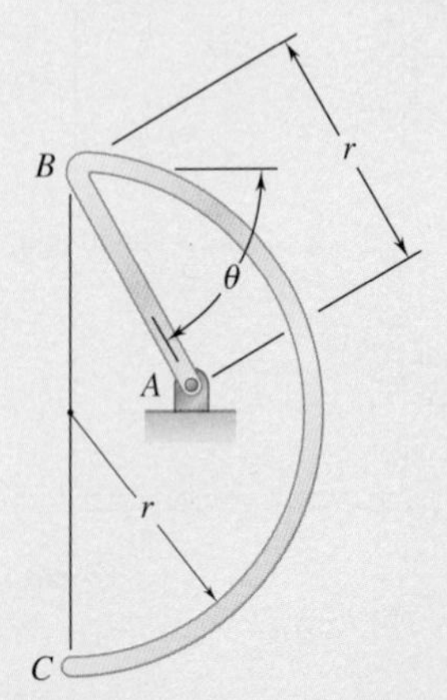

Fig. ***P5.24***

5.2 FURTHER CONSIDERATIONS OF CENTROIDS

The objects we analyzed in Sec. 5.1 were composites of basic geometric shapes like rectangles, triangles, and circles. The same idea of locating a center of gravity or centroid applies for an object with a more complicated shape, but the mathematical techniques for finding the location are a little more difficult.

5.2A Determination of Centroids by Integration

For an area bounded by analytical curves (i.e., curves defined by algebraic equations), we usually determine the centroid by evaluating the integrals in Eqs. (5.3′):

$$\bar{x} = \frac{\int x\,dA}{A} \qquad \bar{y} = \frac{\int y\,dA}{A} \tag{5.3'}$$

If the element of area dA is a small rectangle of sides dx and dy, evaluating each of these integrals requires a *double integration* with respect to x and y. A double integration is also necessary if we use polar coordinates for which dA is a small element with sides dr and $r\,d\theta$.

In most cases, however, it is possible to determine the coordinates of the centroid of an area by performing a single integration. We can achieve this by choosing dA to be a thin rectangle or strip, or it can be a thin sector or pie-shaped element (Fig. 5.12). The centroid of the thin rectangle is located at its center, and the centroid of the thin sector is located at a distance $(2/3)r$ from its vertex (as it is for a triangle). Then we obtain the coordinates of the centroid of the area under consideration

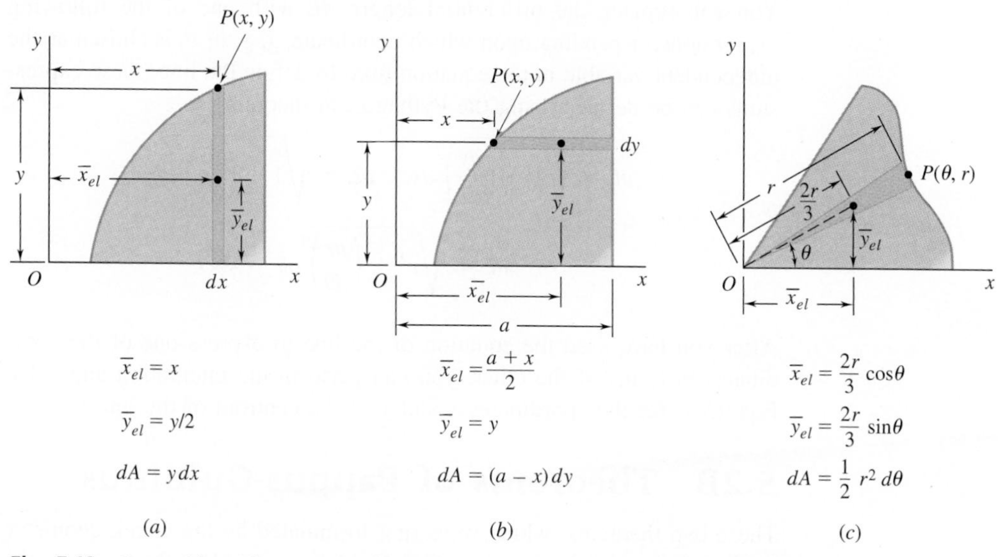

$\bar{x}_{el} = x$

$\bar{y}_{el} = y/2$

$dA = y\,dx$

(*a*)

$\bar{x}_{el} = \dfrac{a + x}{2}$

$\bar{y}_{el} = y$

$dA = (a - x)\,dy$

(*b*)

$\bar{x}_{el} = \dfrac{2r}{3}\cos\theta$

$\bar{y}_{el} = \dfrac{2r}{3}\sin\theta$

$dA = \dfrac{1}{2}r^2\,d\theta$

(*c*)

Fig. 5.12 Centroids and areas of differential elements. (*a*) Vertical rectangular strip; (*b*) horizontal rectangular strip; (*c*) triangular sector.

by setting the first moment of the entire area with respect to each of the coordinate axes equal to the sum (or integral) of the corresponding moments of the elements of the area. Denoting the coordinates of the centroid of the element dA by $\bar{x}_{el}$ and $\bar{y}_{el}$, we have

First moments of area

$$Q_y = \bar{x}A = \int \bar{x}_{el}\, dA$$
$$Q_x = \bar{y}A = \int \bar{y}_{el}\, dA \tag{5.9}$$

If we do not already know the area A, we can also compute it from these elements.

In order to carry out the integration, we need to express the coordinates $\bar{x}_{el}$ and $\bar{y}_{el}$ of the centroid of the element of area dA in terms of the coordinates of a point located on the curve bounding the area under consideration. Also, we should express the area of the element dA in terms of the coordinates of that point and the appropriate differentials. This has been done in Fig. 5.12 for three common types of elements; the pie-shaped element of part (*c*) should be used when the equation of the curve bounding the area is given in polar coordinates. You can substitute the appropriate expressions into formulas (5.9), and then use the equation of the bounding curve to express one of the coordinates in terms of the other. This process reduces the double integration to a single integration. Once you have determined the area and evaluated the integrals in Eqs. (5.9), you can solve these equations for the coordinates $\bar{x}$ and $\bar{y}$ of the centroid of the area.

When a line is defined by an algebraic equation, you can determine its centroid by evaluating the integrals in Eqs. (5.4′):

$$\bar{x} = \frac{\int x\,dL}{L} \qquad \bar{y} = \frac{\int y\,dL}{L} \tag{5.4′}$$

You can replace the differential length dL with one of the following expressions, depending upon which coordinate, x, y, or θ, is chosen as the independent variable in the equation used to define the line (these expressions can be derived using the Pythagorean theorem):

$$dL = \sqrt{1 + \left(\frac{dy}{dx}\right)^2}\,dx \qquad dL = \sqrt{1 + \left(\frac{dx}{dy}\right)^2}\,dy$$

$$dL = \sqrt{r^2 + \left(\frac{dr}{d\theta}\right)^2}\,d\theta$$

After you have used the equation of the line to express one of the coordinates in terms of the other, you can perform the integration and solve Eqs. (5.4) for the coordinates $\bar{x}$ and $\bar{y}$ of the centroid of the line.

5.2B Theorems of Pappus-Guldinus

These two theorems, which were first formulated by the Greek geometer Pappus during the third century C.E. and later restated by the Swiss mathematician Guldinus or Guldin (1577–1643), deal with surfaces and bodies

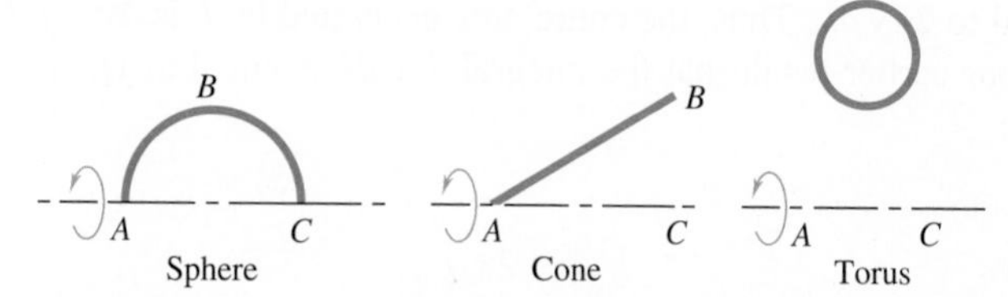

Fig. 5.13 Rotating plane curves about an axis generates surfaces of revolution.

Photo 5.3 The storage tanks shown are bodies of revolution. Thus, their surface areas and volumes can be determined using the theorems of Pappus-Guldinus.

of revolution. A **surface of revolution** is a surface that can be generated by rotating a plane curve about a fixed axis. For example, we can obtain the surface of a sphere by rotating a semicircular arc *ABC* about the diameter *AC* (Fig. 5.13). Similarly, rotating a straight line *AB* about an axis *AC* produces the surface of a cone, and rotating the circumference of a circle about a nonintersecting axis generates the surface of a torus or ring. A **body of revolution** is a body that can be generated by rotating a plane area about a fixed axis. As shown in Fig. 5.14, we can generate a sphere, a cone, and a torus by rotating the appropriate shape about the indicated axis.

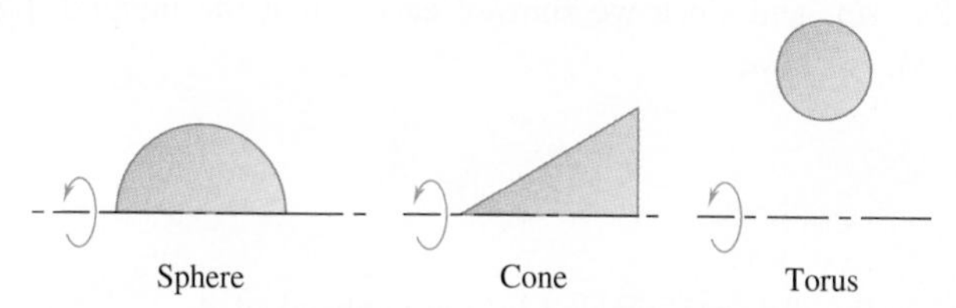

Fig. 5.14 Rotating plane areas about an axis generates volumes of revolution.

Theorem I. *The area of a surface of revolution is equal to the length of the generating curve times the distance traveled by the centroid of the curve while the surface is being generated.*

Proof. Consider an element dL of the line L (Fig. 5.15) that is revolved about the x axis. The circular strip generated by the element dL has an area

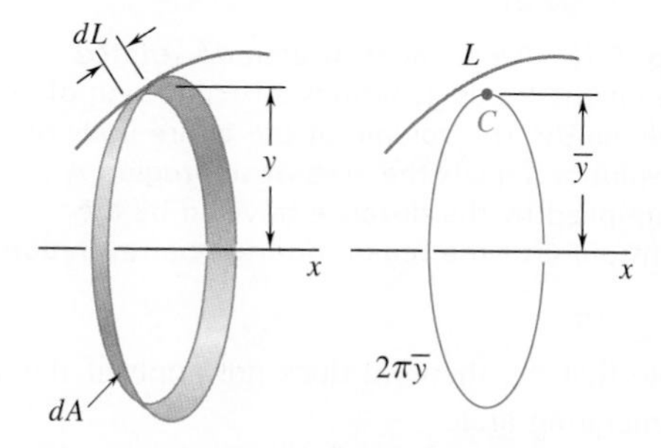

Fig. 5.15 An element of length dL rotated about the x axis generates a circular strip of area dA. The area of the entire surface of revolution equals the length of the line L multiplied by the distance traveled by the centroid C of the line during one revolution.

dA equal to $2\pi y\, dL$. Thus, the entire area generated by L is $A = \int 2\pi y\, dL$. Recall our earlier result that the integral $\int y\, dL$ is equal to $\bar{y}L$. Therefore, we have

$$A = 2\pi \bar{y} L \tag{5.10}$$

Here $2\pi\bar{y}$ is the distance traveled by the centroid C of L (Fig. 5.15). □

Note that the generating curve must not cross the axis about which it is rotated; if it did, the two sections on either side of the axis would generate areas having opposite signs, and the theorem would not apply.

Theorem II. *The volume of a body of revolution is equal to the generating area times the distance traveled by the centroid of the area while the body is being generated.*

Proof. Consider an element dA of the area A that is revolved about the x axis (Fig. 5.16). The circular ring generated by the element dA has a volume dV equal to $2\pi y\, dA$. Thus, the entire volume generated by A is $V = \int 2\pi y\, dA$, and since we showed earlier that the integral $\int y\, dA$ is equal to $\bar{y}A$, we have

$$V = 2\pi \bar{y} A \tag{5.11}$$

Here $2\pi\bar{y}$ is the distance traveled by the centroid of A. □

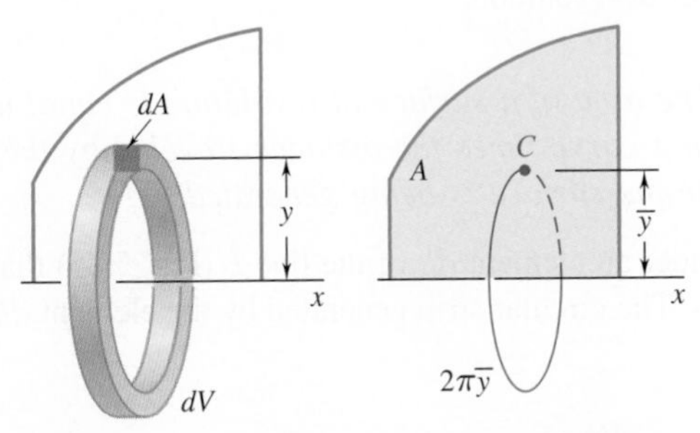

Fig. 5.16 An element of area dA rotated about the x axis generates a circular ring of volume dV. The volume of the entire body of revolution equals the area of the region A multiplied by the distance traveled by the centroid C of the region during one revolution.

Again, note that the theorem does not apply if the axis of rotation intersects the generating area.

The theorems of Pappus-Guldinus offer a simple way to compute the areas of surfaces of revolution and the volumes of bodies of revolution. Conversely, they also can be used to determine the centroid of a plane curve if you know the area of the surface generated by the curve or to determine the centroid of a plane area if you know the volume of the body generated by the area (see Sample Prob. 5.8).

Sample Problem 5.4

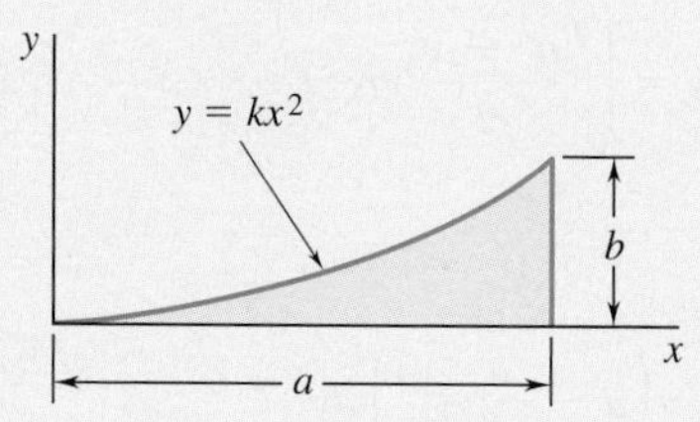

Determine the location of the centroid of a parabolic spandrel by direct integration.

STRATEGY: First express the parabolic curve using the parameters a and b. Then choose a differential element of area and express its area in terms of a, b, x, and y. We illustrate the solution first with a vertical element and then a horizontal element.

MODELING:

Determination of the Constant k. Determine the value of k by substituting $x = a$ and $y = b$ into the given equation. We have $b = ka^2$ or $k = b/a^2$. The equation of the curve is thus

$$y = \frac{b}{a^2}x^2 \qquad \text{or} \qquad x = \frac{a}{b^{1/2}}y^{1/2}$$

ANALYSIS:

Vertical Differential Element. Choosing the differential element shown in Fig. 1, the total area of the region is

$$A = \int dA = \int y\,dx = \int_0^a \frac{b}{a^2}x^2\,dx = \left[\frac{b}{a^2}\frac{x^3}{3}\right]_0^a = \frac{ab}{3}$$

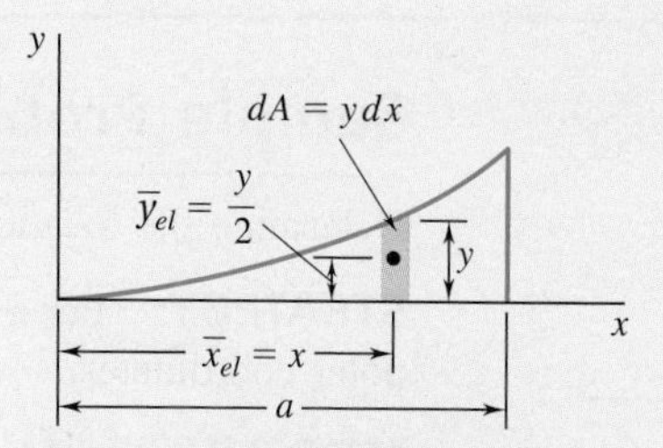

Fig. 1 Vertical differential element used to determine centroid.

The first moment of the differential element with respect to the y axis is $\bar{x}_{el}dA$; hence, the first moment of the entire area with respect to this axis is

$$Q_y = \int \bar{x}_{el}\,dA = \int xy\,dx = \int_0^a x\left(\frac{b}{a^2}x^2\right)dx = \left[\frac{b}{a^2}\frac{x^4}{4}\right]_0^a = \frac{a^2b}{4}$$

Since $Q_y = \bar{x}A$, you have

$$\bar{x}A = \int \bar{x}_{el}\,dA \qquad \bar{x}\frac{ab}{3} = \frac{a^2b}{4} \qquad \bar{x} = \tfrac{3}{4}a \blacktriangleleft$$

Likewise, the first moment of the differential element with respect to the x axis is $\bar{y}_{el}dA$, so the first moment of the entire area about the x axis is

$$Q_x = \int \bar{y}_{el}\,dA = \int \frac{y}{2}y\,dx = \int_0^a \frac{1}{2}\left(\frac{b}{a^2}x^2\right)^2 dx = \left[\frac{b^2}{2a^4}\frac{x^5}{5}\right]_0^a = \frac{ab^2}{10}$$

Since $Q_x = \bar{y}A$, you get

$$\bar{y}A = \int \bar{y}_{el}\,dA \qquad \bar{y}\frac{ab}{3} = \frac{ab^2}{10} \qquad \bar{y} = \tfrac{3}{10}b \blacktriangleleft$$

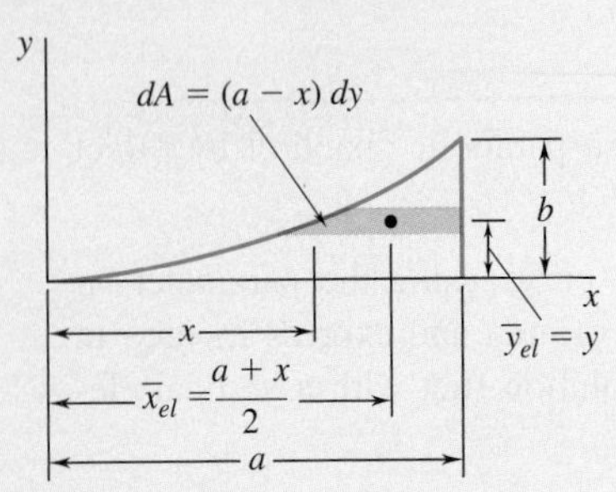

Fig. 2 Horizontal differential element used to determine centroid.

Horizontal Differential Element. You obtain the same results by considering a horizontal element (Fig. 2). The first moments of the area are

$$Q_y = \int \bar{x}_{el}\, dA = \int \frac{a+x}{2}(a-x)\,dy = \int_0^b \frac{a^2 - x^2}{2}\,dy$$

$$= \frac{1}{2}\int_0^b \left(a^2 - \frac{a^2}{b}y\right) dy = \frac{a^2 b}{4}$$

$$Q_x = \int \bar{y}_{el}\, dA = \int y(a-x)\,dy = \int y\left(a - \frac{a}{b^{1/2}}y^{1/2}\right) dy$$

$$= \int_0^b \left(ay - \frac{a}{b^{1/2}}y^{3/2}\right) dy = \frac{ab^2}{10}$$

To determine $\bar{x}$ and $\bar{y}$, again substitute these expressions into the equations defining the centroid of the area.

REFLECT and THINK: You obtain the same results whether you choose a vertical or a horizontal element of area, as you should. You can use both methods as a check against making a mistake in your calculations.

Sample Problem 5.5

Determine the location of the centroid of the circular arc shown.

STRATEGY: For a simple figure with circular geometry, you should use polar coordinates.

MODELING: The arc is symmetrical with respect to the x axis, so $\bar{y} = 0$. Choose a differential element, as shown in Fig. 1.

ANALYSIS: Determine the length of the arc by integration.

$$L = \int dL = \int_{-\alpha}^{\alpha} r\,d\theta = r\int_{-\alpha}^{\alpha} d\theta = 2r\alpha$$

The first moment of the arc with respect to the y axis is

$$Q_y = \int x\,dL = \int_{-\alpha}^{\alpha} (r\cos\theta)(r\,d\theta) = r^2 \int_{-\alpha}^{\alpha} \cos\theta\,d\theta$$

$$= r^2[\sin\theta]_{-\alpha}^{\alpha} = 2r^2 \sin\alpha$$

Since $Q_y = \bar{x}L$, you obtain

$$\bar{x}(2r\alpha) = 2r^2\sin\alpha \qquad \bar{x} = \frac{r\sin\alpha}{\alpha} \blacktriangleleft$$

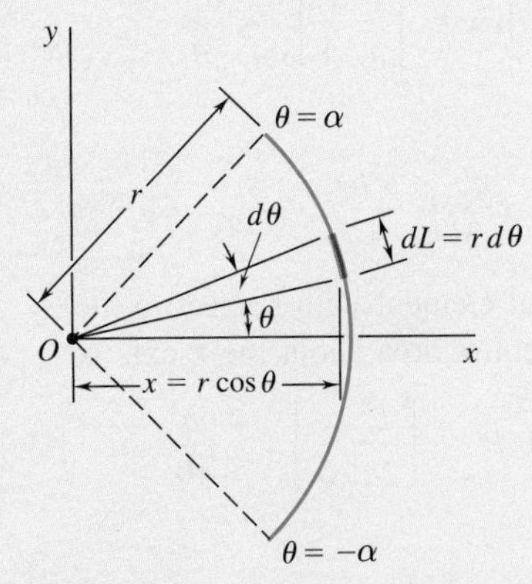

Fig. 1 Differential element used to determine centroid.

REFLECT and THINK: Observe that this result matches that given for this case in Fig. 5.8B.

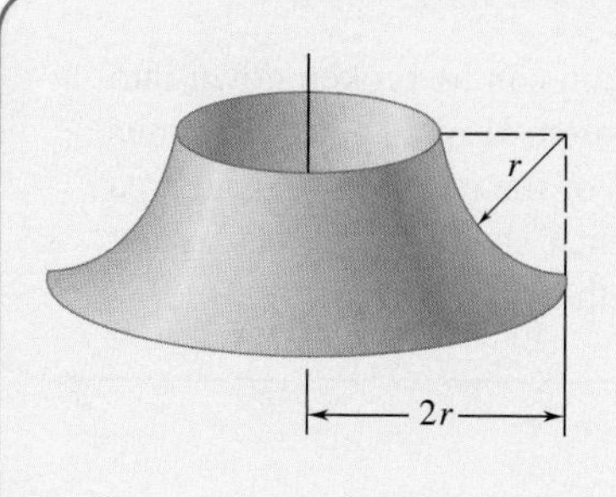

Sample Problem 5.6

Determine the area of the surface of revolution shown that is obtained by rotating a quarter-circular arc about a vertical axis.

STRATEGY: According to the first Pappus-Guldinus theorem, the area of the surface of revolution is equal to the product of the length of the arc and the distance traveled by its centroid.

MODELING and ANALYSIS: Referring to Fig. 5.8B and Fig. 1, you have

$$\bar{x} = 2r - \frac{2r}{\pi} = 2r\left(1 - \frac{1}{\pi}\right)$$

$$A = 2\pi\bar{x}L = 2\pi\left[2r\left(1 - \frac{1}{\pi}\right)\right]\left(\frac{\pi r}{2}\right)$$

$$A = 2\pi r^2(\pi - 1) \blacktriangleleft$$

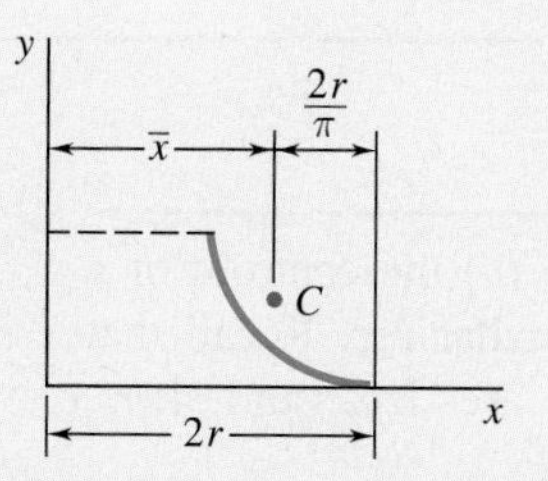

Fig. 1 Centroid location of arc.

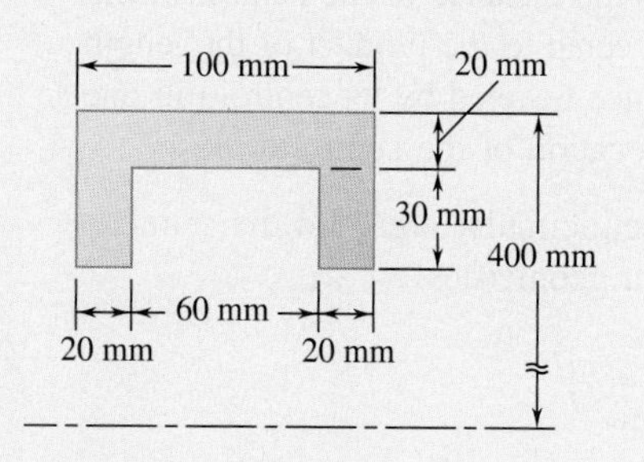

Sample Problem 5.7

The outside diameter of a pulley is 0.8 m, and the cross section of its rim is as shown. Knowing that the pulley is made of steel and that the density of steel is $\rho = 7.85 \times 10^3$ kg/m^3, determine the mass and weight of the rim.

STRATEGY: You can determine the volume of the rim by applying the second Pappus-Guldinus theorem, which states that the volume equals the product of the given cross-sectional area and the distance traveled by its centroid in one complete revolution. However, you can find the volume more easily by observing that the cross section can be formed from rectangle I with a positive area and from rectangle II with a negative area (Fig. 1).

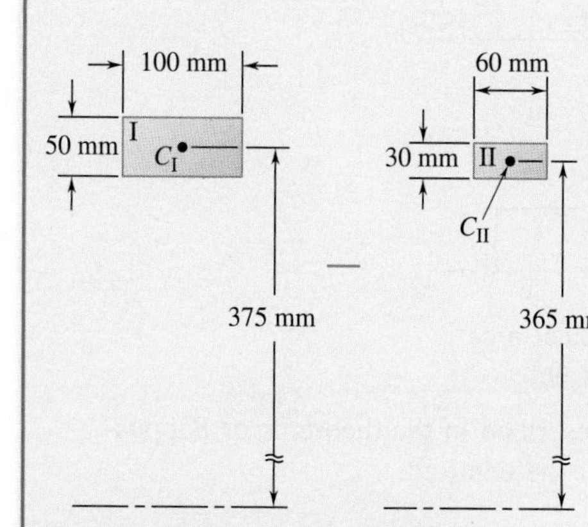

Fig. 1 Modeling the given area by subtracting area II from area I.

MODELING: Use a table to keep track of the data, as you did in Sec. 5.1.

	Area, mm^2	$\bar{y}$, mm	Distance Traveled by C, mm	Volume, mm^3
I	+5000	375	$2\pi(375) = 2356$	$(5000)(2356) = 11.78 \times 10^6$
II	−1800	365	$2\pi(365) = 2293$	$(-1800)(2293) = -4.13 \times 10^6$
				Volume of rim $= 7.65 \times 10^6$

ANALYSIS: Since 1 mm $= 10^{-3}$ m, you have 1 mm$^{-3} = (10^{-3}\text{ m})^3 = 10^{-9}$ m^3. Thus you obtain $V = 7.65 \times 10^6\text{ mm}^3 = (7.65 \times 10^6)(10^{-9}\text{ m}^3) = 7.65 \times 10^{-3}\text{ m}^3$.

$$m = \rho V = (7.85 \times 10^3\text{ kg/m}^3)(7.65 \times 10^{-3}\text{ m}^3) \qquad m = 60.0\text{ kg} \blacktriangleleft$$

$$W = mg = (60.0\text{ kg})(9.81\text{ m/s}^2) = 589\text{ kg}\cdot\text{m/s}^2 \qquad W = 589\text{ N} \blacktriangleleft$$

REFLECT and THINK: When a cross section can be broken down into multiple common shapes, you can apply Theorem II of Pappus–Guldinus in a manner that involves finding the products of the centroid ($\bar{y}$) and area (A), or the first moments of area ($\bar{y}A$), for each shape. Thus, it was not necessary to find the centroid or the area of the overall cross section.

Sample Problem 5.8

Using the theorems of Pappus-Guldinus, determine (*a*) the centroid of a semicircular area and (*b*) the centroid of a semicircular arc. Recall that the volume and the surface area of a sphere are $\frac{4}{3}\pi r^3$ and $4\pi r^2$, respectively.

STRATEGY: The volume of a sphere is equal to the product of the area of a semicircle and the distance traveled by the centroid of the semicircle in one revolution about the *x* axis. Given the volume, you can determine the distance traveled by the centroid and thus the distance of the centroid from the axis. Similarly, the area of a sphere is equal to the product of the length of the generating semicircle and the distance traveled by its centroid in one revolution. You can use this to find the location of the centroid of the arc.

MODELING: Draw diagrams of the semicircular area and the semicircular arc (Fig. 1) and label the important geometries.

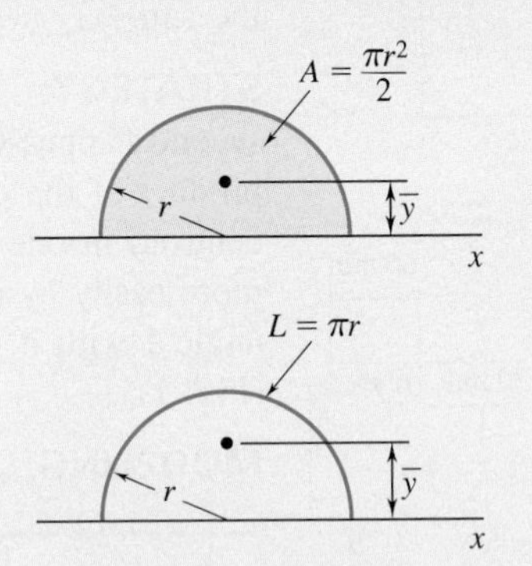

Fig. 1 Semicircular area and semicircular arc.

ANALYSIS: Set up the equalities described in the theorems of Pappus-Guldinus and solve for the location of the centroid.

$$V = 2\pi\bar{y}A \qquad \frac{4}{3}\pi r^3 = 2\pi\bar{y}\left(\frac{1}{2}\pi r^2\right) \qquad \bar{y} = \frac{4r}{3\pi} \quad \blacktriangleleft$$

$$A = 2\pi\bar{y}L \qquad 4\pi r^2 = 2\pi\bar{y}(\pi r) \qquad \bar{y} = \frac{2r}{\pi} \quad \blacktriangleleft$$

REFLECT and THINK: Observe that this result matches those given for these cases in Fig. 5.8.

Problems

5.25 through *5.28* Determine by direct integration the centroid of the area shown.

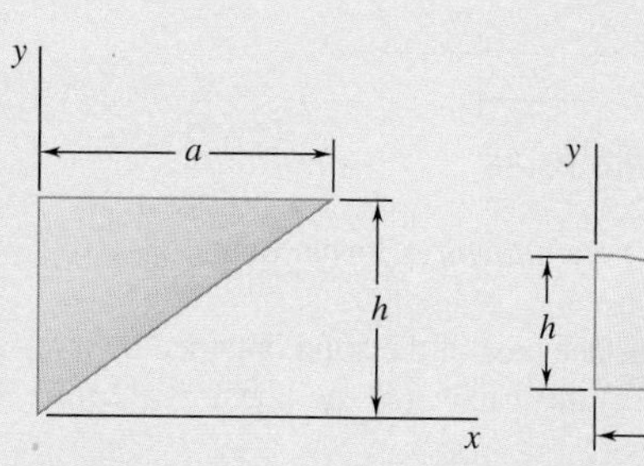

Fig. P5.25

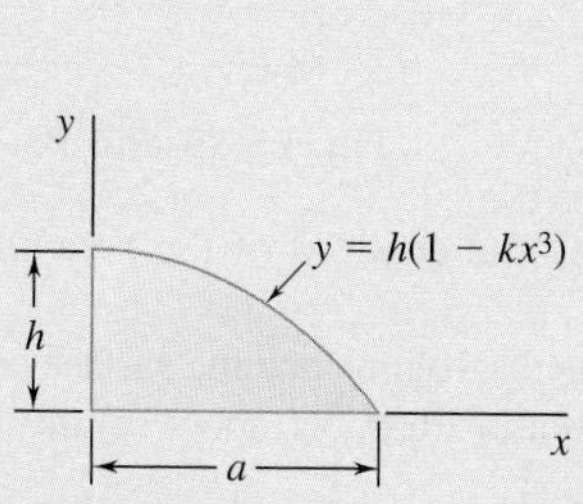

Fig. P5.26

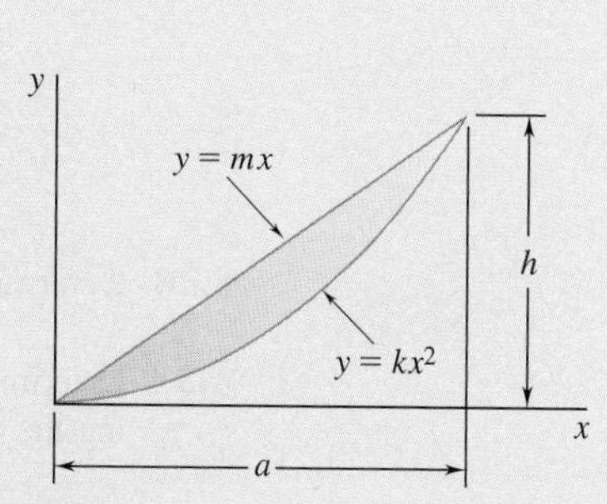

Fig. *5.27*

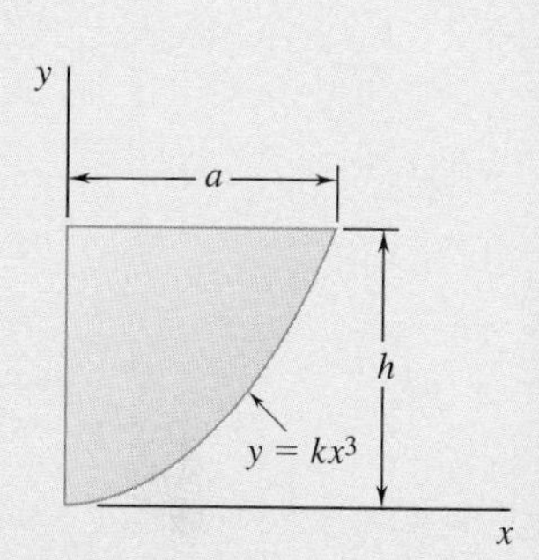

Fig. *P5.28*

5.29 through 5.31 Determine by direct integration the centroid of the area shown.

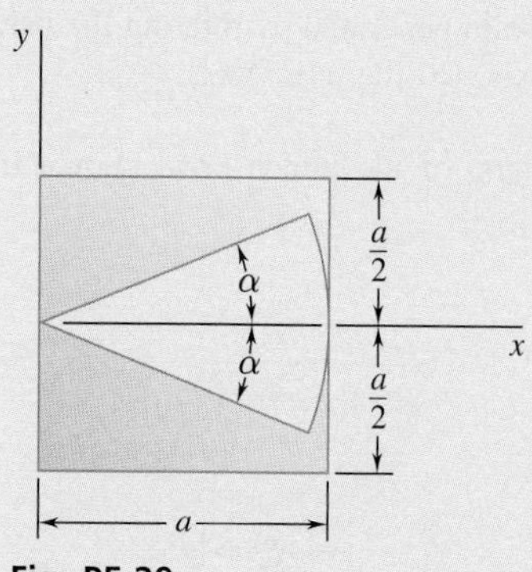

Fig. P5.29

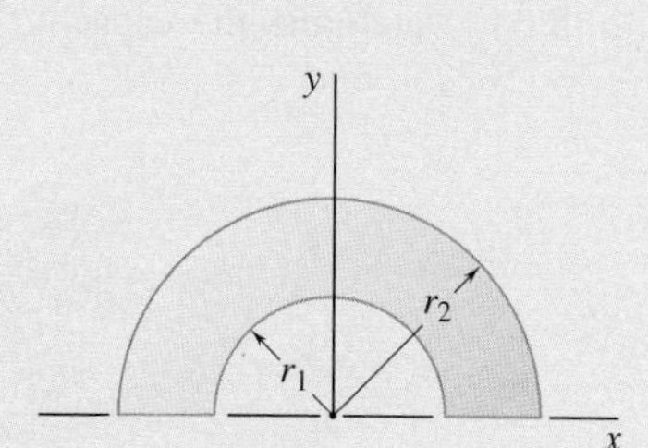

Fig. P5.30

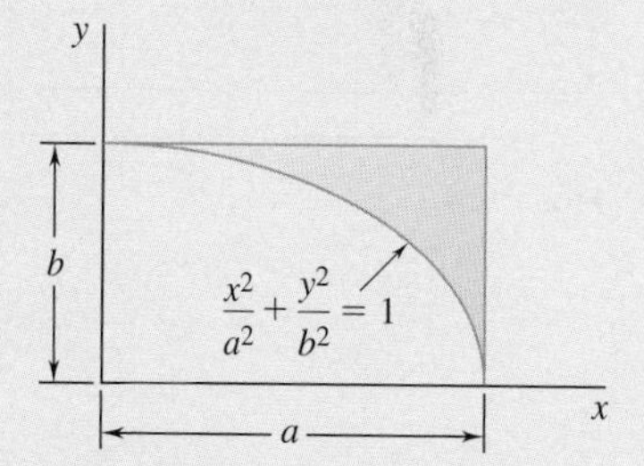

Fig. P5.31

5.32 through 5.34 Determine by direct integration the centroid of the area shown.

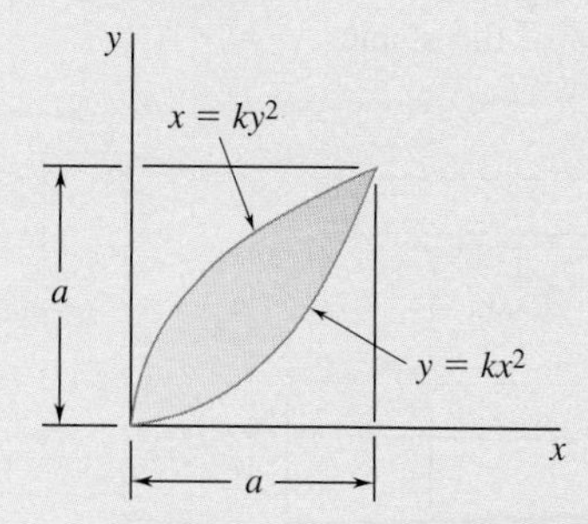

Fig. P5.32

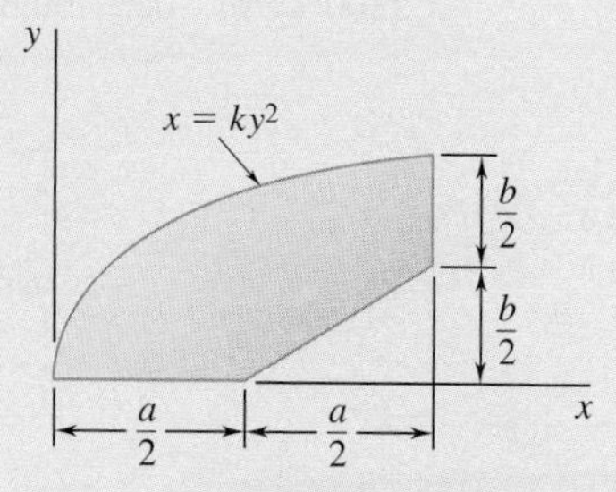

Fig. P5.33

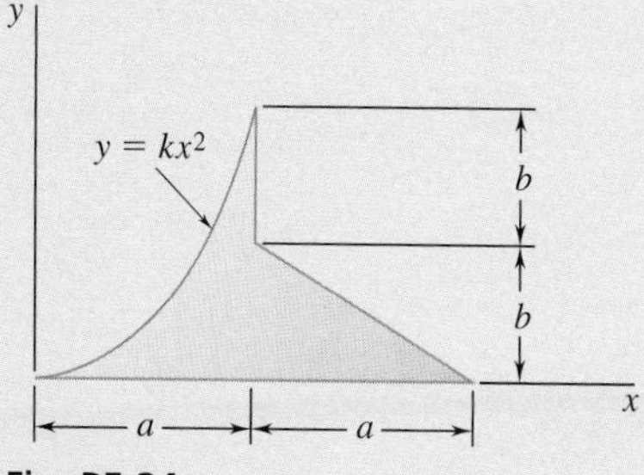

Fig. P5.34

5.35 Determine the centroid of the area shown when $a = 4$ in.

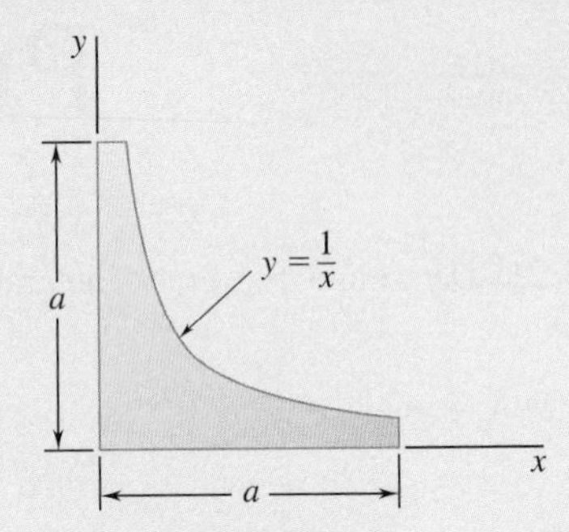

Fig. P5.35 and P5.36

5.36 Determine the centroid of the area shown in terms of a.

5.37 Determine the volume and the surface area of the solid obtained by rotating the area of Prob. 5.1 about (*a*) the x axis, (*b*) the line $x = 72$ mm.

5.38 Determine the volume of the solid obtained by rotating the area of Prob. 5.4 about (*a*) the x axis, (*b*) the y axis.

5.39 Determine the volume and the surface area of the solid obtained by rotating the area of Prob. 5.8 about (*a*) the x axis, (*b*) the y axis.

5.40 Determine the volume of the solid generated by rotating the parabolic area shown about (*a*) the x axis, (*b*) the axis AA'.

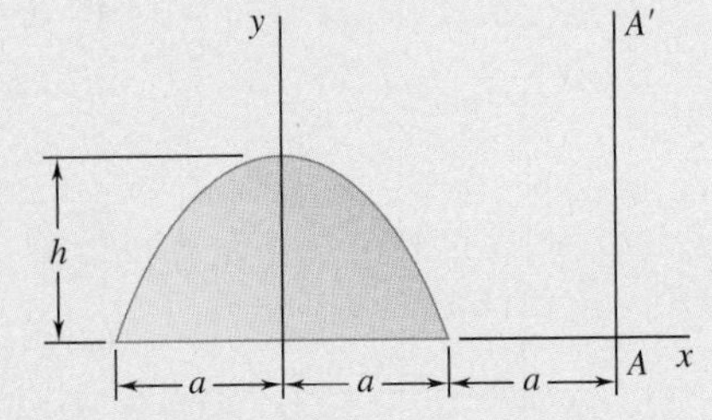

Fig. *P5.40*

5.41 Determine the capacity, in liters, of the punch bowl shown if $R = 250$ mm.

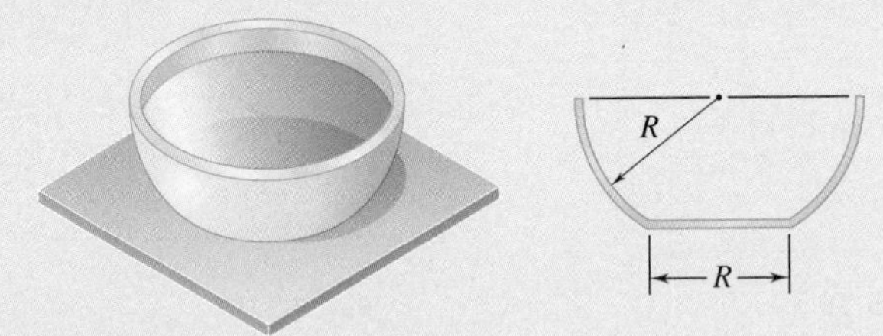

Fig. P5.41

5.42 The aluminum shade for the small high-intensity lamp shown has a uniform thickness of 1 mm. Knowing that the density of aluminum is 2800 kg/m^3, determine the mass of the shade.

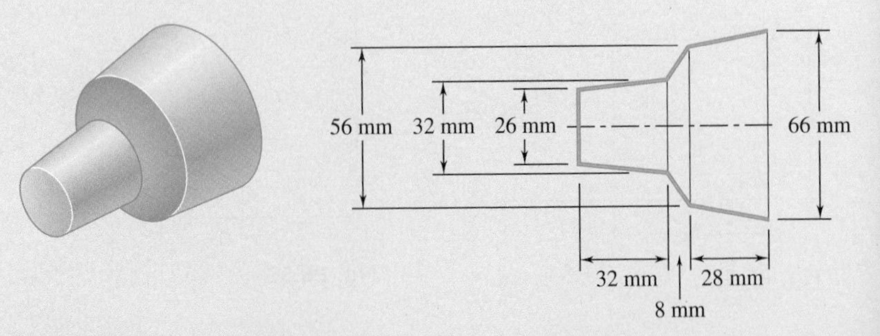

Fig. P5.42

5.43 Knowing that two equal caps have been removed from a 10-in.-diameter wooden sphere, determine the total surface area of the remaining portion.

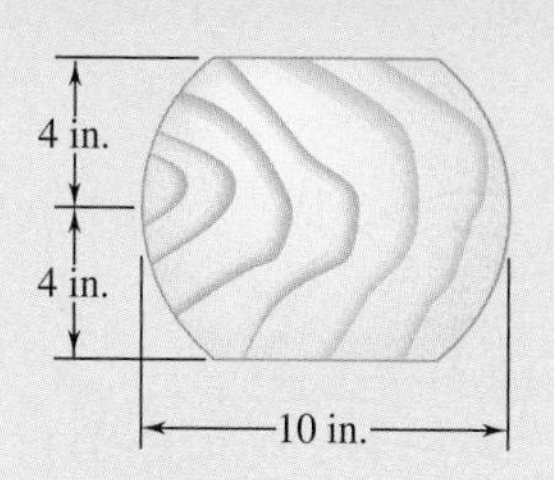

Fig. P5.43

5.44 Three different drive belt profiles are to be studied. If at any given time each belt makes contact with one-half of the circumference of its pulley, determine the *contact area* between the belt and the pulley for each design.

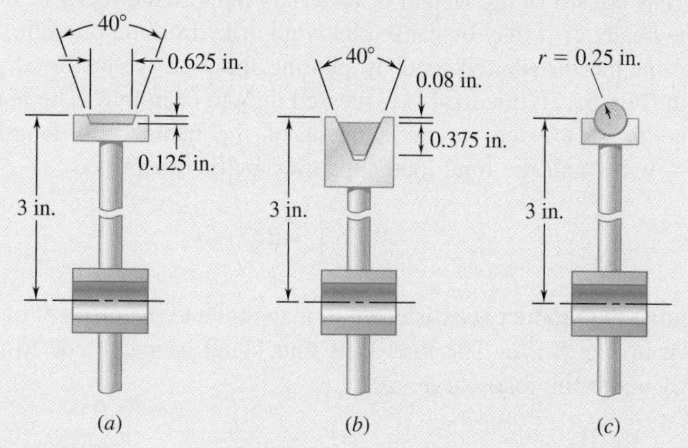

Fig. P5.44

5.45 Determine the volume and weight of the solid brass knob shown, knowing that the specific weight of brass is 0.306 lb/in^3.

5.46 Determine the total surface area of the solid brass knob shown.

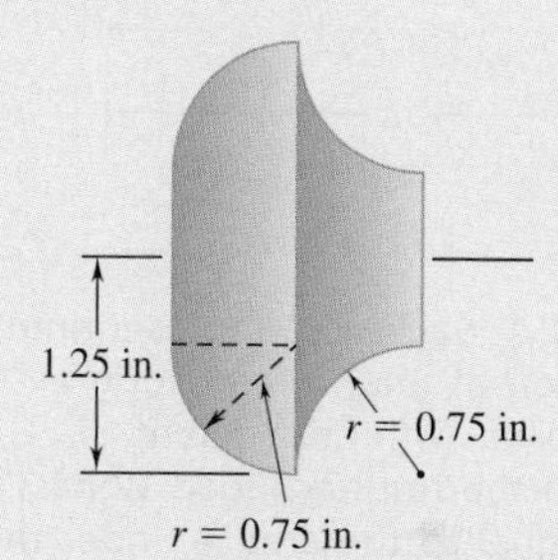

Fig. P5.45 and *P5.46*

5.47 Determine the volume and total surface area of the body shown.

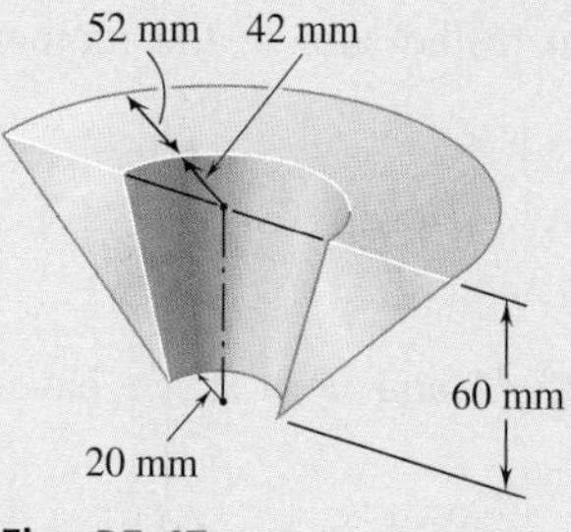

Fig. *P5.47*

5.48 The escutcheon (a decorative plate placed on a pipe where the pipe exits from a wall) shown is cast from brass. Knowing that the density of brass is 8470 kg/m^3, determine the mass of the escutcheon.

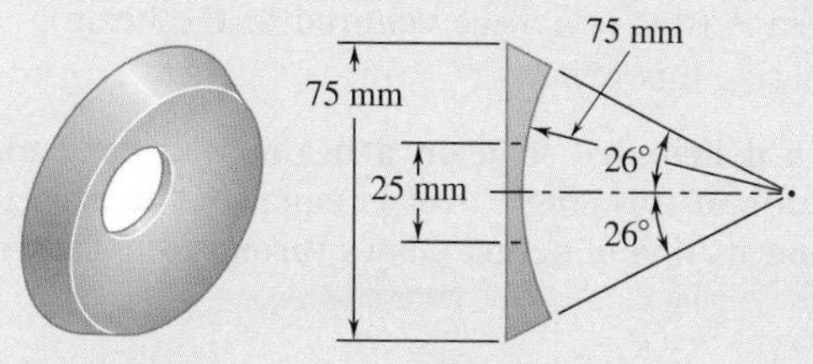

Fig. P5.48

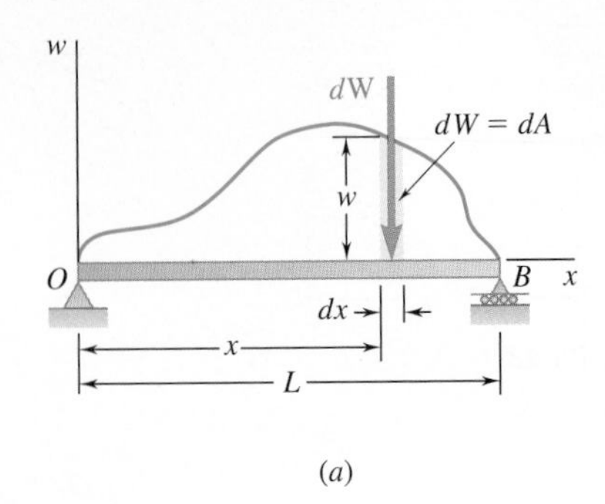

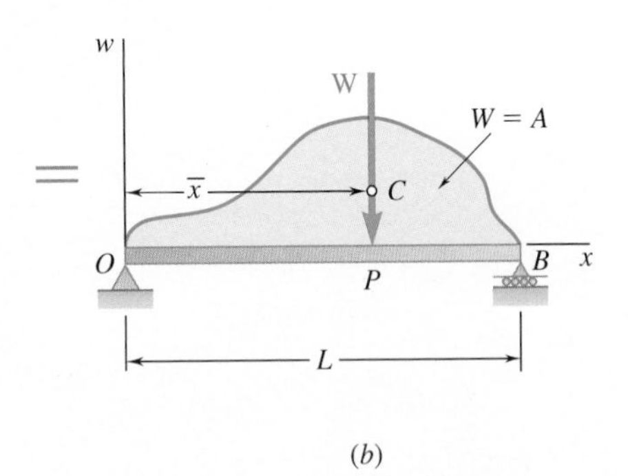

Fig. 5.17 (*a*) A load curve representing the distribution of load forces along a horizontal beam, with an element of length dx; (*b*) the resultant load W has magnitude equal to the area A under the load curve and acts through the centroid of the area.

5.3 DISTRIBUTED LOADS ON BEAMS

We can use the concept of the center of gravity or the centroid of an area to solve other problems besides those dealing with the weights of flat plates. For example, consider a beam supporting a **distributed load**; this load may consist of the weight of materials supported directly or indirectly by the beam, or it may be caused by wind or hydrostatic pressure. We can represent the distributed load by plotting the load w supported per unit length (Fig. 5.17); this load is expressed in N/m or in lb/ft. The magnitude of the force exerted on an element of the beam with length dx is $dW = w\ dx$, and the total load supported by the beam is

$$W = \int_0^L w\,dx$$

Note that the product $w\ dx$ is equal in magnitude to the element of area dA shown in Fig. 5.17*a*. The load W is thus equal in magnitude to the total area A under the load curve, as

$$W = \int dA = A$$

We now want to determine where a *single concentrated load* **W**, of the same magnitude W as the total distributed load, should be applied on the beam if it is to produce the same reactions at the supports (Fig. 5.17*b*). However, this concentrated load **W**, which represents the resultant of the given distributed loading, is equivalent to the loading only when considering the free-body diagram of the entire beam. We obtain the point of application P of the equivalent concentrated load **W** by setting the moment of **W** about point O equal to the sum of the moments of the elemental loads $d\mathbf{W}$ about O. Thus,

$$(OP)W = \int x\,dW$$

Then, since $dW = w\ dx = dA$ and $W = A$, we have

$$(OP)A = \int_0^L x\,dA \tag{5.12}$$

Since this integral represents the first moment with respect to the w axis of the area under the load curve, we can replace it with the product $\bar{x}A$. We therefore have $OP = \bar{x}$, where $\bar{x}$ is the distance from the w axis to the centroid C of the area A (this is *not* the centroid of the beam).

We can summarize this result:

We can replace a distributed load on a beam by a concentrated load; the magnitude of this single load is equal to the area under the load curve, and its line of action passes through the centroid of that area.

Note, however, that the concentrated load is equivalent to the given loading only so far as external forces are concerned. It can be used to determine reactions, but should not be used to compute internal forces and deflections.

Photo 5.4 The roof of the building shown must be able to support not only the total weight of the snow but also the nonsymmetric distributed loads resulting from drifting of the snow.

Sample Problem 5.9

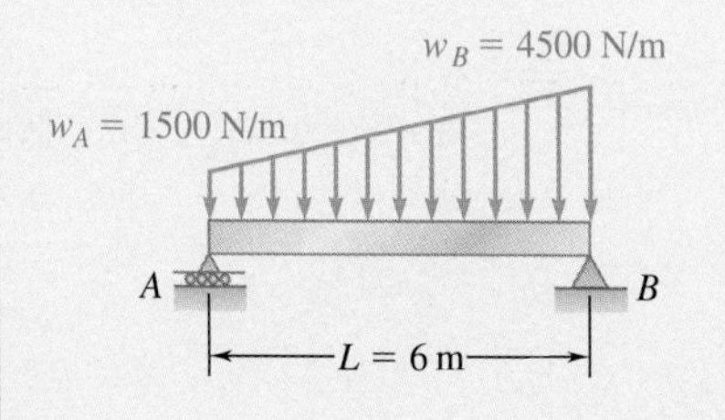

A beam supports a distributed load as shown. (*a*) Determine the equivalent concentrated load. (*b*) Determine the reactions at the supports.

STRATEGY: The magnitude of the resultant of the load is equal to the area under the load curve, and the line of action of the resultant passes through the centroid of the same area. Break down the area into pieces for easier calculation, and determine the resultant load. Then, use the calculated forces or their resultant to determine the reactions.

MODELING and ANALYSIS:

a. *Equivalent Concentrated Load.* Divide the area under the load curve into two triangles (Fig. 1), and construct the table below. To simplify the computations and tabulation, the given loads per unit length have been converted into kN/m.

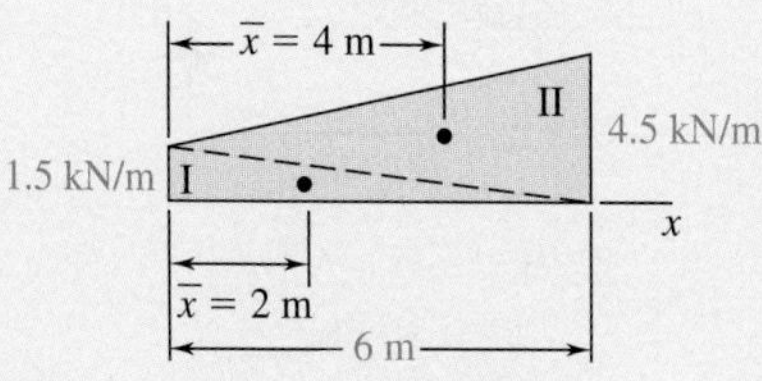

Fig. 1 The load modeled as two triangular areas.

Component	A, kN	$\bar{x}$, m	$\bar{x}A$, kN·m
Triangle I	4.5	2	9
Triangle II	13.5	4	54
	$\Sigma A = 18.0$		$\Sigma \bar{x}A = 63$

Thus, $\bar{X}\Sigma A = \Sigma \bar{x}A$: $\quad \bar{X}(18 \text{ kN}) = 63 \text{ kN·m} \quad \bar{X} = 3.5 \text{ m}$

The equivalent concentrated load (Fig. 2) is

$$\mathbf{W} = 18 \text{ kN} \downarrow \blacktriangleleft$$

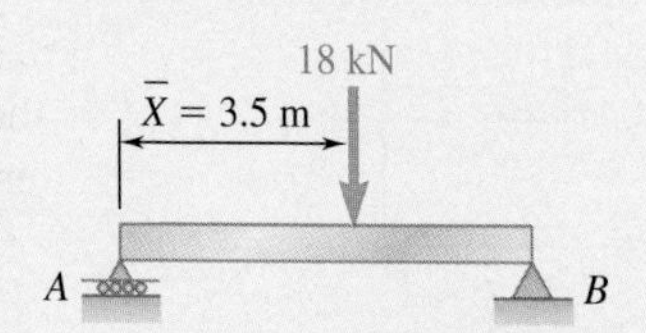

Fig. 2 Equivalent concentrated load.

Its line of action is located at a distance

$$\bar{X} = 3.5 \text{ m to the right of } A \blacktriangleleft$$

b. *Reactions.* The reaction at A is vertical and is denoted by **A**. Represent the reaction at B by its components $\mathbf{B}_x$ and $\mathbf{B}_y$. Consider the given load to be the sum of two triangular loads (see the free-body diagram, Fig. 3). The resultant of each triangular load is equal to the area of the triangle and acts at its centroid.

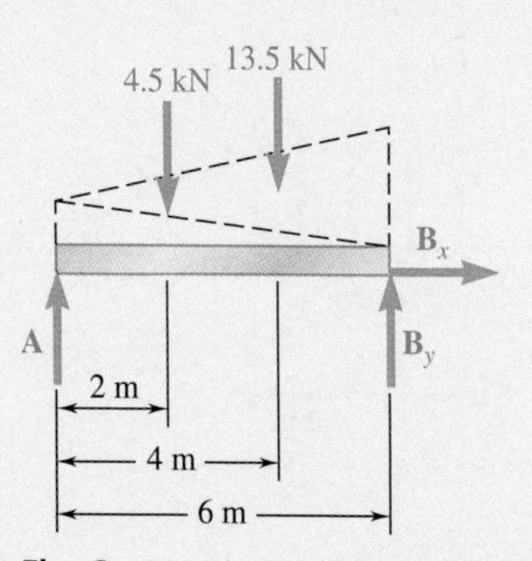

Fig. 3 Free-body diagram of beam.

Write the following equilibrium equations from the free-body diagram:

$\xrightarrow{+} \Sigma F_x = 0$: $\qquad \mathbf{B}_x = 0 \blacktriangleleft$

$+\circlearrowleft \Sigma M_A = 0$: $\qquad -(4.5 \text{ kN})(2 \text{ m}) - (13.5 \text{ kN})(4 \text{ m}) + B_y(6 \text{ m}) = 0$

$$\mathbf{B}_y = 10.5 \text{ kN} \uparrow \blacktriangleleft$$

$+\circlearrowleft \Sigma M_B = 0$: $\qquad +(4.5 \text{ kN})(4 \text{ m}) + (13.5 \text{ kN})(2 \text{ m}) - A(6 \text{ m}) = 0$

$$\mathbf{A} = 7.5 \text{ kN} \uparrow \blacktriangleleft$$

REFLECT and THINK: You can replace the given distributed load by its resultant, which you found in part *a*. Then you can determine the reactions from the equilibrium equations $\Sigma F_x = 0$, $\Sigma M_A = 0$, and $\Sigma M_B = 0$. Again the results are

$$\mathbf{B}_x = 0 \qquad \mathbf{B}_y = 10.5 \text{ kN} \uparrow \qquad \mathbf{A} = 7.5 \text{ kN} \uparrow \blacktriangleleft$$

Problems

5.49 and *5.50* For the beam and loading shown, determine (a) the magnitude and location of the resultant of the distributed load, (b) the reactions at the beam supports.

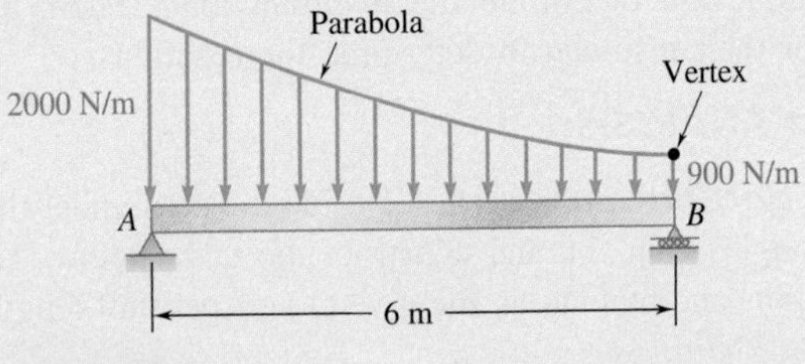

Fig. P5.49

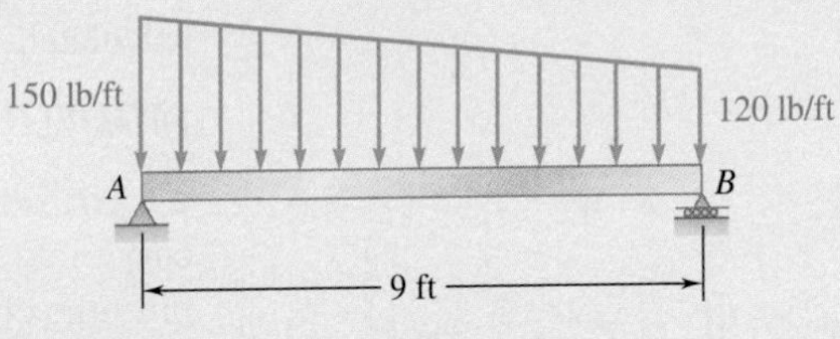

Fig. *P5.50*

5.51 through 5.56 Determine the reactions at the beam supports for the given loading.

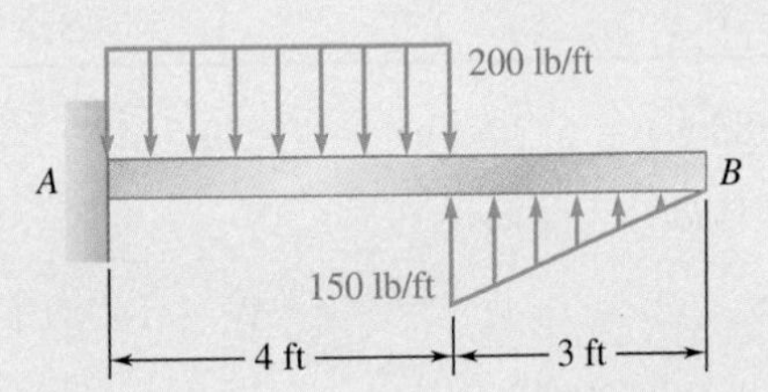

Fig. P5.51

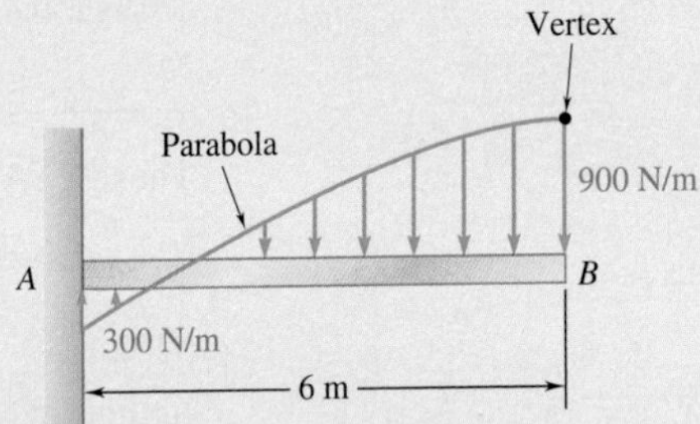

Fig. *P5.52*

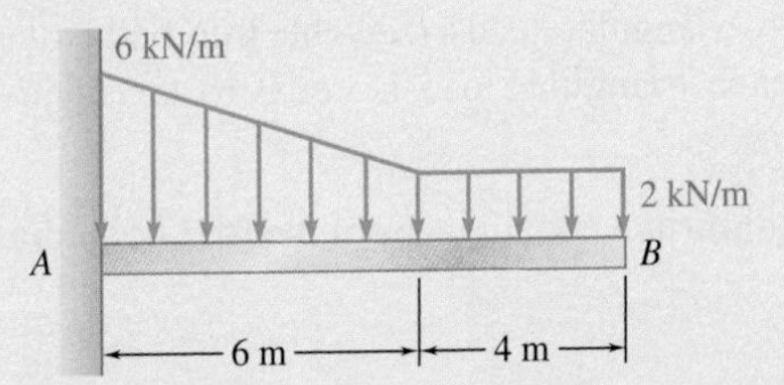

Fig. P5.53

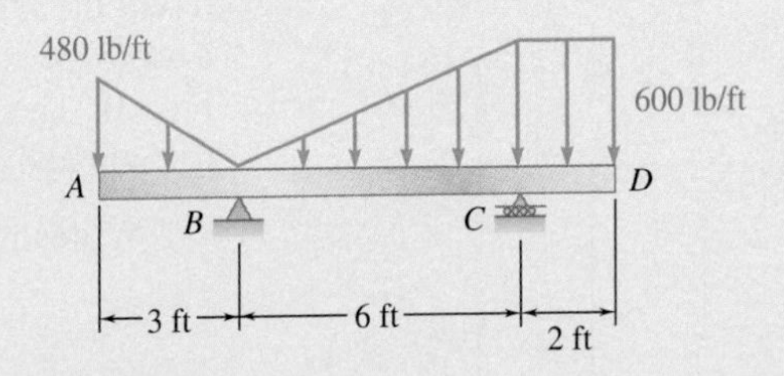

Fig. P5.54

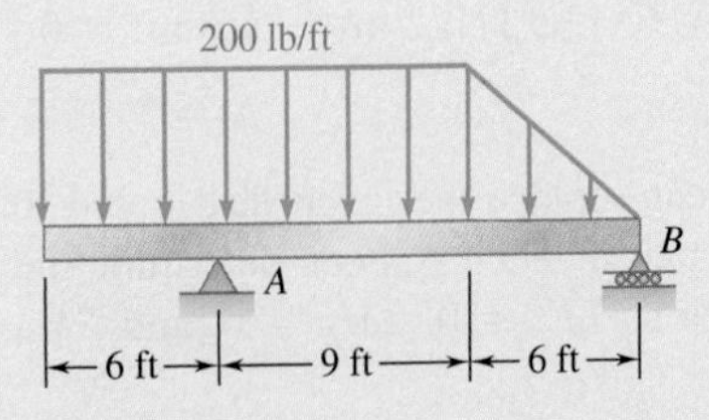

Fig. P5.55

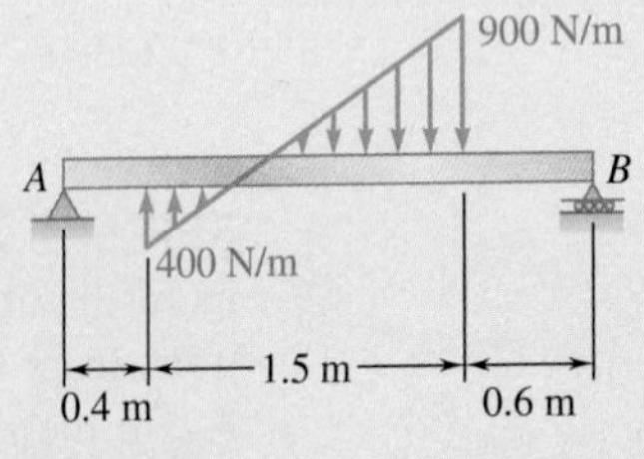

Fig. P5.56

5.4 CENTERS OF GRAVITY AND CENTROIDS OF VOLUMES

So far in this chapter, we have dealt with finding centers of gravity and centroids of two-dimensional areas and objects such as flat plates and plane surfaces. However, the same ideas apply to three-dimensional objects as well. The most general situations require the use of multiple integration for analysis, but we can often use symmetry considerations to simplify the calculations. In this section, we show how to do this.

Photo 5.5 To predict the flight characteristics of the modified Boeing 747 when used to transport a space shuttle, engineers had to determine the center of gravity of each craft.

Carla Thomas/NASA

5.4A Three-Dimensional Centers of Gravity and Centroids

For a three-dimensional body, we obtain the center of gravity G by dividing the body into small elements. The weight $\mathbf{W}$ of the body acting at G is equivalent to the system of distributed forces $\Delta\mathbf{W}$ representing the weights of the small elements. Choosing the y axis to be vertical with positive sense upward (Fig. 5.18) and denoting the position vector of G to be $\bar{\mathbf{r}}$, we set $\mathbf{W}$ equal to the sum of the elemental weights $\Delta\mathbf{W}$ and set its moment about O equal to the sum of the moments about O of the elemental weights. Thus,

$$\begin{aligned} \Sigma\mathbf{F}: &\quad -W\mathbf{j} = \Sigma(-\Delta W\mathbf{j}) \\ \Sigma\mathbf{M}_O: &\quad \bar{\mathbf{r}} \times (-W\mathbf{j}) = \Sigma[\mathbf{r} \times (-\Delta W\mathbf{j})] \end{aligned} \tag{5.13}$$

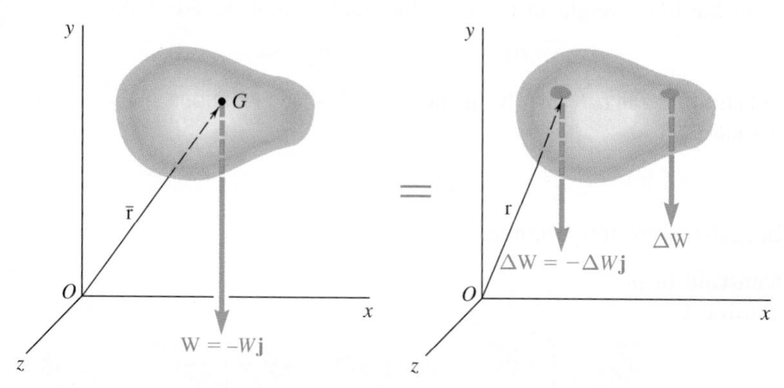

Fig. 5.18 For a three-dimensional body, the weight **W** acting through the center of gravity *G* and its moment about *O* is equivalent to the system of distributed weights acting on all the elements of the body and the sum of their moments about *O*.

We can rewrite the last equation in the form

$$\bar{\mathbf{r}}W \times (-\mathbf{j}) = (\Sigma\mathbf{r}\,\Delta W) \times (-\mathbf{j}) \tag{5.14}$$

From these equations, we can see that the weight $\mathbf{W}$ of the body is equivalent to the system of the elemental weights $\Delta\mathbf{W}$ if the following conditions are satisfied:

$$W = \Sigma\,\Delta W \qquad \bar{\mathbf{r}}W = \Sigma\mathbf{r}\,\Delta W$$

Increasing the number of elements and simultaneously decreasing the size of each element, we obtain in the limit as

Weight, center of gravity of a three-dimensional body

$$W = \int dW \qquad \bar{\mathbf{r}}W = \int \mathbf{r}\, dW \tag{5.15}$$

Note that these relations are independent of the orientation of the body. For example, if the body and the coordinate axes were rotated so that the z axis pointed upward, the unit vector $-\mathbf{j}$ would be replaced by $-\mathbf{k}$ in Eqs. (5.13) and (5.14), but the relations in Eqs. (5.15) would remain unchanged.

Resolving the vectors $\bar{\mathbf{r}}$ and $\mathbf{r}$ into rectangular components, we note that the second of the relations in Eqs. (5.15) is equivalent to the three scalar equations

$$\bar{x}W = \int x\, dW \qquad \bar{y}W = \int y\, dW \qquad \bar{z}W = \int z\, dW \tag{5.16}$$

or

$$\bar{x} = \frac{\int x\, dW}{W} \qquad \bar{y} = \frac{\int y\, dW}{W} \qquad \bar{z} = \frac{\int z\, dW}{W} \tag{5.16$'$}$$

If the body is made of a homogeneous material of specific weight γ, we can express the magnitude dW of the weight of an infinitesimal element in terms of the volume dV of the element and express the magnitude W of the total weight in terms of the total volume V. We obtain

$$dW = \gamma\, dV \qquad W = \gamma V$$

Substituting for dW and W in the second of the relations in Eqs. (5.15), we have

$$\bar{\mathbf{r}}V = \int \mathbf{r}\, dV \tag{5.17}$$

In scalar form, this becomes

Centroid of a volume V

$$\bar{x}V = \int x\, dV \qquad \bar{y}V = \int y\, dV \qquad \bar{z}V = \int z\, dV \tag{5.18}$$

or

$$\bar{x} = \frac{\int x\, dV}{V} \qquad \bar{y} = \frac{\int y\, dV}{V} \qquad \bar{z} = \frac{\int z\, dV}{V} \tag{5.18$'$}$$

The center of gravity of a homogeneous body whose coordinates are $\bar{x}$, $\bar{y}$, $\bar{z}$ is also known as the **centroid C of the volume** V of the body. If the body is not homogeneous, we cannot use Eqs. (5.18) to determine the center of gravity of the body; however, Eqs. (5.18) still define the centroid of the volume.

The integral $\int x\, dV$ is known as the **first moment of the volume with respect to the** yz **plane**. Similarly, the integrals $\int y\, dV$ and $\int z\, dV$ define the first moments of the volume with respect to the zx plane and the xy plane, respectively. You can see from Eqs. (5.18) that if the centroid of a volume is located in a coordinate plane, the first moment of the volume with respect to that plane is zero.

A volume is said to be symmetrical with respect to a given plane if, for every point P of the volume, there exists a point P' of the same volume such that the line PP' is perpendicular to the given plane and is bisected by that plane. We say the plane is a **plane of symmetry** for the given volume. When a volume V possesses a plane of symmetry, the first moment of V with respect to that plane is zero, and the centroid of the volume is located in the plane of symmetry. If a volume possesses two planes of symmetry, the centroid of the volume is located on the line of intersection of the two planes. Finally, if a volume possesses three planes of symmetry that intersect at a well-defined point (i.e., not along a common line), the point of intersection of the three planes coincides with the centroid of the volume. This property enables us to determine immediately the locations of the centroids of spheres, ellipsoids, cubes, rectangular parallelepipeds, etc.

For unsymmetrical volumes or volumes possessing only one or two planes of symmetry, we can determine the location of the centroid by integration.[†] The centroids of several common volumes are shown in Fig. 5.19. Note that, in general, the centroid of a volume of revolution *does not coincide* with the centroid of its cross section. Thus, the centroid of a hemisphere is different from that of a semicircular area, and the centroid of a cone is different from that of a triangle.

5.4B Composite Bodies

If a body can be divided into several of the common shapes shown in Fig. 5.19, we can determine its center of gravity G by setting the moment about O of its total weight equal to the sum of the moments about O of the weights of the various component parts. Proceeding in this way, we obtain the following equations defining the coordinates $\bar{X}, \bar{Y}, \bar{Z}$ of the center of gravity G as

Center of gravity of a body with weight W

$$\bar{X}\Sigma W = \Sigma \bar{x} W \quad \bar{Y}\Sigma W = \Sigma \bar{y} W \quad \bar{Z}\Sigma W = \Sigma \bar{z} W \tag{5.19}$$

or

$$\bar{X} = \frac{\Sigma \bar{x} W}{\Sigma W} \quad \bar{Y} = \frac{\Sigma \bar{y} W}{\Sigma W} \quad \bar{Z} = \frac{\Sigma \bar{z} W}{\Sigma W} \tag{5.19'}$$

If the body is made of a homogeneous material, its center of gravity coincides with the centroid of its volume, and we obtain

Centroid of a volume V

$$\bar{X}\Sigma V = \Sigma \bar{x} V \quad \bar{Y}\Sigma V = \Sigma \bar{y} V \quad \bar{Z}\Sigma V = \Sigma \bar{z} V \tag{5.20}$$

or

$$\bar{X} = \frac{\Sigma \bar{x} V}{\Sigma V} \quad \bar{Y} = \frac{\Sigma \bar{y} V}{\Sigma V} \quad \bar{Z} = \frac{\Sigma \bar{z} V}{\Sigma V} \tag{5.20'}$$

[†]For the determination of centroids of volumes by integration, see Ferdinand P. Beer, E. Russell Johnston, Jr., and David F. Mazurek, *Vector Mechanics for Engineers: Statics,* 11th ed., McGraw-Hill, New York, 2016, Sec. 5.4C.

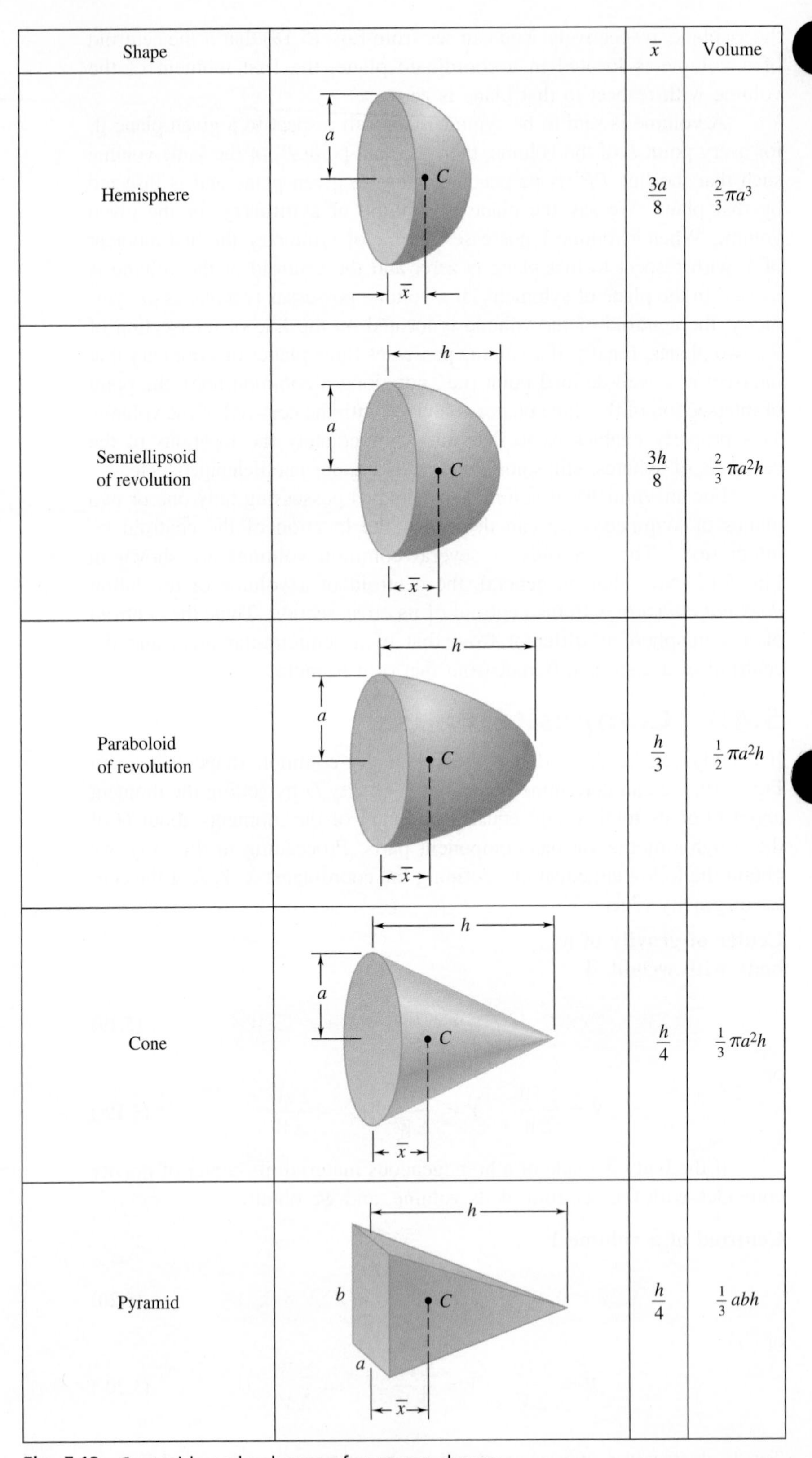

Shape		$\bar{x}$	Volume
Hemisphere		$\frac{3a}{8}$	$\frac{2}{3}\pi a^3$
Semiellipsoid of revolution		$\frac{3h}{8}$	$\frac{2}{3}\pi a^2 h$
Paraboloid of revolution		$\frac{h}{3}$	$\frac{1}{2}\pi a^2 h$
Cone		$\frac{h}{4}$	$\frac{1}{3}\pi a^2 h$
Pyramid		$\frac{h}{4}$	$\frac{1}{3}abh$

Fig. 5.19 Centroids and volumes of common shapes.

Sample Problem 5.10

Determine the location of the center of gravity of the homogeneous body of revolution shown that was obtained by joining a hemisphere and a cylinder and carving out a cone.

STRATEGY: The body is homogeneous, so the center of gravity coincides with the centroid. Since the body was formed from a composite of three simple shapes, you can find the centroid of each shape and combine them using Eq. (5.20).

MODELING: Because of symmetry, the center of gravity lies on the x axis. As shown in Fig. 1, the body is formed by adding a hemisphere to a cylinder and then subtracting a cone. Find the volume and the abscissa of the centroid of each of these components from Fig. 5.19 and enter them in a table (below). Then you can determine the total volume of the body and the first moment of its volume with respect to the yz plane.

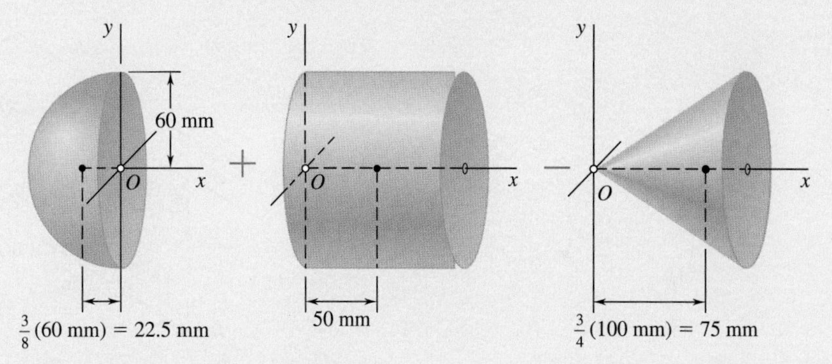

Fig. 1 The given body modeled as the combination of simple geometric shapes.

ANALYSIS: Note that the location of the centroid of the hemisphere is negative because it lies to the left of the origin.

Component	Volume, mm^3	$\bar{x}$, mm	$\bar{x}V$, mm^4
Hemisphere	$\frac{1}{2}\frac{4\pi}{3}(60)^3 = 0.4524 \times 10^6$	-22.5	-10.18×10^6
Cylinder	$\pi(60)^2(100) = 1.1310 \times 10^6$	$+50$	$+56.55 \times 10^6$
Cone	$-\frac{\pi}{3}(60)^2(100) = -0.3770 \times 10^6$	$+75$	-28.28×10^6
	$\Sigma V = 1.206 \times 10^6$		$\Sigma\bar{x}V = +18.09 \times 10^6$

Thus,

$$\bar{X}\Sigma V = \Sigma\bar{x}V: \qquad \bar{X}(1.206 \times 10^6 \text{ mm}^3) = 18.09 \times 10^6 \text{ mm}^4$$

$$\bar{X} = 15 \text{ mm}$$ ◀

REFLECT and THINK: Adding the hemisphere and subtracting the cone have the effect of shifting the centroid of the composite shape to the left of that for the cylinder (50 mm). However, because the first moment of volume for the cylinder is larger than for the hemisphere and cone combined, you should expect the centroid for the composite to still be in the positive x domain. Thus, as a rough visual check, the result of $+15$ mm is reasonable.

Sample Problem 5.11

Locate the center of gravity of the steel machine part shown. The diameter of each hole is 1 in.

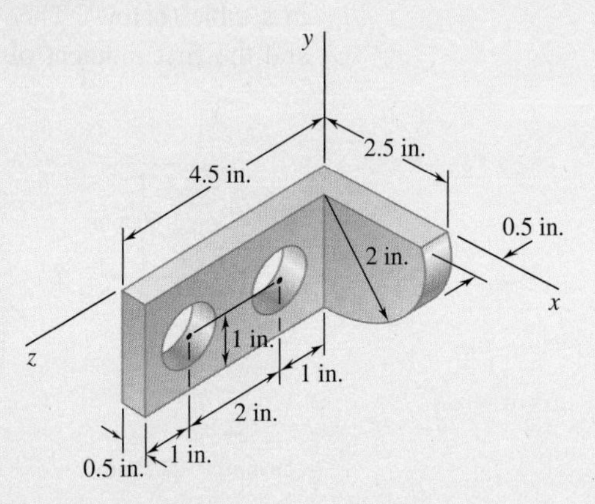

STRATEGY: This part can be broken down into the sum of two volumes minus two smaller volumes (holes). Find the volume and centroid of each volume and combine them using Eq. (5.20) to find the overall centroid.

MODELING: As shown in Fig. 1, the machine part can be obtained by adding a rectangular parallelepiped (I) to a quarter cylinder (II) and then subtracting two 1-in.-diameter cylinders (III and IV). Determine the volume and the coordinates of the centroid of each component and enter them in a table (below). Using the data in the table, determine the total volume and the moments of the volume with respect to each of the coordinate planes.

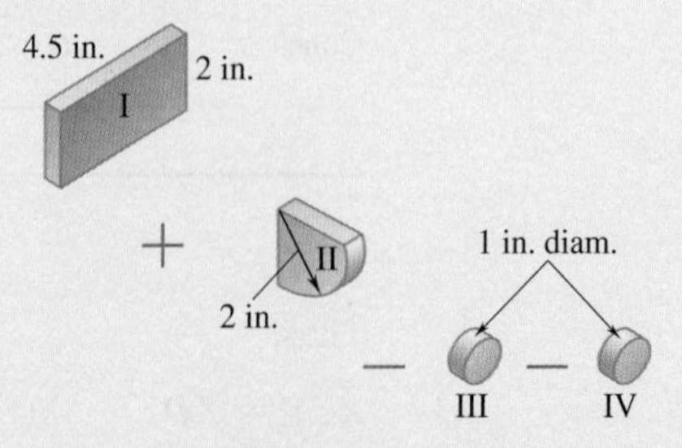

Fig. 1 The given body modeled as the combination of simple geometric shapes.

ANALYSIS: You can treat each component volume as a planar shape using Fig. 5.8A to find the volumes and centroids, but the right-angle joining of components I and II requires calculations in three dimensions. You may find it helpful to draw more detailed sketches of components with the centroids carefully labeled (Fig. 2).

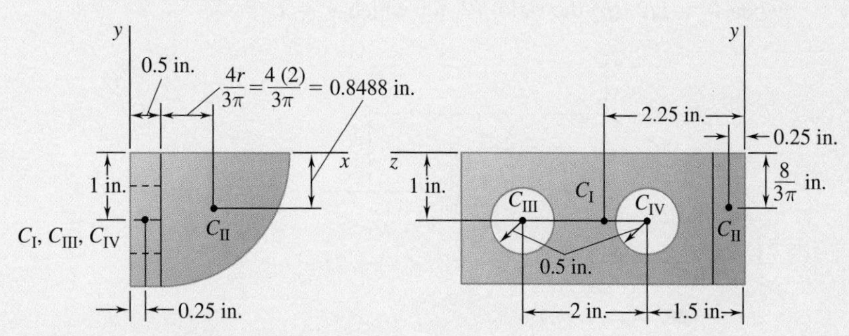

Fig. 2 Centroids of components.

	V, in^3	$\bar{x}$, in.	$\bar{y}$, in.	$\bar{z}$, in.	$\bar{x}V$, in^4	$\bar{y}V$, in^4	$\bar{z}V$, in^4
I	$(4.5)(2)(0.5) = 4.5$	0.25	-1	2.25	1.125	-4.5	10.125
II	$\frac{1}{4}\pi(2)^2(0.5) = 1.571$	1.3488	-0.8488	0.25	2.119	-1.333	0.393
III	$-\pi(0.5)^2(0.5) = -0.3927$	0.25	-1	3.5	-0.098	0.393	-1.374
IV	$-\pi(0.5)^2(0.5) = -0.3927$	0.25	-1	1.5	-0.098	0.393	-0.589
	$\Sigma V = 5.286$				$\Sigma\bar{x}V = 3.048$	$\Sigma\bar{y}V = -5.047$	$\Sigma\bar{z}V = 8.555$

Thus,

$\overline{X}\Sigma V = \Sigma\bar{x}V$: $\quad \overline{X}(5.286 \text{ in}^3) = 3.048 \text{ in}^4 \quad \overline{X} = 0.577$ in. ◀

$\overline{Y}\Sigma V = \Sigma\bar{y}V$: $\quad \overline{Y}(5.286 \text{ in}^3) = -5.047 \text{ in}^4 \quad \overline{Y} = -0.955$ in. ◀

$\overline{Z}\Sigma V = \Sigma\bar{z}V$: $\quad \overline{Z}(5.286 \text{ in}^3) = 8.555 \text{ in}^4 \quad \overline{Z} = 1.618$ in. ◀

REFLECT and THINK: By inspection, you should expect $\overline{X}$ and $\overline{Z}$ to be considerably less than (1/2)(2.5 in.) and (1/2)(4.5 in.), respectively, and $\overline{Y}$ to be slightly less in magnitude than (1/2)(2 in.). Thus, as a rough visual check, the results obtained are as expected.

Problems

5.57 Consider the composite body shown. Determine (*a*) the value of $\bar{x}$ when $h = L/2$, (*b*) the ratio h/L for which $\bar{x} = L$.

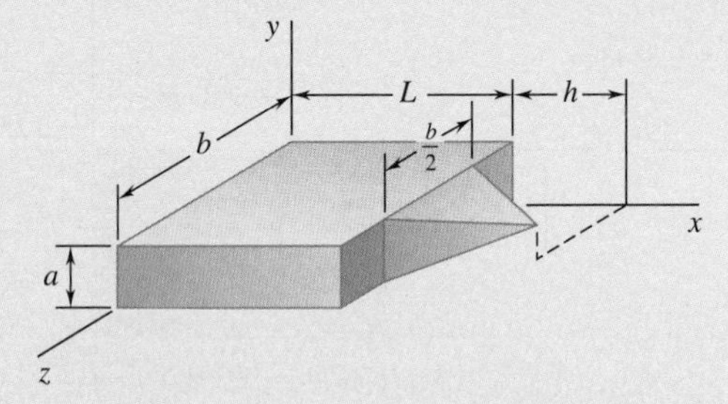

Fig. P5.57

5.58 Determine the location of the centroid of the composite body shown when (*a*) $h = 2b$, (*b*) $h = 2.5b$.

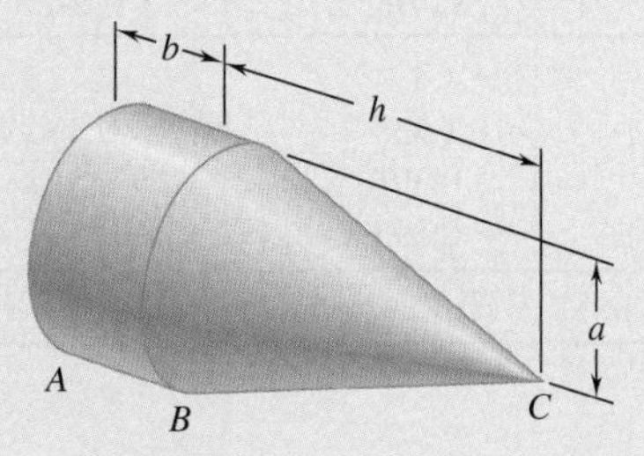

Fig. P5.58

5.59 The composite body shown is formed by removing a semiellipsoid of revolution of semimajor axis h and semiminor axis $a/2$ from a hemisphere of radius a. Determine (*a*) the y coordinate of the centroid when $h = a/2$, (*b*) the ratio h/a for which $\bar{y} = -0.4a$.

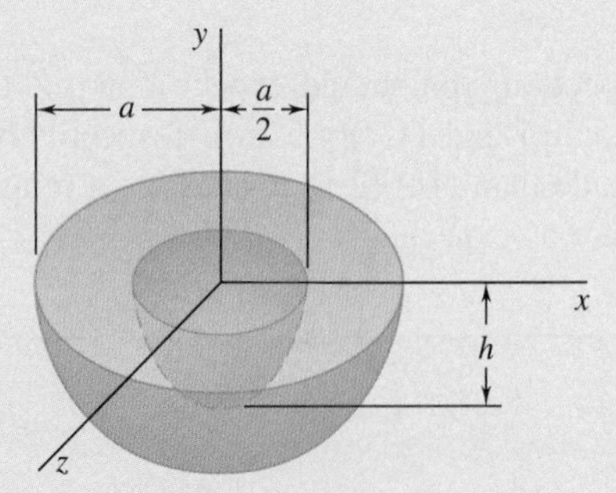

Fig. P5.59

5.60 Locate the centroid of the frustum of a right circular cone when $r_1 = 40$ mm, $r_2 = 50$ mm, and $h = 60$ mm.

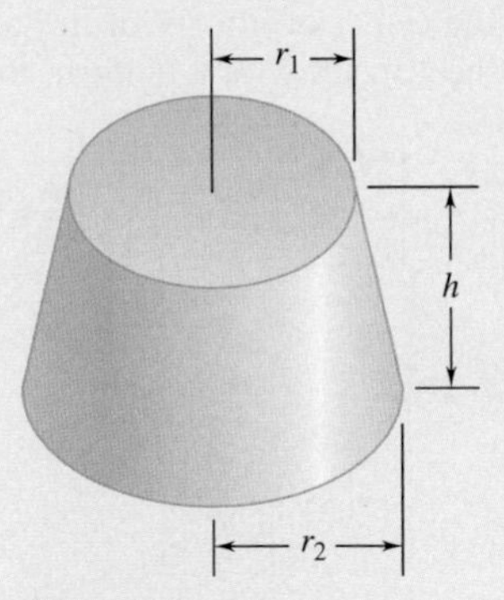

Fig. P5.60

5.61 For the machine element shown, locate the *y* coordinate of the center of gravity.

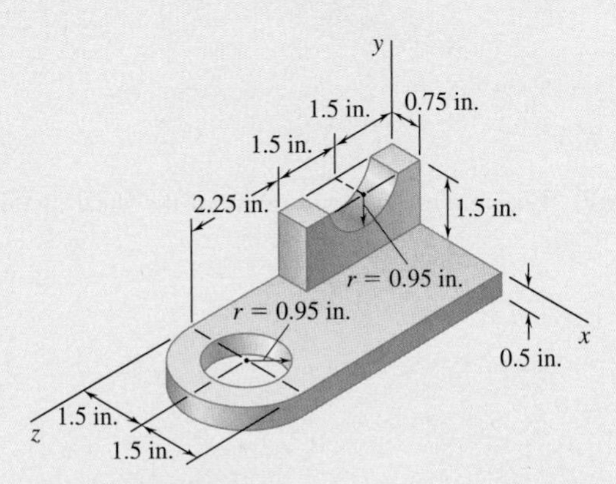

Fig. P5.61 and *P5.62*

5.62 For the machine element shown, locate the *z* coordinate of the center of gravity.

5.63 For the machine element shown, locate the *x* coordinate of the center of gravity.

5.64 For the machine element shown, locate the *y* coordinate of the center of gravity.

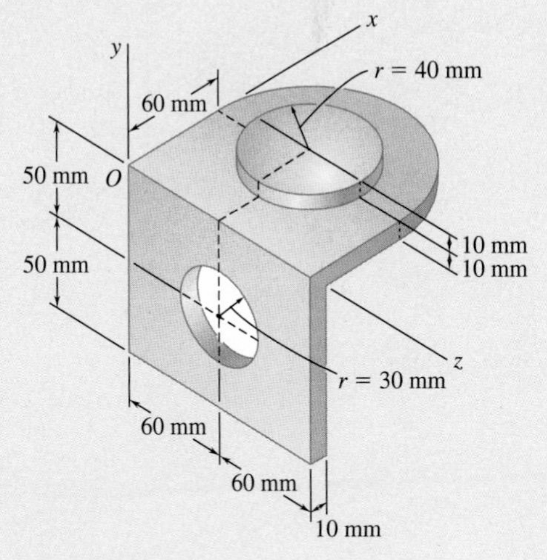

Fig. P5.63 and *P5.64*

5.65 A wastebasket, designed to fit in the corner of a room, is 16 in. high and has a base in the shape of a quarter circle with a radius of 10 in. Locate the center of gravity of the wastebasket, knowing that it is made of sheet metal with a uniform thickness.

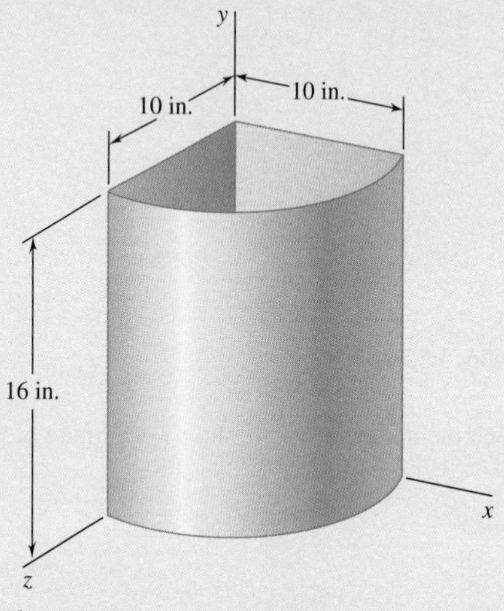

Fig. P5.65

5.66 and *5.67* Locate the center of gravity of the sheet-metal form shown.

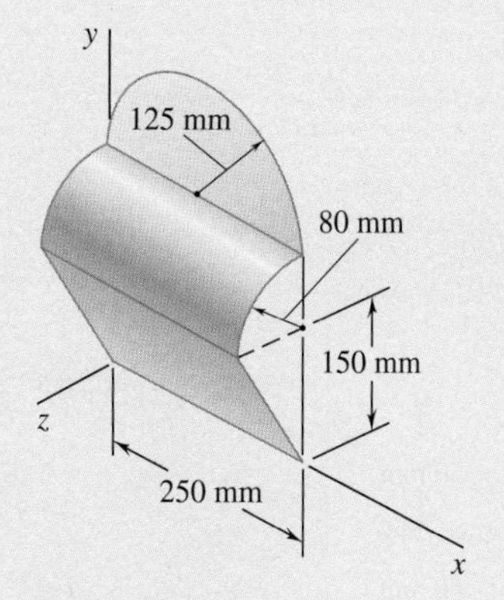

Fig. P5.66

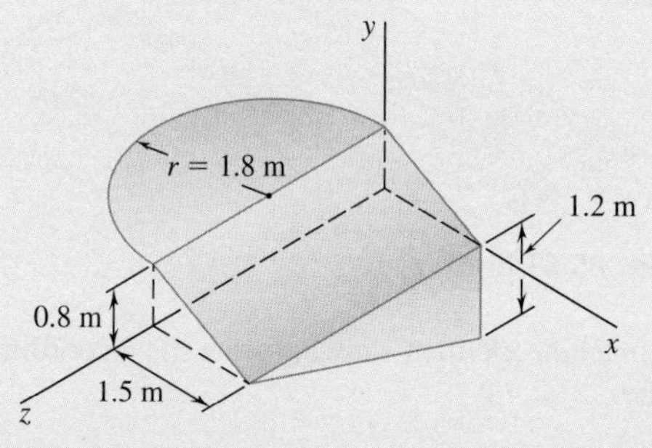

Fig. *P5.67*

5.68 A corner reflector for tracking by radar has two sides in the shape of a quarter circle with a radius of 15 in. and one side in the shape of a triangle. Locate the center of gravity of the reflector, knowing that it is made of sheet metal with a uniform thickness.

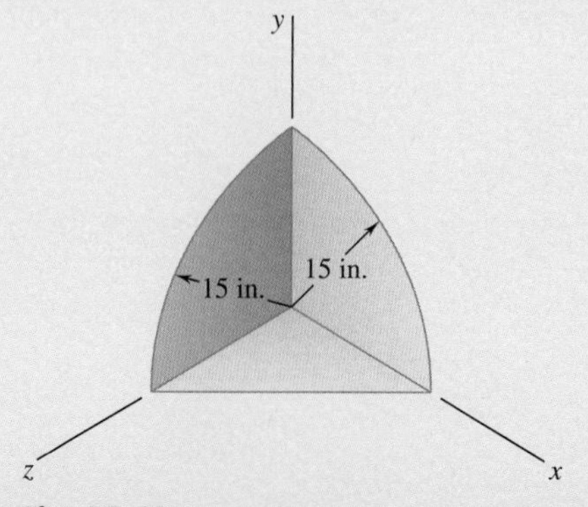

Fig. *P5.68*

5.69 and 5.70 Locate the center of gravity of the figure shown, knowing that it is made of thin brass rods with a uniform diameter.

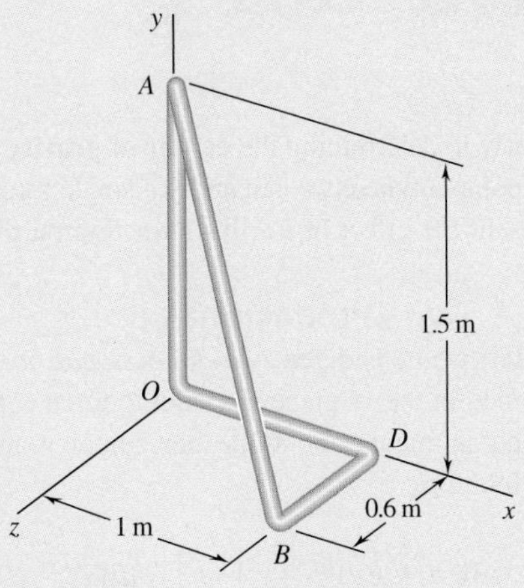

Fig. P5.69

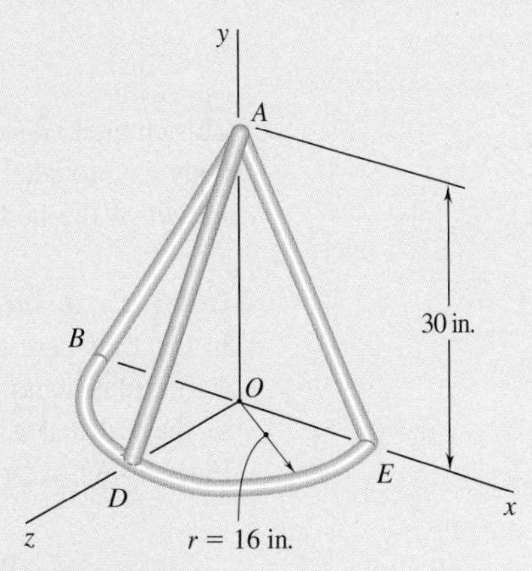

Fig. P5.70

5.71 Three brass plates are brazed to a steel pipe to form the flagpole base shown. Knowing that the pipe has a wall thickness of 8 mm and that each plate is 6 mm thick, determine the location of the center of gravity of the base. (Densities: brass = 8470 kg/m^3, steel = 7860 kg/m^3.)

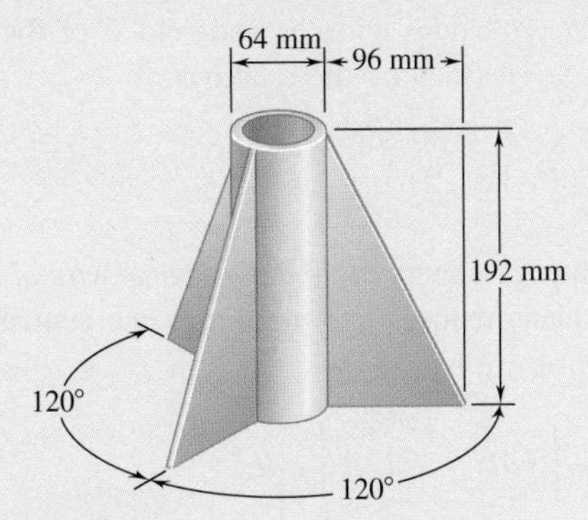

Fig. P5.71

5.72 A brass collar with a length of 2.5 in. is mounted on an aluminum rod with a length of 4 in. Locate the center of gravity of the composite body. (Specific weights: brass = 0.306 lb/in^3, aluminum = 0.101 lb/in^3.)

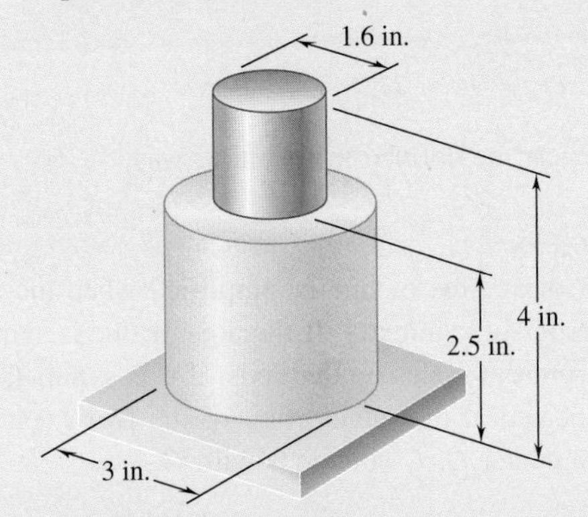

Fig. P5.72

Review and Summary

This chapter was devoted chiefly to determining the **center of gravity** of a rigid body, i.e., to determining the point G where we can apply a single force $\mathbf{W}$—the *weight* of the body—to represent the effect of Earth's attraction on the body.

Center of Gravity of a Two-Dimensional Body

In the first part of this chapter, we considered *two-dimensional bodies,* such as flat plates and wires contained in the xy plane. By adding force components in the vertical z direction and moments about the horizontal y and x axes [Sec. 5.1A], we derived the relations

$$W = \int dW \quad \bar{x}W = \int x\,dW \quad \bar{y}W = \int y\,dW \tag{5.2}$$

These equations define the weight of the body and the coordinates $\bar{x}$ and $\bar{y}$ of its center of gravity.

Centroid of an Area or Line

In the case of a *homogeneous flat* plate of uniform thickness [Sec. 5.1B], the center of gravity G of the plate coincides with the **centroid** C **of the area** A of the plate. The coordinates are defined by the relations

$$\bar{x}A = \int x\,dA \quad \bar{y}A = \int y\,dA \tag{5.3}$$

Similarly, determining the center of gravity of a *homogeneous wire of uniform cross section* contained in a plane reduces to determining the **centroid** C **of the line** L representing the wire; we have

$$\bar{x}L = \int x\,dL \quad \bar{y}L = \int y\,dL \tag{5.4}$$

First Moments

The integrals in Eqs. (5.3) are referred to as the **first moments** of the area A with respect to the y and x axes and are denoted by Q_y and Q_x, respectively [Sec. 5.1C]. We have

$$Q_y = \bar{x}A \quad Q_x = \bar{y}A \tag{5.6}$$

The first moments of a line can be defined in a similar way.

Properties of Symmetry

Determining the centroid C of an area or line is simplified when the area or line possesses certain properties of symmetry. If the area or line is symmetric with respect to an axis, its centroid C lies on that axis; if it is symmetric with respect to two axes, C is located at the intersection of the two axes; if it is symmetric with respect to a center O, C coincides with O.

Center of Gravity of a Composite Body

The areas and the centroids of various common shapes are tabulated in Fig. 5.8. When a flat plate can be divided into several of these shapes, the coordinates $\overline{X}$ and $\overline{Y}$ of its center of gravity G can be determined from the coordinates $\bar{x}_1, \bar{x}_2, \ldots$ and $\bar{y}_1, \bar{y}_2, \ldots$ of the centers of gravity $G_1, G_2, \ldots$ of the various parts [Sec. 5.1D]. Equating moments about the y and x axes, respectively (Fig. 5.20), we have

$$\overline{X}\Sigma W = \Sigma\bar{x}W \qquad \overline{Y}\Sigma W = \Sigma\bar{y}W \tag{5.7}$$

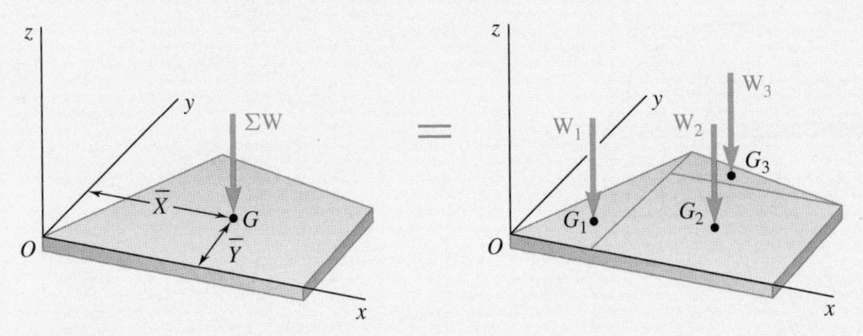

Fig. 5.20

If the plate is homogeneous and of uniform thickness, its center of gravity coincides with the centroid C of the area of the plate, and Eqs. (5.7) reduce to

$$Q_y = \overline{X}\Sigma A = \Sigma\bar{x}A \qquad Q_x = \overline{Y}\Sigma A = \Sigma\bar{y}A \tag{5.8}$$

These equations yield the first moments of the composite area, or they can be solved for the coordinates $\overline{X}$ and $\overline{Y}$ of its centroid [Sample Prob. 5.1]. Determining the center of gravity of a composite wire is carried out in a similar fashion [Sample Prob. 5.2].

Determining a Centroid by Integration

When an area is bounded by analytical curves, you can determine the coordinates of its centroid by *integration* [Sec. 5.2A]. This can be done by evaluating either the double integrals in Eqs. (5.3) or a single integral that uses one of the thin rectangular or pie-shaped elements of area shown in Fig. 5.12. Denoting by $\bar{x}_{el}$ and $\bar{y}_{el}$ the coordinates of the centroid of the element dA, we have

$$Q_y = \bar{x}A = \int \bar{x}_{el}\,dA \qquad Q_x = \bar{y}A = \int \bar{y}_{el}\,dA \tag{5.9}$$

It is advantageous to use the same element of area to compute both of the first moments Q_y and Q_x; we can also use the same element to determine the area A [Sample Prob. 5.4].

Theorems of Pappus–Guldinus

The **theorems of Pappus-Guldinus** relate the area of a surface of revolution or the volume of a body of revolution to the centroid of the generating curve or area [Sec. 5.2B]. The area A of the surface generated by rotating a curve of length L about a fixed axis (Fig. 5.21*a*) is

$$A = 2\pi\bar{y}L \tag{5.10}$$

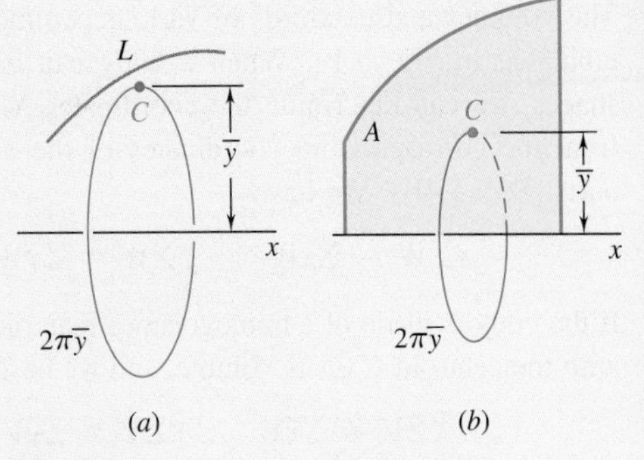

Fig. 5.21

where $\bar{y}$ represents the distance from the centroid C of the curve to the fixed axis. Similarly, the volume V of the body generated by rotating an area A about a fixed axis (Fig. 5.21*b*) is

$$V = 2\pi\bar{y}A \tag{5.11}$$

where $\bar{y}$ represents the distance from the centroid C of the area to the fixed axis.

Distributed Loads

The concept of centroid of an area also can be used to solve problems other than those dealing with the weight of flat plates. For example, to determine the reactions at the supports of a beam [Sec. 5.3], we can replace a **distributed load** w by a concentrated load **W** equal in magnitude to the area A under the load curve and passing through the centroid C of that area (Fig. 5.22).

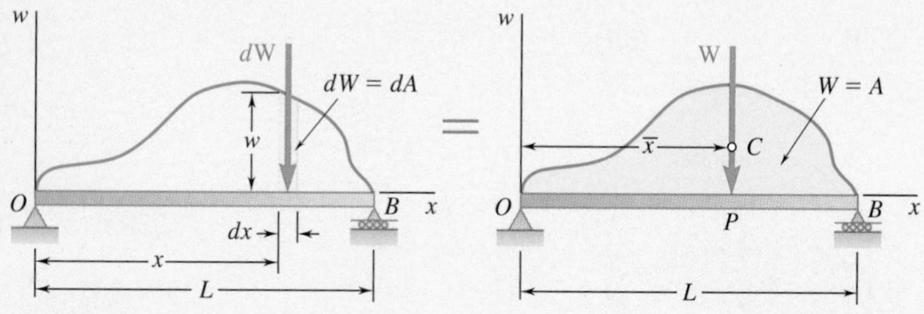

Fig. 5.22

Center of Gravity of a Three-Dimensional Body

The last part of this chapter was devoted to determining the center of gravity G of a three-dimensional body. We defined the coordinates $\bar{x}, \bar{y}, \bar{z}$ of G by the relations

$$\bar{x}W = \int x\,dW \qquad \bar{y}W = \int y\,dW \qquad \bar{z}W = \int z\,dW \tag{5.16}$$

Centroid of a Volume

In the case of a homogeneous body, the center of gravity G coincides with the centroid C of the volume V of the body. The coordinates of C are defined by the relations

$$\bar{x}V = \int x\,dV \qquad \bar{y}V = \int y\,dV \qquad \bar{z}V = \int z\,dV \tag{5.18}$$

If the volume possesses a *plane of symmetry*, its centroid C lies in that plane; if it possesses two planes of symmetry, C is located on the line of intersection of the two planes; if it possesses three planes of symmetry that intersect at only one point, C coincides with that point [Sec. 5.4A].

Center of Gravity of a Composite Body

The volumes and centroids of various common three-dimensional shapes are tabulated in Fig. 5.19. When a body can be divided into several of these shapes, we can determine the coordinates $\overline{X}, \overline{Y}, \overline{Z}$ of its center of gravity G from the corresponding coordinates of the centers of gravity of its various parts [Sec. 5.4B]. We have

$$\overline{X}\Sigma W = \Sigma\bar{x}W \qquad \overline{Y}\Sigma W = \Sigma\bar{y}W \qquad \overline{Z}\Sigma W = \Sigma\bar{z}W \tag{5.19}$$

If the body is made of a homogeneous material, its center of gravity coincides with the centroid C of its volume, and we have [Sample Probs. 5.10 and 5.11]

$$\overline{X}\Sigma V = \Sigma\bar{x}V \qquad \overline{Y}\Sigma V = \Sigma\bar{y}V \qquad \overline{Z}\Sigma V = \Sigma\bar{z}V$$

Review Problems

5.73 and *5.74* Locate the centroid of the plane area shown.

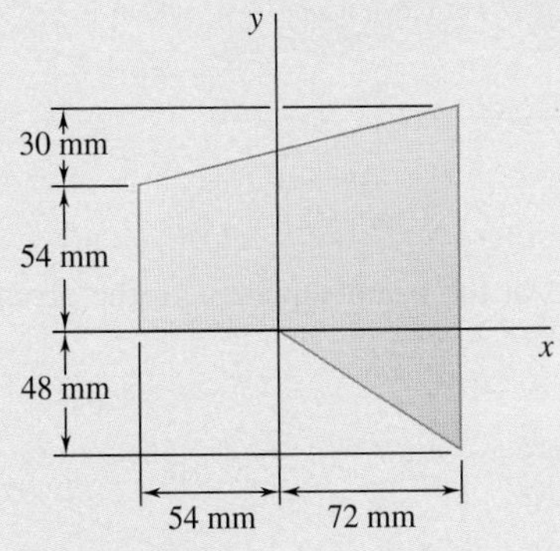

Fig. P5.73

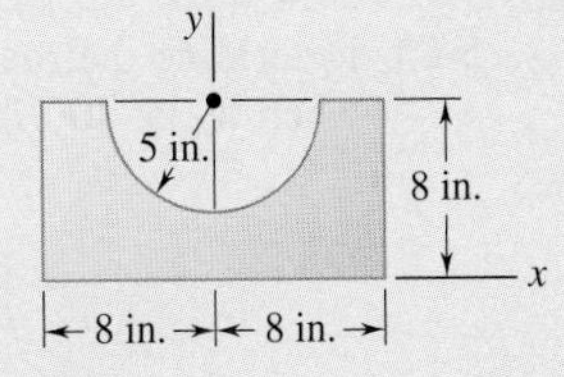

Fig. *P5.74*

5.75 A thin homogeneous wire is bent to form the perimeter of the plane area of Prob. 5.73. Locate the center of gravity of the wire figure thus formed.

5.76 Member *ABCDE* is a component of a mobile and is formed from a single piece of aluminum tubing. Knowing that the member is supported at *C* and that *d* is 0.50 m, determine the length *l* of arm *DE* so that this portion of the member is horizontal.

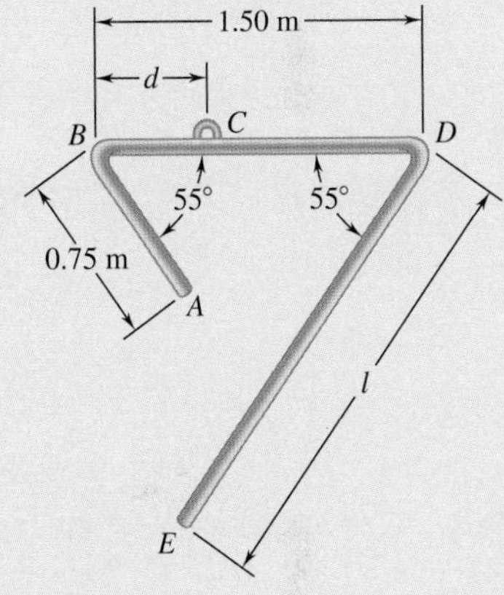

Fig. P5.76

5.77 Determine by direct integration the centroid of the area shown.

5.78 Determine by direct integration the centroid of the area shown.

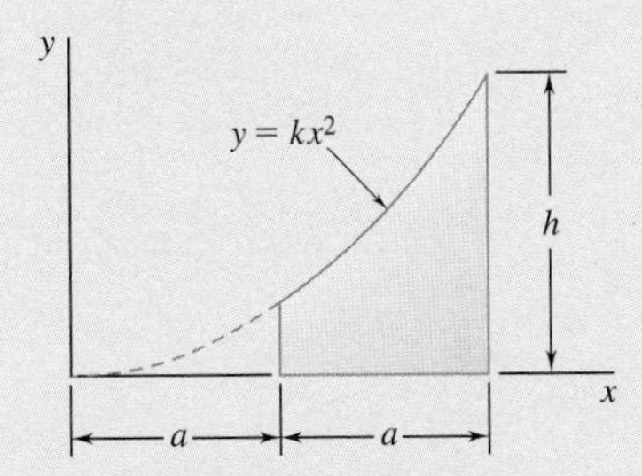

Fig. P5.77

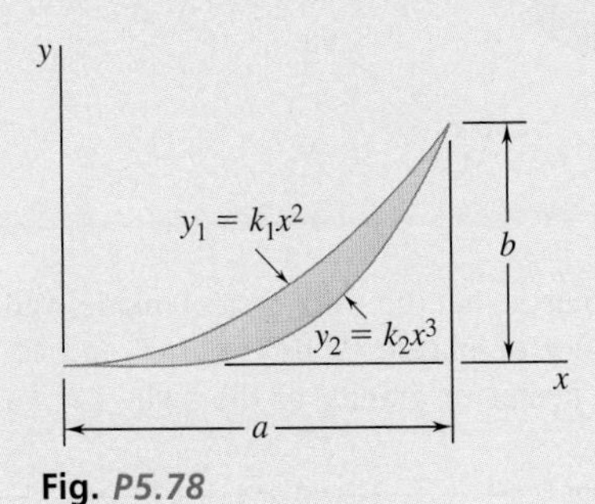

Fig. *P5.78*

5.79 A $\frac{3}{4}$-in.-diameter hole is drilled in a piece of 1-in.-thick steel; the hole is then countersunk as shown. Determine the volume of steel removed during the countersinking process.

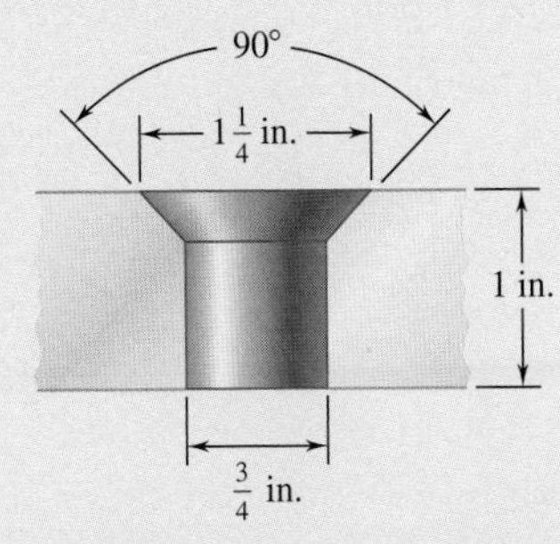

Fig. P5.79

5.80 A manufacturer is planning to produce 20,000 wooden pegs having the shape shown. Determine how many gallons of paint should be ordered, knowing that each peg will be given two coats of paint and that one gallon of paint covers 100 ft^2.

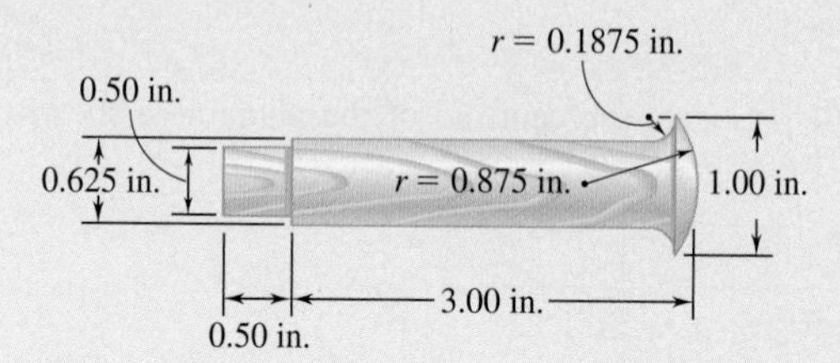

Fig. *P5.80*

5.81 Determine the reactions at the beam supports for the given loading when $w_0 = 400$ lb/ft.

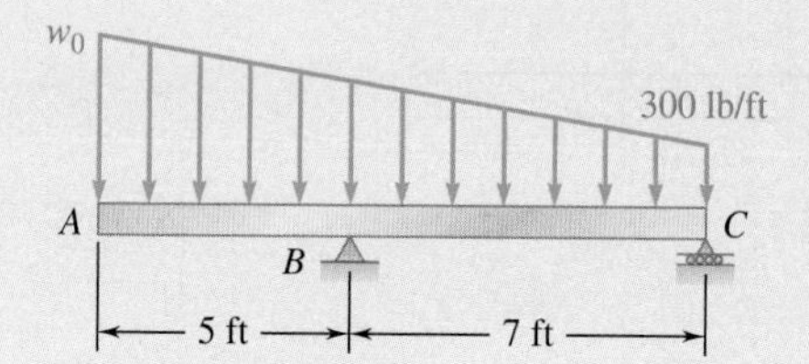

Fig. P5.81 and P5.82

5.82 Determine (*a*) the distributed load w_0 at the end *A* of the beam *ABC* for which the reaction at *C* is zero, (*b*) the corresponding reaction at *B*.

5.83 For the machine element shown, locate the *z* coordinate of the center of gravity.

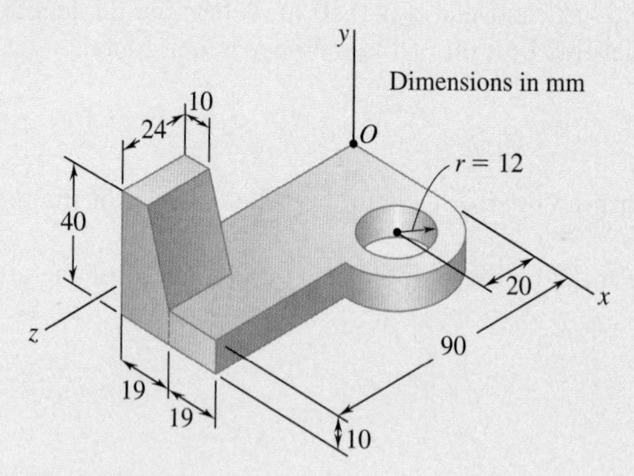

Fig. *P5.83*

5.84 A scratch awl has a plastic handle and a steel blade and shank. Knowing that the density of plastic is 1030 kg/m^3 and of steel is 7860 kg/m^3, locate the center of gravity of the awl.

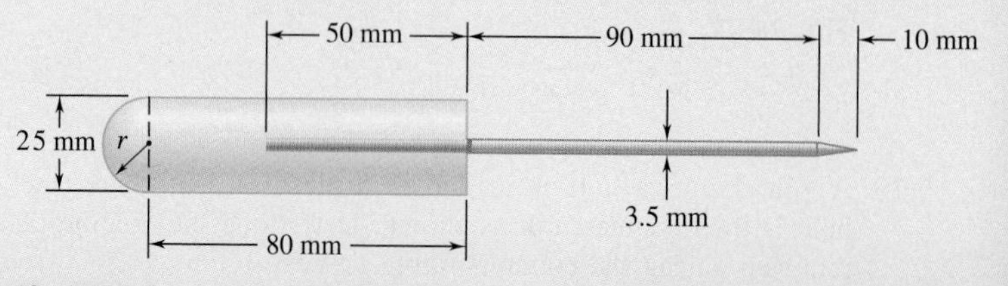

Fig. P5.84

© Lee Rentz/Photoshot

6 Analysis of Structures

Trusses, such as this cantilever arch bridge over Deception Pass in Washington State, provide both a practical and an economical solution to many engineering problems.

Objectives

- **Define** an ideal truss, and consider the attributes of simple trusses.
- **Analyze** plane trusses by the method of joints.
- **Simplify** certain truss analyses by recognizing special loading and geometry conditions.
- **Analyze** trusses by the method of sections.
- **Consider** the characteristics of compound trusses.
- **Analyze** structures containing multiforce members, such as frames and machines.

Introduction

In the preceding chapters, we studied the equilibrium of a single rigid body, where all forces involved were external to the rigid body. We now consider the equilibrium of structures made of several connected parts. This situation calls for determining not only the external forces acting on the structure, but also the forces that hold together the various parts of the structure. From the point of view of the structure as a whole, these forces are **internal forces**.

Consider, for example, the crane shown in Fig. 6.1*a* that supports a load *W*. The crane consists of three beams *AD*, *CF*, and *BE* connected by frictionless pins; it is supported by a pin at *A* and by a cable *DG*. The free-body diagram of the crane is drawn in Fig. 6.1*b*. The external forces shown in the diagram include the weight **W**, the two components $\mathbf{A}_x$ and $\mathbf{A}_y$ of the reaction at *A*, and the force **T** exerted by the cable at *D*. The internal forces holding the various parts of the crane together do not appear in the free-body diagram. If, however, we dismember the crane and draw a free-body diagram for each of its component parts, we can see the forces holding the three beams together, since these forces are external forces from the point of view of each component part (Fig. 6.1*c*).

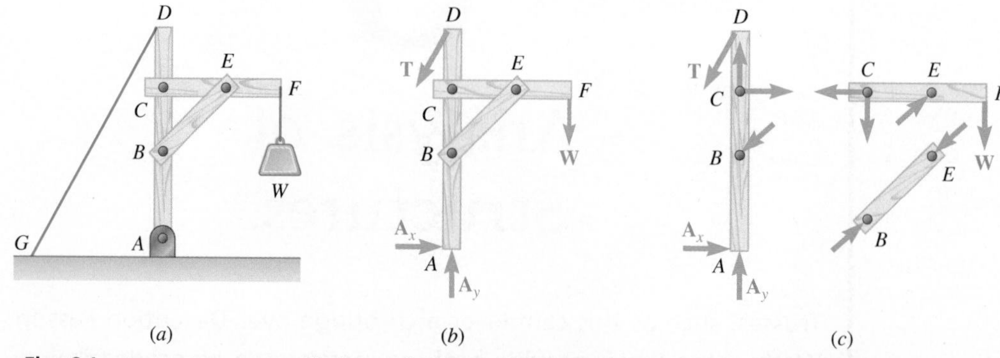

Fig. 6.1 A structure in equilibrium. (*a*) Diagram of a crane supporting a load; (*b*) free-body diagram of the crane; (*c*) free-body diagrams of the components of the crane.

Note that we represent the force exerted at B by member BE on member AD as equal and opposite to the force exerted at the same point by member AD on member BE. Similarly, the force exerted at E by BE on CF is shown equal and opposite to the force exerted by CF on BE, and the components of the force exerted at C by CF on AD are shown equal and opposite to the components of the force exerted by AD on CF. These representations agree with Newton's third law, which states that

The forces of action and reaction between two bodies in contact have the same magnitude, same line of action, and opposite sense.

We pointed out in Chap. 1 that this law, which is based on experimental evidence, is one of the six fundamental principles of elementary mechanics. Its application is essential for solving problems involving connected bodies.

In this chapter, we consider three broad categories of engineering structures:

1. **Trusses**, which are designed to support loads and are usually stationary, fully constrained structures. Trusses consist exclusively of straight members connected at joints located at the ends of each member. Members of a truss, therefore, are **two-force members**, i.e., members acted upon by two equal and opposite forces directed along the member.
2. **Frames**, which are also designed to support loads and are also usually stationary, fully constrained structures. However, like the crane of Fig. 6.1, frames always contain at least one **multi-force member**, i.e., a member acted upon by three or more forces that, in general, are not directed along the member.
3. **Machines**, which are designed to transmit and modify forces and are structures containing moving parts. Machines, like frames, always contain at least one multi-force member.

Two-force member

(*a*) A truss bridge
© *Datacraft Co Ltd/Getty Images RF*

(*b*) A bicycle frame
© *Fuse/Getty Images RF*

(*c*) A hydraulic machine arm
© *Design Pics/Ken Welsh RF*

Photo 6.1 The structures you see around you to support loads or transmit forces are generally trusses, frames, or machines.

6.1 ANALYSIS OF TRUSSES

The truss is one of the major types of engineering structures. It provides a practical and economical solution to many engineering situations, especially in the design of bridges and buildings. In this section, we describe the basic elements of a truss and study a common method for analyzing the forces acting in a truss.

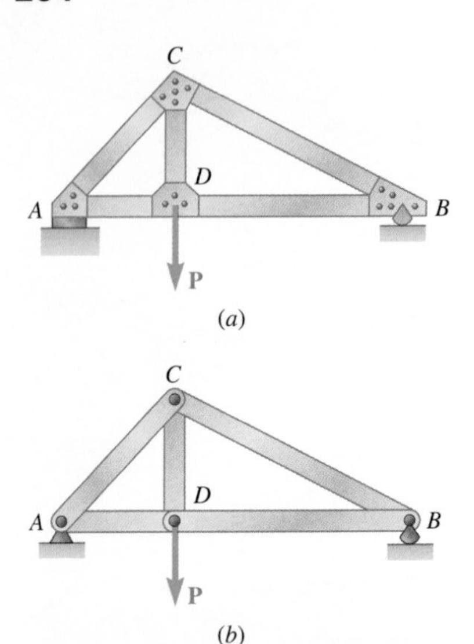

Fig. 6.2 (*a*) A typical truss consists of straight members connected at joints; (*b*) we can model a truss as two-force members connected by pins.

Photo 6.2 Shown is a pin-jointed connection on the former approach span to the San Francisco–Oakland Bay Bridge.

Courtesy of Godden Collection. National Information Service for Earthquake Engineering, University of California, Berkeley

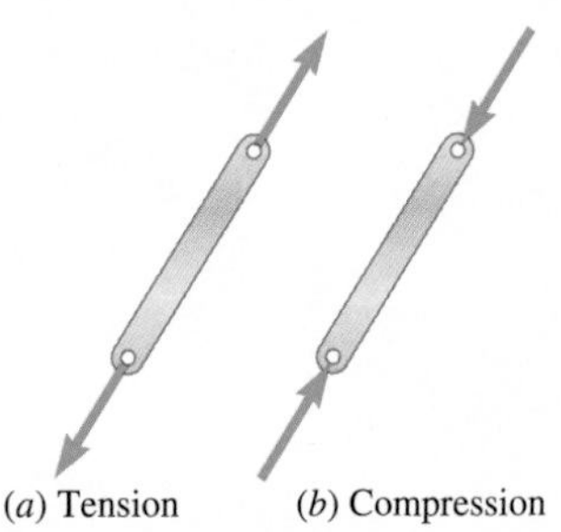

Fig. 6.4 A two-force member of a truss can be in tension or compression.

6.1A Simple Trusses

A truss consists of straight members connected at joints, as shown in Fig. 6.2*a*. Truss members are connected at their extremities only; no member is continuous through a joint. In Fig. 6.2*a*, for example, there is no member *AB*; instead we have two distinct members *AD* and *DB*. Most actual structures are made of several trusses joined together to form a space framework. Each truss is designed to carry those loads that act in its plane and thus may be treated as a two-dimensional structure.

In general, the members of a truss are slender and can support little lateral load; all loads, therefore, must be applied at the various joints and not to the members themselves. When a concentrated load is to be applied between two joints or when the truss must support a distributed load, as in the case of a bridge truss, a floor system must be provided. The floor transmits the load to the joints through the use of stringers and floor beams (Fig. 6.3).

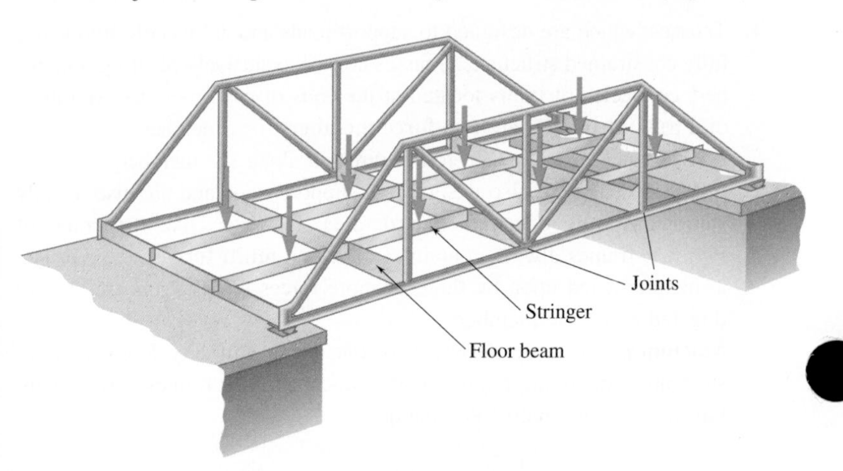

Fig. 6.3 A floor system of a truss uses stringers and floor beams to transmit an applied load to the joints of the truss.

We assume that the weights of the truss members can be applied to the joints, with half of the weight of each member applied to each of the two joints the member connects. Although the members are actually joined together by means of welded, bolted, or riveted connections, it is customary to assume that the members are pinned together; therefore, the forces acting at each end of a member reduce to a single force and no couple. This enables us to model the forces applied to a truss member as a single force at each end of the member. We can then treat each member as a two-force member, and we can consider the entire truss as a group of pins and two-force members (Fig. 6.2*b*). An individual member can be acted upon as shown in either of the two sketches of Fig. 6.4. In Fig. 6.4*a*, the forces tend to pull the member apart, and the member is in tension; in Fig. 6.4*b*, the forces tend to push the member together, and the member is in compression. Some typical trusses are shown in Fig. 6.5.

Consider the truss of Fig. 6.6*a*, which is made of four members connected by pins at *A*, *B*, *C*, and *D*. If we apply a load at *B*, the truss will greatly deform, completely losing its original shape. In contrast, the truss of Fig. 6.6*b*, which is made of three members connected by pins at *A*, *B*, and *C*, will deform only slightly under a load applied at *B*. The only possible

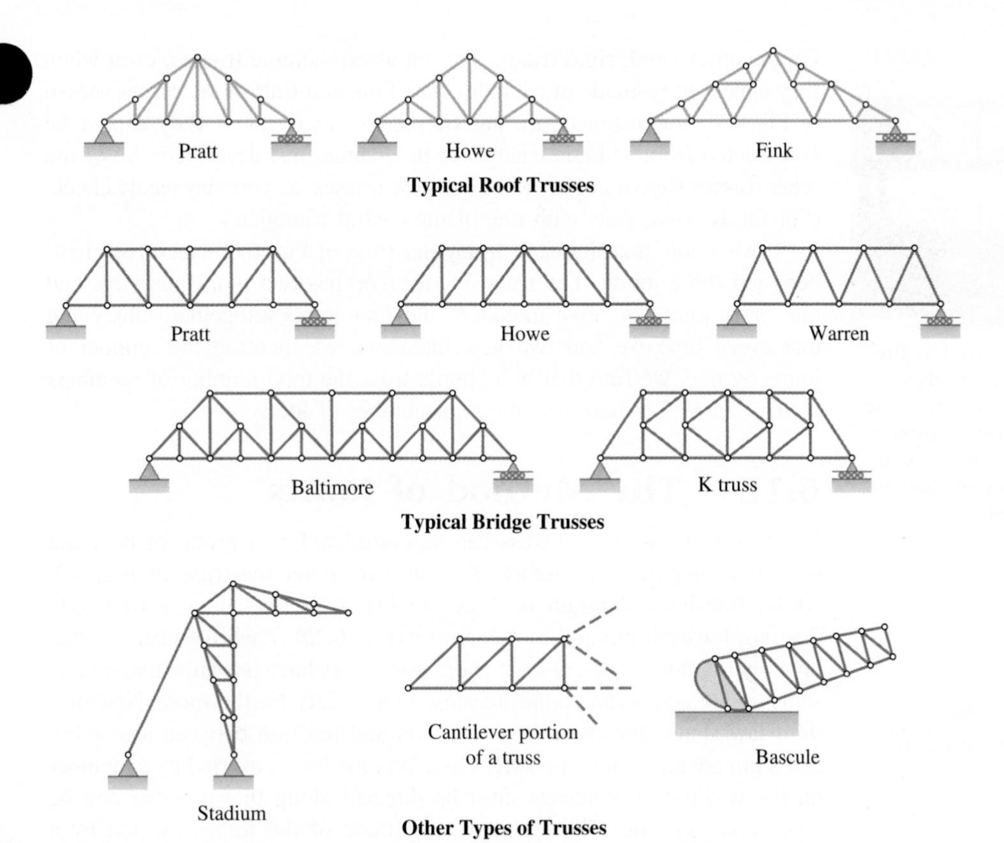

Fig. 6.5 You can often see trusses in the design of a building roof, a bridge, or other other larger structures.

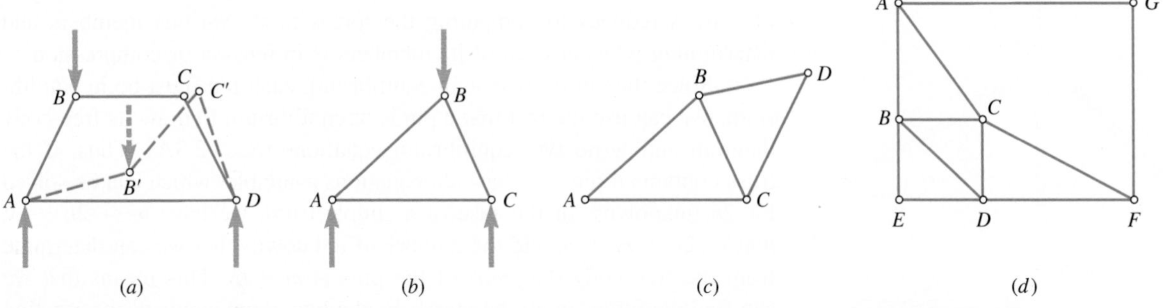

Fig. 6.6 (*a*) A poorly designed truss that cannot support a load; (*b*) the most elementary rigid truss consists of a simple triangle; (*c*) a larger rigid truss built up from the triangle in (*b*); (*d*) a rigid truss not made up of triangles alone.

deformation for this truss is one involving small changes in the length of its members. The truss of Fig. 6.6*b* is said to be a **rigid truss**, the term 'rigid' being used here to indicate that the truss *will not collapse*.

As shown in Fig. 6.6*c*, we can obtain a larger rigid truss by adding two members *BD* and *CD* to the basic triangular truss of Fig. 6.6*b*. We can repeat this procedure as many times as we like, and the resulting truss will be rigid if each time we add two new members they are attached to two existing joints and connected at a new joint. (The three joints must not be in a straight line.) A truss that can be constructed in this manner is called a **simple truss**.

Note that a simple truss is not necessarily made only of triangles. The truss of Fig. 6.6*d*, for example, is a simple truss that we constructed from triangle *ABC* by adding successively the joints *D*, *E*, *F*, and *G*.

Photo 6.3 Two K trusses were used as the main components of the movable bridge shown, which moved above a large stockpile of ore. The bucket below the trusses picked up ore and redeposited it until the ore was thoroughly mixed. The ore was then sent to the mill for processing into steel.

Courtesy of Ferdinand Beer

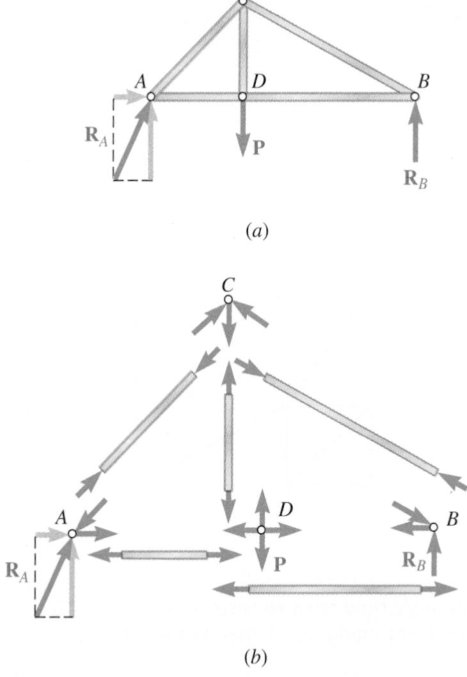

Fig. 6.7 (*a*) Free-body diagram of the truss as a rigid body; (*b*) free-body diagrams of the five members and four pins that make up the truss.

On the other hand, rigid trusses are not always simple trusses, even when they appear to be made of triangles. The Fink and Baltimore trusses shown in Fig. 6.5, for instance, are not simple trusses, because they cannot be constructed from a single triangle in the manner just described. All of the other trusses shown in Fig. 6.5 are simple trusses, as you may easily check. (For the K truss, start with one of the central triangles.)

Also note that the basic triangular truss of Fig. 6.6*b* has three members and three joints. The truss of Fig. 6.6*c* has two more members and one more joint; i.e., five members and four joints altogether. Observing that every time we add two new members, we increase the number of joints by one, we find that in a simple truss the total number of members is $m = 2n - 3$, where n is the total number of joints.

6.1B The Method of Joints

We have just seen that a truss can be considered as a group of pins and two-force members. Therefore, we can dismember the truss of Fig. 6.2, whose free-body diagram is shown in Fig. 6.7*a*, and draw a free-body diagram for each pin and each member (Fig. 6.7*b*). Each member is acted upon by two forces, one at each end; these forces have the same magnitude, same line of action, and opposite sense (Sec. 4.2A). Furthermore, Newton's third law states that the forces of action and reaction between a member and a pin are equal and opposite. Therefore, the forces exerted by a member on the two pins it connects must be directed along that member and be equal and opposite. The common magnitude of the forces exerted by a member on the two pins it connects is commonly referred to as the *force in the member,* even though this quantity is actually a scalar. Since we know the lines of action of all the internal forces in a truss, the analysis of a truss reduces to computing the forces in its various members and determining whether each of its members is in tension or compression.

Since the entire truss is in equilibrium, each pin must be in equilibrium. We can use the fact that a pin is in equilibrium to draw its free-body diagram and write two equilibrium equations (Sec. 2.3A). Thus, if the truss contains n pins, we have $2n$ equations available, which can be solved for $2n$ unknowns. In the case of a simple truss, we have $m = 2n - 3$; that is, $2n = m + 3$, and the number of unknowns that we can determine from the free-body diagrams of the pins is $m + 3$. This means that we can find the forces in all the members, the two components of the reaction $\mathbf{R}_A$, and the reaction $\mathbf{R}_B$ by considering the free-body diagrams of the pins.

We can also use the fact that the entire truss is a rigid body in equilibrium to write three more equations involving the forces shown in the free-body diagram of Fig. 6.7*a*. Since these equations do not contain any new information, they are not independent of the equations associated with the free-body diagrams of the pins. Nevertheless, we can use them to determine the components of the reactions at the supports. The arrangement of pins and members in a simple truss is such that it is always possible to find a joint involving only two unknown forces. We can determine these forces by using the methods of Sec. 2.3C and then transferring their values to the adjacent joints, treating them as known quantities at these joints. We repeat this procedure until we have determined all unknown forces.

As an example, let's analyze the truss of Fig. 6.7 by considering the equilibrium of each pin successively, starting with a joint at which only two forces are unknown. In this truss, all pins are subjected to at least three unknown forces. Therefore, we must first determine the reactions at the supports by considering the entire truss as a free body and using the equations of equilibrium of a rigid body. In this way we find that $\mathbf{R}_A$ is vertical, and we determine the magnitudes of $\mathbf{R}_A$ and $\mathbf{R}_B$.

This reduces the number of unknown forces at joint A to two, and we can determine these forces by considering the equilibrium of pin A. The reaction $\mathbf{R}_A$ and the forces $\mathbf{F}_{AC}$ and $\mathbf{F}_{AD}$ exerted on pin A by members AC and AD, respectively, must form a force triangle. First we draw $\mathbf{R}_A$ (Fig. 6.8); noting that $\mathbf{F}_{AC}$ and $\mathbf{F}_{AD}$ are directed along AC and AD, respectively, we complete the triangle and determine the magnitude and sense of $\mathbf{F}_{AC}$ and $\mathbf{F}_{AD}$. The magnitudes F_{AC} and F_{AD} represent the forces in members AC and AD. Since $\mathbf{F}_{AC}$ is directed down and to the left—that is, *toward* joint A—member AC pushes on pin A and is in compression. (From Newton's third law, pin A pushes *on* member AC.) Since $\mathbf{F}_{AD}$ is directed *away* from joint A, member AD pulls on pin A and is in tension. (From Newton's third law, pin A pulls *away* from member AD.)

Photo 6.4 Because roof trusses, such as those shown, require support only at their ends, it is possible to construct buildings with large, unobstructed interiors.

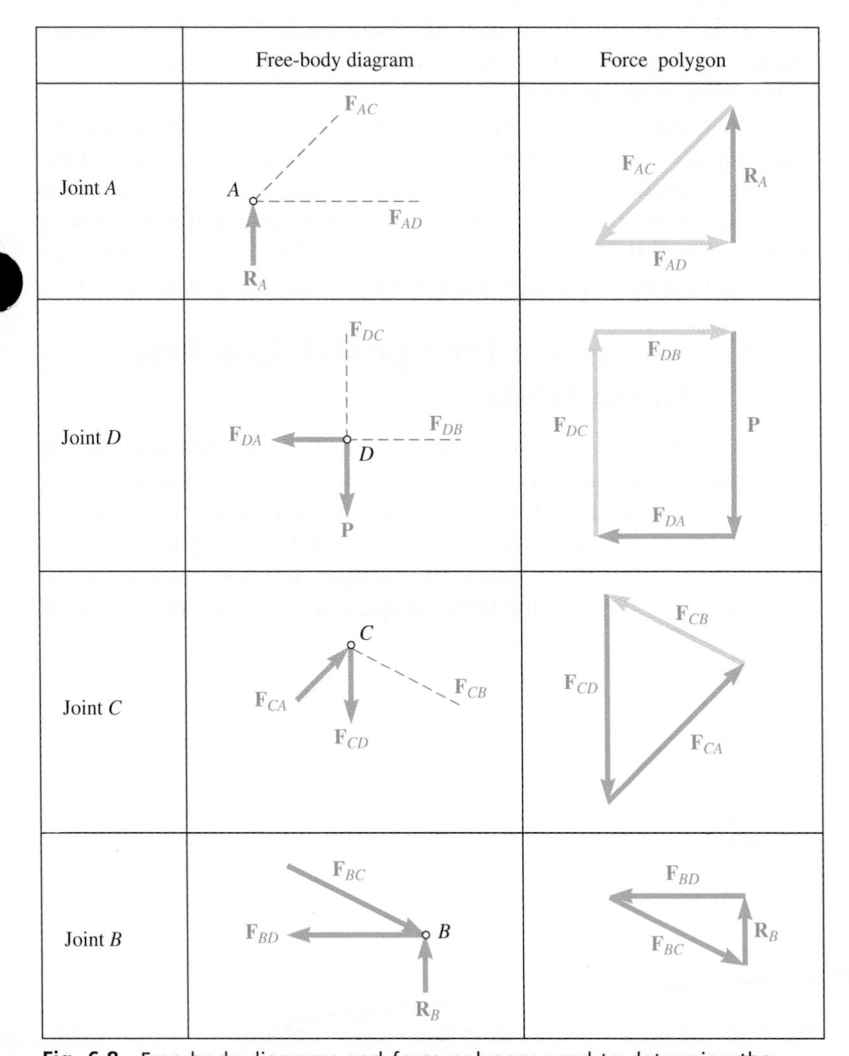

Fig. 6.8 Free-body diagrams and force polygons used to determine the forces on the pins and in the members of the truss in Fig. 6.7.

We can now proceed to joint D, where only two forces, $\mathbf{F}_{DC}$ and $\mathbf{F}_{DB}$, are still unknown. The other forces are the load $\mathbf{P}$, which is given, and the force $\mathbf{F}_{DA}$ exerted on the pin by member AD. As indicated previously, this force is equal and opposite to the force $\mathbf{F}_{AD}$ exerted by the same member on pin A. We can draw the force polygon corresponding to joint D, as shown in Fig. 6.8, and determine the forces $\mathbf{F}_{DC}$ and $\mathbf{F}_{DB}$ from that polygon. However, when more than three forces are involved, it is usually more convenient to solve the equations of equilibrium $\Sigma F_x = 0$ and $\Sigma F_y = 0$ for the two unknown forces. Since both of these forces are directed away from joint D, members DC and DB pull on the pin and are in tension.

Next, we consider joint C; its free-body diagram is shown in Fig. 6.8. Both $\mathbf{F}_{CD}$ and $\mathbf{F}_{CA}$ are known from the analysis of the preceding joints, so only $\mathbf{F}_{CB}$ is unknown. Since the equilibrium of each pin provides sufficient information to determine two unknowns, we can check our analysis at this joint. We draw the force triangle and determine the magnitude and sense of $\mathbf{F}_{CB}$. Since $\mathbf{F}_{CB}$ is directed toward joint C, member CB pushes on pin C and is in compression. The check is obtained by verifying that the force $\mathbf{F}_{CB}$ and member CB are parallel.

Finally, at joint B, we know all of the forces. Since the corresponding pin is in equilibrium, the force triangle must close, giving us an additional check of the analysis.

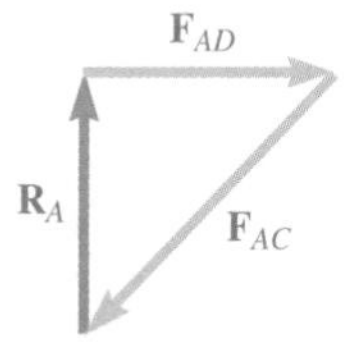

Note that the force polygons shown in Fig. 6.8 are not unique; we could replace each of them by an alternative configuration. For example, the force triangle corresponding to joint A could be drawn as shown in Fig. 6.9. We obtained the triangle actually shown in Fig. 6.8 by drawing the three forces $\mathbf{R}_A$, $\mathbf{F}_{AC}$, and $\mathbf{F}_{AD}$ in tip-to-tail fashion in the order in which we cross their lines of action when moving clockwise around joint A.

Fig. 6.9 Alternative force polygon for joint A in Fig. 6.8.

6.1C Joints Under Special Loading Conditions

Some geometric arrangements of members in a truss are particularly simple to analyze by observation. For example, Fig. 6.10*a* shows a joint connecting four members lying along two intersecting straight lines. The free-body diagram of Fig. 6.10*b* shows that pin A is subjected to two pairs of directly opposite forces. The corresponding force polygon, therefore, must be a parallelogram (Fig. 6.10*c*), and **the forces in opposite members must be equal**.

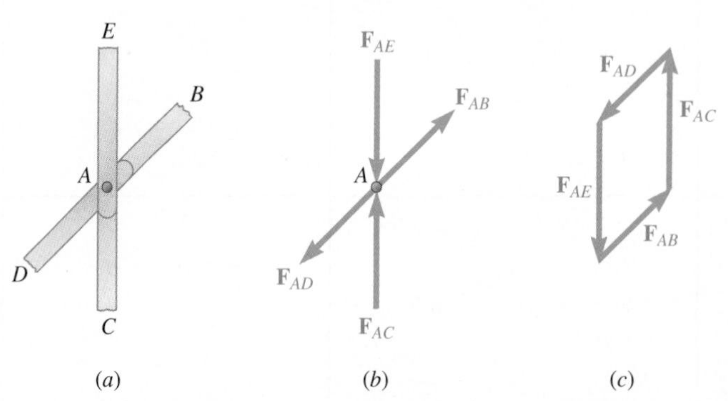

Fig. 6.10 (*a*) A joint A connecting four members of a truss in two straight lines; (*b*) free-body diagram of pin A; (*c*) force polygon (parallelogram) for pin A. Forces in opposite members are equal.

Consider next Fig. 6.11*a*, in which a joint connects three members and supports a load **P**. Two members lie along the same line, and load **P** acts along the third member. The free-body diagram of pin *A* and the corresponding force polygon are the same as in Fig. 6.10*b* and *c*, with $\mathbf{F}_{AE}$ replaced by load **P**. Thus, **the forces in the two opposite members must be equal, and the force in the other member must equal** P. Figure 6.11*b* shows a particular case of special interest. Since, in this case, no external load is applied to the joint, we have $P = 0$, and the force in member *AC* is zero. Member *AC* is said to be a **zero-force member**.

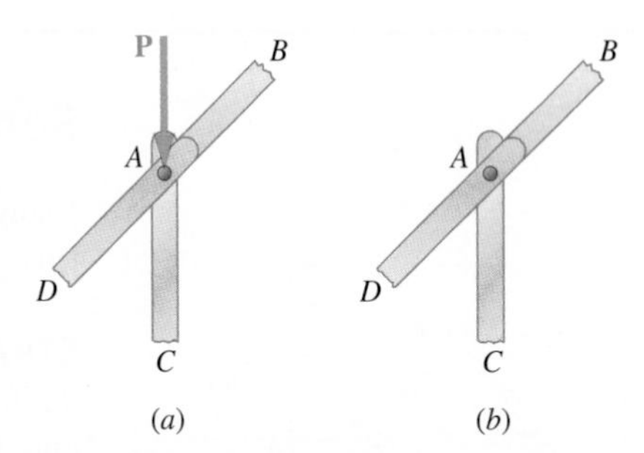

Fig. 6.11 (*a*) Joint *A* in a truss connects three members, two in a straight line and the third along the line of a load. Force in the third member equals the load. (*b*) If the load is zero, the third member is a zero-force member.

Now consider a joint connecting two members only. From Sec. 2.3A, we know that a particle acted upon by two forces is in equilibrium if the two forces have the same magnitude, same line of action, and opposite sense. In the case of the joint of Fig. 6.12*a*, which connects two members *AB* and *AD* lying along the same line, the forces in the two members must be equal for pin *A* to be in equilibrium. In the case of the joint of Fig. 6.12*b*, pin *A* cannot be in equilibrium unless the forces in both members are zero. Members connected as shown in Fig. 6.12*b*, therefore, must be **zero-force members**.

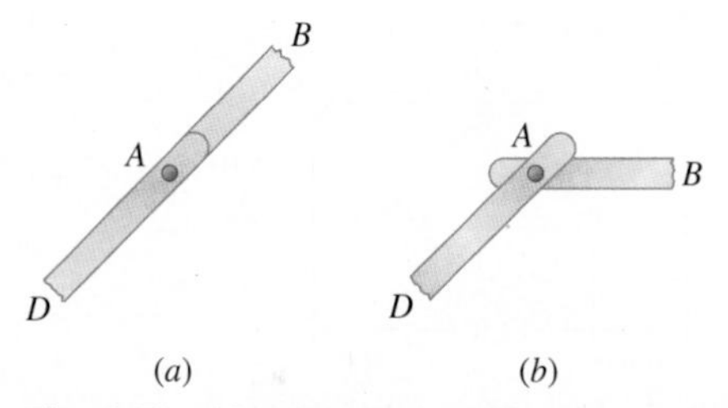

Fig. 6.12 (*a*) A joint in a truss connecting two members in a straight line. Forces in the members are equal. (*b*) If the two members are not in a straight line, they must be zero-force members.

Spotting joints that are under the special loading conditions just described will expedite the analysis of a truss. Consider, for example, a Howe truss loaded as shown in Fig. 6.13. We can recognize all of the members represented by green lines as zero-force members. Joint *C* connects three members, two of which lie in the same line, and is not subjected to any external load; member *BC* is thus a zero-force member. Applying the same reasoning to joint *K*, we find that member *JK* is also a zero-force member. But joint *J* is now in the same situation as joints *C* and *K*, so member *IJ* also must be a zero-force member. Examining joints *C, J,* and *K* also shows that the forces in members *AC* and *CE* are equal, that the forces in members *HJ* and *JL* are equal, and that the forces in members *IK* and *KL* are equal. Turning our attention to joint *I*, where the 20-kN load and member *HI* are collinear, we note that the force in member *HI* is 20 kN (tension) and that the forces in members *GI* and *IK* are equal. Hence, the forces in members *GI, IK,* and *KL* are equal.

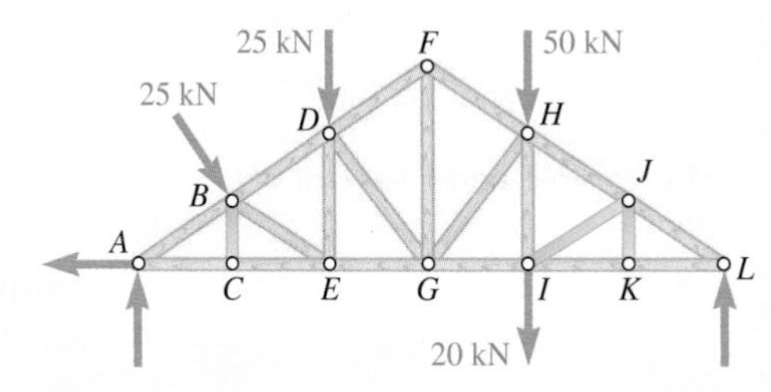

Fig. 6.13 An example of loading on a Howe truss; identifying special loading conditions.

Note that the conditions described here do not apply to joints *B* and *D* in Fig. 6.13, so it is wrong to assume that the force in member *DE* is 25 kN or that the forces in members *AB* and *BD* are equal. To determine the forces in these members and in all remaining members, you need to carry out the analysis of joints *A, B, D, E, F, G, H,* and *L* in the usual manner. Thus, until you have become thoroughly familiar with the conditions under which you can apply the rules described in this section, you should draw the free-body diagrams of all pins and write the corresponding equilibrium equations (or draw the corresponding force polygons) whether or not the joints being considered are under one of these special loading conditions.

A final remark concerning zero-force members: These members are not useless. For example, although the zero-force members of Fig. 6.13 do not carry any loads under the loading conditions shown, the same members would probably carry loads if the loading conditions were changed. Besides, even in the case considered, these members are needed to support the weight of the truss and to maintain the truss in the desired shape.

Sample Problem 6.1

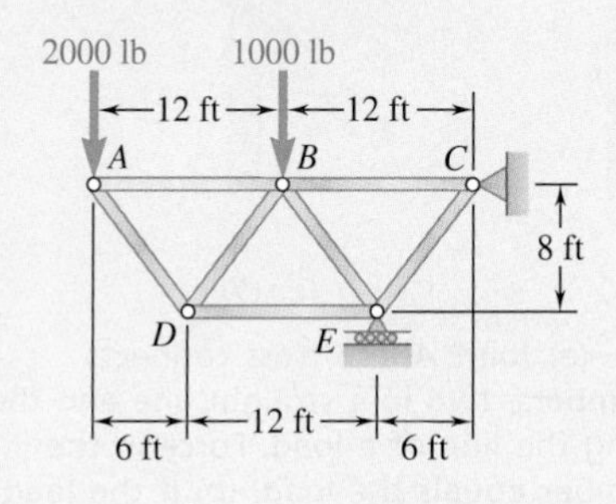

Using the method of joints, determine the force in each member of the truss shown.

STRATEGY: To use the method of joints, you start with an analysis of the free-body diagram of the entire truss. Then look for a joint connecting only two members as a starting point for the calculations. In this example, we start at joint A and proceed through joints D, B, E, and C, but you could also start at joint C and proceed through joints E, B, D, and A.

MODELING and ANALYSIS: You can combine these steps for each joint of the truss in turn. Draw a free-body diagram; draw a force polygon or write the equilibrium equations; and solve for the unknown forces.

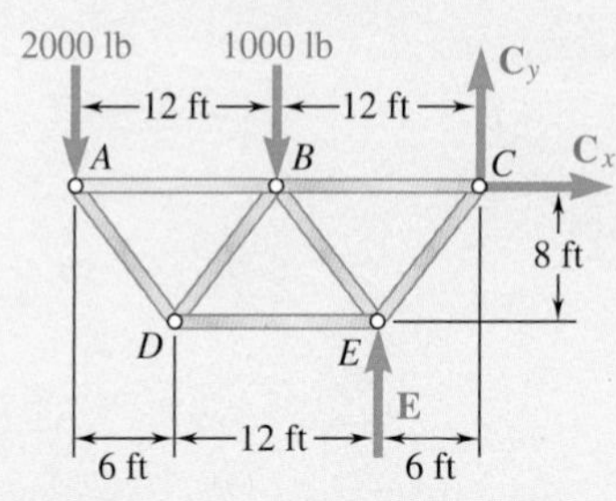

Fig. 1 Free-body diagram of entire truss.

Entire Truss. Draw a free-body diagram of the entire truss (Fig. 1); external forces acting on this free body are the applied loads and the reactions at C and E. Write the equilibrium equations, taking moments about C.

$$+\circlearrowleft \Sigma M_C = 0: \quad (2000 \text{ lb})(24 \text{ ft}) + (1000 \text{ lb})(12 \text{ ft}) - E(6 \text{ ft}) = 0$$

$$E = +10{,}000 \text{ lb} \qquad \mathbf{E} = 10{,}000 \text{ lb}\uparrow$$

$$\xrightarrow{+} \Sigma F_x = 0: \qquad \mathbf{C}_x = 0$$

$$+\uparrow \Sigma F_y = 0: \quad -2000 \text{ lb} - 1000 \text{ lb} + 10{,}000 \text{ lb} + C_y = 0$$

$$C_y = -7000 \text{ lb} \qquad \mathbf{C}_y = 7000 \text{ lb}\downarrow$$

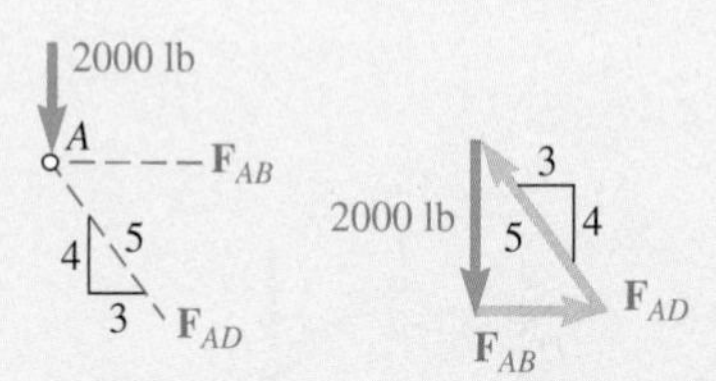

Fig. 2 Free-body diagram of joint A.

Joint A. This joint is subject to only two unknown forces: the forces exerted by AB and those by AD. Use a force triangle to determine $\mathbf{F}_{AB}$ and $\mathbf{F}_{AD}$ (Fig. 2). Note that member AB pulls on the joint so AB is in tension, and member AD pushes on the joint so AD is in compression. Obtain the magnitudes of the two forces from the proportion

$$\frac{2000 \text{ lb}}{4} = \frac{F_{AB}}{3} = \frac{F_{AD}}{5}$$

$$F_{AB} = 1500 \text{ lb } T \blacktriangleleft$$

$$F_{AD} = 2500 \text{ lb } C \blacktriangleleft$$

Joint D. Since you have already determined the force exerted by member AD, only two unknown forces are now involved at this joint. Again, use a force triangle to determine the unknown forces in members DB and DE (Fig. 3).

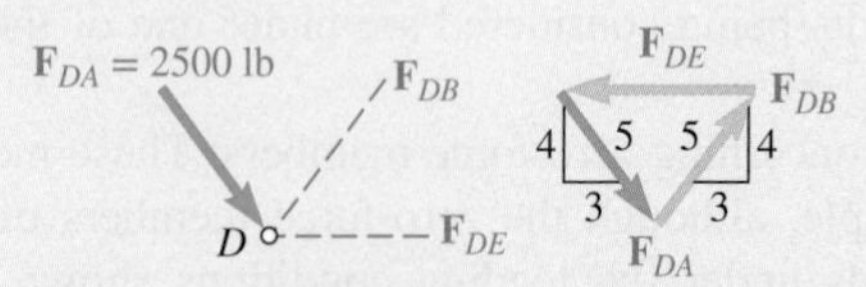

Fig. 3 Free-body diagram of joint D.

$$F_{DB} = F_{DA} \qquad\qquad F_{DB} = 2500 \text{ lb } T \blacktriangleleft$$
$$F_{DE} = 2(\tfrac{3}{5})F_{DA} \qquad\qquad F_{DE} = 3000 \text{ lb } C \blacktriangleleft$$

Joint *B*. Since more than three forces act at this joint (Fig. 4), determine the two unknown forces $\mathbf{F}_{BC}$ and $\mathbf{F}_{BE}$ by solving the equilibrium equations $\Sigma F_x = 0$ and $\Sigma F_y = 0$. Suppose you arbitrarily assume that both unknown forces act away from the joint, i.e., that the members are in tension. The positive value obtained for F_{BC} indicates that this assumption is correct; member BC is in tension. The negative value of F_{BE} indicates that the second assumption is wrong; member BE is in compression.

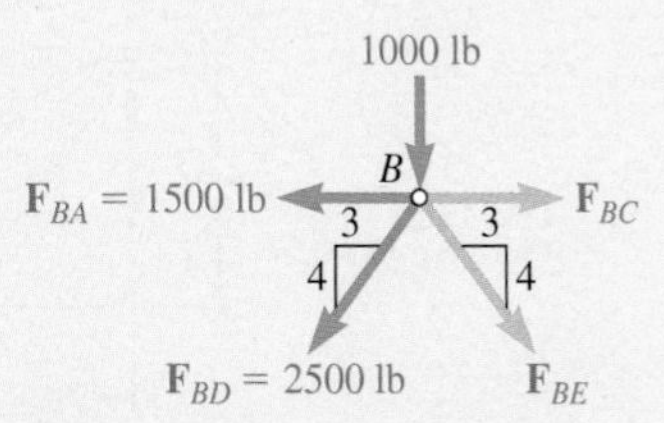

Fig. 4 Free-body diagram of joint *B*.

$$+\uparrow\Sigma F_y = 0: \quad -1000 - \tfrac{4}{5}(2500) - \tfrac{4}{5}F_{BE} = 0$$
$$F_{BE} = -3750 \text{ lb} \qquad\qquad F_{BE} = 3750 \text{ lb } C \blacktriangleleft$$

$$\overset{+}{\rightarrow}\Sigma F_x = 0: \quad F_{BC} - 1500 - \tfrac{3}{5}(2500) - \tfrac{3}{5}(3750) = 0$$
$$F_{BC} = +5250 \text{ lb} \qquad\qquad F_{BC} = 5250 \text{ lb } T \blacktriangleleft$$

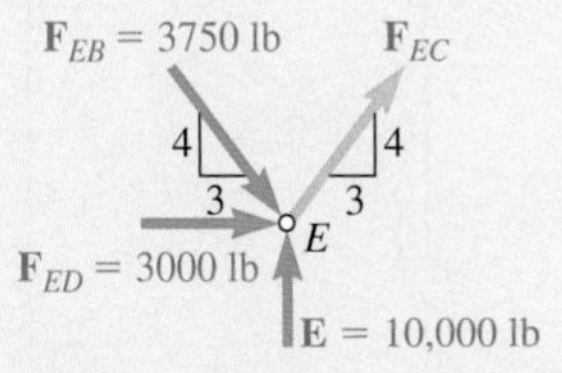

Fig. 5 Free-body diagram of joint *E*.

Joint *E*. Assume the unknown force $\mathbf{F}_{EC}$ acts away from the joint (Fig. 5). Summing x components, you obtain

$$\overset{+}{\rightarrow}\Sigma F_x = 0: \quad \tfrac{3}{5}F_{EC} + 3000 + \tfrac{3}{5}(3750) = 0$$
$$F_{EC} = -8750 \text{ lb} \qquad\qquad F_{EC} = 8750 \text{ lb } C \blacktriangleleft$$

Summing y components, you obtain a check of your computations:

$$+\uparrow\Sigma F_y = 10{,}000 - \tfrac{4}{5}(3750) - \tfrac{4}{5}(8750)$$
$$= 10{,}000 - 3000 - 7000 = 0 \qquad \text{(checks)}$$

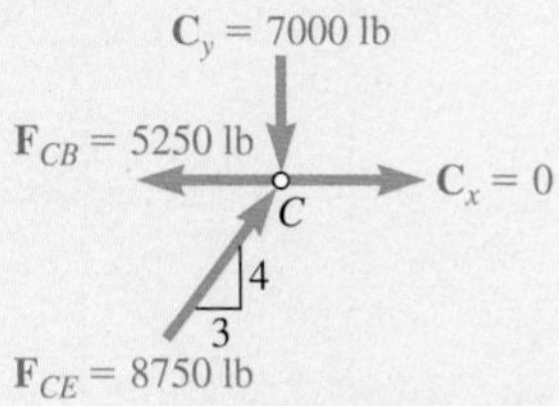

Fig. 6 Free-body diagram of joint *C*.

REFLECT and THINK: Using the computed values of $\mathbf{F}_{CB}$ and $\mathbf{F}_{CE}$, you can determine the reactions $\mathbf{C}_x$ and $\mathbf{C}_y$ by considering the equilibrium of Joint C (Fig. 6). Since these reactions have already been determined from the equilibrium of the entire truss, this provides two checks of your computations. You can also simply use the computed values of all forces acting on the joint (forces in members and reactions) and check that the joint is in equilibrium:

$$\overset{+}{\rightarrow}\Sigma F_x = -5250 + \tfrac{3}{5}(8750) = -5250 + 5250 = 0 \qquad \text{(checks)}$$
$$+\uparrow\Sigma F_y = -7000 + \tfrac{4}{5}(8750) = -7000 + 7000 = 0 \qquad \text{(checks)}$$

Problems

6.1 through *6.14* Using the method of joints, determine the force in each member of the truss shown. State whether each member is in tension or compression.

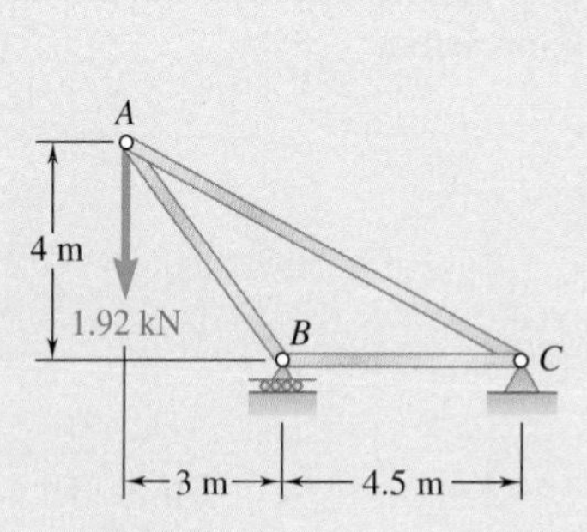

Fig. P6.1

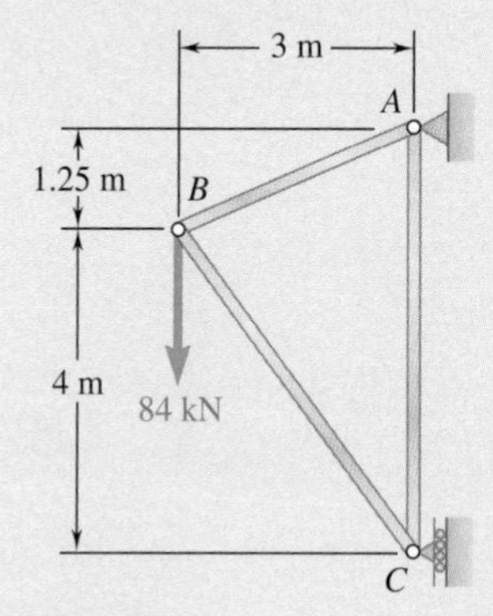

Fig. P6.2

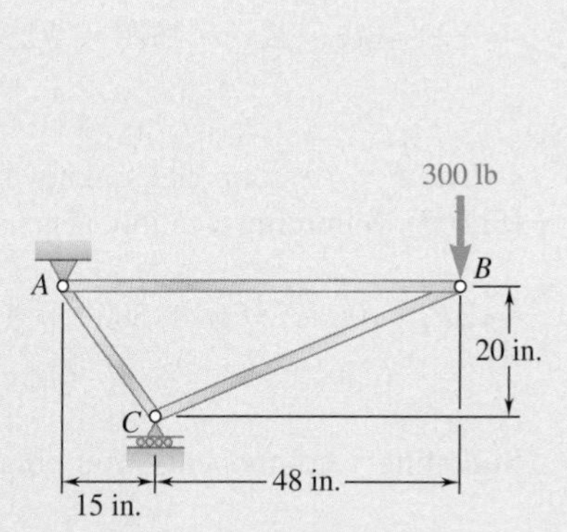

Fig. P6.3

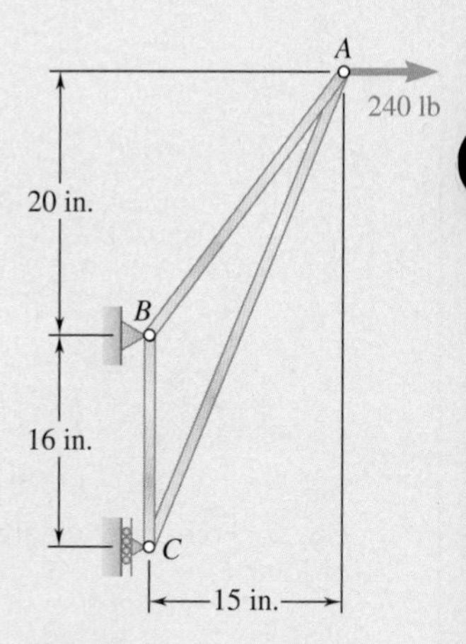

Fig. P6.4

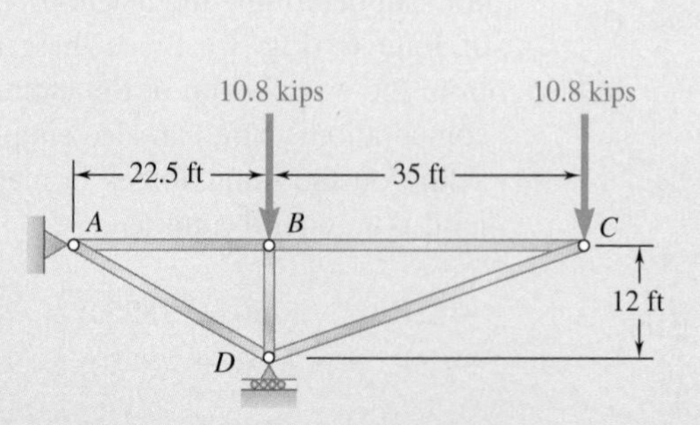

Fig. *P6.5*

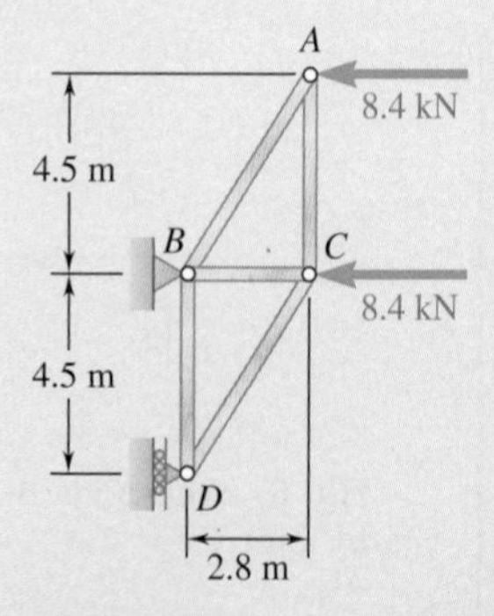

Fig. P6.6

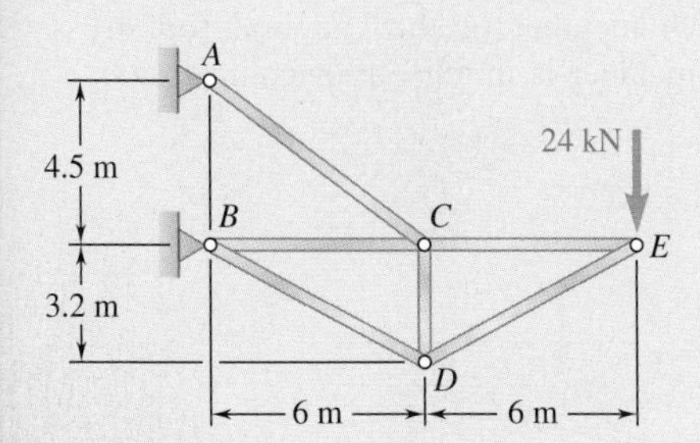

Fig. *P6.7*

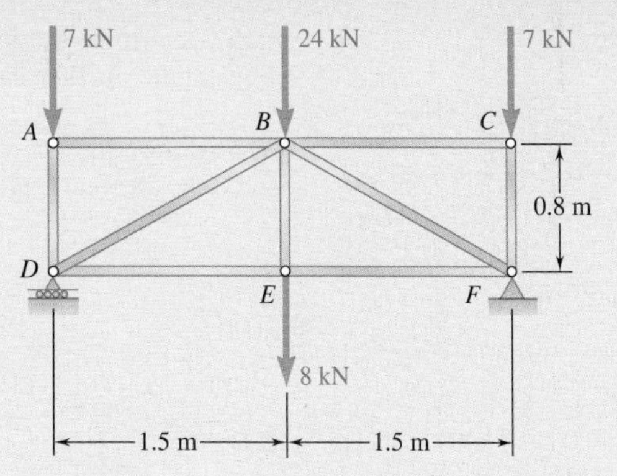

Fig. P6.8

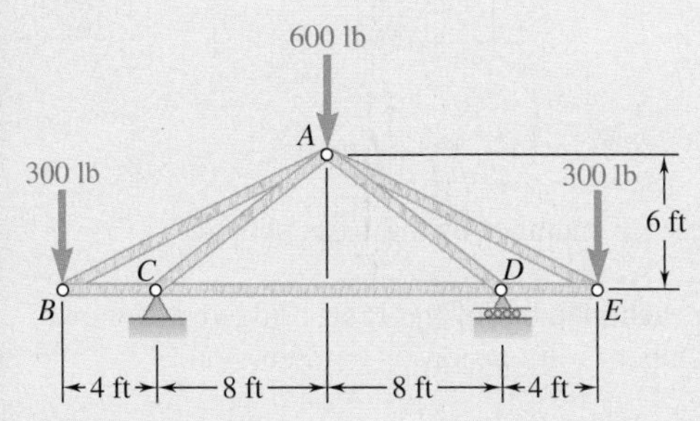

Fig. P6.9

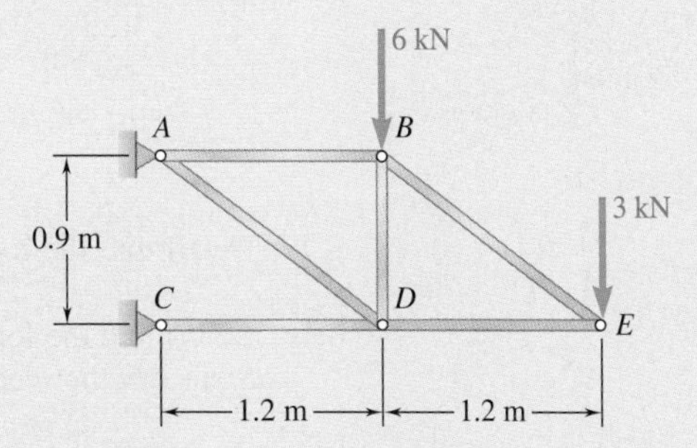

Fig. P6.10

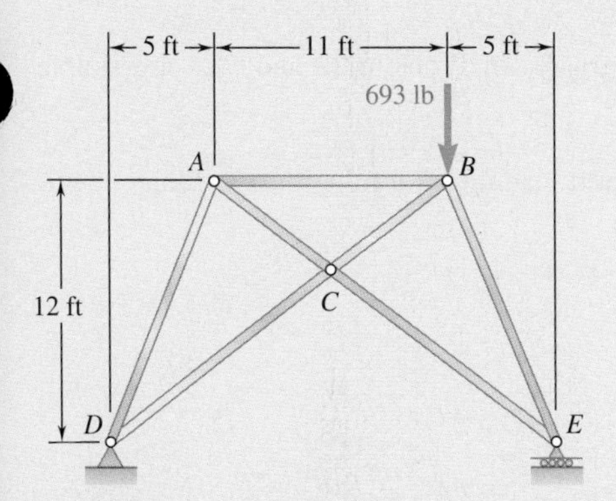

Fig. P6.11

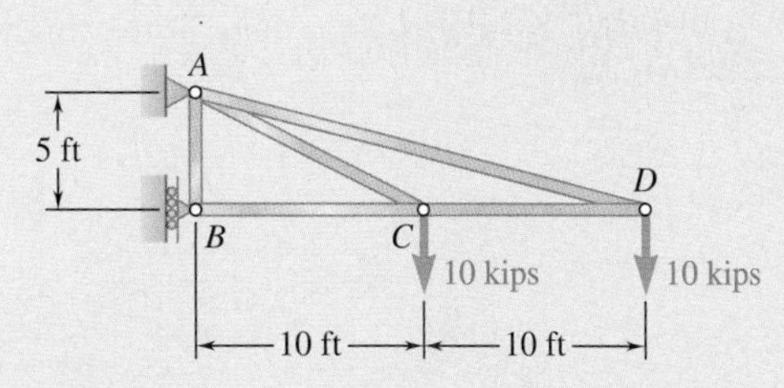

Fig. P6.12

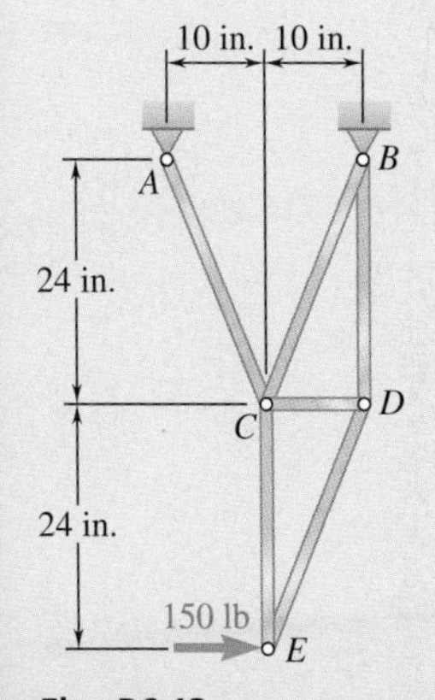

Fig. P6.13

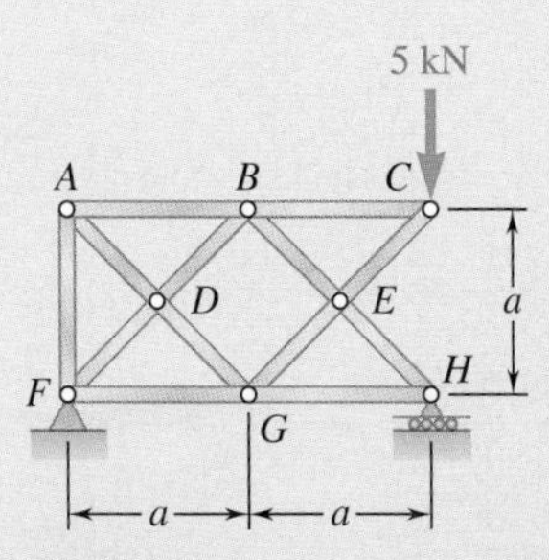

Fig. *P6.14*

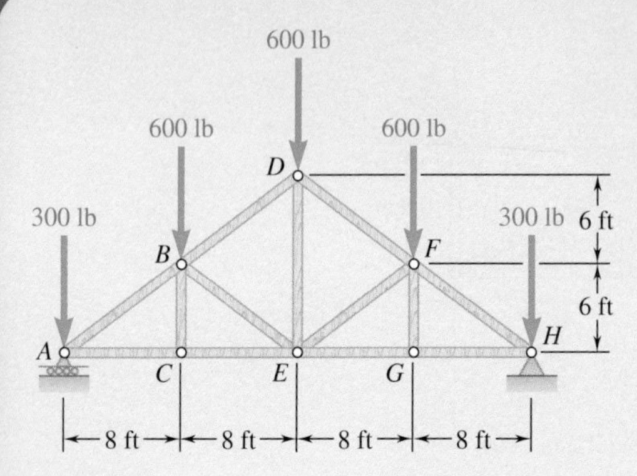

Fig. P6.15

6.15 Determine the force in each member of the Howe roof truss shown. State whether each member is in tension or compression.

6.16 Determine the force in each member of the Gambrel roof truss shown. State whether each member is in tension or compression.

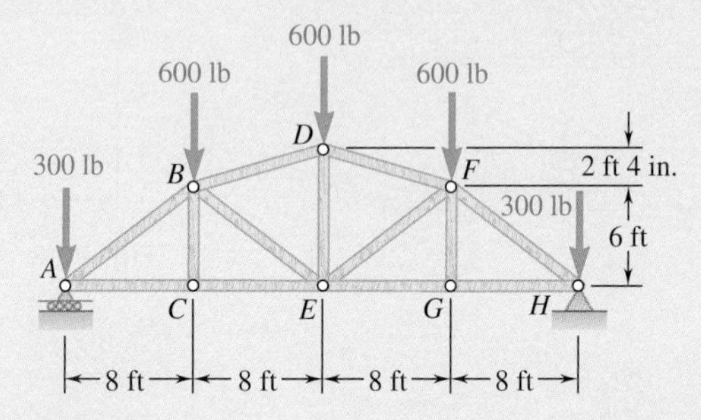

Fig. *P6.16*

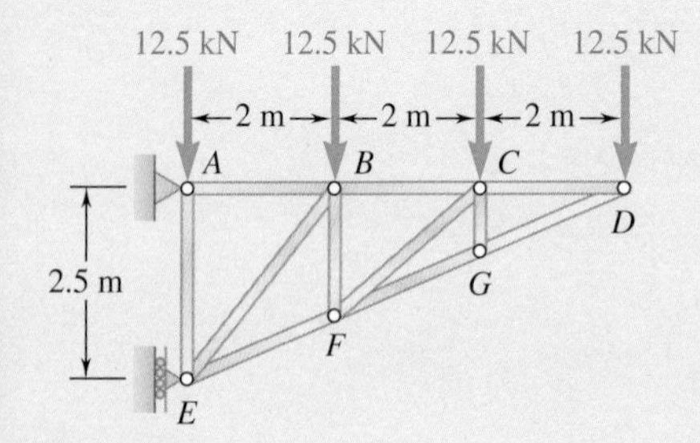

Fig. P6.17

6.17 Determine the force in each member of the truss shown.

6.18 Determine the force in each member of the Pratt bridge truss shown. State whether each member is in tension or compression.

6.19 Determine whether the trusses of Probs. 6.21 and 6.24 are simple trusses.

6.20 Determine whether the trusses of Probs. 6.22 and 6.23 are simple trusses.

6.21 through 6.24 For the given loading, determine the zero-force members in the truss shown.

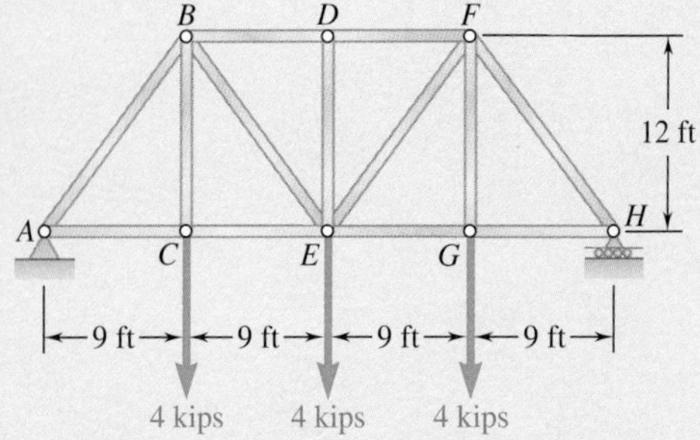

Fig. P6.18

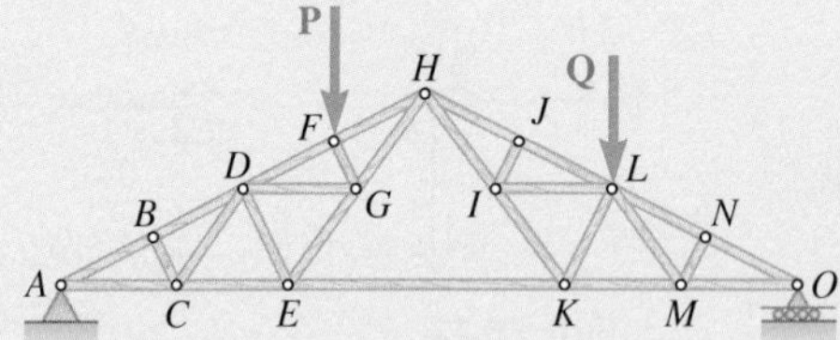

Fig. P6.21

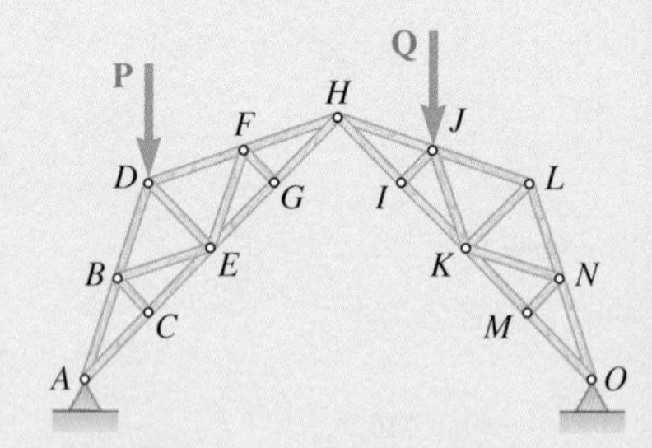

Fig. *P6.22*

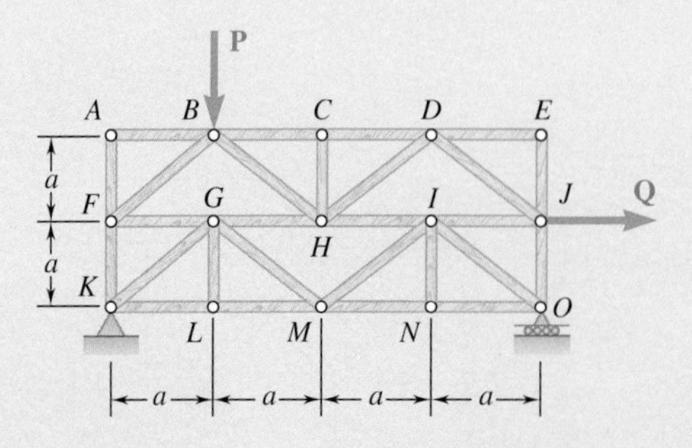

Fig. *P6.23*

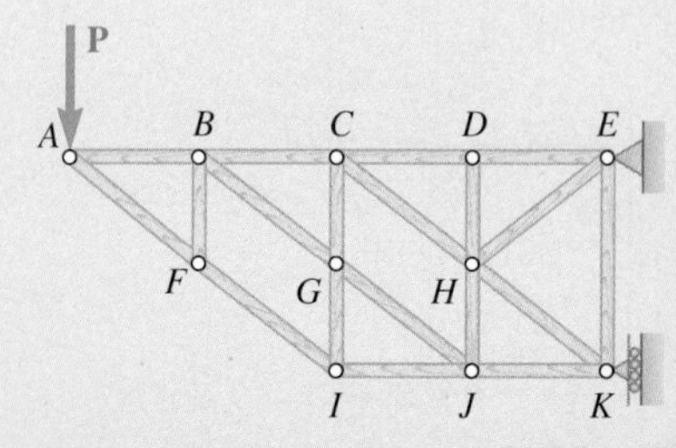

Fig. P6.24

6.2 OTHER TRUSS ANALYSES

The method of joints is most effective when we want to determine the forces in all the members of a truss. If, however, we need to determine the force in only one member or in a very few members, the method of sections is more efficient.

6.2A The Method of Sections

Assume, for example, that we want to determine the force in member *BD* of the truss shown in Fig. 6.14*a*. To do this, we must determine the force with which member *BD* acts on either joint *B* or joint *D*. If we were to use the method of joints, we would choose either joint *B* or joint *D* as a free body. However, we can also choose a larger portion of the truss that is composed of several joints and members, provided that the force we want to find is one of the external forces acting on that portion. If, in addition, we choose the portion of the truss as a free body where a total of only three unknown forces act upon it, we can obtain the desired force by solving the equations of equilibrium for this portion of the truss. In practice, we isolate a portion of the truss by *passing a section* through three members of the truss, one of which is the desired member. That is, we draw a line that divides the truss into two completely separate parts but does not intersect more than three members. We can then use as a free body either of the two portions of the truss obtained after the intersected members have been removed.†

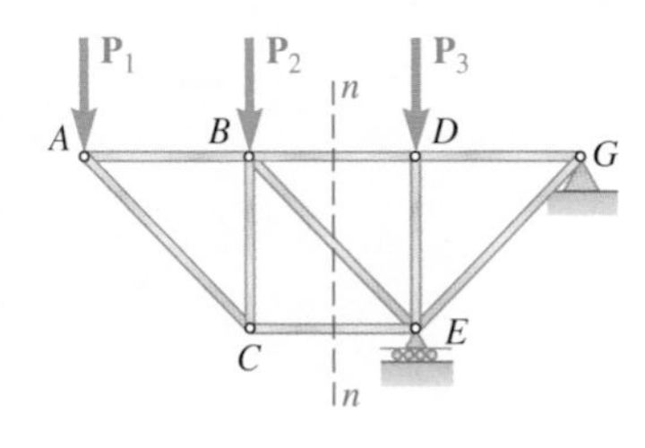

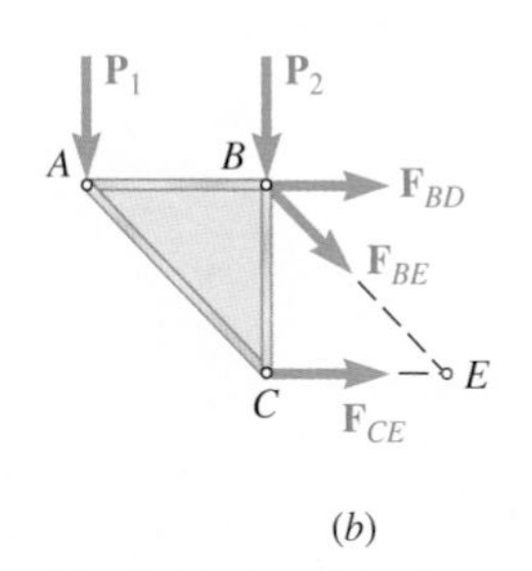

Fig. 6.14 (*a*) We can pass a section *nn* through the truss, dividing the three members *BD*, *BE*, and *CE*. (*b*) Free-body diagram of portion *ABC* of the truss. We assume that members *BD*, *BE*, and *CE* are in tension.

In Fig. 6.14*a*, we have passed the section *nn* through members *BD*, *BE*, and *CE*, and we have chosen the portion *ABC* of the truss as the free body (Fig. 6.14*b*). The forces acting on this free body are the loads $\mathbf{P}_1$ and $\mathbf{P}_2$ at points *A* and *B* and the three unknown forces $\mathbf{F}_{BD}$, $\mathbf{F}_{BE}$, and $\mathbf{F}_{CE}$. Since we do not know whether the members removed are in tension or compression, we have arbitrarily drawn the three forces away from the free body as if the members are in tension.

We use the fact that the rigid body *ABC* is in equilibrium to write three equations that we can solve for the three unknown forces. If we want to determine only force $\mathbf{F}_{BD}$, say, we need write only one equation, provided that the equation does not contain the other unknowns. Thus, the equation $\Sigma M_E = 0$ yields the value of the magnitude F_{BD} (Fig. 6.14*b*). A positive sign in the answer will indicate that our original assumption regarding the sense of $\mathbf{F}_{BD}$ was correct and that member *BD* is in tension; a negative sign will indicate that our assumption was incorrect and that *BD* is in compression.

On the other hand, if we want to determine only force $\mathbf{F}_{CE}$, we need to write an equation that does not involve $\mathbf{F}_{BD}$ or $\mathbf{F}_{BE}$; the appropriate equation is $\Sigma M_B = 0$. Again, a positive sign for the magnitude F_{CE} of the desired force indicates a correct assumption, that is, tension; and a negative sign indicates an incorrect assumption, that is, compression.

If we want to determine only force $\mathbf{F}_{BE}$, the appropriate equation is $\Sigma F_y = 0$. Whether the member is in tension or compression is again determined from the sign of the answer.

†In the analysis of some trusses, we can pass sections through more than three members, provided we can write equilibrium equations involving only one unknown that we can use to determine the forces in one, or possibly two, of the intersected members. See Probs. 6.41 through 6.43.

If we determine the force in only one member, no independent check of the computation is available. However, if we calculate all of the unknown forces acting on the free body, we can check the computations by writing an additional equation. For instance, if we determine $\mathbf{F}_{BD}$, $\mathbf{F}_{BE}$, and $\mathbf{F}_{CE}$ as indicated previously, we can check the work by verifying that $\Sigma F_x = 0$.

6.2B Trusses Made of Several Simple Trusses

Consider two simple trusses *ABC* and *DEF*. If we connect them by three bars *BD, BE,* and *CE* as shown in Fig. 6.15*a*, together they form a rigid truss *ABDF*. We can also combine trusses *ABC* and *DEF* into a single rigid truss by joining joints *B* and *D* at a single joint *B* and connecting joints *C* and *E* by a bar *CE* (Fig. 6.15*b*). This is known as a *Fink truss.* The trusses of Fig. 6.15*a* and *b* are *not* simple trusses; you cannot construct them from a triangular truss by adding successive pairs of members as described in Sec. 6.1A. They are rigid trusses, however, as you can check by comparing the systems of connections used to hold the simple trusses *ABC* and *DEF* together (three bars in Fig. 6.15*a*, one pin and one bar in Fig. 6.15*b*) with the systems of supports discussed in Sec. 4.1. Trusses made of several simple trusses rigidly connected are known as **compound trusses**.

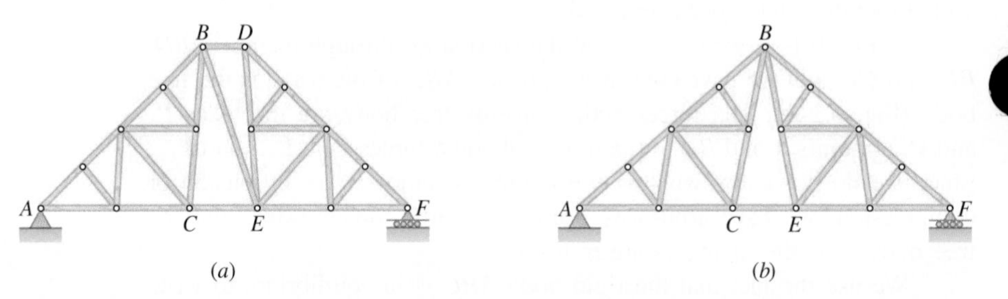

Fig. 6.15 Compound trusses. (*a*) Two simple trusses *ABC* and *DEF* connected by three bars. (*b*) Two simple trusses *ABC* and *DEF* connected by one joint and one bar (a Fink truss).

In a compound truss, the number of members m and the number of joints n are still related by the formula $m = 2n - 3$. You can verify this by observing that if a compound truss is supported by a frictionless pin and a roller (involving three unknown reactions), the total number of unknowns is $m + 3$, and this number must be equal to the number $2n$ of equations obtained by expressing that the n pins are in equilibrium. It follows that $m = 2n - 3$.

Compound trusses supported by a pin and a roller or by an equivalent system of supports are *statically determinate, rigid,* and *completely constrained.* This means that we can determine all of the unknown reactions and the forces in all of the members by using the methods of statics, and the truss will neither collapse nor move. However, the only way to determine all of the forces in the members using the method of joints requires solving a large number of simultaneous equations. In the case of the compound truss of Fig. 6.15*a*, for example, it is more efficient to pass a section through members BD, BE, and CE to determine the forces in these members.

Suppose, now, that the simple trusses ABC and DEF are connected by *four* bars; BD, BE, CD, and CE (Fig. 6.16). The number of members m is now larger than $2n - 3$. This truss is said to be **overrigid**, and one of the four members BD, BE, CD, or CE is **redundant**. If the truss is supported by a pin at A and a roller at F, the total number of unknowns is $m + 3$. Since $m > 2n - 3$, the number $m + 3$ of unknowns is now larger than the number $2n$ of available independent equations; the truss is *statically indeterminate*.

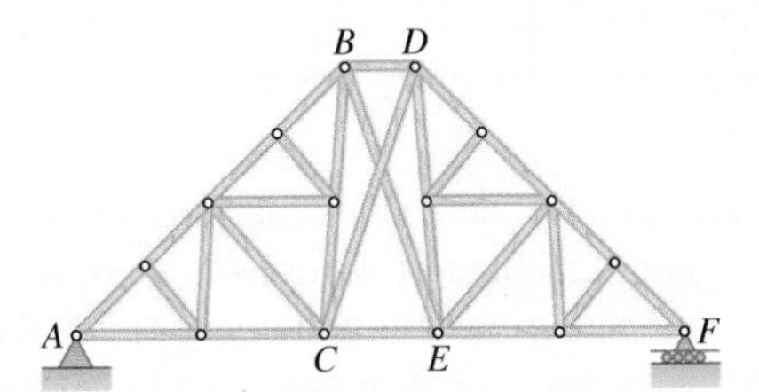

Fig. 6.16 A statically indeterminate, overrigid compound truss, due to a redundant member.

Finally, let us assume that the two simple trusses ABC and DEF are joined by a single pin, as shown in Fig. 6.17*a*. The number of members, m, is now smaller than $2n - 3$. If the truss is supported by a pin at A and a roller at F, the total number of unknowns is $m + 3$. Since $m < 2n - 3$, the number $m + 3$ of unknowns is now smaller than the number $2n$ of equilibrium equations that need to be satisfied. This truss is **nonrigid** and will collapse under its own weight. However, if two pins are used to support it, the truss becomes *rigid* and will not collapse (Fig. 6.17*b*). Note that the total number of unknowns is now $m + 4$ and is equal to the number $2n$ of equations.

More generally, if the reactions at the supports involve r unknowns, the condition for a compound truss to be statically determinate, rigid, and completely constrained is $m + r = 2n$. However, although this condition is necessary, it is not sufficient for the equilibrium of a structure that ceases to be rigid when detached from its supports (see Sec. 6.3B).

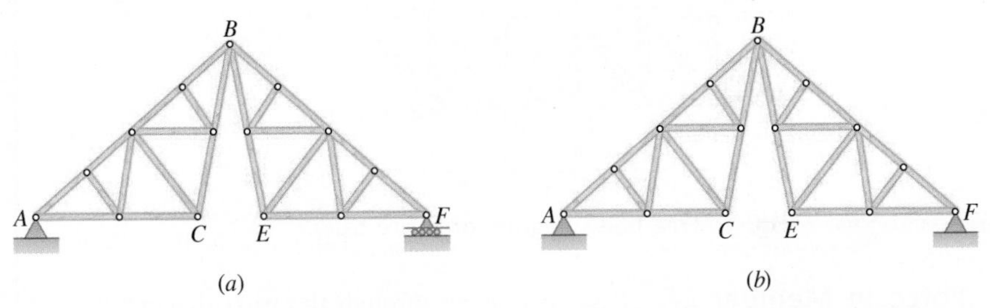

Fig. 6.17 Two simple trusses joined by a pin. (*a*) Supported by a pin and a roller, the truss will collapse under its own weight. (*b*) Supported by two pins, the truss becomes rigid and does not collapse.

Sample Problem 6.2

Determine the forces in members EF and GI of the truss shown.

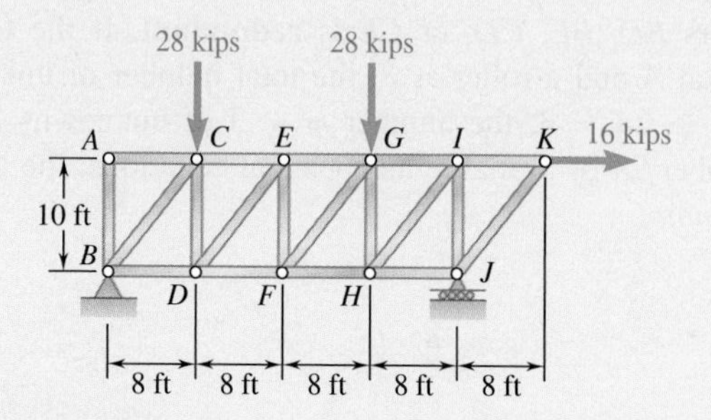

STRATEGY: You are asked to determine the forces in only two of the members in this truss, so the method of sections is more appropriate than the method of joints. You can use a free-body diagram of the entire truss to help determine the reactions, and then pass sections through the truss to isolate parts of it for calculating the desired forces.

MODELING and ANALYSIS: You can go through the steps that follow for the determination of the support reactions, and then for the analysis of portions of the truss.

Free-Body: Entire Truss. Draw a free-body diagram of the entire truss. External forces acting on this free body consist of the applied loads and the reactions at B and J (Fig. 1). Write and solve the following equilibrium equations.

$+\circlearrowleft \Sigma M_B = 0$:

$$-(28 \text{ kips})(8 \text{ ft}) - (28 \text{ kips})(24 \text{ ft}) - (16 \text{ kips})(10 \text{ ft}) + J(32 \text{ ft}) = 0$$

$$J = +33 \text{ kips} \qquad \mathbf{J} = 33 \text{ kips}\uparrow$$

$\xrightarrow{+} \Sigma F_x = 0$: $\quad B_x + 16 \text{ kips} = 0$

$$B_x = -16 \text{ kips} \qquad \mathbf{B}_x = 16 \text{ kips}\leftarrow$$

$+\circlearrowleft \Sigma M_J = 0$:

$$(28 \text{ kips})(24 \text{ ft}) + (28 \text{ kips})(8 \text{ ft}) - (16 \text{ kips})(10 \text{ ft}) - B_y(32 \text{ ft}) = 0$$

$$B_y = +23 \text{ kips} \qquad \mathbf{B}_y = 23 \text{ kips}\uparrow$$

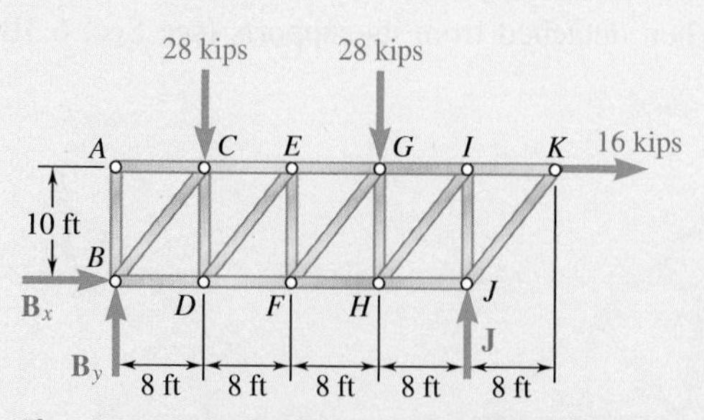

Fig. 1 Free-body diagram of entire truss.

Force in Member *EF*. Pass section nn through the truss diagonally so that it intersects member EF and only two additional members (Fig. 2).

Remove the intersected members and choose the left-hand portion of the truss as a free body (Fig. 3). Three unknowns are involved; to eliminate the two horizontal forces, we write

$$+\uparrow\Sigma F_y = 0: \quad +23 \text{ kips} - 28 \text{ kips} - F_{EF} = 0$$
$$F_{EF} = -5 \text{ kips}$$

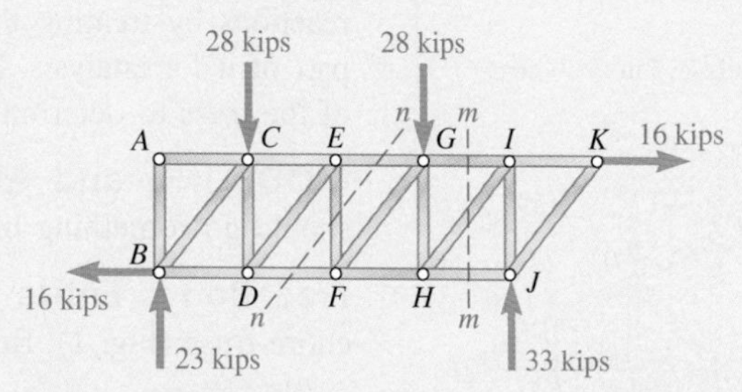

Fig. 2 Sections *nn* and *mm* that will be used to analyze members *EF* and *GI*.

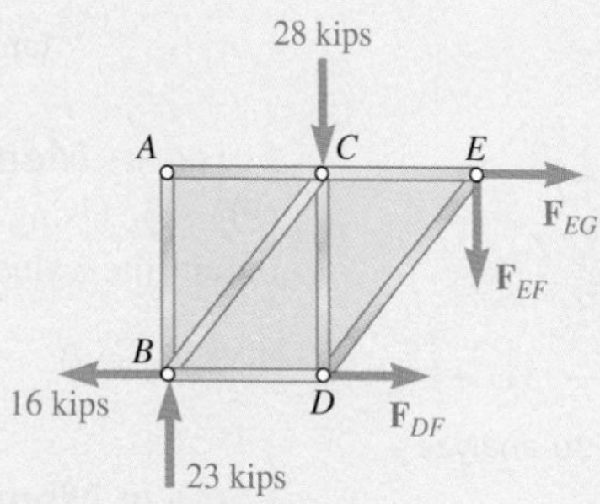

Fig. 3 Free-body diagram to analyze member *EF*.

The sense of $\mathbf{F}_{EF}$ was chosen assuming member *EF* to be in tension; the negative sign indicates that the member is in compression.

$$F_{EF} = 5 \text{ kips } C \blacktriangleleft$$

Force in Member *GI*. Pass section *mm* through the truss vertically so that it intersects member *GI* and only two additional members (Fig. 2). Remove the intersected members and choose the right-hand portion of the truss as a free body (Fig. 4). Again, three unknown forces are involved; to eliminate the two forces passing through point *H*, sum the moments about that point.

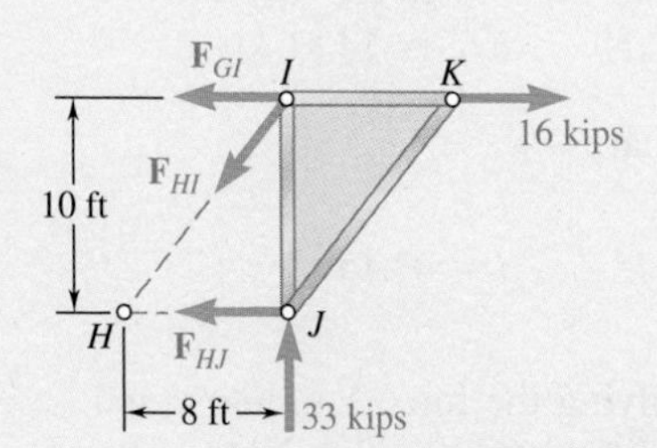

Fig. 4 Free-body diagram to analyze member *GI*.

$$+\circlearrowleft\Sigma M_H = 0: \quad (33 \text{ kips})(8 \text{ ft}) - (16 \text{ kips})(10 \text{ ft}) + F_{GI}(10 \text{ ft}) = 0$$
$$F_{GI} = -10.4 \text{ kips} \qquad F_{GI} = 10.4 \text{ kips } C \blacktriangleleft$$

REFLECT and THINK: Note that a section passed through a truss does not have to be vertical or horizontal; it can be diagonal as well. Choose the orientation that cuts through no more than three members of unknown force and also gives you the simplest part of the truss for which you can write equilibrium equations and determine the unknowns.

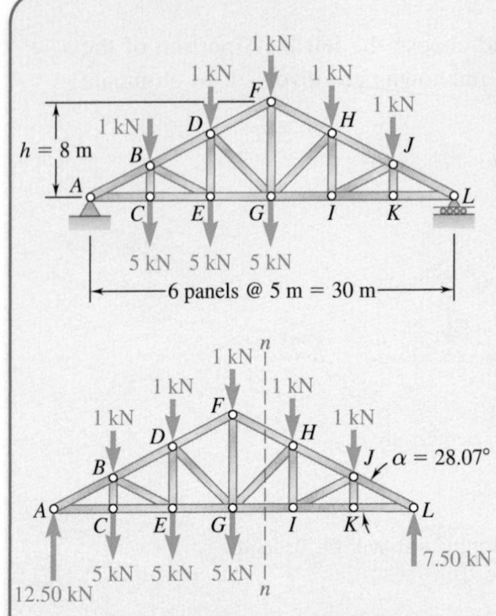

Fig. 1 Free-body diagram of entire truss.

Sample Problem 6.3

Determine the forces in members *FH, GH,* and *GI* of the roof truss shown.

STRATEGY: You are asked to determine the forces in only three members of the truss, so use the method of sections. Determine the reactions by treating the entire truss as a free body and then isolate part of it for analysis. In this case, you can use the same smaller part of the truss to determine all three desired forces.

MODELING and ANALYSIS: Your reasoning and computation should go something like the sequence given here.

Free Body: Entire Truss. From the free-body diagram of the entire truss (Fig. 1), find the reactions at *A* and *L*:

$$\mathbf{A} = 12.50 \text{ kN}\uparrow \qquad \mathbf{L} = 7.50 \text{ kN}\uparrow$$

Note that

$$\tan\alpha = \frac{FG}{GL} = \frac{8\text{ m}}{15\text{ m}} = 0.5333 \qquad \alpha = 28.07°$$

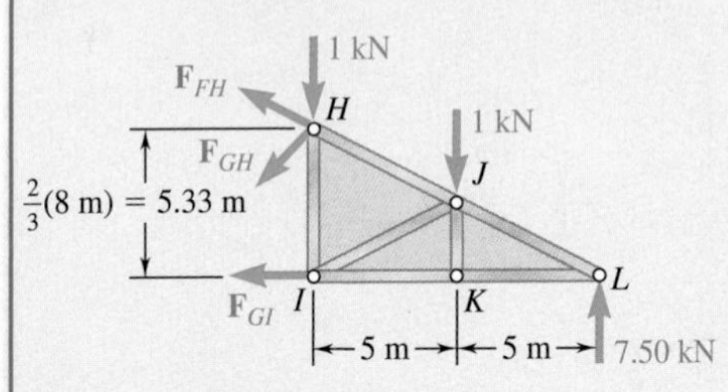

Fig. 2 Free-body diagram to analyze member *GI*.

Force in Member *GI*. Pass section *nn* vertically through the truss (Fig. 1). Using the portion *HLI* of the truss as a free body (Fig. 2), obtain the value of F_{GI}:

$$+\circlearrowleft\Sigma M_H = 0: \qquad (7.50\text{ kN})(10\text{ m}) - (1\text{ kN})(5\text{ m}) - F_{GI}(5.33\text{ m}) = 0$$

$$F_{GI} = +13.13\text{ kN} \qquad F_{GI} = 13.13\text{ kN }T \blacktriangleleft$$

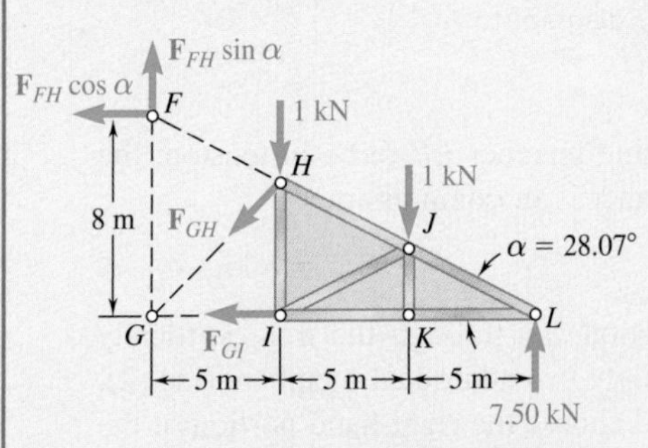

Fig. 3 Simplifying the analysis of member *FH* by first sliding its force to point *F*.

Force in Member *FH*. Determine the value of F_{FH} from the equation $\Sigma M_G = 0$. To do this, move $\mathbf{F}_{FH}$ along its line of action until it acts at point *F*, where you can resolve it into its *x* and *y* components (Fig. 3). The moment of $\mathbf{F}_{FH}$ with respect to point *G* is now $(F_{FH}\cos\alpha)(8\text{ m})$.

$$+\circlearrowleft\Sigma M_G = 0:$$

$$(7.50\text{ kN})(15\text{ m}) - (1\text{ kN})(10\text{ m}) - (1\text{ kN})(5\text{ m}) + (F_{FH}\cos\alpha)(8\text{ m}) = 0$$

$$F_{FH} = -13.81\text{ kN} \qquad F_{FH} = 13.81\text{ kN }C \blacktriangleleft$$

Force in Member *GH*. First note that

$$\tan\beta = \frac{GI}{HI} = \frac{5\text{ m}}{\frac{2}{3}(8\text{ m})} = 0.9375 \qquad \beta = 43.15°$$

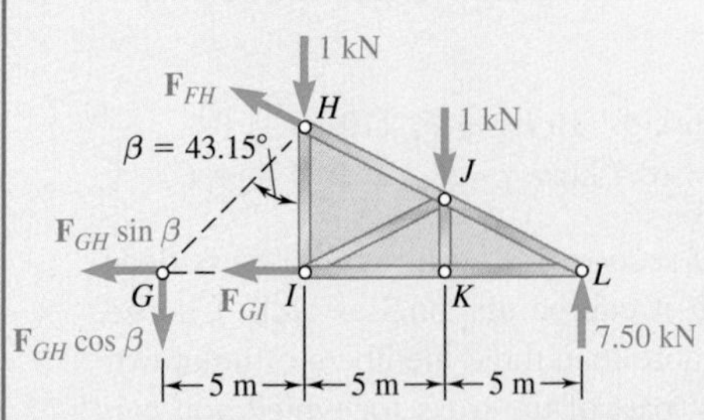

Fig. 4 Simplifying the analysis of member *GH* by first sliding its force to point *G*.

Then determine the value of F_{GH} by resolving the force $\mathbf{F}_{GH}$ into *x* and *y* components at point *G* (Fig. 4) and solving the equation $\Sigma M_L = 0$.

$$+\circlearrowleft\Sigma M_L = 0: \qquad (1\text{ kN})(10\text{ m}) + (1\text{ kN})(5\text{ m}) + (F_{GH}\cos\beta)(15\text{ m}) = 0$$

$$F_{GH} = -1.371\text{ kN} \qquad F_{GH} = 1.371\text{ kN }C \blacktriangleleft$$

REFLECT and THINK: Sometimes you should resolve a force into components to include it in the equilibrium equations. By first sliding this force along its line of action to a more strategic point, you might eliminate one of its components from a moment equilibrium equation.

Problems

6.25 Determine the force in members BD and CD of the truss shown.

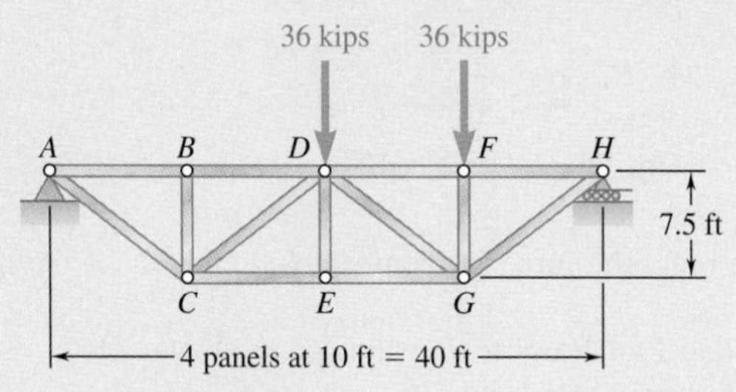

Fig. P6.25 and P6.26

6.26 Determine the force in members DF and DG of the truss shown.

6.27 Determine the force in members BD and DE of the truss shown.

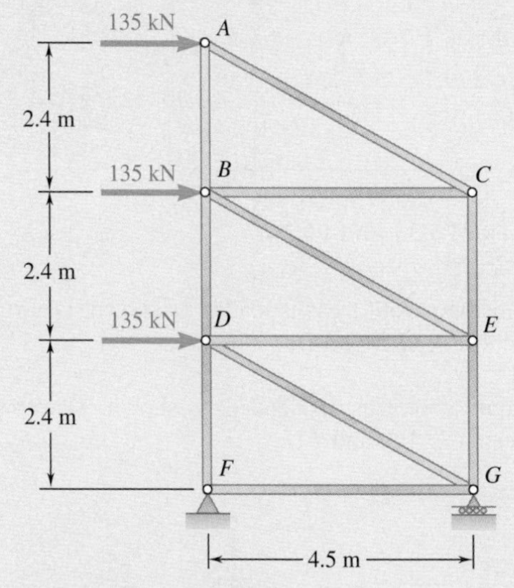

Fig. P6.27 and P6.28

6.28 Determine the force in members DG and EG of the truss shown.

6.29 Determine the force in members DE and DF of the truss shown when $P = 20$ kips.

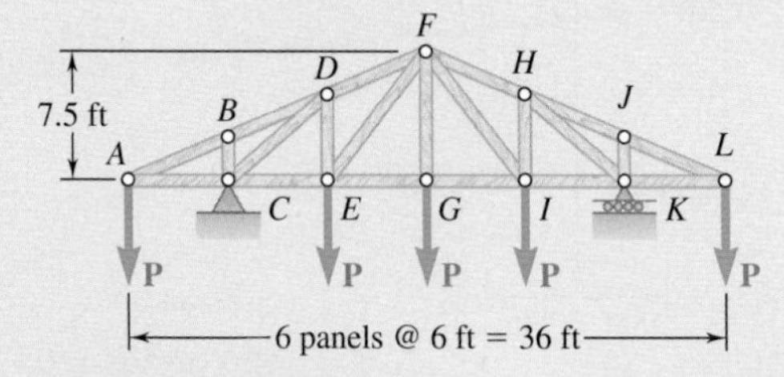

Fig. P6.29 and *P6.30*

6.30 Determine the force in members EG and EF of the truss shown when $P = 20$ kips.

6.31 Determine the force in members DF and DE of the truss shown.

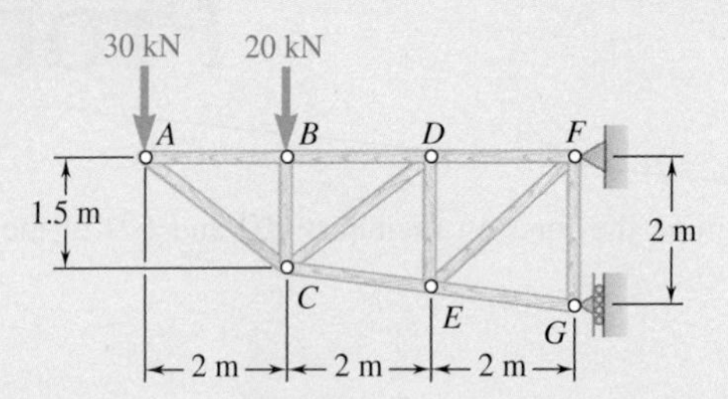

Fig. P6.31 and *P6.32*

6.32 Determine the force in members CD and CE of the truss shown.

6.33 A monosloped roof truss is loaded as shown. Determine the force in members CE, DE, and DF.

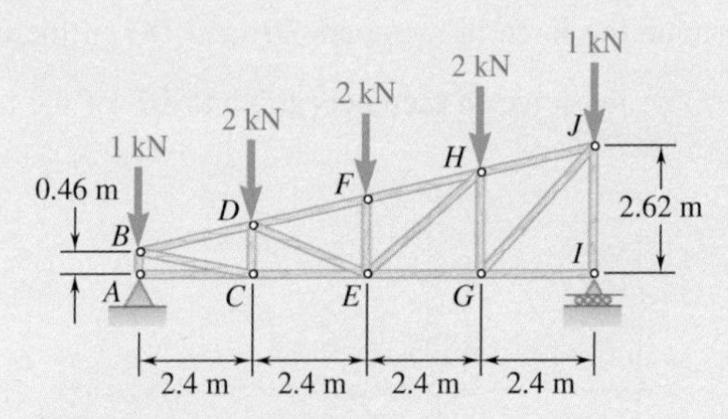

Fig. P6.33 and P6.34

6.34 A monosloped roof truss is loaded as shown. Determine the force in members EG, GH, and HJ.

6.35 A stadium roof truss is loaded as shown. Determine the force in members AB, AG, and FG.

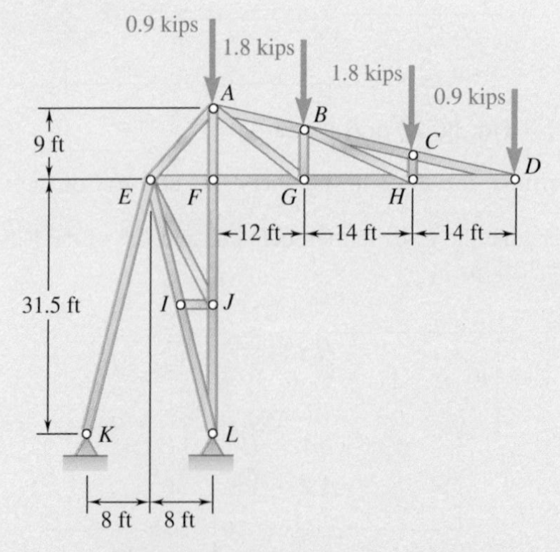

Fig. P6.35 and *P6.36*

6.36 A stadium roof truss is loaded as shown. Determine the force in members AE, EF, and FJ.

6.37 Determine the force in members *DF*, *EF*, and *EG* of the truss shown.

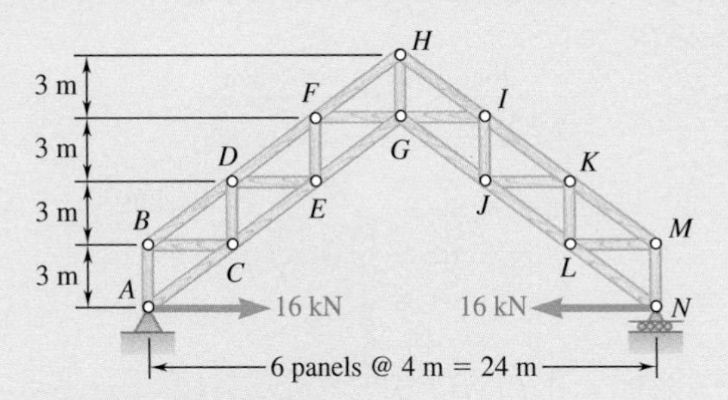

Fig. P6.37 and *P6.38*

6.38 Determine the force in members *GI*, *GJ*, and *HI* of the truss shown.

6.39 Determine the force in members *AD*, *CD*, and *CE* of the truss shown.

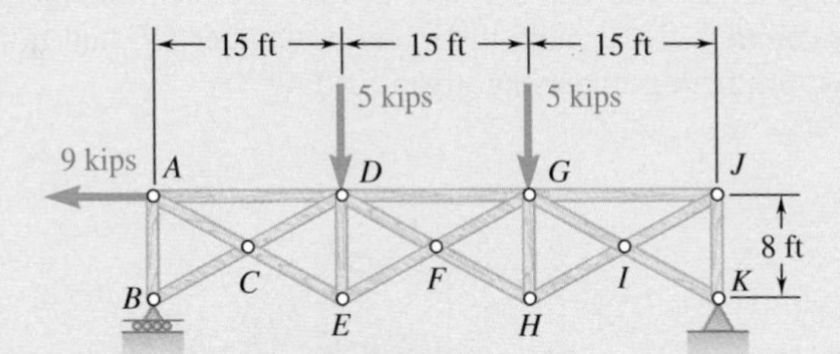

Fig. P6.39 and P6.40

6.40 Determine the force in members *DG*, *FG*, and *FH* of the truss shown.

6.41 Determine the force in members *DG* and *FI* of the truss shown. (*Hint:* Use section *aa*.)

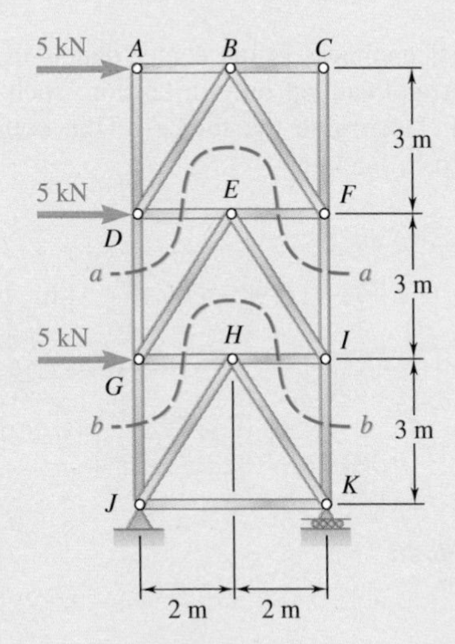

Fig. P6.41 and P6.42

6.42 Determine the force in members *GJ* and *IK* of the truss shown. (*Hint:* Use section *bb*.)

6.43 Determine the force in members AF and EJ of the truss shown when $P = Q = 1.2$ kN. (*Hint:* Use section *aa*.)

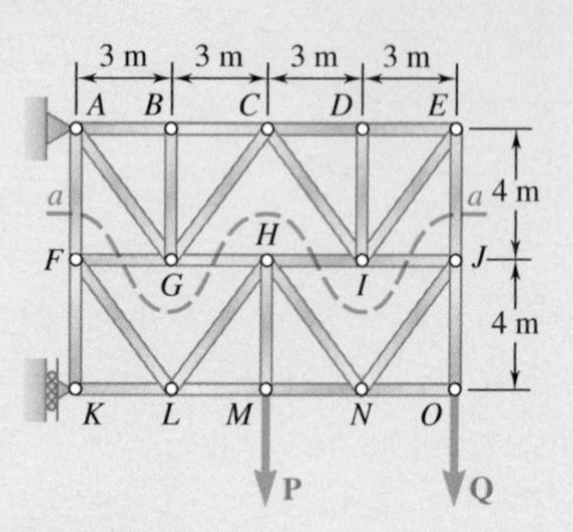

Fig. *P6.43*

6.44 The diagonal members in the center panels of the truss shown are very slender and can act only in tension; such members are known as *counters*. Determine the force in member DE and in the counters that are acting under the given loading.

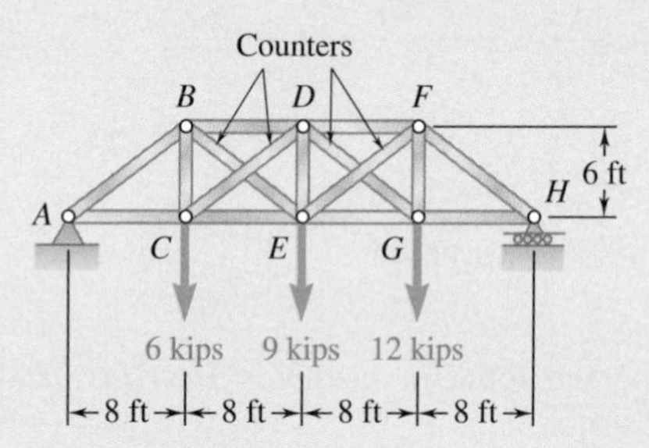

Fig. P6.44

6.45 Solve Prob. 6.44 assuming that the 9-kip load has been removed.

6.46 The diagonal members in the center panels of the truss shown are very slender and can act only in tension; such members are known as *counters*. Determine the forces in the counters that are acting under the given loading.

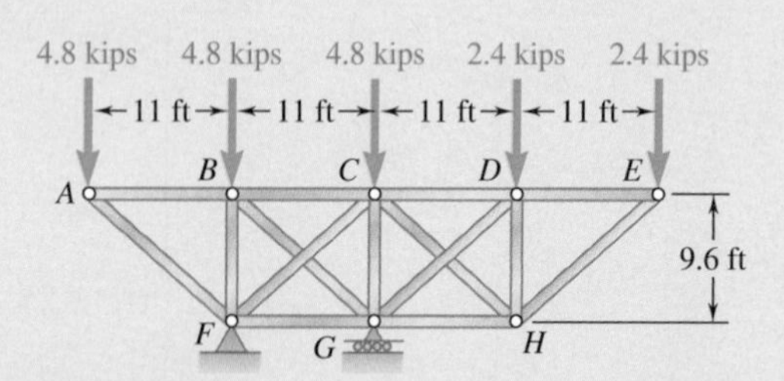

Fig. *P6.46*

6.47 and 6.48 Classify each of the structures shown as completely, partially, or improperly constrained; if completely constrained, further classify as determinate or indeterminate. (All members can act both in tension and in compression.)

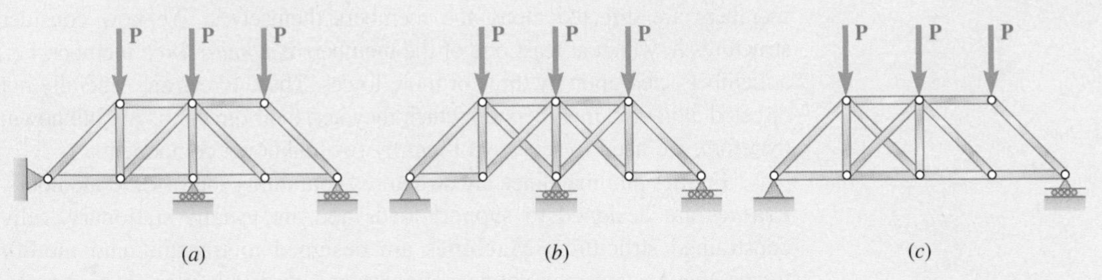

Fig. P6.47

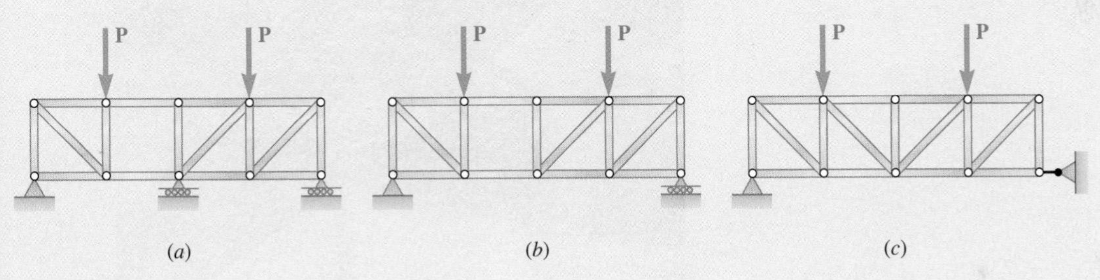

Fig. P6.48

6.3 FRAMES

When we study trusses, we are looking at structures consisting entirely of pins and straight two-force members. The forces acting on the two-force members are directed along the members themselves. We now consider structures in which at least one of the members is a *multi-force* member, i.e., a member acted upon by three or more forces. These forces are generally not directed along the members on which they act; their directions are unknown; therefore, we need to represent them by two unknown components.

Frames and machines are structures containing multi-force members. **Frames** are designed to support loads and are usually stationary, fully constrained structures. **Machines** are designed to transmit and modify forces; they may or may not be stationary and always contain moving parts.

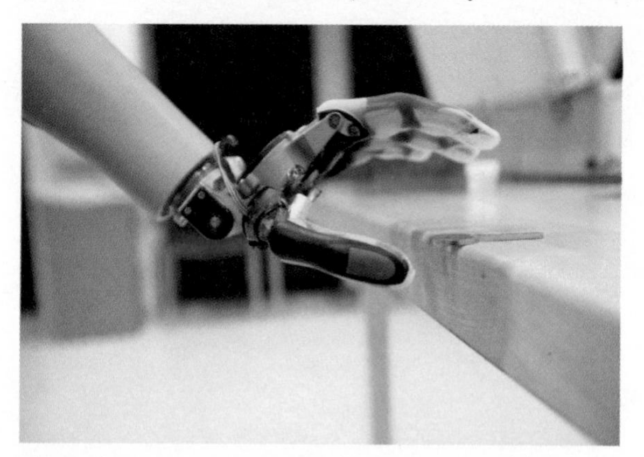

Photo 6.5 Frames and machines contain multi-force members. Frames are fully constrained structures, whereas machines like this prosthetic hand are movable and designed to transmit or modify forces.

6.3A Analysis of a Frame

As the first example of analysis of a frame, we consider again the crane described in Sec. 6.1 that carries a given load W (Fig. 6.18*a*). The free-body diagram of the entire frame is shown in Fig. 6.18*b*. We can use this diagram to determine the external forces acting on the frame. Summing moments about A, we first determine the force $\mathbf{T}$ exerted by the cable;

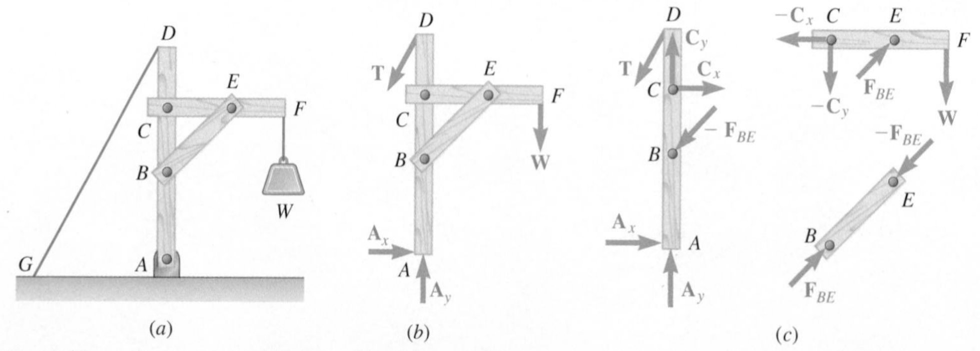

Fig. 6.18 A frame in equilibrium. (*a*) Diagram of a crane supporting a load; (*b*) free-body diagram of the crane; (*c*) free-body diagrams of the components of the crane.

summing x and y components, we then determine the components $\mathbf{A}_x$ and $\mathbf{A}_y$ of the reaction at the pin A.

In order to determine the internal forces holding the various parts of a frame together, we must dismember it and draw a free-body diagram for each of its component parts (Fig. 6.18*c*). First, we examine the two-force members. In this frame, member BE is the only two-force member. The forces acting at each end of this member must have the same magnitude, same line of action, and opposite sense (Sec. 4.2A). They are therefore directed along BE and are denoted, respectively, by $\mathbf{F}_{BE}$ and $-\mathbf{F}_{BE}$. We arbitrarily assume their sense as shown in Fig. 6.18*c*; the sign obtained for the common magnitude F_{BE} of the two forces will confirm or deny this assumption.

Next, we consider the multi-force members, i.e., the members that are acted upon by three or more forces. According to Newton's third law, the force exerted at B by member BE on member AD must be equal and opposite to the force $\mathbf{F}_{BE}$ exerted by AD on BE. Similarly, the force exerted at E by member BE on member CF must be equal and opposite to the force $-\mathbf{F}_{BE}$ exerted by CF on BE. Thus, the forces that the two-force member BE exerts on AD and CF are, respectively, equal to $-\mathbf{F}_{BE}$ and $\mathbf{F}_{BE}$; they have the same magnitude F_{BE}, opposite sense, and should be directed as shown in Fig. 6.18*c*.

Joint C connects two multi-force members. Since neither the direction nor the magnitude of the forces acting at C are known, we represent these forces by their x and y components. The components $\mathbf{C}_x$ and $\mathbf{C}_y$ of the force acting on member AD are arbitrarily directed to the right and upward. Since, according to Newton's third law, the forces exerted by member CF on AD and by member AD on CF are equal and opposite, the components of the force acting on member CF *must* be directed to the left and downward; we denote them, respectively, by $-\mathbf{C}_x$ and $-\mathbf{C}_y$. Whether the force $\mathbf{C}_x$ is actually directed to the right and the force $-\mathbf{C}_x$ is actually directed to the left will be determined later from the sign of their common magnitude C_x with a plus sign indicating that the assumption was correct and a minus sign that it was wrong. We complete the free-body diagrams of the multi-force members by showing the external forces acting at A, D, and F.†

We can now determine the internal forces by considering the free-body diagram of either of the two multi-force members. Choosing the free-body diagram of CF, for example, we write the equations $\Sigma M_C = 0$, $\Sigma M_E = 0$, and $\Sigma F_x = 0$, which yield the values of the magnitudes F_{BE}, C_y, and C_x, respectively. We can check these values by verifying that member AD is also in equilibrium.

Note that we assume the pins in Fig. 6.18 form an integral part of one of the two members they connected, so it is not necessary to show their free-body diagrams. We can always use this assumption to simplify

†It is not strictly necessary to use a minus sign to distinguish the force exerted by one member on another from the equal and opposite force exerted by the second member on the first, since the two forces belong to different free-body diagrams and thus are not easily confused. In the Sample Problems, we use the same symbol to represent equal and opposite forces that are applied to different free bodies. Note that, under these conditions, the sign obtained for a given force component does not directly relate the sense of that component to the sense of the corresponding coordinate axis. Rather, a positive sign indicates that *the sense assumed for that component in the free-body diagram* is correct, and a negative sign indicates that it is wrong.

the analysis of frames and machines. However, when a pin connects three or more members, connects a support and two or more members, or when a load is applied to a pin, we must make a clear decision in choosing the member to which we assume the pin belongs. (If multi-force members are involved, the pin should be attached to one of these members.) We then need to identify clearly the various forces exerted on the pin. This is illustrated in Sample Prob. 6.6.

6.3B Frames That Collapse Without Supports

The crane we just analyzed was constructed so that it could keep the same shape without the help of its supports; we therefore considered it to be a rigid body. Many frames, however, will collapse if detached from their supports; such frames cannot be considered rigid bodies. Consider, for example, the frame shown in Fig. 6.19*a* that consists of two members *AC* and *CB* carrying loads **P** and **Q** at their midpoints. The members are supported by pins at *A* and *B* and are connected by a pin at *C*. If we detach this frame from its supports, it will not maintain its shape. Therefore, we should consider it to be made of *two distinct rigid parts AC* and *CB*.

The equations $\Sigma F_x = 0$, $\Sigma F_y = 0$, and $\Sigma M = 0$ (about any given point) express the conditions for the *equilibrium of a rigid body* (Chap. 4); we should use them, therefore, in connection with the free-body diagrams of members *AC* and *CB* (Fig. 6.19*b*). Since these members are multi-force members and since pins are used at the supports and at the connection, we represent each of the reactions at *A* and *B* and the forces at *C* by two components. In accordance with Newton's third law, we represent the components of the force exerted by *CB* on *AC* and the components of the force exerted by *AC* on *CB* by vectors of the same magnitude and opposite sense. Thus, if the first pair of components consists of $\mathbf{C}_x$ and $\mathbf{C}_y$, the second pair is represented by $-\mathbf{C}_x$ and $-\mathbf{C}_y$.

Note that four unknown force components act on free body *AC*, whereas we need only three independent equations to express that the body is in equilibrium. Similarly, four unknowns, but only three equations, are associated with *CB*. However, only six different unknowns are involved in the analysis of the two members, and altogether, six equations

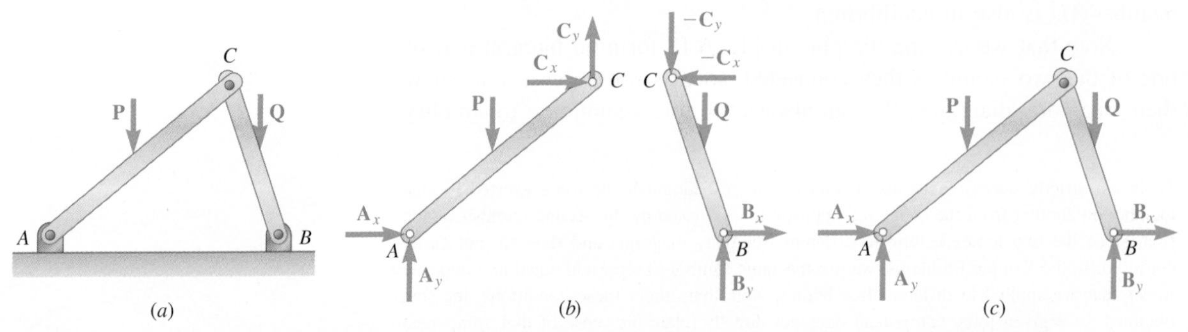

Fig. 6.19 (*a*) A frame of two members supported by two pins and joined together by a third pin. Without the supports, the frame would collapse and is therefore not a rigid body. (*b*) Free-body diagrams of the two members. (*c*) Free-body diagram of the whole frame.

are available to express that the members are in equilibrium. Setting $\Sigma M_A = 0$ for free body AC and $\Sigma M_B = 0$ for CB, we obtain two simultaneous equations that we can solve for the common magnitude C_x of the components $\mathbf{C}_x$ and $-\mathbf{C}_x$ and for the common magnitude C_y of the components $\mathbf{C}_y$ and $-\mathbf{C}_y$. We then have $\Sigma F_x = 0$ and $\Sigma F_y = 0$ for each of the two free bodies, successively obtaining the magnitudes A_x, A_y, B_x, and B_y.

Observe that, since the equations of equilibrium $\Sigma F_x = 0$, $\Sigma F_y = 0$, and $\Sigma M = 0$ (about any given point) are satisfied by the forces acting on free body AC and since they are also satisfied by the forces acting on free body CB, they must be satisfied when the forces acting on the two free bodies are considered simultaneously. Since the internal forces at C cancel each other, we find that the equations of equilibrium must be satisfied by the external forces shown on the free-body diagram of the frame ACB itself (Fig. 6.19*c*), even though the frame is not a rigid body. We can use these equations to determine some of the components of the reactions at A and B. We will find, however, that **the reactions cannot be completely determined from the free-body diagram of the whole frame**. It is thus necessary to dismember the frame and consider the free-body diagrams of its component parts (Fig. 6.19*b*), even when we are interested in determining external reactions only. The reason is that the equilibrium equations obtained for free body ACB *are necessary conditions* for the equilibrium of a nonrigid structure, *but these are not sufficient conditions*.

The method of solution outlined here involved simultaneous equations. We now present a more efficient method that utilizes the free body ACB as well as the free bodies AC and CB. Writing $\Sigma M_A = 0$ and $\Sigma M_B = 0$ for free body ACB, we obtain B_y and A_y. From $\Sigma M_C = 0$, $\Sigma F_x = 0$, and $\Sigma F_y = 0$ for free body AC, we successively obtain A_x, C_x, and C_y. Finally, setting $\Sigma F_x = 0$ for ACB gives us B_x.

We noted previously that the analysis of the frame in Fig. 6.19 involves six unknown force components and six independent equilibrium equations. (The equilibrium equations for the whole frame were obtained from the original six equations and, therefore, are not independent.) Moreover, we checked that all unknowns could be actually determined and that all equations could be satisfied. This frame is **statically determinate and rigid**. (We use the word "rigid" here to indicate that the frame maintains its shape as long as it remains attached to its supports.) In general, to determine whether a structure is statically determinate and rigid, you should draw a free-body diagram for each of its component parts and count the reactions and internal forces involved. You should then determine the number of independent equilibrium equations (excluding equations expressing the equilibrium of the whole structure or of groups of component parts already analyzed). If you have more unknowns than equations, the structure is *statically indeterminate*. If you have fewer unknowns than equations, the structure is *nonrigid*. If you have as many unknowns as equations *and if all unknowns can be determined and all equations satisfied* under general loading conditions, the structure is statically determinate and rigid. If, however, due to an improper arrangement of members and supports, all unknowns cannot be determined and all equations cannot be satisfied, the structure is **statically indeterminate and nonrigid**.

Sample Problem 6.4

In the frame shown, members ACE and BCD are connected by a pin at C and by the link DE. For the loading shown, determine the force in link DE and the components of the force exerted at C on member BCD.

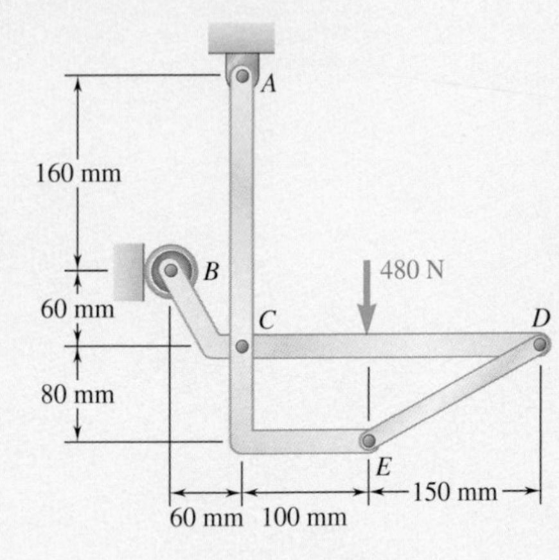

STRATEGY: Follow the general procedure discussed in this section. First treat the entire frame as a free body, which will enable you to find the reactions at A and B. Then dismember the frame and treat each member as a free body, which will give you the equations needed to find the force at C.

MODELING and ANALYSIS: Since the external reactions involve only three unknowns, compute the reactions by considering the free-body diagram of the entire frame (Fig. 1).

$$+\uparrow\Sigma F_y = 0: \quad A_y - 480 \text{ N} = 0 \quad A_y = +480 \text{ N} \quad \mathbf{A}_y = 480 \text{ N}\uparrow$$
$$+\circlearrowleft\Sigma M_A = 0: \quad -(480 \text{ N})(100 \text{ mm}) + B(160 \text{ mm}) = 0$$
$$B = +300 \text{ N} \quad \mathbf{B} = 300 \text{ N}\rightarrow$$
$$\overset{+}{\rightarrow}\Sigma F_x = 0: \quad B + A_x = 0$$
$$300 \text{ N} + A_x = 0 \quad A_x = -300 \text{ N} \quad \mathbf{A}_x = 300 \text{ N}\leftarrow$$

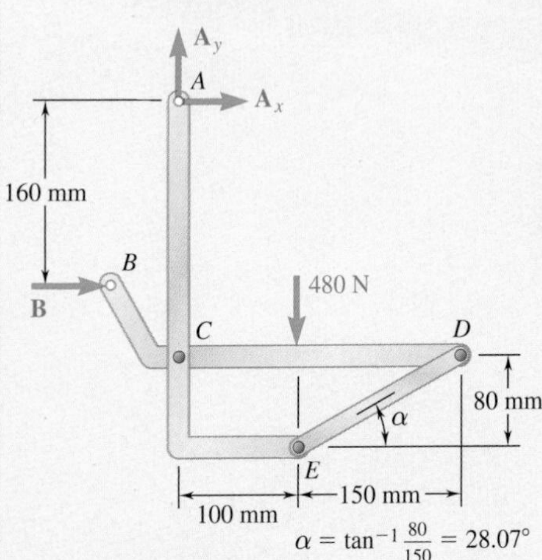

Fig. 1 Free-body diagram of entire frame.

Now dismember the frame (Figs. 2 and 3). Since only two members are connected at C, the components of the unknown forces acting on ACE and BCD are, respectively, equal and opposite. Assume that link DE is in tension (Fig. 3) and exerts equal and opposite forces at D and E, directed as shown.

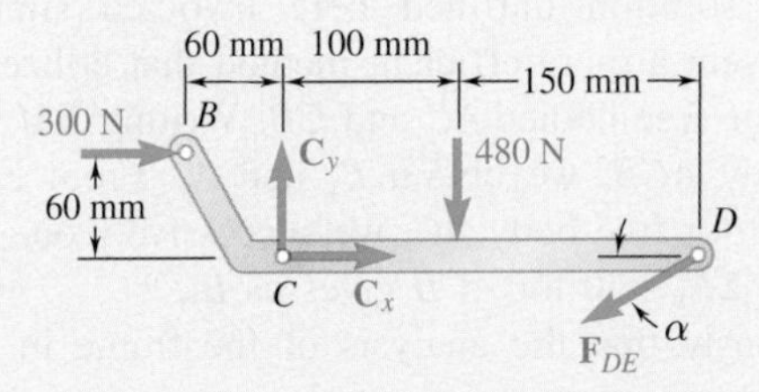

Fig. 2 Free-body diagram of member BCD.

Free Body: Member BCD. Using the free body BCD (Fig. 2), you can write and solve three equilibrium equations:

$$+\circlearrowright\Sigma M_C = 0:$$
$$(F_{DE} \sin \alpha)(250 \text{ mm}) + (300 \text{ N})(80 \text{ mm}) + (480 \text{ N})(100 \text{ mm}) = 0$$
$$F_{DE} = -561 \text{ N} \quad F_{DE} = 561 \text{ N } C \blacktriangleleft$$
$$\overset{+}{\rightarrow}\Sigma F_x = 0: \quad C_x - F_{DE} \cos \alpha + 300 \text{ N} = 0$$
$$C_x - (-561 \text{ N}) \cos 28.07° + 300 \text{ N} = 0 \quad C_x = -795 \text{ N}$$
$$+\uparrow\Sigma F_y = 0: \quad C_y - F_{DE} \sin \alpha - 480 \text{ N} = 0$$
$$C_y - (-561 \text{ N}) \sin 28.07° - 480 \text{ N} = 0 \quad C_y = +216 \text{ N}$$

From the signs obtained for C_x and C_y, the force components $\mathbf{C}_x$ and $\mathbf{C}_y$ exerted on member BCD are directed to the left and up, respectively. Thus, you have

$$\mathbf{C}_x = 795 \text{ N}\leftarrow, \mathbf{C}_y = 216 \text{ N}\uparrow \blacktriangleleft$$

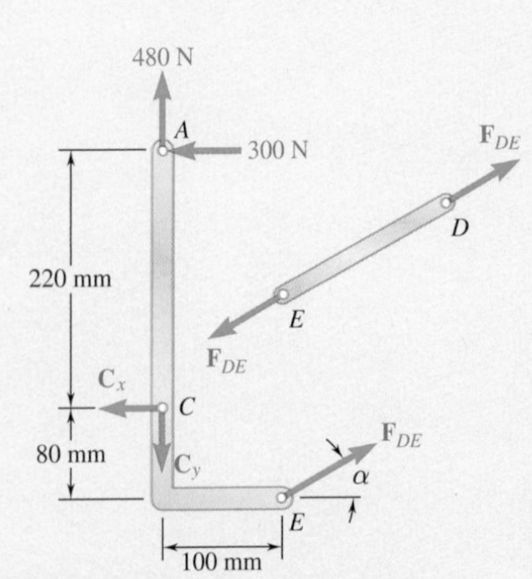

Fig. 3 Free-body diagrams of member ACE and DE.

REFLECT and THINK: Check the computations by considering the free body ACE (Fig. 3). For example,

$$+\circlearrowleft\Sigma M_A = (F_{DE} \cos \alpha)(300 \text{ mm}) + (F_{DE} \sin \alpha)(100 \text{ mm}) - C_x(220 \text{ mm})$$
$$= (-561 \cos \alpha)(300) + (-561 \sin \alpha)(100) - (-795)(220) = 0$$

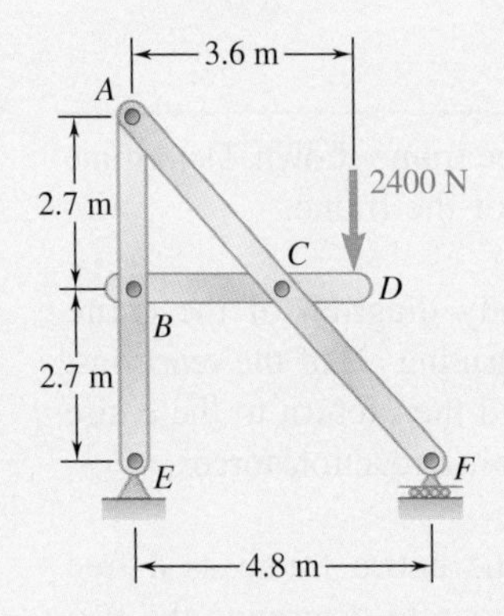

Sample Problem 6.5

Determine the components of the forces acting on each member of the frame shown.

STRATEGY: The approach to this analysis is to consider the entire frame as a free body to determine the reactions, and then consider separate members. However, in this case, you will not be able to determine forces on one member without analyzing a second member at the same time.

MODELING and ANALYSIS: The external reactions involve only three unknowns, so compute the reactions by considering the free-body diagram of the entire frame (Fig. 1).

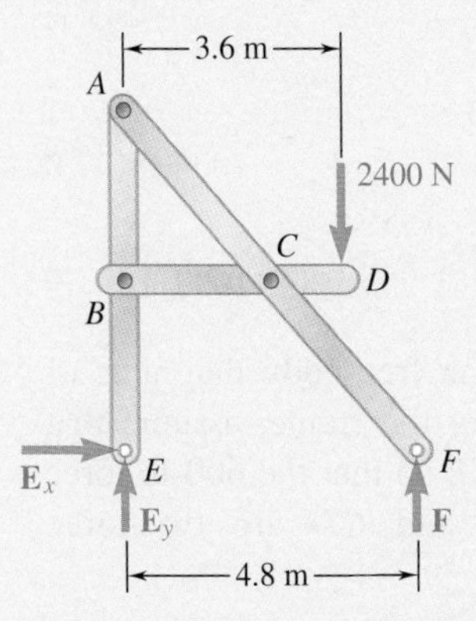

Fig. 1 Free-body diagram of entire frame.

$$+\circlearrowleft \Sigma M_E = 0: \quad -(2400 \text{ N})(3.6 \text{ m}) + F(4.8 \text{ m}) = 0$$
$$F = +1800 \text{ N} \qquad \mathbf{F} = 1800 \text{ N}\uparrow \blacktriangleleft$$
$$+\uparrow \Sigma F_y = 0: \quad -2400 \text{ N} + 1800 \text{ N} + E_y = 0$$
$$E_y = +600 \text{ N} \qquad \mathbf{E}_y = 600 \text{ N}\uparrow \blacktriangleleft$$
$$\xrightarrow{+} \Sigma F_x = 0: \qquad \mathbf{E}_x = 0 \blacktriangleleft$$

Now dismember the frame. Since only two members are connected at each joint, force components are equal and opposite on each member at each joint (Fig. 2).

Free Body: Member BCD.

$$+\circlearrowleft \Sigma M_B = 0: \quad -(2400 \text{ N})(3.6 \text{ m}) + C_y(2.4 \text{ m}) = 0 \qquad C_y = +3600 \text{ N} \blacktriangleleft$$
$$+\circlearrowleft \Sigma M_C = 0: \quad -(2400 \text{ N})(1.2 \text{ m}) + B_y(2.4 \text{ m}) = 0 \qquad B_y = +1200 \text{ N} \blacktriangleleft$$
$$\xrightarrow{+} \Sigma F_x = 0: \quad -B_x + C_x = 0$$

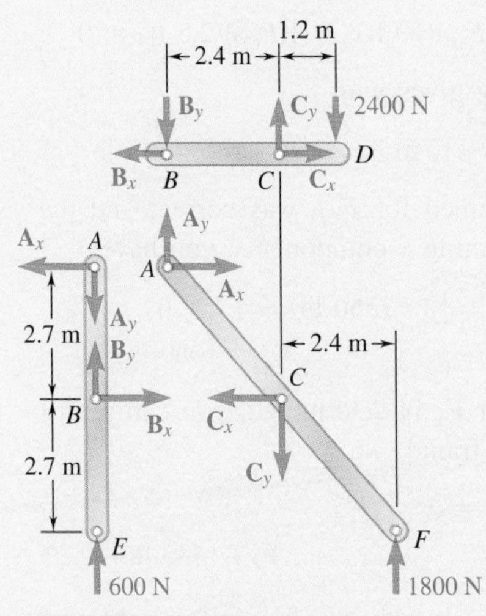

Fig. 2 Free-body diagrams of individual members.

Neither B_x nor C_x can be obtained by considering only member *BCD*; you need to look at member *ABE*. The positive values obtained for B_y and C_y indicate that the force components $\mathbf{B}_y$ and $\mathbf{C}_y$ are directed as assumed.

Free Body: Member ABE.

$$+\circlearrowleft \Sigma M_A = 0: \quad B_x(2.7 \text{ m}) = 0 \qquad B_x = 0 \blacktriangleleft$$
$$\xrightarrow{+} \Sigma F_x = 0: \quad +B_x - A_x = 0 \qquad A_x = 0 \blacktriangleleft$$
$$+\uparrow \Sigma F_y = 0: \quad -A_y + B_y + 600 \text{ N} = 0$$
$$-A_y + 1200 \text{ N} + 600 \text{ N} = 0 \qquad A_y = +1800 \text{ N} \blacktriangleleft$$

Free Body: Member BCD. Returning now to member *BCD*, you have

$$\xrightarrow{+} \Sigma F_x = 0: \quad -B_x + C_x = 0 \qquad 0 + C_x = 0 \qquad C_x = 0 \blacktriangleleft$$

REFLECT and THINK: All unknown components have now been found. To check the results, you can verify that member *ACF* is in equilibrium.

$$+\circlearrowleft \Sigma M_C = (1800 \text{ N})(2.4 \text{ m}) - A_y(2.4 \text{ m}) - A_x(2.7 \text{ m})$$
$$= (1800 \text{ N})(2.4 \text{ m}) - (1800 \text{ N})(2.4 \text{ m}) - 0 = 0 \qquad \text{(checks)}$$

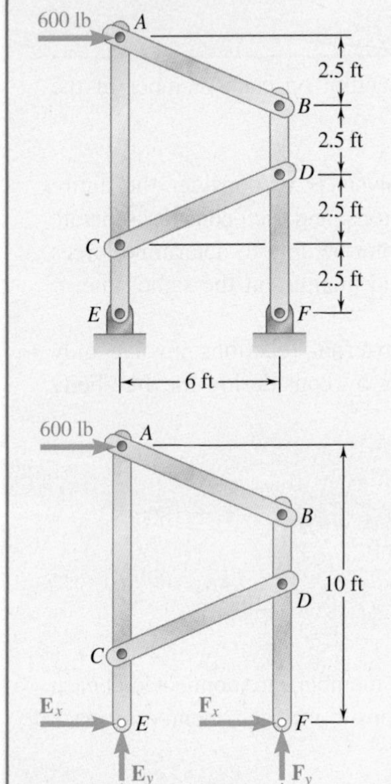

Fig. 1 Free-body diagram of entire frame.

Sample Problem 6.6

A 600-lb horizontal force is applied to pin A of the frame shown. Determine the forces acting on the two vertical members of the frame.

STRATEGY: Begin as usual with a free-body diagram of the entire frame, but this time you will not be able to determine all of the reactions. You will have to analyze a separate member and then return to the entire frame analysis in order to determine the remaining reaction forces.

MODELING and ANALYSIS: Choosing the entire frame as a free body (Fig. 1), you can write equilibrium equations to determine the two force components $\mathbf{E}_y$ and $\mathbf{F}_y$. However, these equations are not sufficient to determine $\mathbf{E}_x$ and $\mathbf{F}_x$.

$$+\circlearrowleft \Sigma M_E = 0: \quad -(600 \text{ lb})(10 \text{ ft}) + F_y(6 \text{ ft}) = 0$$

$$F_y = +1000 \text{ lb} \qquad \mathbf{F}_y = 1000 \text{ lb}\uparrow \blacktriangleleft$$

$$+\uparrow \Sigma F_y = 0: \quad E_y + F_y = 0$$

$$E_y = -1000 \text{ lb} \qquad \mathbf{E}_y = 1000 \text{ lb}\downarrow \blacktriangleleft$$

To proceed with the solution, now consider the free-body diagrams of the various members (Fig. 2). In dismembering the frame, assume that pin A is attached to the multi-force member ACE so that the 600-lb force is applied to that member. Note that AB and CD are two-force members.

Free Body: Member ACE

$$+\uparrow \Sigma F_y = 0: \quad -\tfrac{5}{13}F_{AB} + \tfrac{5}{13}F_{CD} - 1000 \text{ lb} = 0$$

$$+\circlearrowleft \Sigma M_E = 0: \quad -(600 \text{ lb})(10 \text{ ft}) - (\tfrac{12}{13}F_{AB})(10 \text{ ft}) - (\tfrac{12}{13}F_{CD})(2.5 \text{ ft}) = 0$$

Solving these equations simultaneously gives you

$$F_{AB} = -1040 \text{ lb} \qquad F_{CD} = +1560 \text{ lb} \blacktriangleleft$$

The signs indicate that the sense assumed for F_{CD} was correct and the sense for F_{AB} was incorrect. Now summing x components, you have

$$\overset{+}{\rightarrow}\Sigma F_x = 0: \quad 600 \text{ lb} + \tfrac{12}{13}(-1040 \text{ lb}) + \tfrac{12}{13}(+1560 \text{ lb}) + E_x = 0$$

$$E_x = -1080 \text{ lb} \qquad \mathbf{E}_x = 1080 \text{ lb}\leftarrow \blacktriangleleft$$

Free Body: Entire Frame. Now that $\mathbf{E}_x$ is determined, you can return to the free-body diagram of the entire frame.

$$\overset{+}{\rightarrow}\Sigma F_x = 0: \quad 600 \text{ lb} - 1080 \text{ lb} + F_x = 0$$

$$F_x = +480 \text{ lb} \qquad \mathbf{F}_x = 480 \text{ lb}\rightarrow \blacktriangleleft$$

REFLECT and THINK: Check your computations by verifying that the equation $\Sigma M_B = 0$ is satisfied by the forces acting on member BDF.

$$\begin{aligned} +\circlearrowleft \Sigma M_B &= -\left(\tfrac{12}{13}F_{CD}\right)(2.5 \text{ ft}) + (F_x)(7.5 \text{ ft}) \\ &= -\tfrac{12}{13}(1560 \text{ lb})(2.5 \text{ ft}) + (480 \text{ lb})(7.5 \text{ ft}) \\ &= -3600 \text{ lb·ft} + 3600 \text{ lb·ft} = 0 \qquad \text{(checks)} \end{aligned}$$

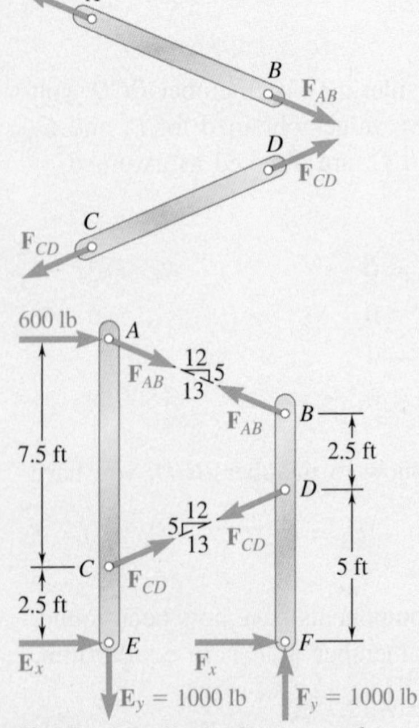

Fig. 2 Free-body diagrams of individual members.

Problems

6.49 through 6.51 Determine the force in member *BD* and the components of the reaction at *C*.

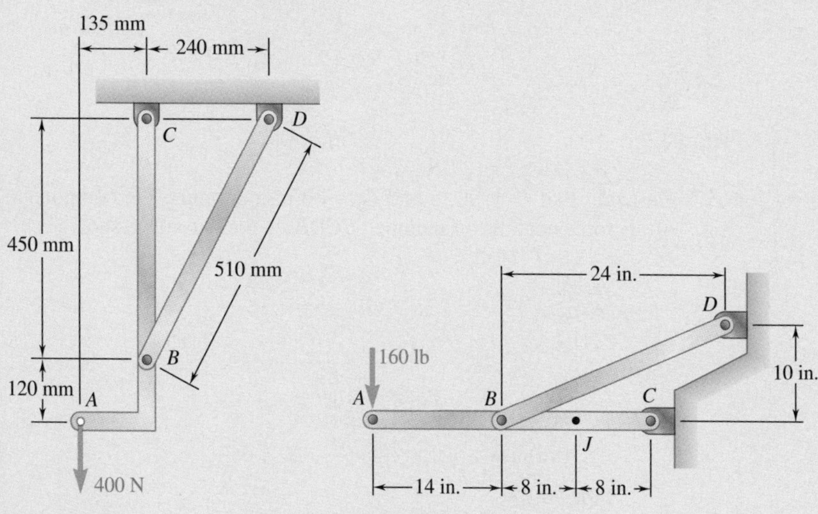

Fig. P6.49

Fig. P6.50

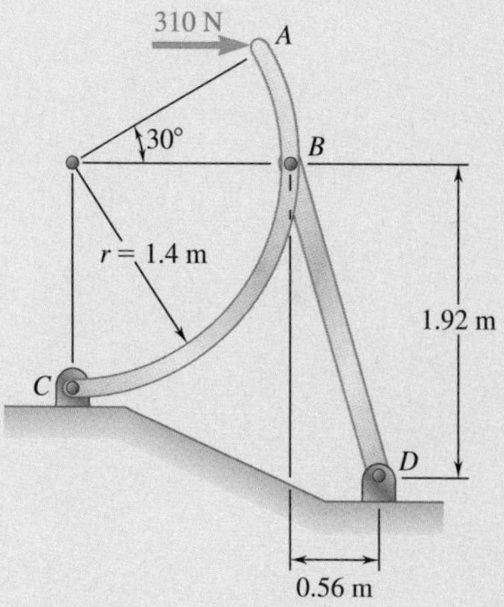

Fig. P6.51

6.52 Determine the components of all forces acting on member *ABCD* of the assembly shown.

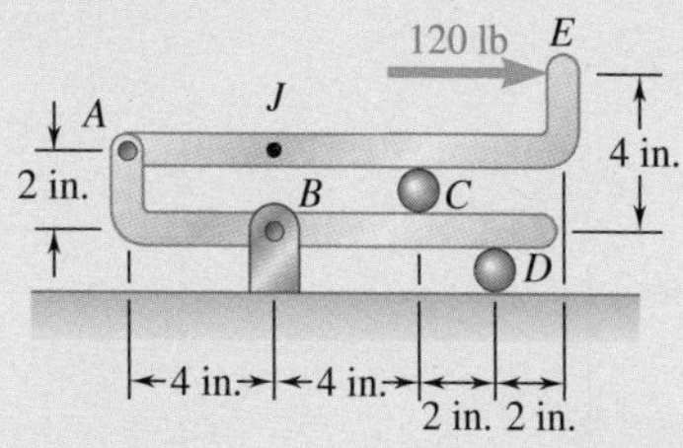

Fig. P6.52

6.53 Determine the components of all forces acting on member *ABCD* when $\theta = 0$.

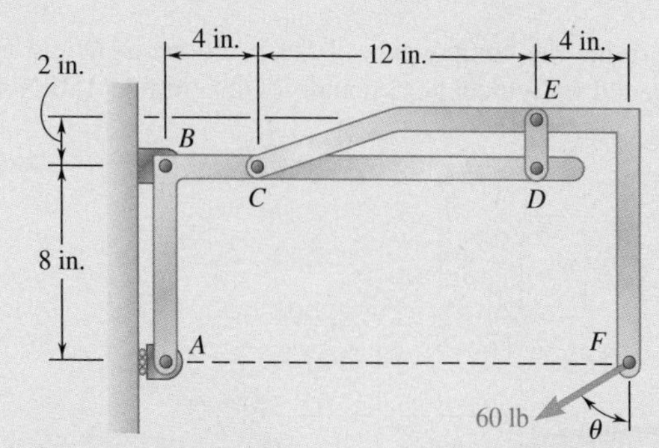

Fig. P6.53 and *P6.54*

6.54 Determine the components of all forces acting on member *ABCD* when $\theta = 90°$.

6.55 and *6.56* Determine the components of the reactions at A and E if a 750-N force directed vertically downward is applied (*a*) at B, (*b*) at D.

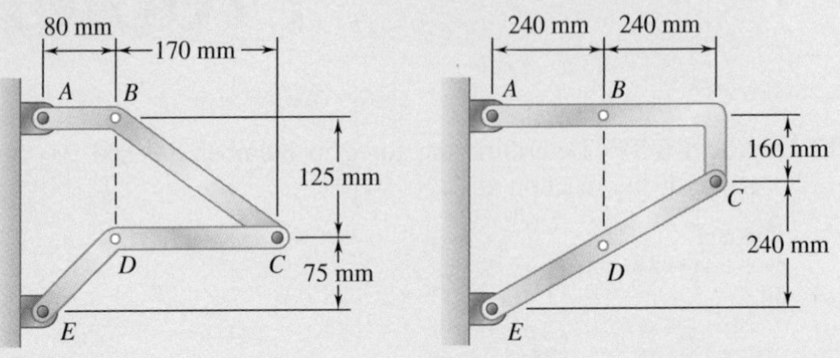

Fig. P6.55

Fig. *P6.56*

6.57 Knowing that $P = 90$ lb and $Q = 60$ lb, determine the components of all forces acting on member $BCDE$ of the assembly shown.

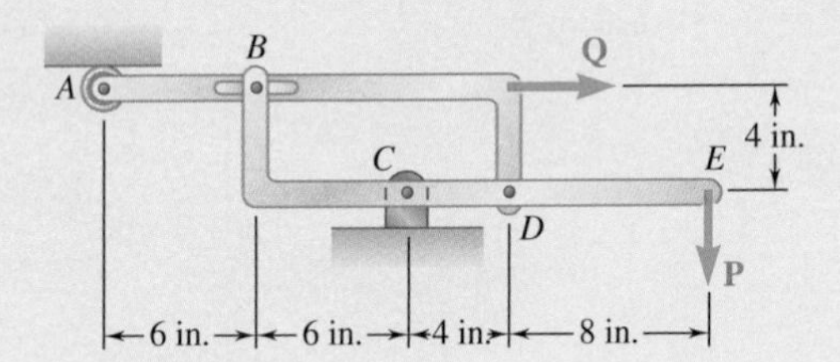

Fig. P6.57

6.58 Determine the components of the reactions at A and E, (*a*) if the 800-N load is applied as shown, (*b*) if the 800-N load is moved along its line of action and is applied at point D.

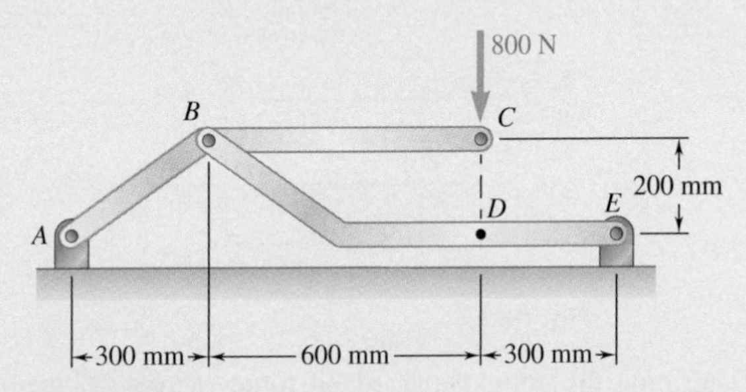

Fig. P6.58

6.59 Determine the components of the reactions at D and E if the frame is loaded by a clockwise couple of magnitude 150 N·m applied (*a*) at A, (*b*) at B.

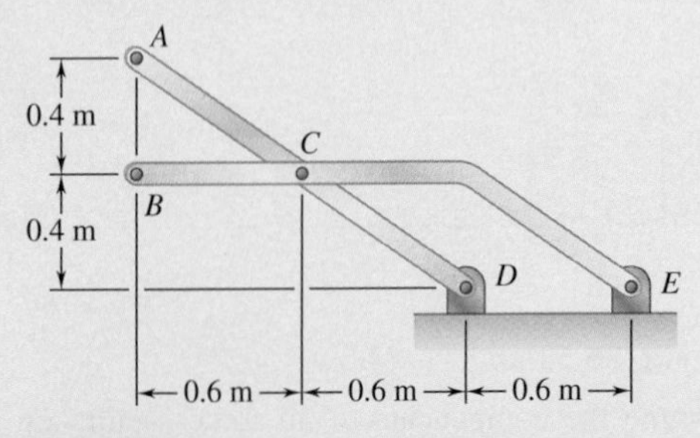

Fig. P6.59

6.60 The 48-lb load can be moved along the line of action shown and applied at A, D, or E. Determine the components of the reactions at B and F if the 48-lb load is applied (*a*) at A, (*b*) at D, (*c*) at E.

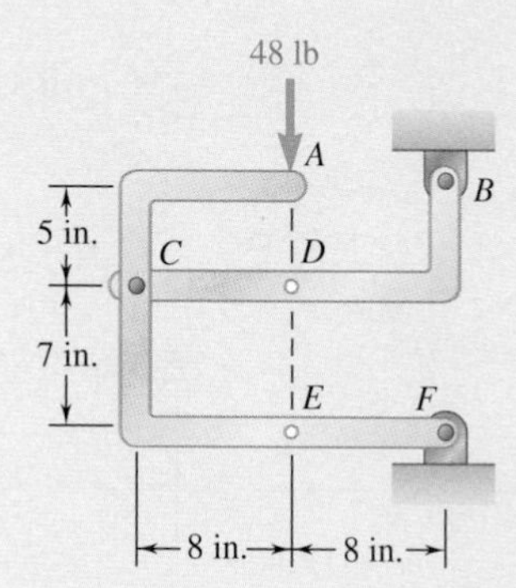

Fig. P6.60 and P6.61

6.61 The 48-lb load is removed and a 288-lb·in. clockwise couple is applied successively at A, D, and E. Determine the components of the reactions at B and F if the couple is applied (*a*) at A, (*b*) at D, (*c*) at E.

6.62 Determine all the forces exerted on member AI if the frame is loaded by a clockwise couple of magnitude 1200 lb·in. applied (*a*) at point D, (*b*) at point E.

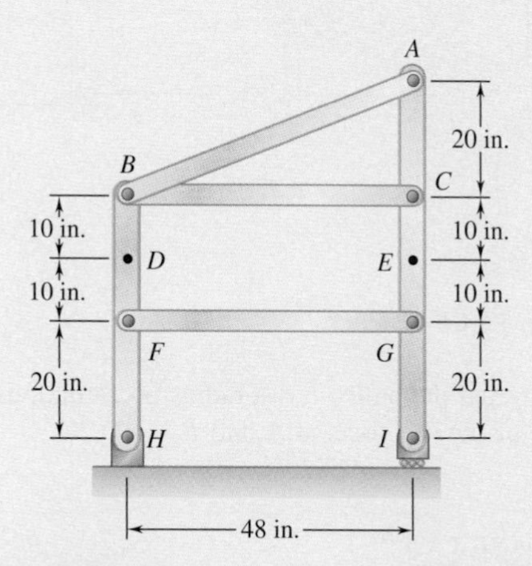

Fig. P6.62

6.63 The hydraulic cylinder CF, which partially controls the position of rod DE, has been locked in the position shown. Knowing that $\theta = 60°$, determine (*a*) the force **P** for which the tension in link AB is 410 N, (*b*) the corresponding force exerted on member BCD at point C.

6.64 The hydraulic cylinder CF, which partially controls the position of rod DE, has been locked in the position shown. Knowing that $P =$ 400 N and $\theta = 75°$, determine (*a*) the force in link AB, (*b*) the corresponding force exerted on member BCD at point C.

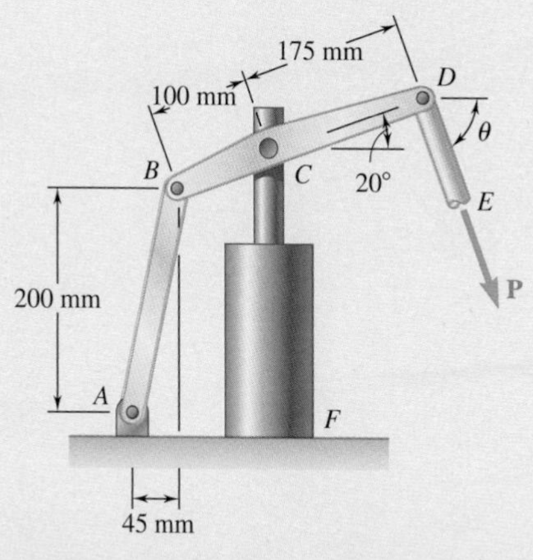

Fig. P6.63 and P6.64

6.65 Two 9-in.-diameter pipes (pipe *1* and pipe *2*) are supported every 7.5 ft by a small frame like that shown. Knowing that the combined weight of each pipe and its contents is 30 lb/ft and assuming frictionless surfaces, determine the components of the reactions at *A* and *G*.

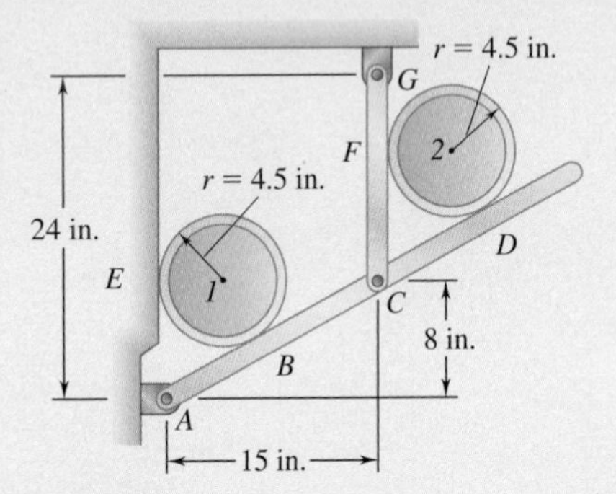

Fig. P6.65

6.66 Solve Prob. 6.65 assuming that pipe *1* is removed and that only pipe *2* is supported by the frames.

6.67 Knowing that each pulley has a radius of 250 mm, determine the components of the reactions at *D* and *E*.

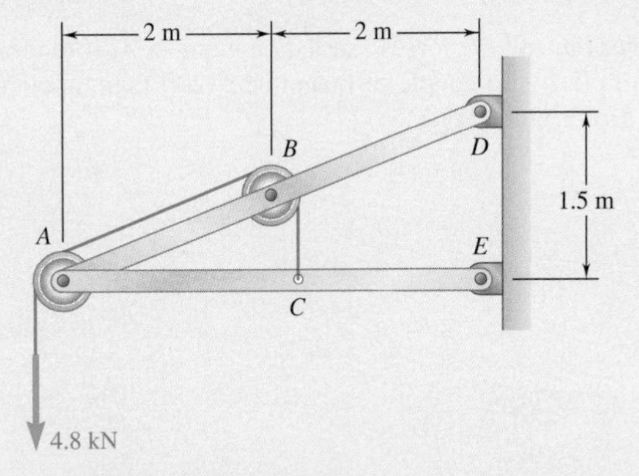

Fig. P6.67

6.68 Knowing that the pulley has a radius of 75 mm, determine the components of the reactions at *A* and *B*.

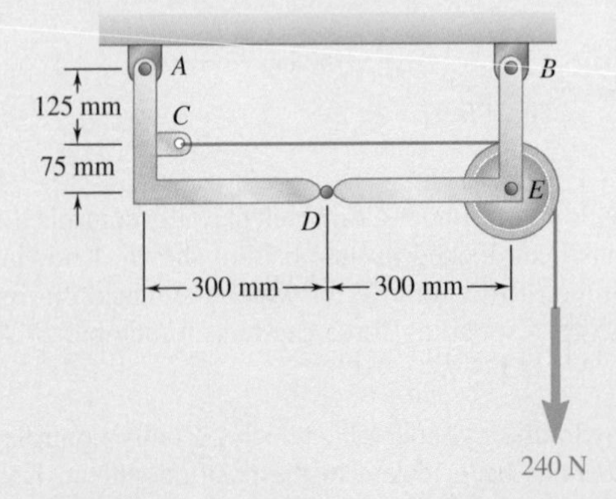

Fig. *P6.68*

6.69 The cab and motor units of the front-end loader shown are connected by a vertical pin located 2 m behind the cab wheels. The distance from C to D is 1 m. The center of gravity of the 300-kN motor unit is located at G_m, while the centers of gravity of the 100-kN cab and 75-kN load are located, respectively, at G_c and G_l. Knowing that the front-end loader is at rest with its brakes released, determine (*a*) the reactions at each of the four wheels, (*b*) the forces exerted on the motor unit at C and D.

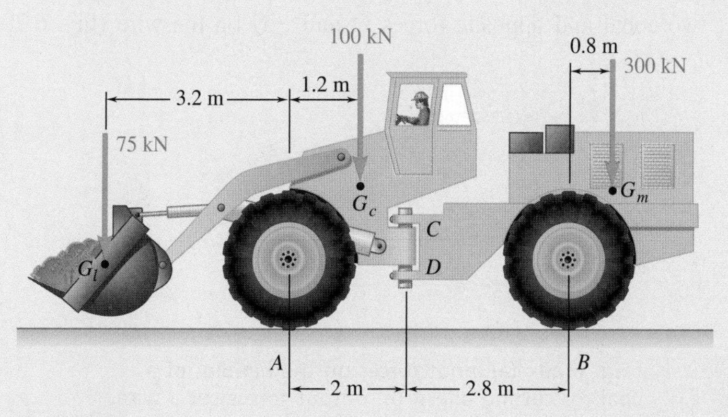

Fig. P6.69

6.70 Solve Prob. 6.69 assuming that the 75-kN load has been removed.

6.71 A trailer weighing 2400 lb is attached to a 2900-lb pickup truck by a ball-and-socket truck hitch at D. Determine (*a*) the reactions at each of the six wheels when the truck and trailer are at rest, (*b*) the additional load on each of the truck wheels due to the trailer.

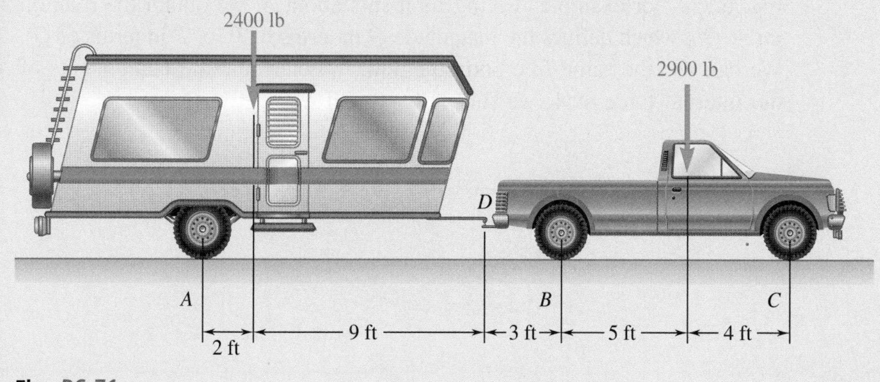

Fig. *P6.71*

6.72 In order to obtain a better weight distribution over the four wheels of the pickup truck of Prob. 6.71, a compensating hitch of the type shown is used to attach the trailer to the truck. The hitch consists of two bar springs (only one is shown in the figure) that fit into bearings inside a support rigidly attached to the truck. The springs are also connected by chains to the trailer frame, and specially designed hooks make it possible to place both chains in tension. (*a*) Determine the tension T required in each of the two chains if the additional load due to the trailer is to be evenly distributed over the four wheels of the truck. (*b*) What are the resulting reactions at each of the six wheels of the trailer-truck combination?

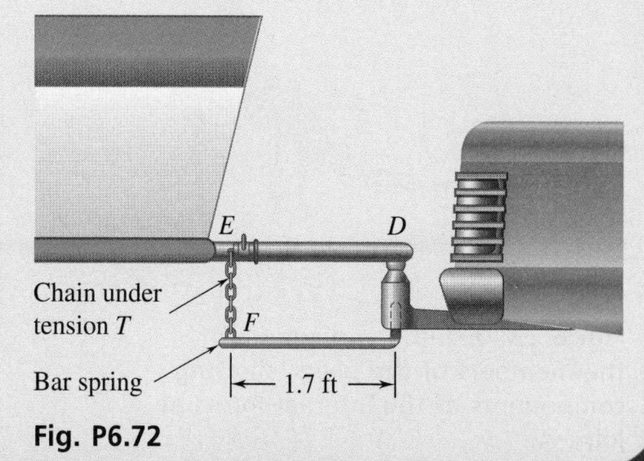

Fig. P6.72

Photo 6.6 This lamp can be placed in many different positions. To determine the forces in the springs and the internal forces at the joints, we need to consider the components of the lamp as free bodies.

6.4 MACHINES

Machines are structures designed to transmit and modify forces. Whether they are simple tools or include complicated mechanisms, their main purpose is to transform **input forces** into **output forces**. Consider, for example, a pair of cutting pliers used to cut a wire (Fig. 6.20*a*). If we apply two equal and opposite forces **P** and −**P** on the handles, the pliers will exert two equal and opposite forces **Q** and −**Q** on the wire (Fig. 6.20*b*).

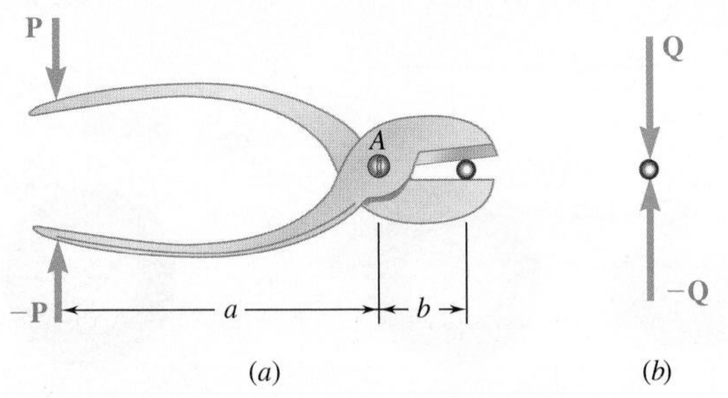

Fig. 6.20 (a) Input forces on the handles of a pair of cutting pliers; (b) output forces cut a wire.

To determine the magnitude Q of the output forces when we know the magnitude P of the input forces (or, conversely, to determine P when Q is known), we draw a free-body diagram of the pliers *alone* (i.e., without the wire), showing the input forces **P** and −**P** and the *reactions* −**Q** and **Q** that the wire exerts on the pliers (Fig. 6.21). However, since a pair of pliers forms a nonrigid structure, we must treat one of the component parts as a free body in order to determine the unknown forces. Consider Fig. 6.22*a*, for example. Taking moments about A, we obtain the relation $Pa = Qb$, which defines the magnitude Q in terms of P (or P in terms of Q). We can use the same free-body diagram to determine the components of the internal force at A; we find $A_x = 0$ and $A_y = P + Q$.

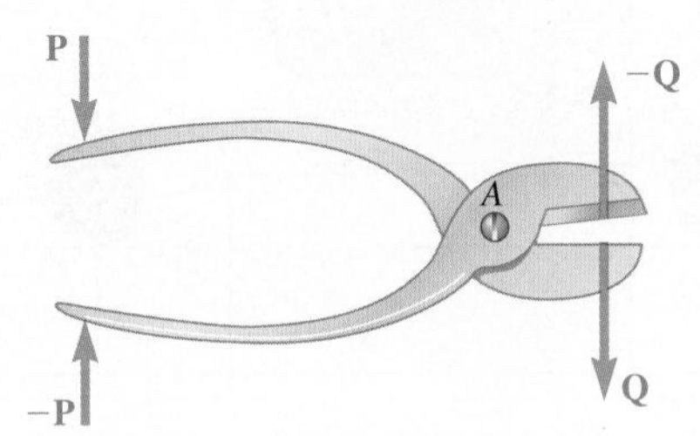

Fig. 6.21 To show a free-body diagram of the pliers in equilibrium, we include the input forces and the reactions to the output forces.

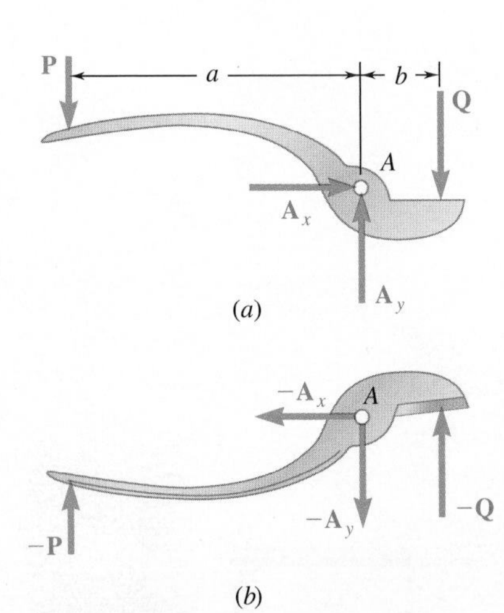

Fig. 6.22 Free-body diagrams of the members of the pliers, showing components of the internal forces at joint *A*.

In the case of more complicated machines, it is generally necessary to use several free-body diagrams and, possibly, to solve simultaneous equations involving various internal forces. You should choose the free bodies to include the input forces and the reactions to the output forces, and the total number of unknown force components involved should not exceed the number of available independent equations. It is advisable, before attempting to solve a problem, to determine whether the structure considered is determinate. There is no point, however, in discussing the rigidity of a machine, since a machine includes moving parts and thus *must* be nonrigid.

Sample Problem 6.7

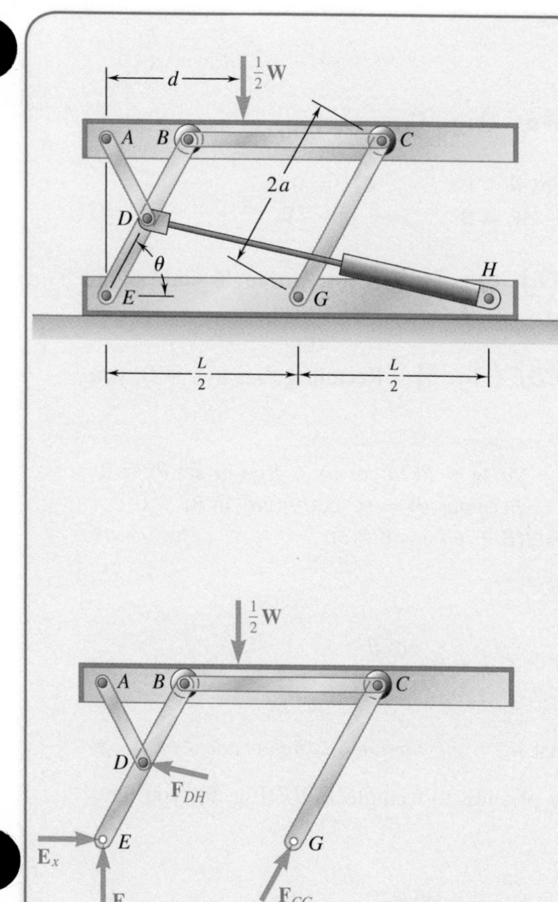

Fig. 1 Free-body diagram of machine.

A hydraulic-lift table is used to raise a 1000-kg crate. The table consists of a platform and two identical linkages on which hydraulic cylinders exert equal forces. (Only one linkage and one cylinder are shown.) Members *EDB* and *CG* are each of length $2a$, and member *AD* is pinned to the midpoint of *EDB*. If the crate is placed on the table so that half of its weight is supported by the system shown, determine the force exerted by each cylinder in raising the crate for $\theta = 60°$, $a = 0.70$ m, and $L = 3.20$ m. Show that the result is independent of the distance d.

STRATEGY: The free-body diagram of the platform and linkage system will involve more than three unknowns, so it alone can not be used to solve this problem. Instead, draw free-body diagrams of each component of the machine and work from them.

MODELING: The machine consists of the platform and the linkage. Its free-body diagram (Fig. 1) includes an input force $\mathbf{F}_{DH}$ exerted by the cylinder; the weight $\mathbf{W}/2$, which is equal and opposite to the output force; and reactions at E and G, which are assumed to be directed as shown. Dismember the mechanism and draw a free-body diagram for each of its component parts (Fig. 2). Note that *AD*, *BC*, and *CG* are two-force members. Member *CG* has already been assumed to be in compression; now assume that *AD* and *BC* are in tension and direct the forces exerted on them as shown. Use equal and opposite vectors to represent the forces exerted by the two-force members on the platform, on member *BDE*, and on roller *C*.

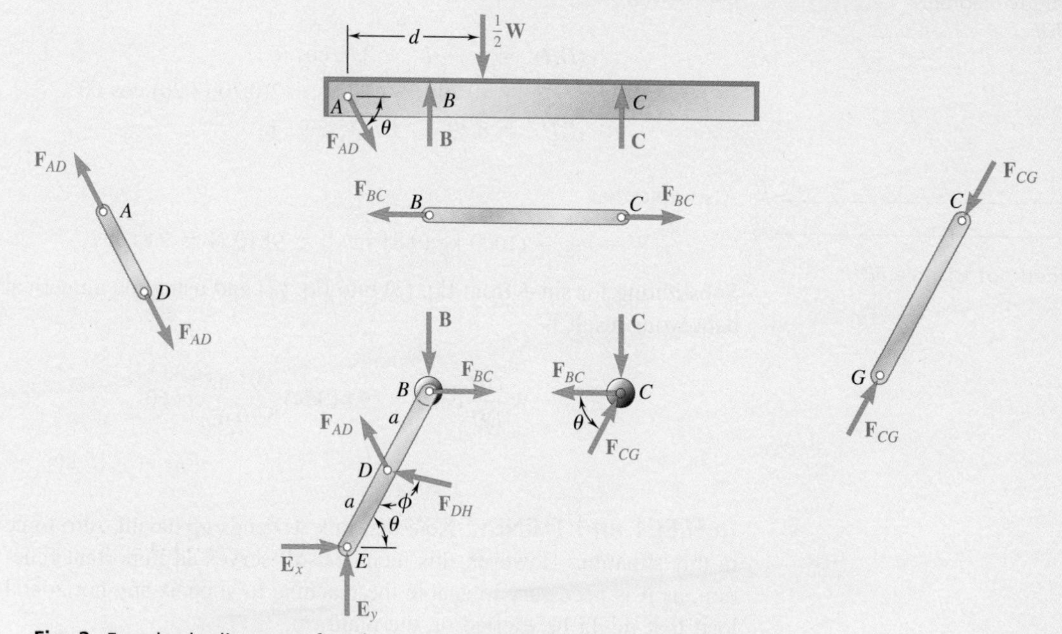

Fig. 2 Free-body diagram of each component part.

(continued)

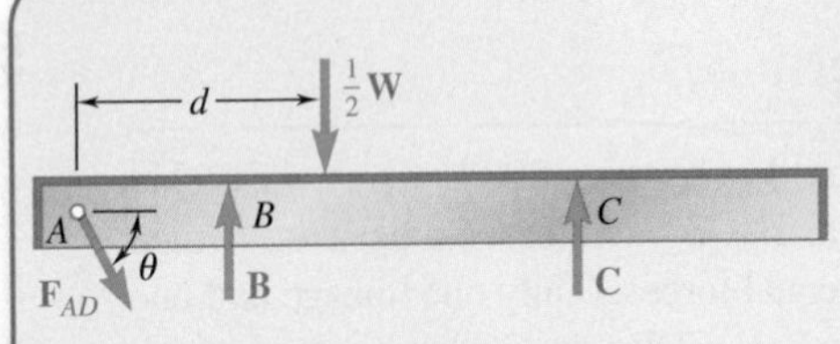

Fig. 3 Free-body diagram of platform *ABC*.

ANALYSIS:

Free Body: Platform *ABC* (Fig. 3).

$$\xrightarrow{+}\Sigma F_x = 0: \quad F_{AD}\cos\theta = 0 \qquad F_{AD} = 0$$

$$+\uparrow \Sigma F_y = 0: \quad B + C - \tfrac{1}{2}W = 0 \qquad B + C = \tfrac{1}{2}W \tag{1}$$

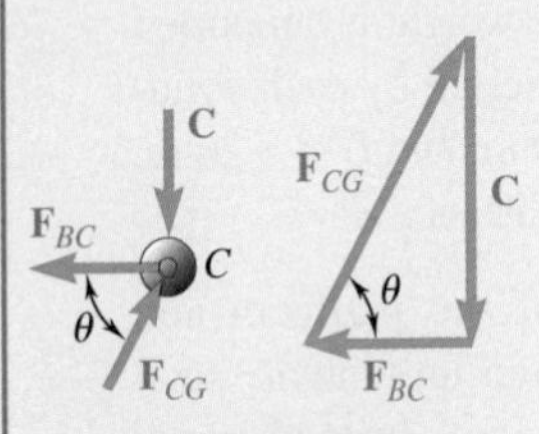

Fig. 4 Free-body diagram of roller *C* and its force triangle.

Free-Body Roller *C* (Fig. 4). Draw a force triangle and obtain $F_{BC} = C\cot\theta$.

Free Body: Member *BDE* (Fig. 5). Recalling that $F_{AD} = 0$, you have

$$+\circlearrowleft\Sigma M_E = 0: \quad F_{DH}\cos(\phi - 90°)a - B(2a\cos\theta) - F_{BC}(2a\sin\theta) = 0$$
$$F_{DH}a\sin\phi - B(2a\cos\theta) - (C\cot\theta)(2a\sin\theta) = 0$$
$$F_{DH}\sin\phi - 2(B + C)\cos\theta = 0$$

From Eq. (1), you obtain

$$F_{DH} = W\frac{\cos\theta}{\sin\phi} \tag{2}$$

Note that *the result obtained is independent of d.* ◀

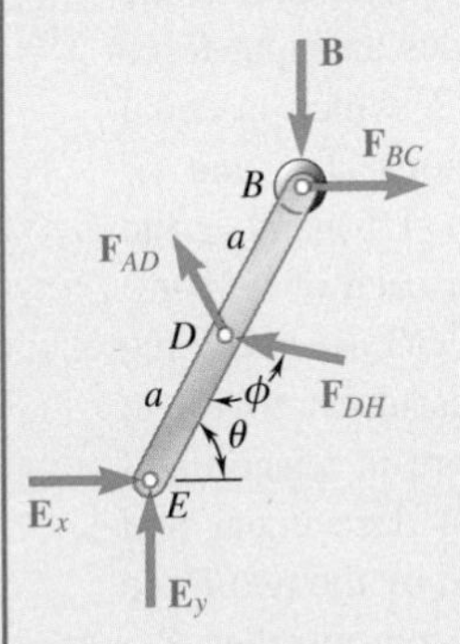

Fig. 5 Free-body diagram of member *BDE*.

Applying first the law of sines to triangle *EDH* (Fig. 6), you have

$$\frac{\sin\phi}{EH} = \frac{\sin\theta}{DH} \qquad \sin\phi = \frac{EH}{DH}\sin\theta \tag{3}$$

Now using the law of cosines, you get

$$(DH)^2 = a^2 + L^2 - 2aL\cos\theta$$
$$= (0.70)^2 + (3.20)^2 - 2(0.70)(3.20)\cos 60°$$
$$(DH)^2 = 8.49 \qquad DH = 2.91 \text{ m}$$

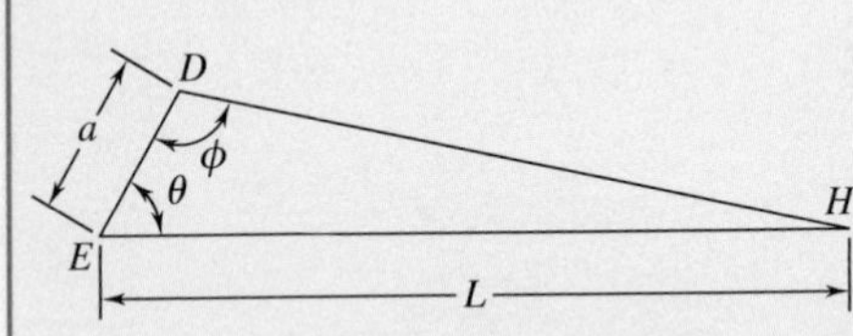

Fig. 6 Geometry of triangle *EDH*.

Also note that

$$W = mg = (1000 \text{ kg})(9.81 \text{ m/s}^2) = 9810 \text{ N} = 9.81 \text{ kN}$$

Substituting for $\sin\phi$ from Eq. (3) into Eq. (2) and using the numerical data, your result is

$$F_{DH} = W\frac{DH}{EH}\cot\theta = (9.81 \text{ kN})\frac{2.91 \text{ m}}{3.20 \text{ m}}\cot 60°$$

$$F_{DH} = 5.15 \text{ kN} \blacktriangleleft$$

REFLECT and THINK: Note that link *AD* ends up having zero force in this situation. However, this member still serves an important function, as it is necessary to enable the machine to support any horizontal load that might be exerted on the platform.

Problems

6.73 A 100-lb force directed vertically downward is applied to the toggle vise at C. Knowing that link BD is 6 in. long and that $a = 4$ in., determine the horizontal force exerted on block E.

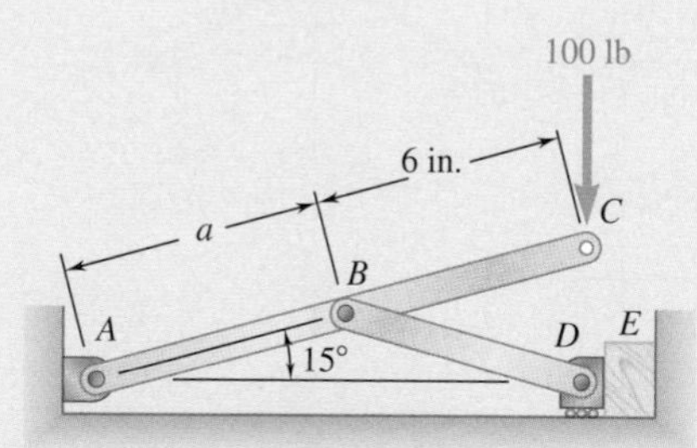

Fig. P6.73 and P6.74

6.74 A 100-lb force directed vertically downward is applied to the toggle vise at C. Knowing that link BD is 6 in. long and that $a = 8$ in., determine the horizontal force exerted on block E.

6.75 The shear shown is used to cut and trim electronic-circuit-board laminates. For the position shown, determine (*a*) the vertical component of the force exerted on the shearing blade at D, (*b*) the reaction at C.

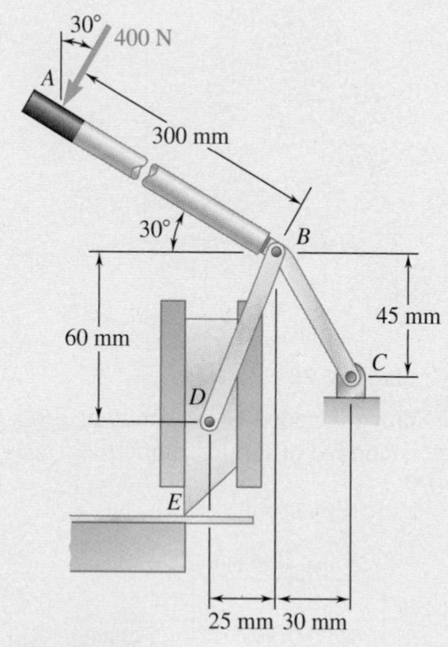

Fig. P6.75

6.76 Water pressure in the supply system exerts a downward force of 135 N on the vertical plug at A. Determine the tension in the fusible link DE and the force exerted on member BCE at B.

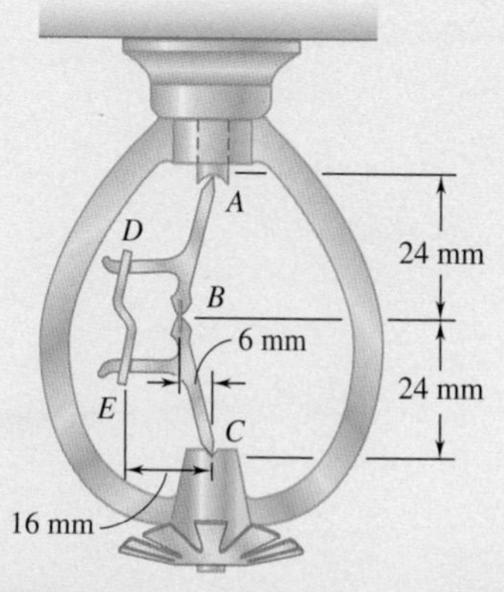

Fig. P6.76

6.77 A 39-ft length of railroad rail of weight 44 lb/ft is lifted by the tongs shown. Determine the forces exerted at D and F on tong BDF.

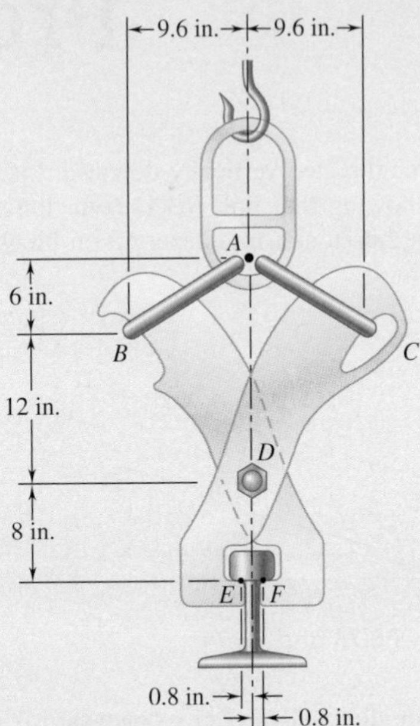

Fig. ***P6.77***

6.78 The tongs shown are used to apply a total upward force of 45 kN on a pipe cap. Determine the forces exerted at D and F on tong ADF.

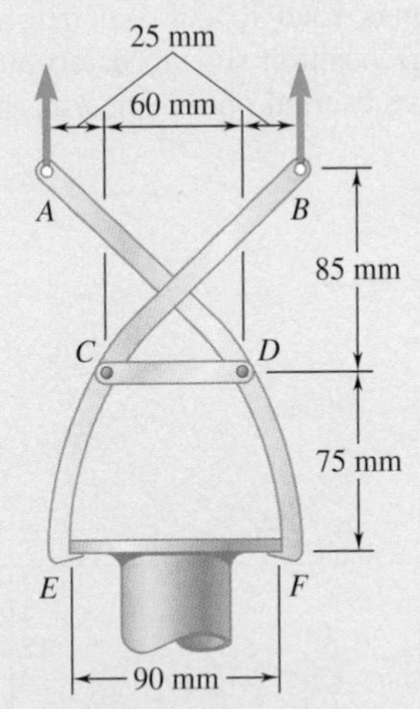

Fig. P6.78

6.79 If the toggle shown is added to the tongs of Prob. 6.78 and a single vertical force is applied at G, determine the forces exerted at D and F on tong ADF.

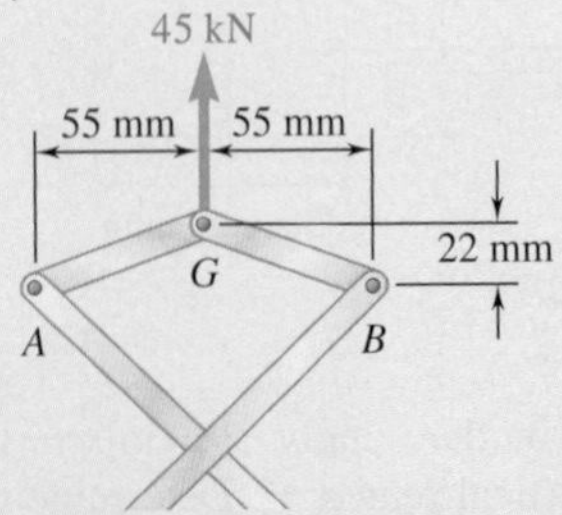

Fig. ***P6.79***

6.80 A small barrel weighing 60 lb is lifted by a pair of tongs as shown. Knowing that $a = 5$ in., determine the forces exerted at B and D on tong ABD.

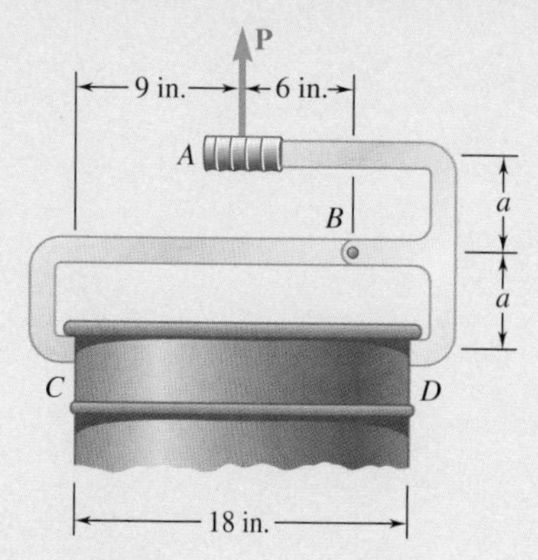

Fig. P6.80

6.81 A force **P** of magnitude 2.4 kN is applied to the piston of the engine system shown. For each of the two positions shown, determine the couple **M** required to hold the system in equilibrium.

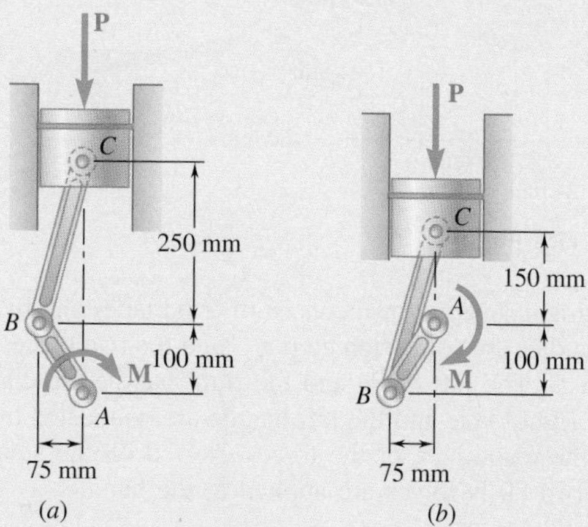

Fig. P6.81 and P6.82

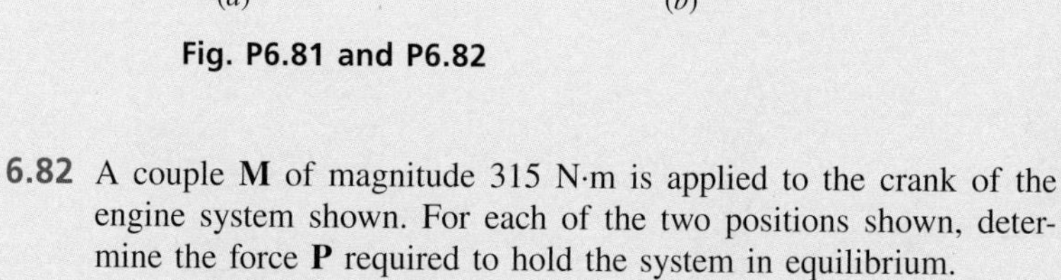

6.82 A couple **M** of magnitude 315 N·m is applied to the crank of the engine system shown. For each of the two positions shown, determine the force **P** required to hold the system in equilibrium.

6.83 and *6.84* Two rods are connected by a frictionless collar B. Knowing that the magnitude of the couple $\mathbf{M}_A$ is 500 lb·in., determine (*a*) the couple $\mathbf{M}_C$ required for equilibrium, (*b*) the corresponding components of the reaction at C.

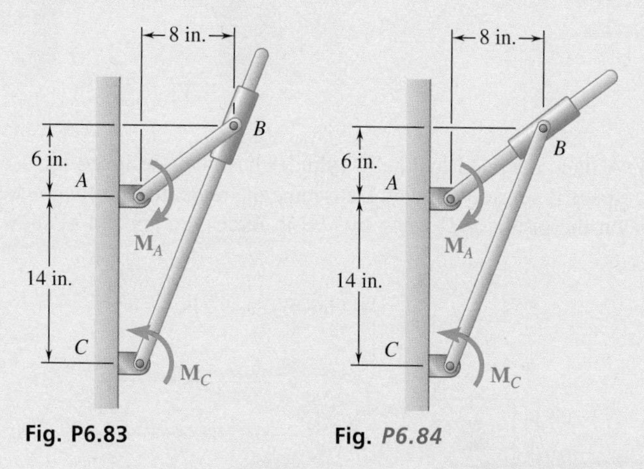

Fig. P6.83 Fig. *P6.84*

6.85 The pliers shown are used to grip a 0.3-in.-diameter rod. Knowing that two 60-lb forces are applied to the handles, determine (*a*) the magnitude of the forces exerted on the rod, (*b*) the force exerted by the pin at A on portion AB of the pliers.

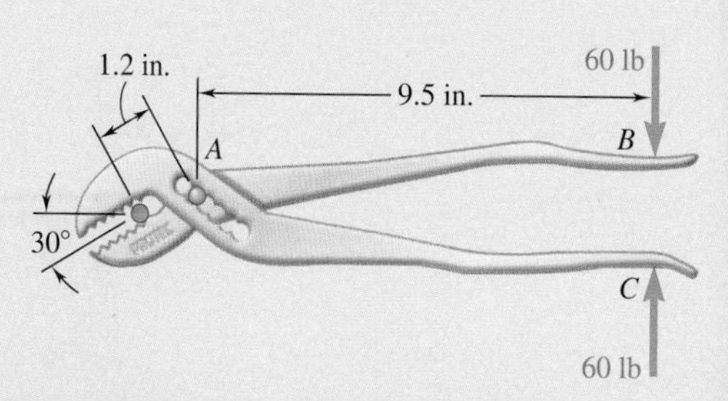

Fig. P6.85

6.86 In using the bolt cutter shown, a worker applies two 300-N forces to the handles. Determine the magnitude of the forces exerted by the cutter on the bolt.

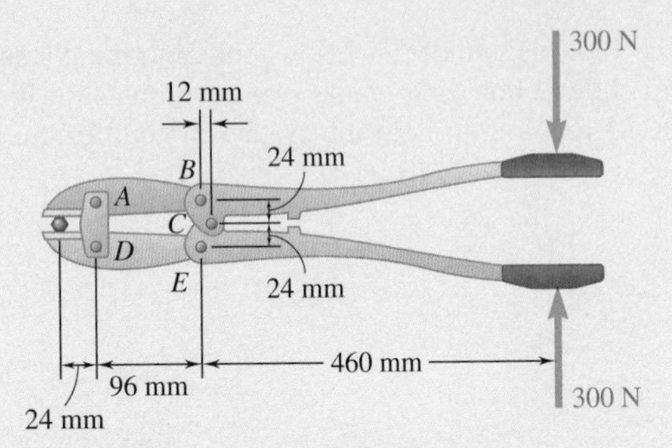

Fig. P6.86

6.87 The garden shears shown consist of two blades and two handles. The two handles are connected by pin C and the two blades are connected by pin D. The left blade and the right handle are connected by pin A; the right blade and the left handle are connected by pin B. Determine the magnitude of the forces exerted on the small branch at E when two 80-N forces are applied to the handles as shown.

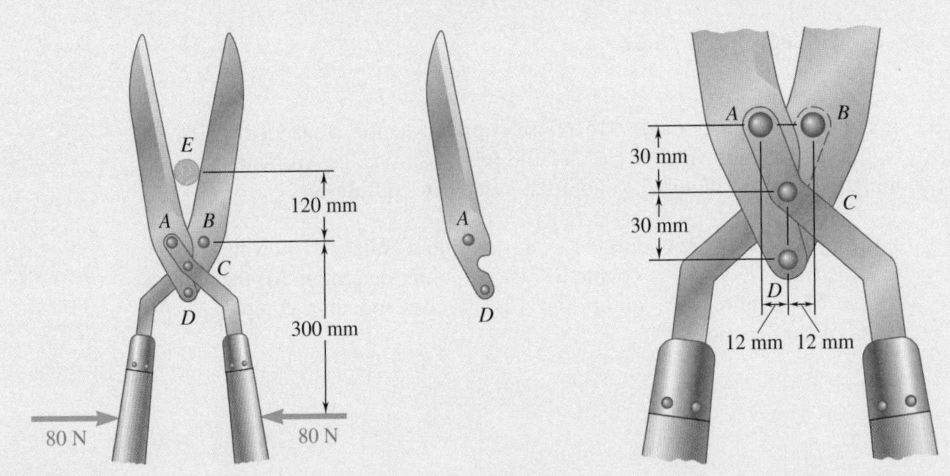

Fig. *P6.87*

6.88 A hand-operated hydraulic cylinder has been designed for use where space is severely limited. Determine the magnitude of the force exerted on the piston at D when two 90-lb forces are applied as shown.

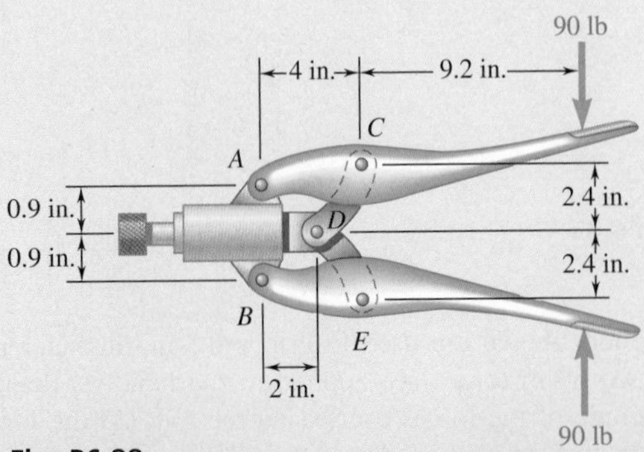

Fig. P6.88

6.89 A 45-lb shelf is held horizontally by a self-locking brace that consists of two parts *EDC* and *CDB* hinged at *C* and bearing against each other at *D*. Determine the force **P** required to release the brace.

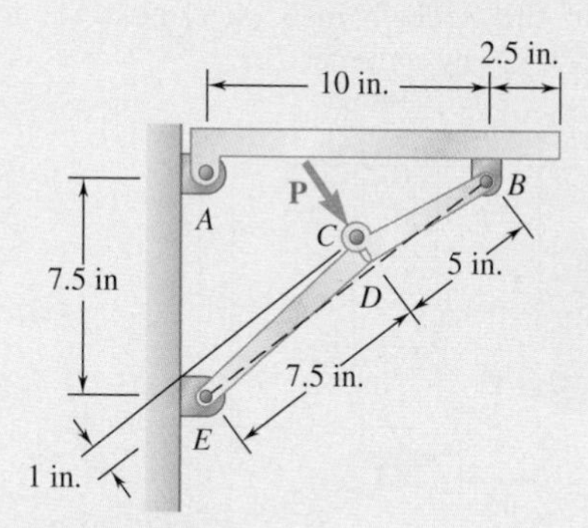

Fig. P6.89

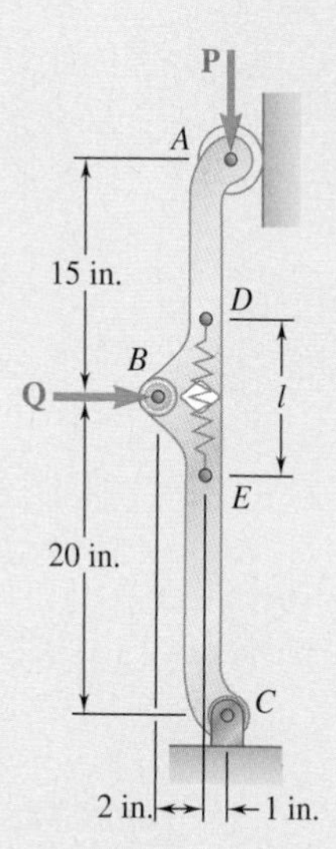

Fig. *P6.90*

6.90 Because the brace shown must remain in position even when the magnitude of **P** is very small, a single safety spring is attached at *D* and *E*. The spring *DE* has a constant of 50 lb/in. and an unstretched length of 7 in. Knowing that $l = 10$ in. and that the magnitude of **P** is 800 lb, determine the force **Q** required to release the brace.

6.91 and 6.92 Determine the force **P** that must be applied to the toggle *CDE* to maintain bracket *ABC* in the position shown.

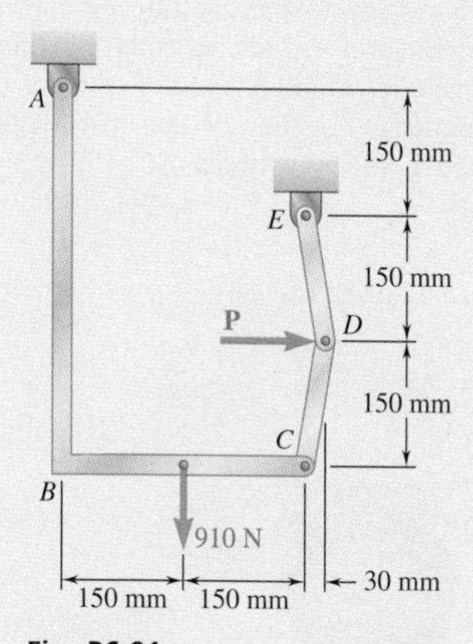

Fig. P6.91

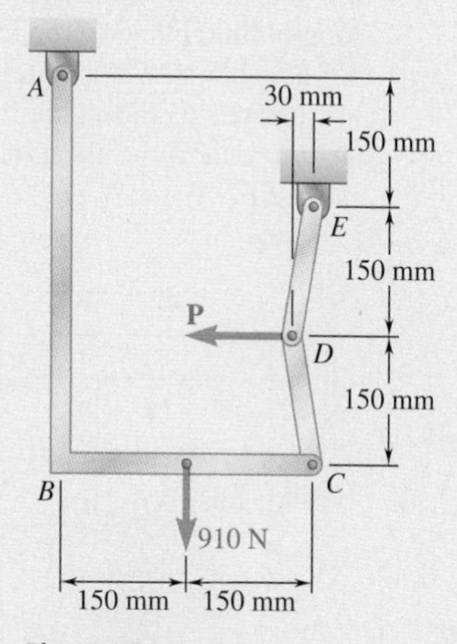

Fig. P6.92

6.93 The telescoping arm *ABC* is used to provide an elevated platform for construction workers. The workers and the platform together have a mass of 200 kg and have a combined center of gravity located directly above *C*. For the position when $\theta = 20°$, determine (*a*) the force exerted at *B* by the single hydraulic cylinder *BD*, (*b*) the force exerted on the supporting carriage at *A*.

6.94 The telescoping arm *ABC* of Prob. 6.93 can be lowered until end *C* is close to the ground, so that workers can easily board the platform. For the position when $\theta = -20°$, determine (*a*) the force exerted at *B* by the single hydraulic cylinder *BD*, (*b*) the force exerted on the supporting carriage at *A*.

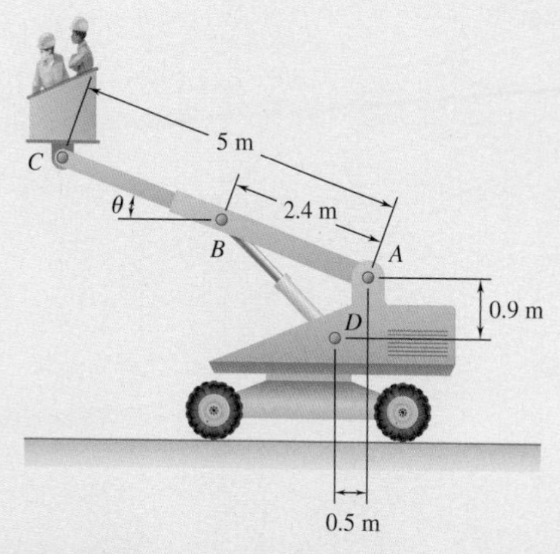

Fig. *P6.93*

6.95 The bucket of the front-end loader shown carries a 3200-lb load. The motion of the bucket is controlled by two identical mechanisms, only one of which is shown. Knowing that the mechanism shown supports one-half of the 3200-lb load, determine the force exerted (*a*) by cylinder *CD*, (*b*) by cylinder *FH*.

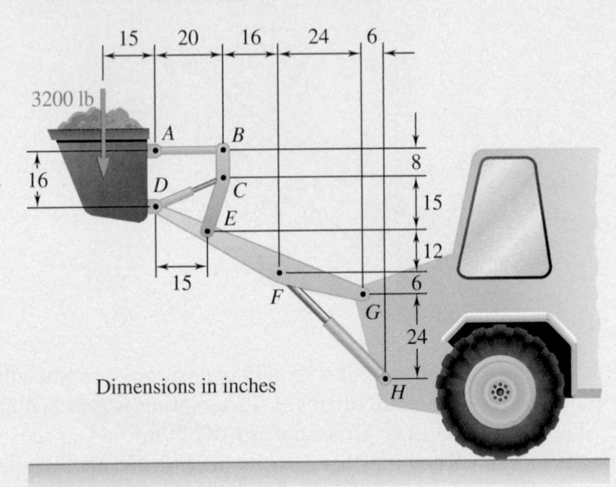

Fig. P6.95

6.96 The motion of the bucket of the front-end loader shown is controlled by two arms and a linkage that are pin-connected at *D*. The arms are located symmetrically with respect to the central vertical and longitudinal plane of the loader; one arm *AFJ* and its control cylinder *EF* are shown. The single linkage *GHDB* and its control cylinder *BC* are located in the plane of symmetry. For the position and loading shown, determine the force exerted (*a*) by cylinder *BC*, (*b*) by cylinder *EF*.

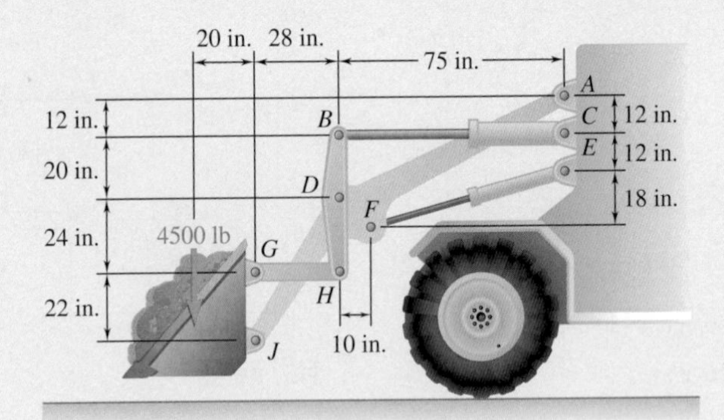

Fig. P6.96

Review and Summary

In this chapter, you studied ways to determine the **internal forces** holding together the various parts of a structure.

Analysis of Trusses

The first half of the chapter presented the analysis of **trusses**, i.e., structures consisting of *straight members connected at their extremities only*. Because the members are slender and unable to support lateral loads, all of the loads must be applied at the joints; thus, we can assume that a truss consists of *pins and two-force members* [Sec. 6.1A].

Simple Trusses

A truss is **rigid** if it is designed in such a way that it does not greatly deform or collapse under a small load. A triangular truss consisting of three members connected at three joints is clearly a rigid truss (Fig. 6.23*a*). The truss obtained by adding two new members to the first one and connecting them at a new joint (Fig. 6.23*b*) is also rigid. Trusses obtained by repeating this procedure are called **simple trusses**. We may check that, in a simple truss, the total number of members is $m = 2n - 3$, where n is the total number of joints [Sec. 6.1A].

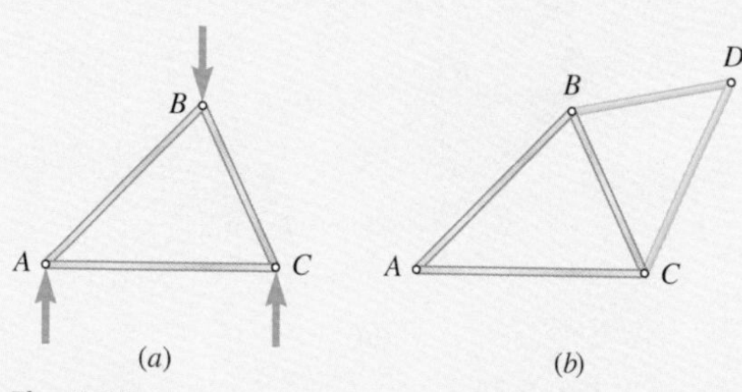

Fig. 6.23

Method of Joints

We can determine the forces in the various members of a simple truss by using the **method of joints** [Sec. 6.1B]. First, we obtain the reactions at the supports by considering the entire truss as a free body. Then we draw the free-body diagram of each pin, showing the forces exerted on the pin by the members or supports it connects. Since the members are straight two-force members, the force exerted by a member on the pin is directed along that member, and only the magnitude of the force is unknown. In the case of a simple truss, it is always possible to draw the free-body diagrams of the pins in such an order that only two unknown forces are included in each diagram. We obtain these forces from the corresponding two equilibrium equations or—if only three forces are involved—from the corresponding force triangle. If the force exerted by a member on a pin is directed toward that pin, the member is in **compression**; if it is directed away from the pin, the member is in **tension** [Sample Prob. 6.1]. The analysis of a truss is sometimes expedited by first recognizing **joints under special loading conditions** [Sec. 6.1C].

Method of Sections

The **method of sections** is usually preferable to the method of joints when we want to determine the force in only one member—or very few members—of a truss [Sec. 6.2A]. To determine the force in member *BD* of the truss of Fig. 6.24*a*, for example, we *pass a section* through members *BD, BE,* and *CE;* remove these members; and use the portion *ABC* of the truss as a free body (Fig. 6.24*b*). Setting $\Sigma M_E = 0$, we determine the magnitude of force $\mathbf{F}_{BD}$ that represents the force in member *BD*. A positive sign indicates that the member is in *tension;* a negative sign indicates that it is in *compression* [Sample Probs. 6.2 and 6.3].

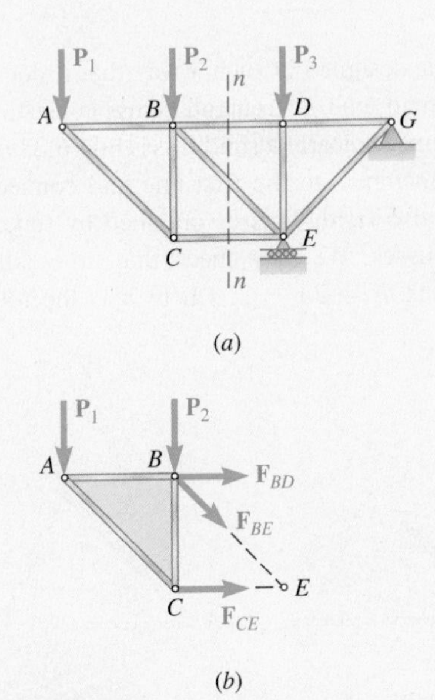

Fig. 6.24

Compound Trusses

The method of sections is particularly useful in the analysis of **compound trusses**, i.e., trusses that cannot be constructed from the basic triangular truss of Fig. 6.23*a* but are built by rigidly connecting several simple trusses [Sec. 6.2B]. If the component trusses are properly connected (e.g., one pin and one link, or three non-concurrent and unparallel links) and if the resulting structure is properly supported (e.g., one pin and one roller), the compound truss is **statically determinate, rigid, and completely constrained**. The following necessary—but not sufficient—condition is then satisfied: $m + r = 2n$, where m is the number of members, r is the number of unknowns representing the reactions at the supports, and n is the number of joints.

Frames and Machines

In the second part of the chapter, we analyzed **frames** and **machines**. These structures contain *multi-force members,* i.e., members acted upon by three or more forces. Frames are designed to support loads and are usually stationary, fully constrained structures. Machines are designed to transmit or modify forces and always contain moving parts [Sec. 6.3].

Analysis of a Frame

To analyze a frame, we first consider the entire frame to be a free body and write three equilibrium equations [Sec. 6.3A]. If the frame remains rigid when detached from its supports, the reactions involve only three unknowns and may be determined from these equations [Sample Probs. 6.4 and 6.5]. On the other hand, if the frame ceases to be rigid when detached from its supports, the reactions involve more than three unknowns, and we cannot determine them completely from the equilibrium equations of the frame [Sec. 6.3B; Sample Prob. 6.6].

Multi-force Members

We then dismember the frame and identify the various members as either two-force members or multi-force members; we assume pins form an integral part of one of the members they connect. We draw the free-body diagram of each of the multi-force members, noting that, when two multi-force members are connected to the same two-force member, they are acted upon by that member with *equal and opposite forces of unknown magnitude but known direction.* When two multi-force members are connected by a pin, they exert on each other *equal and opposite forces of unknown direction* that should be represented by *two unknown components*. We can then solve the equilibrium equations obtained from the free-body diagrams of the multi-force members for the various internal forces [Sample Probs. 6.4 and 6.5]. We also can use the equilibrium equations to complete the determination of the reactions at the supports [Sample Prob. 6.6]. Actually, if the frame is *statically determinate and rigid,* the free-body diagrams of the multi-force members could provide as many equations as there are unknown forces (including the reactions) [Sec. 6.3B]. However, as suggested previously, it is advisable to first consider the free-body diagram of the entire frame to minimize the number of equations that must be solved simultaneously.

Analysis of a Machine

To analyze a machine, we dismember it and, following the same procedure as for a frame, draw the free-body diagram of each multi-force member. The corresponding equilibrium equations yield the **output forces** exerted by the machine in terms of the **input forces** applied to it as well as the **internal forces** at the various connections [Sec. 6.4; Sample Prob. 6.7].

Review Problems

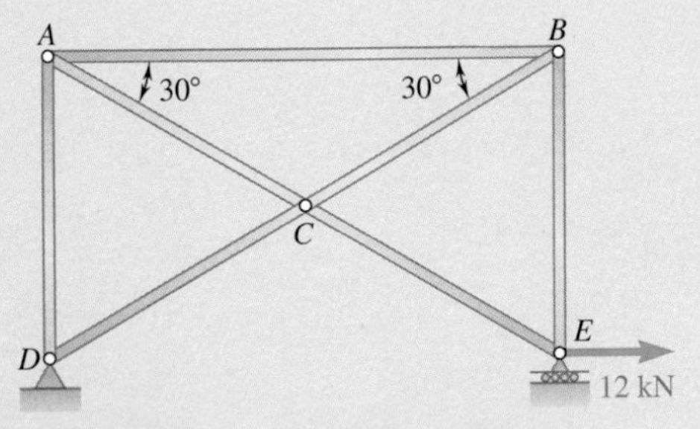

Fig. P6.97

6.97 Determine the force in each member of the truss shown.

6.98 Determine the force in member DE and in each of the members located to the left of DE for the inverted Howe roof truss shown. State whether each member is in tension or compression.

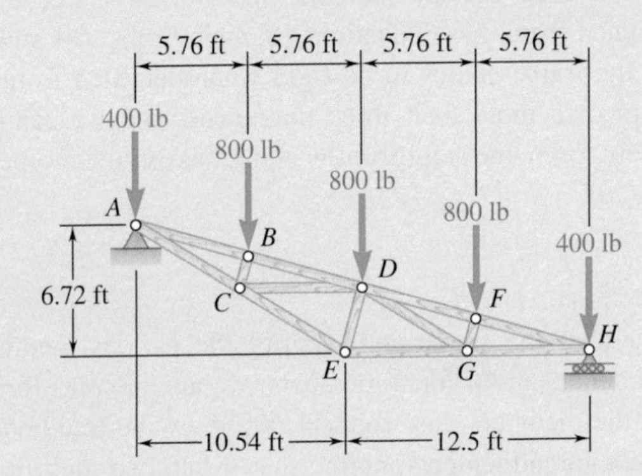

Fig. *P6.98*

6.99 Determine the force in members EH and GI of the truss shown. (*Hint:* Use section *aa*.)

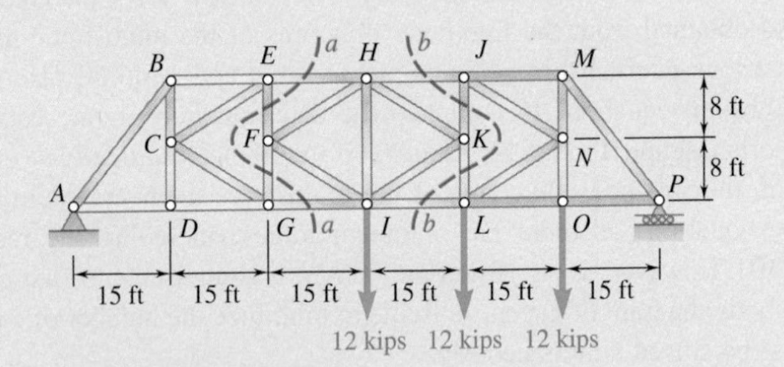

Fig. P6.99 and P6.100

6.100 Determine the force in members HJ and IL of the truss shown. (*Hint:* Use section *bb*.)

6.101 The low-bed trailer shown is designed so that the rear end of the bed can be lowered to ground level in order to facilitate the loading of equipment or wrecked vehicles. A 1400-kg vehicle has been hauled to the position shown by a winch; the trailer is then returned to a traveling position where $\alpha = 0$ and both AB and BE are horizontal. Considering only the weight of the disabled automobile, determine the force that must be exerted by the hydraulic cylinder to maintain a position with $\alpha = 0$.

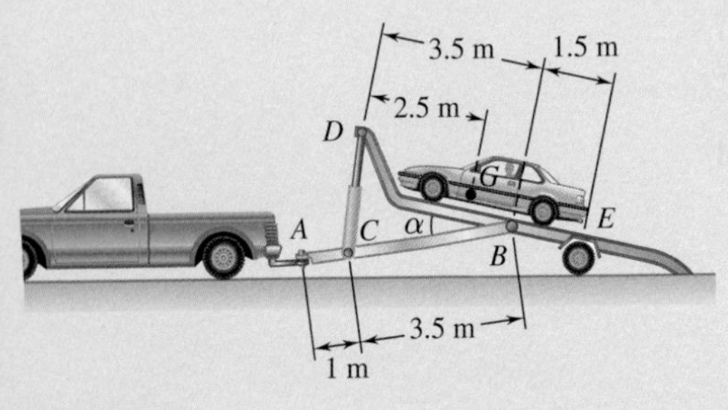

Fig. P6.101

6.102 The axis of the three-hinge arch ABC is a parabola with vertex at B. Knowing that $P = 112$ kN and $Q = 140$ kN, determine (*a*) the components of the reaction at A, (*b*) the components of the force exerted at B on segment AB.

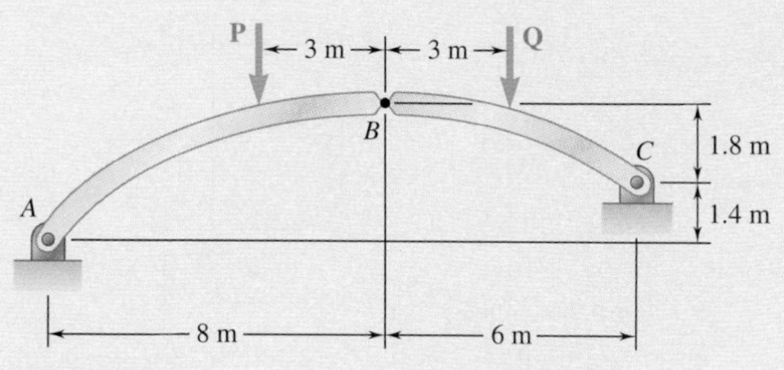

Fig. *P6.102*

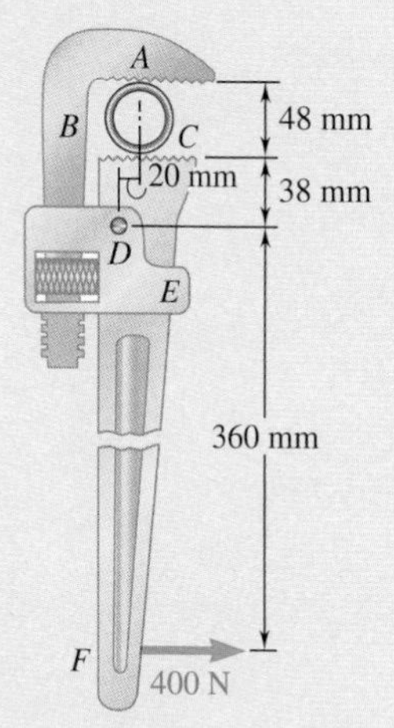

Fig. P6.103

6.103 A 48-mm-diameter pipe is gripped by the Stillson wrench shown. Portions AB and DE of the wrench are rigidly attached to each other and portion CF is connected by a pin at D. Assuming that no slipping occurs between the pipe and the wrench, determine the components of the forces exerted on the pipe at A and C.

6.104 The compound-lever pruning shears shown can be adjusted by placing pin A at various ratchet positions on blade ACE. Knowing that 300-lb vertical forces are required to complete the pruning of a small branch, determine the magnitude P of the forces that must be applied to the handles when the shears are adjusted as shown.

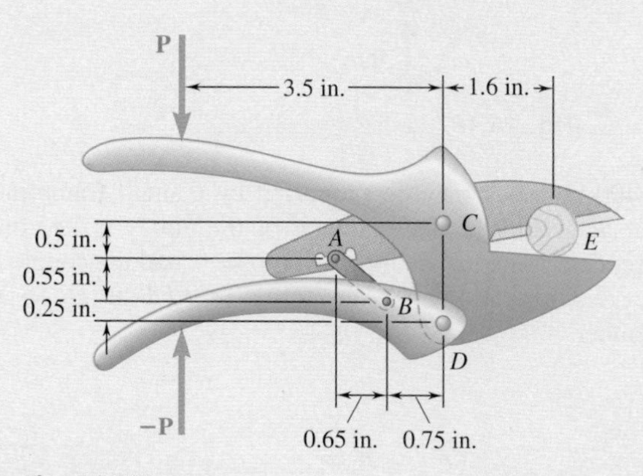

Fig. P6.104

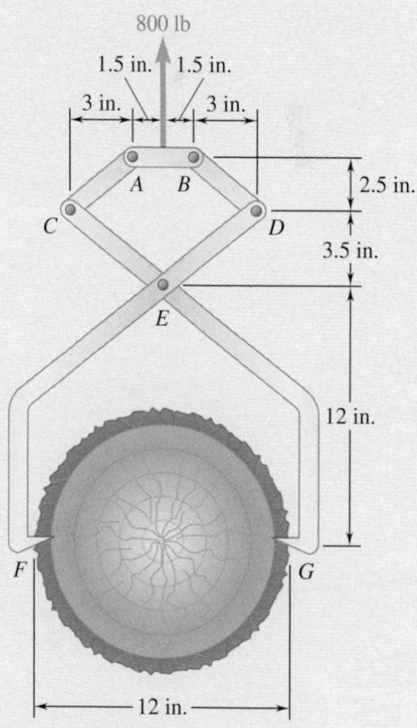

Fig. *P6.105*

6.105 A log weighing 800 lb is lifted by a pair of tongs as shown. Determine the forces exerted at E and F on tong DEF.

6.106 A 3-ft-diameter pipe is supported every 16 ft by a small frame like that shown. Knowing that the combined weight of the pipe and its contents is 500 lb/ft and assuming frictionless surfaces, determine the components (*a*) of the reaction at *E*, (*b*) of the force exerted at *C* on member *CDE*.

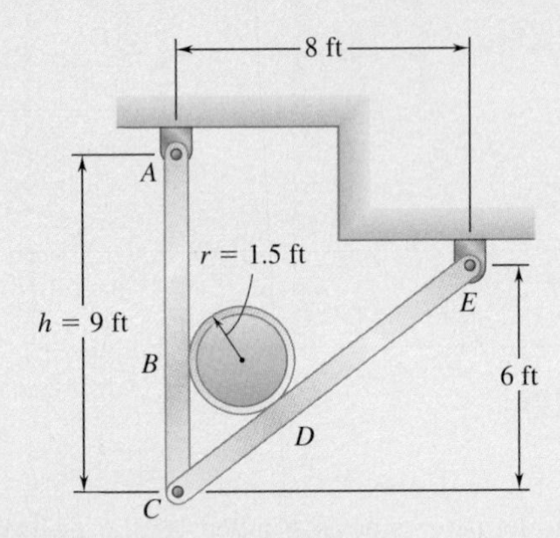

Fig. P6.106

6.107 For the bevel-gear system shown, determine the required value of α if the ratio of M_B to M_A is to be 3.

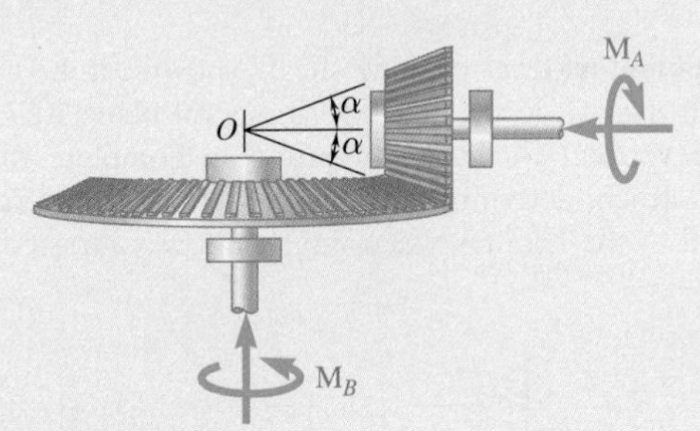

Fig. *P6.107*

6.108 A 400-kg block may be supported by a small frame in each of the four ways shown. The diameter of the pulley is 250 mm. For each case, determine (*a*) the force components and the couple representing the reaction at *A*, (*b*) the force exerted at *D* on the vertical member.

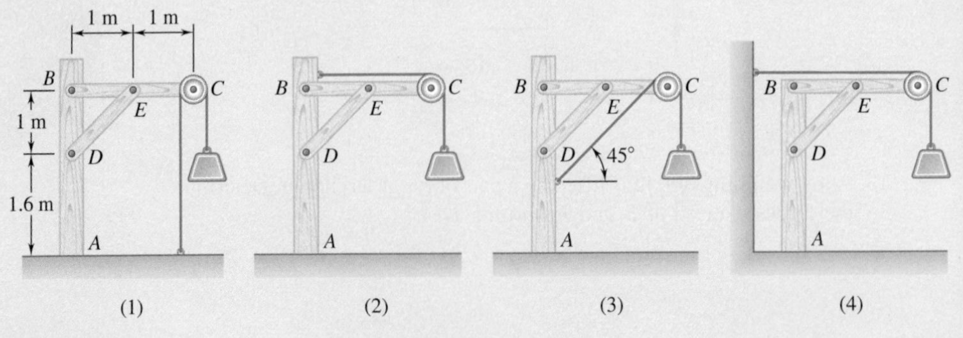

Fig. P6.108

© ConstructionPhotography.com/Photoshot

7

Distributed Forces: Moments of Inertia

The strength of structural members used in the construction of buildings depends to a large extent on the properties of their cross sections. This includes the second moments of area, or moments of inertia, of these cross sections.

Objectives

- **Describe** the second moment, or moment of inertia, of an area.
- **Determine** the rectangular and polar moments of inertia of areas and their corresponding radii of gyration by integration.
- **Develop** the parallel-axis theorem and apply it to determine the moments of inertia of composite areas.

Introduction

In Chap. 5 we analyzed various systems of forces distributed over an area or volume. The three main types of forces considered were (1) weights of homogeneous plates of uniform thickness (Secs. 5.1 and 5.2); (2) distributed loads on beams (Sec. 5.3); and (3) weights of homogeneous three-dimensional bodies (Sec. 5.4). In all of these cases, the distributed forces were proportional to the elemental areas or volumes associated with them. Therefore, we could obtain the resultant of these forces by summing the corresponding areas or volumes, and we determined the moment of the resultant about any given axis by computing the first moments of the areas or volumes about that axis.

In this chapter, we consider distributed forces $\Delta\mathbf{F}$ where the magnitudes depend not only upon the elements of area ΔA on which these forces act but also upon the distance from ΔA to some given axis. More precisely, we assume the magnitude of the force per unit area $\Delta F/\Delta A$ varies linearly with the distance to the axis. Forces of this type arise in the study of the bending of beams.

Starting with the assumption that the elemental forces involved are distributed over an area A and vary linearly with the distance y to the x axis, we will show that the magnitude of their resultant $\mathbf{R}$ depends upon the first moment Q_x of the area A. However, the location of the point where $\mathbf{R}$ is applied depends upon the *second moment,* or *moment of inertia,* I_x of the same area with respect to the x axis. You will see how to compute the moments of inertia of various areas with respect to given x and y axes. We also introduce the *polar moment of inertia* J_O of an area. To facilitate these computations, we establish a relation between the moment of inertia I_x of an area A with respect to a given x axis and the moment of inertia $I_{x'}$ of the same area with respect to the parallel centroidal x' axis (a relation known as the parallel-axis theorem).

7.1 MOMENTS OF INERTIA OF AREAS

In this chapter, we consider distributed forces $\Delta\mathbf{F}$ whose magnitudes ΔF are proportional to the elements of area ΔA on which the forces act and, at the same time, vary linearly with the distance from ΔA to a given axis.

7.1A Second Moment, or Moment of Inertia, of an Area

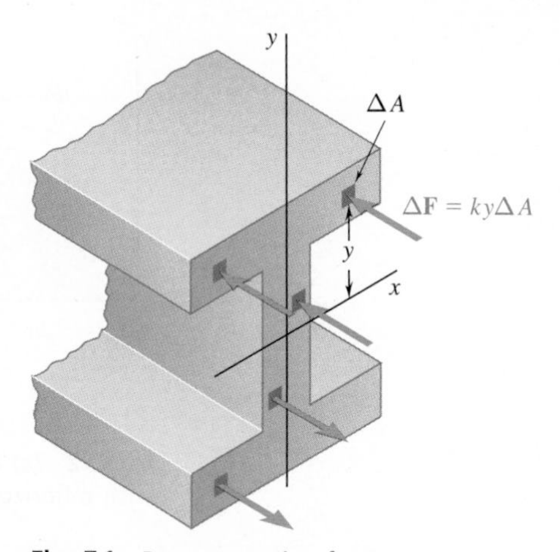

Fig. 7.1 Representative forces on a cross section of a beam subjected to equal and opposite couples at each end.

Consider a beam with a uniform cross section that is subjected to two equal and opposite couples: one applied at each end of the beam. Such a beam is said to be in **pure bending**. The internal forces in any section of the beam are distributed forces whose magnitudes $\Delta F = ky\,\Delta A$ vary linearly with the distance y between the element of area ΔA and an axis passing through the centroid of the section. (This statement can be derived in a course on mechanics of materials.) This axis, represented by the x axis in Fig. 7.1, is known as the **neutral axis** of the section. The forces on one side of the neutral axis are forces of compression, whereas those on the other side are forces of tension. On the neutral axis itself, the forces are zero.

The magnitude of the resultant **R** of the elemental forces **ΔF** that act over the entire section is

$$R = \int ky\,dA = k\int y\,dA$$

You might recognize this last integral as the **first moment** Q_x of the section about the x axis; it is equal to $\bar{y}A$ and is thus equal to zero, since the centroid of the section is located on the x axis. The system of forces **ΔF** thus reduces to a couple. The magnitude M of this couple (bending moment) must be equal to the sum of the moments $\Delta M_x = y\,\Delta F = ky^2\,\Delta A$ of the elemental forces. Integrating over the entire section, we obtain

$$M = \int ky^2\,dA = k\int y^2\,dA$$

This last integral is known as the **second moment**, or **moment of inertia**,† of the beam section with respect to the x axis and is denoted by I_x. We obtain it by multiplying each element of area dA by the *square of its distance* from the x axis and integrating over the beam section. Since each product $y^2\,dA$ is positive, regardless of the sign of y, or zero (if y is zero), the integral I_x is always positive.

7.1B Determining the Moment of Inertia of an Area by Integration

We just defined the second moment, or moment of inertia, I_x of an area A with respect to the x axis. In a similar way, we can also define the moment of inertia I_y of the area A with respect to the y axis (Fig. 7.2*a*):

Moments of inertia of an area

$$I_x = \int y^2\,dA \qquad I_y = \int x^2\,dA \tag{7.1}$$

†The term *second moment* is more proper than the term *moment of inertia,* which logically should be used only to denote integrals of mass. In engineering practice, however, moment of inertia is used in connection with areas as well as masses.

$dA = dx\,dy$ $\quad dI_x = y^2\,dA \quad dI_y = x^2\,dA$

(a)

$dA = (a - x)\,dy$ $\quad dI_x = y^2\,dA$

(b)

$dA = y\,dx$ $\quad dI_y = x^2\,dA$

(c)

Fig. 7.2 (a) Rectangular moments of inertia dI_x and dI_y of an area dA; (b) calculating I_x with a horizontal strip; (c) calculating I_y with a vertical strip.

We can evaluate these integrals, which are known as the **rectangular moments of inertia** of the area A, more easily if we choose dA to be a thin strip parallel to one of the coordinate axes. To compute I_x, we choose the strip parallel to the x axis, so that all points of the strip are at the same distance y from the x axis (Fig. 7.2b). We obtain the moment of inertia dI_x of the strip by multiplying the area dA of the strip by y^2. To compute I_y, we choose the strip parallel to the y axis, so that all points of the strip are at the same distance x from the y axis (Fig. 7.2c). Then the moment of inertia dI_y of the strip is $x^2\,dA$.

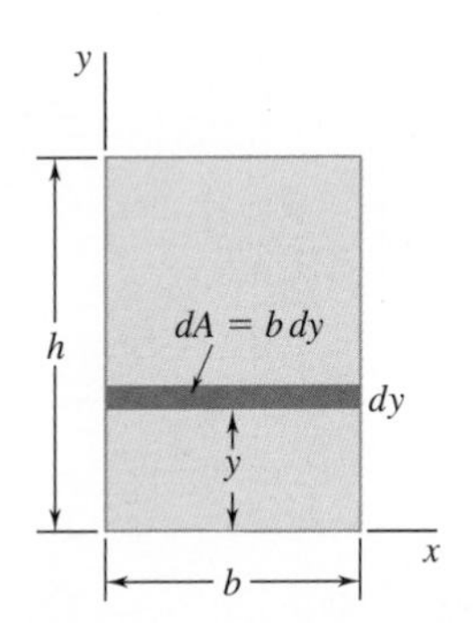

Fig. 7.3 Calculating the moment of inertia of a rectangular area with respect to its base.

Moment of Inertia of a Rectangular Area. As an example, let us determine the moment of inertia of a rectangle with respect to its base (Fig. 7.3). Dividing the rectangle into strips parallel to the x axis, we have

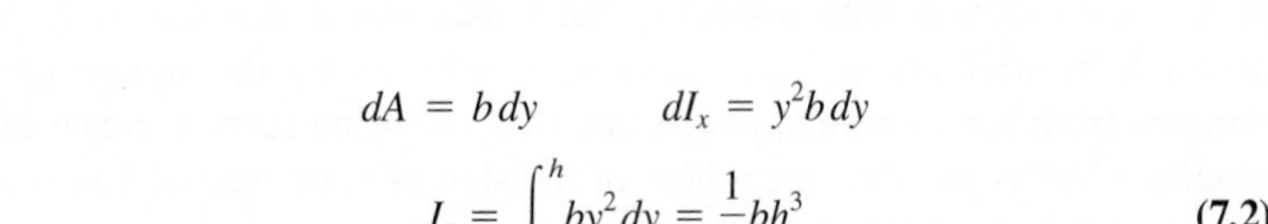

$$dA = b\,dy \qquad dI_x = y^2 b\,dy$$

$$I_x = \int_0^h by^2\,dy = \frac{1}{3}bh^3 \tag{7.2}$$

Computing I_x and I_y Using the Same Elemental Strips. We can use Eq. (7.2) to determine the moment of inertia dI_x with respect to the x axis of a rectangular strip that is parallel to the y axis, such as the strip shown in Fig. 7.2c. Setting $b = dx$ and $h = y$ in formula (7.2), we obtain

$$dI_x = \frac{1}{3}y^3\,dx$$

We also have

$$dI_y = x^2\,dA = x^2 y\,dx$$

Thus, we can use the same element to compute the moments of inertia I_x and I_y of a given area (Fig. 7.4).

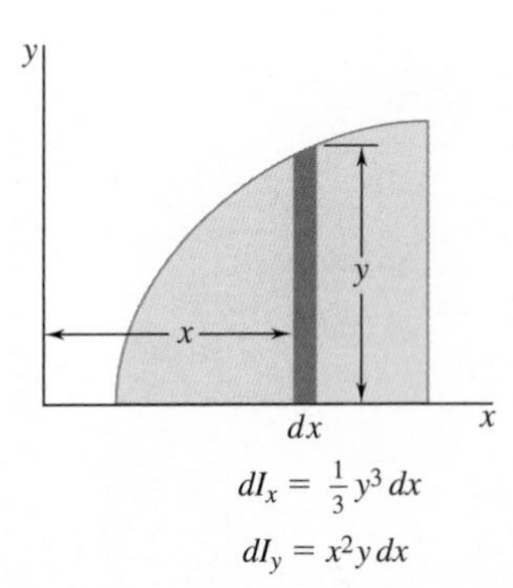

Fig. 7.4 Using the same strip element of a given area to calculate I_x and I_y.

7.1C Polar Moment of Inertia

An integral of great importance in problems concerning the torsion of cylindrical shafts and in problems dealing with the rotation of slabs is

Polar moment of inertia

$$J_O = \int \rho^2 dA \tag{7.3}$$

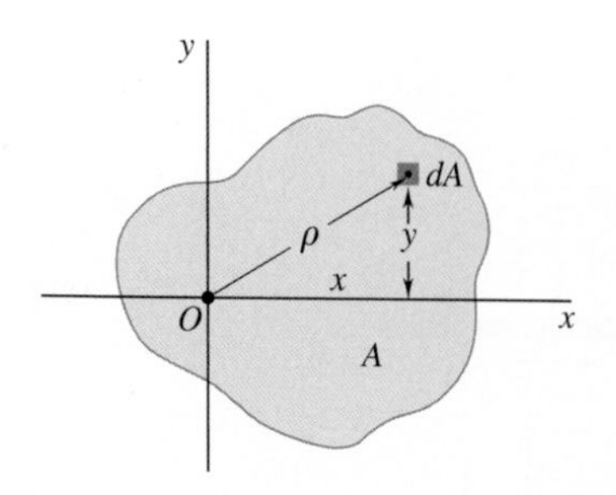

Fig. 7.5 Distance ρ used to evaluate the polar moment of inertia of area A.

where ρ is the distance from O to the element of area dA (Fig. 7.5). This integral is called the **polar moment of inertia** of the area A with respect to the "pole" O.

We can compute the polar moment of inertia of a given area from the rectangular moments of inertia I_x and I_y of the area if these quantities are already known. Indeed, noting that $\rho^2 = x^2 + y^2$, we have

$$J_O = \int \rho^2 dA = \int (x^2 + y^2) dA = \int y^2 dA + \int x^2 dA$$

that is,

$$J_O = I_x + I_y \tag{7.4}$$

7.1D Radius of Gyration of an Area

Consider an area A that has a moment of inertia I_x with respect to the x axis (Fig. 7.6*a*). Imagine that we concentrate this area into a thin strip parallel to the x axis (Fig. 7.6*b*). If the concentrated area A is to have the same moment of inertia with respect to the x axis, the strip should be placed at a distance r_x from the x axis, where r_x is defined by the relation

$$I_x = r_x^2 A$$

Solving for r_x, we have

Radius of gyration

$$r_x = \sqrt{\frac{I_x}{A}} \tag{7.5}$$

The distance r_x is referred to as the **radius of gyration** of the area with respect to the x axis. In a similar way, we can define the radii of gyration r_y and r_O (Fig. 7.6*c* and *d*); we have

$$I_y = r_y^2 A \qquad r_y = \sqrt{\frac{I_y}{A}} \tag{7.6}$$

$$J_O = r_O^2 A \qquad r_O = \sqrt{\frac{J_O}{A}} \tag{7.7}$$

If we rewrite Eq. (7.4) in terms of the radii of gyration, we find that

$$r_O^2 = r_x^2 + r_y^2 \tag{7.8}$$

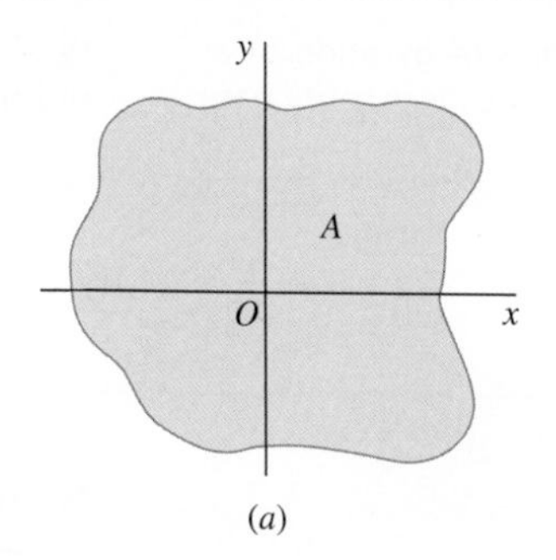

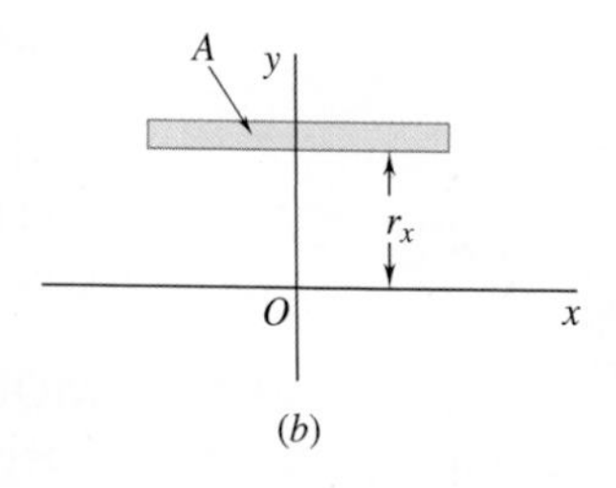

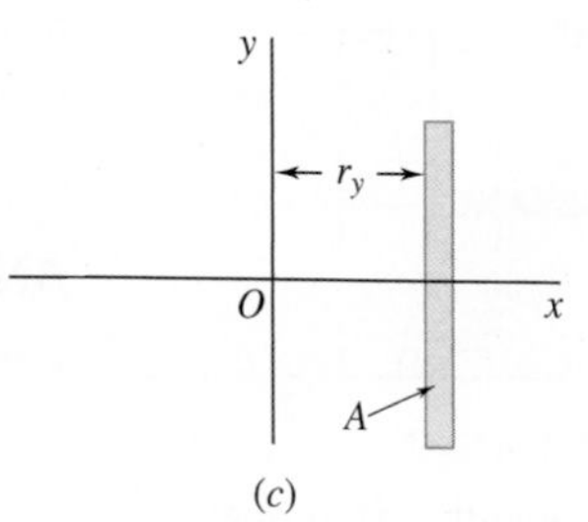

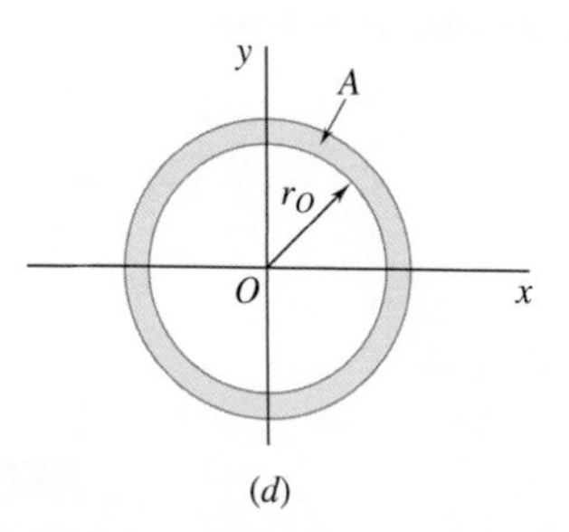

Fig. 7.6 (*a*) Area A with given moment of inertia I_x; (*b*) compressing the area to a horizontal strip with radius of gyration r_x; (*c*) compressing the area to a vertical strip with radius of gyration r_y; (*d*) compressing the area to a circular ring with polar radius of gyration r_O.

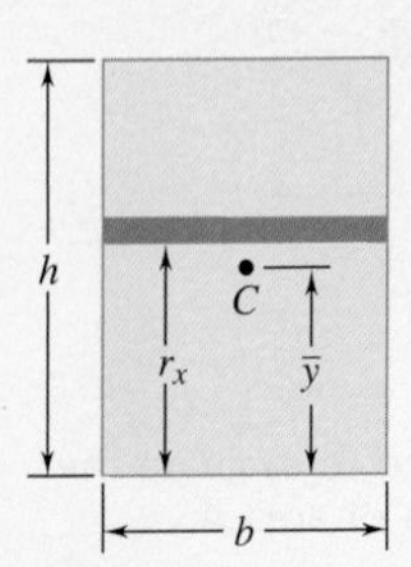

Fig. 7.7 Radius of gyration of a rectangle with respect to its base.

Concept Application 7.1

For the rectangle shown in Fig. 7.7, compute the radius of gyration r_x with respect to its base. Using formulas (7.5) and (7.2), you have

$$r_x^2 = \frac{I_x}{A} = \frac{\frac{1}{3}bh^3}{bh} = \frac{h^2}{3} \qquad r_x = \frac{h}{\sqrt{3}}$$

The radius of gyration r_x of the rectangle is shown in Fig. 7.7. Do not confuse it with the ordinate $\bar{y} = h/2$ of the centroid of the area. The radius of gyration r_x depends upon the *second moment* of the area, whereas the ordinate $\bar{y}$ is related to the *first moment* of the area.

Sample Problem 7.1

Determine the moment of inertia of a triangle with respect to its base.

STRATEGY: To find the moment of inertia with respect to the base, it is expedient to use a differential strip of area parallel to the base. Use the geometry of the situation to carry out the integration.

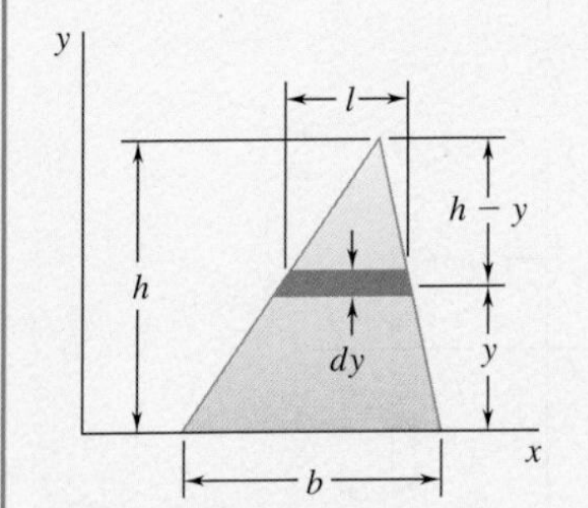

Fig. 1 Triangle with differential strip element parallel to its base.

MODELING: Draw a triangle with a base b and height h, choosing the x axis to coincide with the base (Fig. 1). Choose a differential strip parallel to the x axis to be dA. Since all portions of the strip are at the same distance from the x axis, you have

$$dI_x = y^2\,dA \qquad dA = l\,dy$$

ANALYSIS: Using similar triangles, you have

$$\frac{l}{b} = \frac{h-y}{h} \qquad l = b\frac{h-y}{h} \qquad dA = b\frac{h-y}{h}dy$$

Integrating dI_x from $y = 0$ to $y = h$, you obtain

$$I_x = \int y^2\,dA = \int_0^h y^2 b\frac{h-y}{h}dy = \frac{b}{h}\int_0^h (hy^2 - y^3)\,dy$$

$$= \frac{b}{h}\left[h\frac{y^3}{3} - \frac{y^4}{4}\right]_0^h \qquad I_x = \frac{bh^3}{12} \blacktriangleleft$$

REFLECT and THINK: This problem also could have been solved using a differential strip perpendicular to the base by applying Eq. (7.2) to express the moment of inertia of this strip. However, because of the geometry of this triangle, you would need two integrals to complete the solution.

Sample Problem 7.2

(a) Determine the centroidal polar moment of inertia of a circular area by direct integration. (b) Using the result of part (a), determine the moment of inertia of a circular area with respect to a diameter.

STRATEGY: Since the area is circular, you can evaluate part (a) by using an annular differential area. For part (b), you can use symmetry and Eq. (7.4) to solve for the moment of inertia with respect to a diameter.

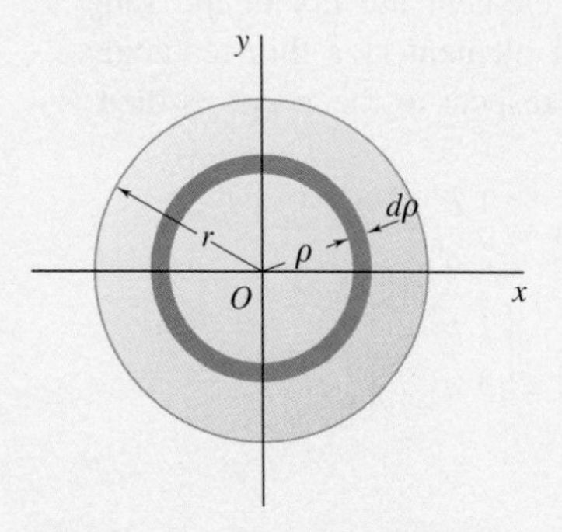

Fig. 1 Circular area with an annular differential element.

MODELING and ANALYSIS:

a. Polar Moment of Inertia. Choose an annular differential element of area to be dA (Fig. 1). Since all portions of the differential area are at the same distance from the origin, you have

$$dJ_O = \rho^2\, dA \qquad dA = 2\pi\rho\, d\rho$$

$$J_O = \int dJ_O = \int_0^r \rho^2(2\pi\rho\, d\rho) = 2\pi \int_0^r \rho^3\, d\rho$$

$$J_O = \frac{\pi}{2} r^4 \blacktriangleleft$$

b. Moment of Inertia with Respect to a Diameter. Because of the symmetry of the circular area, $I_x = I_y$. Then from Eq. (7.4), you have

$$J_O = I_x + I_y = 2I_x \qquad \frac{\pi}{2} r^4 = 2I_x \qquad I_{\text{diameter}} = I_x = \frac{\pi}{4} r^4 \blacktriangleleft$$

REFLECT and THINK: Always look for ways to simplify a problem by the use of symmetry. This is especially true for situations involving circles or spheres.

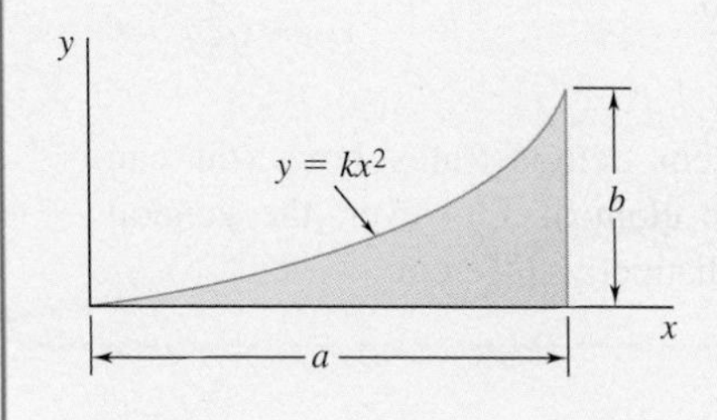

Sample Problem 7.3

(a) Determine the moment of inertia of the shaded region shown with respect to each of the coordinate axes. (Properties of this region were considered in Sample Prob. 5.4.) (b) Using the results of part (a), determine the radius of gyration of the shaded area with respect to each of the coordinate axes.

STRATEGY: You can determine the moments of inertia by using a single differential strip of area; a vertical strip will be more convenient. You can calculate the radii of gyration from the moments of inertia and the area of the region.

MODELING: Referring to Sample Prob. 5.4, you can find the equation of the curve and the total area using

$$y = \frac{b}{a^2}x^2 \qquad A = \tfrac{1}{3}ab$$

ANALYSIS:

a. Moments of Inertia.

Moment of Inertia I_x. Choose a vertical differential element of area for dA (Fig. 1). Since all portions of this element are *not* at the same distance from the x axis, you must treat the element as a thin rectangle. The moment of inertia of the element with respect to the x axis is then

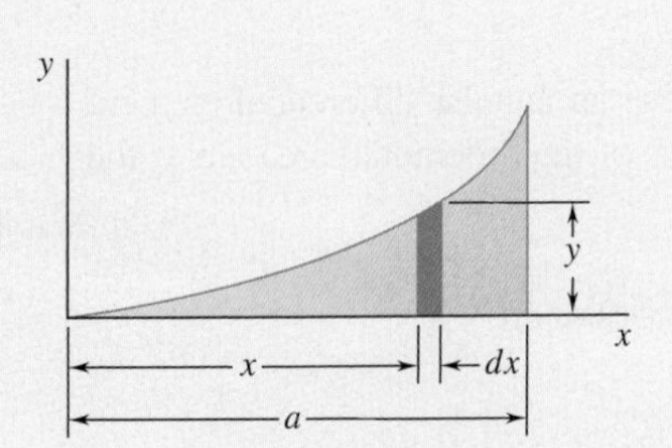

Fig. 1 Subject area with vertical differential strip element.

$$dI_x = \tfrac{1}{3}y^3\,dx = \frac{1}{3}\left(\frac{b}{a^2}x^2\right)^3 dx = \frac{1}{3}\frac{b^3}{a^6}x^6\,dx$$

$$I_x = \int dI_x = \int_0^a \frac{1}{3}\frac{b^3}{a^6}x^6\,dx = \left[\frac{1}{3}\frac{b^3}{a^6}\frac{x^7}{7}\right]_0^a$$

$$I_x = \frac{ab^3}{21} \blacktriangleleft$$

Moment of Inertia I_y. Use the same vertical differential element of area. Since all portions of the element are at the same distance from the y axis, you have

$$dI_y = x^2\,dA = x^2(y\,dx) = x^2\left(\frac{b}{a^2}x^2\right)dx = \frac{b}{a^2}x^4\,dx$$

$$I_y = \int dI_y = \int_0^a \frac{b}{a^2}x^4\,dx = \left[\frac{b}{a^2}\frac{x^5}{5}\right]_0^a$$

$$I_y = \frac{a^3b}{5} \blacktriangleleft$$

b. Radii of Gyration r_x and r_y. From the definition of radius of gyration, you have

$$r_x^2 = \frac{I_x}{A} = \frac{ab^3/21}{ab/3} = \frac{b^2}{7} \qquad r_x = \sqrt{\tfrac{1}{7}}\,b \blacktriangleleft$$

and

$$r_y^2 = \frac{I_y}{A} = \frac{a^3b/5}{ab/3} = \tfrac{3}{5}a^2 \qquad r_y = \sqrt{\tfrac{3}{5}}\,a \blacktriangleleft$$

REFLECT and THINK: This problem demonstrates how you can calculate I_x and I_y using the same strip element. However, the general mathematical approach in each case is distinctly different.

Problems

7.1 through 7.4 Determine by direct integration the moment of inertia of the shaded area with respect to the y axis.

7.5 through *7.8* Determine by direct integration the moment of inertia of the shaded area with respect to the x axis.

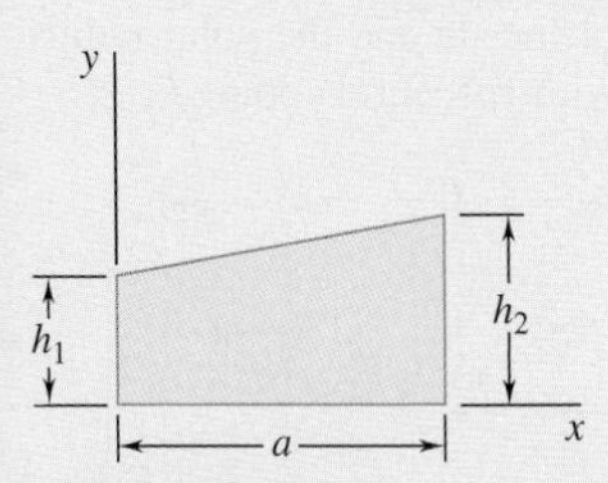

Fig. P7.1 and P7.5

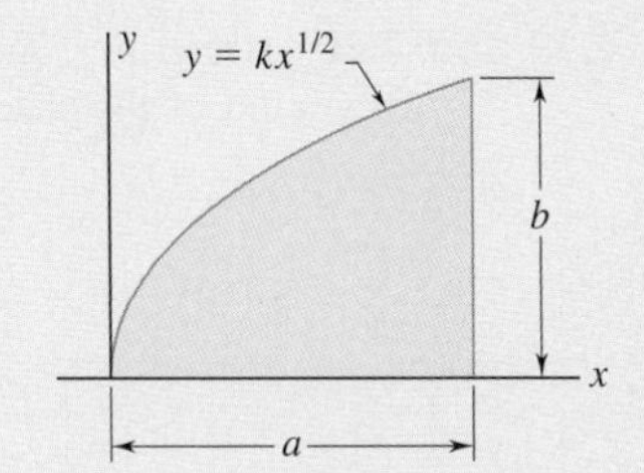

Fig. P7.2 and P7.6

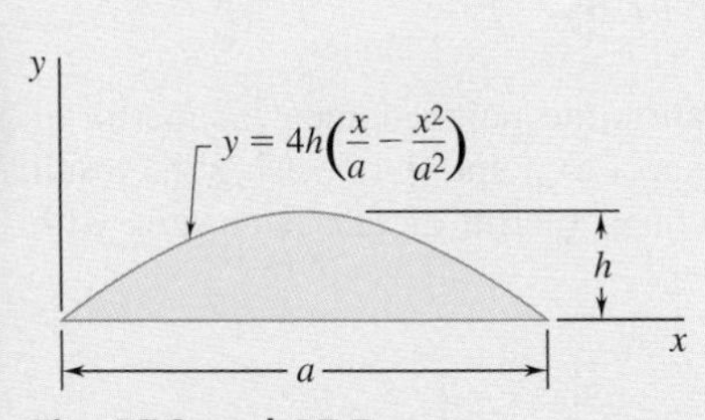

Fig. P7.3 and *P7.7*

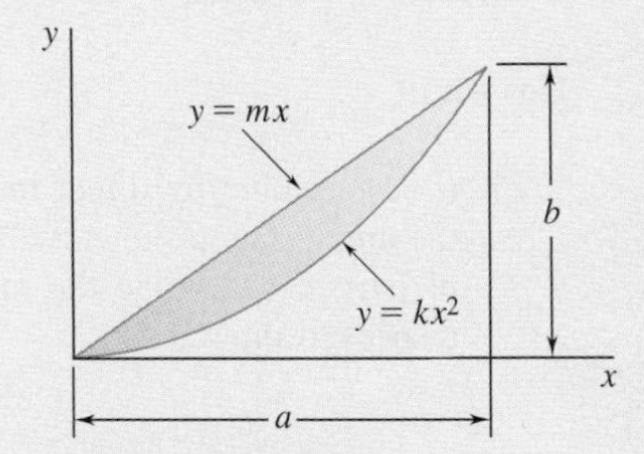

Fig. P7.4 and *P7.8*

7.9 through 7.12 Determine the moment of inertia and radius of gyration of the shaded area shown with respect to the x axis.

7.13 through *7.16* Determine the moment of inertia and radius of gyration of the shaded area shown with respect to the y axis.

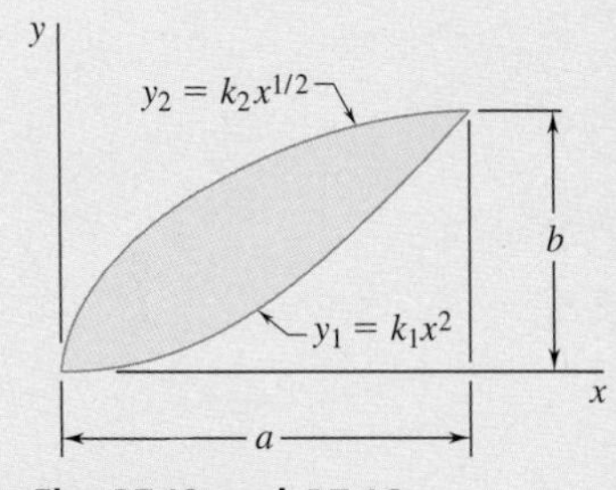

Fig. P7.12 and *P7.16*

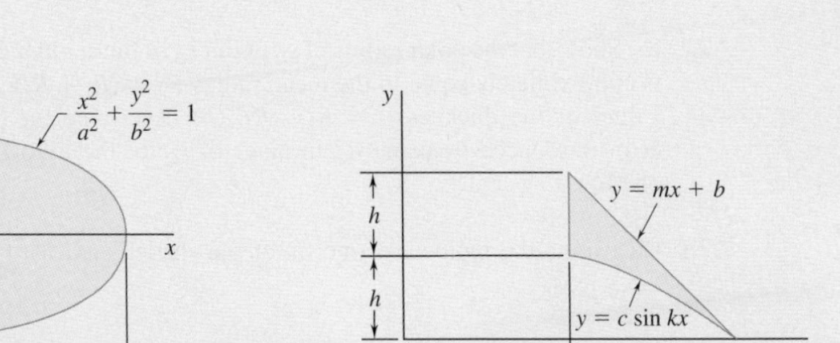

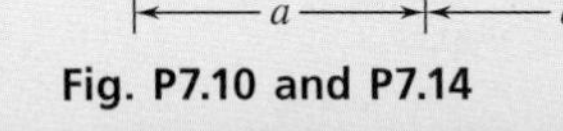

Fig. P7.9 and P7.13

Fig. P7.10 and P7.14

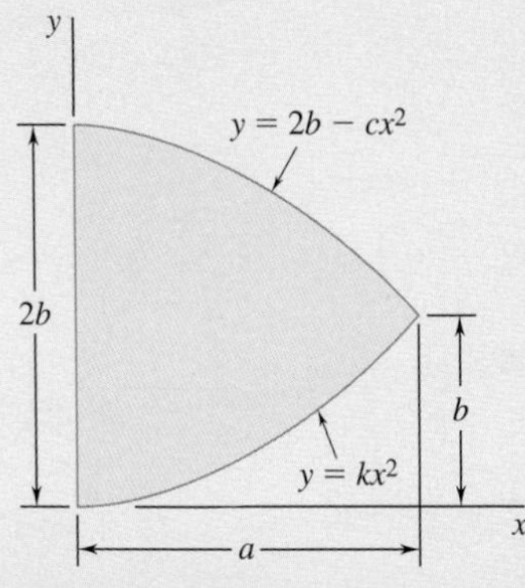

Fig. P7.11 and *P7.15*

7.17 Determine the polar moment of inertia and the polar radius of gyration of the rectangle shown with respect to the midpoint of one of its (*a*) longer sides, (*b*) shorter sides.

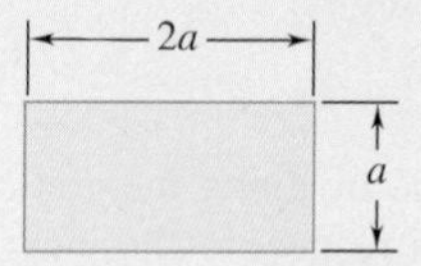

Fig. P7.17 and P7.18

7.18 Determine the polar moment of inertia and the polar radius of gyration of the rectangle shown with respect to one of its corners.

***7.19* and 7.20** Determine the polar moment of inertia and the polar radius of gyration of the shaded area shown with respect to point *P*.

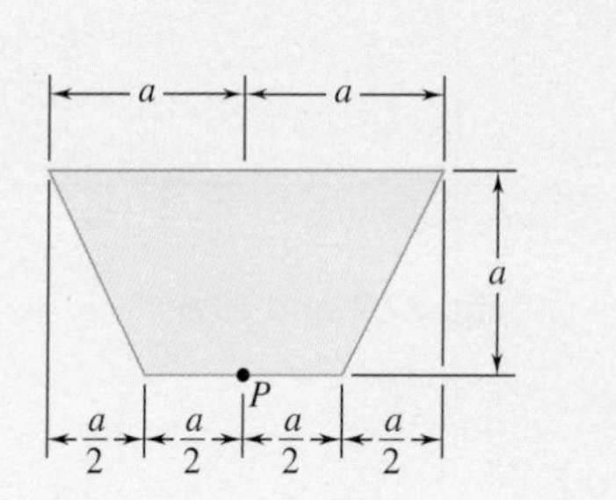

Fig. *P7.19*

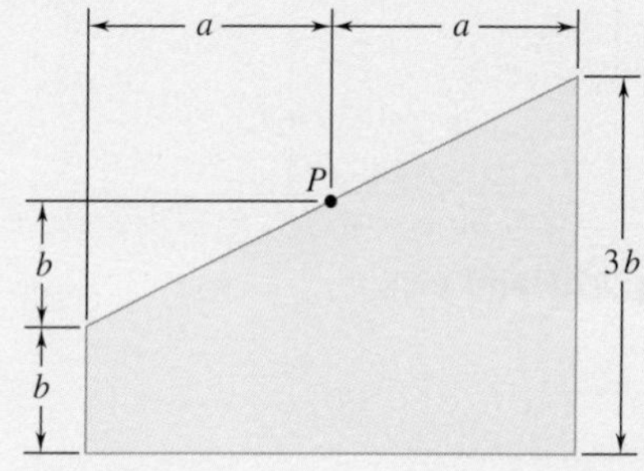

Fig. P7.20

7.21 (*a*) Determine by direct integration the polar moment of inertia of the annular area shown with respect to point *O*. (*b*) Using the result of part *a*, determine the moment of inertia of the given area with respect to the *x* axis.

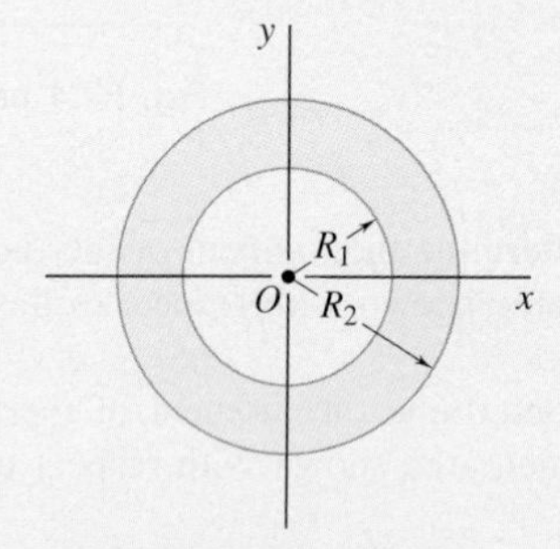

Fig. P7.21 and *P7.22*

7.22 (*a*) Show that the polar radius of gyration r_O of the annular area shown is approximately equal to the mean radius $R_m = (R_1 + R_2)/2$ for small values of the thickness $t = R_2 - R_1$. (*b*) Determine the percentage error introduced by using R_m in place of r_O for the following values of t/R_m: 1, $\frac{1}{2}$, and $\frac{1}{10}$.

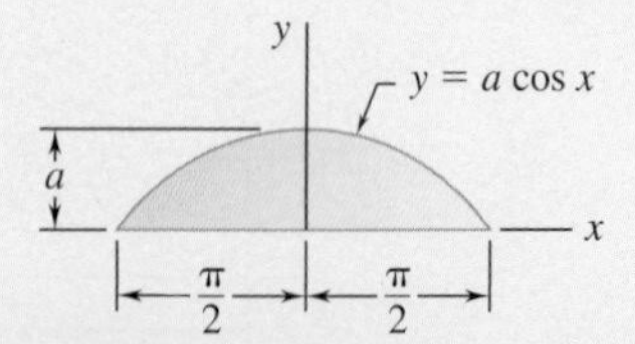

Fig. P7.23 and P7.24

7.23 Determine the moment of inertia of the shaded area with respect to the *x* axis.

7.24 Determine the moment of inertia of the shaded area with respect to the *y* axis.

7.2 PARALLEL-AXIS THEOREM AND COMPOSITE AREAS

In practice, we often need to determine the moment of inertia of a complicated area that can be broken down into a sum of simple areas. However, in doing these calculations, we have to determine the moment of inertia of each simple area with respect to the same axis. In this section, we first derive a formula for computing the moment of inertia of an area with respect to a centroidal axis parallel to a given axis. Then we show how you can use this formula for finding the moment of inertia of a composite area.

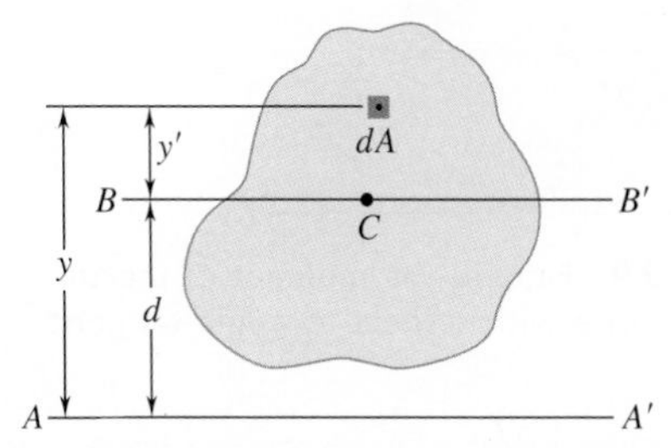

Fig. 7.8 The moment of inertia of an area A with respect to an axis AA' can be determined from its moment of inertia with respect to the centroidal axis BB' by a calculation involving the distance d between the axes.

7.2A The Parallel-Axis Theorem

Consider the moment of inertia I of an area A with respect to an axis AA' (Fig. 7.8). We denote the distance from an element of area dA to AA' by y. This gives us

$$I = \int y^2 dA$$

Let us now draw through the centroid C of the area an axis BB' parallel to AA'; this axis is called a *centroidal axis*. Denoting the distance from the element dA to BB' by y', we have $y = y' + d$, where d is the distance between the axes AA' and BB'. Substituting for y in the previous integral, we obtain

$$I = \int y^2\, dA = \int (y' + d)^2\, dA$$

$$= \int y'^2\, dA + 2d \int y'\, dA + d^2 \int dA$$

The first integral represents the moment of inertia $\bar{I}$ of the area with respect to the centroidal axis BB'. The second integral represents the first moment of the area with respect to BB', but since the centroid C of the area is located on this axis, the second integral must be zero. The last integral is equal to the total area A. Therefore, we have

Parallel-axis theorem

$$I = \bar{I} + Ad^2 \tag{7.9}$$

This formula states that the moment of inertia I of an area with respect to any given axis AA' is equal to the moment of inertia $\bar{I}$ of the area with respect to a centroidal axis BB' parallel to AA' *plus* the product of the area A and the square of the distance d between the two axes. This theorem is known as the **parallel-axis theorem**. Substituting r^2A for I and $\bar{r}^2A$ for $\bar{I}$, we can also express this theorem as

$$r^2 = \bar{r}^2 + d^2 \tag{7.10}$$

A similar theorem relates the polar moment of inertia J_O of an area about a point O to the polar moment of inertia $\bar{J}_C$ of the same area about its centroid C. Denoting the distance between O and C by d, we have

$$J_O = \bar{J}_C + Ad^2 \quad \text{or} \quad r_O^2 = \bar{r}_C^2 + d^2 \tag{7.11}$$

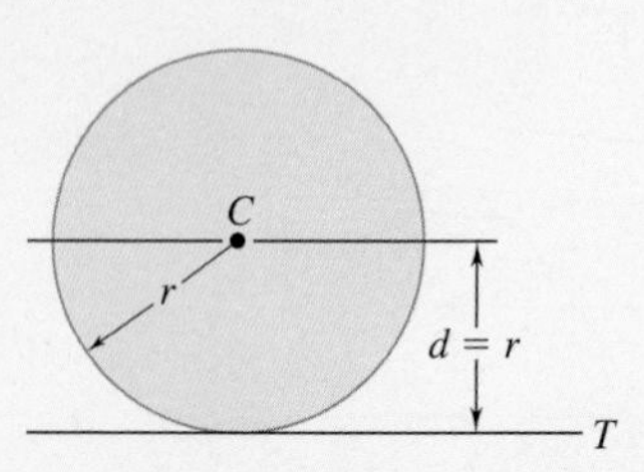

Fig. 7.9 Finding the moment of inertia of a circle with respect to a line tangent to it.

Concept Application 7.2

As an application of the parallel-axis theorem, let us determine the moment of inertia I_T of a circular area with respect to a line tangent to the circle (Fig. 7.9). We found in Sample Prob. 7.2 that the moment of inertia of a circular area about a centroidal axis is $\bar{I} = \frac{1}{4}\pi r^4$. Therefore, we have

$$I_T = \bar{I} + Ad^2 = \tfrac{1}{4}\pi r^4 + (\pi r^2)r^2 = \tfrac{5}{4}\pi r^4$$

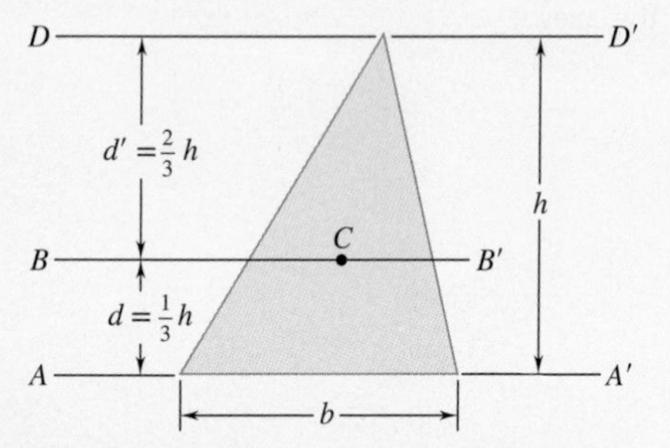

Fig. 7.10 Finding the centroidal moment of inertia of a triangle from the moment of inertia about a parallel axis.

Concept Application 7.3

We can also use the parallel-axis theorem to determine the centroidal moment of inertia of an area when we know the moment of inertia of the area with respect to a parallel axis. Consider, for instance, a triangular area (Fig. 7.10). We found in Sample Prob. 7.1 that the moment of inertia of a triangle with respect to its base AA' is equal to $\frac{1}{12}bh^3$. Using the parallel-axis theorem, we have

$$I_{AA'} = \bar{I}_{BB'} + Ad^2$$
$$\bar{I}_{BB'} = I_{AA'} - Ad^2 = \tfrac{1}{12}bh^3 - \tfrac{1}{2}bh(\tfrac{1}{3}h)^2 = \tfrac{1}{36}bh^3$$

Note that we *subtracted* the product Ad^2 from the given moment of inertia in order to obtain the centroidal moment of inertia of the triangle. That is, this product is *added* when transferring *from* a centroidal axis to a parallel axis, but it is *subtracted* when transferring *to* a centroidal axis. In other words, the moment of inertia of an area is always smaller with respect to a centroidal axis than with respect to any parallel axis.

Returning to Fig. 7.10, we can obtain the moment of inertia of the triangle with respect to the line DD' (which is drawn through a vertex) by writing

$$I_{DD'} = \bar{I}_{BB'} + Ad'^2 = \tfrac{1}{36}bh^3 + \tfrac{1}{2}bh(\tfrac{2}{3}h)^2 = \tfrac{1}{4}bh^3$$

Note that we could not have obtained $I_{DD'}$ directly from $I_{AA'}$. We can apply the parallel-axis theorem only if one of the two parallel axes passes through the centroid of the area.

7.2B Moments of Inertia of Composite Areas

Consider a composite area A made of several component areas $A_1, A_2, A_3, \ldots$. The integral representing the moment of inertia of A can be subdivided into integrals evaluated over $A_1, A_2, A_3, \ldots$. Therefore, we can obtain the moment of inertia of A with respect to a given axis by adding the moments of inertia of the areas $A_1, A_2, A_3, \ldots$ with respect to the same axis.

Figure 7.11 shows several common geometric shapes along with formulas for the moments of inertia of each one. Before adding the moments of inertia of the component areas, however, you may have to use the parallel-axis theorem to transfer each moment of inertia to the desired axis. Sample Probs. 7.4 and 7.5 illustrate the technique.

Shape	Moments of inertia
Rectangle	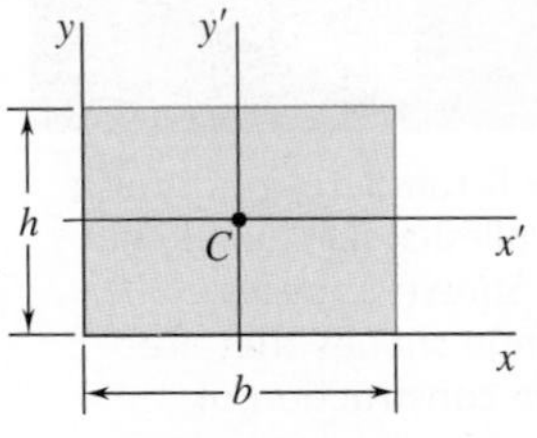 $\bar{I}_{x'} = \frac{1}{12}bh^3$ $\bar{I}_{y'} = \frac{1}{12}b^3h$ $I_x = \frac{1}{3}bh^3$ $I_y = \frac{1}{3}b^3h$ $J_C = \frac{1}{12}bh(b^2 + h^2)$
Triangle	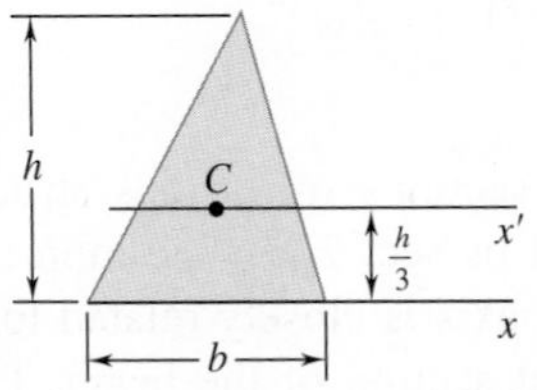 $\bar{I}_{x'} = \frac{1}{36}bh^3$ $I_x = \frac{1}{12}bh^3$
Circle	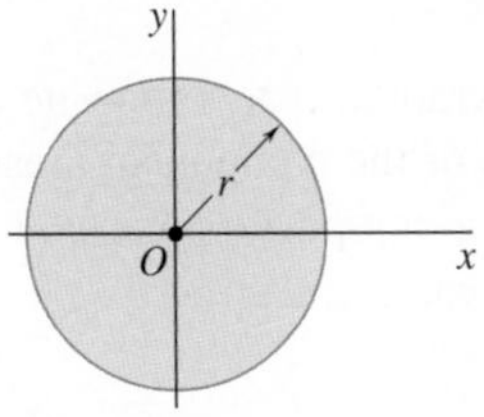 $\bar{I}_x = \bar{I}_y = \frac{1}{4}\pi r^4$ $J_O = \frac{1}{2}\pi r^4$
Semicircle	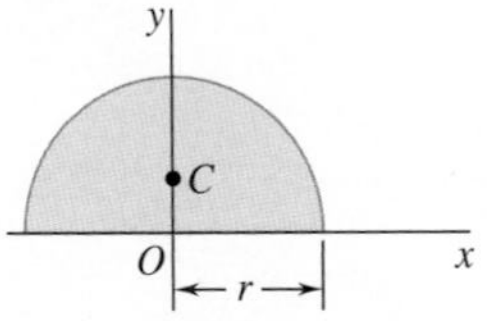 $I_x = I_y = \frac{1}{8}\pi r^4$ $J_O = \frac{1}{4}\pi r^4$
Quarter circle	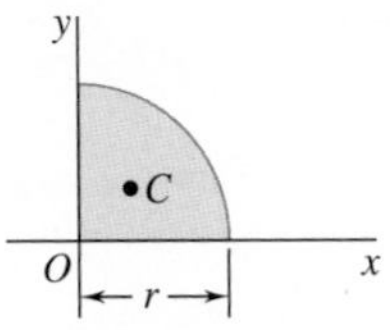 $I_x = I_y = \frac{1}{16}\pi r^4$ $J_O = \frac{1}{8}\pi r^4$
Ellipse	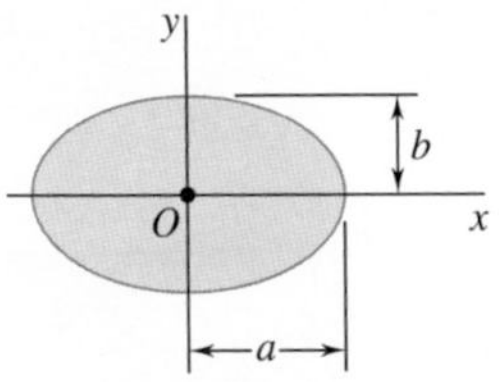 $\bar{I}_x = \frac{1}{4}\pi ab^3$ $\bar{I}_y = \frac{1}{4}\pi a^3b$ $J_O = \frac{1}{4}\pi ab(a^2 + b^2)$

Fig. 7.11 Moments of inertia of common geometric shapes.

Photo 7.1 Appendix B tabulates data for a small sample of the rolled-steel shapes that are readily available. Shown above are examples of wide-flange shapes that are commonly used in the construction of buildings.

© Barry Willis/Getty Images

Properties of the cross sections of various structural shapes are given in App. B. As we noted in Sec. 7.1A, the moment of inertia of a beam section about its neutral axis is closely related to the computation of the bending moment in that section of the beam. Thus, determining moments of inertia is a prerequisite to the analysis and design of structural members.

Note that the radius of gyration of a composite area is *not* equal to the sum of the radii of gyration of the component areas. In order to determine the radius of gyration of a composite area, you must first compute the moment of inertia of the area.

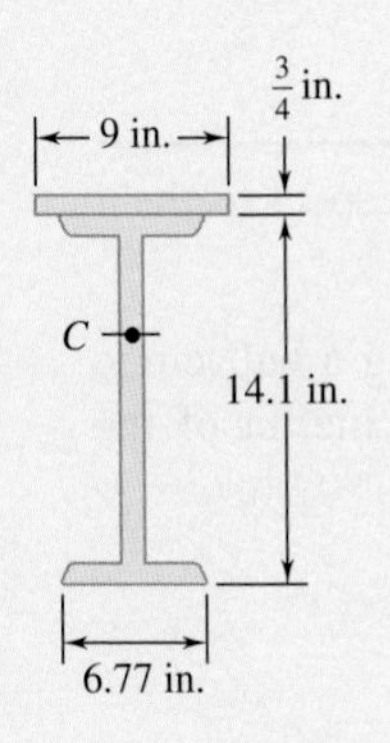

Sample Problem 7.4

The strength of a W14 × 38 rolled-steel beam is increased by attaching a 9 × 3/4-in. plate to its upper flange as shown. Determine the moment of inertia and the radius of gyration of the composite section with respect to an axis that is parallel to the plate and passes through the centroid C of the section.

STRATEGY: This problem involves finding the moment of inertia of a composite area with respect to its centroid. You should first determine the location of this centroid. Then, using the parallel-axis theorem, you can determine the moment of inertia relative to this centroid for the overall section from the centroidal moment of inertia for each component part.

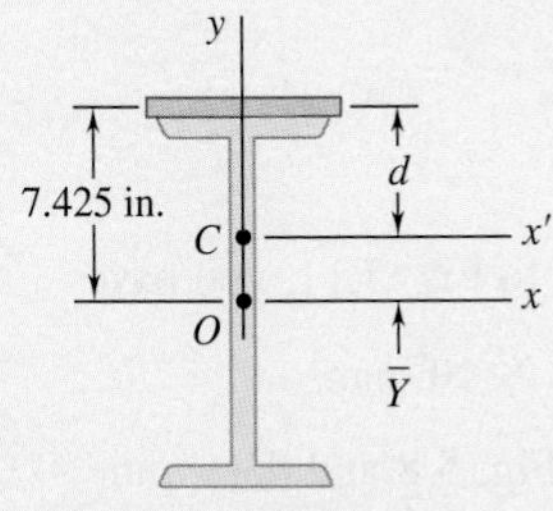

Fig. 1 Origin of coordinates placed at centroid of wide-flange shape.

MODELING and ANALYSIS: Place the origin O of coordinates at the centroid of the wide-flange shape, and compute the distance $\bar{Y}$ to the centroid of the composite section by using the methods of Chap. 5 (Fig. 1). Refer to App. B for the area of the wide-flange shape. The area and the y coordinate of the centroid of the plate are

$$A = (9 \text{ in.})(0.75 \text{ in.}) = 6.75 \text{ in}^2$$
$$\bar{y} = \tfrac{1}{2}(14.1 \text{ in.}) + \tfrac{1}{2}(0.75 \text{ in.}) = 7.425 \text{ in.}$$

Section	Area, in^2	$\bar{y}$, in.	$\bar{y}A$, in^3
Plate	6.75	7.425	50.12
Wide-flange shape	11.2	0	0
	$\Sigma A = 17.95$		$\Sigma \bar{y}A = 50.12$

$$\bar{Y}\Sigma A = \Sigma \bar{y}A \qquad \bar{Y}(17.95) = 50.12 \qquad \bar{Y} = 2.792 \text{ in.}$$

Moment of Inertia. Use the parallel-axis theorem to determine the moments of inertia of the wide-flange shape and the plate with respect to the x' axis. This axis is a centroidal axis for the composite section but *not* for either of the elements considered separately. You can obtain the value of $\bar{I}_x$ for the wide-flange shape from App. B.

For the wide-flange shape,

$$I_{x'} = \bar{I}_x + A\bar{Y}^2 = 385 + (11.2)(2.792)^2 = 472.3 \text{ in}^4$$

For the plate,

$$I_{x'} = \bar{I}_x + Ad^2 = (\tfrac{1}{12})(9)(\tfrac{3}{4})^3 + (6.75)(7.425 - 2.792)^2 = 145.2 \text{ in}^4$$

For the composite area,

$$I_{x'} = 472.3 + 145.2 = 617.5 \text{ in}^4 \qquad I_{x'} = 618 \text{ in}^4 \blacktriangleleft$$

Radius of Gyration. From the moment of inertia and area just calculated, you obtain

$$r_{x'}^2 = \frac{I_{x'}}{A} = \frac{617.5 \text{ in}^4}{17.95 \text{ in}^2} \qquad r_{x'} = 5.87 \text{ in.} \blacktriangleleft$$

REFLECT and THINK: This is a common type of calculation for many different situations. It is often helpful to list data in a table to keep track of the numbers and identify which data you need.

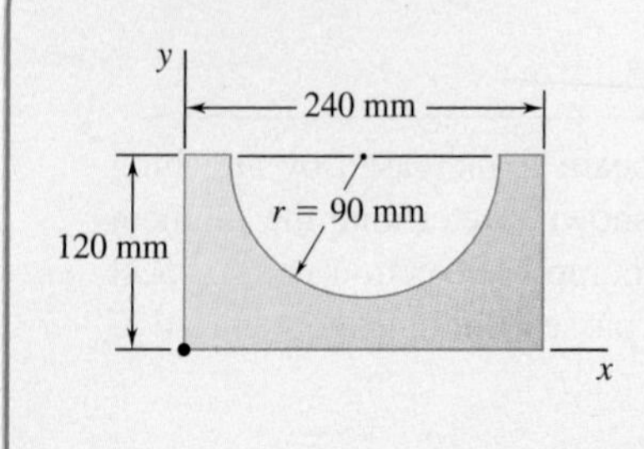

Sample Problem 7.5

Determine the moment of inertia of the shaded area with respect to the x axis.

STRATEGY: You can obtain the given area by subtracting a half circle from a rectangle (Fig. 1). Then compute the moments of inertia of the rectangle and the half circle separately.

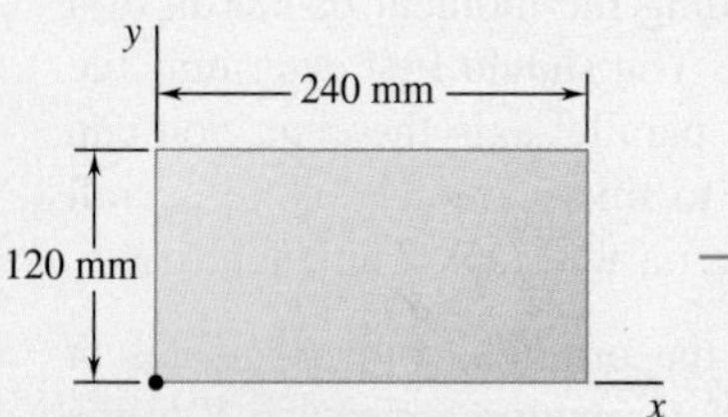

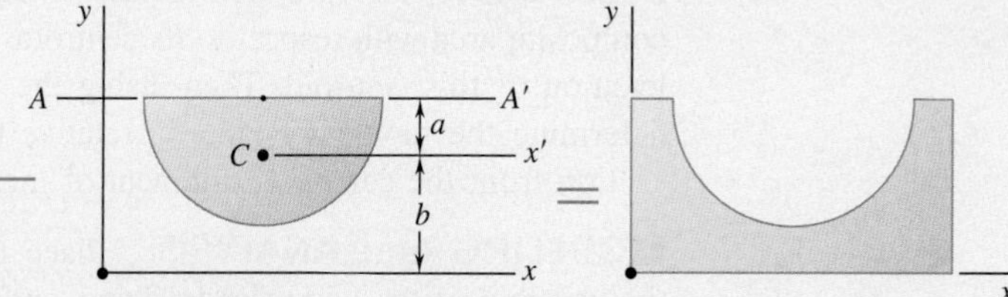

Fig. 1 Modeling given area by subtracting a half circle from a rectangle.

MODELING and ANALYSIS:

Moment of Inertia of Rectangle. Referring to Fig. 7.11, you have

$$I_x = \tfrac{1}{3}bh^3 = \tfrac{1}{3}(240\text{ mm})(120\text{ mm})^3 = 138.2 \times 10^6\text{ mm}^4$$

Moment of Inertia of Half Circle. Refer to Fig. 5.8 and determine the location of the centroid C of the half circle with respect to diameter AA'. As shown in Fig. 2, you have

$$a = \frac{4r}{3\pi} = \frac{(4)(90\text{ mm})}{3\pi} = 38.2\text{ mm}$$

The distance b from the centroid C to the x axis is

$$b = 120\text{ mm} - a = 120\text{ mm} - 38.2\text{ mm} = 81.8\text{ mm}$$

Referring now to Fig. 7.11, compute the moment of inertia of the half circle with respect to diameter AA' and then compute the area of the half circle.

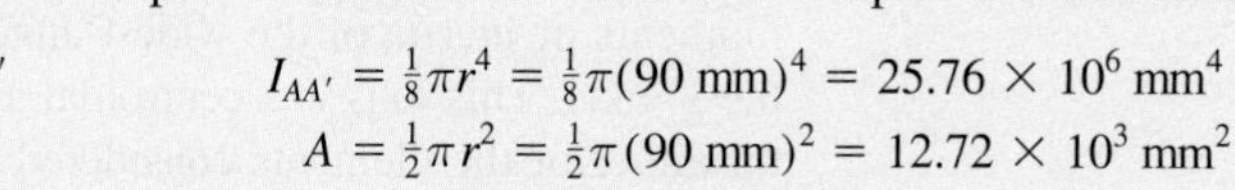

$$I_{AA'} = \tfrac{1}{8}\pi r^4 = \tfrac{1}{8}\pi(90\text{ mm})^4 = 25.76 \times 10^6\text{ mm}^4$$

$$A = \tfrac{1}{2}\pi r^2 = \tfrac{1}{2}\pi(90\text{ mm})^2 = 12.72 \times 10^3\text{ mm}^2$$

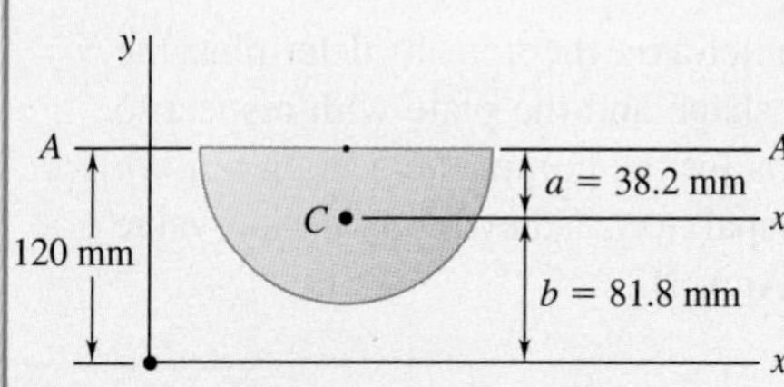

Fig. 2 Centroid location of the half circle.

Next, using the parallel-axis theorem, obtain the value of $\bar{I}_{x'}$ as

$$I_{AA'} = \bar{I}_{x'} + Aa^2$$

$$25.76 \times 10^6\text{ mm}^4 = \bar{I}_{x'} + (12.72 \times 10^3\text{ mm}^2)(38.2\text{ mm})^2$$

$$\bar{I}_{x'} = 7.20 \times 10^6\text{ mm}^4$$

Again using the parallel-axis theorem, obtain the value of I_x as

$$I_x = \bar{I}_{x'} + Ab^2 = 7.20 \times 10^6\text{ mm}^4 + (12.72 \times 10^3\text{ mm}^2)(81.8\text{ mm})^2$$
$$= 92.3 \times 10^6\text{ mm}^4$$

Moment of Inertia of Given Area. Subtracting the moment of inertia of the half circle from that of the rectangle, you obtain

$$I_x = 138.2 \times 10^6\text{ mm}^4 - 92.3 \times 10^6\text{ mm}^4$$

$$I_x = 45.9 \times 10^6\text{ mm}^4 \blacktriangleleft$$

REFLECT and THINK: Figures 5.8 and 7.11 are useful references for locating centroids and moments of inertia of common areas; don't forget to use them.

Problems

7.25 through 7.28 Determine the moment of inertia and the radius of gyration of the shaded area with respect to the x axis.

7.29 through 7.32 Determine the moment of inertia and the radius of gyration of the shaded area with respect to the y axis.

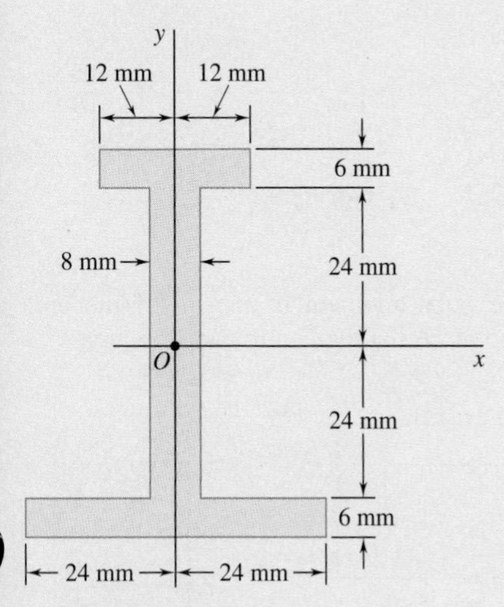

Fig. P7.25 and P7.29

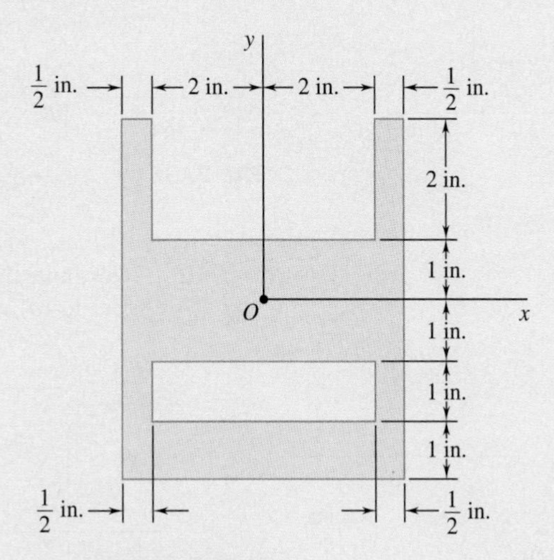

Fig. P7.26 and P7.30

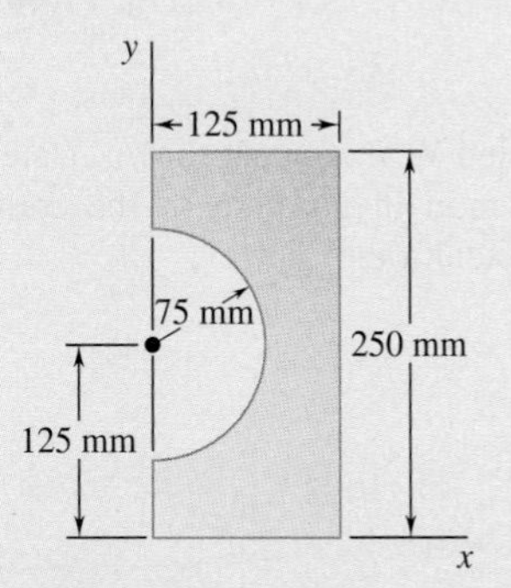

Fig. P7.27 and P7.31

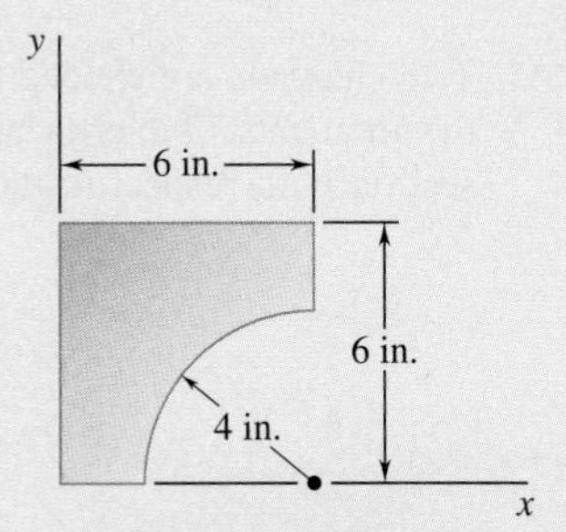

Fig. P7.28 and P7.32

7.33 Determine the shaded area and its moment of inertia with respect to the centroidal axis parallel to AA', knowing that its moments of inertia with respect to AA' and BB' are respectively 2.2×10^6 mm^4 and 4×10^6 mm^4, and that $d_1 = 25$ mm and $d_2 = 10$ mm.

7.34 Knowing that the shaded area is equal to 6000 mm^2 and that its moment of inertia with respect to AA' is 18×10^6 mm^4, determine its moment of inertia with respect to BB' for $d_1 = 50$ mm and $d_2 = 10$ mm.

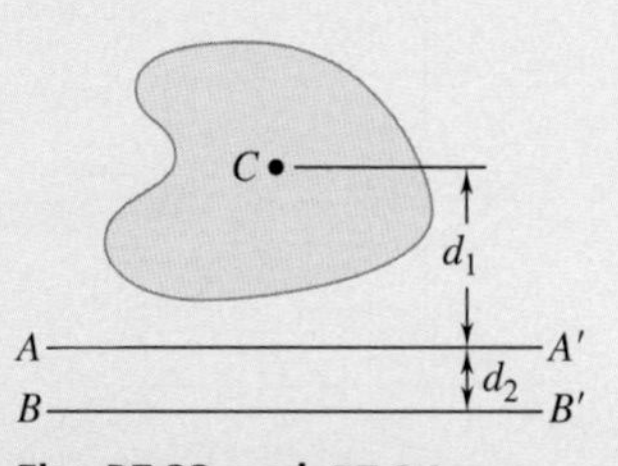

Fig. P7.33 and P7.34

7.35 through *7.37* Determine the moments of inertia $\bar{I}_x$ and $\bar{I}_y$ of the area shown with respect to centroidal axes that are respectively parallel and perpendicular to side AB.

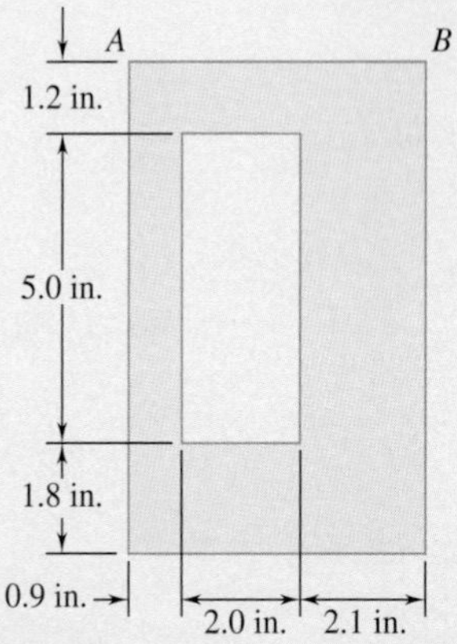

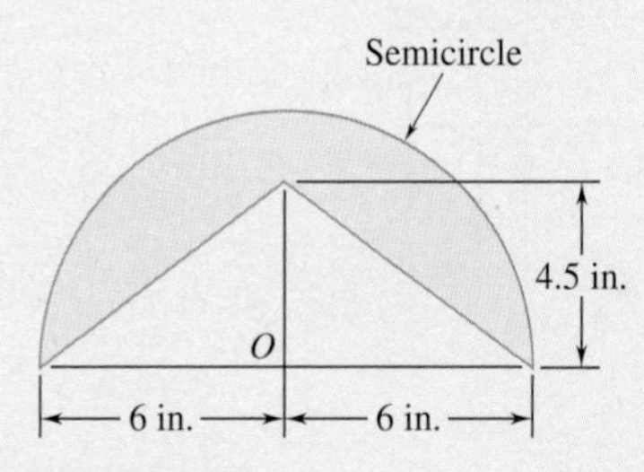

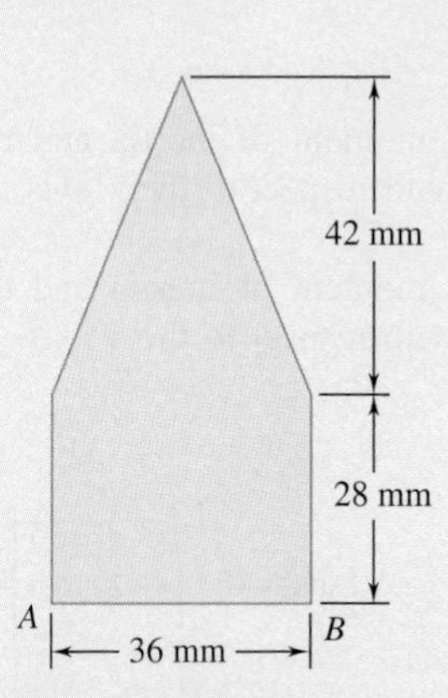

Fig. P7.36

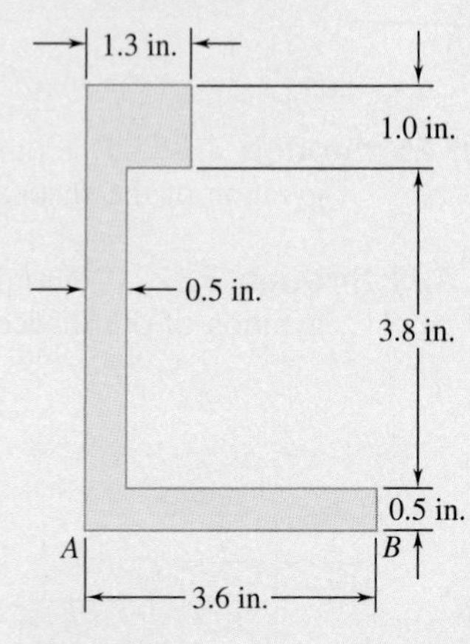

Fig. *P7.37*

7.38 through 7.40 Determine the polar moment of inertia of the area shown with respect to (a) point O, (b) the centroid of the area.

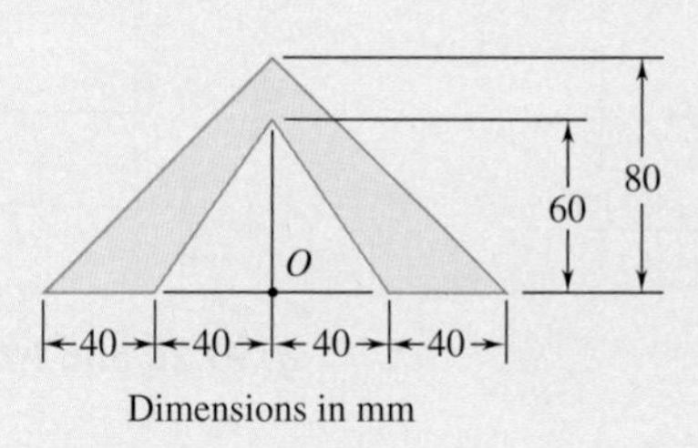

Fig. P7.39

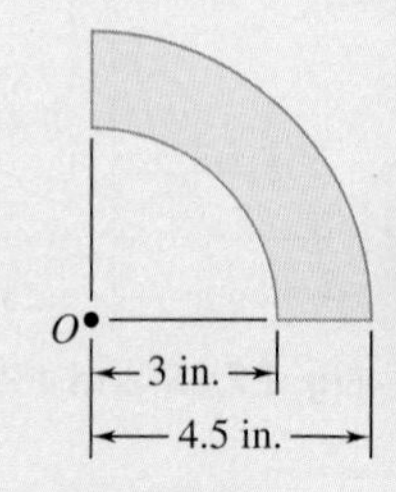

Fig. P7.40

7.41 Two channels are welded to a rolled W section as shown. Determine the moments of inertia and the radii of gyration of the combined section with respect to the centroidal x and y axes.

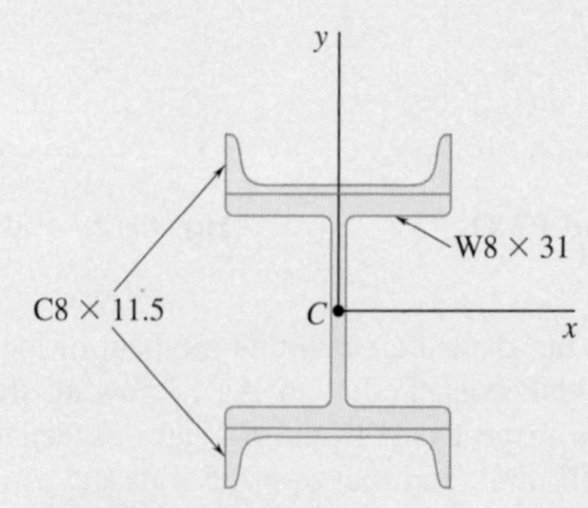

Fig. P7.41

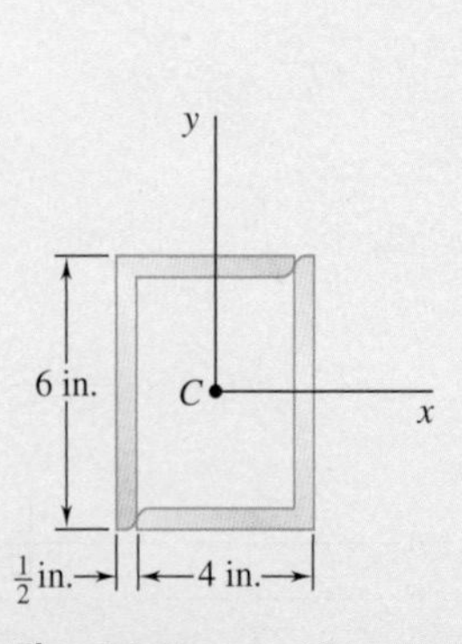

Fig. *P7.42*

7.42 Two L6 × 4 × $\frac{1}{2}$-in. angles are welded together to form the section shown. Determine the moments of inertia and the radii of gyration of the combined section with respect to the centroidal x and y axes.

Fig. P7.35

Semicircle
4.5 in.
O
6 in.
6 in.

Fig. P7.38

7.43 Two channels and two plates are used to form the column section shown. For $b = 200$ mm, determine the moments of inertia and the radii of gyration of the combined section with respect to the centroidal x and y axes.

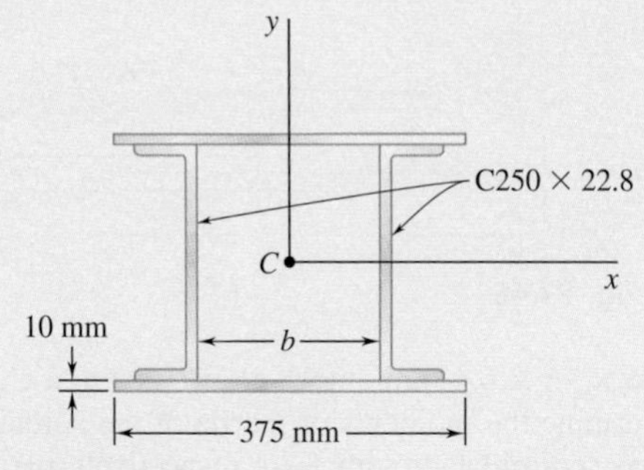

Fig. P7.43

7.44 Two 20-mm steel plates are welded to a rolled S section as shown. Determine the moments of inertia and the radii of gyration of the combined section with respect to the centroidal x and y axes.

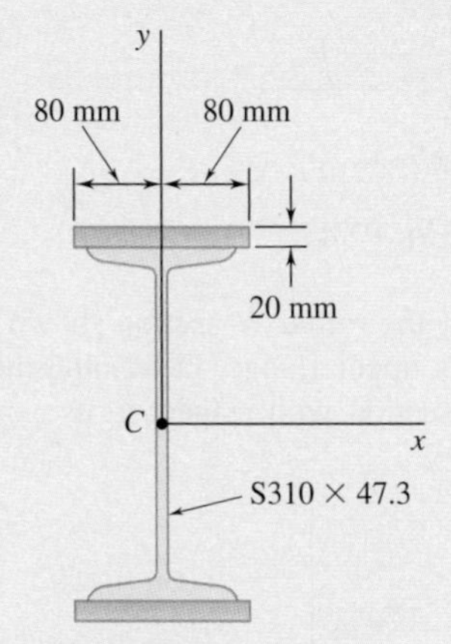

Fig. P7.44

7.45 Two L4 × 4 × $\frac{1}{2}$-in. angles are welded to a steel plate as shown. Determine the moments of inertia of the combined section with respect to the centroidal axes that are respectively parallel and perpendicular to the plate.

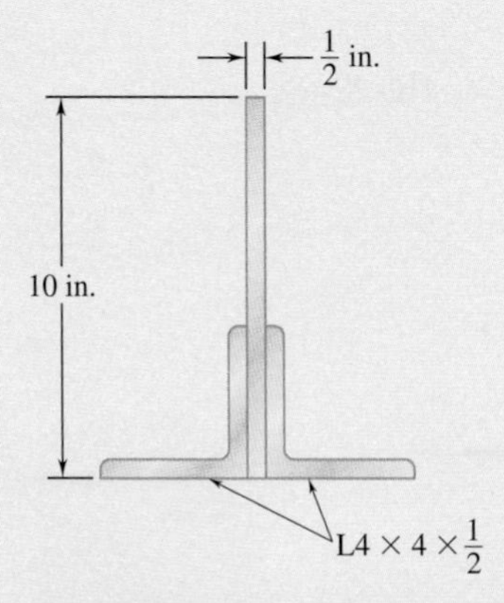

Fig. P7.45

7.46 A channel and a plate are welded together as shown to form a section that is symmetrical with respect to the y axis. Determine the moments of inertia of the combined section with respect to its centroidal x and y axes.

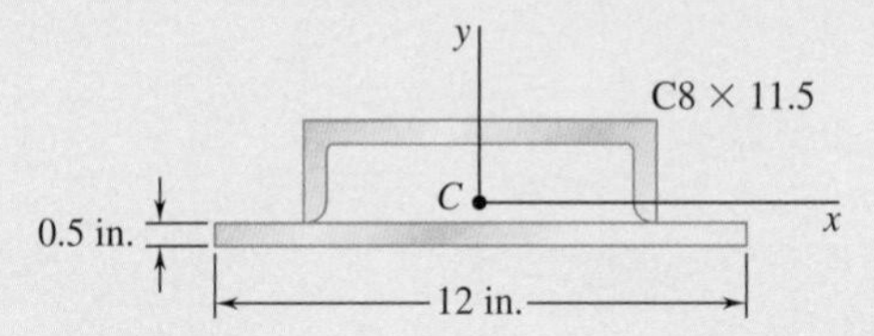

Fig. P7.46

7.47 Two L76 × 76 × 6.4-mm angles are welded to a C250 × 22.8 channel. Determine the moments of inertia of the combined section with respect to centroidal axes that are respectively parallel and perpendicular to the web of the channel.

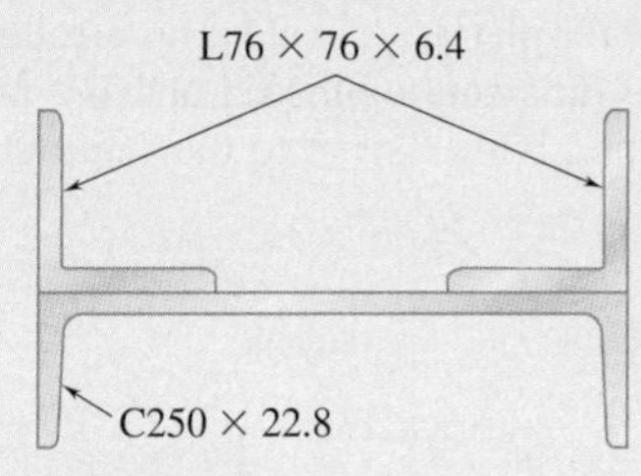

Fig. P7.47

7.48 The strength of the rolled W section shown is increased by welding a channel to its upper flange. Determine the moments of inertia of the combined section with respect to its centroidal x and y axes.

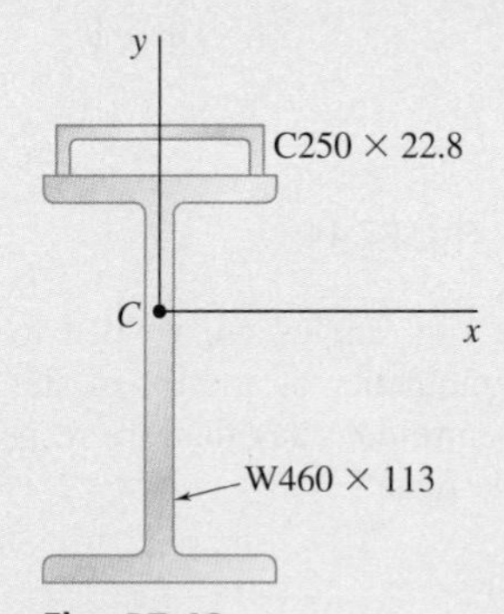

Fig. *P7.48*

Review and Summary

In this chapter, we discussed how to determine the resultant **R** of forces $\Delta\mathbf{F}$ distributed over a plane area A when the magnitudes of these forces are proportional to both the areas ΔA of the elements on which they act and the distances y from these elements to a given x axis; we thus had $\Delta F = ky\,\Delta A$. We found that the magnitude of the resultant **R** is proportional to the first moment $Q_x = \int y\,dA$ of area A, whereas the moment of **R** about the x axis is proportional to the **second moment**, or **moment of inertia**, $I_x = \int y^2\,dA$ of A with respect to the same axis [Sec. 7.1A].

Rectangular Moments of Inertia

The **rectangular moments of inertia I_x and I_y of an area** [Sec. 7.1B] are obtained by evaluating the integrals

$$I_x = \int y^2 dA \qquad I_y = \int x^2 dA \tag{7.1}$$

We can reduce these computations to single integrations by choosing dA to be a thin strip parallel to one of the coordinate axes. We also recall that it is possible to compute I_x and I_y from the same elemental strip (Fig. 7.12) using the formula for the moment of inertia of a rectangular area [Sample Prob. 7.3].

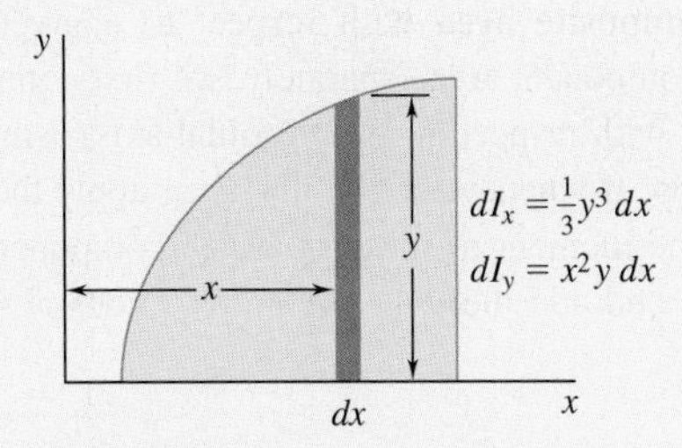

Fig. 7.12

Polar Moment of Inertia

We defined the **polar moment of inertia of an area** A with respect to the pole O [Sec. 7.1C] as

$$J_O = \int \rho^2 dA \tag{7.3}$$

where ρ is the distance from O to the element of area dA (Fig. 7.13). Observing that $\rho^2 = x^2 + y^2$, we established the relation

$$J_O = I_x + I_y \tag{7.4}$$

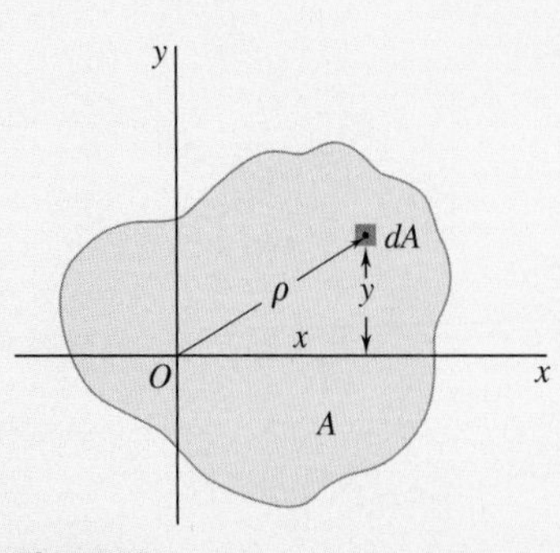

Fig. 7.13

Radius of Gyration

We defined the **radius of gyration of an area** A with respect to the x axis [Sec. 7.1D] as the distance r_x, where $I_x = r_x^2 A$. With similar definitions for the radii of gyration of A with respect to the y axis and with respect to O, we have

$$r_x = \sqrt{\frac{I_x}{A}} \qquad r_y = \sqrt{\frac{I_y}{A}} \qquad r_O = \sqrt{\frac{J_O}{A}} \tag{7.5–7.7}$$

Parallel-Axis Theorem

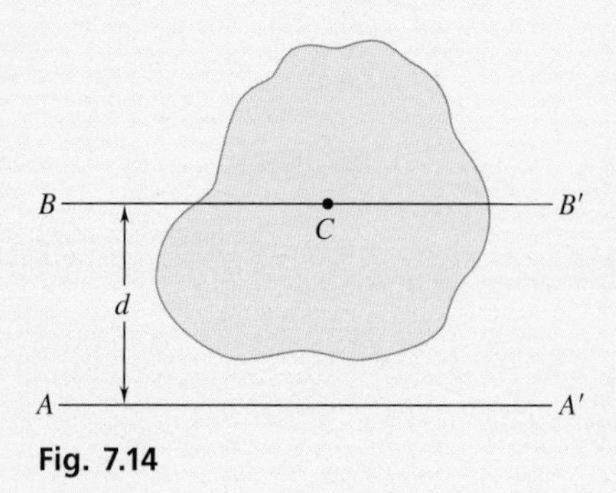

Fig. 7.14

The **parallel-axis theorem**, presented in Sec. 7.2A, states that the moment of inertia I of an area with respect to any given axis AA' (Fig. 7.14) is equal to the moment of inertia $\bar{I}$ of the area with respect to the centroidal axis BB' that is parallel to AA' *plus* the product of the area A and the square of the distance d between the two axes:

$$I = \bar{I} + Ad^2 \tag{7.9}$$

You can use this formula to determine the moment of inertia $\bar{I}$ of an area with respect to a centroidal axis BB' if you know its moment of inertia I with respect to a parallel axis AA'. In this case, however, the product Ad^2 should be *subtracted* from the known moment of inertia I.

A similar relation holds between the polar moment of inertia J_O of an area about a point O and the polar moment of inertia $\bar{J}_C$ of the same area about its centroid C. Letting d be the distance between O and C, we have

$$J_O = \bar{J}_C + Ad^2 \tag{7.11}$$

Composite Areas

The parallel-axis theorem can be used very effectively to compute the **moment of inertia of a composite area** with respect to a given axis [Sec. 7.2B]. Considering each component area separately, we first compute the moment of inertia of each area with respect to its centroidal axis, using the data provided in Fig. 7.11 and App. B whenever possible. Then apply the parallel-axis theorem to determine the moment of inertia of each component area with respect to the desired axis, and add the values [Sample Probs. 7.4 and 7.5].

Review Problems

7.49 Determine by direct integration the moment of inertia of the shaded area with respect to the y axis.

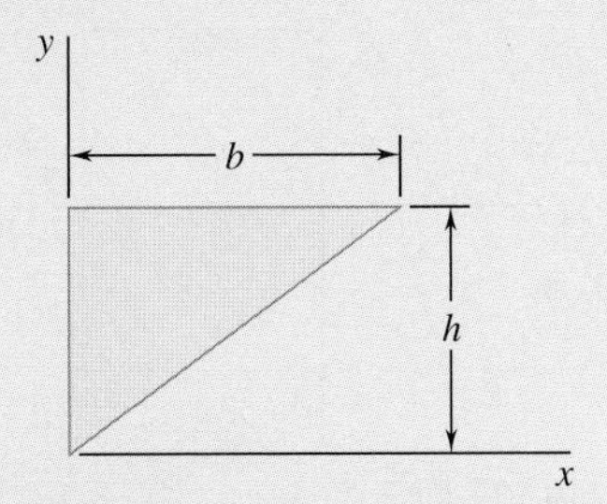

Fig. P7.49 and *P7.50*

7.50 Determine by direct integration the moment of inertia of the shaded area with respect to the x axis.

7.51 Determine the moment of inertia and radius of gyration of the shaded area shown with respect to the x axis.

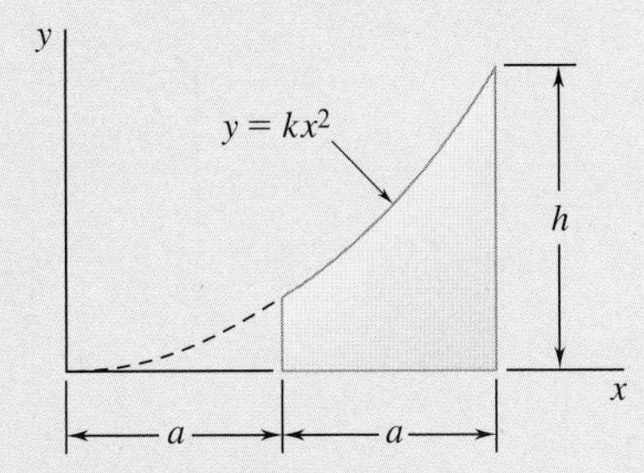

Fig. P7.51 and *P7.52*

7.52 Determine the moment of inertia and radius of gyration of the shaded area shown with respect to the y axis.

7.53 Determine the polar moment of inertia and the polar radius of gyration of the isosceles triangle shown with respect to point O.

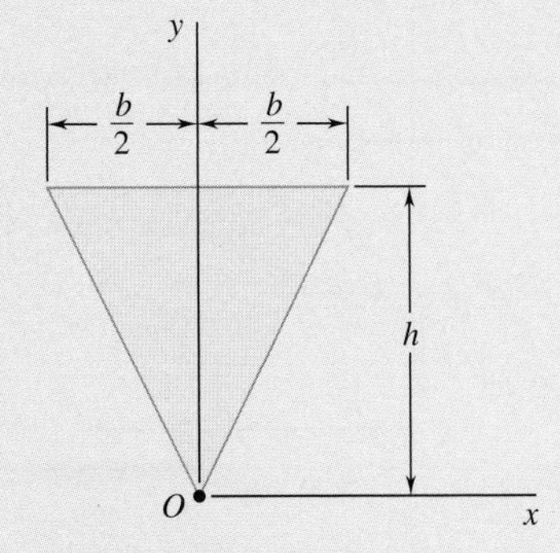

Fig. P7.53

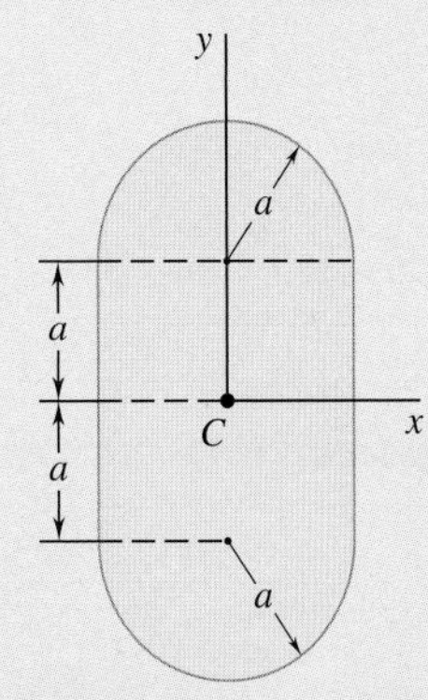

Fig. P7.54

7.54 Determine the moments of inertia of the shaded area shown with respect to the x and y axes when $a = 20$ mm.

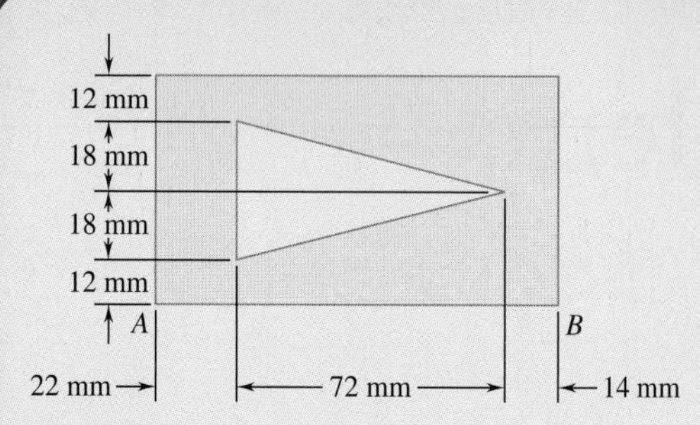

Fig. P7.55

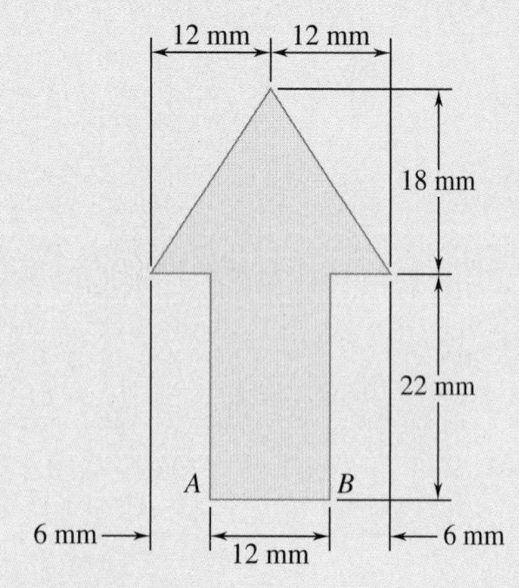

Fig. P7.56

7.55 and 7.56 Determine the moments of inertia $\bar{I}_x$ and $\bar{I}_y$ of the area shown with respect to centroidal axes that are respectively parallel and perpendicular to side AB.

7.57 The shaded area is equal to 50 in^2. Determine its centroidal moments of inertia $\bar{I}_x$ and $\bar{I}_y$, knowing that $\bar{I}_y = 2\bar{I}_x$ and that the polar moment of inertia of the area about point A is $J_A = 2250$ in^4.

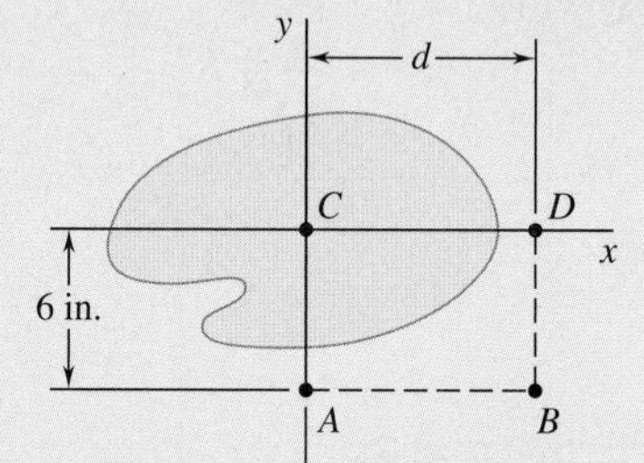

Fig. *P7.57*

7.58 and *7.59* Determine the polar moment of inertia of the area shown with respect to (*a*) point O, (*b*) the centroid of the area.

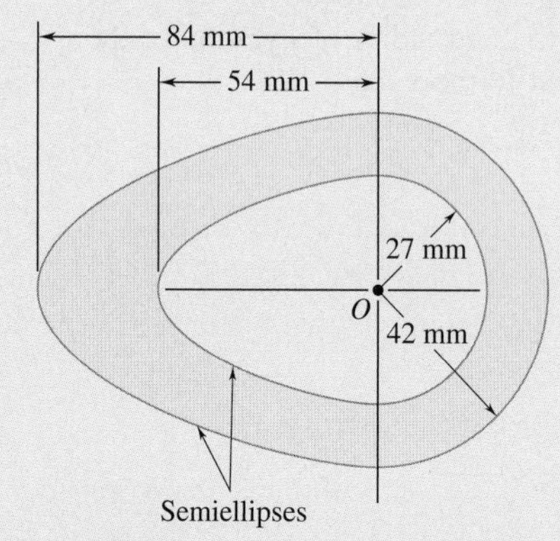

Fig. P7.58

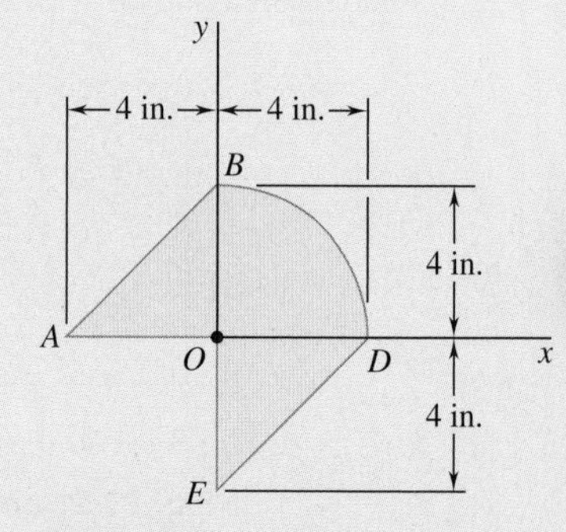

Fig. *P7.59*

7.60 Four L3 × 3 × $\frac{1}{4}$-in. angles are welded to a rolled W section as shown. Determine the moments of inertia and the radii of gyration of the combined section with respect to the centroidal x and y axes.

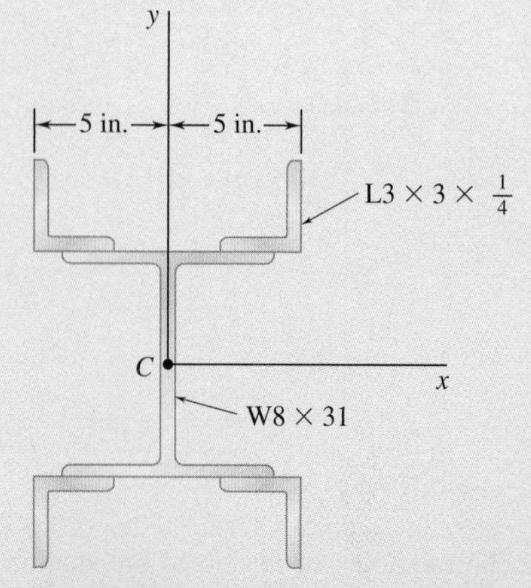

Fig. P7.60

8

Concept of Stress

Stresses occur in all structures subject to loads. This chapter will examine simple states of stress in elements, such as in the two-force members, bolts and pins used in the structure shown.

Objectives

- **Introduce** concept of stress.
- **Define** different stress types: axial normal stress, shearing stress and bearing stress.
- **Discuss** engineer's two principal tasks, namely, the analysis and design of structures and machines.
- **Discuss** the components of stress on different planes and under different loading conditions.
- **Discuss** the many design considerations that an engineer should review before preparing a design.

Introduction

The remainder of this book focuses on *mechanics of materials,* the study of which provides future engineers with the means of analyzing and designing various machines and load-bearing structures involving the determination of *stresses* and *deformations*.

Section 8.1 introduces the concept of *stress* in a member of a structure and how that stress can be determined from the *force* in the member. You will consider the *normal stresses* in a member under axial loading, the *shearing stresses* caused by the application of equal and opposite transverse forces, and the *bearing stresses* created by bolts and pins in the members they connect. Section 8.1 ends with an example showing how the stresses can be determined in a simple two-dimensional structure.

A two-force member under axial loading is observed in Sec. 8.2 where the stresses on an *oblique* plane include both *normal* and *shearing* stresses, while Sec. 8.3 discusses that *six components* are required to describe the state of stress at a point in a body under the most general loading conditions.

Finally, Sec. 8.4 is devoted to the determination of the *ultimate strength* from test specimens and the use of a *factor of safety* to compute the *allowable load* for a structural component made of that material.

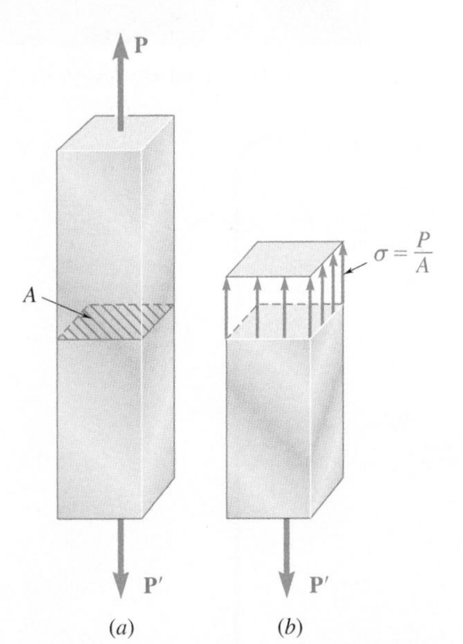

Fig. 8.1 (a) Member with an axial load. (b) Idealized uniform stress distribution at an arbitrary section.

8.1 STRESSES IN THE MEMBERS OF A STRUCTURE

Let us look at the uniformly distributed force using Fig. 8.1. The force per unit area, or intensity of the forces distributed over a given section, is called the *stress* and is denoted by the Greek letter σ (sigma). The stress in a member of cross-sectional area A subjected to an axial load **P** is obtained by dividing the magnitude P of the load by the area A:

$$\sigma = \frac{P}{A} \tag{8.1}$$

A positive sign indicates a tensile stress (member in tension), and a negative sign indicates a compressive stress (member in compression).

When SI metric units are used, P is expressed in newtons (N) and A in square meters (m^2), so the stress σ will be expressed in N/m^2. This unit is called a *pascal* (Pa). However, the pascal is an exceedingly small quantity and often multiples of this unit must be used: the kilopascal (kPa), the megapascal (MPa), and the gigapascal (GPa):

$$1 \text{ kPa} = 10^3 \text{ Pa} = 10^3 \text{ N/m}^2$$

$$1 \text{ MPa} = 10^6 \text{ Pa} = 10^6 \text{ N/m}^2$$

$$1 \text{ GPa} = 10^9 \text{ Pa} = 10^9 \text{ N/m}^2$$

When U.S. customary units are used, force P is usually expressed in pounds (lb) or kilopounds (kip), and the cross-sectional area A is given in square inches (in^2). The stress σ then is expressed in pounds per square inch (psi) or kilopounds per square inch (ksi).[†]

8.1A Axial Stress

The member shown in Fig. 8.1 in the preceding section is subject to forces **P** and **P′** applied at the ends. The forces are directed along the axis of the member, and we say that the member is under *axial loading*. An actual example of structural members under axial loading is provided by the members of the bridge truss shown in Photo 8.1.

Photo 8.1 This bridge truss consists of two-force members that may be in tension or in compression.

As shown in Fig. 8.1, the section through the rod to determine the internal force in the rod and the corresponding stress is perpendicular to the axis of the rod. The corresponding stress is described as a *normal stress*. Thus, Eq. (8.1) gives the *normal stress in a member under axial loading*.

[†]The principal SI and U.S. Customary units used in mechanics are listed in Table 1.3 on p. 12. From the table on the right-hand side, 1 psi is approximately equal to 7 kPa and 1 ksi is approximately equal to 7 MPa.

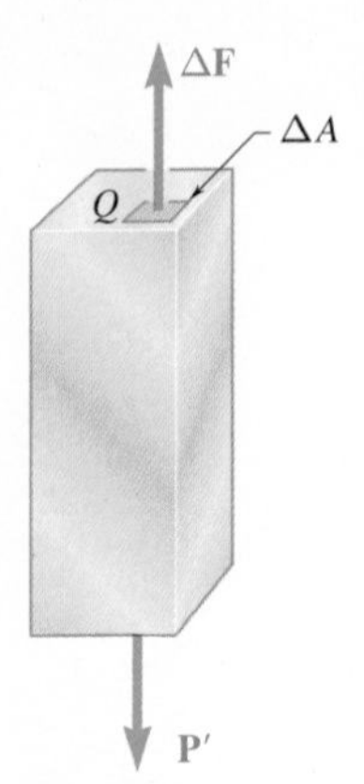

Fig. 8.2 Small area ΔA, at an arbitrary point of the cross section, carries ΔF in this axial member.

Note that in Eq. (8.1), σ represents the *average value* of the stress over the cross section, rather than the stress at a specific point of the cross section. To define the stress at a given point Q of the cross section, consider a small area ΔA (Fig. 8.2). Dividing the magnitude of $\Delta\mathbf{F}$ by ΔA, you obtain the average value of the stress over ΔA. Letting ΔA approach zero, the stress at point Q is

$$\sigma = \lim_{\Delta A \to 0} \frac{\Delta F}{\Delta A} \tag{8.2}$$

In general, the value for the stress σ at a given point Q of the section is different from that for the average stress given by Eq. (8.1), and σ is found to vary across the section. In a slender rod subjected to equal and opposite concentrated loads **P** and **P′** (Fig. 8.3*a*), this variation is small in a section away from the points of application of the concentrated loads (Fig. 8.3*c*), but it is quite noticeable in the neighborhood of these points (Fig. 8.3*b* and *d*).

It follows from Eq. (8.2) that the magnitude of the resultant of the distributed internal forces is

$$\int dF = \int_A \sigma\, dA$$

But the conditions of equilibrium of each of the portions of rod shown in Fig. 8.3 require that this magnitude be equal to the magnitude P of the concentrated loads. Therefore,

$$P = \int dF = \int_A \sigma\, dA \tag{8.3}$$

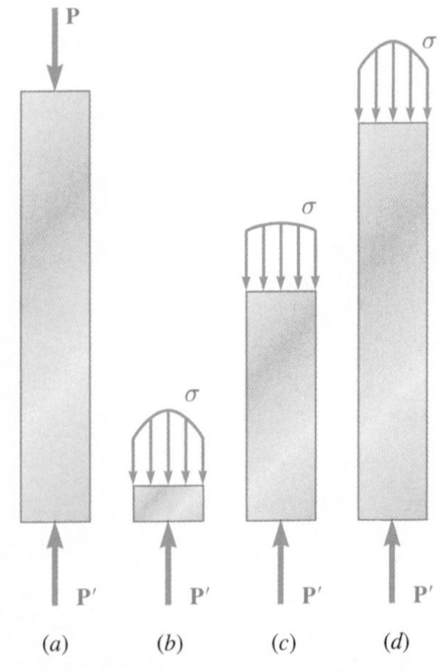

Fig. 8.3 Stress distributions at different sections along axially loaded member.

which means that the volume under each of the stress surfaces in Fig. 8.3 must be equal to the magnitude P of the loads. However, this is the only information derived from statics regarding the distribution of normal stresses in the various sections of the rod. The actual distribution of stresses in any given section is *statically indeterminate*. To learn more about this distribution, it is necessary to consider the deformations resulting from the particular mode of application of the loads at the ends of the rod. This will be discussed further in Chap. 9.

In practice, it is assumed that the distribution of normal stresses in an axially loaded member is uniform, except in the immediate vicinity of the points of application of the loads. The value σ of the stress is then equal to σ_{ave} and can be obtained from Eq. (8.1). However, realize that when we assume a uniform distribution of stresses in the section, it follows from elementary statics[†] that the resultant **P** of the internal forces must be applied at the centroid C of the section (Fig. 8.4). This means that *a uniform distribution of stress is possible only if the line of action of the concentrated loads* **P** *and* **P′** *passes through the centroid of the section considered* (Fig. 8.5). This type of loading is called *centric loading* and will take place in all straight two-force members found in trusses and pin-connected structures.

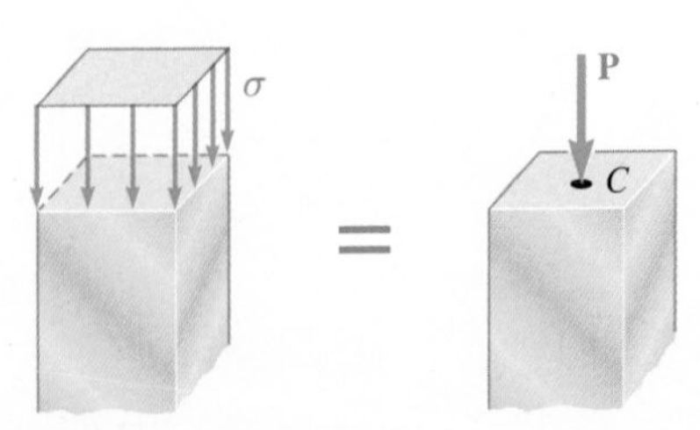

Fig. 8.4 Idealized uniform stress distribution implies the resultant force passes through the cross section's center.

[†]See Ferdinand P. Beer and E. Russell Johnston, Jr., *Mechanics for Engineers,* 5th ed., McGraw-Hill, New York, 2008, or *Vector Mechanics for Engineers,* 11th ed., McGraw-Hill, New York, 2016, Secs. 5.2 and 5.3.

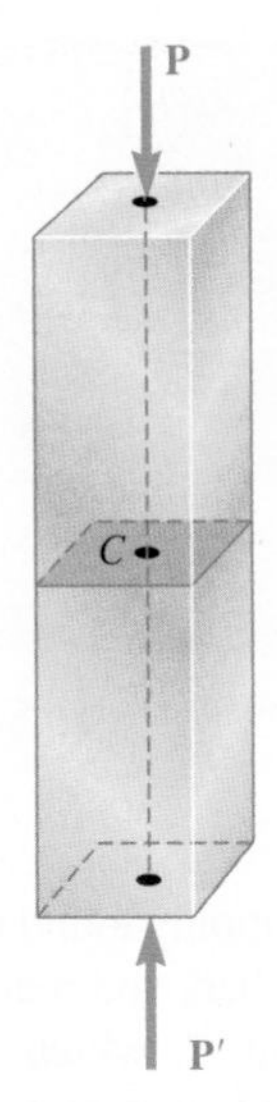

Fig. 8.5 Centric loading having resultant forces passing through the centroid of the section.

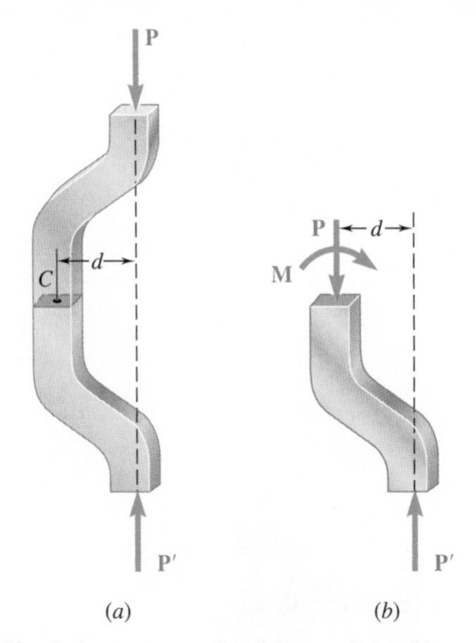

Fig. 8.6 An example of eccentric loading.

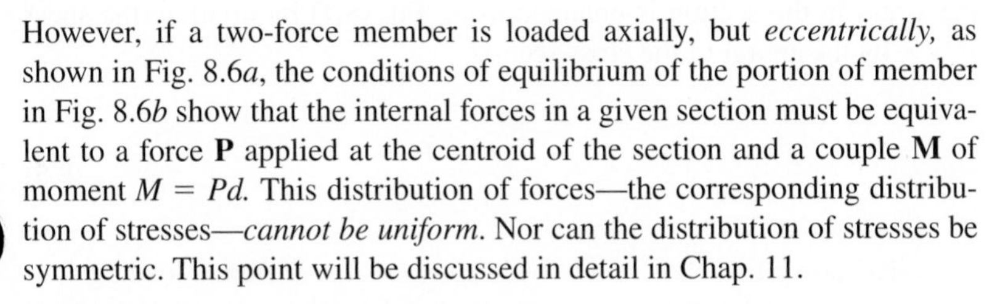

However, if a two-force member is loaded axially, but *eccentrically,* as shown in Fig. 8.6*a*, the conditions of equilibrium of the portion of member in Fig. 8.6*b* show that the internal forces in a given section must be equivalent to a force **P** applied at the centroid of the section and a couple **M** of moment $M = Pd$. This distribution of forces—the corresponding distribution of stresses—*cannot be uniform*. Nor can the distribution of stresses be symmetric. This point will be discussed in detail in Chap. 11.

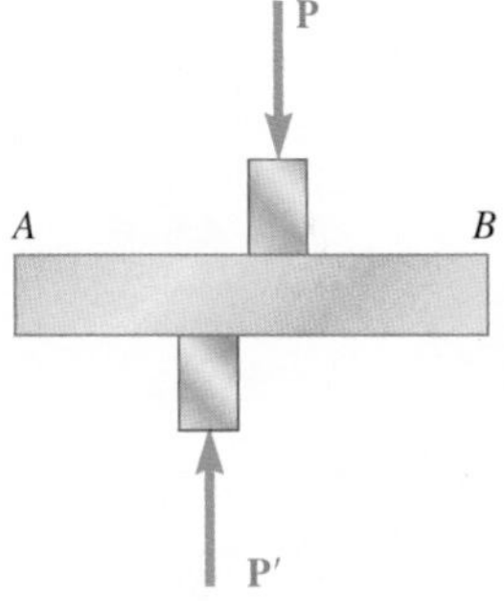

Fig. 8.7 Opposing transverse loads creating shear on member *AB*.

8.1B Shearing Stress

The internal forces and the corresponding stresses discussed in Sec. 8.1A were normal to the section considered. A very different type of stress is obtained when transverse forces **P** and **P′** are applied to a member *AB* (Fig. 8.7). Passing a section at *C* between the points of application of the two forces (Fig. 8.8*a*), you obtain the diagram of portion *AC* shown in Fig. 8.8*b*. Internal forces must exist in the plane of the section, and their resultant is equal to **P**. These elementary internal forces are called *shearing forces,* and the magnitude *P* of their resultant is the *shear* in the section. Dividing the shear *P* by the area *A* of the cross section, you obtain the *average shearing stress* in the section. Denoting the shearing stress by the Greek letter τ (tau), write

$$\tau_{\text{ave}} = \frac{P}{A} \tag{8.4}$$

The value obtained is an average value of the shearing stress over the entire section. Contrary to what was said earlier for normal stresses, the distribution of shearing stresses across the section *cannot* be assumed to be uniform. As you will see in Chap. 13, the actual value τ of the shearing stress varies from zero at the surface of the member to a maximum value τ_{max} that may be much larger than the average value τ_{ave}.

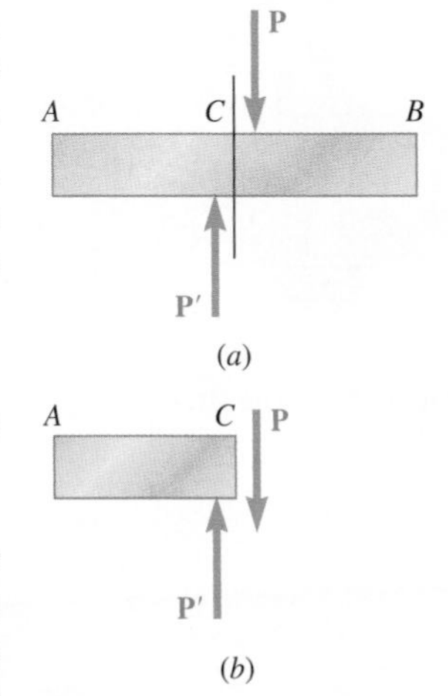

Fig. 8.8 This shows the resulting internal shear force on a section between transverse forces.

Photo 8.2 Cutaway view of a connection with a bolt in shear.
© *John DeWolf*

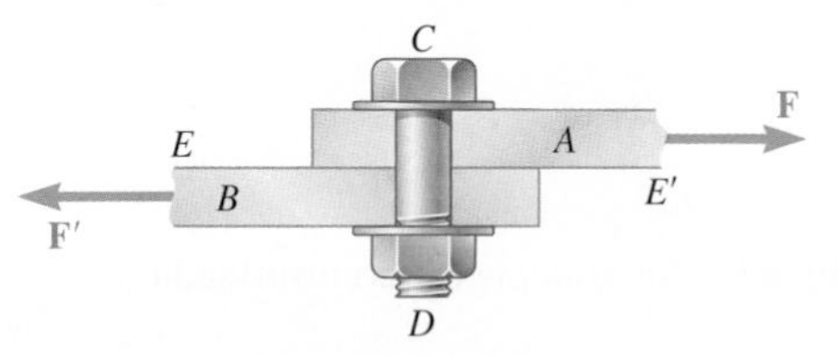

Fig. 8.9 Bolt subject to single shear.

Shearing stresses are commonly found in bolts, pins, and rivets used to connect various structural members and machine components (Photo 8.2). Consider the two plates A and B, which are connected by a bolt CD (Fig. 8.9). If the plates are subjected to tension forces of magnitude F, stresses will develop in the section of bolt corresponding to the plane EE'. Drawing the diagrams of the bolt and of the portion located above the plane EE' (Fig. 8.10), the shear P in the section is equal to F. The average shearing stress in the section is obtained using Eq. (8.4) by dividing the shear $P = F$ by the area A of the cross section:

$$\tau_{\text{ave}} = \frac{P}{A} = \frac{F}{A} \tag{8.5}$$

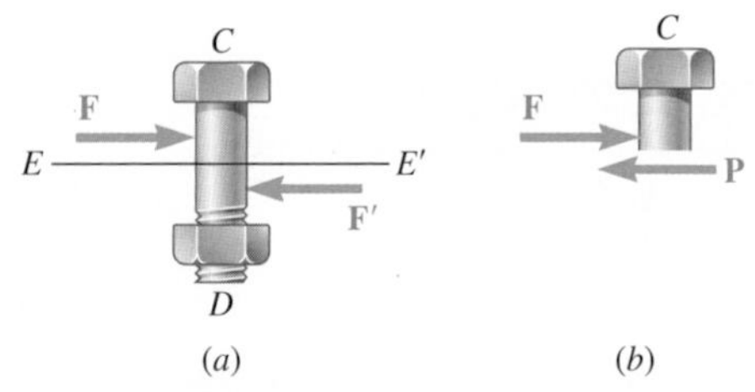

Fig. 8.10 (*a*) Diagram of bolt in single shear; (*b*) section *E-E′* of the bolt.

The previous bolt is said to be in *single shear*. Different loading situations may arise, however. For example, if splice plates C and D are used to connect plates A and B (Fig. 8.11), shear will take place in bolt HJ in each of the two planes KK' and LL' (and similarly in bolt EG). The bolts are said to be in *double shear*. To determine the average shearing stress in each plane, draw free-body diagrams of bolt HJ and of the portion of the bolt located between the two planes (Fig. 8.12). Observing that the shear P in each of the sections is $P = F/2$, the average shearing stress is

$$\tau_{\text{ave}} = \frac{P}{A} = \frac{F/2}{A} = \frac{F}{2A} \tag{8.6}$$

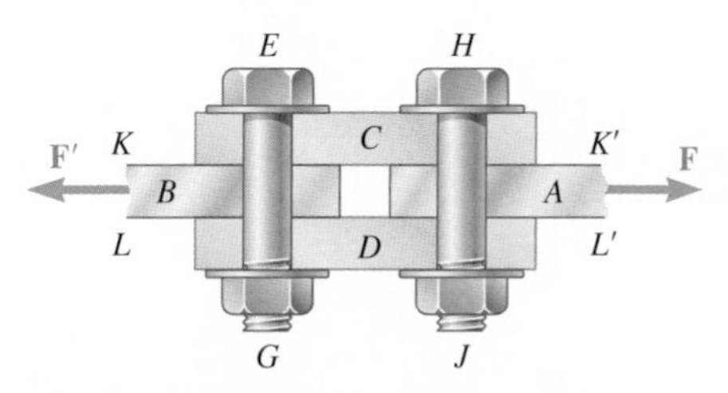

Fig. 8.11 Bolts subject to double shear.

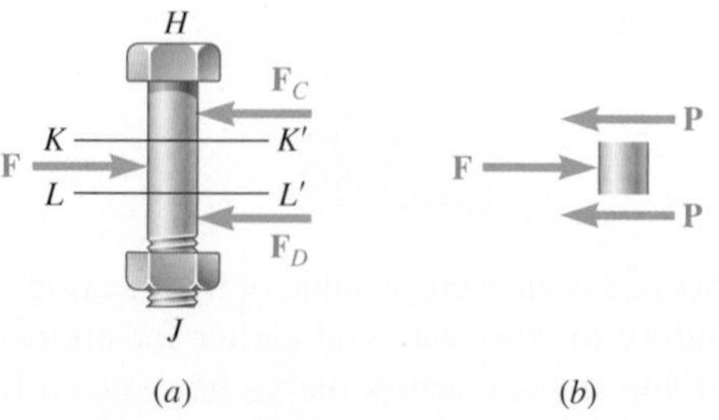

Fig. 8.12 (*a*) Diagram of bolt in double shear; (*b*) section *K-K′* and *L-L′* of the bolt.

8.1C Bearing Stress in Connections

Bolts, pins, and rivets create stresses in the members they connect along the *bearing surface* or surface of contact. For example, consider again the two plates A and B connected by a bolt CD that were discussed in the preceding section (Fig. 8.9). The bolt exerts on plate A a force **P** equal and opposite to the force **F** exerted by the plate on the bolt (Fig. 8.13). The force **P** represents the resultant of elementary forces distributed on the inside surface of a half-cylinder of diameter d and of length t equal to the thickness of the plate. Since the distribution of these forces—and of the corresponding stresses—is quite complicated, in practice one uses an average nominal value σ_b of the stress, called the *bearing stress,* which is obtained by dividing the load P by the area of the rectangle representing the projection of the bolt on the plate section (Fig. 8.14). Since this area is equal to td, where t is the plate thickness and d the diameter of the bolt, we have

$$\sigma_b = \frac{P}{A} = \frac{P}{td} \quad \textbf{(8.7)}$$

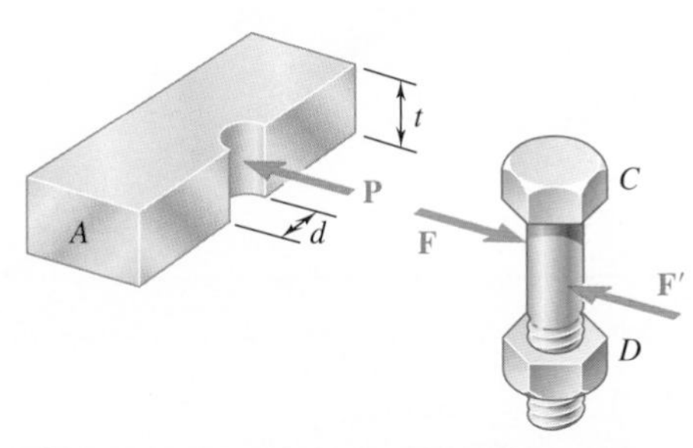

Fig. 8.13 Equal and opposite forces between plate and bolt, exerted over bearing surfaces.

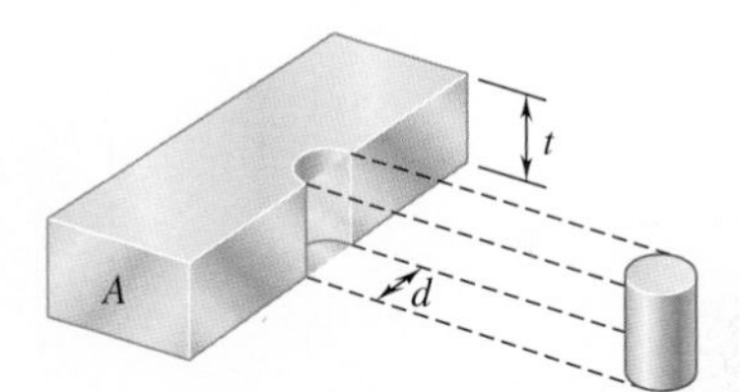

Fig. 8.14 Dimensions for calculating bearing stress area.

8.1D Application to the Analysis and Design of Simple Structures

We are now in a position to determine the stresses in the members and connections of various simple two-dimensional structures and to design such structures. This is illustrated through the following Concept Application.

Photo 8.3 Crane booms used to load and unload ships.

Concept Application 8.1

The structure shown in Fig. 8.15 was designed to support a 30-kN load. It consists of a boom *AB* with a 30 × 50-mm rectangular cross section and a rod *BC* with a 20-mm-diameter circular cross section. The boom and the rod are connected by a pin at *B* and are supported by pins and brackets at *A* and *C*, respectively.

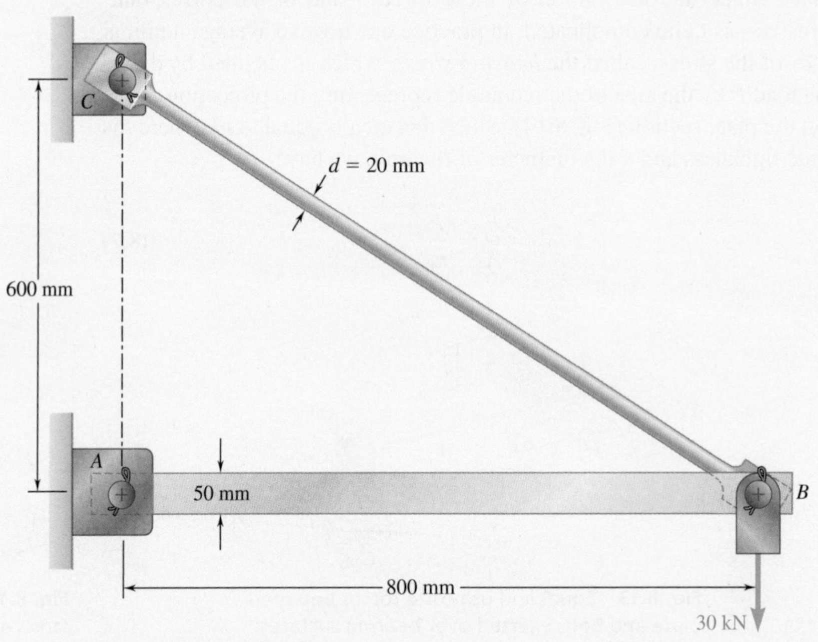

Fig. 8.15 Boom used to support a 30-kN load.

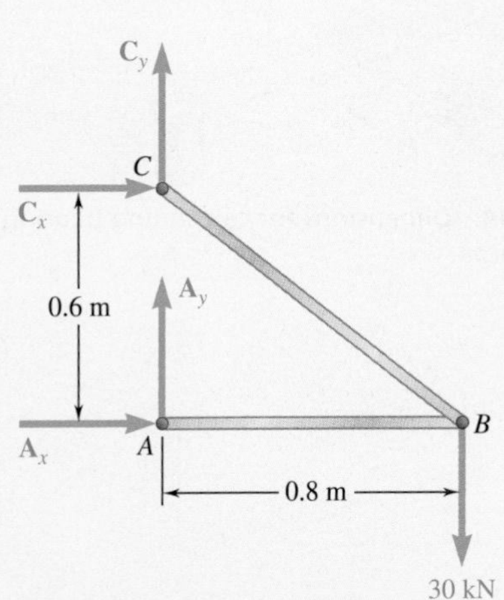

Fig. 8.16 Free-body diagram of boom showing applied load and reaction forces.

We first use the basic methods of statics to find the reactions and then the internal forces in the members. We start by drawing a *free-body diagram* of the structure by detaching it from its supports at *A* and *C*, and showing the reactions that these supports exert on the structure (Fig. 8.16). The reactions are represented by two components $\mathbf{A}_x$ and $\mathbf{A}_y$ at *A*, and $\mathbf{C}_x$ and $\mathbf{C}_y$ at *C*. We write the following three equilibrium equations:

$$+\circlearrowleft \Sigma M_C = 0: \quad A_x(0.6 \text{ m}) - (30 \text{ kN})(0.8 \text{ m}) = 0$$

$$A_x = +40 \text{ kN} \qquad (1)$$

$$\xrightarrow{+} \Sigma F_x = 0: \quad A_x + C_x = 0$$

$$C_x = -A_x \qquad C_x = -40 \text{ kN} \qquad (2)$$

$$+\uparrow \Sigma F_y = 0: \quad A_y + C_y - 30 \text{ kN} = 0$$

$$A_y + C_y = +30 \text{ kN} \qquad (3)$$

We have found two of the four unknowns. We must now dismember the structure. Considering the free-body diagram of the boom *AB* (Fig. 8.17), we write the following equilibrium equation:

$$+\circlearrowleft \Sigma M_B = 0: \quad -A_y(0.8 \text{ m}) = 0 \qquad A_y = 0 \qquad (4)$$

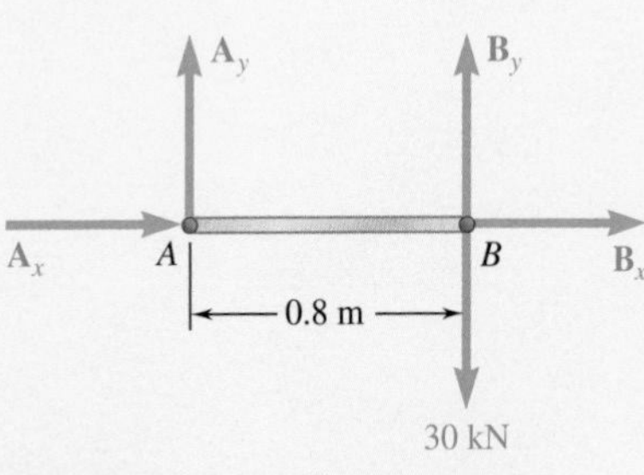

Fig. 8.17 Free-body diagram of member *AB* freed from structure.

Substituting for A_y from (4) into (3), we obtain $C_y = +30$ kN. Expressing the results obtained for the reactions at A and C in vector form, we have

$$\mathbf{A} = 40 \text{ kN} \longrightarrow, \mathbf{C}_x = 40 \text{ kN} \leftarrow, \mathbf{C}_y = 30 \text{ kN}\uparrow$$

We note that the reaction at A is directed along the axis of the boom AB and causes compression in that member. Observing that the components C_x and C_y of the reaction at C are, respectively, proportional to the horizontal and vertical components of the distance from B to C, we conclude that the reaction at C is equal to 50 kN, is directed along the axis of the rod BC, and causes tension in that member.

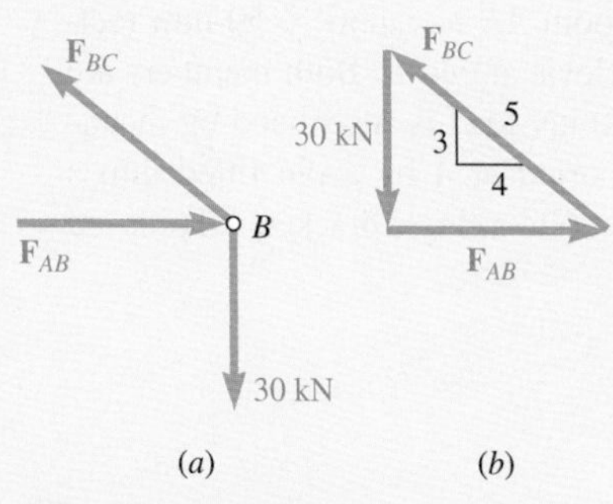

Fig. 8.18 Free-body diagram of boom's joint *B* and associated force triangle.

These results could have been anticipated by recognizing that AB and BC are two-force members, i.e., members that are subjected to forces at only two points, these points being A and B for member AB, and B and C for member BC. Indeed, for a two-force member the lines of action of the resultants of the forces acting at each of the two points are equal and opposite and pass through both points. Using this property, we could have obtained a simpler solution by considering the free-body diagram of pin B. The forces on pin B are the forces $\mathbf{F}_{AB}$ and $\mathbf{F}_{BC}$ exerted, respectively, by members AB and BC, and the 30-kN load (Fig. 8.18*a*). We can express that pin B is in equilibrium by drawing the corresponding force triangle (Fig. 8.18*b*).

Since the force $\mathbf{F}_{BC}$ is directed along member BC, its slope is the same as that of BC, namely, 3/4. We can, therefore, write the proportion

$$\frac{F_{AB}}{4} = \frac{F_{BC}}{5} = \frac{30 \text{ kN}}{3}$$

from which we obtain

$$F_{AB} = 40 \text{ kN} \qquad F_{BC} = 50 \text{ kN}$$

The forces $\mathbf{F}'_{AB}$ and $\mathbf{F}'_{BC}$ exerted by pin B, respectively, on boom AB and rod BC are equal and opposite to $\mathbf{F}_{AB}$ and $\mathbf{F}_{BC}$ (Fig. 8.19).

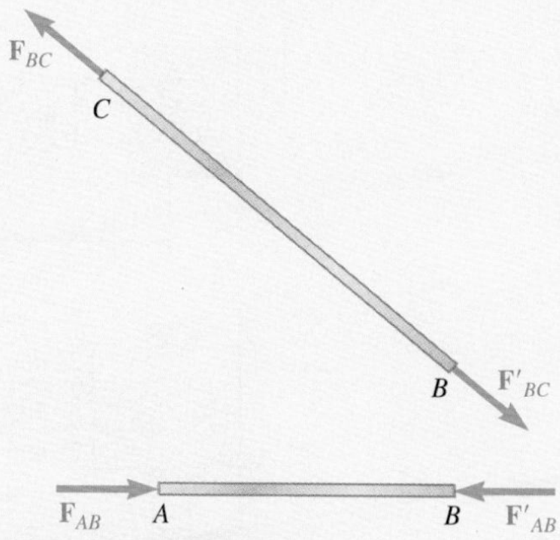

Fig. 8.19 Free-body diagrams of two-force members *AB* and *BC*.

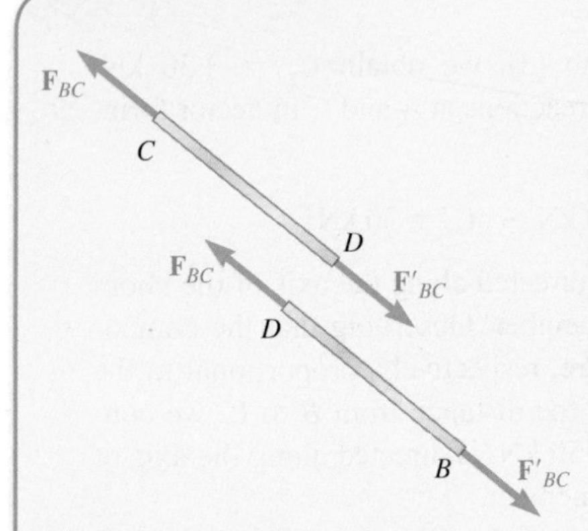

Fig. 8.20 Free-body diagrams of sections of rod *BC*.

Knowing the forces at the ends of each of the members, we can now determine the internal forces in these members. Passing a section at some arbitrary point D of rod BC, we obtain two portions BD and CD (Fig. 8.20). Since 50-kN forces must be applied at D to both portions of the rod to keep them in equilibrium, we conclude that an internal force of 50 kN is produced in rod BC when a 30-kN load is applied at B. We further check from the directions of the forces $\mathbf{F}_{BC}$ and $\mathbf{F}'_{BC}$ in Fig. 8.20 that the rod is in tension. A similar procedure would enable us to determine that the internal force in boom AB is 40 kN and that the boom is in compression.

We now determine the stresses in the members and connections. As shown in Fig. 8.21, the 20-mm-diameter rod BC has flat ends of 20 × 40-mm-rectangular cross section, while boom AB has a 30 × 50-mm rectangular cross section and is fitted with a clevis at end B. Both members are connected at B by a pin from which the 30-kN load is suspended by means of a U-shaped bracket. Boom AB is supported at A by a pin fitted into a double bracket, while rod BC is connected at C to a single bracket. All pins are 25 mm in diameter.

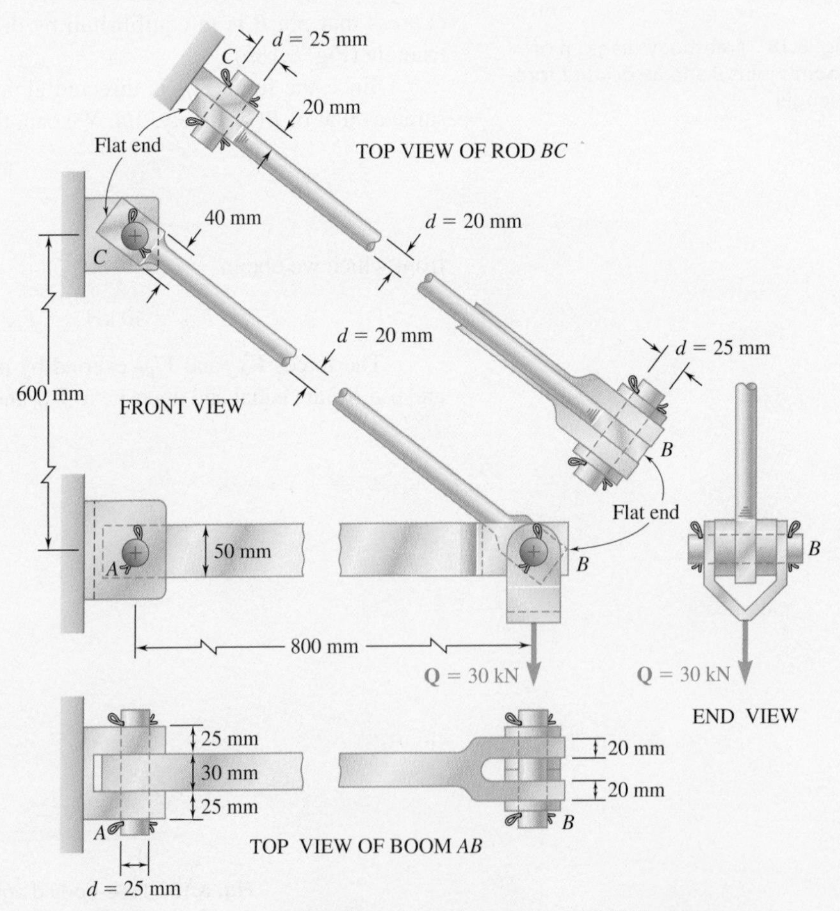

Fig. 8.21 Components of boom used to support 30 kN load.

a. Determination of the Normal Stress in Boom *AB* and Rod *BC*. The force in rod BC is $F_{BC} = 50$ kN (tension). Recalling that the diameter of the rod is 20 mm, we use Eq. (8.1) to determine the stress created in the rod by the given loading. We have

$$P = F_{BC} = +50 \text{ kN} = +50 \times 10^3 \text{ N}$$

$$A = \pi r^2 = \pi\left(\frac{20 \text{ mm}}{2}\right)^2 = \pi(10 \times 10^{-3} \text{ m})^2 = 314 \times 10^{-6} \text{ m}^2$$

$$\sigma_{BC} = \frac{P}{A} = \frac{+50 \times 10^3 \text{ N}}{314 \times 10^{-6} \text{ m}^2} = +159 \times 10^6 \text{ Pa} = +159 \text{ MPa}$$

However, the flat parts of the rod are also under tension and at the narrowest section, where a hole is located, we have

$$A = (20 \text{ mm})(40 \text{ mm} - 25 \text{ mm}) = 300 \times 10^{-6} \text{ m}^2$$

The corresponding average value of the stress, therefore, is

$$(\sigma_{BC})_{\text{end}} = \frac{P}{A} = \frac{50 \times 10^3 \text{ N}}{300 \times 10^{-6} \text{ m}^2} = 167 \text{ MPa}$$

Note that this is an *average value;* close to the hole, the stress will actually reach a much larger value, as you will see in Sec. 9.9. It is clear that, under an increasing load, the rod will fail near one of the holes rather than in its cylindrical portion; its design, therefore, could be improved by increasing the width or the thickness of the flat ends of the rod.

Turning now our attention to boom AB, we recall that the force in the boom is $F_{AB} = 40$ kN (compression). Since the area of the boom's rectangular cross section is $A = 30 \text{ mm} \times 50 \text{ mm} = 1.5 \times 10^{-3} \text{ m}^2$, the average value of the normal stress in the main part of the rod, between pins A and B, is

$$\sigma_{AB} = -\frac{40 \times 10^3 \text{ N}}{1.5 \times 10^{-3} \text{ m}^2} = -26.7 \times 10^6 \text{ Pa} = -26.7 \text{ MPa}$$

Note that the sections of minimum area at A and B are not under stress, since the boom is in compression, and, therefore, *pushes* on the pins (instead of *pulling* on the pins as rod BC does).

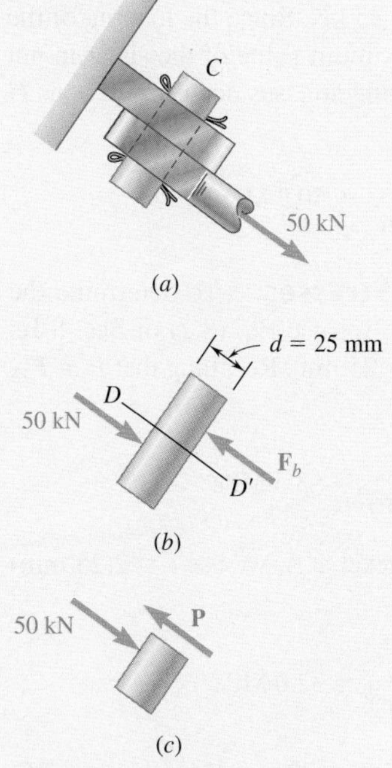

Fig. 8.22 Diagrams of the single shear pin at C.

b. Determination of the Shearing Stress in Various Connections. To determine the shearing stress in a connection such as a bolt, pin, or rivet, we first clearly show the forces exerted by the various members it connects. Thus, in the case of pin C of our example (Fig. 8.22*a*), we draw Fig. 8.22*b*, showing the 50-kN force exerted by member BC on the pin, and the equal and opposite force exerted by the bracket. Drawing now the diagram of the portion of the pin located below the plane DD' where shearing stresses occur (Fig. 8.22*c*), we conclude that the shear in that plane is $P = 50$ kN. Since the cross-sectional area of the pin is

$$A = \pi r^2 = \pi\left(\frac{25 \text{ mm}}{2}\right)^2 = \pi(12.5 \times 10^{-3} \text{ m})^2 = 491 \times 10^{-6} \text{ m}^2$$

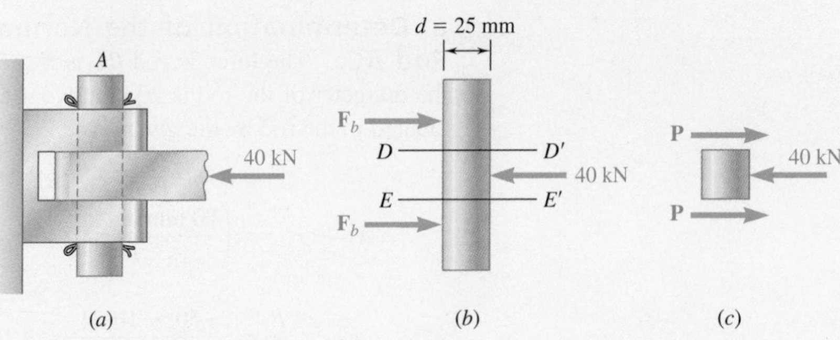

Fig. 8.23 Free-body diagrams of the double shear pin at A.

we find that the average value of the shearing stress in the pin at C is

$$\tau_{ave} = \frac{P}{A} = \frac{50 \times 10^3 \text{ N}}{491 \times 10^{-6} \text{ m}^2} = 102 \text{ MPa}$$

Considering now the pin at A (Fig. 8.23), we note that it is in double shear. Drawing the free-body diagrams of the pin and of the portion of pin located between the planes DD' and EE' where shearing stresses occur, we conclude that $P = 20$ kN and that

$$\tau_{ave} = \frac{P}{A} = \frac{20 \text{ kN}}{491 \times 10^{-6} \text{ m}^2} = 40.7 \text{ MPa}$$

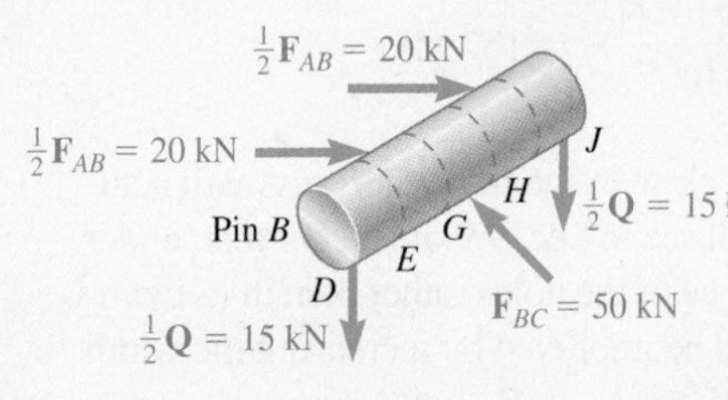

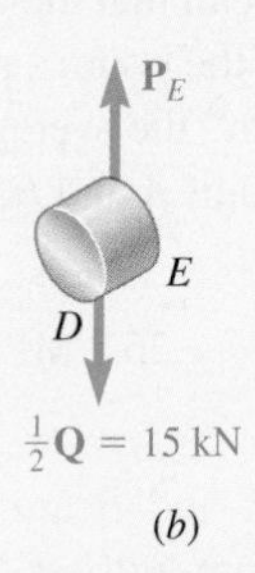

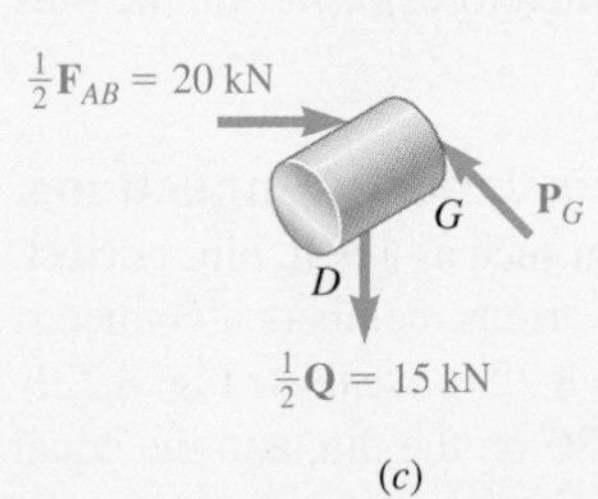

Fig. 8.24 Free-body diagrams for various sections at pin B.

Considering the pin at B (Fig. 8.24*a*), we note that the pin may be divided into five portions which are acted upon by forces exerted by the boom, rod, and bracket. Considering successively the portions DE (Fig. 8.24*b*) and DG (Fig. 8.24*c*), we conclude that the shear in section E is $P_E = 15$ kN, while the shear in section G is $P_G = 25$ kN. Since the loading of the pin is symmetric, we conclude that the maximum value of the shear in pin B is $P_G = 25$ kN, and that the largest shearing stresses occur in sections G and H, where

$$\tau_{ave} = \frac{P_G}{A} = \frac{25 \text{ kN}}{491 \times 10^{-6} \text{ m}^2} = 50.9 \text{ MPa}$$

c. Determination of the Bearing Stresses. To determine the nominal bearing stress at A in member AB, we use Eq. (8.7) of Sec. 8.1c. From Fig. 8.21, we have $t = 30$ mm and $d = 25$ mm. Recalling that $P = F_{AB} = 40$ kN, we have

$$\sigma_b = \frac{P}{td} = \frac{40 \text{ kN}}{(30 \text{ mm})(25 \text{ mm})} = 53.3 \text{ MPa}$$

To obtain the bearing stress in the bracket at A, we use $t = 2(25 \text{ mm}) = 50$ mm and $d = 25$ mm:

$$\sigma_b = \frac{P}{td} = \frac{40 \text{ kN}}{(50 \text{ mm})(25 \text{ mm})} = 32.0 \text{ MPa}$$

The bearing stresses at B in member AB, at B and C in member BC, and in the bracket at C are found in a similar way.

The engineer's role is not limited to the analysis of existing structures and machines subjected to given loading conditions. Of even greater importance to the engineer is the *design* of new structures and machines, that is, the selection of appropriate components to perform a given task.

Considering again the structure of Fig. 8.15, let us assume that rod BC is made of a steel with a maximum allowable stress $\sigma_{all} = 165$ MPa. Can rod BC safely support the load to which it will be subjected? The magnitude of the force F_{BC} in the rod was found earlier to be 50 kN and the stress σ_{BC} was found to be 159 MPa. Since the value obtained is smaller than the value σ_{all} of the allowable stress in the steel used, we conclude that rod BC can safely support the load to which it will be subjected. We should also determine whether the deformations produced by the given loading are acceptable. The study of deformations under axial loads will be the subject of Chap. 9. An additional consideration required for members in compression involves the stability of the member, i.e., its ability to support a given load without experiencing a sudden change in configuration. This will be discussed in Chap. 16.

Concept Application 8.2

As an example of design, let us return to the structure of Fig. 8.15 and assume that aluminum with an allowable stress $\sigma_{all} = 100$ MPa is to be used. Since the force in rod BC is still $P = F_{BC} = 50$ kN under the given loading, from Eq. (8.1), we have

$$\sigma_{all} = \frac{P}{A} \qquad A = \frac{P}{\sigma_{all}} = \frac{50 \times 10^3 \text{ N}}{100 \times 10^6 \text{ Pa}} = 500 \times 10^{-6} \text{ m}^2$$

and since $A = \pi r^2$,

$$r = \sqrt{\frac{A}{\pi}} = \sqrt{\frac{500 \times 10^{-6} \text{ m}^2}{\pi}} = 12.62 \times 10^{-3} \text{ m} = 12.62 \text{ mm}$$

$$d = 2r = 25.2 \text{ mm}$$

Therefore, an aluminum rod 26 mm or more in diameter will be adequate.

Sample Problem 8.1

In the hanger shown, the upper portion of link ABC is $\frac{3}{8}$ in. thick and the lower portions are each $\frac{1}{4}$ in. thick. Epoxy resin is used to bond the upper and lower portions together at B. The pin at A has a $\frac{3}{8}$-in. diameter, while a $\frac{1}{4}$-in.-diameter pin is used at C. Determine (a) the shearing stress in pin A, (b) the shearing stress in pin C, (c) the largest normal stress in link ABC, (d) the average shearing stress on the bonded surfaces at B, and (e) the bearing stress in the link at C.

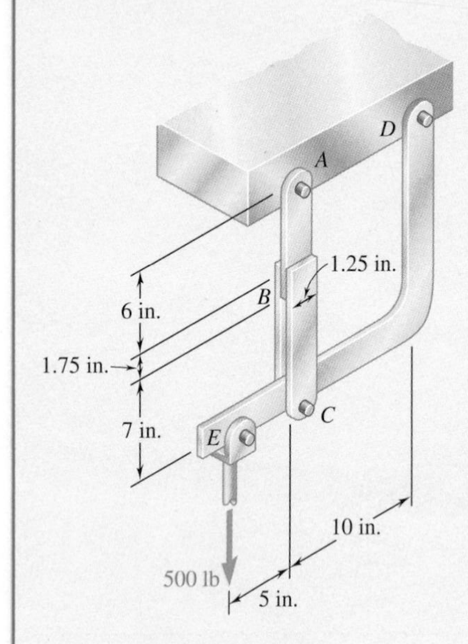

STRATEGY: Consider the free body of the hanger to determine the internal force for member AB and then proceed to determine the shearing and bearing forces applicable to the pins. These forces can then be used to determine the stresses.

MODELING: Draw the free-body diagram of the hanger to determine the support reactions (Fig. 1). Then draw the diagrams of the various components of interest showing the forces needed to determine the desired stresses (Figs. 2-6).

ANALYSIS:

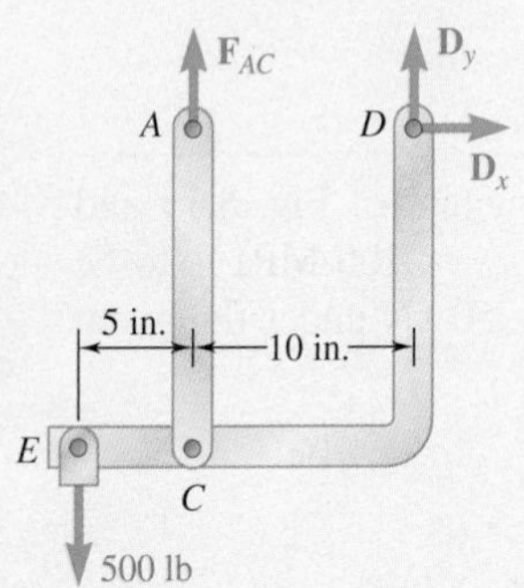

Fig. 1 Free-body diagram of hanger.

Free Body: Entire Hanger. Since the link ABC is a two-force member (Fig. 1), the reaction at A is vertical; the reaction at D is represented by its components D_x and D_y. Thus,

$$+\circlearrowleft \Sigma M_D = 0: \quad (500 \text{ lb})(15 \text{ in.}) - F_{AC}(10 \text{ in.}) = 0$$

$$F_{AC} = +750 \text{ lb} \qquad F_{AC} = 750 \text{ lb} \quad \textit{tension}$$

a. Shearing Stress in Pin A. Since this $\frac{3}{8}$-in.-diameter pin is in single shear (Fig. 2), write

$$\tau_A = \frac{F_{AC}}{A} = \frac{750 \text{ lb}}{\frac{1}{4}\pi(0.375 \text{ in.})^2} \qquad \tau_A = 6790 \text{ psi} \blacktriangleleft$$

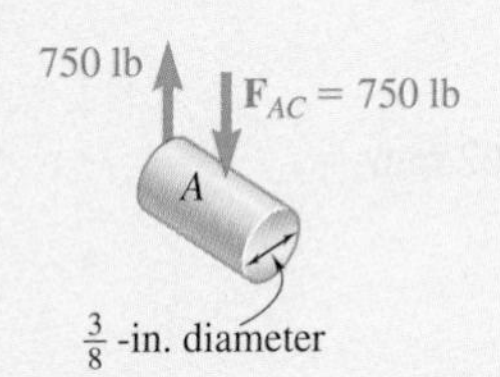

Fig. 2 Pin A.

b. Shearing Stress in Pin C. Since this $\frac{1}{4}$-in.-diameter pin is in double shear (Fig. 3), write

$$\tau_C = \frac{\frac{1}{2}F_{AC}}{A} = \frac{375 \text{ lb}}{\frac{1}{4}\pi(0.25 \text{ in.})^2} \qquad \tau_C = 7640 \text{ psi} \blacktriangleleft$$

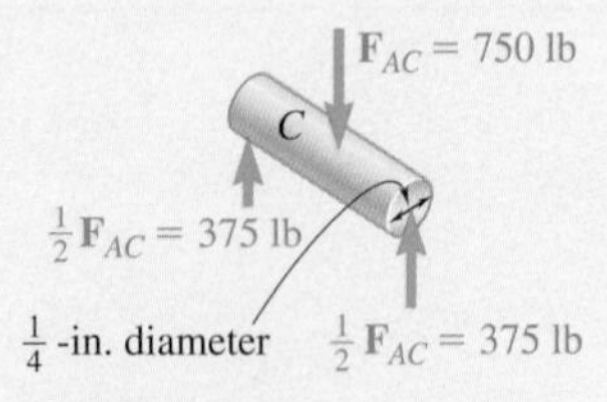

Fig. 3 Pin C.

(continued)

c. Largest Normal Stress in Link *ABC*. The largest stress is found where the area is smallest; this occurs at the cross section at *A* (Fig. 4) where the $\frac{3}{8}$-in. hole is located. We have

$$\sigma_A = \frac{F_{AC}}{A_{net}} = \frac{750 \text{ lb}}{(\frac{3}{8} \text{ in.})(1.25 \text{ in.} - 0.375 \text{ in.})} = \frac{750 \text{ lb}}{0.328 \text{ in}^2} \qquad \sigma_A = 2290 \text{ psi} \blacktriangleleft$$

d. Average Shearing Stress at *B*. We note that bonding exists on both sides of the upper portion of the link (Fig. 5) and that the shear force on each side is $F_1 = (750 \text{ lb})/2 = 375$ lb. The average shearing stress on each surface is

$$\tau_B = \frac{F_1}{A} = \frac{375 \text{ lb}}{(1.25 \text{ in.})(1.75 \text{ in.})} \qquad \tau_B = 171.4 \text{ psi} \blacktriangleleft$$

e. Bearing Stress in Link at *C*. For each portion of the link (Fig. 6), $F_1 = 375$ lb, and the nominal bearing area is (0.25 in.)(0.25 in.) = 0.0625 in^2.

$$\sigma_b = \frac{F_1}{A} = \frac{375 \text{ lb}}{0.0625 \text{ in}^2} \qquad \sigma_b = 6000 \text{ psi} \blacktriangleleft$$

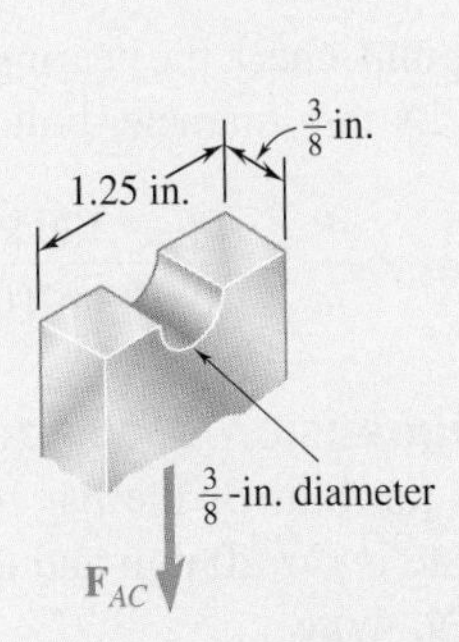

Fig. 4 Link *ABC* section at *A*.

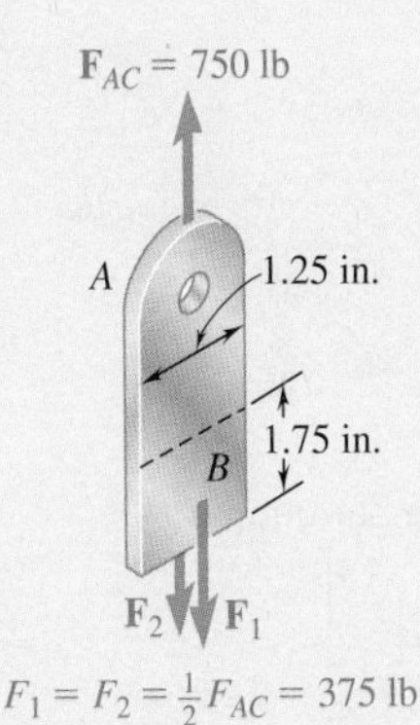

Fig. 5 Element *AB*.

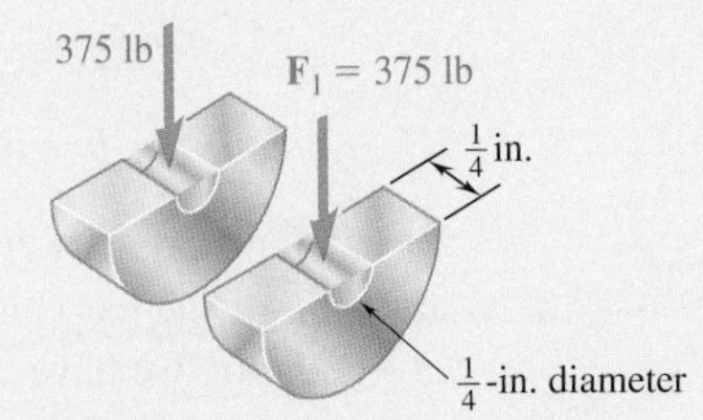

Fig. 6 Link *ABC* section at *C*.

REFLECT and THINK: This sample problem demonstrates the need to draw free-body diagrams of the separate components, carefully considering the behavior in each one. As an example, based on visual inspection of the hanger it is apparent that member *AC* should be in tension for the given load, and the analysis confirms this. Had a compression result been obtained instead, a thorough reexamination of the analysis would have been required.

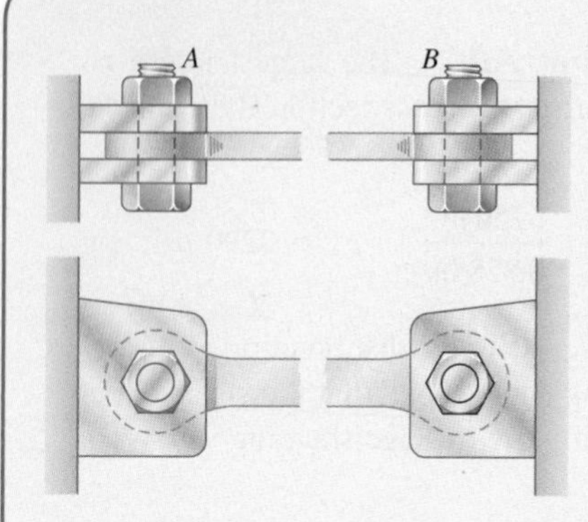

Sample Problem 8.2

The steel tie bar shown is to be designed to carry a tension force of magnitude $P = 120$ kN when bolted between double brackets at A and B. The bar will be fabricated from 20-mm-thick plate stock. For the grade of steel to be used, the maximum allowable stresses are $\sigma = 175$ MPa, $\tau = 100$ MPa, and $\sigma_b = 350$ MPa. Design the tie bar by determining the required values of (*a*) the diameter d of the bolt, (*b*) the dimension b at each end of the bar, and (*c*) the dimension h of the bar.

STRATEGY: Use free-body diagrams to determine the forces needed to obtain the stresses in terms of the design tension force. Setting these stresses equal to the allowable stresses provides for the determination of the required dimensions.

MODELING and ANALYSIS:

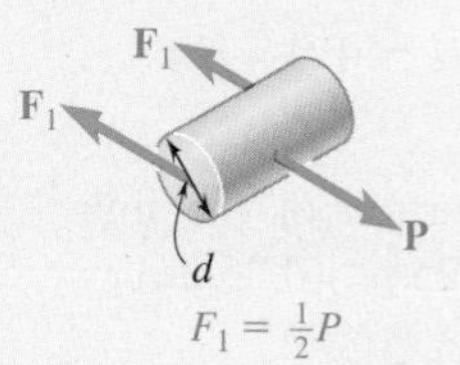

Fig. 1 Sectioned bolt.

a. Diameter of the Bolt. Since the bolt is in double shear (Fig. 1), $F_1 = \frac{1}{2}P = 60$ kN.

$$\tau = \frac{F_1}{A} = \frac{60 \text{ kN}}{\frac{1}{4}\pi d^2} \qquad 100 \text{ MPa} = \frac{60 \text{ kN}}{\frac{1}{4}\pi d^2} \qquad d = 27.6 \text{ mm}$$

$$\text{Use} \qquad d = 28 \text{ mm} \blacktriangleleft$$

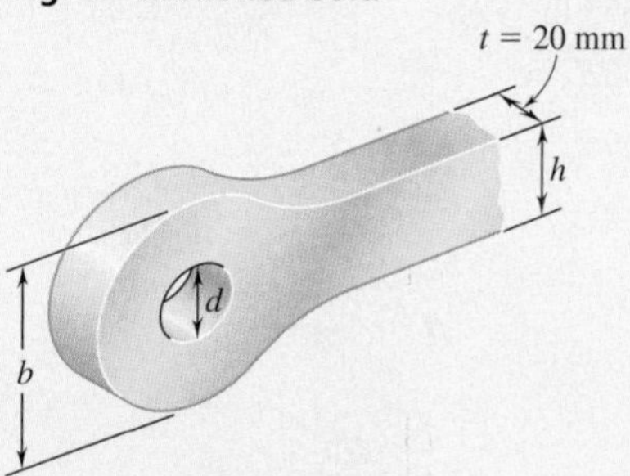

Fig. 2 Tie bar geometry.

At this point, check the bearing stress between the 20-mm-thick plate (Fig. 2) and the 28-mm-diameter bolt.

$$\sigma_b = \frac{P}{td} = \frac{120 \text{ kN}}{(0.020 \text{ m})(0.028 \text{ m})} = 214 \text{ MPa} < 350 \text{ MPa} \qquad \text{OK}$$

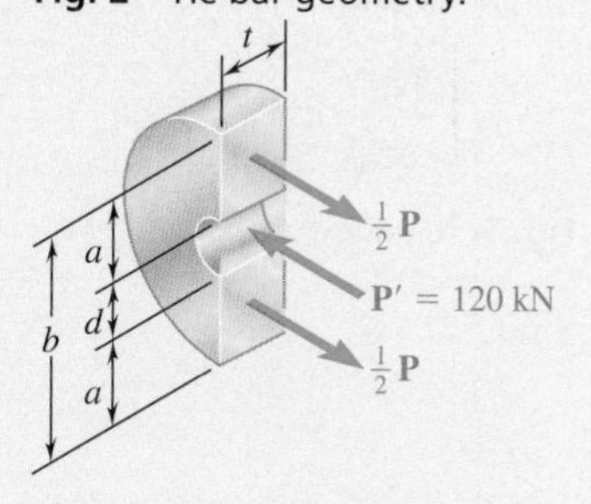

Fig. 3 End section of tie bar.

b. Dimension *b* at Each End of the Bar. We consider one of the end portions of the bar in Fig. 3. Recalling that the thickness of the steel plate is $t = 20$ mm and that the average tensile stress must not exceed 175 MPa, write

$$\sigma = \frac{\frac{1}{2}P}{ta} \qquad 175 \text{ MPa} = \frac{60 \text{ kN}}{(0.02 \text{ m})a} \qquad a = 17.14 \text{ mm}$$

$$b = d + 2a = 28 \text{ mm} + 2(17.14 \text{ mm}) \qquad b = 62.3 \text{ mm} \blacktriangleleft$$

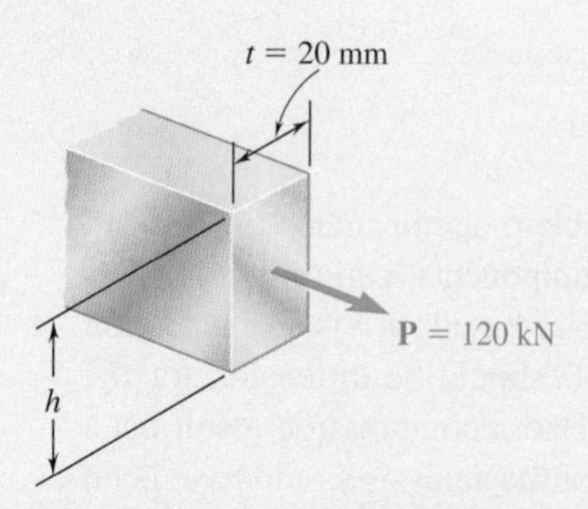

Fig. 4 Mid-body section of tie bar.

c. Dimension *h* of the Bar. We consider a section in the central portion of the bar (Fig. 4). Recalling that the thickness of the steel plate is $t = 20$ mm, we have

$$\sigma = \frac{P}{th} \qquad 175 \text{ MPa} = \frac{120 \text{ kN}}{(0.020 \text{ m})h} \qquad h = 34.3 \text{ mm}$$

$$\text{Use} \qquad h = 35 \text{ mm} \blacktriangleleft$$

REFLECT and THINK: We sized d based on bolt shear, and then checked bearing on the tie bar. Had the maximum allowable bearing stress been exceeded, we would have had to recalculate d based on the bearing criterion.

Problems

8.1 Two solid cylindrical rods AB and BC are welded together at B and loaded as shown. Knowing that $d_1 = 50$ mm and $d_2 = 30$ mm, find the average normal stress at the midsection of (*a*) rod AB, (*b*) rod BC.

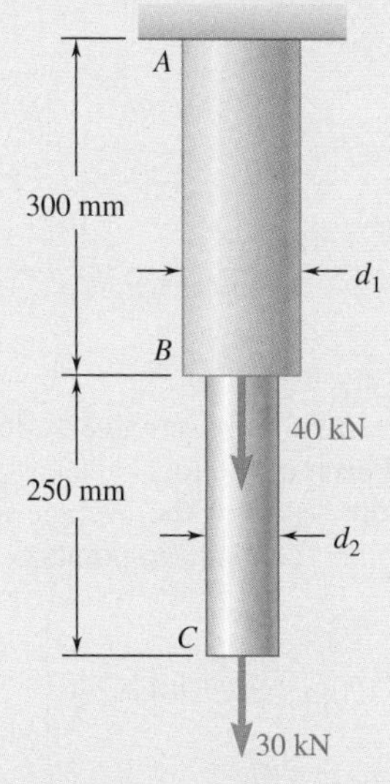

Fig. P8.1 and P8.2

8.2 Two solid cylindrical rods AB and BC are welded together at B and loaded as shown. Knowing that the average normal stress must not exceed 175 MPa in rod AB and 150 MPa in rod BC, determine the smallest allowable values of d_1 and d_2.

8.3 Two solid cylindrical rods AB and BC are welded together at B and loaded as shown. Knowing that $P = 40$ kips, determine the average normal stress at the midsection of (*a*) rod AB, (*b*) rod BC.

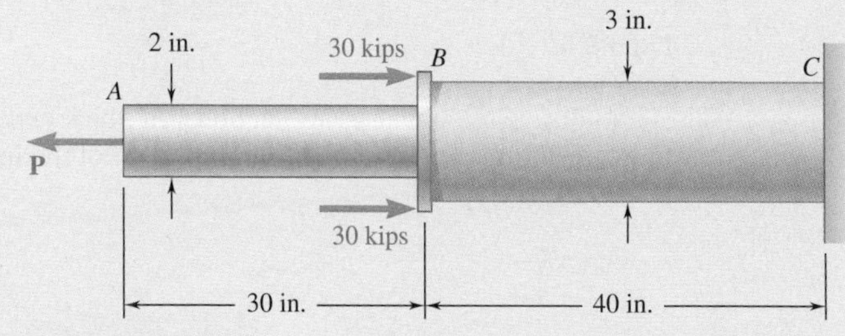

Fig. P8.3 and P8.4

8.4 Two solid cylindrical rods AB and BC are welded together at B and loaded as shown. Determine the magnitude of the force **P** for which the tensile stress in rod AB is twice the magnitude as the compressive stress in rod BC.

8.5 Link *BD* consists of a single bar 36 mm wide and 18 mm thick. Knowing that each pin has a 12-mm diameter, determine the maximum value of the average normal stress in link *BD* if (*a*) $\theta = 0$, (*b*) $\theta = 90°$.

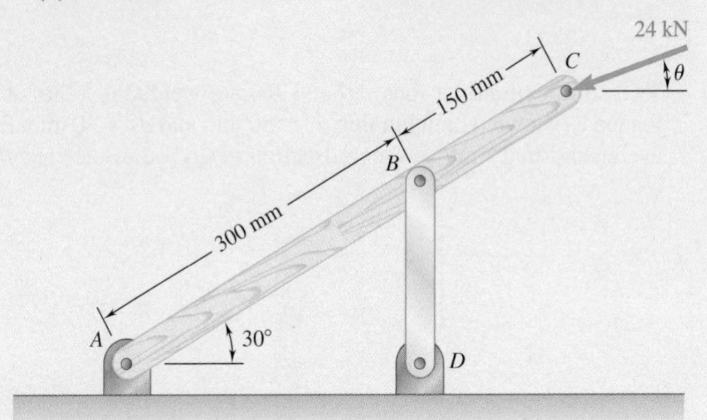

Fig. *P8.5*

8.6 Each of the four vertical links has an 8 × 36-mm uniform rectangular cross section and each of the four pins has a 16-mm diameter. Determine the maximum value of the average normal stress in the links connecting (*a*) points *B* and *D*, (*b*) points *C* and *E*.

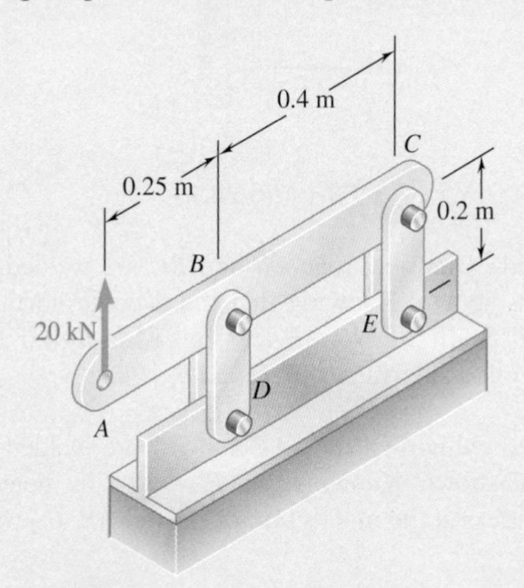

Fig. P8.6

8.7 Link *AC* has a uniform rectangular cross section $\frac{1}{8}$ in. thick and 1 in. wide. Determine the normal stress in the central portion of the link.

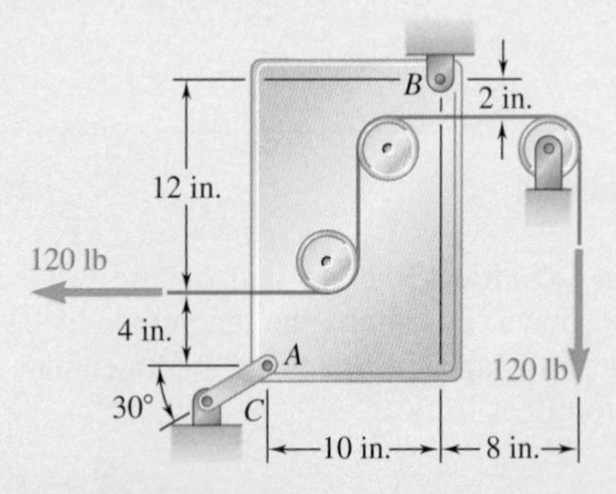

Fig. P8.7

8.8 Two horizontal 5-kip forces are applied to pin B of the assembly shown. Knowing that a pin of 0.8-in. diameter is used at each connection, determine the maximum value of the average normal stress (*a*) in link AB, (*b*) in link BC.

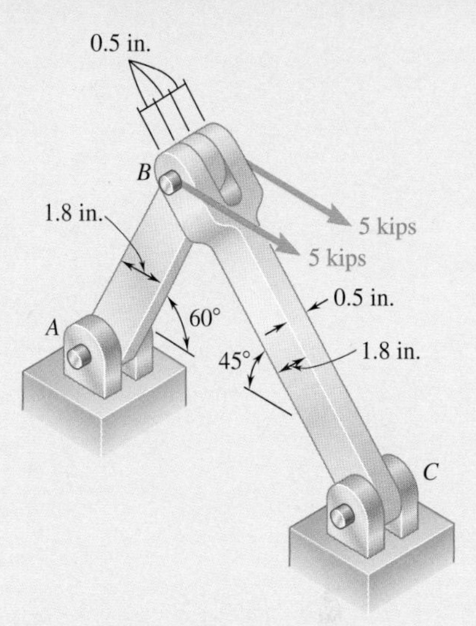

Fig. *P8.8*

8.9 For the Pratt bridge truss and loading shown, determine the average normal stress in member BE, knowing that the cross-sectional area of that member is 5.87 in^2.

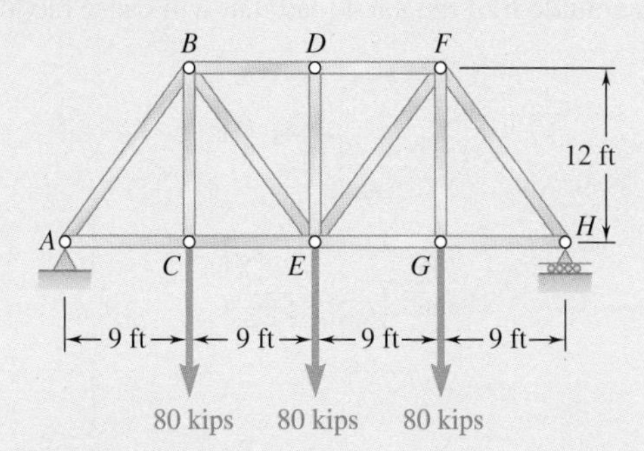

Fig. P8.9 and P8.10

8.10 Knowing that the average normal stress in member CE of the Pratt bridge truss shown must not exceed 21 ksi for the given loading, determine the cross-sectional area of that member that will yield the most economical and safe design. Assume that both ends of the member will be adequately reinforced.

8.11 A couple **M** of magnitude 1500 N·m is applied to the crank of an engine. For the position shown, determine (*a*) the force **P** required to hold the engine system in equilibrium, (*b*) the average normal stress in the connecting rod BC, which has a 450-mm^2 uniform cross section.

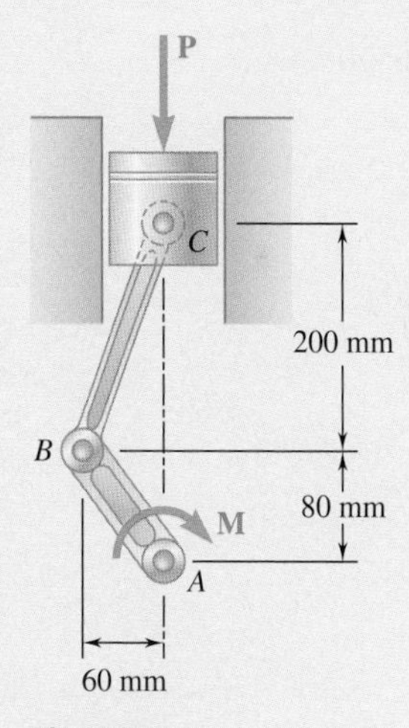

Fig. P8.11

8.12 Two hydraulic cylinders are used to control the position of the robotic arm ABC. Knowing that the control rods attached at A and D each have a 20-mm diameter and happen to be parallel in the position shown, determine the average normal stress in (*a*) member AE, (*b*) member DG.

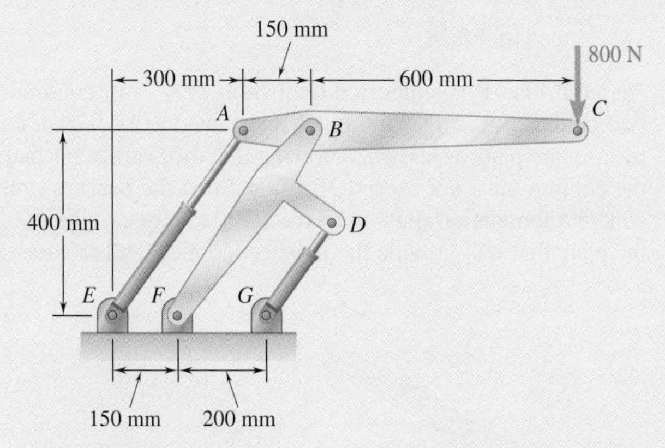

Fig. P8.12

8.13 The wooden members A and B are to be joined by plywood splice plates that will be fully glued on the surfaces in contact. As part of the design of the joint, and knowing that the clearance between the ends of the members is to be 8 mm, determine the smallest allowable length L if the average shearing stress in the glue is not to exceed 800 kPa.

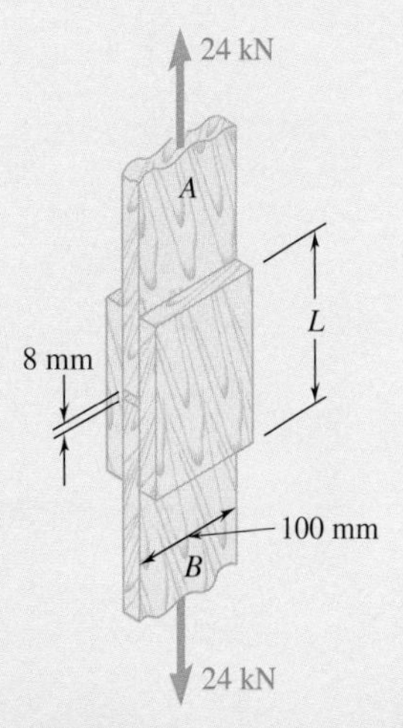

Fig. *P8.13*

8.14 Determine the diameter of the largest circular hole that can be punched into a sheet of polystyrene 6 mm thick, knowing that the force exerted by the punch is 45 kN and that a 55-MPa average shearing stress is required to cause the material to fail.

8.15 Two wooden planks, each $\frac{1}{2}$ in. thick and 9 in. wide, are joined by the dry mortise joint shown. Knowing that the wood used shears off along its grain when the average shearing stress reaches 1.20 ksi, determine the magnitude P of the axial load that will cause the joint to fail.

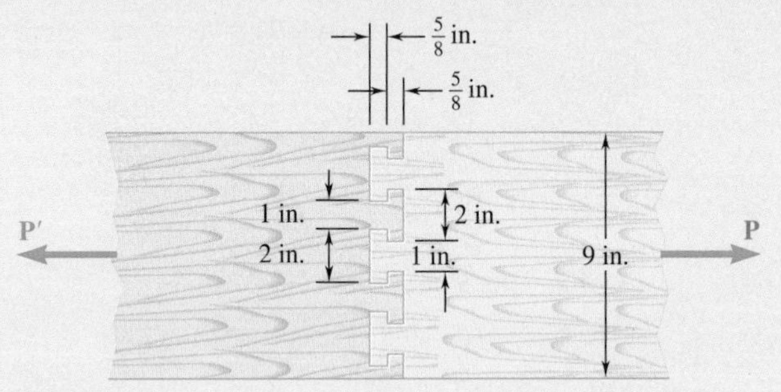

Fig. ***P8.15***

8.16 A load **P** is applied to a steel rod supported as shown by an aluminum plate into which a 0.6-in.-diameter hole has been drilled. Knowing that the shearing stress must not exceed 18 ksi in the steel rod and 10 ksi in the aluminum plate, determine the largest load **P** that may be applied to the rod.

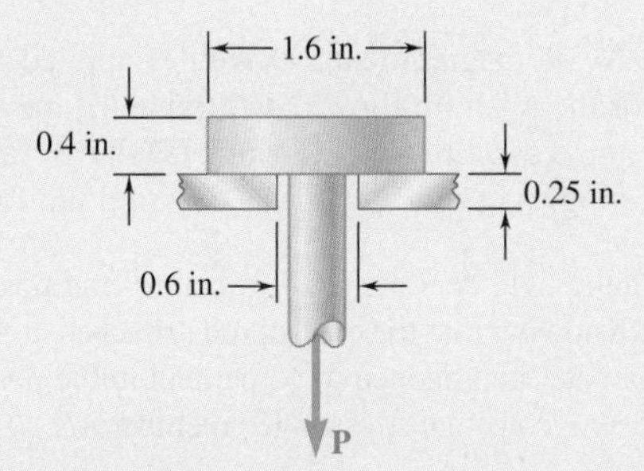

Fig. P8.16

8.17 An axial load **P** is supported by a short W8 × 40 column of cross-sectional area $A = 11.7 \text{ in}^2$ and is distributed to a concrete foundation by a square plate as shown. Knowing that the average normal stress in the column must not exceed 30 ksi and that the bearing stress on the concrete foundation must not exceed 3.0 ksi, determine the side a of the plate that will provide the most economical and safe design.

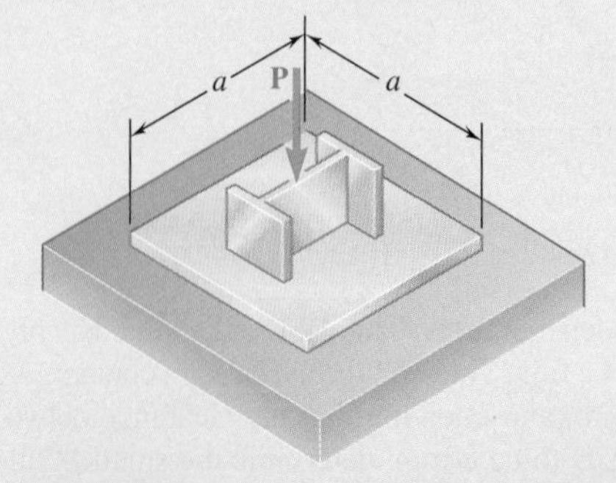

Fig. P8.17

8.18 The axial force in the column supporting the timber beam shown is $P = 20$ kips. Determine the smallest allowable length L of the bearing plate if the bearing stress in the timber is not to exceed 400 psi.

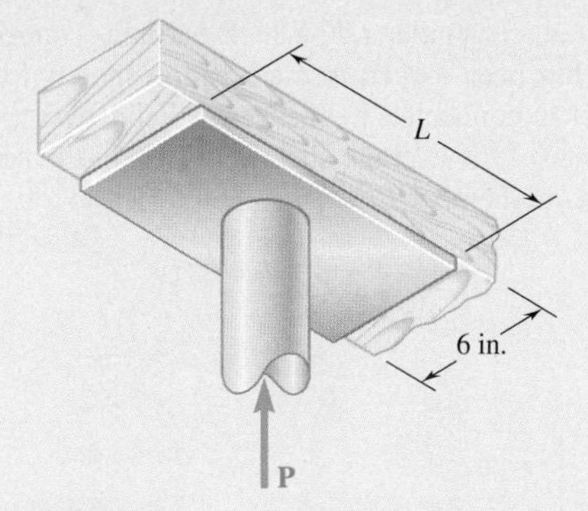

Fig. *P8.18*

8.19 Three wooden planks are fastened together by a series of bolts to form a column. The diameter of each bolt is 12 mm and the inner diameter of each washer is 16 mm, which is slightly larger than the diameter of the holes in the planks. Determine the smallest allowable outer diameter d of the washers, knowing that the average normal stress in the bolts is 36 MPa and that the bearing stress between the washers and the planks must not exceed 8.5 MPa.

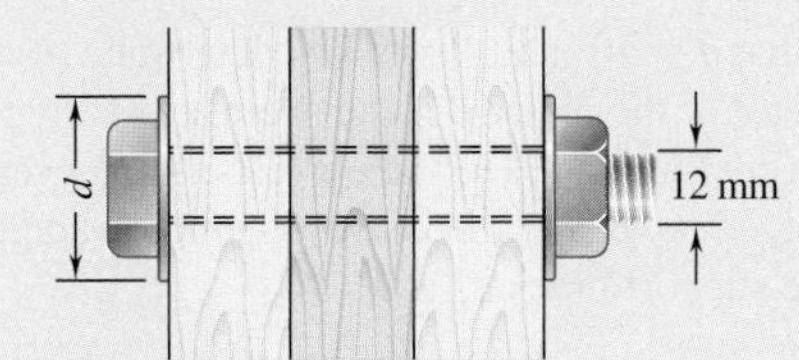

Fig. P8.19

8.20 Link AB, of width $b = 50$ mm and thickness $t = 6$ mm, is used to support the end of a horizontal beam. Knowing that the average normal stress in the link is –140 MPa, and that the average shearing stress in each of the two pins is 80 MPa, determine (*a*) the diameter d of the pins, (*b*) the average bearing stress in the link.

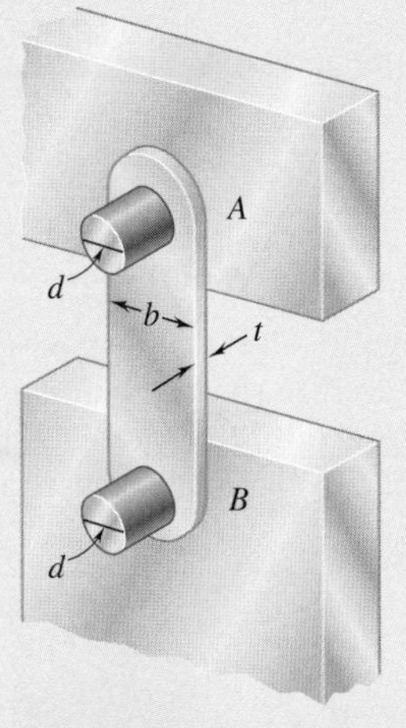

Fig. P8.20

8.21 For the assembly and loading of Prob. 8.8, determine (*a*) the average shearing stress in the pin at C, (*b*) the average bearing stress at C in member BC, (*c*) the average bearing stress at B in member BC.

8.22 The hydraulic cylinder CF, which partially controls the position of rod DE, has been locked in the position shown. Member BD is $\frac{5}{8}$ in. thick and is connected to the vertical rod by a $\frac{3}{8}$-in.-diameter bolt. Determine (*a*) the average shearing stress in the bolt, (*b*) the bearing stress at C in member BD.

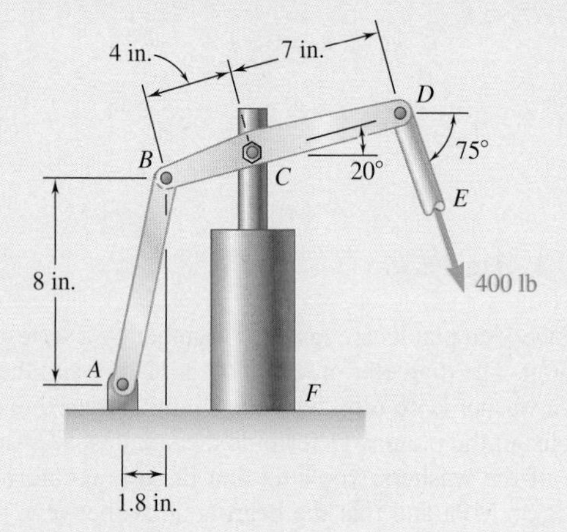

Fig. P8.22

8.23 Knowing that $\theta = 40°$ and $P = 9$ kN, determine (*a*) the smallest allowable diameter of the pin at B if the average shearing stress in the pin is to not exceed 120 MPa, (*b*) the corresponding average bearing stress in member AB at B, (*c*) the corresponding average bearing stress in each of the support brackets at B.

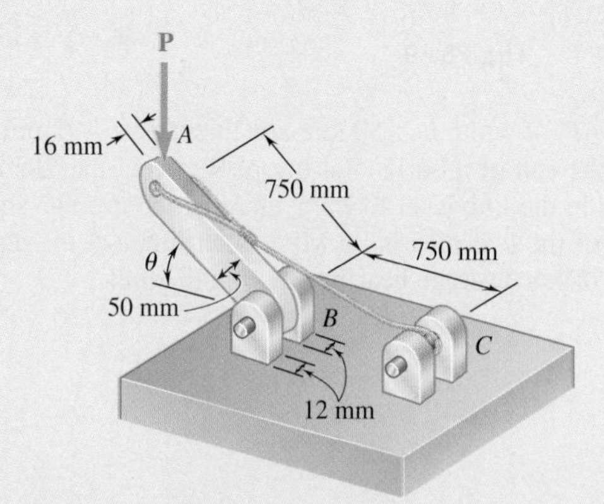

Fig. P8.23 and P8.24

8.24 Determine the largest load **P** that may be applied at A when $\theta = 60°$, knowing that the average shearing stress in the 10-mm-diameter pin at B must not exceed 120 MPa and that the average bearing stress in member AB and in the bracket at B must not exceed 90 MPa.

8.2 STRESS ON AN OBLIQUE PLANE UNDER AXIAL LOADING

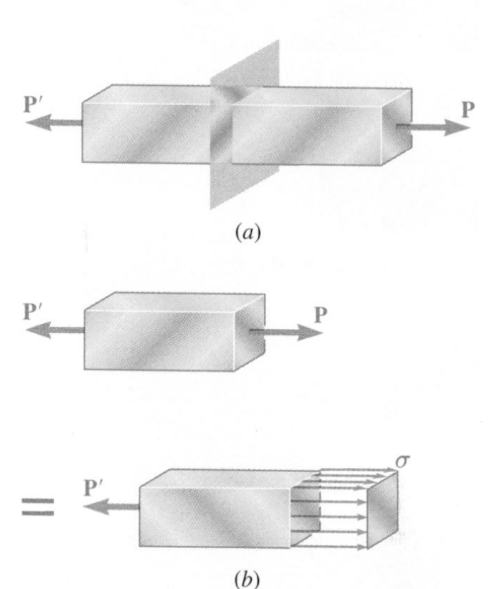

Fig. 8.25 Axial forces on a two-force member. (*a*) Section plane perpendicular to member away from load application. (*b*) Equivalent force diagram models of resultant force acting at centroid and uniform normal stress.

Previously, axial forces exerted on a two-force member (Fig. 8.25*a*) caused normal stresses in that member (Fig. 8.25*b*), while transverse forces exerted on bolts and pins (Fig. 8.26*a*) caused shearing stresses in those connections (Fig. 8.26*b*). Such a relation was observed between axial forces and normal stresses and transverse forces and shearing stresses, because stresses were being determined only on planes perpendicular to the axis of the member or connection. In this section, axial forces cause both normal and shearing stresses on planes that are not perpendicular to the axis of the member. Similarly, transverse forces exerted on a bolt or a pin cause both normal and shearing stresses on planes that are not perpendicular to the axis of the bolt or pin.

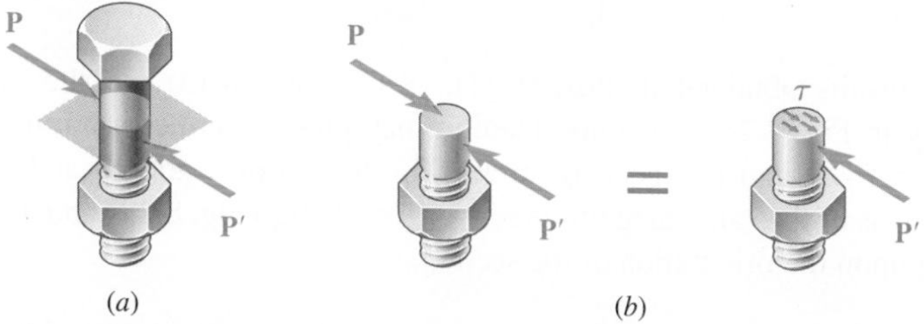

Fig. 8.26 (*a*) Diagram of a bolt from a single-shear joint with a section plane normal to the bolt. (*b*) Equivalent force diagram models of the resultant force acting at the section centroid and the uniform average shear stress.

Consider the two-force member of Fig. 8.25 that is subjected to axial forces **P** and **P′**. If we pass a section forming an angle θ with a normal plane (Fig. 8.27*a*) and draw the free-body diagram of the portion of member located to the left of that section (Fig. 8.27*b*), the equilibrium conditions of the free body show that the distributed forces acting on the section must be equivalent to the force **P.**

Resolving **P** into components **F** and **V**, respectively normal and tangential to the section (Fig. 8.27*c*),

$$F = P\cos\theta \qquad V = P\sin\theta \tag{8.8}$$

Force **F** represents the resultant of normal forces distributed over the section, and force **V** is the resultant of shearing forces (Fig. 8.27*d*). The average values of the corresponding normal and shearing stresses are obtained by dividing F and V by the area A_θ of the section:

$$\sigma = \frac{F}{A_\theta} \qquad \tau = \frac{V}{A_\theta} \tag{8.9}$$

Substituting for F and V from Eq. (8.8) into Eq. (8.9), and observing from Fig. 8.27*c* that $A_0 = A_\theta \cos\theta$ or $A_\theta = A_0/\cos\theta$, where A_0 is the area of a section perpendicular to the axis of the member, we obtain

$$\sigma = \frac{P\cos\theta}{A_0/\cos\theta} \qquad \tau = \frac{P\sin\theta}{A_0/\cos\theta}$$

or

$$\sigma = \frac{P}{A_0}\cos^2\theta \qquad \tau = \frac{P}{A_0}\sin\theta\cos\theta \tag{8.10}$$

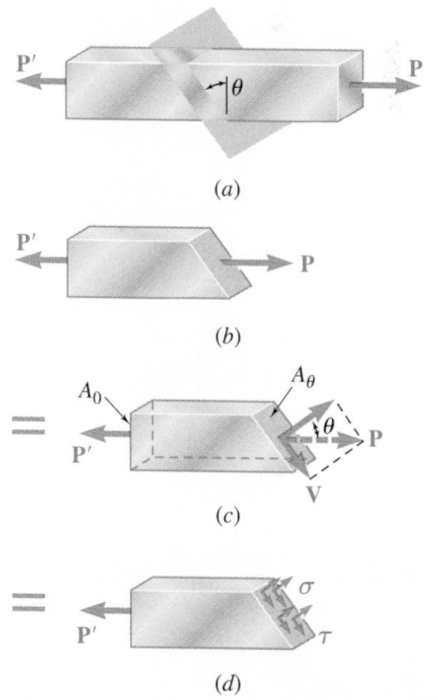

Fig. 8.27 Oblique section through a two-force member. (*a*) Section plane made at an angle θ to the member normal plane, (*b*) Free-body diagram of left section with internal resultant force **P**. (*c*) Free-body diagram of resultant force resolved into components **F** and **V** along the section plane's normal and tangential directions, respectively. (*d*) Free-body diagram with section forces **F** and **V** represented as normal stress, σ, and shearing stress, τ.

(a) Axial loading

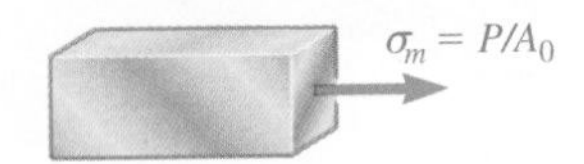

(b) Stresses for $\theta = 0$

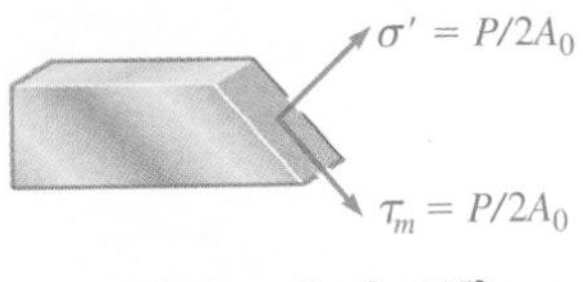

(c) Stresses for $\theta = 45°$

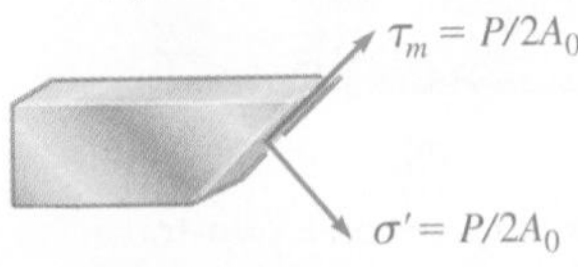

(d) Stresses for $\theta = -45°$

Fig. 8.28 Selected stress results for axial loading.

Note from the first of Eqs. (8.10) that the normal stress σ is maximum when $\theta = 0$ (i.e., the plane of the section is perpendicular to the axis of the member). It approaches zero as θ approaches 90°. We check that the value of σ when $\theta = 0$ is

$$\sigma_m = \frac{P}{A_0} \tag{8.11}$$

The second of Eqs. (8.10) shows that the shearing stress τ is zero for $\theta = 0$ and $\theta = 90°$. For $\theta = 45°$, it reaches its maximum value

$$\tau_m = \frac{P}{A_0} \sin 45° \cos 45° = \frac{P}{2A_0} \tag{8.12}$$

The first of Eqs. (8.10) indicates that, when $\theta = 45°$, the normal stress σ' is also equal to $P/2A_0$:

$$\sigma' = \frac{P}{A_0} \cos^2 45° = \frac{P}{2A_0} \tag{8.13}$$

The results obtained in Eqs. (8.11), (8.12), and (8.13) are shown graphically in Fig. 8.28. The same loading may produce either a normal stress $\sigma_m = P/A_0$ and no shearing stress (Fig. 8.28b) or a normal and a shearing stress of the same magnitude $\sigma' = \tau_m = P/2A_0$ (Fig. 8.28c and d), depending upon the orientation of the section.

8.3 STRESS UNDER GENERAL LOADING CONDITIONS; COMPONENTS OF STRESS

The examples of the previous sections were limited to members under axial loading and connections under transverse loading. Most structural members and machine components are under more involved loading conditions.

Consider a body subjected to several loads $\mathbf{P}_1$, $\mathbf{P}_2$, etc. (Fig. 8.29). To understand the stress condition created by these loads at some point Q within the body, we shall first pass a section through Q, using a plane parallel to the yz plane. The portion of the body to the left of the section is subjected to some of the original loads, and to normal and shearing forces distributed over the section. We shall denote by $\Delta\mathbf{F}^x$ and $\Delta\mathbf{V}^x$, respectively, the normal and the shearing forces acting on a small area ΔA surrounding point Q (Fig. 8.30a). Note that the superscript x is used to indicate that the forces $\Delta\mathbf{F}^x$ and $\Delta\mathbf{V}^x$ act on a surface perpendicular to the x axis. While the normal force $\Delta\mathbf{F}^x$ has a well-defined direction, the shearing force $\Delta\mathbf{V}^x$ may have any direction in the plane of the section. We therefore resolve $\Delta\mathbf{V}^x$ into two component forces, $\Delta\mathbf{V}^x_y$ and $\Delta\mathbf{V}^x_z$, in directions parallel to the y and z axes, respectively (Fig. 8.30b). Dividing the magnitude of each force by the area ΔA and letting ΔA approach zero, we define the three stress components shown in Fig. 8.31:

$$\sigma_x = \lim_{\Delta A \to 0} \frac{\Delta F^x}{\Delta A}$$

$$\tau_{xy} = \lim_{\Delta A \to 0} \frac{\Delta V^x_y}{\Delta A} \qquad \tau_{xz} = \lim_{\Delta A \to 0} \frac{\Delta V^x_z}{\Delta A} \tag{8.14}$$

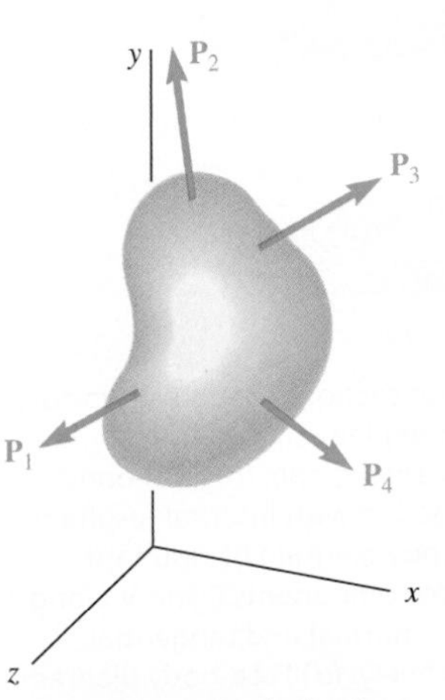

Fig. 8.29 Multiple loads on a general body.

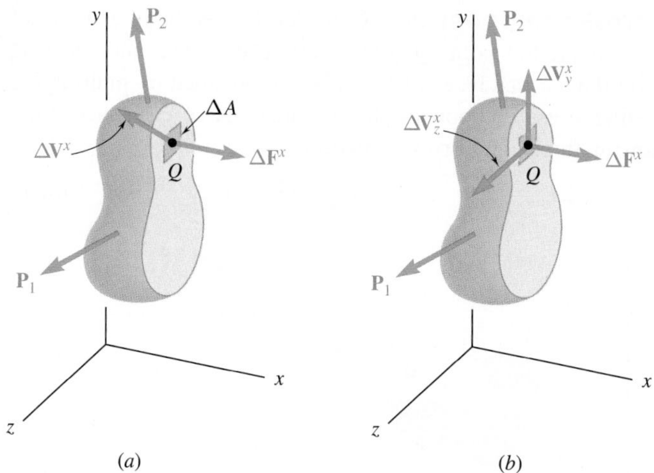

Fig. 8.30 (*a*) Resultant shear and normal forces, $\Delta\mathbf{V}^x$ and $\Delta\mathbf{F}^x$, acting on small area ΔA at point Q. (b) Forces on ΔA resolved into forces in coordinate directions.

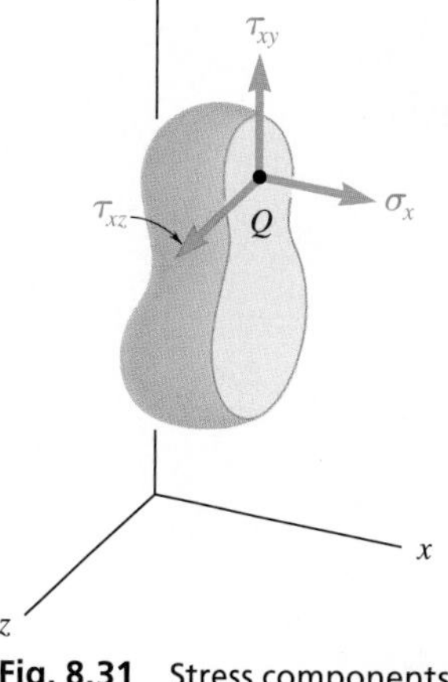

Fig. 8.31 Stress components at point Q on the body to the left of the plane.

Note that the first subscript in σ_x, τ_{xy}, and τ_{xz} is used to indicate that the stresses are exerted *on a surface perpendicular to the x axis.* The second subscript in τ_{xy} and τ_{xz} identifies *the direction of the component.* The normal stress σ_x is positive if the corresponding arrow points in the positive x direction (i.e., if the body is in tension) and negative otherwise. Similarly, the shearing stress components τ_{xy} and τ_{xz} are positive if the corresponding arrows point, respectively, in the positive y and z directions.

This analysis also may be carried out by considering the portion of body located to the right of the vertical plane through Q (Fig. 8.32). The same magnitudes, but opposite directions, are obtained for the normal and shearing forces $\Delta\mathbf{F}^x$, $\Delta\mathbf{V}^x_y$, and $\Delta\mathbf{V}^x_z$. Therefore, the same values are obtained for the corresponding stress components. However as the section in Fig. 8.32 now faces the *negative x axis,* a positive sign for σ_x indicates that the corresponding arrow points *in the negative x direction.* Similarly, positive signs for τ_{xy} and τ_{xz} indicate that the corresponding arrows point in the negative y and z directions, as shown in Fig. 8.32.

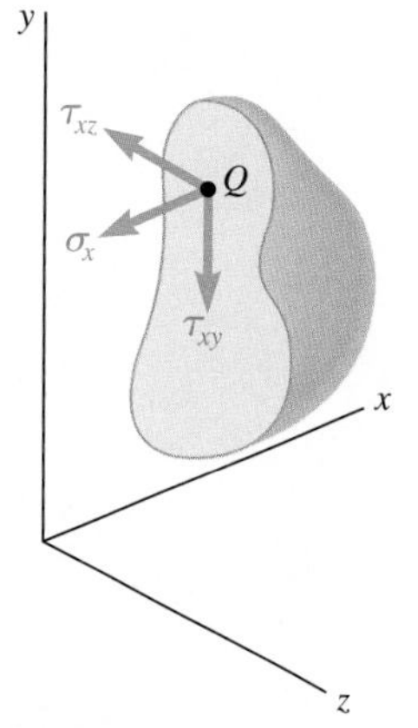

Fig. 8.32 Stress components at point Q on the body to the right of the plane.

Passing a section through Q parallel to the zx plane, we define the stress components, σ_y, τ_{yz}, and τ_{yx}. Then, a section through Q parallel to the xy plane yields the components σ_z, τ_{zx}, and τ_{zy}.

To visualize the stress condition at point Q, consider a small cube of side a centered at Q and the stresses exerted on each of the six faces of the cube (Fig. 8.33). The stress components shown are σ_x, σ_y, and σ_z, which represent the normal stress on faces respectively perpendicular to the x, y, and z axes, and the six shearing stress components τ_{xy}, τ_{xz}, etc. Recall that τ_{xy} represents the y component of the shearing stress exerted on the face perpendicular to the x axis, while τ_{yx} represents the x component of the shearing stress exerted on the face perpendicular to the y axis. Note that only three faces of the cube are actually visible in Fig. 8.33 and that equal and opposite stress components act on the hidden faces. While the stresses acting on the faces of the cube differ slightly from the stresses at Q, the error involved is small and vanishes as side a of the cube approaches zero.

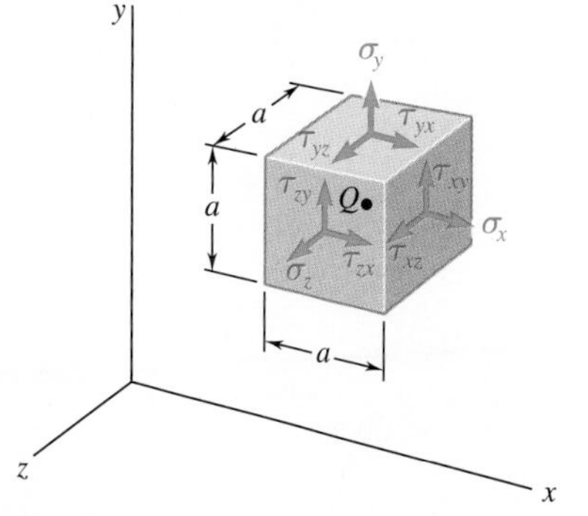

Fig. 8.33 Positive stress components at point Q.

Shearing stress components. Consider the free-body diagram of the small cube centered at point Q (Fig. 8.34). The normal and shearing forces acting on the various faces of the cube are obtained by multiplying the corresponding stress components by the area ΔA of each face. First write the following three equilibrium equations

$$\Sigma F_x = 0 \qquad \Sigma F_y = 0 \qquad \Sigma F_z = 0 \tag{8.15}$$

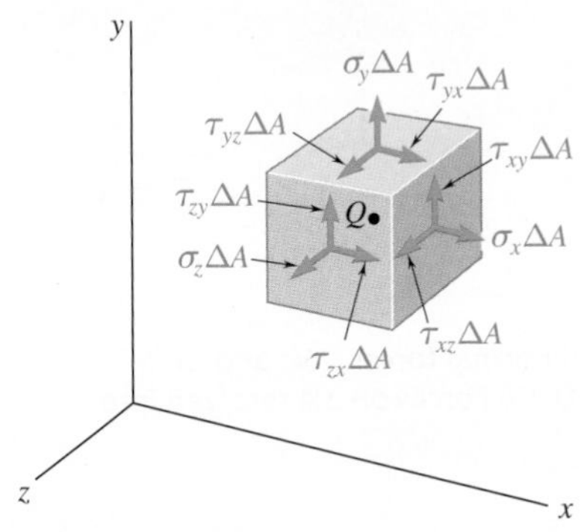

Fig. 8.34 Positive resultant forces on a small element at point Q resulting from a state of general stress.

Since forces equal and opposite to the forces actually shown in Fig. 8.34 are acting on the hidden faces of the cube, Eqs. (8.15) are satisfied. Considering the moments of the forces about axes x', y', and z' drawn from Q in directions respectively parallel to the x, y, and z axes, the three additional equations are

$$\Sigma M_{x'} = 0 \qquad \Sigma M_{y'} = 0 \qquad \Sigma M_{z'} = 0 \tag{8.16}$$

Using a projection on the $x'y'$ plane (Fig. 8.35), note that the only forces with moments about the z axis different from zero are the shearing forces. These forces form two couples: a counterclockwise (positive) moment $(\tau_{xy}\,\Delta A)a$ and a clockwise (negative) moment $-(\tau_{yx}\,\Delta A)a$. The last of the three Eqs. (8.16) yields

$$+\circlearrowleft \Sigma M_z = 0: \qquad (\tau_{xy}\,\Delta A)a - (\tau_{yx}\,\Delta A)a = 0$$

from which

$$\tau_{xy} = \tau_{yx} \tag{8.17}$$

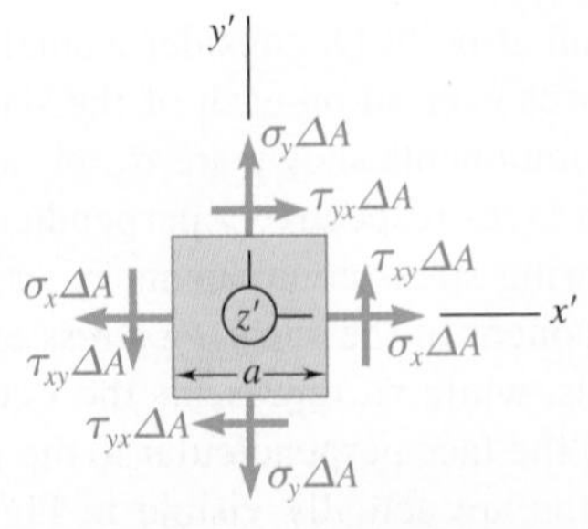

Fig. 8.35 Free-body diagram of small element at Q viewed on projected plane perpendicular to z'-axis. Resultant forces on positive and negative z' faces (not shown) act through the z'-axis, thus do not contribute to the moment about that axis.

This relationship shows that the y component of the shearing stress exerted on a face perpendicular to the x axis is equal to the x component of the shearing stress exerted on a face perpendicular to the y axis. From the remaining parts of Eqs. (8.16), we derive.

$$\tau_{yz} = \tau_{zy} \qquad \tau_{zx} = \tau_{xz} \tag{8.18}$$

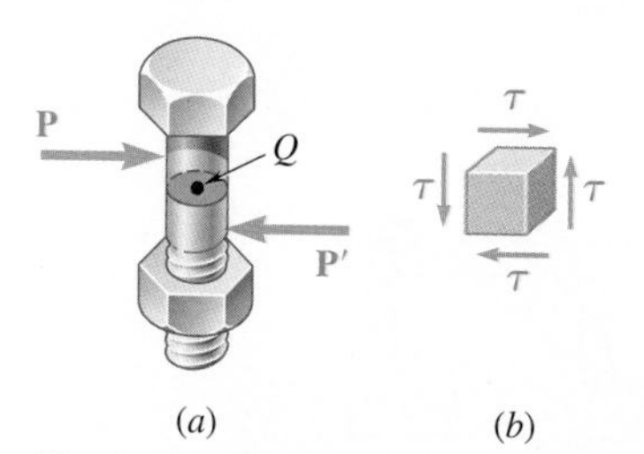

Fig. 8.36 Single-shear bolt with point Q chosen at the center. (b) Pure shear stress element at point Q.

We conclude from Eqs. (8.17) and (8.18), only six stress components are required to define the condition of stress at a given point Q, instead of nine as originally assumed. These components are σ_x, σ_y, σ_z, τ_{xy}, τ_{yz}, and τ_{zx}. Also note that, at a given point, *shear cannot take place in one plane only;* an equal shearing stress must be exerted on another plane perpendicular to the first one. For example, considering the bolt of Fig. 8.26 and a small cube at the center Q (Fig. 8.36*a*), we see that shearing stresses of equal magnitude must be exerted on the two horizontal faces of the cube and on the two faces perpendicular to the forces **P** and **P**′ (Fig. 8.36*b*).

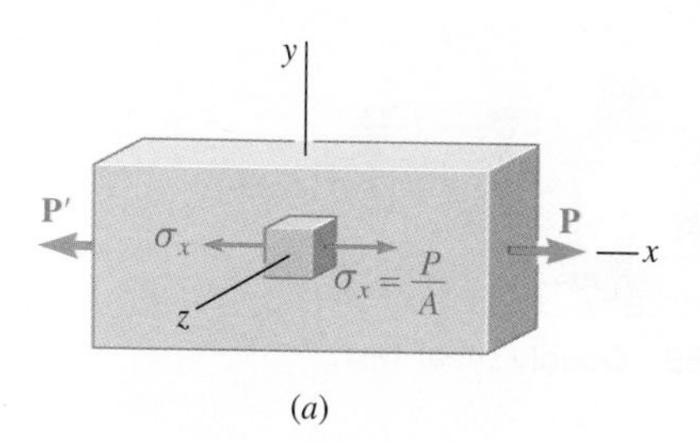

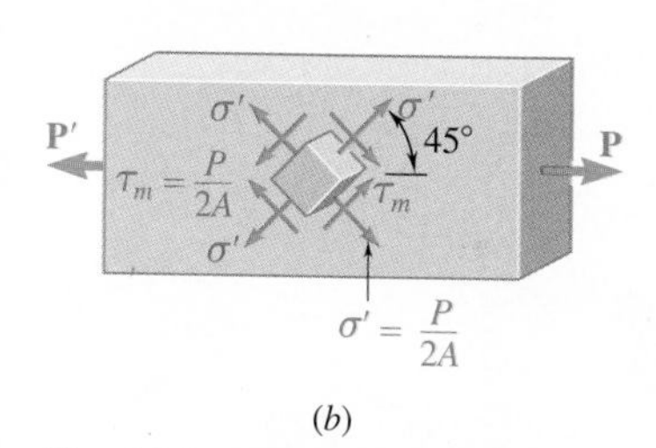

Fig. 8.37 Changing the orientation of the stress element produces different stress components for the same state of stress. This is studied in detail in Chapter 14.

Axial loading. Let us consider again a member under axial loading. If we consider a small cube with faces respectively parallel to the faces of the member and recall the results obtained in Sec. 8.2, the conditions of stress in the member may be described as shown in Fig. 8.37*a*; the only stresses are normal stresses σ_x exerted on the faces of the cube that are perpendicular to the x axis. However, if the small cube is rotated by 45° about the z axis so that its new orientation matches the orientation of the sections considered in Fig. 8.28*c* and *d*, normal and shearing stresses of equal magnitude are exerted on four faces of the cube (Fig. 8.37*b*). Thus, the same loading condition may lead to different interpretations of the stress situation at a given point, depending upon the orientation of the element considered. More will be said about this in Chap. 14: Transformation of Stress and Strain.

8.4 DESIGN CONSIDERATIONS

In engineering applications, the determination of stresses is seldom an end in itself. Rather, the knowledge of stresses is used by engineers to assist in their most important task: the design of structures and machines that will safely and economically perform a specified function.

8.4A Determination of the Ultimate Strength of a Material

An important element to be considered by a designer is how the material will behave under a load. This is determined by performing specific tests on prepared samples of the material. For example, a test specimen of steel may be prepared and placed in a laboratory testing machine to be subjected to a known centric axial tensile force, as described in Sec. 9.1B. As the magnitude of the force is increased, various dimensional changes such as length and diameter are measured. Eventually, the largest force that may be applied to the specimen is reached, and it either breaks or begins to carry

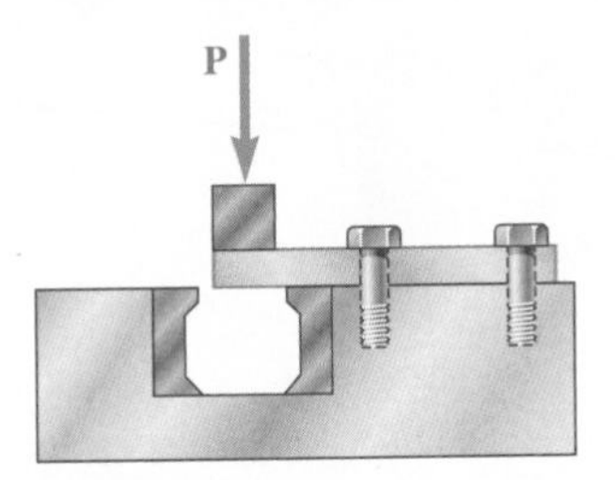

Fig. 8.38 Single shear test.

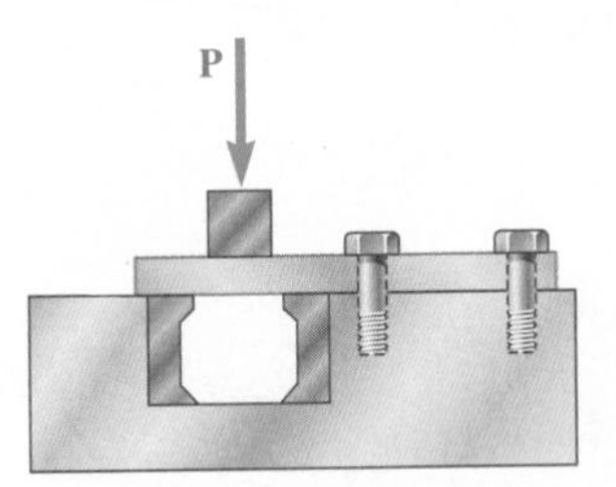

Fig. 8.39 Double shear test.

less load. This largest force is called the *ultimate load* and is denoted by P_U. Since the applied load is centric, the ultimate load is divided by the original cross-sectional area of the rod to obtain the *ultimate normal stress* of the material. This stress, also known as the *ultimate strength in tension,* is

$$\sigma_U = \frac{P_U}{A} \tag{8.19}$$

Several test procedures are available to determine the *ultimate shearing stress* or *ultimate strength in shear*. The one most commonly used involves the twisting of a circular tube (Sec. 10.2). A more direct, if less accurate, procedure clamps a rectangular or round bar in a shear tool (Fig. 8.38) and applies an increasing load P until the ultimate load P_U for single shear is obtained. If the free end of the specimen rests on both of the hardened dies (Fig. 8.39), the ultimate load for double shear is obtained. In either case, the ultimate shearing stress τ_U is

$$\tau_U = \frac{P_U}{A} \tag{8.20}$$

In single shear, this area is the cross-sectional area A of the specimen, while in double shear it is equal to twice the cross-sectional area.

8.4B Allowable Load and Allowable Stress: Factor of Safety

The maximum load that a structural member or a machine component will be allowed to carry under normal conditions is considerably smaller than the ultimate load. This smaller load is the *allowable load* (sometimes called the *working* or *design load*). Thus, only a fraction of the ultimate-load capacity of the member is used when the allowable load is applied. The remaining portion of the load-carrying capacity of the member is kept in reserve to assure its safe performance. The ratio of the ultimate load to the allowable load is used to define the *factor of safety:*†

$$\text{Factor of safety} = F.S. = \frac{\text{ultimate load}}{\text{allowable load}} \tag{8.21}$$

An alternative definition of the factor of safety is based on the use of stresses:

$$\text{Factor of safety} = F.S. = \frac{\text{ultimate stress}}{\text{allowable stress}} \tag{8.22}$$

These two expressions are identical when a linear relationship exists between the load and the stress. In most engineering applications, however,

†In some fields of engineering, notably aeronautical engineering, the *margin of safety* is used in place of the factor of safety. The margin of safety is defined as the factor of safety minus one; that is, margin of safety = $F.S. - 1.00$.

this relationship ceases to be linear as the load approaches its ultimate value, and the factor of safety obtained from Eq. (8.22) does not provide a true assessment of the safety of a given design. Nevertheless, the *allowable-stress method* of design, based on the use of Eq. (8.22), is widely used.

8.4C Factor of Safety Selection

The selection of the factor of safety to be used is one of the most important engineering tasks. If a factor of safety is too small, the possibility of failure becomes unacceptably large. On the other hand, if a factor of safety is unnecessarily large, the result is an uneconomical or nonfunctional design. The choice of the factor of safety for a given design application requires engineering judgment based on many considerations.

1. *Variations that may occur in the properties of the member.* The composition, strength, and dimensions of the member are all subject to small variations during manufacture. In addition, material properties may be altered and residual stresses introduced through heating or deformation that may occur during manufacture, storage, transportation, or construction.
2. *The number of loadings expected during the life of the structure or machine.* For most materials, the ultimate stress decreases as the number of load cycles is increased. This phenomenon is known as *fatigue* and can result in sudden failure if ignored (see Sec. 9.1E).
3. *The type of loadings planned for in the design or that may occur in the future.* Very few loadings are known with complete accuracy—most design loadings are engineering estimates. In addition, future alterations or changes in usage may introduce changes in the actual loading. Larger factors of safety are also required for dynamic, cyclic, or impulsive loadings.
4. *Type of failure.* Brittle materials fail suddenly, usually with no prior indication that collapse is imminent. However, ductile materials, such as structural steel, normally undergo a substantial deformation called *yielding* before failing, providing a warning that overloading exists. Most buckling or stability failures are sudden, whether the material is brittle or not. When the possibility of sudden failure exists, a larger factor of safety should be used than when failure is preceded by obvious warning signs.
5. *Uncertainty due to methods of analysis.* All design methods are based on certain simplifying assumptions that result in calculated stresses being approximations of actual stresses.
6. *Deterioration that may occur in the future because of poor maintenance or unpreventable natural causes.* A larger factor of safety is necessary in locations where conditions such as corrosion and decay are difficult to control or even to discover.
7. *The importance of a given member to the integrity of the whole structure.* Bracing and secondary members in many cases can be designed with a factor of safety lower than that used for primary members.

In addition to these considerations, there is concern of the risk to life and property that a failure would produce. Where a failure would produce no risk to life and only minimal risk to property, the use of a smaller factor of safety can be acceptable. Finally, unless a careful design with a nonexcessive factor of safety is used, a structure or machine might not perform its design function. For example, high factors of safety may have an unacceptable effect on the weight of an aircraft.

For the majority of structural and machine applications, factors of safety are specified by design specifications or building codes written by committees of experienced engineers working with professional societies, industries, or federal, state, or city agencies. Examples of such design specifications and building codes are

1. *Steel:* American Institute of Steel Construction, Specification for Structural Steel Buildings
2. *Concrete:* American Concrete Institute, Building Code Requirement for Structural Concrete
3. *Timber:* American Forest and Paper Association, National Design Specification for Wood Construction
4. *Highway bridges:* American Association of State Highway Officials, Standard Specifications for Highway Bridges

8.4D Load and Resistance Factor Design

The allowable-stress method requires that all the uncertainties associated with the design of a structure or machine element be grouped into a single factor of safety. An alternative method of design makes it possible to distinguish between the uncertainties associated with the structure itself and those associated with the load it is designed to support. Called *Load and Resistance Factor Design (LRFD),* this method allows the designer to distinguish between uncertainties associated with the *live load,* P_L (i.e., the active or time-varying load to be supported by the structure) and the *dead load,* P_D (i.e., the self weight of the structure contributing to the total load).

Using the LRFD method the *ultimate load,* P_U, of the structure (i.e., the load at which the structure ceases to be useful) should be determined. The proposed design is acceptable if the following inequality is satisfied:

$$\gamma_D P_D + \gamma_L P_L \leq \phi P_U \quad (8.23)$$

The coefficient ϕ is the *resistance factor,* which accounts for the uncertainties associated with the structure itself and will normally be less than 1. The coefficients γ_D and γ_L are the *load factors;* they account for the uncertainties associated with the dead and live load and normally will be greater than 1, with γ_L generally larger than γ_D. The allowable-stress method of design will be used in this text.

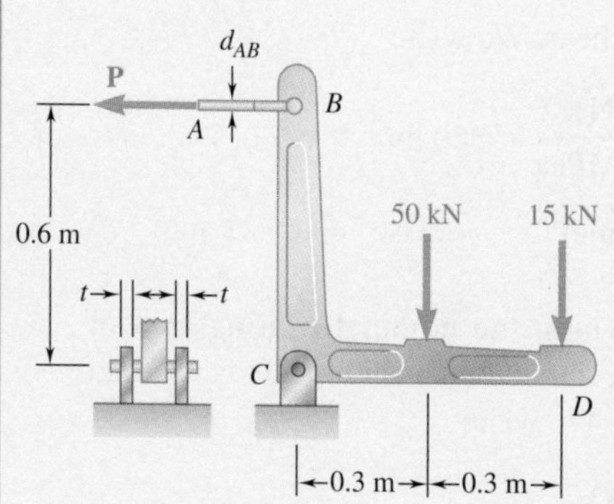

Sample Problem 8.3

Two loads are applied to the bracket *BCD* as shown. (*a*) Knowing that the control rod *AB* is to be made of a steel having an ultimate normal stress of 600 MPa, determine the diameter of the rod for which the factor of safety with respect to failure will be 3.3. (*b*) The pin at *C* is to be made of a steel having an ultimate shearing stress of 350 MPa. Determine the diameter of the pin *C* for which the factor of safety with respect to shear will also be 3.3. (*c*) Determine the required thickness of the bracket supports at *C*, knowing that the allowable bearing stress of the steel used is 300 MPa.

STRATEGY: Consider the free body of the bracket to determine the force **P** and the reaction at *C*. The resulting forces are then used with the allowable stresses, determined from the factor of safety, to obtain the required dimensions.

MODELING: Draw the free-body diagram of the hanger (Fig. 1), and the pin at *C* (Fig. 2).

ANALYSIS:

Free Body: Entire Bracket. Using Fig. 1, the reaction at *C* is represented by its components $\mathbf{C}_x$ and $\mathbf{C}_y$.

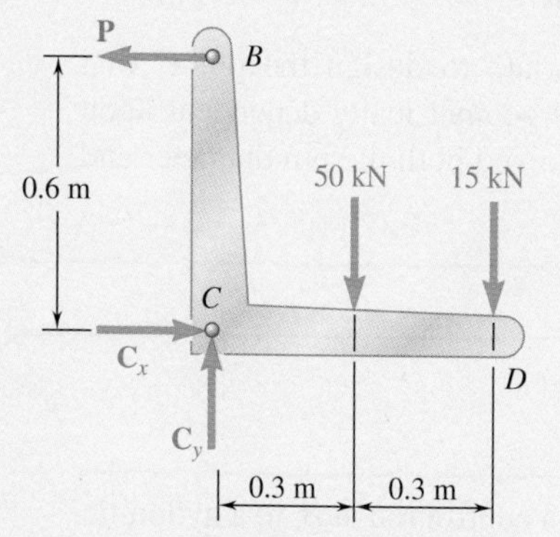

Fig. 1 Free-body diagram of bracket.

$$+\circlearrowleft \Sigma M_C = 0:\quad P(0.6\text{ m}) - (50\text{ kN})(0.3\text{ m}) - (15\text{ kN})(0.6\text{ m}) = 0 \qquad P = 40\text{ kN}$$

$$\Sigma F_x = 0: \qquad C_x = 40\text{ kN}$$

$$\Sigma F_y = 0: \qquad C_y = 65\text{ kN} \qquad C = \sqrt{C_x^2 + C_y^2} = 76.3\text{ kN}$$

a. Control Rod *AB*. Since the factor of safety is 3.3, the allowable stress is

$$\sigma_{\text{all}} = \frac{\sigma_U}{F.S.} = \frac{600\text{ MPa}}{3.3} = 181.8\text{ MPa}$$

For $P = 40$ kN, the cross-sectional area required is

$$A_{\text{req}} = \frac{P}{\sigma_{\text{all}}} = \frac{40\text{ kN}}{181.8\text{ MPa}} = 220 \times 10^{-6}\text{ m}^2$$

$$A_{\text{req}} = \frac{\pi}{4} d_{AB}^2 = 220 \times 10^{-6}\text{ m}^2 \qquad d_{ab} = 16.74\text{ mm} \blacktriangleleft$$

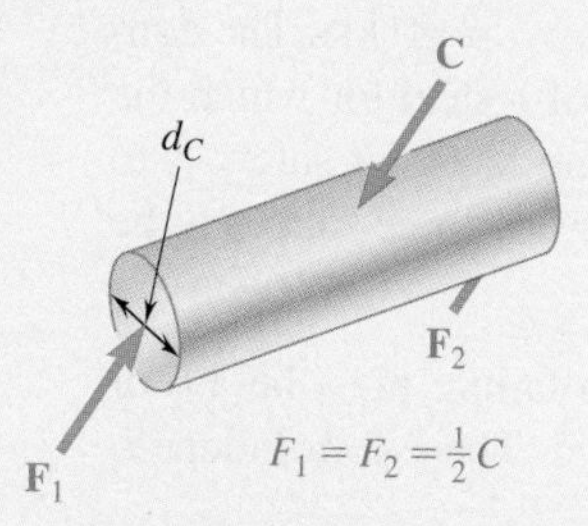

Fig. 2 Free-body diagram of pin at point C.

b. Shear in Pin C. For a factor of safety of 3.3, we have

$$\tau_{\text{all}} = \frac{\tau_U}{F.S.} = \frac{350\text{ MPa}}{3.3} = 106.1\text{ MPa}$$

(continued)

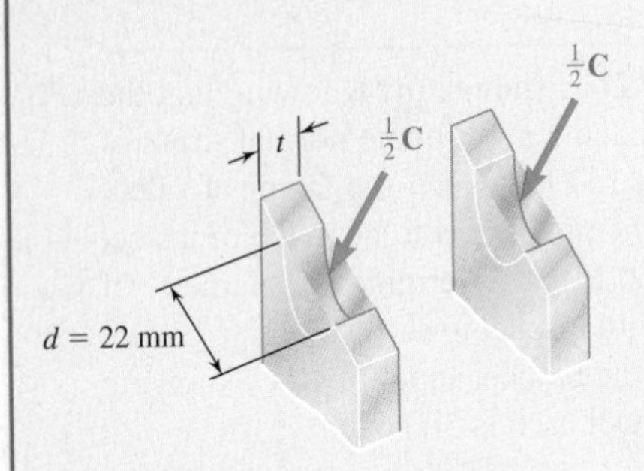

Fig. 3 Bearing loads at bracket support at point C.

As shown in Fig. 2 the pin is in double shear. We write

$$A_{req} = \frac{C/2}{\tau_{all}} = \frac{(76.3 \text{ kN})/2}{106.1 \text{ MPa}} = 360 \text{ mm}^2$$

$$A_{req} = \frac{\pi}{4} d_C^2 = 360 \text{ mm}^2 \qquad d_C = 21.4 \text{ mm} \qquad \text{Use: } d_C = 22 \text{ mm} \blacktriangleleft$$

c. Bearing at C. Using $d = 22$ mm, the nominal bearing area of each bracket is $22t$. From Fig. 3 the force carried by each bracket is $C/2$ and the allowable bearing stress is 300 MPa. We write

$$A_{req} = \frac{C/2}{\sigma_{all}} = \frac{(76.3 \text{ kN})/2}{300 \text{ MPa}} = 127.2 \text{ mm}^2$$

$$\text{Thus, } 22t = 127.2 \qquad t = 5.78 \text{ mm} \qquad \text{Use: } t = 6 \text{ mm} \blacktriangleleft$$

REFLECT and THINK: It was appropriate to design the pin C first and then its bracket, as the pin design was geometrically dependent upon diameter only, while the bracket design involved both the pin diameter and bracket thickness.

Sample Problem 8.4

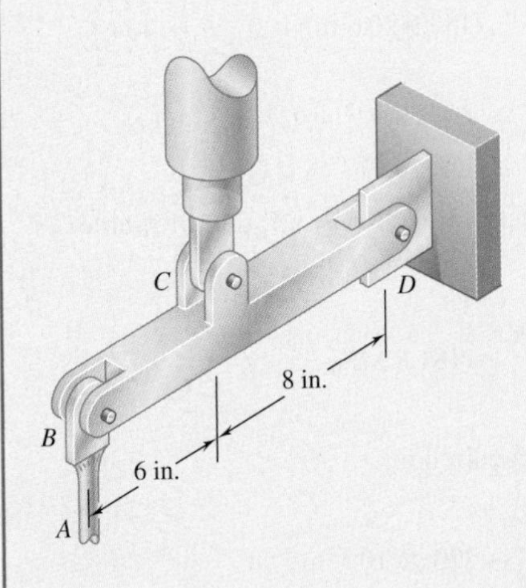

The rigid beam BCD is attached by bolts to a control rod at B, to a hydraulic cylinder at C, and to a fixed support at D. The diameters of the bolts used are: $d_B = d_D = \frac{3}{8}$ in., $d_C = \frac{1}{2}$ in. Each bolt acts in double shear and is made from a steel for which the ultimate shearing stress is $\tau_U = 40$ ksi. The control rod AB has a diameter $d_A = \frac{7}{16}$ in. and is made of a steel for which the ultimate tensile stress is $\sigma_U = 60$ ksi. If the minimum factor of safety is to be 3.0 for the entire unit, determine the largest upward force that may be applied by the hydraulic cylinder at C.

STRATEGY: The factor of safety with respect to failure must be 3.0 or more in each of the three bolts and in the control rod. These four independent criteria need to be considered separately.

MODELING: Draw the free-body diagram of the bar (Fig. 1) and the bolts at B and C (Figs. 2 and 3). Determine the allowable value of the force **C** based on the required design criteria for each part.

ANALYSIS:

Free Body: Beam BCD. Using Fig. 1, first determine the force at C in terms of the force at B and in terms of the force at D.

$$+\circlearrowleft \Sigma M_D = 0: \qquad B(14 \text{ in.}) - C(8 \text{ in.}) = 0 \qquad C = 1.750B \qquad (1)$$

$$+\circlearrowleft \Sigma M_B = 0: \qquad -D(14 \text{ in.}) + C(6 \text{ in.}) = 0 \qquad C = 2.33D \qquad (2)$$

C

B C D

B D

6 in. 8 in.

Fig. 1 Free-body diagram of beam BCD.

(continued)

Control Rod. For a factor of safety of 3.0

$$\sigma_{all} = \frac{\sigma_U}{F.S.} = \frac{60 \text{ ksi}}{3.0} = 20 \text{ ksi}$$

The allowable force in the control rod is

$$B = \sigma_{all}(A) = (20 \text{ ksi})\tfrac{1}{4}\pi\left(\tfrac{7}{16} \text{ in.}\right)^2 = 3.01 \text{ kips}$$

Using Eq. (1), the largest permitted value of C is

$$C = 1.750B = 1.750(3.01 \text{ kips}) \qquad C = 5.27 \text{ kips} \blacktriangleleft$$

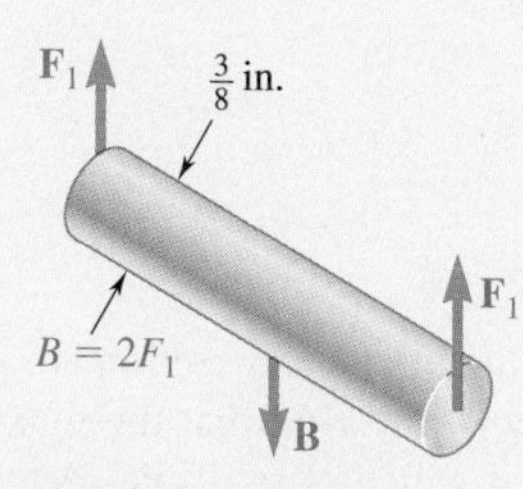

Fig. 2 Free-body diagram of pin at point B.

Bolt at B. $\tau_{all} = \tau_U/F.S. = (40 \text{ ksi})/3 = 13.33$ ksi. Since the bolt is in double shear (Fig. 2), the allowable magnitude of the force **B** exerted on the bolt is

$$B = 2F_1 = 2(\tau_{all}A) = 2(13.33 \text{ ksi})\left(\tfrac{1}{4}\pi\right)\left(\tfrac{3}{8} \text{ in.}\right)^2 = 2.94 \text{ kips}$$

From Eq. (1), $\quad C = 1.750B = 1.750(2.94 \text{ kips}) \qquad C = 5.15 \text{ kips} \blacktriangleleft$

Bolt at D. Since this bolt is the same as bolt B, the allowable force is $D = B = 2.94$ kips. From Eq. (2)

$$C = 2.33D = 2.33(2.94 \text{ kips}) \qquad C = 6.85 \text{ kips} \blacktriangleleft$$

Bolt at C. We again have $\tau_{all} = 13.33$ ksi. Using Fig. 3, we write

$$C = 2F_2 = 2(\tau_{all}A) = 2(13.33 \text{ ksi})\left(\tfrac{1}{4}\pi\right)\left(\tfrac{1}{2} \text{ in.}\right)^2 \qquad C = 5.23 \text{ kips} \blacktriangleleft$$

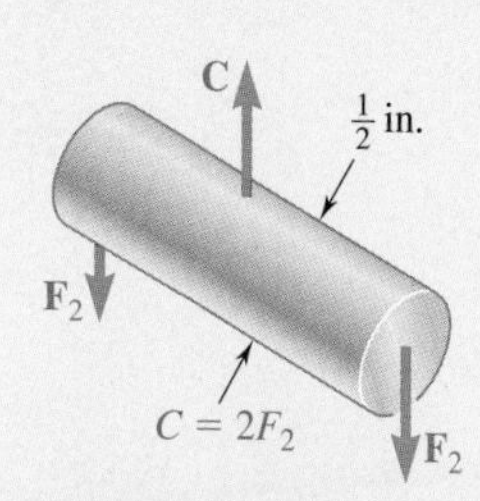

Fig. 3 Free-body diagram of pin at point C.

Summary. We have found separately four maximum allowable values of the force C. In order to satisfy all these criteria, choose the smallest value.

$$C = 5.15 \text{ kips} \blacktriangleleft$$

REFLECT and THINK: This example illustrates that all parts must satisfy the appropriate design criteria, and as a result, some parts have more capacity than needed.

Problems

8.25 Two wooden members of uniform rectangular cross section are joined by the simple glued scarf splice shown. Knowing that $P = 11$ kN, determine the normal and shearing stresses in the glued splice.

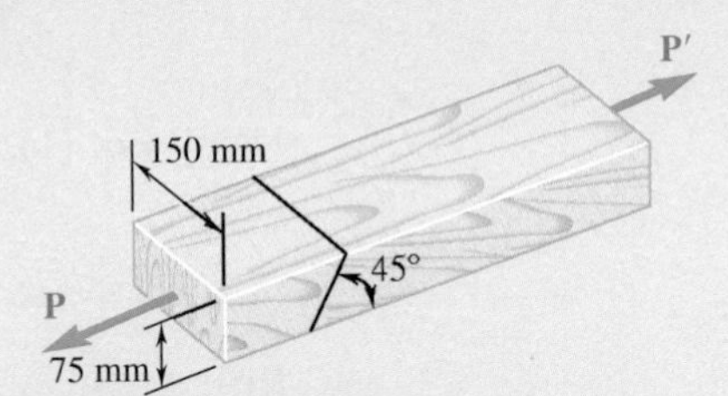

Fig. P8.25 and P8.26

8.26 Two wooden members of uniform rectangular cross section are joined by the simple glued scarf splice shown. Knowing that the maximum allowable shearing stress in the glued splice is 620 kPa, determine (*a*) the largest load **P** that can be safely applied, (*b*) the corresponding tensile stress in the splice.

8.27 The 1.4-kip load **P** is supported by two wooden members of uniform cross section that are joined by the simple glued scarf splice shown. Determine the normal and shearing stresses in the glued splice.

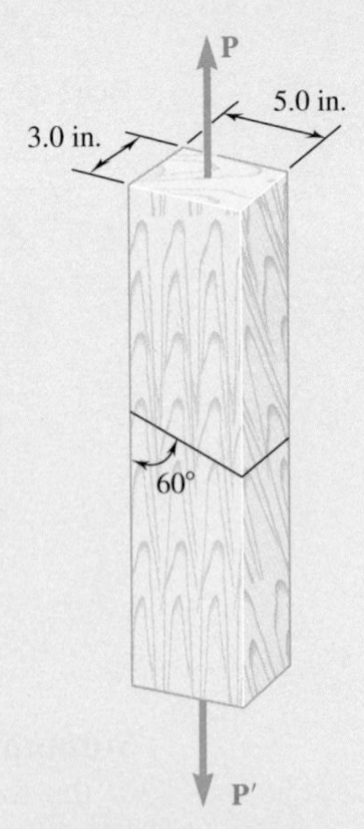

Fig. P8.27 and P8.28

8.28 Two wooden members of uniform cross section are joined by the simple scarf splice shown. Knowing that the maximum allowable tensile stress in the glued splice is 75 psi, determine (*a*) the largest load **P** that can be safely supported, (*b*) the corresponding shearing stress in the splice.

8.29 A 240-kip load **P** is applied to the granite block shown. Determine the resulting maximum value of (*a*) the normal stress, (*b*) the shearing stress. Specify the orientation of the plane on which each of these maximum values occurs.

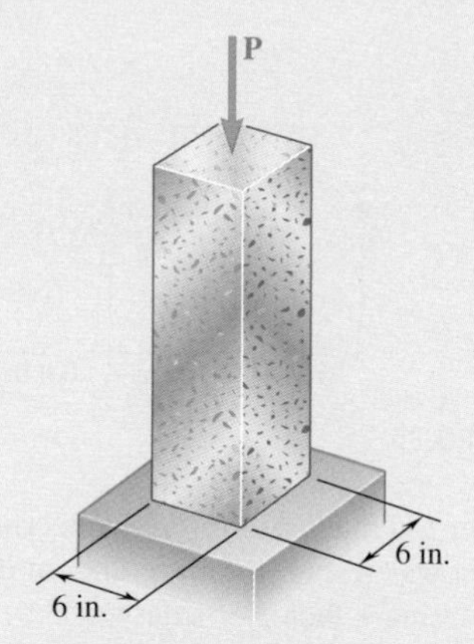

Fig. ***P8.29*** **and P8.30**

8.30 A centric load **P** is applied to the granite block shown. Knowing that the resulting maximum value of the shearing stress in the block is 2.5 ksi, determine (*a*) the magnitude of **P**, (*b*) the orientation of the surface on which the maximum shearing stress occurs, (*c*) the normal stress exerted on the surface, (*d*) the maximum value of the normal stress in the block.

8.31 A steel pipe of 400-mm outer diameter is fabricated from 10-mm-thick plate by welding along a helix that forms an angle of 20° with a plane perpendicular to the axis of the pipe. Knowing that a 300-kN axial force **P** is applied to the pipe, determine the normal and shearing stresses in directions respectively normal and tangential to the weld.

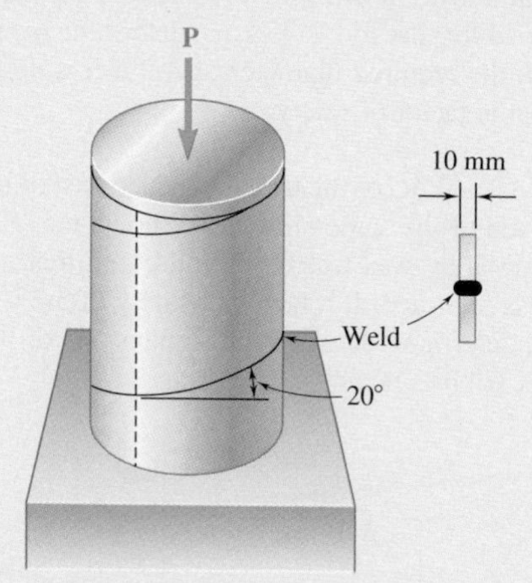

Fig. P8.31 and ***P8.32***

8.32 A steel pipe of 400-mm outer diameter is fabricated from 10-mm-thick plate by welding along a helix that forms an angle of 20° with a plane perpendicular to the axis of the pipe. Knowing that the maximum allowable normal and shearing stresses in the directions respectively normal and tangential to the weld are $\sigma = 60$ MPa and $\tau = 36$ MPa, determine the magnitude P of the largest axial force that can be applied to the pipe.

8.33 Link *AB* is to be made of a steel for which the ultimate normal stress is 450 MPa. Determine the cross-sectional area for *AB* for which the factor of safety will be 3.50. Assume that the link will be adequately reinforced around the pins at *A* and *B*.

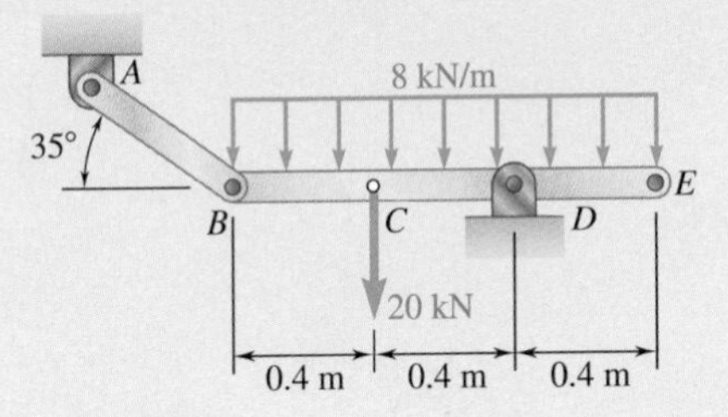

Fig. P8.33

8.34 A $\frac{3}{4}$-in.-diameter rod made of the same material as rods *AC* and *AD* in the truss shown was tested to failure and an ultimate load of 29 kips was recorded. Using a factor of safety of 3.0, determine the required diameter (*a*) of rod *AC*, (*b*) of rod *AD*.

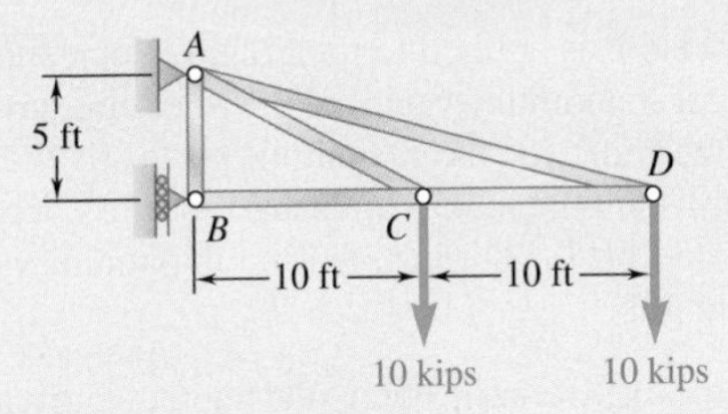

Fig. P8.34 and P8.35

8.35 In the truss shown, members *AC* and *AD* consist of rods made of the same metal alloy. Knowing that *AC* is of 1-in. diameter and that the ultimate load for that rod is 75 kips, determine (*a*) the factor of safety for *AC*, (*b*) the required diameter of *AD* if it is desired that both rods have the same factor of safety.

8.36 Members *AB* and *AC* of the truss shown consist of bars of square cross section made of the same alloy. It is known that a 20-mm-square bar of the same alloy was tested to failure and that an ultimate load of 120 kN was recorded. If a factor of safety of 3.2 is to be achieved for both bars, determine the required dimensions of the cross section of (*a*) bar *AB*, (*b*) bar *AC*.

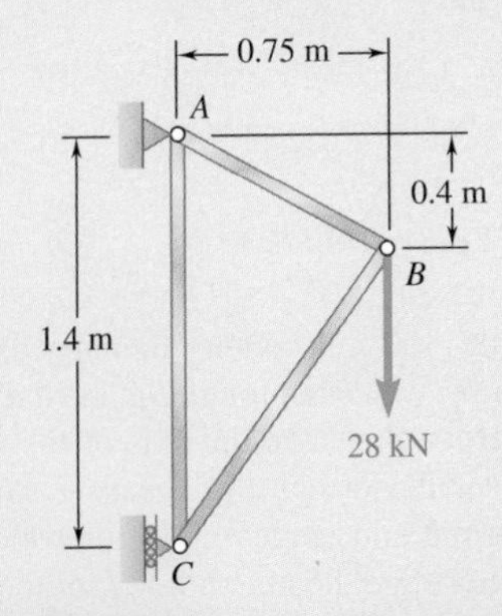

Fig. P8.36

8.37 Three $\frac{3}{4}$-in.-diameter steel bolts are to be used to attach the steel plate shown to a wooden beam. Knowing that the plate will support a load $P = 24$ kips and that the ultimate shearing stress for the steel used is 52 ksi, determine the factor of safety for this design.

Fig. *P8.37*

8.38 Two plates, each $\frac{1}{8}$ in. thick, are used to splice a plastic strip as shown. Knowing that the ultimate shearing stress of the bonding between the surfaces is 130 psi, determine the factor of safety with respect to shear when $P = 325$ lb.

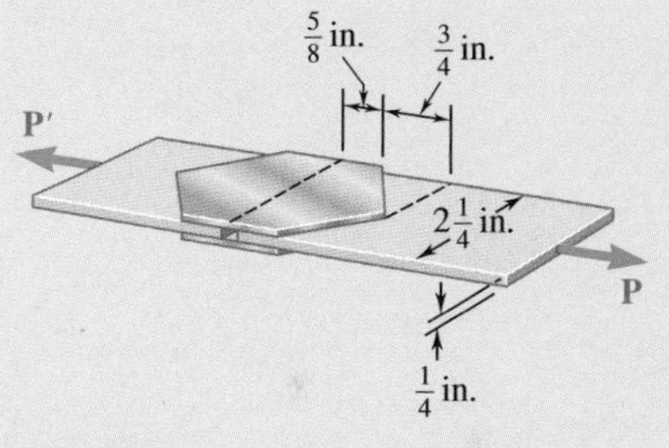

Fig. P8.38

8.39 A load **P** is supported as shown by a steel pin that has been inserted in a short wooden member hanging from the ceiling. The ultimate strength of the wood used is 60 MPa in tension and 7.5 MPa in shear, while the ultimate strength of the steel is 145 MPa in shear. Knowing that $b = 40$ mm, $c = 55$ mm, and $d = 12$ mm, determine the load **P** if an overall factor of safety of 3.2 is desired.

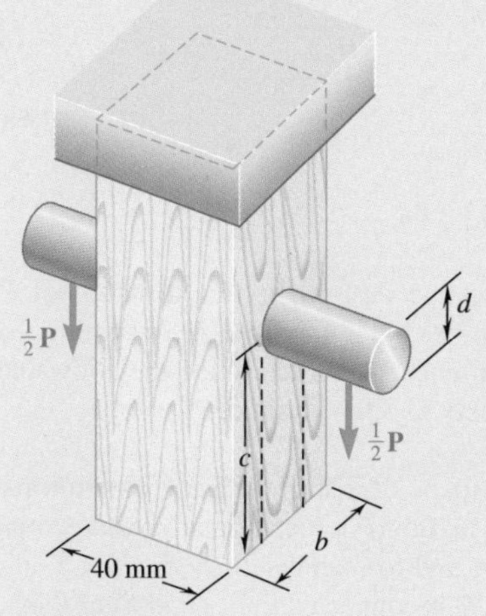

Fig. P8.39

8.40 For the support of Prob. 8.39, knowing that the diameter of the pin is $d = 16$ mm and that the magnitude of the load is $P = 20$ kN, determine (*a*) the factor of safety for the pin, (*b*) the required values of b and c if the factor of safety for the wooden member is the same as that found in part a for the pin.

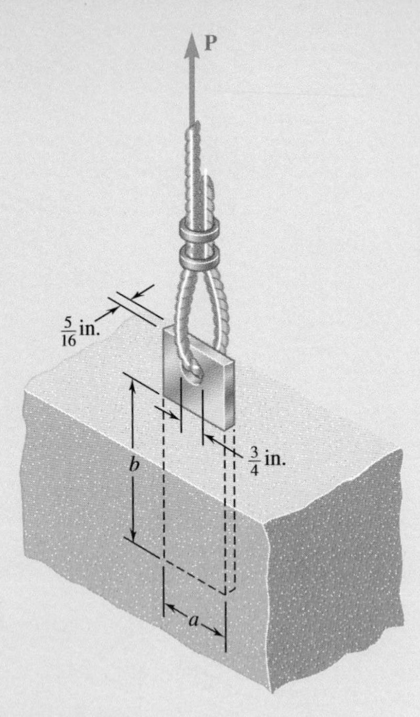

Fig. P8.41

8.41 A steel plate $\frac{5}{16}$ in. thick is embedded in a horizontal concrete slab and is used to anchor a high-strength vertical cable as shown. The diameter of the hole in the plate is $\frac{3}{4}$ in., the ultimate strength of the steel used is 36 ksi, and the ultimate bonding stress between plate and concrete is 300 psi. Knowing that a factor of safety of 3.60 is desired when $P = 2.5$ kips, determine (*a*) the required width a of the plate, (*b*) the minimum depth b to which a plate of that width should be embedded in the concrete slab. (Neglect the normal stresses between the concrete and the lower end of the plate.)

8.42 Determine the factor of safety for the cable anchor in Prob. 8.41 when $P = 3$ kips, knowing that $a = 2$ in. and $b = 7.5$ in.

8.43 In the structure shown, an 8-mm-diameter pin is used at A and 12-mm-diameter pins are used at B and D. Knowing that the ultimate shearing stress is 100 MPa at all connections and the ultimate normal stress is 250 MPa in each of the two links joining B and D, determine the allowable load **P** if an overall factor of safety of 3.0 is desired.

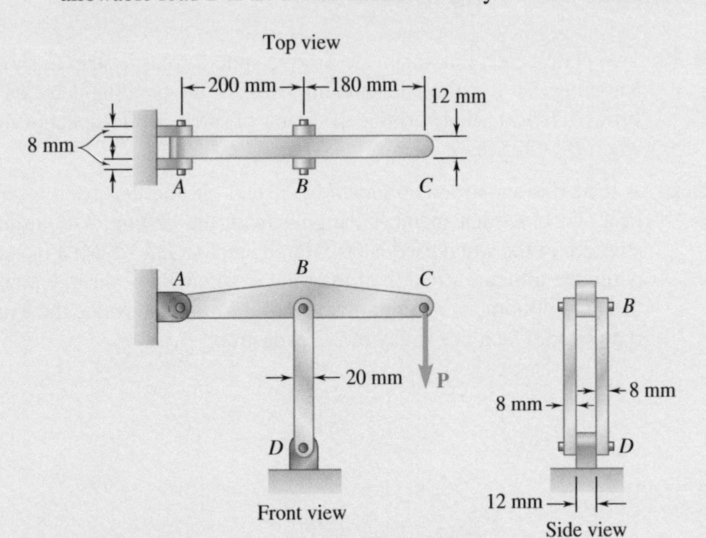

Fig. *P8.43* and P8.44

8.44 In an alternative design for the structure of Prob. 8.43, a pin of 10-mm-diameter is to be used at A. Assuming that all other specifications remain unchanged, determine the allowable load **P** if an overall factor of safety of 3.0 is desired.

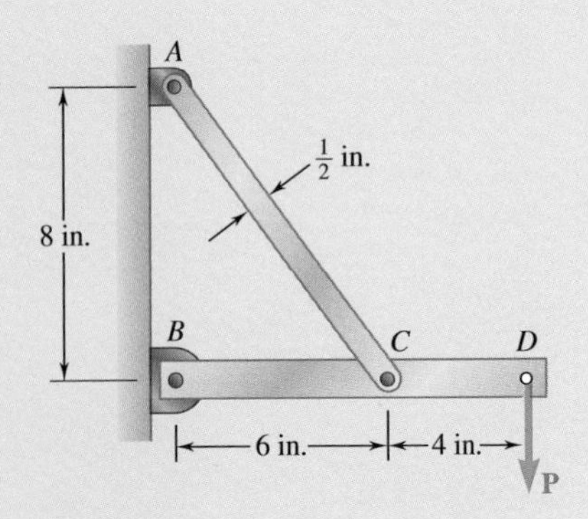

Fig. *P8.45*

8.45 Link AC is made of a steel with a 65-ksi ultimate normal stress and has a $\frac{1}{4} \times \frac{1}{2}$-in. uniform rectangular cross section. It is connected to a support at A and to member BCD at C by $\frac{3}{8}$-in.-diameter pins, while member BCD is connected to its support at B by a $\frac{5}{16}$-in.-diameter pin. All of the pins are made of a steel with a 25-ksi ultimate shearing stress and are in single shear. Knowing that a factor of safety of 3.25 is desired, determine the largest load **P** that can be applied at D. Note that link AC is not reinforced around the pin holes.

8.46 Solve Prob. 8.45, assuming that the structure has been redesigned to use $\frac{5}{16}$-in.-diameter pins at A and C as well as at B and that no other change has been made.

8.47 Each of the two vertical links CF connecting the two horizontal members AD and EG has a 10 × 40-mm uniform rectangular cross section and is made of a steel with an ultimate strength in tension of 400 MPa, while each of the pins at C and F has a 20-mm diameter and is made of a steel with an ultimate strength in shear of 150 MPa. Determine the overall factor of safety for the links CF and the pins connecting them to the horizontal members.

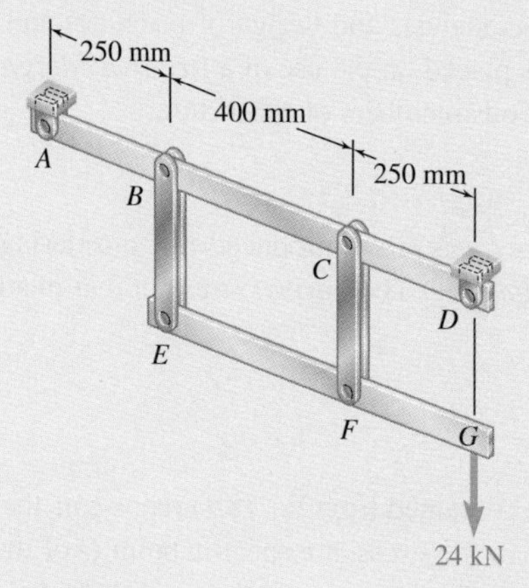

Fig. P8.47

8.48 Solve Prob. 8.47, assuming that the pins at C and F have been replaced by pins with a 30-mm diameter.

Review and Summary

This chapter was devoted to the concept of stress and to an introduction to the methods used for the analysis and design of machines and load-bearing structures. Emphasis was placed on the use of a *free-body diagram* to find the internal forces in the various members of a structure.

Axial Loading: Normal Stress

The concept of *stress* was first introduced by considering a two-force member under an *axial loading*. The *normal stress* in that member (Fig. 8.40) was obtained by

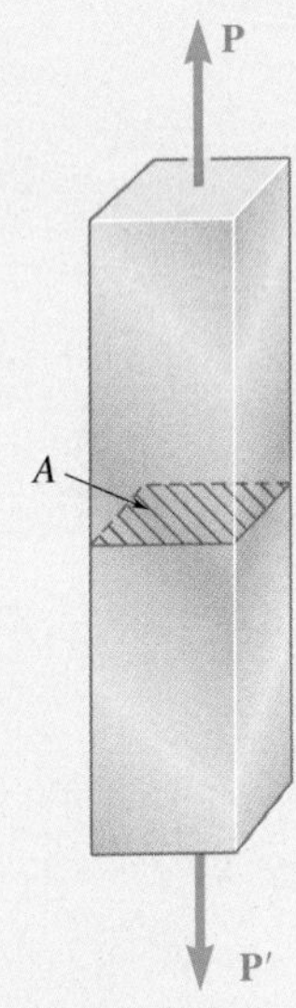

Fig. 8.40

$$\sigma = \frac{P}{A} \tag{8.1}$$

The value of σ obtained from Eq. (8.1) represents the *average stress* over the section rather than the stress at a specific point Q of the section. Considering a small area ΔA surrounding Q and the magnitude ΔF of the force exerted on ΔA, the stress at point Q is

$$\sigma = \lim_{\Delta A \to 0} \frac{\Delta F}{\Delta A} \tag{8.2}$$

In general, the stress σ at point Q in Eq. (8.2) is different from the value of the average stress given by Eq. (8.1) and is found to vary across the section. However, this variation is small in any section away from the points of application of the loads. Therefore, the distribution of the normal stresses in an axially loaded member is assumed to be *uniform,* except in the immediate vicinity of the points of application of the loads.

For the distribution of stresses to be uniform in a given section, the line of action of the loads **P** and **P′** must pass through the centroid C. Such a loading is called a *centric* axial loading. In the case of an *eccentric* axial loading, the distribution of stresses is *not* uniform.

Transverse Forces and Shearing Stress

When equal and opposite *transverse forces* **P** and **P′** of magnitude P are applied to a member AB (Fig. 8.41), *shearing stresses* τ are created over any section located between the points of application of the two forces. These stresses

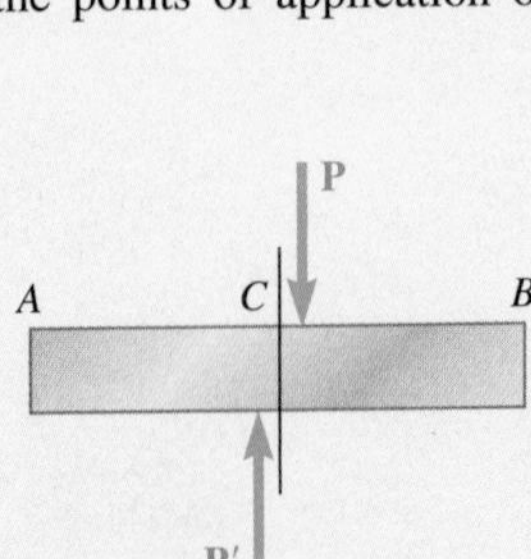

Fig. 8.41

vary greatly across the section and their distribution *cannot* be assumed to be uniform. However, dividing the magnitude P—referred to as the *shear* in the section—by the cross-sectional area A, the *average shearing stress* is:

$$\tau_{\text{ave}} = \frac{P}{A} \tag{8.4}$$

Single and Double Shear

Shearing stresses are found in bolts, pins, or rivets connecting two structural members or machine components. For example, the shearing stress of bolt CD (Fig. 8.42), which is in *single shear*, is written as

$$\tau_{\text{ave}} = \frac{P}{A} = \frac{F}{A} \tag{8.5}$$

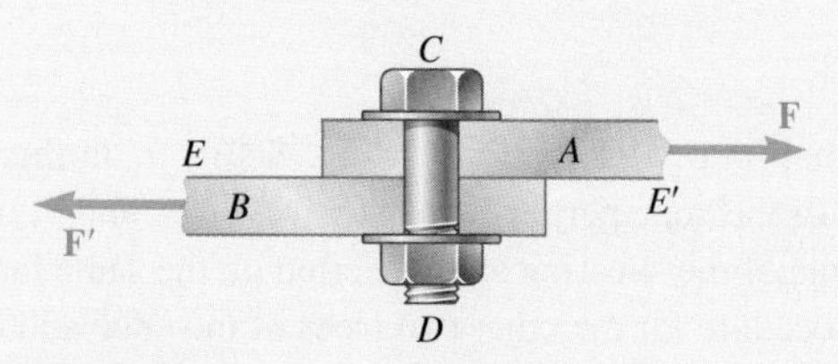

Fig. 8.42

The shearing stresses on bolts EG and HJ (Fig. 8.43), which are both in *double shear*, are written as

$$\tau_{\text{ave}} = \frac{P}{A} = \frac{F/2}{A} = \frac{F}{2A} \tag{8.6}$$

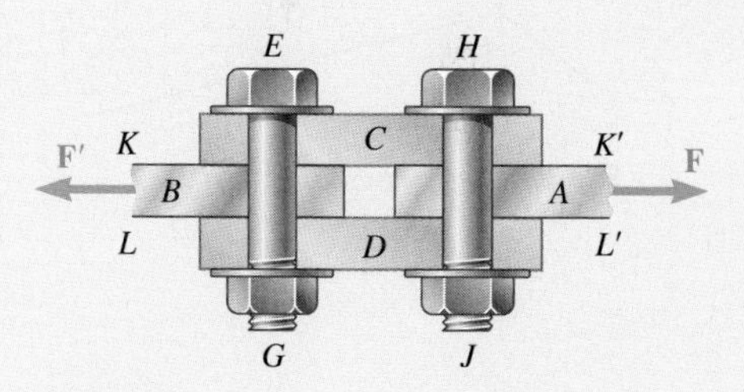

Fig. 8.43

Bearing Stress

Bolts, pins, and rivets also create stresses in the members they connect along the *bearing surface* or surface of contact. Bolt CD of Fig. 8.42 creates stresses on the semicylindrical surface of plate A with which it is in contact (Fig. 8.44). Since the distribution of these stresses is quite complicated, one uses an average nominal value σ_b of the stress, called *bearing stress*.

$$\sigma_b = \frac{P}{A} = \frac{P}{td} \tag{8.7}$$

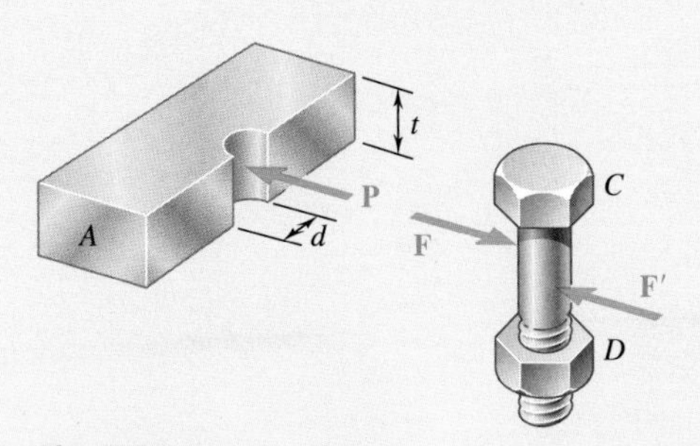

Fig. 8.44

Stresses on an Oblique Section

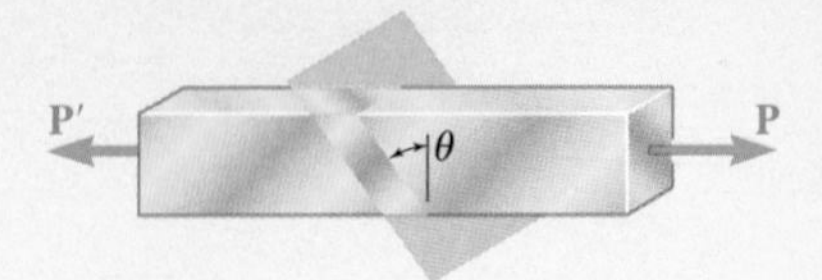

Fig. 8.45

When stresses are created on an *oblique section* in a two-force member under axial loading, both *normal* and *shearing* stresses occur. Denoting by θ the angle formed by the section with a normal plane (Fig. 8.45) and by A_0 the area of a section perpendicular to the axis of the member, the normal stress σ and the shearing stress τ on the oblique section are

$$\sigma = \frac{P}{A_0}\cos^2\theta \qquad \tau = \frac{P}{A_0}\sin\theta\cos\theta \tag{8.10}$$

We observed from these equations that the normal stress is maximum and equal to $\sigma_m = P/A_0$ for $\theta = 0$, while the shearing stress is maximum and equal to $\tau_m = P/2A_0$ for $\theta = 45°$. We also noted that $\tau = 0$ when $\theta = 0$, while $\sigma = P/2A_0$ when $\theta = 45°$.

Stress Under General Loading

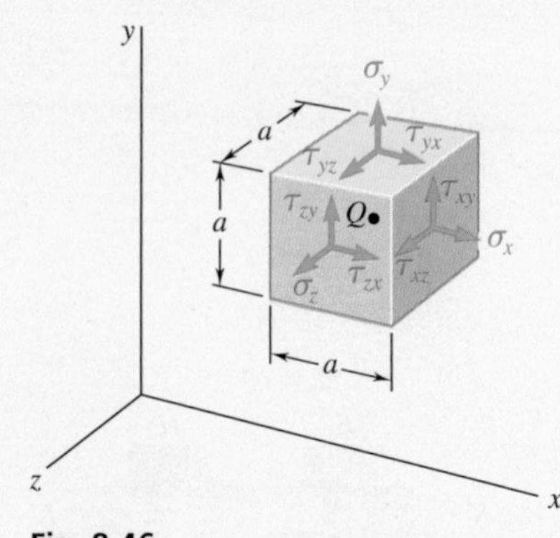

Fig. 8.46

Considering a small cube centered at Q (Fig. 8.46), σ_x is the normal stress exerted on a face of the cube perpendicular to the x axis, and τ_{xy} and τ_{xz} are the y and z components of the shearing stress exerted on the same face of the cube. Repeating this procedure for the other two faces of the cube and observing that $\tau_{xy} = \tau_{yx}$, $\tau_{yz} = \tau_{zy}$, and $\tau_{zx} = \tau_{xz}$, it was determined that *six stress components* are required to define the state of stress at a given point Q, being σ_x, σ_y, σ_z, τ_{xy}, τ_{yz}, and τ_{zx}.

Factor of Safety

The *ultimate load* of a given structural member or machine component is the load at which the member or component is expected to fail. This is computed from the *ultimate stress* or *ultimate strength* of the material used. The ultimate load should be considerably larger than the *allowable load* (i.e., the load that the member or component will be allowed to carry under normal conditions). The ratio of the ultimate load to the allowable load is the *factor of safety:*

$$\text{Factor of safety} = F.S. = \frac{\text{ultimate load}}{\text{allowable load}} \tag{8.21}$$

Load and Resistance Factor Design

Load and Resistance Factor Design (LRFD) allows the engineer to distinguish between the uncertainties associated with the structure and those associated with the load.

Review Problems

8.49 A 40-kN axial load is applied to a short wooden post that is supported by a concrete footing resting on undisturbed soil. Determine (*a*) the maximum bearing stress on the concrete footing, (*b*) the size of the footing for which the average bearing stress in the soil is 145 kPa.

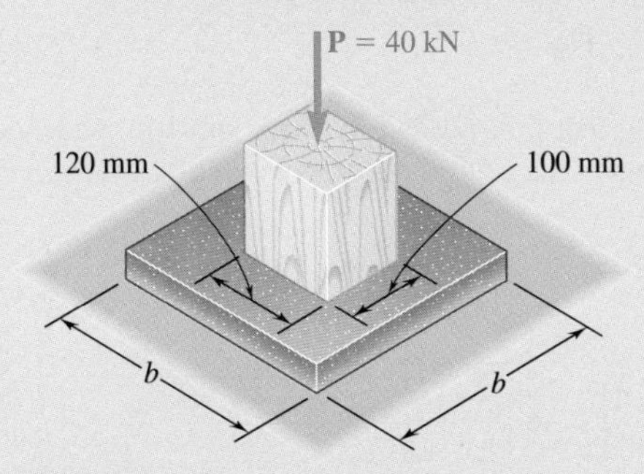

Fig. P8.49

8.50 The frame shown consists of four wooden members, *ABC*, *DEF*, *BE*, and *CF*. Knowing that each member has a 2 × 4-in. rectangular cross section and that each pin has a $\frac{1}{2}$-in. diameter, determine the maximum value of the average normal stress (*a*) in member *BE*, (*b*) in member *CF*.

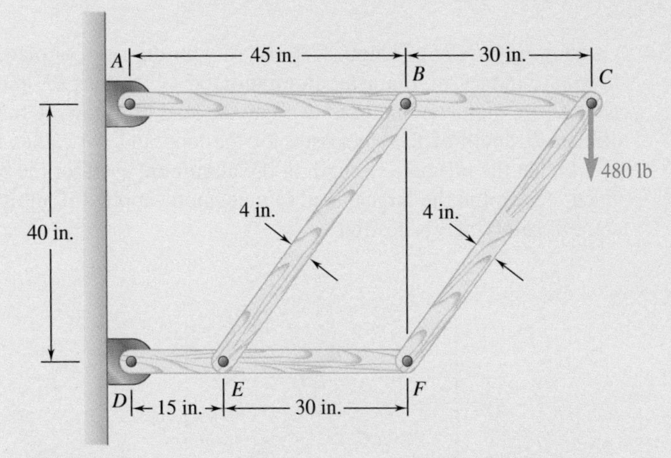

Fig. *P8.50*

8.51 Two steel plates are to be held together by means of 16-mm-diameter high-strength steel bolts fitting snugly inside cylindrical brass spacers. Knowing that the average normal stress must not exceed 200 MPa in the bolts and 130 MPa in the spacers, determine the outer diameter of the spacers that yields the most economical and safe design.

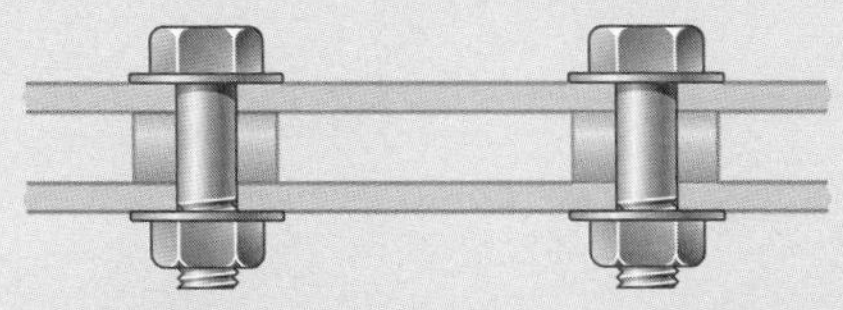

Fig. P8.51

8.52 When the force **P** reached 8 kN, the wooden specimen shown failed in shear along the surface indicated by the dashed line. Determine the average shearing stress along that surface at the time of failure.

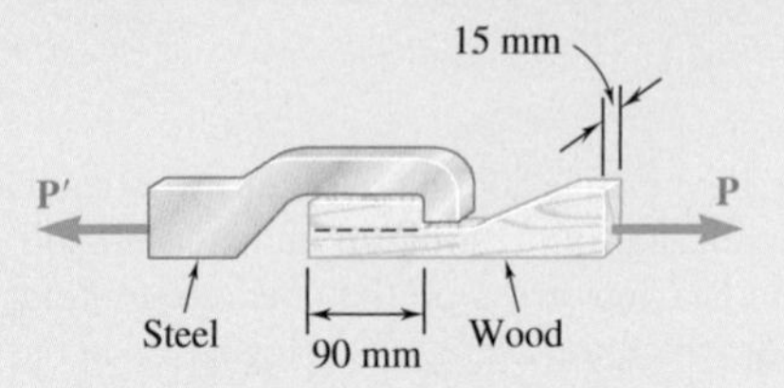

Fig. *P8.52*

8.53 Knowing that link DE is $\frac{1}{8}$ in. thick and 1 in. wide, determine the normal stress in the central portion of that link when (*a*) $\theta = 0$, (*b*) $\theta = 90°$.

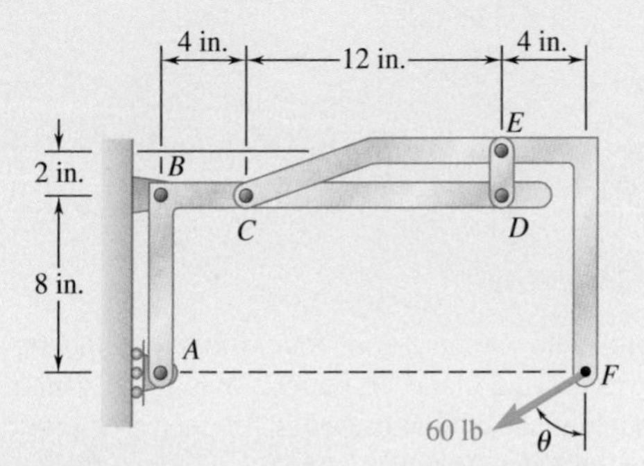

Fig. P8.53

8.54 A steel loop $ABCD$ of length 5 ft and of $\frac{3}{8}$-in. diameter is placed as shown around a 1-in.-diameter aluminum rod AC. Cables BE and DF, each of $\frac{1}{2}$-in. diameter, are used to apply the load **Q**. Knowing that the ultimate strength of the steel used for the loop and the cables is 70 ksi and that the ultimate strength of the aluminum used for the rod is 38 ksi, determine the largest load **Q** that can be applied if an overall factor of safety of 3 is desired.

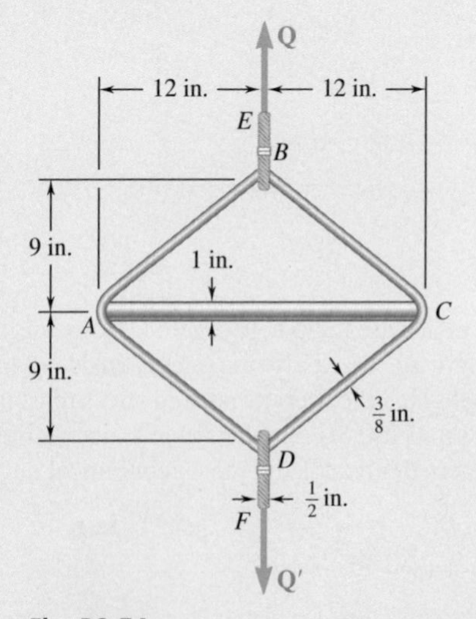

Fig. P8.54

8.55 Two identical linkage-and-hydraulic-cylinder systems control the position of the forks of a fork-lift truck. The load supported by the one system shown is 1500 lb. Knowing that the thickness of member BD is $\frac{5}{8}$ in., determine (*a*) the average shearing stress in the $\frac{1}{2}$-in.-diameter pin at B, (*b*) the bearing stress at B in member BD.

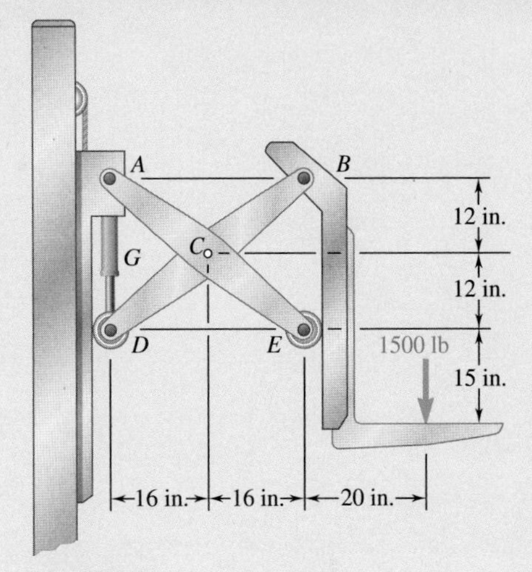

Fig. P8.55

8.56 A $\frac{5}{8}$-in.-diameter steel rod AB is fitted to a round hole near end C of the wooden member CD. For the loading shown, determine (*a*) the maximum average normal stress in the wood, (*b*) the distance b for which the average shearing stress is 100 psi on the surfaces indicated by the dashed lines, (*c*) the average bearing stress on the wood.

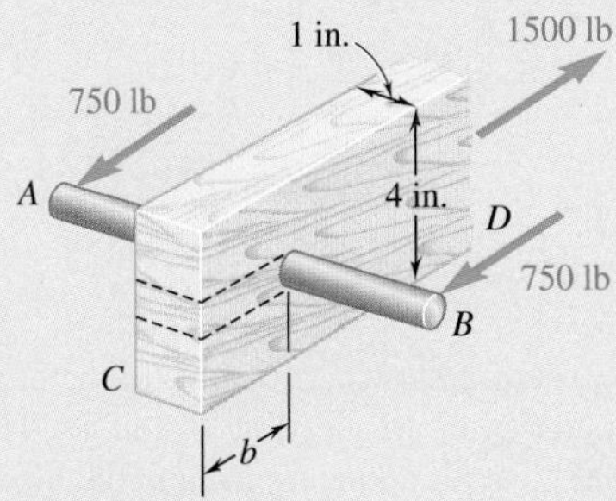

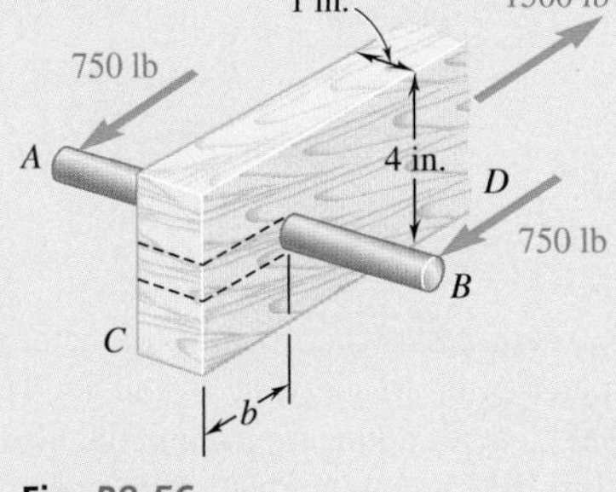

Fig. *P8.56*

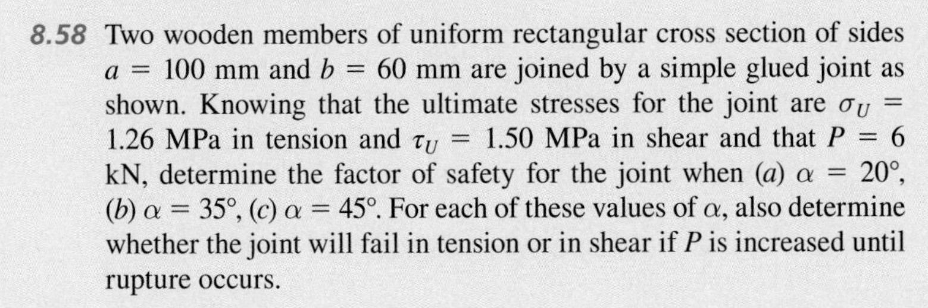

8.57 Member ABC, which is supported by a pin and bracket at C and a cable BD, was designed to support the 16-kN load $\mathbf{P}$ as shown. Knowing that the ultimate load for cable BD is 100 kN, determine the factor of safety with respect to cable failure.

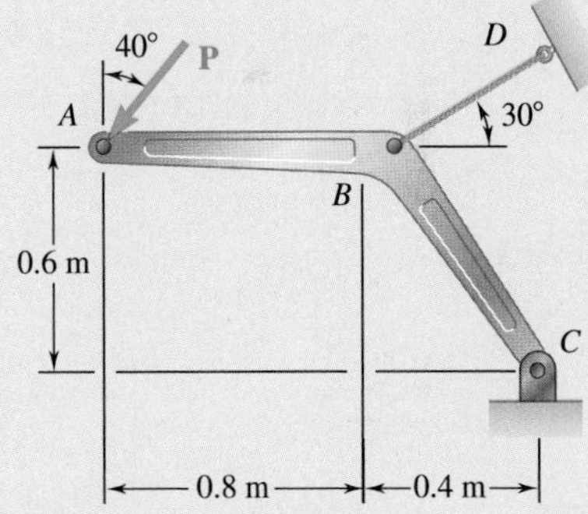

Fig. P8.57

8.58 Two wooden members of uniform rectangular cross section of sides a = 100 mm and b = 60 mm are joined by a simple glued joint as shown. Knowing that the ultimate stresses for the joint are σ_U = 1.26 MPa in tension and τ_U = 1.50 MPa in shear and that P = 6 kN, determine the factor of safety for the joint when (*a*) α = 20°, (*b*) α = 35°, (*c*) α = 45°. For each of these values of α, also determine whether the joint will fail in tension or in shear if P is increased until rupture occurs.

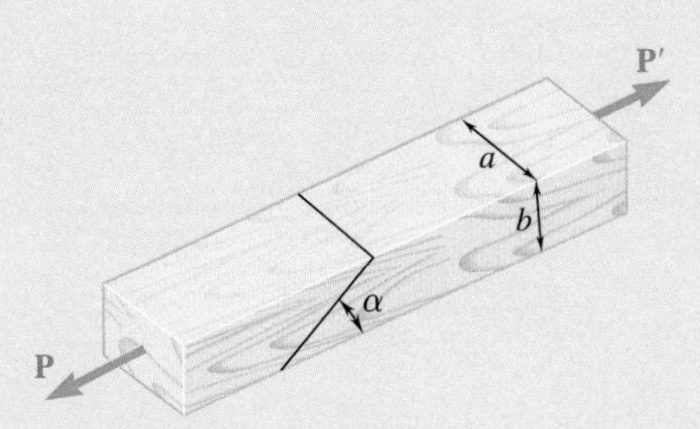

Fig. *P8.58*

8.59 The 2000-lb load may be moved along the beam BD to any position between stops at E and F. Knowing that $\sigma_{all} = 6$ ksi for the steel used in rods AB and CD, determine where the stops should be placed if the permitted motion of the load is to be as large as possible.

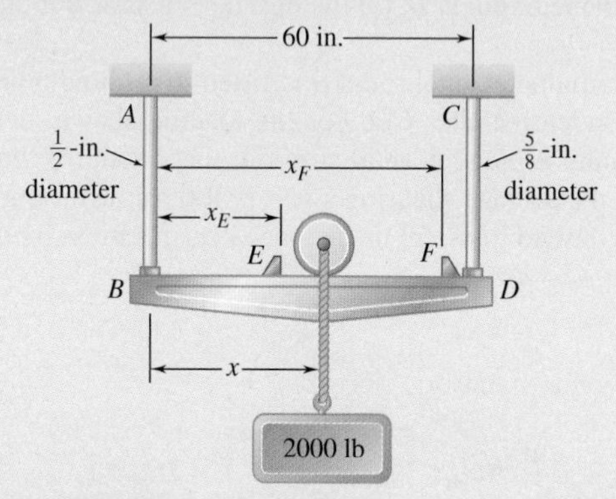

Fig. P8.59

8.60 In the steel structure shown, a 6-mm-diameter pin is used at C and 10-mm-diameter pins are used at B and D. The ultimate shearing stress is 150 MPa at all connections, and the ultimate normal stress is 400 MPa in link BD. Knowing that a factor of safety of 3.0 is desired, determine the largest load $\mathbf{P}$ that can be applied at A. Note that link BD is not reinforced around the pin holes.

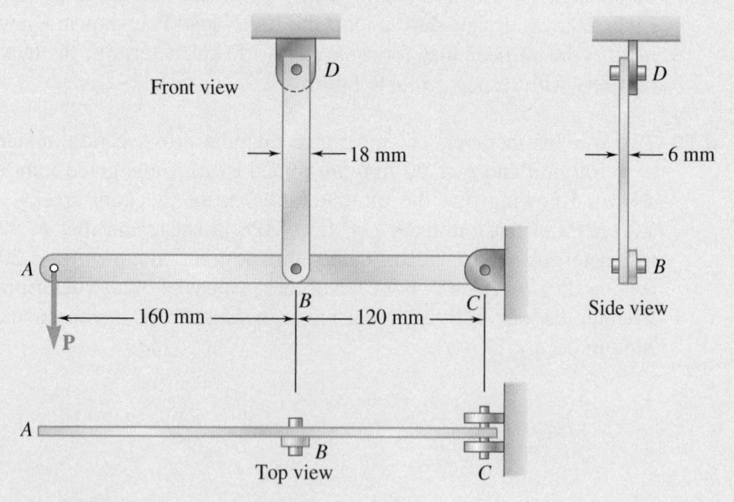

Fig. P8.60

© Sylvain Grandadam/agefotostock

9

Stress and Strain–Axial Loading

This chapter considers deformations occurring in structural components subjected to axial loading. The change in length of the diagonal stays was carefully accounted for in the design of this cable-stayed bridge.

Objectives

In this chapter, we will:

- **Introduce** students to the concept of strain.
- **Discuss** the relationship between stress and strain in different materials.
- **Determine** the deformation of structural components under axial loading.
- **Introduce** Hooke's Law and the modulus of elasticity.
- **Discuss** the concept of lateral strain and Poisson's ratio.
- **Use** axial deformations to solve indeterminate problems.
- **Define** Saint-Venant's principle and the distribution of stresses.
- **Review** stress concentrations and how they are included in design.
- **Define** the difference between elastic and plastic behavior.
- **Look** at specific topics related to fiber-reinforced composite materials, fatigue, multiaxial loading.

Introduction

An important aspect of the analysis and design of structures relates to the *deformations* caused by the loads applied to a structure. It is important to avoid deformations so large that they may prevent the structure from fulfilling the purpose for which it was intended. But the analysis of deformations also helps us to determine stresses. Indeed, it is not always possible to determine the forces in the members of a structure by applying only the principles of statics. This is because statics is based on the assumption of undeformable, rigid structures. By considering engineering structures as *deformable* and analyzing the deformations in their various members, it will be possible for us to compute forces that are *statically indeterminate*. The distribution of stresses in a given member is statically indeterminate, even when the force in that member is known.

In this chapter, you will consider the deformations of a structural member such as a rod, bar, or plate under *axial loading*. First, the *normal strain* ϵ in a member is defined as the *deformation of the member per unit length*. Plotting the stress σ versus the strain ϵ as the load applied to the member is increased produces a *stress-strain diagram* for the material used. From this diagram, some important properties of the material, such as its *modulus of elasticity*, and whether the material is *ductile* or *brittle* can be determined.

From the stress-strain diagram, you also can determine whether the strains in the specimen will disappear after the load has been removed—when the material is said to behave *elastically*—or whether *a permanent set* or *plastic deformation* will result.

You will examine the phenomenon of *fatigue*, which causes structural or machine components to fail after a very large number of repeated loadings, even though the stresses remain in the elastic range.

Sections 9.2 and 9.3 discuss *statically indeterminate problems* in which the reactions and the internal forces *cannot* be determined from statics alone. Here the equilibrium equations derived from the free-body diagram of the member must be complemented by relationships involving deformations that are obtained from the geometry of the problem.

Additional constants associated with isotropic materials—i.e., materials with mechanical characteristics independent of direction—are introduced in Secs. 9.4 through 9.7. They include *Poisson's ratio*, relating lateral and axial strain, and the *modulus of rigidity*, concerning the components of the shearing stress and shearing strain. Stress-strain relationships for an isotropic material under a multiaxial loading also are determined.

In Chap. 8, stresses were assumed uniformly distributed in any given cross section; they were also assumed to remain within the elastic range. The first assumption is discussed in Sec. 9.8, while *stress concentrations* near circular holes and fillets in flat bars are considered in Sec. 9.9.

9.1 BASIC PRINCIPLES OF STRESS AND STRAIN

9.1A Normal Strain Under Axial Loading

Consider a rod BC of length L and uniform cross-sectional area A, which is suspended from B (Fig. 9.1*a*). If you apply a load $\mathbf{P}$ to end C, the rod elongates (Fig. 9.1*b*). Plotting the magnitude P of the load against the deformation δ (Greek letter delta), you obtain a load-deformation diagram (Fig. 9.2). While this diagram contains information useful to the analysis of the rod under consideration, it cannot be used to predict the deformation of a rod of the same material but with different dimensions. Indeed, if a deformation δ is produced in rod BC by a load $\mathbf{P}$, a load $2\mathbf{P}$ is required to cause the same deformation in rod $B'C'$ of the same length L but cross-sectional area $2A$ (Fig. 9.3). Note that in both cases the value of the stress is the same: $\sigma = P/A$. On the other hand, when load $\mathbf{P}$ is applied to a rod $B''C''$ of the same cross-sectional area A but of length $2L$, a deformation 2δ occurs in

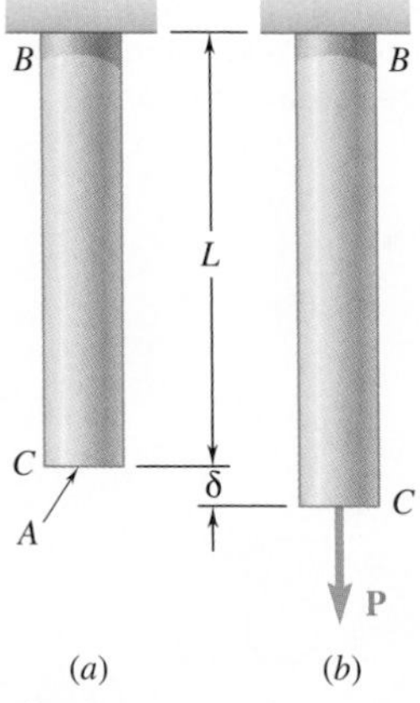

Fig. 9.1 Undeformed and deformed axially-loaded rod.

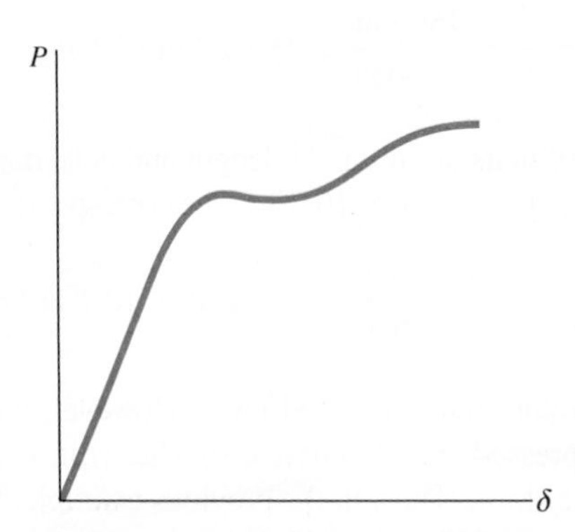

Fig. 9.2 Load-deformation diagram.

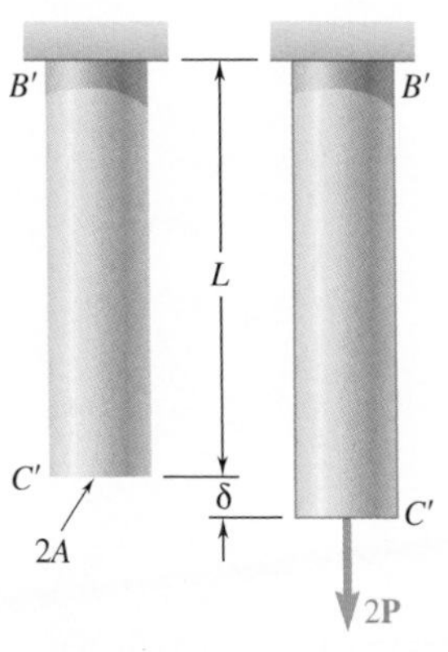

Fig. 9.3 Twice the load is required to obtain the same deformation δ when the cross-sectional area is doubled.

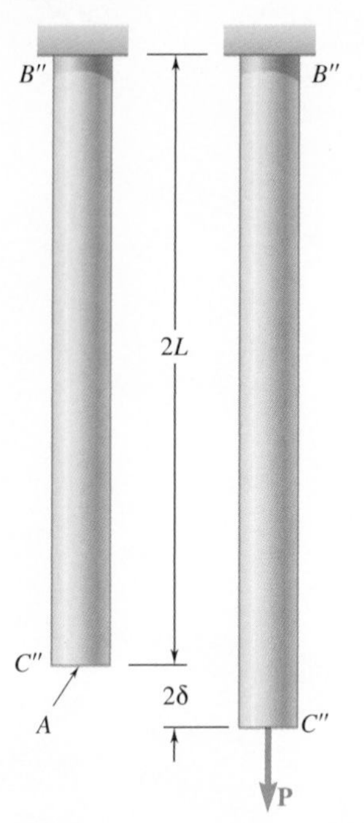

Fig. 9.4 The deformation is doubled when the rod length is doubled while keeping the load **P** and cross-sectional area A the same.

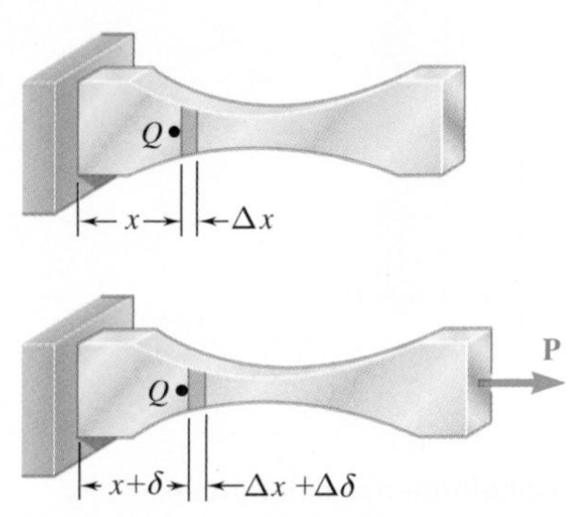

Fig. 9.5 Deformation of axially-loaded member of variable cross-sectional area.

that rod (Fig. 9.4). This is a deformation twice as large as the deformation δ produced in rod BC. In both cases, the ratio of the deformation over the length of the rod is the same at δ/L. This introduces the concept of *strain*. We define the *normal strain* in a rod under axial loading as the *deformation per unit length* of that rod. The normal strain, ϵ (Greek letter epsilon), is

$$\epsilon = \frac{\delta}{L} \tag{9.1}$$

Plotting the stress $\sigma = P/A$ against the strain $\epsilon = \delta/L$ results in a curve that is characteristic of the properties of the material but does not depend upon the dimensions of the specimen used. This curve is called a *stress-strain diagram*.

Since rod BC in Fig. 9.1 has a uniform cross section of area A, the normal stress σ is assumed to have a constant value P/A throughout the rod. The strain ϵ is the ratio of the total deformation δ over the total length L of the rod. It too is consistent throughout the rod. However, for a member of variable cross-sectional area A, the normal stress $\sigma = P/A$ varies along the member, and it is necessary to define the strain at a given point Q by considering a small element of undeformed length Δx (Fig. 9.5). Denoting the deformation of the element under the given loading by $\Delta\delta$, the *normal strain at point Q* is defined as

$$\epsilon = \lim_{\Delta x \to 0} \frac{\Delta\delta}{\Delta x} = \frac{d\delta}{dx} \tag{9.2}$$

Since deformation and length are expressed in the same units, the normal strain ϵ obtained by dividing δ by L (or $d\delta$ by dx) is a *dimensionless quantity*. Thus, the same value is obtained for the normal strain, whether SI metric units or U.S. customary units are used. For instance, consider a bar of length $L = 0.600$ m and uniform cross section that undergoes a deformation $\delta = 150 \times 10^{-6}$ m. The corresponding strain is

$$\epsilon = \frac{\delta}{L} = \frac{150 \times 10^{-6}\text{ m}}{0.600\text{ m}} = 250 \times 10^{-6}\text{ m/m} = 250 \times 10^{-6}$$

Note that the deformation also can be expressed in micrometers: $\delta = 150\ \mu\text{m}$ and the answer written in micros (μ):

$$\epsilon = \frac{\delta}{L} = \frac{150\ \mu\text{m}}{0.600\text{ m}} = 250\ \mu\text{m/m} = 250\ \mu$$

When U.S. customary units are used, the length and deformation of the same bar are $L = 23.6$ in. and $\delta = 5.91 \times 10^{-3}$ in. The corresponding strain is

$$\epsilon = \frac{\delta}{L} = \frac{5.91 \times 10^{-3}\text{ in.}}{23.6\text{ in.}} = 250 \times 10^{-6}\text{ in./in.}$$

which is the same value found using SI units. However, when lengths and deformations are expressed in inches or microinches (μin.), keep the original units obtained for the strain. Thus, in the previous example, the strain would be recorded as either $\epsilon = 250 \times 10^{-6}$ in./in. or $\epsilon = 250\ \mu$in./in.

9.1B Stress-Strain Diagram

Tensile Test. To obtain the stress-strain diagram of a material, a *tensile test* is conducted on a specimen of the material. One type of specimen is shown in Photo 9.1. The cross-sectional area of the cylindrical central portion of the specimen is accurately determined and two gage marks are inscribed on that portion at a distance L_0 from each other. The distance L_0 is known as the *gage length* of the specimen.

The test specimen is then placed in a testing machine (Photo 9.2), which is used to apply a centric load **P**. As load **P** increases, the distance L between the two gage marks also increases (Photo 9.3). The distance L is measured with a dial gage, and the elongation $\delta = L - L_0$ is recorded for each value of P. A second dial gage is often used simultaneously to measure and record the change in diameter of the specimen. From each pair of readings P and δ, the engineering stress σ is

$$\sigma = \frac{P}{A_0} \tag{9.3}$$

and the engineering strain ϵ is

$$\epsilon = \frac{\delta}{L_0} \tag{9.4}$$

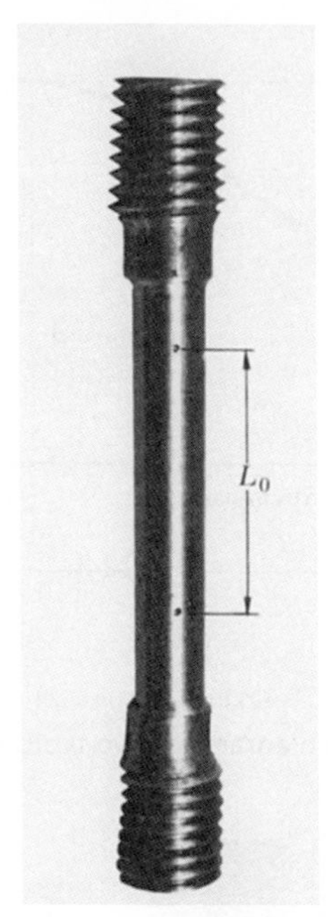

Photo 9.1 Typical tensile-test specimen. Undeformed gage length is L_0.

© *John DeWolf*

Photo 9.2 Universal test machine used to test tensile specimens.

Courtesy of Tinius Olsen Testing Machine Co., Inc.

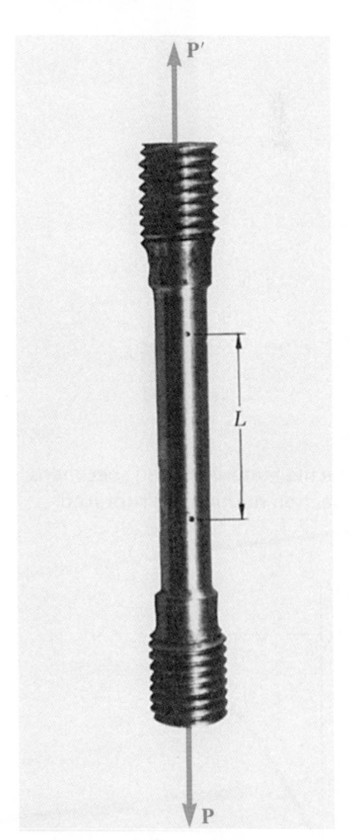

Photo 9.3 Elongated tensile test specimen having load P and deformed length $L > L_0$.

© *John DeWolf*

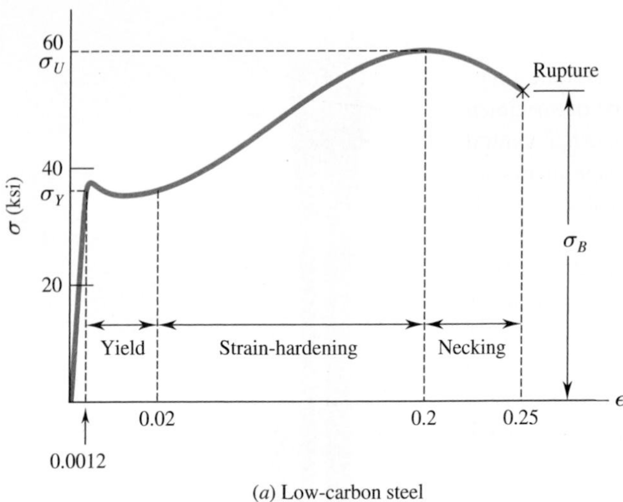

(a) Low-carbon steel

(b) Aluminum alloy

Fig. 9.6 Stress-strain diagrams of two typical ductile materials.

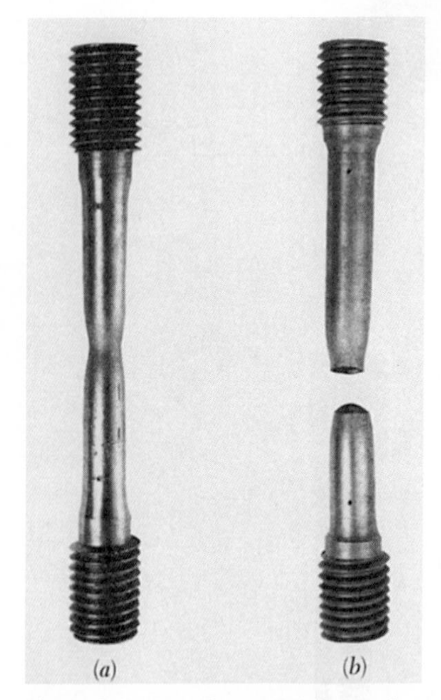

Photo 9.4 Ductile material tested specimens: (a) with cross-section necking, (b) ruptured.

© *John DeWolf*

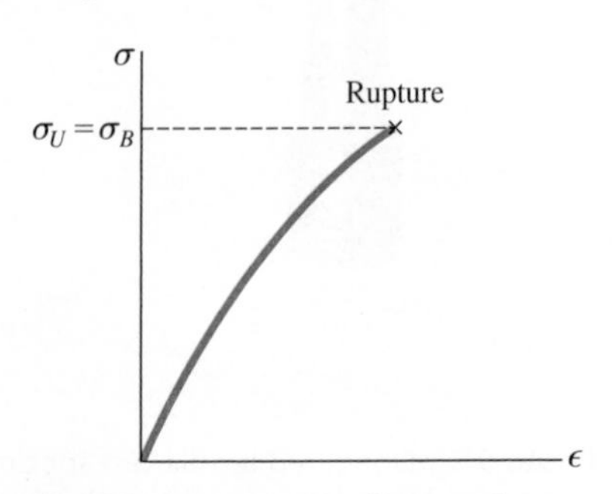

Fig. 9.7 Stress-strain diagram for a typical brittle material.

The stress-strain diagram can be obtained by plotting ϵ as an abscissa and σ as an ordinate.

Stress-strain diagrams of materials vary widely, and different tensile tests conducted on the same material may yield different results, depending upon the temperature of the specimen and the speed of loading. However, some common characteristics can be distinguished from stress-strain diagrams to divide materials into two broad categories: *ductile* and *brittle* materials.

Ductile materials, including structural steel and many alloys of other materials are characterized by their ability to *yield* at normal temperatures. As the specimen is subjected to an increasing load, its length first increases linearly with the load and at a very slow rate. Thus, the initial portion of the stress-strain diagram is a straight line with a steep slope (Fig. 9.6). However, after a critical value σ_Y of the stress has been reached, the specimen undergoes a large deformation with a relatively small increase in the applied load. This deformation is caused by slippage along oblique surfaces and is due primarily to shearing stresses. After a maximum value of the load has been reached, the diameter of a portion of the specimen begins to decrease, due to local instability (Photo 9.4*a*). This phenomenon is known as *necking*. After necking has begun, lower loads are sufficient for specimen to elongate further, until it finally ruptures (Photo 9.4*b*). Note that rupture occurs along a cone-shaped surface that forms an angle of approximately 45° with the original surface of the specimen. This indicates that shear is primarily responsible for the failure of ductile materials, confirming the fact that shearing stresses under an axial load are largest on surfaces forming an angle of 45° with the load (see Sec. 8.2). Note from Fig. 9.6 that the elongation of a ductile specimen after it has ruptured can be 200 times as large as its deformation at yield. The stress σ_Y at which yield is initiated is called the *yield strength* of the material. The stress σ_U corresponding to the maximum load applied is known as the *ultimate strength*. The stress σ_B corresponding to rupture is called the *breaking strength*.

Brittle materials, comprising of cast iron, glass, and stone rupture without any noticeable prior change in the rate of elongation (Fig. 9.7).

Thus, for brittle materials, there is no difference between the ultimate strength and the breaking strength. Also, the strain at the time of rupture is much smaller for brittle than for ductile materials. Note the absence of any necking of the specimen in the brittle material of Photo 9.5 and observe that rupture occurs along a surface perpendicular to the load. Thus, normal stresses are primarily responsible for the failure of brittle materials.†

The stress-strain diagrams of Fig. 9.6 show that while structural steel and aluminum are both ductile, they have different yield characteristics. For structural steel (Fig. 9.6*a*), the stress remains constant over a large range of the strain after the onset of yield. Later, the stress must be increased to keep elongating the specimen until the maximum value σ_U has been reached. This is due to a property of the material known as *strain-hardening*. The *yield strength* of structural steel is determined during the tensile test by watching the load shown on the display of the testing machine. After increasing steadily, the load will suddenly drop to a slightly lower value, which is maintained for a certain period as the specimen keeps elongating. In a very carefully conducted test, one may be able to distinguish between the *upper yield point*, which corresponds to the load reached just before yield starts, and the *lower yield point*, which corresponds to the load required to maintain yield. Since the upper yield point is transient, the lower yield point is used to determine the yield strength of the material.

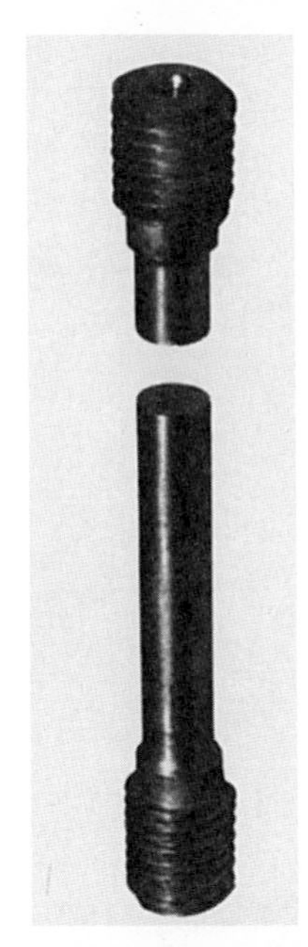

Photo 9.5 Ruptured brittle material specimen.
© *John DeWolf*

For aluminum (Fig. 9.6*b*) and of many other ductile materials, the stress keeps increasing—although not linearly—until the ultimate strength is reached. Necking then begins and eventually ruptures. For such materials, the yield strength σ_Y can be determined using the offset method. For example the yield strength at 0.2% offset is obtained by drawing through the point of the horizontal axis of abscissa $\epsilon = 0.2\%$ (or $\epsilon = 0.002$), which is a line parallel to the initial straight-line portion of the stress-strain diagram (Fig. 9.8). The stress σ_Y corresponding to the point Y is defined as the yield strength at 0.2% offset.

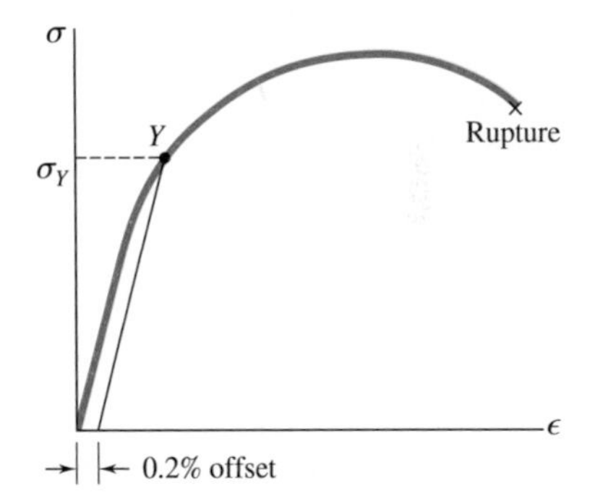

Fig. 9.8 Determination of yield strength by 0.2% offset method.

A standard measure of the ductility of a material is its *percent elongation*:

$$\text{Percent elongation} = 100\frac{L_B - L_0}{L_0}$$

where L_0 and L_B are the initial length of the tensile test specimen and its final length at rupture, respectively. The specified minimum elongation for a 2-in. gage length for commonly used steels with yield strengths up to 50 ksi is 21 percent. This means that the average strain at rupture should be at least 0.21 in./in.

Another measure of ductility that is sometimes used is the *percent reduction in area*:

$$\text{Percent reduction in area} = 100\frac{A_0 - A_B}{A_0}$$

†The tensile tests described in this section were assumed to be conducted at normal temperatures. However, a material that is ductile at normal temperatures may display the characteristics of a brittle material at very low temperatures, while a normally brittle material may behave in a ductile fashion at very high temperatures. At temperatures other than normal, therefore, one should refer to *a material in a ductile state* or to *a material in a brittle state*, rather than to a ductile or brittle material.

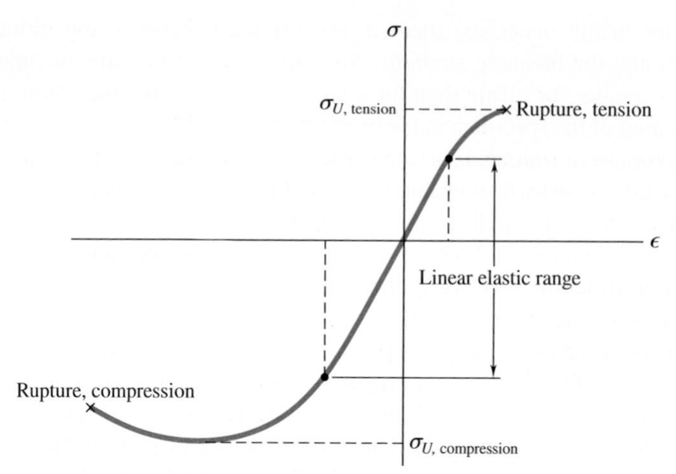

Fig. 9.9 Stress-strain diagram for concrete shows difference in tensile and compression response.

where A_0 and A_B are the initial cross-sectional area of the specimen and its minimum cross-sectional area at rupture, respectively. For structural steel, percent reductions in area of 60 to 70 percent are common.

Compression Test. If a specimen made of a ductile material is loaded in compression instead of tension, the stress-strain curve is essentially the same through its initial straight-line portion and through the beginning of the portion corresponding to yield and strain-hardening. Particularly noteworthy is the fact that for a given steel, the yield strength is the same in both tension and compression. For larger values of the strain, the tension and compression stress-strain curves diverge, and necking does not occur in compression. For most brittle materials, the ultimate strength in compression is much larger than in tension. This is due to the presence of flaws, such as microscopic cracks or cavities that tend to weaken the material in tension, while not appreciably affecting its resistance to compressive failure.

An example of brittle material with different properties in tension and compression is provided by *concrete*, whose stress-strain diagram is shown in Fig. 9.9. On the tension side of the diagram, we first observe a linear elastic range in which the strain is proportional to the stress. After the yield point has been reached, the strain increases faster than the stress until rupture occurs. The behavior of the material in compression is different. First, the linear elastic range is significantly larger. Second, rupture does not occur as the stress reaches its maximum value. Instead, the stress decreases in magnitude while the strain keeps increasing until rupture occurs. Note that the modulus of elasticity, which is represented by the slope of the stress-strain curve in its linear portion, is the same in tension and compression. This is true of most brittle materials.

9.1C Hooke's Law; Modulus of Elasticity

Modulus of Elasticity. Most engineering structures are designed to undergo relatively small deformations, involving only the straight-line portion of the corresponding stress-strain diagram. For that initial portion of the diagram (Fig. 9.6), the stress σ is directly proportional to the strain ϵ:

$$\sigma = E\epsilon \quad (9.5)$$

This is known as *Hooke's law*, after Robert Hooke (1635–1703), an English scientist and one of the early founders of applied mechanics. The coefficient E of the material is the *modulus of elasticity* or *Young's modulus*, after the English scientist Thomas Young (1773–1829). Since the strain ϵ is a dimensionless quantity, E is expressed in the same units as stress σ—in pascals or one of its multiples for SI units and in psi or ksi for U.S. customary units.

The largest value of stress for which Hooke's law can be used for a given material is the *proportional limit* of that material. For ductile materials possessing a well-defined yield point, as in Fig. 9.6*a*, the proportional limit almost coincides with the yield point. For other materials, the proportional limit cannot be determined as easily, since it is difficult to accurately determine the stress σ for which the relation between σ and ϵ ceases to be linear. For such materials, however, using Hooke's law for values of the stress slightly larger than the actual proportional limit will not result in any significant error.

Some physical properties of structural metals, such as strength, ductility, and corrosion resistance, can be greatly affected by alloying, heat treatment, and the manufacturing process used. For example, the stress-strain diagrams of pure iron and three different grades of steel (Fig. 9.10) show that large variations in the yield strength, ultimate strength, and final strain (ductility) exist. All of these metals possess the same modulus of elasticity—their "stiffness," or ability to resist a deformation within the linear range is the same. Therefore, if a high-strength steel is substituted for a lower-strength steel and if all dimensions are kept the same, the structure will have an increased load-carrying capacity, but its stiffness will remain unchanged.

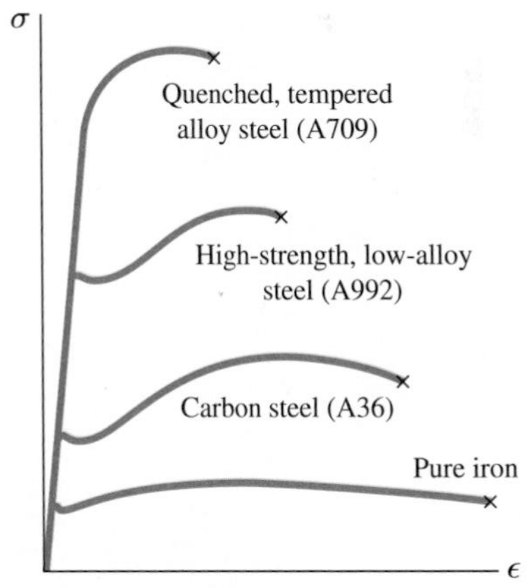

Fig. 9.10 Stress-strain diagrams for iron and different grades of steel.

For the materials considered so far, the relationship between normal stress and normal strain, $\sigma = E\epsilon$, is independent of the direction of loading. This is because the mechanical properties of each material, including its modulus of elasticity E, are independent of the direction considered. Such materials are said to be *isotropic*. Materials whose properties depend upon the direction considered are said to be *anisotropic*.

Fiber-Reinforced Composite Materials. An important class of anisotropic materials consists of *fiber-reinforced composite materials*. These are obtained by embedding fibers of a strong, stiff material into a weaker, softer material, called a *matrix*. Typical materials used as fibers are graphite, glass, and polymers, while various types of resins are used as a matrix. Fig. 9.11 shows a layer, or *lamina*, of a composite material consisting of a large number of parallel fibers embedded in a matrix. An axial load applied to the lamina along the x axis, (in a direction parallel to the fibers) will create a normal stress σ_x in the lamina and a corresponding normal strain ϵ_x, satisfying Hooke's law as the load is increased and as long as the elastic limit of the lamina is not exceeded. Similarly, an axial load applied along the y axis, (in a direction perpendicular to the lamina) will create a normal stress σ_y and a normal strain ϵ_y, and an axial load applied along the z axis will create a normal stress σ_z and a normal strain ϵ_z, all satisfy Hooke's law. However, the moduli of elasticity E_x, E_y, and E_z corresponding, to each of these loadings will be different. Because the fibers are parallel to the

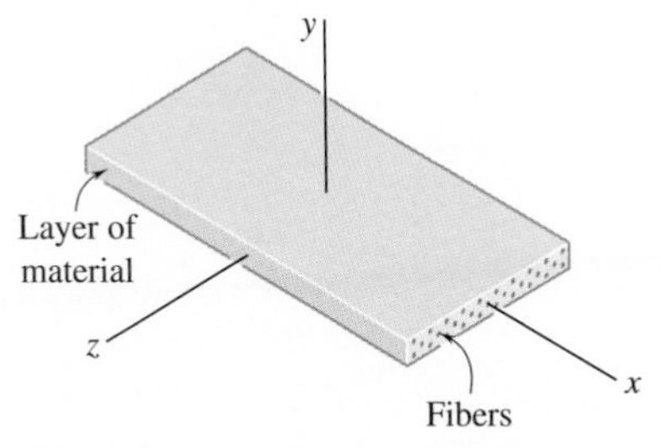

Fig. 9.11 Layer of fiber-reinforced composite material.

x axis, the lamina will offer a much stronger resistance to a load directed along the x axis than to one directed along the y or z axis, and E_x will be much larger than either E_y or E_z.

A flat *laminate* is obtained by superposing a number of layers or laminas. If the laminate is subjected only to an axial load causing tension, the fibers in all layers should have the same orientation as the load in order to obtain the greatest possible strength. But if the laminate is in compression, the matrix material may not be strong enough to prevent the fibers from kinking or buckling. The lateral stability of the laminate can be increased by positioning some of the layers so that their fibers are perpendicular to the load. Positioning some layers so that their fibers are oriented at 30°, 45°, or 60° to the load also can be used to increase the resistance of the laminate to in-plane shear.

*9.1D Elastic Versus Plastic Behavior of a Material

Material behaves *elastically* if the strains in a test specimen from a given load disappear when the load is removed. The largest value of stress causing this elastic behavior is called the *elastic limit* of the material.

If the material has a well-defined yield point as in Fig. 9.6*a*, the elastic limit, the proportional limit, and the yield point are essentially equal. In other words, the material behaves elastically and linearly as long as the stress is kept below the yield point. However, if the yield point is reached, yield takes place as described in Sec. 9.1B. When the load is removed, the stress and strain decrease in a linear fashion along a line CD parallel to the straight-line portion AB of the loading curve (Fig. 9.12). The fact that ϵ does not return to zero after the load has been removed indicates that a *permanent set* or *plastic deformation* of the material has taken place. For most materials, the plastic deformation depends upon both the maximum value reached by the stress and the time elapsed before the load is removed. The stress-dependent part of the plastic deformation is called *slip*, and the time-dependent part—also influenced by the temperature—is *creep*.

When a material does not possess a well-defined yield point, the elastic limit cannot be determined with precision. However, assuming the elastic limit to be equal to the yield strength using the offset method (Sec. 9.1B) results in only a small error. Referring to Fig. 9.8, note that the straight line used to determine point Y also represents the unloading

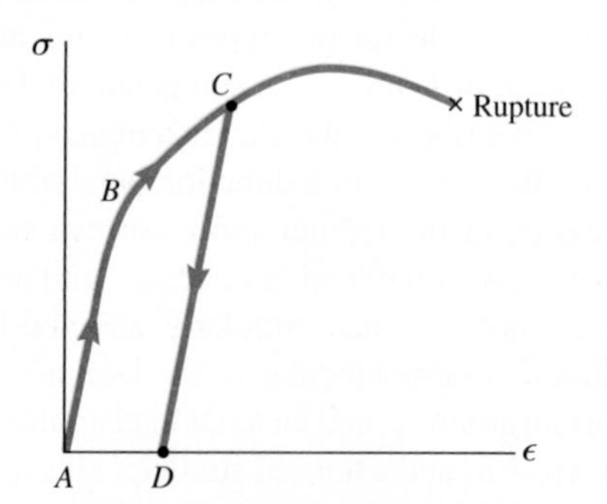

Fig. 9.12 Stress-strain response of ductile material loaded beyond yield and unloaded.

curve after a maximum stress σ_Y has been reached. While the material does not behave truly elastically, the resulting plastic strain is as small as the selected offset.

If, after being loaded and unloaded (Fig. 9.13), the test specimen is loaded again, the new loading curve will follow the earlier unloading curve until it almost reaches point C. Then it will bend to the right and connect with the curved portion of the original stress-strain diagram. This straight-line portion of the new loading curve is longer than the corresponding portion of the initial one. Thus, the proportional limit and the elastic limit have increased as a result of the strain-hardening that occurred during the earlier loading. However, since the point of rupture R remains unchanged, the ductility of the specimen, which should now be measured from point D, has decreased.

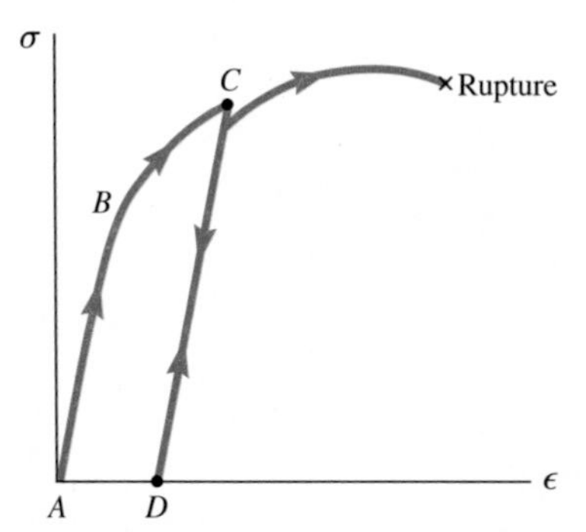

Fig. 9.13 Stress-strain response of ductile material reloaded after prior yielding and unloading.

In previous discussions the specimen was loaded twice in the same direction (i.e., both loads were tensile loads). Now consider that the second load is applied in a direction opposite to that of the first one. Assume the material is mild steel where the yield strength is the same in tension and in compression. The initial load is tensile and is applied until point C is reached on the stress-strain diagram (Fig. 9.14). After unloading (point D), a compressive load is applied, causing the material to reach point H, where the stress is equal to $-\sigma_Y$. Note that portion DH of the stress-strain diagram is curved and does not show any clearly defined yield point. This is referred to as the *Bauschinger effect*. As the compressive load is maintained, the material yields along line HJ.

If the load is removed after point J has been reached, the stress returns to zero along line JK, and the slope of JK is equal to the modulus of elasticity E. The resulting permanent set AK may be positive, negative, or zero, depending upon the lengths of the segments BC and HJ. If a tensile load is applied again to the test specimen, the portion of the stress-strain diagram beginning at K (dashed line) will curve up and to the right until the yield stress σ_Y has been reached.

If the initial loading is large enough to cause strain-hardening of the material (point C'), unloading takes place along line $C'D'$. As the reverse load is applied, the stress becomes compressive, reaching its maximum value at H' and maintaining it as the material yields along line $H'J'$. While the maximum value of the compressive stress is less than σ_Y, the total change in stress between C' and H' is still equal to $2\sigma_Y$.

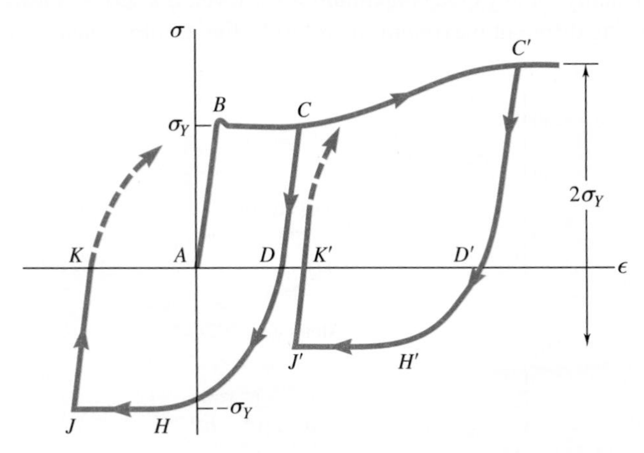

Fig. 9.14 Stress-strain response for mild steel subjected to two cases of reverse loading.

If point K or K' coincides with the origin A of the diagram, the permanent set is equal to zero, and the specimen may appear to have returned to its original condition. However, internal changes will have taken place and, the specimen will rupture without any warning after relatively few repetitions of the loading sequence. Thus, the excessive plastic deformations to which the specimen was subjected caused a radical change in the characteristics of the material. Therefore reverse loadings into the plastic range are seldom allowed, being permitted only under carefully controlled conditions such as in the straightening of damaged material and the final alignment of a structure or machine.

*9.1E Repeated Loadings and Fatigue

You might think that a given load may be repeated many times, provided that the stresses remain in the elastic range. Such a conclusion is correct for loadings repeated a few dozen or even a few hundred times. However, it is not correct when loadings are repeated thousands or millions of times. In such cases, rupture can occur at a stress much lower than the static breaking strength; this phenomenon is known as *fatigue*. A fatigue failure is of a brittle nature, even for materials that are normally ductile.

Fatigue must be considered in the design of all structural and machine components subjected to repeated or fluctuating loads. The number of loading cycles expected during the useful life of a component varies greatly. For example, a beam supporting an industrial crane can be loaded as many as two million times in 25 years (about 300 loadings per working day), an automobile crankshaft is loaded about half a billion times if the automobile is driven 200,000 miles, and an individual turbine blade can be loaded several hundred billion times during its lifetime.

Some loadings are of a fluctuating nature. For example, the passage of traffic over a bridge will cause stress levels that will fluctuate about the stress level due to the weight of the bridge. A more severe condition occurs when a complete reversal of the load occurs during the loading cycle. The stresses in the axle of a railroad car, for example, are completely reversed after each half-revolution of the wheel.

The number of loading cycles required to cause the failure of a specimen through repeated loadings and reverse loadings can be determined experimentally for any given maximum stress level. If a series of tests is conducted using different maximum stress levels, the resulting data is plotted as

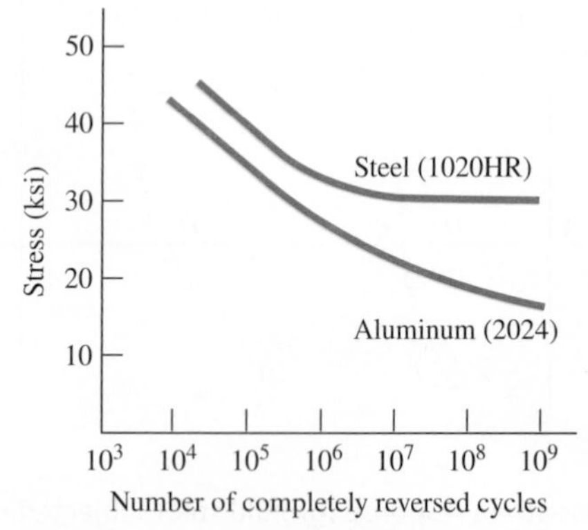

Fig. 9.15 Typical σ-n curves.

a σ-n curve. For each test, the maximum stress σ is plotted as an ordinate and the number of cycles n as an abscissa. Because of the large number of cycles required for rupture, the cycles n are plotted on a logarithmic scale.

A typical σ-n curve for steel is shown in Fig. 9.15. If the applied maximum stress is high, relatively few cycles are required to cause rupture. As the magnitude of the maximum stress is reduced, the number of cycles required to cause rupture increases, until the *endurance limit* is reached. The endurance limit is the stress for which failure does not occur, even for an indefinitely large number of loading cycles. For a low-carbon steel, such as structural steel, the endurance limit is about one-half of the ultimate strength of the steel.

For nonferrous metals, such as aluminum and copper, a typical σ-n curve (Fig. 9.15) shows that the stress at failure continues to decrease as the number of loading cycles is increased. For such metals, the *fatigue limit* is the stress corresponding to failure after a specified number of loading cycles.

Examination of test specimens, shafts, springs, and other components that have failed in fatigue shows that the failure initiated at a microscopic crack or some similar imperfection. At each loading, the crack was very slightly enlarged. During successive loading cycles, the crack propagated through the material until the amount of undamaged material was insufficient to carry the maximum load, and an abrupt, brittle failure occurred. For example, Photo 9.6 shows a progressive fatigue crack in a highway bridge girder that initiated at the irregularity associated with the weld of a cover plate and then propagated through the flange and into the web. Because fatigue failure can be initiated at any crack or imperfection, the surface condition of a specimen has an important effect on the endurance limit obtained in testing. The endurance limit for machined and polished specimens is higher than for rolled or forged components or for components that are corroded. In applications in or near seawater or in other applications where corrosion is expected, a reduction of up to 50 percent in the endurance limit can be expected.

Photo 9.6 Fatigue crack in a steel girder of the Yellow Mill Pond Bridge, Connecticut, prior to repairs.

9.1F Deformations of Members Under Axial Loading

Consider a homogeneous rod BC of length L and uniform cross section of area A subjected to a centric axial load **P** (Fig. 9.16). If the resulting axial stress $\sigma = P/A$ does not exceed the proportional limit of the material, Hooke's law applies and

$$\sigma = E\epsilon \tag{9.5}$$

from which

$$\epsilon = \frac{\sigma}{E} = \frac{P}{AE} \tag{9.6}$$

Recalling that the strain ϵ in Sec. 9.1A is $\epsilon = \delta/L$

$$\delta = \epsilon L \tag{9.7}$$

and substituting for ϵ from Eq. (9.6) into Eq.(9.7):

$$\delta = \frac{PL}{AE} \tag{9.8}$$

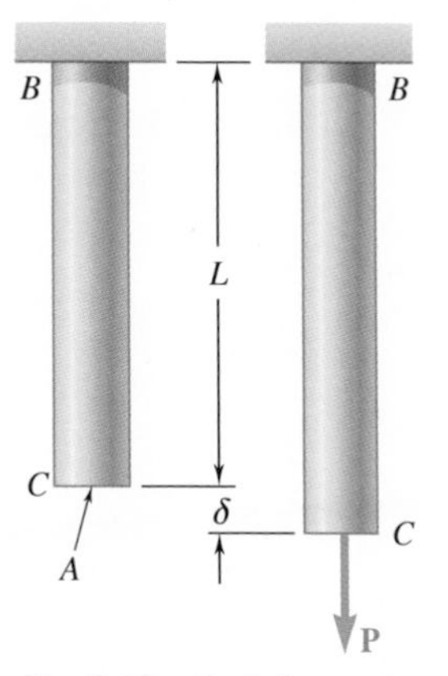

Fig. 9.16 Undeformed and deformed axially-loaded rod.

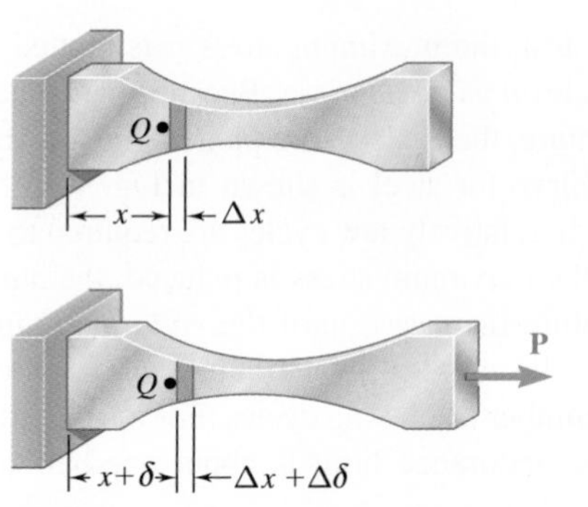

Fig. 9.17 Deformation of axially-loaded member of variable cross-sectional area.

Equation (9.8) can be used only if the rod is homogeneous (constant E), has a uniform cross section of area A, and is loaded at its ends. If the rod is loaded at other points, or consists of several portions of various cross sections and possibly of different materials, it must be divided into component parts that satisfy the required conditions for the application of Eq. (9.8). Using the internal force P_i, length L_i, cross-sectional area A_i, and modulus of elasticity E_i, corresponding to part i, the deformation of the entire rod is

$$\delta = \sum_i \frac{P_i L_i}{A_i E_i} \tag{9.9}$$

In the case of a member of variable cross section (Fig. 9.17), the strain ϵ depends upon the position of the point Q, where it is computed as $\epsilon = d\delta/dx$ (Sec. 9.1A). Solving for $d\delta$ and substituting for ϵ from Eq. (9.6), the deformation of an element of length dx is

$$d\delta = \epsilon\, dx = \frac{P\, dx}{AE}$$

The total deformation δ of the member is obtained by integrating this expression over the length L of the member:

$$\delta = \int_0^L \frac{P\, dx}{AE} \tag{9.10}$$

Equation (9.10) should be used in place of (9.8) when both the cross-sectional area A is a function of x, or when the internal force P depends upon x, as is the case for a rod hanging under its own weight.

Rod BC of Fig. 9.16, used to derive Eq. (9.8), and rod AD of Fig. 9.18 have one end attached to a fixed support. In each case, the deformation δ of the rod was equal to the displacement of its free end. When both ends of a

Concept Application 9.1

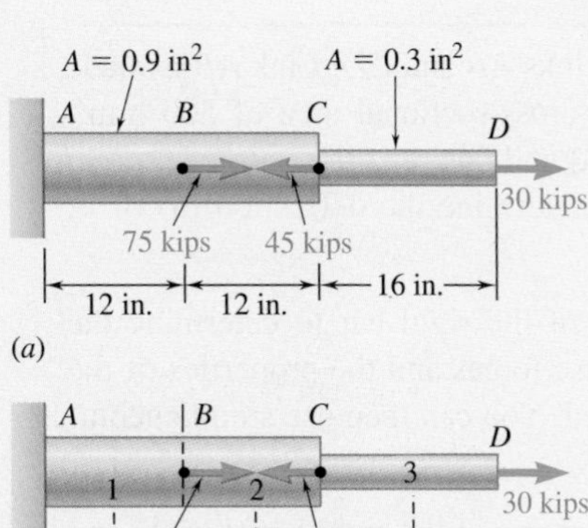

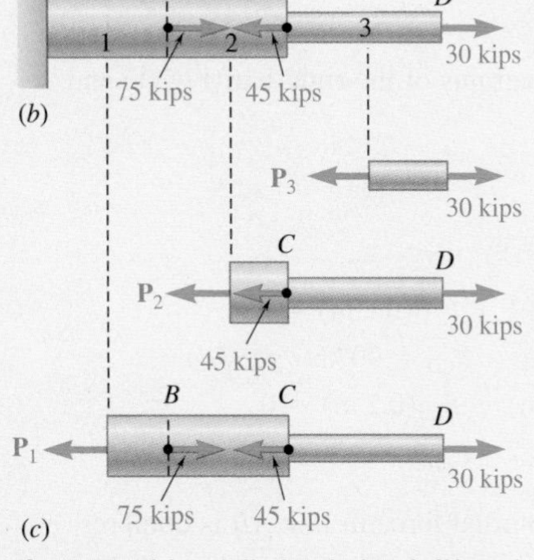

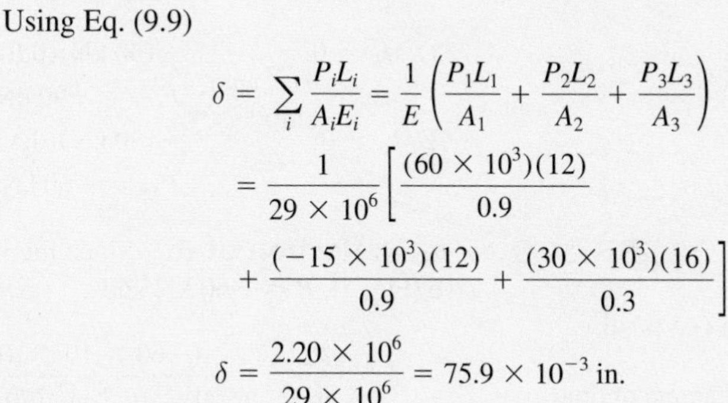

Fig. 9.18 (*a*) Axially-loaded rod. (*b*) Rod divided into three sections. (*c*) Three sectioned free-body diagrams with internal resultant forces P_1, P_2, and P_3.

Determine the deformation of the steel rod shown in Fig. 9.18*a* under the given loads ($E = 29 \times 10^6$ psi).

The rod is divided into three component parts in Fig. 9.18*b*, so

$$L_1 = L_2 = 12 \text{ in.} \qquad L_3 = 16 \text{ in.}$$
$$A_1 = A_2 = 0.9 \text{ in}^2 \qquad A_3 = 0.3 \text{ in}^2$$

To find the internal forces P_1, P_2, and P_3, pass sections through each of the component parts, drawing each time the free-body diagram of the portion of rod located to the right of the section (Fig. 9.18*c*). Each of the free bodies is in equilibrium; thus

$$P_1 = 60 \text{ kips} = 60 \times 10^3 \text{ lb}$$
$$P_2 = -15 \text{ kips} = -15 \times 10^3 \text{ lb}$$
$$P_3 = 30 \text{ kips} = 30 \times 10^3 \text{ lb}$$

Using Eq. (9.9)

$$\delta = \sum_i \frac{P_i L_i}{A_i E_i} = \frac{1}{E}\left(\frac{P_1 L_1}{A_1} + \frac{P_2 L_2}{A_2} + \frac{P_3 L_3}{A_3}\right)$$
$$= \frac{1}{29 \times 10^6}\left[\frac{(60 \times 10^3)(12)}{0.9} + \frac{(-15 \times 10^3)(12)}{0.9} + \frac{(30 \times 10^3)(16)}{0.3}\right]$$
$$\delta = \frac{2.20 \times 10^6}{29 \times 10^6} = 75.9 \times 10^{-3} \text{ in.}$$

rod move, however, the deformation of the rod is measured by the *relative displacement* of one end of the rod with respect to the other. Consider the assembly shown in Fig. 9.19*a*, which consists of three elastic bars of length L connected by a rigid pin at A. If a load **P** is applied at B (Fig. 9.19*b*), each of the three bars will deform. Since the bars AC and AC' are attached to fixed supports at C and C', their common deformation is measured by the displacement δ_A of point A. On the other hand, since both ends of bar AB move, the deformation of AB is measured by the difference between the displacements δ_A and δ_B of points A and B, (i.e., by the relative displacement of B with respect to A). Denoting this relative displacement by $\delta_{B/A}$,

$$\delta_{B/A} = \delta_B - \delta_A = \frac{PL}{AE} \qquad (9.11)$$

where A is the cross-sectional area of AB and E is its modulus of elasticity.

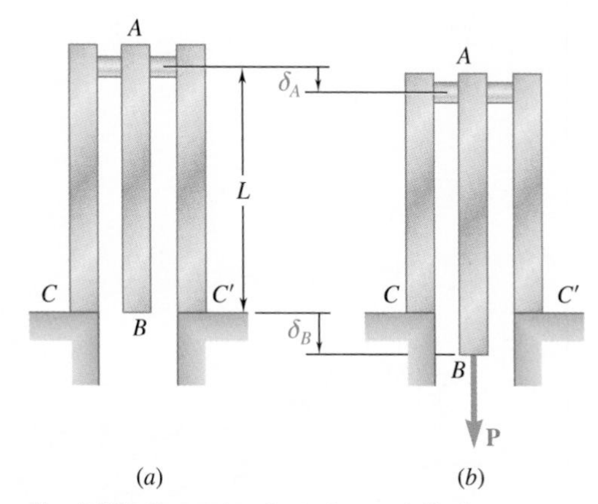

Fig. 9.19 Example of relative end displacement, as exhibited by the middle bar. (*a*) Unloaded. (*b*) Loaded, with deformation.

Sample Problem 9.1

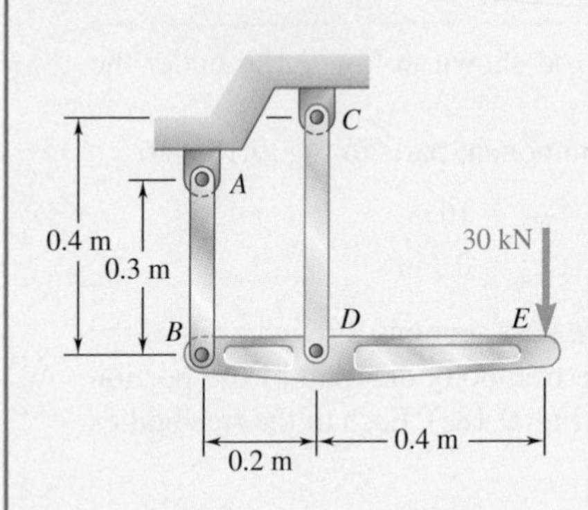

The rigid bar *BDE* is supported by two links *AB* and *CD*. Link *AB* is made of aluminum ($E = 70$ GPa) and has a cross-sectional area of 500 mm². Link *CD* is made of steel ($E = 200$ GPa) and has a cross-sectional area of 600 mm². For the 30-kN force shown, determine the deflection (*a*) of *B*, (*b*) of *D*, and (*c*) of *E*.

STRATEGY: Consider the free body of the rigid bar to determine the internal force of each link. Knowing these forces and the properties of the links, their deformations can be evaluated. You can then use simple geometry to determine the deflection of E.

MODELING: Draw the free body diagrams of the rigid bar (Fig. 1) and the two links (Fig. 2 and 3)

ANALYSIS:

Free Body: Bar *BDE* (Fig. 1)

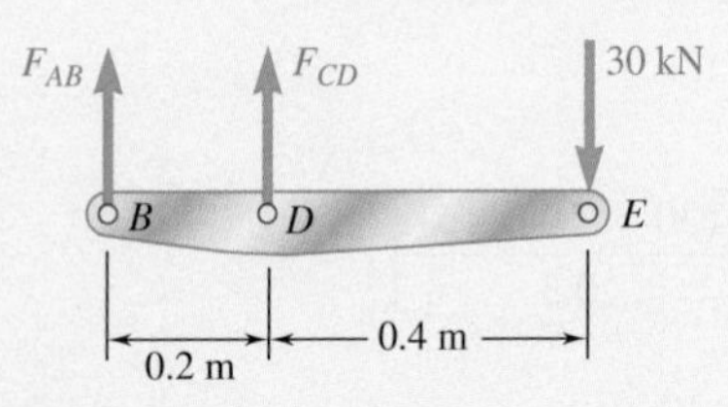

Fig. 1 Free-body diagram of rigid bar *BDE*.

$$+\circlearrowleft\Sigma M_B = 0: \quad -(30\text{ kN})(0.6\text{ m}) + F_{CD}(0.2\text{ m}) = 0$$

$$F_{CD} = +90\text{ kN} \qquad F_{CD} = 90\text{ kN} \quad \textit{tension}$$

$$+\circlearrowleft\Sigma M_D = 0: \quad -(30\text{ kN})(0.4\text{ m}) - F_{AB}(0.2\text{ m}) = 0$$

$$F_{AB} = -60\text{ kN} \qquad F_{AB} = 60\text{ kN} \quad \textit{compression}$$

a. Deflection of *B*. Since the internal force in link *AB* is compressive (Fig. 2), $P = -60$ kN and

$$\delta_B = \frac{PL}{AE} = \frac{(-60 \times 10^3\text{ N})(0.3\text{ m})}{(500 \times 10^{-6}\text{ m}^2)(70 \times 10^9\text{ Pa})} = -514 \times 10^{-6}\text{ m}$$

The negative sign indicates a contraction of member *AB*. Thus, the deflection of end *B* is upward:

$$\delta_B = 0.514\text{ mm} \uparrow \quad ◀$$

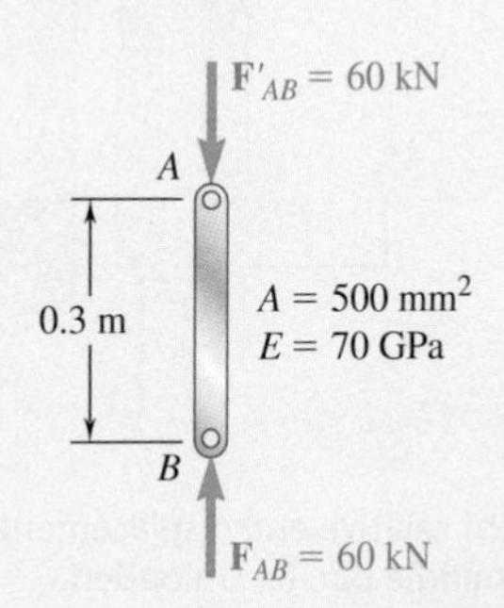

Fig. 2 Free-body diagram of two-force member *AB*.

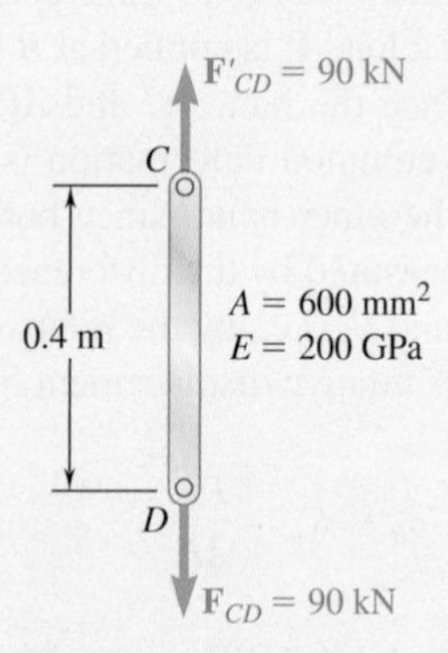

Fig. 3 Free-body diagram of two-force member *CD*.

(continued)

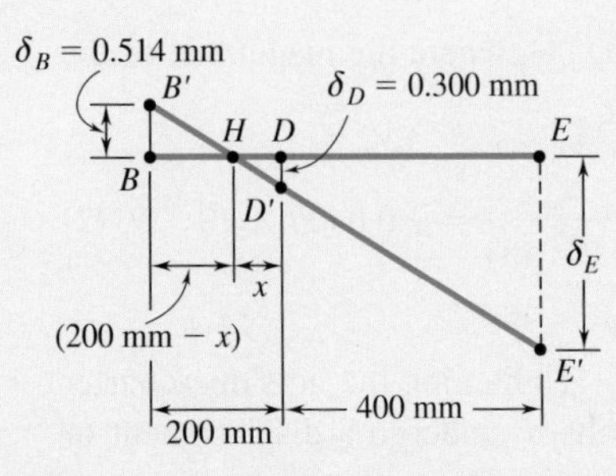

Fig. 4 Deflections at *B* and *D* of rigid bar are used to find δ_E.

b. Deflection of *D*. Since in rod *CD* (Fig. 3), $P = 90$ kN, write

$$\delta_D = \frac{PL}{AE} = \frac{(90 \times 10^3 \text{ N})(0.4 \text{ m})}{(600 \times 10^{-6} \text{ m}^2)(200 \times 10^9 \text{ Pa})}$$

$$= 300 \times 10^{-6} \text{ m} \qquad \delta_D = 0.300 \text{ mm} \downarrow$$

c. Deflection of *E*. Referring to Fig. 4, we denote by *B*′ and *D*′ the displaced positions of points *B* and *D*. Since the bar *BDE* is rigid, points *B*′, *D*′, and *E*′ lie in a straight line. Therefore,

$$\frac{BB'}{DD'} = \frac{BH}{HD} \qquad \frac{0.514 \text{ mm}}{0.300 \text{ mm}} = \frac{(200 \text{ mm}) - x}{x} \qquad x = 73.7 \text{ mm}$$

$$\frac{EE'}{DD'} = \frac{HE}{HD} \qquad \frac{\delta_E}{0.300 \text{ mm}} = \frac{(400 \text{ mm}) + (73.7 \text{ mm})}{73.7 \text{ mm}}$$

$$\delta_E = 1.928 \text{ mm} \downarrow$$

REFLECT and THINK: Comparing the relative magnitude and direction of the resulting deflections, you can see that the answers obtained are consistent with the loading and the deflection diagram of Fig. 4.

Sample Problem 9.2

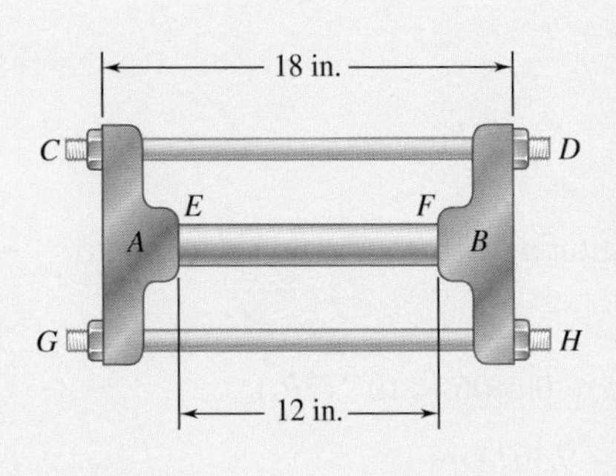

The rigid castings *A* and *B* are connected by two $\frac{3}{4}$-in.-diameter steel bolts *CD* and *GH* and are in contact with the ends of a 1.5-in.-diameter aluminum rod *EF*. Each bolt is single-threaded with a pitch of 0.1 in., and after being snugly fitted, the nuts at *D* and *H* are both tightened one-quarter of a turn. Knowing that *E* is 29×10^6 psi for steel and 10.6×10^6 psi for aluminum, determine the normal stress in the rod.

STRATEGY: The tightening of the nuts causes a displacement of the ends of the bolts relative to the rigid casting that is equal to the difference in displacements between the bolts and the rod. This will give a relation between the internal forces of the bolts and the rod that, when combined with a free body analysis of the rigid casting, will enable you to solve for these forces and determine the corresponding normal stress in the rod.

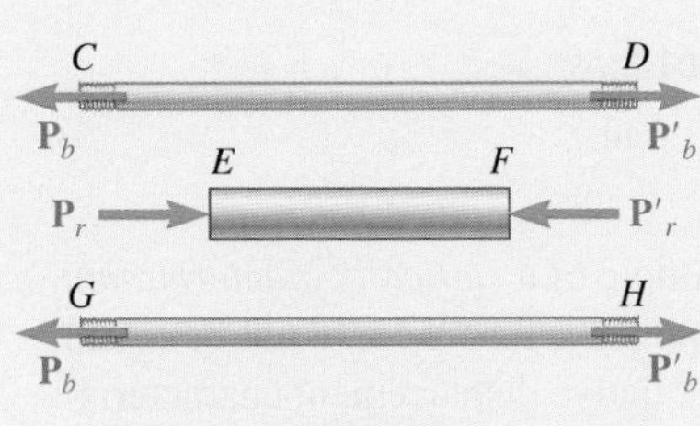

Fig. 1 Free-body diagrams of bolts and aluminum bar.

MODELING: Draw the free body diagrams of the bolts and rod (Fig. 1) and the rigid casting (Fig. 2).

ANALYSIS:

Deformations.

Bolts CD and GH. Tightening the nuts causes tension in the bolts (Fig. 1). Because of symmetry, both are subjected to the same internal force P_b and undergo the same deformation δ_b. Therefore,

$$\delta_b = +\frac{P_b L_b}{A_b E_b} = +\frac{P_b(18 \text{ in.})}{\frac{1}{4}\pi(0.75 \text{ in.})^2(29 \times 10^6 \text{ psi})} = +1.405 \times 10^{-6}\, P_b \quad \textbf{(1)}$$

(continued)

Rod EF. The rod is in compression (Fig. 1), where the magnitude of the force is P_r and the deformation δ_r:

$$\delta_r = -\frac{P_r L_r}{A_r E_r} = -\frac{P_r(12\text{ in.})}{\frac{1}{4}\pi(1.5\text{ in.})^2(10.6 \times 10^6\text{ psi})} = -0.6406 \times 10^{-6} P_r \quad \textbf{(2)}$$

Displacement of D Relative to B. Tightening the nuts one-quarter of a turn causes ends D and H of the bolts to undergo a displacement of $\frac{1}{4}$(0.1 in.) *relative to casting B*. Considering end D,

$$\delta_{D/B} = \tfrac{1}{4}(0.1\text{ in.}) = 0.025\text{ in.} \quad \textbf{(3)}$$

But $\delta_{D/B} = \delta_D - \delta_B$, where δ_D and δ_B represent the displacements of D and B. If casting A is held in a fixed position while the nuts at D and H are being tightened, these displacements are equal to the deformations of the bolts and of the rod, respectively. Therefore,

$$\delta_{D/B} = \delta_b - \delta_r \quad \textbf{(4)}$$

Substituting from Eqs. (1), (2), and (3) into Eq. (4),

$$0.025\text{ in.} = 1.405 \times 10^{-6} P_b + 0.6406 \times 10^{-6} P_r \quad \textbf{(5)}$$

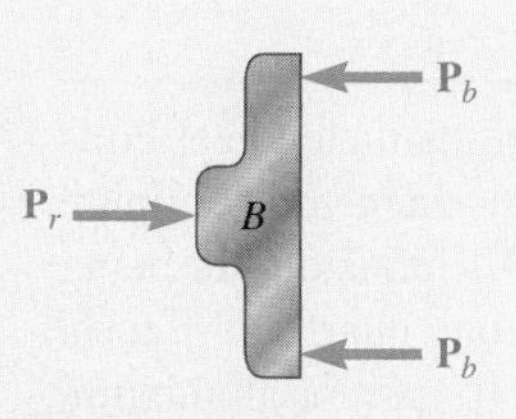

Fig. 2 Free-body diagram of rigid casting.

Free Body: Casting *B* (Fig. 2)

$$\overset{+}{\rightarrow}\Sigma F = 0: \qquad P_r - 2P_b = 0 \qquad P_r = 2P_b \quad \textbf{(6)}$$

Forces in Bolts and Rod Substituting for P_r from Eq. (6) into Eq. (5), we have

$$0.025\text{ in.} = 1.405 \times 10^{-6} P_b + 0.6406 \times 10^{-6}(2P_b)$$

$$P_b = 9.307 \times 10^3\text{ lb} = 9.307\text{ kips}$$

$$P_r = 2P_b = 2(9.307\text{ kips}) = 18.61\text{ kips}$$

Stress in Rod

$$\sigma_r = \frac{P_r}{A_r} = \frac{18.61\text{ kips}}{\frac{1}{4}\pi(1.5\text{ in.})^2} \qquad \sigma_r = 10.53\text{ ksi} \blacktriangleleft$$

REFLECT and THINK: This is an example of a *statically indeterminate* problem, where the determination of the member forces could not be found by equilibrium alone. By considering the relative displacement characteristics of the members, you can obtain additional equations necessary to solve such problems. Situations like this will be examined in more detail in the following section.

Problems

9.1 A 4.8-ft-long steel wire of $\frac{1}{4}$-in. diameter is subjected to a 750-lb tensile load. Knowing that $E = 29 \times 10^6$ psi, determine (*a*) the elongation of the wire, (*b*) the corresponding normal stress.

9.2 Two gage marks are placed exactly 250 mm apart on a 12-mm-diameter aluminum rod with $E = 73$ GPa and an ultimate strength of 140 MPa. Knowing that the distance between the gage marks is 250.28 mm after a load is applied, determine (*a*) the stress in the rod, (*b*) the factor of safety.

9.3 A nylon thread is subjected to a 8.5-N tension force. Knowing that $E = 3.3$ GPa and that the length of the thread increases by 1.1%, determine (*a*) the diameter of the thread, (*b*) the stress in the thread.

9.4 An 18-m-long steel wire of 5-mm diameter is to be used in the manufacture of a prestressed concrete beam. It is observed that the wire stretches 45 mm when a tensile force **P** is applied. Knowing that $E = 200$ GPa, determine (*a*) the magnitude of the force **P**, (*b*) the corresponding normal stress in the wire.

9.5 A steel control rod is 5.5 ft long and must not stretch more than 0.04 in. when a 2-kip tensile load is applied to it. Knowing that $E = 29 \times 10^6$ psi, determine (*a*) the smallest diameter rod that should be used, (*b*) the corresponding normal stress caused by the load.

9.6 A control rod made of yellow brass must not stretch more than 3 mm when the tension in the wire is 4 kN. Knowing that $E = 105$ GPa and that the maximum allowable normal stress is 180 MPa, determine (*a*) the smallest diameter rod that should be used, (*b*) the corresponding maximum length of the rod.

9.7 An aluminum pipe must not stretch more than 0.05 in. when it is subjected to a tensile load. Knowing that $E = 10.1 \times 10^6$ psi and that the maximum allowable normal stress is 14 ksi, determine (*a*) the maximum allowable length of the pipe, (*b*) the required area of the pipe if the tensile load is 127.5 kips.

9.8 A cast-iron tube is used to support a compressive load. Knowing that $E = 10 \times 10^6$ psi and that the maximum allowable change in length is 0.025%, determine (*a*) the maximum normal stress in the tube, (*b*) the minimum wall thickness for a load of 1600 lb if the outside diameter of the tube is 2.0 in.

9.9 A block of 10-in. length and 1.8×1.6-in. cross section is to support a centric compressive load **P**. The material to be used is a bronze for which $E = 14 \times 10^6$ psi. Determine the largest load that can be applied, knowing that the normal stress must not exceed 18 ksi and that the decrease in length of the block should be at most 0.12% of its original length.

9.10 A 4-m-long steel rod must not stretch more than 3 mm and the normal stress must not exceed 150 MPa when the rod is subjected to a 10-kN axial load. Knowing that $E = 200$ GPa, determine the required diameter of the rod.

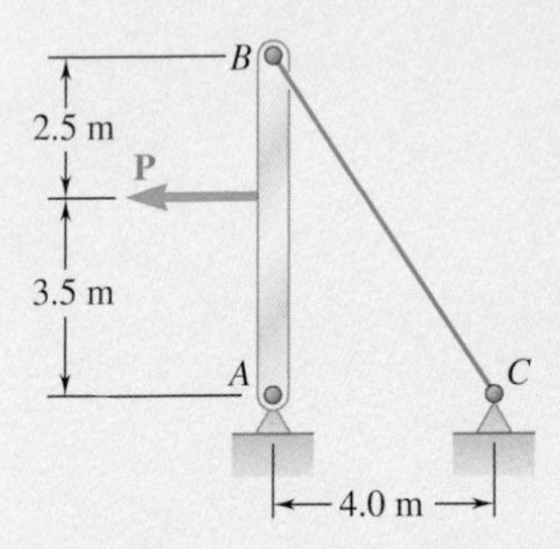

Fig. P9.11

9.11 The 4-mm-diameter cable BC is made of a steel with $E = 200$ GPa. Knowing that the maximum stress in the cable must not exceed 190 MPa and that the elongation of the cable must not exceed 6 mm, find the maximum load **P** that can be applied as shown.

9.12 Rod BD is made of steel ($E = 29 \times 10^6$ psi) and is used to brace the axially compressed member ABC. The maximum force that can be developed in member BD is $0.02P$. If the stress must not exceed 18 ksi and the maximum change in length of BD must not exceed 0.001 times the length of ABC, determine the smallest diameter rod that can be used for member BD.

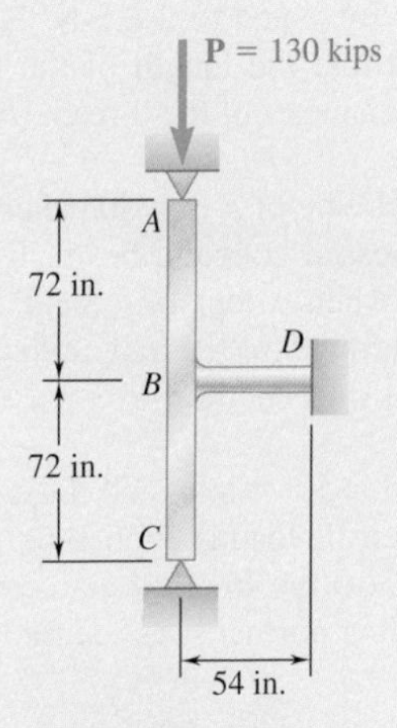

Fig. P9.12

9.13 The specimen shown is made from a 1-in.-diameter cylindrical steel rod with two 1.5-in.-outer-diameter sleeves bonded to the rod as shown. Knowing that $E = 29 \times 10^6$ psi, determine (a) the load **P** so that the total deformation is 0.002 in., (b) the corresponding deformation of the central portion BC.

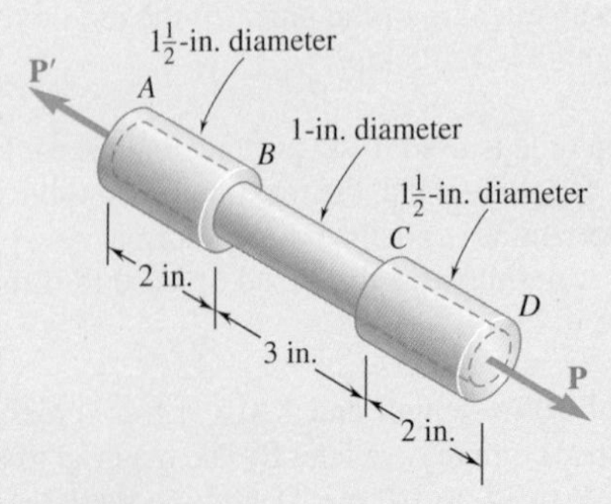

Fig. P9.13

9.14 Both portions of the rod ABC are made of an aluminum for which $E = 70$ GPa. Knowing that the magnitude of **P** is 4 kN, determine (*a*) the value of **Q** so that the deflection at A is zero, (*b*) the corresponding deflection of B.

9.15 The rod ABC is made of an aluminum for which $E = 70$ GPa. Knowing that $P = 6$ kN and $Q = 42$ kN, determine the deflection of (*a*) point A, (*b*) point B.

9.16 Two solid cylindrical rods are joined at B and loaded as shown. Rod AB is made of steel ($E = 29 \times 10^6$ psi), and rod BC of brass ($E = 15 \times 10^6$ psi). Determine (*a*) the total deformation of the composite rod ABC, (*b*) the deflection of point B.

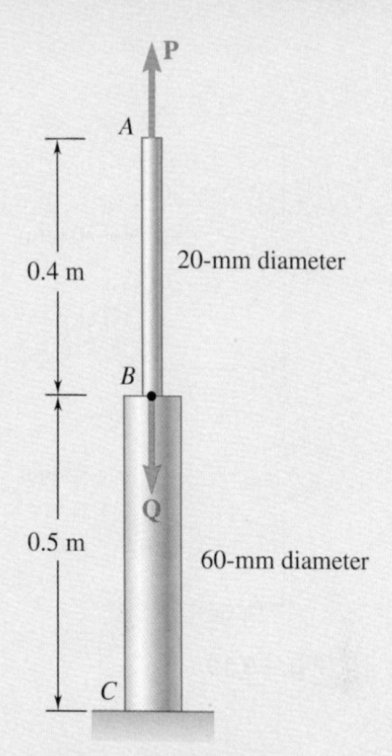

Fig. P9.14 and P9.15

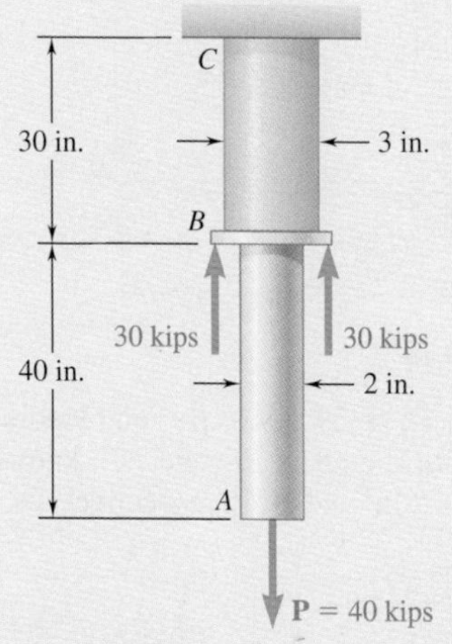

Fig. *P9.16*

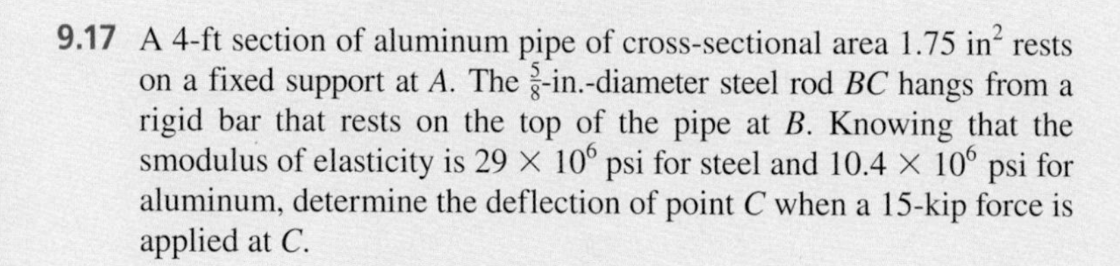

9.17 A 4-ft section of aluminum pipe of cross-sectional area 1.75 in^2 rests on a fixed support at A. The $\frac{5}{8}$-in.-diameter steel rod BC hangs from a rigid bar that rests on the top of the pipe at B. Knowing that the smodulus of elasticity is 29×10^6 psi for steel and 10.4×10^6 psi for aluminum, determine the deflection of point C when a 15-kip force is applied at C.

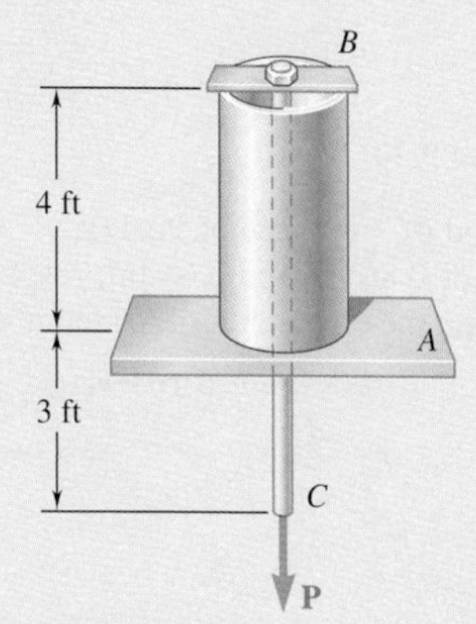

Fig. P9.17

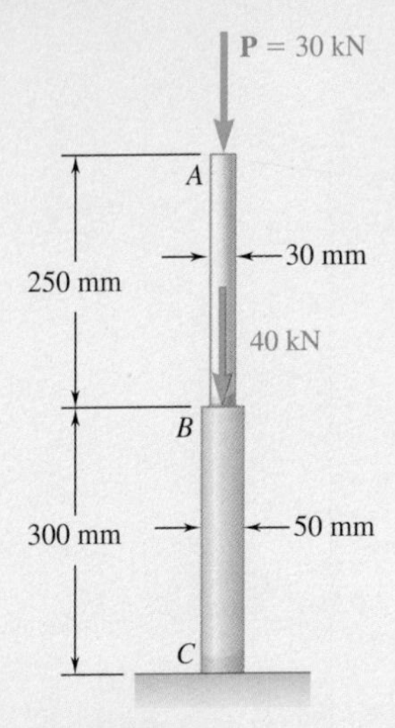

Fig. P9.18

9.18 Two solid cylindrical rods are joined at B and loaded as shown. Rod AB is made of steel (E = 200 GPa) and rod BC of brass (E = 105 GPa). Determine (*a*) the total deformation of the composite rod ABC, (*b*) the deflection of point B.

9.19 The steel frame (E = 200 GPa) shown has a diagonal brace BD with an area of 1920 mm^2. Determine the largest allowable load **P** if the change in length of member BD is not to exceed 1.6 mm.

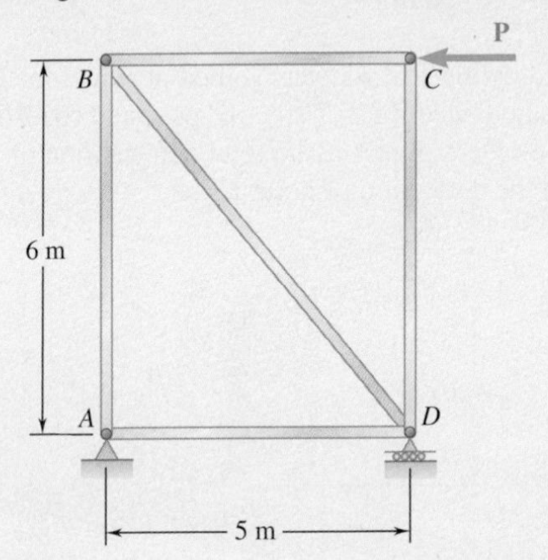

Fig. P9.19

9.20 For the steel truss ($E = 29 \times 10^6$ psi) and loading shown, determine the deformations of members BD and DE, knowing that their cross-sectional areas are 2 in^2 and 3 in^2, respectively.

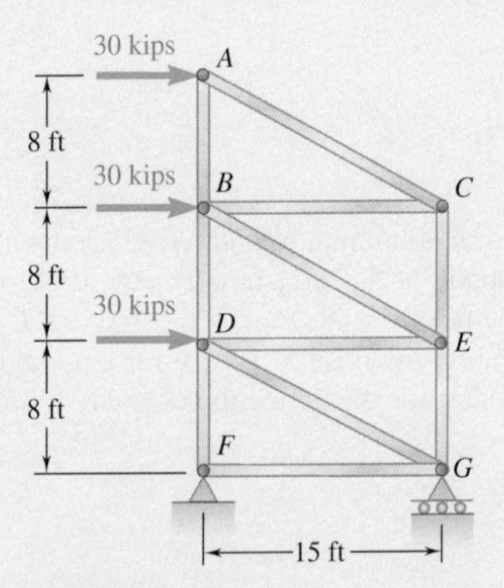

Fig. P9.20

9.21 Members AB and BC are made of steel ($E = 29 \times 10^6$ psi) with cross-sectional areas of 0.80 in^2 and 0.64 in^2, respectively. For the loading shown, determine the elongation of (*a*) member AB, (*b*) member BC.

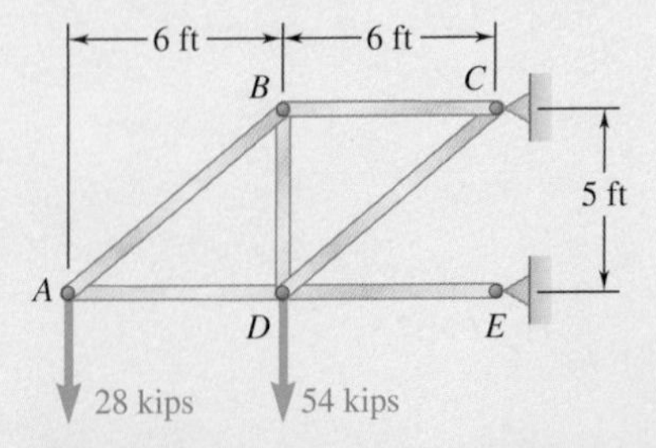

Fig. P9.21

9.22 Members ABC and DEF are joined with steel links ($E = 200$ GPa). Each of the links is made of a pair of 25×35-mm plates. Determine the change in length of (*a*) member BE, (*b*) member CF.

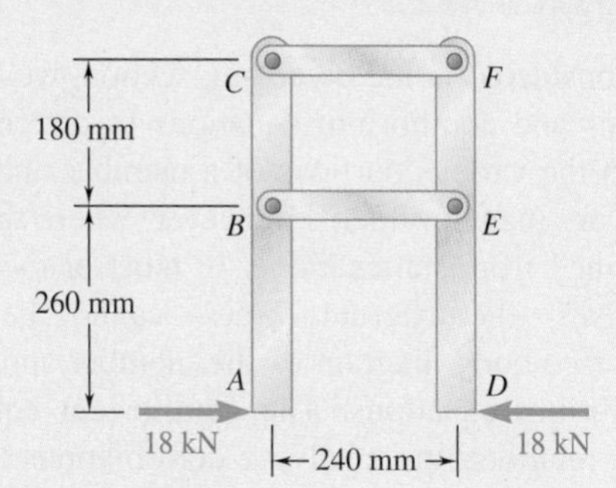

Fig. *P9.22*

9.23 Each of the links AB and CD is made of aluminum ($E = 10.9 \times 10^6$ psi) and has a cross-sectional area of 0.2 in^2. Knowing that they support the rigid member BC, determine the deflection of point E.

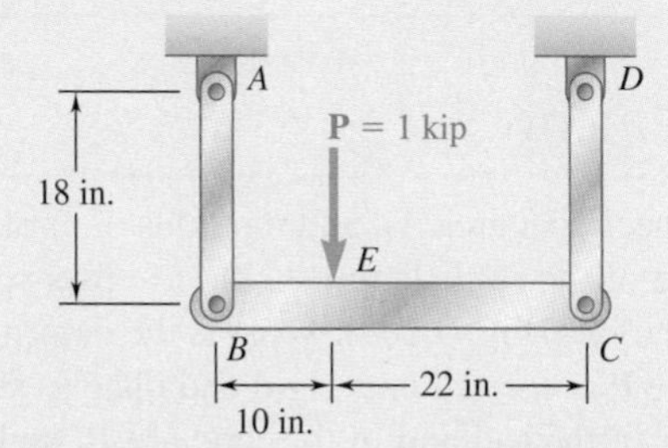

Fig. *P9.23*

9.24 Link BD is made of brass ($E = 105$ GPa) and has a cross-sectional area of 240 mm^2. Link CE is made of aluminum ($E = 72$ GPa) and has a cross-sectional area of 300 mm^2. Knowing that they support rigid member ABC, determine the maximum force **P** that can be applied vertically at point A if the deflection of A is not to exceed 0.35 mm.

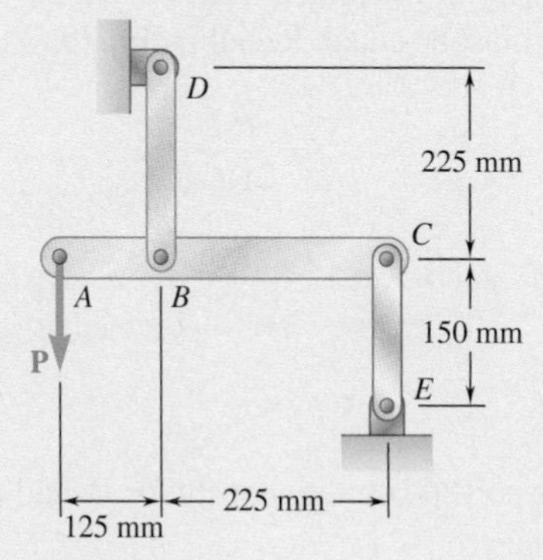

Fig. P9.24

9.2 STATICALLY INDETERMINATE PROBLEMS

In the problems considered in the preceding section, we could always use free-body diagrams and equilibrium equations to determine the internal forces produced in the various portions of a member under given loading conditions. There are many problems, however, where the internal forces cannot be determined from statics alone. In most of these problems, the reactions themselves—the external forces—cannot be determined by simply drawing a free-body diagram of the member and writing the corresponding equilibrium equations. The equilibrium equations must be complemented by relationships involving deformations obtained by considering the geometry of the problem. Because statics is not sufficient to determine either the reactions or the internal forces, problems of this type are called *statically indeterminate*. The following concept applications show how to handle this type of problem.

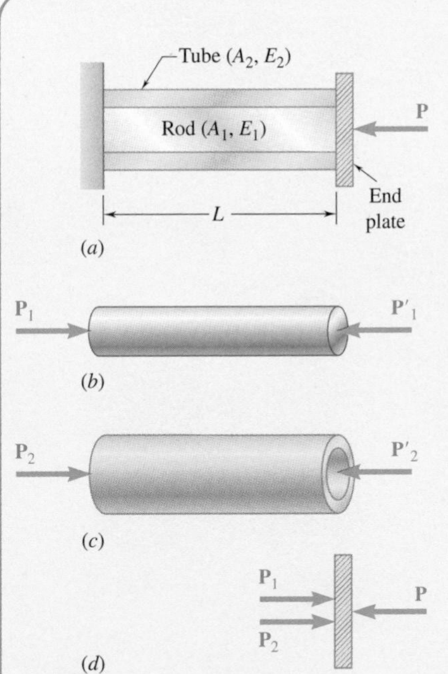

Fig. 9.20 (*a*) Concentric rod and tube, loaded by force *P*. (*b*) Free-body diagram of rod. (*c*) Free-body diagram of tube. (*d*) Free-body diagram of end plate.

Concept Application 9.2

A rod of length L, cross-sectional area A_1, and modulus of elasticity E_1, has been placed inside a tube of the same length L, but of cross-sectional area A_2 and modulus of elasticity E_2 (Fig. 9.20*a*). What is the deformation of the rod and tube when a force **P** is exerted on a rigid end plate as shown?

The axial forces in the rod and in the tube are P_1 and P_2, respectively. Draw free-body diagrams of all three elements (Fig. 9.20*b*, *c*, *d*). Only Fig. 9.20*d* yields any significant information, as:

$$P_1 + P_2 = P \tag{1}$$

Clearly, one equation is not sufficient to determine the two unknown internal forces P_1 and P_2. The problem is statically indeterminate.

However, the geometry of the problem shows that the deformations δ_1 and δ_2 of the rod and tube must be equal. Recalling Eq. (9.8), write

$$\delta_1 = \frac{P_1 L}{A_1 E_1} \qquad \delta_2 = \frac{P_2 L}{A_2 E_2} \tag{2}$$

Equating the deformations δ_1 and δ_2,

$$\frac{P_1}{A_1 E_1} = \frac{P_2}{A_2 E_2} \tag{3}$$

Equations (1) and (3) can be solved simultaneously for P_1 and P_2:

$$P_1 = \frac{A_1 E_1 P}{A_1 E_1 + A_2 E_2} \qquad P_2 = \frac{A_2 E_2 P}{A_1 E_1 + A_2 E_2}$$

Either of Eqs. (2) can be used to determine the common deformation of the rod and tube.

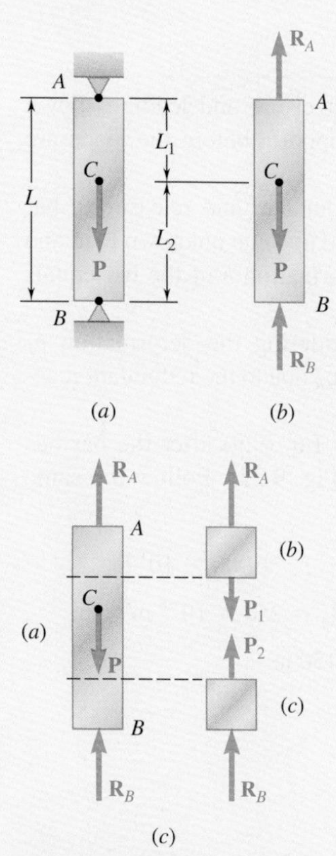

Fig. 9.21 (*a*) Restrained bar with axial load. (*b*) Free-body diagram of bar. (*c*) Free-body diagrams of sections above and below point *C* used to determine internal forces P_1 and P_2.

Concept Application 9.3

A bar *AB* of length *L* and uniform cross section is attached to rigid supports at *A* and *B* before being loaded. What are the stresses in portions *AC* and *BC* due to the application of a load *P* at point *C* (Fig. 9.21*a*)?

Drawing the free-body diagram of the bar (Fig. 9.21*b*), the equilibrium equation is

$$R_A + R_B = P \quad (1)$$

Since this equation is not sufficient to determine the two unknown reactions R_A and R_B, the problem is statically indeterminate.

However, the reactions can be determined if observed from the geometry that the total elongation δ of the bar must be zero. The elongations of the portions *AC* and *BC* are respectively δ_1 and δ_2, so

$$\delta = \delta_1 + \delta_2 = 0$$

Using Eq. (9.8), δ_1 and δ_2 can be expressed in terms of the corresponding internal forces P_1 and P_2,

$$\delta = \frac{P_1 L_1}{AE} + \frac{P_2 L_2}{AE} = 0 \quad (2)$$

Note from the free-body diagrams shown in parts *b* and *c* of Fig. 9.21*c* that $P_1 = R_A$ and $P_2 = -R_B$. Carrying these values into Equation (2),

$$R_A L_1 - R_B L_2 = 0 \quad (3)$$

Equations (1) and (3) can be solved simultaneously for R_A and R_B, as $R_A = PL_2/L$ and $R_B = PL_1/L$. The desired stresses σ_1 in *AC* and σ_2 in *BC* are obtained by dividing $P_1 = R_A$ and $P_2 = -R_B$ by the cross-sectional area of the bar:

$$\sigma_1 = \frac{PL_2}{AL} \qquad \sigma_2 = -\frac{PL_1}{AL}$$

Superposition Method. A structure is statically indeterminate whenever it is held by more supports than are required to maintain its equilibrium. This results in more unknown reactions than available equilibrium equations. It is often convenient to designate one of the reactions as *redundant* and to eliminate the corresponding support. Since the stated conditions of the problem cannot be changed, the redundant reaction must be maintained in the solution. It will be treated as an *unknown load* that, together with the other loads, must produce deformations compatible with the original constraints. The actual solution of the problem considers separately the deformations caused by the given loads and the redundant reaction, and by adding—or *superposing*—the results obtained. The general conditions under which the combined effect of several loads can be obtained in this way are discussed in Sec. 9.5.

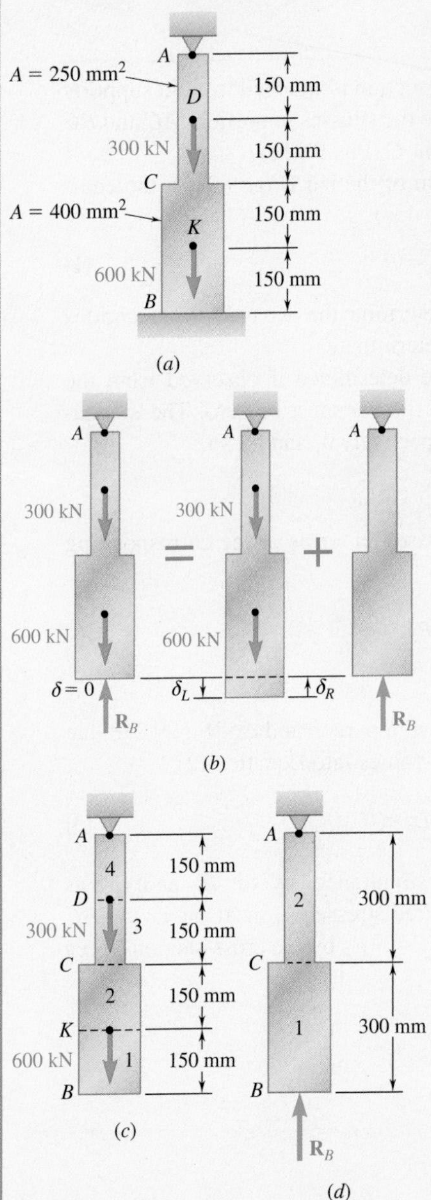

Fig. 9.22 (a) Restrained axially-loaded bar. (b) Reactions will be found by releasing constraint at point B and adding compressive force at point B to enforce zero deformation at point B. (c) Diagram of released structure. (d) Diagram of added reaction force at point B to enforce zero deformation at point B.

Concept Application 9.4

Determine the reactions at A and B for the steel bar and loading shown in Fig. 9.22*a*, assuming a close fit at both supports before the loads are applied.

We consider the reaction at B as redundant and release the bar from that support. The reaction $\mathbf{R}_B$ is considered to be an unknown load and is determined from the condition that the deformation δ of the bar equals zero.

The solution is carried out by considering the deformation δ_L caused by the given loads and the deformation δ_R due to the redundant reaction $\mathbf{R}_B$ (Fig. 9.22*b*).

The deformation δ_L is obtained from Eq. (9.9) after the bar has been divided into four portions, as shown in Fig. 9.22*c*. Follow the same procedure as in Concept Application 9.1:

$$P_1 = 0 \qquad P_2 = P_3 = 600 \times 10^3 \text{ N} \qquad P_4 = 900 \times 10^3 \text{ N}$$
$$A_1 = A_2 = 400 \times 10^{-6} \text{ m}^2 \qquad A_3 = A_4 = 250 \times 10^{-6} \text{ m}^2$$
$$L_1 = L_2 = L_3 = L_4 = 0.150 \text{ m}$$

Substituting these values into Eq. (9.9),

$$\delta_L = \sum_{i=1}^{4} \frac{P_i L_i}{A_i E} = \left(0 + \frac{600 \times 10^3 \text{ N}}{400 \times 10^{-6} \text{ m}^2} + \frac{600 \times 10^3 \text{ N}}{250 \times 10^{-6} \text{ m}^2} + \frac{900 \times 10^3 \text{ N}}{250 \times 10^{-6} \text{ m}^2}\right) \frac{0.150 \text{ m}}{E}$$

$$\delta_L = \frac{1.125 \times 10^9}{E} \tag{1}$$

Considering now the deformation δ_R due to the redundant reaction R_B, the bar is divided into two portions, as shown in Fig. 9.22*d*

$$P_1 = P_2 = -R_B$$
$$A_1 = 400 \times 10^{-6} \text{ m}^2 \qquad A_2 = 250 \times 10^{-6} \text{ m}^2$$
$$L_1 = L_2 = 0.300 \text{ m}$$

Substituting these values into Eq. (9.9),

$$\delta_R = \frac{P_1 L_1}{A_1 E} + \frac{P_2 L_2}{A_2 E} = -\frac{(1.95 \times 10^3) R_B}{E} \tag{2}$$

Express the total deformation δ of the bar as zero:

$$\delta = \delta_L + \delta_R = 0 \tag{3}$$

and, substituting for δ_L and δ_R from Eqs. (1) and (2) into Eqs. (3),

$$\delta = \frac{1.125 \times 10^9}{E} - \frac{(1.95 \times 10^3) R_B}{E} = 0$$

(continued)

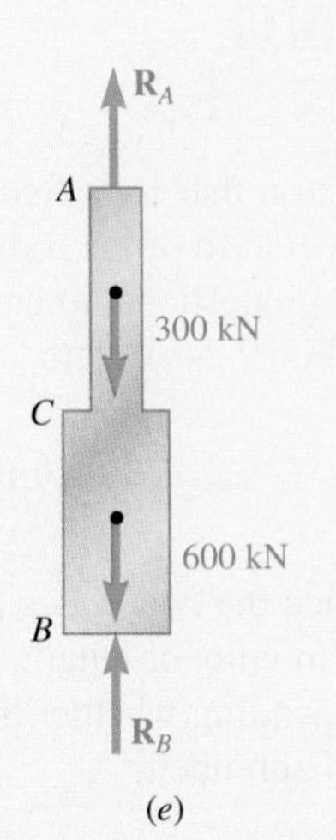

Fig. 9.22 (cont.) (e) Complete free-body diagram of *ACB*.

Solving for R_B,

$$R_B = 577 \times 10^3 \text{ N} = 577 \text{ kN}$$

The reaction R_A at the upper support is obtained from the free-body diagram of the bar (Fig. 9.22*e*),

$$+\uparrow \Sigma F_y = 0: \qquad R_A - 300 \text{ kN} - 600 \text{ kN} + R_B = 0$$

$$R_A = 900 \text{ kN} - R_B = 900 \text{ kN} - 577 \text{ kN} = 323 \text{ kN}$$

Once the reactions have been determined, the stresses and strains in the bar can easily be obtained. Note that, while the total deformation of the bar is zero, each of its component parts *does deform* under the given loading and restraining conditions.

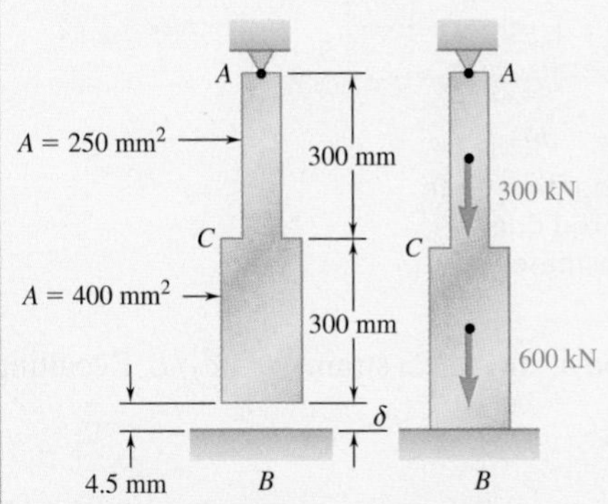

Fig. 9.23 Multi-section bar of Concept Application 9.4 with initial 4.5-mm gap at point *B*. Loading brings bar into contact with constraint.

Concept Application 9.5

Determine the reactions at *A* and *B* for the steel bar and loading of Concept Application 9.4, assuming now that a 4.5-mm clearance exists between the bar and the ground before the loads are applied (Fig. 9.23). Assume $E = 200$ GPa.

Considering the reaction at *B* to be redundant, compute the deformations δ_L and δ_R caused by the given loads and the redundant reaction $\mathbf{R}_B$. However, in this case, the total deformation is $\delta = 4.5$ mm. Therefore,

$$\delta = \delta_L + \delta_R = 4.5 \times 10^{-3} \text{ m} \tag{1}$$

Substituting for δ_L and δ_R into Eq. (1), and recalling that $E = 200 \text{ GPa} = 200 \times 10^9 \text{ Pa}$,

$$\delta = \frac{1.125 \times 10^9}{200 \times 10^9} - \frac{(1.95 \times 10^3)R_B}{200 \times 10^9} = 4.5 \times 10^{-3} \text{ m}$$

Solving for R_B,

$$R_B = 115.4 \times 10^3 \text{ N} = 115.4 \text{ kN}$$

The reaction at *A* is obtained from the free-body diagram of the bar (Fig. 9.22*e*):

$$+\uparrow \Sigma F_y = 0: \qquad R_A - 300 \text{ kN} - 600 \text{ kN} + R_B = 0$$
$$R_A = 900 \text{ kN} - R_B = 900 \text{ kN} - 115.4 \text{ kN} = 785 \text{ kN}$$

9.3 PROBLEMS INVOLVING TEMPERATURE CHANGES

Consider a homogeneous rod AB of uniform cross section that rests freely on a smooth horizontal surface (Fig. 9.24*a*). If the temperature of the rod is raised by ΔT, the rod elongates by an amount δ_T that is proportional to both the temperature change ΔT and the length L of the rod (Fig. 9.24*b*). Here

$$\delta_T = \alpha(\Delta T)L \tag{9.12}$$

where α is a constant characteristic of the material called the *coefficient of thermal expansion*. Since δ_T and L are both expressed in units of length, α represents a quantity *per degree C* or *per degree F*, depending whether the temperature change is expressed in degrees Celsius or Fahrenheit.

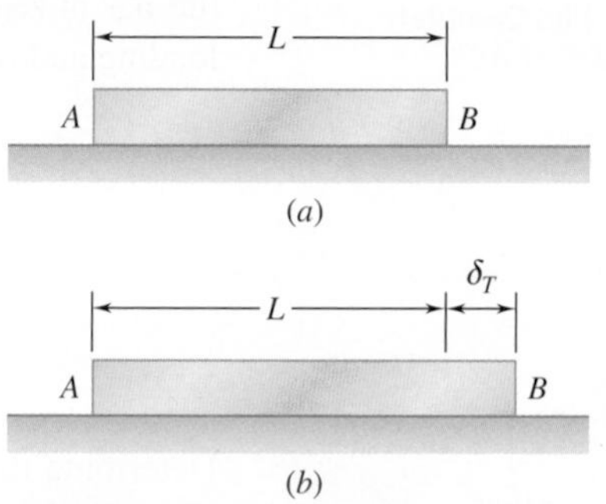

Fig. 9.24 Elongation of an unconstrained rod due to temperature increase.

Associated with deformation δ_T must be a strain $\epsilon_T = \delta_T/L$. Recalling Eq. (9.12),

$$\epsilon_T = \alpha\Delta T \tag{9.13}$$

The strain ϵ_T is called a *thermal strain*, as it is caused by the change in temperature of the rod. However, there is *no stress associated with the strain* ϵ_T.

Assume the same rod AB of length L is placed between two fixed supports at a distance L from each other (Fig. 9.25*a*). Again, there is neither stress nor strain in this initial condition. If we raise the temperature by ΔT, the rod cannot elongate because of the restraints imposed on its ends; the elongation δ_T of the rod is zero. Since the rod is homogeneous and of uniform cross section, the strain ϵ_T at any point is $\epsilon_T = \delta_T/L$ and thus is also zero. However, the supports will exert equal and opposite forces $\mathbf{P}$ and $\mathbf{P}'$ on the rod after the temperature has been raised, to keep it from elongating (Fig. 9.25*b*). It follows that a state of stress (with no corresponding strain) is created in the rod.

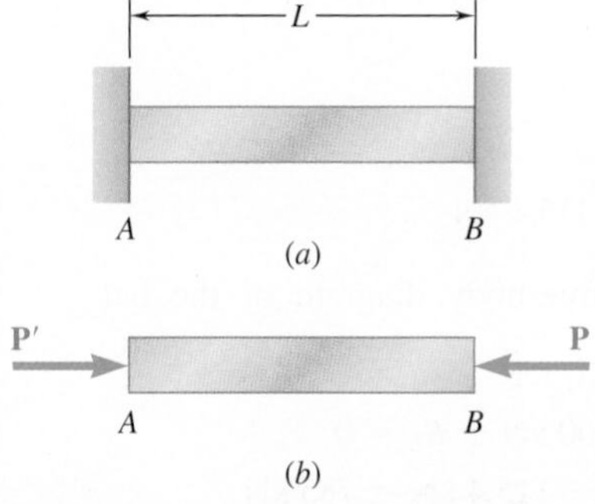

Fig. 9.25 Force P develops when the temperature of the rod increases while ends A and B are restrained.

The problem created by the temperature change ΔT is statically indeterminate. Therefore, the magnitude P of the reactions at the supports is determined from the condition that the elongation of the rod is zero. Using

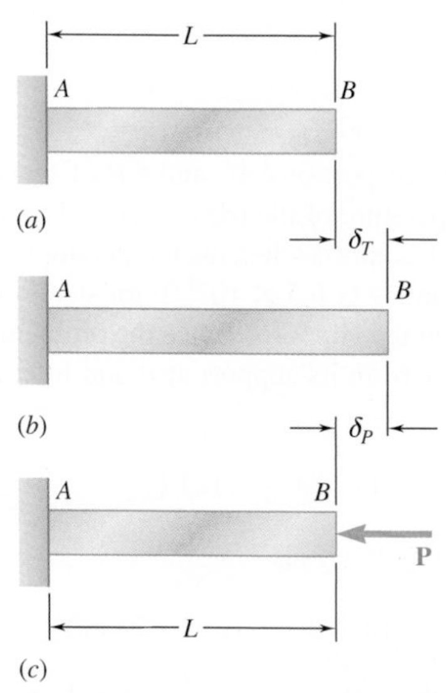

Fig. 9.26 Superposition method to find force at point *B* of restrained rod *AB* undergoing thermal expansion. (*a*) Initial rod length; (*b*) thermally expanded rod length; (*c*) force *P* pushes point *B* back to zero deformation.

the superposition method described in Sec. 9.2, the rod is detached from its support *B* (Fig. 9.26*a*) and elongates freely as it undergoes the temperature change ΔT (Fig. 9.26*b*). According to Eq. (9.12), the corresponding elongation is

$$\delta_T = \alpha(\Delta T)L$$

Applying now to end *B* the force **P** representing the redundant reaction, and recalling Eq. (9.8), a second deformation (Fig. 9.26*c*) is

$$\delta_P = \frac{PL}{AE}$$

Expressing that the total deformation δ must be zero,

$$\delta = \delta_T + \delta_P = \alpha(\Delta T)L + \frac{PL}{AE} = 0$$

from which

$$P = -AE\alpha(\Delta T)$$

The stress in the rod due to the temperature change ΔT is

$$\sigma = \frac{P}{A} = -E\alpha(\Delta T) \qquad \textbf{(9.14)}$$

The absence of any strain in the rod *applies only in the case of a homogeneous rod of uniform cross section*. Any other problem involving a restrained structure undergoing a change in temperature must be analyzed on its own merits. However, the same general approach can be used by considering the deformation due to the temperature change and the deformation due to the redundant reaction separately and superposing the two solutions obtained.

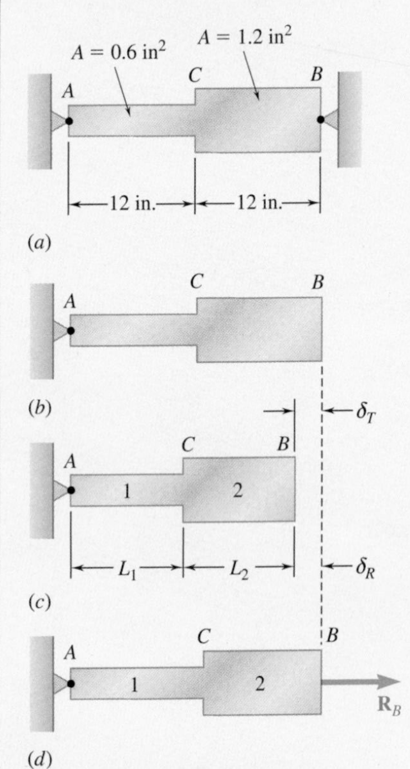

Fig. 9.27 (a) Restrained bar. (b) Bar at +75°F temperature. (c) Bar at lower temperature. (d) Force R_B needed to enforce zero deformation at point B.

Concept Application 9.6

Determine the values of the stress in portions AC and CB of the steel bar shown (Fig. 9.27*a*) when the temperature of the bar is $-50°F$, knowing that a close fit exists at both of the rigid supports when the temperature is $+75°F$. Use the values $E = 29 \times 10^6$ psi and $\alpha = 6.5 \times 10^{-6}/°F$ for steel.

Determine the reactions at the supports. Since the problem is statically indeterminate, detach the bar from its support at B and let it undergo the temperature change

$$\Delta T = (-50°\text{F}) - (75°\text{F}) = -125°\text{F}$$

The corresponding deformation (Fig. 9.27*c*) is

$$\delta_T = \alpha(\Delta T)L = (6.5 \times 10^{-6}/°\text{F})(-125°\text{F})(24 \text{ in.})$$
$$= -19.50 \times 10^{-3} \text{ in.}$$

Applying the unknown force $\mathbf{R}_B$ at end B (Fig. 9.27*d*), use Eq. (9.9) to express the corresponding deformation δ_R. Substituting

$$L_1 = L_2 = 12 \text{ in.}$$
$$A_1 = 0.6 \text{ in}^2 \qquad A_2 = 1.2 \text{ in}^2$$
$$P_1 = P_2 = R_B \qquad E = 29 \times 10^6 \text{ psi}$$

into Eq. (9.9), write

$$\delta_R = \frac{P_1 L_1}{A_1 E} + \frac{P_2 L_2}{A_2 E}$$
$$= \frac{R_B}{29 \times 10^6 \text{ psi}} \left(\frac{12 \text{ in.}}{0.6 \text{ in}^2} + \frac{12 \text{ in.}}{1.2 \text{ in}^2} \right)$$
$$= (1.0345 \times 10^{-6} \text{ in./lb}) R_B$$

Expressing that the total deformation of the bar must be zero as a result of the imposed constraints, write

$$\delta = \delta_T + \delta_R = 0$$
$$= -19.50 \times 10^{-3} \text{ in.} + (1.0345 \times 10^{-6} \text{ in./lb}) R_B = 0$$

from which

$$R_B = 18.85 \times 10^3 \text{ lb} = 18.85 \text{ kips}$$

The reaction at A is equal and opposite.

Noting that the forces in the two portions of the bar are $P_1 = P_2 =$ 18.85 kips, obtain the following values of the stress in portions AC and CB of the bar:

(continued)

$$\sigma_1 = \frac{P_1}{A_1} = \frac{18.85 \text{ kips}}{0.6 \text{ in}^2} = +31.42 \text{ ksi}$$

$$\sigma_2 = \frac{P_2}{A_2} = \frac{18.85 \text{ kips}}{1.2 \text{ in}^2} = +15.71 \text{ ksi}$$

It cannot emphasized too strongly that, while the *total deformation* of the bar must be zero, the deformations of the portions AC and CB *are not zero*. A solution of the problem based on the assumption that these deformations are zero would therefore be wrong. Neither can the values of the strain in AC or CB be assumed equal to zero. To amplify this point, determine the strain ϵ_{AC} in portion AC of the bar. The strain ϵ_{AC} can be divided into two component parts; one is the thermal strain ϵ_T produced in the unrestrained bar by the temperature change ΔT (Fig. 9.27*c*). From Eq. (9.13),

$$\epsilon_T = \alpha\,\Delta T = (6.5 \times 10^{-6}/°\text{F})(-125°\text{F})$$

$$= -812.5 \times 10^{-6} \text{ in./in.}$$

The other component of ϵ_{AC} is associated with the stress σ_1 due to the force $\mathbf{R}_B$ applied to the bar (Fig. 9.27*d*). From Hooke's law, express this component of the strain as

$$\frac{\sigma_1}{E} = \frac{+31.42 \times 10^3 \text{ psi}}{29 \times 10^6 \text{ psi}} = +1083.4 \times 10^{-6} \text{ in./in.}$$

Add the two components of the strain in AC to obtain

$$\epsilon_{AC} = \epsilon_T + \frac{\sigma_1}{E} = -812.5 \times 10^{-6} + 1083.4 \times 10^{-6}$$

$$= +271 \times 10^{-6} \text{ in./in.}$$

A similar computation yields the strain in portion CB of the bar:

$$\epsilon_{CB} = \epsilon_T + \frac{\sigma_2}{E} = -812.5 \times 10^{-6} + 541.7 \times 10^{-6}$$

$$= -271 \times 10^{-6} \text{ in./in.}$$

The deformations δ_{AC} and δ_{CB} of the two portions of the bar are

$$\delta_{AC} = \epsilon_{AC}(AC) = (+271 \times 10^{-6})(12 \text{ in.})$$

$$= +3.25 \times 10^{-3} \text{ in.}$$

$$\delta_{CB} = \epsilon_{CB}(CB) = (-271 \times 10^{-6})(12 \text{ in.})$$

$$= -3.25 \times 10^{-3} \text{ in.}$$

Thus, while the sum $\delta = \delta_{AC} + \delta_{CB}$ of the two deformations is zero, neither of the deformations is zero.

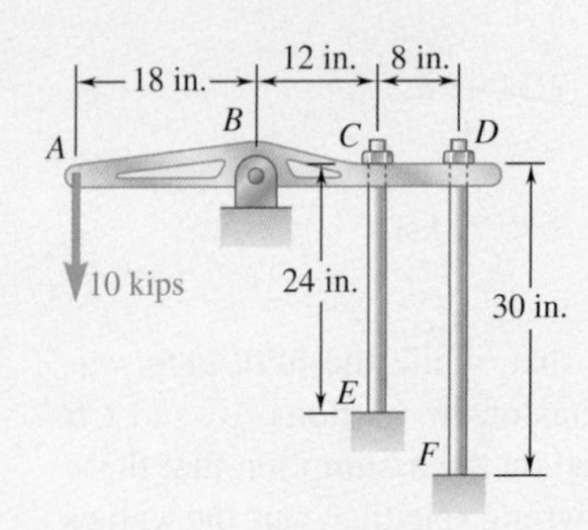

Sample Problem 9.3

The $\frac{1}{2}$-in.-diameter rod *CE* and the $\frac{3}{4}$-in.-diameter rod *DF* are attached to the rigid bar *ABCD* as shown. Knowing that the rods are made of aluminum and using $E = 10.6 \times 10^6$ psi, determine (*a*) the force in each rod caused by the loading shown and (*b*) the corresponding deflection of point *A*.

STRATEGY: To solve this statically indeterminate problem, you must supplement static equilibrium with a relative deflection analysis of the two rods.

MODELING: Draw the free body diagram of the bar (Fig. 1)

ANALYSIS:

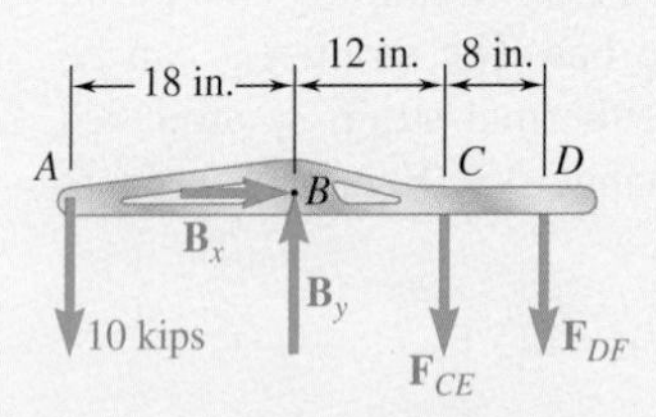

Fig. 1 Free-body diagram of rigid bar *ABCD*.

Statics. Considering the free body of bar *ABCD* in Fig. 1, note that the reaction at *B* and the forces exerted by the rods are indeterminate. However, using statics,

$$+\circlearrowleft \Sigma M_B = 0: \quad (10 \text{ kips})(18 \text{ in.}) - F_{CE}(12 \text{ in.}) - F_{DF}(20 \text{ in.}) = 0$$

$$12F_{CE} + 20F_{DF} = 180 \qquad (1)$$

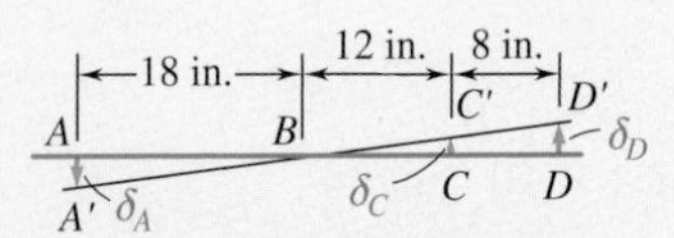

Fig. 2 Linearly proportional displacements along rigid bar *ABCD*.

Geometry. After application of the 10-kip load, the position of the bar is *A'BC'D'* (Fig. 2). From the similar triangles *BAA'*, *BCC'*, and *BDD'*,

$$\frac{\delta_C}{12 \text{ in.}} = \frac{\delta_D}{20 \text{ in.}} \qquad \delta_C = 0.6\delta_D \qquad (2)$$

$$\frac{\delta_A}{18 \text{ in.}} = \frac{\delta_D}{20 \text{ in.}} \qquad \delta_A = 0.9\delta_D \qquad (3)$$

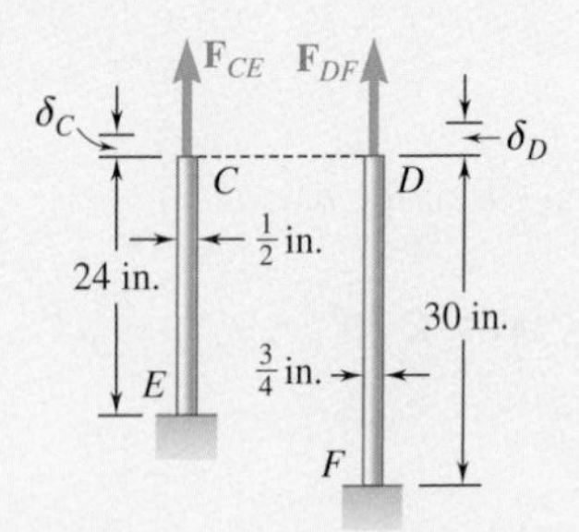

Fig. 3 Forces and deformations in *CE* and *DF*.

Deformations. Using Eq. (9.8), and the data shown in Fig. 3, write

$$\delta_C = \frac{F_{CE}L_{CE}}{A_{CE}E} \qquad \delta_D = \frac{F_{DF}L_{DF}}{A_{DF}E}$$

Substituting for δ_C and δ_D into Eq. (2), write

$$\delta_C = 0.6\delta_D \qquad \frac{F_{CE}L_{CE}}{A_{CE}E} = 0.6\frac{F_{DF}L_{DF}}{A_{DF}E}$$

$$F_{CE} = 0.6\frac{L_{DF}}{L_{CE}}\frac{A_{CE}}{A_{DF}}F_{DF} = 0.6\left(\frac{30 \text{ in.}}{24 \text{ in.}}\right)\left[\frac{\frac{1}{4}\pi(\frac{1}{2}\text{ in.})^2}{\frac{1}{4}\pi(\frac{3}{4}\text{ in.})^2}\right]F_{DF} \quad F_{CE} = 0.333F_{DF}$$

Force in Each Rod. Substituting for F_{CE} into Eq. (1) and recalling that all forces have been expressed in kips,

$$12(0.333F_{DF}) + 20F_{DF} = 180 \qquad F_{DF} = 7.50 \text{ kips} \blacktriangleleft$$

$$F_{CE} = 0.333F_{DF} = 0.333(7.50 \text{ kips}) \qquad F_{CE} = 2.50 \text{ kips} \blacktriangleleft$$

(continued)

Deflections. The deflection of point D is

$$\delta_D = \frac{F_{DF}L_{DF}}{A_{DF}E} = \frac{(7.50 \times 10^3 \text{ lb})(30 \text{ in.})}{\frac{1}{4}\pi(\frac{3}{4} \text{ in.})^2(10.6 \times 10^6 \text{ psi})} \qquad \delta_D = 48.0 \times 10^{-3} \text{ in.}$$

Using Eq. (3),

$$\delta_A = 0.9\delta_D = 0.9(48.0 \times 10^{-3} \text{ in.}) \qquad \delta_A = 43.2 \times 10^{-3} \text{ in.} \blacktriangleleft$$

REFLECT and THINK: You should note that as the rigid bar rotates about B, the deflections at C and D are proportional to their distance from the pivot point B, but *the forces exerted by the rods at these points are not.* Being statically indeterminate, these forces depend upon the deflection attributes of the rods as well as the equilibrium of the rigid bar.

Sample Problem 9.4

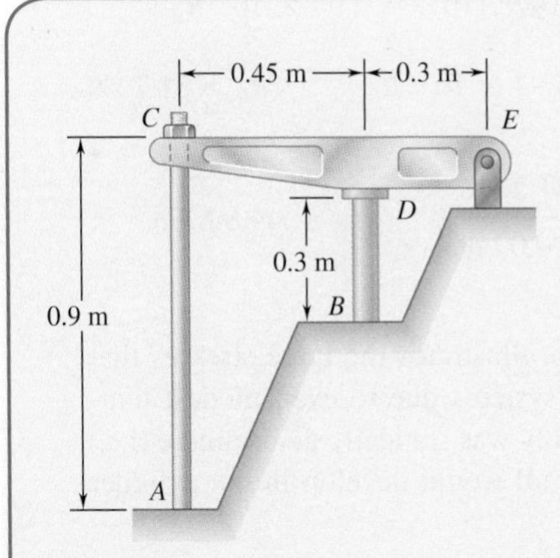

The rigid bar CDE is attached to a pin support at E and rests on the 30-mm-diameter brass cylinder BD. A 22-mm-diameter steel rod AC passes through a hole in the bar and is secured by a nut that is snugly fitted when the temperature of the entire assembly is 20°C. The temperature of the brass cylinder is then raised to 50°C, while the steel rod remains at 20°C. Assuming that no stresses were present before the temperature change, determine the stress in the cylinder.

Rod AC: Steel	Cylinder BD: Brass
$E = 200$ GPa	$E = 105$ GPa
$\alpha = 11.7 \times 10^{-6}/°C$	$\alpha = 20.9 \times 10^{-6}/°C$

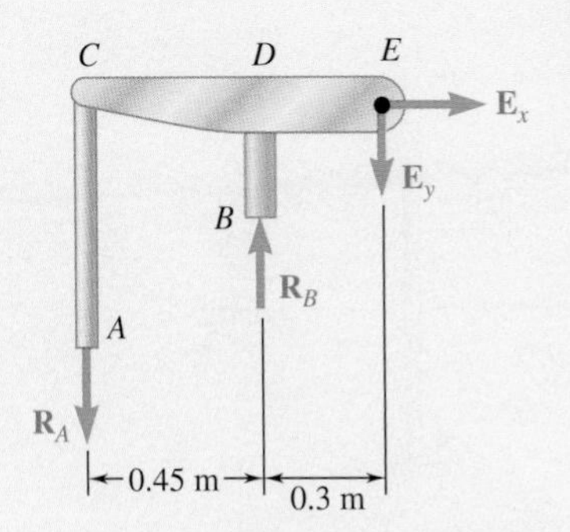

Fig. 1 Free-body diagram of bolt, cylinder and bar.

STRATEGY: You can use the method of superposition, considering $\mathbf{R}_B$ as redundant. With the support at B removed, the temperature rise of the cylinder causes point B to move down through δ_T. The reaction $\mathbf{R}_B$ must cause a deflection δ_1, equal to δ_T so that the final deflection of B will be zero (Fig. 2)

MODELING: Draw the free-body diagram of the entire assembly (Fig. 1).

ANALYSIS:

Statics. Considering the free body of the entire assembly, write

$$+\circlearrowleft\Sigma M_E = 0: \quad R_A(0.75 \text{ m}) - R_B(0.3 \text{ m}) = 0 \quad R_A = 0.4R_B \qquad (1)$$

(continued)

Deflection δ_T. Because of a temperature rise of 50° − 20° = 30°C, the length of the brass cylinder increases by δ_T (Fig. 2*a*).

$$\delta_T = L(\Delta T)\alpha = (0.3 \text{ m})(30°\text{C})(20.9 \times 10^{-6}/°\text{C}) = 188.1 \times 10^{-6} \text{ m} \downarrow$$

Deflection δ_1. From Fig. 2*b*, note that $\delta_D = 0.4\delta_C$ and $\delta_1 = \delta_D + \delta_{B/D}$.

$$\delta_C = \frac{R_A L}{AE} = \frac{R_A(0.9 \text{ m})}{\frac{1}{4}\pi(0.022 \text{ m})^2(200 \text{ GPa})} = 11.84 \times 10^{-9} R_A \uparrow$$

$$\delta_D = 0.40\delta_C = 0.4(11.84 \times 10^{-9} R_A) = 4.74 \times 10^{-9} R_A \uparrow$$

$$\delta_{B/D} = \frac{R_B L}{AE} = \frac{R_B(0.3 \text{ m})}{\frac{1}{4}\pi(0.03 \text{ m})^2(105 \text{ GPa})} = 4.04 \times 10^{-9} R_B \uparrow$$

Recall from Eq. (1) that $R_A = 0.4R_B$, so

$$\delta_1 = \delta_D + \delta_{B/D} = [4.74(0.4R_B) + 4.04R_B]10^{-9} = 5.94 \times 10^{-9} R_B \uparrow$$

But $\delta_T = \delta_1$: $\quad 188.1 \times 10^{-6} \text{ m} = 5.94 \times 10^{-9} R_B \qquad R_B = 31.7 \text{ kN}$

Stress in Cylinder: $\sigma_B = \dfrac{R_B}{A} = \dfrac{31.7 \text{ kN}}{\frac{1}{4}\pi(0.03 \text{ m})^2} \qquad \sigma_B = 44.8 \text{ MPa}$ ◀

REFLECT and THINK: This example illustrates the large stresses that can develop in statically indeterminate systems due to even modest temperature changes. Note that if this assembly was statically determinate (i.e., the steel rod was removed), no stress at all would develop in the cylinder due to the temperature change.

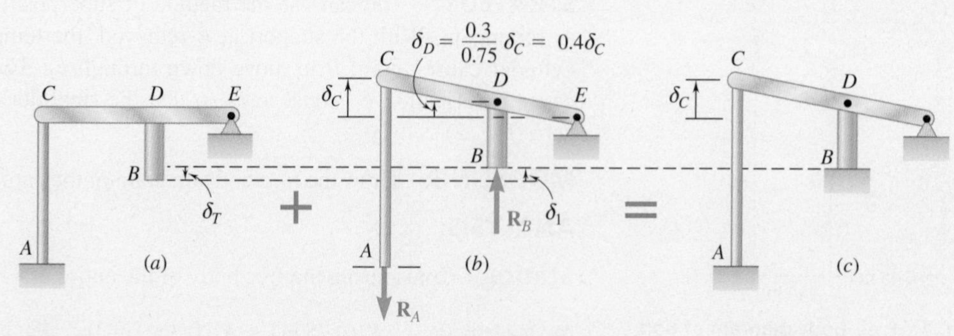

Fig. 2 Superposition of thermal and restraint force deformations (*a*) Support at *B* removed. (*b*) Reaction at *B* applied. (*c*) Final position.

Problems

9.25 An axial force of 200 kN is applied to the assembly shown by means of rigid end plates. Determine (*a*) the normal stress in the aluminum shell, (*b*) the corresponding deformation of the assembly.

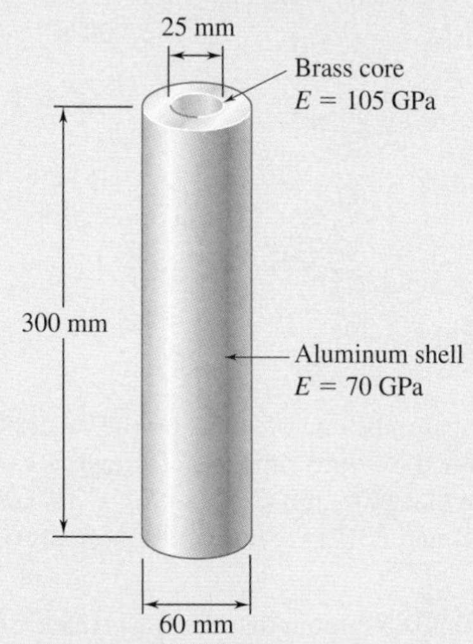

Fig. P9.25 and P9.26

9.26 The length of the assembly shown decreases by 0.40 mm when an axial force is applied by means of rigid end plates. Determine (*a*) the magnitude of the applied force, (*b*) the corresponding stress in the brass core.

9.27 The 4.5-ft concrete post is reinforced with six steel bars, each with a $1\frac{1}{8}$-in. diameter. Knowing that $E_s = 29 \times 10^6$ psi and $E_c = 4.2 \times 10^6$ psi, determine the normal stresses in the steel and in the concrete when a 350-kip axial centric force **P** is applied to the post.

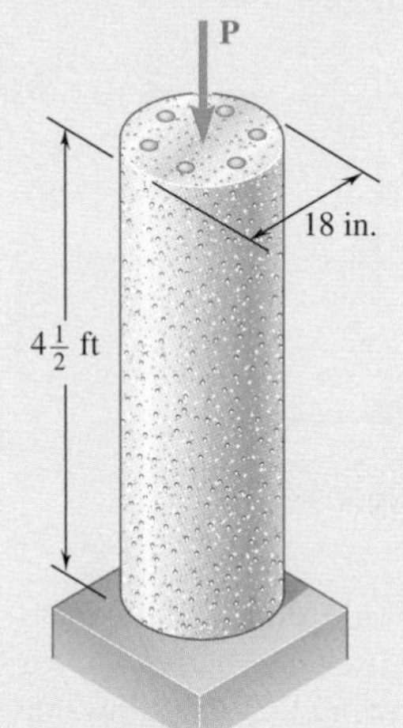

Fig. P9.27

9.28 For the post of Prob. 9.27, determine the maximum centric force that can be applied if the allowable normal stress is 20 ksi in the steel and 2.4 ksi in the concrete.

9.29 Three steel rods ($E = 29 \times 10^6$ psi) support an 8.5-kip load **P**. Each of the rods AB and CD has a 0.32-in^2 cross-sectional area and rod EF has a 1-in^2 cross-sectional area. Neglecting the deformation of bar BED, determine (*a*) the change in length of rod EF, (*b*) the stress in each rod.

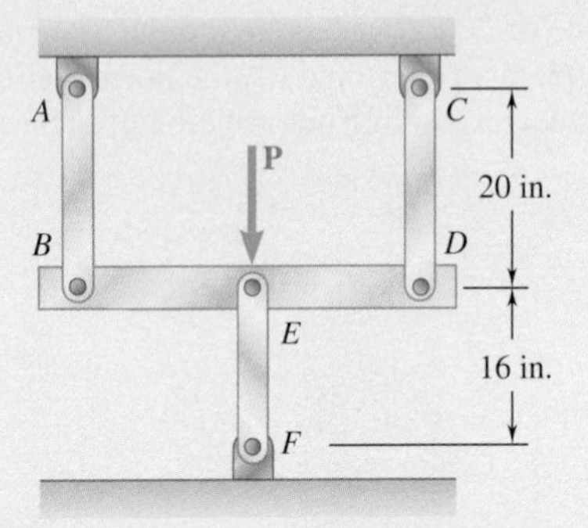

Fig. *P9.29*

9.30 Two cylindrical rods, one of steel and the other of brass, are joined at C and restrained by rigid supports at A and E. For the loading shown and knowing that $E_s = 200$ GPa and $E_b = 105$ GPa, determine (*a*) the reactions at A and E, (*b*) the deflection of point C.

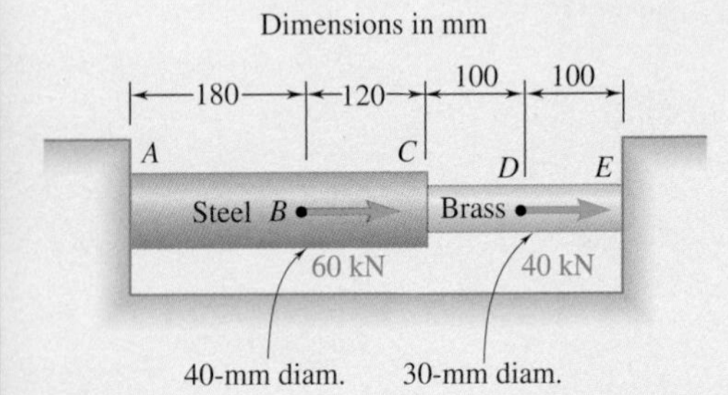

Fig. P9.30

9.31 Solve Prob. 9.30, assuming that rod AC is made of brass and rod CE is made of steel.

9.32 A polystyrene rod consisting of two cylindrical portions AB and BC is restrained at both ends and supports two 6-kip loads as shown. Knowing that $E = 0.45 \times 10^6$ psi, determine (*a*) the reactions at A and C, (*b*) the normal stress in each portion of the rod.

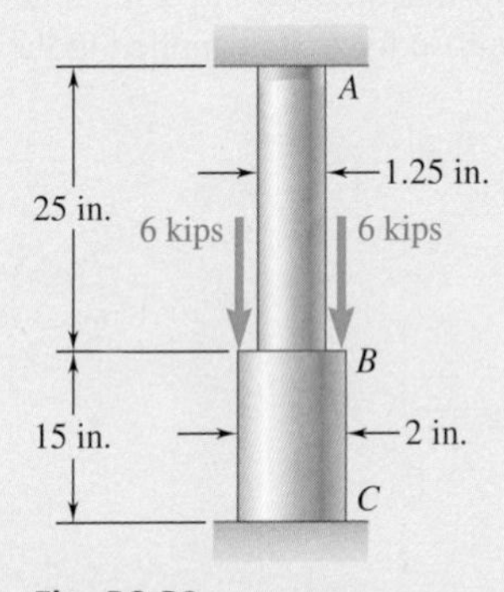

Fig. P9.32

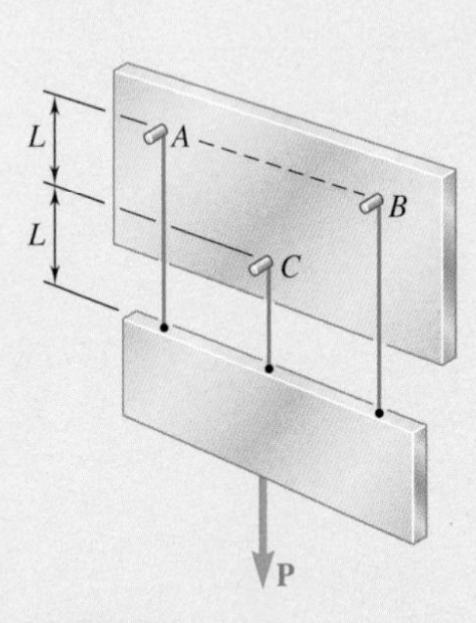

Fig. P9.33

9.33 Three wires are used to suspend the plate shown. Aluminum wires of $\frac{1}{8}$-in. diameter are used at A and B while a steel wire of $\frac{1}{12}$-in. diameter is used at C. Knowing that the allowable stress for aluminum ($E_a = 10.4 \times 10^6$ psi) is 14 ksi and that the allowable stress for steel ($E_s = 29 \times 10^6$ psi) is 18 ksi, determine the maximum load **P** that can be applied.

9.34 The rigid bar AD is supported by two steel wires of $\frac{1}{16}$-in. diameter ($E = 29 \times 10^6$ psi) and a pin and bracket at A. Knowing that the wires were initially taut, determine (*a*) the additional tension in each wire when a 220-lb load **P** is applied at D, (*b*) the corresponding deflection of point D.

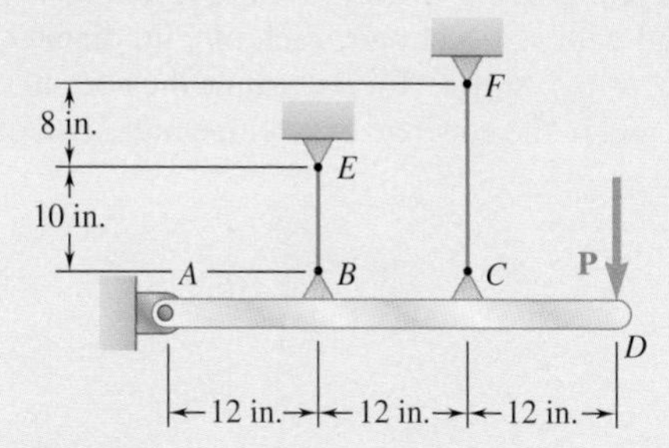

Fig. *P9.34*

9.35 The rigid bar ABC is suspended from three wires of the same material. The cross-sectional area of the wire at B is equal to half of the cross-sectional area of the wires at A and C. Determine the tension in each wire caused by the load **P**.

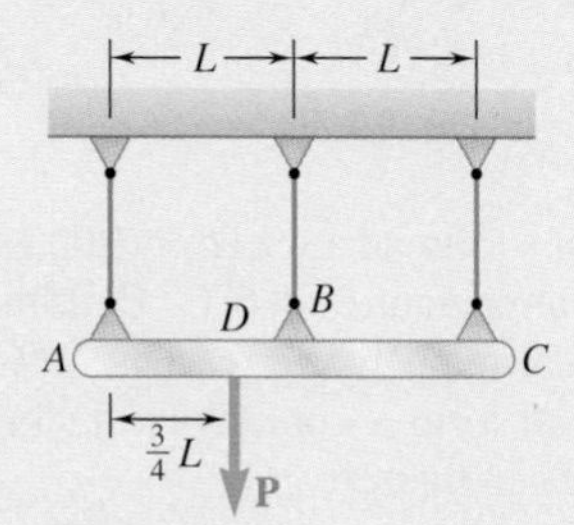

Fig. P9.35

9.36 The rigid bar $ABCD$ is suspended from four identical wires. Determine the tension in each wire caused by the load **P**.

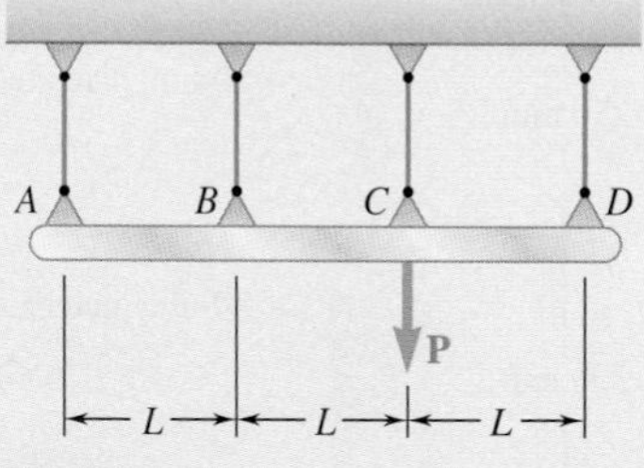

Fig. P9.36

9.37 The brass shell ($\alpha_b = 11.6 \times 10^{-6}$/°F) is fully bonded to the steel core ($\alpha_s = 6.5 \times 10^{-6}$/°F). Determine the largest allowable increase in temperature if the stress in the steel core is not to exceed 8 ksi.

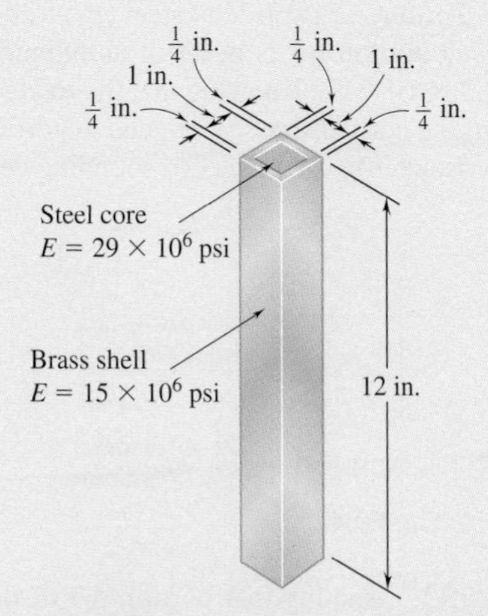

Fig. P9.37

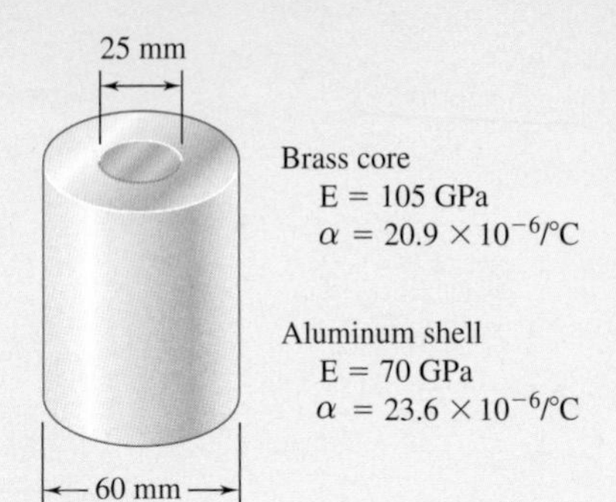

Fig. *P9.38*

9.38 The aluminum shell is fully bonded to the brass core and the assembly is unstressed at a temperature of 15°C. Considering only axial deformations, determine the stress in the aluminum when the temperature reaches 195°C.

9.39 The concrete post (E_c = 3.6 × 10^6 psi and α_c = 5.5 × 10^{-6}/°F) is reinforced with six steel bars, each of $\frac{7}{8}$-in. diameter (E_s = 29 × 10^6 psi and α_s = 6.5 × 10^{-6}/°F). Determine the normal stresses induced in the steel and in the concrete by a temperature rise of 65°F.

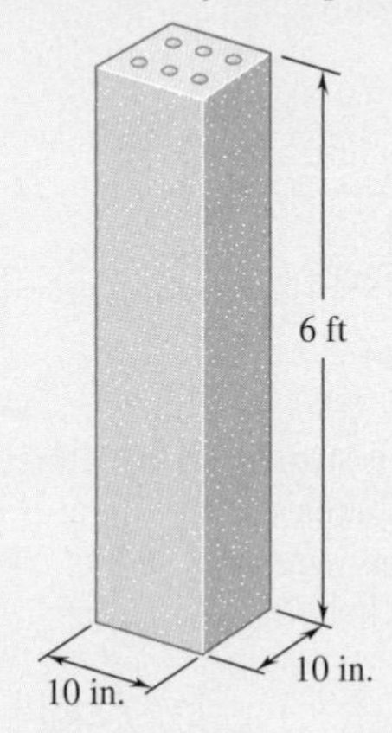

Fig. P9.39

9.40 The steel rails of a railroad track (E_s = 200 GPa, α_s = 11.7 × 10^{-6}/°C) were laid at a temperature of 6°C. Determine the normal stress in the rails when the temperature reaches 48°C, assuming that the rails (*a*) are welded to form a continuous track, (*b*) are 10 m long with 3-mm gaps between them.

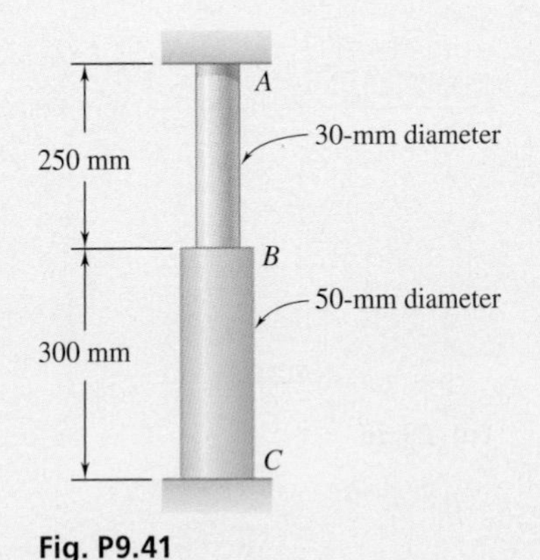

Fig. P9.41

9.41 A rod consisting of two cylindrical portions *AB* and *BC* is restrained at both ends. Portion *AB* is made of steel (E_s = 200 GPa, α_s = 11.7 × 10^{-6}/°C) and portion *BC* is made of brass (E_b = 105 GPa, α_b = 20.9 × 10^{-6}/°C). Knowing that the rod is initially unstressed, determine the compressive force induced in *ABC* when there is a temperature rise of 50°C.

9.42 A rod consisting of two cylindrical portions *AB* and *BC* is restrained at both ends. Portion *AB* is made of steel (E_s = 29 × 10^6 psi, α_s = 6.5 × 10^{-6}/°F) and portion *BC* is made of aluminum (E_a = 10.4 × 10^6 psi, α_a = 13.3 × 10^{-6}/°F). Knowing that the rod is initially unstressed, determine (*a*) the normal stresses induced in portions *AB* and *BC* by a temperature rise of 70°F, (*b*) the corresponding deflection of point *B*.

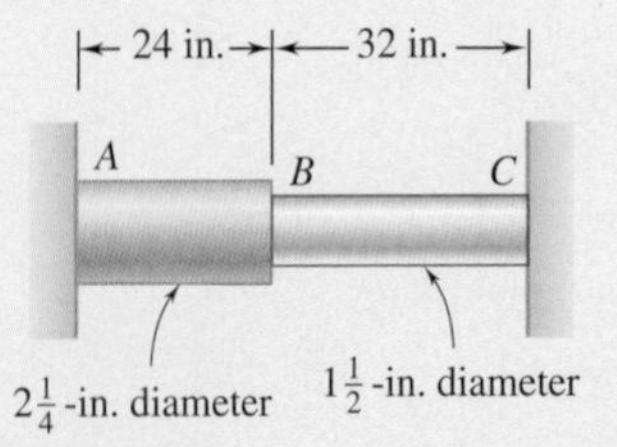

Fig. P9.42

9.43 Solve Prob. 9.42, assuming that portion *AB* of the composite rod is made of aluminum and portion *BC* is made of steel.

9.44 Determine (*a*) the compressive force in the bars shown after a temperature rise of 180°F, (*b*) the corresponding change in length of the bronze bar.

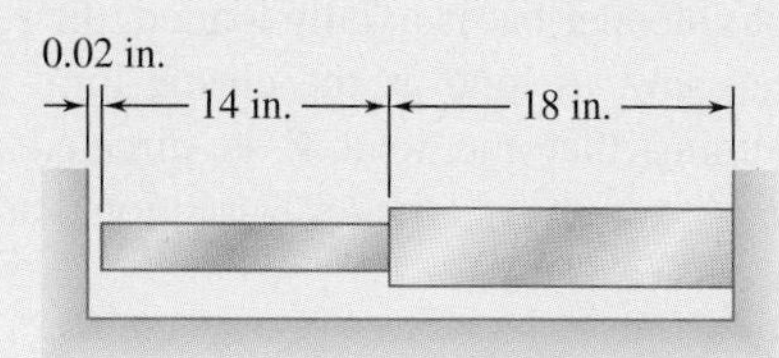

Bronze
$A = 2.4\ \text{in}^2$
$E = 15 \times 10^6$ psi
$\alpha = 12 \times 10^{-6}/°\text{F}$

Aluminum
$A = 2.8\ \text{in}^2$
$E = 10.6 \times 10^6$ psi
$\alpha = 12.9 \times 10^{-6}/°\text{F}$

Fig. P9.44 and P9.45

9.45 Knowing that a 0.02-in. gap exists when the temperature is 75°F, determine (*a*) the temperature at which the normal stress in the aluminum bar will be equal to –11 ksi, (*b*) the corresponding exact length of the aluminum bar.

9.46 At room temperature (20°C) a 0.5-mm gap exists between the ends of the rods shown. At a later time when the temperature has reached 140°C, determine (*a*) the normal stress in the aluminum rod, (*b*) the change in length of the aluminum rod.

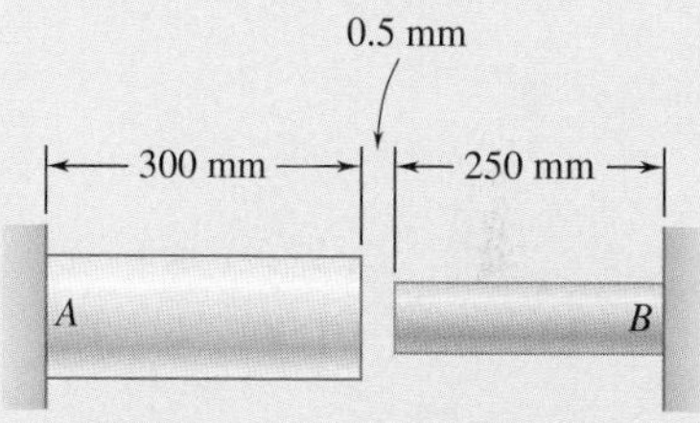

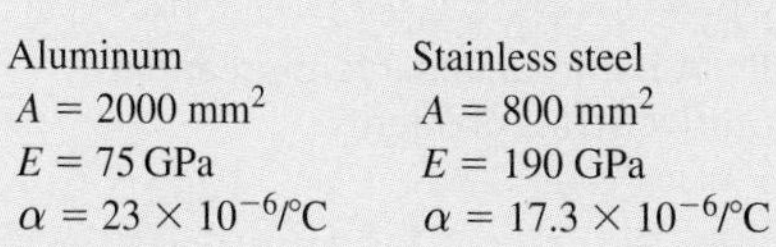

Fig. P9.46

9.47 A brass link ($E_b = 105$ GPa, $\alpha_b = 20.9 \times 10^{-6}/°\text{C}$) and a steel rod ($E_s = 200$ GPa, $\alpha_s = 11.7 \times 10^{-6}/°\text{C}$) have the dimensions shown at a temperature of 20°C. The steel rod is cooled until it fits freely into the link. The temperature of the whole assembly is then raised to 45°C. Determine (*a*) the final normal stress in the steel rod, (*b*) the final length of the steel rod.

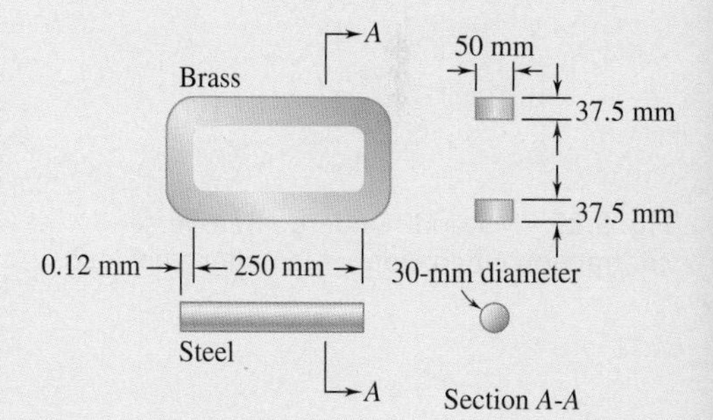

Fig. P9.47

9.48 Two steel bars ($E_s = 200$ GPa and $\alpha_s = 11.7 \times 10^{-6}/°\text{C}$) are used to reinforce a brass bar ($E_b = 105$ GPa, $\alpha_b = 20.9 \times 10^{-6}/°\text{C}$) that is subjected to a load $P = 25$ kN. When the steel bars were fabricated, the distance between the centers of the holes that were to fit on the pins was made 0.5 mm smaller than the 2 m needed. The steel bars were then placed in an oven to increase their length so that they would just fit on the pins. Following fabrication, the temperature in the steel bars dropped back to room temperature. Determine (*a*) the increase in temperature that was required to fit the steel bars on the pins, (*b*) the stress in the brass bar after the load is applied to it.

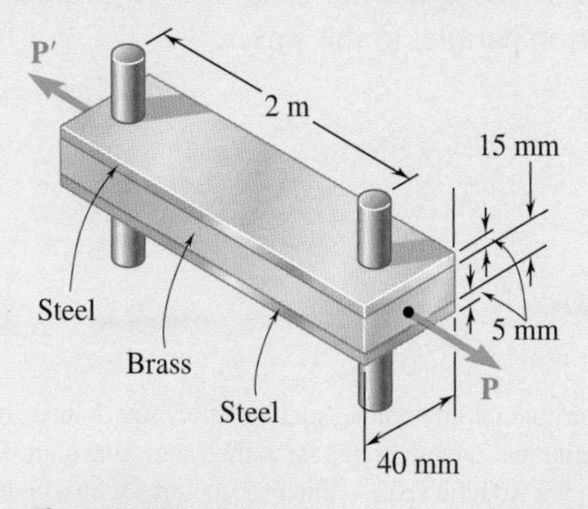

Fig. P9.48

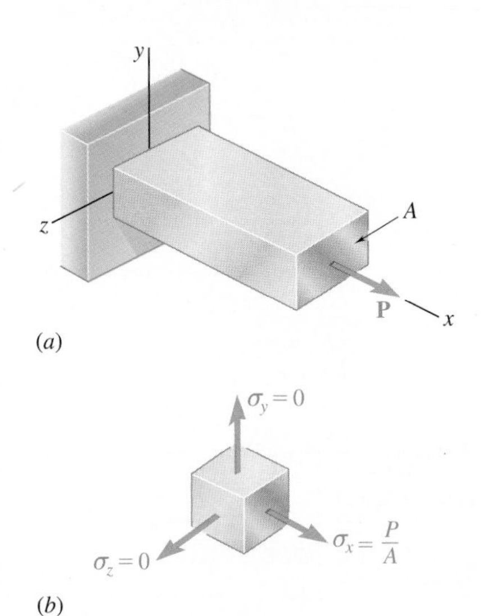

Fig. 9.28 A bar in uniaxial tension and a representative stress element.

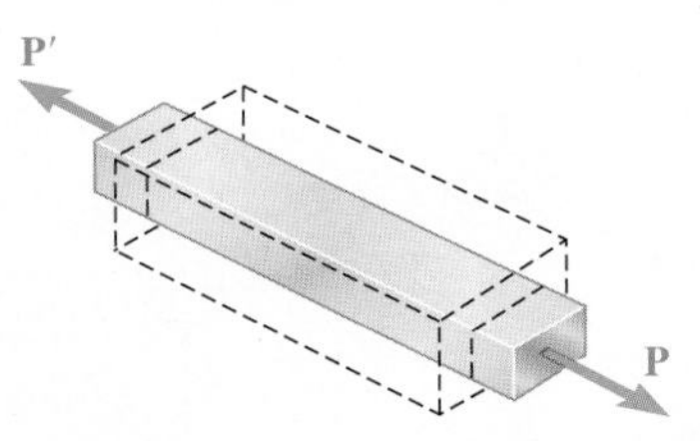

Fig. 9.29 Materials undergo transverse contraction when elongated under axial load.

9.4 POISSON'S RATIO

When a homogeneous slender bar is axially loaded, the resulting stress and strain satisfy Hooke's law, as long as the elastic limit of the material is not exceeded. Assuming that the load **P** is directed along the x axis (Fig. 9.28*a*), $\sigma_x = P/A$, where A is the cross-sectional area of the bar, and from Hooke's law,

$$\epsilon_x = \sigma_x/E \tag{9.15}$$

where E is the modulus of elasticity of the material.

Also, the normal stresses on faces perpendicular to the y and z axes are zero: $\sigma_y = \sigma_z = 0$ (Fig. 9.28*b*). It would be tempting to conclude that the corresponding strains ϵ_y and ϵ_z are also zero. This *is not the case*. In all engineering materials, the elongation produced by an axial tensile force **P** in the direction of the force is accompanied by a contraction in any transverse direction (Fig. 9.29). In this section and the following sections, all materials are assumed to be both *homogeneous* and *isotropic* (i.e., their mechanical properties are independent of both *position* and *direction*). It follows that the strain must have the same value for any transverse direction. Therefore, the loading shown in Fig. 9.28 must have $\epsilon_y = \epsilon_z$. This common value is the *lateral strain*. An important constant for a given material is its *Poisson's ratio*, named after the French mathematician Siméon Denis Poisson (1781–1840) and denoted by the Greek letter ν (nu).

$$\nu = -\frac{\text{lateral strain}}{\text{axial strain}} \tag{9.16}$$

or

$$\nu = -\frac{\epsilon_y}{\epsilon_x} = -\frac{\epsilon_z}{\epsilon_x} \tag{9.17}$$

for the loading condition represented in Fig. 9.28. Note the use of a minus sign in these equations to obtain a positive value for ν, as the axial and lateral strains have opposite signs for all engineering materials.‡ Solving Eq. (9.17) for ϵ_y and ϵ_z, and recalling Eq. (9.15), write the following relationships, which fully describe the condition of strain under an axial load applied in a direction parallel to the x axis:

$$\epsilon_x = \frac{\sigma_x}{E} \qquad \epsilon_y = \epsilon_z = -\frac{\nu\sigma_x}{E} \tag{9.18}$$

‡However, some experimental materials, such as polymer foams, expand laterally when stretched. Since the axial and lateral strains have then the same sign, Poisson's ratio of these materials is negative. (*See* Roderic Lakes, "Foam Structures with a Negative Poisson's Ratio," *Science*, 27 February 1987, Volume 235, pp. 1038–1040.)

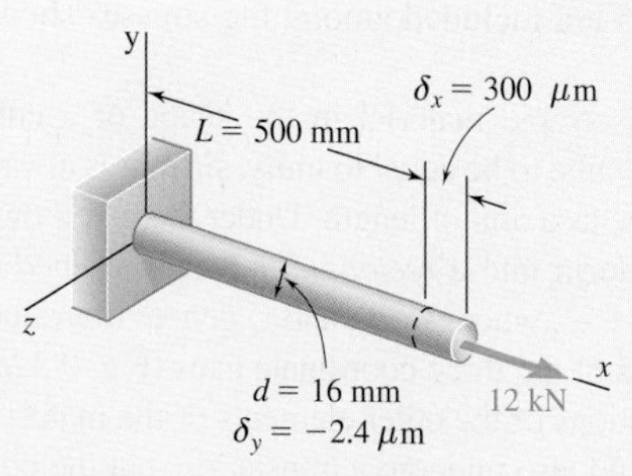

Fig. 9.30 Axially loaded rod.

Concept Application 9.7

A 500-mm-long, 16-mm-diameter rod made of a homogenous, isotropic material is observed to increase in length by 300 μm, and to decrease in diameter by 2.4 μm when subjected to an axial 12-kN load. Determine the modulus of elasticity and Poisson's ratio of the material.

The cross-sectional area of the rod is

$$A = \pi r^2 = \pi(8 \times 10^{-3}\ \text{m})^2 = 201 \times 10^{-6}\ \text{m}^2$$

Choosing the x axis along the axis of the rod (Fig. 9.30), write

$$\sigma_x = \frac{P}{A} = \frac{12 \times 10^3\ \text{N}}{201 \times 10^{-6}\ \text{m}^2} = 59.7\ \text{MPa}$$

$$\epsilon_x = \frac{\delta_x}{L} = \frac{300\ \mu\text{m}}{500\ \text{mm}} = 600 \times 10^{-6}$$

$$\epsilon_y = \frac{\delta_y}{d} = \frac{-2.4\ \mu\text{m}}{16\ \text{mm}} = -150 \times 10^{-6}$$

From Hooke's law, $\sigma_x = E\epsilon_x$,

$$E = \frac{\sigma_x}{\epsilon_x} = \frac{59.7\ \text{MPa}}{600 \times 10^{-6}} = 99.5\ \text{GPa}$$

and from Eq. (9.17),

$$\nu = -\frac{\epsilon_y}{\epsilon_x} = -\frac{-150 \times 10^{-6}}{600 \times 10^{-6}} = 0.25$$

9.5 MULTIAXIAL LOADING: GENERALIZED HOOKE'S LAW

All the examples considered so far in this chapter have dealt with slender members subjected to axial loads, i.e., to forces directed along a single axis. Consider now structural elements subjected to loads acting in the directions of the three coordinate axes and producing normal stresses σ_x, σ_y, and σ_z that are all different from zero (Fig. 9.31). This condition is a *multiaxial*

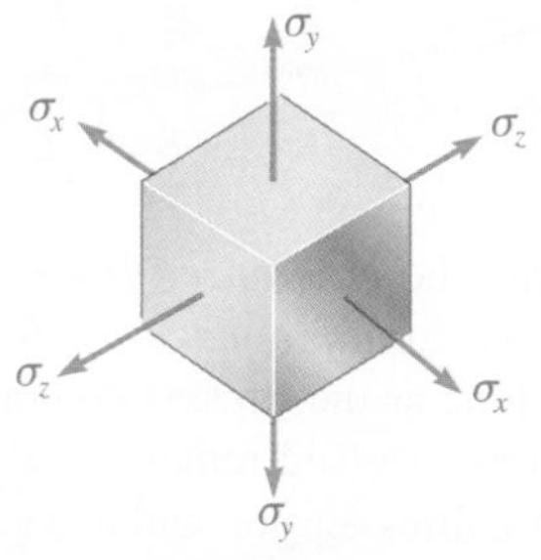

Fig. 9.31 State of stress for multiaxial loading.

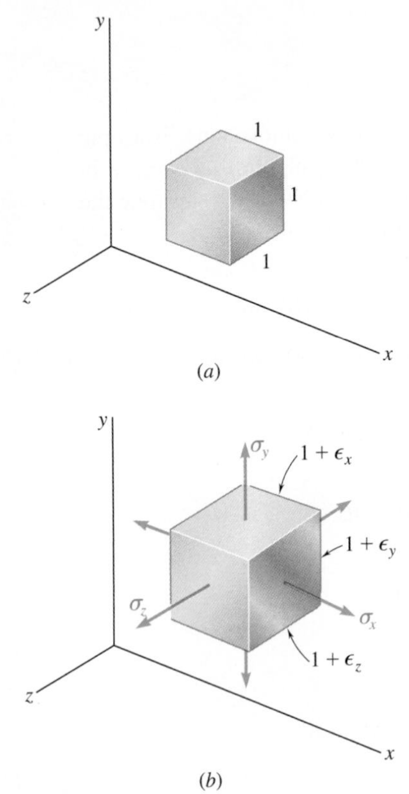

Fig. 9.32 Deformation of unit cube under multiaxial loading: (*a*) unloaded; (*b*) deformed.

loading. Note that this is not the general stress condition described in Sec. 8.2, since no shearing stresses are included among the stresses shown in Fig. 9.31.

Consider an element of an isotropic material in the shape of a cube (Fig. 9.32*a*). Assume the side of the cube to be equal to unity, since it is always possible to select the side of the cube as a unit of length. Under the given multiaxial loading, the element will deform into a *rectangular parallelepiped* of sides equal to $1 + \epsilon_x$, $1 + \epsilon_y$, and $1 + \epsilon_z$, where ϵ_x, ϵ_y, and ϵ_z denote the values of the normal strain in the directions of the three coordinate axes (Fig. 9.32*b*). Note that, as a result of the deformations of the other elements of the material, the element under consideration could also undergo a translation, but the concern here is with the *actual deformation* of the element, not with any possible superimposed rigid-body displacement.

In order to express the strain components ϵ_x, ϵ_y, ϵ_z in terms of the stress components σ_x, σ_y, σ_z, consider the effect of each stress component and combine the results. This approach will be used repeatedly in this text, and is based on the *principle of superposition*. This principle states that the effect of a given combined loading on a structure can be obtained by *determining the effects of the various loads separately and combining the results*, provided that the following conditions are satisfied:

1. Each effect is linearly related to the load that produces it.
2. The deformation resulting from any given load is small and does not affect the conditions of application of the other loads.

For multiaxial loading, the first condition is satisfied if the stresses do not exceed the proportional limit of the material, and the second condition is also satisfied if the stress on any given face does not cause deformations of the other faces that are large enough to affect the computation of the stresses on those faces.

Considering the effect of the stress component σ_x, recall from Sec. 9.4 that σ_x causes a strain equal to σ_x/E in the x direction and strains equal to $-\nu\sigma_x/E$ in each of the y and z directions. Similarly, the stress component σ_y, if applied separately, will cause a strain σ_y/E in the y direction and strains $-\nu\sigma_y/E$ in the other two directions. Finally, the stress component σ_z causes a strain σ_z/E in the z direction and strains $-\nu\sigma_z/E$ in the x and y directions. Combining the results, the components of strain corresponding to the given multiaxial loading are

$$
\begin{aligned}
\epsilon_x &= +\frac{\sigma_x}{E} - \frac{\nu\sigma_y}{E} - \frac{\nu\sigma_z}{E} \\
\epsilon_y &= -\frac{\nu\sigma_x}{E} + \frac{\sigma_y}{E} - \frac{\nu\sigma_z}{E} \\
\epsilon_z &= -\frac{\nu\sigma_x}{E} - \frac{\nu\sigma_y}{E} + \frac{\sigma_z}{E}
\end{aligned}
\qquad (9.19)
$$

Equations (9.19) are the *generalized Hooke's law for the multiaxial loading of a homogeneous isotropic material*. As indicated earlier, these results are valid only as long as the stresses do not exceed the proportional limit and the deformations involved remain small. Also, a positive value for a stress component signifies tension and a negative value compression. Similarly, a positive value for a strain component indicates expansion in the corresponding direction and a negative value contraction.

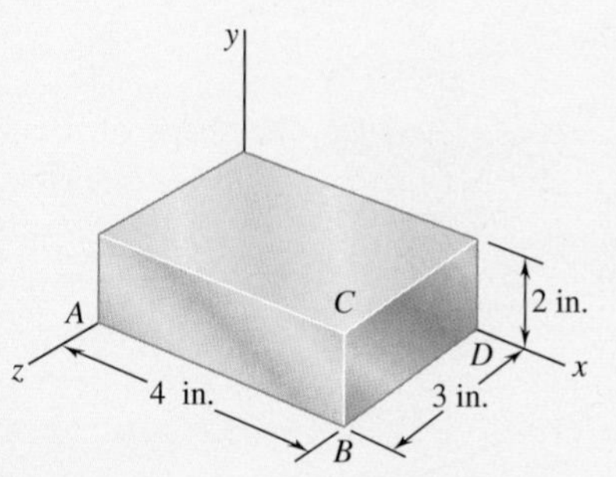

Fig. 9.33 Steel block under uniform pressure p.

Concept Application 9.8

The steel block shown (Fig. 9.33) is subjected to a uniform pressure on all its faces. Knowing that the change in length of edge AB is -1.2×10^{-3} in., determine (*a*) the change in length of the other two edges and (*b*) the pressure p applied to the faces of the block. Assume $E = 29 \times 10^6$ psi and $\nu = 0.29$.

a. Change in Length of Other Edges. Substituting $\sigma_x = \sigma_y = \sigma_z = -p$ into Eqs. (9.19), the three strain components have the common value

$$\epsilon_x = \epsilon_y = \epsilon_z = -\frac{p}{E}(1 - 2\nu) \tag{1}$$

Since

$$\epsilon_x = \delta_x/AB = (-1.2 \times 10^{-3}\text{ in.})/(4\text{ in.})$$
$$= -300 \times 10^{-6}\text{ in./in.}$$

obtain

$$\epsilon_y = \epsilon_z = \epsilon_x = -300 \times 10^{-6}\text{ in./in.}$$

from which

$$\delta_y = \epsilon_y(BC) = (-300 \times 10^{-6})(2\text{ in.}) = -600 \times 10^{-6}\text{ in.}$$
$$\delta_z = \epsilon_z(BD) = (-300 \times 10^{-6})(3\text{ in.}) = -900 \times 10^{-6}\text{ in.}$$

b. Pressure. Solving Eq. (1) for p,

$$p = -\frac{E\epsilon_x}{1 - 2\nu} = -\frac{(29 \times 10^6\text{ psi})(-300 \times 10^{-6})}{1 - 0.58}$$
$$p = 20.7\text{ ksi}$$

9.6 SHEARING STRAIN

When we derived in Sec. 9.5 the relations (9.19) between normal stresses and normal strains in a homogeneous isotropic material, we assumed that no shearing stresses were involved. In the more general stress situation represented in Fig. 9.34, shearing stresses τ_{xy}, τ_{yz}, and τ_{zx} are present (as well as the corresponding shearing stresses τ_{yx}, τ_{zy}, and τ_{xz}). These stresses have no direct effect on the normal strains and, as long as all the deformations involved remain small, they will not affect the derivation nor the validity of Eqs. (9.19). The shearing stresses, however, tend to deform a cubic element of material into an *oblique* parallelepiped.

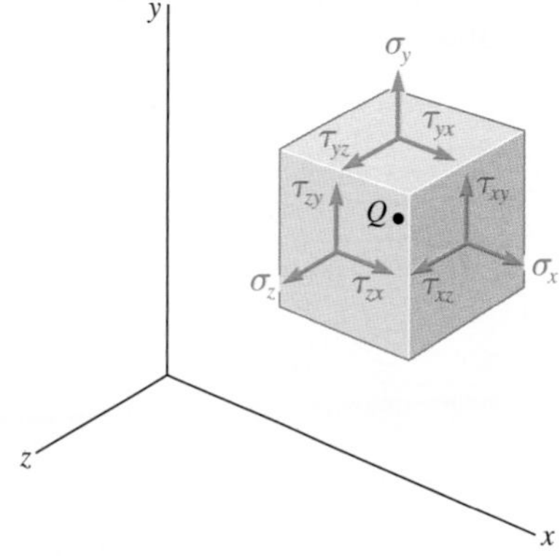

Fig. 9.34 Positive stress components at point Q for a general state of stress.

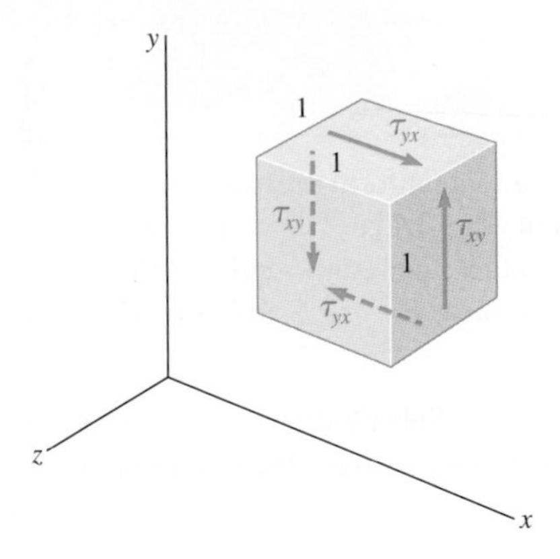

Fig. 9.35 Unit cubic element subjected to shearing stress.

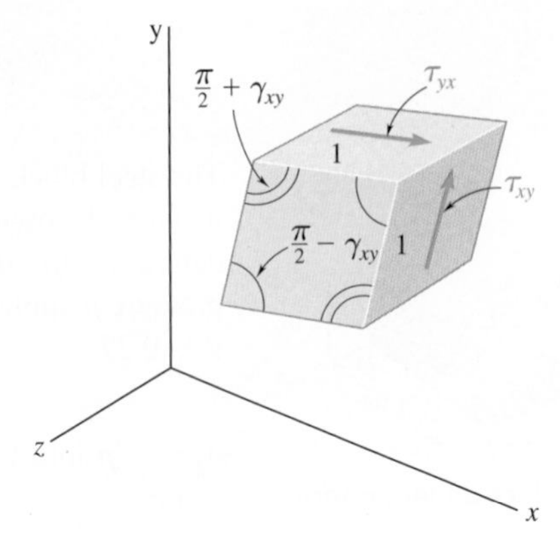

Fig. 9.36 Deformation of unit cubic element due to shearing stress.

Consider a cubic element (Fig. 9.35) subjected to only the shearing stresses τ_{xy} and τ_{yx} applied to faces of the element respectively perpendicular to the x and y axes. (Recall from Sec. 8.3 that $\tau_{xy} = \tau_{yx}$.) The cube is observed to deform into a rhomboid of sides equal to one (Fig. 9.36). Two of the angles formed by the four faces under stress are reduced from $\frac{\pi}{2}$ to $\frac{\pi}{2} - \gamma_{xy}$, while the other two are increased from $\frac{\pi}{2}$ to $\frac{\pi}{2} + \gamma_{xy}$. The small angle γ_{xy} (expressed in radians) defines the *shearing strain* corresponding to the x and y directions. When the deformation involves a *reduction* of the angle formed by the two faces oriented toward the positive x and y axes (as shown in Fig. 9.36), the shearing strain γ_{xy} is *positive*; otherwise, it is negative.

As a result of the deformations of the other elements of the mate rial, the element under consideration also undergoes an overall rotation. The concern here is with the *actual deformation* of the element, not with any possible superimposed rigid-body displacement.[†]

Plotting successive values of τ_{xy} against the corresponding values of γ_{xy}, the shearing stress-strain diagram is obtained for the material. (This can be accomplished by carrying out a torsion test, as you will see in Chap. 10.) This diagram is similar to the normal stress-strain diagram from the tensile test described earlier; however, the values for the yield strength, ultimate strength, etc., are about half as large in shear as they are in tension. As for normal stresses and strains, the initial portion of the shearing stress-strain diagram is a straight line. For values of the shearing stress that do not exceed the proportional limit in shear, it can be written for any homogeneous isotropic material that

$$\tau_{xy} = G\gamma_{xy} \tag{9.20}$$

This relationship is *Hooke's law for shearing stress and strain*, and the constant G is called the *modulus of rigidity* or *shear modulus* of the material.

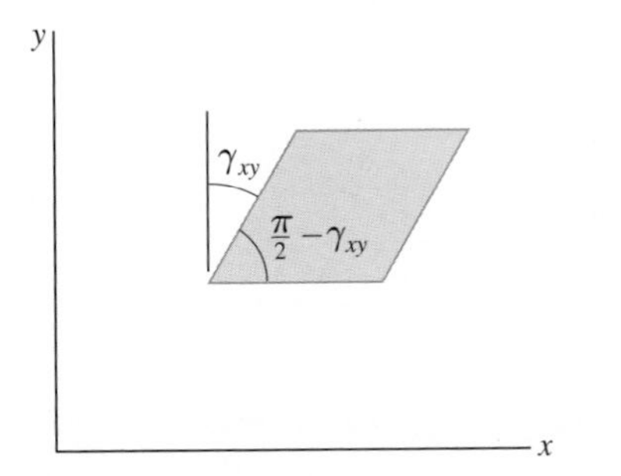

Fig. 9.37 Cubic element as viewed in *xy*-plane after rigid rotation.

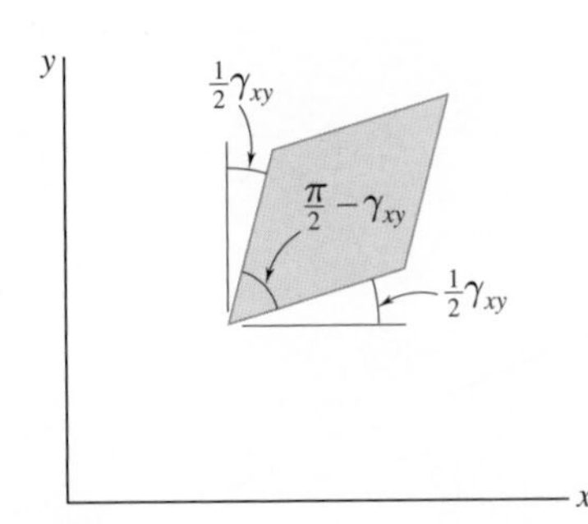

Fig. 9.38 Cubic element as viewed in *xy*-plane with equal rotation of *x* and *y* faces.

[†] In defining the strain γ_{xy}, some authors arbitrarily assume that the actual deformation of the element is accompanied by a rigid-body rotation where the horizontal faces of the element do not rotate. The strain γ_{xy} is then represented by the angle through which the other two faces have rotated (Fig. 9.37). Others assume a rigid-body rotates where the horizontal faces rotate through $\frac{1}{2}\gamma_{xy}$ counterclockwise and the vertical faces through $\frac{1}{2}\gamma_{xy}$ clockwise (Fig. 9.38). Since both assumptions are unnecessary and may lead to confusion, in this text you will associate the shearing strain γ_{xy} with the *change in the angle* formed by the two faces, rather than with the *rotation of a given face* under restrictive conditions.

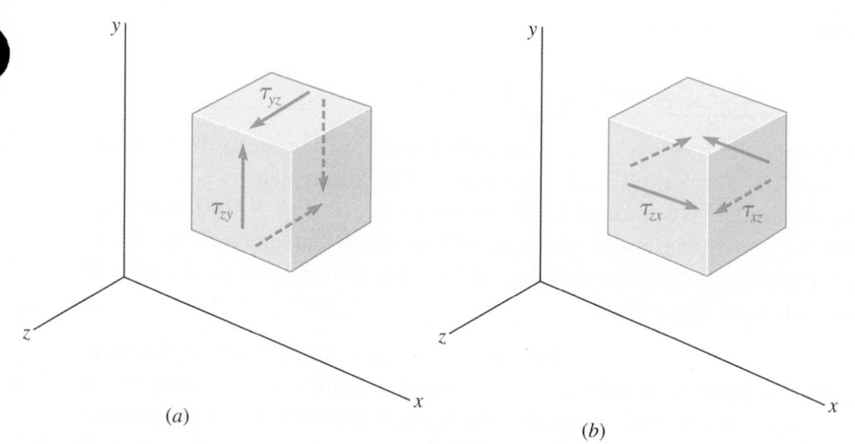

Fig. 9.39 States of pure shear in: (*a*) *yz*-plane; (*b*) *xz*-plane.

Since the strain γ_{xy} is defined as an angle in radians, it is dimensionless, and the modulus G is expressed in the same units as τ_{xy} in pascals or in psi. The modulus of rigidity G of any given material is less than one-half, but more than one-third of the modulus of elasticity E of that material.

Now consider a small element of material subjected to shearing stresses τ_{yz} and τ_{zy} (Fig. 9.39*a*), where the shearing strain γ_{yz} is the change in the angle formed by the faces under stress. The shearing strain γ_{zx} is found in a similar way by considering an element subjected to shearing stresses τ_{zx} and τ_{xz} (Fig. 9.39*b*). For values of the stress that do not exceed the proportional limit, you can write two additional relationships:

$$\tau_{yz} = G\gamma_{yz} \qquad \tau_{zx} = G\gamma_{zx} \tag{9.21}$$

where the constant G is the same as in Eq. (9.20).

For the general stress condition represented in Fig. 9.34, and as long as none of the stresses involved exceeds the corresponding proportional limit, you can apply the principle of superposition and combine the results. The generalized Hooke's law for a homogeneous isotropic material under the most general stress condition is

$$\begin{aligned}
\epsilon_x &= +\frac{\sigma_x}{E} - \frac{\nu\sigma_y}{E} - \frac{\nu\sigma_z}{E} \\
\epsilon_y &= -\frac{\nu\sigma_x}{E} + \frac{\sigma_y}{E} - \frac{\nu\sigma_z}{E} \\
\epsilon_z &= -\frac{\nu\sigma_x}{E} - \frac{\nu\sigma_y}{E} + \frac{\sigma_z}{E} \\
\gamma_{xy} = \frac{\tau_{xy}}{G} \qquad \gamma_{yz} &= \frac{\tau_{yz}}{G} \qquad \gamma_{zx} = \frac{\tau_{zx}}{G}
\end{aligned} \tag{9.22}$$

An examination of Eqs. (9.22) leads us to three distinct constants, E, ν, and G, which are used to predict the deformations caused in a given material by an arbitrary combination of stresses. Only two of these constants need be determined experimentally for any given material. The next section explains that the third constant can be obtained through a very simple computation.

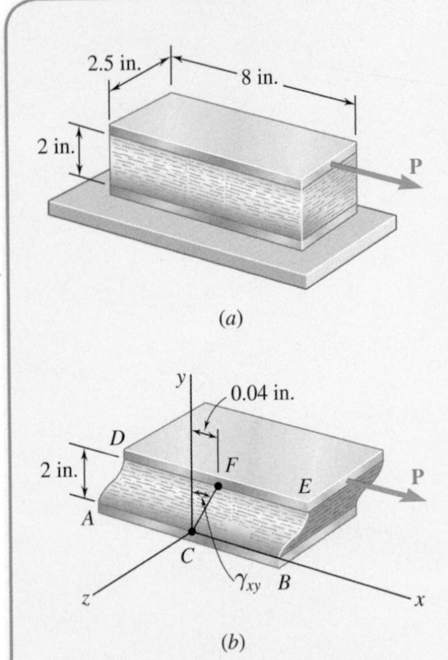

Fig. 9.40 (a) Rectangular block loaded in shear. (b) Deformed block showing the shearing strain.

Concept Application 9.9

A rectangular block of a material with a modulus of rigidity $G = 90$ ksi is bonded to two rigid horizontal plates. The lower plate is fixed, while the upper plate is subjected to a horizontal force **P** (Fig. 9.40*a*). Knowing that the upper plate moves through 0.04 in. under the action of the force, determine (*a*) the average shearing strain in the material and (*b*) the force **P** exerted on the upper plate.

a. Shearing Strain. The coordinate axes are centered at the midpoint *C* of edge *AB* and directed as shown (Fig. 9.40*b*). The shearing strain γ_{xy} is equal to the angle formed by the vertical and the line *CF* joining the midpoints of edges *AB* and *DE*. Noting that this is a very small angle and recalling that it should be expressed in radians, write

$$\gamma_{xy} \approx \tan \gamma_{xy} = \frac{0.04 \text{ in.}}{2 \text{ in.}} \qquad \gamma_{xy} = 0.020 \text{ rad}$$

b. Force Exerted on Upper Plate. Determine the shearing stress τ_{xy} in the material. Using Hooke's law for shearing stress and strain,

$$\tau_{xy} = G\gamma_{xy} = (90 \times 10^3 \text{ psi})(0.020 \text{ rad}) = 1800 \text{ psi}$$

The force exerted on the upper plate is

$$P = \tau_{xy} A = (1800 \text{ psi})(8 \text{ in.})(2.5 \text{ in.}) = 36.0 \times 10^3 \text{ lb}$$

$$P = 36.0 \text{ kips}$$

*9.7 DEFORMATIONS UNDER AXIAL LOADING–RELATION BETWEEN E, ν, AND G

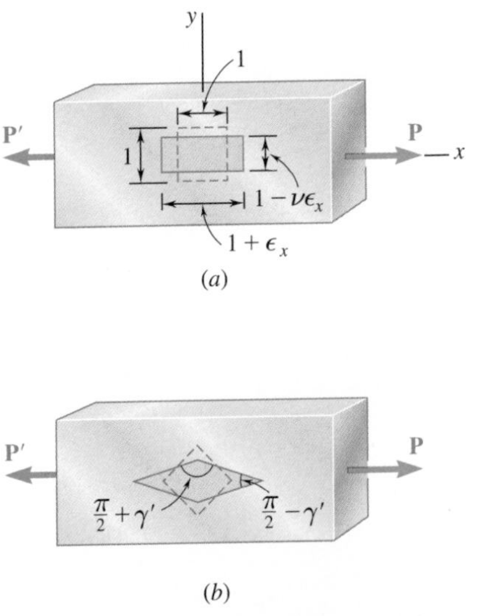

Section 9.4 showed that a slender bar subjected to an axial tensile load **P** directed along the *x* axis will elongate in the *x* direction and contract in both of the transverse *y* and *z* directions. If ϵ_x denotes the axial strain, the lateral strain is expressed as $\epsilon_y = \epsilon_z = -\nu\epsilon_x$, where ν is Poisson's ratio. Thus, an element in the shape of a cube of side equal to one and oriented as shown in Fig. 9.41*a* will deform into a rectangular parallelepiped of sides $1 + \epsilon_x$, $1 - \nu\epsilon_x$, and $1 - \nu\epsilon_x$. (Note that only one face of the element is shown in the figure.) On the other hand, if the element is oriented at 45° to the axis of the load (Fig. 9.41*b*), the face shown deforms into a rhombus. Therefore, the axial load **P** causes a shearing strain γ' equal to the amount by which each of the angles shown in Fig. 9.41*b* increases or decreases.

The fact that shearing strains, as well as normal strains, result from an axial loading is not a surprise, since it was observed at the end of Sec. 8.3 that an axial load **P** causes normal and shearing stresses of equal magnitude on four of the faces of an element oriented at 45° to the axis of the member. This was illustrated in Fig. 8.37, which has been repeated here. It was also shown in Sec. 8.2 that the shearing stress is maximum on a plane

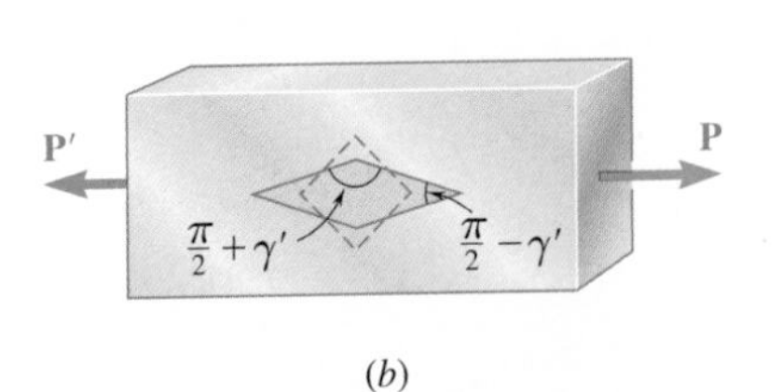

Fig. 9.41 Representations of strain in an axially-loaded bar: (a) cubic strain element faces aligned with coordinate axes; (b) cubic strain element faces rotated 45° about z-axis.

forming an angle of 45° with the axis of the load. It follows from Hooke's law for shearing stress and strain that the shearing strain γ' associated with the element of Fig. 9.41*b* is also maximum: $\gamma' = \gamma_m$.

We will now derive a relationship between the maximum shearing strain $\gamma' = \gamma_m$ associated with the element of Fig. 9.41*b* and the normal strain ϵ_x in the direction of the load. Consider the prismatic element obtained by intersecting the cubic element of Fig. 9.41*a* by a diagonal plane (Fig. 9.42*a* and *b*). Referring to Fig. 9.41*a*, this new element will deform into that shown in Fig. 9.42*c*, which has horizontal and vertical sides equal to $1 + \epsilon_x$ and $1 - \nu\epsilon_x$. But the angle formed by the oblique and horizontal faces of Fig. 9.42*b* is precisely half of one of the right angles of the cubic element in Fig. 9.41*b*. The angle β into which this angle deforms must be equal to half of $\pi/2 - \gamma_m$. Therefore,

$$\beta = \frac{\pi}{4} - \frac{\gamma_m}{2}$$

Applying the formula for the tangent of the difference of two angles,

$$\tan\beta = \frac{\tan\frac{\pi}{4} - \tan\frac{\gamma_m}{2}}{1 + \tan\frac{\pi}{4}\tan\frac{\gamma_m}{2}} = \frac{1 - \tan\frac{\gamma_m}{2}}{1 + \tan\frac{\gamma_m}{2}}$$

or since $\gamma_m/2$ is a very small angle,

$$\tan\beta = \frac{1 - \frac{\gamma_m}{2}}{1 + \frac{\gamma_m}{2}} \tag{9.23}$$

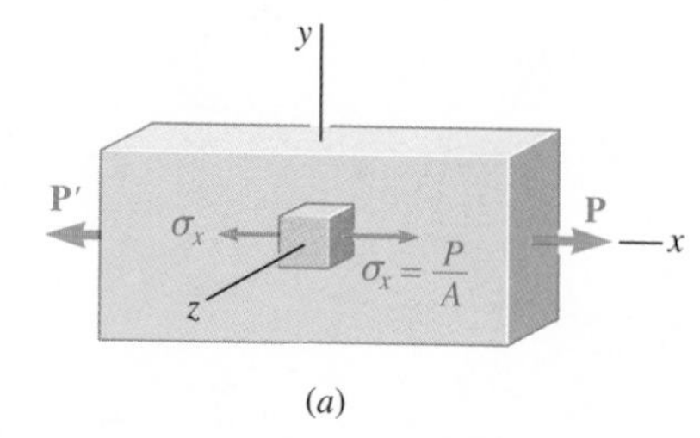

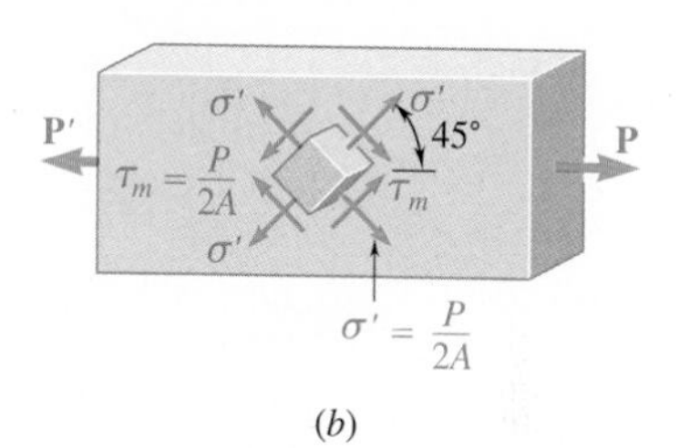

Fig. 8.37 (repeated)

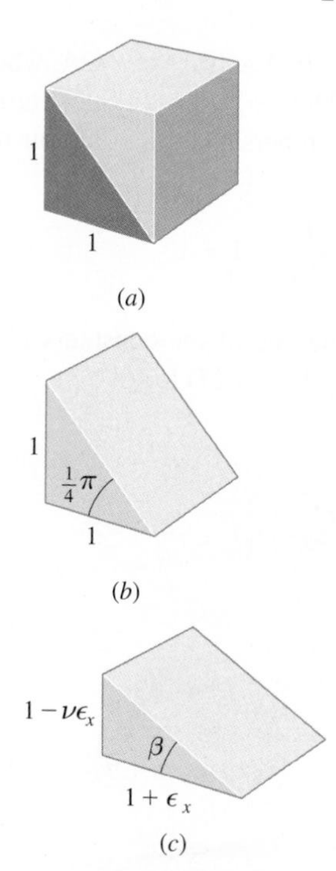

Fig. 9.42 (*a*) Cubic strain unit element, to be sectioned on a diagonal plane. (*b*) Undeformed section of unit element. (*c*) Deformed section of unit element.

From Fig. 9.42*c*, observe that

$$\tan \beta = \frac{1 - \nu\epsilon_x}{1 + \epsilon_x} \tag{9.24}$$

Equating the right-hand members of Eqs. (9.23) and (9.24) and solving for γ_m, results in

$$\gamma_m = \frac{(1 + \nu)\epsilon_x}{1 + \dfrac{1 - \nu}{2}\epsilon_x}$$

Since $\epsilon_x \ll 1$, the denominator in the expression obtained can be assumed equal to one. Therefore,

$$\gamma_m = (1 + \nu)\epsilon_x \tag{9.25}$$

which is the desired relation between the maximum shearing strain γ_m and the axial strain ϵ_x.

To obtain a relation among the constants E, ν, and G, we recall that, by Hooke's law, $\gamma_m = \tau_m/G$, and for an axial loading, $\epsilon_x = \sigma_x/E$. Equation (9.25) can be written as

$$\frac{\tau_m}{G} = (1 + \nu)\frac{\sigma_x}{E}$$

or

$$\frac{E}{G} = (1 + \nu)\frac{\sigma_x}{\tau_m} \tag{9.26}$$

Recall from Fig. 8.37 that $\sigma_x = P/A$ and $\tau_m = P/2A$, where A is the cross-sectional area of the member. Thus, $\sigma_x/\tau_m = 2$. Substituting this value into Eq. (9.26) and dividing both members by 2, the relationship is

$$\frac{E}{2G} = 1 + \nu \tag{9.27}$$

which can be used to determine one of the constants E, ν, or G from the other two. For example, solving Eq. (9.27) for G,

$$G = \frac{E}{2(1 + \nu)} \tag{9.28}$$

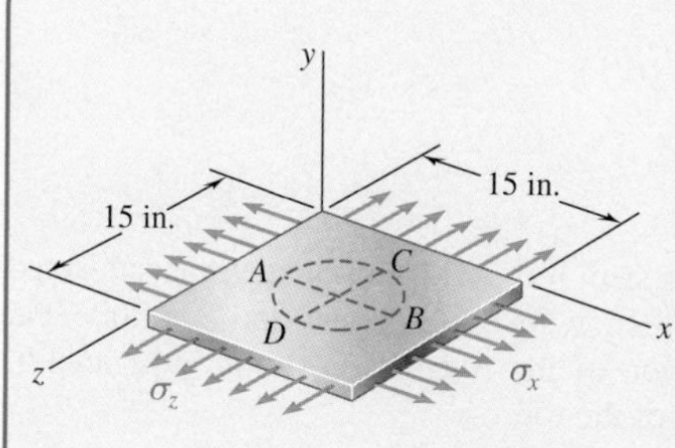

Sample Problem 9.5

A circle of diameter $d = 9$ in. is scribed on an unstressed aluminum plate of thickness $t = \frac{3}{4}$ in. Forces acting in the plane of the plate later cause normal stresses $\sigma_x = 12$ ksi and $\sigma_z = 20$ ksi. For $E = 10 \times 10^6$ psi and $\nu = \frac{1}{3}$, determine the change in (*a*) the length of diameter *AB*, (*b*) the length of diameter *CD*, (*c*) the thickness of the plate.

STRATEGY: You can use the generalized Hooke's Law to determine the components of strain. These strains can then be used to evaluate the various dimensional changes to the plate.

ANALYSIS:

Hooke's Law. Note that $\sigma_y = 0$. Using Eqs. (9.19), find the strain in each of the coordinate directions.

$$\epsilon_x = +\frac{\sigma_x}{E} - \frac{\nu\sigma_y}{E} - \frac{\nu\sigma_z}{E}$$

$$= \frac{1}{10 \times 10^6 \text{ psi}}\left[(12 \text{ ksi}) - 0 - \frac{1}{3}(20 \text{ ksi})\right] = +0.533 \times 10^{-3} \text{ in./in.}$$

$$\epsilon_y = -\frac{\nu\sigma_x}{E} + \frac{\sigma_y}{E} - \frac{\nu\sigma_z}{E}$$

$$= \frac{1}{10 \times 10^6 \text{ psi}}\left[-\frac{1}{3}(12 \text{ ksi}) + 0 - \frac{1}{3}(20 \text{ ksi})\right] = -1.067 \times 10^{-3} \text{ in./in.}$$

$$\epsilon_z = -\frac{\nu\sigma_x}{E} - \frac{\nu\sigma_y}{E} + \frac{\sigma_z}{E}$$

$$= \frac{1}{10 \times 10^6 \text{ psi}}\left[-\frac{1}{3}(12 \text{ ksi}) - 0 + (20 \text{ ksi})\right] = +1.600 \times 10^{-3} \text{ in./in.}$$

a. Diameter *AB*. The change in length is $\delta_{B/A} = \epsilon_x d$.

$$\delta_{B/A} = \epsilon_x d = (+0.533 \times 10^{-3} \text{ in./in.})(9 \text{ in.})$$

$$\delta_{B/A} = +4.8 \times 10^{-3} \text{ in.} \blacktriangleleft$$

b. Diameter *CD*.

$$\delta_{C/D} = \epsilon_z d = (+1.600 \times 10^{-3} \text{ in./in.})(9 \text{ in.})$$

$$\delta_{C/D} = +14.4 \times 10^{-3} \text{ in.} \blacktriangleleft$$

c. Thickness. Recalling that $t = \frac{3}{4}$ in.,

$$\delta_t = \epsilon_y t = (-1.067 \times 10^{-3} \text{ in./in.})(\tfrac{3}{4} \text{ in.})$$

$$\delta_t = -0.800 \times 10^{-3} \text{ in.} \blacktriangleleft$$

Problems

9.49 In a standard tensile test, a steel rod of 22-mm diameter is subjected to a tension force of 75 kN. Knowing that $\nu = 0.30$ and $E = 200$ GPa, determine (*a*) the elongation of the rod in a 200-mm gage length, (*b*) the change in diameter of the rod.

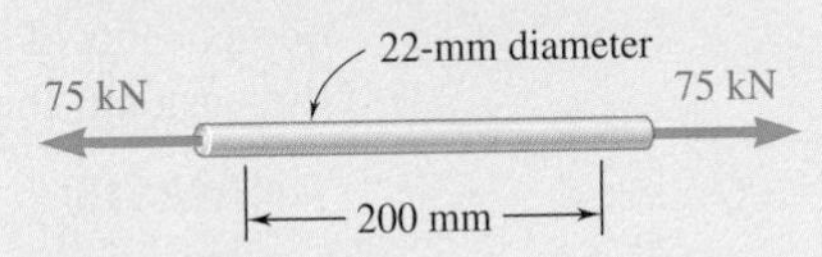

Fig. P9.49

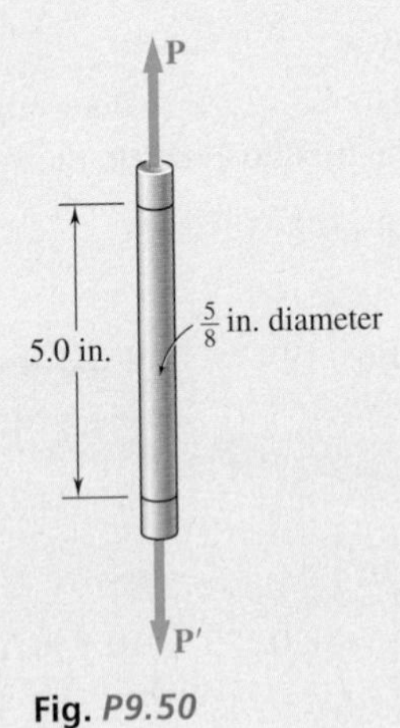

Fig. *P9.50*

9.50 A standard tension test is used to determine the properties of an experimental plastic. The test specimen is a $\frac{5}{8}$-in.-diameter rod and it is subjected to an 800-lb tensile force. Knowing that an elongation of 0.45 in. and a decrease in diameter of 0.025 in. are observed in a 5-in. gage length, determine the modulus of elasticity, the modulus of rigidity, and Poisson's ratio for the material.

9.51 A 2-m length of an aluminum pipe of 240-mm outer diameter and 10-mm wall thickness is used as a short column to carry a 640-kN centric axial load. Knowing that $E = 73$ GPa and $\nu = 0.33$, determine (*a*) the change in length of the pipe, (*b*) the change in its outer diameter, (*c*) the change in its wall thickness.

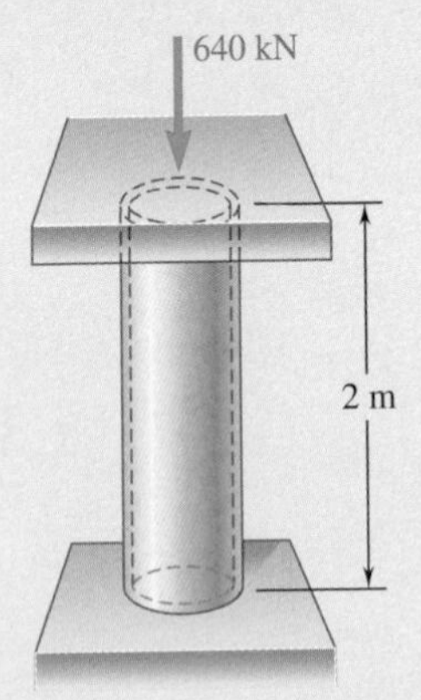

Fig. *P9.51*

9.52 The change in diameter of a large steel bolt is carefully measured as the nut is tightened. Knowing that $E = 29 \times 10^6$ psi and $\nu = 0.30$, determine the internal force in the bolt if the diameter is observed to decrease by 0.5×10^{-3} in.

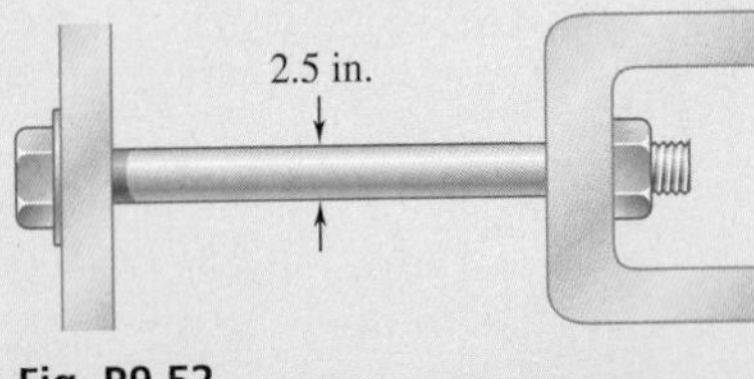

Fig. P9.52

9.53 An aluminum plate ($E = 74$ GPa, $\nu = 0.33$) is subjected to a centric axial load that causes a normal stress σ. Knowing that, before loading, a line of slope 2:1 is scribed on the plate, determine the slope of the line when $\sigma = 125$ MPa.

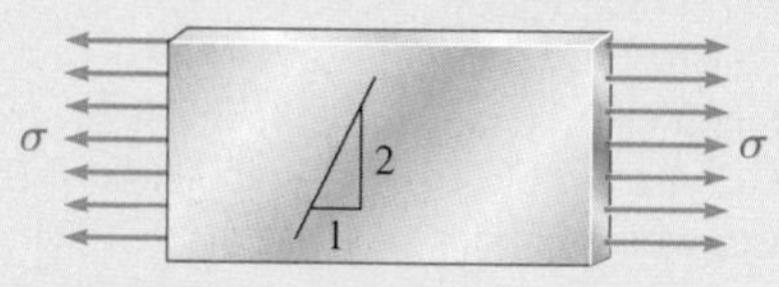

Fig. P9.53

9.54 A 2.75-kN tensile load is applied to a test coupon made from 1.6-mm flat steel plate ($E = 200$ GPa, $\nu = 0.30$). Determine the resulting change (*a*) in the 50-mm gage length, (*b*) in the width of portion *AB* of the test coupon, (*c*) in the thickness of portion *AB*, (*d*) in the cross-sectional area of portion *AB*.

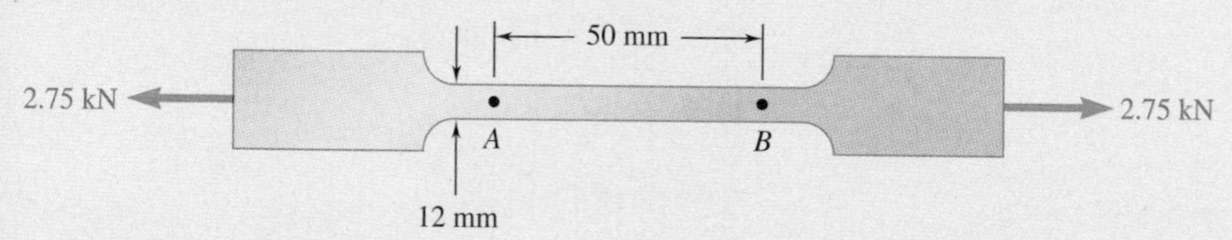

Fig. P9.54

9.55 The aluminum rod *AD* is fitted with a jacket that is used to apply a hydrostatic pressure of 6000 psi to the 12-in. portion *BC* of the rod. Knowing that $E = 10.1 \times 10^6$ psi and $\nu = 0.36$, determine (*a*) the change in the total length *AD*, (*b*) the change in diameter at the middle of the rod.

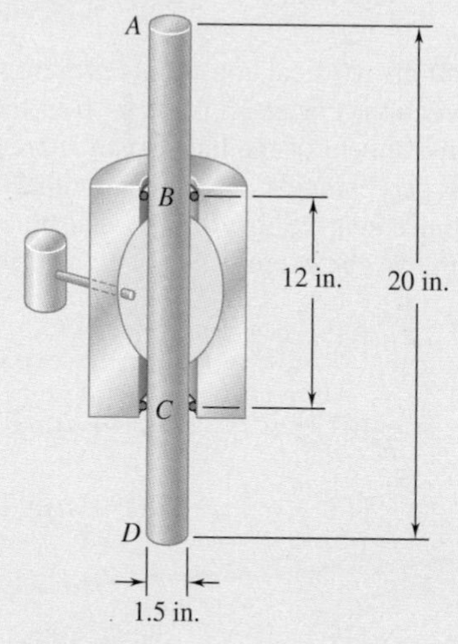

Fig. P9.55

9.56 For the rod of Prob. 9.55, determine the forces that should be applied to the ends *A* and *D* of the rod (*a*) if the axial strain in portion *BC* of the rod is to remain zero as the hydrostatic pressure is applied, (*b*) if the total length *AD* of the rod is to remain unchanged.

9.57 A 30-mm square has been scribed on the side of a large steel pressure vessel. After pressurization, the biaxial stress condition at the square is as shown. Knowing that $E = 200$ GPa and $\nu = 0.30$, determine the change in length of (*a*) side *AB*, (*b*) side *BC*, (*c*) diagonal *AC*.

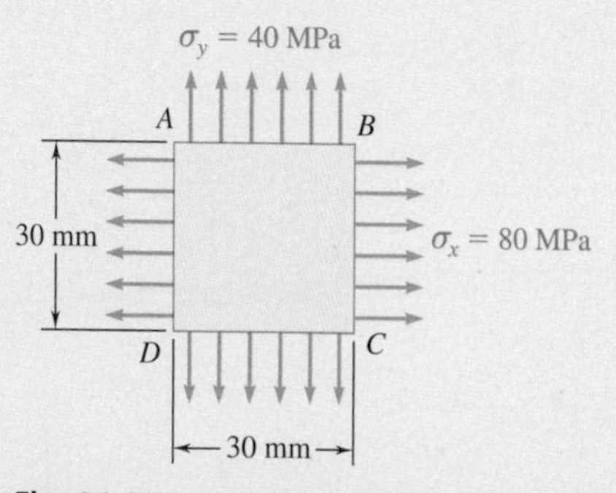

Fig. P9.57

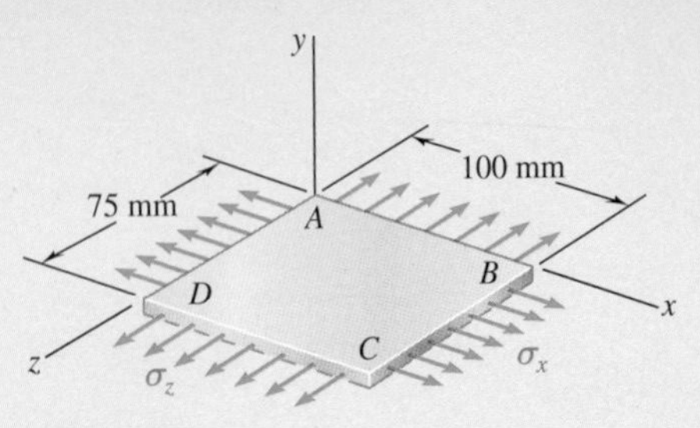

Fig. P9.58

9.58 A fabric used in air-inflated structures is subjected to a biaxial loading that results in normal stresses $\sigma_x = 120$ MPa and $\sigma_z = 160$ MPa. Knowing that the properties of the fabric can be approximated as $E = 87$ GPa and $\nu = 0.34$, determine the change in length of (*a*) side *AB*, (*b*) side *BC*, (*c*) diagonal *AC*.

9.59 In many situations it is known that the normal stress in a given direction is zero. For example, $\sigma_z = 0$ in the case of the thin plate shown. For this case, which is known as *plane stress*, show that if the strains ϵ_x and ϵ_y have been determined experimentally, we can express σ_x, σ_y and ϵ_z as follows:

$$\sigma_x = E\frac{\epsilon_x + \nu\epsilon_y}{1 - \nu^2}$$

$$\sigma_y = E\frac{\epsilon_y + \nu\epsilon_x}{1 - \nu^2}$$

$$\epsilon_z = -\frac{\nu}{1 - \nu}(\epsilon_x + \epsilon_y)$$

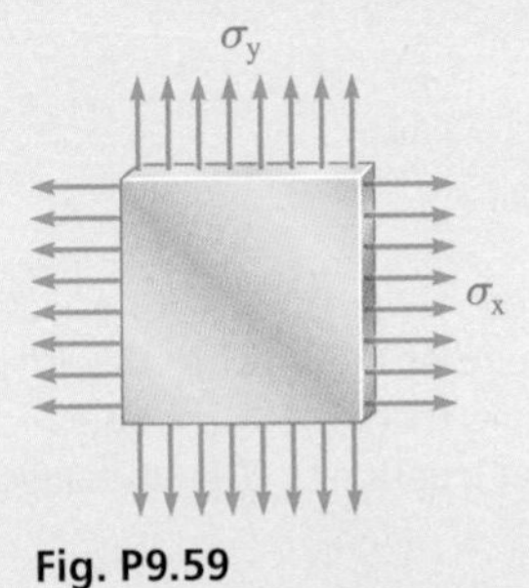

Fig. P9.59

9.60 In many situations physical constraints prevent strain from occurring in a given direction. For example, $\epsilon_z = 0$ in the case shown, where longitudinal movement of the long prism is prevented at every point. Plane sections perpendicular to the longitudinal axis remain plane and the same distance apart. Show that for this situation, which is known as *plane strain*, we can express σ_z, ϵ_x, and ϵ_y as follows:

$$\sigma_z = \nu(\sigma_x + \sigma_y)$$

$$\epsilon_x = \frac{1}{E}[(1 - \nu^2)\sigma_x - \nu(1 + \nu)\sigma_y]$$

$$\epsilon_y = \frac{1}{E}[(1 - \nu^2)\sigma_y - \nu(1 + \nu)\sigma_x]$$

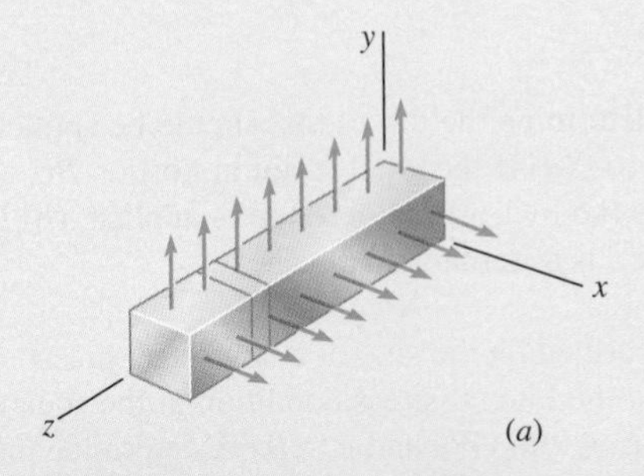

σ_y σ_x σ_z (*b*)

Fig. P9.60

9.61 The plastic block shown is bonded to a rigid support and to a vertical plate to which a 55-kip load **P** is applied. Knowing that for the plastic used $G = 150$ ksi, determine the deflection of the plate.

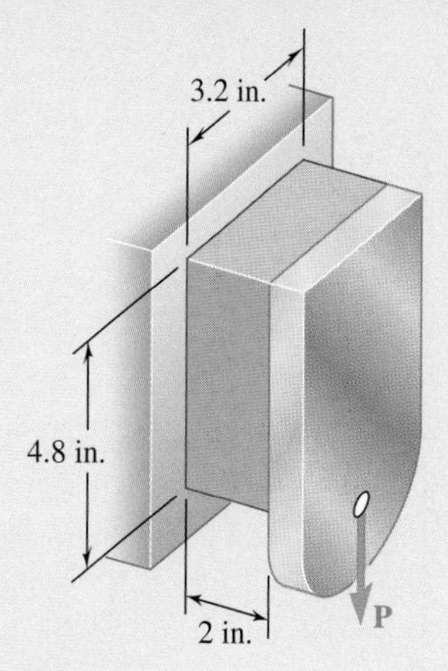

Fig. P9.61

9.62 A vibration isolation unit consists of two blocks of hard rubber bonded to a plate AB and to rigid supports as shown. Knowing that a force of magnitude $P = 25$ kN causes a deflection $\delta = 1.5$ mm of plate AB, determine the modulus of rigidity of the rubber used.

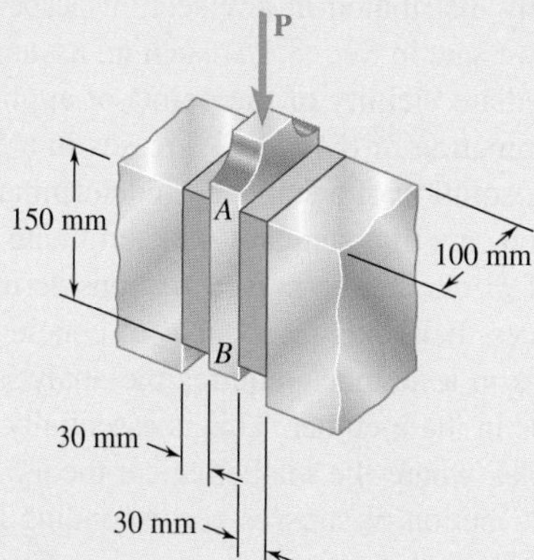
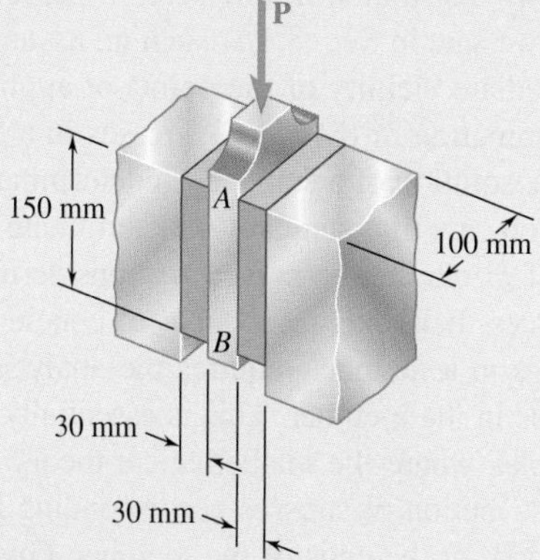

Fig. P9.62 and P9.63

9.63 A vibration isolation unit consists of two blocks of hard rubber with a modulus of rigidity $G = 19$ MPa bonded to a plate AB and to rigid supports as shown. Denoting by P the magnitude of the force applied to the plate and by δ the corresponding deflection, determine the effective spring constant, $k = P/\delta$, of the system.

9.64 An elastomeric bearing ($G = 130$ psi) is used to support a bridge girder as shown to provide flexibility during earthquakes. The beam must not displace more than $\frac{3}{8}$ in. when a 5-kip lateral load is applied as shown. Knowing that the maximum allowable shearing stress is 60 psi, determine (*a*) the smallest allowable dimension b, (*b*) the smallest required thickness a.

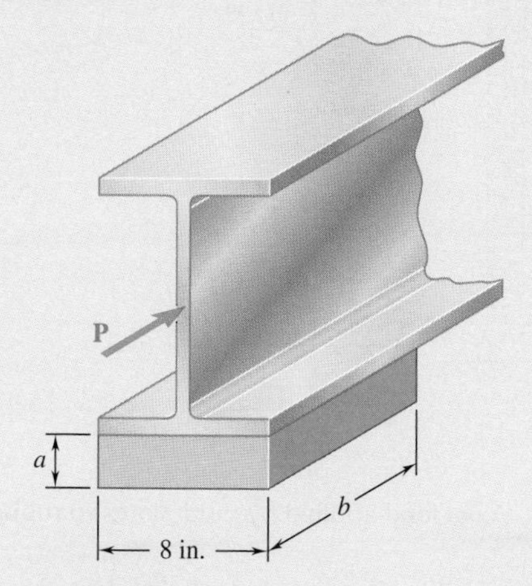

Fig. P9.64

Fig. 9.43 Axial load applied by rigid plates.

9.8 STRESS AND STRAIN DISTRIBUTION UNDER AXIAL LOADING: SAINT-VENANT'S PRINCIPLE

We have assumed so far that, in an axially loaded member, the normal stresses are uniformly distributed in any section perpendicular to the axis of the member. As we saw in Sec. 8.1A, such an assumption may be quite in error in the immediate vicinity of the points of application of the loads. However, the determination of the actual stresses in a given section of the member requires the solution of a statically indeterminate problem.

In Sec. 9.2, you saw that statically indeterminate problems involving the determination of *forces* can be solved by considering the *deformations* caused by these forces. It is thus reasonable to conclude that the determination of the *stresses* in a member requires the analysis of the strains produced by the stresses in the member. This is essentially the approach found in advanced textbooks, where the mathematical theory of elasticity is used to determine the distribution of stresses corresponding to various modes of application of the loads at the ends of the member. Given the more limited mathematical means at our disposal, our analysis of stresses will be restricted to the particular case when two rigid plates are used to transmit the loads to a member made of a homogeneous isotropic material (Fig. 9.43).

If the loads are applied at the center of each plate,[†] the plates will move toward each other without rotating, causing the member to get shorter, while increasing in width and thickness. It is assumed that the member will remain straight, plane sections will remain plane, and all elements of the member will deform in the same way, since this assumption is compatible with the given end conditions. Fig. 9.44 shows a rubber model before and after loading.[‡] Now, if all elements deform in the same way, the distribution of strains

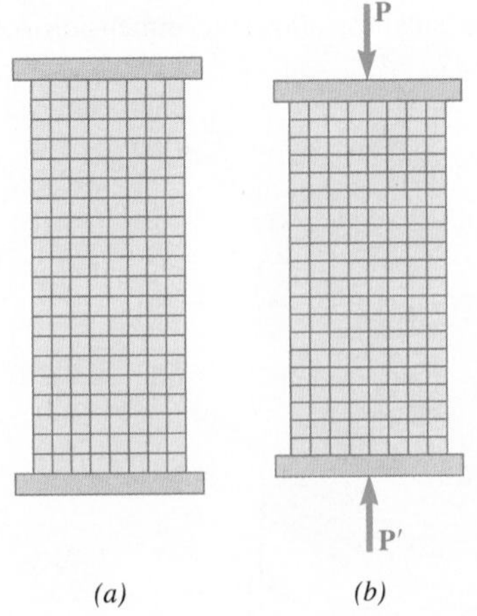

Fig. 9.44 Axial load applied by rigid plates to rubber model.

[†]More precisely, the common line of action of the loads should pass through the centroid of the cross section (cf. Sec. 8.1A).

[‡]Note that for long, slender members, another configuration is possible and will prevail if the load is sufficiently large; the member *buckles* and assumes a curved shape. This will be discussed in Chap. 16.

throughout the member must be uniform. In other words, the axial strain ϵ_y and the lateral strain $\epsilon_x = -\nu\epsilon_y$ are constant. But, if the stresses do not exceed the proportional limit, Hooke's law applies, and $\sigma_y = E\epsilon_y$, so the normal stress σ_y is also constant. Thus, the distribution of stresses is uniform throughout the member, and at any point,

$$\sigma_y = (\sigma_y)_{\text{ave}} = \frac{P}{A}$$

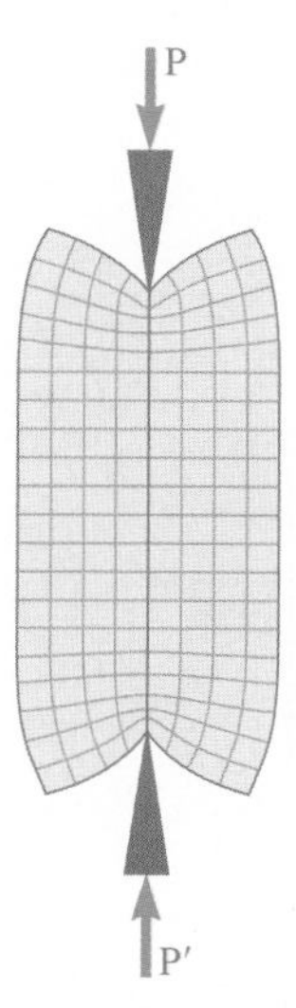

Fig. 9.45 Concentrated axial load applied to rubber model.

If the loads are concentrated, as in Fig. 9.45, the elements in the immediate vicinity of the points of application of the loads are subjected to very large stresses, while other elements near the ends of the member are unaffected by the loading. This results in large deformations, strains, and stresses near the points of application of the loads, while no deformation takes place at the corners. Considering elements farther and farther from the ends, a progressive equalization of the deformations and a more uniform distribution of the strains and stresses are seen across a section of the member. Using the mathematical theory of elasticity found in advanced textbooks, Fig. 9.46 shows the resulting distribution of stresses across various sections of a thin rectangular plate subjected to concentrated loads. Note that at a

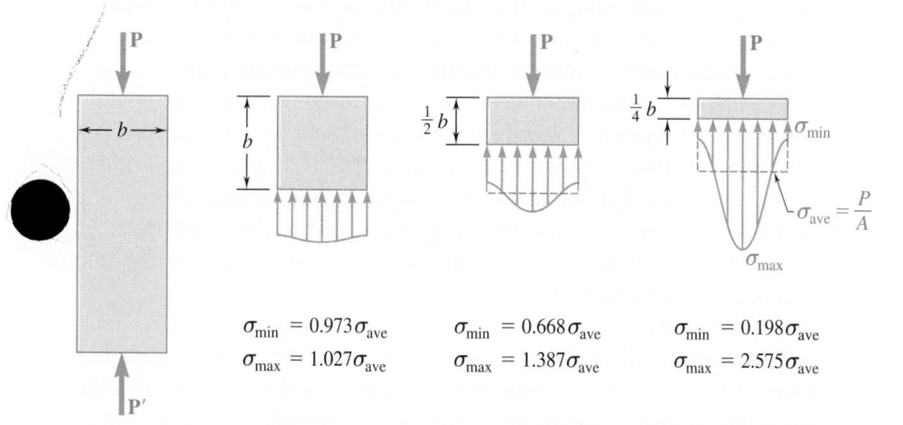

Fig. 9.46 Stress distributions in a plate under concentrated axial loads.

distance b from either end, where b is the width of the plate, the stress distribution is nearly uniform across the section, and the value of the stress σ_y at any point of that section can be assumed to be equal to the average value P/A. Thus, at a distance equal to or greater than the width of the member, the distribution of stresses across a section is the same, whether the member is loaded as shown in Fig. 9.43 or Fig. 9.45. In other words, except in the immediate vicinity of the points of application of the loads, the stress distribution is assumed independent of the actual mode of application of the loads. This statement, which applies to axial loadings and to practically any type of load, is known as *Saint-Venant's principle*, after the French mathematician and engineer Adhémar Barré de Saint-Venant (1797–1886).

While Saint-Venant's principle makes it possible to replace a given loading by a simpler one to compute the stresses in a structural member, keep in mind two important points when applying this principle:

1. The actual loading and the loading used to compute the stresses must be *statically equivalent*.

2. Stresses cannot be computed in this manner in the immediate vicinity of the points of application of the loads. Advanced theoretical or experimental methods must be used to determine the distribution of stresses in these areas.

You should also observe that the plates used to obtain a uniform stress distribution in the member of Fig. 9.44 must allow the member to freely expand laterally. Thus, the plates cannot be rigidly attached to the member; assume them to be just in contact with the member and smooth enough not to impede lateral expansion. While such end conditions can be achieved for a member in compression, they cannot be physically realized in the case of a member in tension. It does not matter, whether or not an actual fixture can be realized and used to load a member so that the distribution of stresses in the member is uniform. The important thing is to *imagine a model* that will allow such a distribution of stresses and to keep this model in mind so that it can be compared with the actual loading conditions.

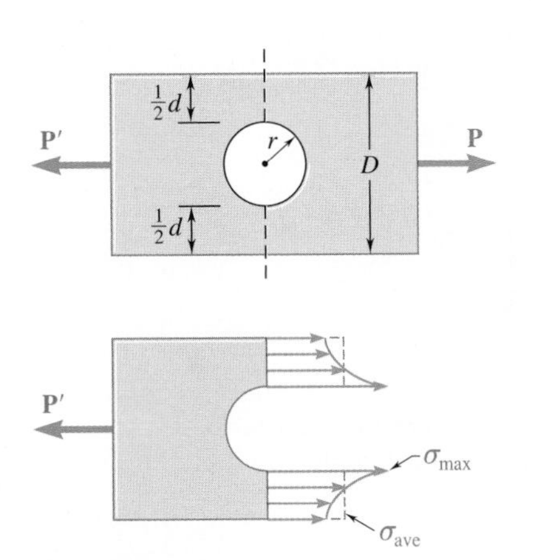

Fig. 9.47 Stress distribution near circular hole in flat bar under axial loading.

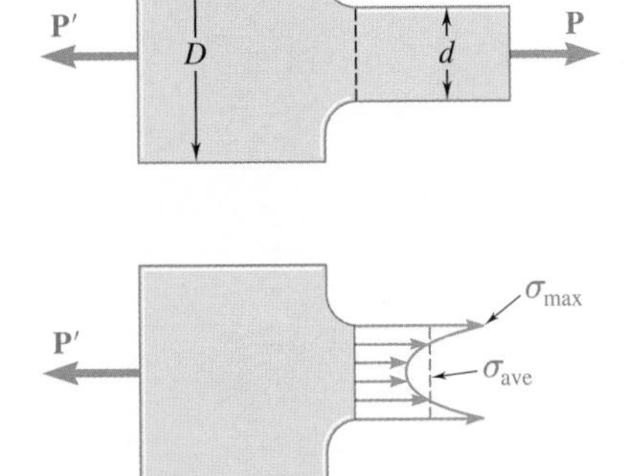

Fig. 9.48 Stress distribution near fillets in flat bar under axial loading.

9.9 STRESS CONCENTRATIONS

As you saw in the preceding section, the stresses near the points of application of concentrated loads can reach values much larger than the average value of the stress in the member. When a structural member contains a discontinuity, such as a hole or a sudden change in cross section, high localized stresses can occur. Figs. 9.47 and 9.48 show the distribution of stresses in critical sections corresponding to two situations. Fig. 9.47 shows a flat bar with a *circular hole* and shows the stress distribution in a section passing through the center of the hole. Fig. 9.48 shows a flat bar consisting of two portions of different widths connected by *fillets*; here the stress distribution is in the narrowest part of the connection, where the highest stresses occur.

These results were obtained experimentally through the use of a photoelastic method. Fortunately for the engineer, these results are independent of the size of the member and of the material used; they depend only upon the ratios of the geometric parameters involved (i.e., the ratio $2r/D$ for a circular hole, and the ratios r/d and D/d for fillets). Furthermore, the designer is more interested in the *maximum value* of the stress in a given section than the actual distribution of stresses. The main concern is to determine *whether* the allowable stress will be exceeded under a given loading, not *where* this value will be exceeded. Thus, the ratio

$$K = \frac{\sigma_{max}}{\sigma_{ave}} \quad (9.29)$$

is computed in the critical (narrowest) section of the discontinuity. This ratio is the *stress-concentration factor* of the discontinuity. Stress-concentration factors can be computed in terms of the ratios of the geometric parameters involved, and the results can be expressed in tables or graphs, as shown in Fig. 9.49. To determine the maximum stress occurring near a discontinuity in a given member subjected to a given axial load P, the designer needs to compute the average stress $\sigma_{ave} = P/A$ in the critical section and multiply the result obtained by the appropriate value of the stress-concentration factor K. Note that this procedure is valid only as long as σ_{max} does not exceed the proportional limit of the material, since the values of K plotted in Fig. 9.49 were obtained by assuming a linear relation between stress and strain.

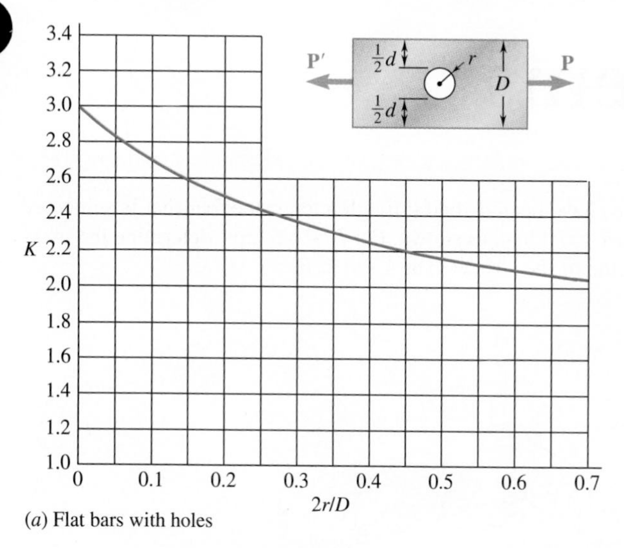

(a) Flat bars with holes

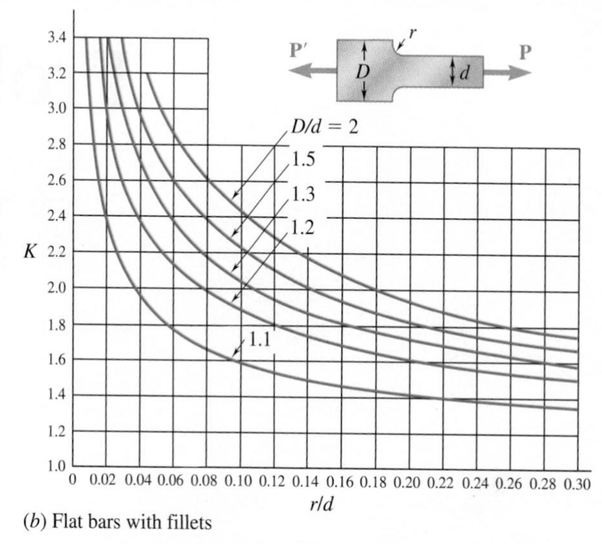

(b) Flat bars with fillets

Fig. 9.49 Stress concentration factors for flat bars under axial loading. Note that the average stress must be computed across the narrowest section: $\sigma_{ave} = P/td$, where t is the thickness of the bar. (Source: W. D. Pilkey and D.F. Pilkey, *Peterson's Stress Concentration Factors*, 3rd ed., John Wiley & Sons, New York, 2008.)

Concept Application 9.10

Determine the largest axial load **P** that can be safely supported by a flat steel bar consisting of two portions, both 10 mm thick and, respectively, 40 and 60 mm wide, connected by fillets of radius $r = 8$ mm. Assume an allowable normal stress of 165 MPa.

First compute the ratios

$$\frac{D}{d} = \frac{60\text{ mm}}{40\text{ mm}} = 1.50 \qquad \frac{r}{d} = \frac{8\text{ mm}}{40\text{ mm}} = 0.20$$

Using the curve in Fig. 9.49*b* corresponding to $D/d = 1.50$, the value of the stress-concentration factor corresponding to $r/d = 0.20$ is

$$K = 1.82$$

Then carrying this value into Eq. (9.29) and solving for σ_{ave},

$$\sigma_{ave} = \frac{\sigma_{max}}{1.82}$$

But σ_{max} cannot exceed the allowable stress $\sigma_{all} = 165$ MPa. Substituting this value for σ_{max}, the average stress in the narrower portion ($d = 40$ mm) of the bar should not exceed the value

$$\sigma_{ave} = \frac{165\text{ MPa}}{1.82} = 90.7\text{ MPa}$$

Recalling that $\sigma_{ave} = P/A$,

$$P = A\sigma_{ave} = (40\text{ mm})(10\text{ mm})(90.7\text{ MPa}) = 36.3 \times 10^3\text{ N}$$

$$P = 36.3\text{ kN}$$

Problems

9.65 Two holes have been drilled through a long steel bar that is subjected to a centric axial load as shown. For $P = 6.5$ kips, determine the maximum value of the stress (*a*) at *A*, (*b*) at *B*.

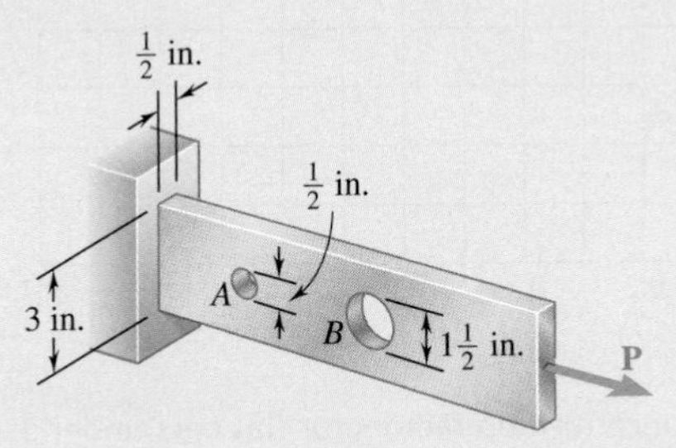

Fig. P9.65 and *P9.66*

9.66 Knowing that $\sigma_{all} = 16$ ksi, determine the maximum allowable value of the centric axial load **P**.

9.67 Knowing that, for the plate shown, the allowable stress is 125 MPa, determine the maximum allowable value of P when (*a*) $r = 12$ mm, (*b*) $r = 18$ mm.

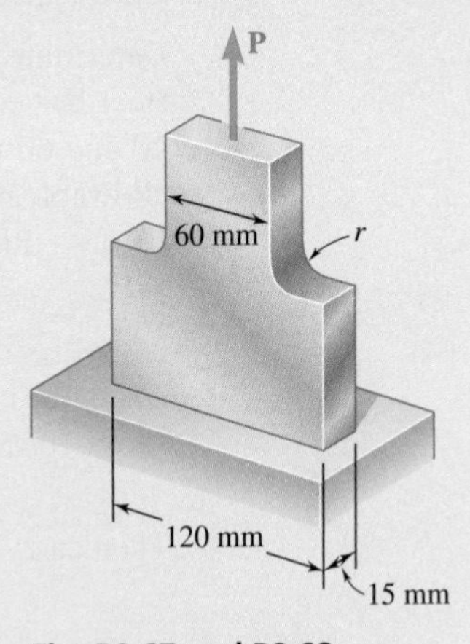

Fig. P9.67 and P9.68

9.68 Knowing that $P = 38$ kN, determine the maximum stress when (*a*) $r = 10$ mm, (*b*) $r = 16$ mm, (*c*) $r = 18$ mm.

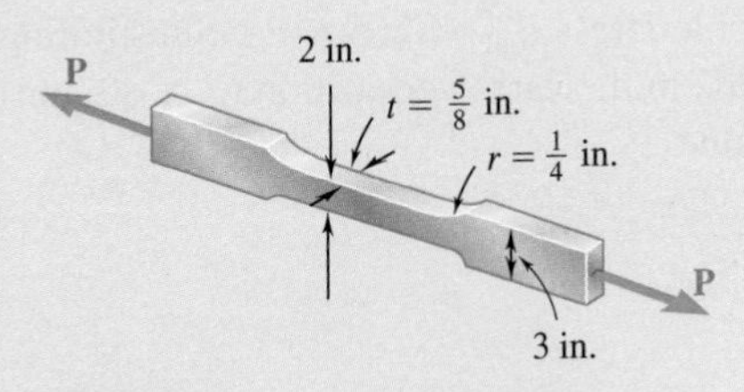

Fig. P9.69

9.69 (*a*) Knowing that the allowable stress is 20 ksi, determine the maximum allowable magnitude of the centric load **P**. (*b*) Determine the percent change in the maximum allowable magnitude of **P** if the raised portions are removed at the ends of the specimen.

9.70 A centric axial force is applied to the steel bar shown. Knowing that $\sigma_{all} = 20$ ksi, determine the maximum allowable load **P**.

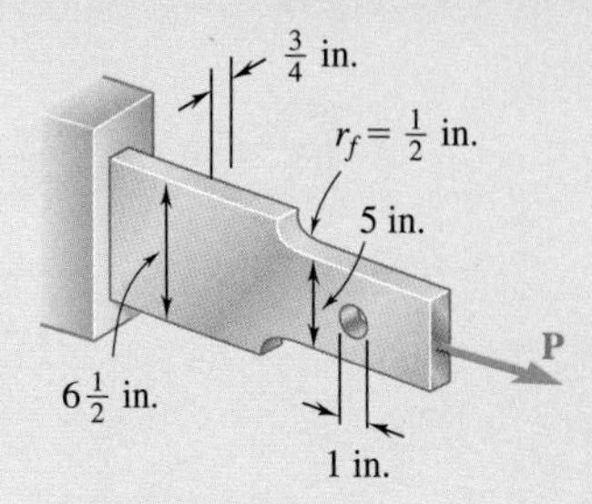

Fig. P9.70

9.71 Knowing that the hole has a diameter of 9 mm, determine (*a*) the radius r_f of the fillets for which the same maximum stress occurs at the hole *A* and at the fillets, (*b*) the corresponding maximum allowable load **P** if the allowable stress is 100 MPa.

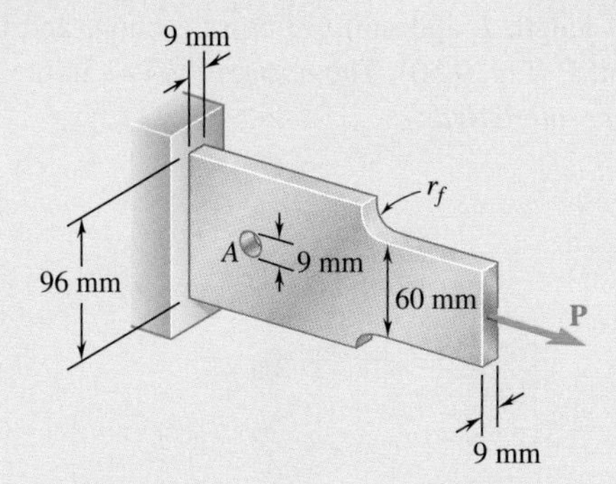

Fig. *P9.71*

9.72 For $P = 100$ kN, determine the minimum plate thickness t required if the allowable stress is 125 MPa.

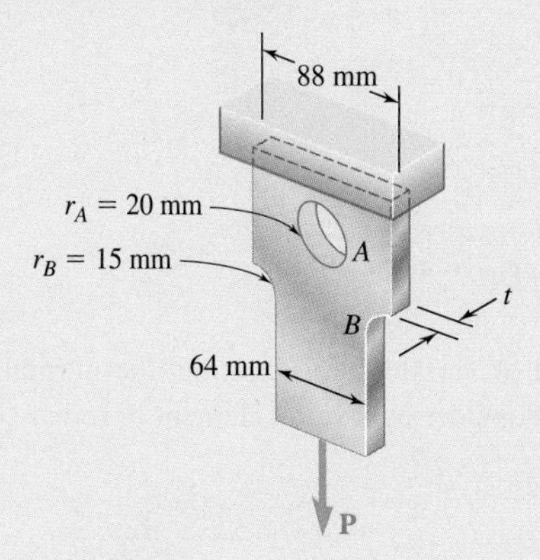

Fig. P9.72

Review and Summary

Normal Strain

Consider a rod of length L and uniform cross section, and its deformation δ under an axial load **P** (Fig. 9.50). The *normal strain* ϵ in the rod is defined as the *deformation per unit length*:

$$\epsilon = \frac{\delta}{L} \tag{9.1}$$

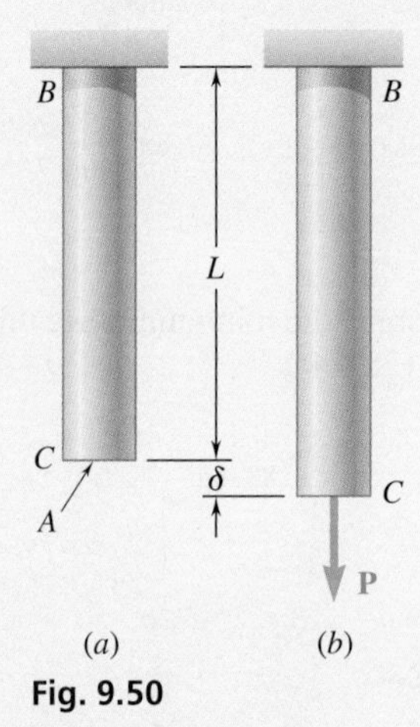

Fig. 9.50

In the case of a rod of variable cross section, the normal strain at any given point Q is found by considering a small element of rod at Q:

$$\epsilon = \lim_{\Delta x \longrightarrow 0} \frac{\Delta \delta}{\Delta x} = \frac{d\delta}{dx} \tag{9.2}$$

Stress-Strain Diagram

A *stress-strain diagram* is obtained by plotting the stress σ versus the strain ϵ as the load increases. These diagrams can be used to distinguish between *brittle* and *ductile* materials. A brittle material ruptures without any noticeable prior change in the rate of elongation (Fig. 9.51), while a ductile material *yields* after a critical

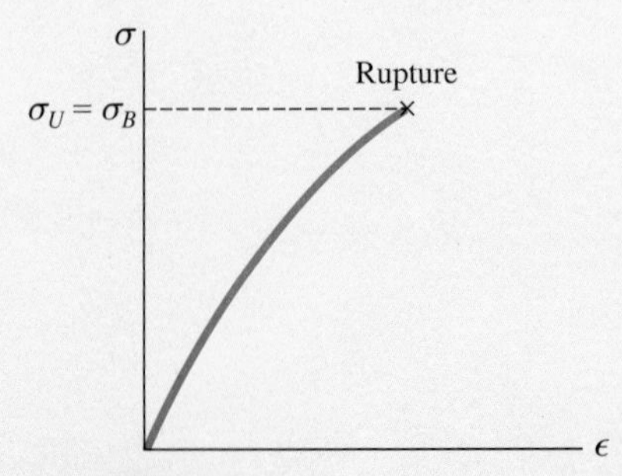

Fig. 9.51

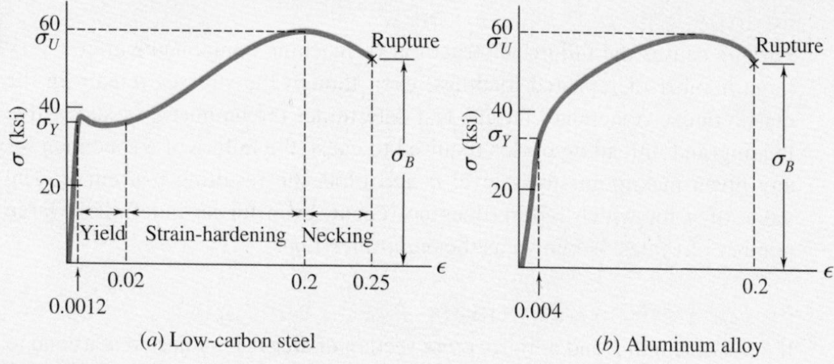

Fig. 9.52

stress σ_Y (the *yield strength*) has been reached (Fig. 9.52). The specimen undergoes a large deformation before rupturing, with a relatively small increase in the applied load. An example of brittle material with different properties in tension and compression is *concrete*.

Hooke's Law and Modulus of Elasticity

The initial portion of the stress-strain diagram is a straight line. Thus, for small deformations, the stress is directly proportional to the strain:

$$\sigma = E\epsilon \tag{9.5}$$

This relationship is *Hooke's law*, and the coefficient E is the *modulus of elasticity* of the material. The *proportional limit* is the largest stress for which Eq. (9.5) applies.

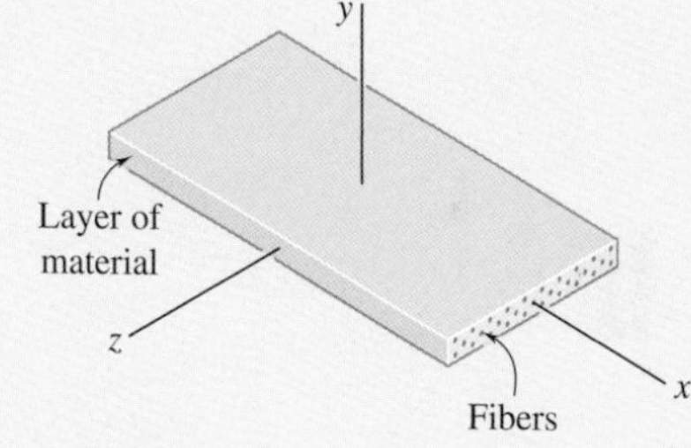

Fig. 9.53

Properties of *isotropic* materials are independent of direction, while properties of *anisotropic* materials depend upon direction. *Fiber-reinforced composite materials* are made of fibers of a strong, stiff material embedded in layers of a weaker, softer material (Fig. 9.53).

Elastic Limit and Plastic Deformation

If the strains caused in a test specimen by the application of a given load disappear when the load is removed, the material is said to behave *elastically*. The largest stress for which this occurs is called the *elastic limit* of the material. If the elastic limit is exceeded, the stress and strain decrease in a linear fashion when the load is removed, and the strain does not return to zero (Fig. 9.54), indicating that a *permanent set* or *plastic deformation* of the material has taken place.

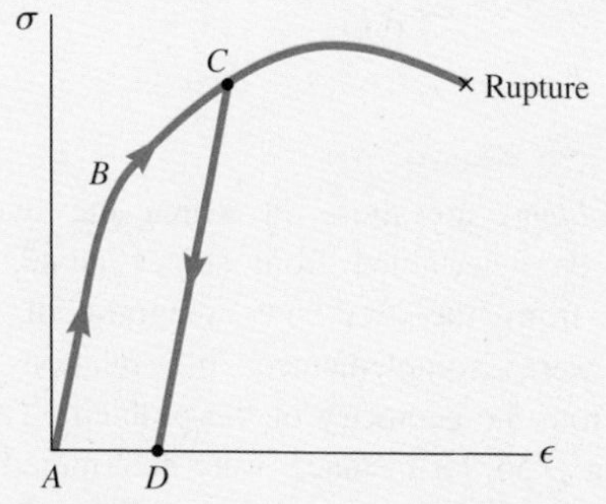

Fig. 9.54

Fatigue and Endurance Limit

Fatigue causes the failure of structural or machine components after a very large number of repeated loadings, even though the stresses remain in the elastic range. A standard fatigue test determines the number n of successive loading-and-unloading cycles required to cause the failure of a specimen for any given maximum stress level σ and plots the resulting σ-n curve. The value of σ for which failure does not occur, even for an indefinitely large number of cycles, is known as the *endurance limit*.

Elastic Deformation Under Axial Loading

If a rod of length L and uniform cross section of area A is subjected at its end to a centric axial load **P** (Fig. 9.55), the corresponding deformation is

$$\delta = \frac{PL}{AE} \tag{9.8}$$

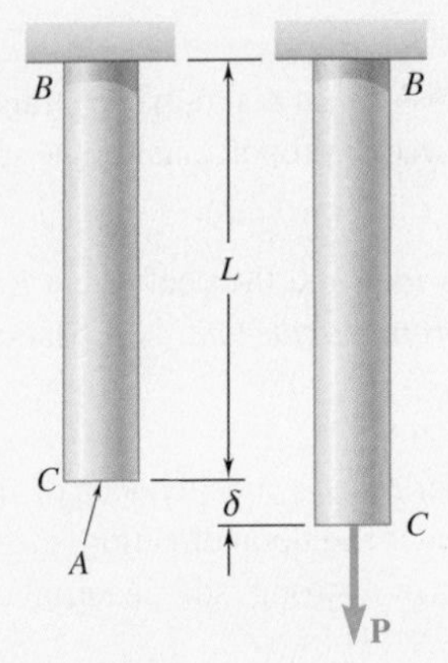

Fig. 9.55

If the rod is loaded at several points or consists of several parts of various cross sections and possibly of different materials, the deformation δ of the rod must be expressed as the sum of the deformations of its component parts:

$$\delta = \sum_i \frac{P_i L_i}{A_i E_i} \tag{9.9}$$

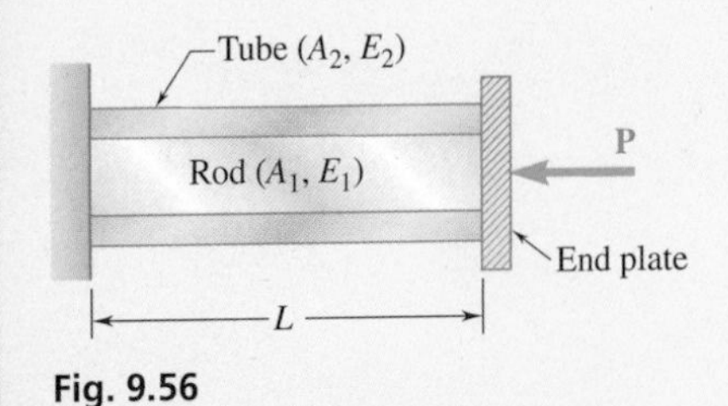

Fig. 9.56

Statically Indeterminate Problems

Statically indeterminate problems are those in which the reactions and the internal forces *cannot* be determined from statics alone. The equilibrium equations derived from the free-body diagram of the member under consideration were complemented by relations involving deformations and obtained from the geometry of the problem. The forces in the rod and in the tube of Fig. 9.56, for instance, were determined by observing that their sum is equal to P, and that they cause equal deformations in the rod and in the tube. Similarly, the reactions at the supports of the bar of

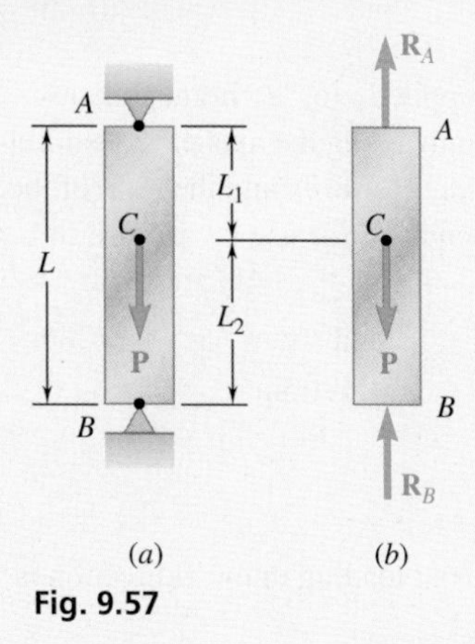

Fig. 9.57

Fig. 9.57 could not be obtained from the free-body diagram of the bar alone, but they could be determined by expressing that the total elongation of the bar must be equal to zero.

Problems with Temperature Changes

When the temperature of an *unrestrained rod AB* of length L is increased by ΔT, its elongation is

$$\delta_T = \alpha(\Delta T)L \tag{9.12}$$

where α is the *coefficient of thermal expansion* of the material. The corresponding strain, called *thermal strain*, is

$$\epsilon_T = \alpha \Delta T \tag{9.13}$$

and *no stress* is associated with this strain. However, if rod AB is *restrained* by fixed supports (Fig. 9.58), stresses develop in the rod as the temperature increases, because of the reactions at the supports. To determine the magnitude P of the reactions, the rod is first detached from its support at B (Fig. 9.59*a*). The deformation δ_T of the rod occurs as it expands due to of the temperature change (Fig. 9.59*b*). The deformation δ_P caused by the force **P** is required to bring it back to its original length, so that it may be reattached to the support at B (Fig. 9.59*c*).

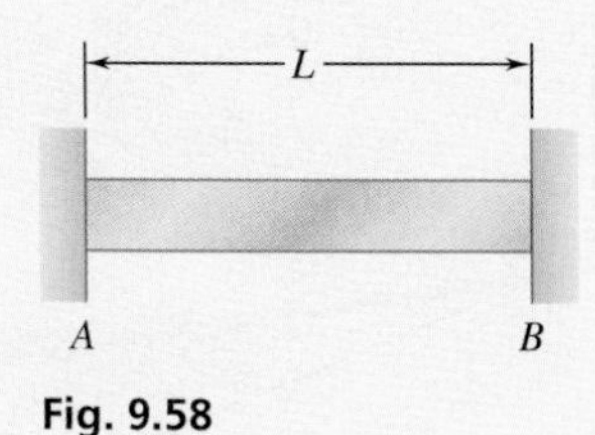

Fig. 9.58

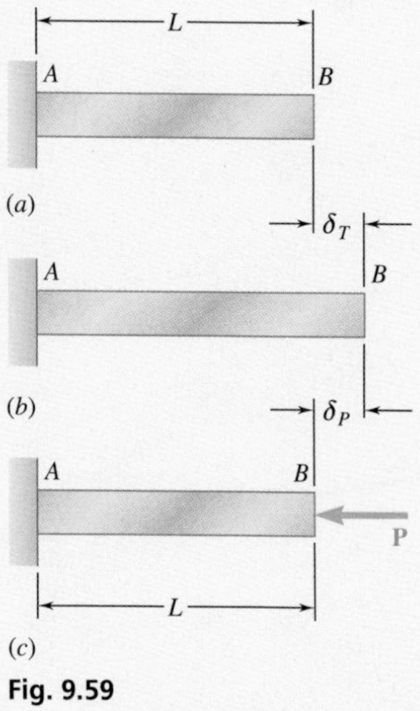

Fig. 9.59

Lateral Strain and Poisson's Ratio

When an axial load **P** is applied to a homogeneous, slender bar (Fig. 9.60), it causes a strain, not only along the axis of the bar but in any transverse direction. This strain is the *lateral strain*, and the ratio of the lateral strain over the axial strain is called *Poisson's ratio*:

$$\nu = -\frac{\text{lateral strain}}{\text{axial strain}} \tag{9.16}$$

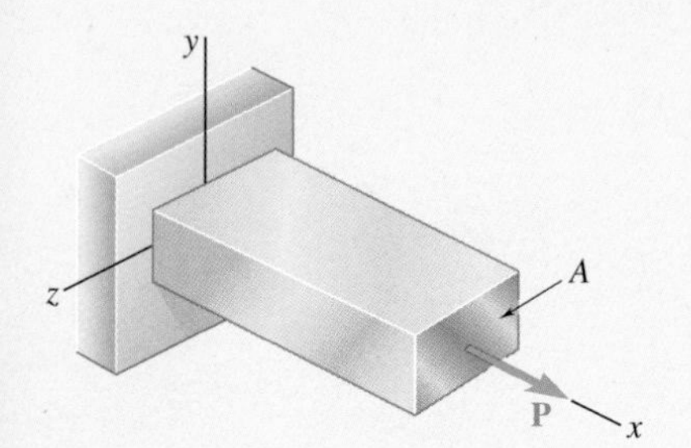

Fig. 9.60

Multiaxial Loading

The condition of strain under an axial loading in the x direction is

$$\epsilon_x = \frac{\sigma_x}{E} \qquad \epsilon_y = \epsilon_z = -\frac{\nu\sigma_x}{E} \tag{9.18}$$

A *multiaxial loading* causes the state of stress shown in Fig. 9.61. The resulting strain condition was described by the *generalized Hooke's law* for a multiaxial loading.

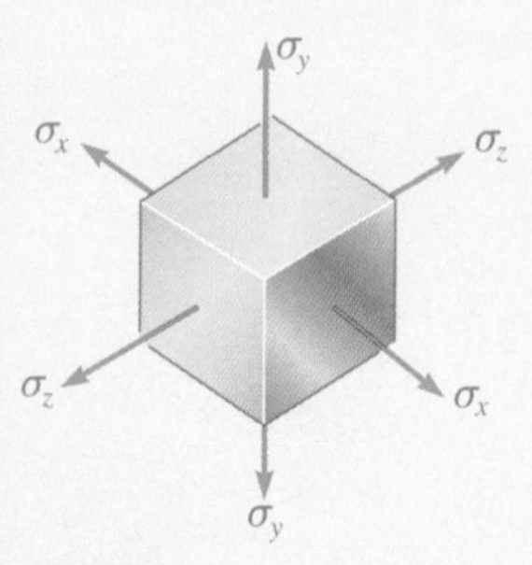

Fig. 9.61

$$\begin{aligned}
\epsilon_x &= +\frac{\sigma_x}{E} - \frac{\nu\sigma_y}{E} - \frac{\nu\sigma_z}{E} \\
\epsilon_y &= -\frac{\nu\sigma_x}{E} + \frac{\sigma_y}{E} - \frac{\nu\sigma_z}{E} \\
\epsilon_z &= -\frac{\nu\sigma_x}{E} - \frac{\nu\sigma_y}{E} + \frac{\sigma_z}{E}
\end{aligned} \tag{9.19}$$

Shearing Strain: Modulus of Rigidity

The state of stress in a material under the most general loading condition involves shearing stresses, as well as normal stresses (Fig. 9.62). The shearing stresses tend to deform a cubic element of material into an oblique parallelepiped. The

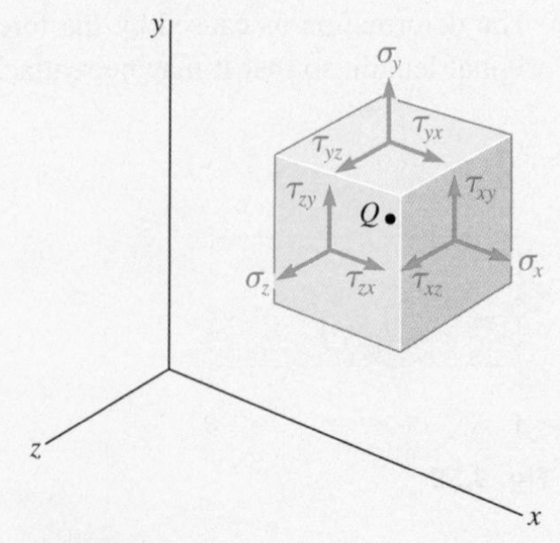

Fig. 9.62

stresses τ_{xy} and τ_{yx} shown in Fig. 9.63 cause the angles formed by the faces on which they act to either increase or decrease by a small angle γ_{xy}. This angle defines the *shearing strain* corresponding to the x and y directions. Defining in a similar way the shearing strains γ_{yz} and γ_{zx}, the following relations were written:

$$\tau_{xy} = G\gamma_{xy} \qquad \tau_{yz} = G\gamma_{yz} \qquad \tau_{zx} = G\gamma_{zx} \qquad \textbf{(9.20, 21)}$$

which are valid for any homogeneous isotropic material within its proportional limit in shear. The constant G is the *modulus of rigidity* of the material, and the relationships obtained express *Hooke's law for shearing stress and strain.* Together with Eqs. (9.19), they form a group of equations representing the generalized Hooke's law for a homogeneous isotropic material under the most general stress condition.

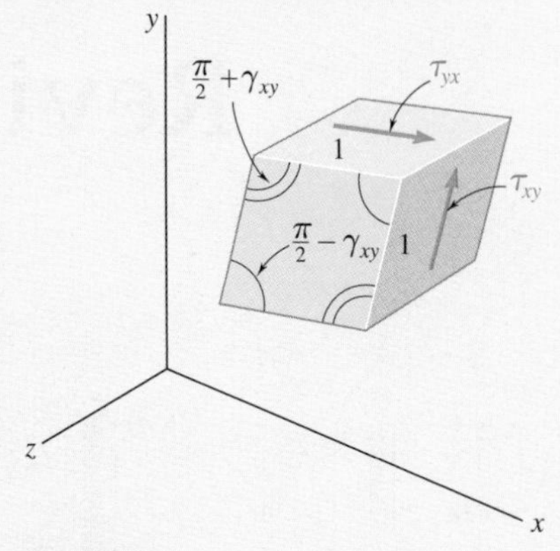

Fig. 9.63

While an axial load exerted on a slender bar produces only normal strains—both axial and transverse—on an element of material oriented along the axis of the bar, it will produce both normal and shearing strains on an element rotated through 45° (Fig. 9.64). The three constants E, ν, and G are not independent. They satisfy the relation

$$\frac{E}{2G} = 1 + \nu \qquad \textbf{(9.27)}$$

This equation can be used to determine any of the three constants in terms of the other two.

Saint-Venant's Principle

Saint-Venant's principle states that except in the immediate vicinity of the points of application of the loads, the distribution of stresses in a given member is independent of the actual mode of application of the loads. This principle makes it possible to assume a uniform distribution of stresses in a member subjected to concentrated axial loads, except close to the points of application of the loads, where stress concentrations will occur.

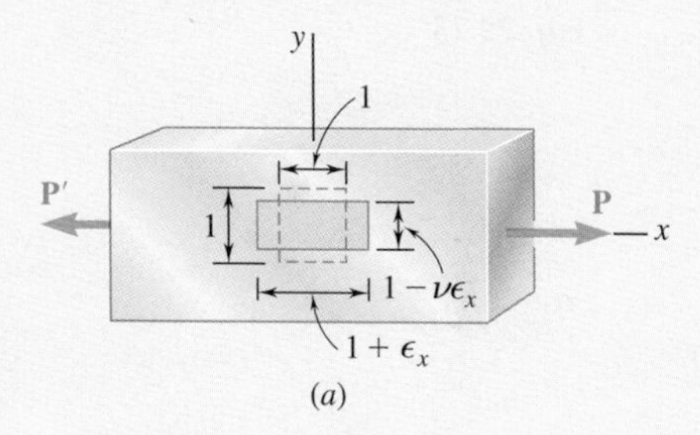

(a)

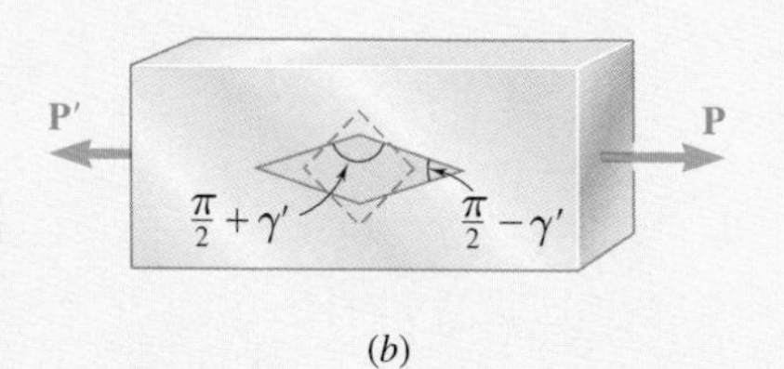

(b)

Fig. 9.64

Stress Concentrations

Stress concentrations will also occur in structural members near a discontinuity, such as a hole or a sudden change in cross section. The ratio of the maximum value of the stress occurring near the discontinuity over the average stress computed in the critical section is referred to as the *stress-concentration factor* of the discontinuity:

$$K = \frac{\sigma_{\max}}{\sigma_{\text{ave}}} \qquad \textbf{(9.29)}$$

Review Problems

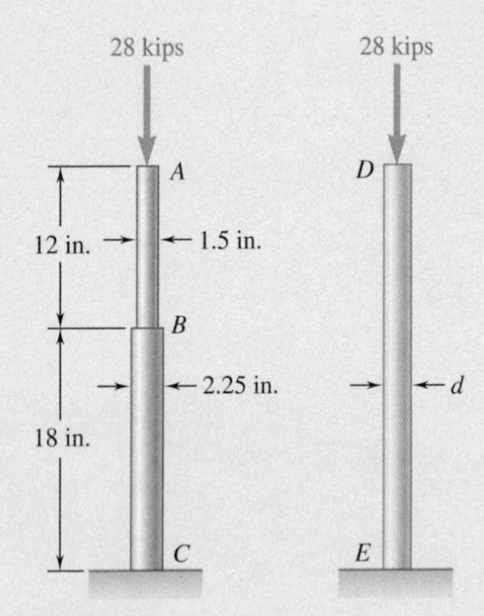

Fig. P9.73

9.73 The aluminum rod ABC ($E = 10.1 \times 10^6$ psi), which consists of two cylindrical portions AB and BC, is to be replaced with a cylindrical steel rod DE ($E = 29 \times 10^6$ psi) of the same overall length. Determine the minimum required diameter d of the steel rod if its vertical deformation is not to exceed the deformation of the aluminum rod under the same load and if the allowable stress in the steel rod is not to exceed 24 ksi.

9.74 The brass tube AB ($E = 105$ GPa) has a cross-sectional area of 140 mm^2 and is fitted with a plug at A. The tube is attached at B to a rigid plate that is itself attached at C to the bottom of an aluminum cylinder ($E = 72$ GPa) with a cross-sectional area of 250 mm^2. The cylinder is then hung from a support at D. In order to close the cylinder, the plug must move down through 1 mm. Determine the force **P** that must be applied to the cylinder.

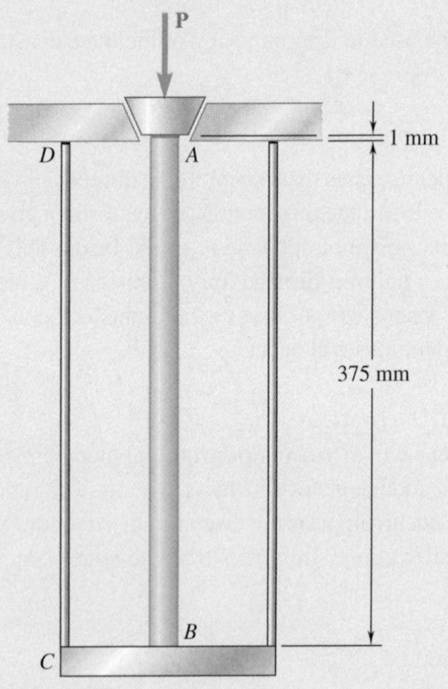

Fig. P9.74

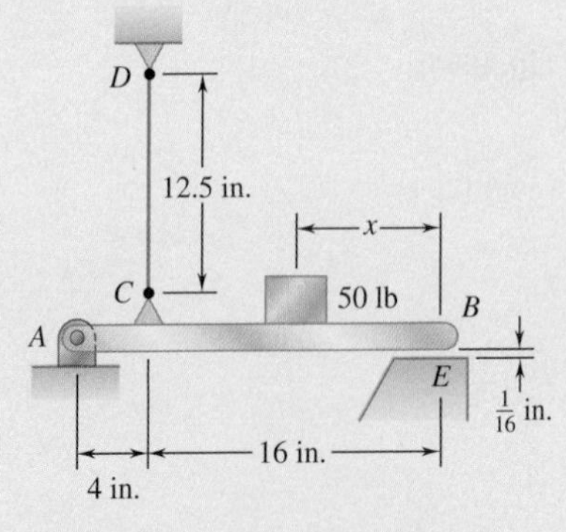

Fig. *P9.75*

9.75 The length of the $\frac{3}{32}$-in.-diameter steel wire CD has been adjusted so that with no load applied, a gap of $\frac{1}{16}$ in. exists between the end B of the rigid beam ACB and a contact point E. Knowing that $E = 29 \times 10^6$ psi, determine where a 50-lb block should be placed on the beam in order to cause contact between B and E.

9.76 Each of the four vertical links connecting the two rigid horizontal members is made of aluminum ($E = 70$ GPa) and has a uniform rectangular cross section of 10×40 mm. For the loading shown, determine the deflection of (*a*) point *E*, (*b*) point *F*, (*c*) point *G*.

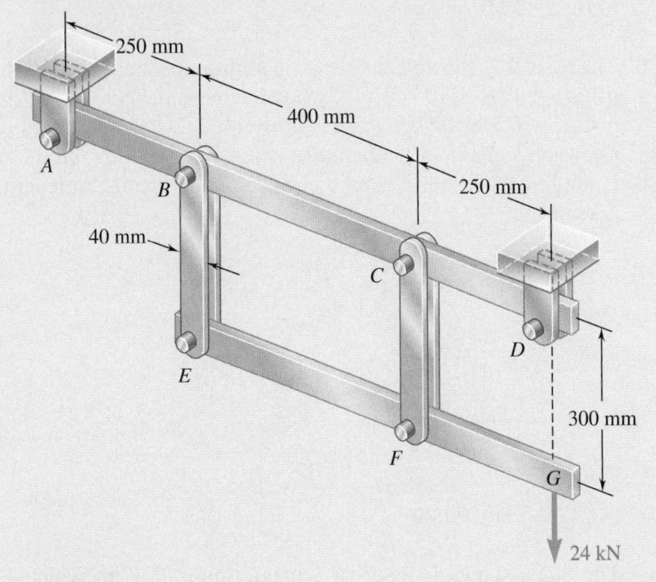

Fig. P9.76

9.77 Each of the rods *BD* and *CE* is made of brass ($E = 105$ GPa) and has a cross-sectional area of 200 mm^2. Determine the deflection of end *A* of the rigid member *ABC* caused by the 2-kN load.

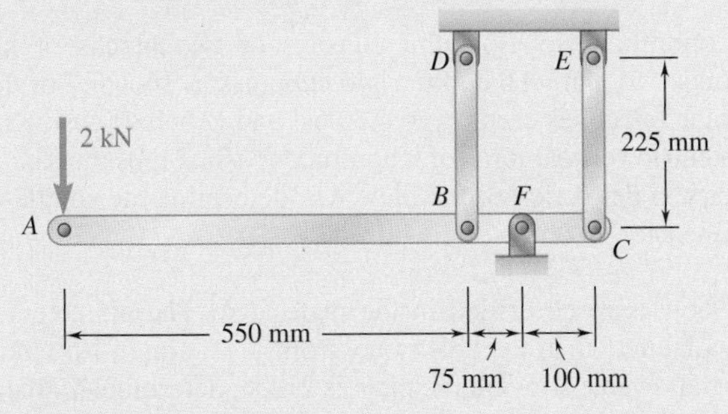

Fig. P9.77

9.78 The brass strip *AB* has been attached to a fixed support at *A* and rests on a rough support at *B*. Knowing that the coefficient of friction is 0.60 between the strip and the support at *B*, determine the decrease in temperature for which slipping will impend.

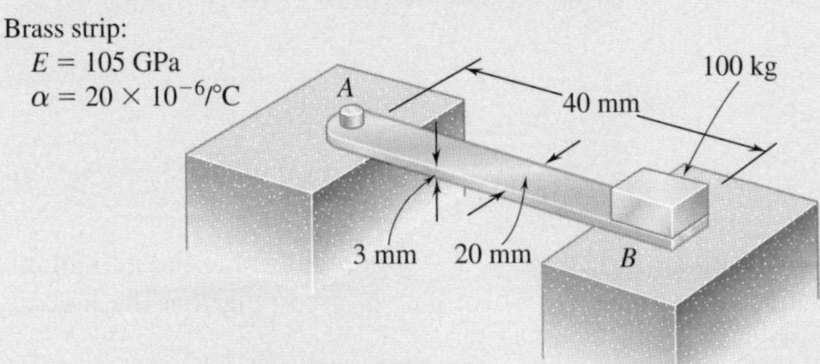

Fig. P9.78

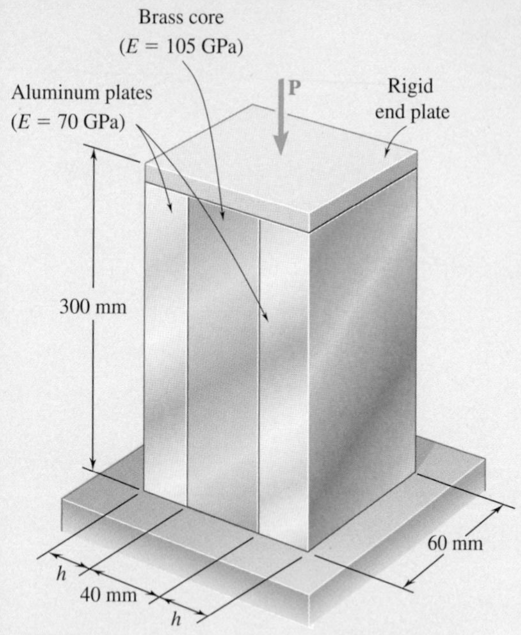

Fig. *P9.79*

9.79 An axial centric force P is applied to the composite block shown by means of a rigid end plate. Determine (*a*) the value of h if the portion of the load carried by the aluminum plates is half the portion of the load carried by the brass core, (*b*) the total load if the stress in the brass is 80 MPa.

9.80 The assembly shown consists of an aluminum shell ($E_a = 10.6 \times 10^6$ psi, $\alpha_a = 12.9 \times 10^{-6}$/°F) fully bonded to a steel core ($E_s = 29 \times 10^6$ psi, $\alpha_s = 6.5 \times 10^{-6}$/°F) and is unstressed. Determine (*a*) the largest allowable change in temperature if the stress in the aluminum shell is not to exceed 6 ksi, (*b*) the corresponding change in length of the assembly.

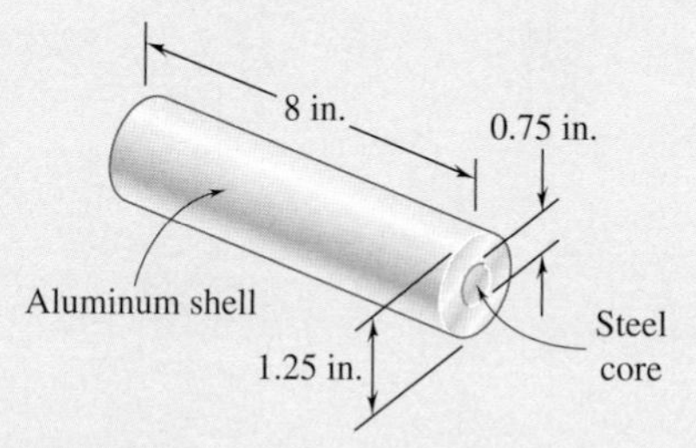

Fig. P9.80

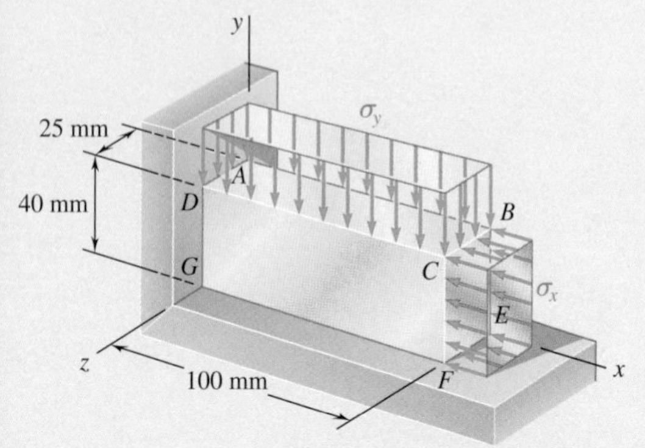

Fig. P9.81

9.81 The block shown is made of a magnesium alloy for which $E = 45$ GPa and $\nu = 0.35$. Knowing that $\sigma_x = -180$ MPa, determine (*a*) the magnitude of σ_y for which the change in the height of the block will be zero, (*b*) the corresponding change in the area of the face *ABCD*, (*c*) the corresponding change in the volume of the block.

9.82 A vibration isolation unit consists of two blocks of hard rubber bonded to plate *AB* and to rigid supports as shown. For the type and grade of rubber used, $\tau_{all} = 220$ psi and $G = 1800$ psi. Knowing that a centric vertical force of magnitude $P = 3.2$ kips must cause a 0.1-in. vertical deflection of the plate *AB*, determine the smallest allowable dimensions *a* and *b* of the block.

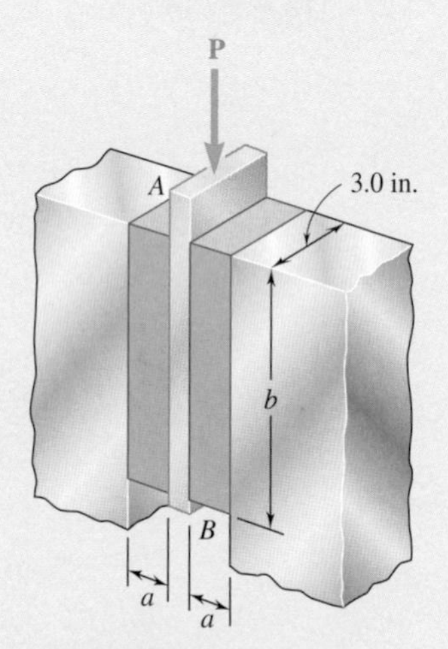

Fig. P9.82

9.83 A hole is to be drilled in the plate at *A*. The diameters of the bits available to drill the hole range from $\frac{1}{2}$ to $1\frac{1}{2}$ in. in $\frac{1}{4}$-in. increments. If the allowable stress in the plate is 21 ksi, determine (*a*) the diameter *d* of the largest bit that can be used if the allowable load **P** at the hole is to exceed that at the fillets, (*b*) the corresponding allowable load **P**.

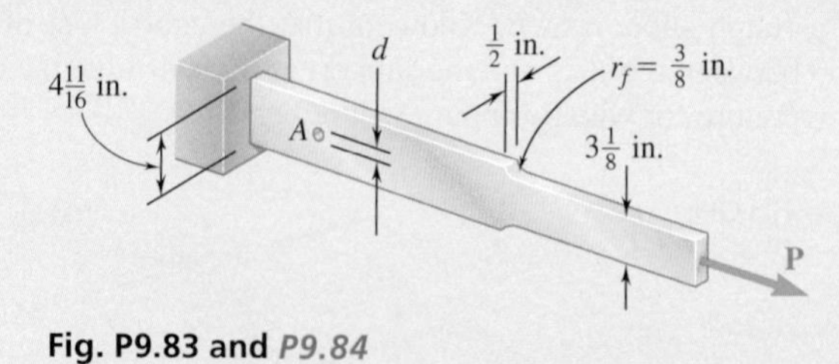

Fig. P9.83 and *P9.84*

9.84 (*a*) For $P = 13$ kips and $d = \frac{1}{2}$ in., determine the maximum stress in the plate shown. (*b*) Solve part *a*, assuming that the hole at *A* is not drilled.

10

Torsion

In the part of the jet engine shown here, the central shaft links the components of the engine to develop the thrust that propels the aircraft.

Objectives

In this chapter, you will:

- **Consider** the concept of torsion in structural members and machine parts
- **Define** shearing stresses and strains in a circular shaft subject to torsion
- **Define angle** of twist in terms of the applied torque, geometry of the shaft, and material
- **Use** torsional deformations to solve indeterminate problems

Introduction

In this chapter, structural members and machine parts that are in *torsion* will be analyzed, where the stresses and strains in members of circular cross section are subjected to twisting couples, or *torques*, **T** and **T**′ (Fig. 10.1). These couples have a common magnitude T, and opposite senses. They are vector quantities and can be represented either by curved arrows (Fig. 10.1*a*) or by couple vectors (Fig. 10.1*b*).

Members in torsion are encountered in many engineering applications. The most common application is provided by *transmission shafts*, which are used to transmit power from one point to another (Photo 10.1). These shafts can be either solid, as shown in Fig. 10.1, or hollow.

Photo 10.1 In this automotive power train, the shaft transmits power from the engine to the rear wheels.

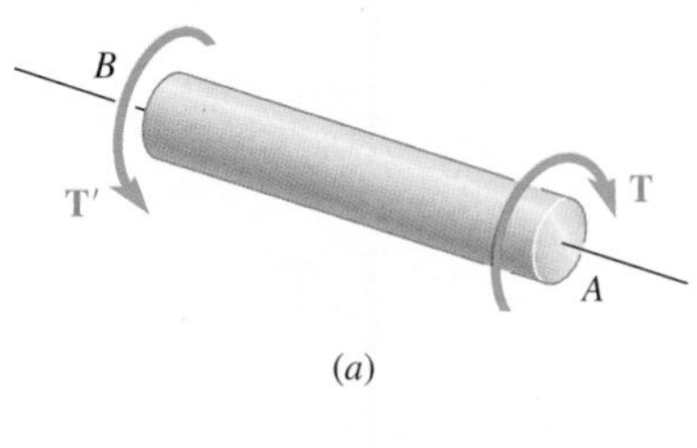

(*a*)

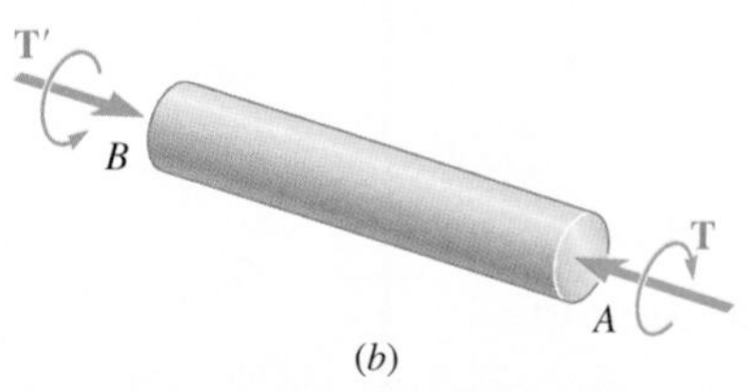

(*b*)

Fig. 10.1 Two equivalent ways to represent a torque in a free-body diagram.

The system shown in Fig. 10.2*a* consists of a turbine A and an electric generator B connected by a transmission shaft AB. Breaking the system into its three component parts (Fig. 10.2*b*), the turbine exerts a

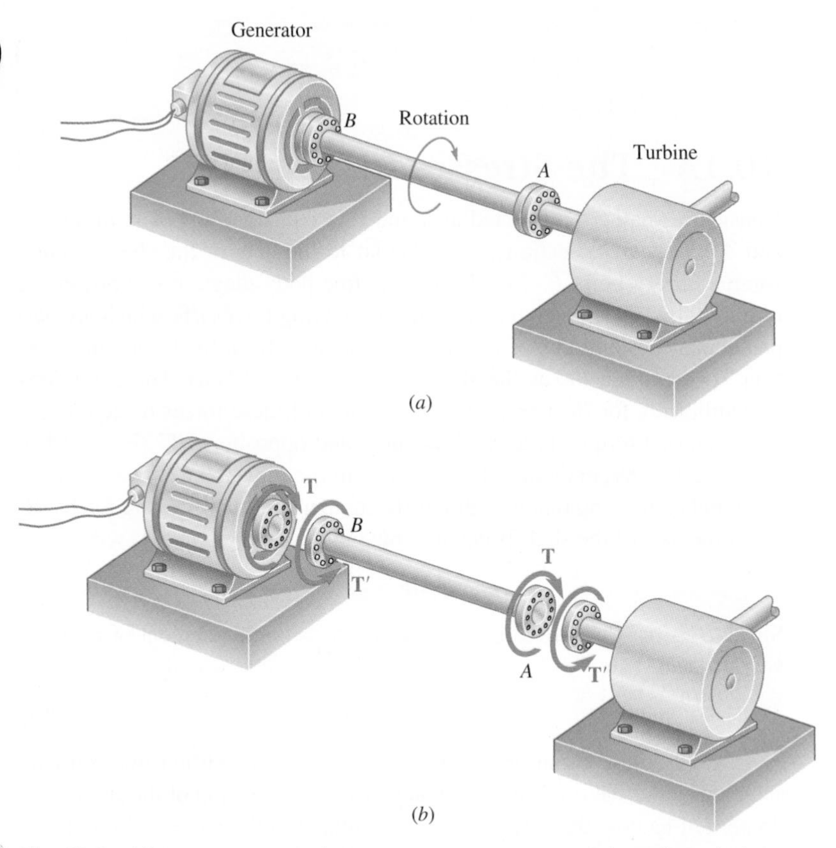

Fig. 10.2 (*a*) A generator receives power at a constant number of revolutions per minute from a turbine through shaft *AB*. (*b*) Free-body diagram of shaft *AB* along with the driving and reacting torques on the generator and turbine, respectively.

twisting couple or torque **T** on the shaft, which then exerts an equal torque on the generator. The generator reacts by exerting the equal and opposite torque **T**′ on the shaft, and the shaft reacts by exerting the torque **T**′ on the turbine.

First the stresses and deformations that take place in circular shafts will be analyzed. Then an important property of circular shafts is demonstrated: *When a circular shaft is subjected to torsion, every cross section remains plane and undistorted*. Therefore, while the various cross sections along the shaft rotate through different angles, each cross section rotates as a solid rigid slab. This property helps to determine the *distribution of shearing strains in a circular shaft and to conclude that the shearing strain varies linearly with the distance from the axis of the shaft*.

Deformations in the *elastic range* and Hooke's law for shearing stress and strain are used to determine the *distribution of shearing stresses* in a circular shaft and derive the *elastic torsion formulas*.

In Sec. 10.2, the *angle of twist* of a circular shaft is found when subjected to a given torque, assuming elastic deformations. The solution of problems involving *statically indeterminate shafts* is discussed in Sec. 10.3.

10.1 CIRCULAR SHAFTS IN TORSION

10.1A The Stresses in a Shaft

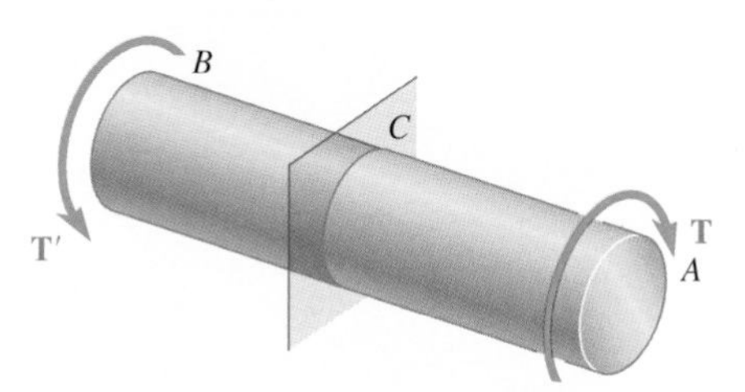

Fig. 10.3 Shaft subject to torques, with a section plane at *C*.

Consider a shaft *AB* subjected at *A* and *B* to equal and opposite torques **T** and **T′**. We pass a section perpendicular to the axis of the shaft through some arbitrary point *C* (Fig. 10.3). The free-body diagram of portion *BC* of the shaft must include the elementary shearing forces $d\mathbf{F}$, which are perpendicular to the radius of the shaft. These arise from the torque that portion *AC* exerts on *BC* as the shaft is twisted (Fig. 10.4*a*). The conditions of equilibrium for *BC* require that the system of these forces be equivalent to an internal torque **T**, as well as equal and opposite to **T′** (Fig. 10.4*b*). Denoting the perpendicular distance ρ from the force $d\mathbf{F}$ to the axis of the shaft and expressing that the sum of the moments of the shearing forces $d\mathbf{F}$ about the axis of the shaft is equal in magnitude to the torque **T**, write

$$\int \rho dF = T$$

Since $dF = \tau dA$, where τ is the shearing stress on the element of area dA, you also can write

$$\int \rho(\tau dA) = T \quad \textbf{(10.1)}$$

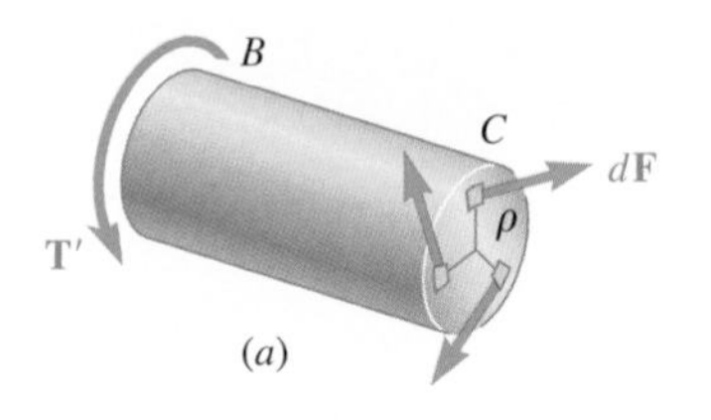

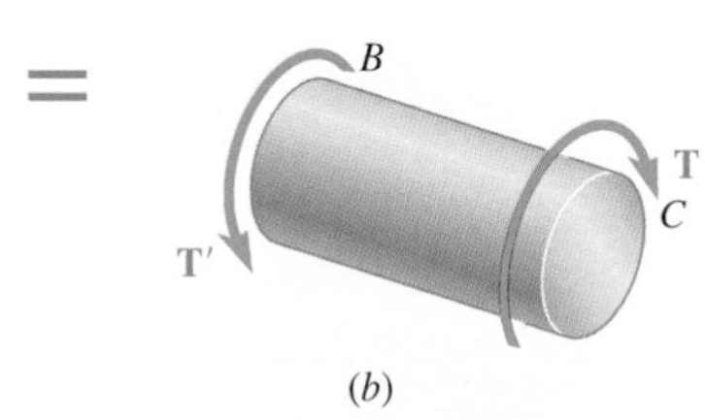

Fig. 10.4 (*a*) Free body diagram of section *BC* with torque at *C* represented by the contributions of small elements of area carrying forces *dF* at a radius ρ from the section center. (*b*) Free-body diagram of section *BC* having all the small area elements summed, resulting in torque *T*.

While these equations express an important condition that must be satisfied by the shearing stresses in any given cross section of the shaft, they do *not* tell us how these stresses are distributed in the cross section. Thus, the actual distribution of stresses under a given load is *statically indeterminate* (i.e., this distribution *cannot be determined by the methods of statics*). However, it was assumed in Sec. 8.1A that the normal stresses produced by an axial centric load were uniformly distributed, and this assumption was justified in Sec. 9.8, except in the neighborhood of concentrated loads. A similar assumption with respect to the distribution of shearing stresses in an elastic shaft *would be wrong*. Withhold any judgment until the *deformations* that are produced in the shaft have been analyzed. This will be done in the next section.

As indicated in Sec. 8.3, shear cannot take place in one plane only. Consider the very small element of shaft shown in Fig. 10.5. The torque applied to the shaft produces shearing stresses τ on the faces perpendicular to the axis of the shaft. However, the conditions of equilibrium (Sec. 8.3) require the existence of equal stresses on the faces formed by the two planes containing the axis of the shaft. That such shearing stresses actually occur in torsion can be demonstrated by considering a

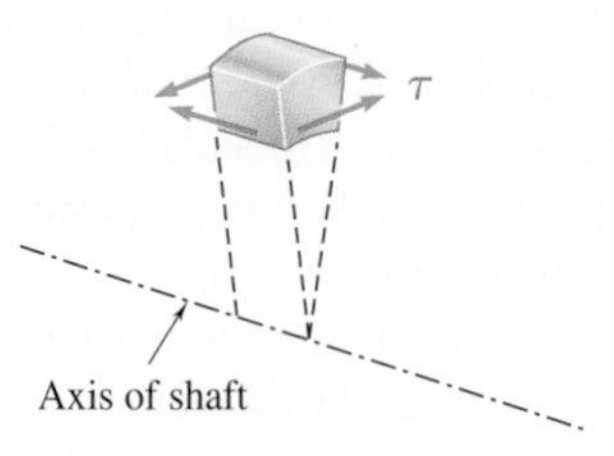

Fig. 10.5 Small element in shaft showing how shearing stress components act.

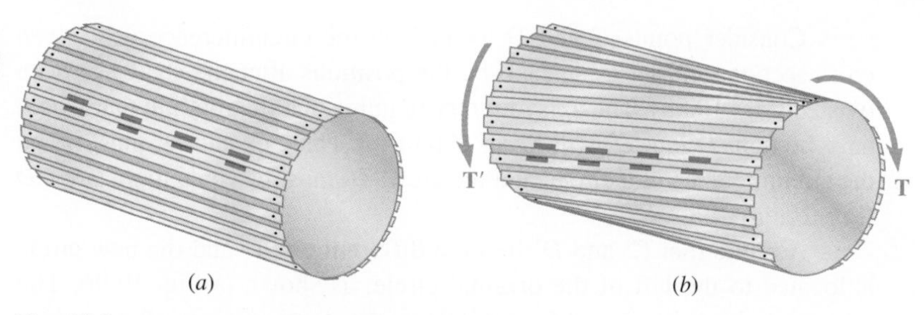

Fig. 10.6 Demonstration of shear in a shaft (*a*) undeformed; (*b*) loaded and deformed.

"shaft" made of separate slats pinned at both ends to disks, as shown in Fig. 10.6*a*. If markings have been painted on two adjoining slats, it is observed that the slats will slide with respect to each other when equal and opposite torques are applied to the ends of the "shaft" (Fig. 10.6*b*). While sliding will not actually take place in a shaft made of a homogeneous and cohesive material, the tendency for sliding will exist, showing that stresses occur on longitudinal planes as well as on planes perpendicular to the axis of the shaft.[†]

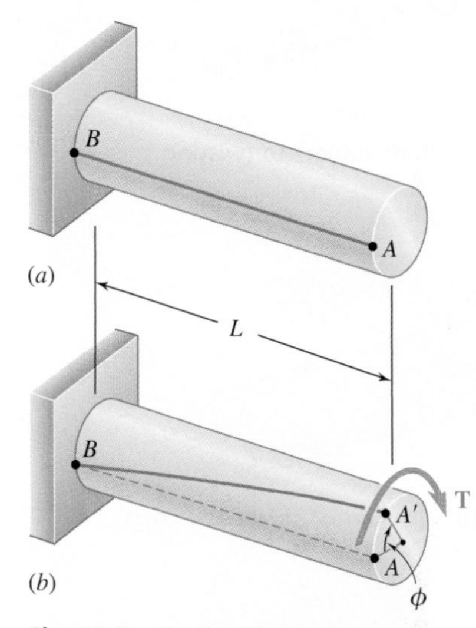

Fig. 10.7 Shaft with fixed support and line *AB* drawn showing deformation under torsion loading: (*a*) unloaded; (*b*) loaded.

10.1B Deformations in a Circular Shaft

Deformation Characteristics. Consider a circular shaft attached to a fixed support at one end (Fig. 10.7*a*). If a torque **T** is applied to the other end, the shaft will twist, with its free end rotating through an angle ϕ called *the angle of twist* (Fig. 10.7*b*). Within a certain range of values of T, the angle of twist ϕ is proportional to T. Also, ϕ is proportional to the length L of the shaft. In other words, the angle of twist for a shaft of the same material and same cross section, but twice as long, will be twice as large under the same torque **T**.

When a circular shaft is subjected to torsion, *every cross section remains plane and undistorted.* In other words, while the various cross sections along the shaft rotate through different amounts, each cross section rotates as a solid rigid slab. This is illustrated in Fig. 10.8*a*, which shows the deformations in a rubber model subjected to torsion. This property is characteristic of circular shafts, whether solid or hollow—but not of members with noncircular cross section. For example, when a bar of square cross section is subjected to torsion, its various cross sections warp and do not remain plane (Fig. 10.8*b*).

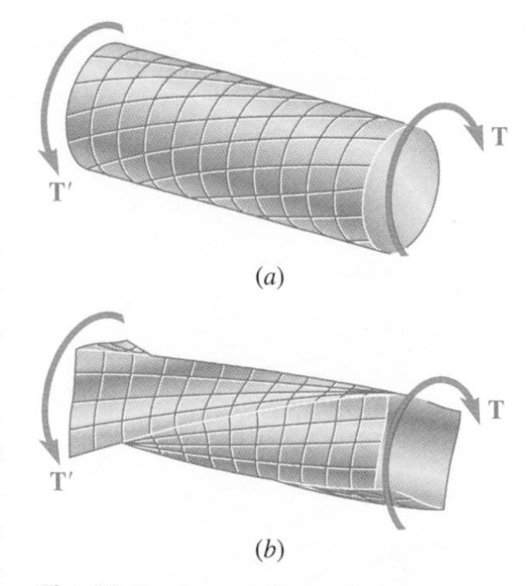

Fig. 10.8 Comparison of deformations in (*a*) circular and (*b*) square shafts.

The cross sections of a circular shaft remain plane and undistorted because a circular shaft is *axisymmetric* (i.e., its appearance remains the same when it is viewed from a fixed position and rotated about its axis through an arbitrary angle). Square bars, on the other hand, retain the same appearance only if they are rotated through 90° or 180°. Theoretically the axisymmetry of circular shafts can be used to prove that their cross sections remain plane and undistorted.

[†]The twisting of a cardboard tube that has been slit lengthwise provides another demonstration of the existence of shearing stresses on longitudinal planes.

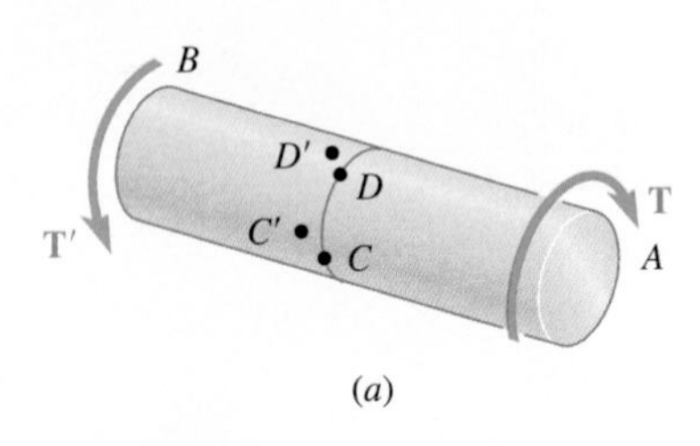

(a)

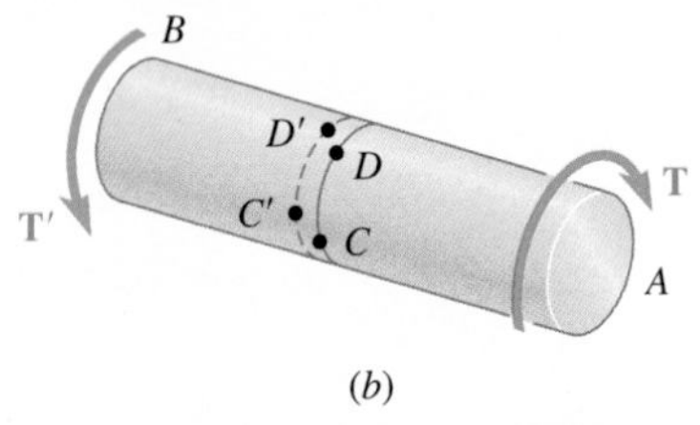

(b)

Fig. 10.9 Shaft subject to twisting.

Consider points C and D located on the circumference of a given cross section, and let C' and D' be the positions after the shaft has been twisted (Fig. 10.9*a*). The axisymmetry requires that the rotation that would have brought D into D' will bring C into C'. Thus, C' and D' must lie on the circumference of a circle, and the arc $C'D'$ must be equal to the arc CD (Fig. 10.9*b*).

Assume that C' and D' lie on a different circle, and the new circle is located to the left of the original circle, as shown in Fig. 10.9*b*. The same situation will prevail for any other cross section, since all cross sections of the shaft are subjected to the same internal torque T, and looking at the shaft from its end A shows that the loading causes any given circle drawn on the shaft to move *away*. But viewed from B, the given load looks the same (a clockwise couple in the foreground and a counterclockwise couple in the background), where the circle moves *toward* you. This contradiction proves that C' and D' lie on the same circle as C and D. Thus, as the shaft is twisted, the original circle just rotates in its own plane. Since the same reasoning can be applied to any smaller, concentric circle located in the cross section, the entire cross section remains plane (Fig. 10.10).

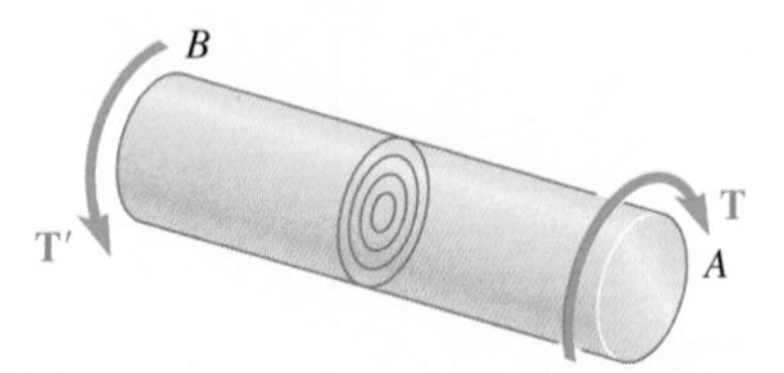

Fig. 10.10 Concentric circles at a cross section.

This argument does not preclude the possibility for the various concentric circles of Fig. 10.10 to rotate by different amounts when the shaft is twisted. But if that were so, a given diameter of the cross section would be distorted into a curve, as shown in Fig. 10.11*a*. Looking at this curve from A, the outer layers of the shaft get more twisted than the inner ones, while looking from B reveals the opposite (Fig. 10.11*b*). This inconsistency indicates that any diameter of a given cross section remains straight (Fig. 10.11*c*); therefore, any given cross section of a circular shaft remains plane and undistorted.

Now consider the mode of application of the twisting couples **T** and **T**′. If *all* sections of the shaft, from one end to the other, are to remain plane and undistorted, the couples are applied so the ends of the shaft remain plane and undistorted. This can be accomplished by applying the couples **T** and **T**′ to rigid plates that are solidly attached to the ends of the shaft (Fig. 10.12*a*). All sections will remain plane and undistorted when the loading is applied, and the resulting deformations will be uniform throughout the entire length of the shaft. All of the equally spaced circles shown in Fig. 10.12*a* will rotate by the same amount relative to their neighbors, and each of the straight lines will be transformed into a curve (helix) intersecting the various circles at the same angle (Fig. 10.12*b*).

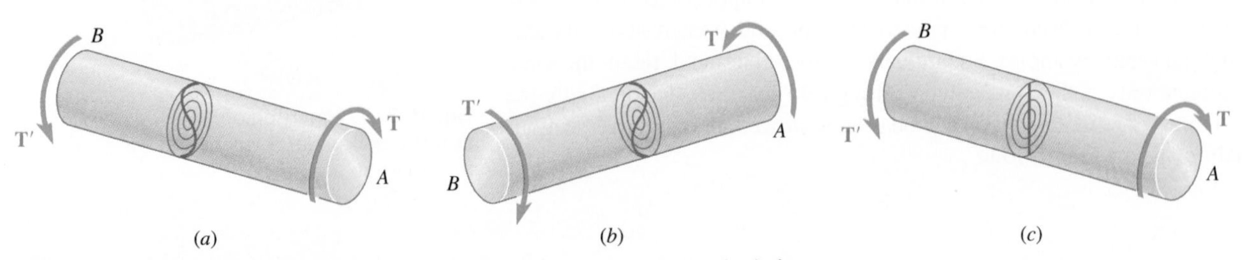

Fig. 10.11 Potential deformations of diameter lines if section's concentric circles rotate different amounts (*a*, *b*) or the same amount (*c*).

Shearing Strains. The examples given in this and the following sections are based on the assumption of rigid end plates. However, loading conditions may differ from those corresponding to the model of Fig. 10.12. This model helps to define a torsion problem for which we can obtain an exact solution. By use of Saint-Venant's principle, the results obtained for this idealized model may be extended to most engineering applications.

Now we will determine the distribution of *shearing strains* in a circular shaft of length L and radius c that has been twisted through an angle ϕ (Fig. 10.13*a*). Detaching from the shaft a cylinder of radius ρ, consider the small square element formed by two adjacent circles and two adjacent straight lines traced on the surface before any load is applied (Fig. 10.13*b*). As the shaft is subjected to a torsional load, the element deforms into a rhombus (Fig. 10.13*c*). Here the shearing strain γ in a given element is measured by the change in the angles formed by the sides of that element (Sec. 9.6). Since the circles defining two of the sides remain unchanged, the shearing strain γ must be equal to the angle between lines AB and $A'B$.

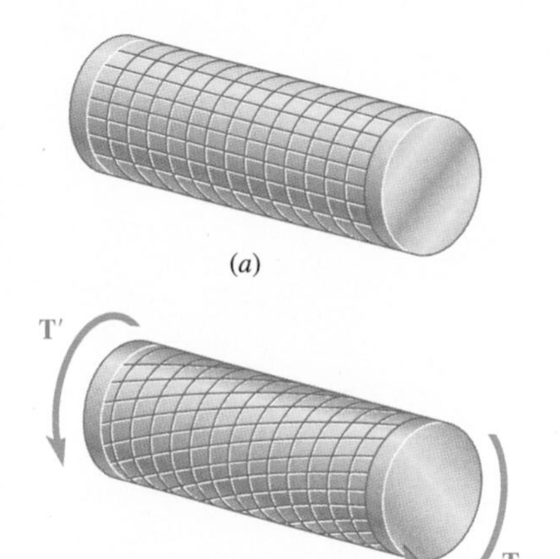

Fig. 10.12 Visualization of deformation resulting from twisting couples: (*a*) undeformed, (*b*) deformed.

Fig. 10.13*c* shows that, for small values of γ, the arc length AA' is expressed as $AA' = L\gamma$. But since $AA' = \rho\phi$, it follows that $L\gamma = \rho\phi$, or

$$\gamma = \frac{\rho\phi}{L} \tag{10.2}$$

where γ and ϕ are in radians. This equation shows that the shearing strain γ at a given point of a shaft in torsion is proportional to the angle of twist ϕ. It also shows that γ is proportional to the distance ρ from the axis of the shaft to that point. Thus, *the shearing strain in a circular shaft varies linearly with the distance from the axis of the shaft.*

From Eq. (10.2), the shearing strain is maximum on the surface of the shaft, where $\rho = c$.

$$\gamma_{\max} = \frac{c\phi}{L} \tag{10.3}$$

Eliminating ϕ from Eqs. (10.2) and (10.3), the shearing strain γ at a distance ρ from the axis of the shaft is

$$\gamma = \frac{\rho}{c}\gamma_{\max} \tag{10.4}$$

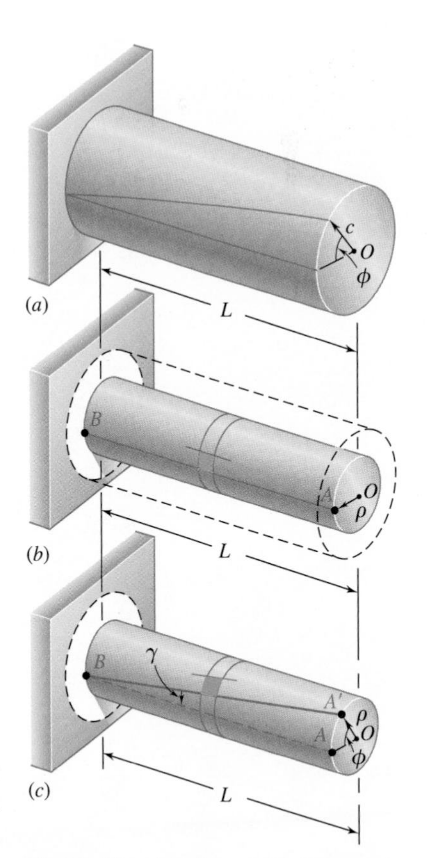

Fig. 10.13 Shearing strain deformation. (*a*) The angle of twist ϕ. (*b*) Undeformed portion of shaft of radius ρ. (*c*) Deformed portion of shaft; angle of twist ϕ and shearing strain γ share the same arc length AA'.

10.1C Stresses in the Elastic Range

When the torque **T** is such that all shearing stresses in the shaft remain below the yield strength τ_Y, the stresses in the shaft will remain below both the proportional limit and the elastic limit. Thus, Hooke's law will apply, and there will be no permanent deformation.

Recalling Hooke's law for shearing stress and strain from Sec. 9.6, write

$$\tau = G\gamma \tag{10.5}$$

where G is the modulus of rigidity or shear modulus of the material. Multiplying both members of Eq. (10.4) by G, write

$$G\gamma = \frac{\rho}{c} G\gamma_{\max}$$

or, making use of Eq. (10.5),

$$\tau = \frac{\rho}{c} \tau_{\max} \qquad \textbf{(10.6)}$$

This equation shows that, as long as the yield strength (or proportional limit) is not exceeded in any part of a circular shaft, *the shearing stress in the shaft varies linearly with the distance ρ from the axis of the shaft.* Fig. 10.14*a* shows the stress distribution in a solid circular shaft of radius c. A hollow circular shaft of inner radius c_1 and outer radius c_2 is shown in Fig. 10.14*b*. From Eq. (10.6),

$$\tau_{\min} = \frac{c_1}{c_2} \tau_{\max} \qquad \textbf{(10.7)}$$

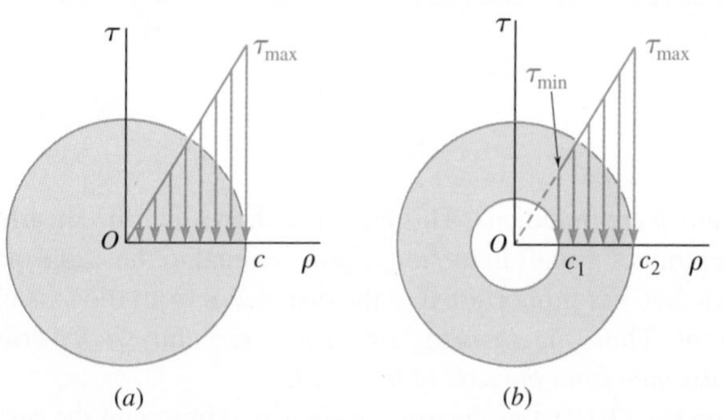

Fig. 10.14 Distribution of shearing stresses in a torqued shaft: (*a*) solid shaft, (*b*) hollow shaft.

Recall from Sec. 10.1A that the sum of the moments of the elementary forces exerted on any cross section of the shaft must be equal to the magnitude T of the torque exerted on the shaft:

$$\int \rho(\tau\, dA) = T \qquad \textbf{(10.1)}$$

Substituting for τ from Eq. (10.6) into Eq. (10.1),

$$T = \int \rho\tau\, dA = \frac{\tau_{\max}}{c} \int \rho^2\, dA$$

The integral in the last part represents the polar moment of inertia J of the cross section with respect to its center O. Therefore,

$$T = \frac{\tau_{\max} J}{c} \qquad \textbf{(10.8)}$$

or solving for $\tau_{\max}$,

$$\tau_{\max} = \frac{Tc}{J} \qquad \textbf{(10.9)}$$

Substituting for τ_{max} from Eq. (10.9) into Eq. (10.6), the shearing stress at any distance ρ from the axis of the shaft is

$$\tau = \frac{T\rho}{J} \tag{10.10}$$

Eqs. (10.9) and (10.10) are known as the *elastic torsion formulas*. Recall from statics that the polar moment of inertia of a circle of radius c is $J = \frac{1}{2}\pi c^4$. For a hollow circular shaft of inner radius c_1 and outer radius c_2, the polar moment of inertia is

$$J = \tfrac{1}{2}\pi c_2^4 - \tfrac{1}{2}\pi c_1^4 = \tfrac{1}{2}\pi(c_2^4 - c_1^4) \tag{10.11}$$

When SI metric units are used in Eq. (10.9) or (10.10), T is given in N·m, c or ρ in meters, and J in m^4. The resulting shearing stress is given in N/m^2, that is, pascals (Pa). When U.S. customary units are used, T is given in lb·in., c or ρ in inches, and J in in^4. The resulting shearing stress is given in psi.

Concept Application 10.1

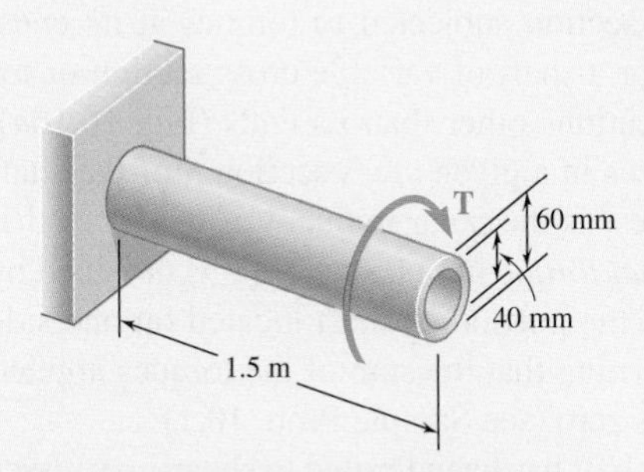

Fig. 10.15 Hollow, fixed-end shaft having torque T applied at end.

A hollow cylindrical steel shaft is 1.5 m long and has inner and outer diameters respectively equal to 40 and 60 mm (Fig. 10.15). (*a*) What is the largest torque that can be applied to the shaft if the shearing stress is not to exceed 120 MPa? (*b*) What is the corresponding minimum value of the shearing stress in the shaft?

The largest torque **T** that can be applied to the shaft is the torque for which $\tau_{max} = 120$ MPa. Since this is less than the yield strength for any steel, use Eq. (10.9). Solving this equation for T,

$$T = \frac{J\tau_{max}}{c} \tag{1}$$

Recalling that the polar moment of inertia J of the cross section is given by Eq. (10.11), where $c_1 = \frac{1}{2}(40 \text{ mm}) = 0.02$ m and $c_2 = \frac{1}{2}(60 \text{ mm}) = 0.03$ m, write

$$J = \tfrac{1}{2}\pi(c_2^4 - c_1^4) = \tfrac{1}{2}\pi(0.03^4 - 0.02^4) = 1.021 \times 10^{-6} \text{ m}^4$$

Substituting for J and τ_{max} into Eq. (1) and letting $c = c_2 = 0.03$ m,

$$T = \frac{J\tau_{max}}{c} = \frac{(1.021 \times 10^{-6} \text{ m}^4)(120 \times 10^6 \text{ Pa})}{0.03 \text{ m}} = 4.08 \text{ kN·m}$$

The minimum shearing stress occurs on the inner surface of the shaft. Eq. (10.7) expresses that τ_{min} and τ_{max} are respectively proportional to c_1 and c_2:

$$\tau_{min} = \frac{c_1}{c_2}\tau_{max} = \frac{0.02 \text{ m}}{0.03 \text{ m}}(120 \text{ MPa}) = 80 \text{ MPa}$$

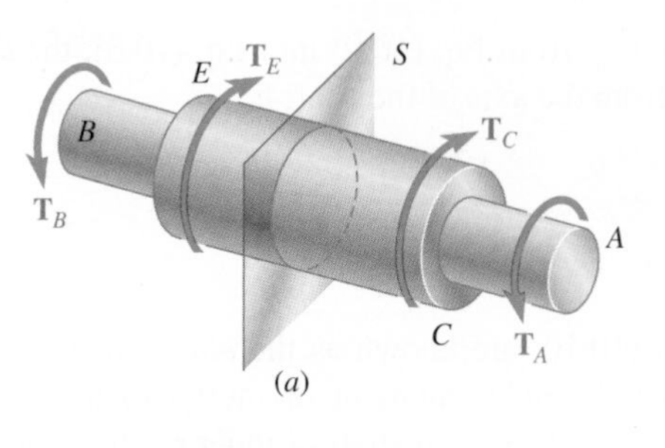

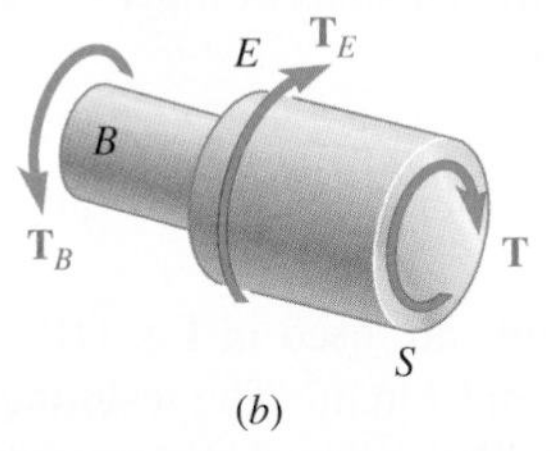

Fig. 10.16 Shaft with variable cross section. (*a*) With applied torques and section *S*. (*b*) Free-body diagram of sectioned shaft.

The torsion formulas of Eqs. (10.9) and (10.10) were derived for a shaft of uniform circular cross section subjected to torques at its ends. However, they also can be used for a shaft of variable cross section or for a shaft subjected to torques at locations other than its ends (Fig. 10.16*a*). The distribution of shearing stresses in a given cross section S of the shaft is obtained from Eq. (10.9), where J is the polar moment of inertia of that section and T represents the *internal torque* in that section. T is obtained by drawing the free-body diagram of the portion of shaft located on one side of the section (Fig. 10.16*b*) and writing that the sum of the torques applied (including the internal torque **T**) is zero (see Sample Prob. 10.1).

Our analysis of stresses in a shaft has been limited to shearing stresses due to the fact that the element selected was oriented so that its faces were either parallel or perpendicular to the axis of the shaft (Fig. 10.5). Now consider two elements a and b located on the surface of a circular shaft subjected to torsion (Fig. 10.17). Since the faces of element a are respectively parallel and perpendicular to the axis of the shaft, the only stresses on the element are the shearing stresses

$$\tau_{\max} = \frac{Tc}{J} \tag{10.9}$$

On the other hand, the faces of element b, which form arbitrary angles with the axis of the shaft, are subjected to a combination of normal and shearing stresses. Consider the stresses and resulting forces on faces that

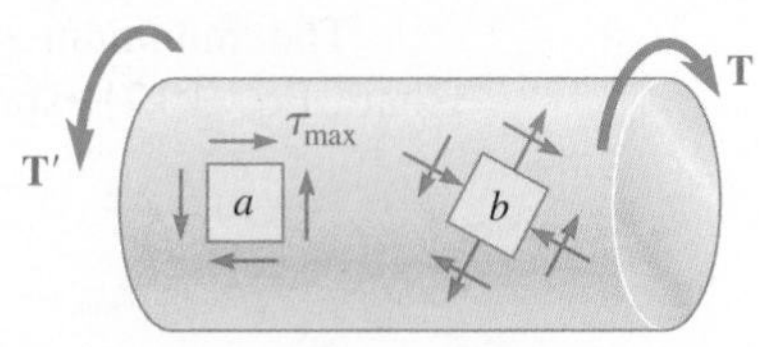

Fig. 10.17 Circular shaft with stress elements at different orientations.

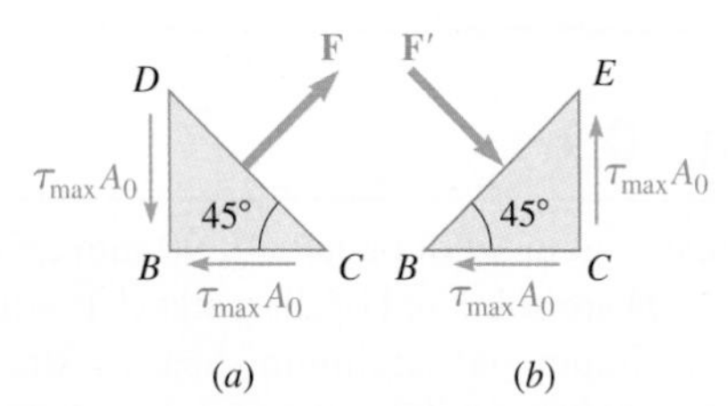

Fig. 10.18 Forces on faces at 45° to shaft axis.

are at 45° to the axis of the shaft. The free-body diagrams of the two triangular elements are shown in Fig. 10.18. From Fig. 10.18*a*, the stresses exerted on the faces *BC* and *BD* are the shearing stresses $\tau_{max} = Tc/J$. The magnitude of the corresponding shear forces is $\tau_{max}A_0$, where A_0 is the area of the face. Observing that the components along *DC* of the two shear forces are equal and opposite, the force **F** exerted on *DC* must be perpendicular to that face and is a tensile force. Its magnitude is

$$F = 2(\tau_{max}A_0)\cos 45° = \tau_{max}A_0\sqrt{2} \quad \textbf{(10.12)}$$

The corresponding stress is obtained by dividing the force *F* by the area *A* of face *DC*. Observing that $A = A_0\sqrt{2}$,

$$\sigma = \frac{F}{A} = \frac{\tau_{max}A_0\sqrt{2}}{A_0\sqrt{2}} = \tau_{max} \quad \textbf{(10.13)}$$

A similar analysis of the element of Fig. 10.18*b* shows that the stress on the face *BE* is $\sigma = -\tau_{max}$. Therefore, the stresses exerted on the faces of an element *c* at 45° to the axis of the shaft (Fig. 10.19) are normal stresses equal to $\pm\tau_{max}$. Thus, while element *a* in Fig. 10.19 is in pure shear, element *c* in the same figure is subjected to a tensile stress on two of its faces and a compressive stress on the other two. Also note that all of the stresses involved have the same magnitude, *Tc*/*J*.[†]

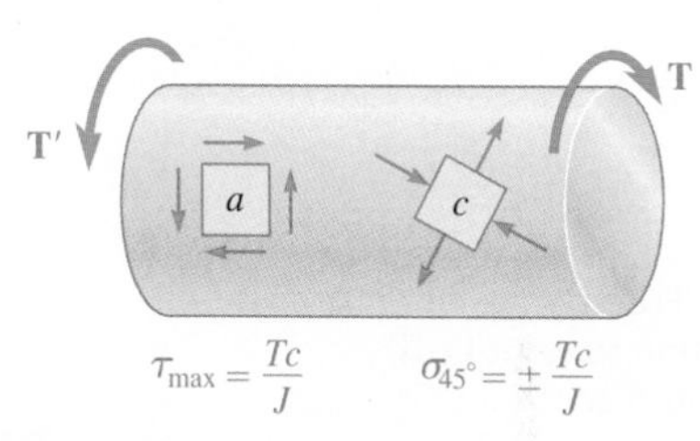

Fig. 10.19 Shaft elements with only shearing stresses or normal stresses.

Because ductile materials generally fail in shear, a specimen subjected to torsion breaks along a plane perpendicular to its longitudinal axis (Photo 10.2*a*). On the other hand, brittle materials are weaker in tension than in shear. Thus, when subjected to torsion, a brittle material tends to break along surfaces perpendicular to the direction in which tension is maximum, forming a 45° angle with the longitudinal axis of the specimen (Photo 10.2*b*).

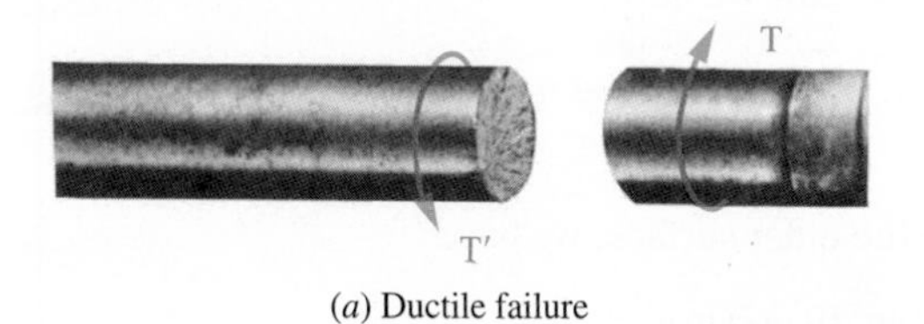

(*a*) Ductile failure

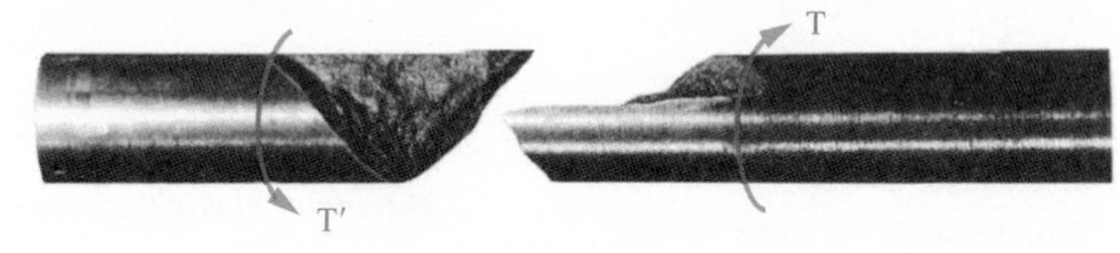

(*b*) Brittle failure

Photo 10.2 Shear failure of shaft subject to torque.

[†]Stresses on elements of arbitrary orientation, such as in Fig. 10.18*b*, will be discussed in Chap. 14.

Sample Problem 10.1

Shaft BC is hollow with inner and outer diameters of 90 mm and 120 mm, respectively. Shafts AB and CD are solid and of diameter d. For the loading shown, determine (*a*) the maximum and minimum shearing stress in shaft BC, (*b*) the required diameter d of shafts AB and CD if the allowable shearing stress in these shafts is 65 MPa.

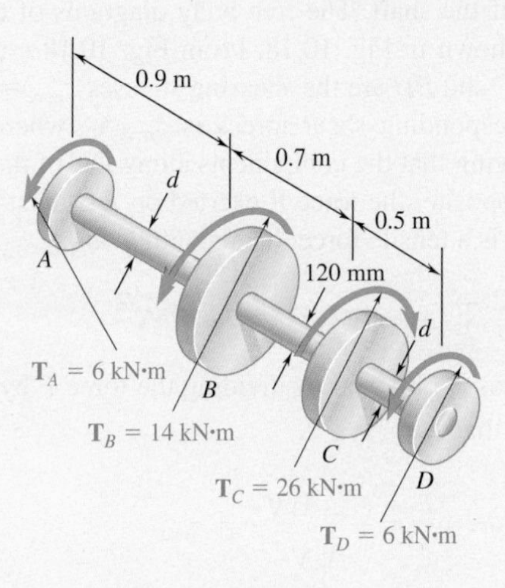

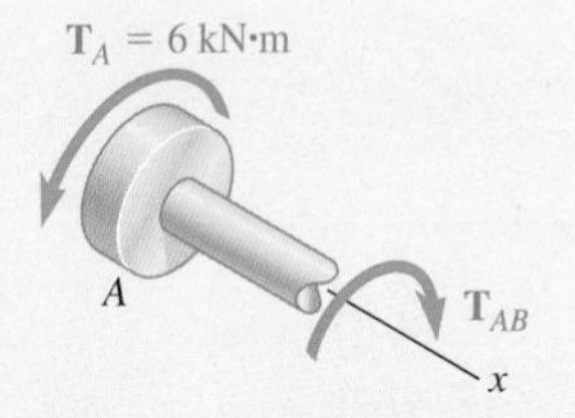

Fig. 1 Free-body diagram for section to left of cut between *A* and *B*.

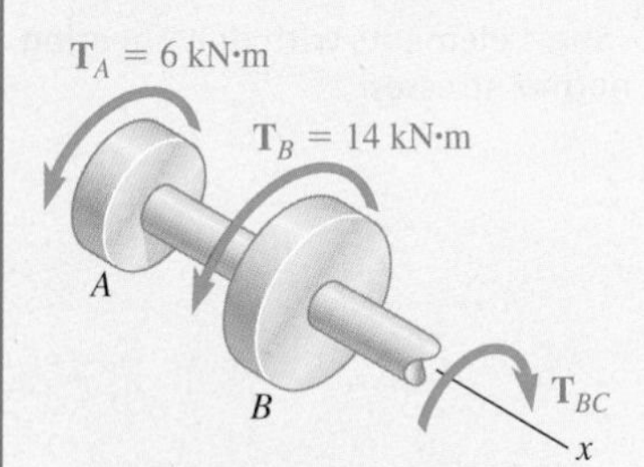

Fig. 2 Free-body diagram for section to left of cut between *B* and *C*.

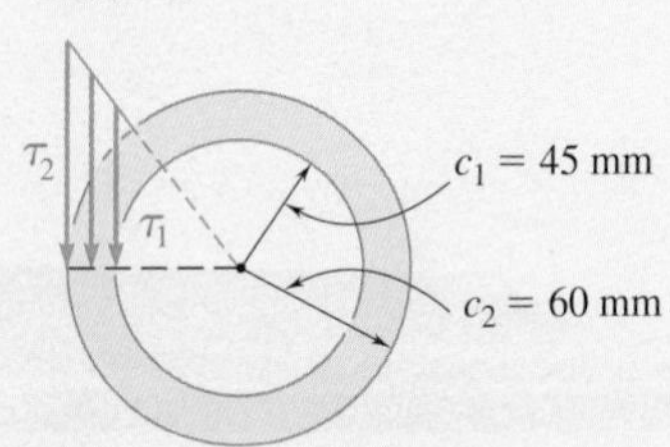

Fig. 3 Shearing stress distribution on cross section.

STRATEGY: Use free-body diagrams to determine the torque in each shaft. The torques can then be used to find the stresses for shaft BC and the required diameters for shafts AB and CD.

MODELING: Denoting by $\mathbf{T}_{AB}$ the torque in shaft AB (Fig. 1), we pass a section through shaft AB and, for the free body shown, we write

$$\Sigma M_x = 0: \qquad (6\text{ kN}\cdot\text{m}) - T_{AB} = 0 \qquad T_{AB} = 6\text{ kN}\cdot\text{m}$$

We now pass a section through shaft BC (Fig. 2) and, for the free body shown, we have

$$\Sigma M_x = 0: \quad (6\text{ kN}\cdot\text{m}) + (14\text{ kN}\cdot\text{m}) - T_{BC} = 0 \qquad T_{BC} = 20\text{ kN}\cdot\text{m}$$

ANALYSIS:

a. Shaft *BC*. For this hollow shaft we have

$$J = \frac{\pi}{2}(c_2^4 - c_1^4) = \frac{\pi}{2}[(0.060)^4 - (0.045)^4] = 13.92 \times 10^{-6}\text{ m}^4$$

Maximum Shearing Stress. On the outer surface, we have

$$\tau_{\max} = \tau_2 = \frac{T_{BC}c_2}{J} = \frac{(20\text{ kN}\cdot\text{m})(0.060\text{ m})}{13.92 \times 10^{-6}\text{ m}^4} \qquad \tau_{\max} = 86.2\text{ MPa} ◀$$

(continued)

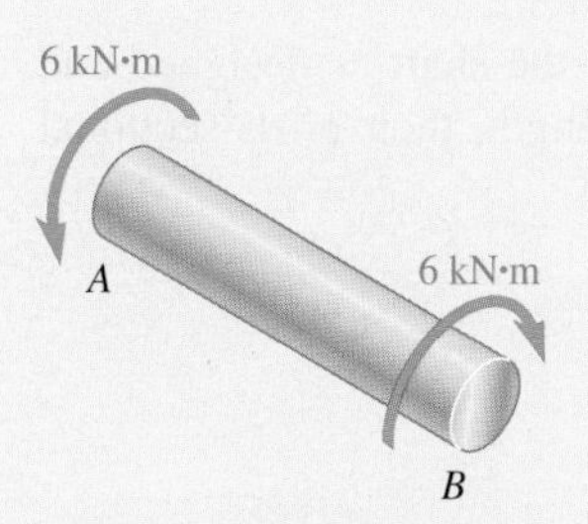

Fig. 4 Free-body diagram of shaft portion AB.

Minimum Shearing Stress. As shown in Fig. 3 the stresses are proportional to the distance from the axis of the shaft.

$$\frac{\tau_{min}}{\tau_{max}} = \frac{c_1}{c_2} \qquad \frac{\tau_{min}}{86.2 \text{ MPa}} = \frac{45 \text{ mm}}{60 \text{ mm}} \qquad \tau_{min} = 64.7 \text{ MPa} \blacktriangleleft$$

b. Shafts *AB* and *CD*. We note that both shafts have the same torque $T = 6$ kN·m (Fig. 4). Denoting the radius of the shafts by c and knowing that $\tau_{all} = 65$ MPa, we write

$$\tau = \frac{Tc}{J} \qquad 65 \text{ MPa} = \frac{(6 \text{ kN·m})c}{\frac{\pi}{2}c^4}$$

$$c^3 = 58.8 \times 10^{-6} \text{ m}^3 \qquad c = 38.9 \times 10^{-3} \text{ m}$$

$$d = 2c = 2(38.9 \text{ mm}) \qquad d = 77.8 \text{ mm} \blacktriangleleft$$

Sample Problem 10.2

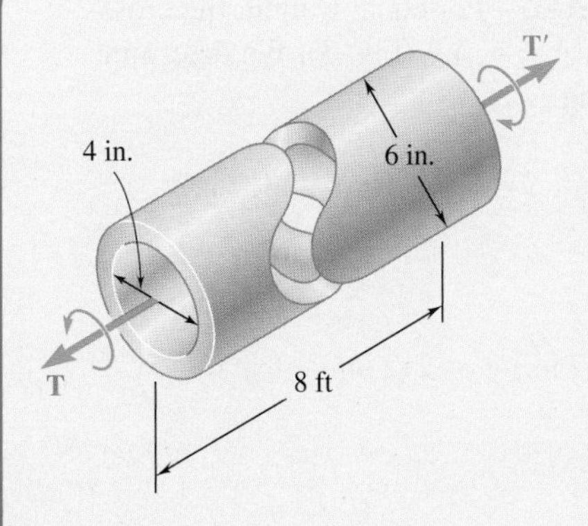

The preliminary design of a motor-to-generator connection calls for the use of a large hollow shaft with inner and outer diameters of 4 in. and 6 in., respectively. Knowing that the allowable shearing stress is 12 ksi, determine the maximum torque that can be transmitted by (*a*) the shaft as designed, (*b*) a solid shaft of the same weight, and (*c*) a hollow shaft of the same weight and an 8-in. outer diameter.

STRATEGY: Use Eq. (10.9) to determine the maximum torque using the allowable stress.

MODELING and ANALYSIS:

a. Hollow Shaft as Designed. Using Fig. 1 and setting $\tau_{all} = 12$ ksi, we write

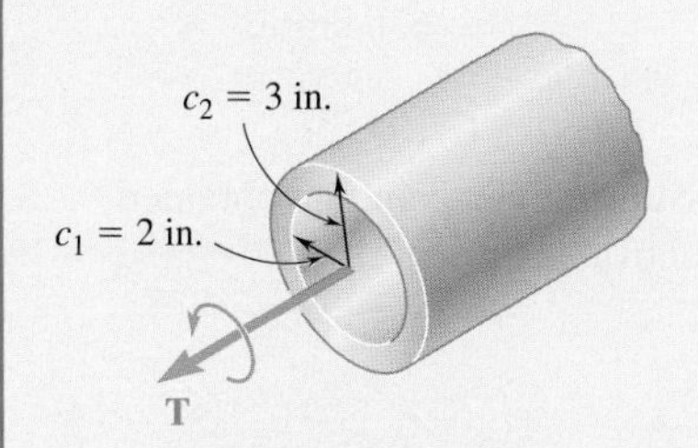

Fig. 1 Shaft as designed.

$$J = \frac{\pi}{2}(c_2^4 - c_1^4) = \frac{\pi}{2}[(3 \text{ in.})^4 - (2 \text{ in.})^4] = 102.1 \text{ in}^4$$

Using Eq. (10.9), we write

$$\tau_{max} = \frac{Tc_2}{J} \qquad 12 \text{ ksi} = \frac{T(3 \text{ in.})}{102.1 \text{ in}^4} \qquad T = 408 \text{ kip·in.} \blacktriangleleft$$

(continued)

b. Solid Shaft of Equal Weight. For the shaft as designed and this solid shaft to have the same weight and length, their cross-sectional areas must be equal, i.e. $A_{(a)} = A_{(b)}$.

$$\pi[(3 \text{ in.})^2 - (2 \text{ in.})^2] = \pi c_3^2 \qquad c_3 = 2.24 \text{ in.}$$

Using Fig. 2 and setting $\tau_{all} = 12$ ksi, we write

$$\tau_{max} = \frac{Tc_3}{J} \qquad 12 \text{ ksi} = \frac{T(2.24 \text{ in.})}{\frac{\pi}{2}(2.24 \text{ in.})^4} \qquad T = 211 \text{ kip·in.} \blacktriangleleft$$

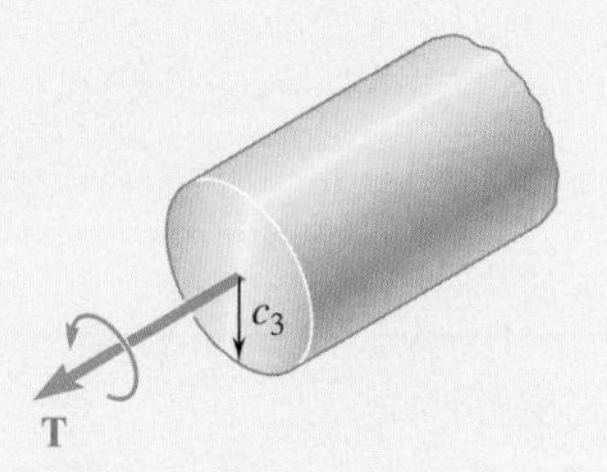

Fig. 2 Solid shaft having equal weight.

c. Hollow Shaft of 8-in. Diameter. For equal weight, the cross-sectional areas again must be equal, i.e., $A_{(a)} = A_{(c)}$ (Fig. 3). We determine the inside diameter of the shaft by writing

$$\pi[(3 \text{ in.})^2 - (2 \text{ in.})^2] = \pi[(4 \text{ in.})^2 - c_5^2] \qquad c_5 = 3.317 \text{ in.}$$

For $c_5 = 3.317$ in. and $c_4 = 4$ in.,

$$J = \frac{\pi}{2}[(4 \text{ in.})^4 - (3.317 \text{ in.})^4] = 212 \text{ in}^4$$

With $\tau_{all} = 12$ ksi and $c_4 = 4$ in.,

$$\tau_{max} = \frac{Tc_4}{J} \qquad 12 \text{ ksi} = \frac{T(4 \text{ in.})}{212 \text{ in}^4} \qquad T = 636 \text{ kip·in.} \blacktriangleleft$$

REFLECT and THINK: This example illustrates the advantage obtained when the shaft material is further from the centroidal axis.

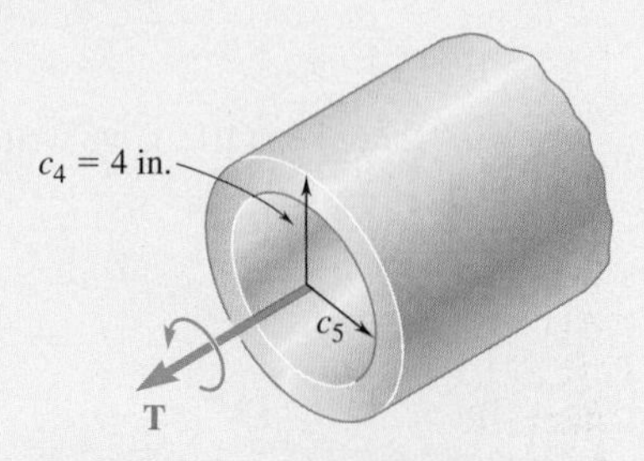

Fig. 3 Hollow shaft with an 8-in. outer diameter, having equal weight.

Problems

10.1 Determine the torque **T** that causes a maximum shearing stress of 70 MPa in the steel cylindrical shaft shown.

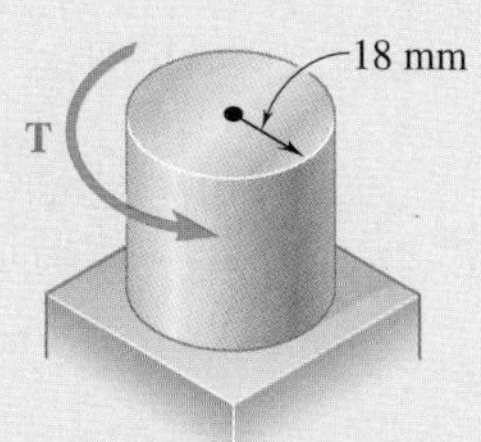

Fig. P10.1 and P10.2

10.2 Determine the maximum shearing stress caused by a torque of magnitude $T = 800$ N·m.

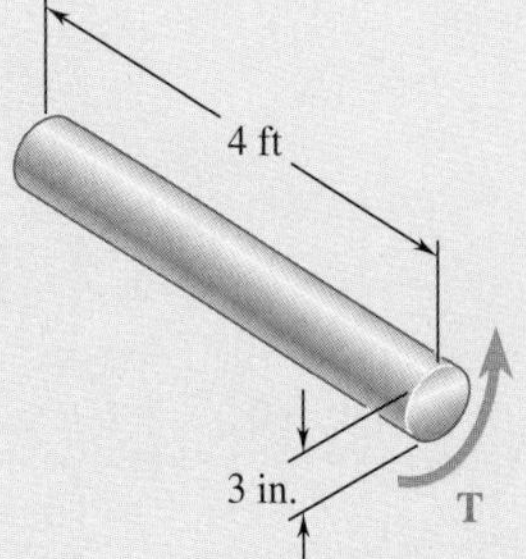

Fig. P10.3

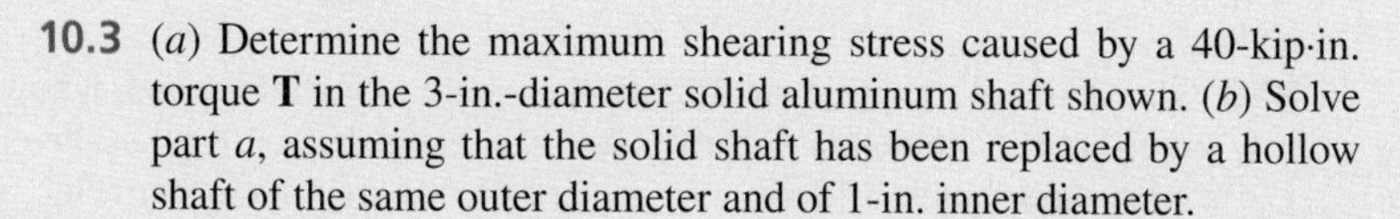

10.3 (*a*) Determine the maximum shearing stress caused by a 40-kip·in. torque **T** in the 3-in.-diameter solid aluminum shaft shown. (*b*) Solve part *a*, assuming that the solid shaft has been replaced by a hollow shaft of the same outer diameter and of 1-in. inner diameter.

10.4 (*a*) Determine the torque that can be applied to a solid shaft of 20-mm diameter without exceeding an allowable shearing stress of 80 MPa. (*b*) Solve part *a*, assuming that the solid shaft has been replaced by a hollow shaft of the same cross-sectional area and with an inner diameter equal to half of its outer diameter.

10.5 A torque $T = 3$ kN·m is applied to the solid bronze cylinder shown. Determine (*a*) the maximum shearing stress, (*b*) the shearing stress at point *D* that lies on a 15-mm radius circle drawn on the end of the cylinder, (*c*) the percent of the torque carried by the portion of the cylinder within the 15 mm radius.

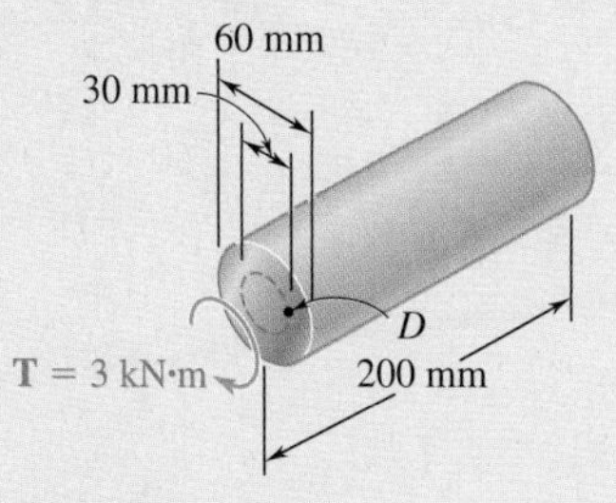

Fig. *P10.5*

10.6 (*a*) For the 3-in.-diameter solid cylinder and loading shown, determine the maximum shearing stress. (*b*) Determine the inner diameter of the 4-in.-diameter hollow cylinder shown, for which the maximum stress is the same as in part *a*.

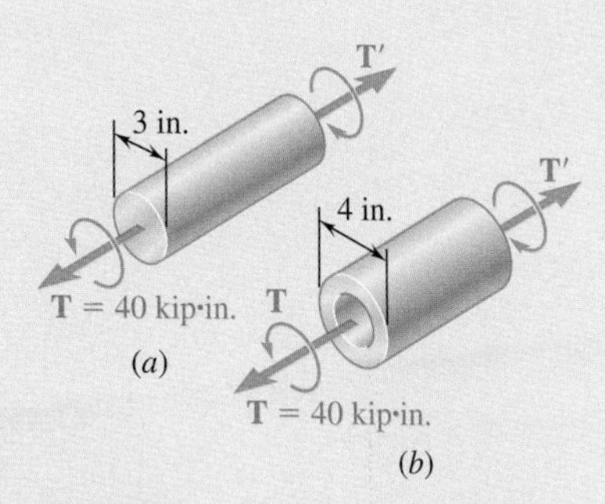

Fig. P10.6

10.7 The torques shown are exerted on pulleys *A*, *B*, and *C*. Knowing that both shafts are solid, determine the maximum shearing stress in (*a*) shaft *AB*, (*b*) shaft *BC*.

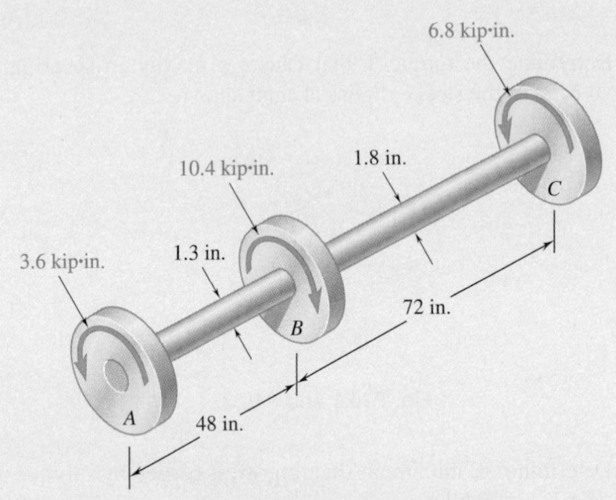

Fig. *P10.7* and P10.8

10.8 The shafts of the pulley assembly shown are to be redesigned. Knowing that the allowable shearing stress in each shaft is 8.5 ksi, determine the smallest allowable diameter of (*a*) shaft *AB*, (*b*) shaft *BC*.

10.9 Under normal operating conditions, the electric motor exerts a 12-kip·in. torque at *E*. Knowing that each shaft is solid, determine the maximum shearing stress in (*a*) shaft *BC*, (*b*) shaft *CD*, (*c*) shaft *DE*.

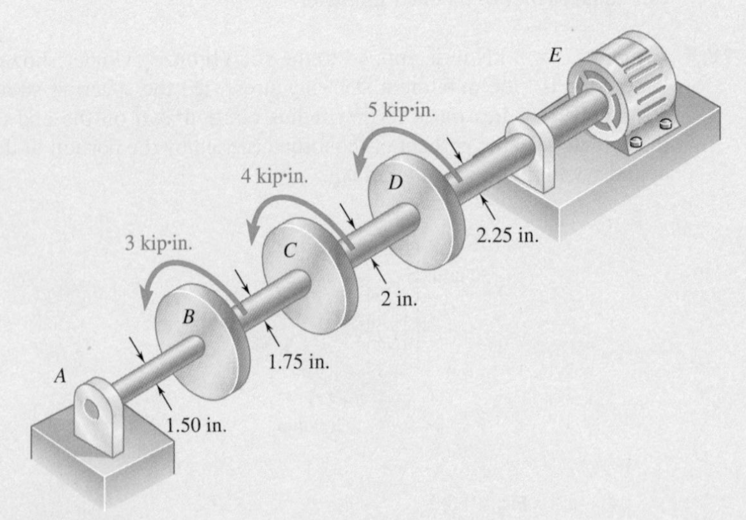

Fig. P10.9

10.10 Solve Prob. 10.9, assuming that a 1-in.-diameter hole has been drilled into each shaft.

10.11 Under normal operating conditions, the electric motor exerts a torque of 2.4 kN·m on shaft AB. Knowing that each shaft is solid, determine the maximum shearing stress (a) in shaft AB, (b) in shaft BC, (c) in shaft CD.

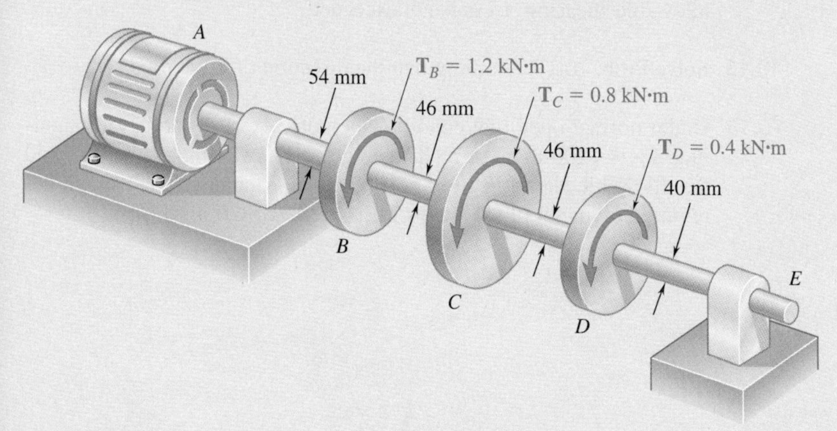

Fig. ***P10.11***

10.12 In order to reduce the total mass of the assembly of Prob. 10.11, a new design is being considered in which the diameter of shaft BC will be smaller. Determine the smallest diameter of shaft BC for which the maximum value of the shearing stress in the assembly will not be increased.

10.13 The allowable shearing stress is 15 ksi in the 1.5-in.-diameter steel rod AB and 8 ksi in the 1.8-in.-diameter rod BC. Neglecting the effect of stress concentrations, determine the largest torque that can be applied at A.

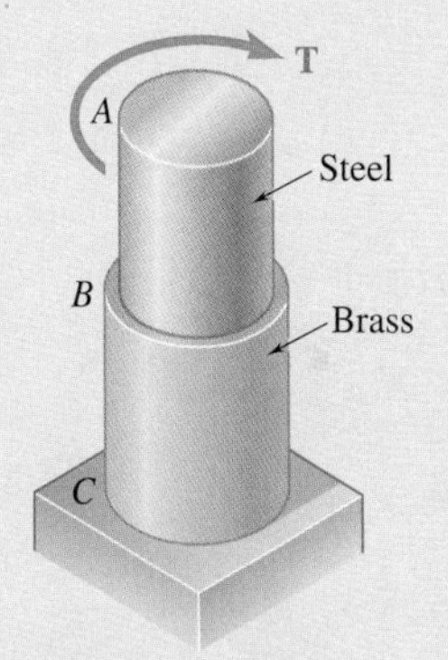

Fig. P10.13 and ***P10.14***

10.14 The allowable shearing stress is 15 ksi in the steel rod AB and 8 ksi in the brass rod BC. Knowing that a torque of magnitude T = 10 kip·in. is applied at A and neglecting the effect of stress concentrations, determine the required diameter of (a) rod AB, (b) rod BC.

10.15 The solid rod AB has a diameter d_{AB} = 60 mm and is made of a steel for which the allowable shearing stress is 85 MPa. The pipe CD, which has an outer diameter of 90 mm and a wall thickness of 6 mm, is made of an aluminum for which the allowable shearing stress is 54 MPa. Determine the largest torque **T** that can be applied at A.

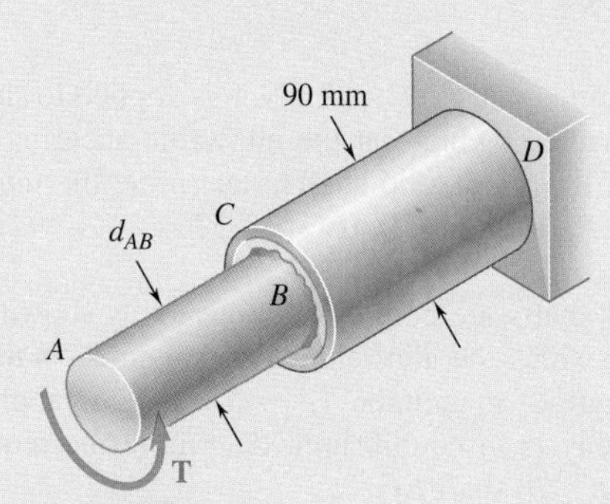

Fig. P10.15

10.16 The allowable shearing stress is 50 MPa in the brass rod AB and 25 MPa in the aluminum rod BC. Knowing that a torque of magnitude T = 1250 N·m is applied at A, determine the required diameter of (a) rod AB, (b) rod BC.

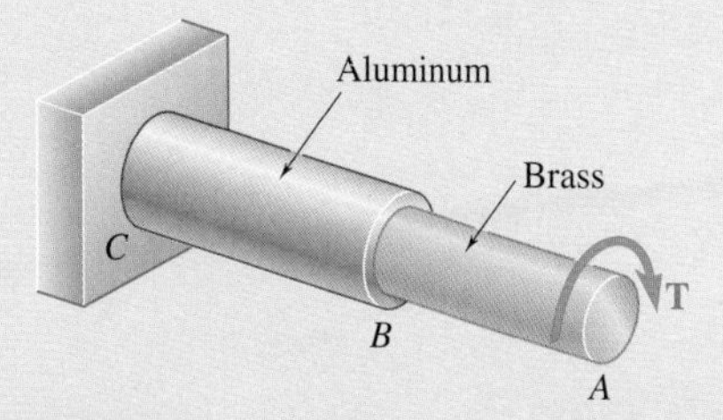

Fig. P10.16

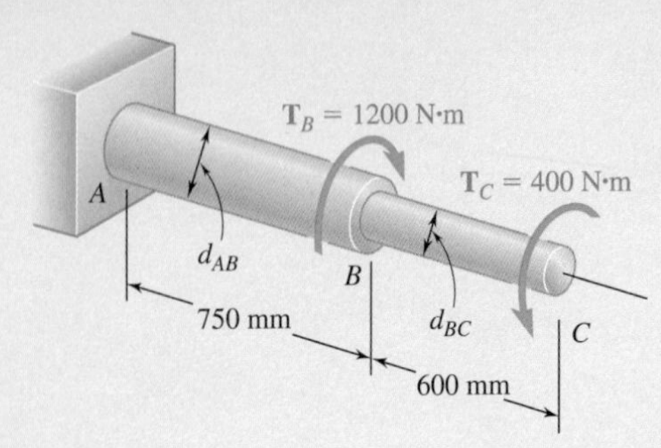

Fig. P10.17

10.17 The solid shaft shown is formed of a brass for which the allowable shearing stress is 55 MPa. Neglecting the effect of stress concentrations, determine the smallest diameters d_{AB} and d_{BC} for which the allowable shearing stress is not exceeded.

10.18 Solve Prob. 10.17 assuming that the direction of $\mathbf{T}_C$ is reversed.

10.19 Under normal operating conditions a motor exerts a torque of magnitude $T_F = 1200$ lb·in. at F. Knowing that $r_D = 8$ in., $r_G = 3$ in., and the allowable shearing stress is 10.5 ksi in each shaft, determine the required diameter of (*a*) shaft *CDE*, (*b*) shaft *FGH*.

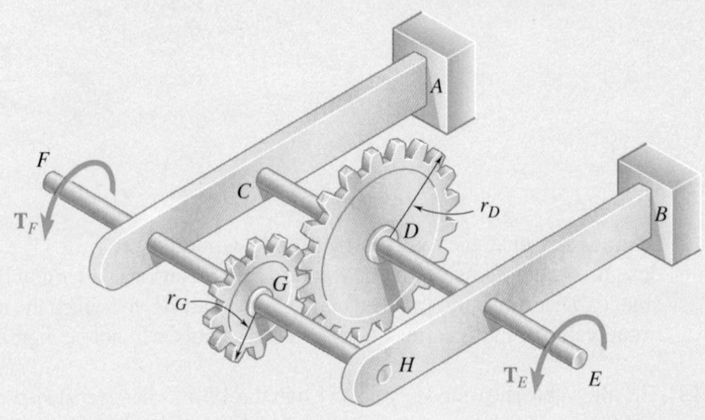

Fig. *P10.19* and P10.20

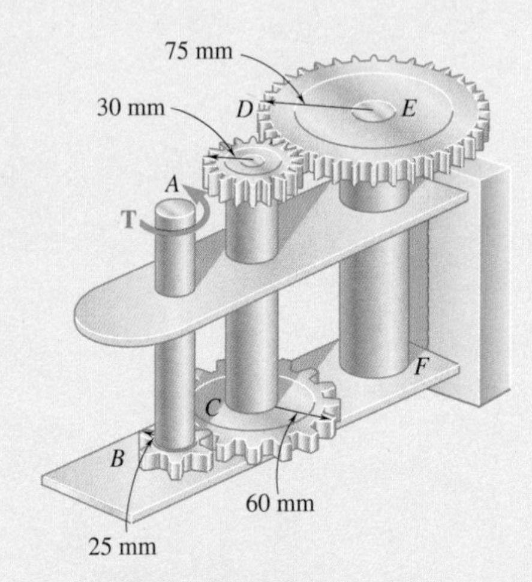

Fig. P10.21 and *P10.22*

10.20 Under normal operating conditions a motor exerts a torque of magnitude T_F at F. The shafts are made of a steel for which the allowable shearing stress is 12 ksi and have diameters $d_{CDE} = 0.900$ in. and $d_{FGH} = 0.800$ in. Knowing that $r_D = 6.5$ in. and $r_G = 4.5$ in., determine the largest allowable value of T_F.

10.21 A torque of magnitude $T = 100$ N·m is applied to shaft *AB* of the gear train shown. Knowing that the diameters of the three solid shafts are, respectively, $d_{AB} = 21$ mm, $d_{CD} = 30$ mm, and $d_{EF} = 40$ mm, determine the maximum shearing stress in (*a*) shaft *AB*, (*b*) shaft *CD*, (*c*) shaft *EF*.

10.22 A torque of magnitude $T = 120$ N·m is applied to shaft *AB* of the gear train shown. Knowing that the allowable shearing stress is 75 MPa in each of the three solid shafts, determine the required diameter of (*a*) shaft *AB*, (*b*) shaft *CD*, (*c*) shaft *EF*.

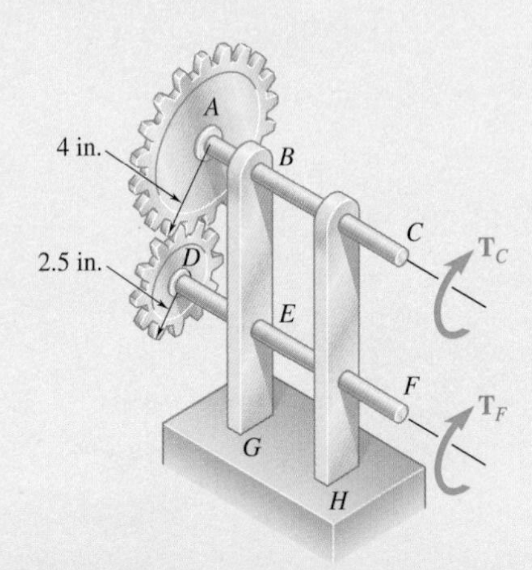

Fig. P10.23 and P10.24

10.23 Two solid shafts are connected by gears as shown and are made of a steel for which the allowable shearing stress is 8500 psi. Knowing that a torque of magnitude $T_C = 5$ kip·in. is applied at C and that the assembly is in equilibrium, determine the required diameter of (*a*) shaft *BC*, (*b*) shaft *EF*.

10.24 Two solid shafts are connected by gears as shown and are made of a steel for which the allowable shearing stress is 7000 psi. Knowing that the diameters of the two shafts are, respectively $d_{BC} = 1.6$ in. and $d_{EF} = 1.25$ in., determine the largest torque $\mathbf{T}_C$ that can be applied at C.

10.2 ANGLE OF TWIST IN THE ELASTIC RANGE

In this section, a relationship will be determined between the angle of twist ϕ of a circular shaft and the torque **T** exerted on the shaft. The entire shaft is assumed to remain elastic. Considering first the case of a shaft of length L with a uniform cross section of radius c subjected to a torque **T** at its free end (Fig. 10.20), recall that the angle of twist ϕ and the maximum shearing strain γ_{max} are related as

$$\gamma_{max} = \frac{c\phi}{L} \tag{10.3}$$

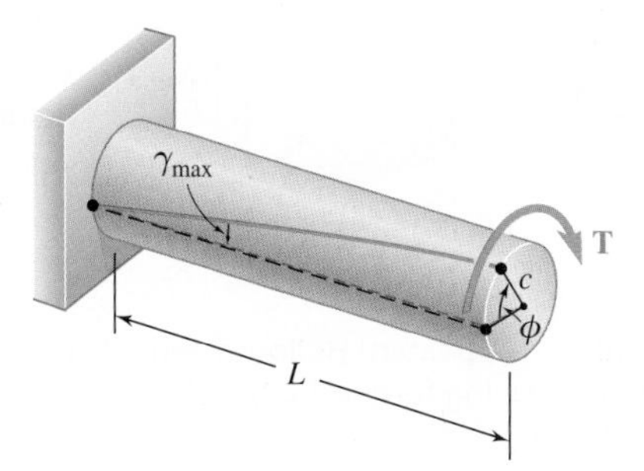

Fig. 10.20 Torque applied to free end of shaft resulting in angle of twist ϕ.

But in the elastic range, the yield stress is not exceeded anywhere in the shaft. Hooke's law applies, and $\gamma_{max} = \tau_{max}/G$. Recalling Eq. (10.9),

$$\gamma_{max} = \frac{\tau_{max}}{G} = \frac{Tc}{JG} \tag{10.14}$$

Equating the right-hand members of Eqs. (10.3) and (10.14) and solving for ϕ, write

$$\phi = \frac{TL}{JG} \tag{10.15}$$

where ϕ is in radians. The relationship obtained shows that, within the elastic range, *the angle of twist ϕ is proportional to the torque T applied to the shaft*. This agrees with the discussion at the beginning of Sec. 10.1B.

Eq. (10.15) provides a convenient method to determine the modulus of rigidity. A cylindrical rod of a material is placed in a *torsion testing machine* (Photo 10.3). Torques of increasing magnitude T are applied to the specimen, and the corresponding values of the angle of twist ϕ in a length L of the specimen are recorded. As long as the yield stress of the material is not exceeded, the points obtained by plotting ϕ against T fall on a straight line. The slope of this line represents the quantity JG/L, from which the modulus of rigidity G can be computed.

Photo 10.3 Tabletop torsion testing machine.
Courtesy of Tinius Olsen Testing Machine Co., Inc.

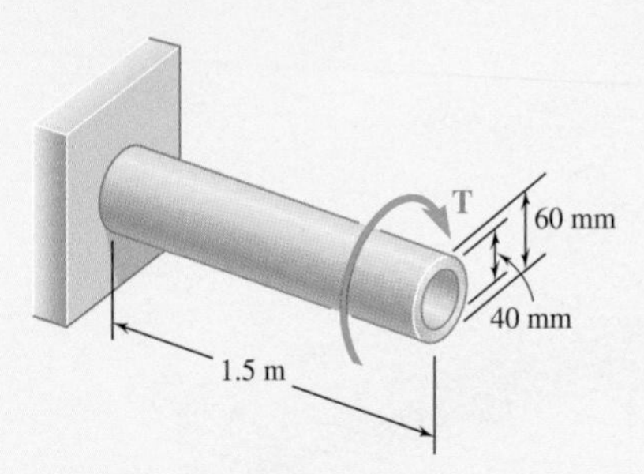

Fig. 10.15 (repeated) Hollow, fixed-end shaft having torque T applied at end.

Concept Application 10.2

What torque should be applied to the end of the shaft of Concept Application 10.1 to produce a twist of 2°? Use the value $G = 77$ GPa for the modulus of rigidity of steel.

Solving Eq. (10.15) for T, write

$$T = \frac{JG}{L}\phi$$

Substituting the given values

$$G = 77 \times 10^9 \text{ Pa} \qquad L = 1.5 \text{ m}$$

$$\phi = 2°\left(\frac{2\pi \text{ rad}}{360°}\right) = 34.9 \times 10^{-3} \text{ rad}$$

and recalling that, for the given cross section,

$$J = 1.021 \times 10^{-6} \text{ m}^4$$

we have

$$T = \frac{JG}{L}\phi = \frac{(1.021 \times 10^{-6} \text{ m}^4)(77 \times 10^9 \text{ Pa})}{1.5 \text{ m}}(34.9 \times 10^{-3} \text{ rad})$$

$$T = 1.829 \times 10^3 \text{ N·m} = 1.829 \text{ kN·m}$$

Concept Application 10.3

What angle of twist will create a shearing stress of 70 MPa on the inner surface of the hollow steel shaft of Concept Applications 10.1 and 10.2?

One method for solving this problem is to use Eq. (10.10) to find the torque T corresponding to the given value of τ and Eq. (10.15) to determine the angle of twist ϕ corresponding to the value of T just found.

A more direct solution is to use Hooke's law to compute the shearing strain on the inner surface of the shaft:

$$\gamma_{\min} = \frac{\tau_{\min}}{G} = \frac{70 \times 10^6 \text{ Pa}}{77 \times 10^9 \text{ Pa}} = 909 \times 10^{-6}$$

Recalling Eq. (10.2), which was obtained by expressing the length of arc AA' in Fig. 10.13c in terms of both γ and ϕ, we have

$$\phi = \frac{L\gamma_{\min}}{c_1} = \frac{1500 \text{ mm}}{20 \text{ mm}}(909 \times 10^{-6}) = 68.2 \times 10^{-3} \text{ rad}$$

To obtain the angle of twist in degrees, write

$$\phi = (68.2 \times 10^{-3} \text{ rad})\left(\frac{360°}{2\pi \text{ rad}}\right) = 3.91°$$

Eq. (10.15) can be used for the angle of twist only if the shaft is homogeneous (constant G), has a uniform cross section, and is loaded only at its ends. If the shaft is subjected to torques at locations other than its ends or if it has several portions with various cross sections and possibly of different materials, it must be divided into parts that satisfy the required conditions for Eq. (10.15). For shaft AB shown in Fig. 10.21, four different parts should be considered: AC, CD, DE, and EB. The total angle of twist of the shaft (i.e., the angle through which end A rotates with respect to end B) is obtained by *algebraically* adding the angles of twist of each component part. Using the internal torque T_i, length L_i, cross-sectional polar moment of inertia J_i, and modulus of rigidity G_i, corresponding to part i, the total angle of twist of the shaft is

$$\phi = \sum_i \frac{T_i L_i}{J_i G_i} \tag{10.16}$$

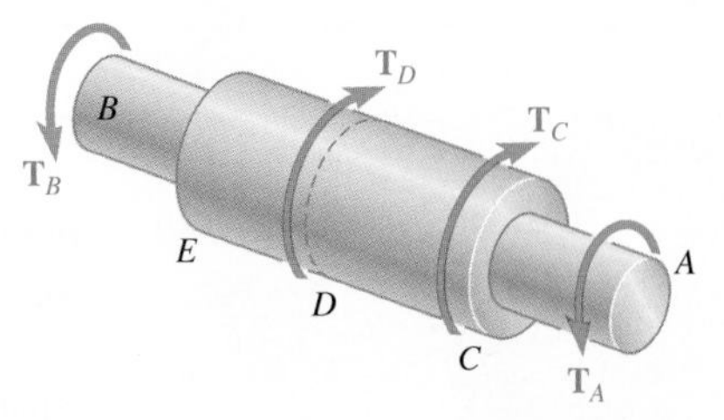

Fig. 10.21 Shaft with multiple cross-section dimensions and multiple loads.

The internal torque T_i in any given part of the shaft is obtained by passing a section through that part and drawing the free-body diagram of the portion of shaft located on one side of the section. This procedure is applied in Sample Prob. 10.3.

For a shaft with a variable circular cross section, as shown in Fig. 10.22, Eq. (10.15) is applied to a disk of thickness dx. The angle by which one face of the disk rotates with respect to the other is

$$d\phi = \frac{T\,dx}{JG}$$

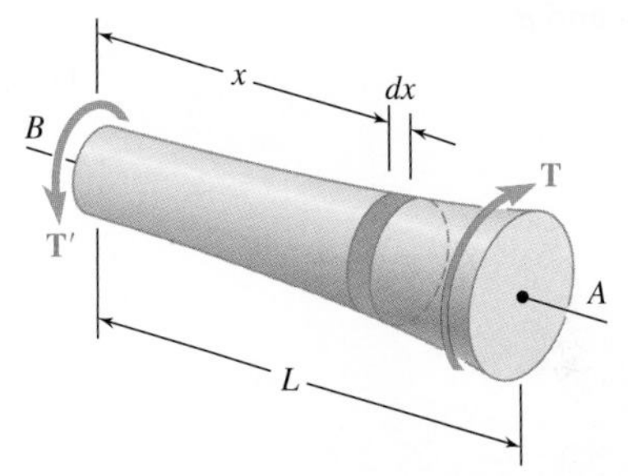

Fig. 10.22 Torqued shaft with variable cross section.

where J is a function of x. Integrating in x from 0 to L, the total angle of twist of the shaft is

$$\phi = \int_0^L \frac{T\,dx}{JG} \tag{10.17}$$

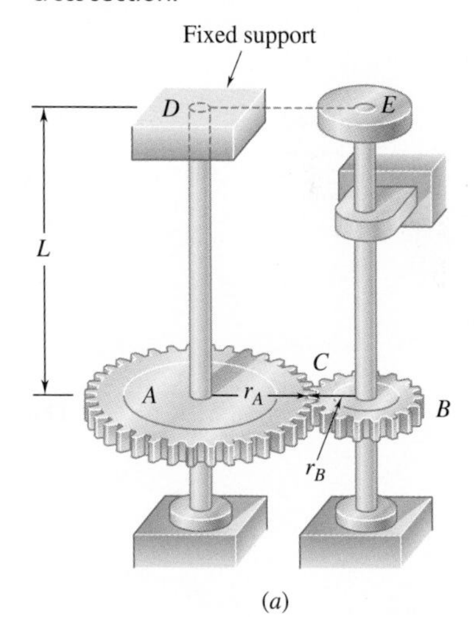

The shafts shown in Figs. 10.15 and 10.20 both had one end attached to a fixed support. In each case, the angle of twist ϕ was equal to the angle of rotation of its free end. When both ends of a shaft rotate, however, the angle of twist of the shaft is equal to the angle through which one end of the shaft rotates *with respect to the other*. For example, consider the assembly shown in Fig. 10.23*a*, consisting of two elastic shafts AD and BE, each of length L, radius c, modulus of rigidity G, and attached to gears meshed at C. If a torque **T** is applied at E (Fig. 10.23*b*), both shafts will be twisted. Since the end D of shaft AD is fixed, the angle of twist of AD is measured by the angle of rotation ϕ_A of end A. On the other hand, since both ends of shaft BE rotate, the angle of twist of BE is equal to the difference between the angles of rotation ϕ_B and ϕ_E (i.e., the angle of twist is equal to the angle through which end E rotates with respect to end B). This relative angle of rotation, $\phi_{E/B}$, is

$$\phi_{E/B} = \phi_E - \phi_B = \frac{TL}{JG}$$

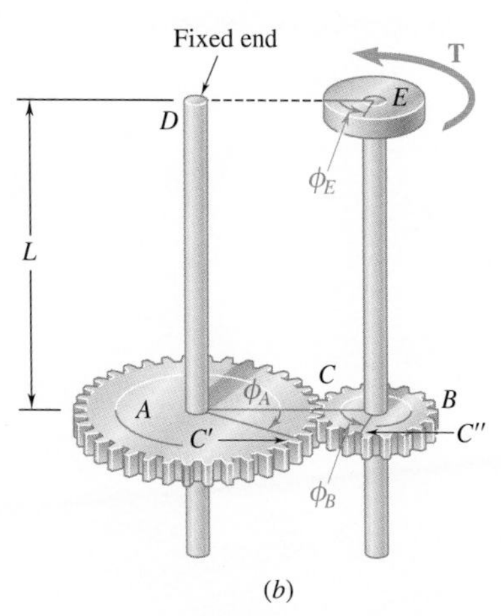

Fig. 10.23 (*a*) Gear assembly for transmitting torque from point *E* to point *D*. (*b*) Angles of twist at disk *E*, gear *B*, and gear *A*.

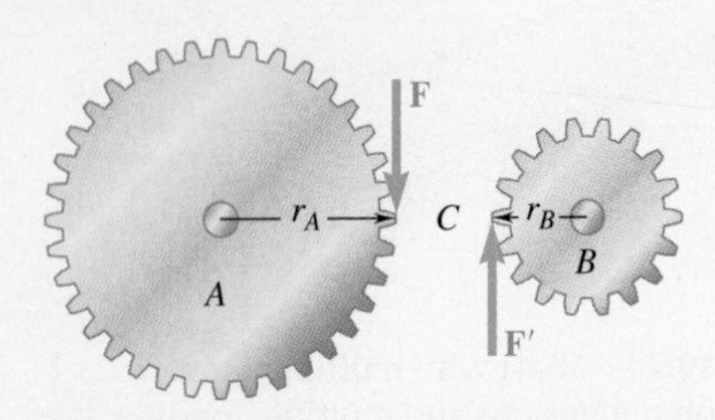

Fig. 10.24 Gear teeth forces for gears *A* and *B*.

Concept Application 10.4

For the assembly of Fig. 10.23, knowing that $r_A = 2r_B$, determine the angle of rotation of end *E* of shaft *BE* when the torque **T** is applied at *E*.

First determine the torque $\mathbf{T}_{AD}$ exerted on shaft *AD*. Observing that equal and opposite forces **F** and **F′** are applied on the two gears at *C* (Fig. 10.24) and recalling that $r_A = 2r_B$, the torque exerted on shaft *AD* is twice as large as the torque exerted on shaft *BE*. Thus, $T_{AD} = 2T$.

Since the end *D* of shaft *AD* is fixed, the angle of rotation ϕ_A of gear *A* is equal to the angle of twist of the shaft and is

$$\phi_A = \frac{T_{AD}L}{JG} = \frac{2TL}{JG}$$

Since the arcs CC' and CC'' in Fig. 10.23*b* must be equal, $r_A\phi_A = r_B\phi_B$. So,

$$\phi_B = (r_A/r_B)\phi_A = 2\phi_A$$

Therefore,

$$\phi_B = 2\phi_A = \frac{4TL}{JG}$$

Next, consider shaft *BE*. The angle of twist of the shaft is equal to the angle $\phi_{E/B}$ through which end *E* rotates with respect to end *B*. Thus,

$$\phi_{E/B} = \frac{T_{BE}L}{JG} = \frac{TL}{JG}$$

The angle of rotation of end *E* is obtained by

$$\phi_E = \phi_B + \phi_{E/B}$$

$$= \frac{4TL}{JG} + \frac{TL}{JG} = \frac{5TL}{JG}$$

10.3 STATICALLY INDETERMINATE SHAFTS

There are situations where the internal torques cannot be determined from statics alone. In such cases, the external torques (i.e., those exerted on the shaft by the supports and connections) cannot be determined from the free-body diagram of the entire shaft. The equilibrium equations must be complemented by relations involving the deformations of the shaft and obtained by the geometry of the problem. Because statics is not sufficient to determine external and internal torques, the shafts are *statically indeterminate*. The following Concept Application as well as Sample Prob. 10.5 show how to analyze statically indeterminate shafts.

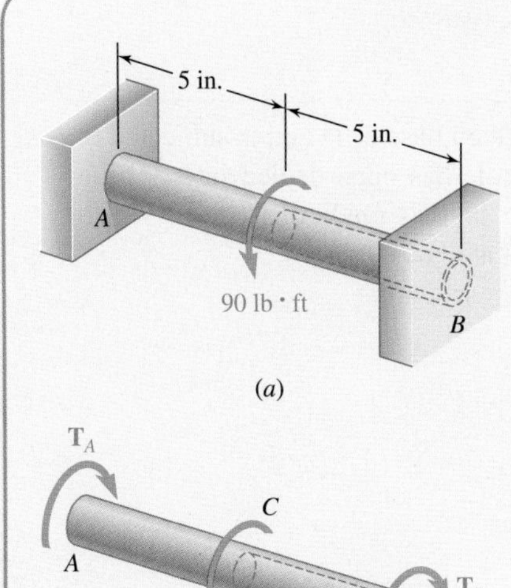

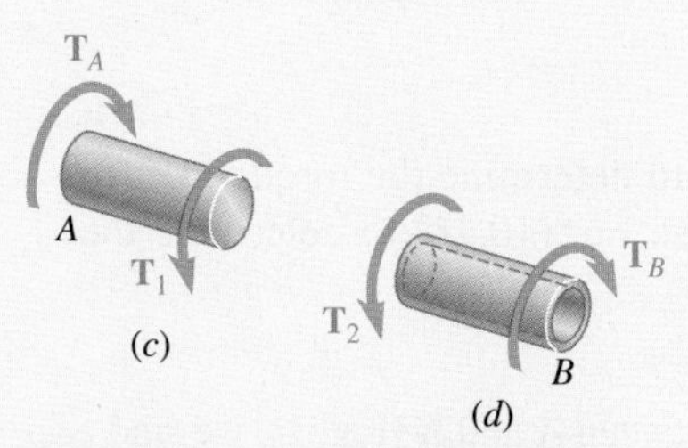

Fig. 10.25 (*a*) Shaft with central applied torque and fixed ends. (*b*) Free-body diagram of shaft *AB*. (*c*) Free-body diagrams for solid and hollow segments.

Concept Application 10.5

A circular shaft *AB* consists of a 10-in.-long, $\frac{7}{8}$-in.-diameter steel cylinder, in which a 5-in.-long, $\frac{5}{8}$-in.-diameter cavity has been drilled from end *B*. The shaft is attached to fixed supports at both ends, and a 90 lb·ft torque is applied at its midsection (Fig. 10.25*a*). Determine the torque exerted on the shaft by each of the supports.

Drawing the free-body diagram of the shaft and denoting by $\mathbf{T}_A$ and $\mathbf{T}_B$ the torques exerted by the supports (Fig. 10.25*b*), the equilibrium equation is

$$T_A + T_B = 90 \text{ lb·ft}$$

Since this equation is not sufficient to determine the two unknown torques $\mathbf{T}_A$ and $\mathbf{T}_B$, the shaft is statically indeterminate.

However, $\mathbf{T}_A$ and $\mathbf{T}_B$ can be determined if we observe that the total angle of twist of shaft *AB* must be zero, since both of its ends are restrained. Denoting by ϕ_1 and ϕ_2, respectively, the angles of twist of portions *AC* and *CB*, we write

$$\phi = \phi_1 + \phi_2 = 0$$

From the free-body diagram of a small portion of shaft including end *A* (Fig. 10.25*c*), we note that the internal torque T_1 in *AC* is equal to T_A; from the free-body diagram of a small portion of shaft including end *B* (Fig. 10.25*d*), we note that the internal torque T_2 in *CB* is equal to T_B. Recalling Eq. (10.15) and observing that portions *AC* and *CB* of the shaft are twisted in opposite senses, write

$$\phi = \phi_1 + \phi_2 = \frac{T_A L_1}{J_1 G} - \frac{T_B L_2}{J_2 G} = 0$$

Solving for T_B,

$$T_B = \frac{L_1 J_2}{L_2 J_1} T_A$$

Substituting the numerical data gives

$$L_1 = L_2 = 5 \text{ in.}$$

$$J_1 = \tfrac{1}{2}\pi(\tfrac{7}{16} \text{ in.})^4 = 57.6 \times 10^{-3} \text{ in}^4$$

$$J_2 = \tfrac{1}{2}\pi[(\tfrac{7}{16} \text{ in.})^4 - (\tfrac{5}{16} \text{ in.})^4] = 42.6 \times 10^{-3} \text{ in}^4$$

Therefore,

$$T_B = 0.740\, T_A$$

Substitute this expression into the original equilibrium equation:

$$1.740\, T_A = 90 \text{ lb·ft}$$

$$T_A = 51.7 \text{ lb·ft} \qquad T_B = 38.3 \text{ lb·ft}$$

Sample Problem 10.3

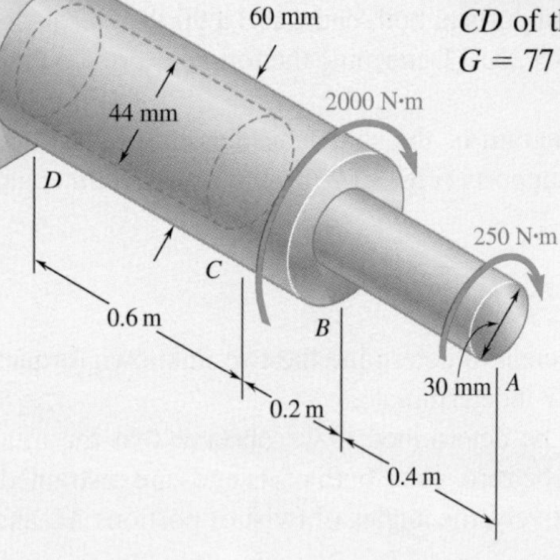

The horizontal shaft AD is attached to a fixed base at D and is subjected to the torques shown. A 44-mm-diameter hole has been drilled into portion CD of the shaft. Knowing that the entire shaft is made of steel for which $G = 77$ GPa, determine the angle of twist at end A.

STRATEGY: Use free-body diagrams to determine the torque in each shaft segment AB, BC, and CD. Then use Eq. (10.16) to determine the angle of twist at end A.

MODELING:

Passing a section through the shaft between A and B (Fig. 1), we find

$$\Sigma M_x = 0: \qquad (250 \text{ N·m}) - T_{AB} = 0 \qquad T_{AB} = 250 \text{ N·m}$$

Passing now a section between B and C (Fig. 2) we have

$$\Sigma M_x = 0: (250 \text{ N·m}) + (2000 \text{ N·m}) - T_{BC} = 0 \qquad T_{BC} = 2250 \text{ N·m}$$

Since no torque is applied at C,

$$T_{CD} = T_{BC} = 2250 \text{ N·m}$$

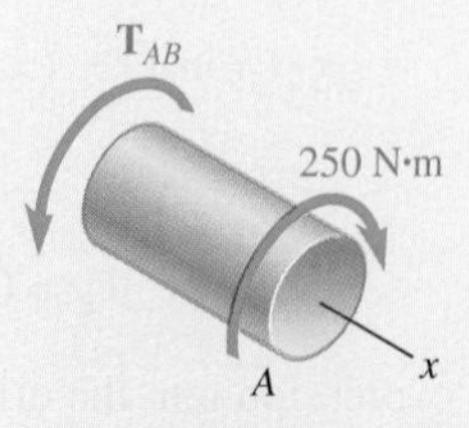

Fig. 1 Free-body diagram for finding internal torque in segment *AB*.

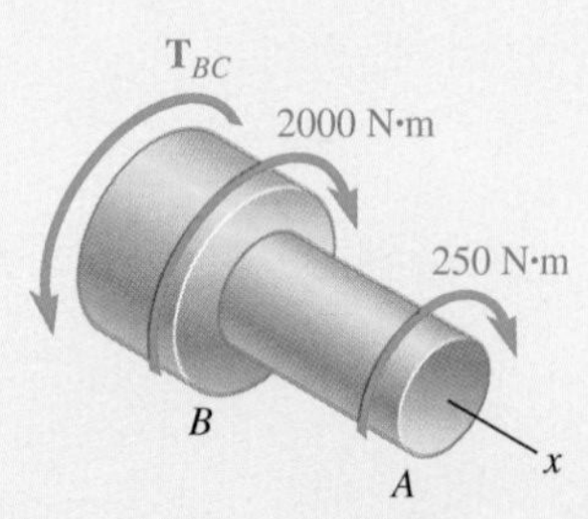

Fig. 2 Free-body diagram for finding internal torque in segment *BC*.

(continued)

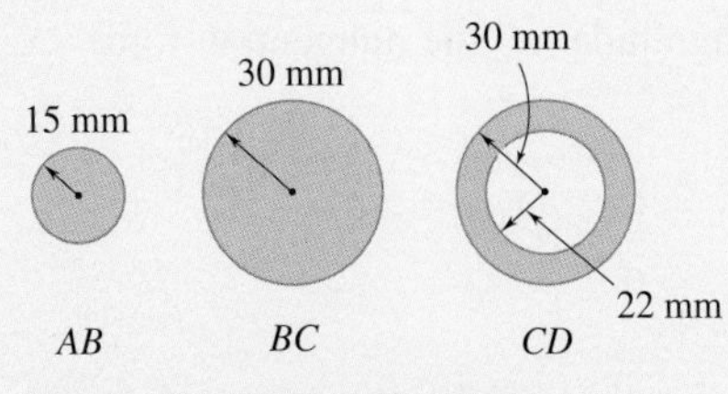

Fig. 3 Dimensions for three cross sections of shaft.

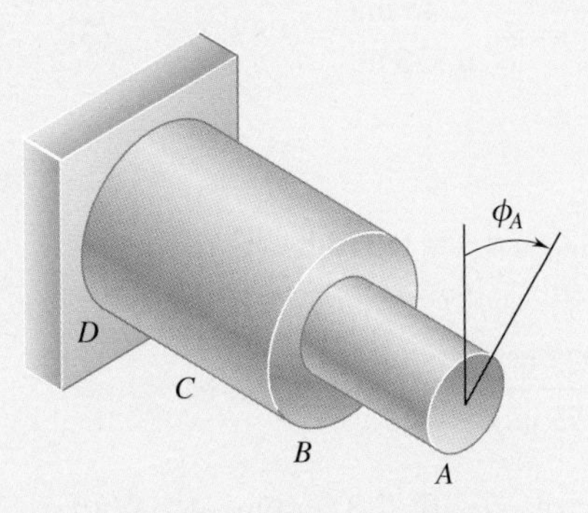

Fig. 4 Representation of angle of twist at end *A*.

ANALYSIS:

Polar of Moments of Inertia.
Using Fig. 3

$$J_{AB} = \frac{\pi}{2}c^4 = \frac{\pi}{2}(0.015 \text{ m})^4 = 0.0795 \times 10^{-6} \text{ m}^4$$

$$J_{BC} = \frac{\pi}{2}c^4 = \frac{\pi}{2}(0.030 \text{ m})^4 = 1.272 \times 10^{-6} \text{ m}^4$$

$$J_{CD} = \frac{\pi}{2}(c_2^4 - c_1^4) = \frac{\pi}{2}[(0.030 \text{ m})^4 - (0.022 \text{ m})^4] = 0.904 \times 10^{-6} \text{m}^4$$

Angle of Twist. From Fig. 4, using Eq. (10.16) and recalling that $G = 77$ GPa for the entire shaft, we have

$$\phi_A = \sum_i \frac{T_i L_i}{J_i G} = \frac{1}{G}\left(\frac{T_{AB}L_{AB}}{J_{AB}} + \frac{T_{BC}L_{BC}}{J_{BC}} + \frac{T_{CD}L_{CD}}{J_{CD}}\right)$$

$$\phi_A = \frac{1}{77 \text{ GPa}}\left[\frac{(250 \text{ N}\cdot\text{m})(0.4 \text{ m})}{0.0795 \times 10^{-6} \text{ m}^4} + \frac{(2250)(0.2)}{1.272 \times 10^{-6}} + \frac{(2250)(0.6)}{0.904 \times 10^{-6}}\right]$$

$$= 0.01634 + 0.00459 + 0.01939 = 0.0403 \text{ rad}$$

$$\phi_A = (0.0403 \text{ rad})\frac{360°}{2\pi \text{ rad}} \qquad \phi_A = 2.31° \blacktriangleleft$$

Sample Problem 10.4

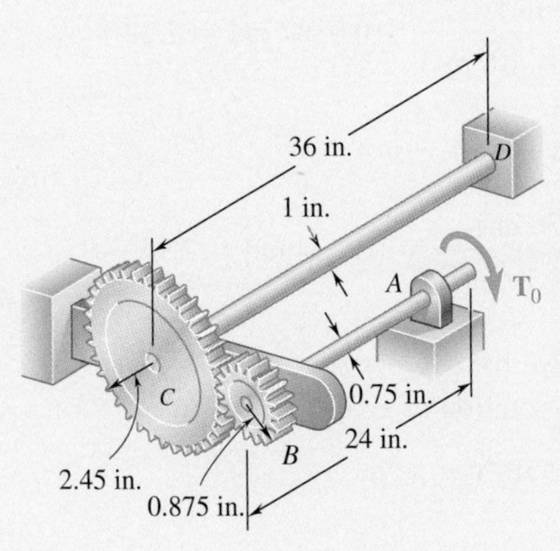

Two solid steel shafts are connected by the gears shown. Knowing that for each shaft $G = 11.2 \times 10^6$ psi and the allowable shearing stress is 8 ksi, determine (*a*) the largest torque $\mathbf{T}_0$ that may be applied to end *A* of shaft *AB* and (*b*) the corresponding angle through which end *A* of shaft *AB* rotates.

STRATEGY: Use the free-body diagrams and kinematics to determine the relation between the torques and twist in each shaft segment, *AB* and *CD*. Then use the allowable stress to determine the torque that can be applied and Eq. (10.15) to determine the angle of twist at end *A*.

(continued)

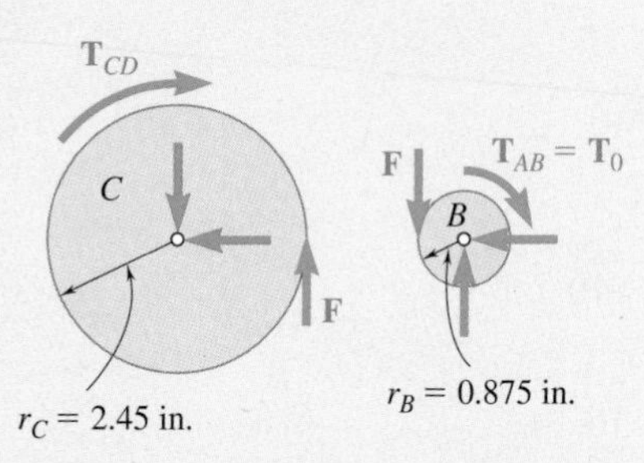

Fig. 1 Free-body diagrams of gears *B* and *C*.

MODELING: Denoting by F the magnitude of the tangential force between gear teeth (Fig. 1), we have

$$\textbf{\textit{Gear B.}}\quad \Sigma M_B = 0: \qquad F(0.875\text{ in.}) - T_0 = 0 \qquad T_{CD} = 2.8T_0 \tag{1}$$

$$\textbf{\textit{Gear C.}}\quad \Sigma M_C = 0: \qquad F(2.45\text{ in.}) - T_{CD} = 0$$

Using kinematics with Fig. 2, we see that the peripheral motions of the gears are equal and write

$$r_B\phi_B = r_C\phi_C \qquad \phi_B = \phi_C\frac{r_C}{r_B} = \phi_C\frac{2.45\text{ in.}}{0.875\text{ in.}} = 2.8\phi_C \tag{2}$$

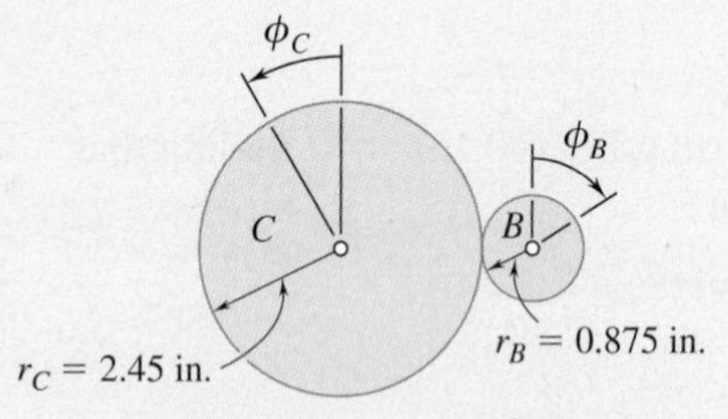

Fig. 2 Angle of twist for gears *B* and *C*.

ANALYSIS:

a. Torque T_0. For shaft AB, $T_{AB} = T_0$ and $c = 0.375$ in. (Fig. 3); considering maximum permissible shearing stress, we write

$$\tau = \frac{T_{AB}c}{J} \qquad 8000\text{ psi} = \frac{T_0(0.375\text{ in.})}{\frac{1}{2}\pi(0.375\text{ in.})^4} \qquad T_0 = 663\text{ lb}\cdot\text{in.} \blacktriangleleft$$

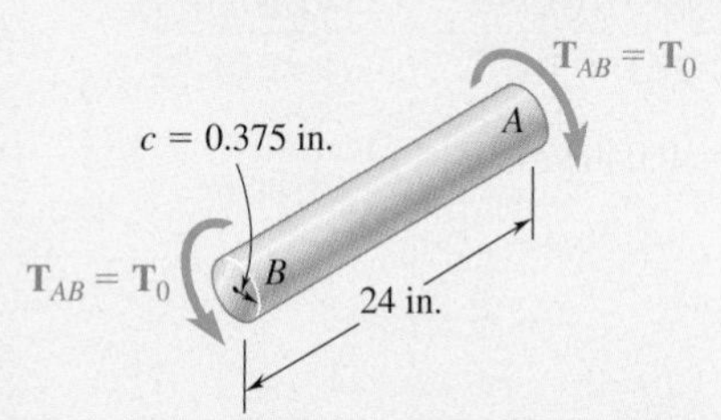

Fig. 3 Free-body diagram of shaft *AB*.

For shaft CD using Eq. (1) we have $T_{CD} = 2.8T_0$ (Fig. 4). With $c = 0.5$ in. and $\tau_{\text{all}} = 8000$ psi, we write

$$\tau = \frac{T_{CD}c}{J} \qquad 8000\text{ psi} = \frac{2.8T_0(0.5\text{ in.})}{\frac{1}{2}\pi(0.5\text{ in.})^4} \qquad T_0 = 561\text{ lb}\cdot\text{in.} \blacktriangleleft$$

The maximum permissible torque is the smaller value obtained for T_0.

$$T_0 = 561\text{ lb}\cdot\text{in.} \blacktriangleleft$$

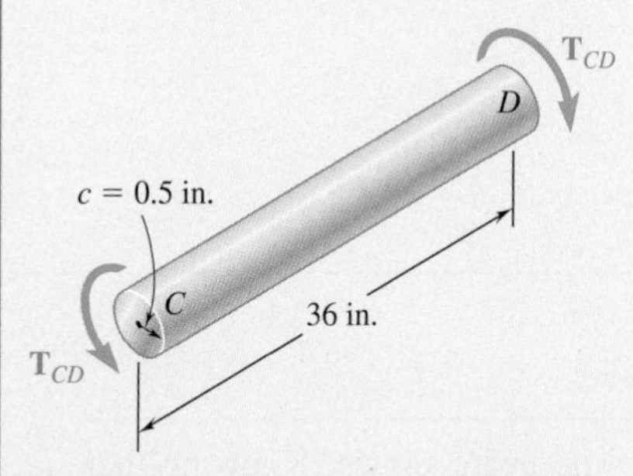

Fig. 4 Free-body diagram of shaft *CD*.

b. Angle of Rotation at End *A*. We first compute the angle of twist for each shaft.

Shaft AB. For $T_{AB} = T_0 = 561$ lb·in., we have

$$\phi_{A/B} = \frac{T_{AB}L}{JG} = \frac{(561\text{ lb}\cdot\text{in.})(24\text{ in.})}{\frac{1}{2}\pi(0.375\text{ in.})^4(11.2\times10^6\text{ psi})} = 0.0387\text{ rad} = 2.22^\circ$$

Shaft CD. $T_{CD} = 2.8T_0 = 2.8(561\text{ lb}\cdot\text{in.})$

$$\phi_{C/D} = \frac{T_{CD}L}{JG} = \frac{2.8(561\text{ lb}\cdot\text{in.})(36\text{ in.})}{\frac{1}{2}\pi(0.5\text{ in.})^4(11.2\times10^6\text{ psi})} = 0.0514\text{ rad} = 2.95^\circ$$

Since end D of shaft CD is fixed, we have $\phi_C = \phi_{C/D} = 2.95^\circ$. Using Eq. (2) with Fig. 5, we find the angle of rotation of gear B is

$$\phi_B = 2.8\phi_C = 2.8(2.95^\circ) = 8.26^\circ$$

For end A of shaft AB, we have

$$\phi_A = \phi_B + \phi_{A/B} = 8.26^\circ + 2.22^\circ \qquad \phi_A = 10.48^\circ \blacktriangleleft$$

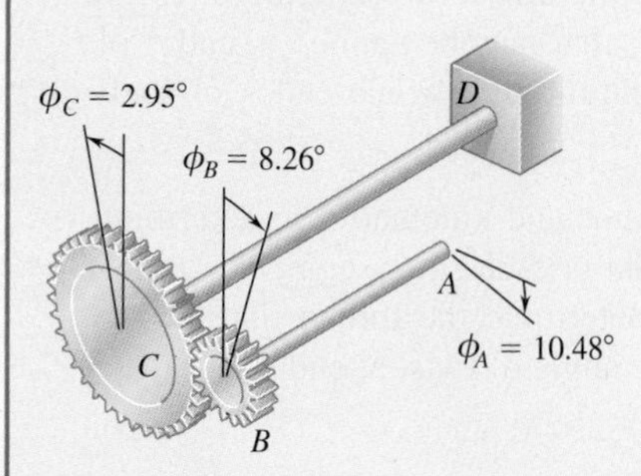

Fig. 5 Angle of twist results.

Sample Problem 10.5

A steel shaft and an aluminum tube are connected to a fixed support and to a rigid disk as shown in the cross section. Knowing that the initial stresses are zero, determine the maximum torque $\mathbf{T}_0$ that can be applied to the disk if the allowable stresses are 120 MPa in the steel shaft and 70 MPa in the aluminum tube. Use $G = 77$ GPa for steel and $G = 27$ GPa for aluminum.

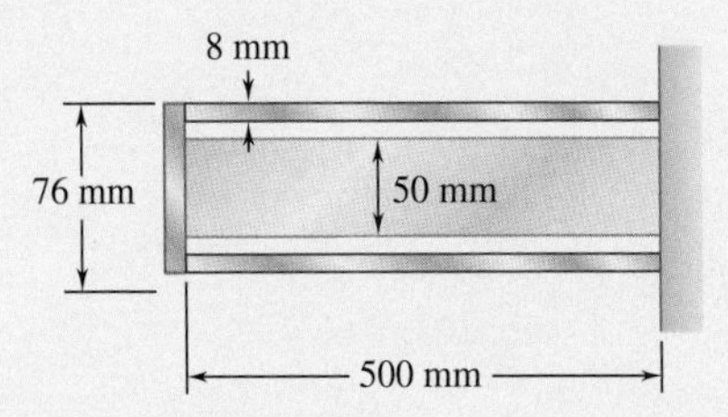

STRATEGY: We know that the applied load is resisted by both the shaft and the tube, but we do not know the portion carried by each part. Thus we need to look at the deformations. We know that both the shaft and tube are connected to the rigid disk and that the angle of twist is therefore the same for each. Once we know the portion of the torque carried by each part, we can use the allowable stress for each to determine which one governs and use this to determine the maximum torque.

MODELING:

We first draw a free-body diagram of the disk (Fig. 1) and find

$$T_0 = T_1 + T_2 \tag{1}$$

Knowing that the angle of twist is the same for the shaft and tube, we write

$$\phi_1 = \phi_2: \qquad \frac{T_1 L_1}{J_1 G_1} = \frac{T_2 L_2}{J_2 G_2}$$

$$\frac{T_1 (0.5 \text{ m})}{(2.003 \times 10^{-6} \text{ m}^4)(27 \text{ GPa})} = \frac{T_2 (0.5 \text{ m})}{(0.614 \times 10^{-6} \text{ m}^4)(77 \text{ GPa})}$$

$$T_2 = 0.874 T_1 \tag{2}$$

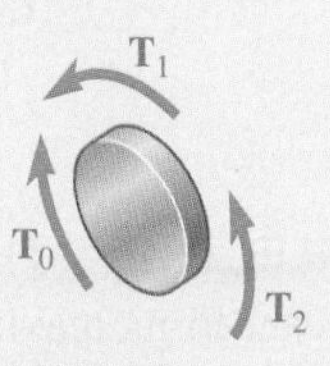

Fig. 1 Free-body diagram of end cap.

(continued)

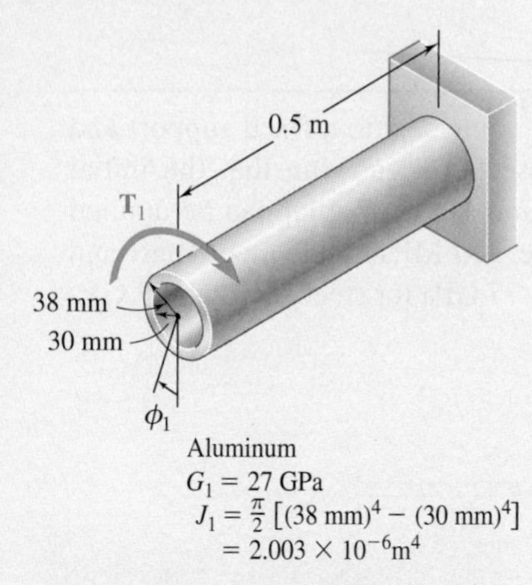

Fig. 2 Torque and angle of twist for hollow shaft.

ANALYSIS: We need to determine which part reaches its allowable stress first, and so we arbitrarily assume that the requirement $\tau_{alum} \leq$ 70 MPa is critical. For the aluminum tube in Fig. 2, we have

$$T_1 = \frac{\tau_{alum} J_1}{c_1} = \frac{(70 \text{ MPa})(2.003 \times 10^{-6} \text{ m}^4)}{0.038 \text{ m}} = 3690 \text{ N·m}$$

Using Eq. (2), compute the corresponding value T_2 and then find the maximum shearing stress in the steel shaft of Fig. 3.

$$T_2 = 0.874 T_1 = 0.874(3690) = 3225 \text{ N·m}$$

$$\tau_{steel} = \frac{T_2 c_2}{J_2} = \frac{(3225 \text{ N·m})(0.025 \text{ m})}{0.614 \times 10^{-6} \text{ m}^4} = 131.3 \text{ MPa}$$

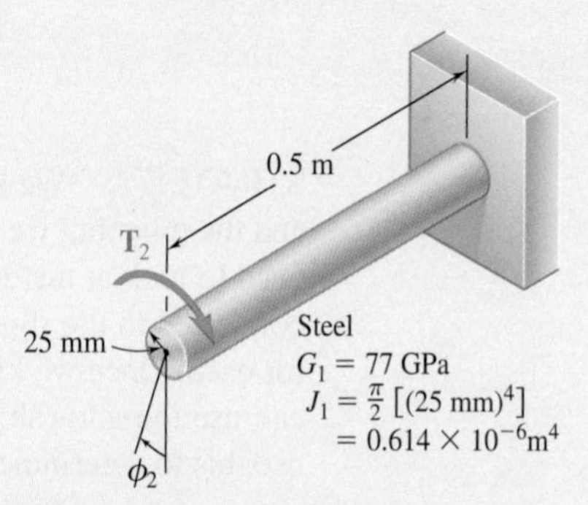

Fig. 3 Torque and angle of twist for solid shaft.

Note that the allowable steel stress of 120 MPa is exceeded; the assumption was *wrong*. Thus, the maximum torque $\mathbf{T}_0$ will be obtained by making $\tau_{steel} = 120$ MPa. Determine the torque $\mathbf{T}_2$:

$$T_2 = \frac{\tau_{steel} J_2}{c_2} = \frac{(120 \text{ MPa})(0.614 \times 10^{-6} \text{ m}^4)}{0.025 \text{ m}} = 2950 \text{ N·m}$$

From Eq. (2), we have

$$2950 \text{ N·m} = 0.874 T_1 \qquad T_1 = 3375 \text{ N·m}$$

Using Eq. (1), we obtain the maximum permissible torque:

$$T_0 = T_1 + T_2 = 3375 \text{ N·m} + 2950 \text{ N·m}$$

$$T_0 = 6.325 \text{ kN·m} \blacktriangleleft$$

REFLECT and THINK: This example illustrates that each part must not exceed its maximum allowable stress. Since the steel shaft reaches its allowable stress level first, the maximum stress in the aluminum shaft is below its maximum.

Problems

10.25 For the aluminum shaft shown ($G = 3.9 \times 10^6$ psi), determine (*a*) the torque **T** that causes an angle of twist of 5°, (*b*) the angle of twist caused by the same torque **T** in a solid cylindrical shaft of the same length and cross-sectional area.

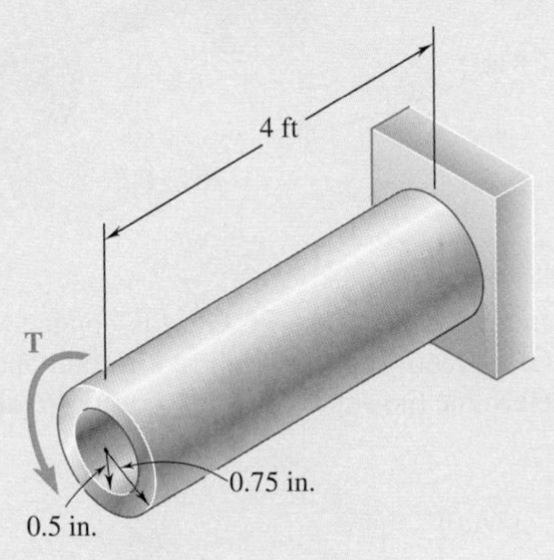

Fig. P10.25

10.26 (*a*) For the solid steel shaft shown, determine the angle of twist at *A*. Use $G = 11.2 \times 10^6$ psi. (*b*) Solve part *a*, assuming that the steel shaft is hollow with a 1.5-in. outer radius and a 0.75-in. inner radius.

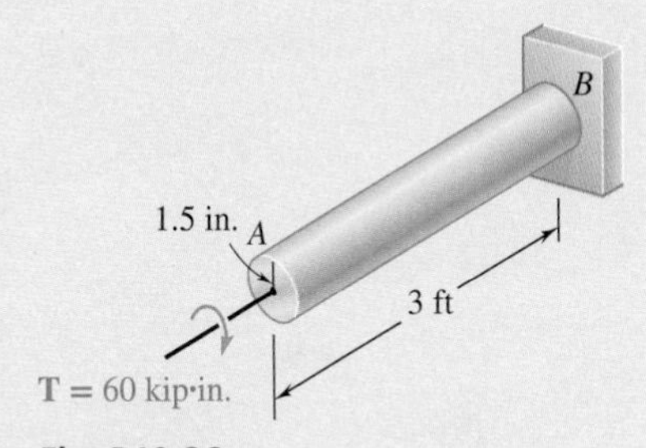

Fig. P10.26

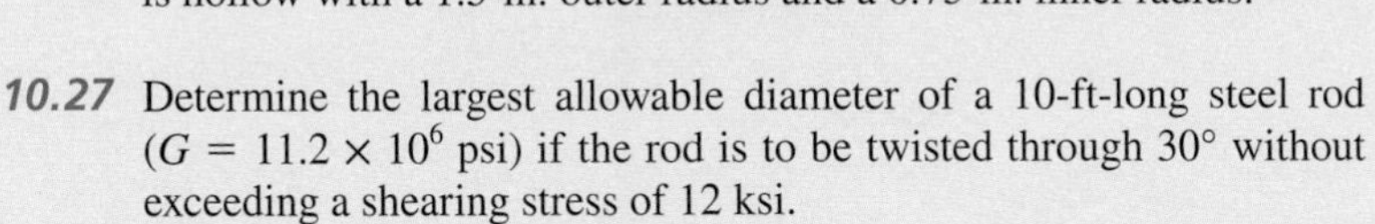

10.27 Determine the largest allowable diameter of a 10-ft-long steel rod ($G = 11.2 \times 10^6$ psi) if the rod is to be twisted through 30° without exceeding a shearing stress of 12 ksi.

10.28 The ship at *A* has just started to drill for oil on the ocean floor at a depth of 5000 ft. Knowing that the top of the 8-in.-diameter steel drill pipe ($G = 11.2 \times 10^6$ psi) rotates through two complete revolutions before the drill bit at *B* starts to operate, determine the maximum shearing stress caused in the pipe by torsion.

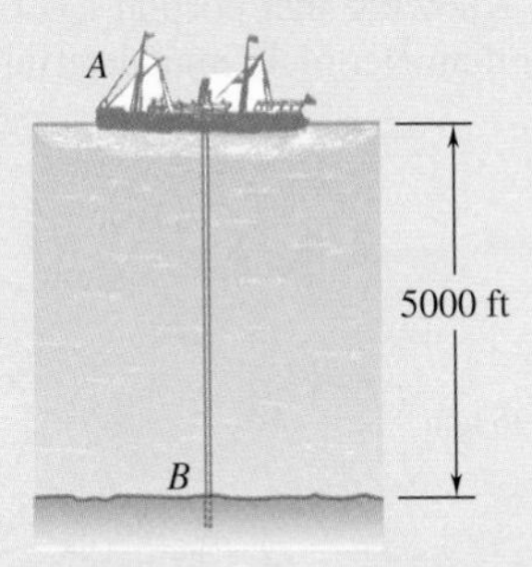

Fig. P10.28

10.29 The torques shown are exerted on pulleys *A* and *B*. Knowing that the shafts are solid and made of steel ($G = 77$ GPa), determine the angle of twist between (*a*) *A* and *B*, (*b*) *A* and *C*.

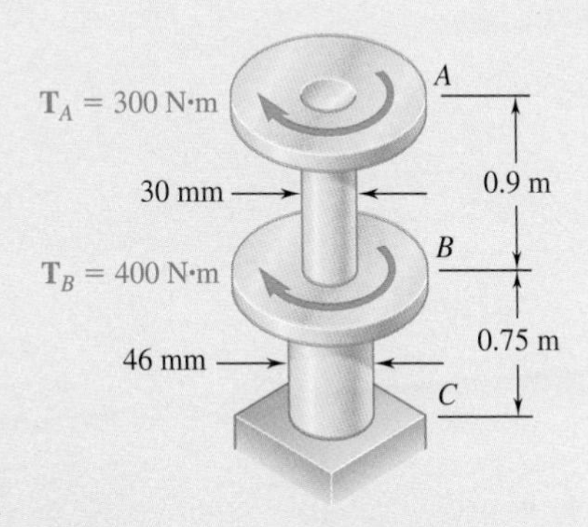

Fig. *P10.29*

10.30 The torques shown are exerted on pulleys B, C, and D. Knowing that the entire shaft is made of aluminum (G = 27 GPa), determine the angle of twist between (*a*) C and B, (*b*) D and B.

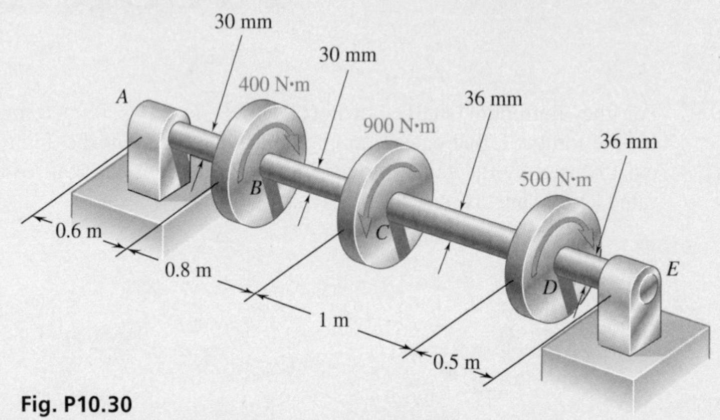

Fig. P10.30

10.31 The aluminum rod BC (G = 26 GPa) is bonded to the brass rod AB (G = 39 GPa). Knowing that each rod is solid and has a diameter of 12 mm, determine the angle of twist (*a*) at B, (*b*) at C.

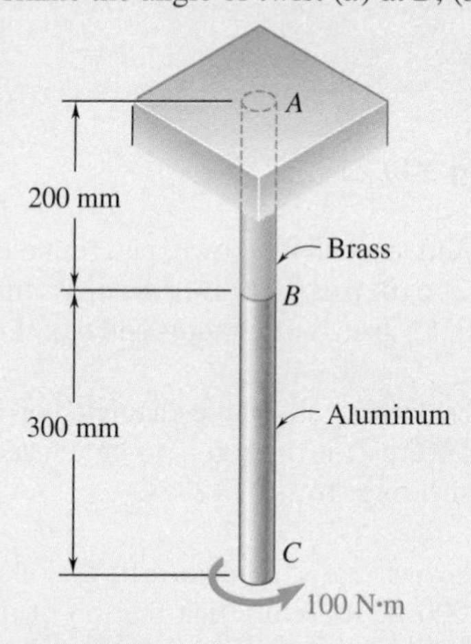

Fig. P10.31

10.32 The aluminum rod AB (G = 27 GPa) is bonded to the brass rod BD (G = 39 GPa). Knowing that portion CD of the brass rod is hollow and has an inner diameter of 40 mm, determine the angle of twist at A.

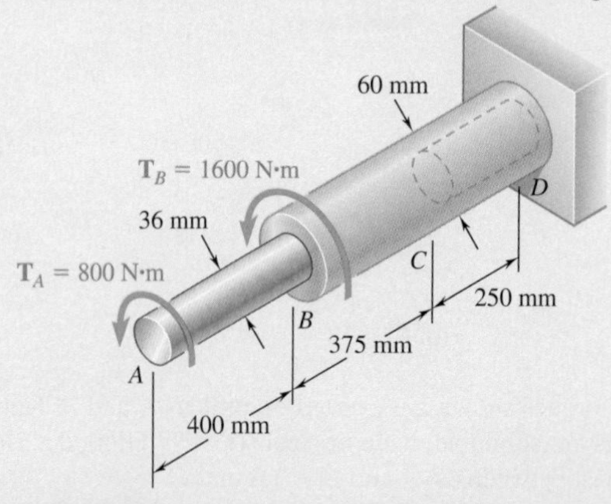

Fig. P10.32

10.33 Two solid shafts are connected by the gears shown. Knowing that $G = 77$ GPa for each shaft, determine the angle through which end A rotates when $T_A = 1200$ N·m.

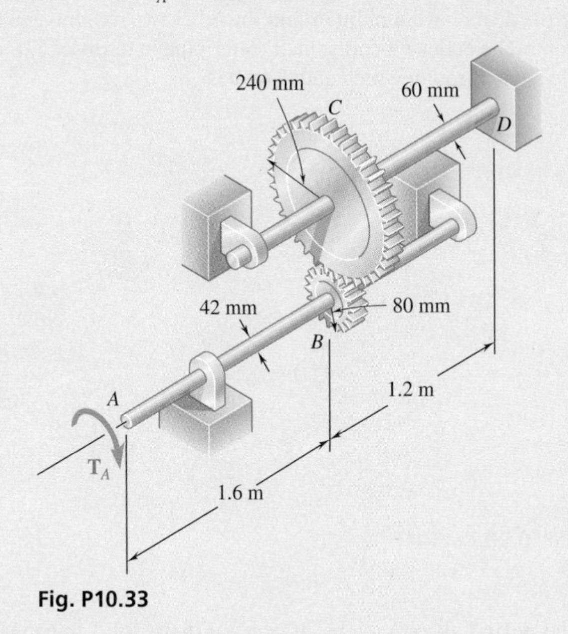

Fig. P10.33

10.34 Two solid steel shafts, each of 30-mm diameter, are connected by the gears shown. Knowing that $G = 77.2$ GPa, determine the angle through which end A rotates when a torque of magnitude $T = 200$ N·m is applied at A.

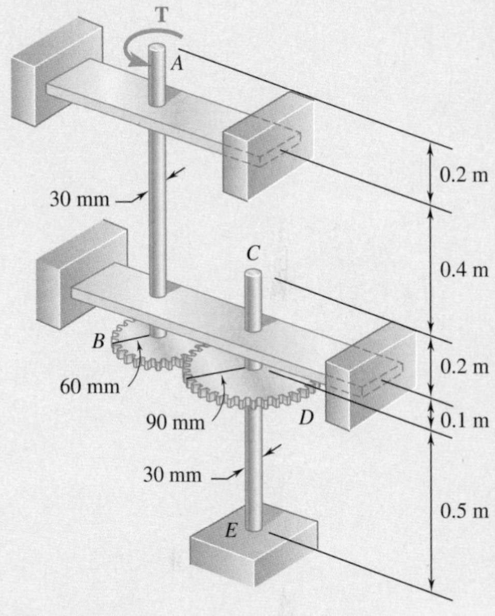

Fig. P10.34

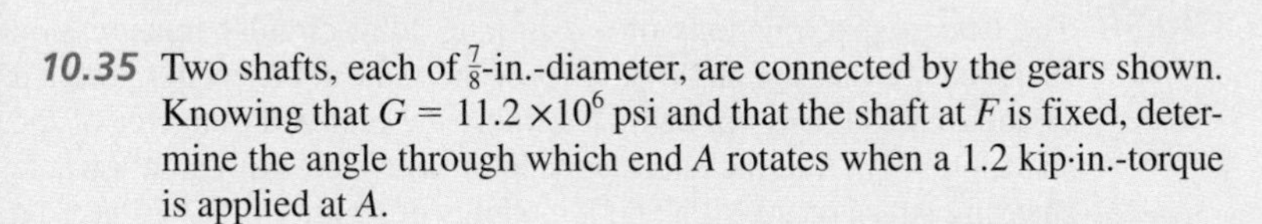
10.35 Two shafts, each of $\frac{7}{8}$-in.-diameter, are connected by the gears shown. Knowing that $G = 11.2 \times 10^6$ psi and that the shaft at F is fixed, determine the angle through which end A rotates when a 1.2 kip·in.-torque is applied at A.

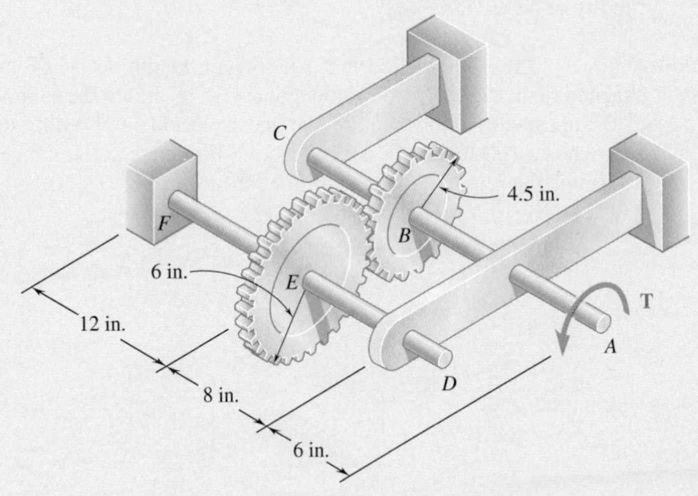

Fig. ***P10.35***

10.36 A coder F, used to record in digital form the rotation of shaft A, is connected to the shaft by means of the gear train shown, which consists of four gears and three solid steel shafts each of diameter d. Two of the gears have a radius r and the other two a radius nr. If the rotation of the coder F is prevented, determine in terms of T, l, G, J, and n the angle through which end A rotates.

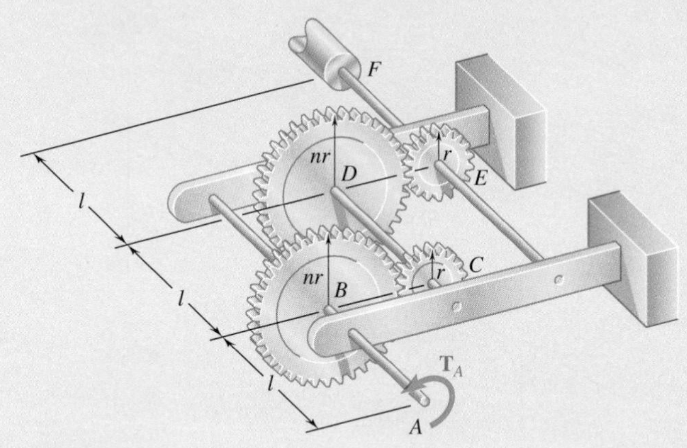

Fig. P10.36

10.37 The design specifications of a 1.2-m-long solid transmission shaft require that the angle of twist of the shaft not exceed 4° when a torque of 750 N·m is applied. Determine the required diameter of the shaft, knowing that the shaft is made of a steel with an allowable shearing stress of 90 MPa and a modulus of rigidity of 77.2 GPa.

10.38 The design specifications of a 2-m-long solid circular transmission shaft require that the angle of twist of the shaft not exceed 3° when a torque of 9 kN·m is applied. Determine the required diameter of the shaft, knowing that the shaft is made of (*a*) a steel with an allowable shearing stress of 90 MPa and a modulus of rigidity of 77 GPa, (*b*) a bronze with an allowable shearing stress of 35 MPa and a modulus of rigidity of 42 GPa.

10.39 and 10.40 The solid cylindrical rod BC of length L = 24 in. is attached to the rigid lever AB of length a = 15 in. and to the support at C. Design specifications require that the displacement of A not exceed 1 in. when a 100-lb force **P** is applied at A. For the material indicated, determine the required diameter of the rod.

10.39 *Steel*: τ_{all} = 15 ksi, $G = 11.2 \times 10^6$ psi.
10.40 *Aluminum*: τ_{all} = 10 ksi, $G = 3.9 \times 10^6$ psi.

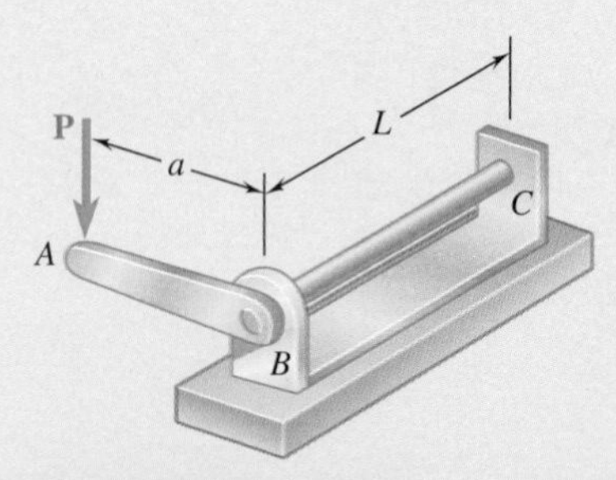

Fig. P10.39 and P10.40

10.41 A torque of magnitude $T = 4$ kN·m is applied at end A of the composite shaft shown. Knowing that the modulus of rigidity is 77 GPa for the steel and 27 GPa for the aluminum, determine (*a*) the maximum shearing stress in the steel core, (*b*) the maximum shearing stress in the aluminum jacket, (*c*) the angle of twist at A.

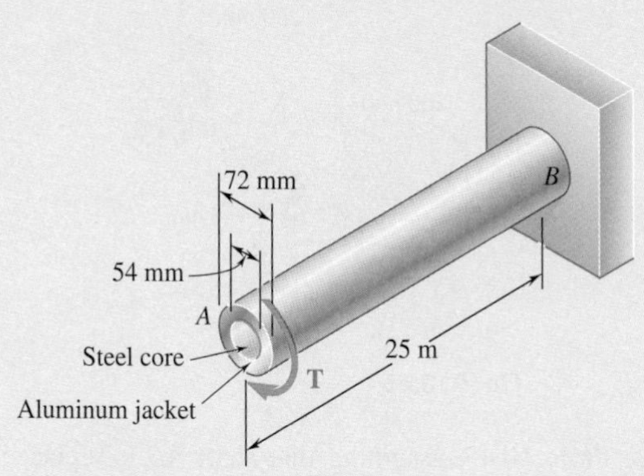

Fig. P10.41 and *P10.42*

10.42 The composite shaft shown is to be twisted by applying a torque **T** at end A. Knowing that the modulus of rigidity is 77 GPa for the steel and 27 GPa for the aluminum, determine the largest angle through which end A can be rotated if the following allowable stresses are not be exceeded: $\tau_{\text{steel}} = 60$ MPa and $\tau_{\text{aluminum}} = 45$ MPa.

10.43 The composite shaft shown consists of a 0.2-in.-thick brass jacket ($G_{\text{brass}} = 5.6 \times 10^6$ psi) bonded to a 1.2-in.-diameter steel core ($G_{\text{steel}} = 11.2 \times 10^6$ psi). Knowing that the shaft is subjected to a 5-kip·in. torque, determine (*a*) the maximum shearing stress in the steel core, (*b*) the angle of twist of B relative to end A.

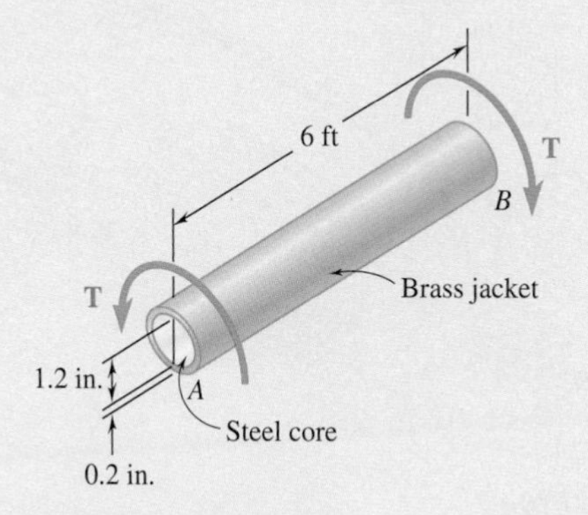

Fig. P10.43 and P10.44

10.44 The composite shaft shown consists of a 0.2-in.-thick brass jacket ($G_{\text{brass}} = 5.6 \times 10^6$ psi) bonded to a 1.2-in.-diameter steel core ($G_{\text{steel}} = 11.2 \times 10^6$ psi). Knowing that the shaft is being subjected to the torques shown, determine the largest angle through which it can be twisted if the following allowable stresses are not to be exceeded: $\tau_{\text{steel}} = 15$ ksi and $\tau_{\text{brass}} = 8$ ksi.

10.45 Two solid steel shafts (G = 77.2 GPa) are connected to a coupling disk B and to fixed supports at A and C. For the loading shown, determine (*a*) the reaction at each support, (*b*) the maximum shearing stress in shaft AB, (*c*) the maximum shearing stress in shaft BC.

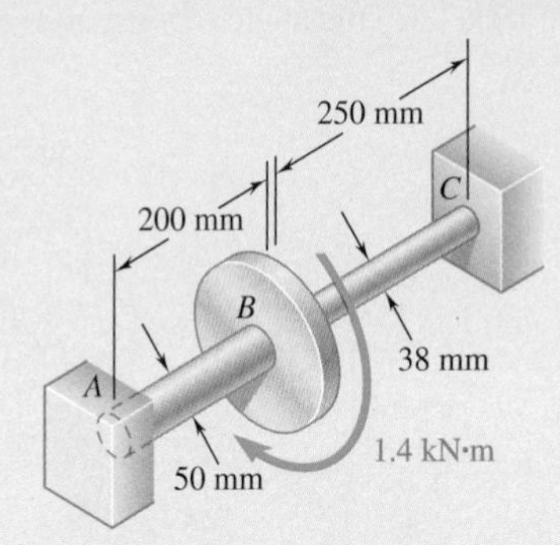

Fig. P10.45

10.46 Solve Prob. 10.45, assuming that shaft AB is replaced by a hollow shaft of the same outer diameter and of 25-mm inner diameter.

10.47 The design specifications for the gear-and-shaft system shown require that the same diameter be used for both shafts and that the angle through which pulley A will rotate when subjected to a 2-kip·in. torque $\mathbf{T}_A$ while pulley D is held fixed will not exceed 7.5°. Determine the required diameter of the shafts if both shafts are made of a steel with $G = 11.2 \times 10^6$ psi and τ_{all} = 12 ksi.

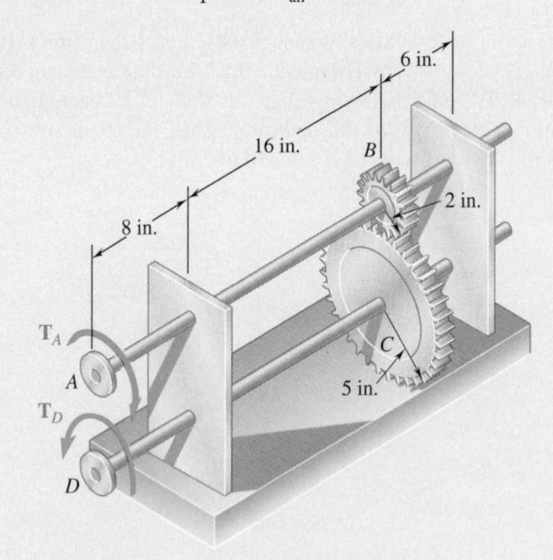

Fig. *P10.47*

10.48 Solve Prob. 10.47, assuming that both shafts are made of a brass with $G = 5.6 \times 10^6$ psi and τ_{all} = 8 ksi.

Review and Summary

This chapter was devoted to the analysis and design of *shafts* subjected to twisting couples, or *torques*. Except for the last two sections of the chapter, our discussion was limited to *circular shafts*.

Deformations in Circular Shafts

The distribution of stresses in the cross section of a circular shaft is *statically indeterminate*. The determination of these stresses requires a prior analysis of the *deformations* occurring in the shaft [Sec. 10.1B]. In a circular shaft subjected to torsion, *every cross section remains plane and undistorted*. The *shearing strain* in a small element with sides parallel and perpendicular to the axis of the shaft and at a distance ρ from that axis is

$$\gamma = \frac{\rho\phi}{L} \qquad \textbf{(10.2)}$$

where ϕ is the angle of twist for a length L of the shaft (Fig. 10.26). Eq. (10.2) shows that the *shearing strain in a circular shaft varies linearly with the distance from the axis of the shaft*. It follows that the strain is maximum at the surface of the shaft, where ρ is equal to the radius c of the shaft:

$$\gamma_{\max} = \frac{c\phi}{L} \qquad \gamma = \frac{\rho}{c}\gamma_{\max} \qquad \textbf{(10.3, 4)}$$

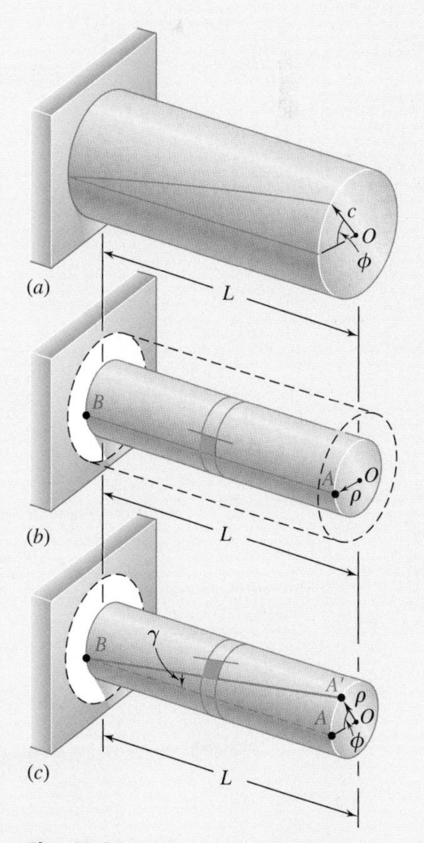

Fig. 10.26

Shearing Stresses in Elastic Range

The relationship between *shearing stresses* in a circular shaft within the elastic range [Sec. 10.1C] and Hooke's law for shearing stress and strain, $\tau = G\gamma$, is

$$\tau = \frac{\rho}{c}\tau_{\max} \qquad \textbf{(10.6)}$$

which shows that within the elastic range, the *shearing stress τ in a circular shaft also varies linearly with the distance from the axis of the shaft*. Equating the sum of the moments of the elementary forces exerted on any section of the shaft to the magnitude T of the torque applied to the shaft, the *elastic torsion formulas* are

$$\tau_{\max} = \frac{Tc}{J} \qquad \tau = \frac{T\rho}{J} \qquad \textbf{(10.9, 10)}$$

where c is the radius of the cross section and J its centroidal polar moment of inertia. $J = \frac{1}{2}\pi c^4$ for a solid shaft, and $J = \frac{1}{2}\pi(c_2^4 - c_1^4)$ for a hollow shaft of inner radius c_1 and outer radius c_2.

We noted that while the element a in Fig. 10.27 is in pure shear, the element c in the same figure is subjected to normal stresses of the same magnitude, Tc/J, with two of the normal stresses being tensile and two compressive. This explains why in a torsion test ductile materials, which generally fail in shear, will break along a plane perpendicular to the axis of the specimen, while brittle materials, which are weaker in tension than in shear, will break along surfaces forming a 45° angle with that axis.

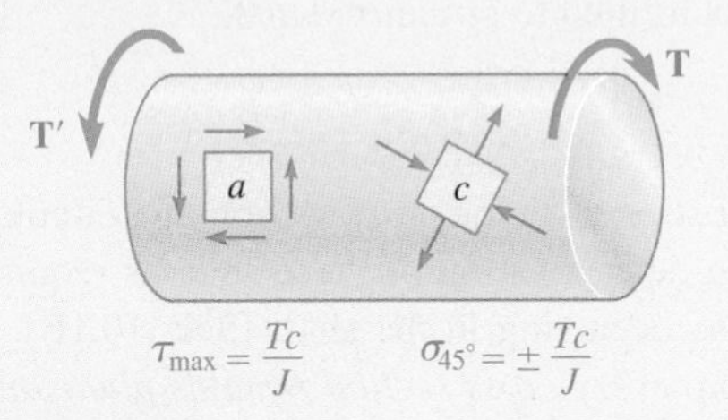

Fig. 10.27

Angle of Twist

Within the elastic range, the angle of twist ϕ of a circular shaft is proportional to the torque T applied to it (Fig. 10.28).

$$\phi = \frac{TL}{JG} \quad \text{(units of radians)} \tag{10.15}$$

where L = length of shaft
J = polar moment of inertia of cross section
G = modulus of rigidity of material

If the shaft is subjected to torques at locations other than its ends or consists of several parts of various cross sections and possibly of different materials, the angle of twist of the shaft must be expressed as the *algebraic sum* of the angles of twist of its component parts:

$$\phi = \sum_i \frac{T_i L_i}{J_i G_i} \tag{10.16}$$

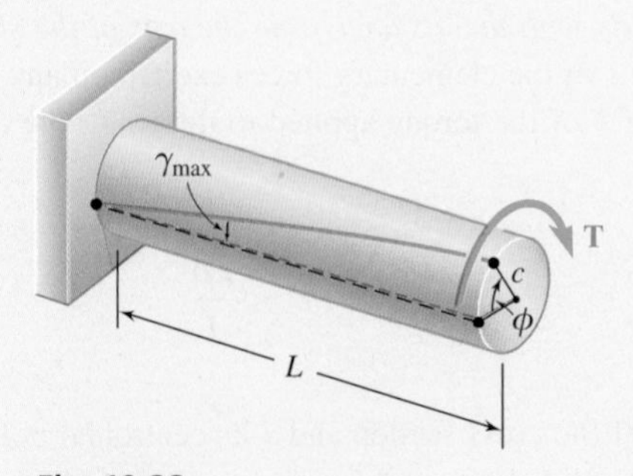

Fig. 10.28

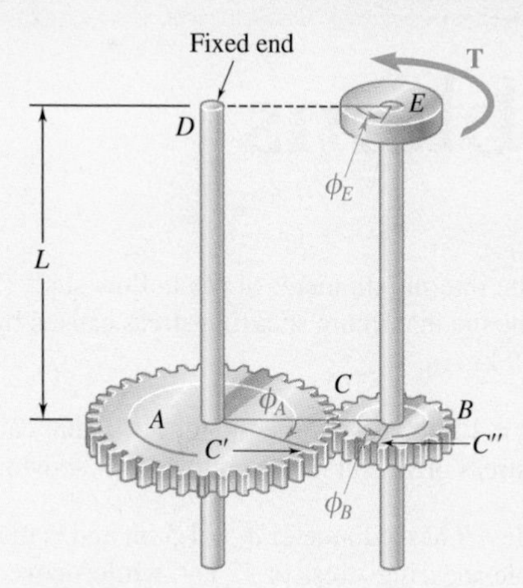

Fig. 10.29

When both ends of a shaft BE rotate (Fig. 10.29), the angle of twist is equal to the *difference* between the angles of rotation ϕ_B and ϕ_E of its ends. When two shafts AD and BE are connected by gears A and B, the torques applied by gear A on shaft AD and gear B on shaft BE are *directly proportional* to the radii r_A and r_B of the two gears—since the forces applied on each other by the gear teeth at C are equal and opposite. On the other hand, the angles ϕ_A and ϕ_B are *inversely proportional* to r_A and r_B—since the arcs CC' and CC'' described by the gear teeth are equal.

Statically Indeterminate Shafts

If the reactions at the supports of a shaft or the internal torques cannot be determined from statics alone, the shaft is said to be *statically indeterminate.* The equilibrium equations obtained from free-body diagrams must be complemented by relationships involving deformations of the shaft and obtained from the geometry of the problem.

Review Problems

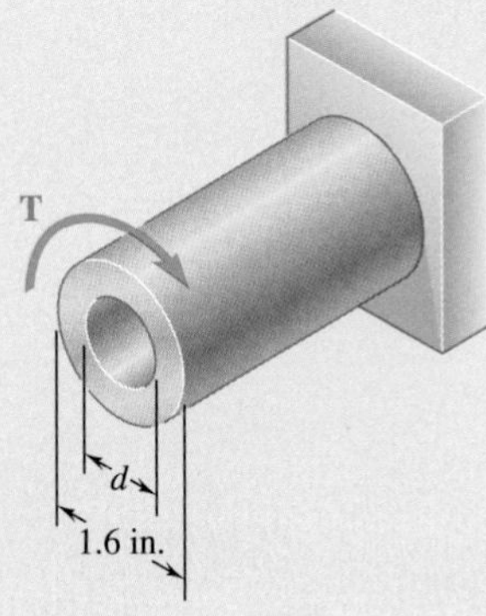

Fig. P10.49 and *P10.50*

10.49 Knowing that the internal diameter of the hollow shaft shown is $d =$ 0.9 in., determine the maximum shearing stress caused by a torque of magnitude $T = 9$ kip·in.

10.50 Knowing that $d = 1.2$ in., determine the torque **T** that causes a maximum shearing stress of 7.5 ksi in the hollow shaft shown.

10.51 The solid spindle AB has a diameter $d_s = 1.5$ in. and is made of a steel with an allowable shearing stress of 12 ksi, while sleeve CD is made of a brass with an allowable shearing stress of 7 ksi. Determine the largest torque **T** that can be applied at A.

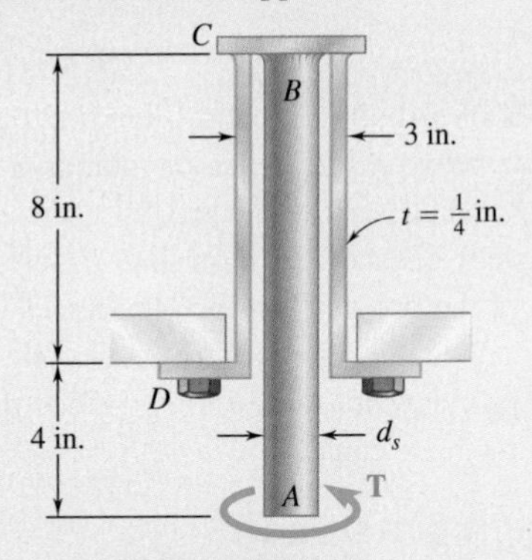

Fig. P10.51 and P10.52

10.52 The solid spindle AB is made of a steel with an allowable shearing stress of 12 ksi, while sleeve CD is made of a brass with an allowable shearing stress of 7 ksi. Determine (*a*) the largest torque **T** that can be applied at A if the allowable shearing stress is not to be exceeded in sleeve CD, (*b*) the corresponding required value of the diameter d_s of spindle AB.

10.53 A steel pipe of 12-in. outer diameter is fabricated from $\frac{1}{4}$-in.-thick plate by welding along a helix that forms an angle of 45° with a plane parallel to the axis of the pipe. Knowing that the maximum allowable tensile stress in the weld is 12 ksi, determine the largest torque that can be applied to the pipe.

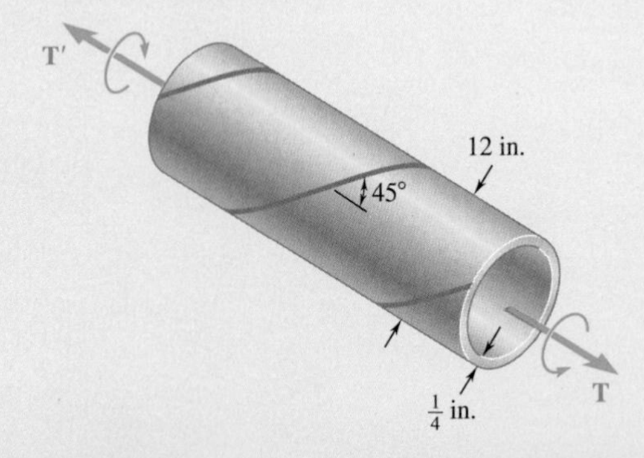

Fig. *P10.53*

10.54 Two solid brass rods AB and CD are brazed to a brass sleeve EF. Determine the ratio d_2/d_1 for which the same maximum shearing stress occurs in the rods and in the sleeve.

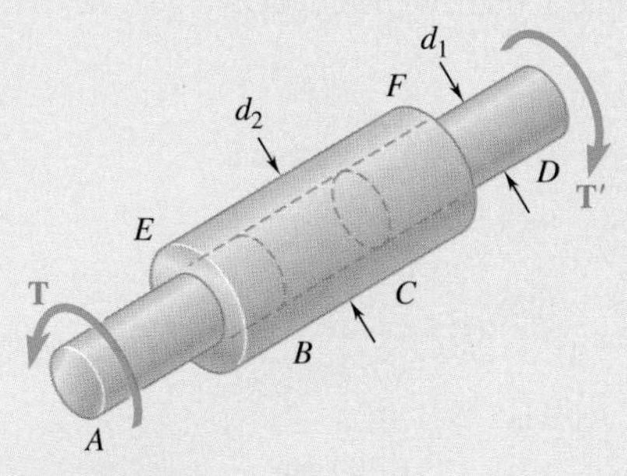

Fig. P10.54

10.55 The design of the gear-and-shaft system shown requires that steel shafts of the same diameter be used for both AB and CD. It is further required that $\tau_{max} \leq 60$ MPa and that the angle ϕ_D through which end D of shaft CD rotates not exceed 1.5°. Knowing that $G = 77.2$ GPa, determine the required diameter of the shafts.

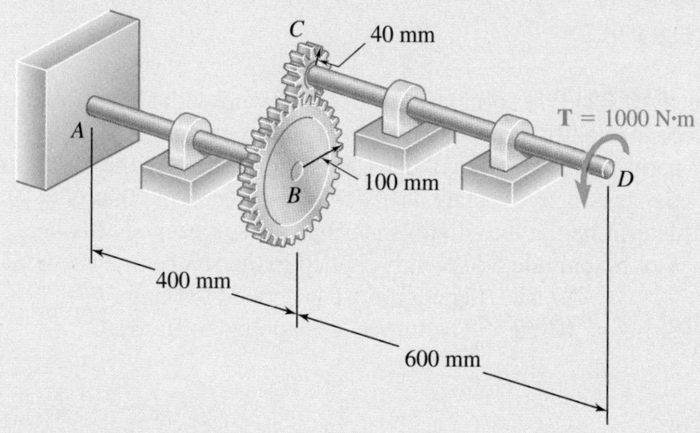

Fig. *P10.55*

10.56 In the bevel-gear system shown, $\alpha = 18.43°$. Knowing that the allowable shearing stress is 8 ksi in each shaft and that the system is in equilibrium, determine the largest torque $\mathbf{T}_A$ that can be applied at A.

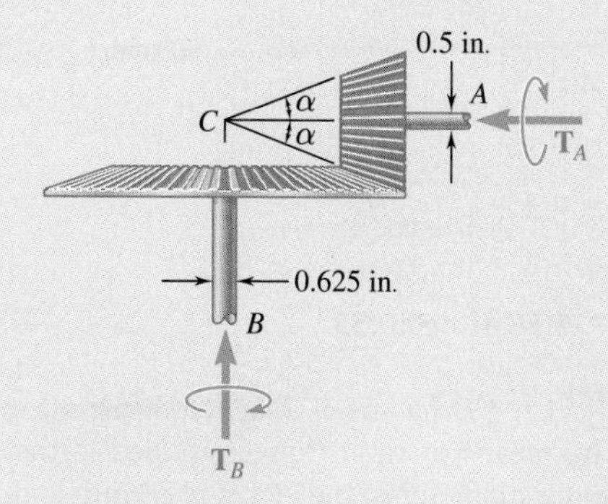

Fig. P10.56

10.57 Ends A and D of the two solid steel shafts AB and CD are fixed, while ends B and C are connected to gears as shown. Knowing that the allowable shearing stress is 50 MPa in each shaft, determine the largest torque **T** that can be applied to gear B.

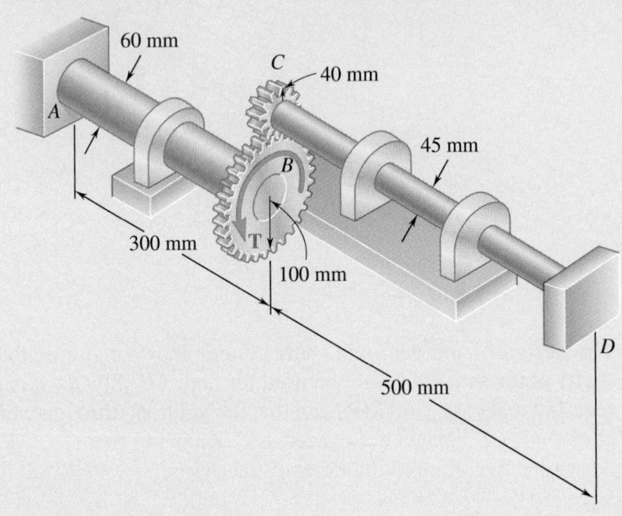

Fig. *P10.57*

10.58 and 10.59 Two solid steel shafts are fitted with flanges that are then connected by bolts as shown. The bolts are slightly undersized and permit a 1.5° rotation of one flange with respect to the other before the flanges begin to rotate as a single unit. Knowing that $G = 77.2$ GPa, determine the maximum shearing stress in each shaft when a torque **T** of magnitude 500 N·m is applied to the flange indicated.

10.58 The torque **T** is applied to flange B.

10.59 The torque **T** is applied to flange C.

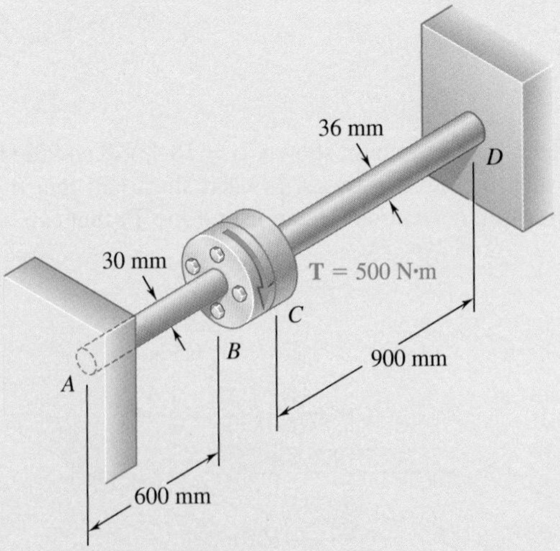

Fig. P10.58 and P10.59

10.60 The steel jacket CD has been attached to the 40-mm-diameter steel shaft AE by means of rigid flanges welded to the jacket and to the rod. The outer diameter of the jacket is 80 mm and its wall thickness is 4 mm. If 500 N·m torques are applied as shown, determine the maximum shearing stress in the jacket.

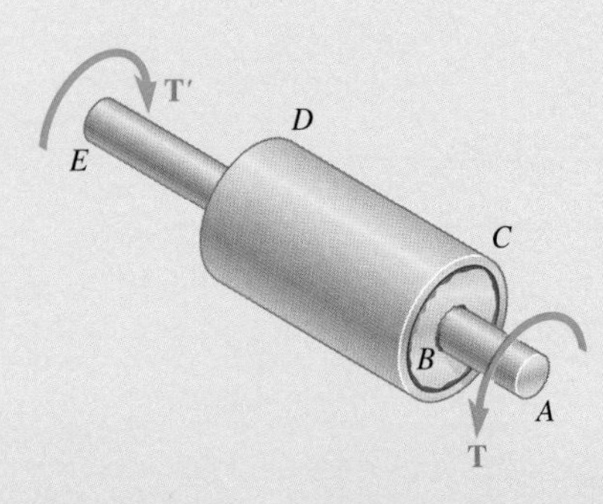

Fig. P10.60

11

Pure Bending

The normal stresses and the curvature resulting from pure bending, such as those developed in the center portion of the barbell shown, will be studied in this chapter.

Objectives

In this chapter, you will:

- **Consider** the general principles of bending behavior
- **Define** the deformations, strains, and normal stresses in beams subject to pure bending
- **Describe** the behavior of composite beams made of more than one material
- **Analyze** members subject to eccentric axial loading, involving both axial stresses and bending stresses
- **Review** beams subject to unsymmetric bending, i.e., where bending does not occur in a plane of symmetry

Introduction

This chapter and the following two analyze the stresses and strains in prismatic members subjected to *bending*. Bending is a major concept used in the design of many machine and structural components, such as beams and girders.

This chapter is devoted to the analysis of prismatic members subjected to equal and opposite couples **M** and **M′** acting in the same longitudinal plane. Such members are said to be in *pure bending*. The members are assumed to possess a plane of symmetry with the couples **M** and **M′** acting in that plane (Fig. 11.1).

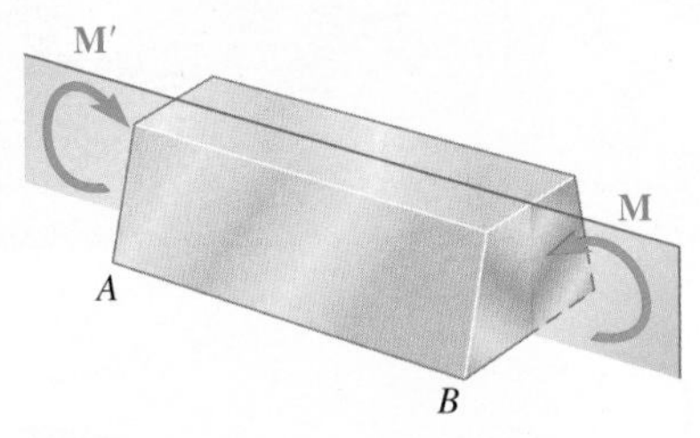

Fig. 11.1 Member in pure bending.

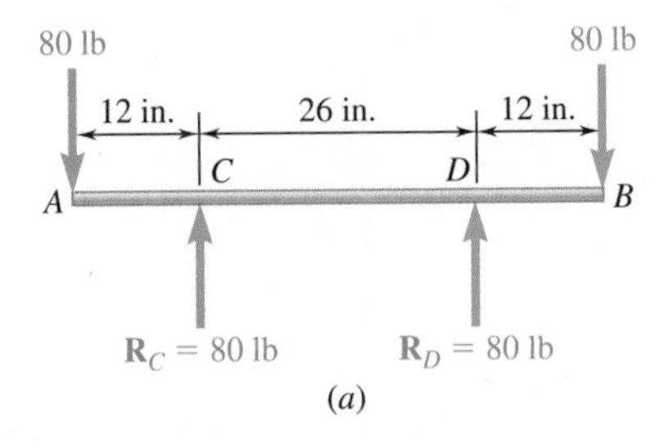

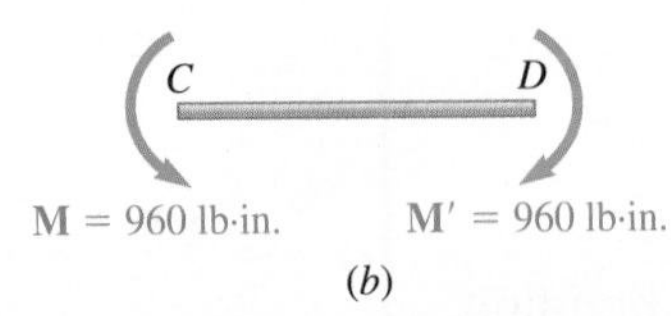

Fig. 11.2 (*a*) Free-body diagram of the barbell pictured in the chapter opening photo and (*b*) free-body diagram of the center portion of the bar, which is in pure bending.

An example of pure bending is provided by the bar of a typical barbell as it is held overhead by a weight lifter as shown in the opening photo for this chapter. The bar carries equal weights at equal distances from the hands of the weight lifter. Because of the symmetry of the free-body diagram of the bar (Fig. 11.2*a*), the reactions at the hands must be equal and opposite to the weights. Therefore, as far as the middle portion *CD* of the bar is concerned, the weights and the reactions can be replaced by two equal and opposite 960-lb·in. couples (Fig. 11.2*b*), showing that the middle portion of the bar is in pure bending. A similar analysis of a small sport buggy (Photo 11.1) shows that the axle is in pure bending between the two points where it is attached to the frame.

The results obtained from the direct applications of pure bending will be used in the analysis of other types of loadings, such as *eccentric axial loadings* and *transverse loadings*.

Photo 11.1 The center portion of the rear axle of the sport buggy is in pure bending.

Courtesy of Flexifoil

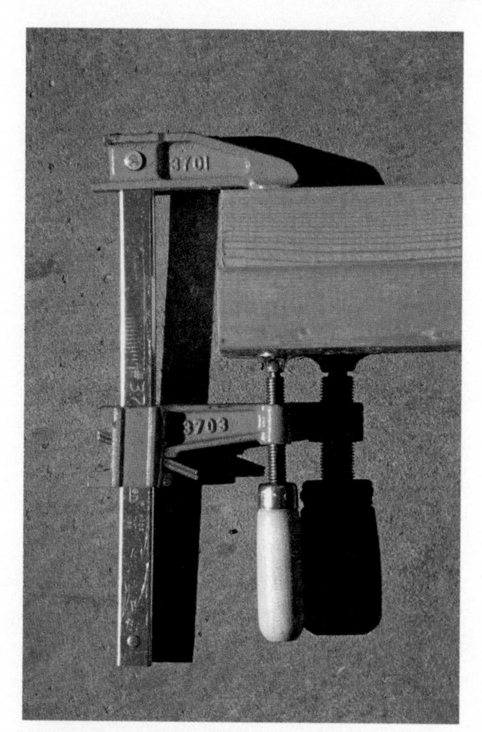

Photo 11.2 Clamp used to glue lumber pieces together.

Photo 11.2 shows a 12-in. steel bar clamp used to exert 150-lb forces on two pieces of lumber as they are being glued together. Fig. 11.3*a* shows the equal and opposite forces exerted by the lumber on the clamp. These forces result in an *eccentric loading* of the straight portion of the clamp. In Fig. 11.3*b*, a section CC' has been passed through the clamp and a free-body diagram has been drawn of the upper half of the clamp. The internal forces in the section are equivalent to a 150-lb axial tensile force **P** and a 750-lb·in. couple **M**. By combining our knowledge of the stresses under a *centric* load and the results of an analysis of stresses in pure bending, the distribution of stresses under an *eccentric* load is obtained. This is discussed in Sec. 11.4.

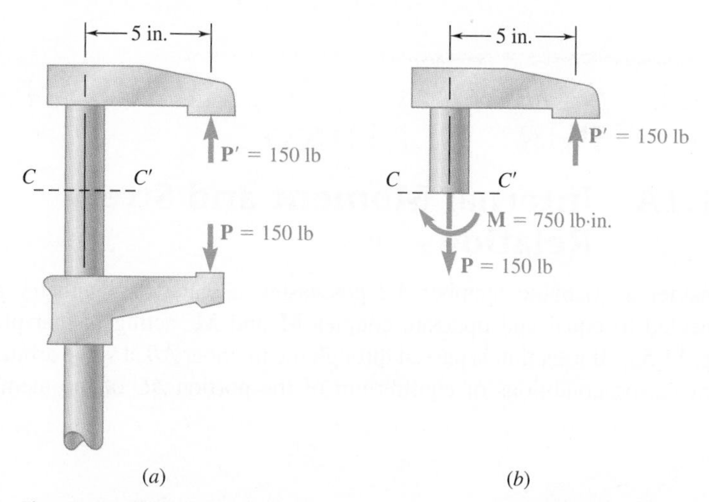

Fig. 11.3 (*a*) Free-body diagram of a clamp, (*b*) free-body diagram of the upper portion of the clamp.

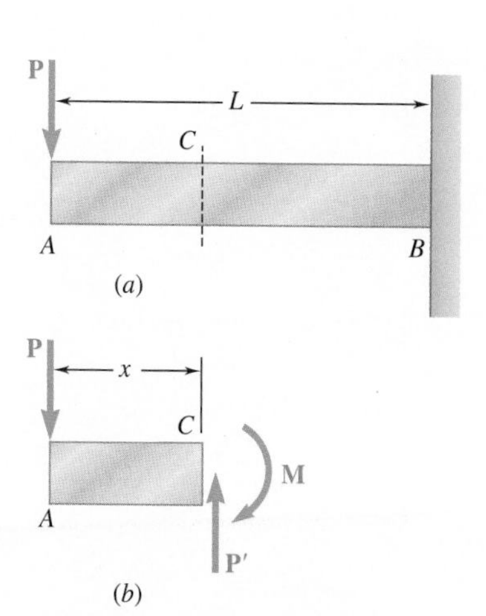

Fig. 11.4 (*a*) Cantilevered beam with end loading. (*b*) As portion *AC* shows, beam is not in pure bending.

The study of pure bending plays an essential role in the study of beams (i.e., prismatic members) subjected to various types of *transverse loads*. Consider a cantilever beam *AB* supporting a concentrated load **P** at its free end (Fig. 11.4*a*). If a section is passed through *C* at a distance *x* from *A*, the free-body diagram of *AC* (Fig. 11.4*b*) shows that the internal forces in the section consist of a force $\mathbf{P}'$ equal and opposite to $\mathbf{P}$ and a couple $\mathbf{M}$ of magnitude $M = Px$. The distribution of normal stresses in the section can be obtained from the couple $\mathbf{M}$ as if the beam were in pure bending. The shearing stresses in the section depend on the force $\mathbf{P}'$, and their distribution over a given section is discussed in Chap. 13.

The first part of this chapter covers the analysis of stresses and deformations caused by pure bending in a homogeneous member possessing a plane of symmetry and made of a material following Hooke's law. The methods of statics are used in Sec. 11.1A to derive three fundamental equations which must be satisfied by the normal stresses in any given cross section of the member. In Sec. 11.1B, it will be proved that *transverse sections remain plane* in a member subjected to pure bending, while in Sec. 11.2, formulas are developed to determine the *normal stresses* and *radius of curvature* for that member within the elastic range.

Sec. 11.3 covers the stresses and deformations in *composite members* made of more than one material, such as reinforced-concrete beams, which utilize the best features of steel and concrete and are extensively used in the construction of buildings and bridges. You will learn to draw a *transformed section* representing a member made of a homogeneous material that undergoes the same deformations as the composite member under the same loading. The transformed section is used to find the stresses and deformations in the original composite member.

In Sec. 11.4, you will analyze an *eccentric axial loading* in a plane of symmetry (Fig. 11.3) by superposing the stresses due to pure bending and a centric axial loading.

The study of the bending of prismatic members concludes with the analysis of *unsymmetric bending* (Sec. 11.5), and the study of the general case of *eccentric axial loading* (Sec. 11.6).

11.1 SYMMETRIC MEMBERS IN PURE BENDING

11.1A Internal Moment and Stress Relations

Consider a prismatic member AB possessing a plane of symmetry and subjected to equal and opposite couples $\mathbf{M}$ and $\mathbf{M}'$ acting in that plane (Fig. 11.5*a*). If a section is passed through the member AB at some arbitrary point C, the conditions of equilibrium of the portion AC of the member

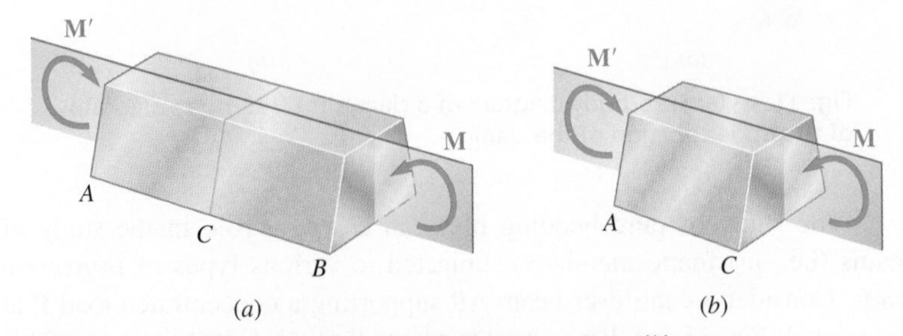

Fig. 11.5 (*a*) A member in a state of pure bending. (*b*) Any intermediate portion of AB will also be in pure bending.

require the internal forces in the section to be equivalent to the couple **M** (Fig. 11.5*b*). The moment M of that couple is the *bending moment* in the section. Following the usual convention, a positive sign is assigned to M when the member is bent as shown in Fig. 11.5*a* (i.e., when the concavity of the beam faces upward) and a negative sign otherwise.

Denoting by σ_x the normal stress at a given point of the cross section and by τ_{xy} and τ_{xz} the components of the shearing stress, we express that the system of the elementary internal forces exerted on the section is equivalent to the couple **M** (Fig. 11.6).

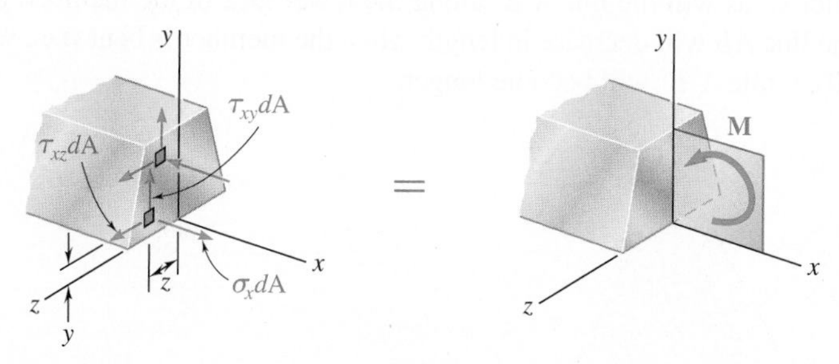

Fig. 11.6 Stresses resulting from pure bending moment M.

Recall from statics that a couple **M** actually consists of two equal and opposite forces. The sum of the components of these forces in any direction is therefore equal to zero. Moreover, the moment of the couple is the same about *any* axis perpendicular to its plane and is zero about any axis contained in that plane. Selecting arbitrarily the z axis shown in Fig. 11.6, the equivalence of the elementary internal forces and the couple **M** is expressed by writing that the sums of the components and moments of the forces are equal to the corresponding components and moments of the couple **M**:

$$x \text{ components:} \quad \int \sigma_x \, dA = 0 \tag{11.1}$$

$$\text{Moments about } y \text{ axis:} \quad \int z\sigma_x \, dA = 0 \tag{11.2}$$

$$\text{Moments about } z \text{ axis:} \quad \int (-y\sigma_x \, dA) = M \tag{11.3}$$

Three additional equations could be obtained by setting equal to zero the sums of the y components, z components, and moments about the x axis, but these equations would involve only the components of the shearing stress and, as you will see in the next section, the components of the shearing stress are both equal to zero.

Two remarks should be made at this point:

1. The minus sign in Eq. (11.3) is due to the fact that a tensile stress ($\sigma_x > 0$) leads to a negative moment (clockwise) of the normal force $\sigma_x \, dA$ about the z axis.
2. Eq. (11.2) could have been anticipated, since the application of couples in the plane of symmetry of member AB result in a distribution of normal stresses symmetric about the y axis.

Once more, note that the actual distribution of stresses in a given cross section cannot be determined from statics alone. It is *statically indeterminate* and may be obtained only by analyzing the *deformations* produced in the member.

11.1B Deformations

We will now analyze the deformations of a prismatic member possessing a plane of symmetry. Its ends are subjected to equal and opposite couples **M** and **M′** acting in the plane of symmetry. The member will bend under the action of the couples, but will remain symmetric with respect to that plane (Fig. 11.7). Moreover, since the bending moment M is the same in any cross section, the member will bend uniformly. Thus, the line AB along the upper face of the member intersecting the plane of the couples will have a constant curvature. In other words, the line AB will be transformed into a circle of center C, as will the line $A'B'$ along the lower face of the member. Note that the line AB will decrease in length when the member is bent (i.e., when $M > 0$), while $A'B'$ will become longer.

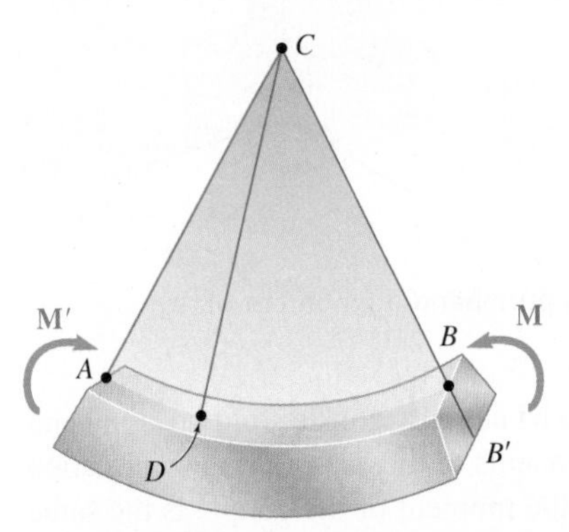

Fig. 11.7 Initially straight members in pure bending deform into a circular arc.

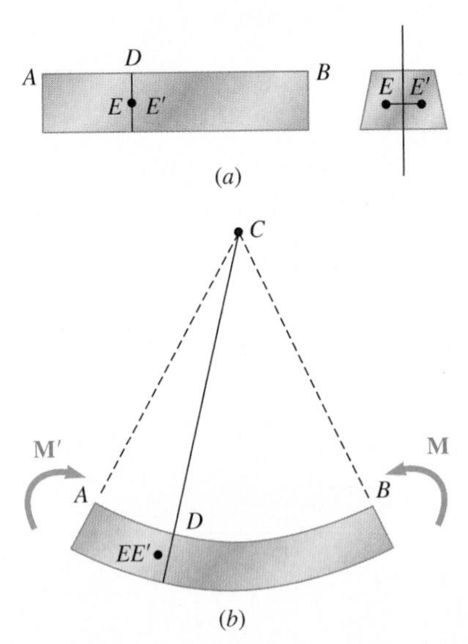

Fig. 11.8 (a) Two points in a cross section at D that is perpendicular to the member's axis. (b) Considering the possibility that these points do not remain in the cross section after bending.

Next we will prove that any cross section perpendicular to the axis of the member remains plane, and that the plane of the section passes through C. If this were not the case, we could find a point E of the original section through D (Fig. 11.8a) which, after the member has been bent, would *not* lie in the plane perpendicular to the plane of symmetry that contains line CD (Fig. 11.8b). But, because of the symmetry of the member, there would be another point E' that would be transformed exactly in the same way. Let us assume that, after the beam has been bent, both points would be located to the left of the plane defined by CD, as shown in Fig. 11.8b. Since the bending moment M is the same throughout the member, a similar situation would prevail in any other cross section, and the points corresponding to E and E' would also move to the left. Thus, an observer at A would conclude that the loading causes the points E and E' in the various cross sections to move forward (toward the observer). But an observer at B, to whom the loading looks the same, and who observes the points E and E' in the same positions (except that they are now inverted) would reach the opposite conclusion. This inconsistency leads us to conclude that E and E' will lie in the plane defined by CD and, therefore, that the section remains plane and passes through C. We should note, however, that this discussion does not rule out the possibility of deformations within the plane of the section.

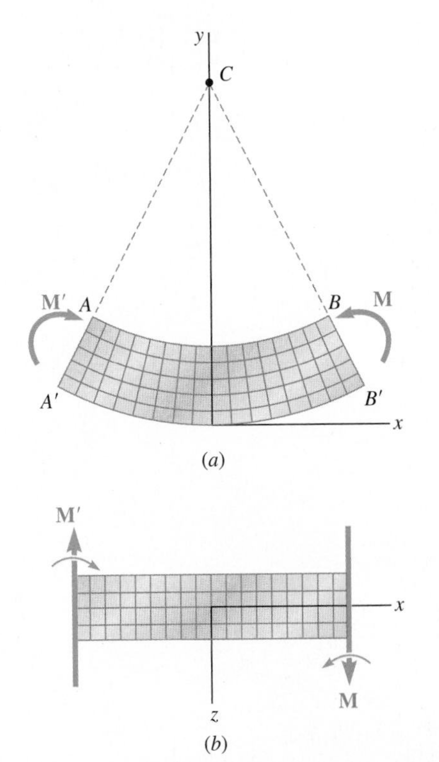

Fig. 11.9 Member subject to pure bending shown in two views. (a) Longitudinal, vertical section (plane of symmetry). (b) Longitudinal, horizontal section.

Suppose that the member is divided into a large number of small cubic elements with faces respectively parallel to the three coordinate planes. The property we have established requires that these elements be transformed as shown in Fig. 11.9 when the member is subjected to the couples **M** and **M′**. Since all the faces represented in the two projections of

Fig. 11.9 are at 90° to each other, we conclude that $\gamma_{xy} = \gamma_{zx} = 0$ and, thus, that $\tau_{xy} = \tau_{xz} = 0$. Regarding the three stress components that we have not yet discussed, namely, σ_y, σ_z, and τ_{yz}, we note that they must be zero on the surface of the member. Since, on the other hand, the deformations involved do not require any interaction between the elements of a given transverse cross section, we can assume that these three stress components are equal to zero throughout the member. This assumption is verified, both from experimental evidence and from the theory of elasticity, for slender members undergoing small deformations. We conclude that the only nonzero stress component exerted on any of the small cubic elements considered here is the normal component σ_x. Thus, at any point of a slender member in pure bending, we have a state of *uniaxial stress*. Recalling that, for $M > 0$, lines AB and $A'B'$ are observed, respectively, to decrease and increase in length, we note that the strain ϵ_x and the stress σ_x are negative in the upper portion of the member (*compression*) and positive in the lower portion (*tension*).

It follows from above that a surface parallel to the upper and lower faces of the member must exist where ϵ_x and σ_x are zero. This surface is called the *neutral surface*. The neutral surface intersects the plane of symmetry along an arc of circle DE (Fig. 11.10*a*), and it intersects a transverse section along a straight line called the *neutral axis* of the section (Fig. 11.10*b*). The origin of coordinates is now selected on the neutral surface—rather than on the lower face of the member—so that the distance from any point to the neutral surface is measured by its coordinate y.

Denoting by ρ the radius of arc DE (Fig. 11.10*a*), by θ the central angle corresponding to DE, and observing that the length of DE is equal to the length L of the undeformed member, we write

$$L = \rho\theta \tag{11.4}$$

Considering the arc JK located at a distance y above the neutral surface, its length L' is

$$L' = (\rho - y)\theta \tag{11.5}$$

Since the original length of arc JK was equal to L, the deformation of JK is

$$\delta = L' - L \tag{11.6}$$

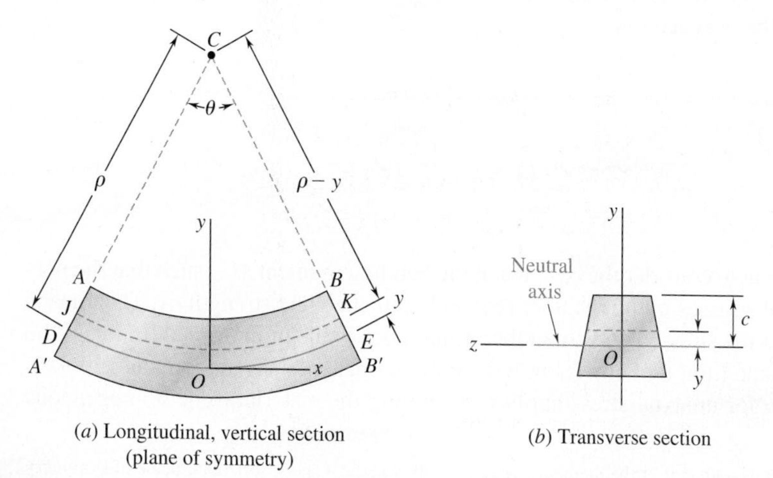

Fig. 11.10 Establishment of neutral axis. (*a*) Longitudinal-vertical view. (*b*) Transverse section at origin.

or, substituting from Eqs. (11.4) and (11.5) into Eq. (11.6),

$$\delta = (\rho - y)\theta - \rho\theta = -y\theta \quad \textbf{(11.7)}$$

The longitudinal strain ϵ_x in the elements of JK is obtained by dividing δ by the original length L of JK. Write

$$\epsilon_x = \frac{\delta}{L} = \frac{-y\theta}{\rho\theta}$$

or

$$\epsilon_x = -\frac{y}{\rho} \quad \textbf{(11.8)}$$

The minus sign is due to the fact that it is assumed the bending moment is positive, and thus the beam is concave upward.

Because of the requirement that transverse sections remain plane, identical deformations occur in all planes parallel to the plane of symmetry. Thus, the value of the strain given by Eq. (11.8) is valid anywhere, and the *longitudinal normal strain* ϵ_x *varies linearly with the distance* y *from the neutral surface*.

The strain ϵ_x reaches its maximum absolute value when y is largest. Denoting the largest distance from the neutral surface as c (corresponding to either the upper or the lower surface of the member) and the *maximum absolute value* of the strain as ϵ_m, we have

$$\epsilon_m = \frac{c}{\rho} \quad \textbf{(11.9)}$$

Solving Eq. (11.9) for ρ and substituting into Eq. (11.8),

$$\epsilon_x = -\frac{y}{c}\epsilon_m \quad \textbf{(11.10)}$$

To compute the strain or stress at a given point of the member, we must first locate the neutral surface in the member. To do this, we must specify the stress-strain relation of the material used, as will be considered in the next section.[†]

11.2 STRESSES AND DEFORMATIONS IN THE ELASTIC RANGE

We now consider the case when the bending moment M is such that the normal stresses in the member remain below the yield strength σ_Y. This means that the stresses in the member remain below the proportional limit and the elastic limit as well. There will be no permanent deformation, and Hooke's law for uniaxial stress applies. Assuming the material to be homogeneous

[†]Let us note that, if the member possesses both a vertical and a horizontal plane of symmetry (e.g., a member with a rectangular cross section) and the stress-strain curve is the same in tension and compression, the neutral surface will coincide with the plane of symmetry.

and denoting its modulus of elasticity by E, the normal stress in the longitudinal x direction is

$$\sigma_x = E\epsilon_x \tag{11.11}$$

Recalling Eq. (11.10) and multiplying both members by E, we write

$$E\epsilon_x = -\frac{y}{c}(E\epsilon_m)$$

or using Eq. (11.11),

$$\sigma_x = -\frac{y}{c}\sigma_m \tag{11.12}$$

where σ_m denotes the *maximum absolute value* of the stress. This result shows that, *in the elastic range, the normal stress varies linearly with the distance from the neutral surface* (Fig. 11.11).

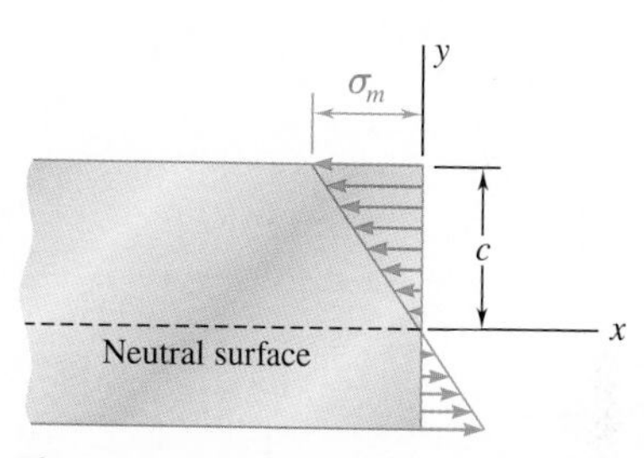

Fig. 11.11 Bending stresses vary linearly with distance from the neutral axis.

Note that neither the location of the neutral surface nor the maximum value σ_m of the stress have yet to be determined. Both can be found using Eqs. (11.1) and (11.3). Substituting for σ_x from Eq. (11.12) into Eq. (11.1), write

$$\int \sigma_x\, dA = \int\left(-\frac{y}{c}\sigma_m\right) dA = -\frac{\sigma_m}{c}\int y\, dA = 0$$

from which

$$\int y\, dA = 0 \tag{11.13}$$

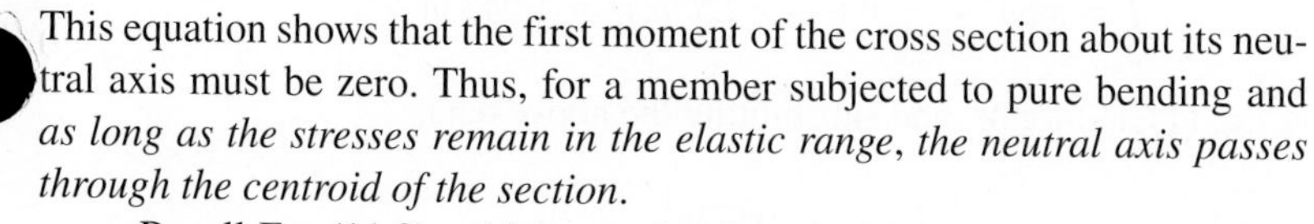

This equation shows that the first moment of the cross section about its neutral axis must be zero. Thus, for a member subjected to pure bending and *as long as the stresses remain in the elastic range, the neutral axis passes through the centroid of the section.*

Recall Eq. (11.3), which was developed with respect to an *arbitrary* horizontal z axis:

$$\int(-y\sigma_x\, dA) = M \tag{11.3}$$

Specifying that the z axis coincides with the neutral axis of the cross section, substitute σ_x from Eq. (11.12) into Eq. (11.3):

$$\int(-y)\left(-\frac{y}{c}\sigma_m\right) dA = M$$

or

$$\frac{\sigma_m}{c}\int y^2\, dA = M \tag{11.14}$$

Recall that for pure bending the neutral axis passes through the centroid of the cross section and I is the moment of inertia or second moment of area of the cross section with respect to a centroidal axis perpendicular to the plane of the couple **M**. Solving Eq. (11.14) for σ_m,†

$$\sigma_m = \frac{Mc}{I} \tag{11.15}$$

†Recall that the bending moment is assumed to be positive. If the bending moment is negative, M should be replaced in Eq. (11.15) by its absolute value $|M|$.

Substituting for σ_m from Eq. (11.15) into Eq. (11.12), we obtain the normal stress σ_x at any distance y from the neutral axis:

$$\sigma_x = -\frac{My}{I} \tag{11.16}$$

Eqs. (11.15) and (11.16) are called the *elastic flexure formulas*, and the normal stress σ_x caused by the bending or "flexing" of the member is often referred to as the *flexural stress*. The stress is compressive ($\sigma_x < 0$) above the neutral axis ($y > 0$) when the bending moment M is positive and tensile ($\sigma_x > 0$) when M is negative.

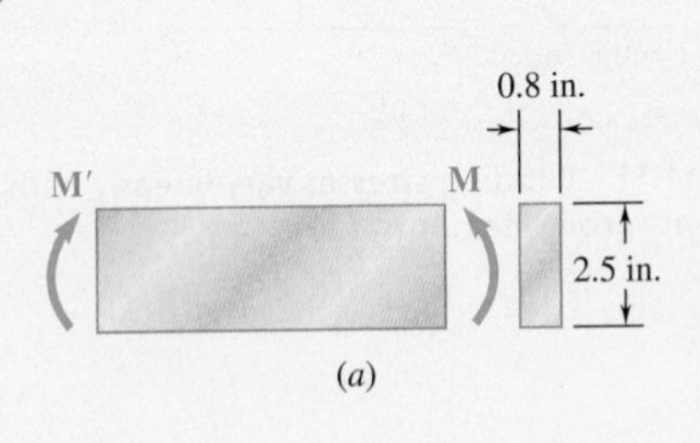

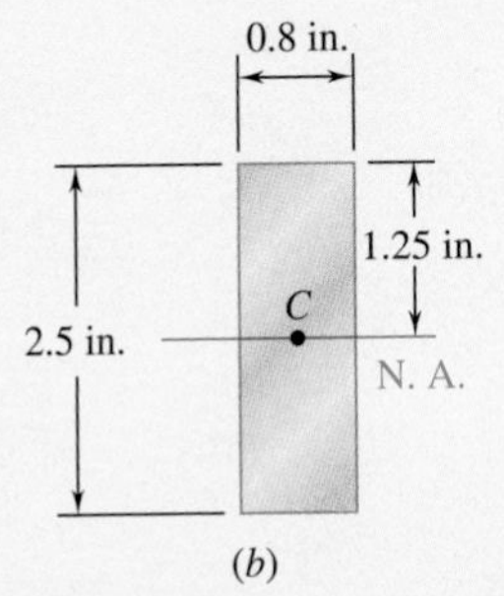

Fig. 11.12 (*a*) Bar of rectangular cross-section in pure bending. (*b*) Centroid and dimensions of cross section.

Concept Application 11.1

A steel bar of 0.8 × 2.5-in. rectangular cross section is subjected to two equal and opposite couples acting in the vertical plane of symmetry of the bar (Fig. 11.12*a*). Determine the value of the bending moment M that causes the bar to yield. Assume $\sigma_Y = 36$ ksi.

Since the neutral axis must pass through the centroid C of the cross section, $c = 1.25$ in. (Fig. 11.12*b*). On the other hand, the centroidal moment of inertia of the rectangular cross section is

$$I = \tfrac{1}{12}bh^3 = \tfrac{1}{12}(0.8 \text{ in.})(2.5 \text{ in.})^3 = 1.042 \text{ in}^4$$

Solving Eq. (11.15) for M, and substituting the above data,

$$M = \frac{I}{c}\sigma_m = \frac{1.042 \text{ in}^4}{1.25 \text{ in.}}(36 \text{ ksi})$$

$$M = 30 \text{ kip·in.}$$

Returning to Eq. (11.15), the ratio I/c depends only on the geometry of the cross section. This ratio is defined as the *elastic section modulus S*, where

$$\text{Elastic section modulus} = S = \frac{I}{c} \tag{11.17}$$

Substituting S for I/c into Eq. (11.15), this equation in alternative form is

$$\sigma_m = \frac{M}{S} \tag{11.18}$$

Since the maximum stress σ_m is inversely proportional to the elastic section modulus S, beams should be designed with as large a value of S as is

Photo 11.3 Wide-flange steel beams are used in the frame of this building.
© *Hisham Ibrahim/Getty Images RF*

practical. For example, a wooden beam with a rectangular cross section of width b and depth h has

$$S = \frac{I}{c} = \frac{\frac{1}{12}bh^3}{h/2} = \tfrac{1}{6}bh^2 = \tfrac{1}{6}Ah \tag{11.19}$$

where A is the cross-sectional area of the beam. For two beams with the same cross-sectional area A (Fig. 11.13), the beam with the larger depth h will have the larger section modulus and will be the more effective in resisting bending.[†]

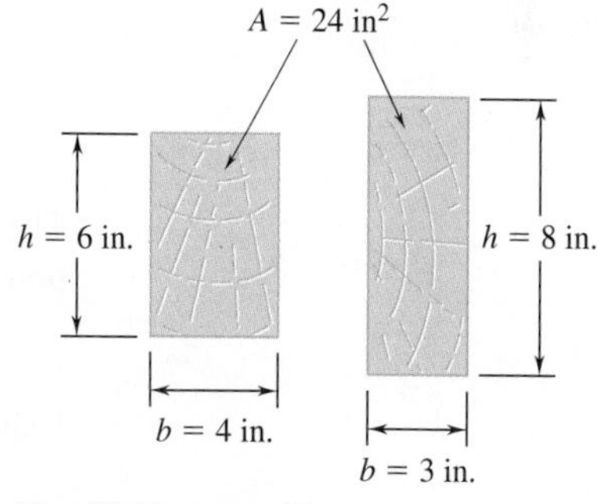

Fig. 11.13 Wood beam cross sections.

In the case of structural steel (Photo 11.3), American standard beams (S-beams) and wide-flange beams (W-beams) are preferred to other shapes because a large portion of their cross section is located far from the neutral axis (Fig. 11.14). Thus, for a given cross-sectional area and a given depth, their design provides large values of I and S. Values of the elastic section modulus of commonly manufactured beams can be obtained from tables listing the various geometric properties of such beams. To determine the maximum stress σ_m in a given section of a standard beam, the engineer needs only to read the value of the elastic section modulus S in such a table and divide the bending moment M in the section by S.

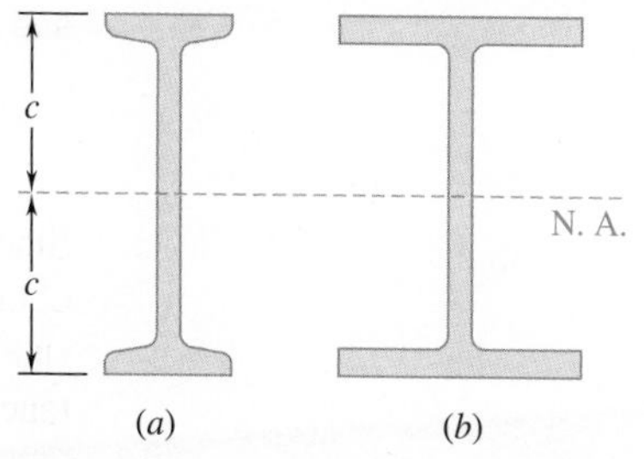

Fig. 11.14 Two types of steel beam cross sections: (*a*) American Standard beam (*S*) (*b*) wide-flange beam (*W*).

The deformation of the member caused by the bending moment M is measured by the *curvature* of the neutral surface. The curvature is defined as the reciprocal of the radius of curvature ρ and can be obtained by solving Eq. (11.9) for $1/\rho$:

$$\frac{1}{\rho} = \frac{\epsilon_m}{c} \tag{11.20}$$

[†]However, large values of the ratio h/b could result in lateral instability of the beam.

In the elastic range, $\epsilon_m = \sigma_m/E$. Substituting for ϵ_m into Eq. (11.20) and recalling Eq. (11.15), write

$$\frac{1}{\rho} = \frac{\sigma_m}{Ec} = \frac{1}{Ec}\frac{Mc}{I}$$

or

$$\frac{1}{\rho} = \frac{M}{EI} \tag{11.21}$$

Concept Application 11.2

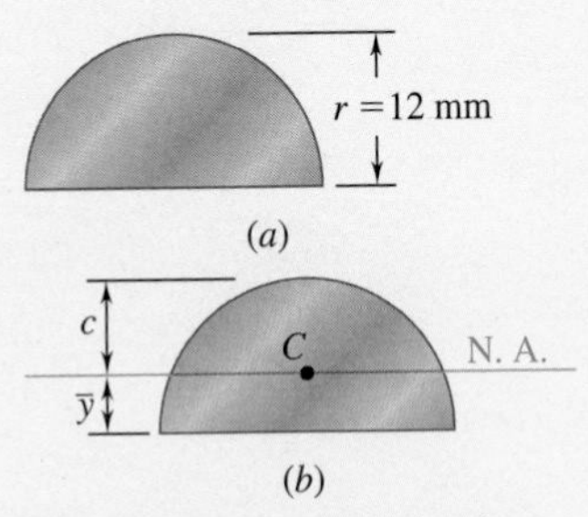

Fig. 11.15 (a) Semi-circular section of rod in pure bending. (b) Centroid and neutral axis of cross section.

An aluminum rod with a semicircular cross section of radius $r = 12$ mm (Fig. 11.15*a*) is bent into the shape of a circular arc of mean radius $\rho =$ 2.5 m. Knowing that the flat face of the rod is turned toward the center of curvature of the arc, determine the maximum tensile and compressive stress in the rod. Use $E = 70$ GPa.

We can use Eq. (11.21) to determine the bending moment M corresponding to the given radius of curvature ρ and then Eq. (11.15) to determine σ_m. However, it is simpler to use Eq. (11.9) to determine ϵ_m and Hooke's law to obtain σ_m.

The ordinate $\bar{y}$ of the centroid C of the semicircular cross section is

$$\bar{y} = \frac{4r}{3\pi} = \frac{4(12 \text{ mm})}{3\pi} = 5.093 \text{ mm}$$

The neutral axis passes through C (Fig. 11.15*b*), and the distance c to the point of the cross section farthest away from the neutral axis is

$$c = r - \bar{y} = 12 \text{ mm} - 5.093 \text{ mm} = 6.907 \text{ mm}$$

Using Eq. (11.9),

$$\epsilon_m = \frac{c}{\rho} = \frac{6.907 \times 10^{-3} \text{ m}}{2.5 \text{ m}} = 2.763 \times 10^{-3}$$

and applying Hooke's law,

$$\sigma_m = E\epsilon_m = (70 \times 10^9 \text{ Pa})(2.763 \times 10^{-3}) = 193.4 \text{ MPa}$$

Since this side of the rod faces away from the center of curvature, the stress obtained is a tensile stress. The maximum compressive stress occurs on the flat side of the rod. Using the fact that the stress is proportional to the distance from the neutral axis, write

$$\sigma_{comp} = -\frac{\bar{y}}{c}\sigma_m = -\frac{5.093 \text{ mm}}{6.907 \text{ mm}}(193.4 \text{ MPa})$$

$$= -142.6 \text{ MPa}$$

Sample Problem 11.1

The rectangular tube shown is extruded from an aluminum alloy for which $\sigma_Y = 40$ ksi, $\sigma_U = 60$ ksi, and $E = 10.6 \times 10^6$ psi. Neglecting the effect of fillets, determine (*a*) the bending moment M for which the factor of safety will be 3.00 and (*b*) the corresponding radius of curvature of the tube.

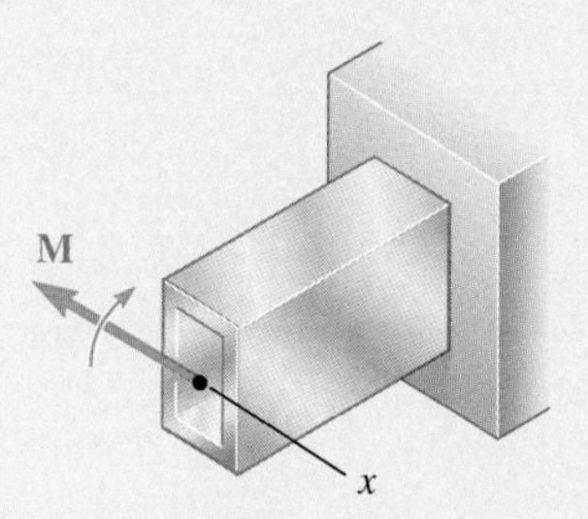

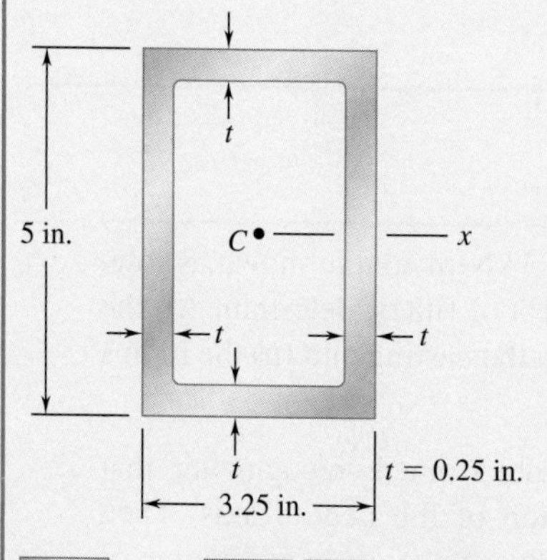

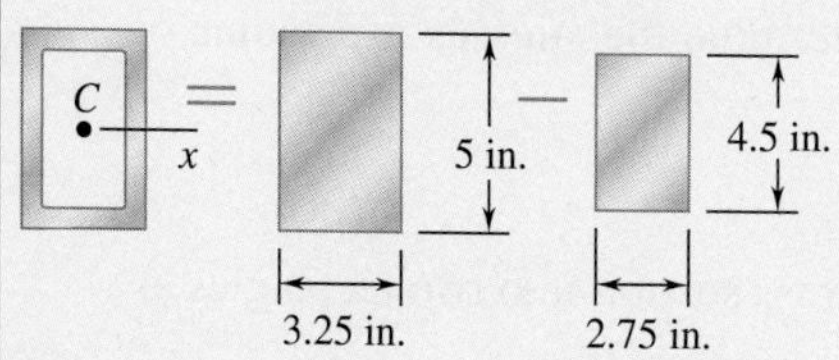

Fig. 1 Superposition for calculating moment of inertia.

STRATEGY: Use the factor of safety to determine the allowable stress. Then calculate the bending moment and radius of curvature using Eqs. (11.15) and (11.21).

MODELING and ANALYSIS:

Moment of Inertia. Considering the cross-sectional area of the tube as the difference between the two rectangles shown in Fig. 1 and recalling the formula for the centroidal moment of inertia of a rectangle, write

$$I = \tfrac{1}{12}(3.25)(5)^3 - \tfrac{1}{12}(2.75)(4.5)^3 \qquad I = 12.97 \text{ in}^4$$

Allowable Stress. For a factor of safety of 3.00 and an ultimate stress of 60 ksi, we have

$$\sigma_{all} = \frac{\sigma_U}{F.S.} = \frac{60 \text{ ksi}}{3.00} = 20 \text{ ksi}$$

Since $\sigma_{all} < \sigma_Y$, the tube remains in the elastic range and we can apply the results of Sec. 11.2.

a. Bending Moment. With $c = \frac{1}{2}(5 \text{ in.}) = 2.5$ in., we write

$$\sigma_{all} = \frac{Mc}{I} \qquad M = \frac{I}{c}\sigma_{all} = \frac{12.97 \text{ in}^4}{2.5 \text{ in.}}(20 \text{ ksi}) \qquad M = 103.8 \text{ kip·in.} \blacktriangleleft$$

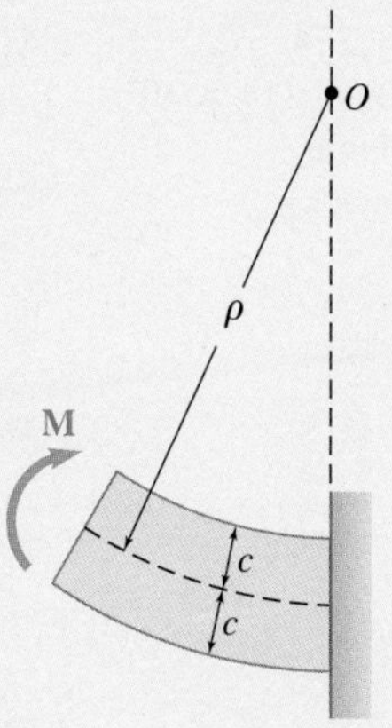

Fig. 2 Deformed shape of beam.

b. Radius of Curvature. Using Fig. 2 and recalling that $E = 10.6 \times 10^6$ psi, we substitute this value and the values obtained for I and M into Eq. (11.21) and find

$$\frac{1}{\rho} = \frac{M}{EI} = \frac{103.8 \times 10^3 \text{ lb·in.}}{(10.6 \times 10^6 \text{ psi})(12.97 \text{ in}^4)} = 0.755 \times 10^{-3} \text{ in}^{-1}$$

$$\rho = 1325 \text{ in.} \qquad \rho = 110.4 \text{ ft} \blacktriangleleft$$

(continued)

REFLECT and THINK: Alternatively, we can calculate the radius of curvature using Eq. (11.9). Since we know that the maximum stress is $\sigma_{all} = 20$ ksi, the maximum strain ϵ_m can be determined, and Eq. (11.9) gives

$$\epsilon_m = \frac{\sigma_{all}}{E} = \frac{20 \text{ ksi}}{10.6 \times 10^6 \text{ psi}} = 1.887 \times 10^{-3} \text{ in./in.}$$

$$\epsilon_m = \frac{c}{\rho} \qquad \rho = \frac{c}{\epsilon_m} = \frac{2.5 \text{ in.}}{1.887 \times 10^{-3} \text{ in./in.}}$$

$$\rho = 1325 \text{ in.} \qquad \rho = 110.4 \text{ ft} \blacktriangleleft$$

Sample Problem 11.2

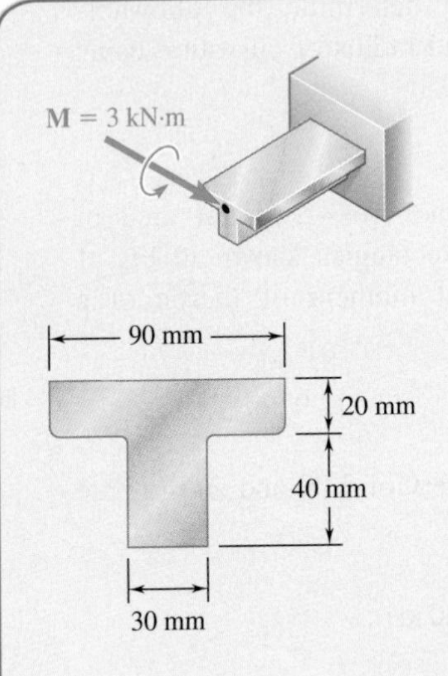

A cast-iron machine part is acted upon by the 3 kN·m couple shown. Knowing that $E = 165$ GPa and neglecting the effect of fillets, determine (*a*) the maximum tensile and compressive stresses in the casting and (*b*) the radius of curvature of the casting.

STRATEGY: The moment of inertia is determined, recognizing that it is first necessary to determine the location of the neutral axis. Then Eqs. (11.15) and (11.21) are used to determine the stresses and radius of curvature.

MODELING and ANALYSIS:

Centroid. Divide the T-shaped cross section into two rectangles as shown in Fig. 1 and write

	Area, mm^2	$\bar{y}$, mm	$\bar{y}A$, mm^3
1	$(20)(90) = 1800$	50	90×10^3
2	$(40)(30) = 1200$	20	24×10^3
	$\Sigma A = 3000$		$\Sigma\bar{y}A = 114 \times 10^3$

$\bar{Y}\Sigma A = \Sigma\bar{y}A$

$\bar{Y}(3000) = 114 \times 10^6$

$\bar{Y} = 38$ mm

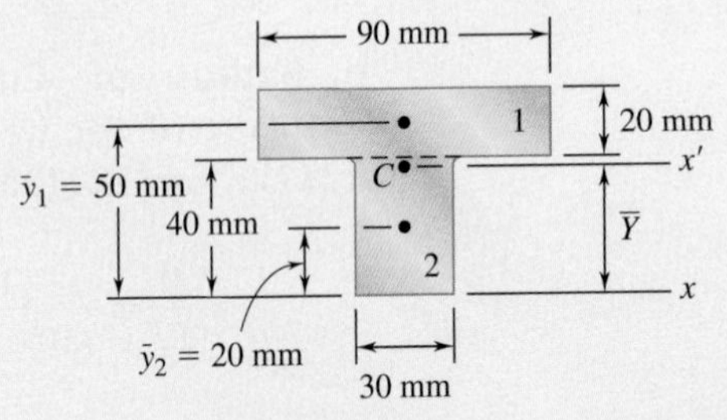

Fig. 1 Composite areas for calculating centroid.

(continued)

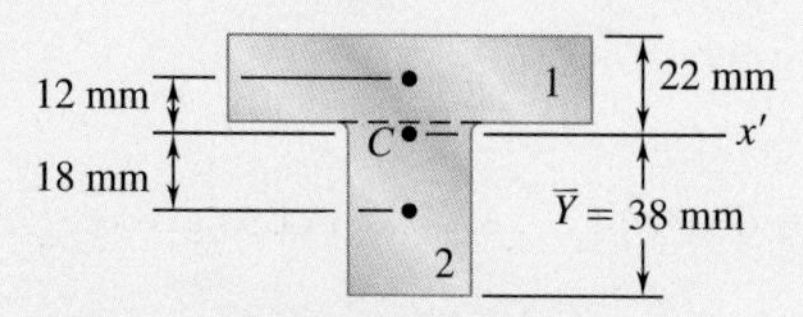

Fig. 2 Composite areas for calculating moment of inertia.

Centroidal Moment of Inertia. The parallel-axis theorem is used to determine the moment of inertia of each rectangle (Fig. 2) with respect to the axis x' that passes through the centroid of the composite section. Adding the moments of inertia of the rectangles, write

$$I_{x'} = \Sigma(\bar{I} + Ad^2) = \Sigma(\tfrac{1}{12}bh^3 + Ad^2)$$

$$= \tfrac{1}{12}(90)(20)^3 + (90 \times 20)(12)^2 + \tfrac{1}{12}(30)(40)^3 + (30 \times 40)(18)^2$$

$$= 868 \times 10^3 \text{ mm}^4$$

$$I = 868 \times 10^{-9} \text{ m}^4$$

a. Maximum Tensile Stress. Since the applied couple bends the casting downward, the center of curvature is located below the cross section. The maximum tensile stress occurs at point A (Fig. 3), which is farthest from the center of curvature.

$$\sigma_A = \frac{Mc_A}{I} = \frac{(3 \text{ kN·m})(0.022 \text{ m})}{868 \times 10^{-9} \text{ m}^4} \qquad \sigma_A = +76.0 \text{ MPa} \blacktriangleleft$$

Maximum Compressive Stress. This occurs at point B (Fig. 3):

$$\sigma_B = -\frac{Mc_B}{I} = -\frac{(3 \text{ kN·m})(0.038 \text{ m})}{868 \times 10^{-9} \text{ m}^4} \qquad \sigma_B = -131.3 \text{ MPa} \blacktriangleleft$$

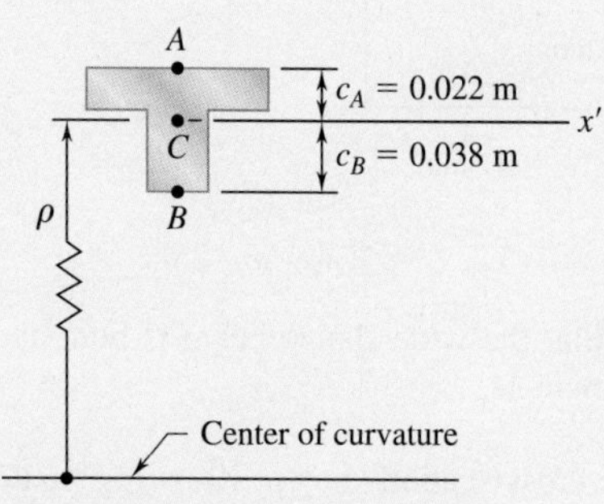

Fig. 3 Radius of curvature is measured to the centroid of the cross section.

b. Radius of Curvature. From Eq. (11.21), using Fig. 3, we have

$$\frac{1}{\rho} = \frac{M}{EI} = \frac{3 \text{ kN·m}}{(165 \text{ GPa})(868 \times 10^{-9} \text{ m}^4)}$$

$$= 20.95 \times 10^{-3} \text{ m}^{-1} \qquad \rho = 47.7 \text{ m} \blacktriangleleft$$

REFLECT and THINK: Note the T-section has a vertical plane of symmetry, with the applied moment in that plane. Thus the couple of this applied moment lies in the plane of symmetry, resulting in symmetrical bending. Had the couple been in another plane, we would have unsymmetric bending and thus would need to apply the principles of Sec. 11.5.

Problems

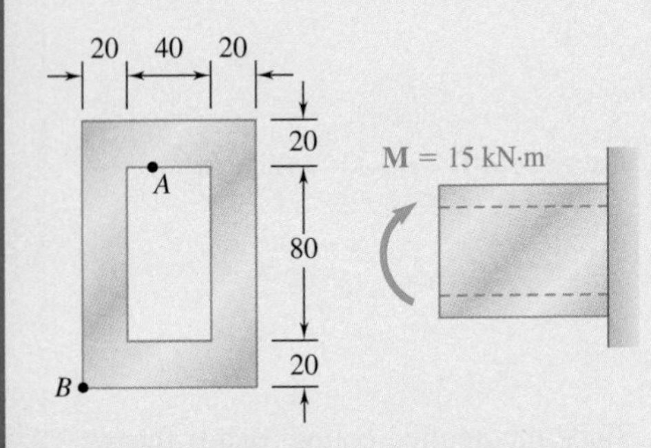

Fig. P11.1

11.1 and 11.2 Knowing that the couple shown acts in a vertical plane, determine the stress at (*a*) point *A*, (*b*) point *B*.

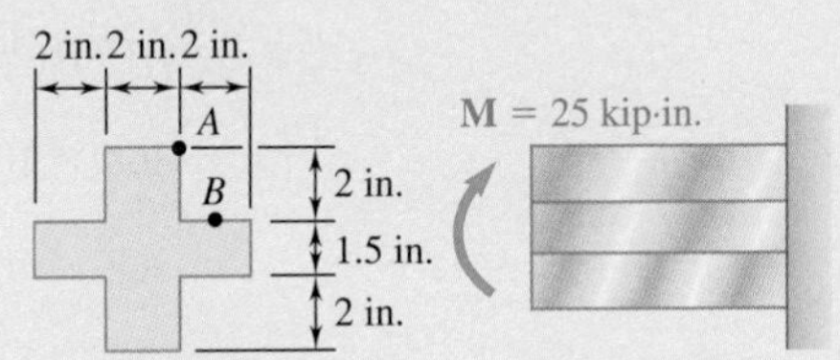

Fig. P11.2

11.3 Using an allowable stress of 155 MPa, determine the largest bending moment M_x that can be applied to the wide-flange beam shown. Neglect the effect of the fillets.

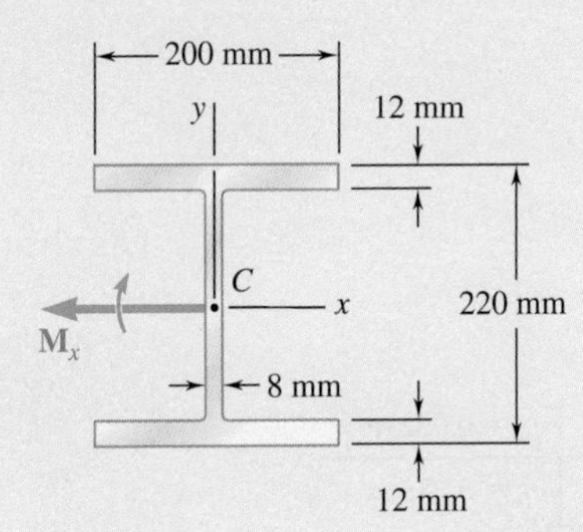

Fig. P11.3

11.4 Solve Prob. 11.3, assuming that the wide-flange beam is bent about the *y* axis by a couple of moment M_y.

11.5 A nylon spacing bar has the cross section shown. Knowing that the allowable stress for the grade of nylon used is 24 MPa, determine the largest couple $\mathbf{M}_z$ that can be applied to the bar.

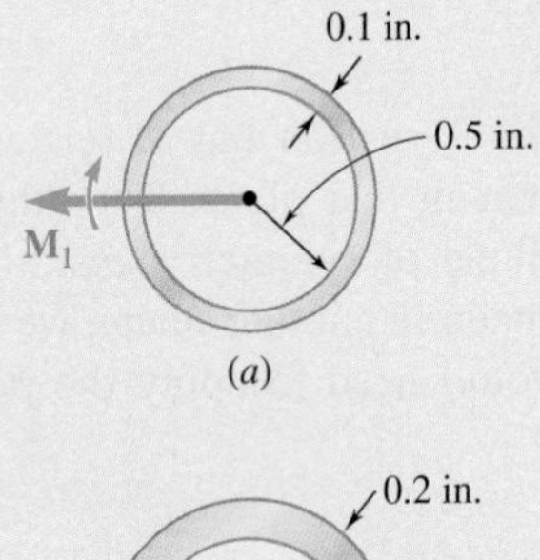

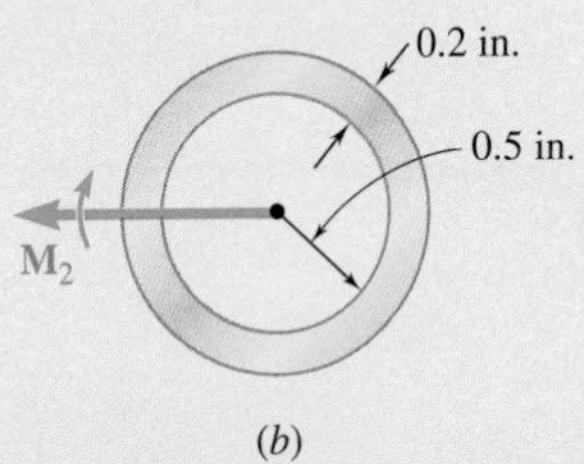

Fig. P11.6

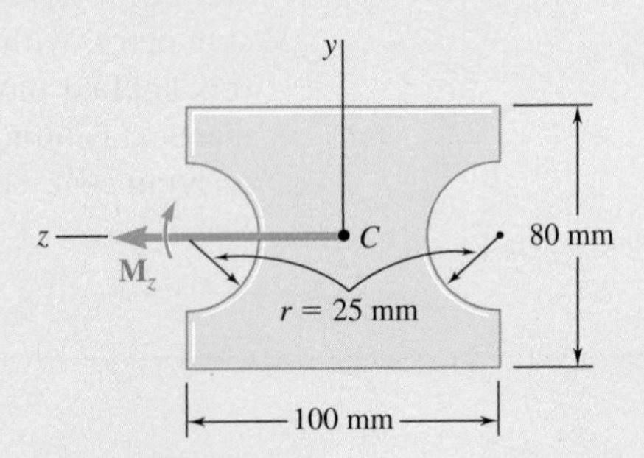

Fig. *P11.5*

11.6 Using an allowable stress of 16 ksi, determine the largest couple that can be applied to each pipe.

11.7 and *11.8* Two W4 × 13 rolled sections are welded together as shown. Knowing that for the steel alloy used $\sigma_U = 58$ ksi and using a factor of safety of 3.0, determine the largest couple that can be applied when the assembly is bent about the z axis.

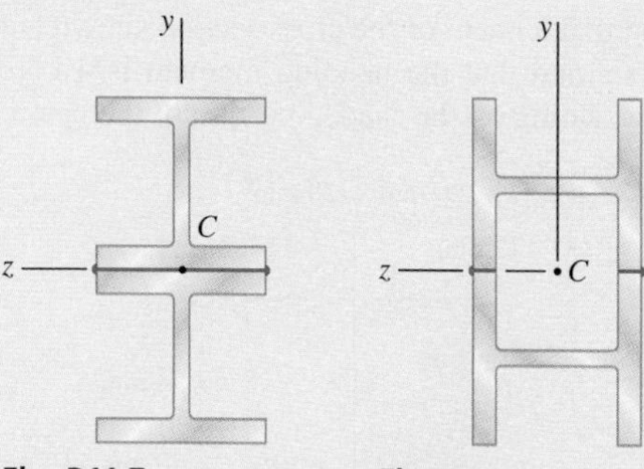

Fig. P11.7 **Fig. *P11.8***

11.9 through *11.11* Two vertical forces are applied to the beam of the cross section shown. Determine the maximum tensile and compressive stresses in portion BC of the beam.

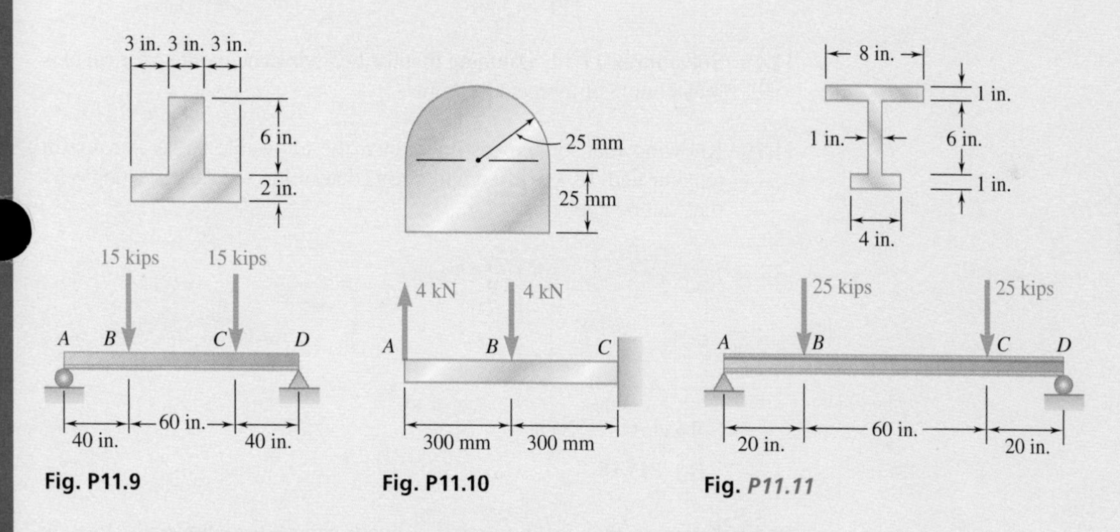

Fig. P11.9 **Fig. P11.10** **Fig. *P11.11***

11.12 Knowing that for the extruded beam shown the allowable stress is 120 MPa in tension and 150 MPa in compression, determine the largest couple **M** that can be applied.

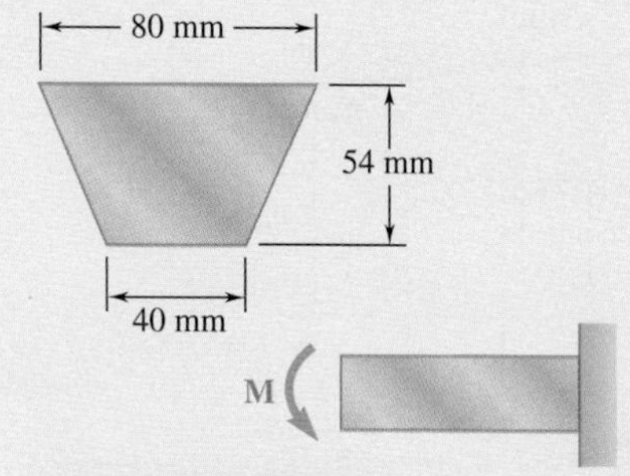

Fig. P11.12

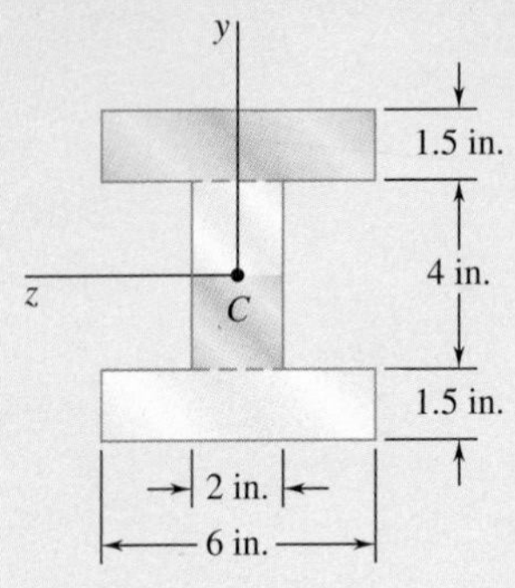

Fig. P11.13

11.13 Knowing that a beam of the cross section shown is bent about a horizontal axis and that the bending moment is 50 kip·in., determine the total force acting (*a*) on the top flange (*b*) on the shaded portion of the web.

11.14 Knowing that a beam of the cross section shown is bent about a horizontal axis and that the bending moment is 4 kN·m, determine the total force acting on the shaded portion of the beam.

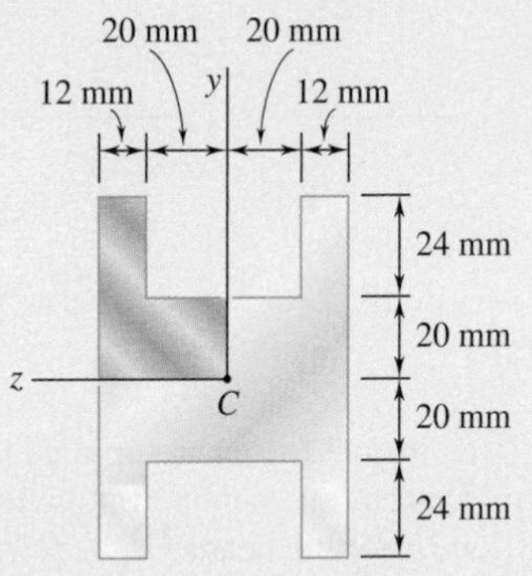

Fig. *P11.14*

11.15 Solve Prob. 11.14, assuming that the beam is bent about a vertical axis by a couple of moment 4 kN·m.

11.16 Knowing that for the casting shown the allowable stress is 5 ksi in tension and 18 ksi in compression, determine the largest couple **M** that can be applied.

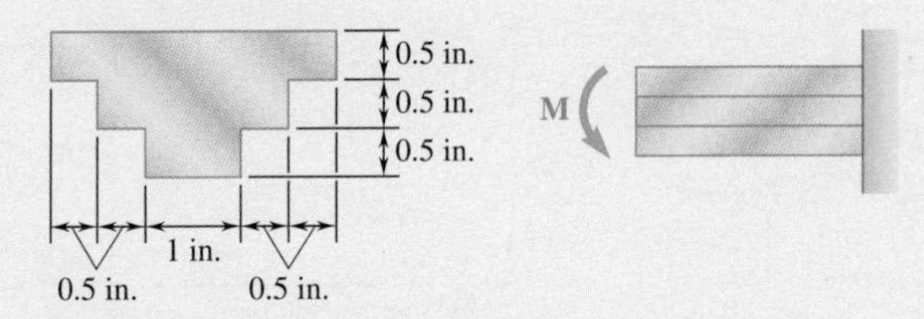

Fig. P11.16

11.17 Knowing that for the extruded beam shown the allowable stress is 120 MPa in tension and 150 MPa in compression, determine the largest couple **M** that can be applied.

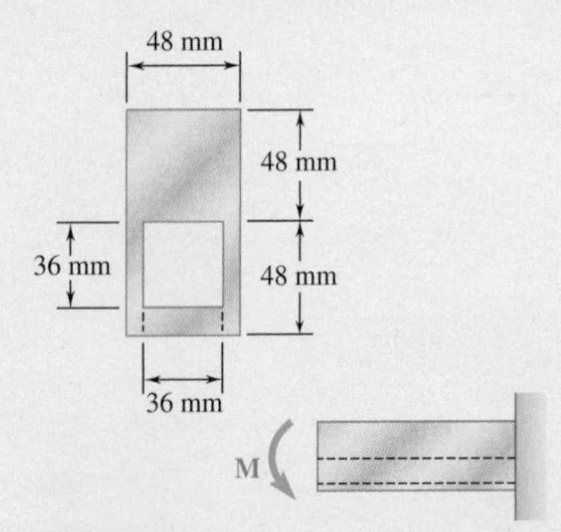

Fig. P11.17

11.18 Knowing that for the extruded beam shown the allowable stress is 12 ksi in tension and 16 ksi in compression, determine the largest couple **M** that can be applied.

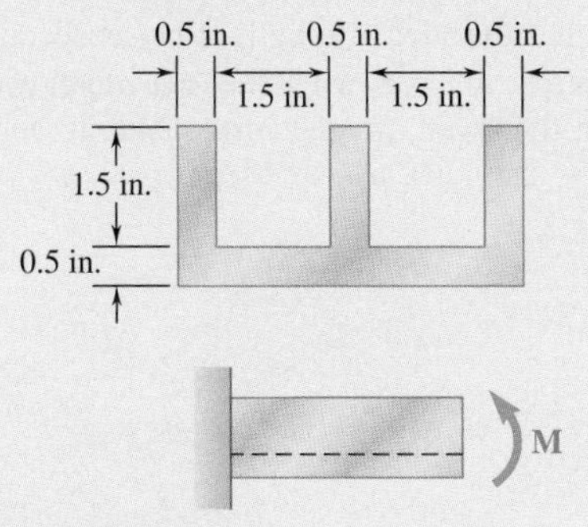

Fig. P11.18

11.19 The beam shown is made of a nylon for which the allowable stress is 24 MPa in tension and 30 MPa in compression. Determine the largest couple **M** that can be applied to the beam.

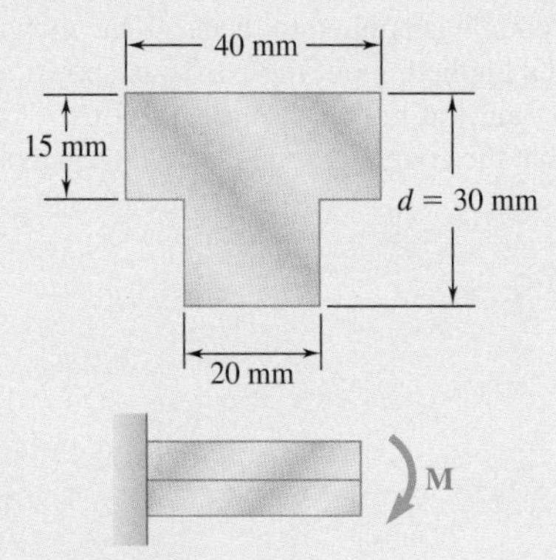

Fig. P11.19

11.20 Solve Prob. 11.19, assuming that $d = 40$ mm.

11.21 Knowing that for the beam shown the allowable stress is 12 ksi in tension and 16 ksi in compression, determine the largest couple **M** that can be applied.

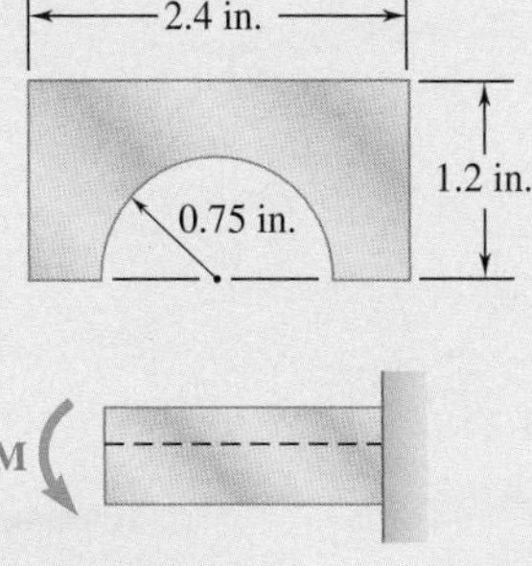

Fig. P11.21

11.22 Straight rods of 0.30-in. diameter and 200-ft length are sometimes used to clear underground conduits of obstructions or to thread wires through a new conduit. The rods are made of high-strength steel and, for storage and transportation, are wrapped on spools of 5-ft diameter. Assuming that the yield strength is not exceeded, determine (*a*) the maximum stress in a rod, when the rod, which was initially straight, is wrapped on the spool, (*b*) the corresponding bending moment in the rod. Use $E = 29 \times 10^6$ psi.

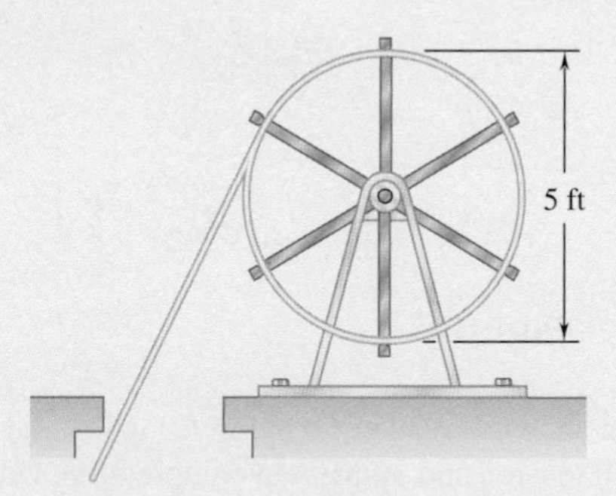

Fig. *P11.22*

11.23 A 60-N·m couple is applied to the steel bar shown. (*a*) Assuming that the couple is applied about the *z* axis as shown, determine the maximum stress and the radius of curvature of the bar. (*b*) Solve part *a*, assuming that the couple is applied about the *y* axis. Use $E = 200$ GPa.

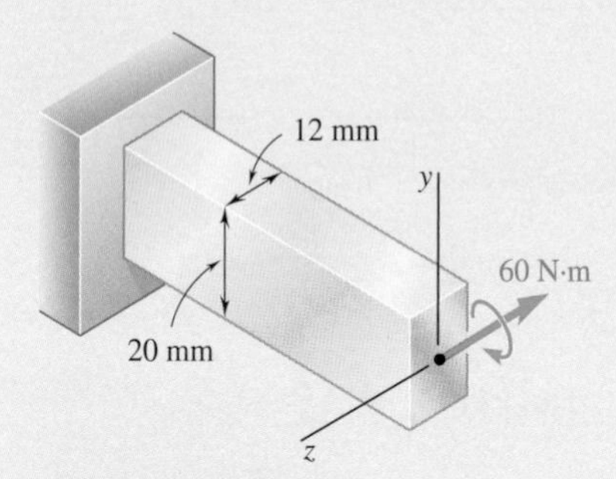

Fig. P11.23

11.24 A couple of magnitude *M* is applied to a square bar of side *a*. For each of the orientations shown, determine the maximum stress and the curvature of the bar.

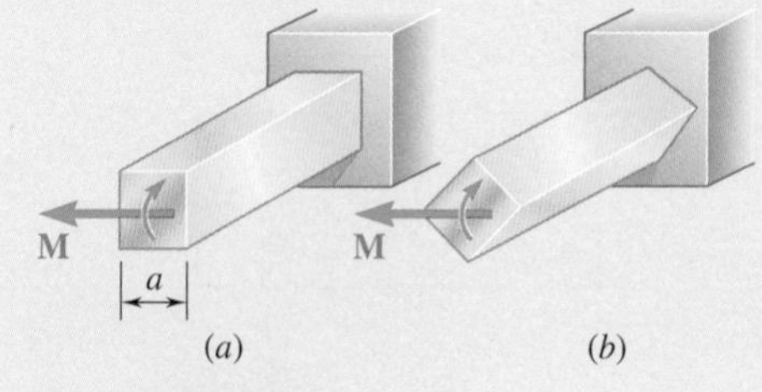

Fig. P11.24

11.3 MEMBERS MADE OF COMPOSITE MATERIALS

The derivations given in Sec. 11.2 are based on the assumption of a homogeneous material with a given modulus of elasticity E. If the member is made of two or more materials with different moduli of elasticity, the member is a composite member.

Consider a bar consisting of two portions of different materials bonded together as shown in Fig. 11.16. This composite bar will deform as described in Sec. 11.1B, since its cross section remains the same throughout its entire length, and since no assumption was made in Sec. 11.1B regarding the stress-strain relationship of the material or materials involved. Thus, the normal strain ϵ_x still varies linearly with the distance y from the neutral axis of the section (Fig. 11.17*a* and *b*), and formula (11.8) holds:

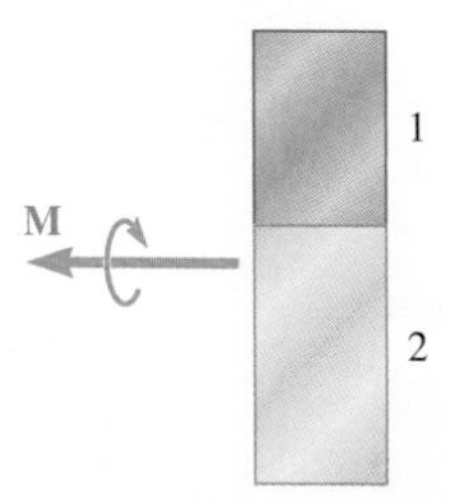

Fig. 11.16 Cross section made with different materials.

$$\epsilon_x = -\frac{y}{\rho} \tag{11.8}$$

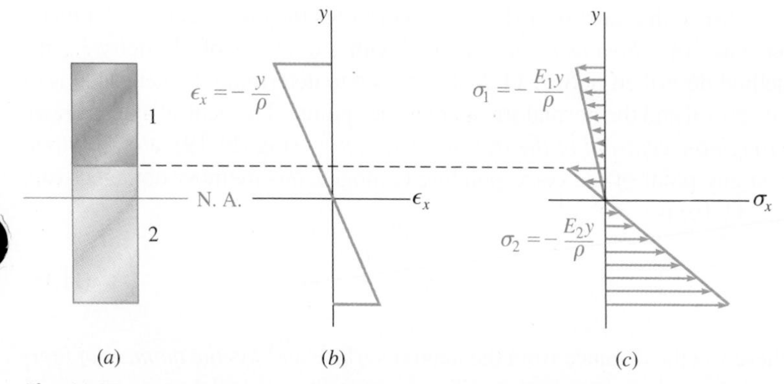

Fig. 11.17 Stress and strain distributions in bar made of two materials. (*a*) Neutral axis shifted from centroid. (*b*) Strain distribution. (*c*) Corresponding stress distribution.

However, it cannot be assumed that the neutral axis passes through the centroid of the composite section, and one of the goals of this analysis is to determine the location of this axis.

Since the moduli of elasticity E_1 and E_2 of the two materials are different, the equations for the normal stress in each material are

$$\sigma_1 = E_1\epsilon_x = -\frac{E_1 y}{\rho}$$

$$\sigma_2 = E_2\epsilon_x = -\frac{E_2 y}{\rho} \tag{11.22}$$

A stress-distribution curve is obtained that consists of two segments with straight lines as shown in Fig. 11.17*c*. It follows from Eqs. (11.22) that the force dF_1 exerted on an element of area dA of the upper portion of the cross section is

$$dF_1 = \sigma_1\, dA = -\frac{E_1 y}{\rho}\, dA \tag{11.23}$$

while the force dF_2 exerted on an element of the same area dA of the lower portion is

$$dF_2 = \sigma_2 dA = -\frac{E_2 y}{\rho} dA \tag{11.24}$$

Denoting the ratio E_2/E_1 of the two moduli of elasticity by n, we can write

$$dF_2 = -\frac{(nE_1)y}{\rho} dA = -\frac{E_1 y}{\rho}(n\, dA) \tag{11.25}$$

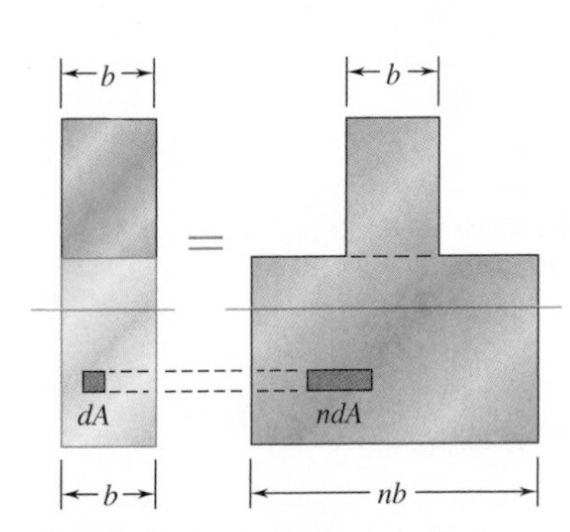

Fig. 11.18 Transformed section based on replacing lower material with that used on top.

Comparing Eqs. (11.23) and (11.25), we note that the same force dF_2 would be exerted on an element of area $n\, dA$ of the first material. Thus, the resistance to bending of the bar would remain the same if both portions were made of the first material, provided that the width of each element of the lower portion were multiplied by the factor n. Note that this widening (if $n > 1$) or narrowing (if $n < 1$) must be *in a direction parallel to the neutral axis of the section*, since it is essential that the distance y of each element from the neutral axis remain the same. This new cross section is called the *transformed section* of the member (Fig. 11.18).

Since the transformed section represents the cross section of a member made of a *homogeneous material* with a modulus of elasticity E_1, the method described in Sec. 11.2 can be used to determine the neutral axis of the section and the normal stress at various points. The neutral axis is drawn *through the centroid of the transformed section* (Fig. 11.19), and the stress σ_x at any point of the corresponding homogeneous member obtained from Eq. (11.16) is

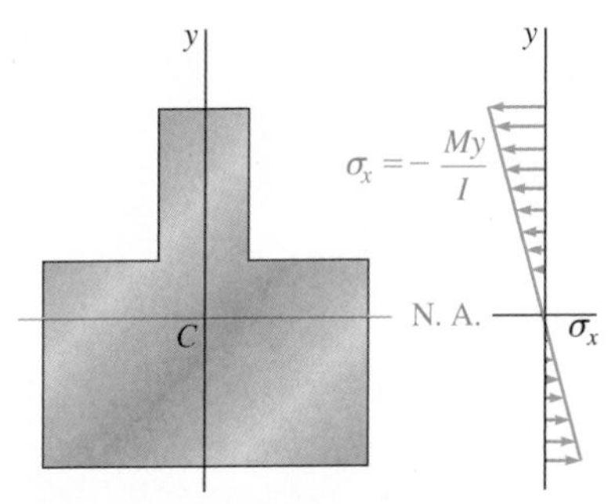

Fig. 11.19 Distribution of stresses in transformed section.

$$\sigma_x = -\frac{My}{I} \tag{11.16}$$

where y is the distance from the neutral surface and I is *the moment of inertia of the transformed section* with respect to its centroidal axis.

To obtain the stress σ_1 at a point located in the upper portion of the cross section of the original composite bar, compute the stress σ_x at the corresponding point of the transformed section. However, to obtain the stress σ_2 at a point in the lower portion of the cross section, we must *multiply by* n the stress σ_x computed at the corresponding point of the transformed section. Indeed, the same elementary force dF_2 is applied to an element of area $n\, dA$ of the transformed section and to an element of area dA of the original section. Thus, the stress σ_2 at a point of the original section must be n times larger than the stress at the corresponding point of the transformed section.

The deformations of a composite member can also be determined by using the transformed section. We recall that the transformed section represents the cross section of a member, made of a homogeneous material of modulus E_1, which deforms in the same manner as the composite member. Therefore, using Eq. (11.21), we write that the curvature of the composite member is

$$\frac{1}{\rho} = \frac{M}{E_1 I}$$

where I is the moment of inertia of the transformed section with respect to its neutral axis.

Concept Application 11.3

A bar obtained by bonding together pieces of steel ($E_s = 29 \times 10^6$ psi) and brass ($E_b = 15 \times 10^6$ psi) has the cross section shown (Fig. 11.20*a*). Determine the maximum stress in the steel and in the brass when the bar is in pure bending with a bending moment $M = 40$ kip·in.

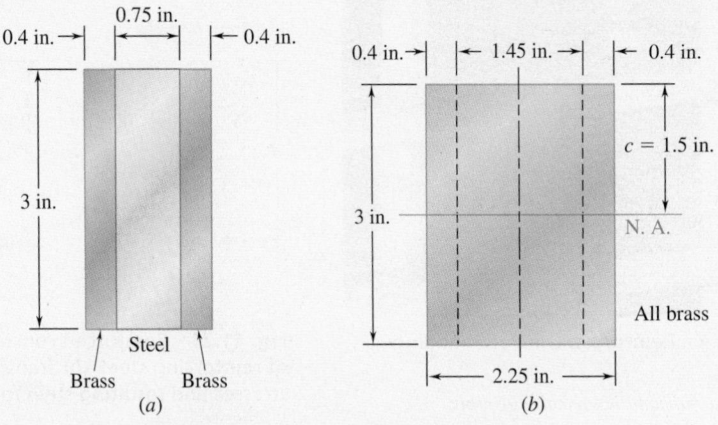

Fig. 11.20 (*a*) Composite bar. (*b*) Transformed section.

The transformed section corresponding to an equivalent bar made entirely of brass is shown in Fig. 11.20*b*. Since

$$n = \frac{E_s}{E_b} = \frac{29 \times 10^6 \text{ psi}}{15 \times 10^6 \text{ psi}} = 1.933$$

the width of the central portion of brass, which replaces the original steel portion, is obtained by multiplying the original width by 1.933:

$$(0.75 \text{ in.})(1.933) = 1.45 \text{ in.}$$

Note that this change in dimension occurs in a direction parallel to the neutral axis. The moment of inertia of the transformed section about its centroidal axis is

$$I = \tfrac{1}{12} bh^3 = \tfrac{1}{12}(2.25 \text{ in.})(3 \text{ in.})^3 = 5.063 \text{ in}^4$$

and the maximum distance from the neutral axis is $c = 1.5$ in. Using Eq. (11.15), the maximum stress in the transformed section is

$$\sigma_m = \frac{Mc}{I} = \frac{(40 \text{ kip·in.})(1.5 \text{ in.})}{5.063 \text{ in}^4} = 11.85 \text{ ksi}$$

This value also represents the maximum stress in the brass portion of the original composite bar. The maximum stress in the steel portion, however, will be larger than for the transformed section, since the area of the central portion must be reduced by the factor $n = 1.933$. Thus,

$$(\sigma_{\text{brass}})_{\max} = 11.85 \text{ ksi}$$

$$(\sigma_{\text{steel}})_{\max} = (1.933)(11.85 \text{ ksi}) = 22.9 \text{ ksi}$$

Photo 11.4 Reinforced concrete building frame.

An important example of structural members made of two different materials is furnished by *reinforced concrete beams* (Photo 11.4). These beams, when subjected to positive bending moments, are reinforced by steel rods placed a short distance above their lower face (Fig. 11.21*a*). Since concrete is very weak in tension, it cracks below the neutral surface, and the steel rods carry the entire tensile load, while the upper part of the concrete beam carries the compressive load.

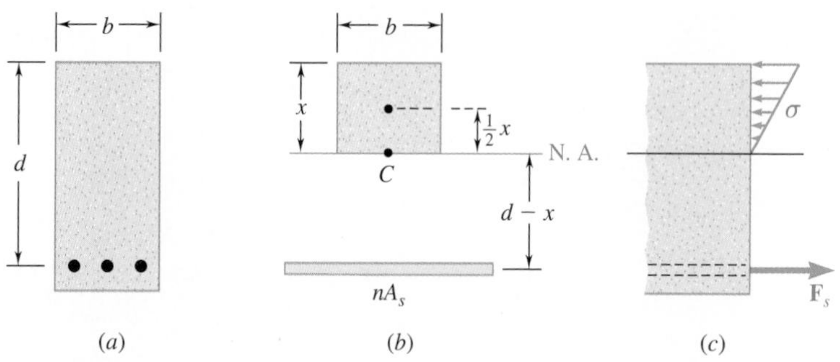

Fig. 11.21 Reinforced concrete beam: (*a*) Cross section showing location of reinforcing steel. (*b*) Transformed section of all concrete. (*c*) Concrete stresses and resulting steel force.

To obtain the transformed section of a reinforced concrete beam, we replace the total cross-sectional area A_s of the steel bars by an equivalent area nA_s, where n is the ratio E_s/E_c of the moduli of elasticity of steel and concrete (Fig. 11.21*b*). Since the concrete in the beam acts effectively only in compression, only the portion located above the neutral axis should be used in the transformed section.

The position of the neutral axis is obtained by determining the distance x from the upper face of the beam to the centroid C of the transformed section. Using the width of the beam b and the distance d from the upper face to the center line of the steel rods, the first moment of the transformed section with respect to the neutral axis must be zero. Since the first moment of each portion of the transformed section is obtained by multiplying its area by the distance of its own centroid from the neutral axis,

$$(bx)\frac{x}{2} - nA_s(d - x) = 0$$

or

$$\frac{1}{2}bx^2 + nA_s x - nA_s d = 0 \tag{11.26}$$

Solving this quadratic equation for x, both the position of the neutral axis in the beam and the portion of the cross section of the concrete beam that is effectively used are obtained.

The stresses in the transformed section are determined as explained earlier in this section (see Sample Prob. 11.4). The distribution of the compressive stresses in the concrete and the resultant $\mathbf{F}_s$ of the tensile forces in the steel rods are shown in Fig. 11.21*c*.

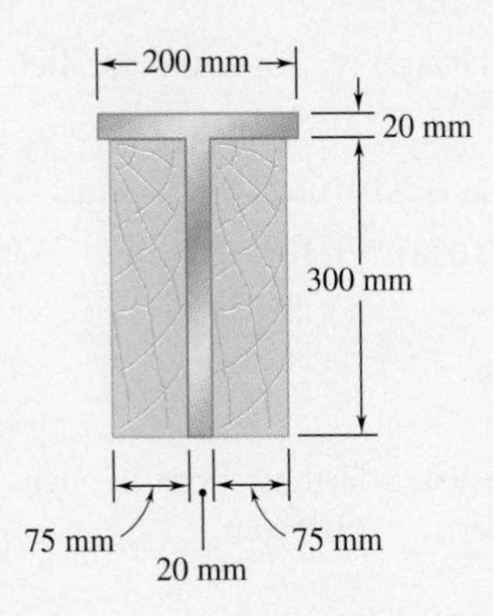

Sample Problem 11.3

Two steel plates have been welded together to form a beam in the shape of a T that has been strengthened by securely bolting to it the two oak timbers shown in the figure. The modulus of elasticity is 12.5 GPa for the wood and 200 GPa for the steel. Knowing that a bending moment $M = 50$ kN·m is applied to the composite beam, determine (*a*) the maximum stress in the wood and (*b*) the stress in the steel along the top edge.

STRATEGY: The beam is first transformed to a beam made of a single material (either steel or wood). The moment of inertia is then determined for the transformed section, and this is used to determine the required stresses, remembering that the actual stresses must be based on the original material.

MODELING:

Transformed Section. First compute the ratio

$$n = \frac{E_s}{E_w} = \frac{200 \text{ GPa}}{12.5 \text{ GPa}} = 16$$

Multiplying the horizontal dimensions of the steel portion of the section by $n = 16$, a transformed section made entirely of wood is obtained.

Neutral Axis. Fig. 1 shows the transformed section. The neutral axis passes through the centroid of the transformed section. Since the section consists of two rectangles,

$$\overline{Y} = \frac{\Sigma \overline{y} A}{\Sigma A} = \frac{(0.160 \text{ m})(3.2 \text{ m} \times 0.020 \text{ m}) + 0}{3.2 \text{ m} \times 0.020 \text{ m} + 0.470 \text{ m} \times 0.300 \text{ m}} = 0.050 \text{ m}$$

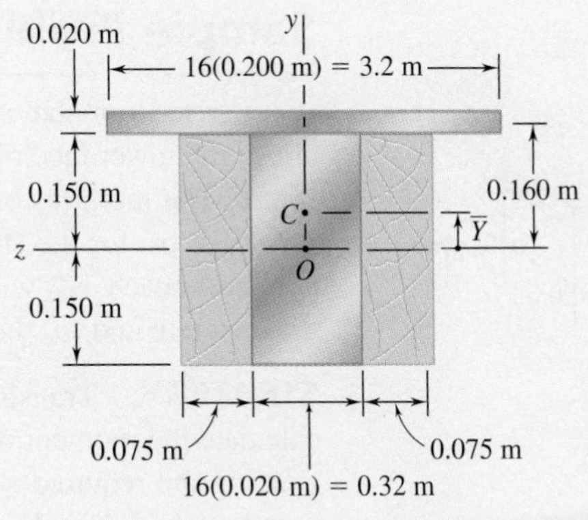

Fig. 1 Transformed cross section.

(continued)

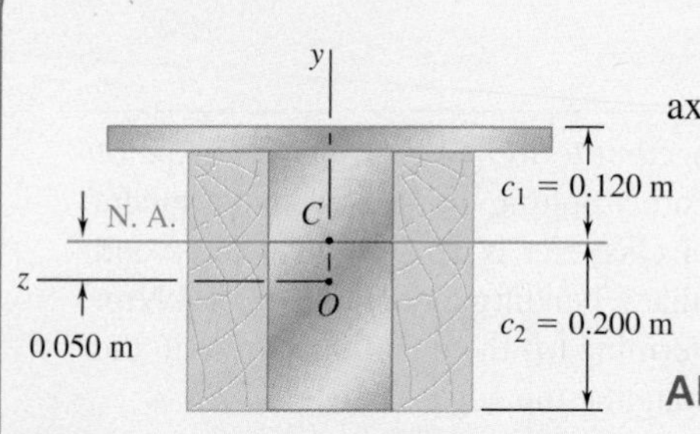

Fig. 2 Transformed section showing neutral axis and distances to extreme fibers.

Centroidal Moment of Inertia. Using Fig. 2 and the parallel-axis theorem,

$$I = \tfrac{1}{12}(0.470)(0.300)^3 + (0.470 \times 0.300)(0.050)^2$$
$$+\tfrac{1}{12}(3.2)(0.020)^3 + (3.2 \times 0.020)(0.160 - 0.050)^2$$
$$I = 2.19 \times 10^{-3}\ \text{m}^4$$

ANALYSIS:

a. Maximum Stress in Wood. The wood farthest from the neutral axis is located along the bottom edge, where $c_2 = 0.200$ m.

$$\sigma_w = \frac{Mc_2}{I} = \frac{(50 \times 10^3\ \text{N·m})(0.200\ \text{m})}{2.19 \times 10^{-3}\ \text{m}^4}$$

$$\sigma_w = 4.57\ \text{MPa} \blacktriangleleft$$

b. Stress in Steel. Along the top edge, $c_1 = 0.120$ m. From the transformed section we obtain an equivalent stress in wood, which must be multiplied by n to obtain the stress in steel.

$$\sigma_s = n\frac{Mc_1}{I} = (16)\frac{(50 \times 10^3\ \text{N·m})(0.120\ \text{m})}{2.19 \times 10^{-3}\ \text{m}^4}$$

$$\sigma_s = 43.8\ \text{MPa} \blacktriangleleft$$

REFLECT and THINK: Since the transformed section was based on a beam made entirely of wood, it was necessary to use n to get the actual stress in the steel. Furthermore, at any common distance from the neutral axis, the stress in the steel will be substantially greater than that in the wood, reflective of the much larger modulus of elasticity for the steel.

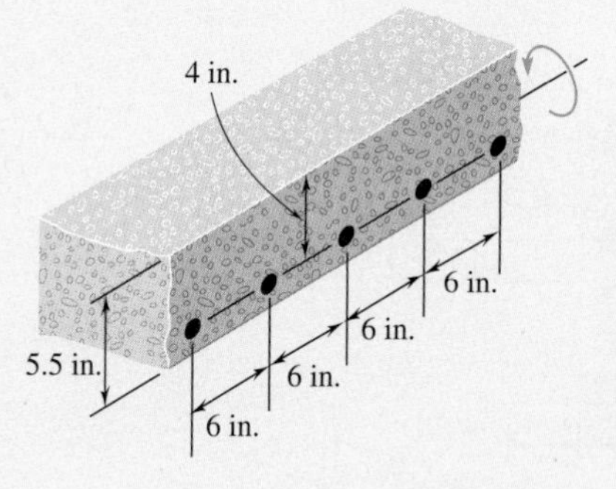

Sample Problem 11.4

A concrete floor slab is reinforced by $\frac{5}{8}$-in.-diameter steel rods placed 1.5 in. above the lower face of the slab and spaced 6 in. on centers, as shown in the figure. The modulus of elasticity is 3.6×10^6 psi for the concrete used and 29×10^6 psi for the steel. Knowing that a bending moment of 40 kip·in. is applied to each 1-ft width of the slab, determine (*a*) the maximum stress in the concrete and (*b*) the stress in the steel.

STRATEGY: Transform the section to a single material, concrete, and then calculate the moment of inertia for the transformed section. Continue by calculating the required stresses, remembering that the actual stresses must be based on the original material.

(continued)

MODELING:

Transformed Section. Consider a portion of the slab 12 in. wide, in which there are two $\frac{5}{8}$-in.-diameter rods having a total cross-sectional area

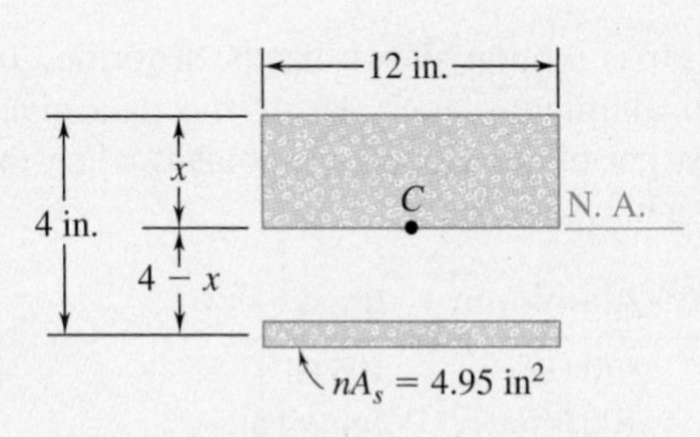

Fig. 1 Transformed section.

$$A_s = 2\left[\frac{\pi}{4}\left(\frac{5}{8}\text{ in.}\right)^2\right] = 0.614 \text{ in}^2$$

Since concrete acts only in compression, all the tensile forces are carried by the steel rods, and the transformed section (Fig. 1) consists of the two areas shown. One is the portion of concrete in compression (located above the neutral axis), and the other is the transformed steel area nA_s. We have

$$n = \frac{E_s}{E_c} = \frac{29 \times 10^6 \text{ psi}}{3.6 \times 10^6 \text{ psi}} = 8.06$$

$$nA_s = 8.06(0.614 \text{ in}^2) = 4.95 \text{ in}^2$$

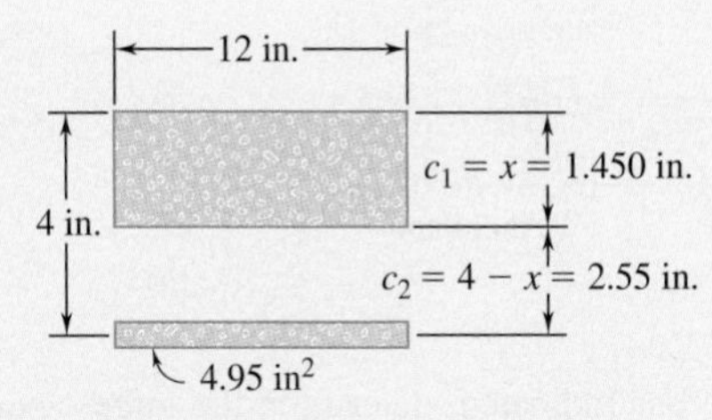

Fig. 2 Dimensions of transformed section used to calculate moment of inertia.

Neutral Axis. The neutral axis of the slab passes through the centroid of the transformed section. Summing moments of the transformed area about the neutral axis, write

$$12x\left(\frac{x}{2}\right) - 4.95(4 - x) = 0 \qquad x = 1.450 \text{ in.}$$

Moment of Inertia. Using Fig. 2, the centroidal moment of inertia of the transformed area is

$$I = \tfrac{1}{3}(12)(1.450)^3 + 4.95(4 - 1.450)^2 = 44.4 \text{ in}^4$$

ANALYSIS:

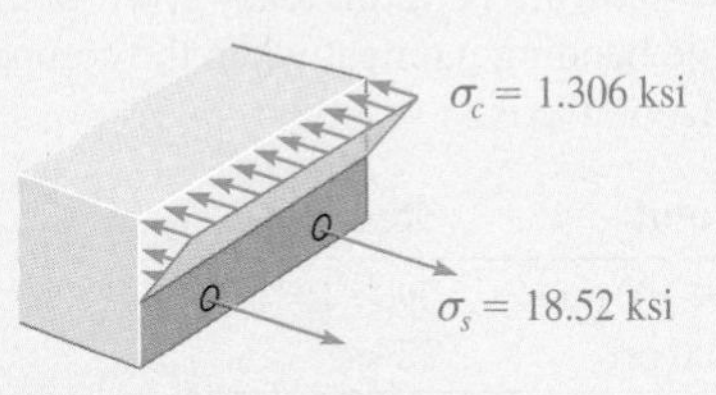

Fig. 3 Stress diagram.

a. Maximum Stress in Concrete. Fig. 3 shows the stresses on the cross section. At the top of the slab, we have $c_1 = 1.450$ in. and

$$\sigma_c = \frac{Mc_1}{I} = \frac{(40 \text{ kip·in.})(1.450 \text{ in.})}{44.4 \text{ in}^4} \qquad \sigma_c = 1.306 \text{ ksi} \blacktriangleleft$$

b. Stress in Steel. For the steel, we have $c_2 = 2.55$ in., $n = 8.06$ and

$$\sigma_s = n\frac{Mc_2}{I} = 8.06\frac{(40 \text{ kip·in.})(2.55 \text{ in.})}{44.4 \text{ in}^4} \qquad \sigma_s = 18.52 \text{ ksi} \blacktriangleleft$$

REFLECT and THINK: Since the transformed section was based on a beam made entirely of concrete, it was necessary to use n to get the actual stress in the steel. The difference in the resulting stresses reflects the large differences in the moduli of elasticity.

Problems

11.25 and 11.26 A bar having the cross section shown has been formed by securely bonding brass and aluminum stock. Using the data given below, determine the largest permissible bending moment when the composite bar is bent about a horizontal axis.

	Aluminum	Brass
Modulus of elasticity	70 GPa	105 GPa
Allowable stress	100 MPa	160 MPa

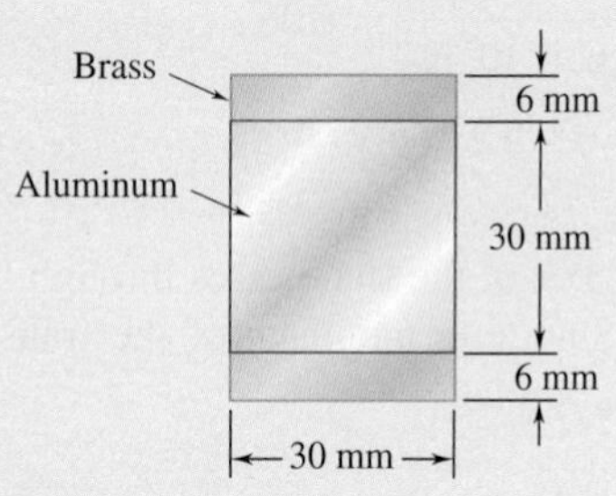

Fig. P11.25

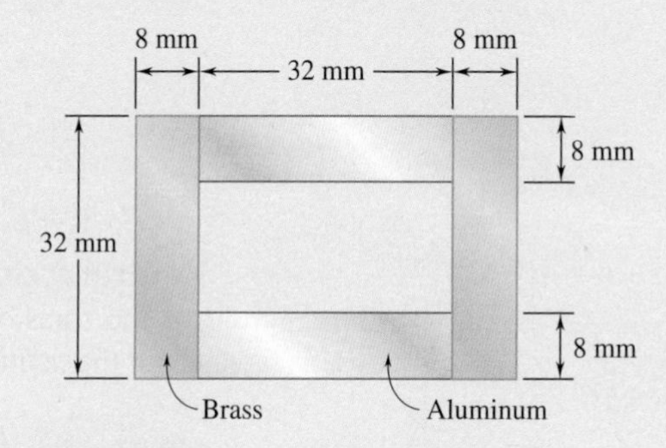

Fig. P11.26

11.27 and *11.28* For the composite bar indicated, determine the largest permissible bending moment when the bar is bent about a vertical axis.

11.27 Bar of Prob. 11.25.

11.28 Bar of Prob. 11.26.

11.29 and 11.30 Wooden beams and steel plates are securely bolted together to form the composite member shown. Using the data given below, determine the largest permissible bending moment when the composite member is bent about a horizontal axis.

	Wood	Steel
Modulus of elasticity	2×10^6 psi	29×10^6 psi
Allowable stress	2000 psi	22 ksi

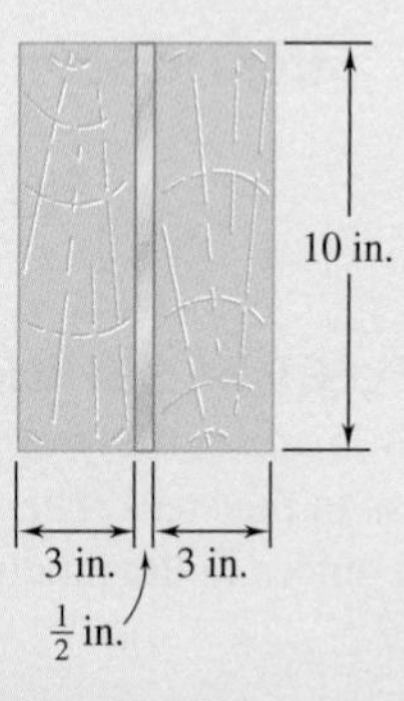

Fig. P11.29

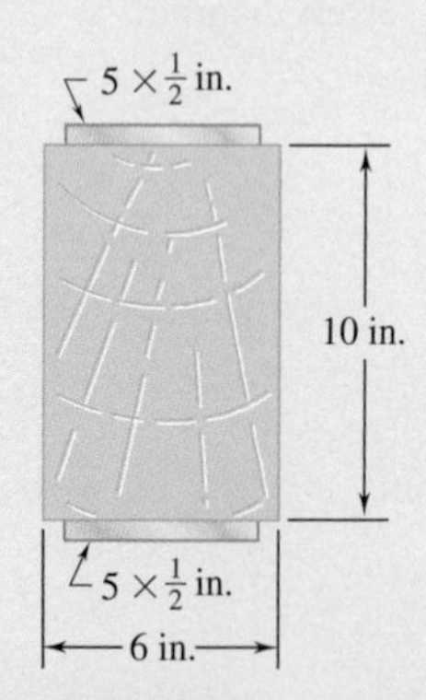

Fig. P11.30

11.31 and 11.32 A steel bar and an aluminum bar are bonded together to form the composite beam shown. The modulus of elasticity for aluminum is 70 GPa and for steel is 200 GPa. Knowing that the beam is bent about a horizontal axis by a couple of moment $M = 1500$ N·m, determine the maximum stress in (*a*) the aluminum, (*b*) the steel.

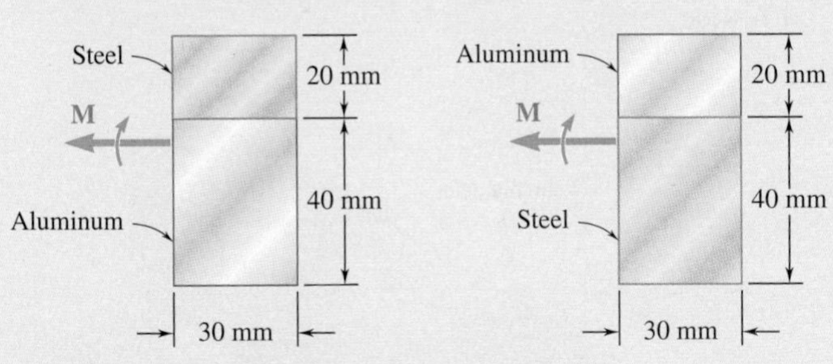

Fig. P11.31 **Fig. P11.32**

11.33 and *11.34* The 6 × 12-in. timber beam has been strengthened by bolting to it the steel reinforcement shown. The modulus of elasticity for wood is 1.8×10^6 psi and for steel 29×10^6 psi. Knowing that the beam is bent about a horizontal axis by a couple of moment 450 kip·in., determine the maximum stress in (*a*) the wood, (*b*) the steel.

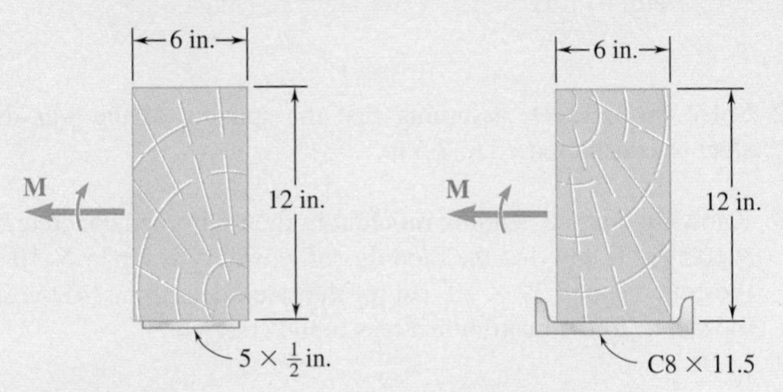

Fig. P11.33 **Fig. *P11.34***

***11.35* and 11.36** For the composite bar indicated, determine the radius of curvature caused by the couple of moment 1500 N·m.

11.35 Bar of Prob. 11.31.

11.36 Bar of Prob. 11.32.

11.37 and 11.38 For the composite beam indicated, determine the radius of curvature caused by the couple of moment 450 kip·in.

11.37 Bar of Prob. 11.33.

11.38 Bar of Prob. 11.34.

11.39 The reinforced concrete beam shown is subjected to a positive bending moment of 175 kN·m. Knowing that the modulus of elasticity is 25 GPa for the concrete and 200 GPa for the steel, determine (*a*) the stress in the steel, (*b*) the maximum stress in the concrete.

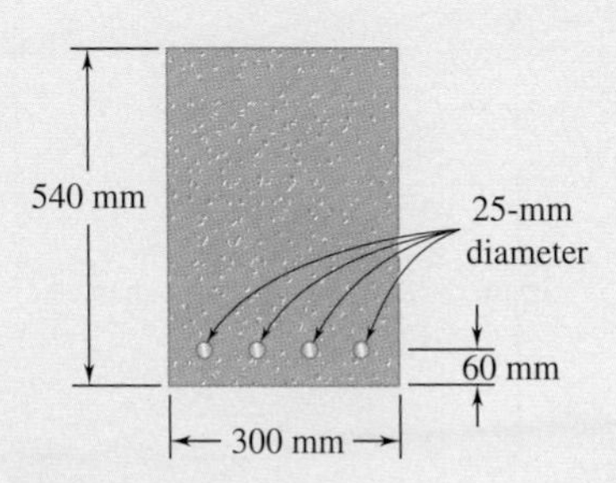

Fig. P11.39

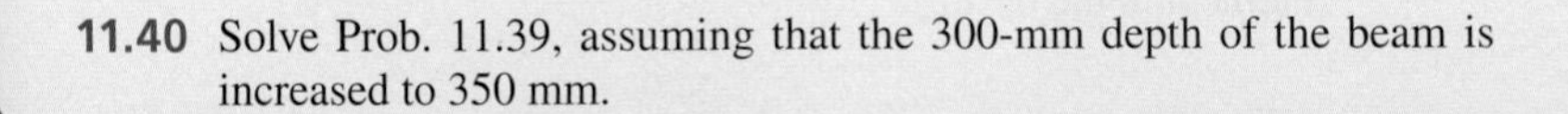

11.40 Solve Prob. 11.39, assuming that the 300-mm depth of the beam is increased to 350 mm.

11.41 A concrete slab is reinforced by $\frac{5}{8}$-in.-diameter steel rods placed on 5.5-in. centers as shown. The modulus of elasticity is 3×10^6 psi for the concrete and 29×10^6 psi for the steel. Using an allowable stress of 1400 psi for the concrete and 20 psi for the steel, determine the largest allowable positive bending moment in a portion of the slab 1 ft wide.

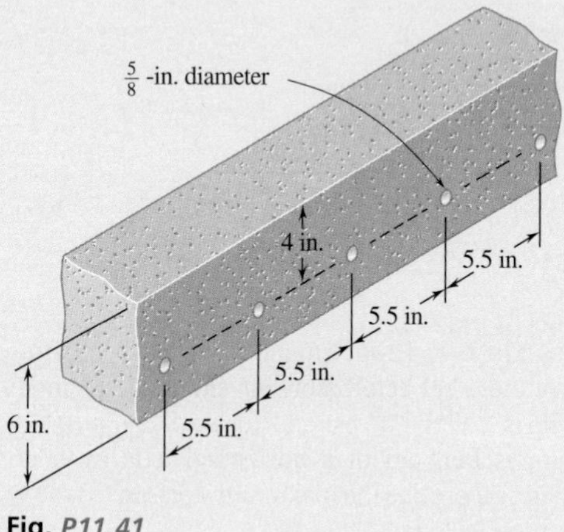

Fig. P11.41

11.42 Solve Prob. 11.41, assuming that the spacing of the $\frac{5}{8}$-in.-diameter steel rods is increased to 7.5 in.

11.43 Knowing that the bending moment in the reinforced concrete beam is +100 kip·ft and that the modulus of elasticity is 3.625×10^6 psi for the concrete and 29×10^6 psi for the steel, determine (*a*) the stress in the steel, (*b*) the maximum stress in the concrete.

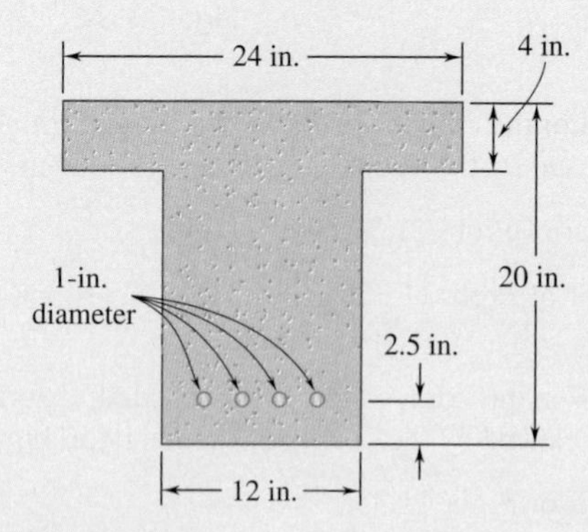

Fig. P11.43

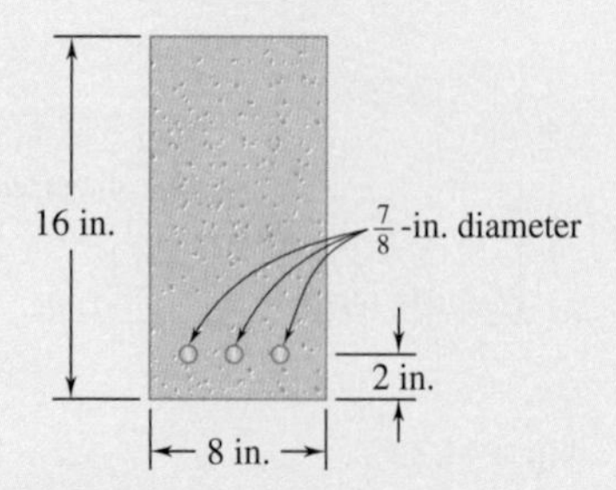

Fig. P11.44

11.44 A concrete beam is reinforced by three steel rods placed as shown. The modulus of elasticity is 3×10^6 psi for the concrete and 30×10^6 psi for the steel. Using an allowable stress of 1350 psi for the concrete and 20 ksi for the steel, determine the largest permissible positive bending moment in the beam.

11.45 and 11.46 Five metal strips, each of 0.5×1.5-in. cross section, are bonded together to form the composite beam shown. The modulus of elasticity is 30×10^6 psi for the steel, 15×10^6 psi for the brass, 10×10^6 psi for the aluminum. Knowing that the beam is bent about a horizontal axis by a couple of moment 12 kip·in., determine (*a*) the maximum stress in each of the three metals, (*b*) the radius of curvature of the composite beam.

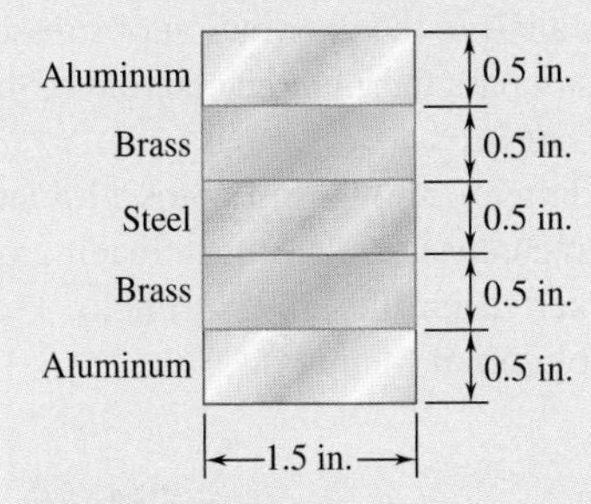

Fig. ***P11.45***

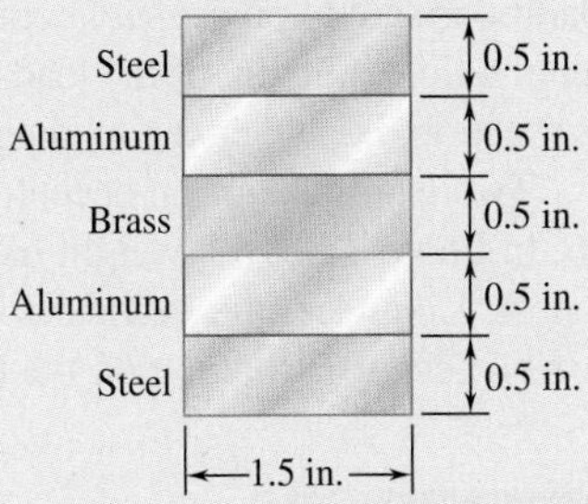

Fig. P11.46

11.47 The composite beam shown is formed by bonding together a brass rod and an aluminum rod of semicircular cross sections. The modulus of elasticity is 105 GPa for the brass and 70 GPa for the aluminum. Knowing that the composite beam is bent about a horizontal axis by couples of moment 900 N·m, determine the maximum stress (*a*) in the brass, (*b*) in the aluminum.

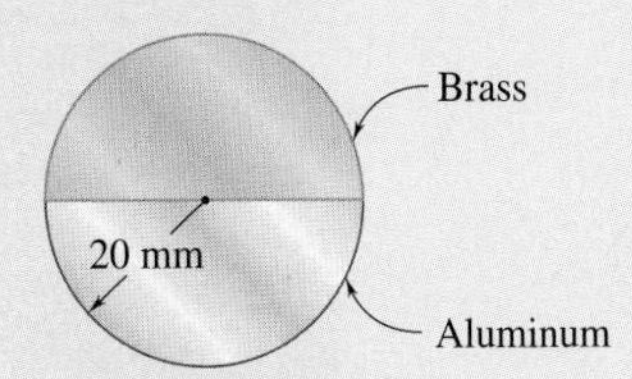

Fig. ***P11.47***

11.48 A steel pipe and an aluminum pipe are securely bonded together to form the composite beam shown. The modulus of elasticity is 200 GPa for the steel and 70 GPa for the aluminum. Knowing that the composite beam is bent by a couple of moment 500 N·m, determine the maximum stress (*a*) in the aluminum, (*b*) in the steel.

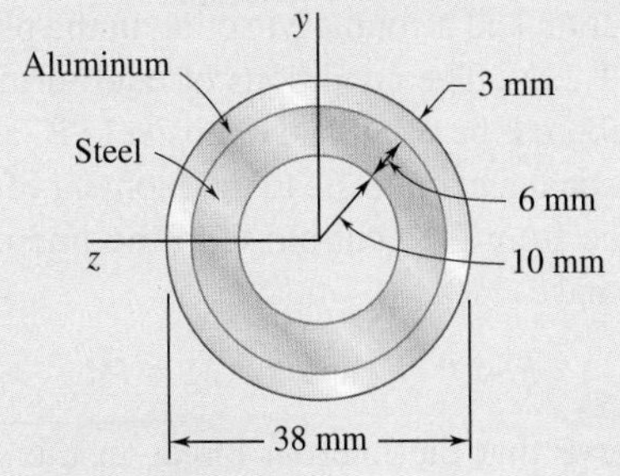

Fig. P11.48

11.4 ECCENTRIC AXIAL LOADING IN A PLANE OF SYMMETRY

We saw in Sec. 8.2A that the distribution of stresses in the cross section of a member under axial loading can be assumed uniform only if the line of action of the loads **P** and **P′** passes through the centroid of the cross section. Such a loading is said to be *centric*. Let us now analyze the distribution of stresses when the line of action of the loads does *not* pass through the centroid of the cross section, i.e., when the loading is *eccentric*.

Two examples of an eccentric loading are shown in Photos 11.5 and 11.6. In Photo 11.5, the weight of the lamp causes an eccentric loading on the post. Likewise, the vertical forces exerted on the press in Photo. 11.6 cause an eccentric loading on the back column of the press.

Photo 11.5 Walkway light.

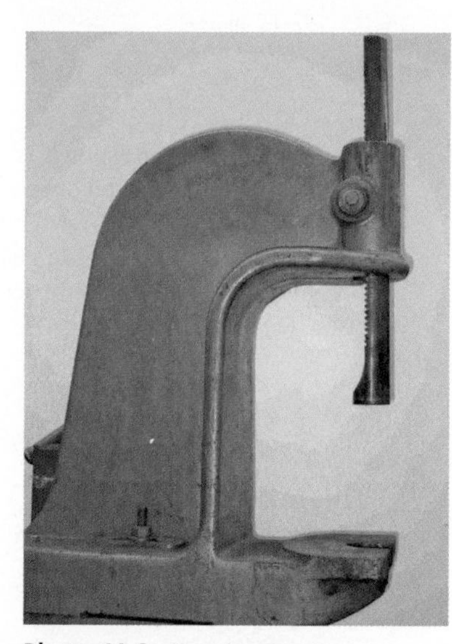

Photo 11.6 Bench press.

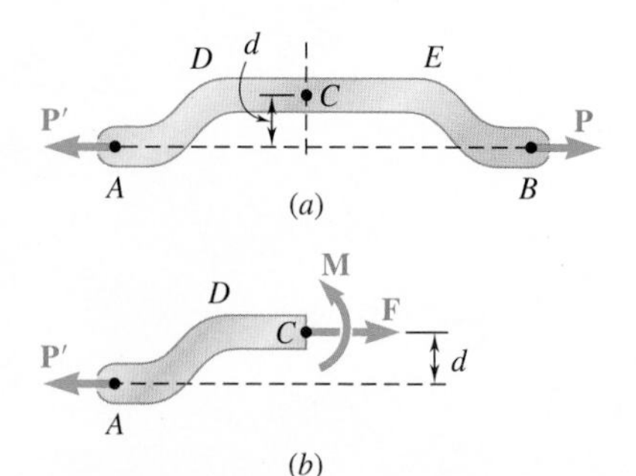

Fig. 11.22 (*a*) Member with eccentric loading. (*b*) Free-body diagram of the member with internal forces at section *C*.

In this section, our analysis will be limited to members that possess a plane of symmetry, and it will be assumed that the loads are applied in the plane of symmetry of the member (Fig. 11.22*a*). The internal forces acting on a given cross section may then be represented by a force **F** applied at the centroid *C* of the section and a couple **M** acting in the plane of symmetry of the member (Fig. 11.22*b*). The conditions of equilibrium of the free body *AC* require that the force **F** be equal and opposite to **P′** and that the moment of the couple **M** be equal and opposite to the moment of **P′** about *C*. Denoting by *d* the distance from the centroid *C* to the line of action *AB* of the forces **P** and **P′**, we have

$$F = P \qquad \text{and} \qquad M = Pd \tag{11.27}$$

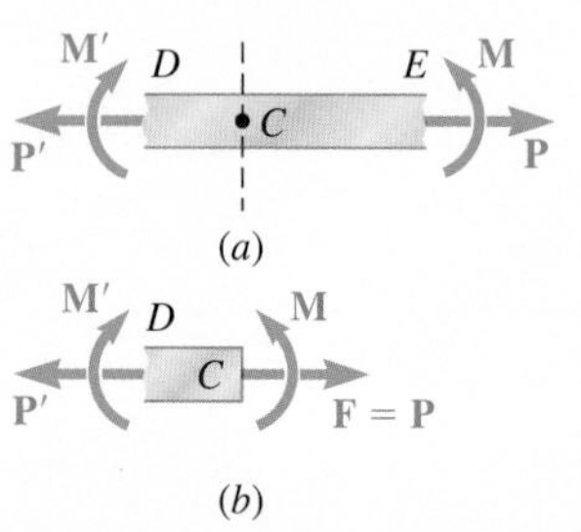

Fig. 11.23 (*a*) Free-body diagram of straight portion *DE*. (*b*) Free-body diagram of portion *CD*.

We now observe that the internal forces in the section would have been represented by the same force and couple if the straight portion *DE* of member *AB* had been detached from *AB* and subjected simultaneously to the centric loads **P** and **P′** and to the bending couples **M** and **M′** (Fig. 11.23). Thus, the stress distribution due to the original eccentric loading can be

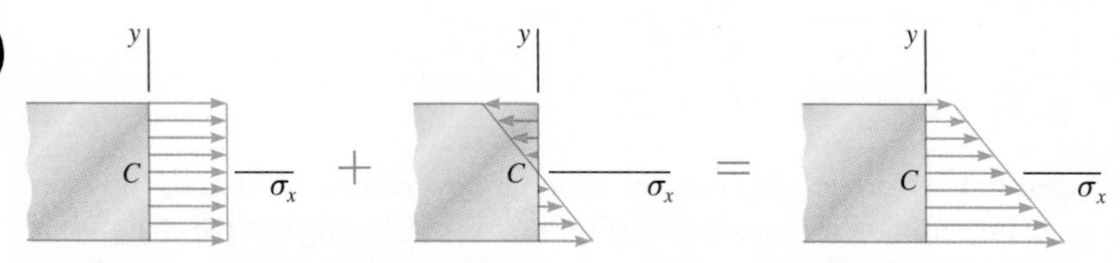

Fig. 11.24 Stress distribution for eccentric loading is obtained by superposing the axial and pure bending distributions.

obtained by superposing the uniform stress distribution corresponding to the centric loads **P** and **P′** and the linear distribution corresponding to the bending couples **M** and **M′** (Fig. 11.24). Write

$$\sigma_x = (\sigma_x)_{\text{centric}} + (\sigma_x)_{\text{bending}}$$

or recalling Eqs. (8.5) and (11.16),

$$\sigma_x = \frac{P}{A} - \frac{My}{I} \tag{11.28}$$

where A is the area of the cross section and I its centroidal moment of inertia and y is measured from the centroidal axis of the cross section. This relationship shows that the distribution of stresses across the section is *linear but not uniform.* Depending upon the geometry of the cross section and the eccentricity of the load, the combined stresses may all have the same sign, as shown in Fig. 11.24, or some may be positive and others negative, as shown in Fig. 11.25. In the latter case, there will be a line in the section, along which $\sigma_x = 0$. This line represents the *neutral axis* of the section. We note that the neutral axis does *not* coincide with the centroidal axis of the section, since $\sigma_x \neq 0$ for $y = 0$.

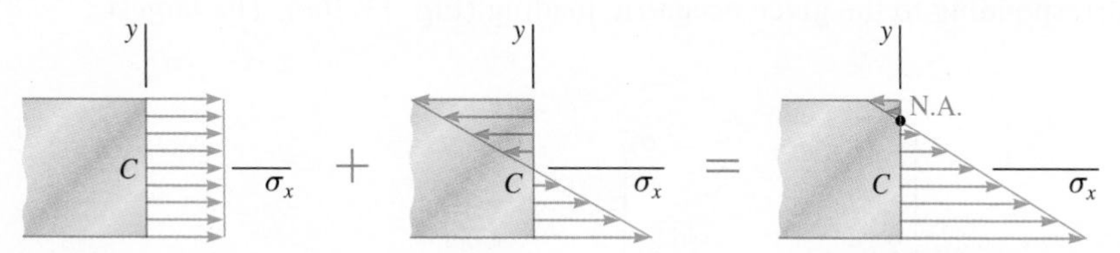

Fig. 11.25 Alternative stress distribution for eccentric loading that results in zones of tension and compression.

The results obtained are valid only to the extent that the conditions of applicability of the superposition principle (Sec. 9.5) and of Saint-Venant's principle (Sec. 9.8) are met. This means that the stresses involved must not exceed the proportional limit of the material. The deformations due to bending must not appreciably affect the distance d in Fig. 11.22a, and the cross section where the stresses are computed must not be too close to points D or E. The first of these requirements clearly shows that the superposition method cannot be applied to plastic deformations.

Concept Application 11.4

An open-link chain is obtained by bending low-carbon steel rods of 0.5-in. diameter into the shape shown (Fig. 11.26*a*). Knowing that the chain carries a load of 160 lb, determine (*a*) the largest tensile and compressive stresses in the straight portion of a link, (*b*) the distance between the centroidal and the neutral axis of a cross section.

a. Largest Tensile and Compressive Stresses. The internal forces in the cross section are equivalent to a centric force **P** and a bending couple **M** (Fig. 11.26*b*) of magnitudes

$$P = 160 \text{ lb}$$

$$M = Pd = (160 \text{ lb})(0.65 \text{ in.}) = 104 \text{ lb}\cdot\text{in.}$$

The corresponding stress distributions are shown in Fig. 11.26*c* and *d*. The distribution due to the centric force P is uniform and equal to $\sigma_0 = P/A$. We have

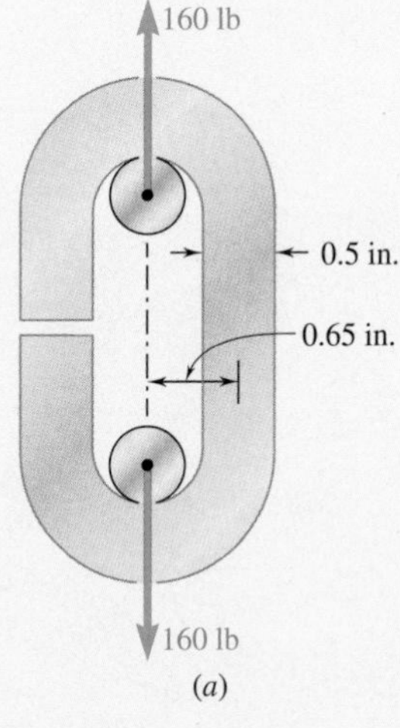

$$A = \pi c^2 = \pi(0.25 \text{ in.})^2 = 0.1963 \text{ in}^2$$

$$\sigma_0 = \frac{P}{A} = \frac{160 \text{ lb}}{0.1963 \text{ in}^2} = 815 \text{ psi}$$

The distribution due to the bending couple **M** is linear with a maximum stress $\sigma_m = Mc/I$. We write

$$I = \tfrac{1}{4}\pi c^4 = \tfrac{1}{4}\pi(0.25 \text{ in.})^4 = 3.068 \times 10^{-3} \text{ in}^4$$

$$\sigma_m = \frac{Mc}{I} = \frac{(104 \text{ lb}\cdot\text{in.})(0.25 \text{ in.})}{3.068 \times 10^{-3} \text{ in}^4} = 8475 \text{ psi}$$

Superposing the two distributions, we obtain the stress distribution corresponding to the given eccentric loading (Fig. 11.26*e*). The largest

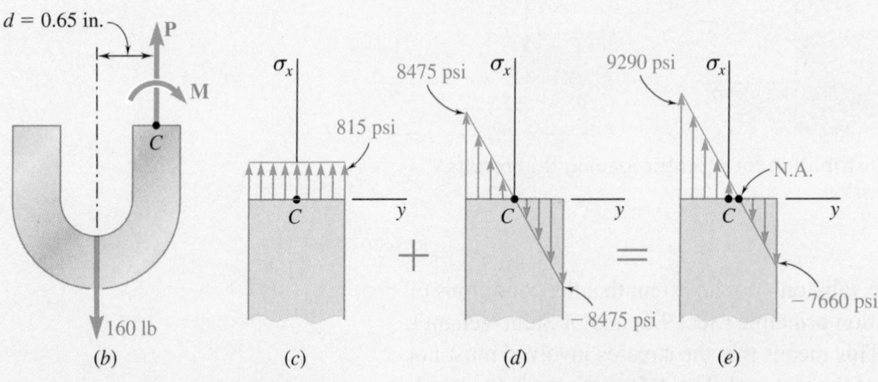

Fig. 11.26 (*a*) Open chain link under loading. (*b*) Free-body diagram for section at *C*. (*c*) Axial stress at section *C*. (*d*) Bending stress at *C*. (*e*) Superposition of stresses.

(continued)

tensile and compressive stresses in the section are found to be, respectively,

$$\sigma_t = \sigma_0 + \sigma_m = 815 + 8475 = 9290 \text{ psi}$$

$$\sigma_c = \sigma_0 - \sigma_m = 815 - 8475 = -7660 \text{ psi}$$

b. Distance Between Centroidal and Neutral Axes. The distance y_0 from the centroidal to the neutral axis of the section is obtained by setting $\sigma_x = 0$ in Eq. (11.28) and solving for y_0:

$$0 = \frac{P}{A} - \frac{My_0}{I}$$

$$y_0 = \left(\frac{P}{A}\right)\left(\frac{I}{M}\right) = (815 \text{ psi})\frac{3.068 \times 10^{-3} \text{ in}^4}{104 \text{ lb·in.}}$$

$$y_0 = 0.0240 \text{ in.}$$

Sample Problem 11.5

Knowing that for the cast iron link shown the allowable stresses are 30 MPa in tension and 120 MPa in compression, determine the largest force **P** which can be applied to the link. (*Note*: The T-shaped cross section of the link has previously been considered in Sample Prob. 11.2.)

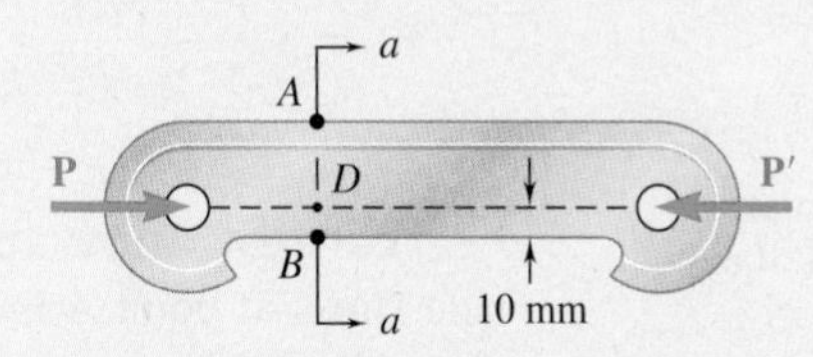

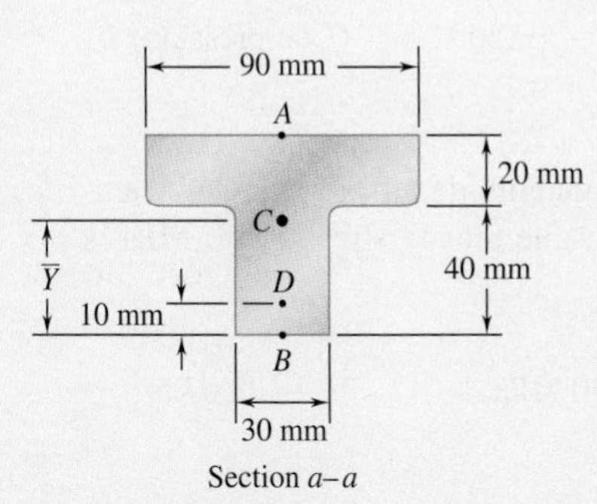

Fig. 1 Section geometry to find centroid location.

STRATEGY: The stresses due to the axial load and the couple resulting from the eccentricity of the axial load with respect to the neutral axis are superposed to obtain the maximum stresses. The cross section is singly symmetric, so it is necessary to determine both the maximum compression stress and the maximum tension stress and compare each to the corresponding allowable stress to find **P**.

MODELING and ANALYSIS:

Properties of Cross Section. The cross section is shown in Fig. 1. From Sample Prob. 11.2, we have

$$A = 3000 \text{ mm}^2 = 3 \times 10^{-3} \text{ m}^2 \qquad \bar{Y} = 38 \text{ mm} = 0.038 \text{ m}$$

$$I = 868 \times 10^{-9} \text{ m}^4$$

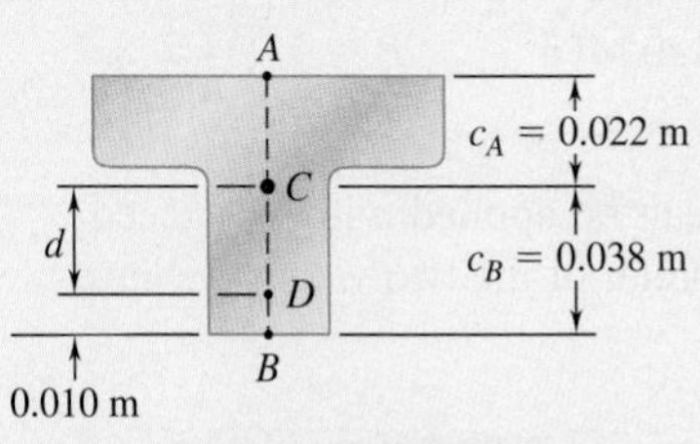

Fig. 2 Dimensions for finding *d*.

We now write (Fig. 2): $\quad d = (0.038 \text{ m}) - (0.010 \text{ m}) = 0.028 \text{ m}$

(continued)

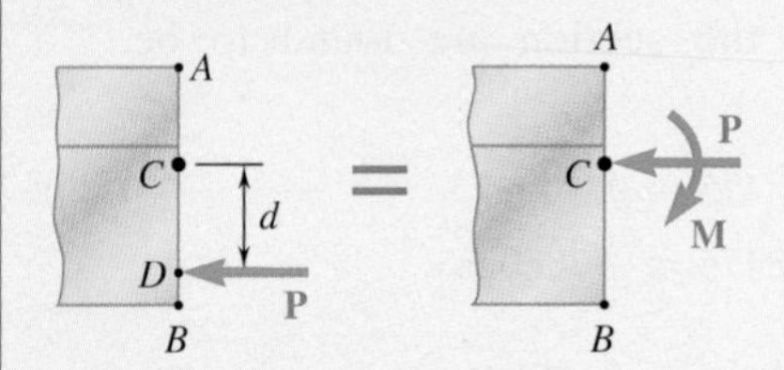

Fig. 3 Equivalent force-couple system at centroid C.

Force and Couple at C. Using Fig. 3, we replace **P** by an equivalent force-couple system at the centroid C.

$$P = P \qquad M = P(d) = P(0.028 \text{ m}) = 0.028P$$

The force **P** acting at the centroid causes a uniform stress distribution (Fig. 4*a*). The bending couple **M** causes a linear stress distribution (Fig. 4*b*).

$$\sigma_0 = \frac{P}{A} = \frac{P}{3 \times 10^{-3}} = 333P \qquad \text{(Compression)}$$

$$\sigma_1 = \frac{Mc_A}{I} = \frac{(0.028P)(0.022)}{868 \times 10^{-9}} = 710P \qquad \text{(Tension)}$$

$$\sigma_2 = \frac{Mc_B}{I} = \frac{(0.028P)(0.038)}{868 \times 10^{-9}} = 1226P \qquad \text{(Compression)}$$

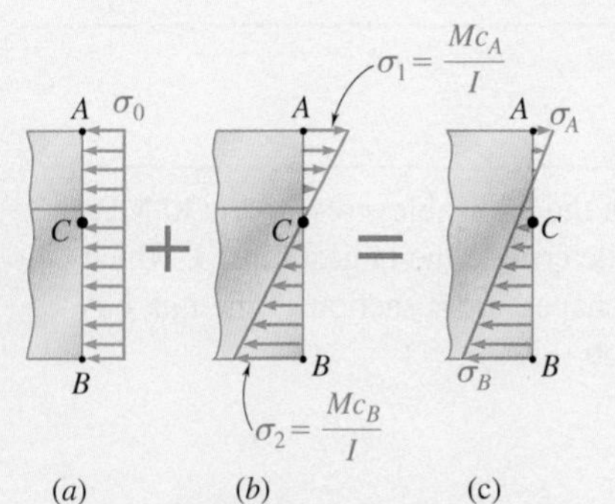

Fig. 4 Stress distribution at section C is superposition of axial and bending distributions.

Superposition. The total stress distribution (Fig. 4*c*) is found by superposing the stress distributions caused by the centric force **P** and by the couple **M**. Since tension is positive, and compression negative, we have

$$\sigma_A = -\frac{P}{A} + \frac{Mc_A}{I} = -333P + 710P = +377P \qquad \text{(Tension)}$$

$$\sigma_B = -\frac{P}{A} - \frac{Mc_B}{I} = -333P - 1226P = -1559P \qquad \text{(Compression)}$$

Largest Allowable Force. The magnitude of **P** for which the tensile stress at point A is equal to the allowable tensile stress of 30 MPa is found by writing

$$\sigma_A = 377P = 30 \text{ MPa} \qquad P = 79.6 \text{ kN} \blacktriangleleft$$

We also determine the magnitude of **P** for which the stress at B is equal to the allowable compressive stress of 120 MPa.

$$\sigma_B = -1559P = -120 \text{ MPa} \qquad P = 77.0 \text{ kN} \blacktriangleleft$$

The magnitude of the largest force **P** that can be applied without exceeding either of the allowable stresses is the smaller of the two values we have found.

$$P = 77.0 \text{ kN} \blacktriangleleft$$

Problems

11.49 Two forces **P** can be applied separately or at the same time to a plate that is welded to a solid circular bar of radius r. Determine the largest compressive stress in the circular bar (*a*) when both forces are applied, (*b*) when only one of the forces is applied.

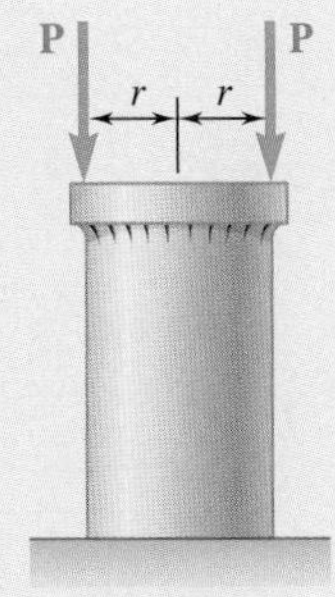

Fig. P11.49

11.50 As many as three axial loads each of magnitude $P = 10$ kips can be applied to the end of a W8 × 21 rolled-steel shape. Determine the stress at point A, (*a*) for the loading shown, (*b*) if loads are applied at points 1 and 2 only.

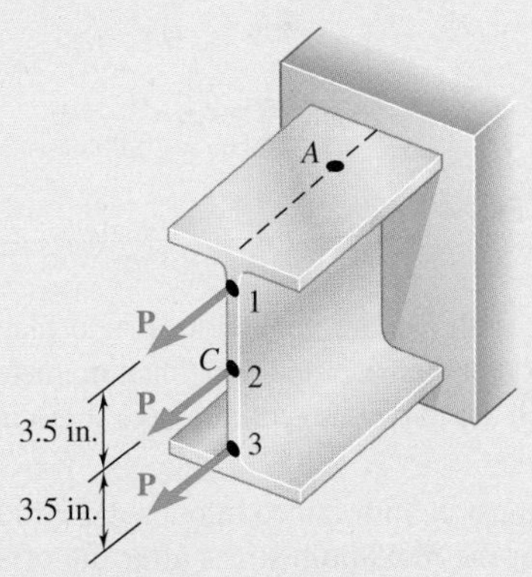

Fig. P11.50

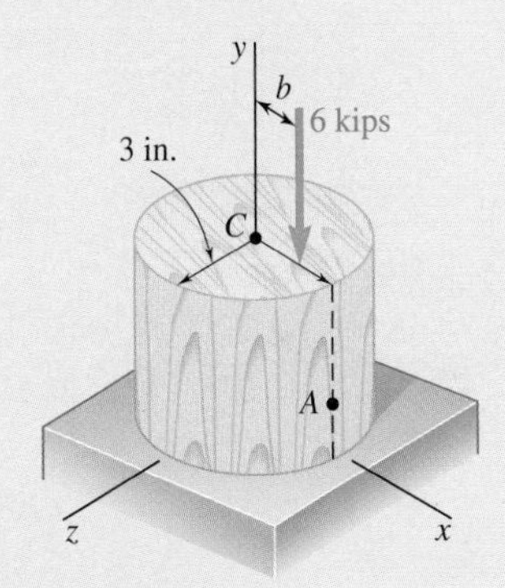

Fig. P11.51

11.51 A short wooden post supports a 6-kip axial load as shown. Determine the stress at point A when (*a*) $b = 0$, (*b*) $b = 1.5$ in., (*c*) $b = 3$ in.

11.52 Knowing that the magnitude of the horizontal force **P** is 8 kN, determine the stress at (*a*) point A, (*b*) point B.

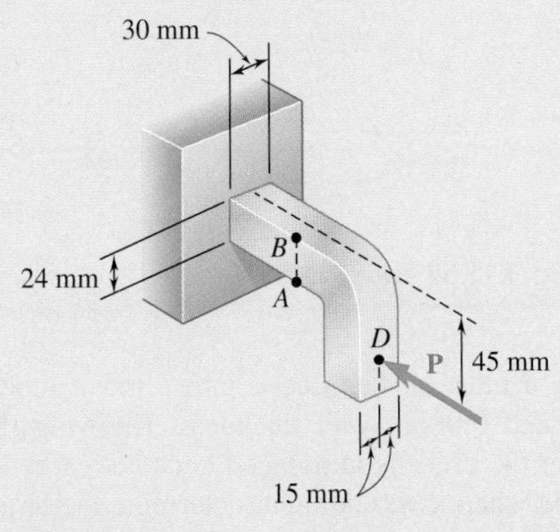

Fig. P11.52

11.53 The vertical portion of the press shown consists of a rectangular tube having a wall thickness $t = 10$ mm. Knowing that the press has been tightened on wooden planks being glued together until $P = 20$ kN, determine the stress (*a*) at point A, (*b*) point B.

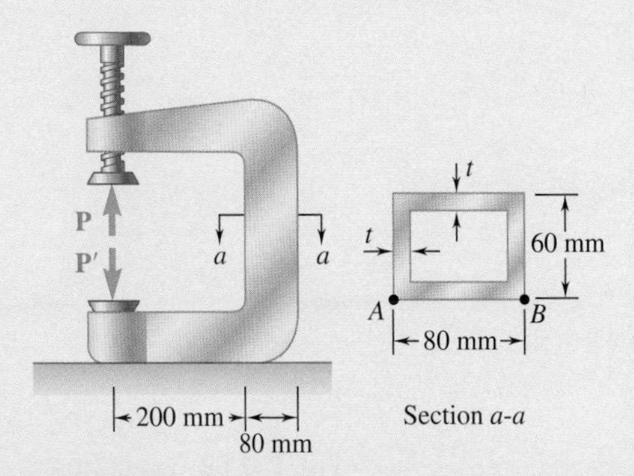

Fig. P11.53

11.54 Solve Prob. 11.53, assuming that $t = 8$ mm.

11.55 Determine the stress at points A and B, (a) for the loading shown, (b) if the 60-kN loads are applied at points 1 and 2 only.

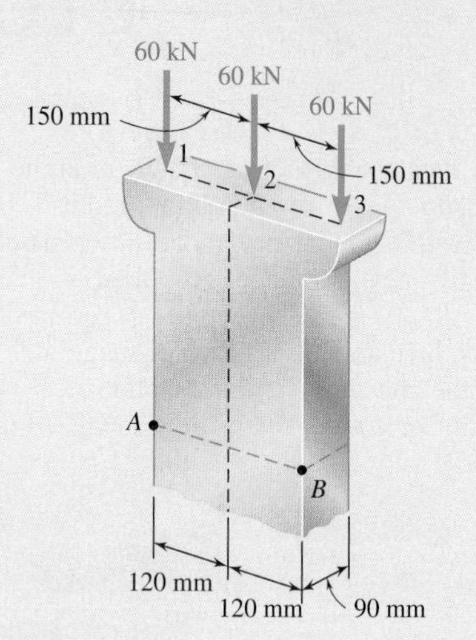

Fig. *P11.55*

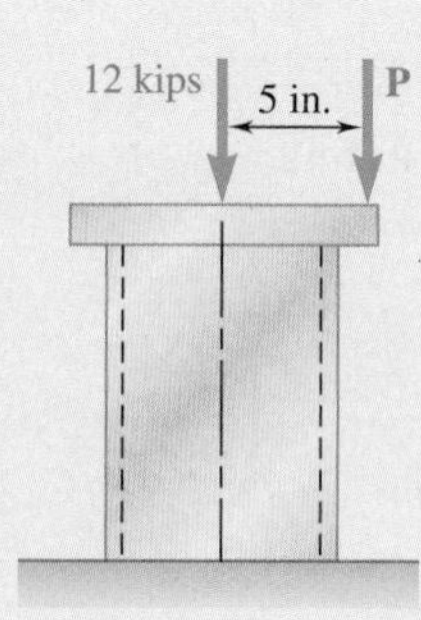

Fig. *P11.56*

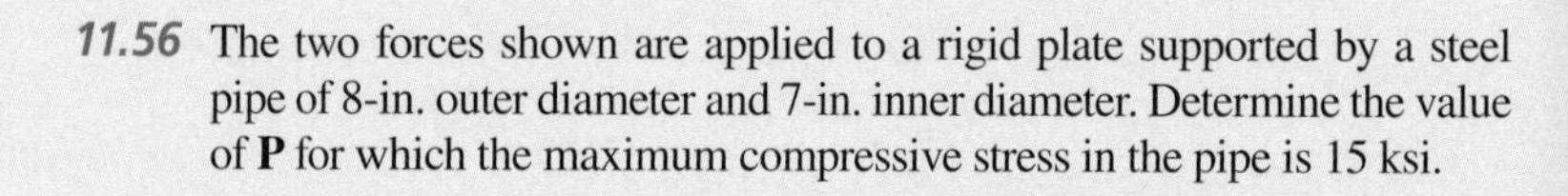

11.56 The two forces shown are applied to a rigid plate supported by a steel pipe of 8-in. outer diameter and 7-in. inner diameter. Determine the value of **P** for which the maximum compressive stress in the pipe is 15 ksi.

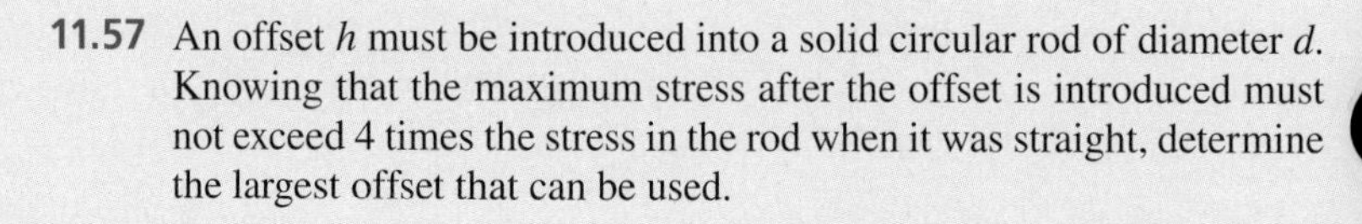

11.57 An offset h must be introduced into a solid circular rod of diameter d. Knowing that the maximum stress after the offset is introduced must not exceed 4 times the stress in the rod when it was straight, determine the largest offset that can be used.

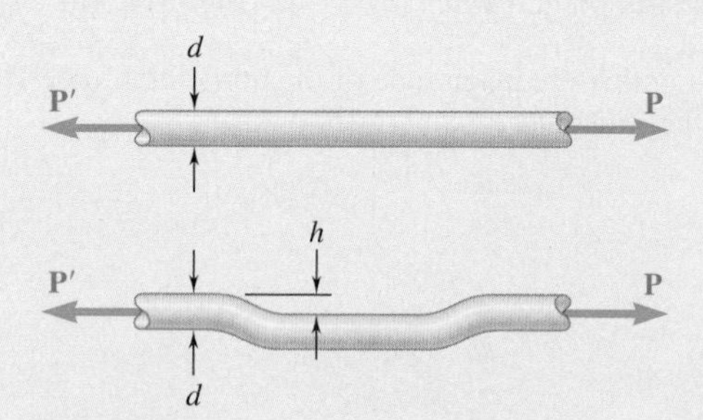

Fig. P11.57 and P11.58

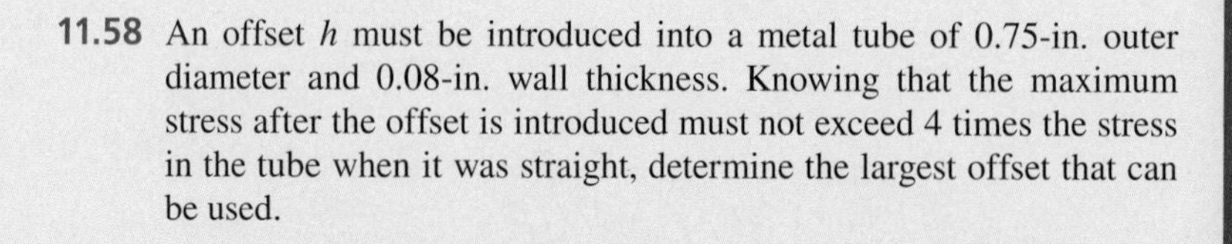

11.58 An offset h must be introduced into a metal tube of 0.75-in. outer diameter and 0.08-in. wall thickness. Knowing that the maximum stress after the offset is introduced must not exceed 4 times the stress in the tube when it was straight, determine the largest offset that can be used.

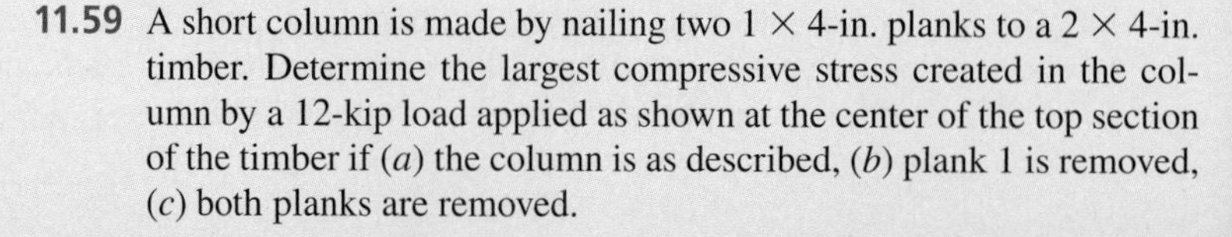

11.59 A short column is made by nailing two 1 × 4-in. planks to a 2 × 4-in. timber. Determine the largest compressive stress created in the column by a 12-kip load applied as shown at the center of the top section of the timber if (a) the column is as described, (b) plank 1 is removed, (c) both planks are removed.

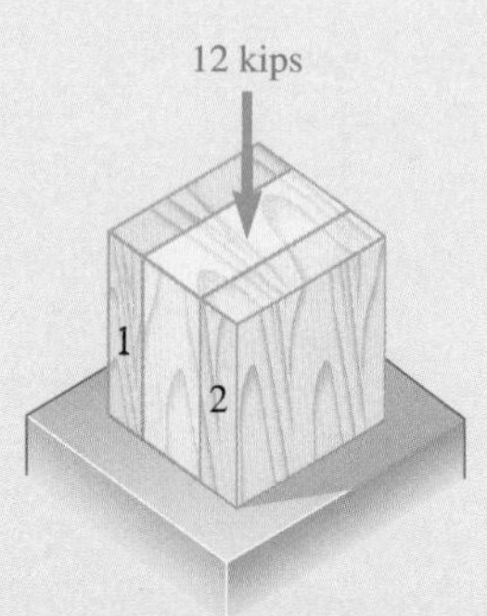

Fig. P11.59

11.60 Knowing that the allowable stress in section ABD is 10 ksi, determine the largest force **P** that can be applied to the bracket shown.

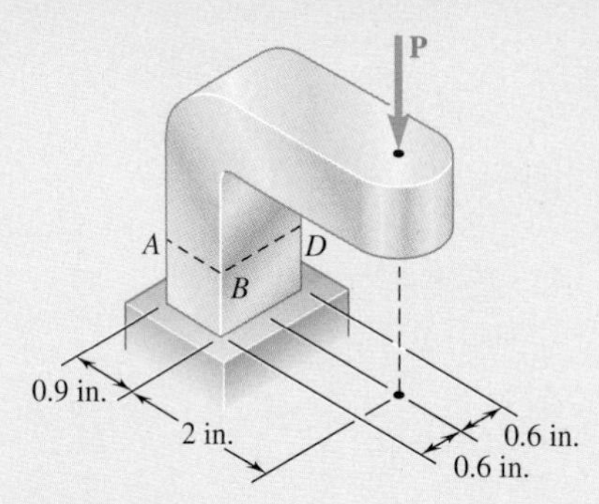

Fig. P11.60

11.61 A milling operation was used to remove a portion of a solid bar of square cross section. Knowing that $a = 30$ mm, $d = 20$ mm, and $\sigma_{all} = 60$ MPa, determine the magnitude P of the largest forces that can be safely applied at the centers of the ends of the bar.

11.62 A milling operation was used to remove a portion of a solid bar of square cross section. Forces of magnitude $P = 18$ kN are applied at the center of the ends of the bar. Knowing that $a = 30$ mm and $\sigma_{all} = 135$ MPa, determine the smallest allowable depth d of the milled portion of the rod.

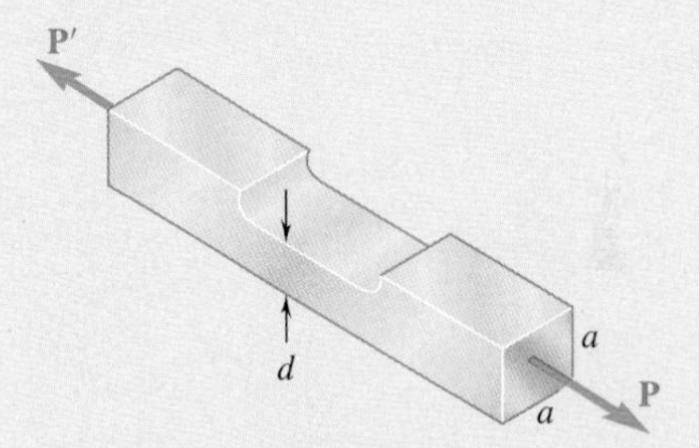

Fig. *P11.61* and P11.62

11.63 A vertical rod is attached at point A to the cast iron hanger shown. Knowing that the allowable stresses in the hanger are $\sigma_{all} = +5$ ksi and $\sigma_{all} = -12$ ksi, determine the largest downward force and the largest upward force that can be exerted by the rod.

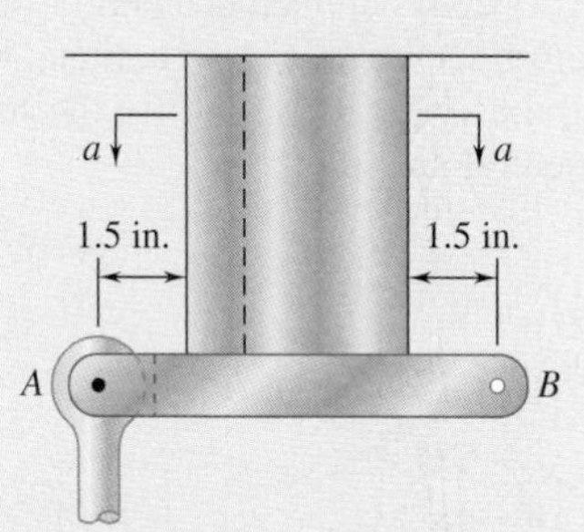

Fig. *P11.63*

11.64 A steel rod is welded to a steel plate to form the machine element shown. Knowing that the allowable stress is 135 MPa, determine (*a*) the largest force **P** that can be applied to the element, (*b*) the corresponding location of the neutral axis. *Given*: The centroid of the cross section is at C and $I_z = 4195$ mm^4.

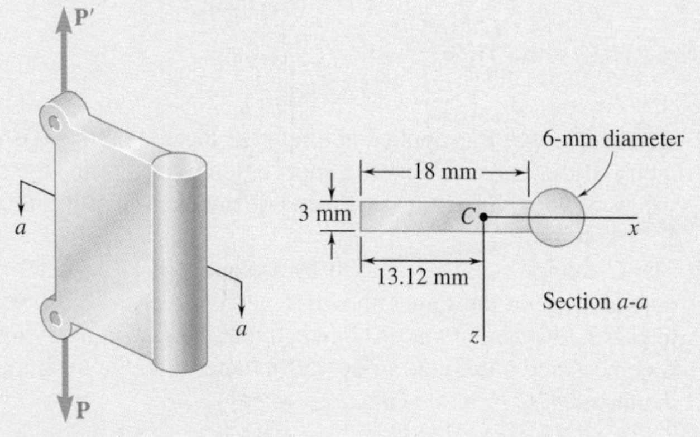

Fig. P11.64

11.65 The shape shown was formed by bending a thin steel plate. Assuming that the thickness t is small compared to the length a of a side of the shape, determine the stress (*a*) at A, (*b*) at B, (*c*) at C.

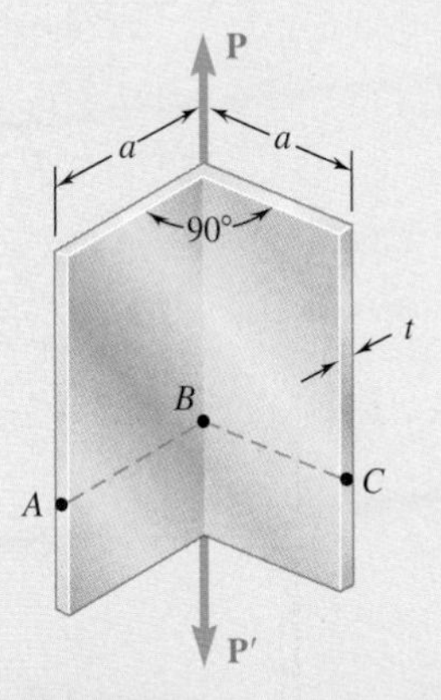

Fig. P11.65

11.66 Knowing that the clamp shown has been tightened until $P = 400$ N, determine (*a*) the stress at point *A*, (*b*) the stress at point *B*, (*c*) the location of the neutral axis of section *a-a*.

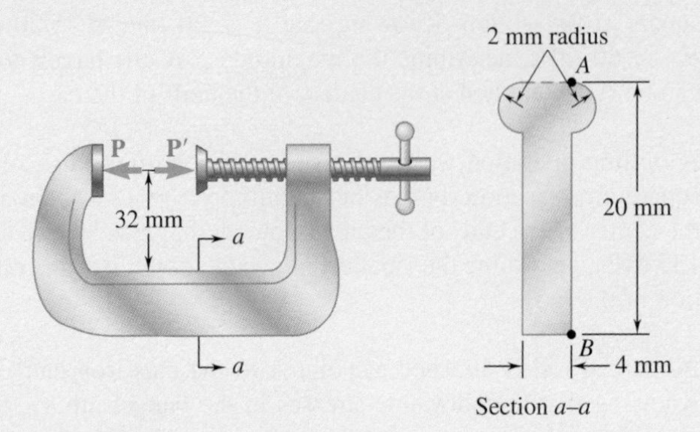

Fig. P11.66

11.67 A vertical force **P** of magnitude 20 kips is applied at a point *C* located on the axis of symmetry of the cross section of a short column. Knowing that $y = 5$ in., determine (*a*) the stress at point *A*, (*b*) the stress at point *B*, (*c*) the location of the neutral axis.

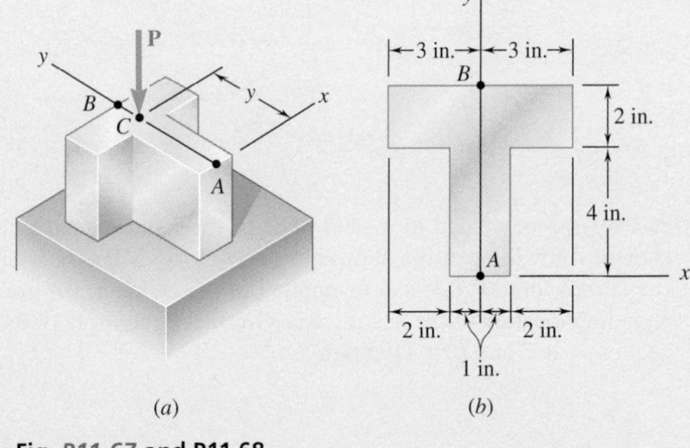

Fig. ***P11.67*** **and P11.68**

11.68 A vertical force **P** is applied at a point *C* located on the axis of symmetry of the cross section of a short column. Determine the range of values of *y* for which tensile stresses do not occur in the column.

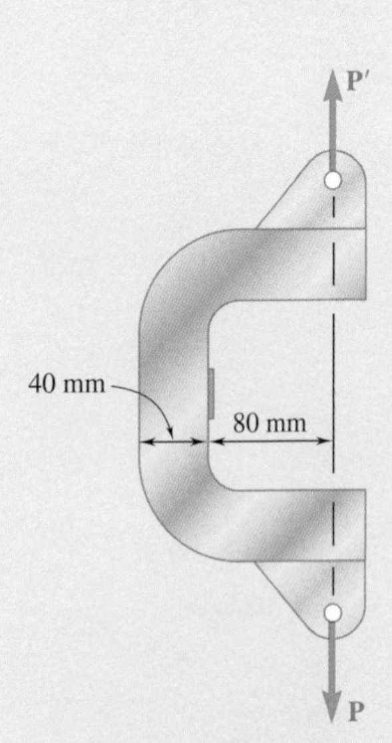

Fig. ***P11.69***

11.69 The C-shaped steel bar is used as a dynamometer to determine the magnitude *P* of the forces shown. Knowing that the cross section of the bar is a square of side 40 mm and that the strain on the inner edge was measured and found to be 450 μ, determine the magnitude *P* of the forces. Use $E = 200$ GPa.

11.70 A short length of a rolled-steel column supports a rigid plate on which two loads **P** and **Q** are applied as shown. The strains at two points A and B on the center lines of the outer faces of the flanges have been measured and found to be

$$\epsilon_A = -400 \times 10^{-6} \text{ in./in.} \qquad \epsilon_B = -300 \times 10^{-6} \text{ in./in.}$$

Knowing that $E = 29 \times 10^6$ psi, determine the magnitude of each load.

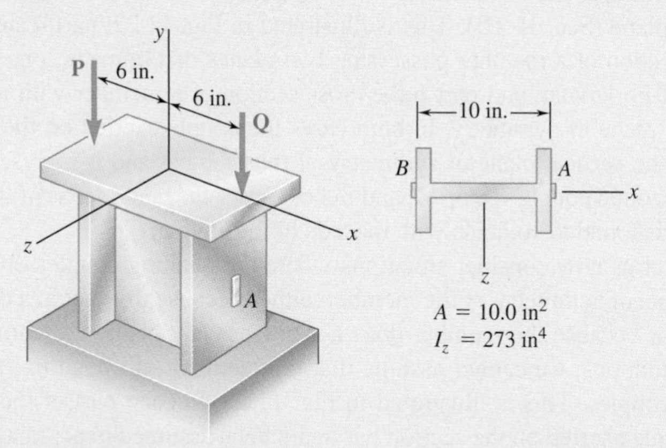

Fig. P11.70

11.71 An eccentric force **P** is applied as shown to a steel bar of 25 × 90-mm cross section. The strains at A and B have been measured and found to be

$$\epsilon_A = +350 \ \mu \qquad \epsilon_B = -70 \ \mu$$

Knowing that $E = 200$ GPa, determine (*a*) the distance d, (*b*) the magnitude of the force **P**.

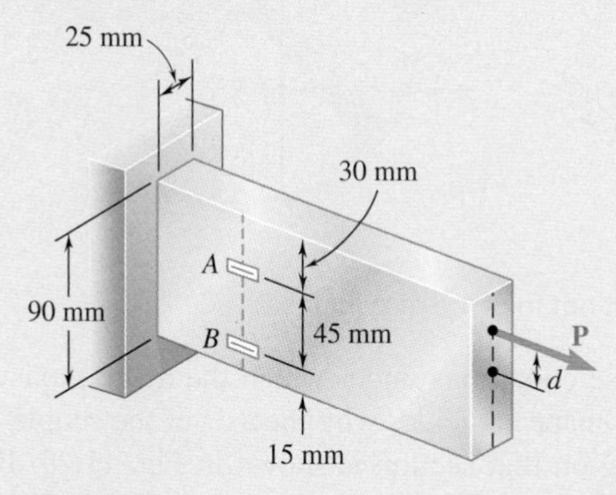

Fig. P11.71

11.72 Solve Prob. 11.71, assuming that the measured strains are

$$\epsilon_A = +600 \ \mu \qquad \epsilon_B = +420 \ \mu$$

11.5 UNSYMMETRIC BENDING ANALYSIS

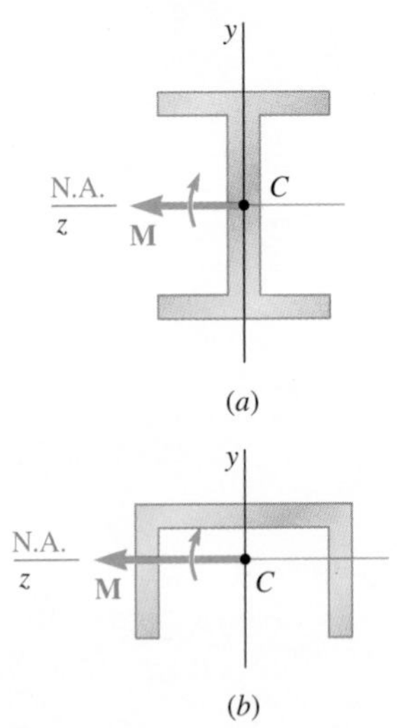

Fig. 11.27 Moment in plane of symmetry.

Our analysis of pure bending has been limited so far to members possessing at least one plane of symmetry and subjected to couples acting in that plane. Because of the symmetry of such members and of their loadings, the members remain symmetric with respect to the plane of the couples and thus bend in that plane (Sec. 11.1B). This is illustrated in Fig. 11.27; part *a* shows the cross section of a member possessing two planes of symmetry, one vertical and one horizontal, and part *b* the cross section of a member with a single, vertical plane of symmetry. In both cases the couple exerted on the section acts in the vertical plane of symmetry of the member and is represented by the horizontal couple vector **M**, and in both cases the neutral axis of the cross section is found to coincide with the axis of the couple.

Let us now consider situations where the bending couples do *not* act in a plane of symmetry of the member, either because they act in a different plane, or because the member does not possess any plane of symmetry. In such situations, we cannot assume that the member will bend in the plane of the couples. This is illustrated in Fig. 11.28. In each part of the figure, the couple exerted on the section has again been assumed to act in a vertical plane and has been represented by a horizontal couple vector **M**. However, since the vertical plane is not a plane of symmetry, *we cannot expect the member to bend in that plane or the neutral axis of the section to coincide with the axis of the couple.*

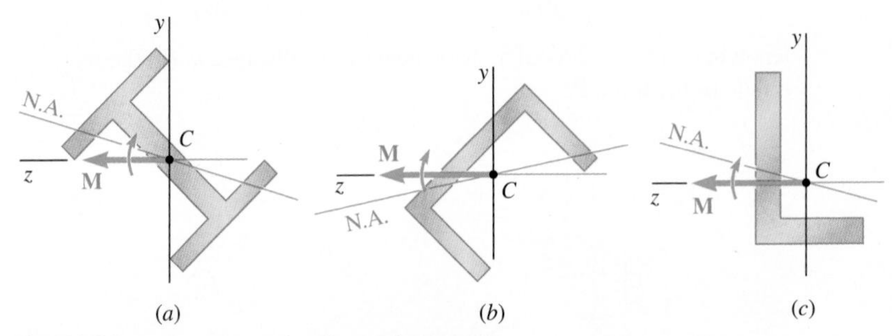

Fig. 11.28 Moment not in plane of symmetry.

The precise conditions under which the neutral axis of a cross section of arbitrary shape coincides with the axis of the couple **M** representing the forces acting on that section is shown in Fig. 11.29. Both the couple

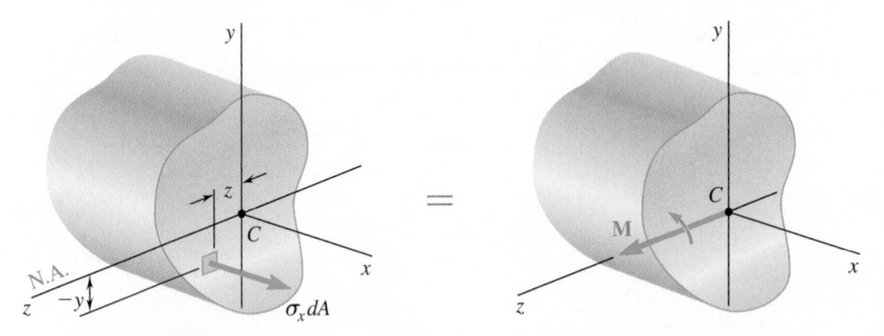

Fig. 11.29 Section of arbitrary shape where the neutral axis coincides with the axis of couple **M**.

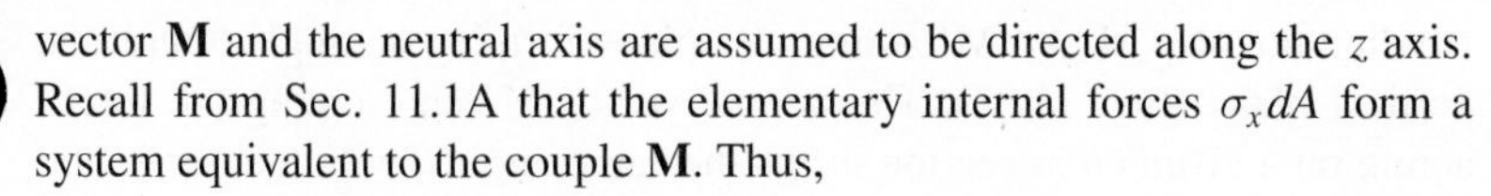

vector **M** and the neutral axis are assumed to be directed along the z axis. Recall from Sec. 11.1A that the elementary internal forces $\sigma_x dA$ form a system equivalent to the couple **M**. Thus,

$$x \text{ components:} \qquad \int \sigma_x dA = 0 \qquad \textbf{(11.1)}$$

$$\text{moments about } y \text{ axis:} \qquad \int z\sigma_x dA = 0 \qquad \textbf{(11.2)}$$

$$\text{moments about } z \text{ axis:} \qquad \int (-y\sigma_x dA) = M \qquad \textbf{(11.3)}$$

When all of the stresses are within the proportional limit, the first of these equations leads to the requirement that the neutral axis be a centroidal axis, and the last to the fundamental relation $\sigma_x = -My/I$. Since we had assumed in Sec. 11.1A that the cross section was symmetric with respect to the y axis, Eq. (11.2) was dismissed as trivial at that time. Now that we are considering a cross section of arbitrary shape, Eq. (11.2) becomes highly significant. Assuming the stresses to remain within the proportional limit of the material, $\sigma_x = -\sigma_m y/c$ is substituted into Eq. (11.2) for

$$\int z\left(-\frac{\sigma_m y}{c}\right) dA = 0 \qquad \text{or} \qquad \int yz\,dA = 0 \qquad \textbf{(11.29)}$$

The integral $\int yz\,dA$ represents the product of inertia I_{yz} of the cross section with respect to the y and z axes, and will be zero if these axes are the *principal centroidal axes of the cross section.*[†] Thus the neutral axis of the cross section coincides with the axis of the couple **M** representing the forces acting on that section *if, and only if, the couple vector* **M** *is directed along one of the principal centroidal axes of the cross section.*

Note that the cross sections shown in Fig. 11.27 are symmetric with respect to at least one of the coordinate axes. In each case, the y and z axes are the principal centroidal axes of the section. Since the couple vector **M** is directed along one of the principal centroidal axes, the neutral axis coincides with the axis of the couple. Also, if the cross sections are rotated through 90° (Fig. 11.30), the couple vector **M** is still directed along a principal centroidal axis, and the neutral axis again coincides with the axis of the couple, even though in case b the couple does *not* act in a plane of symmetry of the member.

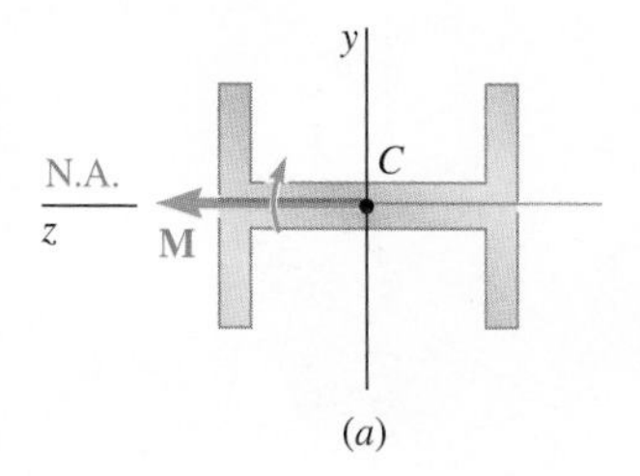

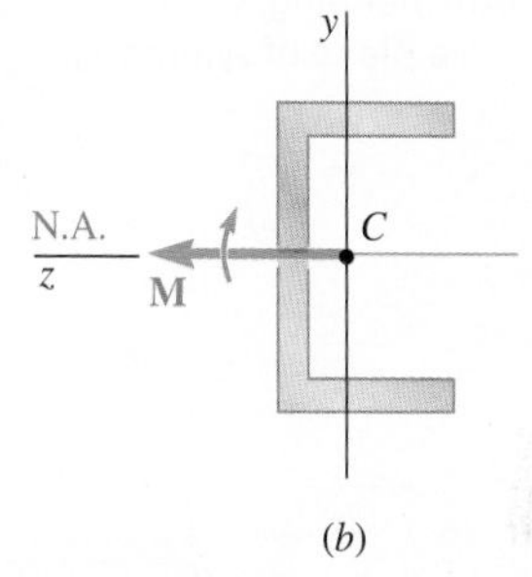

Fig. 11.30 Moment aligned with principal centroidal axis.

In Fig. 11.28, neither of the coordinate axes is an axis of symmetry for the sections shown, and the coordinate axes are not principal axes. Thus, the couple vector **M** is not directed along a principal centroidal axis, and the neutral axis does not coincide with the axis of the couple. However, any given section possesses principal centroidal axes, even if it is unsymmetric, as the section shown in Fig. 11.28c, and these axes may be determined analytically or by using Mohr's circle.[†] If the couple vector **M** is directed along one of the principal centroidal axes of the section, the neutral axis will coincide with the axis of the couple (Fig. 11.31), and the equations derived for symmetric members can be used to determine the stresses.

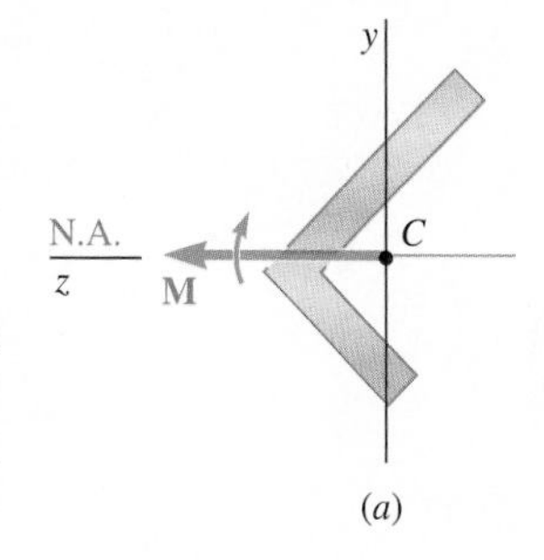

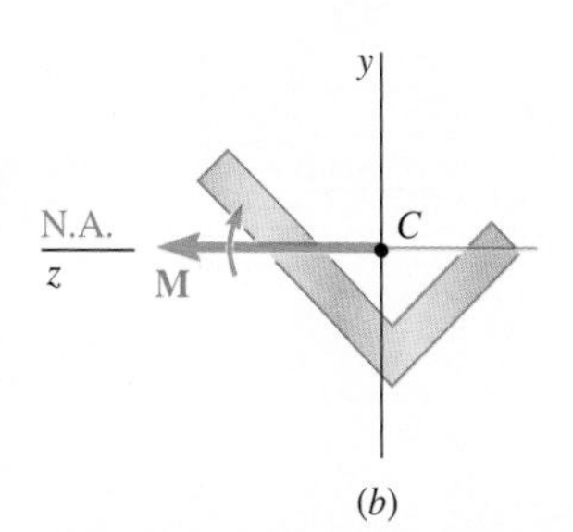

Fig. 11.31 Moment aligned with principal centroidal axis of an unsymmetric shape.

As you will see presently, the principle of superposition can be used to determine stresses in the most general case of unsymmetric bending. Consider first a member with a vertical plane of symmetry subjected to

[†]See Ferdinand P. Beer and E. Russell Johnston, Jr., *Mechanics for Engineers*, 5th ed., McGraw-Hill, New York, 2008, or *Vector Mechanics for Engineers*, 11th ed., McGraw-Hill, New York, 2016, Secs. 9.3–9.4.

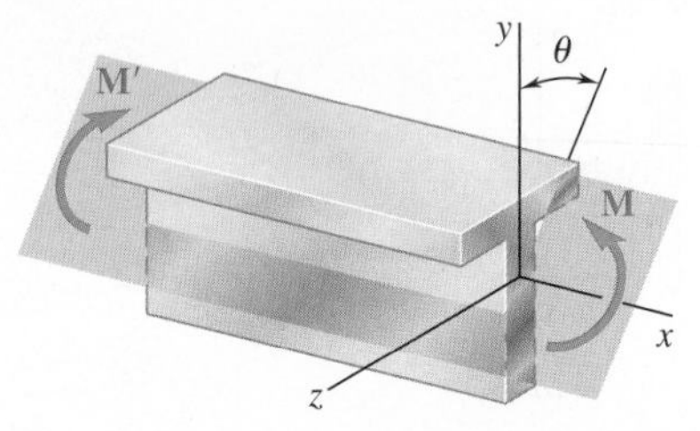

Fig. 11.32 Unsymmetric bending, with bending moment not in a plane of symmetry.

bending couples **M** and **M′** acting in a plane forming an angle θ with the vertical plane (Fig. 11.32). The couple vector **M** representing the forces acting on a given cross section forms the same angle θ with the horizontal z axis (Fig. 11.33). Resolving the vector **M** into component vectors $\mathbf{M}_z$ and $\mathbf{M}_y$ along the z and y axes, respectively, gives

$$M_z = M \cos \theta \qquad M_y = M \sin \theta \tag{11.30}$$

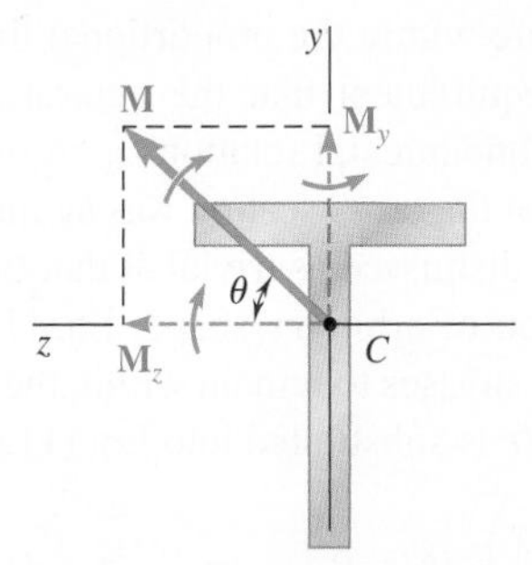

Fig. 11.33 Applied moment is resolved into y and z components.

Since the y and z axes are the principal centroidal axes of the cross section, Eq. (11.16) determines the stresses resulting from the application of either of the couples represented by $\mathbf{M}_z$ and $\mathbf{M}_y$. The couple $\mathbf{M}_z$ acts in a vertical plane and bends the member in that plane (Fig. 11.34). The resulting stresses are

$$\sigma_x = -\frac{M_z y}{I_z} \tag{11.31}$$

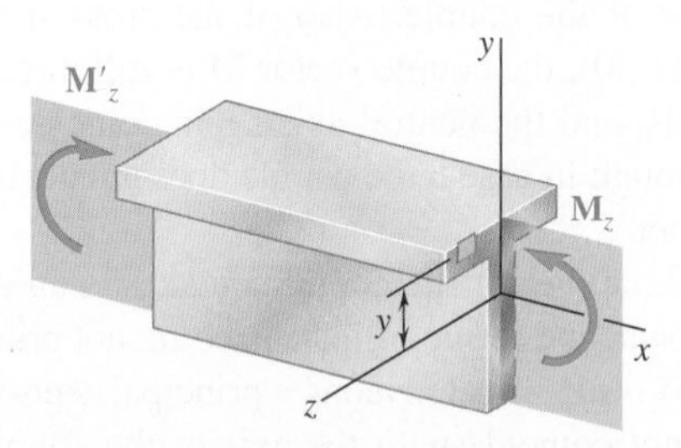

Fig. 11.34 $\mathbf{M}_z$ acts in a plane that includes a principal centroidal axis, bending the member in the vertical plane.

where I_z is the moment of inertia of the section about the principal centroidal z axis. The negative sign is due to the compression above the xz plane ($y > 0$) and tension below ($y < 0$). The couple $\mathbf{M}_y$ acts in a horizontal plane and bends the member in that plane (Fig. 11.35). The resulting stresses are

$$\sigma_x = +\frac{M_y z}{I_y} \tag{11.32}$$

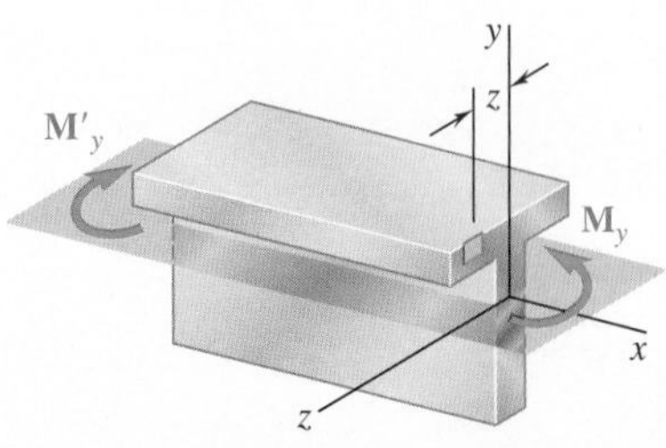

Fig. 11.35 $\mathbf{M}_y$ acts in a plane that includes a principal centroidal axis, bending the member in the horizontal plane.

where I_y is the moment of inertia of the section about the principal centroidal y axis, and where the positive sign is due to the fact that we have tension to the left of the vertical xy plane ($z > 0$) and compression to its right ($z < 0$). The distribution of the stresses caused by the original couple **M** is obtained by superposing the stress distributions defined by Eqs. (11.31) and (11.32), respectively. We have

$$\sigma_x = -\frac{M_z y}{I_z} + \frac{M_y z}{I_y} \qquad (11.33)$$

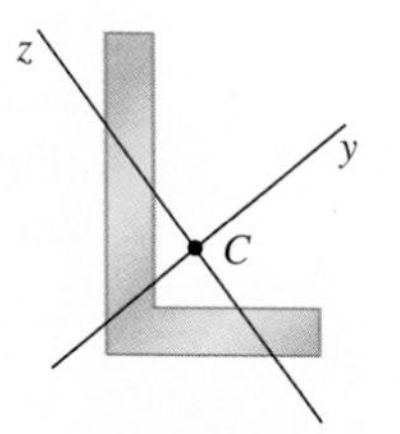

Fig. 11.36 Unsymmetric cross section with principal axes.

Note that the expression obtained can also be used to compute the stresses in an unsymmetric section, as shown in Fig. 11.36, once the principal centroidal y and z axes have been determined. However, Eq. (11.33) is valid only if the conditions of applicability of the principle of superposition are met. It should not be used if the combined stresses exceed the proportional limit of the material or if the deformations caused by one of the couples appreciably affect the distribution of the stresses due to the other.

Eq. (11.33) shows that the distribution of stresses caused by unsymmetric bending is linear. However, the neutral axis of the cross section will not, in general, coincide with the axis of the bending couple. Since the normal stress is zero at any point of the neutral axis, the equation defining that axis is obtained by setting $\sigma_x = 0$ in Eq. (11.33).

$$-\frac{M_z y}{I_z} + \frac{M_y z}{I_y} = 0$$

Solving for y and substituting for M_z and M_y from Eqs. (11.30) gives

$$y = \left(\frac{I_z}{I_y} \tan \theta\right) z \qquad (11.34)$$

This equation is for a straight line of slope $m = (I_z/I_y) \tan \theta$. Thus, the angle ϕ that the neutral axis forms with the z axis (Fig. 11.37) is defined by the relation

$$\tan \phi = \frac{I_z}{I_y} \tan \theta \qquad (11.35)$$

where θ is the angle that the couple vector **M** forms with the same axis. Since I_z and I_y are both positive, ϕ and θ have the same sign. Furthermore, $\phi > \theta$ when $I_z > I_y$, and $\phi < \theta$ when $I_z < I_y$. Thus, the neutral axis is always located between the couple vector **M** and the principal axis corresponding to the minimum moment of inertia.

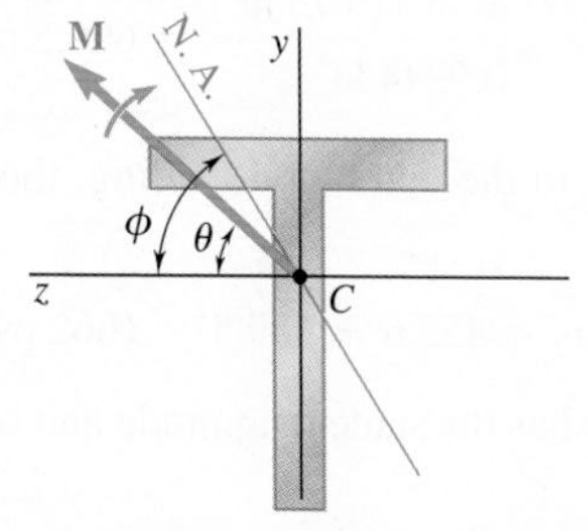

Fig. 11.37 Neutral axis for unsymmetric bending.

Concept Application 11.5

A 1600-lb·in. couple is applied to a wooden beam, of rectangular cross section 1.5 by 3.5 in., in a plane forming an angle of 30° with the vertical (Fig. 11.38*a*). Determine (*a*) the maximum stress in the beam and (*b*) the angle that the neutral surface forms with the horizontal plane.

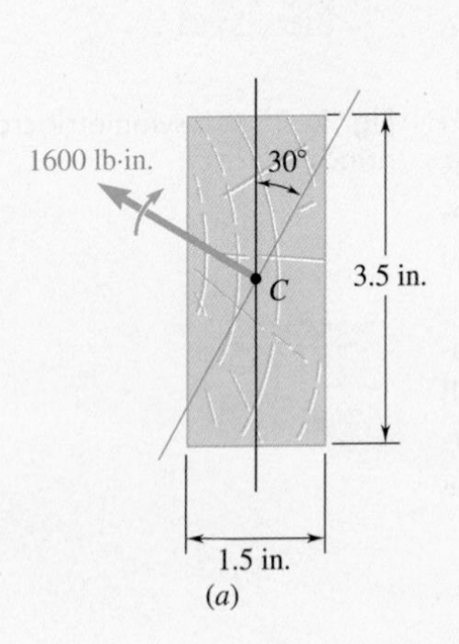

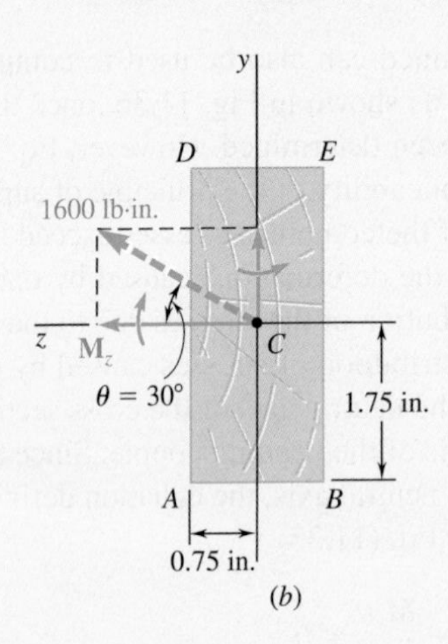

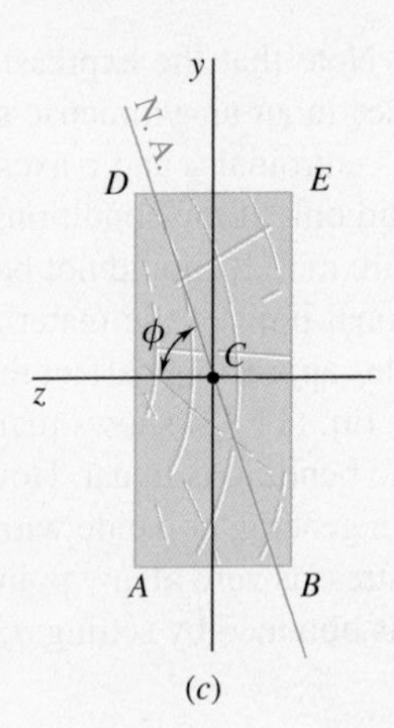

a. Maximum Stress. The components $\mathbf{M}_z$ and $\mathbf{M}_y$ of the couple vector are first determined (Fig. 11.38*b*):

$$M_z = (1600 \text{ lb·in.}) \cos 30° = 1386 \text{ lb·in.}$$

$$M_y = (1600 \text{ lb·in.}) \sin 30° = 800 \text{ lb·in.}$$

Compute the moments of inertia of the cross section with respect to the *z* and *y* axes:

$$I_z = \tfrac{1}{12}(1.5 \text{ in.})(3.5 \text{ in.})^3 = 5.359 \text{ in}^4$$

$$I_y = \tfrac{1}{12}(3.5 \text{ in.})(1.5 \text{ in.})^3 = 0.9844 \text{ in}^4$$

The largest tensile stress due to $\mathbf{M}_z$ occurs along *AB* and is

$$\sigma_1 = \frac{M_z y}{I_z} = \frac{(1386 \text{ lb·in.})(1.75 \text{ in.})}{5.359 \text{ in}^4} = 452.6 \text{ psi}$$

The largest tensile stress due to $\mathbf{M}_y$ occurs along *AD* and is

$$\sigma_2 = \frac{M_y z}{I_y} = \frac{(800 \text{ lb·in.})(0.75 \text{ in.})}{0.9844 \text{ in}^4} = 609.5 \text{ psi}$$

The largest tensile stress due to the combined loading, therefore, occurs at *A* and is

$$\sigma_{max} = \sigma_1 + \sigma_2 = 452.6 + 609.5 = 1062 \text{ psi}$$

The largest compressive stress has the same magnitude and occurs at *E*.

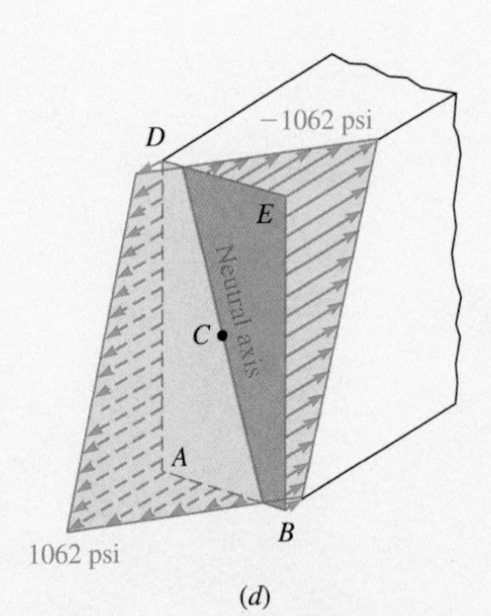

Fig. 11.38 (*a*) Rectangular wood beam subject to unsymmetric bending. (*b*) Bending moment resolved into components. (*c*) Cross section with neutral axis. (*d*) Stress distribution.

(continued)

b. Angle of Neutral Surface with Horizontal Plane. The angle ϕ that the neutral surface forms with the horizontal plane (Fig. 11.38*c*) is obtained from Eq. (11.35):

$$\tan \phi = \frac{I_z}{I_y} \tan \theta = \frac{5.359 \text{ in}^4}{0.9844 \text{ in}^4} \tan 30° = 3.143$$

$$\phi = 72.4°$$

The distribution of the stresses across the section is shown in Fig. 11.38*d*.

11.6 GENERAL CASE OF ECCENTRIC AXIAL LOADING ANALYSIS

In Sec. 11.4 we analyzed the stresses produced in a member by an eccentric axial load applied in a plane of symmetry of the member. We will now study the more general case when the axial load is not applied in a plane of symmetry.

Consider a straight member AB subjected to equal and opposite eccentric axial forces **P** and **P**′ (Fig. 11.39*a*), and let a and b be the distances from the line of action of the forces to the principal centroidal axes of the cross section of the member. The eccentric force **P** is statically equivalent to the system consisting of a centric force **P** and of the two couples $\mathbf{M}_y$ and $\mathbf{M}_z$ of moments $M_y = Pa$ and $M_z = Pb$ in Fig. 11.39*b*. Similarly, the eccentric force **P**′ is equivalent to the centric force **P**′ and the couples $\mathbf{M}'_y$ and $\mathbf{M}'_z$.

By virtue of Saint-Venant's principle (Sec. 9.8), replace the original loading of Fig. 11.39*a* by the statically equivalent loading of Fig. 11.39*b* to determine the distribution of stresses in section S of the member (as long as that section is not too close to either end). The stresses due to the loading of Fig. 11.39*b* can be obtained by superposing the stresses corresponding to the centric axial load **P** and to the bending couples $\mathbf{M}_y$ and $\mathbf{M}_z$, as long as the conditions of the principle of superposition are satisfied (Sec. 9.5). The stresses due to the centric load **P** are given by Eq. (8.1), and the stresses due to the bending couples by Eq. (11.33). Therefore,

$$\sigma_x = \frac{P}{A} - \frac{M_z y}{I_z} + \frac{M_y z}{I_y} \quad \textbf{(11.36)}$$

where y and z are measured from the principal centroidal axes of the section. This relationship shows that the distribution of stresses across the section is *linear*.

In computing the combined stress σ_x from Eq. (11.36), be sure to correctly determine the sign of each of the three terms in the right-hand member, since each can be positive or negative, depending upon the sense

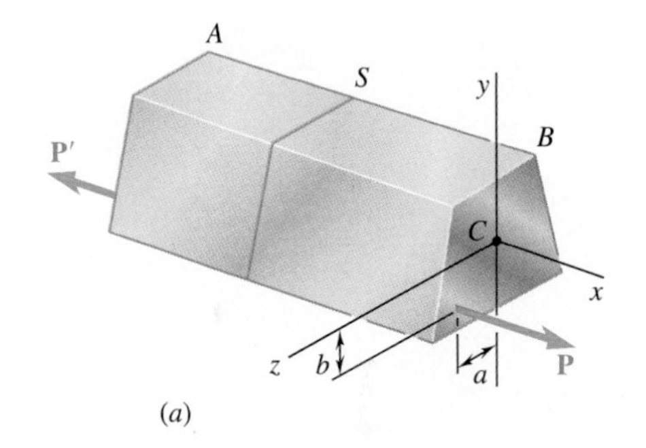

(*a*)

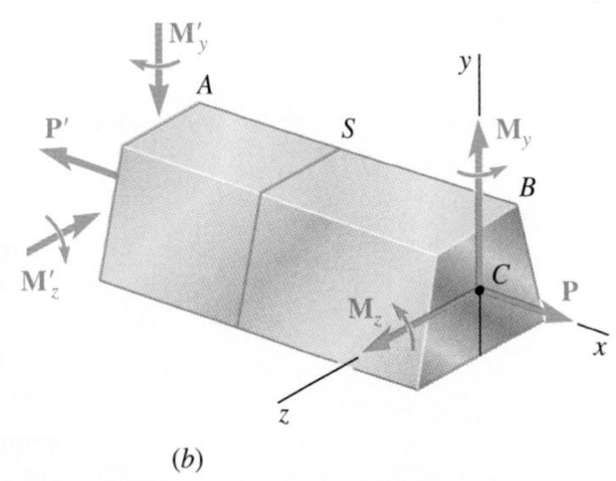

(*b*)

Fig. 11.39 Eccentric axial loading. (*a*) Axial force applied away from section centroid. (*b*) Equivalent force-couple system acting at centroid.

of the loads **P** and **P′** and the location of their line of action with respect to the principal centroidal axes of the cross section. The combined stresses σ_x obtained from Eq. (11.36) at various points of the section may all have the same sign, or some may be positive and others negative. In the latter case, there will be a line in the section along which the stresses are zero. Setting $\sigma_x = 0$ in Eq. (11.36), the equation of a straight line representing the *neutral axis* of the section is

$$\frac{M_z}{I_z}y - \frac{M_y}{I_y}z = \frac{P}{A}$$

Concept Application 11.6

A vertical 4.80-kN load is applied as shown on a wooden post of rectangular cross section, 80 by 120 mm (Fig. 11.40*a*). (*a*) Determine the stress at points *A*, *B*, *C*, and *D*. (*b*) Locate the neutral axis of the cross section.

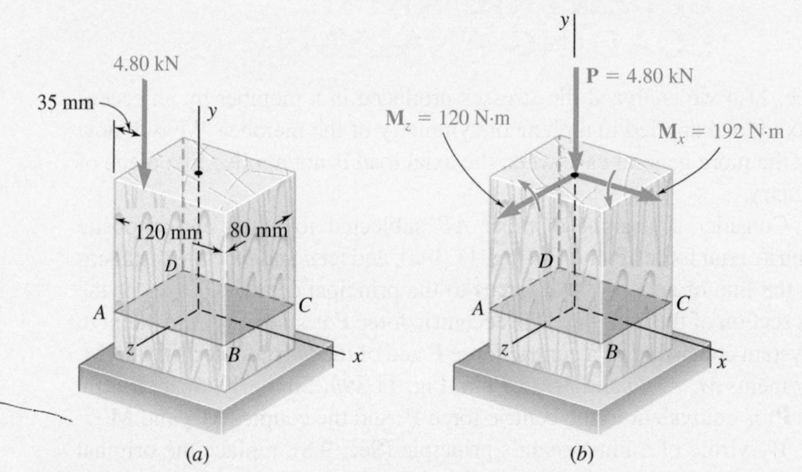

Fig. 11.40 (a) Eccentric load on a rectangular wood column. (*b*) Equivalent force-couple system for eccentric load.

a. Stresses. The given eccentric load is replaced by an equivalent system consisting of a centric load **P** and two couples $\mathbf{M}_x$ and $\mathbf{M}_z$ represented by vectors directed along the principal centroidal axes of the section (Fig. 11.40*b*). Thus

$$M_x = (4.80 \text{ kN})(40 \text{ mm}) = 192 \text{ N·m}$$

$$M_z = (4.80 \text{ kN})(60 \text{ mm} - 35 \text{ mm}) = 120 \text{ N·m}$$

Compute the area and the centroidal moments of inertia of the cross section:

$$A = (0.080 \text{ m})(0.120 \text{ m}) = 9.60 \times 10^{-3} \text{ m}^2$$

$$I_x = \tfrac{1}{12}(0.120 \text{ m})(0.080 \text{ m})^3 = 5.12 \times 10^{-6} \text{ m}^4$$

$$I_z = \tfrac{1}{12}(0.080 \text{ m})(0.120 \text{ m})^3 = 11.52 \times 10^{-6} \text{ m}^4$$

(continued)

The stress σ_0 due to the centric load **P** is negative and uniform across the section:

$$\sigma_0 = \frac{P}{A} = \frac{-4.80 \text{ kN}}{9.60 \times 10^{-3} \text{ m}^2} = -0.5 \text{ MPa}$$

The stresses due to the bending couples $\mathbf{M}_x$ and $\mathbf{M}_z$ are linearly distributed across the section with maximum values equal to

$$\sigma_1 = \frac{M_x z_{\max}}{I_x} = \frac{(192 \text{ N·m})(40 \text{ mm})}{5.12 \times 10^{-6} \text{ m}^4} = 1.5 \text{ MPa}$$

$$\sigma_2 = \frac{M_z x_{\max}}{I_z} = \frac{(120 \text{ N·m})(60 \text{ mm})}{11.52 \times 10^{-6} \text{ m}^4} = 0.625 \text{ MPa}$$

The stresses at the corners of the section are

$$\sigma_y = \sigma_0 \pm \sigma_1 \pm \sigma_2$$

where the signs must be determined from Fig. 11.40*b*. Noting that the stresses due to $\mathbf{M}_x$ are positive at C and D and negative at A and B, and the stresses due to $\mathbf{M}_z$ are positive at B and C and negative at A and D, we obtain

$$\sigma_A = -0.5 - 1.5 - 0.625 = -2.625 \text{ MPa}$$

$$\sigma_B = -0.5 - 1.5 + 0.625 = -1.375 \text{ MPa}$$

$$\sigma_C = -0.5 + 1.5 + 0.625 = +1.625 \text{ MPa}$$

$$\sigma_D = -0.5 + 1.5 - 0.625 = +0.375 \text{ MPa}$$

b. Neutral Axis. The stress will be zero at a point G between B and C, and at a point H between D and A (Fig. 11.40*c*). Since the stress distribution is linear,

$$\frac{BG}{80 \text{ mm}} = \frac{1.375}{1.625 + 1.375} \qquad BG = 36.7 \text{ mm}$$

$$\frac{HA}{80 \text{ mm}} = \frac{2.625}{2.625 + 0.375} \qquad HA = 70 \text{ mm}$$

The neutral axis can be drawn through points G and H (Fig. 11.40*d*). The distribution of the stresses across the section is shown in Fig. 11.40*e*.

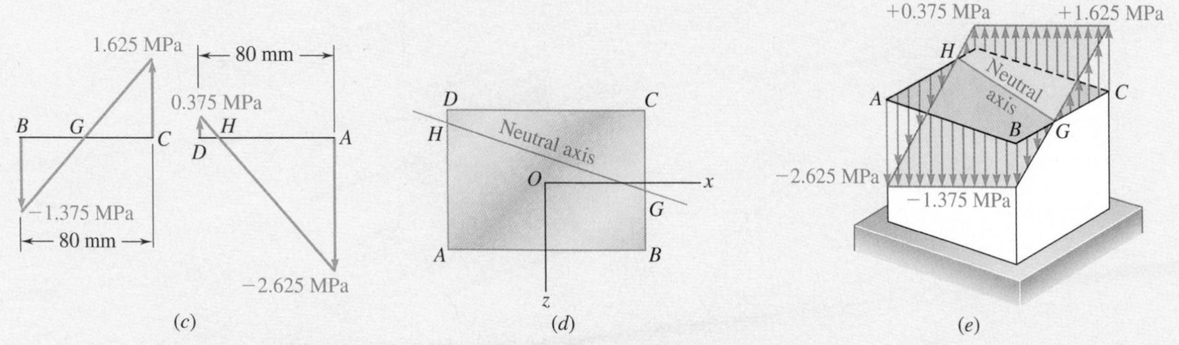

Fig. 11.40 (*cont.*) (*c*) Stress distributions along edges *BC* and *AD*. (*d*) Neutral axis is line through points *G* and *H*. (*e*) Stress distribution for eccentric load.

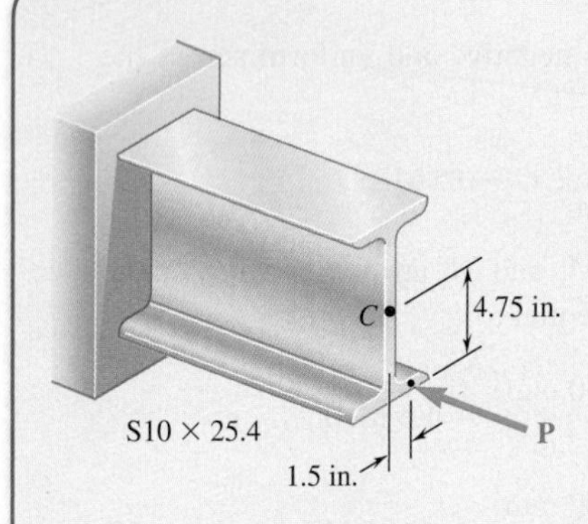

Sample Problem 11.6

A horizontal load **P** is applied as shown to a short section of an S10 × 25.4 rolled-steel member. Knowing that the compressive stress in the member is not to exceed 12 ksi, determine the largest permissible load **P**.

STRATEGY: The load is applied eccentrically with respect to both centroidal axes of the cross section. The load is replaced with an equivalent force-couple system at the centroid of the cross section. The stresses due to the axial load and the two couples are then superposed to determine the maximum stresses on the cross section.

MODELING and ANALYSIS:

Properties of Cross Section. The cross section is shown in Fig. 1, and the following data are taken from Appendix B.

$$\textit{Area: } A = 7.46 \text{ in}^2$$
$$\textit{Section moduli: } S_x = 24.7 \text{ in}^3 \qquad S_y = 2.91 \text{ in}^3$$

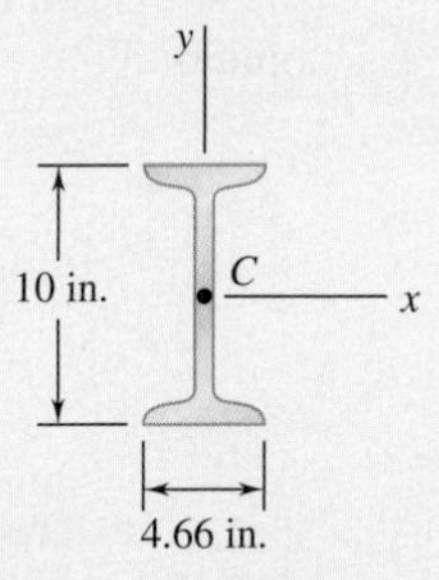

Fig. 1 Rolled-steel member

Force and Couple at C. Using Fig. 2, we replace **P** by an equivalent force-couple system at the centroid *C* of the cross section.

$$M_x = (4.75 \text{ in.})P \qquad M_y = (1.5 \text{ in.})P$$

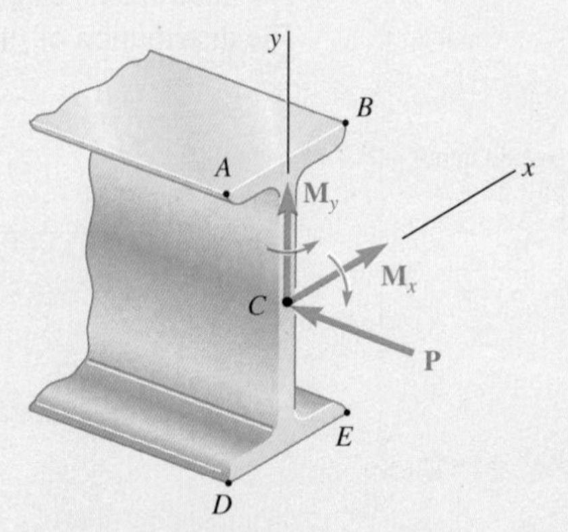

Fig. 2 Equivalent force-couple system at section centroid.

(continued)

Note that the couple vectors $\mathbf{M}_x$ and $\mathbf{M}_y$ are directed along the principal axes of the cross section.

Normal Stresses. The absolute values of the stresses at points A, B, D, and E due, respectively, to the centric load $\mathbf{P}$ and to the couples $\mathbf{M}_x$ and $\mathbf{M}_y$ are

$$\sigma_1 = \frac{P}{A} = \frac{P}{7.46 \text{ in}^2} = 0.1340P$$

$$\sigma_2 = \frac{M_x}{S_x} = \frac{4.75P}{24.7 \text{ in}^3} = 0.1923P$$

$$\sigma_3 = \frac{M_y}{S_y} = \frac{1.5P}{2.91 \text{ in}^3} = 0.5155P$$

Superposition. The total stress at each point is found by superposing the stresses due to $\mathbf{P}$, $\mathbf{M}_x$, and $\mathbf{M}_y$. We determine the sign of each stress by carefully examining the sketch of the force-couple system.

$$\sigma_A = -\sigma_1 + \sigma_2 + \sigma_3 = -0.1340P + 0.1923P + 0.5155P = +0.574P$$

$$\sigma_B = -\sigma_1 + \sigma_2 - \sigma_3 = -0.1340P + 0.1923P - 0.5155P = -0.457P$$

$$\sigma_D = -\sigma_1 - \sigma_2 + \sigma_3 = -0.1340P - 0.1923P + 0.5155P = +0.189P$$

$$\sigma_E = -\sigma_1 - \sigma_2 - \sigma_3 = -0.1340P - 0.1923P - 0.5155P = -0.842P$$

Largest Permissible Load. The maximum compressive stress occurs at point E. Recalling that $\sigma_{\text{all}} = -12$ ksi, we write

$$\sigma_{\text{all}} = \sigma_E \qquad -12 \text{ ksi} = -0.842P \qquad P = 14.3 \text{ kips} \blacktriangleleft$$

Problems

11.73 through 11.78 The couple **M** is applied to a beam of the cross section shown in a plane forming an angle β with the vertical. Determine the stress at (*a*) point *A*, (*b*) point *B*, (*c*) point *D*.

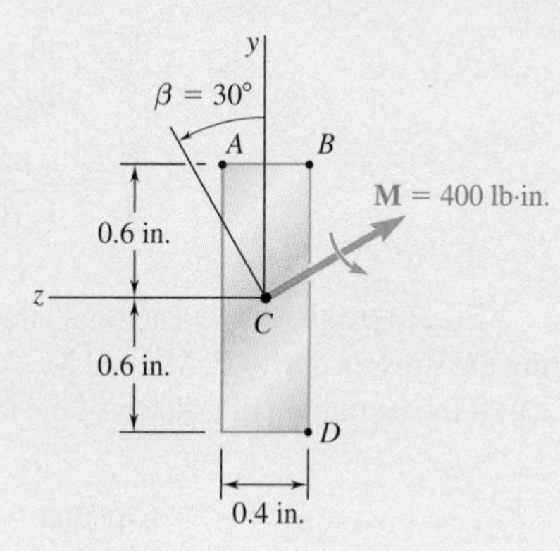

Fig. P11.73

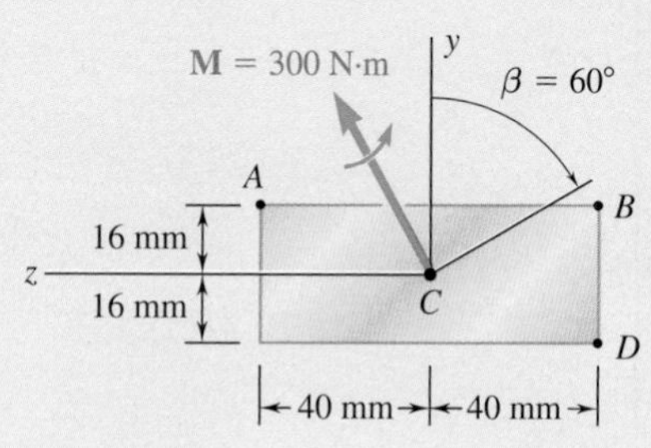

Fig. P11.74

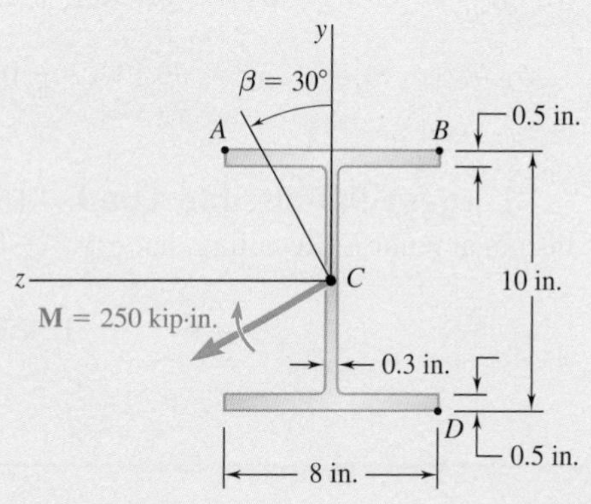

Fig. P11.75

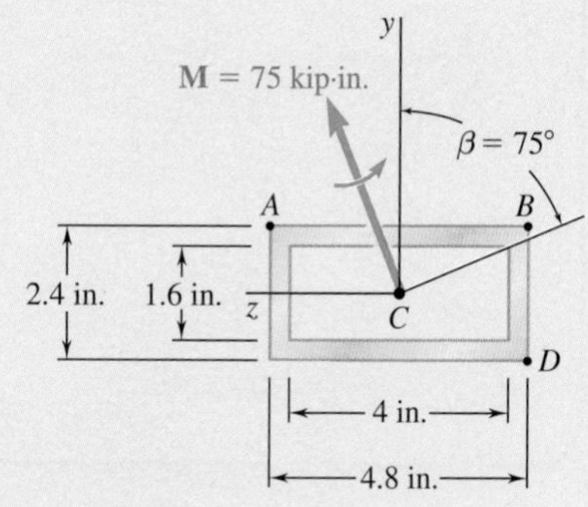

Fig. P11.76

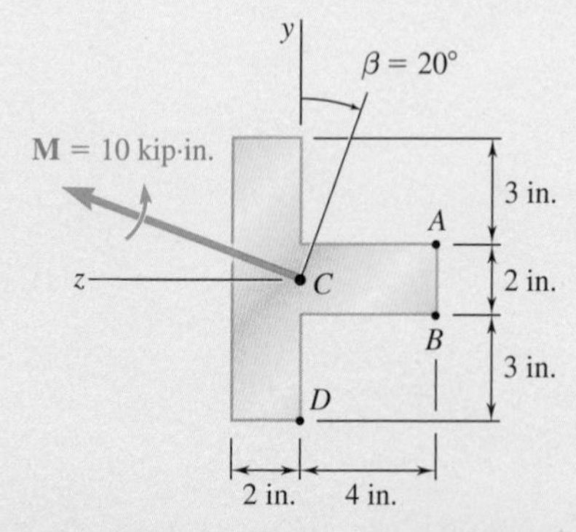

Fig. *P11.77*

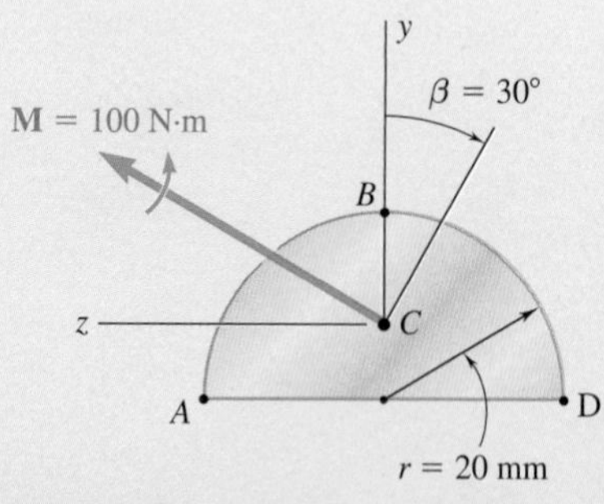

Fig. P11.78

11.79 through 11.84 The couple **M** acts in a vertical plane and is applied to a beam oriented as shown. Determine (*a*) the angle that the neutral axis forms with the horizontal plane, (*b*) the maximum tensile stress in the beam.

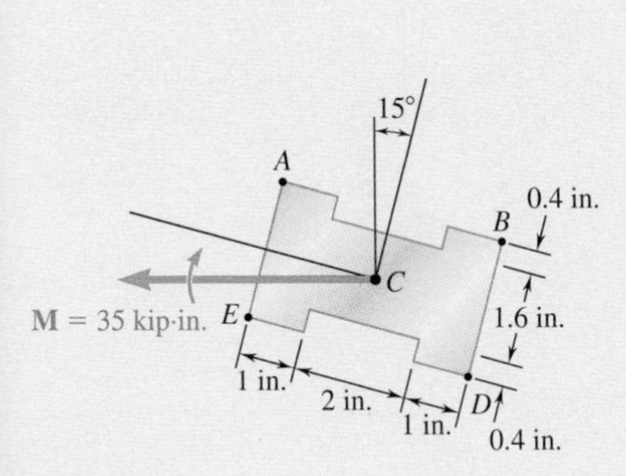

Fig. P11.79

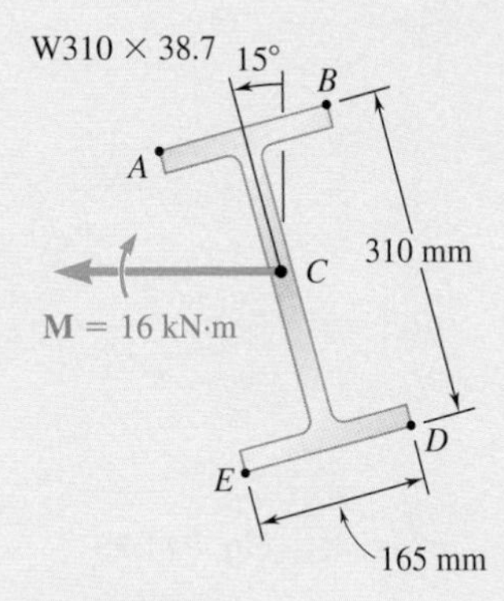

Fig. P11.80

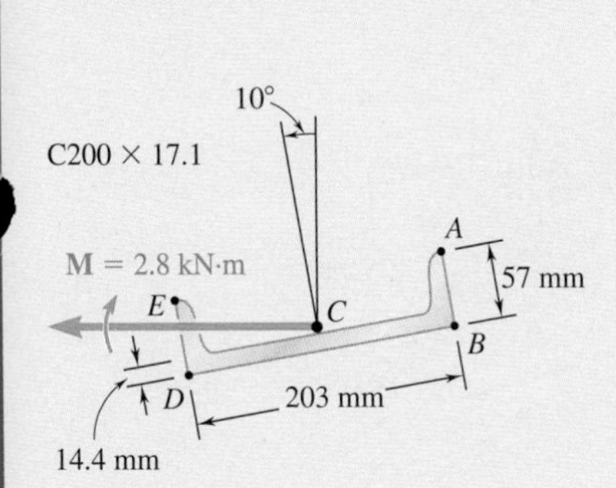

Fig. P11.81

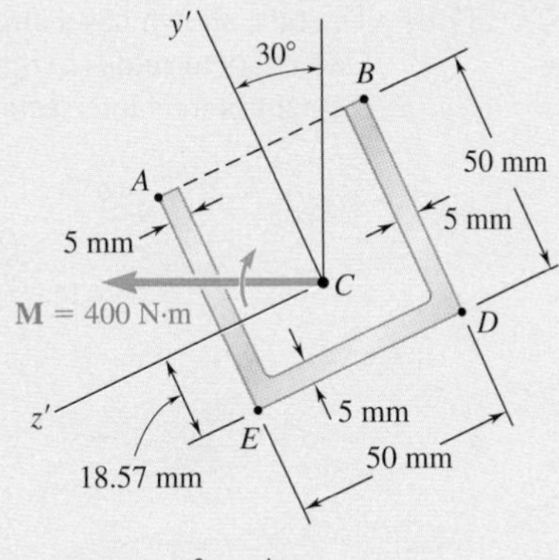

$I_{y'} = 281 \times 10^3 \text{ mm}^4$

$I_{z'} = 176.9 \times 10^3 \text{ mm}^4$

Fig. P11.82

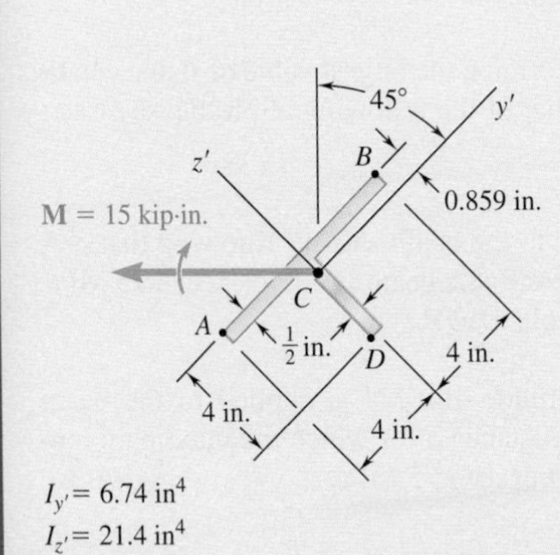

$I_{y'} = 6.74 \text{ in}^4$

$I_{z'} = 21.4 \text{ in}^4$

Fig. P11.83

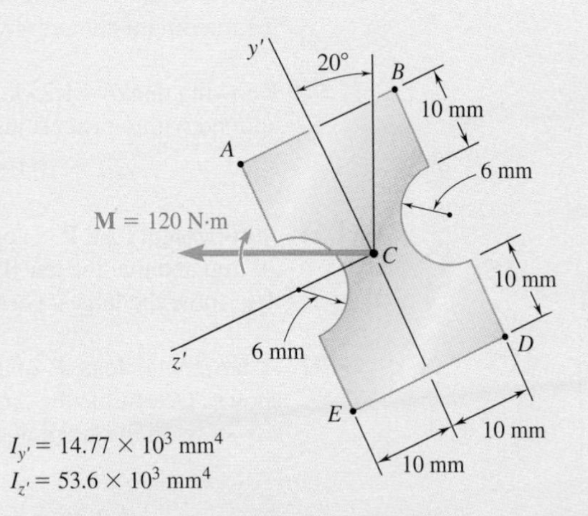

$I_{y'} = 14.77 \times 10^3 \text{ mm}^4$

$I_{z'} = 53.6 \times 10^3 \text{ mm}^4$

Fig. P11.84

11.85 For the loading shown, determine (*a*) the stress at points *A* and *B*, (*b*) the point where the neutral axis intersects line *ABD*.

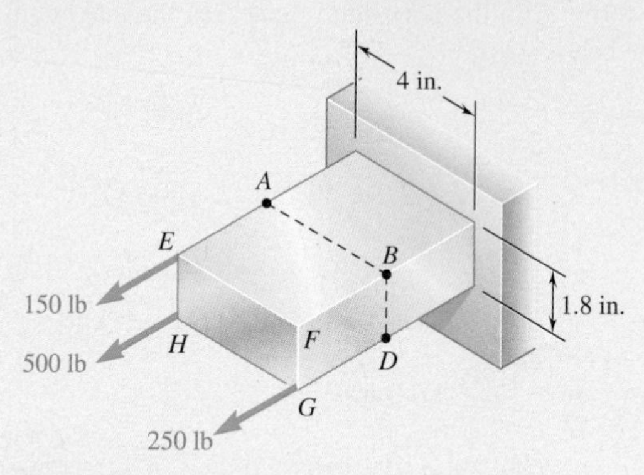

Fig. P11.85

11.86 Solve Prob. 11.85, assuming that the magnitude of the force applied at *G* is increased from 250 lb to 400 lb.

11.87 The tube shown has a uniform wall thickness of 12 mm. For the given loading, determine (*a*) the stress at points *A* and *B*, (*b*) the point where the neutral axis intersects line *ABD*.

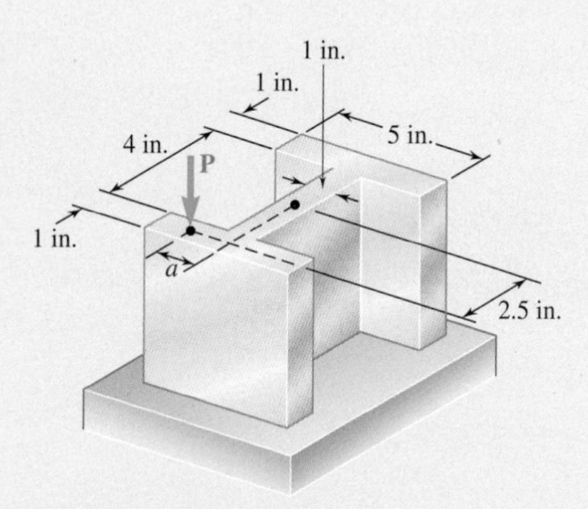

Fig. P11.89 and *P11.90*

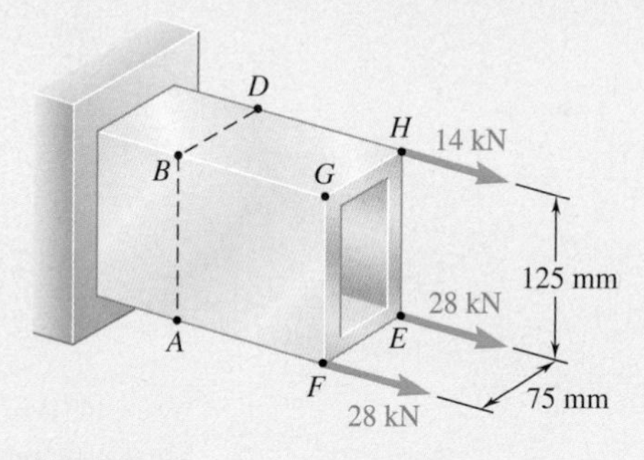

Fig. P11.87

11.88 Solve Prob. 11.87, assuming that the 28-kN force at point *E* is removed.

11.89 Knowing that $P = 90$ kips, determine the largest distance *a* for which the maximum compressive stress does not exceed 18 ksi.

11.90 Knowing that $a = 1.25$ in., determine the largest value of **P** that can be applied without exceeding either of the following allowable stresses:

$$\sigma_{ten} = 10 \text{ ksi} \qquad \sigma_{comp} = 18 \text{ ksi}$$

11.91 A horizontal load **P** is applied to the beam shown. Knowing that $a =$ 20 mm and that the tensile stress in the beam is not to exceed 75 MPa, determine the largest permissible load **P**.

11.92 A horizontal load **P** of magnitude 100 kN is applied to the beam shown. Determine the largest distance *a* for which the maximum tensile stress in the beam does not exceed 75 MPa.

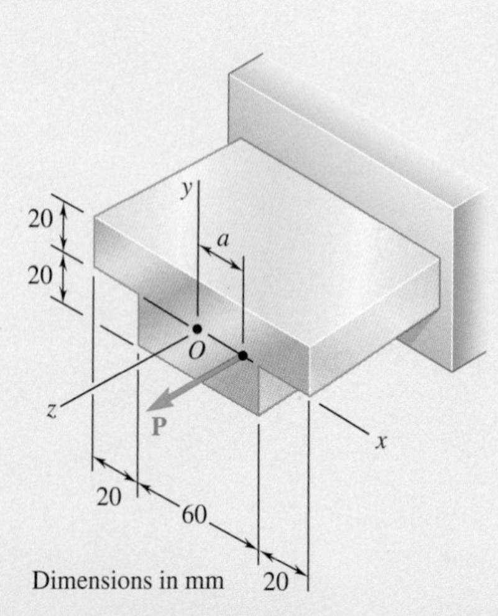

Fig. P11.91 and *P11.92*

Review and Summary

This chapter was devoted to the analysis of members in *pure bending*. The stresses and deformation in members subjected to equal and opposite couples **M** and **M**′ acting in the same longitudinal plane (Fig. 11.41) were studied.

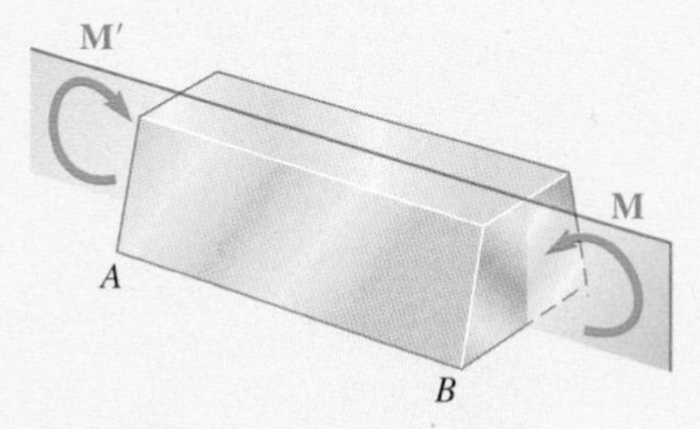

Fig. 11.41

Normal Strain in Bending

In members possessing a plane of symmetry and subjected to couples acting in that plane, it was proven that *transverse sections remain plane* as a member is deformed. A member in pure bending also has a *neutral surface* along which normal strains and stresses are zero. The longitudinal *normal strain* ϵ_x *varies linearly* with the distance y from the neutral surface:

$$\epsilon_x = -\frac{y}{\rho} \quad \textbf{(11.8)}$$

where ρ is the *radius of curvature* of the neutral surface (Fig. 11.42). The intersection of the neutral surface with a transverse section is known as the *neutral axis* of the section.

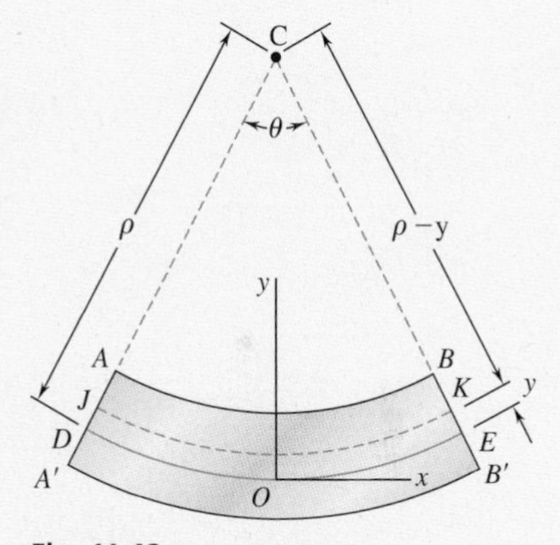

Fig. 11.42

Normal Stress in Elastic Range

For members made of a material that follows Hooke's law, the *normal stress* σ_x *varies linearly* with the distance from the neutral axis (Fig. 11.43). Using the maximum stress σ_m, the normal stress is

$$\sigma_x = -\frac{y}{c}\sigma_m \quad \textbf{(11.12)}$$

where c is the largest distance from the neutral axis to a point in the section.

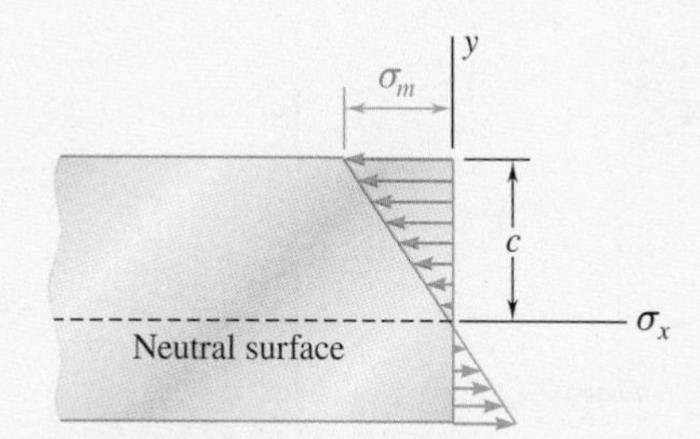

Fig. 11.43

Elastic Flexure Formula

By setting the sum of the elementary forces $\sigma_x\, dA$ equal to zero, we proved that the *neutral axis passes through the centroid* of the cross section of a member in pure bending. Then by setting the sum of the moments of the elementary forces equal to the bending moment, the *elastic flexure formula* is

$$\sigma_m = \frac{Mc}{I} \quad \textbf{(11.15)}$$

where I is the moment of inertia of the cross section with respect to the neutral axis. The normal stress at any distance y from the neutral axis is

$$\sigma_x = -\frac{My}{I} \quad \textbf{(11.16)}$$

Elastic Section Modulus

Noting that I and c depend only on the geometry of the cross section we introduced the *elastic section modulus*

$$S = \frac{I}{c} \tag{11.17}$$

Use the section modulus to write an alternative expression for the maximum normal stress:

$$\sigma_m = \frac{M}{S} \tag{11.18}$$

Curvature of Member

The *curvature* of a member is the reciprocal of its radius of curvature, and may be found by

$$\frac{1}{\rho} = \frac{M}{EI} \tag{11.21}$$

Members Made of Several Materials

We considered the bending of members made of several materials with *different moduli of elasticity*. While transverse sections remain plane, the *neutral axis does not pass through the centroid* of the composite cross section (Fig. 11.44). Using the ratio of the moduli of elasticity of the materials, we obtained a *transformed section* corresponding to an equivalent member made entirely of one material. The methods previously developed are used to determine the stresses in this equivalent homogeneous member (Fig. 11.45), and the ratio of the moduli of elasticity is used to determine the stresses in the composite beam.

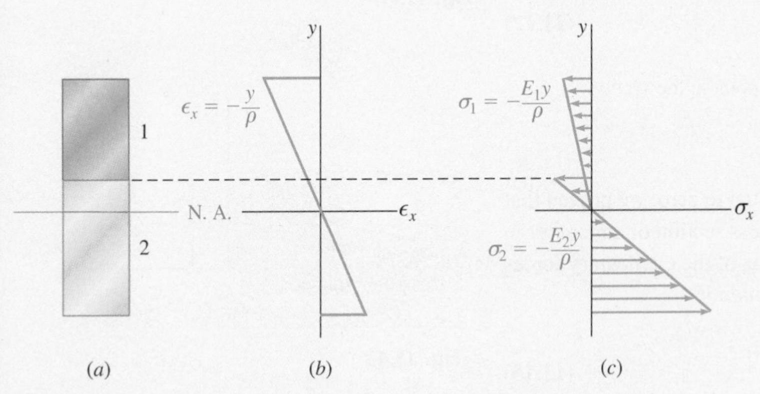

Fig. 11.44

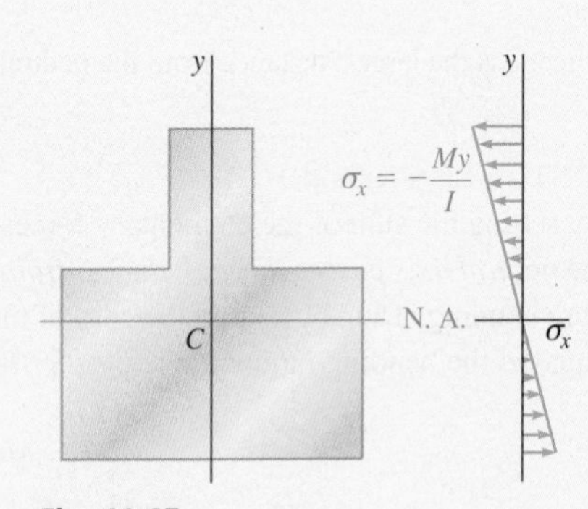

Fig. 11.45

Eccentric Axial Loading

When a member is loaded *eccentrically in a plane of symmetry*, the *eccentric load* is replaced with a force-couple system located at the centroid of the cross section (Fig. 11.46). The stresses from the centric load and the bending couple are superposed (Fig. 11.47):

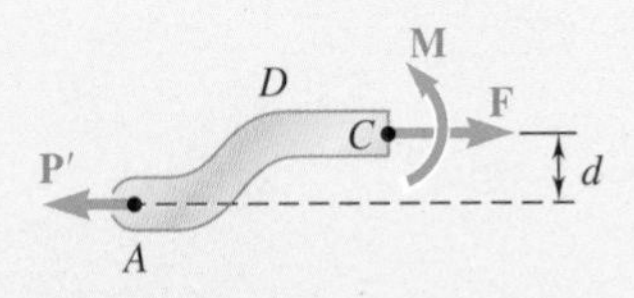

Fig. 11.46

$$\sigma_x = \frac{P}{A} - \frac{My}{I} \tag{11.28}$$

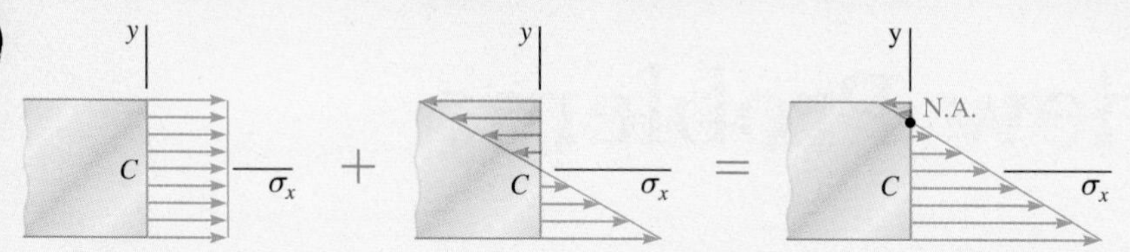

Fig. 11.47

Unsymmetric Bending

For bending of members of *unsymmetric cross section*, the flexure formula may be used, provided that the couple vector **M** is directed along one of the principal centroidal axes of the cross section. When necessary, **M** can be resolved into components along the principal axes, and the stresses superposed due to the component couples (Figs. 11.48 and 11.49).

$$\sigma_x = -\frac{M_z y}{I_z} + \frac{M_y z}{I_y} \tag{11.33}$$

For the couple **M** shown in Fig. 11.50, the orientation of the neutral axis is defined by

$$\tan \phi = \frac{I_z}{I_y} \tan \theta \tag{11.35}$$

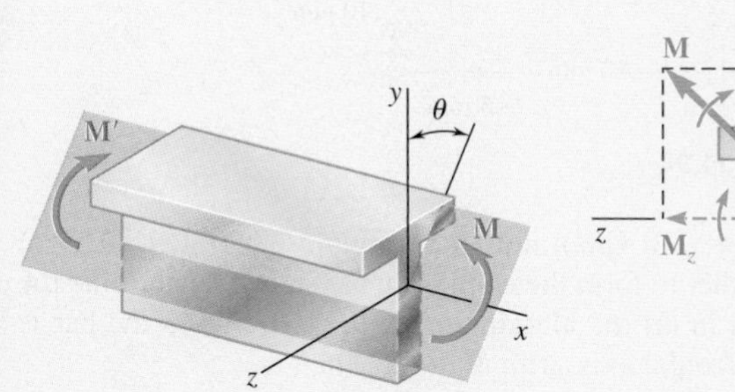

Fig. 11.48

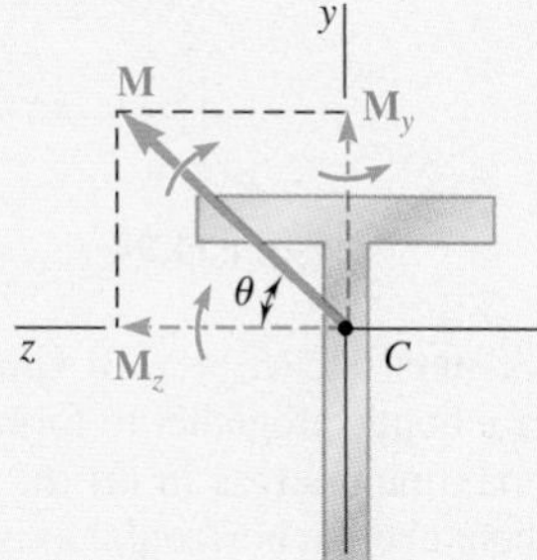

Fig. 11.49

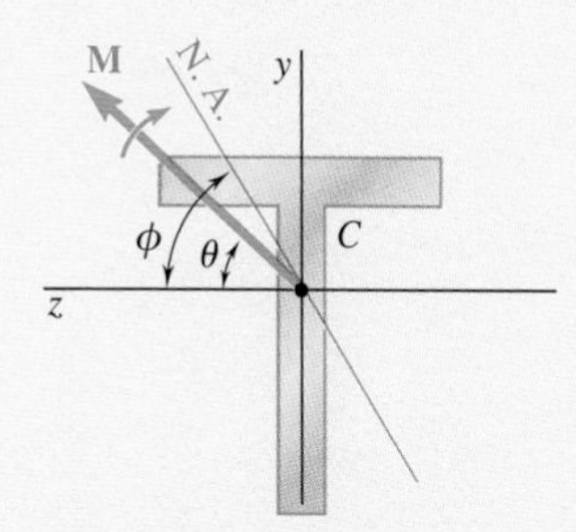

Fig. 11.50

General Eccentric Axial Loading

For the general case of *eccentric axial loading*, the load is replaced by a force-couple system located at the centroid. The stresses are superposed due to the centric load and the two component couples directed along the principal axes:

$$\sigma_x = \frac{P}{A} - \frac{M_z y}{I_z} + \frac{M_y z}{I_y} \tag{11.36}$$

Review Problems

11.93 Knowing that the couple shown acts in a vertical plane, determine the stress at (*a*) point *A*, (*b*) point *B*.

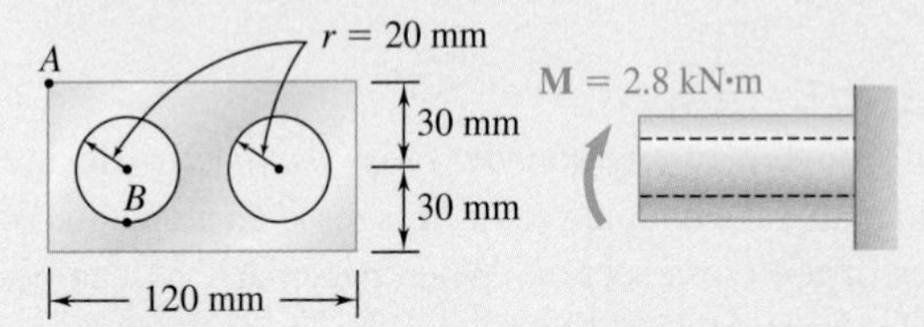

Fig. P11.93

11.94 (*a*) Using an allowable stress of 120 MPa, determine the largest couple **M** that can be applied to a beam of the cross section shown. (*b*) Solve part *a*, assuming that the cross section of the beam is an 80-mm square.

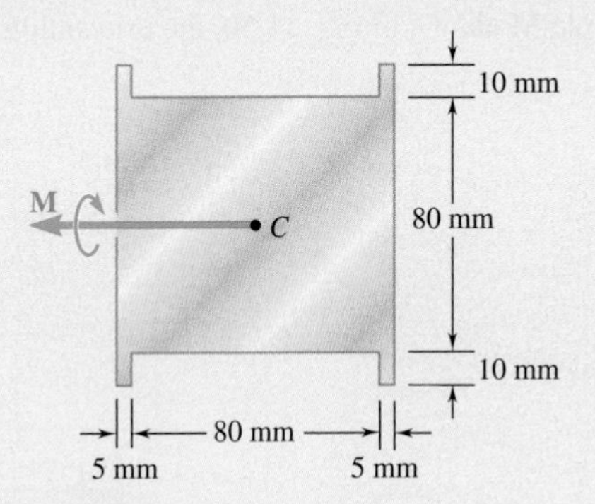

Fig. P11.94

11.95 A steel bar (E_s = 210 GPa) and an aluminum bar (E_a = 70 GPa) are bonded together to form the composite bar shown. Determine the maximum stress in (*a*) the aluminum, (*b*) the steel, when the bar is bent about a horizontal axis, with M = 60 N·m.

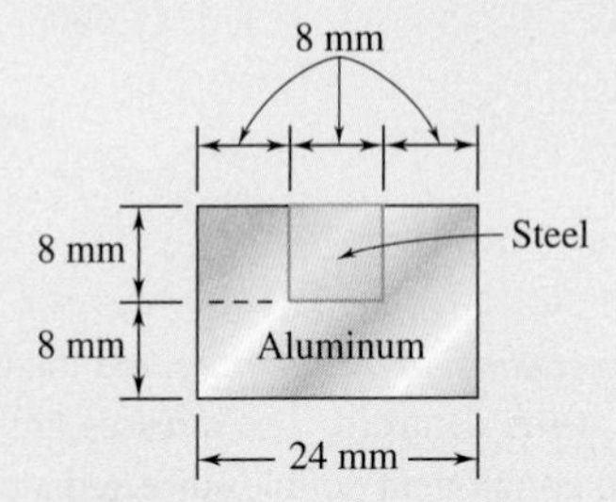

Fig. *P11.95*

11.96 A single vertical force **P** is applied to a short steel post as shown. Gages located at A, B, and C indicate the following strains:

$$\epsilon_A = -500\ \mu \qquad \epsilon_B = -1000\ \mu \qquad \epsilon_C = -200\ \mu$$

Knowing that $E = 29 \times 10^6$ psi, determine (*a*) the magnitude of **P**, (*b*) the line of action of **P**, (*c*) the corresponding strain at the hidden edge of the post, where $x = -2.5$ in. and $z = -1.5$ in.

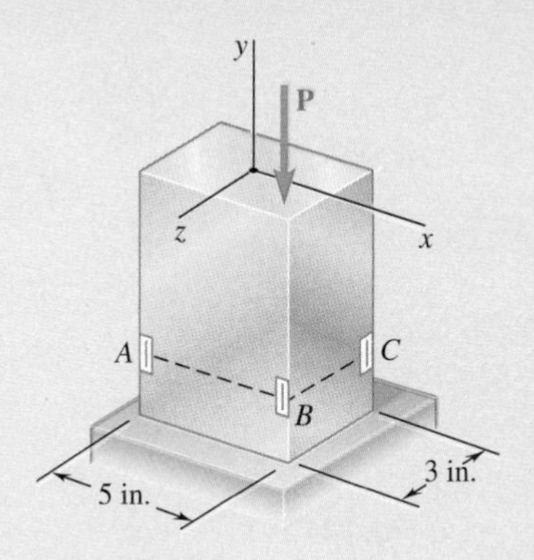

Fig. P11.96

11.97 Two vertical forces are applied to a beam of the cross section shown. Determine the maximum tensile and compressive stresses in portion BC of the beam.

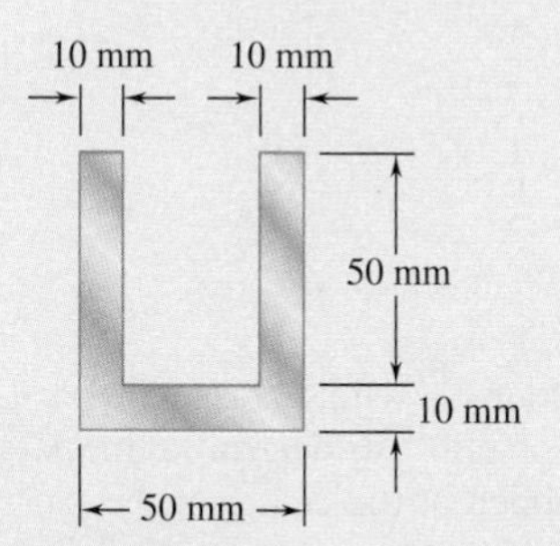

Fig. P11.97

11.98 In order to increase corrosion resistance, a 2-mm-thick cladding of aluminum has been added to a steel bar as shown. The modulus of elasticity is $E = 200$ GPa for steel and $E = 70$ GPa for aluminum. For a bending moment of 300 N·m, determine (*a*) the maximum stress in the steel, (*b*) the maximum stress in the aluminum, (*c*) the radius of curvature of the bar.

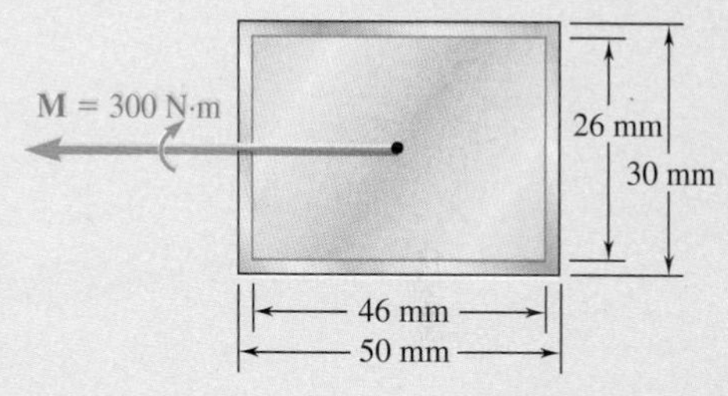

Fig. P11.98

11.99 A 6 × 10-in. timber beam has been strengthened by bolting to it the steel straps shown. The modulus of elasticity is $E = 1.5 \times 10^6$ psi for the wood and $E = 30 \times 10^6$ psi for the steel. Knowing that the beam is bent about a horizontal axis by a couple of moment 200 kip·in., determine the maximum stress in (*a*) the wood, (*b*) the steel.

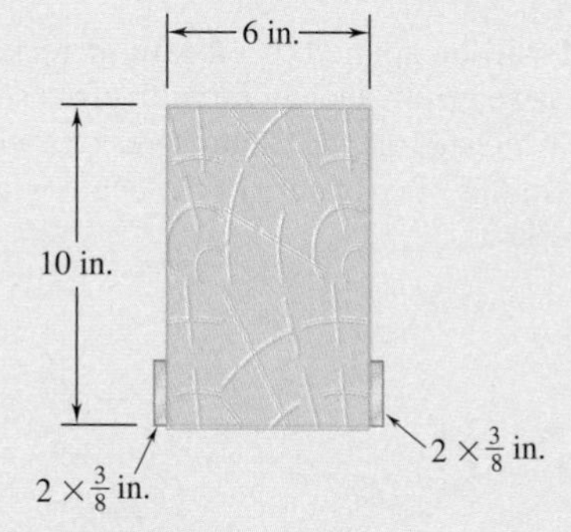

Fig. P11.99

11.100 The four forces shown are applied to a rigid plate supported by a solid steel post of radius a. Knowing that $P = 24$ kips and $a = 1.6$ in., determine the maximum stress in the post when (*a*) the force at D is removed, (*b*) the forces at C and D are removed.

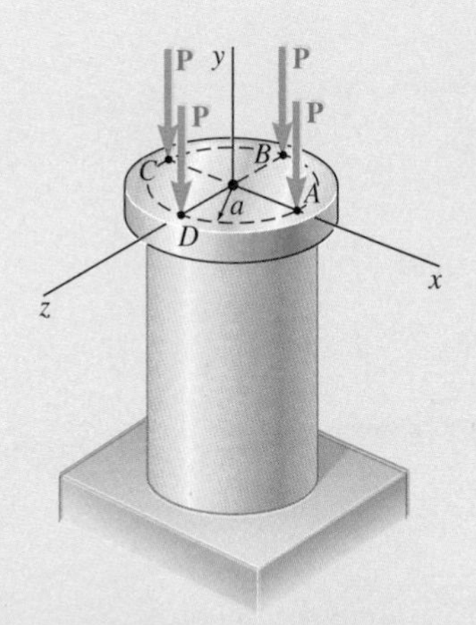

Fig. P11.100

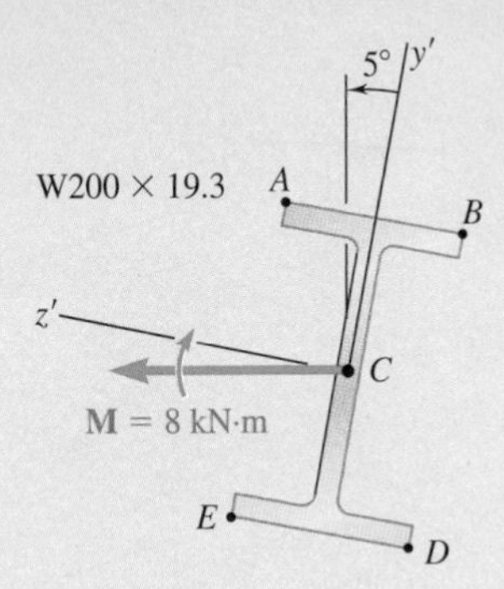

Fig. P11.101

11.101 A couple **M** of moment 8 kN·m acting in a vertical plane is applied to a W200 × 19.3 rolled-steel beam as shown. Determine (*a*) the angle that the neutral axis forms with the horizontal plane, (*b*) the maximum stress in the beam.

11.102 Portions of a $\frac{1}{2} \times \frac{1}{2}$ -in. square bar have been bent to form the two machine components shown. Knowing that the allowable stress is 15 ksi, determine the maximum load that can be applied to each component.

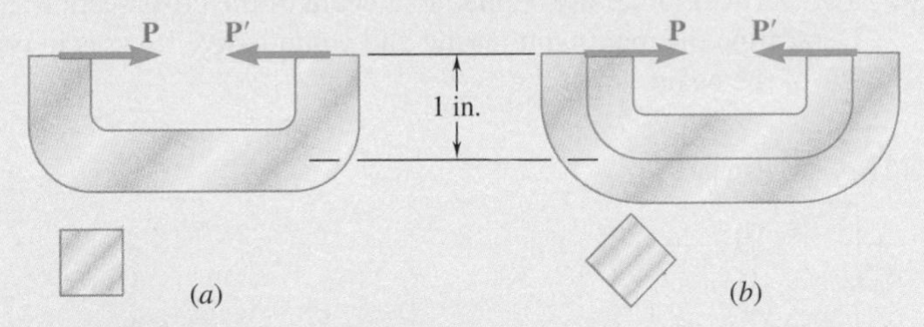

Fig. P11.102

11.103 A short column is made by nailing four 1 × 4-in. planks to a 4 × 4-in. timber. Using an allowable stress of 600 psi, determine the largest compressive load **P** that can be applied at the center of the top section of the timber column as shown if (*a*) the column is as described, (*b*) plank 1 is removed, (*c*) planks 1 and 2 are removed, (*d*) planks 1, 2, and 3 are removed, (*e*) all planks are removed.

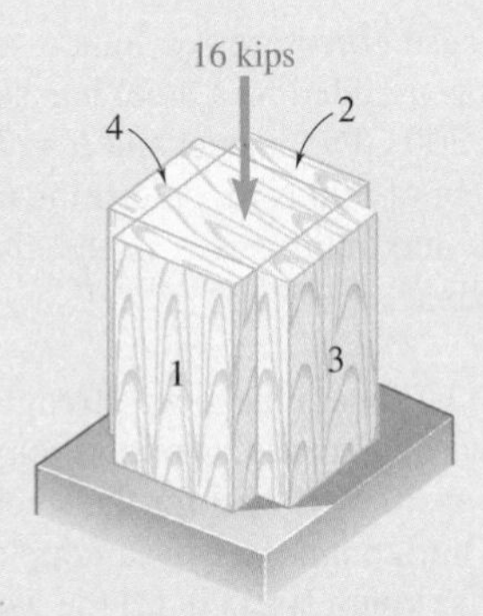

Fig. *P11.103*

11.104 A couple **M** will be applied to a beam of rectangular cross section that is to be sawed from a log of circular cross section. Determine the ratio *d*/*b* for which (*a*) the maximum stress σ_m will be as small as possible, (*b*) the radius of curvature of the beam will be maximum.

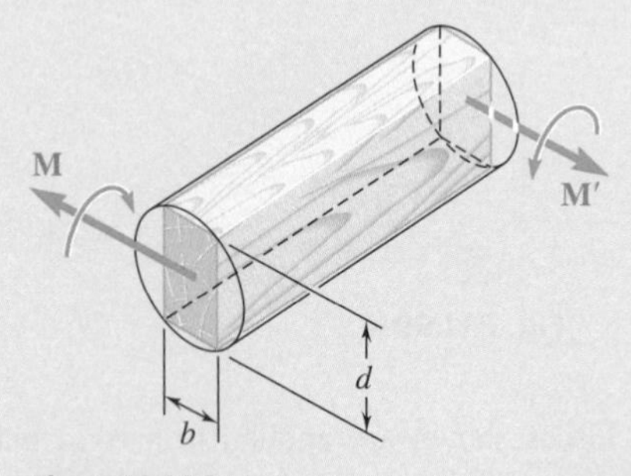

Fig. P11.104

© Digital Vision/Getty Images RF

12

Analysis and Design of Beams for Bending

The beams supporting the overhead crane system are subject to transverse loads, causing the beams to bend. The normal stresses resulting from such loadings will be determined in this chapter.

Objectives

In this chapter, you will:

- **Draw** shear and bending-moment diagrams using static equilibrium applied to sections.
- **Describe** the relationships between applied loads, shear, and bending moments throughout a beam.
- **Use** section modulus to design beams.

Introduction

This chapter and most of the next one are devoted to the analysis and the design of *beams*, which are structural members supporting loads applied at various points along the member. Beams are usually long, straight prismatic members. Steel and aluminum beams play an important part in both structural and mechanical engineering. Timber beams are widely used in home construction (Photo 12.1). In most cases, the loads are perpendicular to the axis of the beam. This *transverse loading* causes only bending and shear in the beam. When the loads are not at a right angle to the beam, they also produce axial forces in the beam.

Photo 12.1 Timber beams used in a residential dwelling.
© *Huntstock/agefotostock RF*

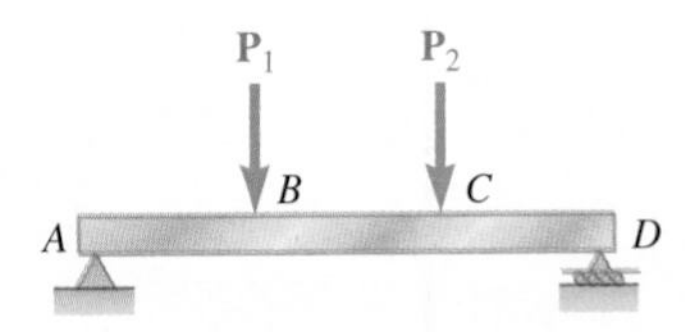

(*a*) Concentrated loads

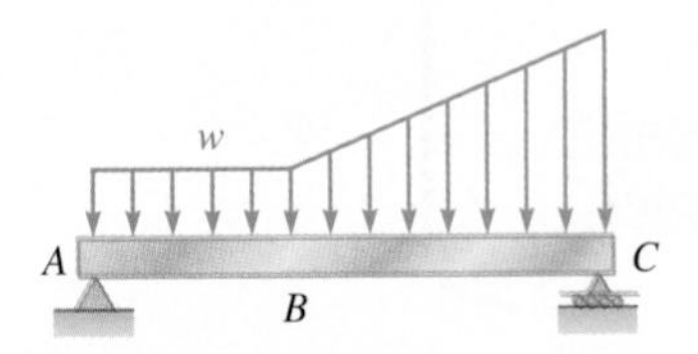

(*b*) Distributed loads

Fig. 12.1 Transversely loaded beams.

The transverse loading of a beam may consist of *concentrated loads* $\mathbf{P}_1, \mathbf{P}_2, \ldots$ expressed in newtons, pounds, or their multiples of kilonewtons and kips (Fig. 12.1*a*); of a *distributed load* w expressed in N/m, kN/m, lb/ft, or kips/ft (Fig. 12.1*b*); or of a combination of both. When the load w per unit length has a constant value over part of the beam (as between A and B in Fig. 12.1*b*), the load is *uniformly distributed.*

Beams are classified according to the way they are supported, as shown in Fig. 12.2. The distance L is called the *span*. Note that the reactions at the supports of the beams in Fig. 12.2 *a*, *b*, and *c* involve a total of only three unknowns and can be determined by the methods of statics. Such beams are said to be *statically determinate*. On the other hand, the reactions at the supports of the beams in Fig. 12.2 *d*, *e*, and *f* involve more than three unknowns and cannot be determined by the methods of statics alone. The properties of the beams with regard to their resistance to deformations must

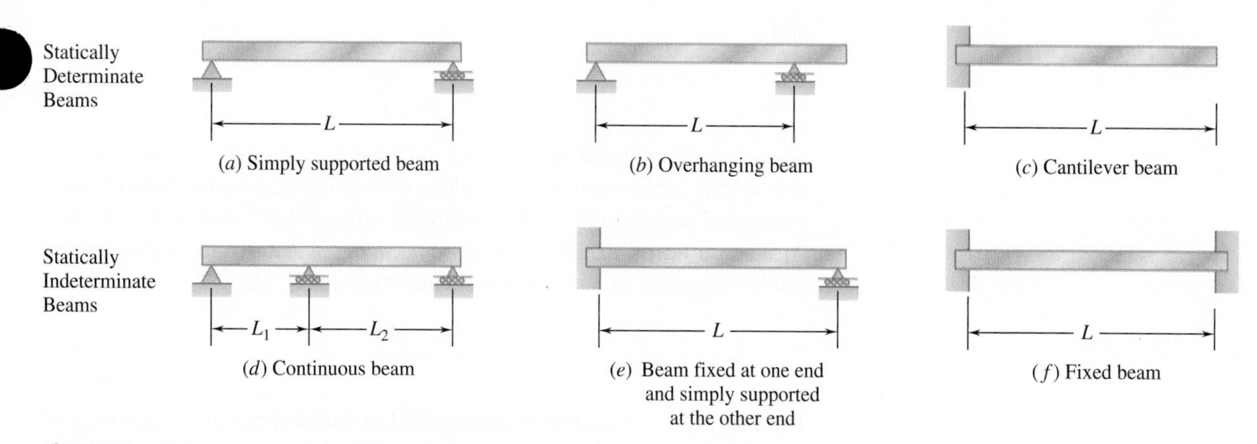

Fig. 12.2 Common beam support configurations.

be taken into consideration. Such beams are said to be *statically indeterminate*, and their analysis will be discussed in Chap. 15.

Sometimes two or more beams are connected by hinges to form a single continuous structure. Two examples of beams hinged at a point H are shown in Fig. 12.3. Note that the reactions at the supports involve four unknowns and cannot be determined from the free-body diagram of the two-beam system. They can be determined by recognizing that the internal moment at the hinge is zero. Then, after considering the free-body diagram of each beam separately, six unknowns are involved (including two force components at the hinge), and six equations are available.

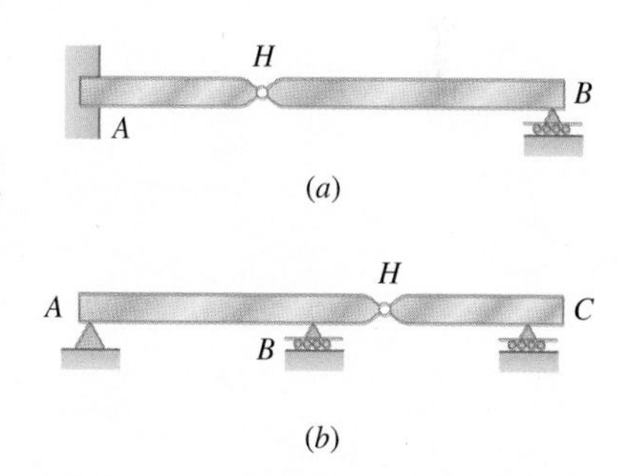

Fig. 12.3 Beams connected by hinges.

When a beam is subjected to transverse loads, the internal forces in any section of the beam consist of a shear force **V** and a bending couple **M**. For example, a simply supported beam AB is carrying two concentrated loads and a uniformly distributed load (Fig. 12.4*a*). To determine the internal forces in a section through point C, draw the free-body diagram of the entire beam to obtain the reactions at the supports (Fig. 12.4*b*). Passing a section through C, then draw the free-body diagram of AC (Fig. 12.4*c*), from which the shear force **V** and the bending couple **M** are found.

The bending couple **M** creates *normal stresses* in the cross section, while the shear force **V** creates *shearing stresses*. In most cases, the dominant criterion in the design of a beam for strength is the maximum value of the normal stress in the beam. The normal stresses in a beam are the subject of this chapter, while shearing stresses are discussed in Chap. 13.

Since the distribution of the normal stresses in a given section depends only upon the bending moment M and the geometry of the section,[†] the elastic flexure formulas derived in Sec. 11.2 are used to determine the maximum stress, σ_m, as well as the stress at any given point on the cross section, σ_x:[‡]

$$\sigma_m = \frac{|M|c}{I} \tag{12.1}$$

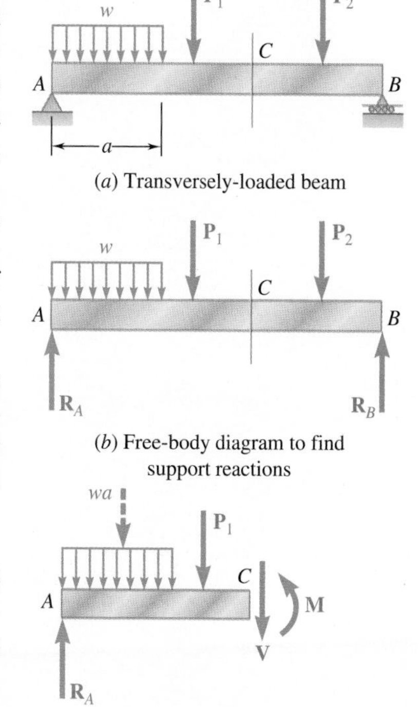

Fig. 12.4 Analysis of a simply supported beam.

[†]It is assumed that the distribution of the normal stresses in a given cross section is not affected by the deformations caused by the shearing stresses.

[‡]Recall from Sec. 11.1 that M can be positive or negative, depending upon whether the concavity of the beam at the point considered faces upward or downward. Thus, in a transverse loading the sign of M can vary along the beam. On the other hand, since σ_m is a positive quantity, the absolute value of M is used in Eq. (12.1).

and

$$\sigma_x = -\frac{My}{I} \tag{12.2}$$

where I is the moment of inertia of the cross section with respect to a centroidal axis perpendicular to the plane of the couple, y is the distance from the neutral surface, and c is the maximum value of that distance (Fig. 11.11). Also recall from Sec. 11.2 that the maximum value σ_m of the normal stress can be expressed in terms of the section modulus S. Thus

$$\sigma_m = \frac{|M|}{S} \tag{12.3}$$

The fact that σ_m is inversely proportional to S underlines the importance of selecting beams with a large section modulus. Section moduli of various rolled-steel shapes are given in Appendix B, while the section modulus of a rectangular shape is

$$S = \tfrac{1}{6}bh^2 \tag{12.4}$$

where b and h are, respectively, the width and the depth of the cross section.

Eq. (12.3) also shows that for a beam of uniform cross section, σ_m is proportional to $|M|$. Thus, the maximum value of the normal stress in the beam occurs in the section where $|M|$ is largest. One of the most important parts of the design of a beam for a given loading condition is the determination of the location and magnitude of the largest bending moment.

This task is made easier if a *bending-moment diagram* is drawn, where the bending moment M is determined at various points of the beam and plotted against the distance x measured from one end. It is also easier if a *shear diagram* is drawn by plotting the shear V against x. The sign convention used to record the values of the shear and bending moment is discussed in Sec. 12.1.

In Sec. 12.2 relationships between load, shear, and bending moments are derived and used to obtain the shear and bending-moment diagrams. This approach facilitates the determination of the largest absolute value of the bending moment and the maximum normal stress in the beam.

In Sec. 12.3 beams are designed for bending such that the maximum normal stress in these beams will not exceed their allowable values.

12.1 SHEAR AND BENDING-MOMENT DIAGRAMS

The maximum absolute values of the shear and bending moment in a beam are easily found if V and M are plotted against the distance x measured from one end of the beam. Besides, as you will see in Chap. 15, the knowledge of M as a function of x is essential to determine the deflection of a beam.

In this section, the shear and bending-moment diagrams are obtained by determining the values of V and M at selected points of the beam. These values are found by passing a section through the point to be determined (Fig. 12.5*a*) and considering the equilibrium of the portion of beam located on either side of the section (Fig. 12.5*b*). Since the shear forces **V** and **V′**

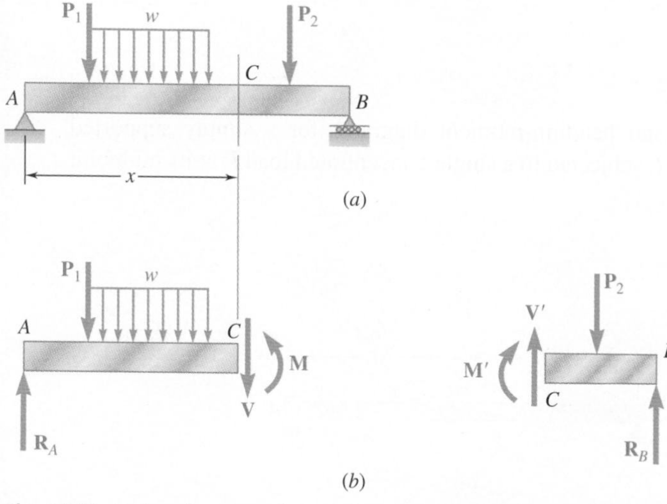

Fig. 12.5 Determination of shear force, V, and bending moment, M, at a given section. (*a*) Loaded beam with section indicated at arbitrary position x. (*b*) Free-body diagrams drawn to the left and right of the section at C.

have opposite senses, recording the shear at point C with an up or down arrow is meaningless, unless it is indicated at the same time which of the free bodies AC and CB is being considered. For this reason, the shear V is recorded with a *plus sign* if the shear forces are directed as in Fig. 12.5*b* and a *minus sign* otherwise. A similar convention is applied for the bending moment M.† Summarizing the sign conventions:

The shear V and the bending moment M at a given point of a beam are positive when the internal forces and couples acting on each portion of the beam are directed as shown in Fig. 12.6a.

1. *The shear at any given point of a beam is positive when the* **external** *forces (loads and reactions) acting on the beam tend to shear off the beam at that point as indicated in Fig. 12.6b.*
2. *The bending moment at any given point of a beam is positive when the* **external** *forces acting on the beam tend to bend the beam at that point as indicated in Fig. 12.6c.*

It is helpful to note that the values of the shear and of the bending moment are positive in the left half of a simply supported beam carrying a single concentrated load at its midpoint, as is discussed in the following Concept Application.

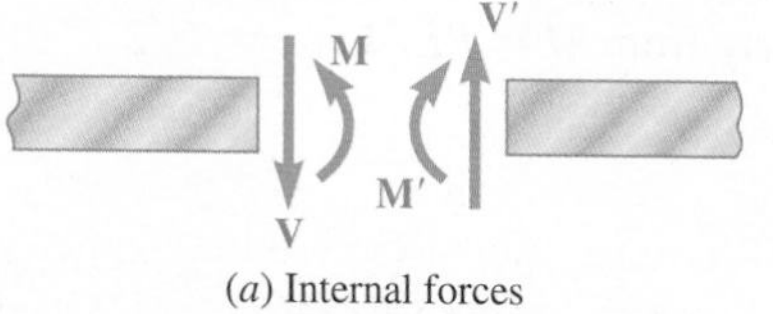

(*a*) Internal forces
(positive shear and positive bending moment)

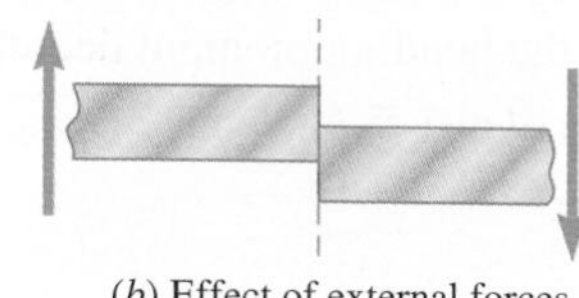

(*b*) Effect of external forces
(positive shear)

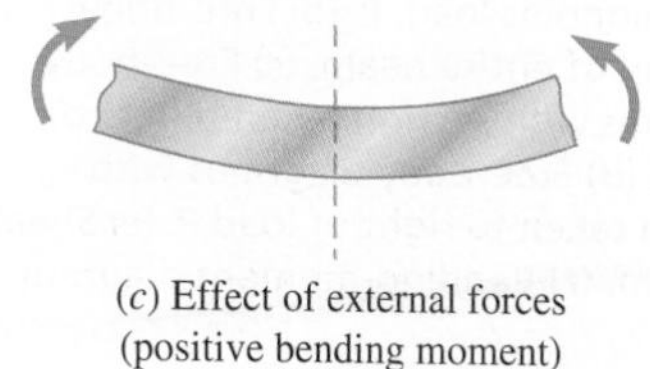

(*c*) Effect of external forces
(positive bending moment)

Fig. 12.6 Sign convention for shear and bending moment.

†This convention is the same that we used earlier in Sec. 11.1.

Concept Application 12.1

Draw the shear and bending-moment diagrams for a simply supported beam AB of span L subjected to a single concentrated load **P** at its midpoint C (Fig. 12.7*a*).

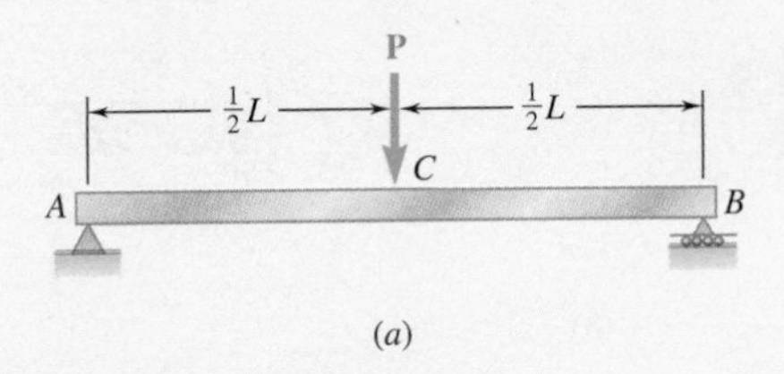

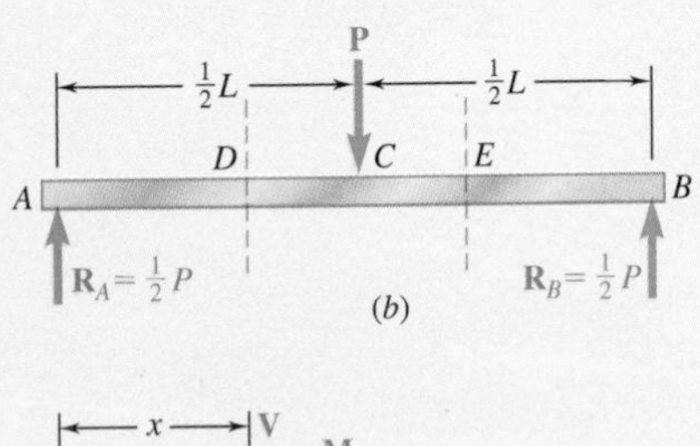

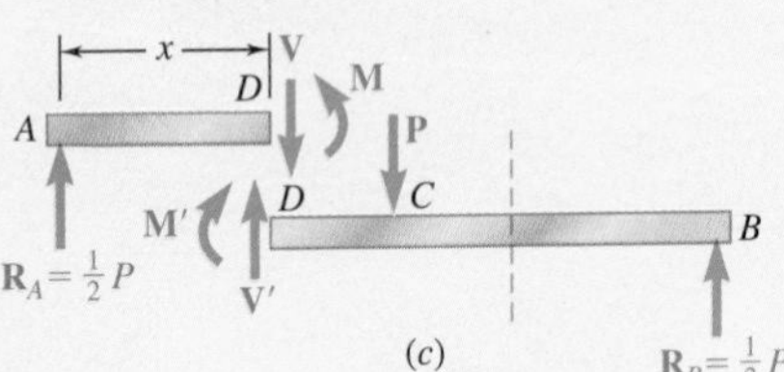

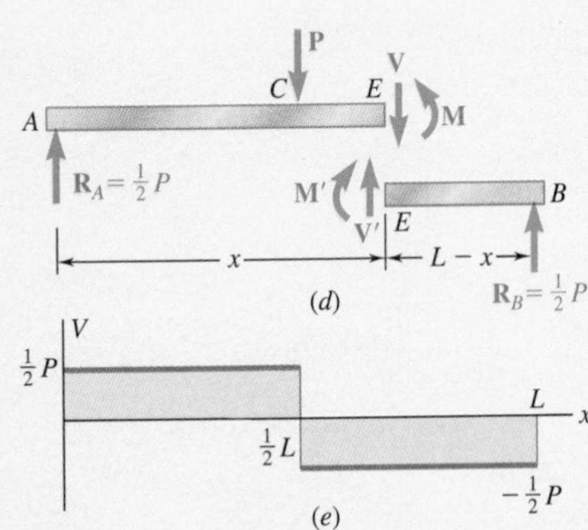

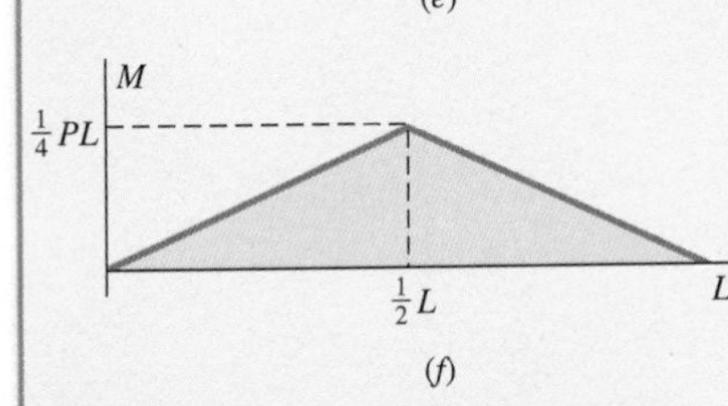

Fig. 12.7 (*a*) Simply supported beam with midpoint load, **P.** (*b*) Free-body diagram of entire beam. (*c*) Free-body diagrams with section taken to left of load **P.** (*d*) Free-body diagrams with section taken to right of load **P.** (*e*) Shear diagram. (*f*) Bending-moment diagram.

Determine the reactions at the supports from the free-body diagram of the entire beam (Fig. 12.7*b*). The magnitude of each reaction is equal to $P/2$.

Next cut the beam at a point D between A and C and draw the free-body diagrams of AD and DB (Fig. 12.7*c*). *Assuming that the shear and bending moment are positive*, we direct the internal forces **V** and **V′** and the internal couples **M** and **M′** as in Fig. 12.6*a*. Consider the free body AD. The sum of the vertical components and the sum of the moments about D of the forces acting on the free body are zero, so $V = +P/2$ and $M = +Px/2$. Both the shear and the bending moment are positive. This is checked by observing that the reaction at A tends to shear off and bend the beam at D as indicated in Figs. 12.6*b* and *c*. We plot V and M between A and C (Figs. 12.7*d* and *e*). The shear has a constant value $V = P/2$, while the bending moment increases linearly from $M = 0$ at $x = 0$ to $M = PL/4$ at $x = L/2$.

Cutting the beam at a point E between C and B and considering the free body EB (Fig. 12.7*d*), the sum of the vertical components and the sum of the moments about E of the forces acting on the free body are zero. Obtain $V = -P/2$ and $M = P(L - x)/2$. Therefore, the shear is negative, and the bending moment positive. This is checked by observing that the reaction at B bends the beam at E as in Fig. 12.6*c* but tends to shear it off in a manner opposite to that shown in Fig. 12.6*b*. The shear and bending-moment diagrams of Figs. 12.7*e* and *f* are completed by showing the shear with a constant value $V = -P/2$ between C and B, while the bending moment decreases linearly from $M = PL/4$ at $x = L/2$ to $M = 0$ at $x = L$.

Note from the previous Concept Application that when a beam is subjected only to concentrated loads, the shear is constant between loads and the bending moment varies linearly between loads. In such situations, the shear and bending-moment diagrams can be drawn easily once the values of V and M have been obtained at sections selected just to the left and just to the right of the points where the loads and reactions are applied (see Sample Prob. 12.1).

Concept Application 12.2

Draw the shear and bending-moment diagrams for a cantilever beam AB of span L supporting a uniformly distributed load w (Fig. 12.8a).

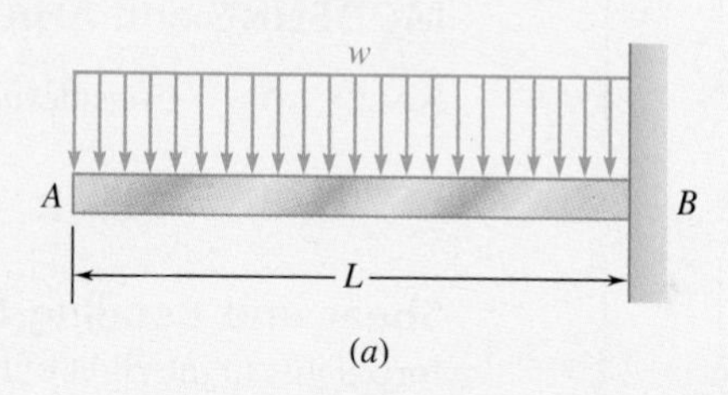

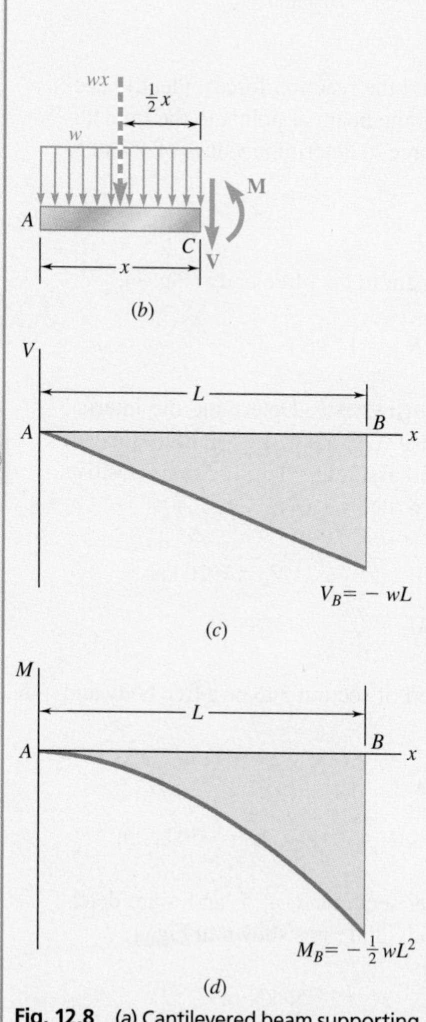

Fig. 12.8 (a) Cantilevered beam supporting a uniformly distributed load. (b) Free-body diagram of section AC. (c) Shear diagram. (d) Bending-moment diagram.

Cut the beam at a point C, located between A and B, and draw the free-body diagram of AC (Fig. 12.8b), directing **V** and **M** as in Fig. 12.6a. Using the distance x from A to C and replacing the distributed load over AC by its resultant wx applied at the midpoint of AC, write

$$+\uparrow \Sigma F_y = 0: \qquad -wx - V = 0 \qquad V = -wx$$

$$+\circlearrowleft \Sigma M_C = 0: \qquad wx\left(\frac{x}{2}\right) + M = 0 \qquad M = -\frac{1}{2}wx^2$$

Note that the shear diagram is represented by an oblique straight line (Fig. 12.8c) and the bending-moment diagram by a parabola (Fig. 12.8d). The maximum values of V and M both occur at B, where

$$V_B = -wL \qquad M_B = -\tfrac{1}{2}wL^2$$

Sample Problem 12.1

For the timber beam and loading shown, draw the shear and bending-moment diagrams and determine the maximum normal stress due to bending.

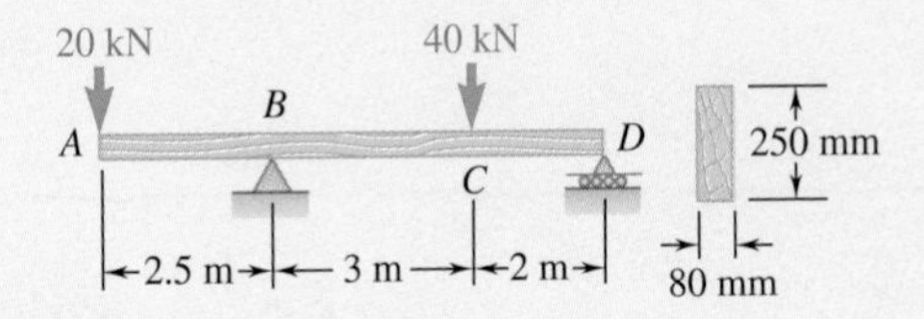

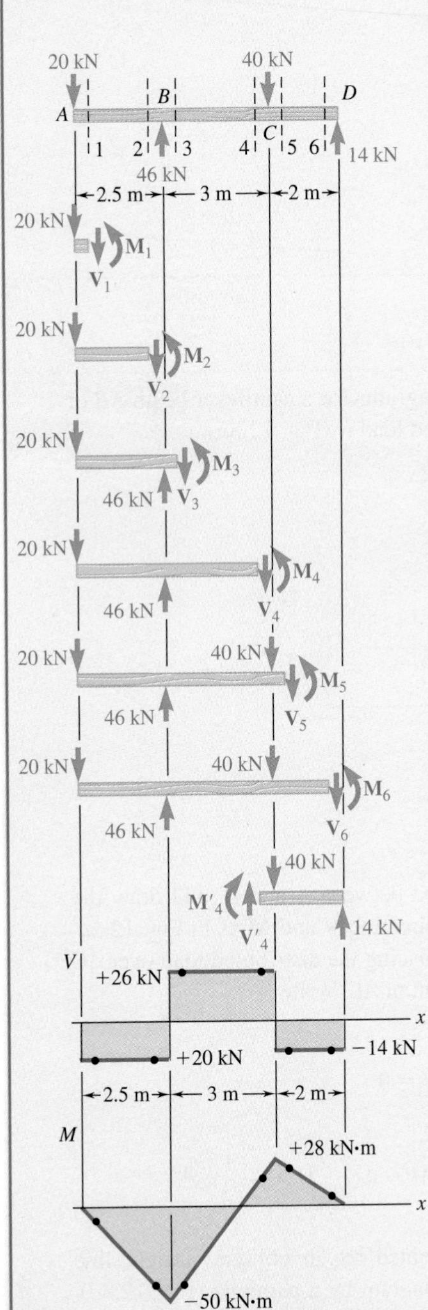

Fig. 1 Free-body diagram of beam, free-body diagrams of sections to left of cut, shear diagram, bending-moment diagram.

STRATEGY: After using statics to find the reaction forces, identify sections to be analyzed. You should section the beam at points to the immediate left and right of each concentrated force to determine values of V and M at these points.

MODELING and ANALYSIS:

Reactions. Considering the entire beam to be a free body (Fig. 1),

$$\mathbf{R}_B = 40 \text{ kN}\uparrow \qquad \mathbf{R}_D = 14 \text{ kN}\uparrow$$

Shear and Bending-Moment Diagrams. Determine the internal forces just to the right of the 20-kN load at A. Considering the stub of beam to the left of section *1* as a free body and assuming V and M to be positive (according to the standard convention), write

$$+\uparrow\Sigma F_y = 0: \quad -20 \text{ kN} - V_1 = 0 \qquad V_1 = -20 \text{ kN}$$

$$+\circlearrowleft\Sigma M_1 = 0: \quad (20 \text{ kN})(0 \text{ m}) + M_1 = 0 \qquad M_1 = 0$$

Next consider the portion to the left of section *2* to be a free body and write

$$+\uparrow\Sigma F_y = 0: \quad -20 \text{ kN} - V_2 = 0 \qquad V_2 = -20 \text{ kN}$$

$$+\circlearrowleft\Sigma M_2 = 0: \quad (20 \text{ kN})(2.5 \text{ m}) + M_2 = 0 \qquad M_2 = -50 \text{ kN·m}$$

The shear and bending moment at sections *3*, *4*, *5*, and *6* are determined in a similar way from the free-body diagrams shown in Fig. 1:

$$V_3 = +26 \text{ kN} \qquad M_3 = -50 \text{ kN·m}$$

$$V_4 = +26 \text{ kN} \qquad M_4 = +28 \text{ kN·m}$$

$$V_5 = -14 \text{ kN} \qquad M_5 = +28 \text{ kN·m}$$

$$V_6 = -14 \text{ kN} \qquad M_6 = 0$$

(continued)

For several of the latter sections, the results may be obtained more easily by considering the portion to the right of the section to be a free body. For example, for the portion of beam to the right of section 4,

$$+\uparrow\Sigma F_y = 0: \qquad V_4 - 40\text{ kN} + 14\text{ kN} = 0 \qquad V_4 = +26\text{ kN}$$

$$+\circlearrowleft\Sigma M_4 = 0: \qquad -M_4 + (14\text{ kN})(2\text{ m}) = 0 \qquad M_4 = +28\text{ kN·m}$$

Now plot the six points shown on the shear and bending-moment diagrams. As indicated earlier, the shear is of constant value between concentrated loads, and the bending moment varies linearly.

Maximum Normal Stress. This occurs at B, where $|M|$ is largest. Use Eq. (12.4) to determine the section modulus of the beam:

$$S = \tfrac{1}{6}bh^2 = \tfrac{1}{6}(0.080\text{ m})(0.250\text{ m})^2 = 833.33 \times 10^{-6}\text{ m}^3$$

Substituting this value and $|M| = |M_B| = 50 \times 10^3$ N·m into Eq. (12.3) gives

$$\sigma_m = \frac{|M_B|}{S} = \frac{(50 \times 10^3\text{ N·m})}{833.33 \times 10^{-6}} = 60.00 \times 10^6\text{ Pa}$$

Maximum normal stress in the beam = 60.0 MPa ◀

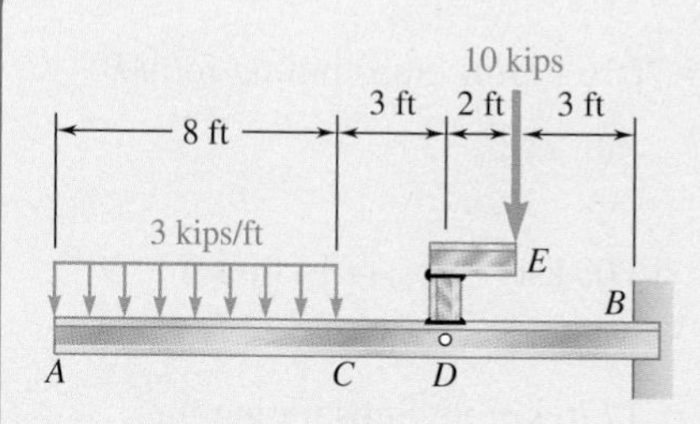

Sample Problem 12.2

The structure shown consists of a W10 × 112 rolled-steel beam AB and two short members welded together and to the beam. (*a*) Draw the shear and bending-moment diagrams for the beam and the given loading. (*b*) Determine the maximum normal stress in sections just to the left and just to the right of point D.

STRATEGY: You should first replace the 10-kip load with an equivalent force-couple system at D. You can section the beam within each region of continuous load (including regions of no load) and find equations for the shear and bending moment.

MODELING and ANALYSIS:

Equivalent Loading of Beam. The 10-kip load is replaced by an equivalent force-couple system at D. The reaction at B is determined by considering the beam to be free body (Fig. 1).

(continued)

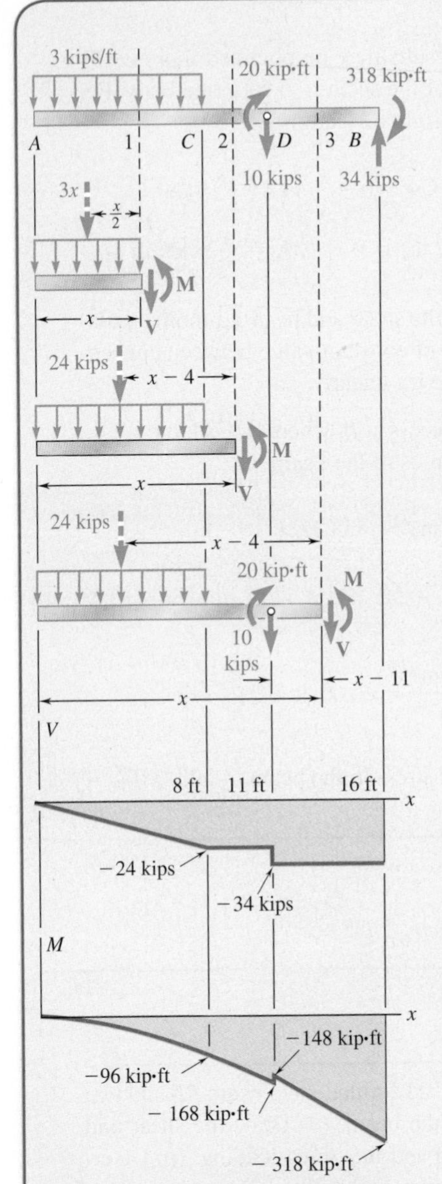

Fig. 1 Free-body diagram of beam, free-body diagrams of sections to left of cut, shear diagram, bending-moment diagram.

a. Shear and Bending-Moment Diagrams

From A to C. Determine the internal forces at a distance x from point A by considering the portion of beam to the left of section *1*. That part of the distributed load acting on the free body is replaced by its resultant, and

$$+\uparrow\Sigma F_y = 0: \qquad -3x - V = 0 \qquad V = -3x \text{ kips}$$

$$+\circlearrowleft\Sigma M_1 = 0: \qquad 3x(\tfrac{1}{2}x) + M = 0 \qquad M = -1.5x^2 \text{ kip·ft}$$

Since the free-body diagram shown in Fig. 1 can be used for all values of x smaller than 8 ft, the expressions obtained for V and M are valid in the region $0 < x < 8$ ft.

From C to D. Considering the portion of beam to the left of section 2 and again replacing the distributed load by its resultant,

$$+\uparrow\Sigma F_y = 0: \qquad -24 - V = 0 \qquad V = -24 \text{ kips}$$

$$+\circlearrowleft\Sigma M_2 = 0: \qquad 24(x - 4) + M = 0 \qquad M = 96 - 24x \quad \text{kip·ft}$$

These expressions are valid in the region 8 ft $< x <$ 11 ft.

From D to B. Using the position of beam to the left of section *3*, the region 11 ft $< x <$ 16 ft is

$$V = -34 \text{ kips} \qquad M = 226 - 34x \quad \text{kip·ft}$$

The shear and bending-moment diagrams for the entire beam now can be plotted. Note that the couple of moment 20 kip·ft applied at point D introduces a discontinuity into the bending-moment diagram.

b. Maximum Normal Stress to the Left and Right of Point D.

From Appendix B for the W10 × 112 rolled-steel shape, $S = 126 \text{ in}^3$ about the X-X axis.

To the left of D: $|M| = 168$ kip·ft $= 2016$ kip·in. Substituting for $|M|$ and S into Eq. (12.3), write

$$\sigma_m = \frac{|M|}{S} = \frac{2016 \text{ kip·in.}}{126 \text{ in}^3} = 16.00 \text{ ksi} \qquad \sigma_m = 16.00 \text{ ksi} \blacktriangleleft$$

To the right of D: $|M| = 148$ kip·ft $= 1776$ kip·in. Substituting for $|M|$ and S into Eq. (12.3), write

$$\sigma_m = \frac{|M|}{S} = \frac{1776 \text{ kip·in.}}{126 \text{ in}^3} = 14.10 \text{ ksi} \qquad \sigma_m = 14.10 \text{ ksi} \blacktriangleleft$$

REFLECT and THINK: It was not necessary to determine the reactions at the right end to draw the shear and bending-moment diagrams. However, having determined these at the start of the solution, they can be used as checks of the values at the right end of the shear and bending-moment diagrams.

Problems

12.1 through 12.4 For the beam and loading shown, (*a*) draw the shear and bending-moment diagrams, (*b*) determine the equations of the shear and bending-moment curves.

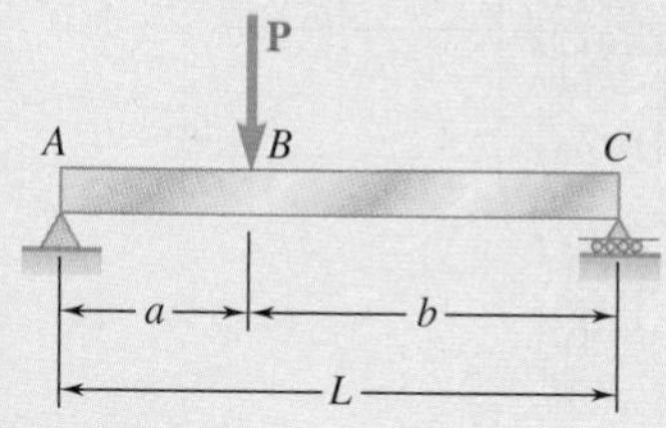

Fig. P12.1

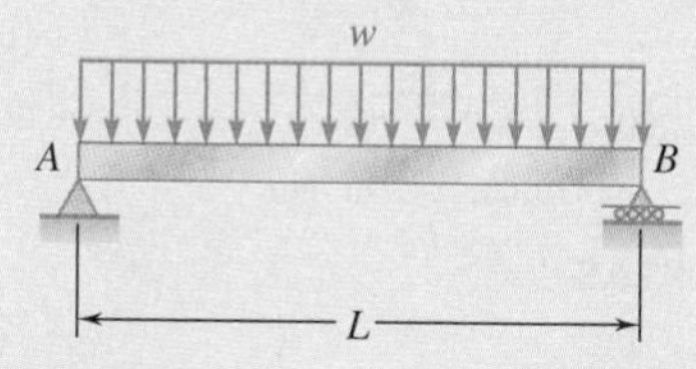

Fig. P12.2

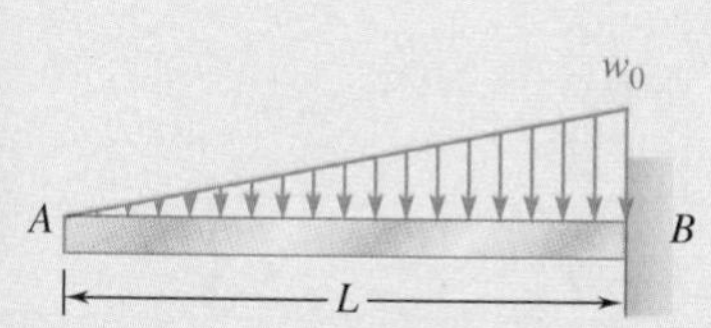

Fig. P12.3

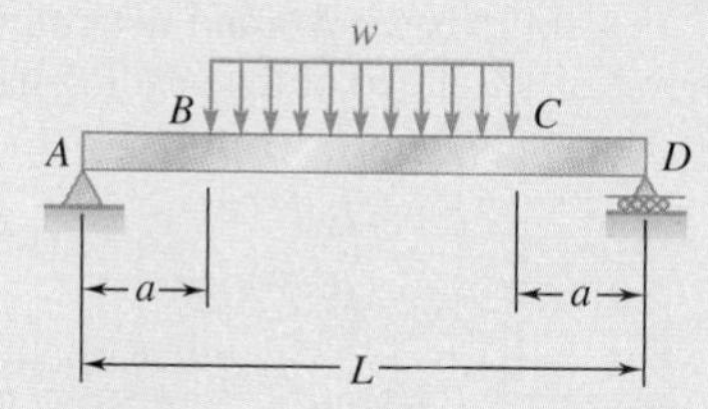

Fig. P12.4

12.5 and *12.6* Draw the shear and bending-moment diagrams for the beam and loading shown, and determine the maximum absolute value (*a*) of the shear, (*b*) of the bending moment.

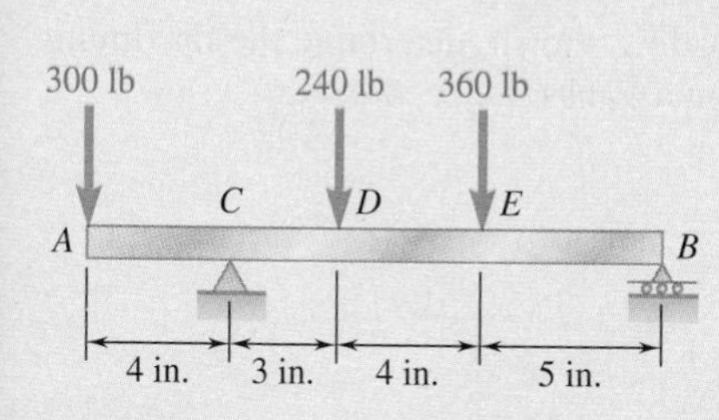

Fig. P12.5

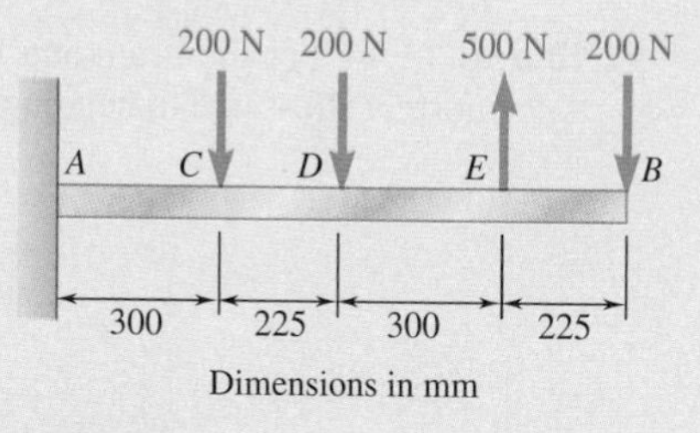

Fig. *P12.6*

12.7 and *12.8* Draw the shear and bending-moment diagrams for the beam and loading shown, and determine the maximum absolute value (*a*) of the shear, (*b*) of the bending moment.

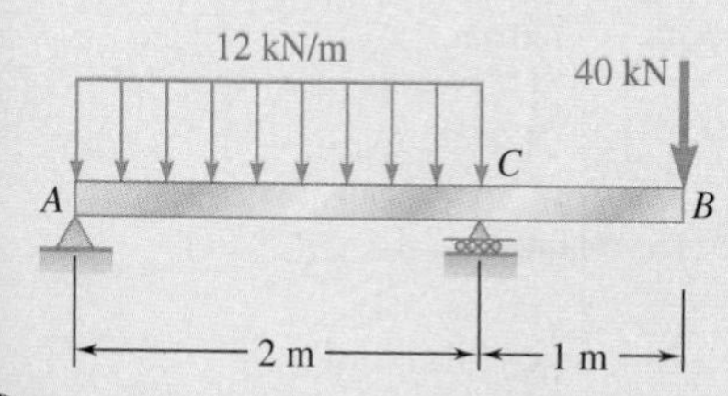

Fig. P12.7

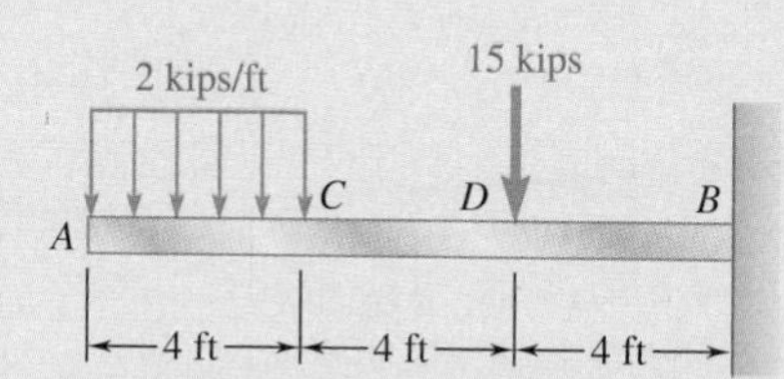

Fig. *P12.8*

12.9 and 12.10 Draw the shear and bending-moment diagrams for the beam and loading shown, and determine the maximum absolute value (*a*) of the shear, (*b*) of the bending moment.

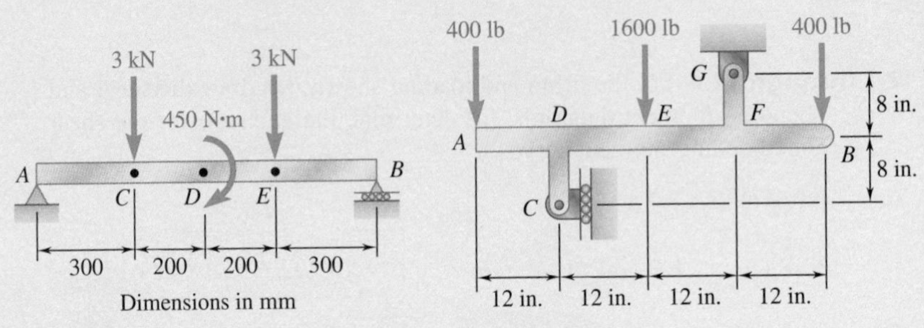

Fig. P12.9

Fig. P12.10

12.11 and 12.12 Assuming that the upward reaction of the ground is uniformly distributed, draw the shear and bending-moment diagrams for the beam *AB* and determine the maximum absolute value (*a*) of the shear, (*b*) of the bending moment.

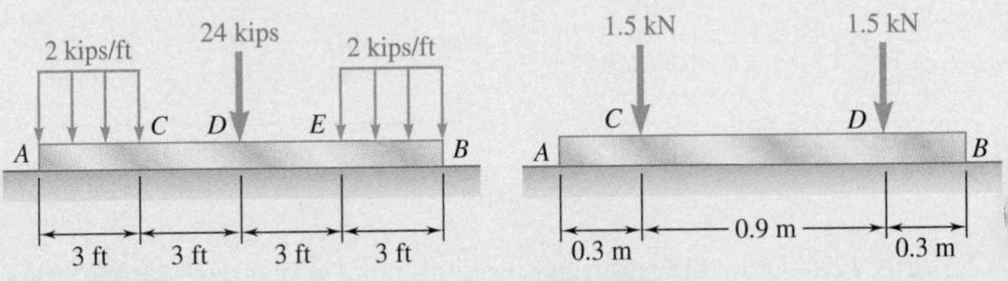

Fig. P12.11

Fig. P12.12

12.13 and 12.14 For the beam and loading shown, determine the maximum normal stress due to bending on a transverse section at *C*.

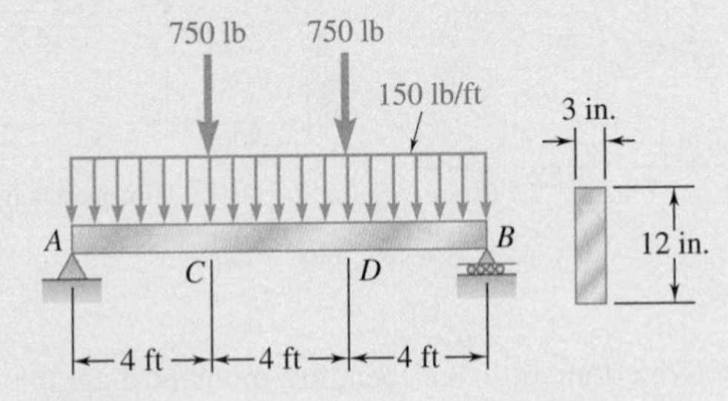

Fig. P12.13

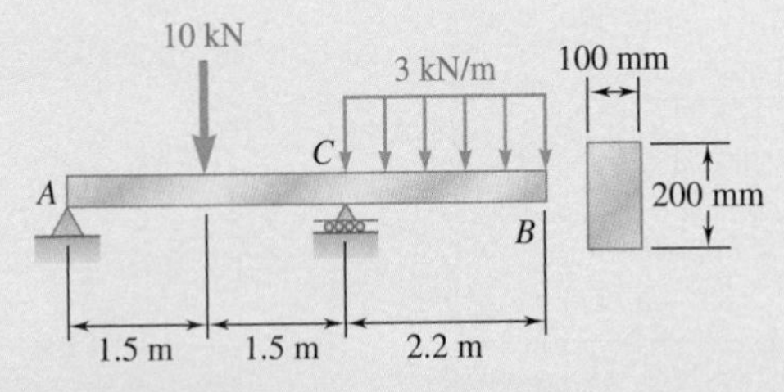

Fig. P12.14

12.15 For the beam and loading shown, determine the maximum normal stress due to bending on section *a-a*.

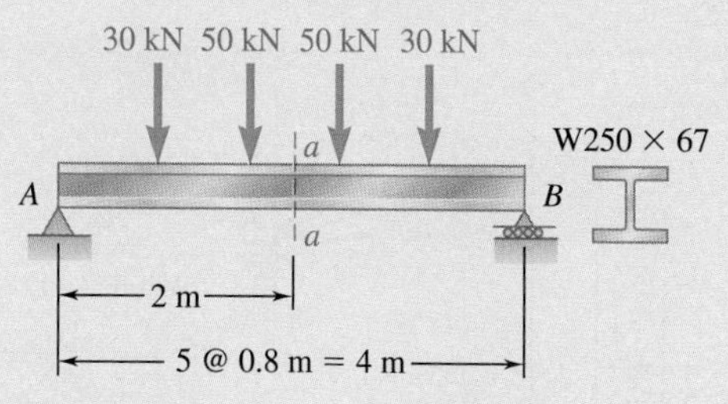

Fig. P12.15

12.16 For the beam and loading shown, determine the maximum normal stress due to bending on a transverse section at *C*.

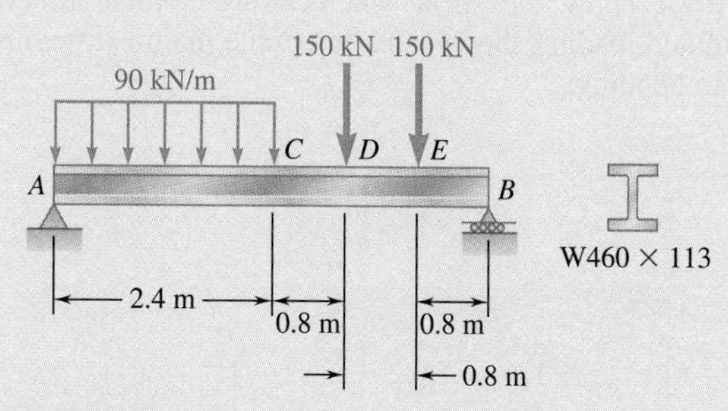

Fig. P12.16

12.17 and *12.18* For the beam and loading shown, determine the maximum normal stress due to bending on a transverse section at *C*.

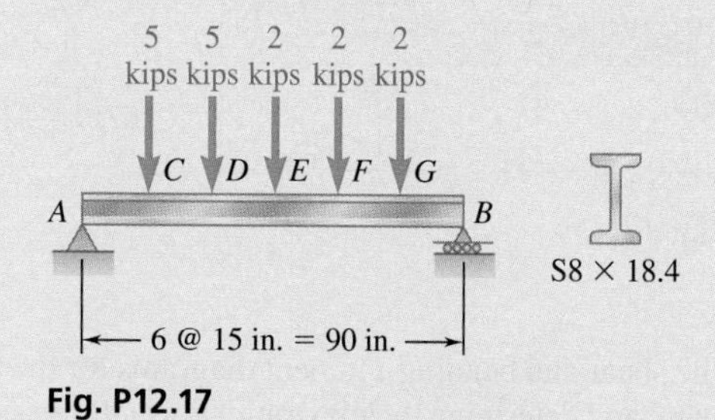

Fig. P12.17

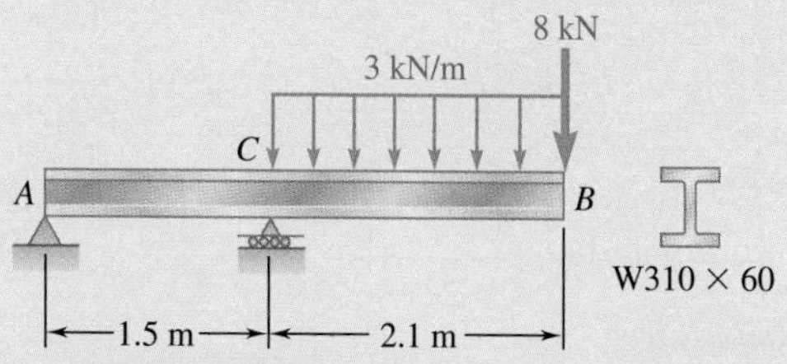

Fig. *P12.18*

12.19 and 12.20 Draw the shear and bending-moment diagrams for the beam and loading shown, and determine the maximum normal stress due to bending.

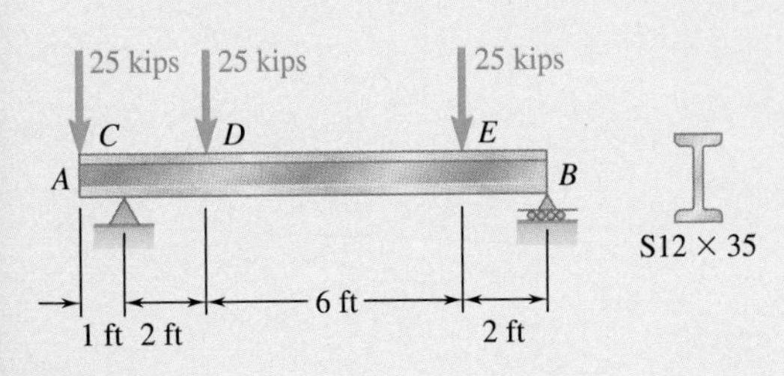

Fig. P12.19

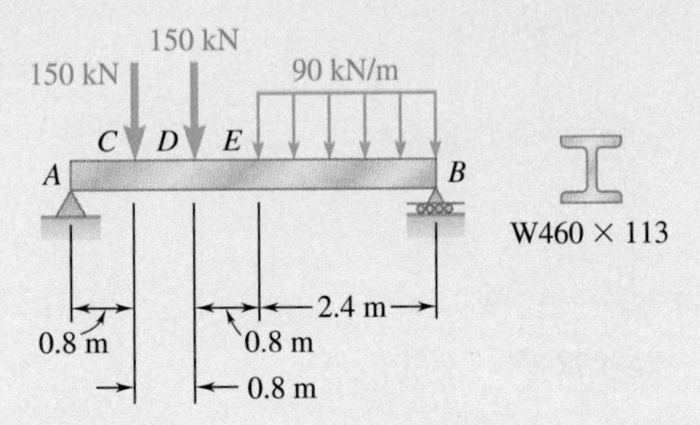

Fig. P12.20

12.21 and 12.22 Draw the shear and bending-moment diagrams for the beam and loading shown, and determine the maximum normal stress due to bending.

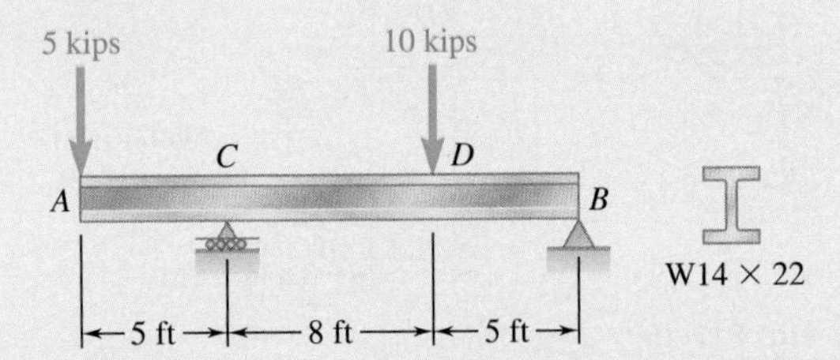

Fig. P12.21

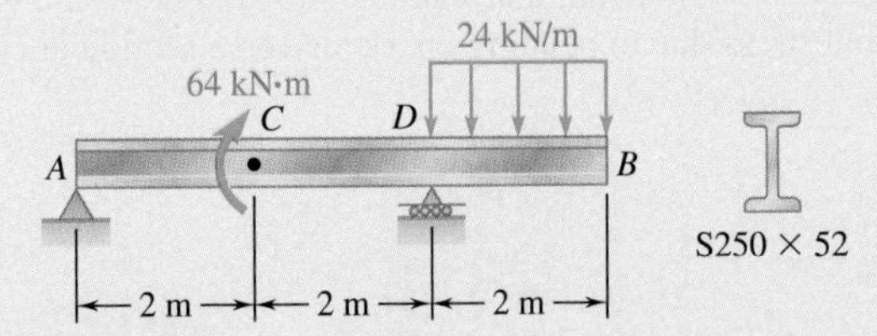

Fig. P12.22

12.23 Draw the shear and bending-moment diagrams for the beam and loading shown, and determine the maximum normal stress due to bending.

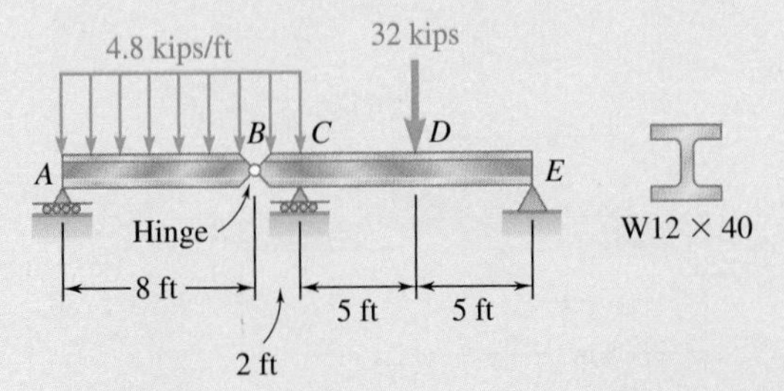

Fig. P12.23

12.24 Knowing that $W = 3$ kips, draw the shear and bending-moment diagrams for beam AB, and determine the maximum normal stress due to bending.

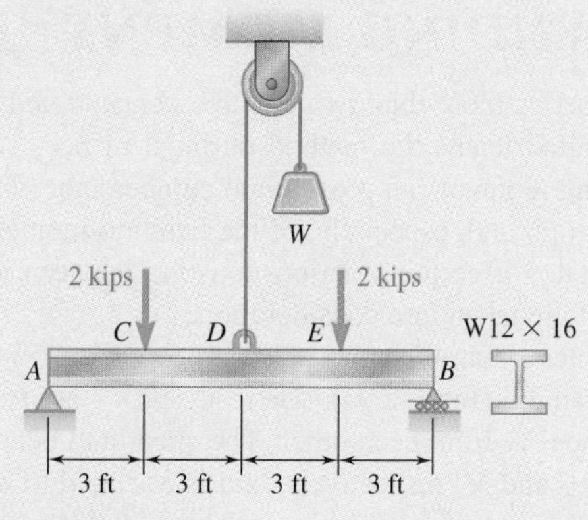

Fig. P12.24

12.25 Knowing that $P = Q = 480$ N, determine (*a*) the distance a for which the absolute value of the bending moment in the beam is as small as possible, (*b*) the corresponding maximum normal stress due to bending. (*Hint*: Draw the bending-moment diagram and equate the absolute values of the largest positive and negative bending moments obtained.)

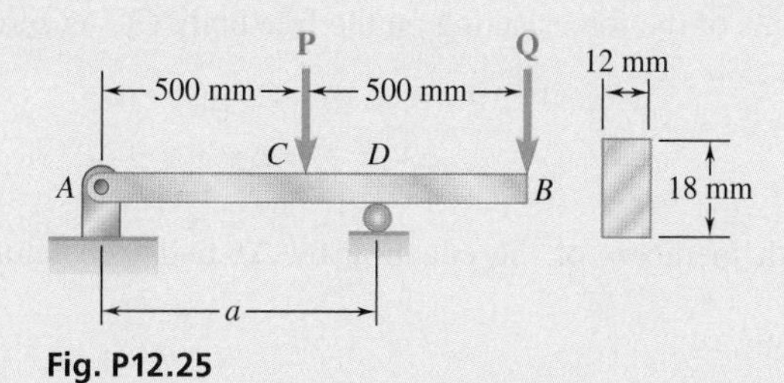

Fig. P12.25

12.26 Solve Prob. 12.25, assuming that P = 480 N and Q = 320 N.

12.27 Determine (*a*) the distance a for which the maximum absolute value of the bending moment in the beam is as small as possible, (*b*) the corresponding maximum normal stress due to bending. (See hint for Prob. 12.25.)

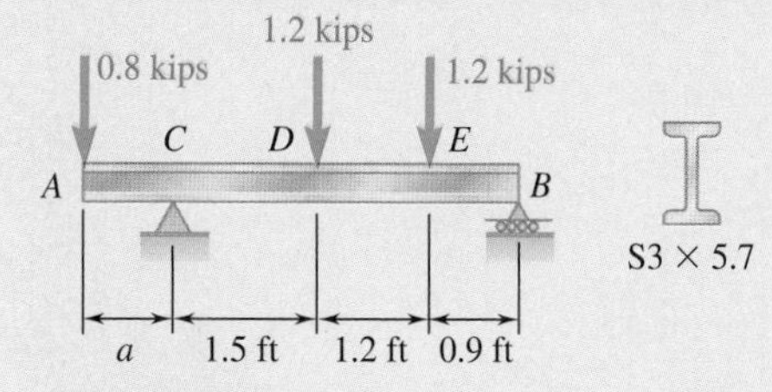

Fig. P12.27

12.28 A solid steel rod of diameter d is supported as shown. Knowing that for steel $\gamma = 490$ lb/ft^3, determine the smallest diameter d that can be used if the normal stress due to bending is not to exceed 4 ksi.

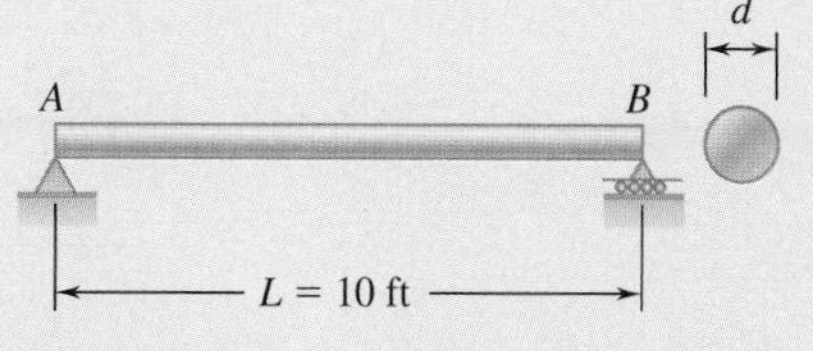

Fig. ***P12.28***

12.2 RELATIONSHIPS BETWEEN LOAD, SHEAR, AND BENDING MOMENT

When a beam carries more than two or three concentrated loads, or when it carries distributed loads, the method outlined in Sec. 12.1 for plotting shear and bending moment can prove quite cumbersome. The construction of the shear diagram and, especially, of the bending-moment diagram will be greatly facilitated if certain relations existing between load, shear, and bending moment are taken into consideration.

For example, a simply supported beam AB is carrying a distributed load w per unit length (Fig. 12.9a), where C and C' are two points of the beam at a distance Δx from each other. The shear and bending moment at C is denoted by V and M, respectively, and is assumed to be positive. The shear and bending moment at C' is denoted by $V + \Delta V$ and $M + \Delta M$.

Detach the portion of beam CC' and draw its free-body diagram (Fig. 12.9b). The forces exerted on the free body include a load of magnitude $w\,\Delta x$ and internal forces and couples at C and C'. Since shear and bending moment are assumed to be positive, the forces and couples are directed as shown.

Relationships between Load and Shear. The sum of the vertical components of the forces acting on the free body CC' is zero, so

$$+\uparrow \Sigma F_y = 0: \qquad V - (V + \Delta V) - w\,\Delta x = 0$$

$$\Delta V = -w\,\Delta x$$

Dividing both members of the equation by Δx and then letting Δx approach zero,

$$\frac{dV}{dx} = -w \tag{12.5}$$

Eq. (12.5) indicates that, for a beam loaded as shown in Fig. 12.9a, the slope dV/dx of the shear curve is negative. The magnitude of the slope at any point is equal to the load per unit length at that point.

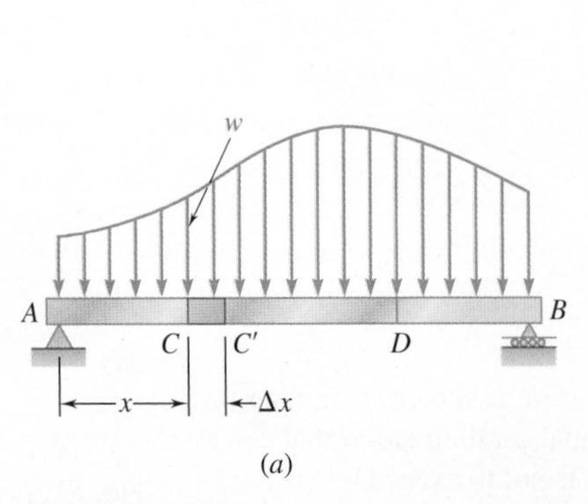

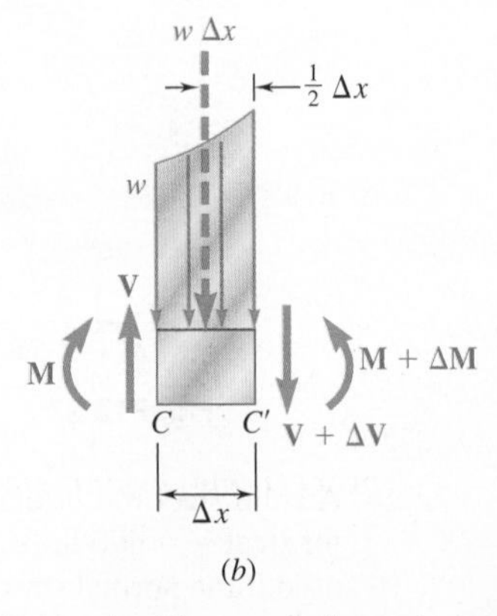

Fig. 12.9 (*a*) Simply supported beam subjected to a distributed load, with a small element between *C* and *C'*, (*b*) free-body diagram of the element.

Integrating Eq. (12.5) between points C and D,

$$V_D - V_C = -\int_{x_C}^{x_D} w\,dx \tag{12.6a}$$

$$V_D - V_C = -(\text{area under load curve between } C \text{ and } D) \tag{12.6b}$$

This result is illustrated in Fig. 12.10*b*. Note that this result could be obtained by considering the equilibrium of the portion of beam CD, since the area under the load curve represents the total load applied between C and D.

Also, Eq. (12.5) is not valid at a point where a concentrated load is applied; the shear curve is discontinuous at such a point, as seen in Sec. 12.1. Similarly, Eqs. (12.6a) and (12.6b) are not valid when concentrated loads are applied between C and D, since they do not take into account the sudden change in shear caused by a concentrated load. Eqs. (12.6a) and (12.6b), should be applied only between successive concentrated loads.

Relationships between Shear and Bending Moment. Returning to the free-body diagram of Fig. 12.9*b* and writing that the sum of the moments about C' is zero, we have

$$+\circlearrowleft \Sigma M_{C'} = 0: \qquad (M + \Delta M) - M - V\,\Delta x + w\,\Delta x \frac{\Delta x}{2} = 0$$

$$\Delta M = V\,\Delta x - \frac{1}{2} w\,(\Delta x)^2$$

Dividing both members by Δx and then letting Δx approach zero,

$$\frac{dM}{dx} = V \tag{12.7}$$

Eq. (12.7) indicates that the slope dM/dx of the bending-moment curve is equal to the value of the shear. This is true at any point where the shear has a well-defined value (i.e., no concentrated load is applied). Eq. (12.7) also shows that $V = 0$ at points where M is maximum. This property facilitates the determination of the points where the beam is likely to fail under bending.

Integrating Eq. (12.7) between points C and D,

$$M_D - M_C = \int_{x_C}^{x_D} V\,dx \tag{12.8a}$$

$$M_D - M_C = \text{area under shear curve between } C \text{ and } D \tag{12.8b}$$

This result is illustrated in Fig. 12.10*c*. Note that the area under the shear curve is positive where the shear is positive and negative where the shear is negative. Eqs. (12.8a) and (12.8*b*) are valid even when concentrated loads are applied between C and D, as long as the shear curve has been drawn correctly. The equations are not valid if a couple is applied at a point between C and D, since they do not take into account the sudden change in bending moment caused by a couple (see Sample Prob. 12.6).

In most engineering applications, one needs to know the value of the bending moment at only a few specific points. Once the shear diagram has been drawn and after M has been determined at one of the ends of the beam, the value of the bending moment can be obtained at any given point by computing the area under the shear curve and using Eq. (12.8*b*).

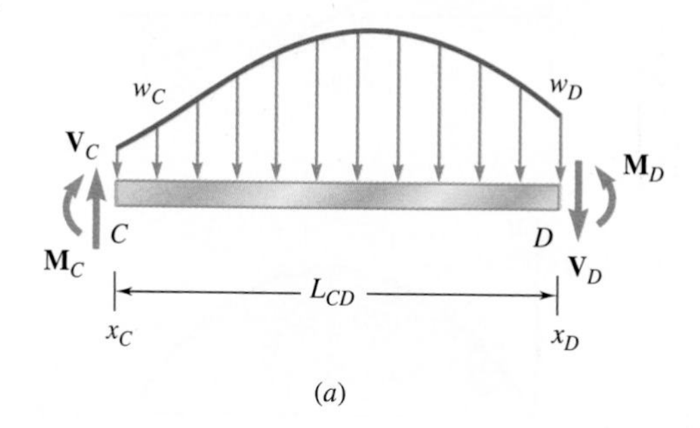

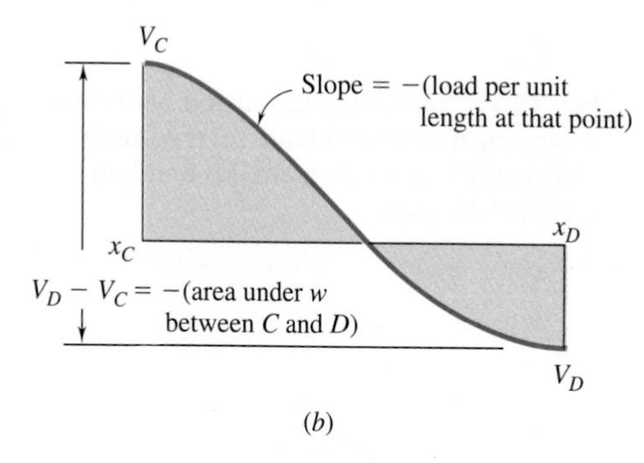

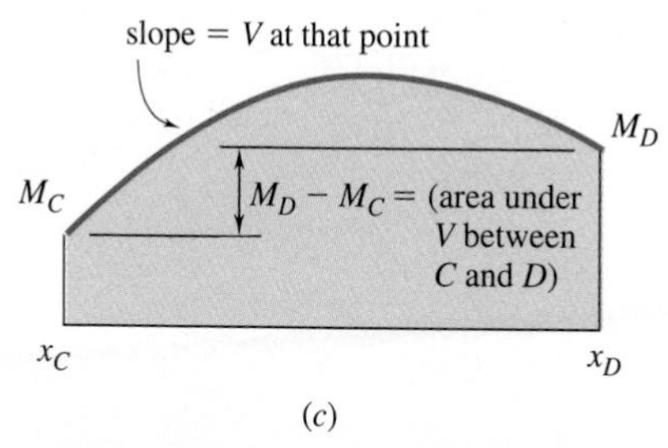

Fig. 12.10 Relationships between load, shear, and bending moment. (*a*) Section of loaded beam. (*b*) Shear curve for section. (*c*) Bending-moment curve for section.

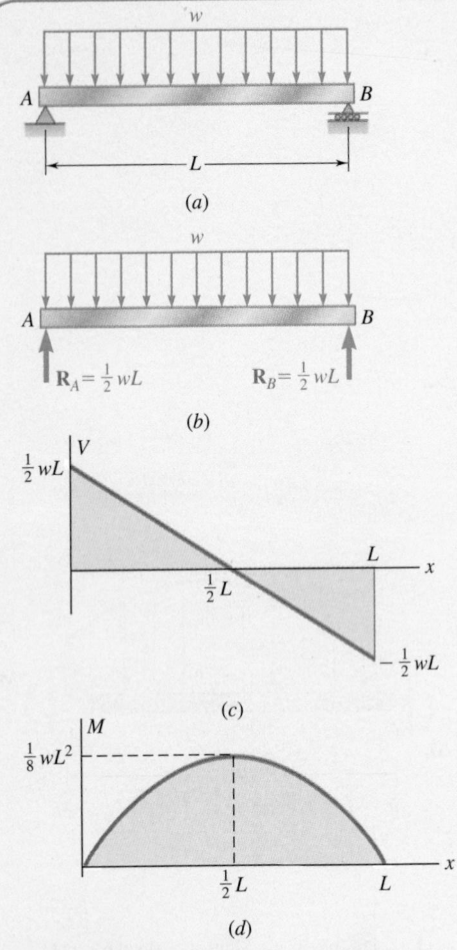

Fig. 12.11 (*a*) Simply supported beam with uniformly distributed load. (*b*) Free-body diagram. (*c*) Shear diagram. (*d*) Bending-moment diagram.

Concept Application 12.3

Draw the shear and bending-moment diagrams for the simply supported beam shown in Fig. 12.11*a* and determine the maximum value of the bending moment.

From the free-body diagram of the entire beam (Fig. 12.11*b*), we determine the magnitude of the reactions at the supports:

$$R_A = R_B = \tfrac{1}{2}wL$$

Next, draw the shear diagram. Close to the end *A* of the beam, the shear is equal to R_A, (that is, to $\frac{1}{2}wL$) which can be checked by considering as a free body a very small portion of the beam. Using Eq. (12.6*a*), the shear *V* at any distance *x* from *A* is

$$V - V_A = -\int_0^x w\,dx = -wx$$

$$V = V_A - wx = \tfrac{1}{2}wL - wx = w(\tfrac{1}{2}L - x)$$

Thus the shear curve is an oblique straight line that crosses the *x* axis at $x = L/2$ (Fig. 12.11*c*). Considering the bending moment, observe that $M_A = 0$. The value *M* of the bending moment at any distance *x* from *A* is obtained from Eq. (12.8*a*):

$$M - M_A = \int_0^x V\,dx$$

$$M = \int_0^x w(\tfrac{1}{2}L - x)dx = \tfrac{1}{2}w(Lx - x^2)$$

The bending-moment curve is a parabola. The maximum value of the bending moment occurs when $x = L/2$, since *V* (and thus dM/dx) is zero for this value of *x*. Substituting $x = L/2$ in the last equation, $M_{max} = wL^2/8$ (Fig. 12.11*d*).

For instance, since $M_A = 0$ for the beam of Concept Application 12.3, the maximum value of the bending moment for that beam is obtained simply by measuring the area of the shaded triangle of the positive portion of the shear diagram of Fig. 12.11*c*. So,

$$M_{max} = \frac{1}{2}\frac{L}{2}\frac{wL}{2} = \frac{wL^2}{8}$$

Note that the load curve is a horizontal straight line, the shear curve an oblique straight line, and the bending-moment curve a parabola. If the load curve had been an oblique straight line (first degree), the shear curve would have been a parabola (second degree), and the bending-moment curve a cubic (third degree). The shear and bending-moment curves are always one and two degrees higher than the load curve, respectively. With this in mind, the shear and bending-moment diagrams can be drawn without actually determining the functions $V(x)$ and $M(x)$. The sketches will be more accurate if we make use of the fact that at any point where the curves are continuous, the slope of the shear curve is equal to $-w$ and the slope of the bending-moment curve is equal to *V*.

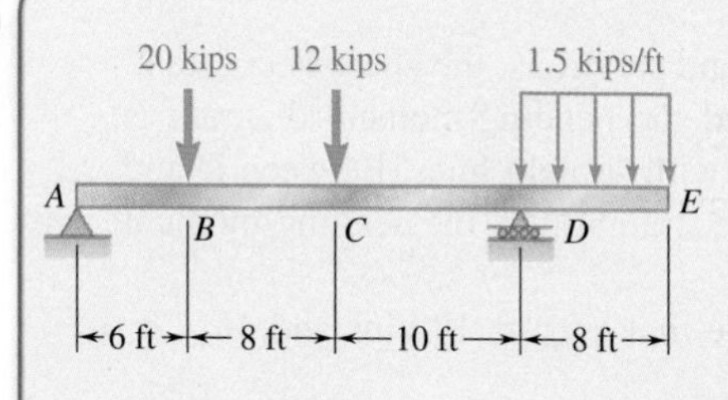

Sample Problem 12.3

Draw the shear and bending-moment diagrams for the beam and loading shown.

STRATEGY: The beam supports two concentrated loads and one distributed load. You can use the equations in this section between these loads and under the distributed load, but you should expect changes in the diagrams at the concentrated load points.

MODELING and ANALYSIS:

Reactions. Consider the entire beam as a free body as shown in Fig. 1.

$+\circlearrowleft \Sigma M_A = 0$:

$$D(24\text{ ft}) - (20\text{ kips})(6\text{ ft}) - (12\text{ kips})(14\text{ ft}) - (12\text{ kips})(28\text{ ft}) = 0$$

$$D = +26\text{ kips} \qquad \mathbf{D} = 26\text{ kips}\uparrow$$

$+\uparrow \Sigma F_y = 0$: $\quad A_y - 20\text{ kips} - 12\text{ kips} + 26\text{ kips} - 12\text{ kips} = 0$

$$A_y = +18\text{ kips} \qquad \mathbf{A}_y = 18\text{ kips}\uparrow$$

$\overset{+}{\rightarrow} \Sigma F_x = 0$: $\quad A_x = 0 \qquad \mathbf{A}_x = 0$

Note that at both A and E the bending moment is zero. Thus, two points (indicated by dots) are obtained on the bending-moment diagram.

Shear Diagram. Since $dV/dx = -w$, between concentrated loads and reactions the slope of the shear diagram is zero (i.e., the shear is constant). The shear at any point is determined by dividing the beam into two parts and considering either part to be a free body. For example, using the portion of beam to the left of section *1*, the shear between B and C is

$+\uparrow \Sigma F_y = 0$: $\quad +18\text{ kips} - 20\text{ kips} - V = 0 \qquad V = -2\text{ kips}$

Also, the shear is $+12$ kips just to the right of D and zero at end E. Since the slope $dV/dx = -w$ is constant between D and E, the shear diagram between these two points is a straight line.

Bending-Moment Diagram. Recall that the area under the shear curve between two points is equal to the change in bending moment between the same two points. For convenience, the area of each portion of the shear diagram is computed and indicated in parentheses on the diagram in Fig. 1. Since the bending moment M_A at the left end is known to be zero,

$$M_B - M_A = +108 \qquad M_B = +108\text{ kip}\cdot\text{ft}$$
$$M_C - M_B = -16 \qquad M_C = +92\text{ kip}\cdot\text{ft}$$
$$M_D - M_C = -140 \qquad M_D = -48\text{ kip}\cdot\text{ft}$$
$$M_E - M_D = +48 \qquad M_E = 0$$

Since M_E is known to be zero, a check of the computations is obtained.

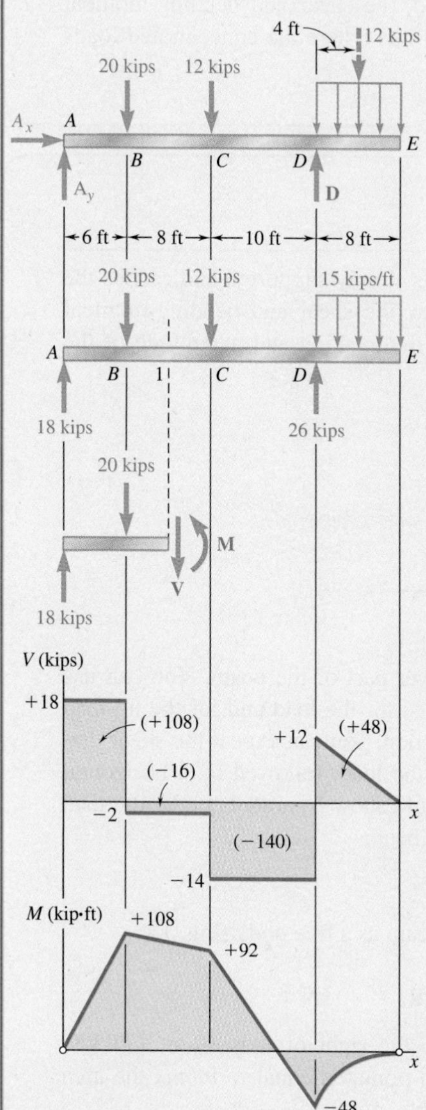

Fig. 1 Free-body diagrams of beam, free-body diagram of section to left of cut, shear diagram, bending-moment diagram.

(continued)

Between the concentrated loads and reactions, the shear is constant. Thus, the slope dM/dx is constant, and the bending-moment diagram is drawn by connecting the known points with straight lines. Between D and E where the shear diagram is an oblique straight line, the bending-moment diagram is a parabola.

From the V and M diagrams, note that $V_{max} = 18$ kips and $M_{max} = 108$ kip · ft.

REFLECT and THINK: As expected, the shear and bending-moment diagrams show abrupt changes at the points where the concentrated loads act.

Sample Problem 12.4

The W360 × 79 rolled-steel beam AC is simply supported and carries the uniformly distributed load shown. Draw the shear and bending-moment diagrams for the beam, and determine the location and magnitude of the maximum normal stress due to bending.

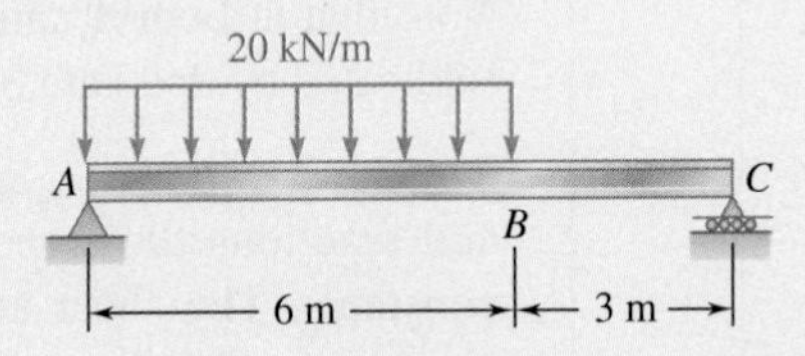

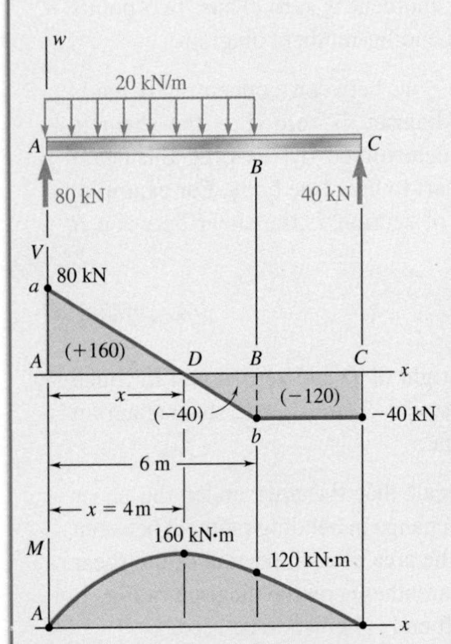

Fig. 1 Free-body diagram, shear diagram, bending-moment diagram.

STRATEGY: A load is distributed over part of the beam. You can use the equations in this section in two parts: for the load and for the no-load regions. From the discussion in this section, you can expect the shear diagram will show an oblique line under the load, followed by a horizontal line. The bending-moment diagram should show a parabola under the load and an oblique line under the rest of the beam.

MODELING and ANALYSIS:

Reactions. Considering the entire beam as a free body (Fig. 1),

$$\mathbf{R}_A = 80 \text{ kN} \uparrow \qquad \mathbf{R}_C = 40 \text{ kN} \uparrow$$

Shear Diagram. The shear just to the right of A is $V_A = +80$ kN. Since the change in shear between two points is equal to *minus* the area under the load curve between the same two points, V_B is

$$V_B - V_A = -(20 \text{ kN/m})(6 \text{ m}) = -120 \text{ kN}$$
$$V_B = -120 + V_A = -120 + 80 = -40 \text{ kN}$$

(continued)

The slope $dV/dx = -w$ is constant between A and B, and the shear diagram between these two points is represented by a straight line. Between B and C, the area under the load curve is zero; therefore,

$$V_C - V_B = 0 \qquad V_C = V_B = -40 \text{ kN}$$

and the shear is constant between B and C.

Bending-Moment Diagram. Note that the bending moment at each end is zero. In order to determine the maximum bending moment, locate the section D of the beam where $V = 0$.

$$V_D - V_A = -wx$$

$$0 - 80 \text{ kN} = -(20 \text{ kN/m})\, x$$

Solving for x, $x = 4$ m ◀

The maximum bending moment occurs at point D, where $dM/dx = V = 0$. The areas of various portions of the shear diagram are computed and given (in parentheses). The area of the shear diagram between two points is equal to the change in bending moment between the same two points, giving

$$M_D - M_A = +\ 160 \text{ kN·m} \qquad M_D = +160 \text{ kN·m}$$

$$M_B - M_D = -\ \ 40 \text{ kN·m} \qquad M_B = +120 \text{ kN·m}$$

$$M_C - M_B = -\ 120 \text{ kN·m} \qquad M_C = 0$$

The bending-moment diagram consists of an arc of parabola followed by a segment of straight line. The slope of the parabola at A is equal to the value of V at that point.

Maximum Normal Stress. This occurs at D, where $|M|$ is largest. From Appendix B, for a W360 × 79 rolled-steel shape, $S = 1270 \text{ mm}^3$ about a horizontal axis. Substituting this and $|M| = |M_D| = 160 \times 10^3$ N·m into Eq. (12.3),

$$\sigma_m = \frac{|M_D|}{S} = \frac{160 \times 10^3 \text{ N·m}}{1270 \times 10^{-6} \text{ m}^3} = 126.0 \times 10^6 \text{ Pa}$$

Maximum normal stress in the beam = 126.0 MPa ◀

Sample Problem 12.5

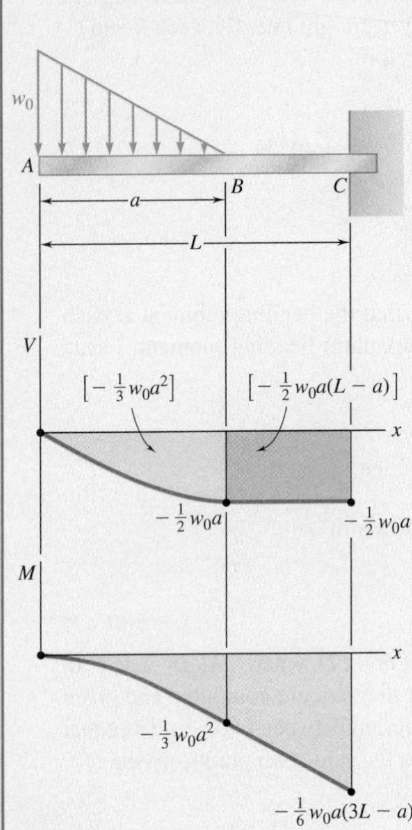

Fig. 1 Beam with load, shear diagram, bending-moment diagram.

Sketch the shear and bending-moment diagrams for the cantilever beam shown in Fig. 1.

STRATEGY: Because there are no support reactions until the right end of the beam, you can rely solely on the equations from this section without needing to use free-body diagrams and equilibrium equations. Due to the non-uniform distributed load, you should expect the results to involve equations of higher degree, with a parabolic curve in the shear diagram and a cubic curve in the bending-moment diagram.

MODELING and ANALYSIS:

Shear Diagram. At the free end of the beam, $V_A = 0$. Between A and B, the area under the load curve is $\frac{1}{2}w_0a$. Thus,

$$V_B - V_A = -\tfrac{1}{2}w_0a \qquad V_B = -\tfrac{1}{2}w_0a$$

Between B and C, the beam is not loaded, so $V_C = V_B$. At A, $w = w_0$. According to Eq. (12.5), the slope of the shear curve is $dV/dx = -w_0$, while at B the slope is $dV/dx = 0$. Between A and B, the loading decreases linearly, and the shear diagram is parabolic. Between B and C, $w = 0$, and the shear diagram is a horizontal line.

Bending-Moment Diagram. The bending moment M_A at the free end of the beam is zero. Compute the area under the shear curve to obtain.

$$M_B - M_A = -\tfrac{1}{3}w_0a^2 \qquad M_B = -\tfrac{1}{3}w_0a^2$$

$$M_C - M_B = -\tfrac{1}{2}w_0a(L-a)$$

$$M_C = -\tfrac{1}{6}w_0a(3L-a)$$

The sketch of the bending-moment diagram is completed by recalling that $dM/dx = V$. Between A and B, the diagram is represented by a cubic curve with zero slope at A and between B and C by a straight line.

REFLECT and THINK: Although not strictly required for the solution of this problem, determination of the support reactions would serve as an excellent check of the final values of the shear and bending-moment diagrams.

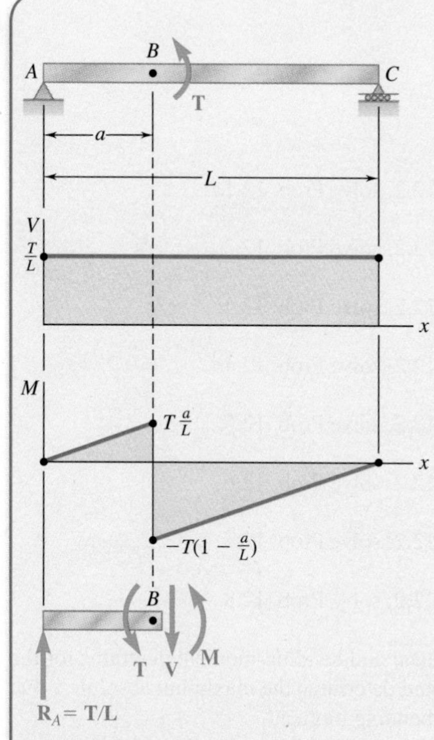

Fig. 1 Beam with load, shear diagram, bending-moment diagram, free-body diagram of section to left of B.

Sample Problem 12.6

The simple beam AC in Fig. 1 is loaded by a couple of moment T applied at point B. Draw the shear and bending-moment diagrams of the beam.

STRATEGY: The load supported by the beam is a concentrated couple. Since the only vertical forces are those associated with the support reactions, you should expect the shear diagram to be of constant value. However, the bending-moment diagram will have a discontinuity at B due to the couple.

MODELING and ANALYSIS:

The entire beam is taken as a free body.

$$\mathbf{R}_A = \frac{T}{L}\uparrow \qquad \mathbf{R}_C = \frac{T}{L}\downarrow$$

The shear at any section is constant and equal to T/L. Since a couple is applied at B, the bending-moment diagram is discontinuous at B. It is represented by two oblique straight lines and decreases suddenly at B by an amount equal to T. This discontinuity can be verified by equilibrium analysis. For example, considering the free body of the portion of the beam from A to just beyond the right of B as shown in Fig. 1, M is

$$+\circlearrowleft\Sigma M_B = 0: \qquad -\frac{T}{L}a + T + M = 0 \qquad M = -T\left(1 - \frac{a}{L}\right)$$

REFLECT and THINK: Notice that the applied couple results in a sudden change to the moment diagram at the point of application in the same way that a concentrated force results in a sudden change to the shear diagram.

Problems

12.29 Using the method of Sec. 12.2, solve Prob. 12.1*a*.

12.30 Using the method of Sec. 12.2, solve Prob. 12.2*a*.

12.31 Using the method of Sec. 12.2, solve Prob. 12.3*a*.

12.32 Using the method of Sec. 12.2, solve Prob. 12.4*a*.

12.33 Using the method of Sec. 12.2, solve Prob. 12.5.

12.34 Using the method of Sec. 12.2, solve Prob. 12.6.

12.35 Using the method of Sec. 12.2, solve Prob. 12.7.

12.36 Using the method of Sec. 12.2, solve Prob. 12.8.

12.37 through 12.40 Draw the shear and bending-moment diagrams for the beam and loading shown, and determine the maximum absolute value (*a*) of the shear, (*b*) of the bending moment.

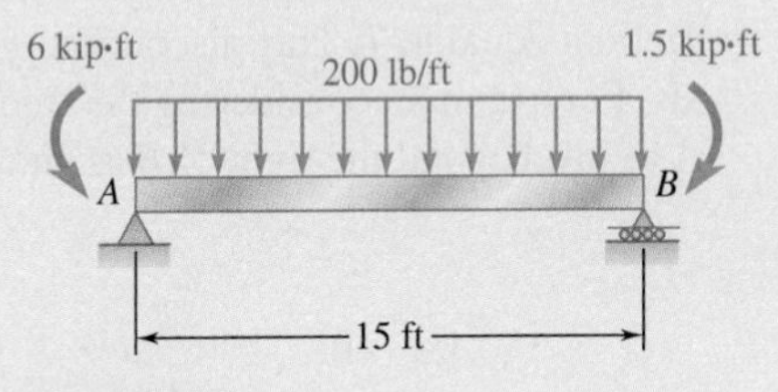

Fig. P12.37

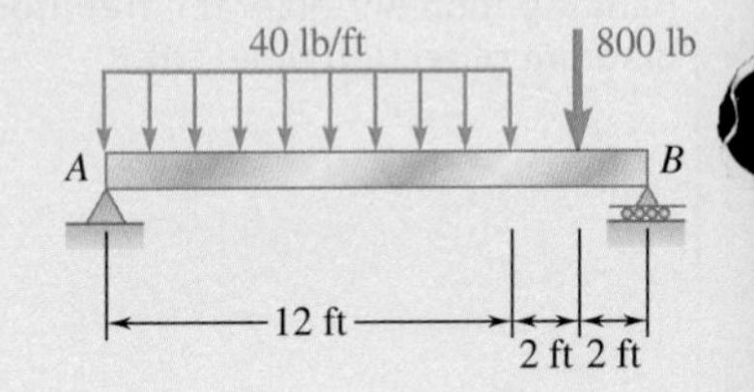

Fig. P12.38

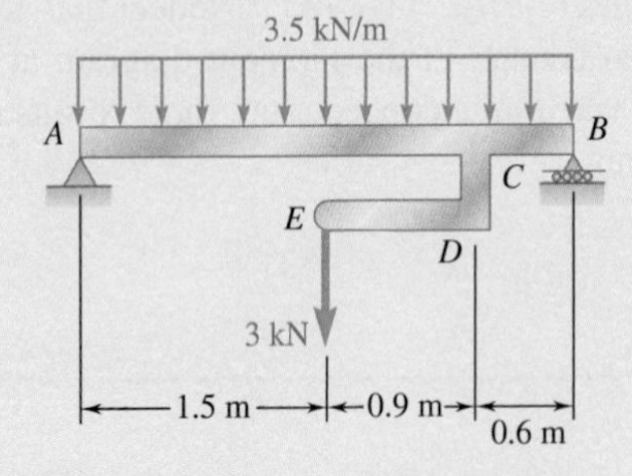

Fig. P12.39

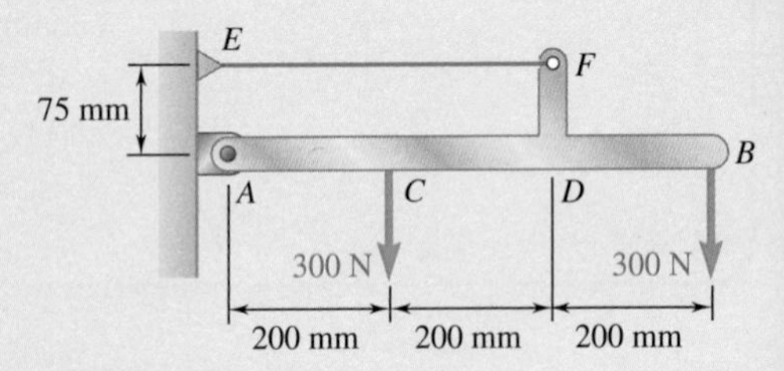

Fig. P12.40

12.41 Using the method of Sec. 12.2, solve Prob. 12.13.

12.42 Using the method of Sec. 12.2, solve Prob. 12.14.

12.43 Using the method of Sec. 12.2, solve Prob. 12.15.

12.44 Using the method of Sec. 12.2, solve Prob. 12.17.

12.45 and *12.46* Determine (*a*) the equations of the shear and bending-moment curves for the beam and loading shown, (*b*) the maximum absolute value of the bending moment in the beam.

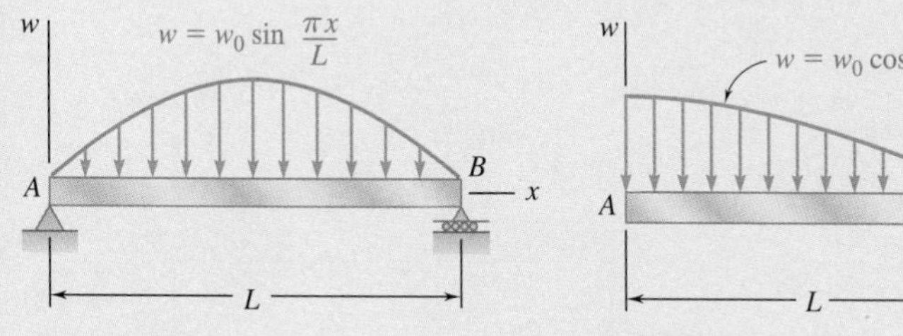

Fig. P12.45 **Fig. *P12.46***

12.47 Determine (*a*) the equations of the shear and bending-moment curves for the beam and loading shown, (*b*) the maximum absolute value of the bending moment in the beam.

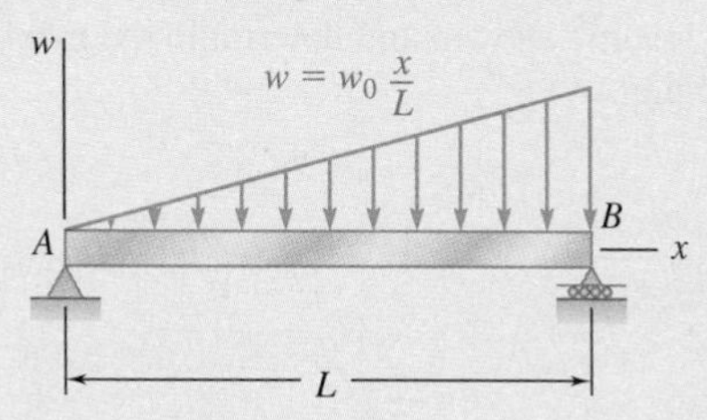

Fig. P12.47

12.48 For the beam and loading shown, determine the equations of the shear and bending-moment curves and the maximum absolute value of the bending moment in the beam, knowing that (*a*) $k = 1$, (*b*) $k = 0.5$.

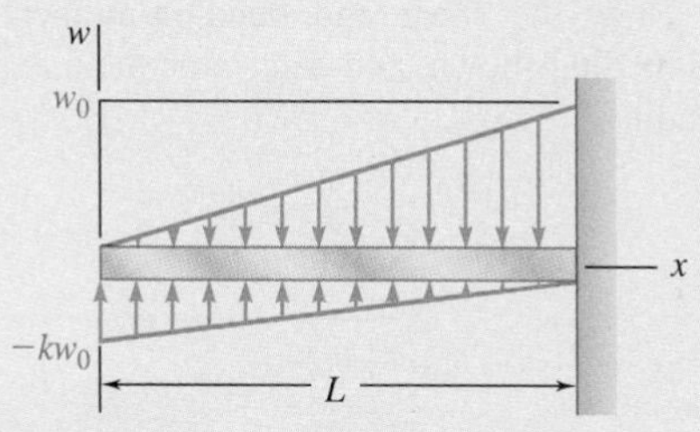

Fig. *P12.48*

12.49 and 12.50 Draw the shear and bending-moment diagrams for the beam and loading shown, and determine the maximum normal stress due to bending.

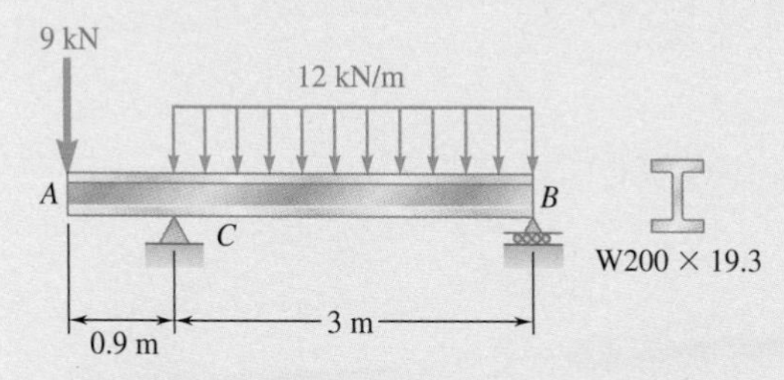

Fig. P12.49

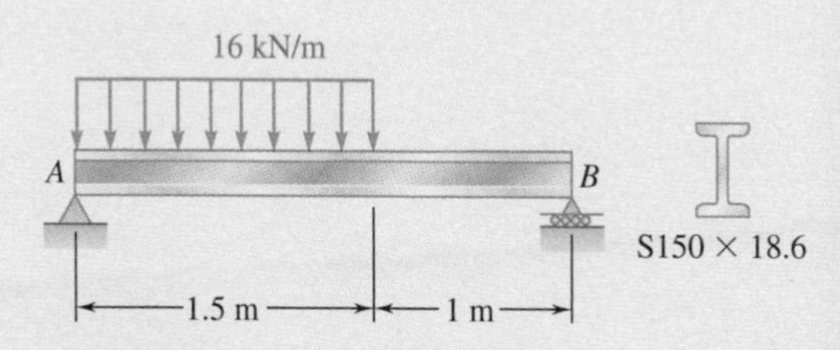

Fig. P12.50

12.51 and 12.52 Draw the shear and bending-moment diagrams for the beam and loading shown, and determine the maximum normal stress due to bending.

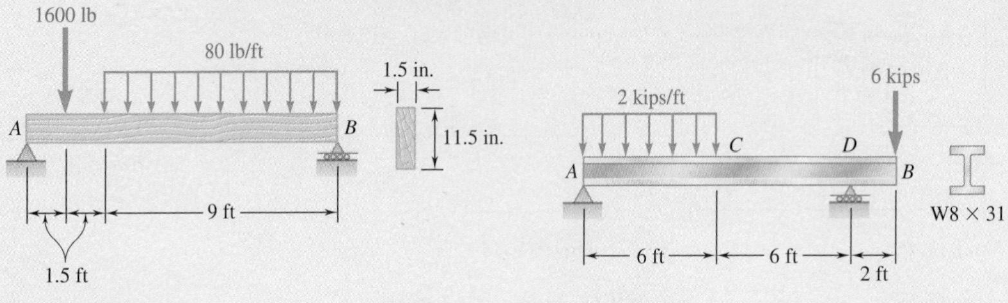

Fig. P12.51

Fig. P12.52

12.53 **and 12.54** Draw the shear and bending-moment diagrams for the beam and loading shown, and determine the maximum normal stress due to bending.

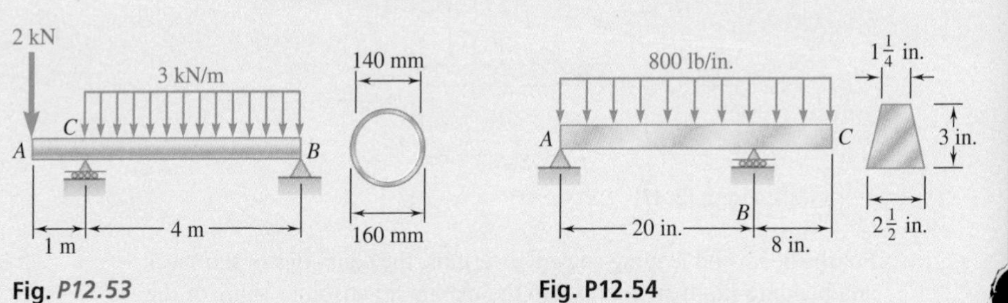

Fig. *P12.53*

Fig. P12.54

12.55 and *12.56* Draw the shear and bending-moment diagrams for the beam and loading shown, and determine the maximum normal stress due to bending.

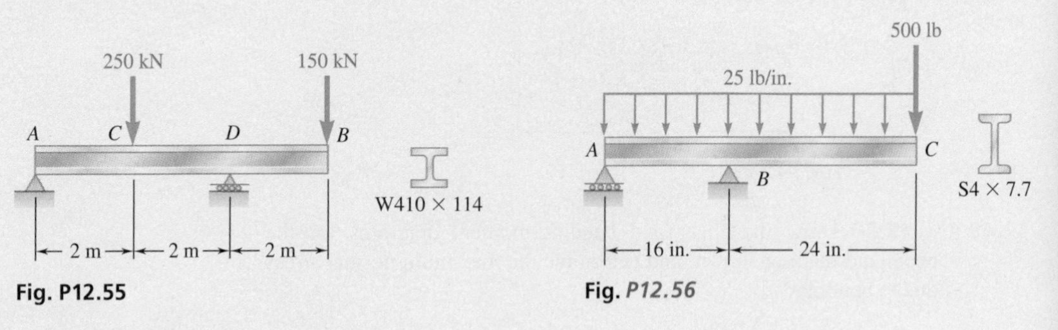

Fig. P12.55

Fig. *P12.56*

12.3 DESIGN OF PRISMATIC BEAMS FOR BENDING

The design of a beam is usually controlled by the maximum absolute value $|M|_{max}$ of the bending moment that occurs in the beam. The largest normal stress σ_m in the beam is found at the surface of the beam in the critical section where $|M|_{max}$ occurs and is obtained by substituting $|M|_{max}$ for $|M|$ in Eq. (12.1) or Eq. (12.3).†

$$\sigma_m = \frac{|M|_{max}c}{I} \quad (12.1a)$$

$$\sigma_m = \frac{|M|_{max}}{S} \quad (12.3a)$$

A safe design requires that $\sigma_m \leq \sigma_{all}$, where σ_{all} is the allowable stress for the material used. Substituting σ_{all} for σ_m in (12.3*a*) and solving for S yields the minimum allowable value of the section modulus for the beam being designed:

$$S_{min} = \frac{|M|_{max}}{\sigma_{all}} \quad (12.9)$$

The design of common types of beams, such as timber beams of rectangular cross section and rolled-steel members of various cross-sectional shapes, is discussed in this section. A proper procedure should lead to the most economical design. This means that among beams of the same type and same material, and other things being equal, the beam with the smallest weight per unit length—and, thus, the smallest cross-sectional area—should be selected, since this beam will be the least expensive.

The design procedure generally includes the following steps:‡

Step 1. First determine the value of σ_{all} for the material selected from a table of properties of materials or from design specifications. You also can compute this value by dividing the ultimate strength σ_U of the material by an appropriate factor of safety (Sec. 8.4C). Assuming that the value of σ_{all} is the same in tension and in compression, proceed as follows.

Step 2. Draw the shear and bending-moment diagrams corresponding to the specified loading conditions, and determine the maximum absolute value $|M|_{max}$ of the bending moment in the beam.

Step 3. Determine from Eq. (12.9) the minimum allowable value S_{min} of the section modulus of the beam.

Step 4. For a timber beam, the depth h of the beam, its width b, or the ratio h/b characterizing the shape of its cross section probably will have been specified. The unknown dimensions can be selected by using Eq. (11.19), so b and h satisfy the relation $\frac{1}{6}bh^2 = S \geq S_{min}$.

†For beams that are not symmetrical with respect to their neutral surface, the largest of the distances from the neutral surface to the surfaces of the beam should be used for c in Eq. (12.1) and in the computation of the section modulus $S = I/c$.

‡ It is assumed that all beams considered in this chapter are adequately braced to prevent lateral buckling and bearing plates are provided under concentrated loads applied to rolled-steel beams to prevent local buckling (crippling) of the web.

Step 5. For a rolled-steel beam, consult the appropriate table in Appendix B. Of the available beam sections, consider only those with a section modulus $S \geq S_{min}$ and select the section with the smallest weight per unit length. This is the most economical of the sections for which $S \geq S_{min}$. Note that this is not necessarily the section with the smallest value of S (see Concept Application 12.4). In some cases, the selection of a section may be limited by considerations such as the allowable depth of the cross section or the allowable deflection of the beam (see Chap. 15).

Fig. 12.12 Cantilevered wide-flange beam with end load.

Concept Application 12.4

Select a wide-flange beam to support the 15-kip load as shown in Fig. 12.12. The allowable normal stress for the steel used is 24 ksi.

1. The allowable normal stress is given: $\sigma_{all} = 24$ ksi.
2. The shear is constant and equal to 15 kips. The bending moment is maximum at B.

$$|M|_{max} = (15 \text{ kips})(8 \text{ ft}) = 120 \text{ kip·ft} = 1440 \text{ kip·in.}$$

3. The minimum allowable section modulus is

$$S_{min} = \frac{|M|_{max}}{\sigma_{all}} = \frac{1440 \text{ kip·in.}}{24 \text{ ksi}} = 60.0 \text{ in}^3$$

4. Referring to the table of *Properties of Rolled-Steel Shapes* in Appendix B, note that the shapes are arranged in groups of the same depth and are listed in order of decreasing weight. Choose the lightest beam in each group having a section modulus $S = I/c$ at least as large as S_{min} and record the results in the following table.

Shape	S, in^3
W21 × 44	81.6
W18 × 50	88.9
W16 × 40	64.7
W14 × 43	62.6
W12 × 50	64.2
W10 × 54	60.0

The most economical is the W16 × 40 shape since it weighs only 40 lb/ft, even though it has a larger section modulus than two of the other shapes. The total weight of the beam will be (8 ft) × (40 lb) = 320 lb. This weight is small compared to the 15,000-1b load and thus can be neglected in our analysis.

The previous discussion was limited to materials for which σ_{all} is the same in tension and compression. If σ_{all} is different, make sure to select the beam section where $\sigma_m \leq \sigma_{all}$ for both tensile and compressive stresses. If the cross section is not symmetric about its neutral axis, the largest tensile and the largest compressive stresses will not necessarily occur in the section where $|M|$ is maximum (one may occur where M is maximum and the other where M is minimum). Thus, step 2 should include the determination of both M_{max} and M_{min}, and step 3 should take into account both tensile and compressive stresses.

Finally, the design procedure described in this section takes into account only the normal stresses occurring on the surface of the beam. Short beams, especially those made of timber, may fail in shear under a transverse loading. The determination of shearing stresses in beams will be discussed in Chap. 13.

Sample Problem 12.7

A 12-ft-long overhanging timber beam AC with an 8-ft span AB is to be designed to support the distributed and concentrated loads shown. Knowing that timber of 4-in. nominal width (3.5-in. actual width) with a 1.75-ksi allowable stress is to be used, determine the minimum required depth h of the beam.

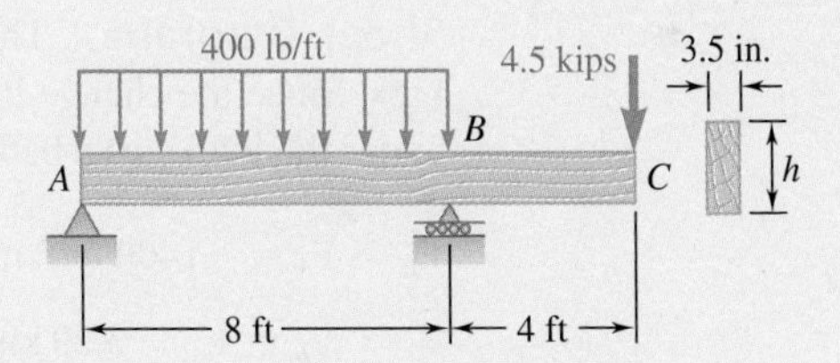

STRATEGY: Draw the bending-moment diagram to find the absolute maximum bending-moment. Then, using this bending-moment, you can determine the required section properties that satisfy the given allowable stress.

MODELING and ANALYSIS:

Reactions. Consider the entire beam to be a free body (Fig. 1).

$$+\circlearrowleft \Sigma M_A = 0: \quad B(8 \text{ ft}) - (3.2 \text{ kips})(4 \text{ ft}) - (4.5 \text{ kips})(12 \text{ ft}) = 0$$

$$B = 8.35 \text{ kips} \qquad \mathbf{B} = 8.35 \text{ kips} \uparrow$$

$$\xrightarrow{+} \Sigma F_x = 0: \quad A_x = 0$$

$$+\uparrow \Sigma F_y = 0: A_y + 8.35 \text{ kips} - 3.2 \text{ kips} - 4.5 \text{ kips} = 0$$

$$A_y = -0.65 \text{ kips} \qquad \mathbf{A} = 0.65 \text{ kips} \downarrow$$

(continued)

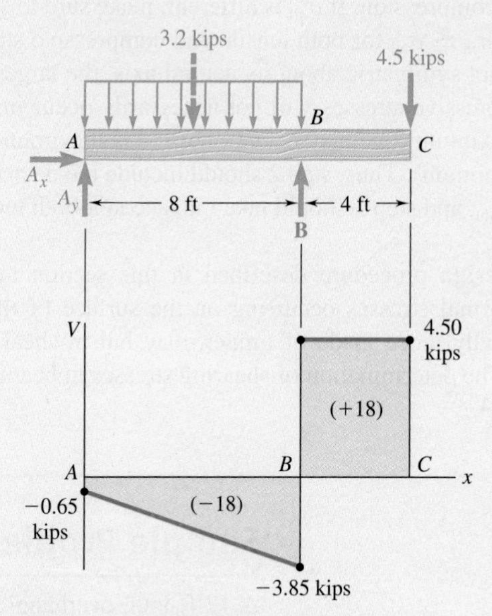

Fig. 1 Free-body diagram of beam and its shear diagram.

Shear Diagram. The shear just to the right of A is $V_A = A_y = -0.65$ kips. Since the change in shear between A and B is equal to *minus* the area under the load curve between these two points, V_B is obtained by

$$V_B - V_A = -(400 \text{ lb/ft})(8 \text{ ft}) = -3200 \text{ lb} = -3.20 \text{ kips}$$

$$V_B = V_A - 3.20 \text{ kips} = -0.65 \text{ kips} - 3.20 \text{ kips} = -3.85 \text{ kips.}$$

The reaction at B produces a sudden increase of 8.35 kips in V, resulting in a shear equal to 4.50 kips to the right of B. Since no load is applied between B and C, the shear remains constant between these two points.

Determination of $|M|_{max}$. Observe that the bending moment is equal to zero at both ends of the beam: $M_A = M_C = 0$. Between A and B, the bending moment decreases by an amount equal to the area under the shear curve, and between B and C it increases by a corresponding amount. Thus, the maximum absolute value of the bending moment is $|M|_{max} = 18.00$ kip·ft.

Minimum Allowable Section Modulus. Substituting the values of σ_{all} and $|M|_{max}$ into Eq. (12.9) gives

$$S_{min} = \frac{|M|_{max}}{\sigma_{all}} = \frac{(18 \text{ kip·ft})(12 \text{ in./ft})}{1.75 \text{ ksi}} = 123.43 \text{ in}^3$$

(continued)

Minimum Required Depth of Beam. Recalling the formula developed in step 4 of the design procedure and substituting the values of b and S_{min}, we have

$$\tfrac{1}{6}bh^2 \geq S_{min} \qquad \tfrac{1}{6}(3.5\text{ in.})h^2 \geq 123.43\text{ in}^3 \qquad h \geq 14.546\text{ in.}$$

The minimum required depth of the beam is $h = 14.55$ in. ◀

REFLECT and THINK: In practice, standard wood shapes are specified by nominal dimensions that are slightly larger than actual. In this case, specify a 4-in. × 16-in. member with the actual dimensions of 3.5 in. × 15.25 in.

Sample Problem 12.8

A 5-m-long, simply supported steel beam AD is to carry the distributed and concentrated loads shown. Knowing that the allowable normal stress for the grade of steel is 160 MPa, select the wide-flange shape to be used.

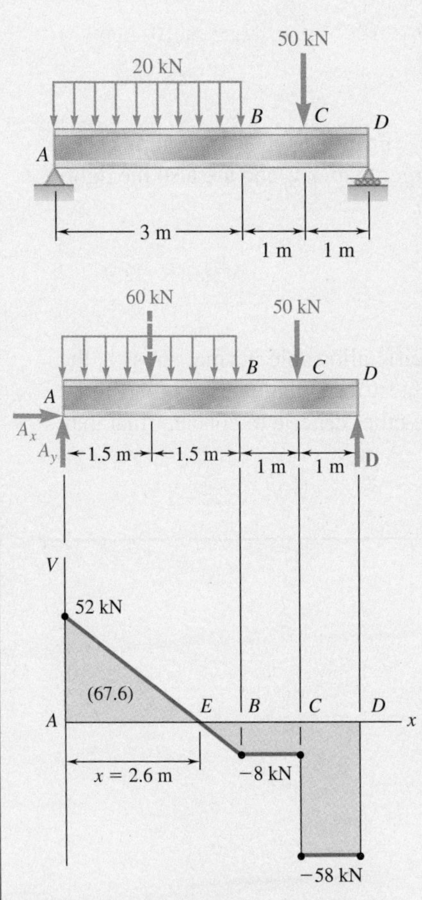

Fig. 1 Free-body diagram of beam and its shear diagram.

STRATEGY: Draw the bending-moment diagram to find the absolute maximum bending moment. Then, using this moment, you can determine the required section modulus that satisfies the given allowable stress.

MODELING and ANALYSIS:

Reactions. Consider the entire beam to be a free body (Fig. 1).

$$+\circlearrowleft\Sigma M_A = 0:\ D(5\text{ m}) - (60\text{ kN})(1.5\text{ m}) - (50\text{ kN})(4\text{ m}) = 0$$

$$D = 58.0\text{ kN} \qquad \mathbf{D} = 58.0\text{ kN}\uparrow$$

$$\overset{+}{\rightarrow}\Sigma F_x = 0: \qquad A_x = 0$$

$$+\uparrow\Sigma F_y = 0:\ A_y + 58.0\text{ kN} - 60\text{ kN} - 50\text{ kN} = 0$$

$$A_y = 52.0\text{ kN} \qquad \mathbf{A} = 52.0\text{ kN}\uparrow$$

Shear Diagram. The shear just to the right of A is $V_A = A_y = +52.0$ kN. Since the change in shear between A and B is equal to *minus* the area under the load curve between these two points,

$$V_B = 52.0\text{ kN} - 60\text{ kN} = -8\text{ kN}$$

(continued)

The shear remains constant between B and C, where it drops to -58 kN, and keeps this value between C and D. Locate the section E of the beam where $V = 0$ by

$$V_E - V_A = -wx$$
$$0 - 52.0 \text{ kN} = -(20 \text{ kN/m})\,x$$

So, $x = 2.60$ m.

Determination of $|M|_{max}$. The bending moment is maximum at E, where $V = 0$. Since M is zero at the support A, its maximum value at E is equal to the area under the shear curve between A and E. Therefore, $|M|_{max} = M_E = 67.6$ kN·m.

Minimum Allowable Section Modulus. Substituting the values of σ_{all} and $|M|_{max}$ into Eq. (12.9) gives

$$S_{min} = \frac{|M|_{max}}{\sigma_{all}} = \frac{67.6 \text{ kN·m}}{160 \text{ MPa}} = 422.5 \times 10^{-6} \text{ m}^3 = 422.5 \times 10^3 \text{ mm}^3$$

Selection of Wide-Flange Shape. From Appendix B, compile a list of shapes that have a section modulus larger than S_{min} and are also the lightest shape in a given depth group (Fig. 2).

Shape	S, mm^3
W410 × 38.8	629
W360 × 32.9	475
W310 × 38.7	547
W250 × 44.8	531
W200 × 46.1	451

Fig. 2 Lightest shape in each depth group that provide the required section modulus.

The lightest shape available is W360 × 32.9 ◀

REFLECT and THINK: When a specific allowable normal stress is the sole design criterion for beams, the lightest acceptable shapes tend to be deeper sections. In practice, there will be other criteria to consider that may alter the final shape selection.

Problems

12.57 and 12.58 For the beam and loading shown, design the cross section of the beam knowing that the grade of timber used has an allowable normal stress of 12 MPa.

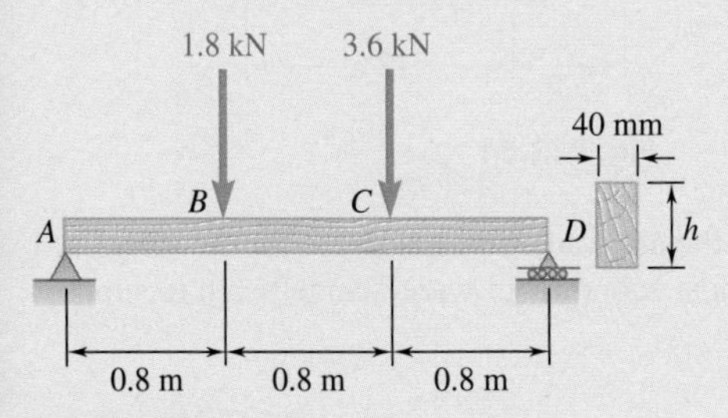

Fig. P12.57

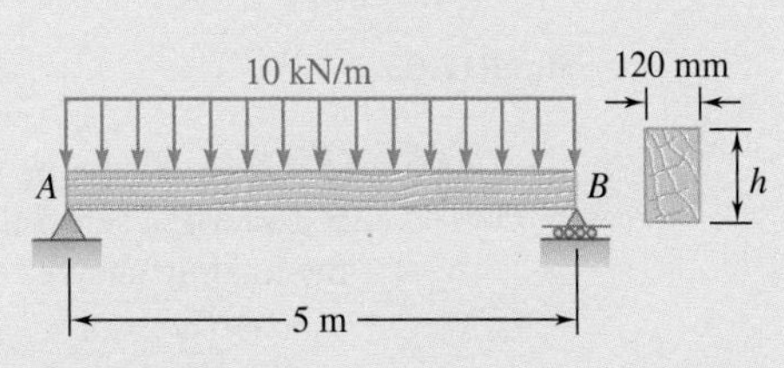

Fig. P12.58

***12.59* and 12.60** For the beam and loading shown, design the cross section of the beam knowing that the grade of timber used has an allowable normal stress of 1750 psi.

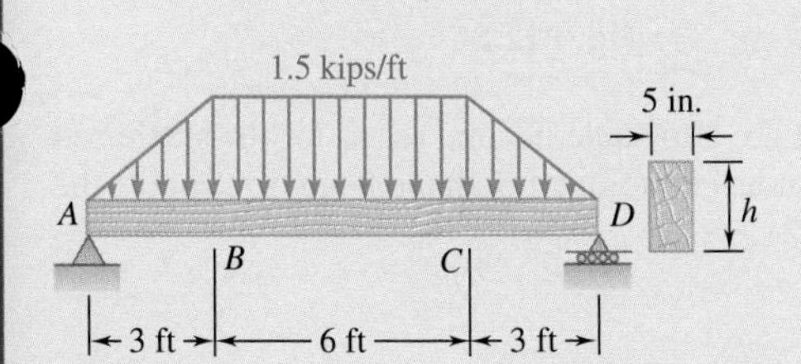

Fig. *P12.59*

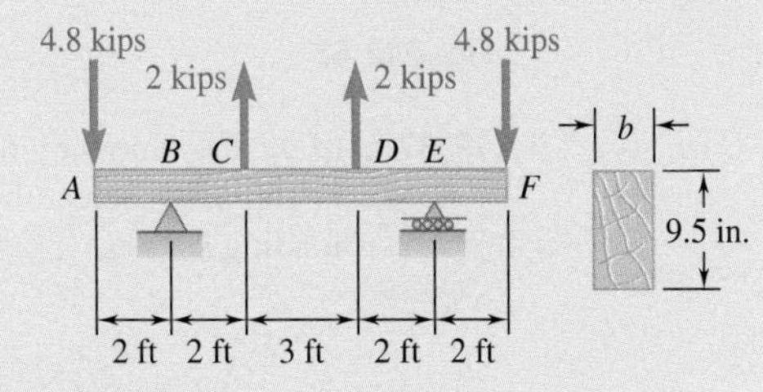

Fig. P12.60

12.61 For the beam and loading shown, design the cross section of the beam knowing that the grade of timber used has an allowable normal stress of 12 MPa.

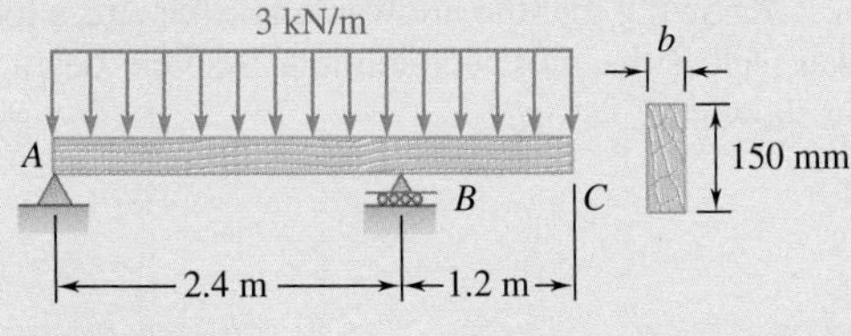

Fig. *P12.61*

12.62 For the beam and loading shown, design the cross section of the beam knowing that the grade of timber used has an allowable normal stress of 1750 psi.

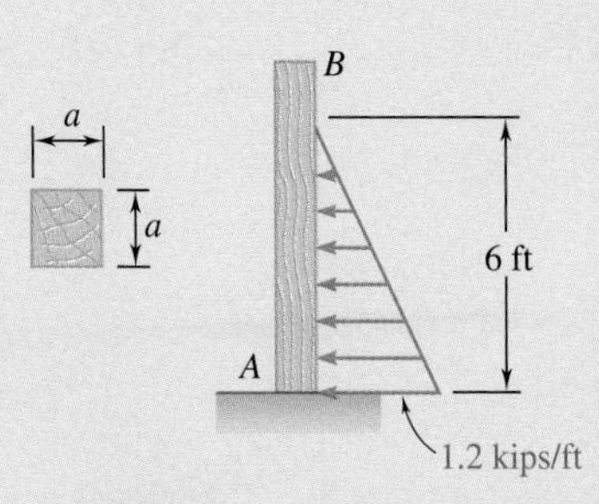

Fig. P12.62

12.63 and 12.64 Knowing that the allowable normal stress for the steel used is 24 ksi, select the most economical wide-flange beam to support the loading shown.

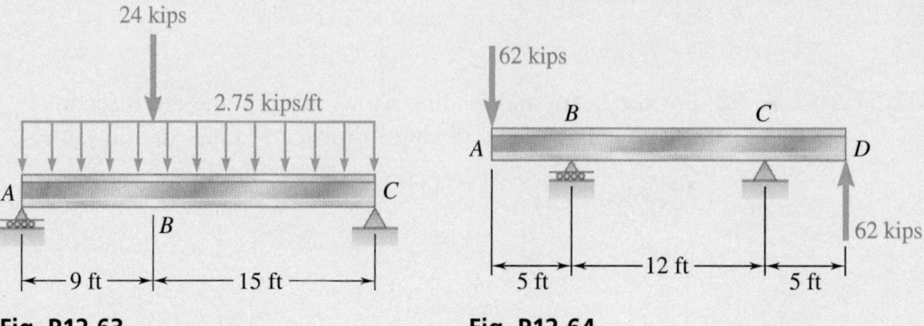

Fig. P12.63

Fig. P12.64

12.65 and 12.66 Knowing that the allowable normal stress for the steel used is 160 MPa, select the most economical wide-flange beam to support the loading shown.

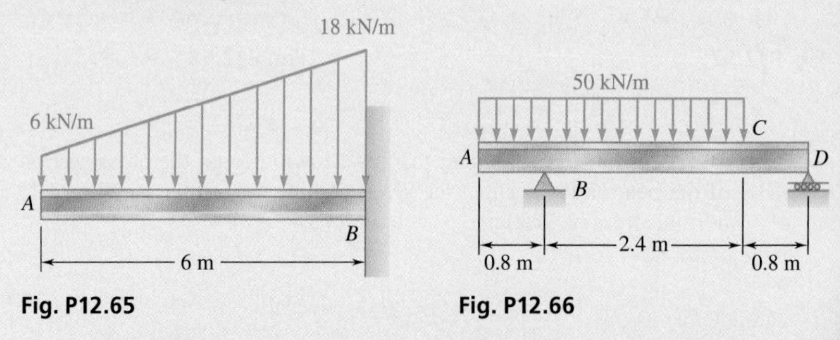

Fig. P12.65

Fig. P12.66

12.67 and *12.68* Knowing that the allowable normal stress for the steel used is 160 MPa, select the most economical S-shape beam to support the loading shown.

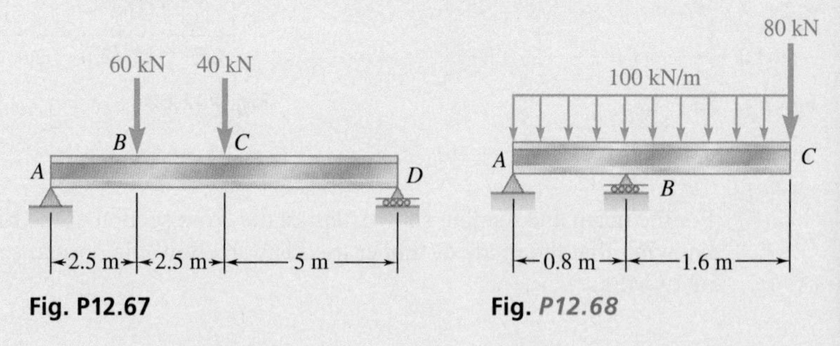

Fig. P12.67

Fig. *P12.68*

12.69 and *12.70* Knowing that the allowable normal stress for the steel used is 24 ksi, select the most economical S-shape beam to support the loading shown.

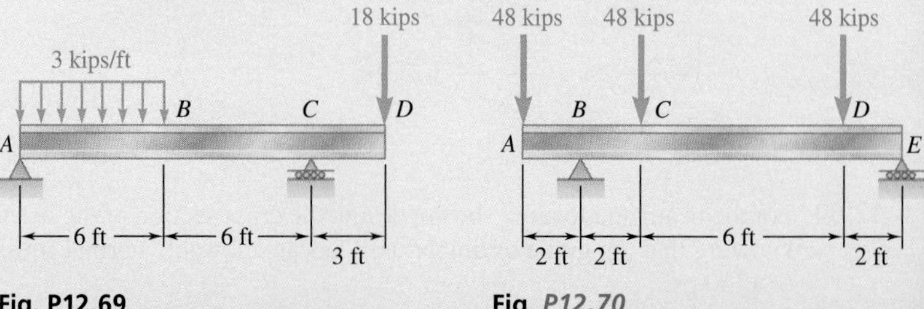

Fig. P12.69

Fig. *P12.70*

12.71 Two rolled-steel channels are to be welded back to back and used to support the loading shown. Knowing that the allowable normal stress for the steel used is 30 ksi, determine the most economical channels that can be used.

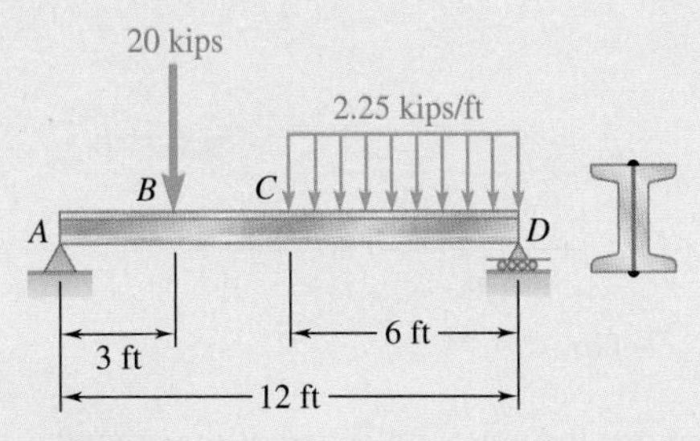

Fig. P12.71

12.72 Two metric rolled-steel channels are to be welded along their edges and used to support the loading shown. Knowing that the allowable normal stress for the steel used is 150 MPa, determine the most economical channels that can be used.

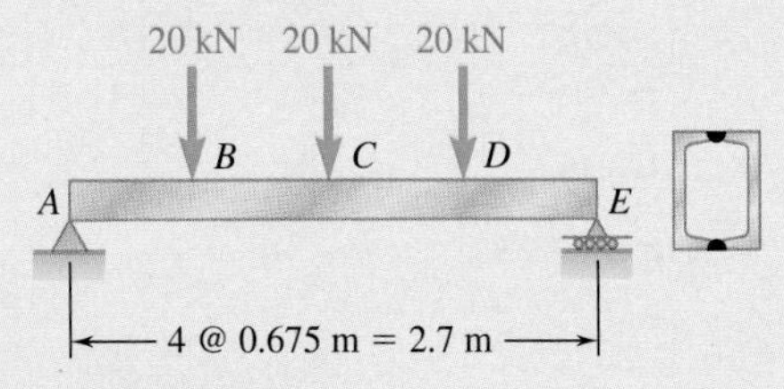

Fig. P12.72

12.73 Two L4 × 3 rolled-steel angles are bolted together and used to support the loading shown. Knowing that the allowable normal stress for the steel used is 24 ksi, determine the minimum angle thickness that can be used.

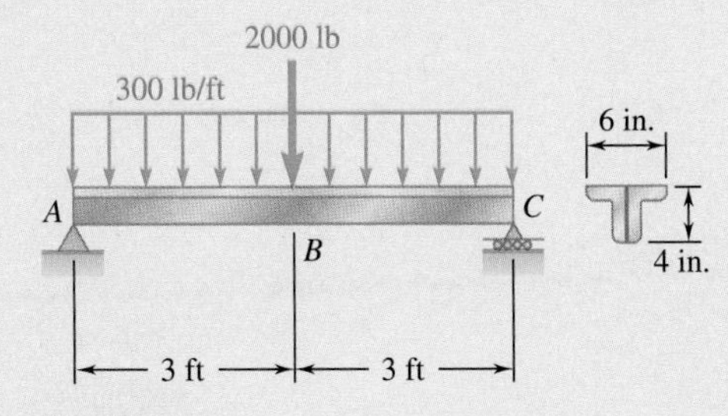

Fig. P12.73

12.74 A steel pipe of 100-mm diameter is to support the loading shown. Knowing that the stock of pipes available has thicknesses varying from 6 mm to 24 mm in 3-mm increments and that the allowable normal stress for the steel used is 150 MPa, determine the minimum wall thickness t that can be used.

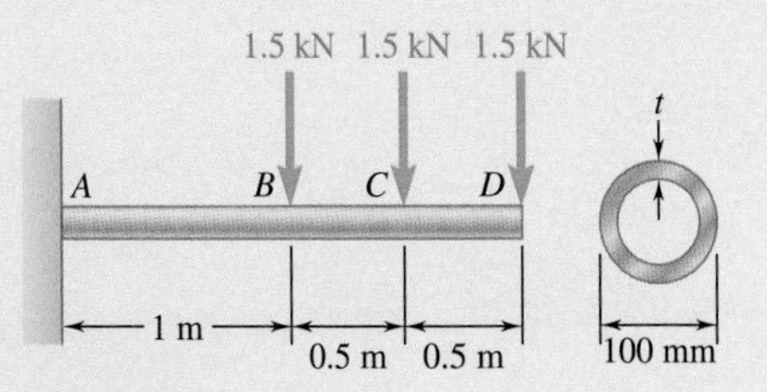

Fig. P12.74

12.75 Assuming that the upward reaction of the ground is uniformly distributed and knowing that the allowable normal stress for the steel used is 170 MPa, select the most economical wide-flange beam to support the loading shown.

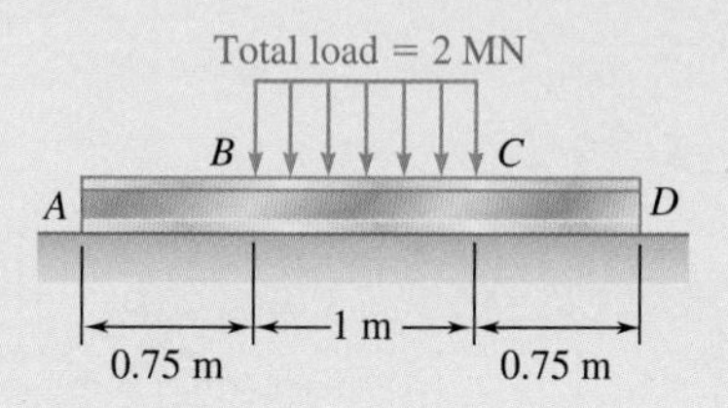

Fig. ***P12.75***

12.76 Assuming that the upward reaction of the ground is uniformly distributed and knowing that the allowable normal stress for the steel used is 24 ksi, select the most economical wide-flange beam to support the loading shown.

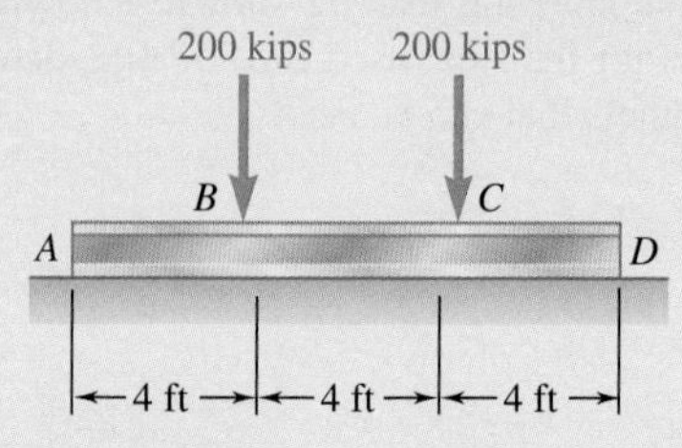

Fig. ***P12.76***

Review and Summary

Design of Prismatic Beams

This chapter was devoted to the analysis and design of beams under transverse loadings consisting of concentrated or distributed loads. The beams are classified according to the way they are supported (Fig. 12.13). Only *statically determinate* beams were considered, where all support reactions can be determined by statics.

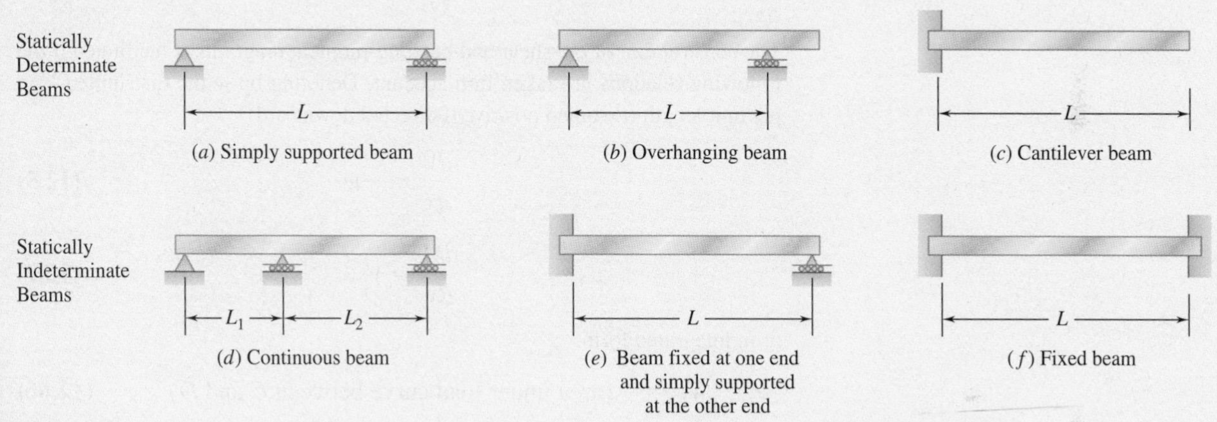

Fig. 12.13

Normal Stresses Due to Bending

While transverse loadings cause both bending and shear in a beam, the normal stresses caused by bending are the dominant criterion in the design of a beam for strength [Sec. 12.1]. Therefore, this chapter dealt only with the determination of the normal stresses in a beam, the effect of shearing stresses being examined in the next one.

The flexure formula for the determination of the maximum value σ_m of the normal stress in a given section of the beam is

$$\sigma_m = \frac{|M|c}{I} \tag{12.1}$$

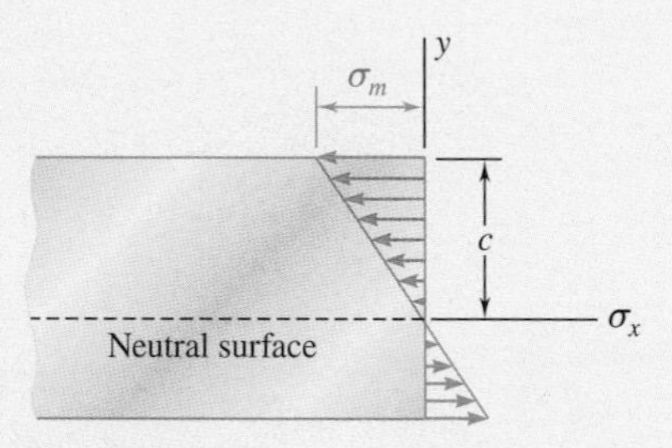

Fig. 12.14

where I is the moment of inertia of the cross section with respect to a centroidal axis perpendicular to the plane of the bending couple **M** and c is the maximum distance from the neutral surface (Fig. 12.14). Introducing the elastic section modulus $S = I/c$ of the beam, the maximum value σ_m of the normal stress in the section can be expressed also as

$$\sigma_m = \frac{|M|}{S} \tag{12.3}$$

Shear and Bending-Moment Diagrams

From Eq. (12.1) it is seen that the maximum normal stress occurs in the section where $|M|$ is largest and at the point farthest from the neutral axis.

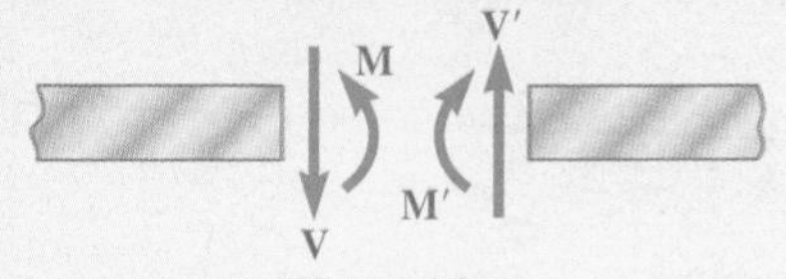

(*a*) Internal forces
(positive shear and positive bending moment)

Fig. 12.15

The determination of the maximum value of $|M|$ and of the critical section of the beam in which it occurs is simplified if *shear diagrams* and *bending-moment diagrams are drawn.* These diagrams represent the variation of the shear and of the bending moment along the beam and are obtained by determining the values of V and M at selected points of the beam. These values are found by passing a section through the point and drawing the free-body diagram of either of the portions of beam. To avoid any confusion regarding the sense of the shearing force **V** and of the bending couple **M** (which act in opposite sense on the two portions of the beam), we follow the sign convention adopted earlier, as illustrated in Fig. 12.15.

Relationships Between Load, Shear, and Bending Moment

The construction of the shear and bending-moment diagrams is facilitated if the following relations are taken into account. Denoting by w the distributed load per unit length (assumed positive if directed downward)

$$\frac{dV}{dx} = -w \tag{12.5}$$

$$\frac{dM}{dx} = V \tag{12.7}$$

or in integrated form,

$$V_D - V_C = -(\text{area under load curve between } C \text{ and } D) \tag{12.6b}$$

$$M_D - M_C = \text{area under shear curve between } C \text{ and } D \tag{12.8b}$$

Eq. (12.6b) makes it possible to draw the shear diagram of a beam from the curve representing the distributed load on that beam and V at one end of the beam. Similarly, Eq. (2.8b) makes it possible to draw the bending-moment diagram from the shear diagram and M at one end of the beam. However, concentrated loads introduce discontinuities in the shear diagram and concentrated couples in the bending-moment diagram, none of which is accounted for in these equations. The points of the beam where the bending moment is maximum or minimum are also the points where the shear is zero (Eq. 12.7).

Design of Prismatic Beams

Having determined σ_{all} for the material used and assuming that the design of the beam is controlled by the maximum normal stress in the beam, the minimum allowable value of the section modulus is

$$S_{min} = \frac{|M|_{max}}{\sigma_{all}} \tag{12.9}$$

For a timber beam of rectangular cross section, $S = \frac{1}{6}bh^2$, where b is the width of the beam and h its depth. The dimensions of the section, therefore, must be selected so that $\frac{1}{6}bh^2 \geq S_{min}$.

For a rolled-steel beam, consult the appropriate table in Appendix B. Of the available beam sections, consider only those with a section modulus $S \geq S_{min}$. From this group we normally select the section with the smallest weight per unit length.

Review Problems

12.77 and 12.78 Draw the shear and bending-moment diagrams for the beam and loading shown, and determine the maximum absolute value (*a*) of the shear, (*b*) of the bending moment.

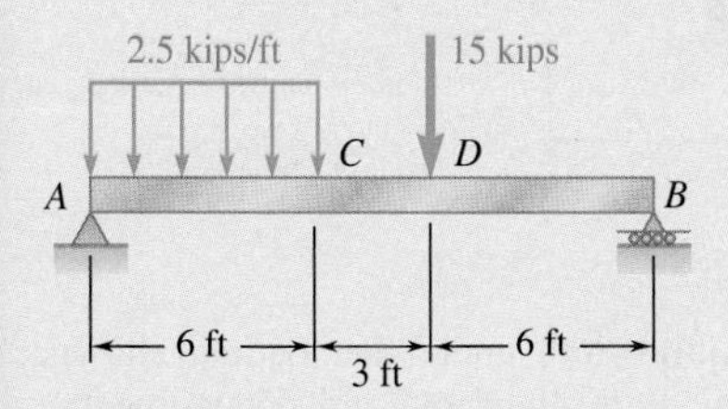

Fig. P12.77

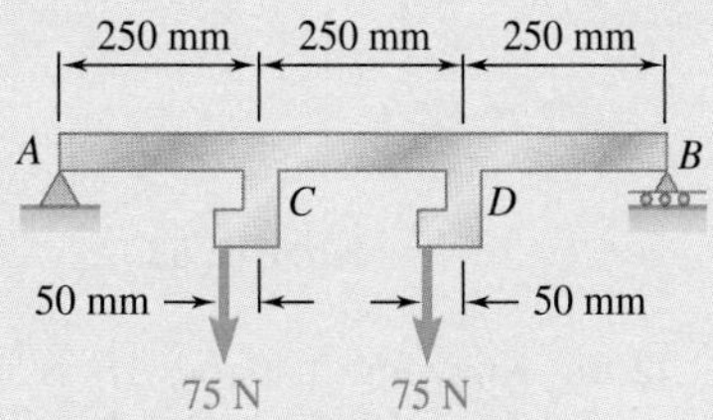

250 mm 250 mm 250 mm

A B C D

50 mm 50 mm

75 N 75 N

Fig. P12.78

12.79 Determine (*a*) the equations of the shear and bending-moment curves for the beam and loading shown, (*b*) the maximum absolute value of the bending moment in the beam.

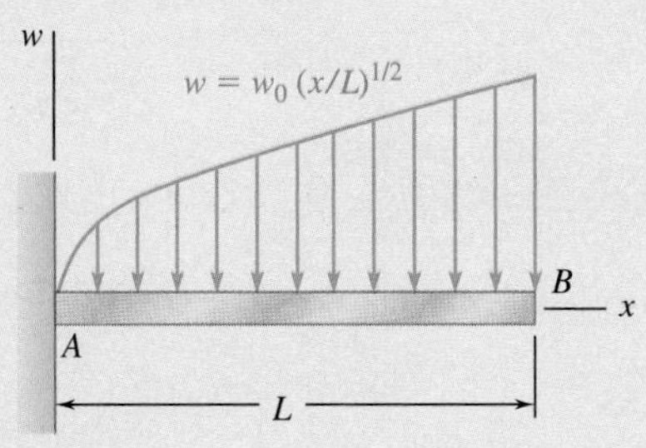

Fig. *P12.79*

12.80 and 12.81 Draw the shear and bending-moment diagrams for the beam and loading shown and determine the maximum normal stress due to bending.

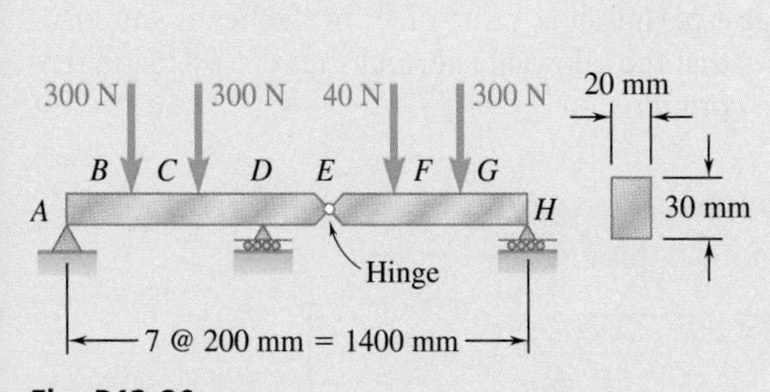

Fig. P12.80

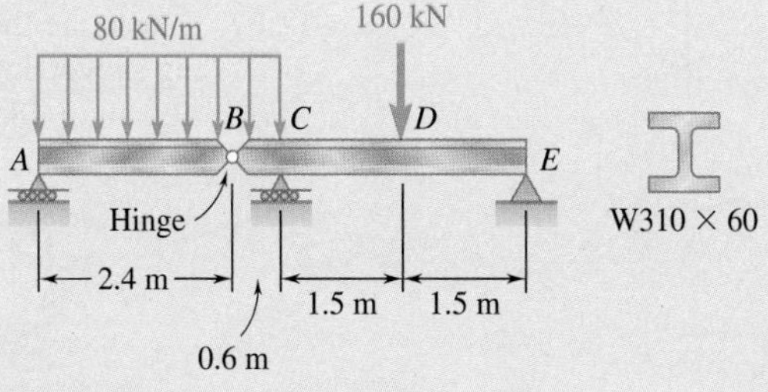

Fig. P12.81

12.82 Determine (*a*) the distance *a* for which the absolute value of the bending moment in the beam is as small as possible, (*b*) the corresponding maximum normal stress due to bending. (*Hint*: Draw the bending-moment diagram and equate the absolute values of the largest positive and negative bending moments obtained.)

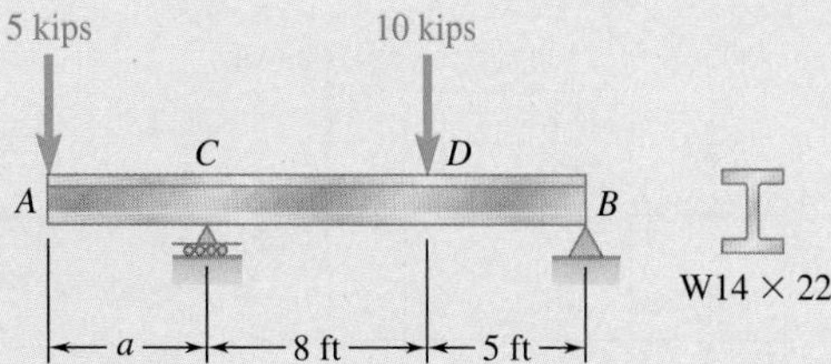

Fig. *P12.82*

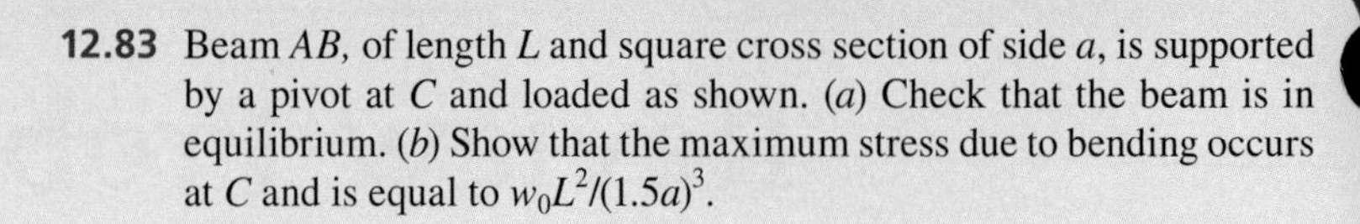

12.83 Beam AB, of length L and square cross section of side a, is supported by a pivot at C and loaded as shown. (*a*) Check that the beam is in equilibrium. (*b*) Show that the maximum stress due to bending occurs at C and is equal to $w_0L^2/(1.5a)^3$.

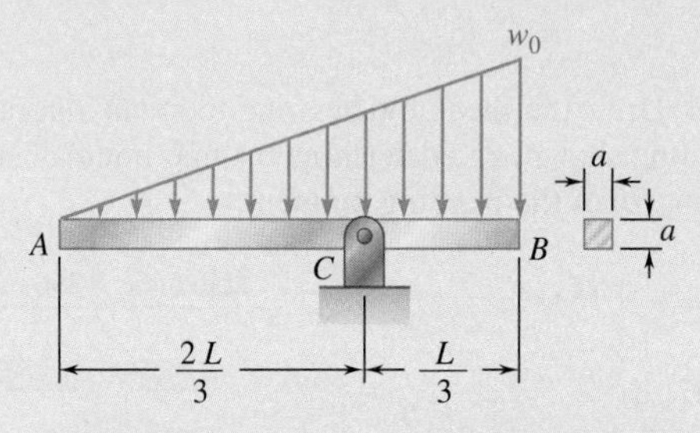

Fig. P12.83

12.84 Knowing that rod AB is in equilibrium under the loading shown, draw the shear and bending-moment diagrams and determine the maximum normal stress due to bending.

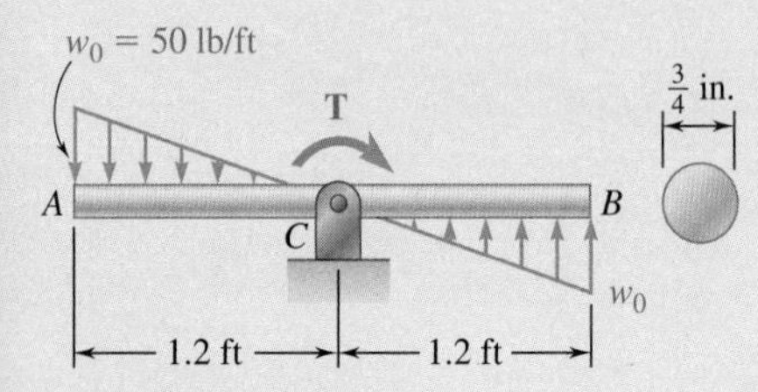

Fig. P12.84

12.85 Three steel plates are welded together to form the beam shown. Knowing that the allowable normal stress for the steel used is 22 ksi, determine the minimum flange width b that can be used.

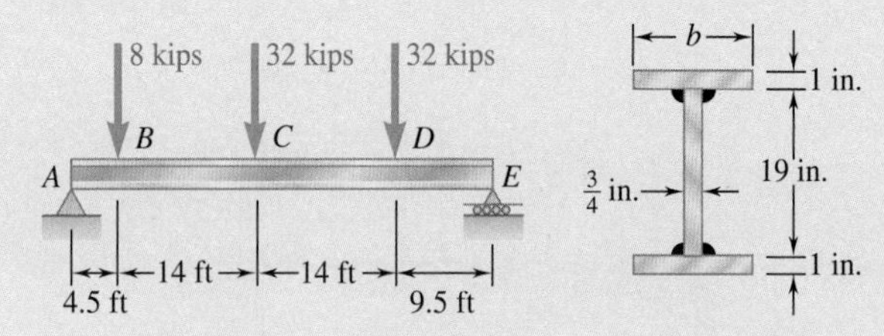

Fig. P12.85

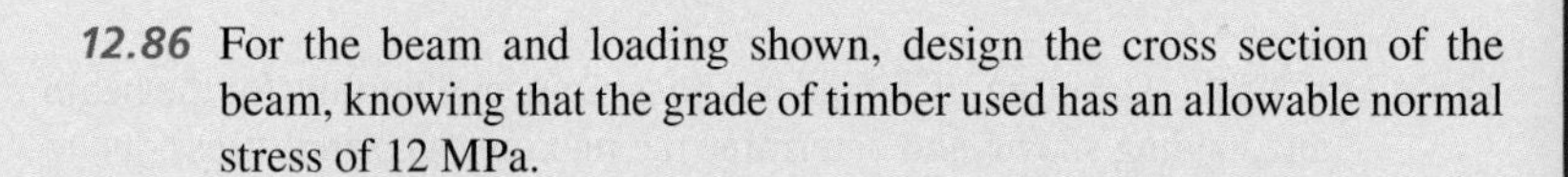

12.86 For the beam and loading shown, design the cross section of the beam, knowing that the grade of timber used has an allowable normal stress of 12 MPa.

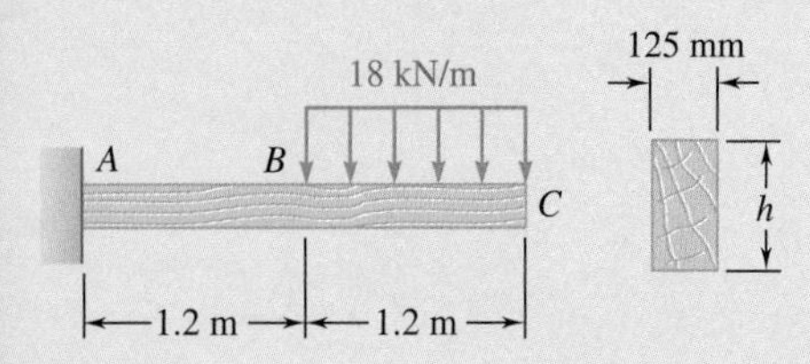

Fig. *P12.86*

12.87 Determine the largest permissible value of **P** for the beam and loading shown, knowing that the allowable normal stress is +8 ksi in tension and −18 ksi in compression.

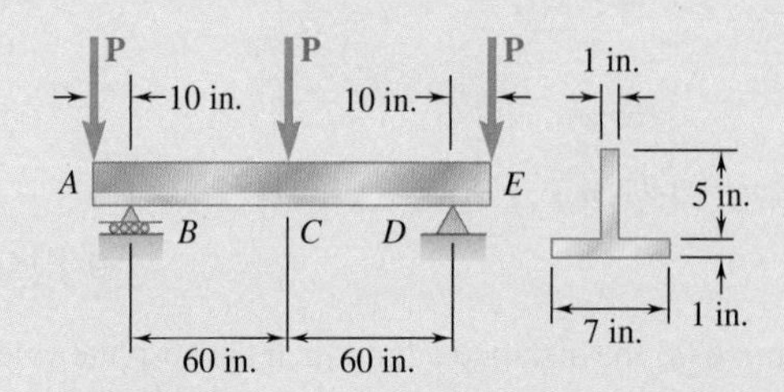

Fig. P12.87

12.88 Knowing that the allowable normal stress for the steel used is 24 ksi, select the most economical wide-flange beam to support the loading shown.

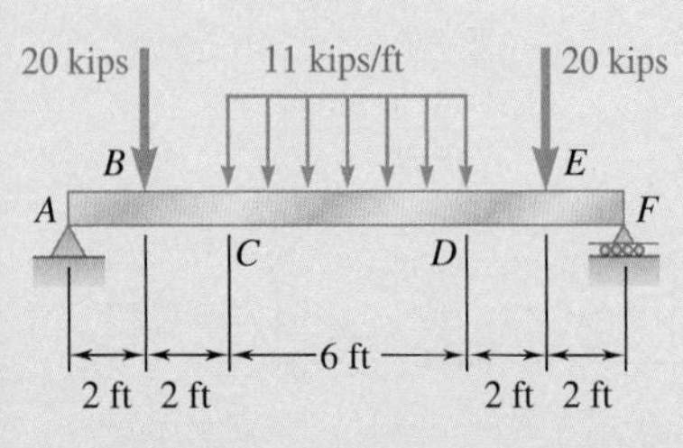

Fig. *P12.88*

© *Aurora Photos/Alamy*

13

Shearing Stresses in Beams and Thin-Walled Members

A reinforced concrete deck will be attached to each of the thin-walled steel sections to form a composite box girder bridge. In this chapter, shearing stresses will be determined in various types of beams and girders.

Objectives

In this chapter, you will:

- **Demonstrate** how transverse loads on a beam generate shearing stresses.
- **Determine** the stresses and shear flow on a horizontal section in a beam.
- **Determine** the shearing stresses in a thin-walled beam.

Introduction

Shearing stresses are important, particularly in the design of short, stubby beams. Their analysis is the subject of this chapter.

Fig. 13.1 graphically expresses the elementary normal and shearing forces exerted on a transverse section of a prismatic beam with a vertical plane of symmetry that are equivalent to the bending couple **M** and the shearing force **V**. Six equations can be written to express this. Three of these equations involve only the normal forces $\sigma_x\, dA$ and have been discussed in Sec. 11.1. These are Eqs. (11.1), (11.2), and (11.3), which express that the sum of the normal forces is zero and that the sums of their moments about the y and z axes are equal to zero and M, respectively. Three more equations involving the shearing forces $\tau_{xy}\, dA$ and $\tau_{xz}\, dA$ now can be written. One equation expresses that the sum of the moments of the shearing forces about the x axis is zero and can be dismissed as trivial in view of the symmetry of the beam with respect to the xy plane. The other two involve the y and z components of the elementary forces and are

$$y \text{ components:} \quad \int \tau_{xy}\, dA = -V \tag{13.1}$$

$$z \text{ components:} \quad \int \tau_{xz}\, dA = 0 \tag{13.2}$$

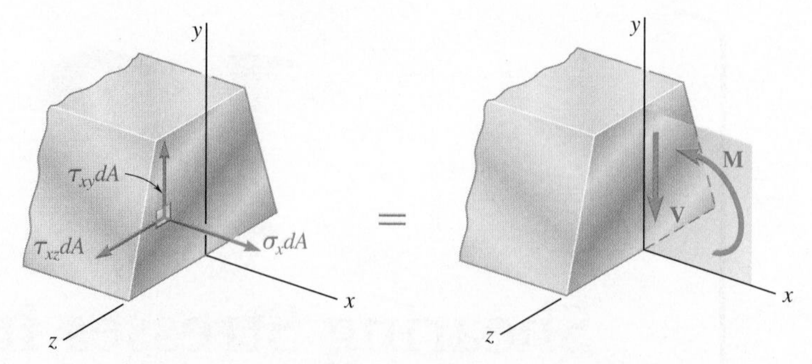

Fig. 13.1 All the stresses on elemental areas (left) sum to give the resultant shear V and bending moment M.

Eq. (13.1) shows that vertical shearing stresses must exist in a transverse section of a beam under transverse loading. Eq. (13.2) indicates that the average lateral shearing stress in any section is zero. However, this does not mean that the shearing stress τ_{xz} is zero everywhere.

Now consider a small cubic element located in the vertical plane of symmetry of the beam (where τ_{xz} must be zero) and examine the stresses exerted on its faces (Fig. 13.2). A normal stress σ_x and a shearing stress τ_{xy} are exerted on each of the two faces perpendicular to the x axis. But we know from Chapter 8 that when shearing stresses τ_{xy} are exerted on the vertical faces of an element, equal stresses must be exerted on the horizontal faces of the same element. Thus, the longitudinal shearing stresses must exist in any member subjected to a transverse loading. This is verified by considering a cantilever beam made of separate planks clamped together at the fixed end (Fig. 13.3*a*). When a transverse load **P** is applied to the free end of this composite beam, the planks slide with respect to each other (Fig. 13.3*b*). In contrast, if a couple **M** is applied to the free end of the same composite beam (Fig. 13.3*c*), the various planks bend into circular concentric arcs and do not slide with respect to each other. This verifies the fact that shear does not occur in a beam subjected to pure bending (see Sec. 11.3).

While sliding does not actually take place when a transverse load **P** is applied to a beam made of a homogeneous and cohesive material such as steel, the tendency to slide exists, showing that stresses occur on horizontal longitudinal planes as well as on vertical transverse planes. In timber beams, whose resistance to shear is weaker between fibers, failure due to shear occurs along a longitudinal plane rather than a transverse plane (Photo 13.1).

In Sec. 13.1A, a beam element of length Δx is considered that is bounded by one horizontal and two transverse planes. The shearing force $\Delta\mathbf{H}$ exerted on its horizontal face will be determined, as well as the shear per unit length q, which is known as *shear flow*. An equation for the shearing stress in a beam with a vertical plane of symmetry is obtained in Sec. 13.1B and used in Sec. 13.1C to determine the shearing stresses in common types of beams.

The method in Sec. 13.1 is extended in Sec. 13.2 to cover the case of a beam element bounded by two transverse planes and a curved surface. This allows us to determine the shearing stresses at any point of a symmetric thin-walled member, such as the flanges of wide-flange beams and box beams in Sec. 13.3.

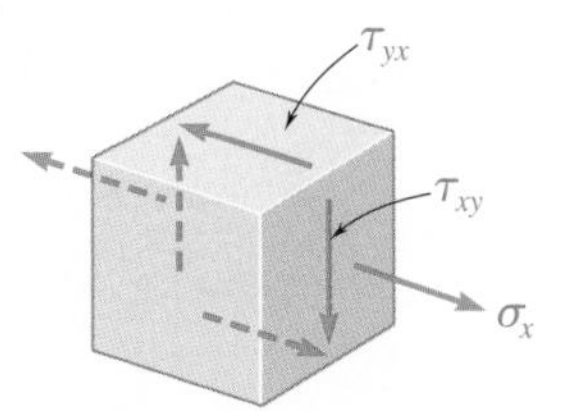

Fig. 13.2 Stress element from section of a transversely loaded beam.

(*a*)
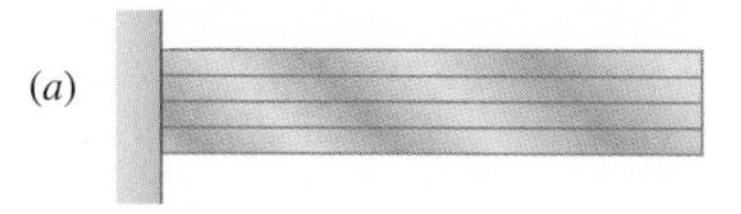

(*b*)
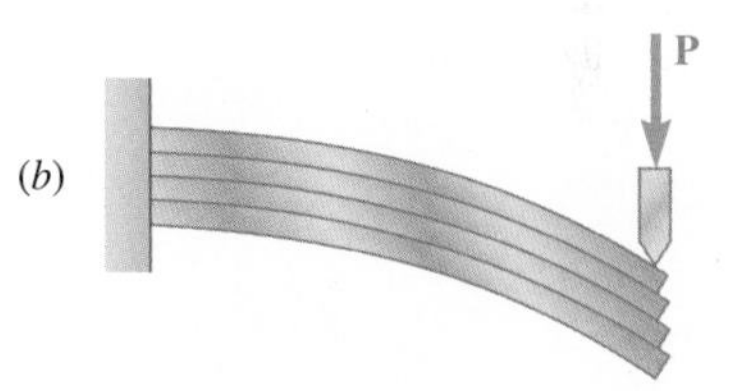

(*c*)
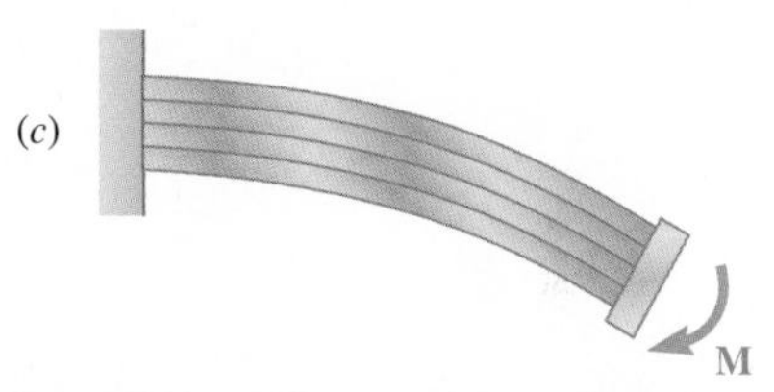

Fig. 13.3 (*a*) Beam made of planks to illustrate the role of shearing stresses. (*b*) Beam planks slide relative to each other when transversely loaded. (*c*) Bending moment causes deflection without sliding.

Photo 13.1 Longitudinal shear failure in timber beam loaded in the laboratory.

13.1 HORIZONTAL SHEARING STRESS IN BEAMS

13.1A Shear on the Horizontal Face of a Beam Element

Consider a prismatic beam AB with a vertical plane of symmetry that supports various concentrated and distributed loads (Fig. 13.4). At a distance x from end A, we detach from the beam an element $CDD'C'$ with length of Δx extending across the width of the beam from the upper surface to a

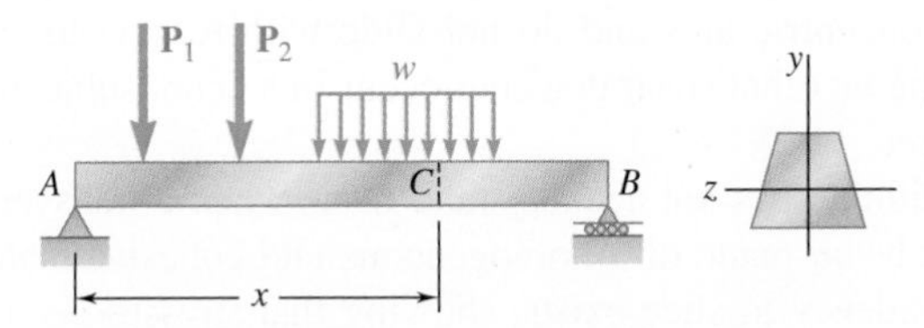

Fig. 13.4 Transversely loaded beam with vertical plane of symmetry.

horizontal plane located at a distance y_1 from the neutral axis (Fig. 13.5). The forces exerted on this element consist of vertical shearing forces $\mathbf{V}'_C$ and $\mathbf{V}'_D$, a horizontal shearing force $\Delta\mathbf{H}$ exerted on the lower face of the element, elementary horizontal normal forces $\sigma_C dA$ and $\sigma_D dA$, and possibly a load $w\,\Delta x$ (Fig. 13.6). The equilibrium equation for horizontal forces is

$$\xrightarrow{+} \Sigma F_x = 0: \qquad \Delta H + \int_{\mathcal{a}} (\sigma_C - \sigma_D)\, dA = 0$$

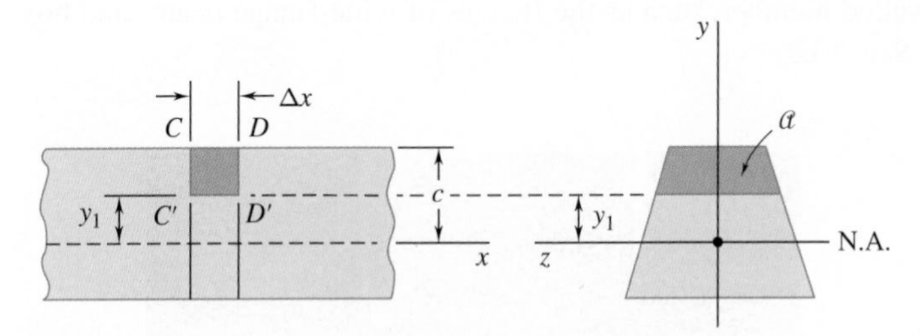

Fig. 13.5 Short segment of beam with stress element $CDD'C'$ defined.

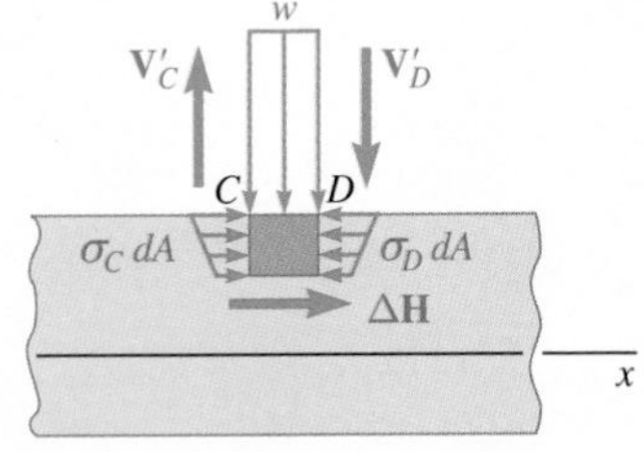

Fig. 13.6 Forces exerted on element $CCD'C'$.

where the integral extends over the shaded area $\mathcal{a}$ of the section located above the line $y = y_1$. Solving this equation for ΔH and using Eq. (12.2), $\sigma = My/I$, to express the normal stresses in terms of the bending moments at C and D, provides

$$\Delta H = \frac{M_D - M_C}{I} \int_{\mathcal{a}} y\, dA \tag{13.3}$$

The integral in Eq. (13.3) represents the *first moment* with respect to the neutral axis of the portion $\mathcal{A}$ of the cross section of the beam that is located above the line $y = y_1$ and will be denoted by Q. On the other hand, recalling Eq. (12.7), the increment $M_D - M_C$ of the bending moment is

$$M_D - M_C = \Delta M = (dM/dx)\,\Delta x = V\,\Delta x$$

Substituting into Eq. (13.3), the horizontal shear exerted on the beam element is

$$\Delta H = \frac{VQ}{I}\Delta x \qquad \textbf{(13.4)}$$

The same result is obtained if a free body the lower element $C'D'D''C''$ is used instead of the upper element $CDD'C'$ (Fig. 13.7), since the shearing forces $\Delta\mathbf{H}$ and $\Delta\mathbf{H}'$ exerted by the two elements on each other are equal and opposite. This leads us to observe that the first moment Q of the portion $\mathcal{A}'$ of the cross section located below the line $y = y_1$ (Fig. 13.7) is equal in magnitude and opposite in sign to the first moment of the portion $\mathcal{A}$ located above that line (Fig. 13.5). Indeed, the sum of these two moments is equal to the moment of the area of the entire cross section with respect to its centroidal axis and, thus must be zero. This property is sometimes used to simplify the computation of Q. Also note that Q is maximum for $y_1 = 0$, since the elements of the cross section located above the neutral axis contribute positively to the integral in Eq. (13.3) that defines Q, while the elements located below that axis contribute negatively.

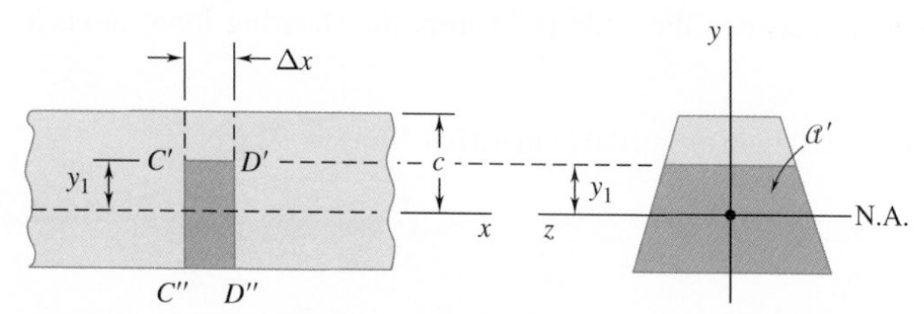

Fig. 13.7 Short segment of beam with stress element $C'D'D''C''$ defined.

The *horizontal shear per unit length*, which will be denoted by q, is obtained by dividing both members of Eq. (13.4) by Δx:

$$q = \frac{\Delta H}{\Delta x} = \frac{VQ}{I} \qquad \textbf{(13.5)}$$

Recall that Q is the first moment with respect to the neutral axis of the portion of the cross section located either above or below the point at which q is being computed and that I is the centroidal moment of inertia of the *entire* cross-sectional area. The horizontal shear per unit length q is also called the *shear flow* and will be discussed in Sec. 13.3.

Concept Application 13.1

A beam is made of three planks, 20 by 100 mm in cross section, and nailed together (Fig. 13.8*a*). Knowing that the spacing between nails is 25 mm and the vertical shear in the beam is $V = 500$ N, determine the shearing force in each nail.

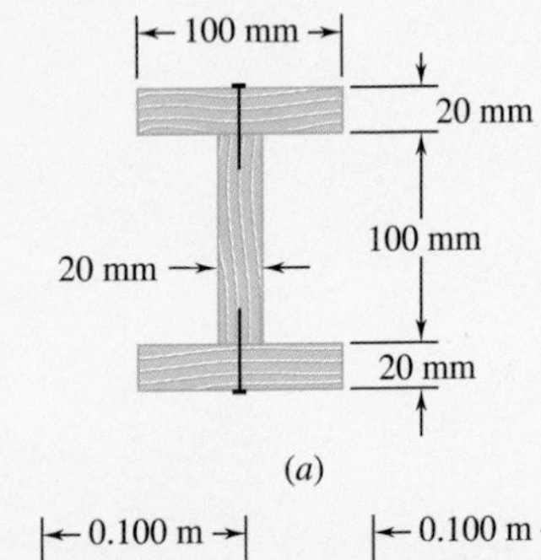

Determine the horizontal force per unit length q exerted on the lower face of the upper plank. Use Eq. (13.5), where Q represents the first moment with respect to the neutral axis of the shaded area A shown in Fig. 13.8*b*, and I is the moment of inertia about the same axis of the entire cross-sectional area (Fig. 13.8*c*). Recalling that the first moment of an area with respect to a given axis is equal to the product of the area and of the distance from its centroid to the axis,†

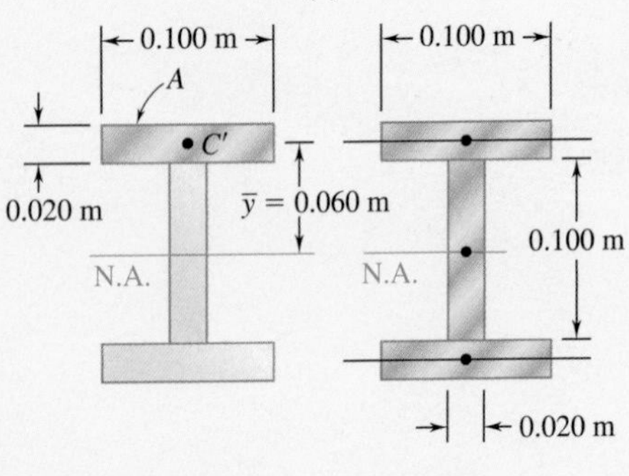

Fig. 13.8 (*a*) Beam made of three boards nailed together. (*b*) Cross section for computing Q. (*c*) Cross section for computing moment of inertia.

$$Q = A\bar{y} = (0.020 \text{ m} \times 0.100 \text{ m})(0.060 \text{ m})$$
$$= 120 \times 10^{-6} \text{ m}^3$$
$$I = \tfrac{1}{12}(0.020 \text{ m})(0.100 \text{ m})^3$$
$$+2[\tfrac{1}{12}(0.100 \text{ m})(0.020 \text{ m})^3$$
$$+(0.020 \text{ m} \times 0.100 \text{ m})(0.060 \text{ m})^2]$$
$$= 1.667 \times 10^{-6} + 2(0.0667 + 7.2)10^{-6}$$
$$= 16.20 \times 10^{-6} \text{ m}^4$$

Substituting into Eq. (13.5),

$$q = \frac{VQ}{I} = \frac{(500 \text{ N})(120 \times 10^{-6} \text{ m}^3)}{16.20 \times 10^{-6} \text{ m}^4} = 3704 \text{ N/m}$$

Since the spacing between the nails is 25 mm, the shearing force in each nail is

$$F = (0.025 \text{ m})q = (0.025 \text{ m})(3704 \text{ N/m}) = 92.6 \text{ N}$$

13.1B Shearing Stresses in a Beam

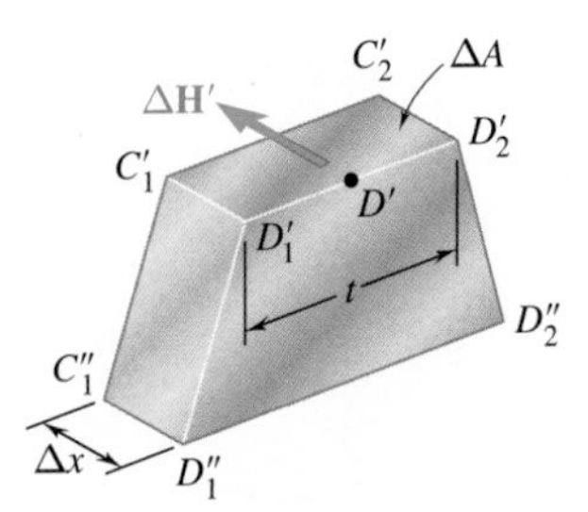

Fig. 13.9 Stress element $C'D'D''C''$ showing the shear force on a horizontal plane.

Consider again a beam with a vertical plane of symmetry that is subjected to various concentrated or distributed loads applied in that plane. If, through two vertical cuts and one horizontal cut, an element of length Δx is detached from the beam (Fig. 13.9), the magnitude ΔH of the shearing force exerted on the horizontal face of the element can be obtained from Eq. (13.4). The *average shearing stress* τ_{ave} on that face of the element is obtained by dividing ΔH by the area ΔA of the face. Observing that $\Delta A = t\,\Delta x$, where t is the width of the element at the cut, we write

$$\tau_{ave} = \frac{\Delta H}{\Delta A} = \frac{VQ}{I}\frac{\Delta x}{t\,\Delta x}$$

†See Sec. 5.1C.

or

$$\tau_{ave} = \frac{VQ}{It} \tag{13.6}$$

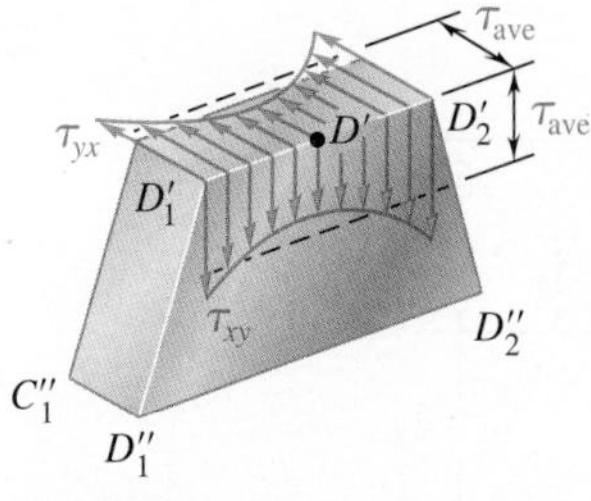

Fig. 13.10 Stress element $C'D'D''C''$ showing the shearing stress distribution along $D'_1 D'_2$.

Note that since the shearing stresses τ_{xy} and τ_{yx} exerted on a transverse and a horizontal plane through D' are equal, the expression also represents the average value of τ_{xy} along the line $D'_1 D'_2$ (Fig. 13.10).

Observe that $\tau_{yx} = 0$ on the upper and lower faces of the beam, since no forces are exerted on these faces. It follows that $\tau_{xy} = 0$ along the upper and lower edges of the transverse section (Fig. 13.11). Also note that while Q is maximum for $y = 0$ (see Sec. 13.1A), τ_{ave} may not be maximum along the neutral axis, since τ_{ave} depends upon the width t of the section as well as upon Q.

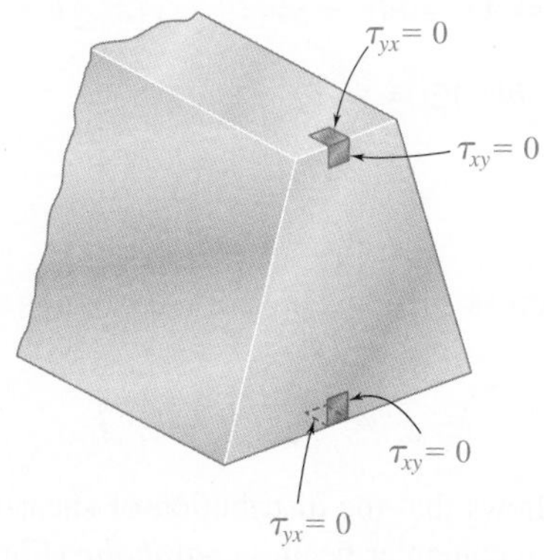

Fig. 13.11 Beam cross section showing that the shearing stress is zero at the top and bottom of the beam.

As long as the width of the beam cross section remains small compared to its depth, the shearing stress varies only slightly along the line $D'_1 D'_2$ (Fig. 13.10), and Eq. (13.6) can be used to compute τ_{xy} at any point along $D'_1 D'_2$. Actually, τ_{xy} is larger at points D'_1 and D'_2 than at D', but the theory of elasticity shows† that, for a beam of rectangular section of width b and depth h, and as long as $b \leq h/4$, the value of the shearing stress at points C_1 and C_2 (Fig. 13.12) does not exceed by more than 0.8% the average value of the stress computed along the neutral axis.

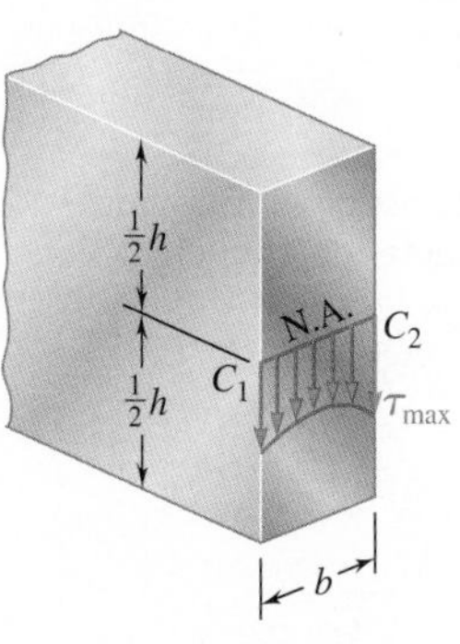

Fig. 13.12 Shearing stress distribution along neutral axis of rectangular beam cross section.

On the other hand, for large values of b/h, τ_{max} of the stress at C_1 and C_2 may be many times larger then the average value τ_{ave} computed along the neutral axis, as shown in the following table.

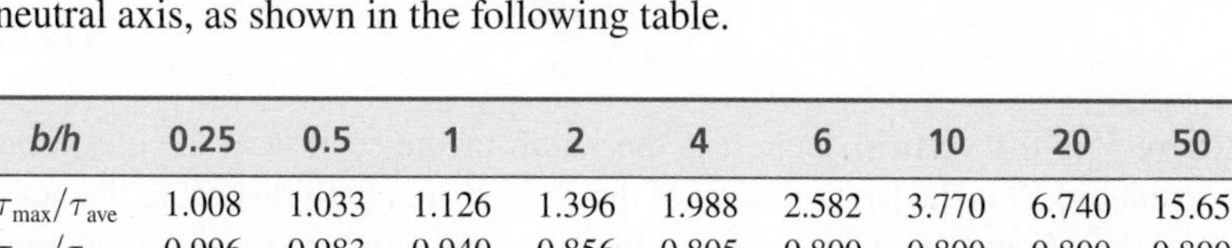

b/h	0.25	0.5	1	2	4	6	10	20	50
τ_{max}/τ_{ave}	1.008	1.033	1.126	1.396	1.988	2.582	3.770	6.740	15.65
τ_{min}/τ_{ave}	0.996	0.983	0.940	0.856	0.805	0.800	0.800	0.800	0.800

†See S. P. Timoshenko and J. N. Goodier, *Theory of Elasticity*, McGraw-Hill, New York, 3d ed., 1970, sec. 124.

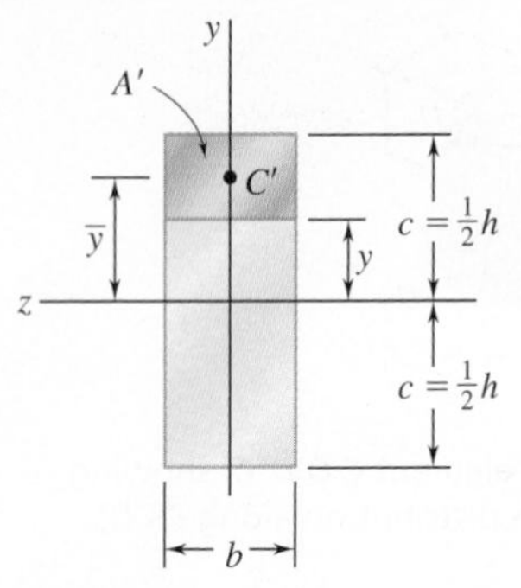

Fig. 13.13 Geometric terms for rectangular section used to calculate shearing stress.

13.1C Shearing Stresses τ_{xy} In Common Beam Types

In the preceding section for a *narrow rectangular beam* (i.e., a beam of rectangular section of width b and depth h with $b \leq \frac{1}{4}h$), the variation of the shearing stress τ_{xy} across the width of the beam is less than 0.8% of τ_{ave}. Therefore, Eq. (13.6) is used in practical applications to determine the shearing stress at any point of the cross section of a narrow rectangular beam, and

$$\tau_{xy} = \frac{VQ}{It} \tag{13.7}$$

where t is equal to the width b of the beam and Q is the first moment with respect to the neutral axis of the shaded area A (Fig. 13.13).

Observing that the distance from the neutral axis to the centroid C' of A is $\bar{y} = \frac{1}{2}(c + y)$ and recalling that $Q = A\bar{y}$,

$$Q = A\bar{y} = b(c - y)\tfrac{1}{2}(c + y) = \tfrac{1}{2}b(c^2 - y^2) \tag{13.8}$$

Recalling that $I = bh^3/12 = \frac{2}{3}bc^3$,

$$\tau_{xy} = \frac{VQ}{Ib} = \frac{3}{4}\frac{c^2 - y^2}{bc^3}V$$

or noting that the cross-sectional area of the beam is $A = 2bc$,

$$\tau_{xy} = \frac{3}{2}\frac{V}{A}\left(1 - \frac{y^2}{c^2}\right) \tag{13.9}$$

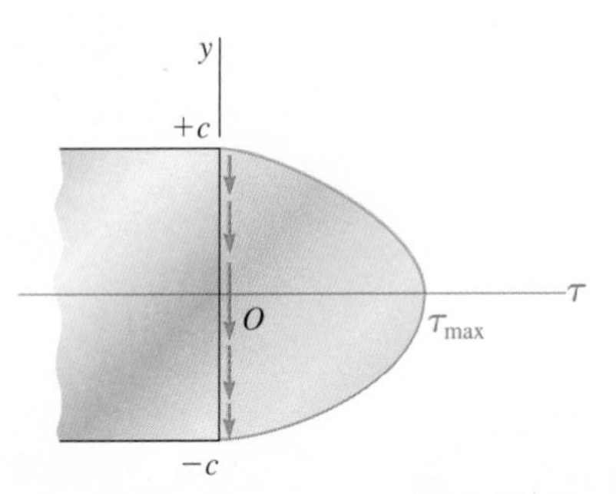

Fig. 13.14 Shearing stress distribution on transverse section of rectangular beam.

Eq. (13.9) shows that the distribution of shearing stresses in a transverse section of a rectangular beam is *parabolic* (Fig. 13.14). As observed in the preceding section, the shearing stresses are zero at the top and bottom of the cross section ($y = \pm c$). Making $y = 0$ in Eq. (13.9), the value of the maximum shearing stress in a given section of a *narrow rectangular beam* is

$$\tau_{max} = \frac{3}{2}\frac{V}{A} \tag{13.10}$$

This relationship shows that the maximum value of the shearing stress in a beam of rectangular cross section is 50% larger than the value V/A obtained by wrongly assuming a uniform stress distribution across the entire cross section.

In an *American standard beam* (S-beam) or a *wide-flange beam* (W-beam), Eq. (13.6) can be used to determine the average value of the shearing stress τ_{xy} over a section aa' or bb' of the transverse cross section of the beam (Figs. 13.15*a* and *b*). So

$$\tau_{ave} = \frac{VQ}{It} \tag{13.6}$$

where V is the vertical shear, t is the width of the section at the elevation considered, Q is the first moment of the shaded area with respect to the neutral axis cc', and I is the moment of inertia of the entire cross-sectional area about cc'. Plotting τ_{ave} against the vertical distance y provides the curve shown in Fig. 13.15*c*. Note the discontinuities existing in this curve, which reflect the difference between the values of t corresponding respectively to the flanges $ABGD$ and $A'B'G'D'$ and to the web $EFF'E'$.

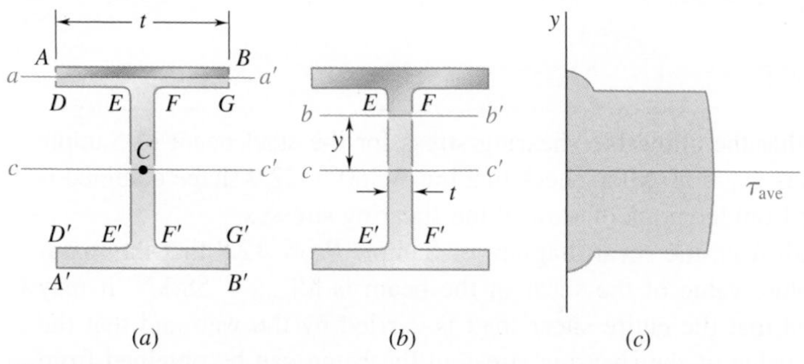

Fig. 13.15 Wide-flange beam. (*a*) Area for finding first moment of area in flange. (*b*) Area for finding first moment of area in web. (*c*) Shearing stress distribution.

In the web, the shearing stress τ_{xy} varies only very slightly across the section bb' and is assumed to be equal to its average value τ_{ave}. This is not true, however, for the flanges. For example, considering the horizontal line $DEFG$, note that τ_{xy} is zero between D and E and between F and G, since these two segments are part of the free surface of the beam. However, the value of τ_{xy} between E and F is non-zero and can be obtained by making $t = EF$ in Eq. (13.6). In practice, one usually assumes that the entire shear load is carried by the web and that a good approximation of the maximum value of the shearing stress in the cross section can be obtained by dividing V by the cross-sectional area of the web.

$$\tau_{max} = \frac{V}{A_{web}} \tag{13.11}$$

However, while the vertical component τ_{xy} of the shearing stress in the flanges can be neglected, its horizontal component τ_{xz} has a significant value that will be determined in Sec. 13.3.

Concept Application 13.2

Knowing that the allowable shearing stress for the timber beam of Sample Prob. 12.7 is $\tau_{all} = 0.250$ ksi, check that the design is acceptable from the point of view of the shearing stresses.

Recall from the shear diagram of Sample Prob. 12.7 that $V_{max} = 4.50$ kips. The actual width of the beam was given as $b = 3.5$ in., and the value obtained for its depth was $h = 14.55$ in. Using Eq. (13.10) for the maximum shearing stress in a narrow rectangular beam,

$$\tau_{max} = \frac{3}{2}\frac{V}{A} = \frac{3}{2}\frac{V}{bh} = \frac{3(4.50 \text{ kips})}{2(3.5 \text{ in.})(14.55 \text{ in.})} = 0.1325 \text{ ksi}$$

Since $\tau_{max} < \tau_{all}$, the design obtained in Sample Prob. 12.7 is acceptable.

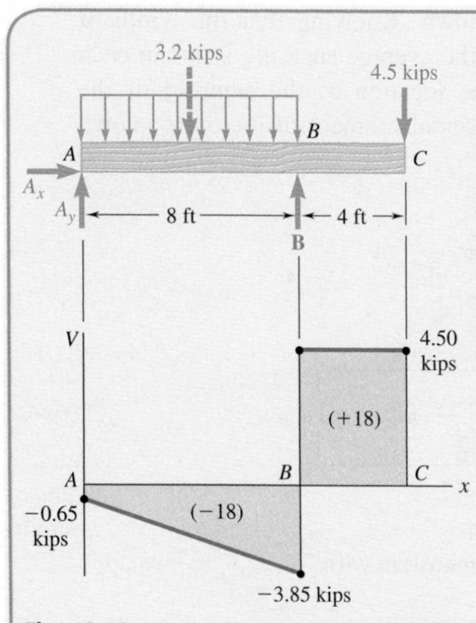

Fig. 13.16 Shear diagram for beam of Sample Problem 12.7.

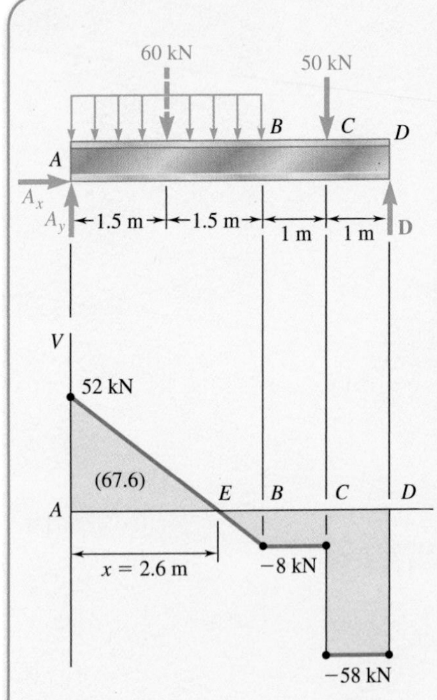

Fig. 13.17 Shear diagram for beam of Sample Problem 12.8.

Concept Applications 13.3

Knowing that the allowable shearing stress for the steel beam of Sample Prob. 12.8 is $\tau_{all} = 90$ MPa, check that the W360 × 32.9 shape obtained is acceptable from the point of view of the shearing stresses.

Recall from the shear diagram of Sample Prob. 12.8 that the maximum absolute value of the shear in the beam is $|V|_{max} = 58$ kN. It may be assumed that the entire shear load is carried by the web and that the maximum value of the shearing stress in the beam can be obtained from Eq. (13.11). From Appendix B, for a W360 × 32.9 shape, the depth of the beam and the thickness of its web are $d = 348$ mm and $t_w = 5.84$ mm. Thus,

$$A_{web} = d\,t_w = (348\text{ mm})(5.84\text{ mm}) = 2032\text{ mm}^2$$

Substituting $|V|_{max}$ and A_{web} into Eq. (13.11),

$$\tau_{max} = \frac{|V|_{max}}{A_{web}} = \frac{58\text{ kN}}{2032\text{ mm}^2} = 28.5\text{ MPa}$$

Since $\tau_{max} < \tau_{all}$, the design obtained in Sample Prob. 12.8 is acceptable.

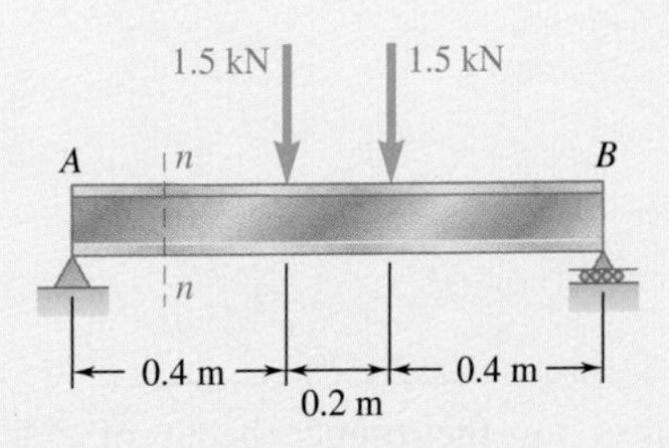

Sample Problem 13.1

Beam *AB* is made of three plates glued together and is subjected, in its plane of symmetry, to the loading shown. Knowing that the width of each glued joint is 20 mm, determine the average shearing stress in each joint at section *n–n* of the beam. The location of the centroid of the section is given in Fig. 1 and the centroidal moment of inertia is known to be $I = 8.63 \times 10^{-6}$ m^4.

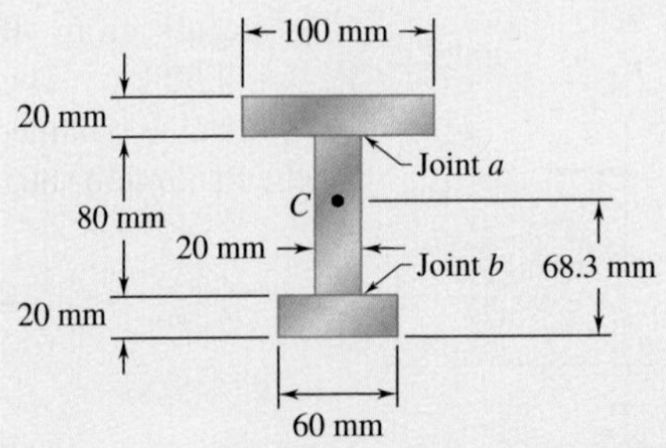

Fig. 1 Cross section dimensions with location of centroid.

(continued)

STRATEGY: A free-body diagram is first used to determine the shear at the required section. Eq. (13.7) is then used to determine the average shearing stress in each joint.

MODELING:

Vertical Shear at Section *n–n*. As shown in the free-body diagram in Fig. 2, the beam and loading are both symmetric with respect to the center of the beam. Thus, we have **A** = **B** = 1.5 kN ↑.

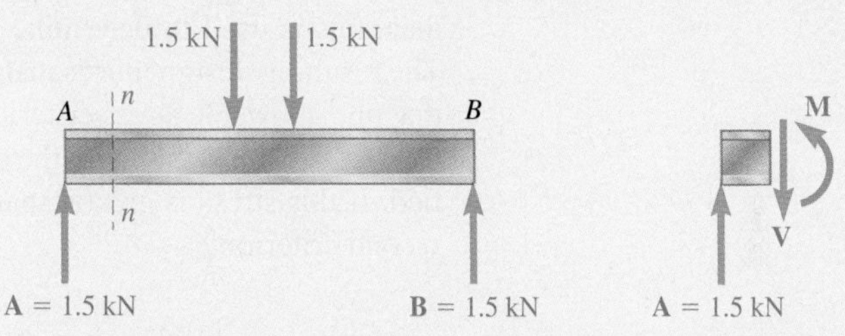

Fig. 2 Free-body diagram of beam and segment of beam to left of section *n–n*.

Drawing the free-body diagram of the portion of the beam to the left of section *n–n* (Fig. 2), we write

$$+\uparrow \Sigma F_y = 0: \qquad 1.5 \text{ kN} - V = 0 \qquad V = 1.5 \text{ kN}$$

ANALYSIS:

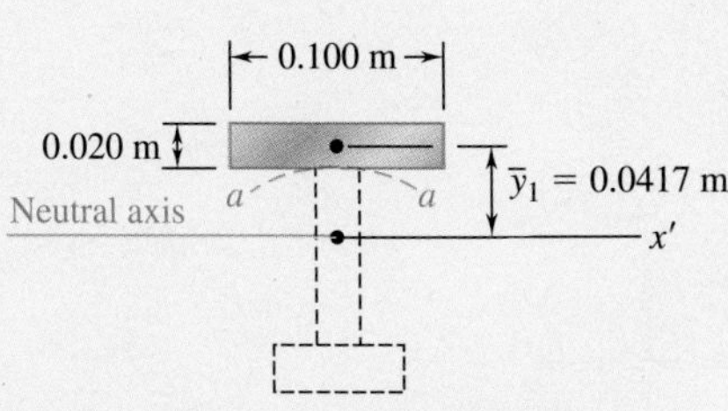

Fig. 3 Using area above section *a–a* to find *Q*.

Shearing Stress in Joint *a*. Using Fig. 3, pass the section *a–a* through the glued joint and separate the cross-sectional area into two parts. We choose to determine Q by computing the first moment with respect to the neutral axis of the area above section *a–a*.

$$Q = A\bar{y}_1 = [(0.100 \text{ m})(0.020 \text{ m})](0.0417 \text{ m}) = 83.4 \times 10^{-6} \text{ m}^3$$

Recalling that the width of the glued joint is $t = 0.020$ m, we use Eq. (13.7) to determine the average shearing stress in the joint.

$$\tau_{\text{ave}} = \frac{VQ}{It} = \frac{(1500 \text{ N})(83.4 \times 10^{-6} \text{ m}^3)}{(8.63 \times 10^{-6} \text{ m}^4)(0.020 \text{ m})} \qquad \tau_{\text{ave}} = 725 \text{ kPa} \blacktriangleleft$$

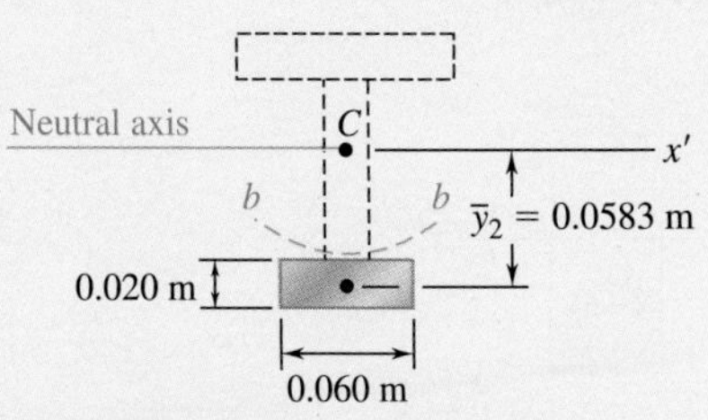

Fig. 4 Using area below section *b–b* to find *Q*.

Shearing Stress in Joint *b*. Using Fig. 4, now pass section *b–b* and compute Q by using the area below the section.

$$Q = A\bar{y}_2 = [(0.060 \text{ m})(0.020 \text{ m})](0.0583 \text{ m}) = 70.0 \times 10^{-6} \text{ m}^3$$

$$\tau_{\text{ave}} = \frac{VQ}{It} = \frac{(1500 \text{ N})(70.0 \times 10^{-6} \text{ m}^3)}{(8.63 \times 10^{-6} \text{ m}^4)(0.020 \text{ m})} \qquad \tau_{\text{ave}} = 608 \text{ kPa} \blacktriangleleft$$

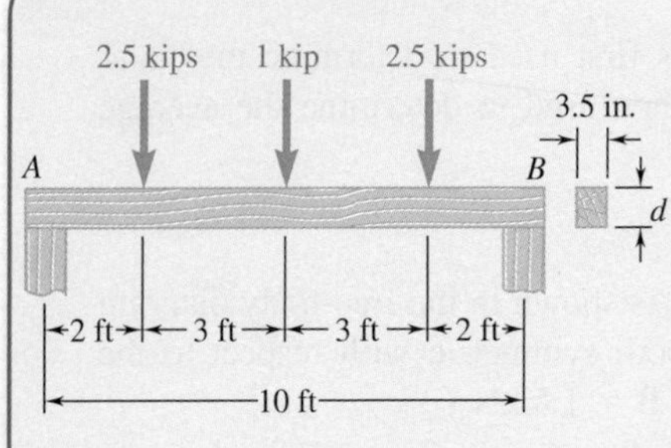

Sample Problem 13.2

A timber beam AB of span 10 ft and nominal width 4 in. (actual width = 3.5 in.) is to support the three concentrated loads shown. Knowing that for the grade of timber used $\sigma_{all} = 1800$ psi and $\tau_{all} = 120$ psi, determine the minimum required depth d of the beam.

STRATEGY: A free-body diagram with the shear and bending-moment diagrams is used to determine the maximum shear and bending moment. The resulting design must satisfy both allowable stresses. Start by assuming that one allowable stress criterion governs, and solve for the required depth d. Then use this depth with the other criterion to determine if it is also satisfied. If this stress is greater than the allowable, revise the design using the second criterion.

MODELING:

Maximum Shear and Bending Moment. The free-body diagram is used to determine the reactions and draw the shear and bending-moment diagrams in Fig. 1. We note that

$$M_{max} = 7.5 \text{ kip·ft} = 90 \text{ kip·in.}$$

$$V_{max} = 3 \text{ kips}$$

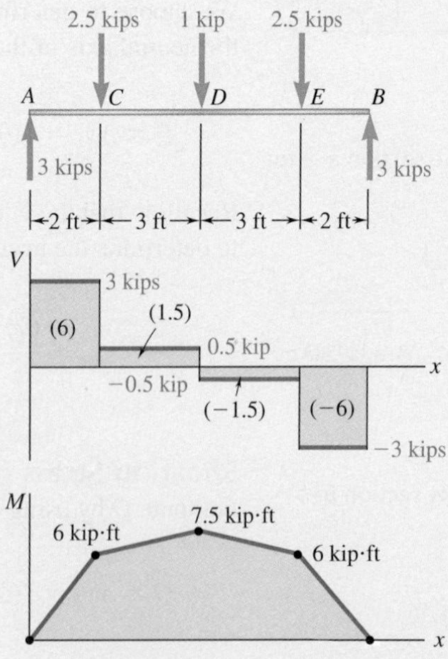

Fig. 1 Free-body diagram of beam with shear and bending-moment diagrams.

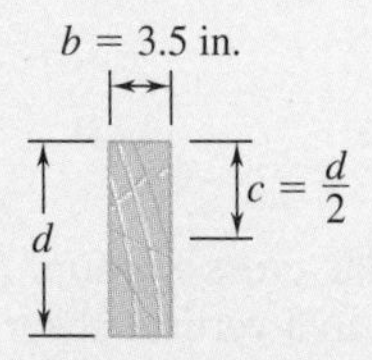

Fig. 2 Section of beam having depth *d*.

ANALYSIS:

Design Based on Allowable Normal Stress. We first express the elastic section modulus S in terms of the depth d (Fig. 2). We have

$$I = \frac{1}{12}bd^3 \qquad S = \frac{1}{c} = \frac{1}{6}bd^2 = \frac{1}{6}(3.5)d^2 = 0.5833d^2$$

For $M_{max} = 90$ kip·in. and $\sigma_{all} = 1800$ psi, we write

$$S = \frac{M_{max}}{\sigma_{all}} \qquad 0.5833d^2 = \frac{90 \times 10^3 \text{ lb·in.}}{1800 \text{ psi}}$$

$$d^2 = 85.7 \qquad d = 9.26 \text{ in.}$$

We have satisfied the requirement that $\sigma_m \leq 1800$ psi.

Check Shearing Stress. For $V_{max} = 3$ kips and $d = 9.26$ in., we find

$$\tau_m = \frac{3}{2}\frac{V_{max}}{A} = \frac{3}{2}\frac{3000 \text{ lb}}{(3.5 \text{ in.})(9.26 \text{ in.})} \qquad \tau_m = 138.8 \text{ psi}$$

Since $\tau_{all} = 120$ psi, the depth $d = 9.26$ in. is *not* acceptable and we must redesign the beam on the basis of the requirement that $\tau_m \leq 120$ psi.

Design Based on Allowable Shearing Stress. Since we now know that the allowable shearing stress controls the design, we write

$$\tau_m = \tau_{all} = \frac{3}{2}\frac{V_{max}}{A} \qquad 120 \text{ psi} = \frac{3}{2}\frac{3000 \text{ lb}}{(3.5 \text{ in.})d}$$

$$d = 10.71 \text{ in.} \blacktriangleleft$$

The normal stress is, of course, less than $\sigma_{all} = 1800$ psi, and the depth of 10.71 in. is fully acceptable.

REFLECT and THINK: Since timber is normally available in nominal depth increments of 2 in., a 4 × 12-in. standard size timber should be used. The actual cross section would then be 3.5 × 11.25 in. (Fig. 3).

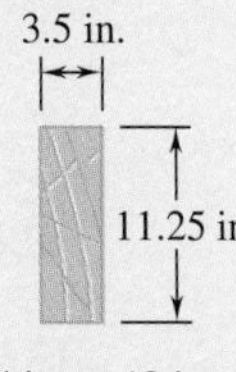

Fig. 3 Design cross section.

Problems

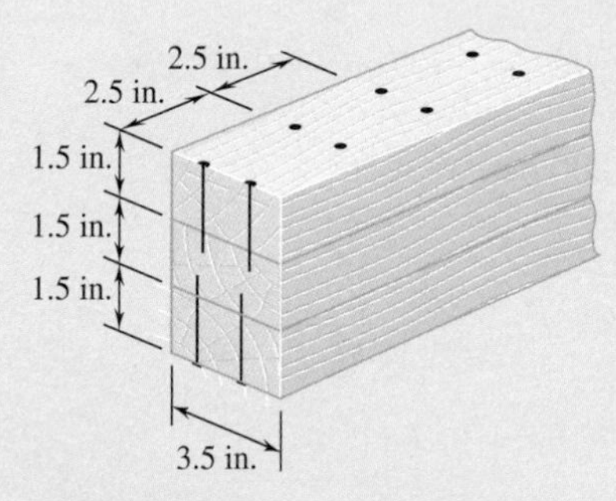

Fig. P13.1

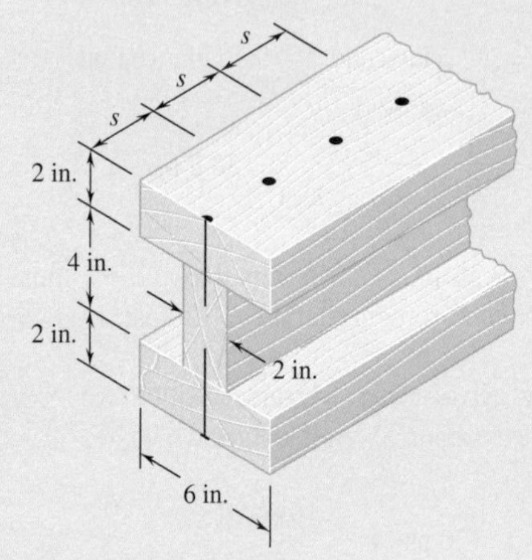

Fig. P13.2

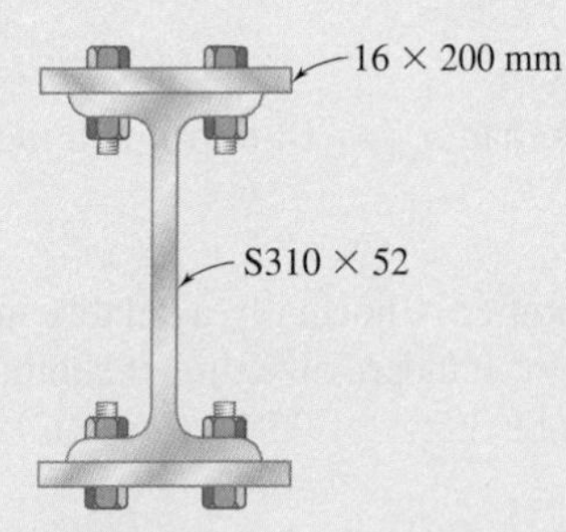

Fig. P13.5

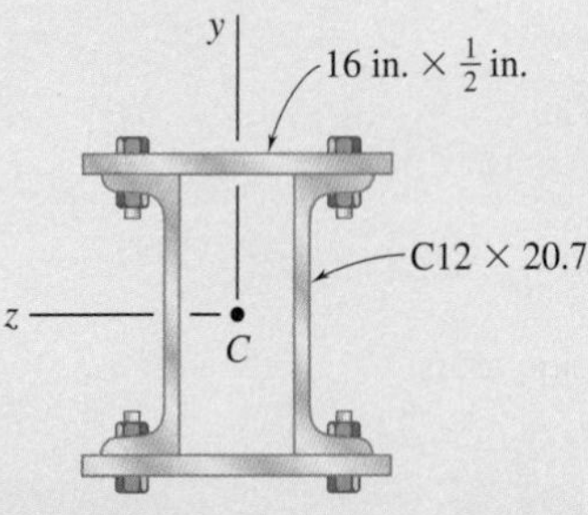

Fig. P13.7

13.1 Three boards, each of 1.5×3.5-in. rectangular cross section, are nailed together to form a beam that is subjected to a vertical shear of 250 lb. Knowing that the spacing between each pair of nails is 2.5 in., determine the shearing force in each nail.

13.2 Three boards, each 2 in. thick, are nailed together to form a beam that is subjected to a vertical shear. Knowing that the allowable shearing force in each nail is 150 lb, determine the allowable shear if the spacing s between the nails is 3 in.

13.3 Three boards are nailed together to form a beam shown, which is subjected to a vertical shear. Knowing that the spacing between the nails is $s = 75$ mm and that the allowable shearing force in each nail is 400 N, determine the allowable shear when $w = 120$ mm.

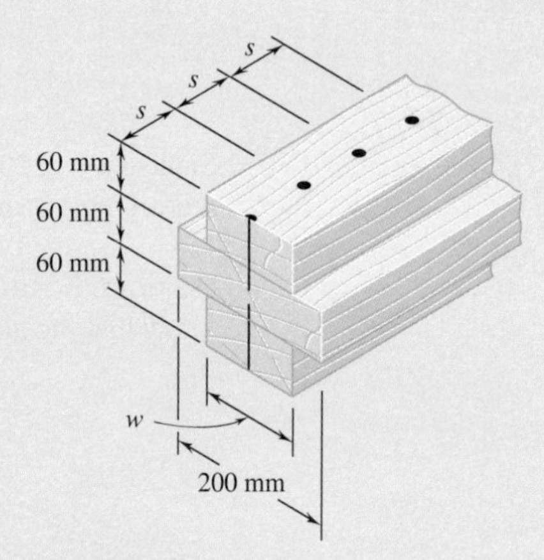

Fig. P13.3

13.4 Solve Prob. 13.3, assuming that the width of the top and bottom boards is changed to $w = 100$ mm.

13.5 The American Standard rolled-steel beam shown has been reinforced by attaching to it two 16×200-mm plates, using 18-mm-diameter bolts spaced longitudinally every 120 mm. Knowing that the average allowable shearing stress in the bolts is 90 MPa, determine the largest permissible vertical shearing force.

13.6 Solve Prob. 13.5, assuming that the reinforcing plates are only 12 mm thick.

13.7 The beam shown is fabricated by connecting two channel shapes and two plates, using bolts of $\frac{3}{4}$-in. diameter spaced longitudinally every 7.5 in. Determine the average shearing stress in the bolts caused by a shearing force of 25 kips parallel to the y axis.

13.8 A beam is fabricated by connecting the rolled-steel members shown by bolts of $\frac{3}{4}$-in. diameter spaced longitudinally every 5 in. Determine the average shearing stress in the bolts caused by a shearing force of 30 kips parallel to the y axis.

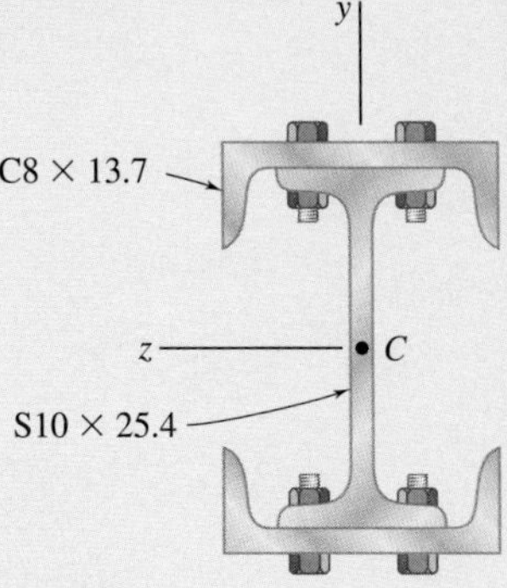

Fig. P13.8

13.9 through 13.12 For the beam and loading shown, consider section n–n and determine (*a*) the largest shearing stress in that section, (*b*) the shearing stress at point a.

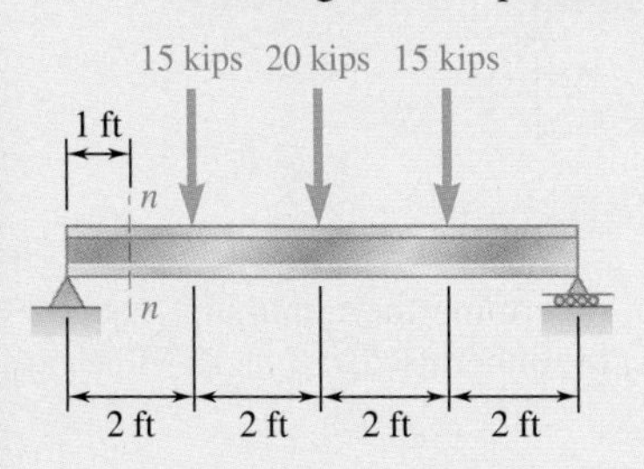

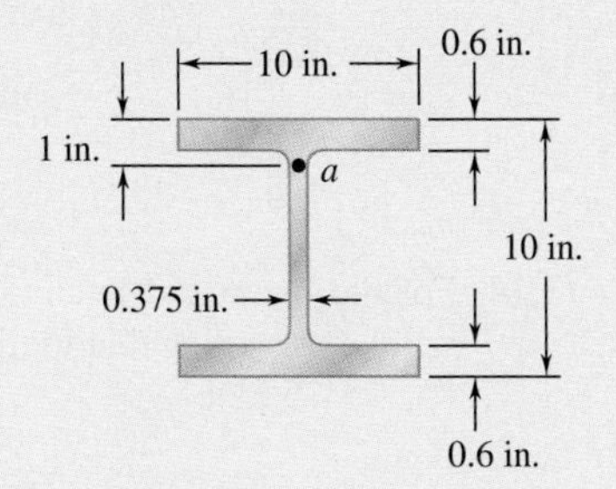

Fig. P13.9

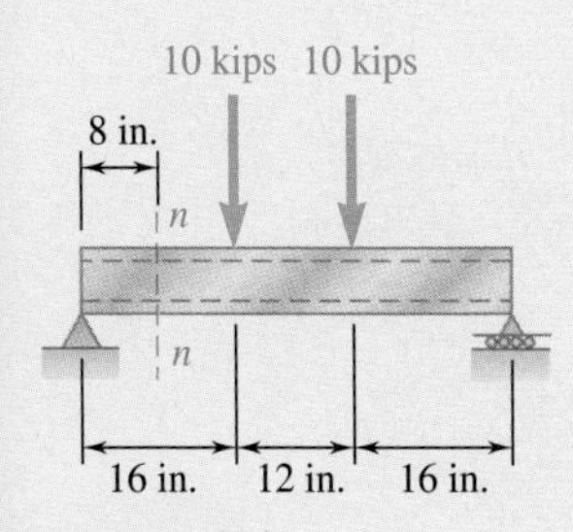

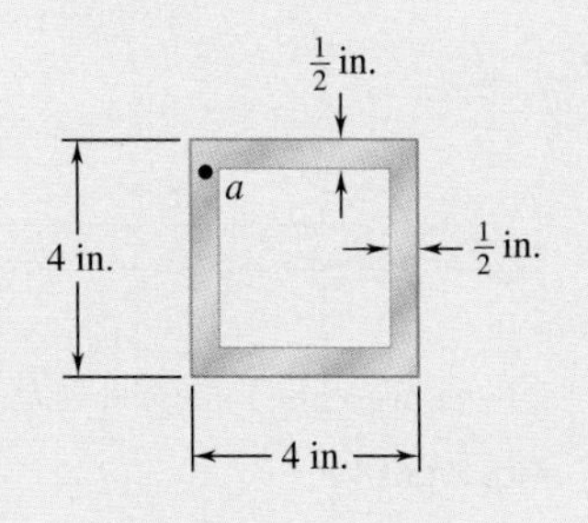

Fig. P13.10

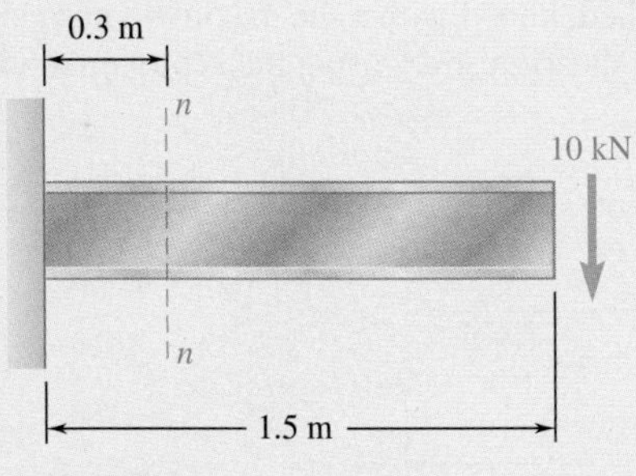

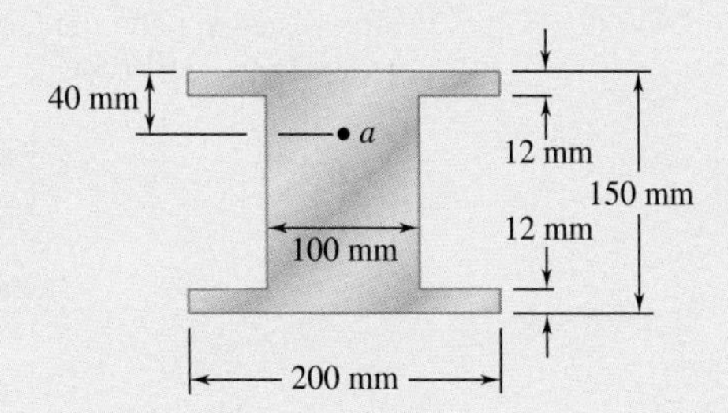

Fig. P13.11

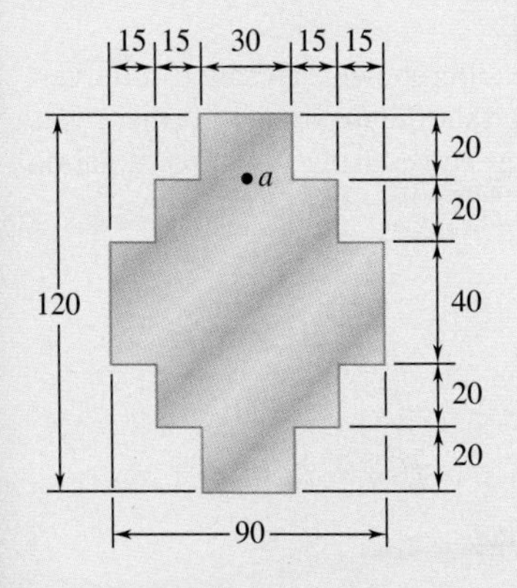

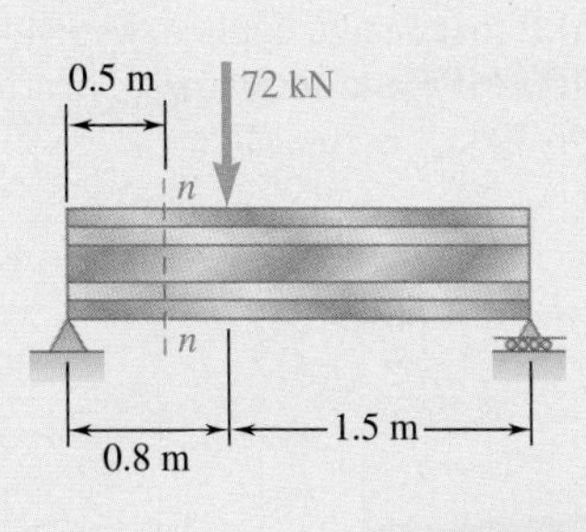

Fig. P13.12

13.13 For the beam and loading shown, determine the minimum required depth h, knowing that for the grade of timber used, $\sigma_{all} = 1750$ psi and $\tau_{all} = 130$ psi.

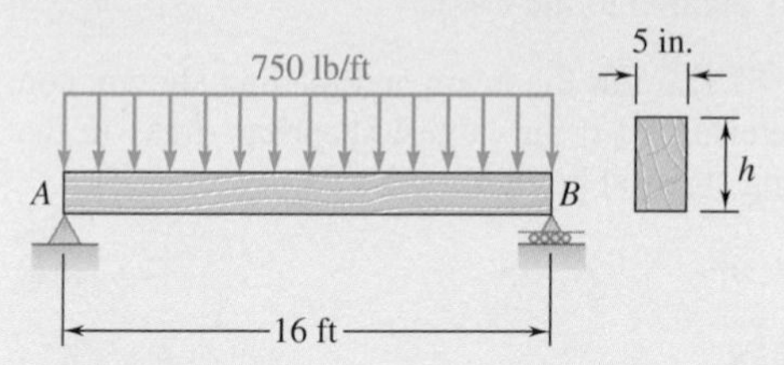

Fig. P13.13

13.14 For the beam and loading shown, determine the minimum required width b, knowing that for the grade of timber used, $\sigma_{all} = 12$ MPa and $\tau_{all} = 825$ kPa.

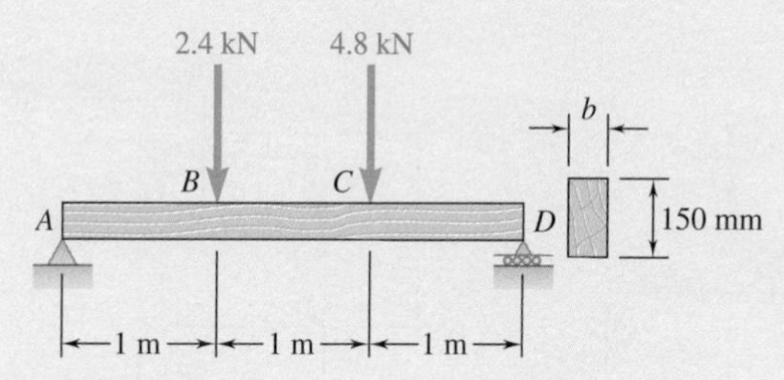

Fig. P13.14

13.15 For the wide-flange beam with the loading shown, determine the largest load **P** that can be applied, knowing that the maximum normal stress is 24 ksi and the largest shearing stress using the approximation $\tau_m = V/A_{web}$ is 14.5 ksi.

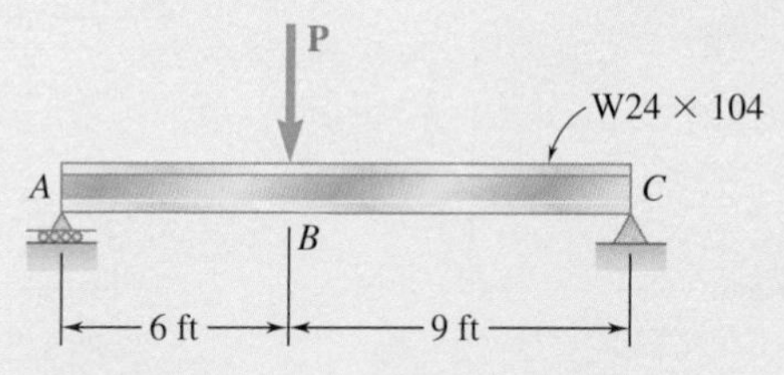

Fig. ***P13.15***

13.16 For the wide-flange beam with the loading shown, determine the largest load **P** that can be applied, knowing that the maximum normal stress is 160 MPa and the largest shearing stress using the approximation $\tau_m = V/A_{web}$ is 100 MPa.

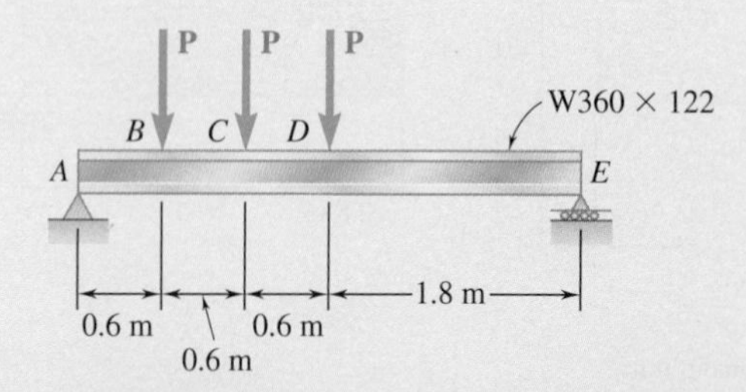

Fig. ***P13.16***

13.17 and 13.18 For the beam and loading shown, consider section n–n and determine the shearing stress at (*a*) point *a*, (*b*) point *b*.

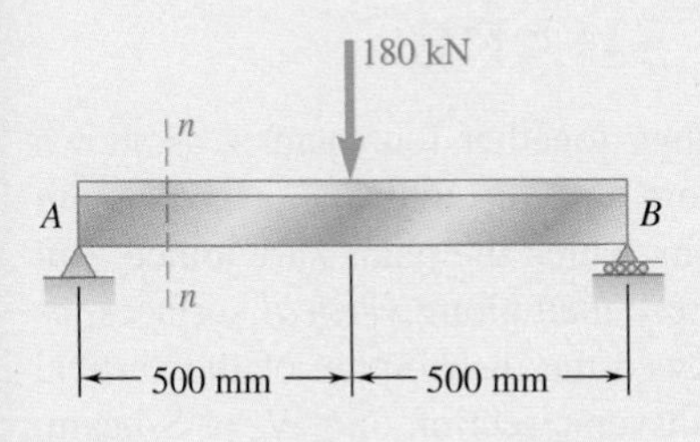

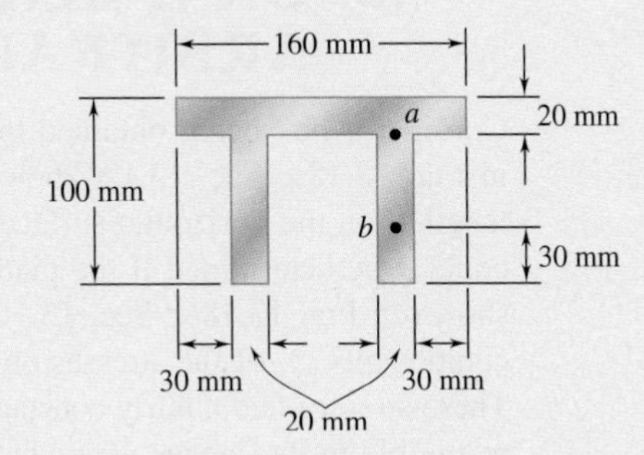

Fig. P13.17 and P13.19

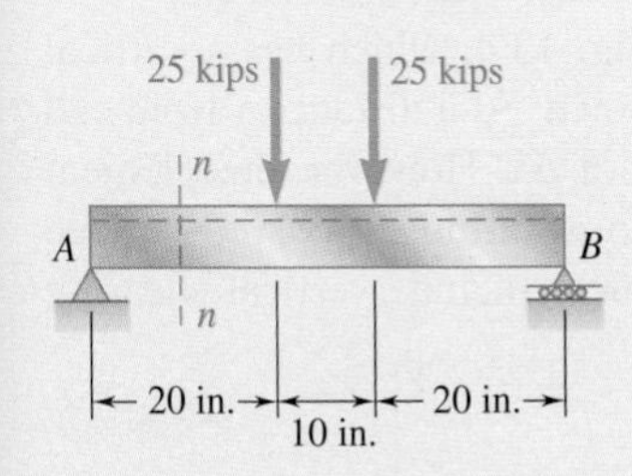

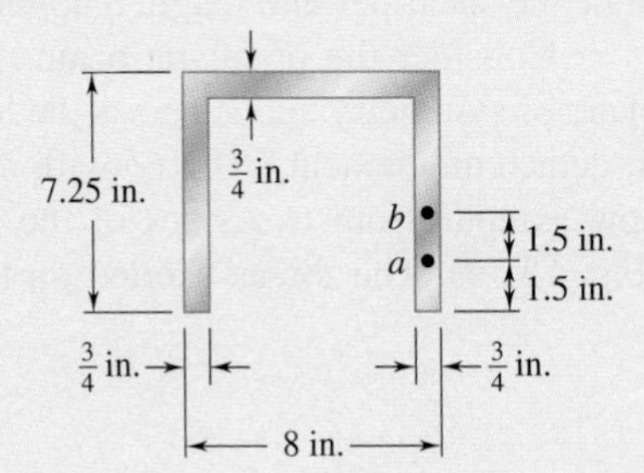

Fig. P13.18 and P13.20

13.19 and 13.20 For the beam and loading shown, determine the largest shearing stress in section n–n.

***13.21* through 13.24** A beam having the cross section shown is subjected to a vertical shear **V**. Determine (*a*) the horizontal line along which the shearing stress is maximum, (*b*) the constant k in the following expression for the maximum shearing stress

$$\tau_{max} = k\frac{V}{A}$$

where A is the cross-sectional area of the beam.

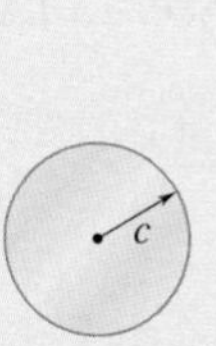

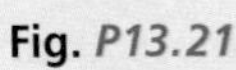

Fig. *P13.21*

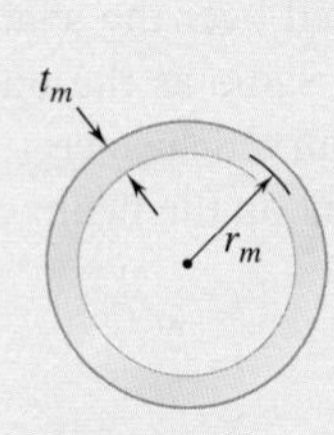

Fig. P13.22

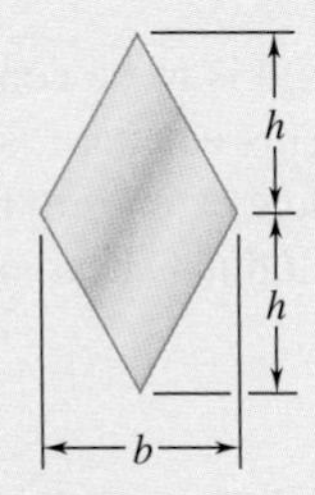

Fig. *P13.23*

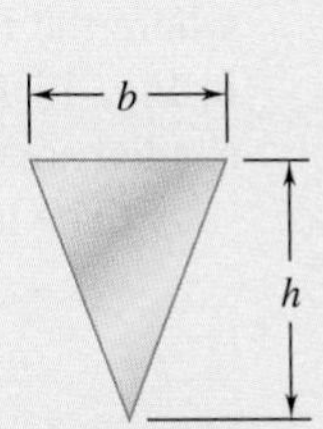

Fig. P13.24

13.2 LONGITUDINAL SHEAR ON A BEAM ELEMENT OF ARBITRARY SHAPE

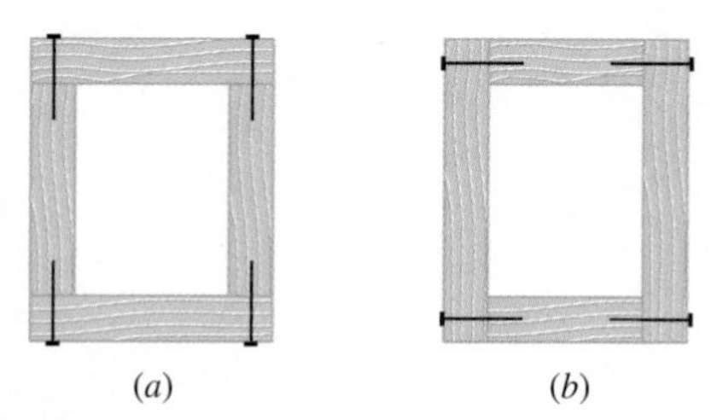

Fig. 13.18 Box beam formed by nailing planks together.

Consider a box beam obtained by nailing together four planks, as shown in Fig. 13.18*a*. Sec. 13.1A showed how to determine the shear per unit length q on the horizontal surfaces along which the planks are joined. But could q be determined if the planks are joined along *vertical* surfaces, as shown in Fig. 13.18*b*? Sec. 13.1C showed the distribution of the vertical components τ_{xy} of the stresses on a transverse section of a W- or S-beam. These stresses had a fairly constant value in the web of the beam and were negligible in its flanges. But what about the *horizontal* components τ_{xz} of the stresses in the flanges? The procedure developed in Sec. 13.1A to determine the shear per unit length q applies to the cases just described.

Consider the prismatic beam AB of Fig. 13.4, which has a vertical plane of symmetry and supports the loads shown. At a distance x from end A, detach an element $CDD'C'$ with a length of Δx. However, this element now extends from two sides of the beam to an arbitrary curved surface (Fig. 13.19). The forces exerted on the element include vertical shearing

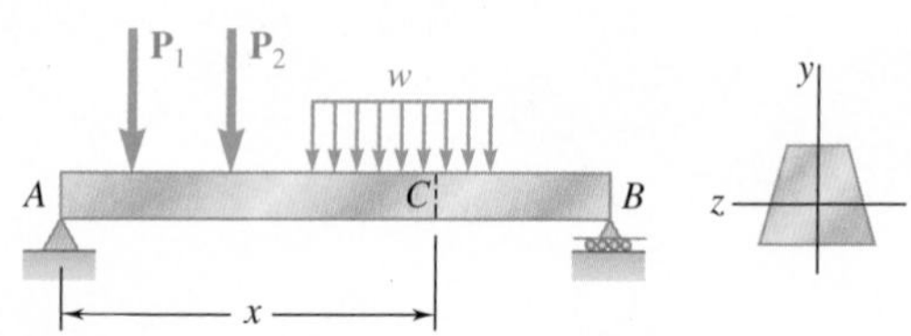

Fig. 13.4 (*repeated*) Beam example.

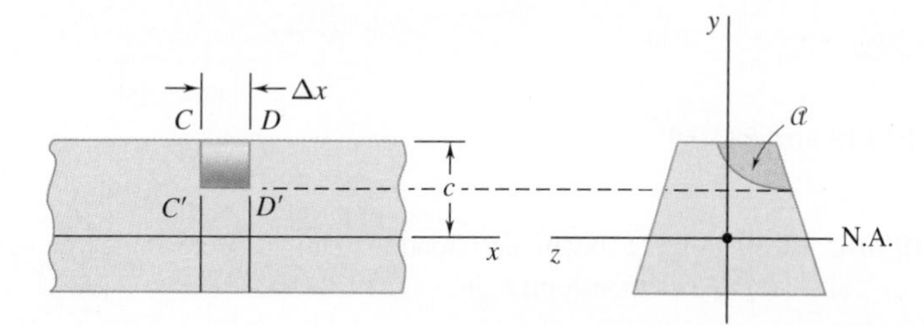

Fig. 13.19 Short segment of beam with element $CDD'C'$ of length Δx.

forces $\mathbf{V}'_C$ and $\mathbf{V}'_D$, elementary horizontal normal forces $\sigma_C dA$ and $\sigma_D dA$, possibly a load $w\,\Delta x$, and a longitudinal shearing force $\Delta\mathbf{H}$, which represent the resultant of the elementary longitudinal shearing forces exerted on the curved surface (Fig. 13.20). The equilibrium equation is

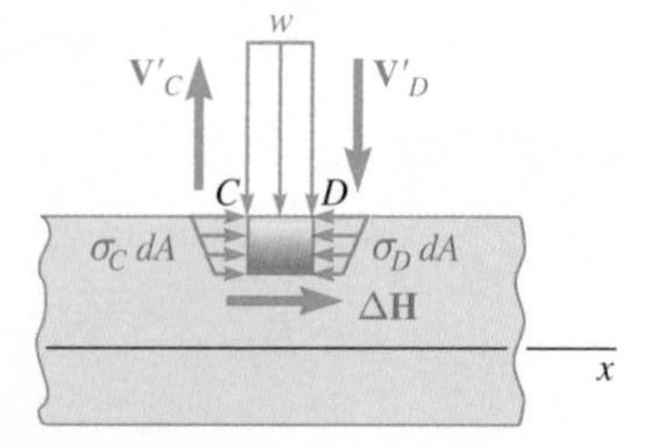

Fig. 13.20 Forces exerted on element $CDD'C'$.

$$\overset{+}{\rightarrow}\Sigma F_x = 0: \qquad \Delta H + \int_{\mathcal{a}} (\sigma_C - \sigma_D)\, dA = 0$$

where the integral is to be computed over the shaded area $\mathcal{a}$ of the section in Fig. 13.19. This equation is the same as the one in Sec. 13.1A, but the shaded area $\mathcal{a}$ now extends to the curved surface.

The longitudinal shear exerted on the beam element is

$$\Delta H = \frac{VQ}{I}\Delta x \tag{13.4}$$

where I is the centroidal moment of inertia of the entire section, Q is the first moment of the shaded area $\mathcal{a}$ with respect to the neutral axis, and V is the vertical shear in the section. Dividing both members of Eq. (13.4) by Δx, the horizontal shear per unit length or shear flow is

$$q = \frac{\Delta H}{\Delta x} = \frac{VQ}{I} \tag{13.5}$$

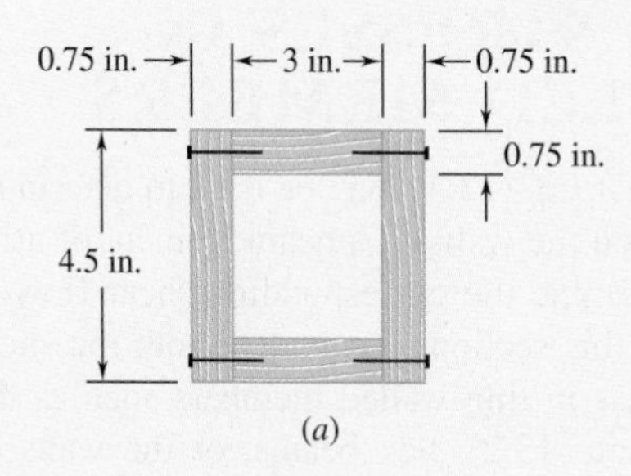

Concept Application 13.4

A square box beam is made of two 0.75 × 3-in. planks and two 0.75 × 4.5-in. planks nailed together, as shown (Fig. 13.21*a*). Knowing that the spacing between nails is 1.75 in. and that the beam is subjected to a vertical shear with a magnitude of $V = 600$ lb, determine the shearing force in each nail.

Isolate the upper plank and consider the total force per unit length q exerted on its two edges. Use Eq. (13.5), where Q represents the first moment with respect to the neutral axis of the shaded area A' shown in Fig. 13.21*b* and I is the moment of inertia about the same axis of the entire cross-sectional area of the box beam (Fig. 13.21*c*).

$$Q = A'\bar{y} = (0.75 \text{ in.})(3 \text{ in.})(1.875 \text{ in.}) = 4.22 \text{ in}^3$$

Recalling that the moment of inertia of a square of side a about a centroidal axis is $I = \frac{1}{12}a^4$,

$$I = \tfrac{1}{12}(4.5 \text{ in.})^4 - \tfrac{1}{12}(3 \text{ in.})^4 = 27.42 \text{ in}^4$$

Substituting into Eq. (13.5),

$$q = \frac{VQ}{I} = \frac{(600 \text{ lb})(4.22 \text{ in}^3)}{27.42 \text{ in}^4} = 92.3 \text{ lb/in.}$$

Because both the beam and the upper plank are symmetric with respect to the vertical plane of loading, equal forces are exerted on both edges of the plank. The force per unit length on each of these edges is thus $\frac{1}{2}q = \frac{1}{2}(92.3) = 46.15$ lb/in. Since the spacing between nails is 1.75 in., the shearing force in each nail is

$$F = (1.75 \text{ in.})(46.15 \text{ lb/in.}) = 80.8 \text{ lb}$$

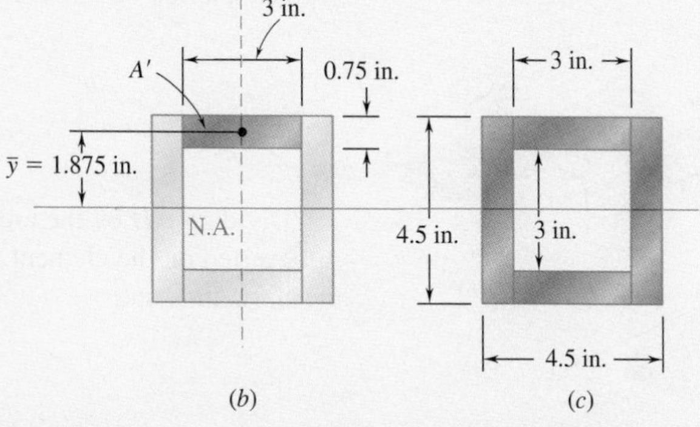

Fig. 13.21 (*a*) Box beam made from planks nailed together. (*b*) Geometry for finding first moment of area of top plank. (*c*) Geometry for finding the moment of inertia of entire cross section.

13.3 SHEARING STRESSES IN THIN-WALLED MEMBERS

We saw in the preceding section that Eq. (13.4) may be used to determine the longitudinal shear $\Delta\mathbf{H}$ exerted on the walls of a beam element of arbitrary shape and Eq. (13.5) to determine the corresponding shear flow q. Eqs. (13.4) and (13.5) are used in this section to calculate both the shear flow and the average shearing stress in thin-walled members such as the flanges of wide-flange beams (Photo 13.2), box beams, or the walls of structural tubes (Photo 13.3).

Photo 13.2 Wide-flange beams.

Photo 13.3 Structural tubes.

Consider a segment of length Δx of a wide-flange beam (Fig. 13.22*a*) where $\mathbf{V}$ is the vertical shear in the transverse section shown. Detach an element $ABB'A'$ of the upper flange (Fig. 13.22*b*). The longitudinal shear $\Delta\mathbf{H}$ exerted on that element can be obtained from Eq. (13.4):

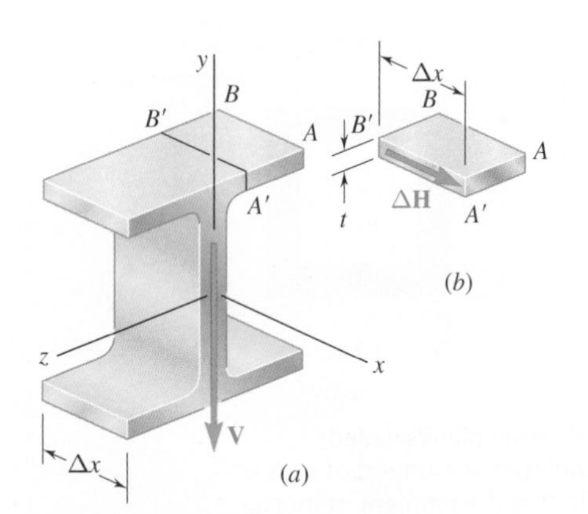

Fig. 13.22 (*a*) Wide-flange beam section with vertical shear V. (*b*) Segment of flange with longitudinal shear ΔH.

$$\Delta H = \frac{VQ}{I}\,\Delta x \tag{13.4}$$

Dividing ΔH by the area $\Delta A = t\,\Delta x$ of the cut, the average shearing stress exerted on the element is the same expression obtained in Sec. 13.1B for a horizontal cut:

$$\tau_{\text{ave}} = \frac{VQ}{It} \tag{13.6}$$

Note that τ_{ave} now represents the average value of the shearing stress τ_{zx} over a vertical cut, but since the thickness t of the flange is small, there is very little variation of τ_{zx} across the cut. Recalling that $\tau_{xz} = \tau_{zx}$

(Fig. 13.23), the horizontal component τ_{xz} of the shearing stress at any point of a transverse section of the flange can be obtained from Eq. (13.6), where Q is the first moment of the shaded area about the neutral axis (Fig. 13.24*a*). A similar result was obtained for the vertical component τ_{xy} of the shearing stress in the web (Fig. 13.24*b*). Eq. (13.6) can be used to determine shearing stresses in box beams (Fig. 13.25), half pipes (Fig. 13.26), and other thin-walled members, as long as the loads are applied in a plane of symmetry. In each case, the cut must be perpendicular to the surface of the member, and Eq. (13.6) will yield the component of the shearing stress in the direction tangent to that surface. (The other component is assumed to be equal to zero, because of the proximity of the two free surfaces.)

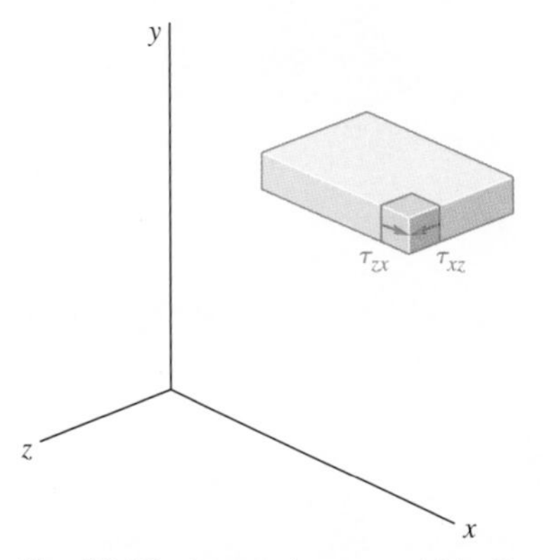

Fig. 13.23 Stress element within flange segment.

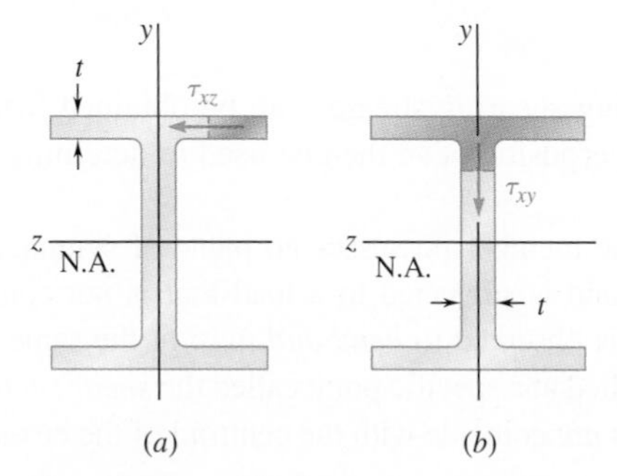

Fig. 13.24 Wide-flange beam sections showing shearing stress (*a*) in flange and (*b*) in web. The shaded area is that used for calculating the first moment of area.

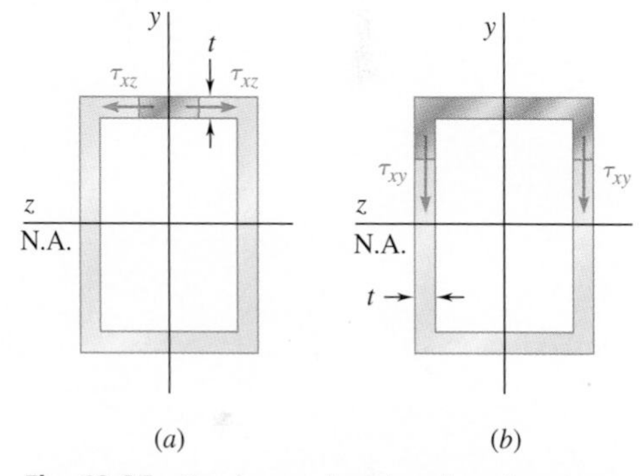

Fig. 13.25 Box beam showing shearing stress (*a*) in flange, (*b*) in web. Shaded area is that used for calculating the first moment of area.

Comparing Eqs. (13.5) and (13.6), the product of the shearing stress τ at a given point of the section and the thickness t at that point is equal to q. Since V and I are constant, q depends only upon the first moment Q and easily can be sketched on the section. For a box beam (Fig. 13.27), q grows smoothly from zero at A to a maximum value at C and C' on the neutral axis and decreases back to zero as E is reached. There is no sudden variation in the magnitude of q as it passes a corner at B, D, B', or D', and the sense of q in the horizontal portions of the section is easily obtained from its sense in the vertical portions (the sense of the shear $\mathbf{V}$). In a wide-flange section (Fig. 13.28), the values of q in portions AB and $A'B$ of the upper flange are distributed symmetrically. At B in the web, q corresponds to the two halves of the flange, which must be combined to obtain the value of q at the top of the web. After reaching a maximum value at C on the neutral axis, q decreases and splits into two equal parts at D, which corresponds at D to the two halves of the lower flange. The shear per unit length q is commonly called the *shear flow* and reflects the similarity between the properties of q just described and some of the characteristics of a fluid flow through an open channel or pipe.

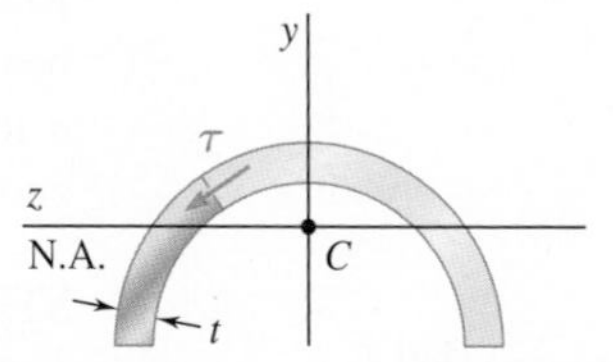

Fig. 13.26 Half pipe section showing shearing stress, and shaded area for calculating first moment of area.

So far, all of the loads were applied in a plane of symmetry of the member. In the case of members possessing two planes of symmetry (Fig. 13.24 or 13.27), any load applied through the centroid of a given cross section can be resolved into components along the two axes of symmetry. Each component will cause the member to bend in a plane of symmetry,

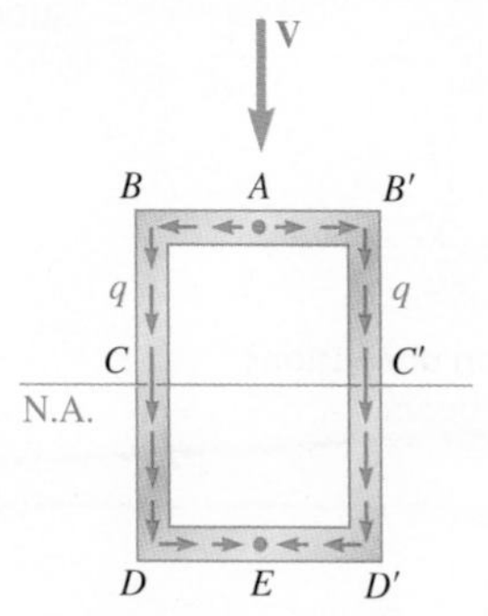

Fig. 13.27 Shear flow, *q*, in a box beam section.

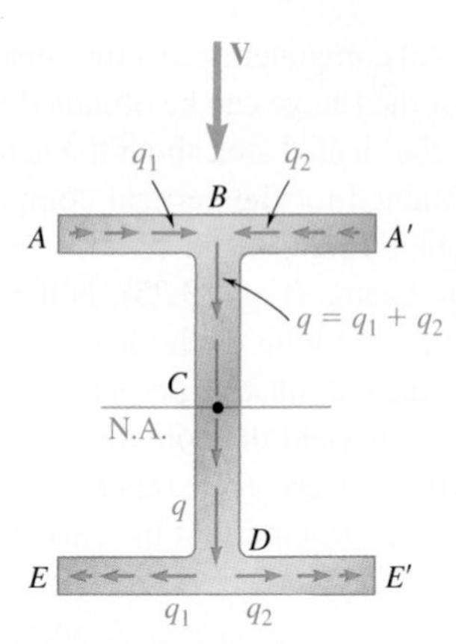

Fig. 13.28 Shear flow, q, in a wide-flange beam section.

and the corresponding shearing stresses can be obtained from Eq. (13.6). The principle of superposition can then be used to determine the resulting stresses.

However, if the member possesses no plane of symmetry or a single plane of symmetry and is subjected to a load that is not contained in that plane, that member is observed to *bend and twist* at the same time—except when the load is applied at a specific point called the *shear center*.† The shear center normally does *not* coincide with the centroid of the cross section.

Sample Problem 13.3

Knowing that the vertical shear is 50 kips in a W10 × 68 rolled-steel beam, determine the horizontal shearing stress in the top flange at a point a located 4.31 in. from the edge of the beam. The dimensions and other geometric data of the rolled-steel section are given in Appendix B.

STRATEGY: Determine the horizontal shearing stress at the required section.

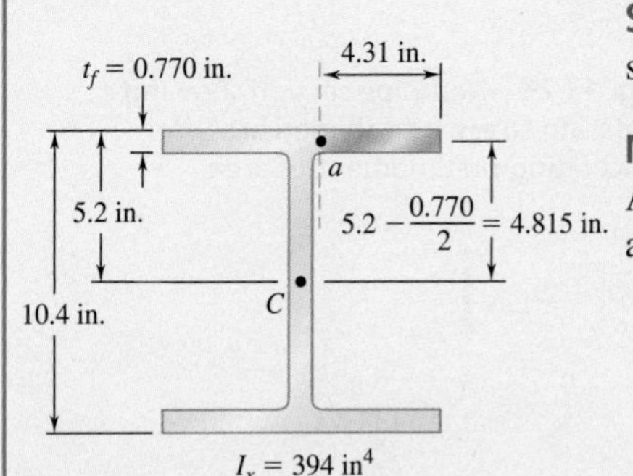

Fig. 1 Cross section dimensions for W10 × 68 steel beam.

MODELING and ANALYSIS:

As shown in Fig. 1, we isolate the shaded portion of the flange by cutting along the dashed line that passes through point a.

$$Q = (4.31 \text{ in.})(0.770 \text{ in.})(4.815 \text{ in.}) = 15.98 \text{ in}^3$$

$$\tau = \frac{VQ}{It} = \frac{(50 \text{ kips})(15.98 \text{ in}^3)}{(394 \text{ in}^4)(0.770 \text{ in.})} \qquad \tau = 2.63 \text{ ksi} \blacktriangleleft$$

†See Ferdinand P. Beer, E. Russell Johnston Jr., John T. DeWolf, and David F. Mazurek, *Mechanics of Materials*, 7th ed., McGraw-Hill, New York, 2015, sec. 6.6.

Sample Problem 13.4

Solve Sample Prob. 13.3, assuming that 0.75×12-in. plates have been attached to the flanges of the W10 × 68 beam by continuous fillet welds as shown.

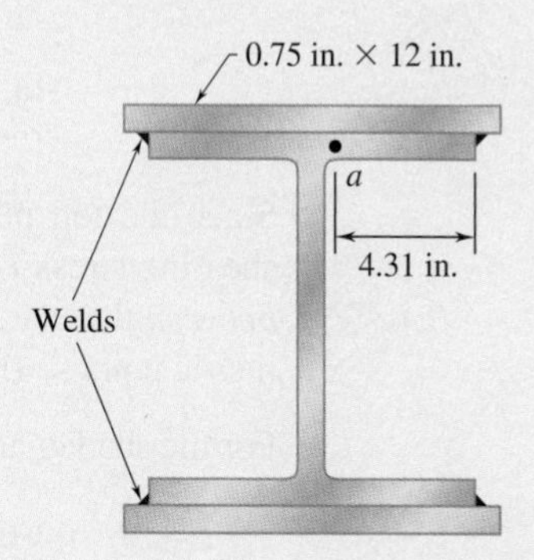

STRATEGY: Calculate the properties for the composite beam and then determine the shearing stress at the required section.

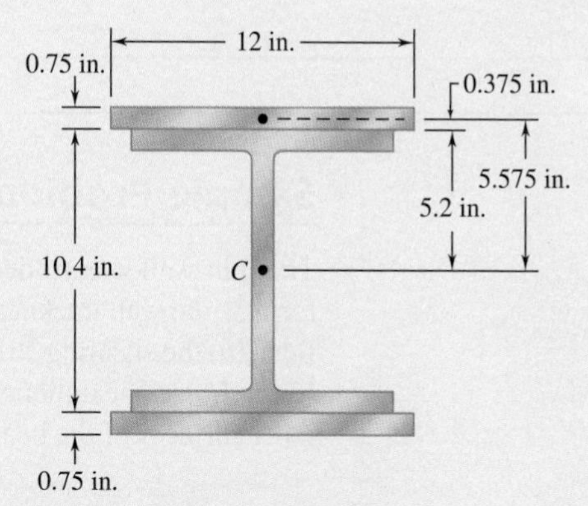

Fig. 1 Cross section dimensions for calculating moment of inertia.

MODELING and ANALYSIS:

For the composite beam shown in Fig. 1, the centroidal moment of inertia is

$$I = 394 \text{ in}^4 + 2\left[\tfrac{1}{12}(12 \text{ in.})(0.75 \text{ in.})^3 + (12 \text{ in.})(0.75 \text{ in.})(5.575 \text{ in.})^2\right]$$
$$I = 954 \text{ in}^4$$

(continued)

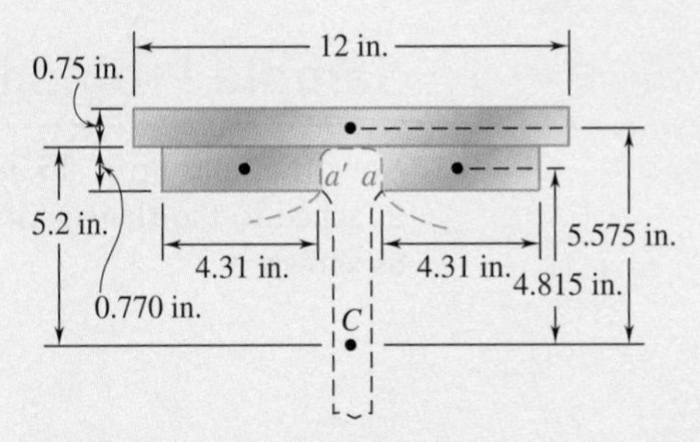

Fig. 2 Dimensions used to find first moment of area and shearing stress at flange-web junction.

Since the top plate and the flange are connected only at the welds, the shearing stress is found at a by passing a section through the flange at a, *between* the plate and the flange, and again through the flange at the symmetric point a' (Fig. 2).

For the shaded area,

$$t = 2t_f = 2(0.770 \text{ in.}) = 1.540 \text{ in.}$$

$$Q = 2[(4.31 \text{ in.})(0.770 \text{ in.})(4.815 \text{ in.})] + (12 \text{ in.})(0.75 \text{ in.})(5.575 \text{ in.})$$

$$Q = 82.1 \text{ in}^3$$

$$\tau = \frac{VQ}{It} = \frac{(50 \text{ kips})(82.1 \text{ in}^3)}{(954 \text{ in}^4)(1.540 \text{ in.})} \qquad \tau = 2.79 \text{ ksi}$$

Sample Problem 13.5

The thin-walled extruded beam shown is made of aluminum and has a uniform 3-mm wall thickness. Knowing that the shear in the beam is 5 kN, determine (*a*) the shearing stress at point A, (*b*) the maximum shearing stress in the beam. *Note*: The dimensions given are to lines midway between the outer and inner surfaces of the beam.

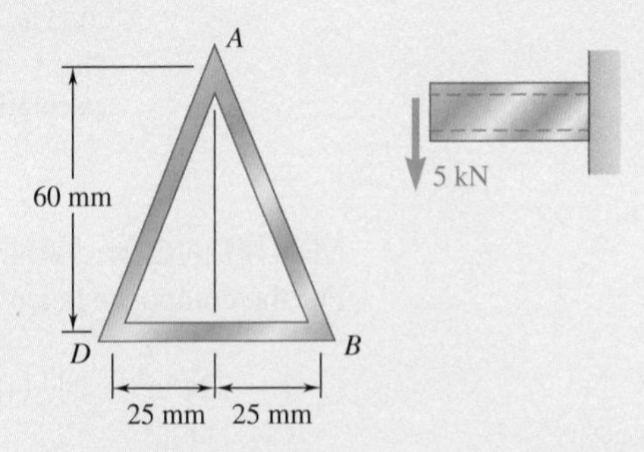

STRATEGY: Determine the location of the centroid and then calculate the moment of inertia. Calculate the two required stresses.

(continued)

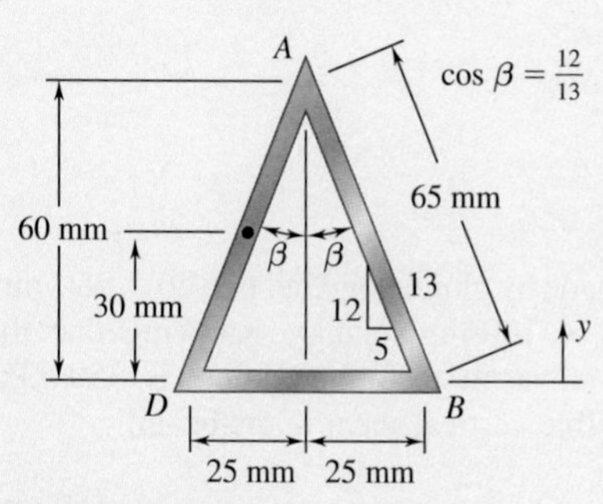

Fig. 1 Section dimensions for finding centroid.

MODELING and ANALYSIS:

Centroid. Using Fig. 1, we note that $AB = AD = 65$ mm.

$$\overline{Y} = \frac{\Sigma\, \overline{y}A}{\Sigma\, A} = \frac{2[(65 \text{ mm})(3 \text{ mm})(30 \text{ mm})]}{2[(65 \text{ mm})(3 \text{ mm})] + (50 \text{ mm})(3 \text{ mm})}$$

$$\overline{Y} = 21.67 \text{ mm}$$

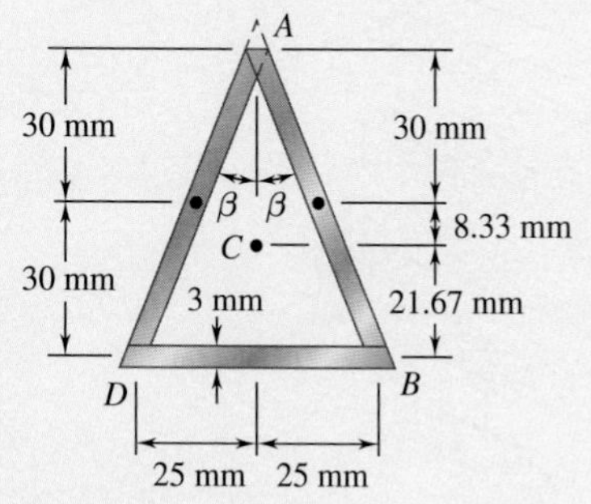

Fig. 2 Dimensions locating centroid.

Centroidal Moment of Inertia. Each side of the thin-walled beam can be considered as a parallelogram (Fig. 2), and we recall that for the case shown $I_{nn} = bh^3/12$, where b is measured parallel to the axis nn. Using Fig. 3 we write

$$b = (3 \text{ mm})/\cos \beta = (3 \text{ mm})/(12/13) = 3.25 \text{ mm}$$

$$I = \Sigma(\overline{I} + Ad^2) = 2[\tfrac{1}{12}(3.25 \text{ mm})(60 \text{ mm})^3$$
$$+ (3.25 \text{ mm})(60 \text{ mm})(8.33 \text{ mm})^2] + [\tfrac{1}{12}(50 \text{ mm})(3 \text{ mm})^3$$
$$+ (50 \text{ mm})(3 \text{ mm})(21.67 \text{ mm})^2]$$

$$I = 214.6 \times 10^3 \text{ mm}^4 \qquad I = 0.2146 \times 10^{-6} \text{ m}^4$$

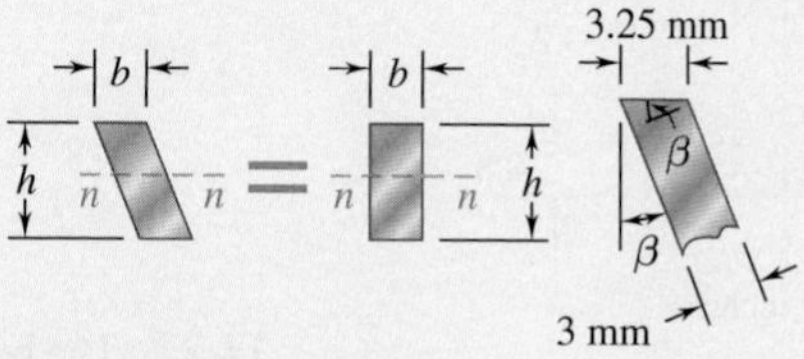

Fig. 3 Determination of horizontal width for side elements.

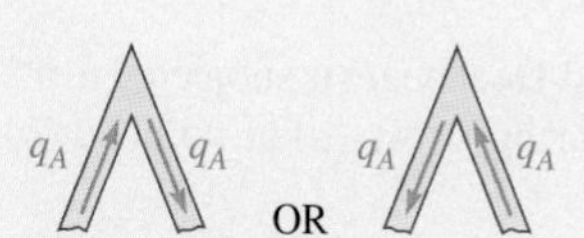

Fig. 4 Possible directions for shear flow at A.

a. Shearing Stress at A. If a shearing stress τ_A occurs at A, the shear flow will be $q_A = \tau_A t$ and must be directed in one of the two ways shown in Fig 4. But the cross section and the loading are symmetric about a vertical line through A, and thus the shear flow must also be symmetric. Since neither of the possible shear flows is symmetric, we conclude that

$$\tau_A = 0. \quad ◀$$

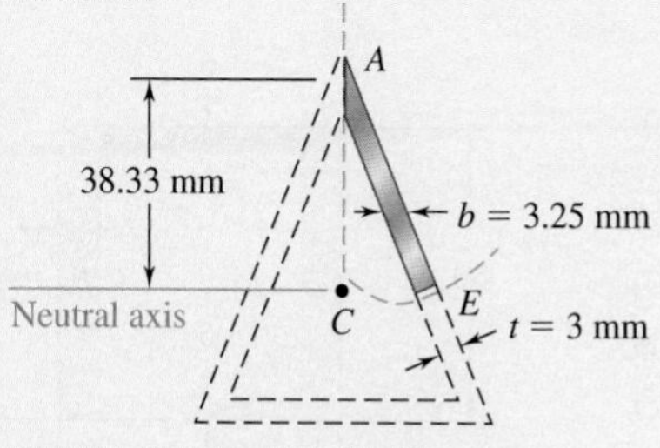

Fig. 5 Section for finding the maximum shearing stress.

b. Maximum Shearing Stress. Since the wall thickness is constant, the maximum shearing stress occurs at the neutral axis, where Q is maximum. Since we know that the shearing stress at A is zero, we cut the section along the dashed line shown and isolate the shaded portion of the beam (Fig. 5). In order to obtain the largest shearing stress, the cut at the neutral axis is made perpendicular to the sides and is of length $t = 3$ mm.

$$Q = [(3.25 \text{ mm})(38.33 \text{ mm})]\left(\frac{38.33 \text{ mm}}{2}\right) = 2387 \text{ mm}^3$$

$$Q = 2.387 \times 10^{-6} \text{ m}^3$$

$$\tau_E = \frac{VQ}{It} = \frac{(5 \text{ kN})(2.387 \times 10^{-6} \text{ m}^3)}{(0.2146 \times 10^{-6} \text{ m}^4)(0.003 \text{ m})} \qquad \tau_{max} = \tau_E = 18.54 \text{ MPa} \quad ◀$$

Problems

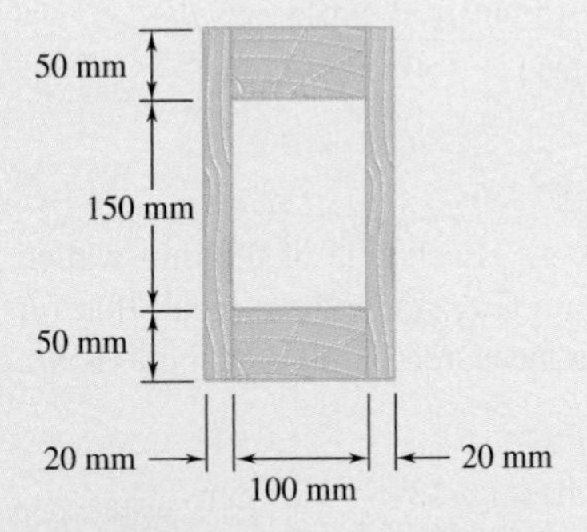

Fig. P13.25

13.25 The built-up beam shown is made by gluing together two 20 × 250-mm plywood strips and two 50 × 100-mm planks. Knowing that the allowable average shearing stress in the glued joints is 350 kPa, determine the largest permissible vertical shear in the beam.

13.26 The built-up timber beam is subjected to a vertical shear of 1200 lb. Knowing that the allowable shearing force in the nails is 75 lb, determine the largest permissible spacing s of the nails.

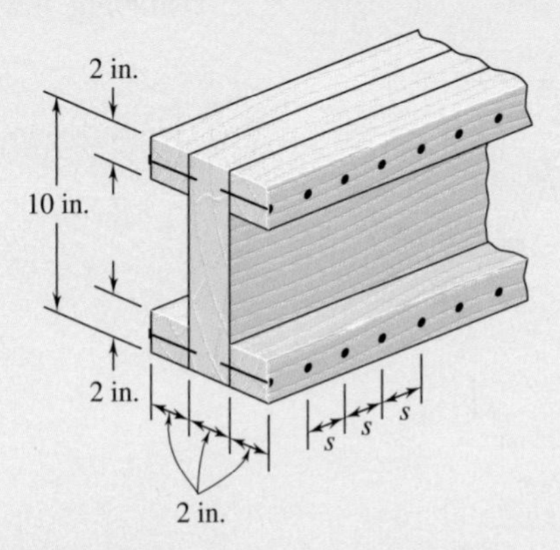

Fig. P13.26

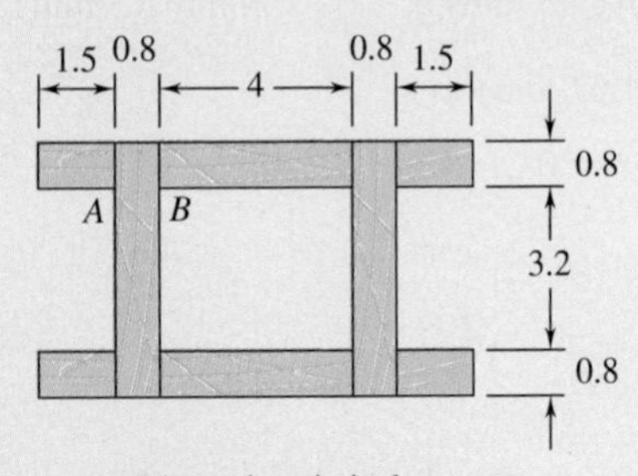

Fig. P13.27

13.27 The built-up beam was made by gluing together several wooden planks. Knowing that the beam is subjected to a 1200-lb vertical shear, determine the average shearing stress in the glued joint (*a*) at *A*, (*b*) at *B*.

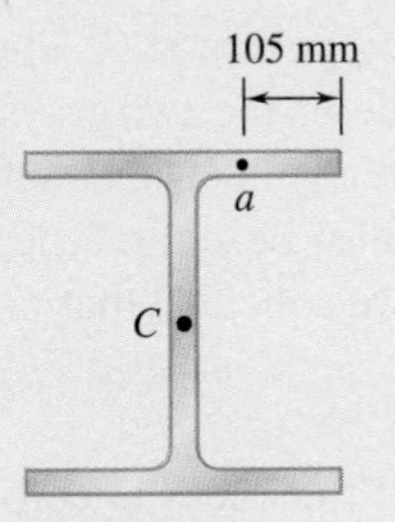

Fig. P13.28

13.28 Knowing that a W360 × 122 rolled-steel beam is subjected to a 250-kN vertical shear, determine the shearing stress (*a*) at point *a*, (*b*) at the centroid *C* of the section.

13.29 and *13.30* An extruded aluminum beam has the cross section shown. Knowing that the vertical shear in the beam is 150 kN, determine the corresponding shearing stress at (*a*) point *a*, (*b*) point *b*.

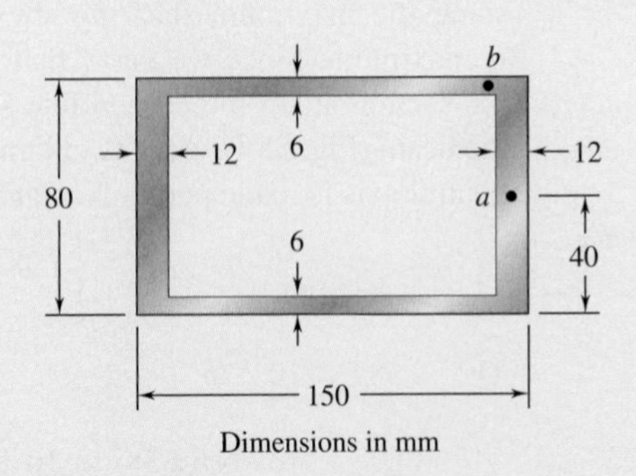

Fig. P13.29

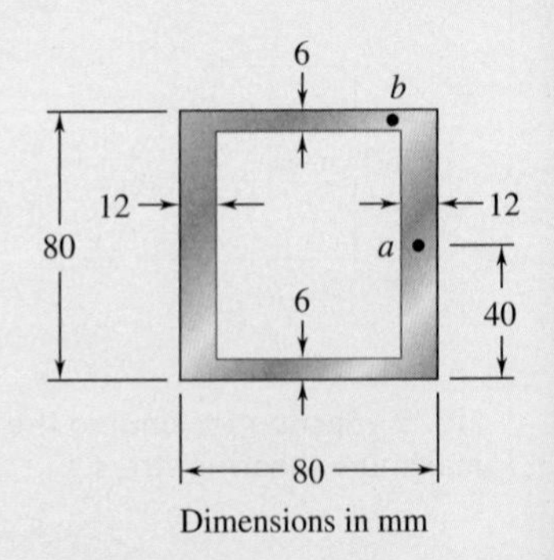

Fig. *P13.30*

13.31 and *13.32* The extruded aluminum beam shown has a uniform wall thickness of $\frac{1}{8}$ in. Knowing that the vertical shear in the beam is 2 kips, determine the corresponding shearing stress at each of the five points indicated.

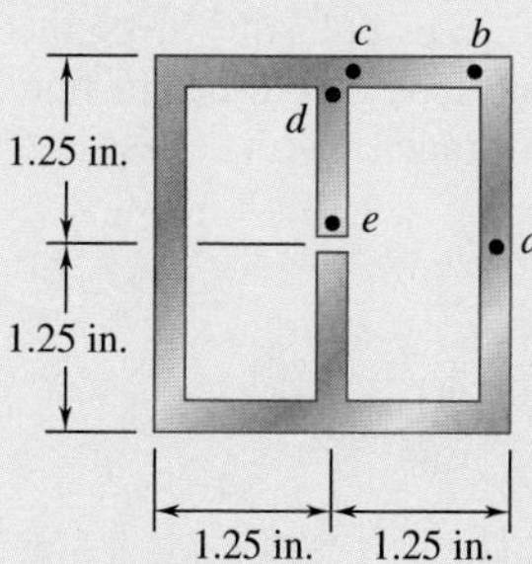

Fig. P13.31

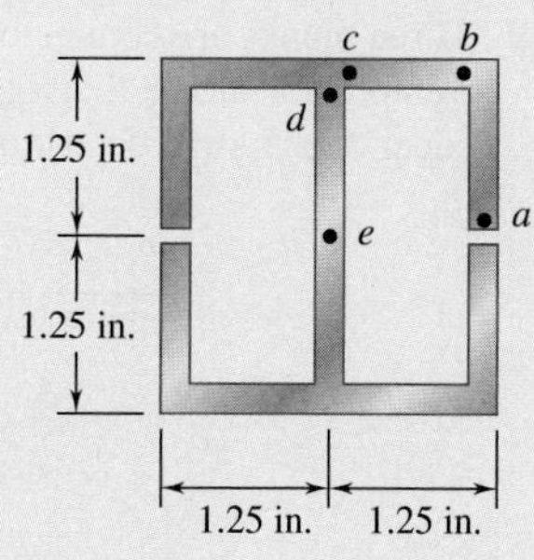

Fig. *P13.32*

13.33 Knowing that a given vertical shear **V** causes a maximum shearing stress of 75 MPa in the hat-shaped extrusion shown, determine the corresponding shearing stress at (*a*) point *a*, (*b*) point *b*.

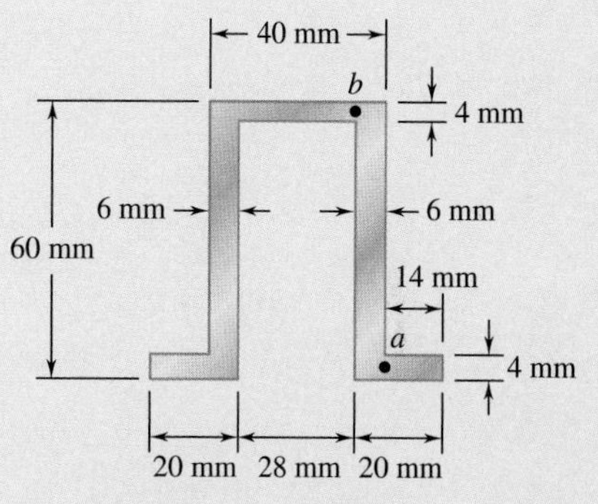

Fig. P13.33

13.34 Knowing that a given vertical shear **V** causes a maximum shearing stress of 50 MPa in a thin-walled member having the cross section shown, determine the corresponding shearing stress at (*a*) point *a*, (*b*) point *b*, (*c*) point *c*.

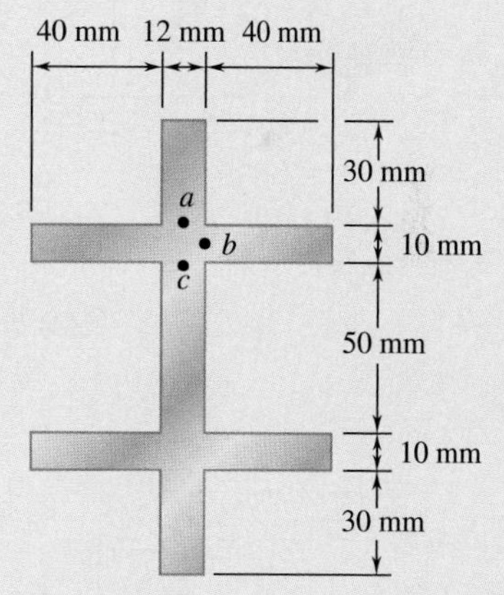

Fig. P13.34

13.35 The vertical shear is 1200 lb in a beam having the cross section shown. Knowing that $d = 4$ in., determine the shearing stress at (*a*) point *a*, (*b*) point *b*.

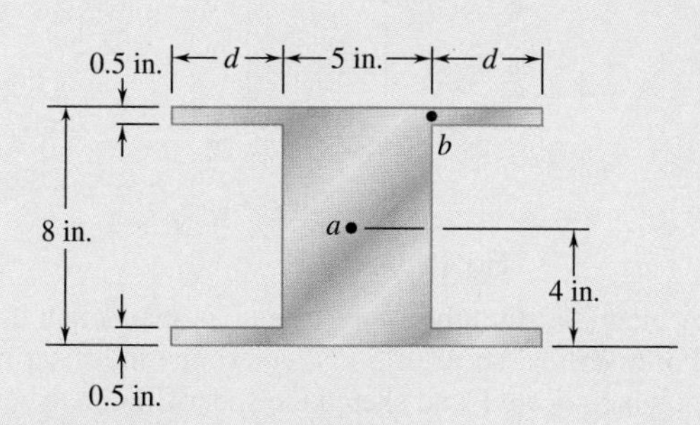

Fig. P13.35 and P13.36

13.36 The vertical shear is 1200 lb in a beam having the cross section shown. Determine (*a*) the distance *d* for which $\tau_a = \tau_b$, (*b*) the corresponding shearing stress at points *a* and *b*.

13.37 A beam consists of three planks connected by steel bolts with a longitudinal spacing of 225 mm. Knowing that the shear in the beam is vertical and equal to 6 kN and that the allowable average shearing stress in each bolt is 60 MPa, determine the smallest permissible bolt diameter that can be used.

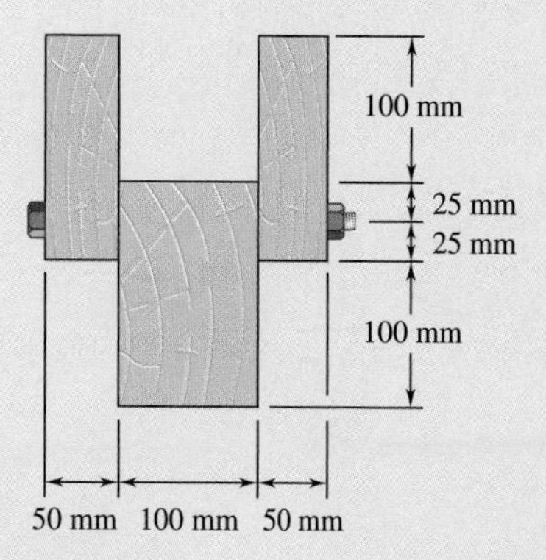

Fig. P13.37

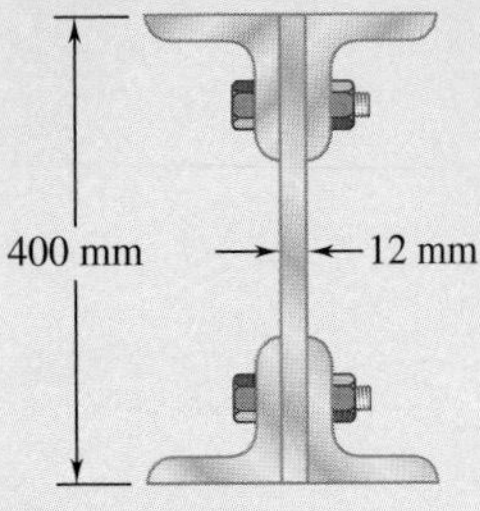

Fig. P13.38

13.38 Four L102 × 102 × 9.5 steel angle shapes and a 12 × 400-mm steel plate are bolted together to form a beam with the cross section shown. The bolts are of 22-mm diameter and are spaced longitudinally every 120 mm. Knowing that the beam is subjected to a vertical shear of 240 kN, determine the average shearing stress in each bolt.

13.39 Three planks are connected as shown by bolts of $\frac{3}{8}$-in. diameter spaced every 6 in. along the longitudinal axis of the beam. For a vertical shear of 2.5 kips, determine the average shearing stress in the bolts.

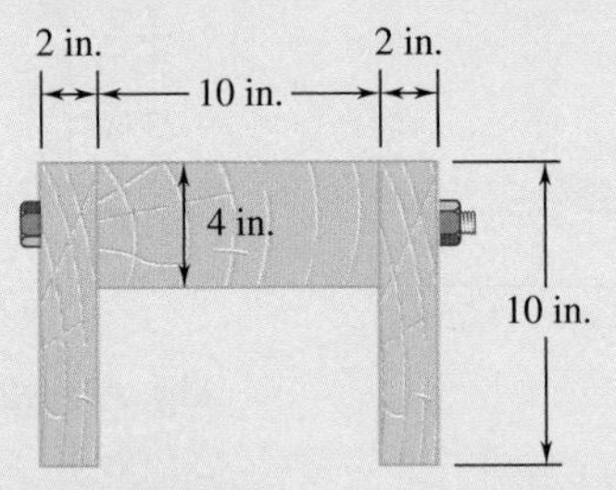

Fig. P13.39

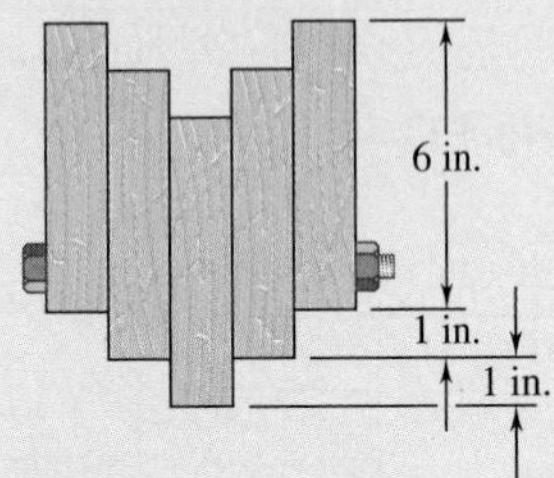

Fig. P13.40

13.40 A beam consists of five planks of 1.5 × 6-in. cross section connected by steel bolts with a longitudinal spacing of 9 in. Knowing that the shear in the beam is vertical and equal to 2000 lb and that the allowable average shearing stress in each bolt is 7500 psi, determine the smallest permissible bolt diameter that can be used.

13.41 A plate of 4-mm thickness is bent as shown and then used as a beam. For a vertical shear of 12 kN, determine (*a*) the shearing stress at point *A*, (*b*) the maximum shearing stress in the beam. Also sketch the shear flow in the cross section.

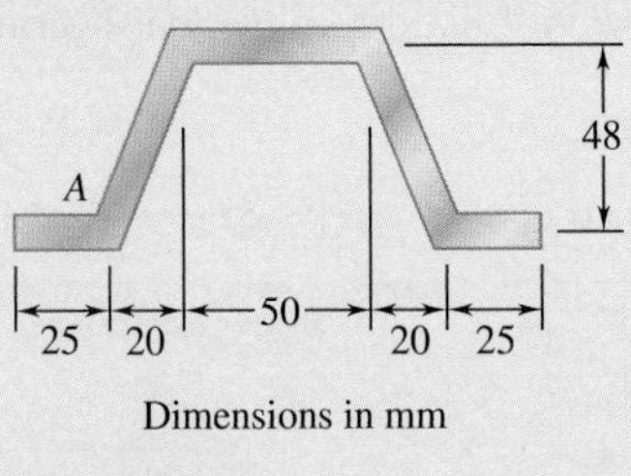

Fig. P13.41

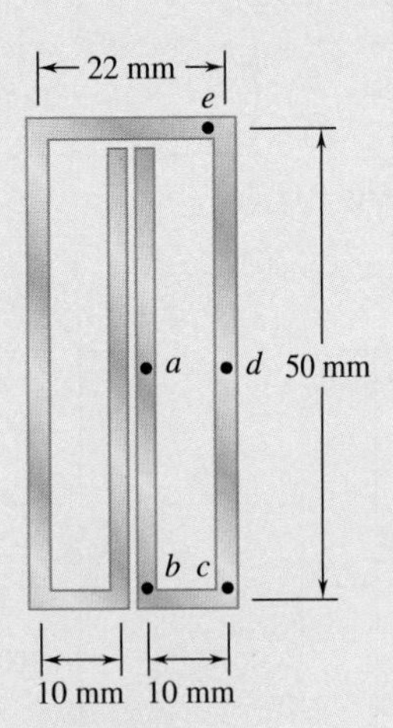

Fig. P13.42

13.42 A plate of 2-mm thickness is bent as shown and then used as a beam. For a vertical shear of 5 kN, determine the shearing stress at the five points indicated, and sketch the shear flow in the cross section.

13.43 A plate of $\frac{1}{4}$-in. thickness is corrugated as shown and then used as a beam. For a vertical shear of 1.2 kips, determine (*a*) the maximum shearing stress in the section, (*b*) the shearing stress at point *B*. Also sketch the shear flow in the cross section.

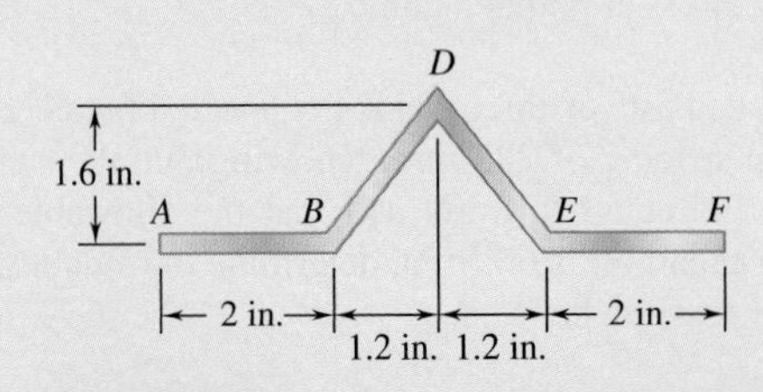

Fig. P13.43

13.44 A plate of thickness t is bent as shown and then used as a beam. For a vertical shear of 600 lb, determine (*a*) the thickness t for which the maximum shearing stress is 300 psi, (*b*) the corresponding shearing stress at point E. Also sketch the shear flow in the cross section.

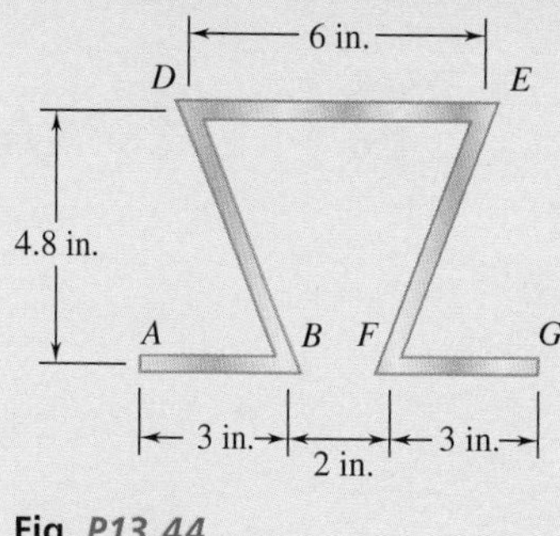

Fig. *P13.44*

13.45 For a beam made of two or more materials, with each material having a different modulus of elasticity, show that Eq. (13.6)

$$\tau_{ave} = \frac{VQ}{It}$$

remains valid provided that both Q and I are computed by using the transformed section of the beam (see Sec. 11.3) and provided further that t is the actual width of the beam where τ_{ave} is computed.

13.46 and *13.47* A composite beam is made by attaching the timber and steel portions shown with bolts of 12-mm diameter spaced longitudinally every 200 mm. The modulus of elasticity is 10 GPa for the wood and 200 GPa for the steel. For a vertical shear of 4 kN, determine (*a*) the average shearing stress in the bolts, (*b*) the shearing stress at the center of the cross section. (*Hint*: Use the method indicated in Prob. 13.45.)

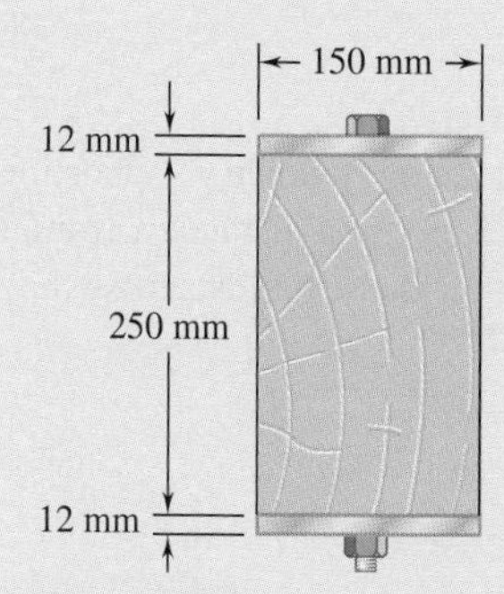

Fig. P13.46

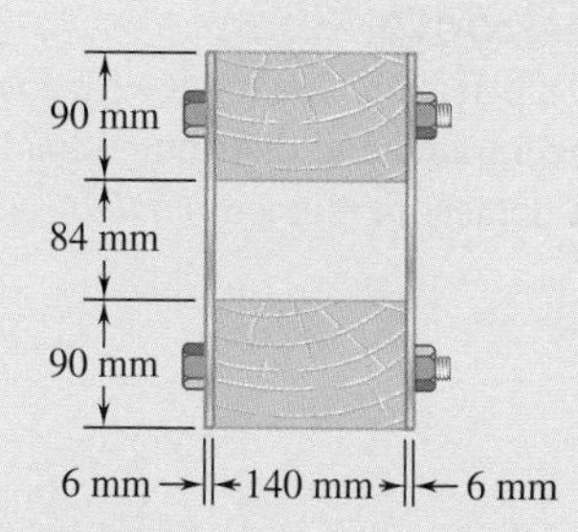

Fig. *P13.47*

13.48 and 13.49 A steel bar and an aluminum bar are bonded together as shown to form a composite beam. Knowing that the vertical shear in the beam is 4 kips and that the modulus of elasticity is 29×10^6 psi for the steel and 10.6×10^6 psi for the aluminum, determine (*a*) the average shearing stress at the bonded surface, (*b*) the maximum shearing stress in the beam. (*Hint*: Use the method indicated in Prob. 13.45.)

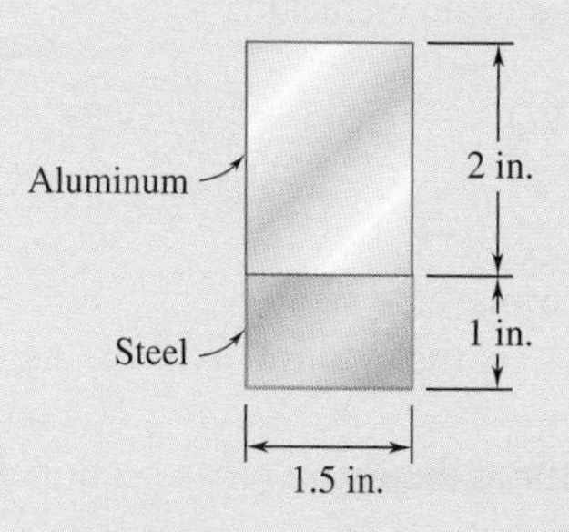

Fig. P13.48

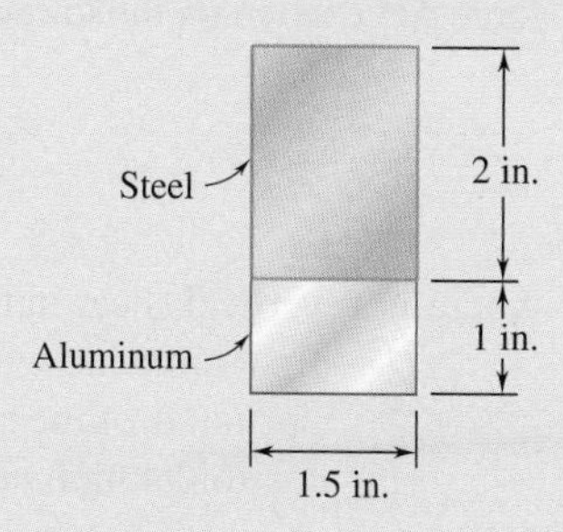

Fig. P13.49

Review and Summary

Stresses on a Beam Element

A small element located in the vertical plane of symmetry of a beam under a transverse loading was considered (Fig. 13.29), and it was found that normal stresses σ_x and shearing stresses τ_{xy} are exerted on the transverse faces of that element, while shearing stresses τ_{yx}, equal in magnitude to τ_{xy}, are exerted on its horizontal faces.

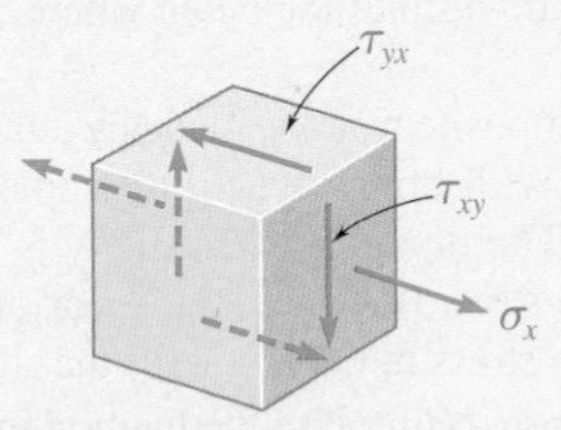

Fig. 13.29

Horizontal Shear

For a prismatic beam AB with a vertical plane of symmetry supporting various concentrated and distributed loads (Fig. 13.30), at a distance x from end A we can detach an element $CDD'C'$ of length Δx that extends across the width of

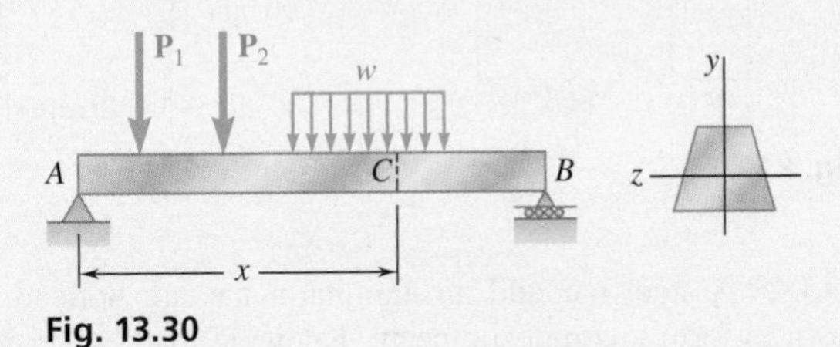

Fig. 13.30

the beam from the upper surface of the beam to a horizontal plane located at a distance y_1 from the neutral axis (Fig. 13.31). The magnitude of the shearing force $\Delta \mathbf{H}$ exerted on the lower face of the beam element is

$$\Delta H = \frac{VQ}{I} \Delta x \tag{13.4}$$

where V = vertical shear in the given transverse section

Q = first moment with respect to the neutral axis of the shaded portion $\mathcal{A}$ of the section

I = centroidal moment of inertia of the entire cross-sectional area

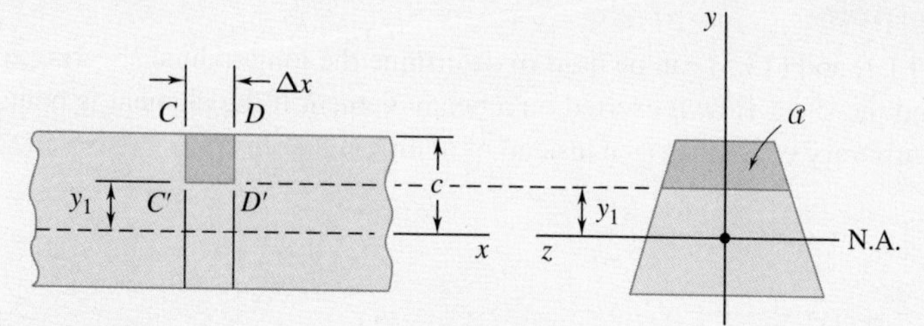

Fig. 13.31

Shear Flow

The *horizontal shear per unit length* or *shear flow*, denoted by the letter q, is obtained by dividing both members of Eq. (13.4) by Δx:

$$q = \frac{\Delta H}{\Delta x} = \frac{VQ}{I} \tag{13.5}$$

Shearing Stresses in a Beam

Dividing both members of Eq. (13.4) by the area ΔA of the horizontal face of the element and observing that $\Delta A = t\,\Delta x$, where t is the width of the element at the cut, the *average shearing stress* on the horizontal face of the element is

$$\tau_{\text{ave}} = \frac{VQ}{It} \tag{13.6}$$

Since the shearing stresses τ_{xy} and τ_{yx} are exerted on a transverse and a horizontal plane through D' and are equal, Eq. (13.6) also represents the average value of τ_{xy} along the line $D'_1D'_2$ (Fig. 13.32).

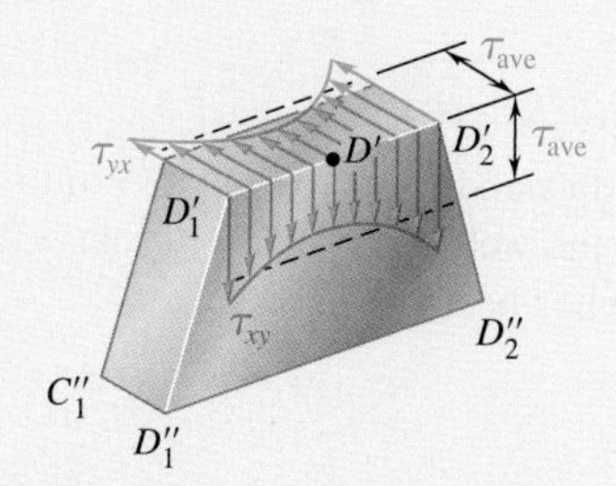

Fig. 13.32

Shearing Stresses in a Beam of Rectangular Cross Section

The distribution of shearing stresses in a beam of rectangular cross section was found to be parabolic, and the maximum stress, which occurs at the center of the section, is

$$\tau_{\max} = \frac{3}{2}\frac{V}{A} \tag{13.10}$$

where A is the area of the rectangular section. For wide-flange beams, a good approximation of the maximum shearing stress is obtained by dividing the shear V by the cross-sectional area of the web.

Longitudinal Shear on Curved Surface

Eqs. (13.4) and (13.5) can be used to determine the longitudinal shearing force $\Delta\mathbf{H}$ and the shear flow q exerted on a beam element if the element is bounded by an arbitrary curved surface instead of a horizontal plane (Fig. 13.33).

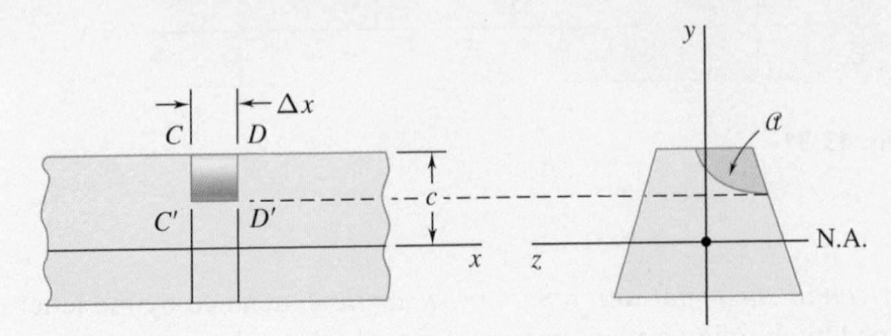

Fig. 13.33

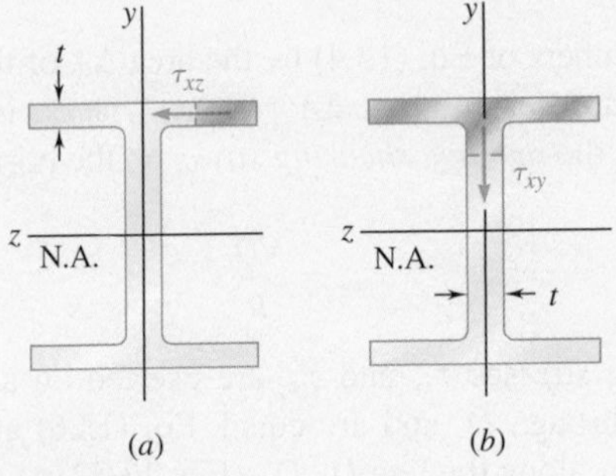

Fig. 13.34

Shearing Stresses in Thin-Walled Members

We found that we could extend the use of Eq. (13.6) to determine the average shearing stress in both the webs and flanges of thin-walled members, such as wide-flange beams and box beams (Fig. 13.34).

Review Problems

13.50 A square box beam is made of two 20×80-mm planks and two 20×120-mm planks nailed together as shown. Knowing that the spacing between the nails is $s = 30$ mm and that the vertical shear in the beam is $V = 1200$ N, determine (*a*) the shearing force in each nail, (*b*) the maximum shearing stress in the beam.

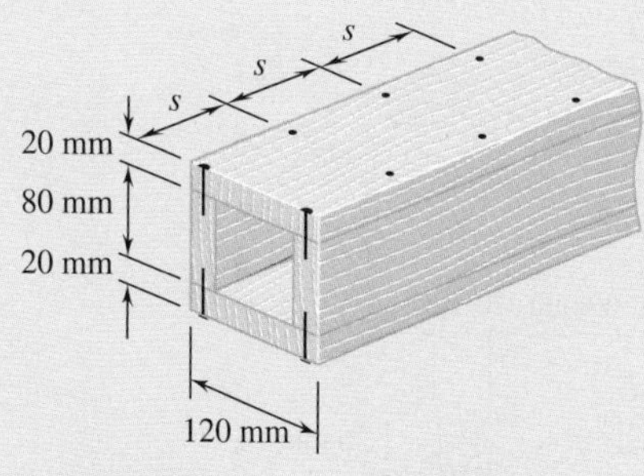

Fig. P13.50

13.51 The composite beam shown is fabricated by connecting two W6 × 20 rolled-steel members, using bolts of $\frac{5}{8}$-in. diameter spaced longitudinally every 6 in. Knowing that the average allowable shearing stress in the bolts is 10.5 ksi, determine the largest allowable vertical shear in the beam.

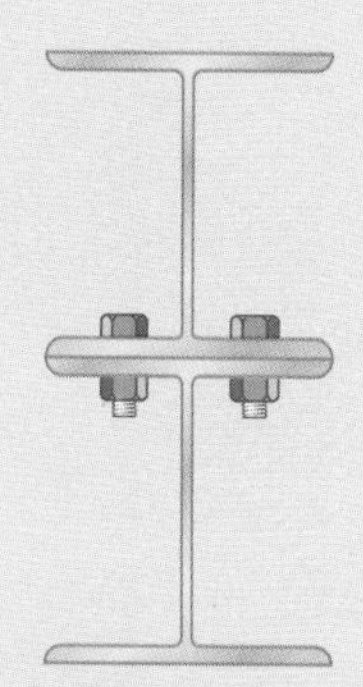

Fig. P13.51

13.52 For the beam and loading shown, consider section *n–n* and determine (*a*) the largest shearing stress in that section, (*b*) the shearing stress at point *a*.

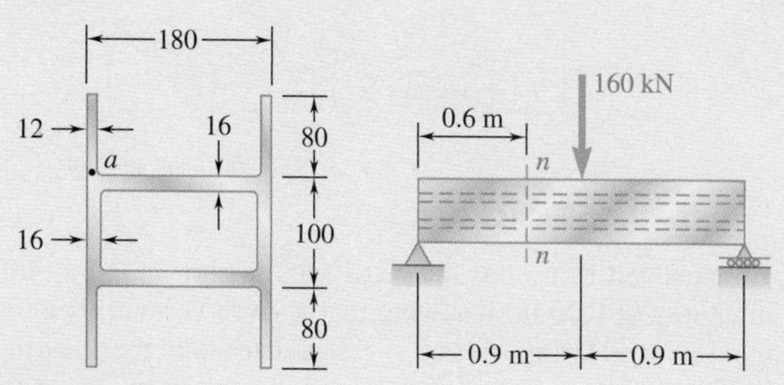

Dimensions in mm

Fig. *P13.52*

13.53 A timber beam *AB* of length *L* and rectangular cross section carries a uniformly distributed load *w* and is supported as shown. (*a*) Show that the ratio τ_m/σ_m of the maximum values of the shearing and normal stresses in the beam is equal to $2h/L$, where *h* and *L* are, respectively, the depth and the length of the beam. (*b*) Determine the depth *h* and the width *b* of the beam, knowing that $L = 5$ m, $w = 8$ kN/m, $\tau_m = 1.08$ MPa, and $\sigma_m = 12$ MPa.

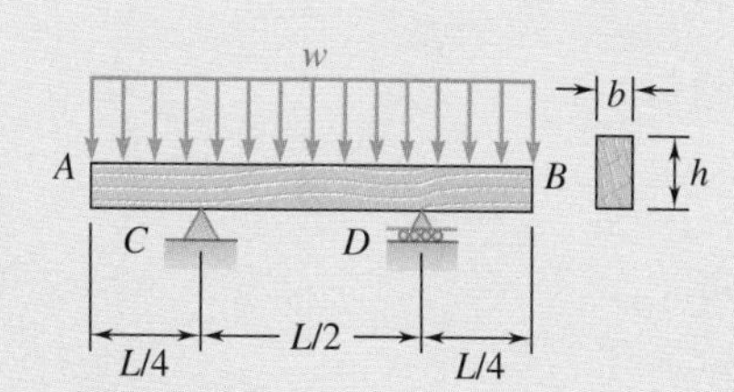

Fig. P13.53

13.54 For the beam and loading shown, consider section *n–n* and determine the shearing stress at (*a*) point *a*, (*b*) point *b*.

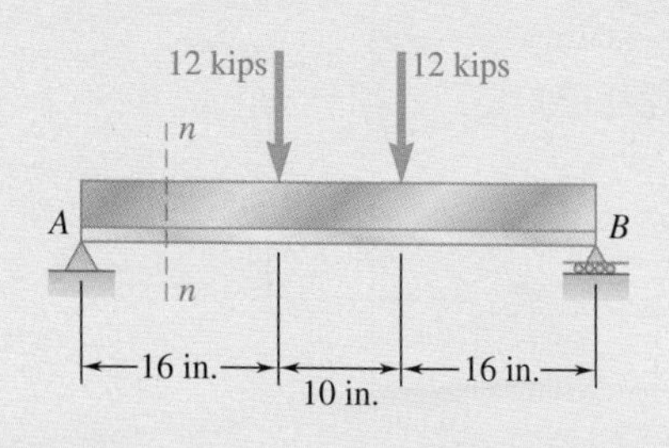

Fig. *P13.54*

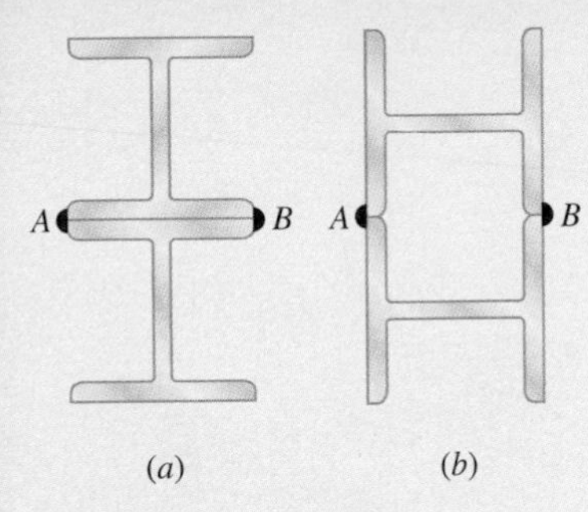

Fig. P13.55

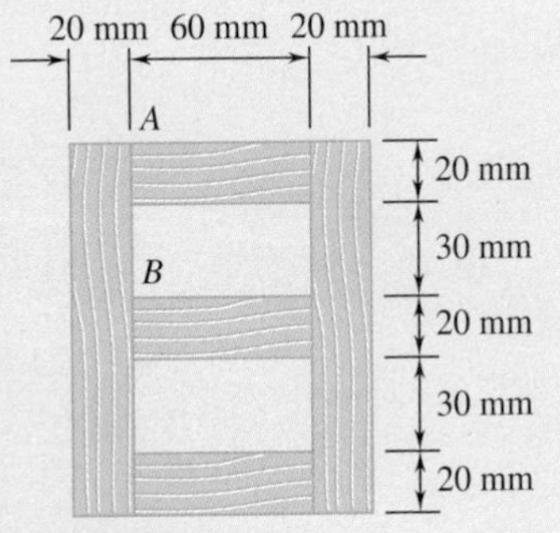

Fig. P13.56

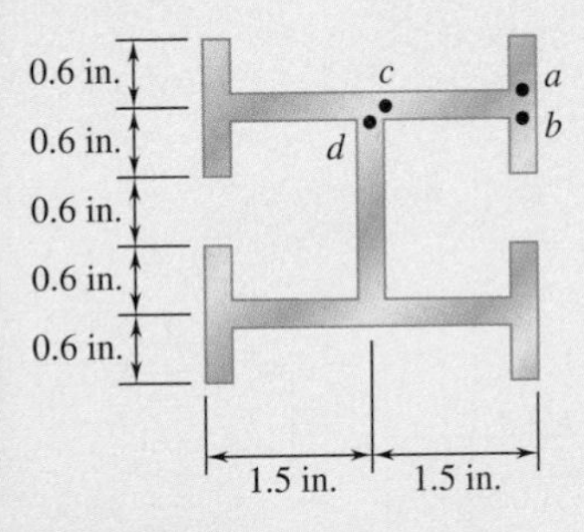

Fig. P13.58

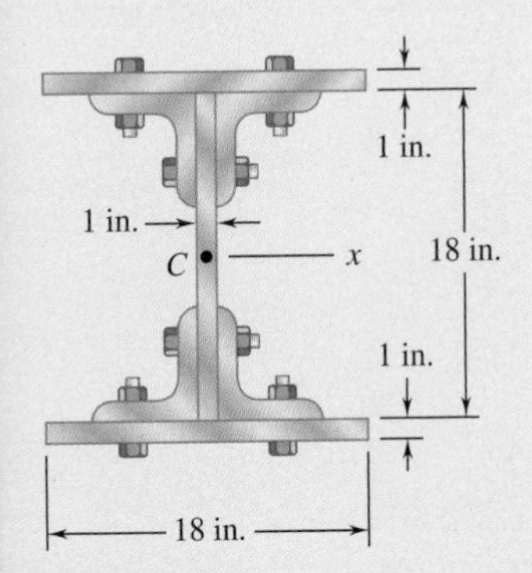

Fig. P13.60

13.55 Two W8 × 31 rolled sections can be welded at A and B in either of the two ways shown in order to form a composite beam. Knowing that for each weld the allowable horizontal shearing force is 3000 lb per inch of weld, determine the maximum allowable vertical shear in the composite beam for each of the two arrangements shown.

13.56 Several wooden planks are glued together to form the box beam shown. Knowing that the beam is subjected to a vertical shear of 3 kN, determine the average shearing stress in the glued joint (*a*) at A, (*b*) at B.

13.57 The built-up wooden beam shown is subjected to a vertical shear of 8 kN. Knowing that the nails are spaced longitudinally every 60 mm at A and every 25 mm at B, determine the shearing force in the nails (*a*) at A, (*b*) at B. (*Given*: $I_x = 1.504 \times 10^9$ mm^4)

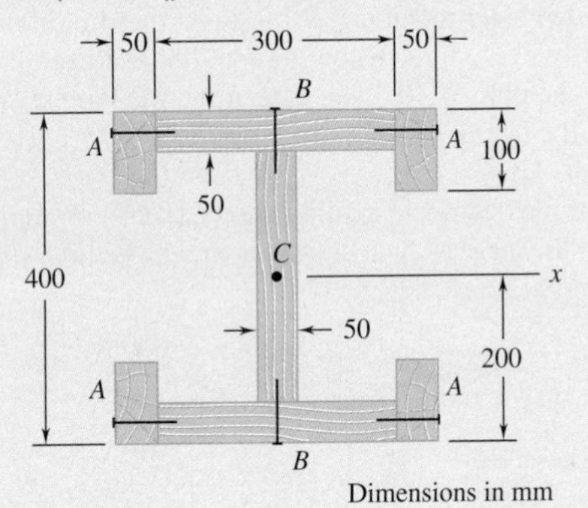

Fig. P13.57

13.58 An extruded beam has the cross section shown and a uniform wall thickness of 0.20 in. Knowing that a given vertical shear **V** causes a maximum shearing stress $\tau = 9$ ksi, determine the shearing stress at the four points indicated.

13.59 Solve Prob. 13.58 assuming that the beam is subjected to a horizontal shear **V**.

13.60 Three 1 × 18-in. steel plates are bolted to four L6 × 6 × 1 angles to form a beam with the cross section shown. The bolts have a $\frac{7}{8}$-in. diameter and are spaced longitudinally every 5 in. Knowing that the allowable average shearing stress in the bolts is 12 ksi, determine the largest permissible vertical shear in the beam. (*Given*: $I_x = 6123$ in^4.)

13.61 An extruded beam has the cross section shown and a uniform wall thickness of 3 mm. For a vertical shear of 10 kN, determine (*a*) the shearing stress at point A, (*b*) the maximum shearing stress in the beam. Also sketch the shear flow in the cross section.

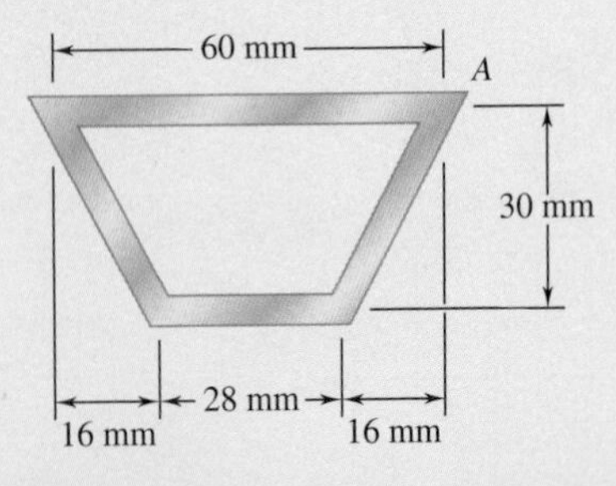

Fig. *P13.61*

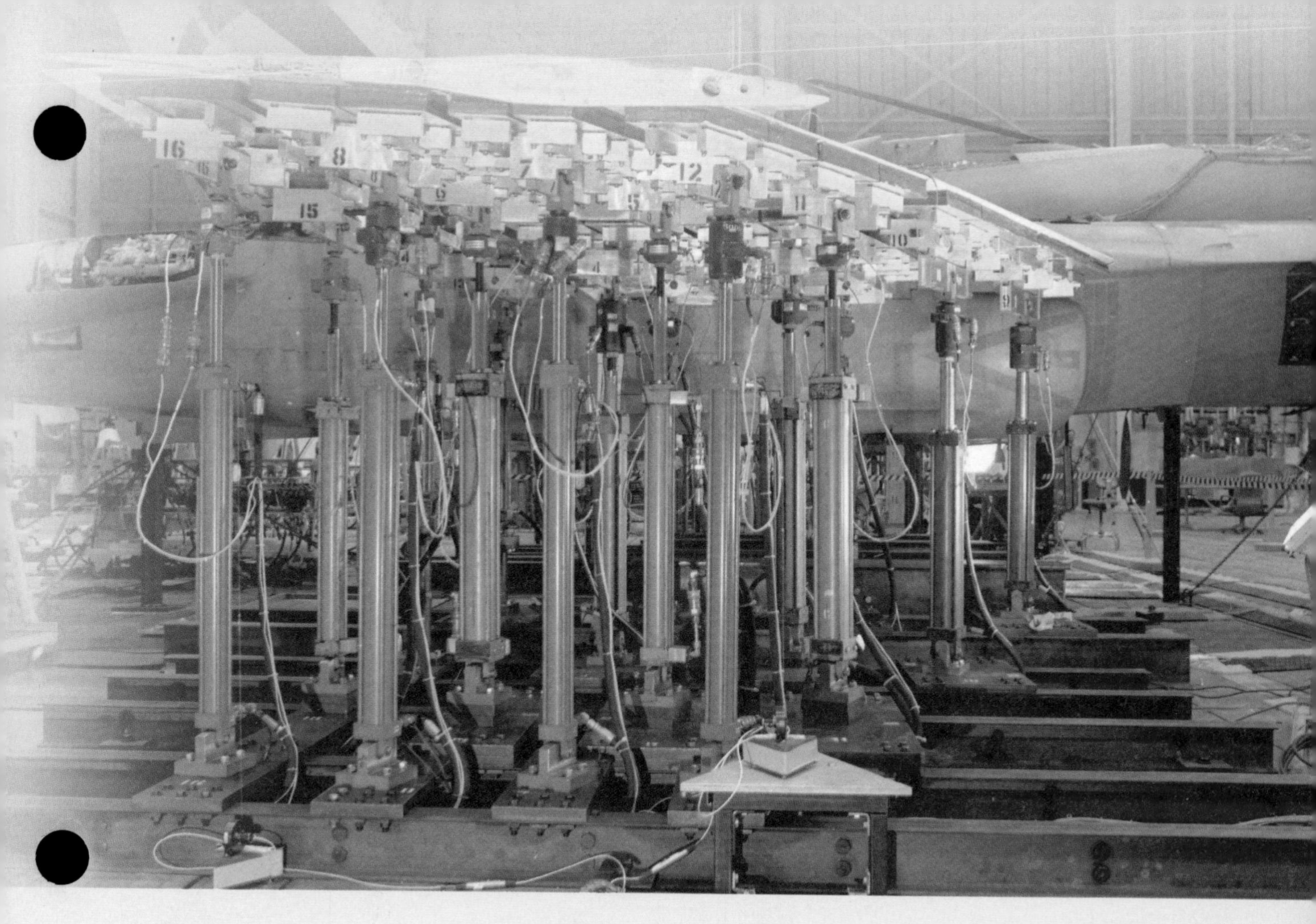

NASA

14

Transformations of Stress

The aircraft wing shown is being tested to determine how forces due to lift are distributed through the wing. This chapter will examine methods for determining the maximum stresses at any point in a structure.

Introduction

Objectives

In this chapter, you will:

- **Apply** stress transformation equations to plane stress situations to determine any stress component at a point.
- **Apply** the alternative Mohr's circle approach to perform plane stress transformations.
- **Use** transformation techniques to identify key components of stress, such as principal stresses.
- **Analyze** plane stress states in thin-walled pressure vessels.

Introduction

The most general state of stress at a given point Q is represented by six components (Sec. 8.3). Three of these components, σ_x, σ_y, and σ_z, are the normal stresses exerted on the faces of a small cubic element centered at Q with the same orientation as the coordinate axes (Fig. 14.1*a*). The other three, τ_{xy}, τ_{yz}, and τ_{zx},† are the components of the shearing stresses on the same element. The same state of stress will be represented by a different set of components if the coordinate axes are rotated (Fig. 14.1*b*). The first part of this chapter determines how the components of stress are transformed under a rotation of the coordinate axes.

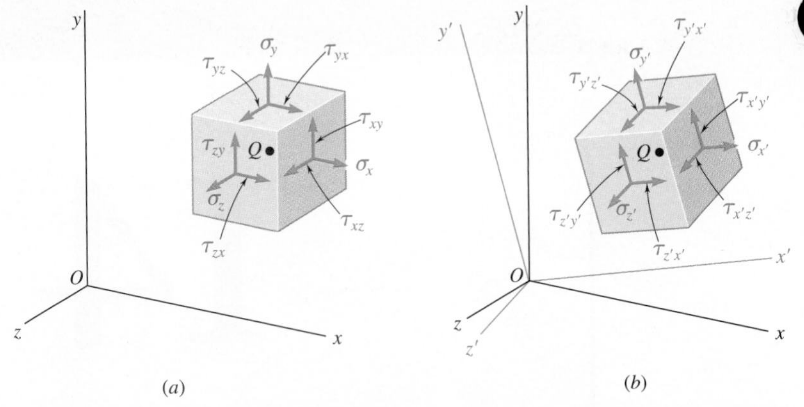

Fig. 14.1 General state of stress at a point: (*a*) referred to {*xyz*}, (*b*) referred to {*x'y'z'*}.

Our discussion of the transformation of stress will deal mainly with *plane stress*, i.e., with a situation in which two of the faces of the cubic element are free of any stress. If the z axis is chosen perpendicular to these faces, $\sigma_z = \tau_{zx} = \tau_{zy} = 0$, and the only remaining stress components are σ_x, σ_y, and τ_{xy} (Fig. 14.2). This situation occurs in a thin plate subjected to forces acting in the midplane of the plate (Fig. 14.3). It also occurs on the free surface of a structural element or machine component where any point of the surface of that element or component is not subjected to an external force (Fig. 14.4).

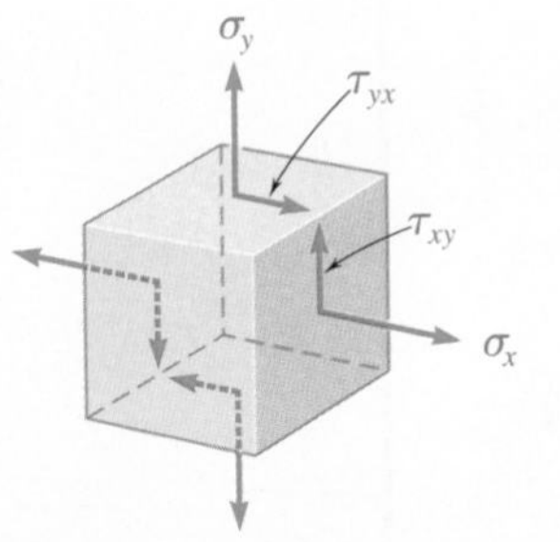

Fig. 14.2 Non-zero stress components for state of plane stress.

†Recall that $\tau_{yx} = \tau_{xy}$, $\tau_{zy} = \tau_{yz}$, and $\tau_{xz} = \tau_{zx}$ (Sec. 8.3).

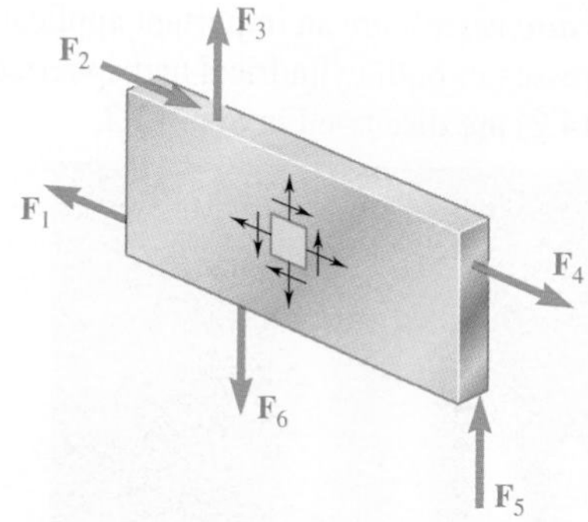

Fig. 14.3 Example of plane stress: thin plate subjected to only in-plane loads.

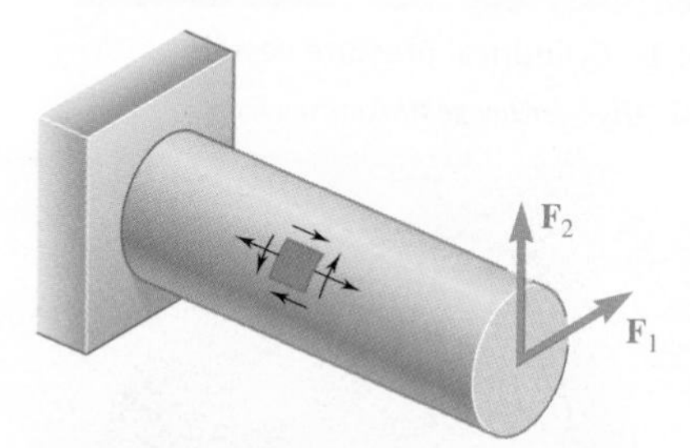

Fig. 14.4 Example of plane stress: free surface of a structural component.

In Sec. 14.1A, a state of plane stress at a given point Q is characterized by the stress components σ_x, σ_y, and τ_{xy} associated with the element shown in Fig. 14.5*a*. Components $\sigma_{x'}$, $\sigma_{y'}$, and $\tau_{x'y'}$ associated with that element after it has been rotated through an angle θ about the z axis (Fig. 14.5*b*) will then be determined. In Sec. 14.1B, the value θ_p of θ will be found, where the stresses $\sigma_{x'}$ and $\sigma_{y'}$ are the maximum and minimum stresses. These values of the normal stress are the *principal stresses* at point Q, and the faces of the corresponding element define the *principal planes of stress* at that point. The angle of rotation θ_s for which the shearing stress is maximum also is discussed.

In Sec. 14.2, an alternative method to solve problems involving the transformation of plane stress, based on the use of *Mohr's circle*, is presented.

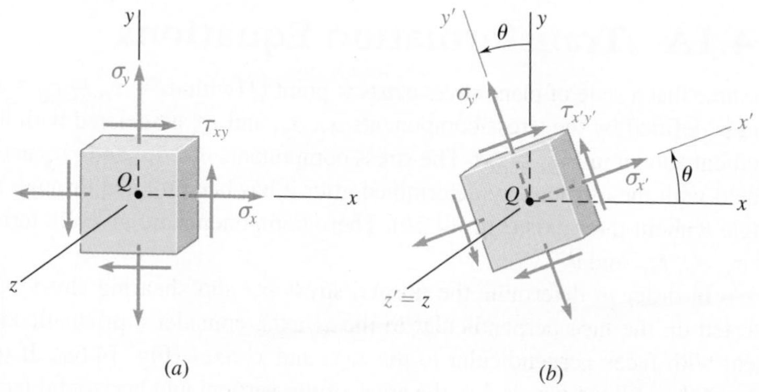

Fig. 14.5 State of plane stress: (*a*) referred to {*xyz*}, (*b*) referred to {*x'y'z'*}.

Thin-walled pressure vessels are an important application of the analysis of plane stress. Stresses in both cylindrical and spherical pressure vessels (Photos 14.1 and 14.2) are discussed in Sec. 14.3.

Photo 14.1 Cylindrical pressure vessels.

Photo 14.2 Spherical pressure vessel.

14.1 TRANSFORMATION OF PLANE STRESS

14.1A Transformation Equations

Assume that a state of plane stress exists at point Q (with $\sigma_z = \tau_{zx} = \tau_{zy} = 0$) and is defined by the stress components σ_x, σ_y, and τ_{xy} associated with the element shown in Fig. 14.5*a*. The stress components $\sigma_{x'}$, $\sigma_{y'}$, and $\tau_{x'y'}$ associated with the element are determined after it has been rotated through an angle θ about the z axis (Fig. 14.5*b*). These components are given in terms of σ_x, σ_y, τ_{xy}, and θ.

In order to determine the normal stress $\sigma_{x'}$ and shearing stress $\tau_{x'y'}$ exerted on the face perpendicular to the x' axis, consider a prismatic element with faces perpendicular to the x, y, and x' axes (Fig. 14.6*a*). If the area of the oblique face is ΔA, the areas of the vertical and horizontal faces are equal to $\Delta A \cos\theta$ and $\Delta A \sin\theta$, respectively. The *forces* exerted on

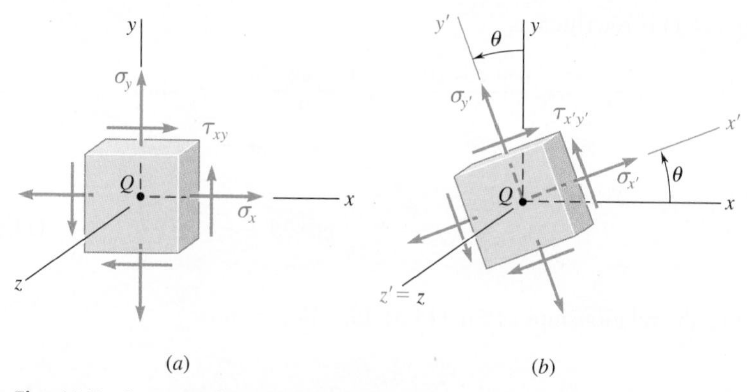

Fig. 14.5 (repeated) State of plane stress: (*a*) referred to {*xyz*}, (*b*) referred to {*x*′*y*′*z*′}.

the three faces are as shown in Fig. 14.6*b*. (No forces are exerted on the triangular faces of the element, since the corresponding normal and shearing stresses are assumed equal to zero.) Using components along the x' and y' axes, the equilibrium equations are

$$\Sigma F_{x'} = 0: \quad \sigma_{x'}\,\Delta A - \sigma_x(\Delta A \cos\theta)\cos\theta - \tau_{xy}(\Delta A\cos\theta)\sin\theta$$
$$-\sigma_y(\Delta A\sin\theta)\sin\theta - \tau_{xy}(\Delta A\sin\theta)\cos\theta = 0$$

$$\Sigma F_{y'} = 0: \quad \tau_{x'y'}\,\Delta A + \sigma_x(\Delta A\cos\theta)\sin\theta - \tau_{xy}(\Delta A\cos\theta)\cos\theta$$
$$-\sigma_y(\Delta A\sin\theta)\cos\theta + \tau_{xy}(\Delta A\sin\theta)\sin\theta = 0$$

Solving the first equation for $\sigma_{x'}$ and the second for $\tau_{x'y'}$,

$$\sigma_{x'} = \sigma_x\cos^2\theta + \sigma_y\sin^2\theta + 2\tau_{xy}\sin\theta\cos\theta \tag{14.1}$$

$$\tau_{x'y'} = -(\sigma_x - \sigma_y)\sin\theta\cos\theta + \tau_{xy}(\cos^2\theta - \sin^2\theta) \tag{14.2}$$

Recalling the trigonometric relations

$$\sin 2\theta = 2\sin\theta\cos\theta \qquad \cos 2\theta = \cos^2\theta - \sin^2\theta \tag{14.3}$$

and

$$\cos^2\theta = \frac{1 + \cos 2\theta}{2} \qquad \sin^2\theta = \frac{1 - \cos 2\theta}{2} \tag{14.4}$$

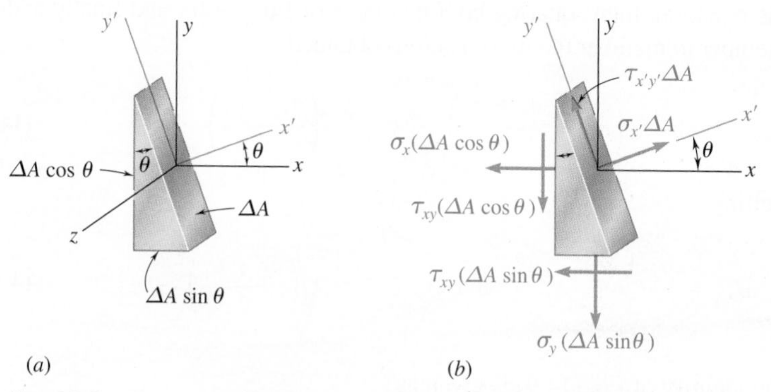

Fig. 14.6 Stress transformation equations are determined by considering an arbitrary prismatic wedge element. (*a*) Geometry of the element. (*b*) Free-body diagram.

Eq. (14.1) is rewritten as

$$\sigma_{x'} = \sigma_x \frac{1 + \cos 2\theta}{2} + \sigma_y \frac{1 - \cos 2\theta}{2} + \tau_{xy} \sin 2\theta$$

or

$$\sigma_{x'} = \frac{\sigma_x + \sigma_y}{2} + \frac{\sigma_x - \sigma_y}{2} \cos 2\theta + \tau_{xy} \sin 2\theta \qquad (14.5)$$

Using the relationships of Eq. (14.3), Eq. (14.2) is now

$$\tau_{x'y'} = -\frac{\sigma_x - \sigma_y}{2} \sin 2\theta + \tau_{xy} \cos 2\theta \qquad (14.6)$$

The normal stress $\sigma_{y'}$ is obtained by replacing θ in Eq. (14.5) by the angle $\theta + 90°$ that the y' axis forms with the x axis. Since $\cos(2\theta + 180°) = -\cos 2\theta$ and $\sin(2\theta + 180°) = -\sin 2\theta$,

$$\sigma_{y'} = \frac{\sigma_x + \sigma_y}{2} - \frac{\sigma_x - \sigma_y}{2} \cos 2\theta - \tau_{xy} \sin 2\theta \qquad (14.7)$$

Adding Eqs. (14.5) and (14.7) member to member,

$$\sigma_{x'} + \sigma_{y'} = \sigma_x + \sigma_y \qquad (14.8)$$

Since $\sigma_z = \sigma_{z'} = 0$, we thus verify for plane stress that the sum of the normal stresses exerted on a cubic element of material is independent of the orientation of that element.

14.1B Principal Stresses and Maximum Shearing Stress

Eqs. (14.5) and (14.6) are the parametric equations of a circle. This means that, if a set of rectangular axes is used to plot a point M of abscissa $\sigma_{x'}$ and ordinate $\tau_{x'y'}$ for any given parameter θ, all of the points obtained will lie on a circle. To establish this property, we eliminate θ from Eqs. (14.5) and (14.6) by first transposing $(\sigma_x + \sigma_y)/2$ in Eq. (14.5) and squaring both members of the equation, then squaring both members of Eq. (14.6), and finally adding member to member the two equations obtained:

$$\left(\sigma_{x'} - \frac{\sigma_x + \sigma_y}{2}\right)^2 + \tau_{x'y'}^2 = \left(\frac{\sigma_x - \sigma_y}{2}\right)^2 + \tau_{xy}^2 \qquad (14.9)$$

Setting

$$\sigma_{\text{ave}} = \frac{\sigma_x + \sigma_y}{2} \quad \text{and} \quad R = \sqrt{\left(\frac{\sigma_x - \sigma_y}{2}\right)^2 + \tau_{xy}^2} \qquad (14.10)$$

the identity of Eq. (14.9) is given as

$$(\sigma_{x'} - \sigma_{\text{ave}})^2 + \tau_{x'y'}^2 = R^2 \qquad (14.11)$$

which is the equation of a circle of radius R centered at the point C of abscissa σ_{ave} and ordinate 0 (Fig. 14.7). Due to the symmetry of the circle about the horizontal axis, the same result is obtained if a point N of abscissa $\sigma_{x'}$ and ordinate $-\tau_{x'y'}$ is plotted instead of M. (Fig. 14.8). This property will be used in Sec. 14.2.

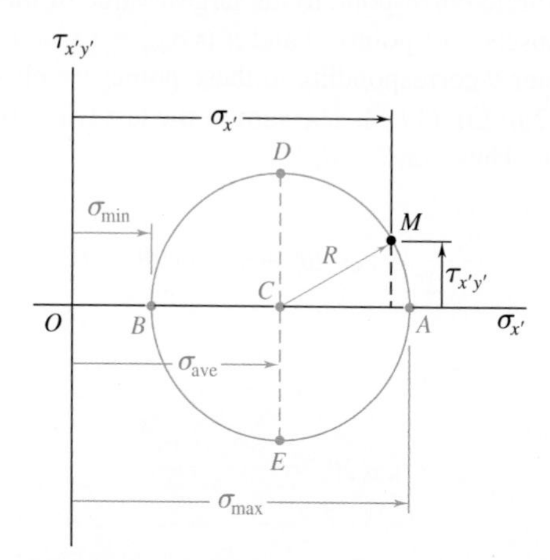

Fig. 14.7 Circular relationship of transformed stresses.

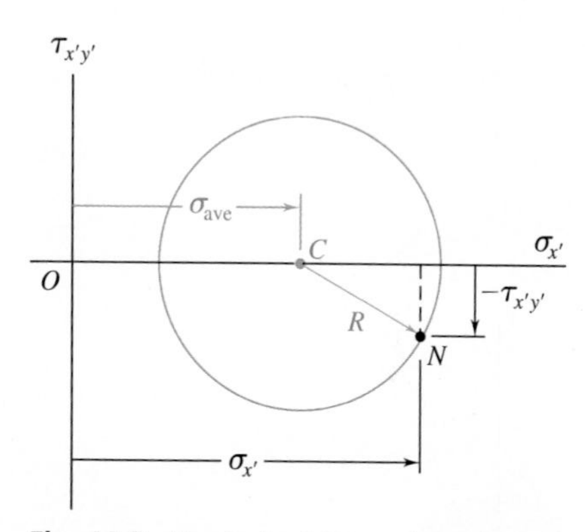

Fig. 14.8 Equivalent formation of stress transformation circle.

The points A and B where the circle of Fig. 14.7 intersects the horizontal axis are of special interest: point A corresponds to the maximum value of the normal stress $\sigma_{x'}$, while point B corresponds to its minimum value. Both points also correspond to a zero value of the shearing stress $\tau_{x'y'}$. Thus, the values θ_p of the parameter θ which correspond to points A and B can be obtained by setting $\tau_{x'y'} = 0$ in Eq. (14.6).[†]

$$\tan 2\theta_p = \frac{2\tau_{xy}}{\sigma_x - \sigma_y} \quad \textbf{(14.12)}$$

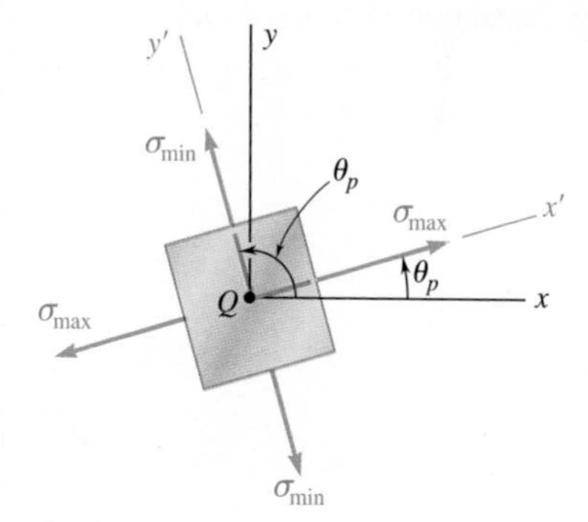

Fig. 14.9 Principal stresses.

This equation defines two values $2\theta_p$ that are 180° apart and thus two values θ_p that are 90° apart. Either value can be used to determine the orientation of the corresponding element (Fig. 14.9). The planes containing the faces of the element obtained in this way are the *principal planes of stress* at point Q, and the corresponding values σ_{max} and σ_{min} exerted on these planes are the *principal stresses* at Q. Since both values θ_p defined by Eq. (14.12) are obtained by setting $\tau_{x'y'} = 0$ in Eq. (14.6), it is clear that no shearing stress is exerted on the principal planes.

From Fig. 14.7,

$$\sigma_{max} = \sigma_{ave} + R \quad \text{and} \quad \sigma_{min} = \sigma_{ave} - R \quad \textbf{(14.13)}$$

Substituting for σ_{ave} and R from Eq. (14.10),

$$\sigma_{max,\,min} = \frac{\sigma_x + \sigma_y}{2} \pm \sqrt{\left(\frac{\sigma_x - \sigma_y}{2}\right)^2 + \tau_{xy}^2} \quad \textbf{(14.14)}$$

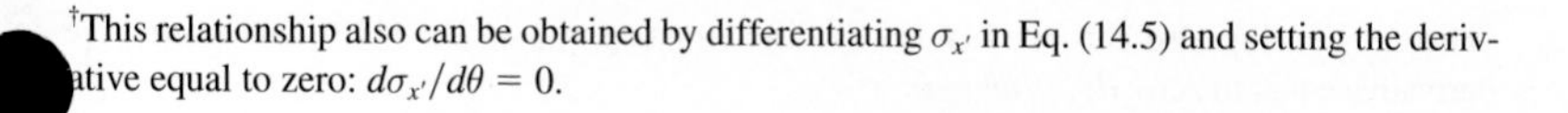

[†]This relationship also can be obtained by differentiating $\sigma_{x'}$ in Eq. (14.5) and setting the derivative equal to zero: $d\sigma_{x'}/d\theta = 0$.

Unless it is possible to tell by inspection which of these principal planes is subjected to σ_{max} and which is subjected to σ_{min}, it is necessary to substitute one of the values θ_p into Eq. (14.5) in order to determine which corresponds to the maximum value of the normal stress.

Referring again to Fig. 14.7, points D and E located on the vertical diameter of the circle correspond to the largest value of the shearing stress $\tau_{x'y'}$. Since the abscissa of points D and E is $\sigma_{ave} = (\sigma_x + \sigma_y)/2$, the values θ_s of the parameter θ corresponding to these points are obtained by setting $\sigma_{x'} = (\sigma_x + \sigma_y)/2$ in Eq. (14.5). The sum of the last two terms in that equation must be zero. Thus, for $\theta = \theta_s$,[†]

$$\frac{\sigma_x - \sigma_y}{2} \cos 2\theta_s + \tau_{xy} \sin 2\theta_s = 0$$

or

$$\tan 2\theta_s = -\frac{\sigma_x - \sigma_y}{2\tau_{xy}} \tag{14.15}$$

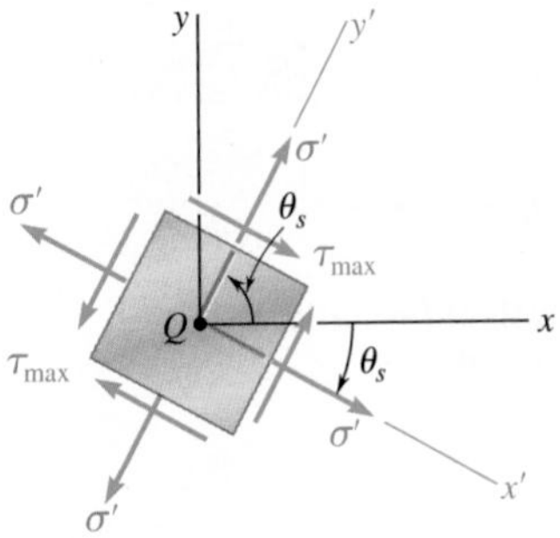

Fig. 14.10 Maximum shearing stress.

This equation defines two values $2\theta_s$ that are 180° apart, and thus two values θ_s that are 90° apart. Either of these values can be used to determine the orientation of the element corresponding to the maximum shearing stress (Fig. 14.10). Fig. 14.7 shows that the maximum value of the shearing stress is equal to the radius R of the circle. Recalling the second of Eqs. (14.10),

$$\tau_{max} = \sqrt{\left(\frac{\sigma_x - \sigma_y}{2}\right)^2 + \tau_{xy}^2} \tag{14.16}$$

As observed earlier, the normal stress corresponding to the condition of maximum shearing stress is

$$\sigma' = \sigma_{ave} = \frac{\sigma_x + \sigma_y}{2} \tag{14.17}$$

Comparing Eqs. (14.12) and (14.15), $\tan 2\theta_s$ is the negative reciprocal of $\tan 2\theta_p$. Thus, angles $2\theta_s$ and $2\theta_p$ are 90° apart, and therefore angles θ_s and θ_p are 45° apart. Thus, *the planes of maximum shearing stress are at 45° to the principal planes*. This confirms the results found in Sec. 8.3 for a centric axial load (Fig. 8.37) and in Sec. 10.1C for a torsional load (Fig. 10.17).

Be aware that the analysis of the transformation of plane stress has been limited to rotations *in the plane of stress*. If the cubic element of Fig. 14.5 is rotated about an axis other than the z axis, its faces may be subjected to shearing stresses larger than defined by Eq. (14.16). In these cases, the value given by Eq. (14.16) is referred to as the maximum *in-plane* shearing stress.

[†]This relationship also can be obtained by differentiating $\tau_{x'y'}$ in Eq. (14.6) and setting the derivative equal to zero: $d\tau_{x'y'}/d\theta = 0$.

Concept Application 14.1

For the state of plane stress shown in Fig. 14.11*a*, determine (*a*) the principal planes, (*b*) the principal stresses, (*c*) the maximum shearing stress and the corresponding normal stress.

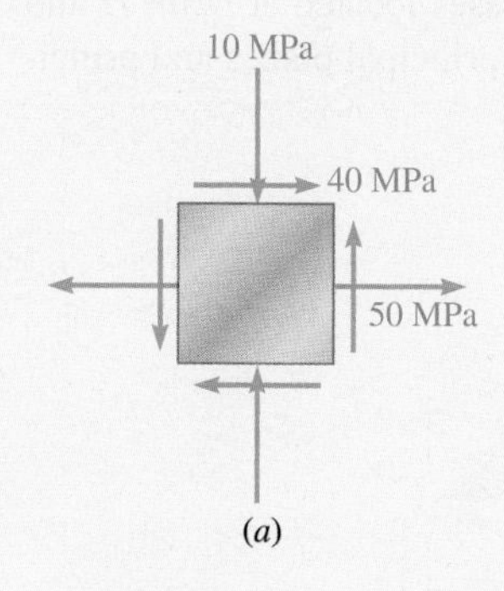

a. Principal Planes. Following the usual sign convention, the stress components are

$$\sigma_x = +50 \text{ MPa} \qquad \sigma_y = -10 \text{ MPa} \qquad \tau_{xy} = +40 \text{ MPa}$$

Substituting into Eq. (14.12),

$$\tan 2\theta_p = \frac{2\tau_{xy}}{\sigma_x - \sigma_y} = \frac{2(+40)}{50 - (-10)} = \frac{80}{60}$$

$$2\theta_p = 53.1^\circ \quad \text{and} \quad 180^\circ + 53.1^\circ = 233.1^\circ$$

$$\theta_p = 26.6^\circ \quad \text{and} \quad 116.6^\circ$$

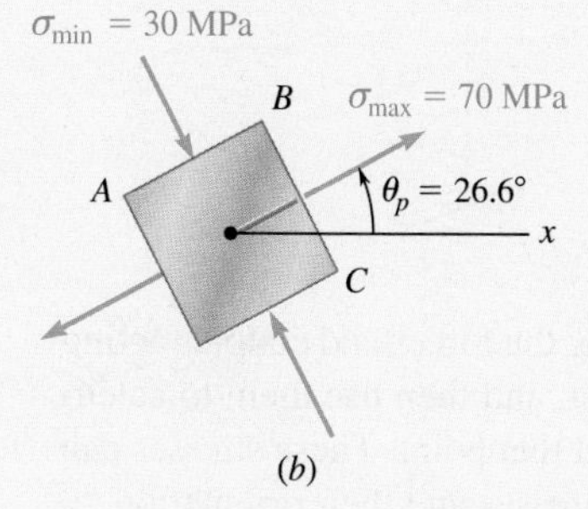

b. Principal Stresses. Eq. (14.14) yields

$$\sigma_{\text{max, min}} = \frac{\sigma_x + \sigma_y}{2} \pm \sqrt{\left(\frac{\sigma_x - \sigma_y}{2}\right)^2 + \tau_{xy}^2}$$

$$= 20 \pm \sqrt{(30)^2 + (40)^2}$$

$$\sigma_{\max} = 20 + 50 = 70 \text{ MPa}$$

$$\sigma_{\min} = 20 - 50 = -30 \text{ MPa}$$

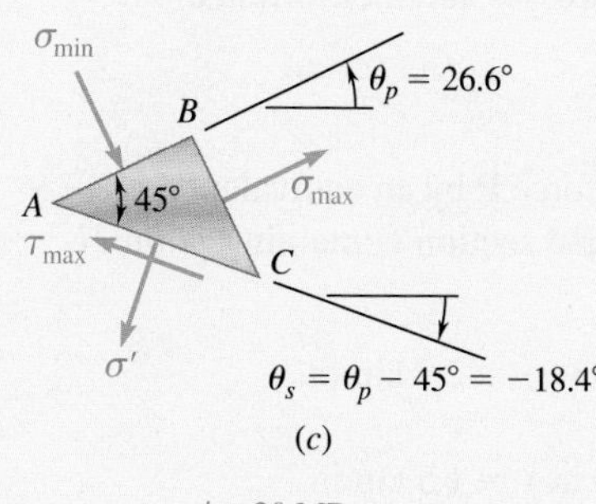

The principal planes and principal stresses are shown in Fig. 14.11*b*. Making $2\theta = 53.1^\circ$ in Eq. (14.5), it is confirmed that the normal stress exerted on face *BC* of the element is the maximum stress:

$$\sigma_{x'} = \frac{50 - 10}{2} + \frac{50 + 10}{2} \cos 53.1^\circ + 40 \sin 53.1^\circ$$

$$= 20 + 30 \cos 53.1^\circ + 40 \sin 53.1^\circ = 70 \text{ MPa} = \sigma_{\max}$$

c. Maximum Shearing Stress. Eq. (14.16) yields

$$\tau_{\max} = \sqrt{\left(\frac{\sigma_x - \sigma_y}{2}\right)^2 + \tau_{xy}^2} = \sqrt{(30)^2 + (40)^2} = 50 \text{ MPa}$$

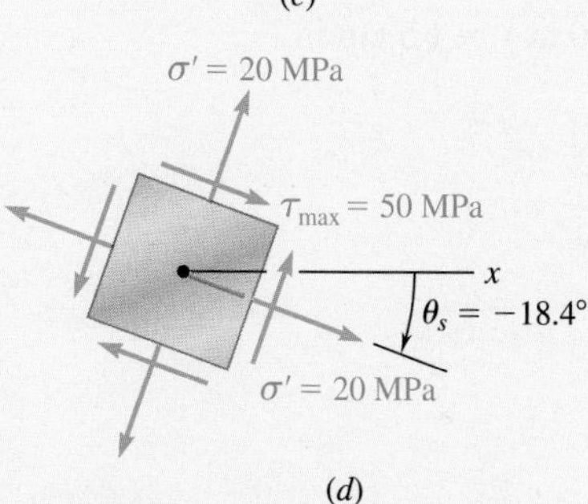

Fig. 14.11 (*a*) Plane stress element. (*b*) Plane stress element oriented in principal directions. (*c*) Plane stress element showing principal and maximum shear planes. (*d*) Plane stress element showing maximum shear orientation.

Since $\sigma_{\max}$ and $\sigma_{\min}$ have opposite signs, $\tau_{\max}$ actually represents the maximum value of the shearing stress at the point. The orientation of the planes of maximum shearing stress and the sense of the shearing stresses are determined by passing a section along the diagonal plane *AC* of the element of Fig. 14.11*b*. Since the faces *AB* and *BC* of the element are in the principal planes, the diagonal plane *AC* must be one of the planes of maximum shearing stress (Fig. 14.11*c*). Furthermore, the equilibrium conditions for the prismatic element *ABC* require that the shearing stress exerted on *AC* be directed as shown. The cubic element corresponding to the maximum shearing stress is shown in Fig. 14.11*d*. The normal stress on each of the four faces of the element is given by Eq. (14.17):

$$\sigma' = \sigma_{\text{ave}} = \frac{\sigma_x + \sigma_y}{2} = \frac{50 - 10}{2} = 20 \text{ MPa}$$

Sample Problem 14.1

A single horizontal force **P** with a magnitude of 150 lb is applied to end D of lever ABD. Knowing that portion AB of the lever has a diameter of 1.2 in., determine (*a*) the normal and shearing stresses located at point H and having sides parallel to the x and y axes, (*b*) the principal planes and principal stresses at point H.

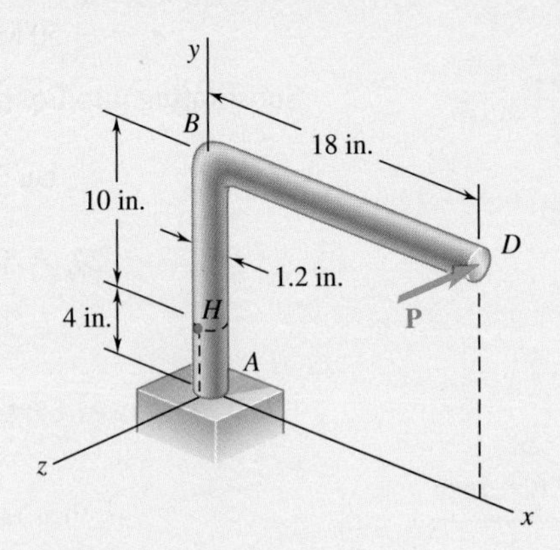

STRATEGY: You can begin by determining the forces and couples acting on the section containing the point of interest, and then use them to calculate the normal and shearing stresses acting at that point. These stresses can then be transformed to obtain the principal stresses and their orientation.

MODELING and ANALYSIS:

Force-Couple System. We replace the force **P** by an equivalent force-couple system at the center C of the transverse section containing point H (Fig.1):

$$P = 150 \text{ lb} \qquad T = (150 \text{ lb})(18 \text{ in.}) = 2.7 \text{ kip·in.}$$

$$M_x = (150 \text{ lb})(10 \text{ in.}) = 1.5 \text{ kip·in.}$$

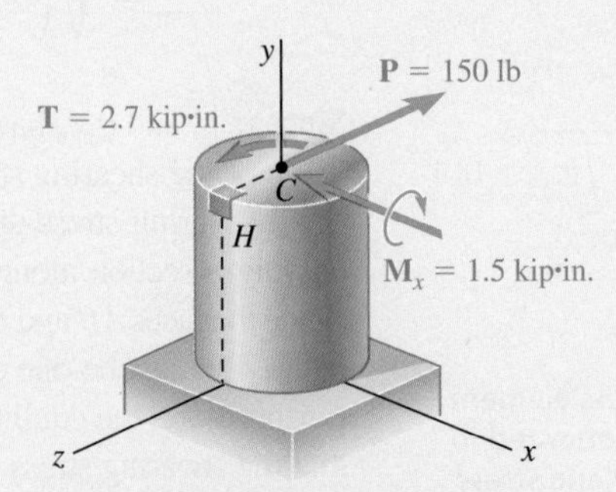

Fig. 1 Equivalent force-couple system acting on transverse section containing point H.

(continued)

a. Stresses σ_x, σ_y, τ_{xy} at Point *H*. Using the sign convention shown in Fig. 14.2, the sense and the sign of each stress component are found by carefully examining the force-couple system at point *C* (Fig. 1):

$$\sigma_x = 0 \quad \sigma_y = +\frac{Mc}{I} = +\frac{(1.5 \text{ kip·in.})(0.6 \text{ in.})}{\frac{1}{4}\pi\,(0.6 \text{ in.})^4} \qquad \sigma_y = +8.84 \text{ ksi} \blacktriangleleft$$

$$\tau_{xy} = +\frac{Tc}{J} = +\frac{(2.7 \text{ kip·in.})(0.6 \text{ in.})}{\frac{1}{2}\pi\,(0.6 \text{ in.})^4} \qquad \tau_{xy} = +7.96 \text{ ksi} \blacktriangleleft$$

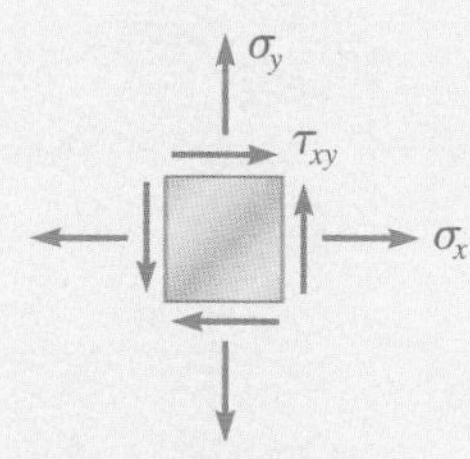

Fig. 2 General plane stress element (showing positive directions).

We note that the shearing force **P** does not cause any shearing stress at point *H*. The general plane stress element (Fig. 2) is completed to reflect these stress results (Fig. 3).

b. Principal Planes and Principal Stresses. Substituting the values of the stress components into Eq. (14.12), the orientation of the principal planes is

$$\tan 2\theta_p = \frac{2\tau_{xy}}{\sigma_x - \sigma_y} = \frac{2(7.96)}{0 - 8.84} = -1.80$$

$$2\theta_p = -61.0^\circ \quad \text{and} \quad 180^\circ - 61.0^\circ = +119^\circ$$

$$\theta_p = -30.5^\circ \quad \text{and} \quad +59.5^\circ \blacktriangleleft$$

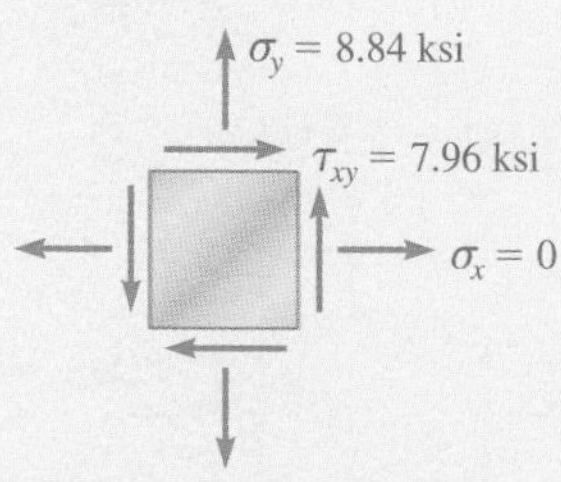

Fig. 3 Stress element at point *H*.

Substituting into Eq. (14.14), the magnitudes of the principal stresses are

$$\sigma_{\max,\min} = \frac{\sigma_x + \sigma_y}{2} \pm \sqrt{\left(\frac{\sigma_x - \sigma_y}{2}\right)^2 + \tau_{xy}^2}$$

$$= \frac{0 + 8.84}{2} \pm \sqrt{\left(\frac{0 - 8.84}{2}\right)^2 + (7.96)^2} = +4.42 \pm 9.10$$

$$\sigma_{\max} = +13.52 \text{ ksi} \blacktriangleleft$$

$$\sigma_{\min} = -4.68 \text{ ksi} \blacktriangleleft$$

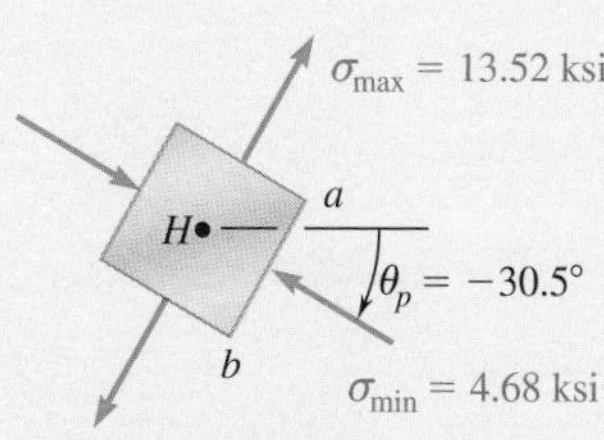

Fig. 4 Stress element at point *H* oriented in principal directions.

Considering face *ab* of the element shown, $\theta_p = -30.5^\circ$ in Eq. (14.5) and $\sigma_{x'} = -4.68$ ksi. The principal stresses are as shown in Fig. 4.

Problems

14.1 through 14.4 For the given state of stress, determine the normal and shearing stresses exerted on the oblique face of the shaded triangular element shown. Use a method of analysis based on the equilibrium of that element, as was done in the derivations of Sec. 14.1A.

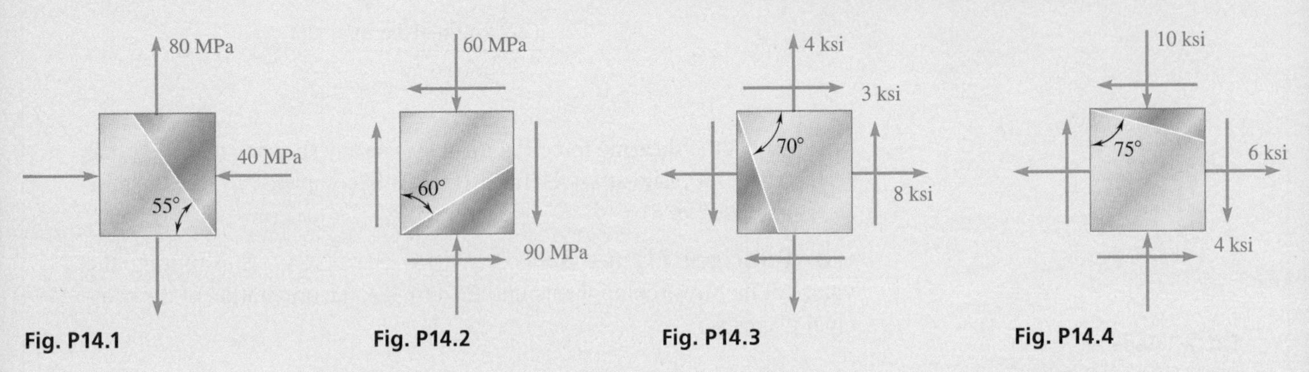

Fig. P14.1 **Fig. P14.2** **Fig. P14.3** **Fig. P14.4**

14.5 through *14.8* For the given state of stress, determine (*a*) the principal planes, (*b*) the principal stresses.

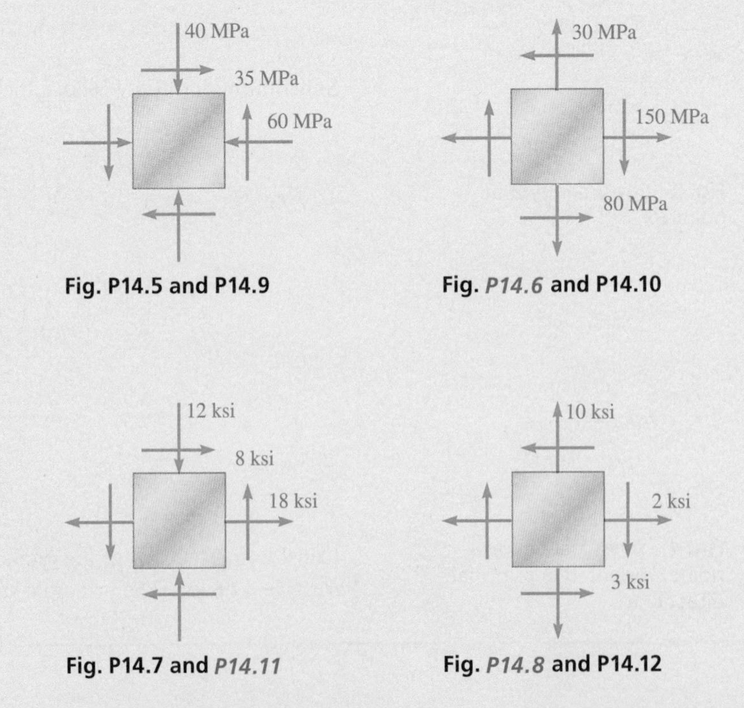

Fig. P14.5 and P14.9

Fig. *P14.6* and P14.10

Fig. P14.7 and *P14.11*

Fig. *P14.8* and P14.12

14.9 through 14.12 For the given state of stress, determine (*a*) the orientation of the planes of maximum in-plane shearing stress, (*b*) the maximum in-plane shearing stress, (*c*) the corresponding normal stress.

14.13 through 14.16 For the given state of stress, determine the normal and shearing stresses after the element shown has been rotated through (*a*) 25° clockwise, (*b*) 10° counterclockwise.

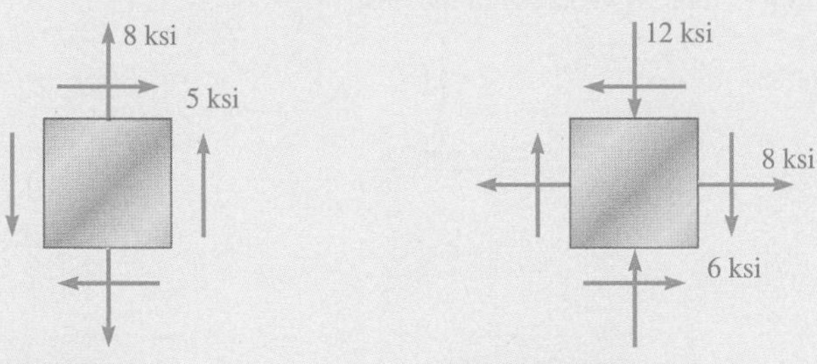

Fig. P14.13 Fig. P14.14

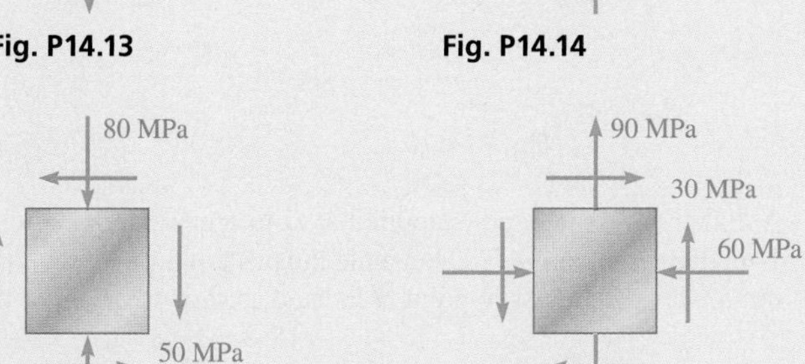

Fig. P14.15 Fig. P14.16

14.17 and 14.18 The grain of a wooden member forms an angle of 15° with the vertical. For the state of stress shown, determine (*a*) the in-plane shearing stress parallel to the grain, (*b*) the normal stress perpendicular to the grain.

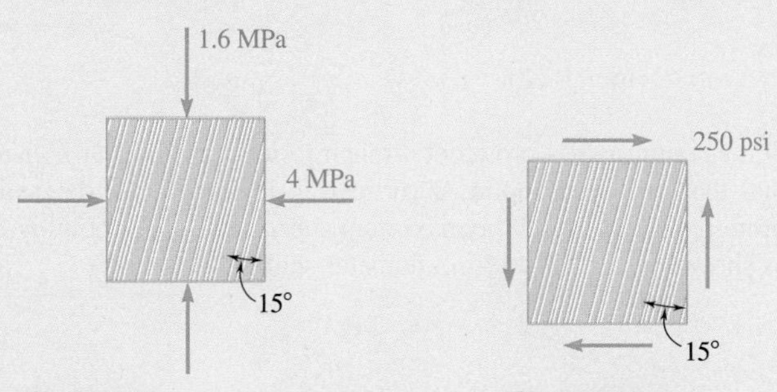

Fig. P14.17 Fig. P14.18

14.19 Two steel plates of uniform cross section 10 × 80 mm are welded together as shown. Knowing that centric 100-kN forces are applied to the welded plates and that $\beta = 25°$, determine (*a*) the in-plane shearing stress parallel to the weld, (*b*) the normal stress perpendicular to the weld.

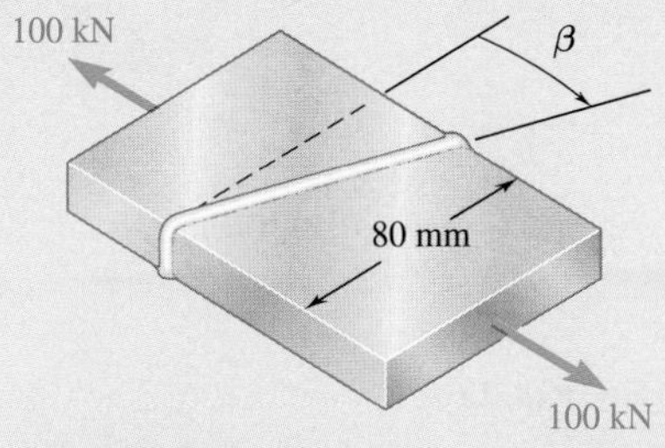

Fig. P14.19

14.20 The centric force **P** is applied to a short post as shown. Knowing that the stresses on plane *a–a* are $\sigma = -15$ ksi and $\tau = 5$ ksi, determine (*a*) the angle β that plane *a–a* forms with the horizontal, (*b*) the maximum compressive stress in the post.

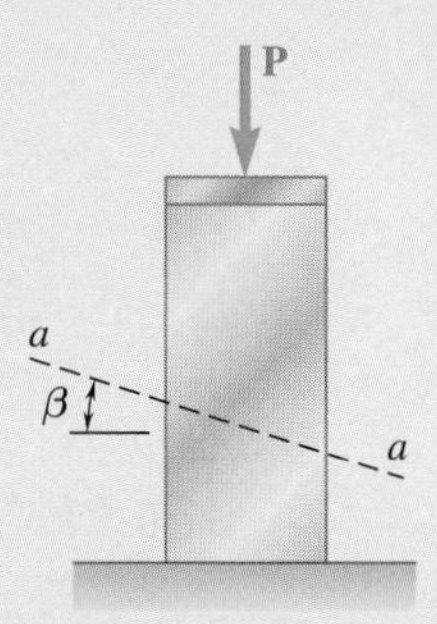

Fig. P14.20

14.21 A 400-lb vertical force is applied at *D* to a gear attached to the solid 1-in. diameter shaft *AB*. Determine the principal stresses and the maximum shearing stress at point *H* located as shown on top of the shaft.

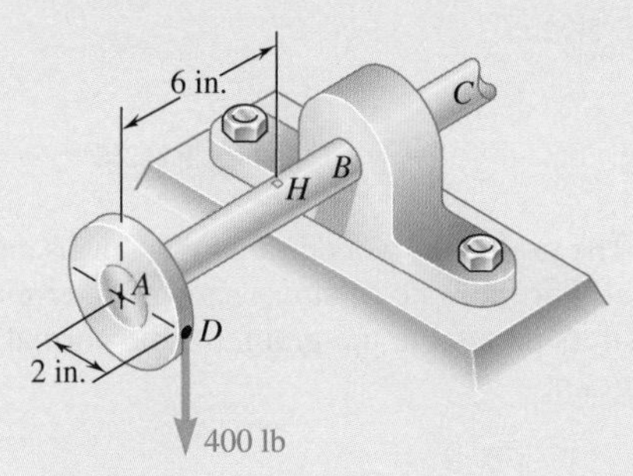

Fig. *P14.21*

14.22 A mechanic uses a crowfoot wrench to loosen a bolt at *E*. Knowing that the mechanic applies a vertical 24-lb force at *A*, determine the principal stresses and the maximum shearing stress at point *H* located as shown on top of the $\frac{3}{4}$-in.-diameter shaft.

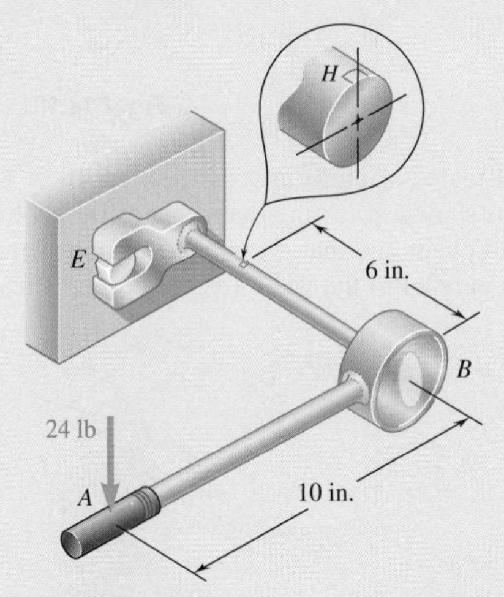

Fig. P14.22

14.23 The steel pipe AB has a 102-mm outer diameter and a 6-mm wall thickness. Knowing that arm CD is rigidly attached to the pipe, determine the principal stresses and the maximum shearing stress at point K.

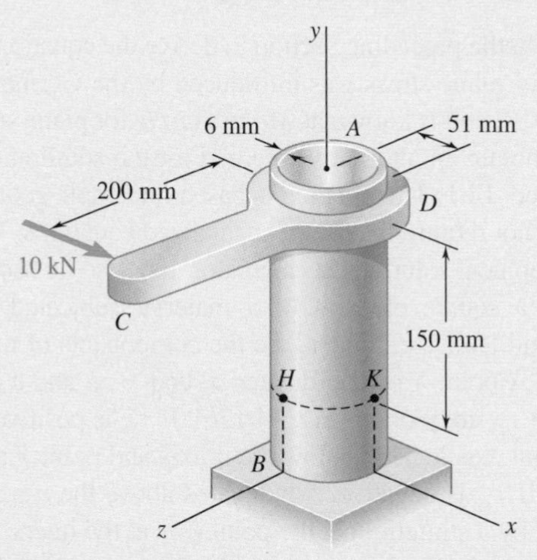

Fig. P14.23

14.24 For the state of plane stress shown, determine the largest value of σ_y for which the maximum in-plane shearing stress is equal to or less than 75 MPa.

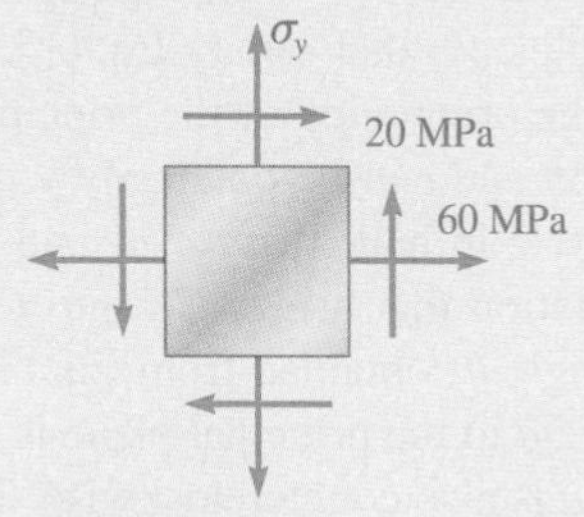

Fig. P14.24

14.2 MOHR'S CIRCLE FOR PLANE STRESS

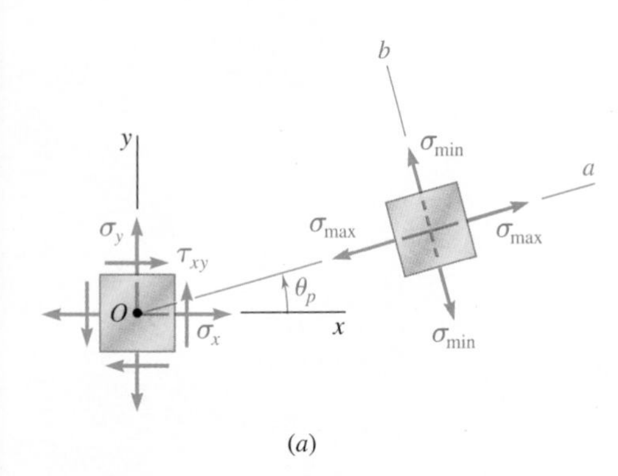

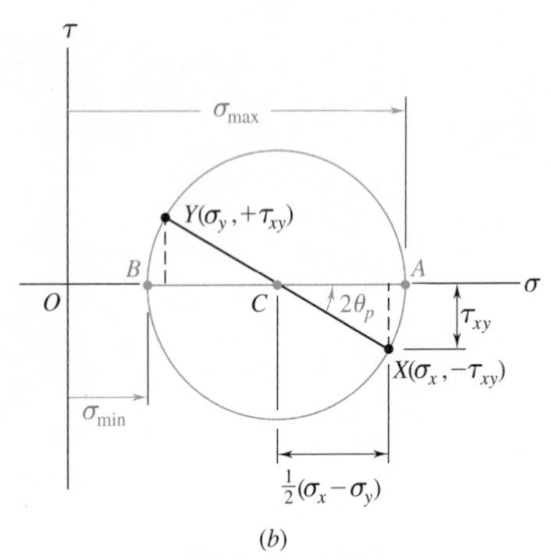

Fig. 14.12 (*a*) Plane stress element and the orientation of principal planes. (*b*) Corresponding Mohr's circle.

The circle used in the preceding section to derive the equations relating to the transformation of plane stress was introduced by the German engineer Otto Mohr (1835–1918) and is known as *Mohr's circle* for plane stress. This circle can be used to obtain an alternative method for the solution of the problems considered in Sec. 14.1. This method is based on simple geometric considerations and does not require the use of specialized equations. While originally designed for graphical solutions, a calculator may also be used.

Consider a square element of a material subjected to plane stress (Fig. 14.12*a*), and let σ_x, σ_y, and τ_{xy} be the components of the stress exerted on the element. A point X of coordinates σ_x and $-\tau_{xy}$ and a point Y of coordinates σ_y and $+\tau_{xy}$ are plotted (Fig. 14.12*b*). If τ_{xy} is positive, as assumed in Fig. 14.12*a*, point X is located below the σ axis and point Y above, as shown in Fig. 14.12*b*. If τ_{xy} is negative, X is located above the σ axis and Y below. Joining X and Y by a straight line, the point C is at the intersection of line XY with the σ axis, and the circle is drawn with its center at C and having a diameter XY. The abscissa of C and the radius of the circle are respectively equal to σ_{ave} and R in Eqs. (14.10). The circle obtained is Mohr's circle for plane stress. Thus, the abscissas of points A and B where the circle intersects the σ axis represent the principal stresses σ_{max} and σ_{min} at the point considered.

Since tan $(XCA) = 2\tau_{xy}/(\sigma_x - \sigma_y)$, the angle XCA is equal in magnitude to one of the angles $2\theta_p$ that satisfy Eq. (14.12). Thus, the angle θ_p in Fig. 14.12*a* defines the orientation of the principal plane corresponding to point A in Fig. 14.12*b* and can be obtained by dividing the angle XCA measured on Mohr's circle in half. If $\sigma_x > \sigma_y$ and $\tau_{xy} > 0$, as in the case considered here, the rotation that brings CX into CA is counterclockwise. But, in that case, the angle θ_p obtained from Eq. (14.12) and defining the direction of the normal Oa to the principal plane is positive; thus, the rotation bringing Ox into Oa is also counterclockwise. Therefore, the senses of rotation in both parts of Fig. 14.12 are the same. So, if a counterclockwise rotation through $2\theta_p$ is required to bring CX into CA on Mohr's circle, a counterclockwise rotation through θ_p will bring Ox into Oa in Fig. 14.12*a*.[†]

Since Mohr's circle is uniquely defined, the same circle can be obtained from the stress components $\sigma_{x'}$, $\sigma_{y'}$, and $\tau_{x'y'}$, which correspond to the x' and y' axes shown in Fig. 14.13*a*. Point X' of coordinates $\sigma_{x'}$ and $-\tau_{x'y'}$ and point Y' of coordinates $\sigma_{y'}$ and $+\tau_{x'y'}$ are located on Mohr's circle, and the angle $X'CA$ in Fig. 14.13*b* must be equal to twice the angle $x'Oa$ in Fig. 14.13*a*. Since the angle XCA is twice the angle xOa, the angle XCX' in Fig. 14.13*b* is twice the angle xOx' in Fig. 14.13*a*. Thus the diameter $X'Y'$ defining the normal and shearing stresses $\sigma_{x'}$, $\sigma_{y'}$, and $\tau_{x'y'}$ is obtained by rotating the diameter XY through an angle equal to twice the angle θ formed by the $x'$and x axes in Fig. 14.13*a*. The rotation that brings the diameter XY into the diameter $X'Y'$ in Fig. 14.13*b* has the same sense as the rotation that brings the xy axes into the $x'y'$ axes in Fig. 14.13*a*.

This property can be used to verify that planes of maximum shearing stress are at 45° to the principal planes. Indeed, points D and E on

[†]This is due to the fact that we are using the circle of Fig 14.8 rather than the circle of Fig. 14.7 as Mohr's circle.

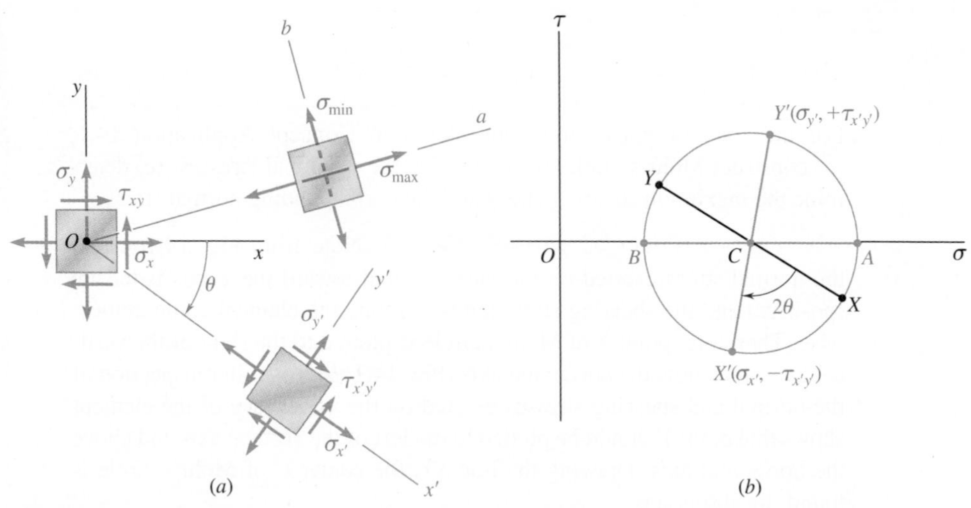

Fig. 14.13 (*a*) Stress element referenced to *xy* axes, transformed to obtain components referenced to *x'y'* axes. (*b*) Corresponding Mohr's circle.

Mohr's circle correspond to the planes of maximum shearing stress, while *A* and *B* correspond to the principal planes (Fig. 14.14*b*). Since the diameters *AB* and *DE* of Mohr's circle are at 90° to each other, the faces of the corresponding elements are at 45° to each other (Fig. 14.14*a*).

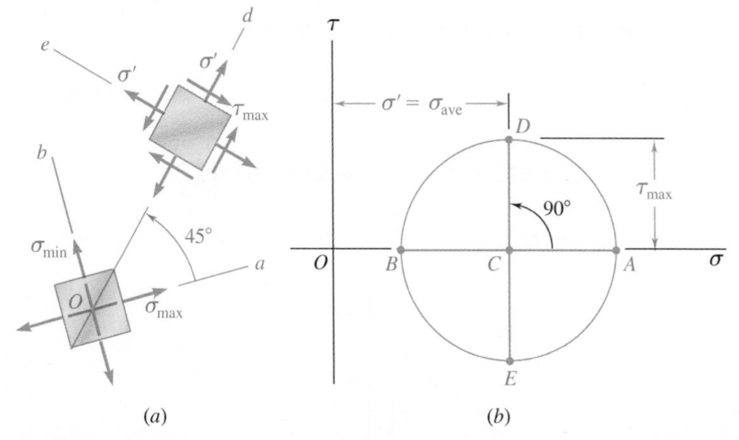

Fig. 14.14 (*a*) Stress elements showing orientation of planes of maximum shearing stress relative to principal planes. (*b*) Corresponding Mohr's circle.

The construction of Mohr's circle for plane stress is simplified if each face of the element used to define the stress components is considered separately. From Figs. 14.12 and 14.13, when the shearing stress exerted *on a given face* tends to rotate the element *clockwise*, the point on Mohr's circle corresponding to that face is located *above* the σ axis. When the shearing stress on a given face tends to rotate the element *counterclockwise*, the point corresponding to that face is located *below* the σ axis (Fig. 14.15).[†] As far as the normal stresses are concerned, the usual convention holds, so that a tensile stress is positive and is plotted to the right, while a compressive stress is considered negative and is plotted to the left.

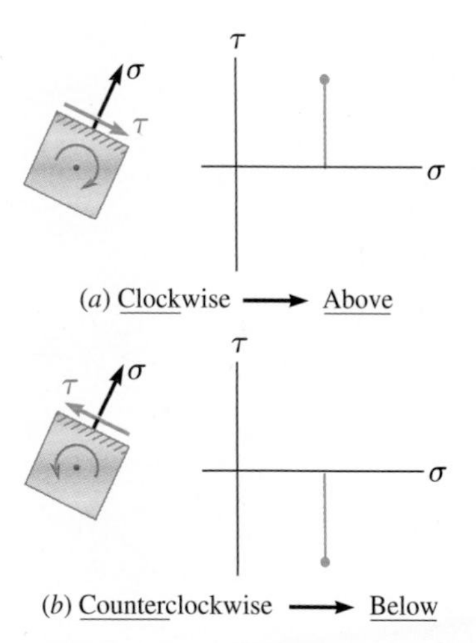

Fig. 14.15 Convention for plotting shearing stress on Mohr's circle.

[†]To remember this convention, think "In the kitchen, the *clock* is above, and the *counter* is below."

Concept Application 14.2

For the state of plane stress considered in Concept Application 14.1, (*a*) construct Mohr's circle, (*b*) determine the principal stresses, (*c*) determine the maximum shearing stress and the corresponding normal stress.

a. Construction of Mohr's Circle. Note from Fig. 14.16*a* that the normal stress exerted on the face oriented toward the x axis is tensile (positive) and the shearing stress tends to rotate the element counterclockwise. Therefore, point X of Mohr's circle is plotted to the right of the vertical axis and below the horizontal axis (Fig. 14.16*b*). A similar inspection of the normal and shearing stresses exerted on the upper face of the element shows that point Y should be plotted to the left of the vertical axis and above the horizontal axis. Drawing the line XY, the center C of Mohr's circle is found. Its abscissa is

$$\sigma_{\text{ave}} = \frac{\sigma_x + \sigma_y}{2} = \frac{50 + (-10)}{2} = 20 \text{ MPa}$$

Since the sides of the shaded triangle are

$$CF = 50 - 20 = 30 \text{ MPa} \qquad \text{and} \qquad FX = 40 \text{ MPa}$$

the radius of the circle is

$$R = CX = \sqrt{(30)^2 + (40)^2} = 50 \text{ MPa}$$

b. Principal Planes and Principal Stresses. The principal stresses are

$$\sigma_{\max} = OA = OC + CA = 20 + 50 = 70 \text{ MPa}$$

$$\sigma_{\min} = OB = OC - BC = 20 - 50 = -30 \text{ MPa}$$

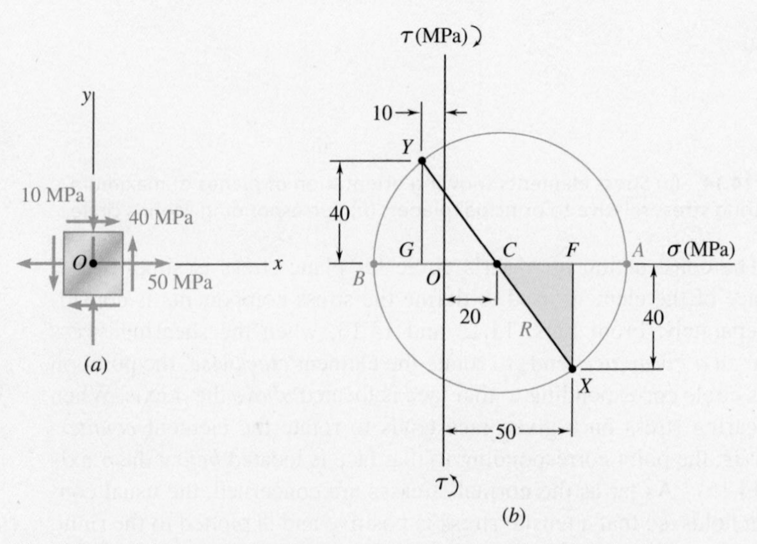

Fig. 14.16 (*a*) Plane stress element. (*b*) Corresponding Mohr's circle.

(continued)

Recalling that the angle ACX represents $2\theta_p$ (Fig. 14.16*b*),

$$\tan 2\theta_p = \frac{FX}{CF} = \frac{40}{30}$$

$$2\theta_p = 53.1° \qquad \theta_p = 26.6°$$

Since the rotation that brings CX into CA in Fig. 14.16*d* is counterclockwise, the rotation that brings Ox into the axis Oa corresponding to σ_{max} in Fig. 14.16*c* is also counterclockwise.

c. Maximum Shearing Stress. Since a further rotation of 90° counterclockwise brings CA into CD in Fig. 14.16*d*, a further rotation of 45° counterclockwise will bring the axis Oa into the axis Od corresponding to the maximum shearing stress in Fig. 14.16*d*. Note from Fig. 14.16*d* that $\tau_{max} = R = 50$ MPa and the corresponding normal stress is $\sigma' = \sigma_{ave} = 20$ MPa. Since point D is located above the σ axis in Fig. 14.16*c*, the shearing stresses exerted on the faces perpendicular to Od in Fig. 14.16*d* must be directed so that they will tend to rotate the element clockwise.

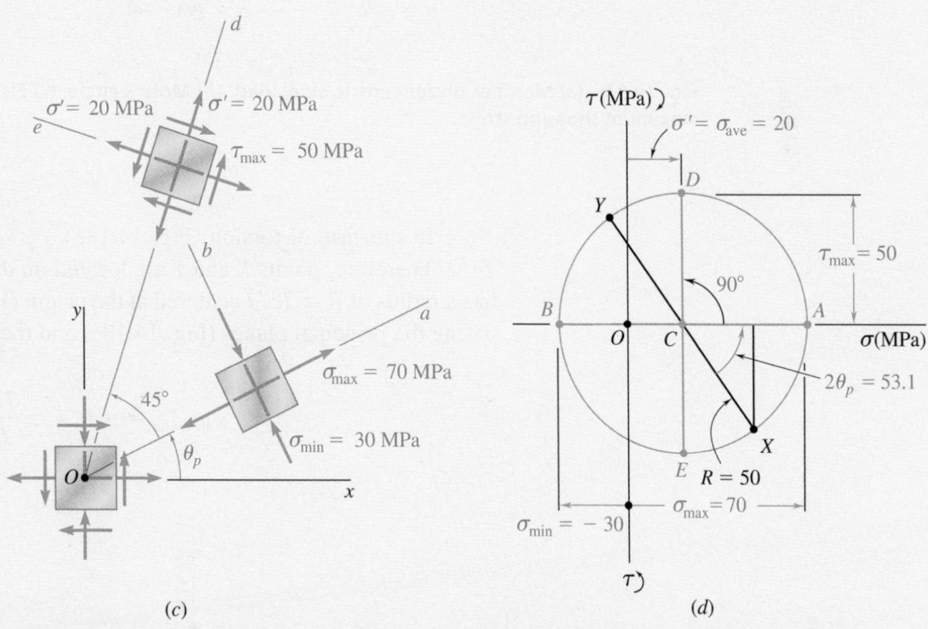

Fig. 14.16 (*cont.*) (*c*) Stress element orientations for principal and maximum shearing stresses. (*d*) Mohr's circle used to determine principal and maximum shearing stresses.

Mohr's circle provides a convenient way of checking the results obtained earlier for stresses under a centric axial load (Sec. 8.3) and under a torsional load (Sec. 10.1c). In the first case (Fig. 14.17*a*), $\sigma_x = P/A$, $\sigma_y = 0$, and $\tau_{xy} = 0$. The corresponding points X and Y define a circle of radius $R = P/2A$ that passes through the origin of coordinates (Fig. 14.17*b*). Points D and E yield the orientation of the planes of maximum shearing stress (Fig. 14.17*c*), as well as τ_{max} and the corresponding normal stresses σ':

$$\tau_{max} = \sigma' = R = \frac{P}{2A} \tag{14.18}$$

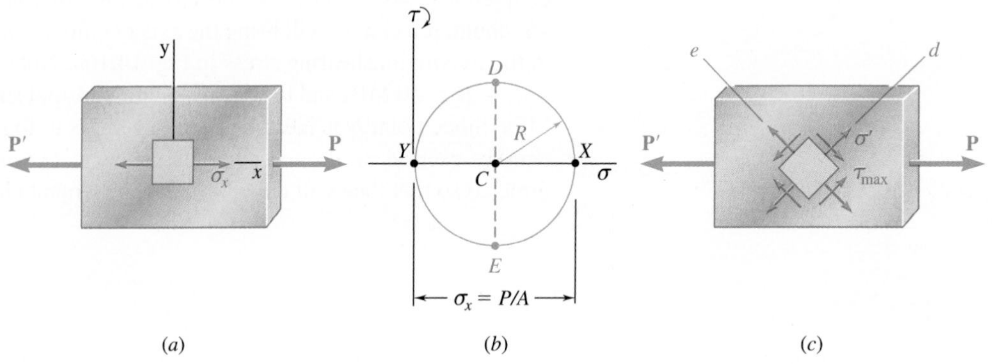

Fig. 14.17 (*a*) Member under centric axial load. (*b*) Mohr's circle. (*c*) Element showing planes of maximum shearing stress.

In the case of torsion (Fig. 14.18*a*), $\sigma_x = \sigma_y = 0$ and $\tau_{xy} = \tau_{max} = Tc/J$. Therefore, points X and Y are located on the τ axis, and Mohr's circle has a radius of $R = Tc/J$ centered at the origin (Fig. 14.18*b*). Points A and B define the principal planes (Fig. 14.18*c*) and the principal stresses:

$$\sigma_{max,\,min} = \pm R = \pm \frac{Tc}{J} \tag{14.19}$$

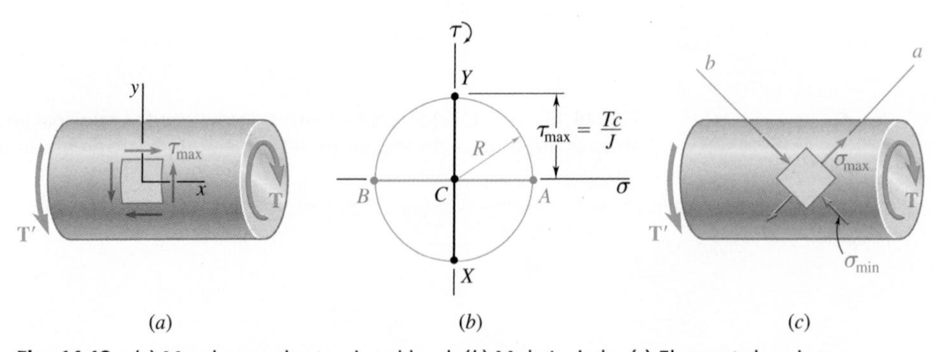

Fig. 14.18 (*a*) Member under torsional load. (*b*) Mohr's circle. (*c*) Element showing orientation of principal stresses.

Sample Problem 14.2

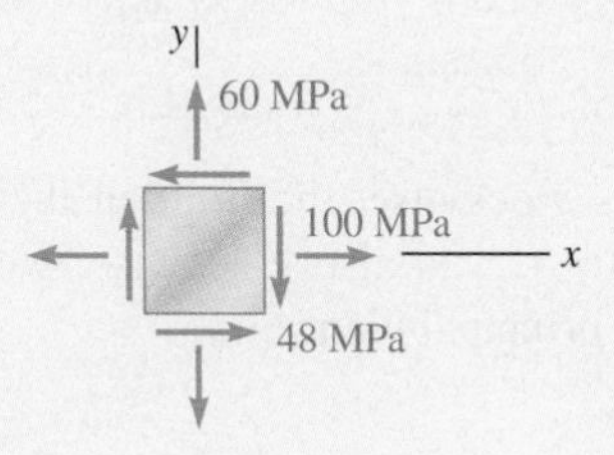

For the state of plane stress shown determine (*a*) the principal planes and the principal stresses, (*b*) the stress components exerted on the element obtained by rotating the given element counterclockwise through 30°.

STRATEGY: Since the given state of stress represents two points on Mohr's circle, you can use these points to generate the circle. The state of stress on any other plane, including the principal planes, can then be readily determined through the geometry of the circle.

MODELING and ANALYSIS:

Construction of Mohr's Circle (Fig 1). On a face perpendicular to the x axis, the normal stress is tensile, and the shearing stress tends to rotate the element clockwise. Thus, X is plotted at a point 100 units to the right of the vertical axis and 48 units above the horizontal axis. By examining the stress components on the upper face, point $Y(60, -48)$ is plotted. Join points X and Y by a straight line to define the center C of Mohr's circle. The abscissa of C, which represents σ_{ave}, and the radius R of the circle, can be measured directly or calculated as

$$\sigma_{ave} = OC = \tfrac{1}{2}(\sigma_x + \sigma_y) = \tfrac{1}{2}(100 + 60) = 80 \text{ MPa}$$

$$R = \sqrt{(CF)^2 + (FX)^2} = \sqrt{(20)^2 + (48)^2} = 52 \text{ MPa}$$

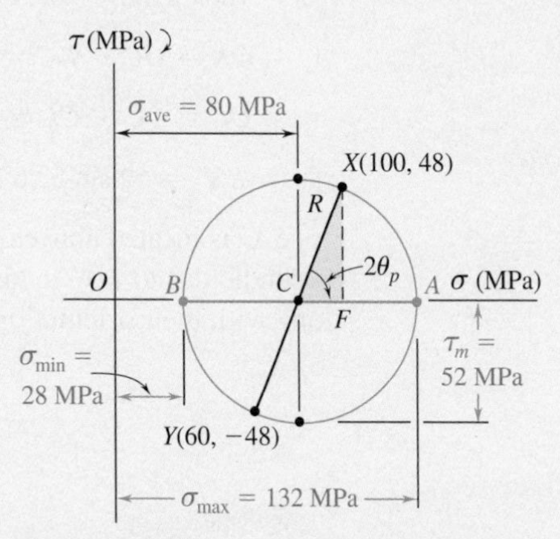

Fig. 1 Mohr's circle for given stress state.

a. Principal Planes and Principal Stresses. We rotate the diameter XY clockwise through $2\theta_p$ until it coincides with the diameter AB. Thus,

$$\tan 2\theta_p = \frac{XF}{CF} = \frac{48}{20} = 2.4 \quad 2\theta_p = 67.4° \downarrow \quad \theta_p = 33.7° \downarrow \blacktriangleleft$$

(continued)

The principal stresses are represented by the abscissas of points A and B:

$$\sigma_{max} = OA = OC + CA = 80 + 52 \qquad \sigma_{max} = +132 \text{ MPa}$$

$$\sigma_{min} = OB = OC - BC = 80 - 52 \qquad \sigma_{min} = +28 \text{ MPa}$$

Since the rotation that brings XY into AB is clockwise, the rotation that brings Ox into the axis Oa corresponding to σ_{max} is also clockwise; we obtain the orientation shown in Fig. 2 for the principal planes.

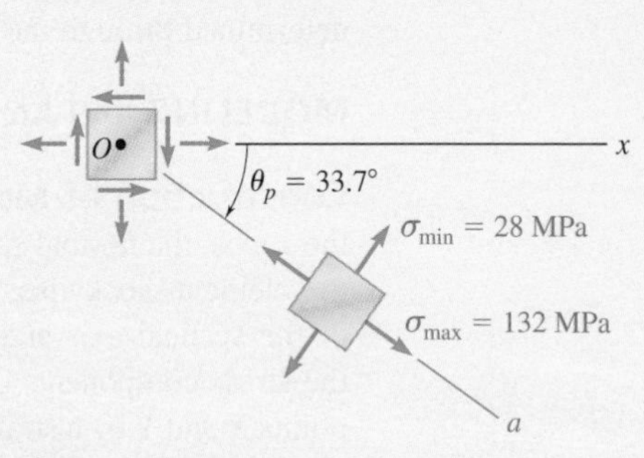

Fig. 2 Orientation of principal stress element.

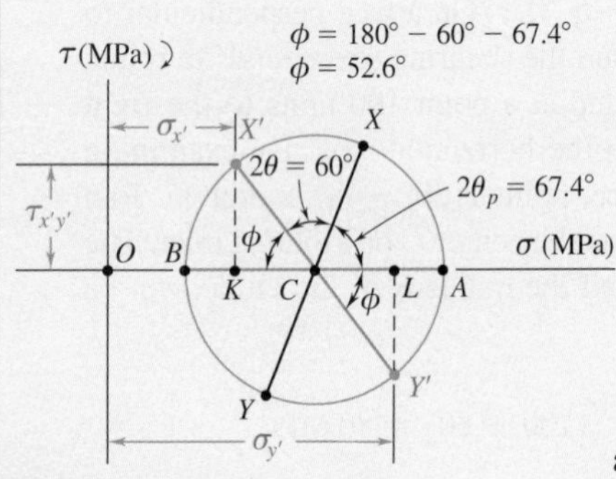

Fig. 3 Mohr's circle analysis for element rotation of 30° counterclockwise.

b. Stress Components on Element Rotated 30°↺.

Points X' and Y' on Mohr's circle that correspond to the stress components on the rotated element are obtained by rotating XY counterclockwise through $2\theta = 60°$ (Fig. 3). We find

$$\phi = 180° - 60° - 67.4° \qquad \phi = 52.6°$$

$$\sigma_{x'} = OK = OC - KC = 80 - 52 \cos 52.6° \qquad \sigma_{x'} = +48.4 \text{ MPa}$$

$$\sigma_{y'} = OL = OC + CL = 80 + 52 \cos 52.6° \qquad \sigma_{y'} = +111.6 \text{ MPa}$$

$$\tau_{x'y'} = KX' = 52 \sin 52.6° \qquad \tau_{x'y'} = 41.3 \text{ MPa}$$

Since X' is located above the horizontal axis, the shearing stress on the face perpendicular to Ox' tends to rotate the element clockwise. The stresses, along with their orientation, are shown in Fig. 4.

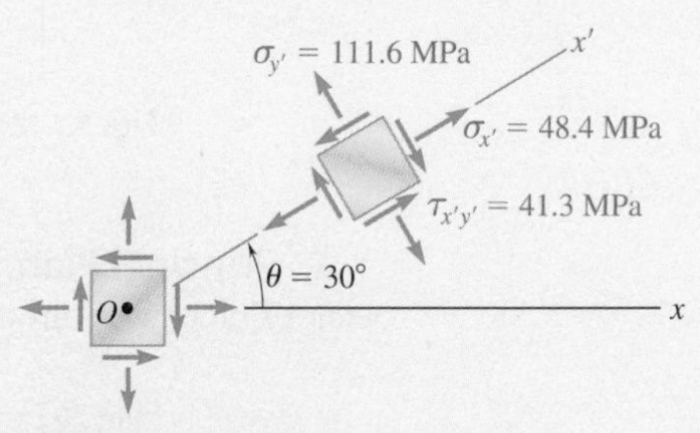

Fig. 4 Stress components obtained by rotating original element 30° counterclockwise.

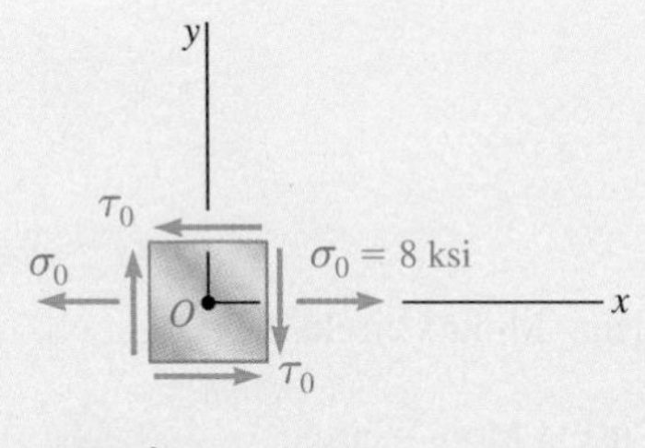

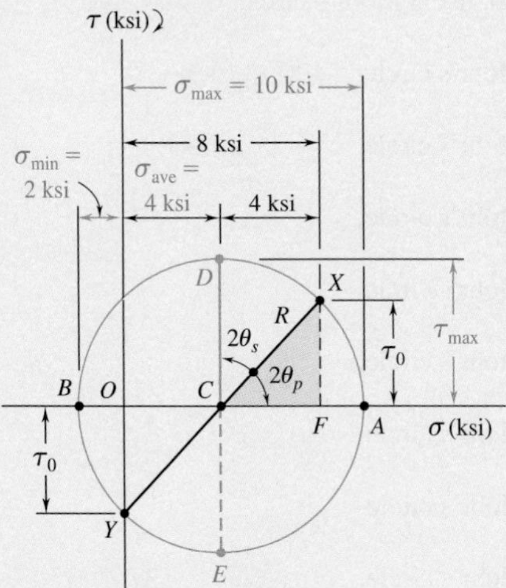

Fig. 1 Mohr's circle for given state of stress.

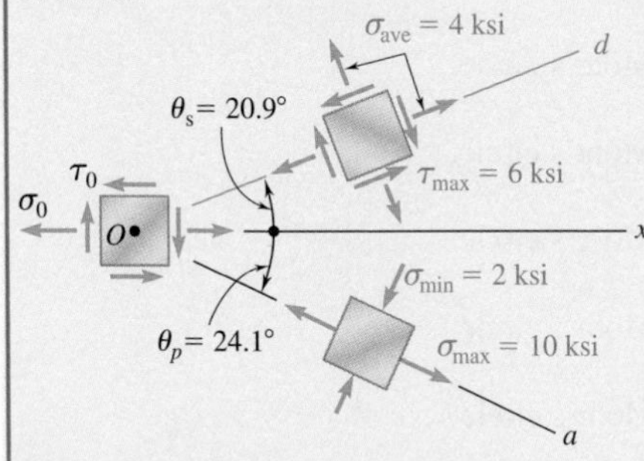

Fig. 2 Orientation of principal and maximum shearing stress planes for assumed sense of τ_0.

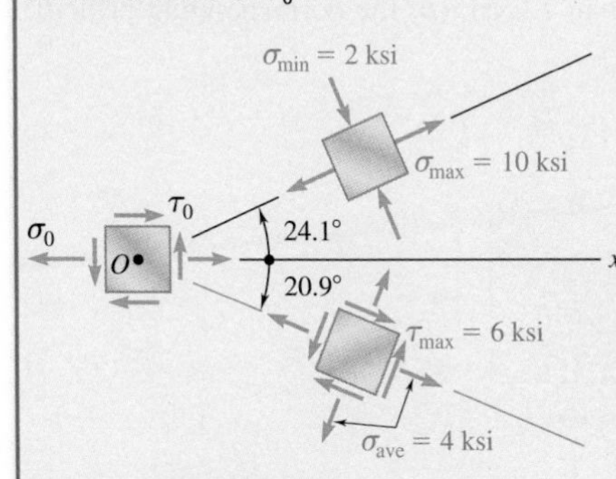

Fig. 3 Orientation of principal and maximum shearing stress planes for opposite sense of τ_0.

Sample Problem 14.3

A state of plane stress consists of a tensile stress $\sigma_0 = 8$ ksi exerted on vertical surfaces and of unknown shearing stresses. Determine (*a*) the magnitude of the shearing stress τ_0 for which the largest normal stress is 10 ksi, (*b*) the corresponding maximum shearing stress.

STRATEGY: You can use the normal stresses on the given element to determine the average normal stress, thereby establishing the center of Mohr's circle. Knowing that the given maximum normal stress is also a principal stress, you can use this to complete the construction of the circle.

MODELING and ANALYSIS:

Construction of Mohr's Circle (Fig.1). Assume that the shearing stresses act in the senses shown. Thus, the shearing stress τ_0 on a face perpendicular to the x axis tends to rotate the element clockwise, and point X of coordinates 8 ksi and τ_0 is plotted above the horizontal axis. Considering a horizontal face of the element, $\sigma_y = 0$ and τ_0 tends to rotate the element counterclockwise. Thus, Y is plotted at a distance τ_0 below O.

The abscissa of the center C of Mohr's circle is

$$\sigma_{ave} = \tfrac{1}{2}(\sigma_x + \sigma_y) = \tfrac{1}{2}(8 + 0) = 4 \text{ ksi}$$

The radius R of the circle is found by observing that $\sigma_{max} = 10$ ksi and is represented by the abscissa of point A:

$$\sigma_{max} = \sigma_{ave} + R$$

$$10 \text{ ksi} = 4 \text{ ksi} + R \qquad R = 6 \text{ ksi}$$

a. Shearing Stress τ_0. Considering the right triangle CFX,

$$\cos 2\theta_p = \frac{CF}{CX} = \frac{CF}{R} = \frac{4 \text{ ksi}}{6 \text{ ksi}} \qquad 2\theta_p = 48.2° \qquad \theta_p = 24.1° \downarrow$$

$$\tau_0 = FX = R \sin 2\theta_p = (6 \text{ ksi}) \sin 48.2° \qquad \tau_0 = 4.47 \text{ ksi} \blacktriangleleft$$

b. Maximum Shearing Stress. The coordinates of point D of Mohr's circle represent the maximum shearing stress and the corresponding normal stress.

$$\tau_{max} = R = 6 \text{ ksi} \qquad \tau_{max} = 6 \text{ ksi} \blacktriangleleft$$

$$2\theta_s = 90° - 2\theta_p = 90° - 48.2° = 41.8° \circlearrowleft \qquad \theta_x = 20.9° \circlearrowleft$$

The maximum shearing stress is exerted on an element that is oriented as shown in Fig. 2. (The element upon which the principal stresses are exerted is also shown.)

REFLECT and THINK. If our original assumption regarding the sense of τ_0 was reversed, we would obtain the same circle and the same answers, but the orientation of the elements would be as shown in Fig. 3.

Problems

14.25 Solve Probs. 14.5 and 14.9, using Mohr's circle.

14.26 Solve Probs. 14.6 and 14.10, using Mohr's circle.

14.27 Solve Prob. 14.11, using Mohr's circle.

14.28 Solve Prob. 14.12, using Mohr's circle.

14.29 Solve Prob. 14.13, using Mohr's circle.

14.30 Solve Prob. 14.14, using Mohr's circle

14.31 Solve Prob. 14.15, using Mohr's circle.

14.32 Solve Prob. 14.16, using Mohr's circle.

14.33 Solve Prob. 14.17, using Mohr's circle.

14.34 Solve Prob. 14.18, using Mohr's circle.

14.35 Solve Prob. 14.19, using Mohr's circle.

14.36 Solve Prob. 14.20, using Mohr's circle.

14.37 Solve Prob. 14.21, using Mohr's circle.

14.38 Solve Prob. 14.22, using Mohr's circle.

14.39 Solve Prob. 14.23, using Mohr's circle.

14.40 Solve Prob. 14.24, using Mohr's circle.

14.41 For the state of plane stress shown, use Mohr's circle to determine (*a*) the largest value of τ_{xy} for which the maximum in-plane shearing stress is equal to or less than 12 ksi, (*b*) the corresponding principal stresses.

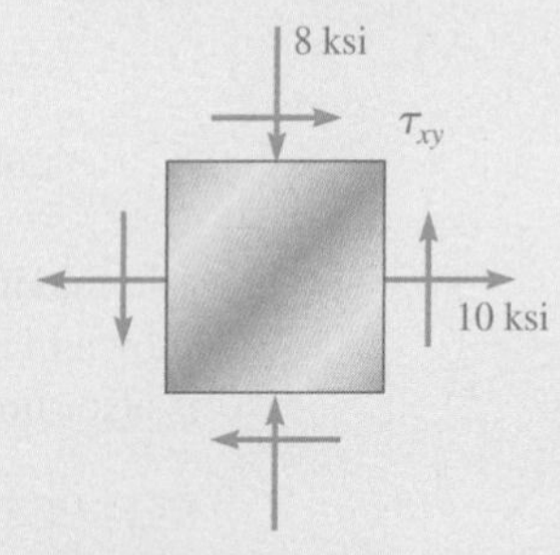

Fig. P14.41

14.42 For the element shown, determine the range of values of τ_{xy} for which the maximum in-plane shearing stress is equal to or less than 150 MPa.

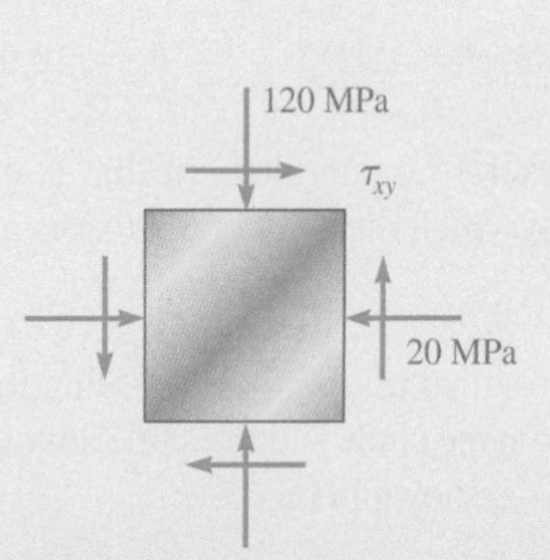

Fig. ***P14.42***

14.43 For the state of plane stress shown, use Mohr's circle to determine (*a*) the value of τ_{xy} for which the in-plane shearing stress parallel to the weld is zero, (*b*) the corresponding principal stresses.

14.44 Solve Prob. 14.43, assuming that the weld forms an angle of 60° with the horizontal.

***14.45* through 14.48** Determine the principal planes and the principal stresses for the state of plane stress resulting from the superposition of the two states of stress shown.

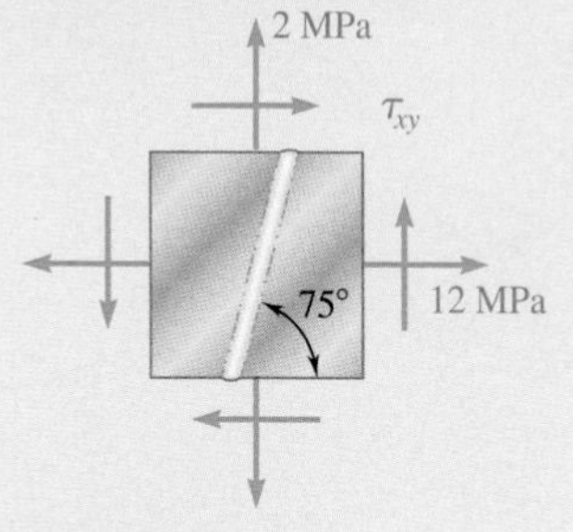

Fig. P14.43

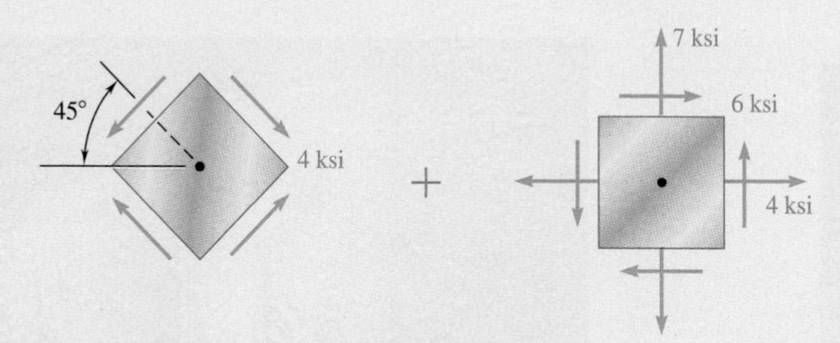

Fig. *P14.45*

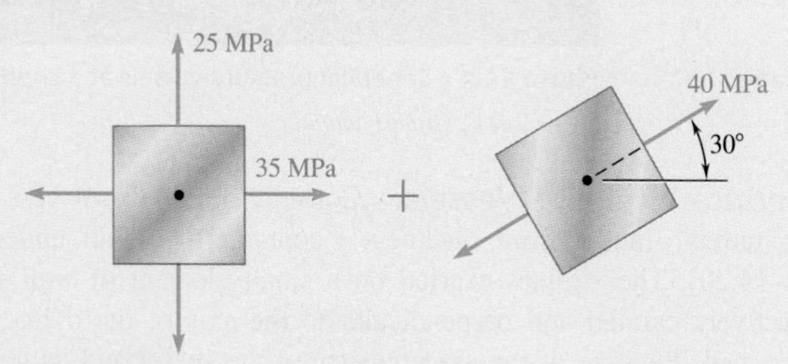

Fig. P14.46

Fig. P14.47

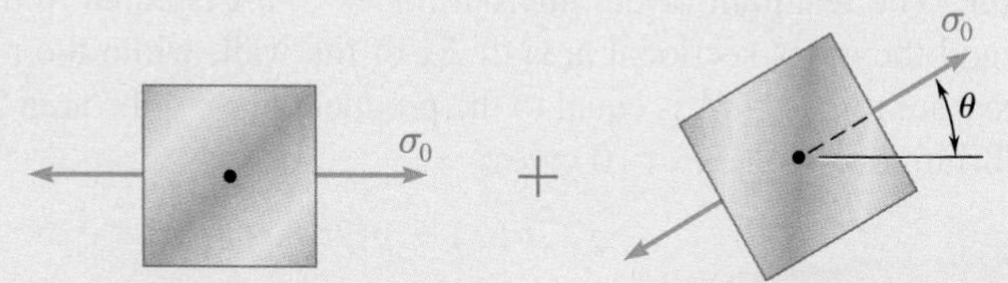

Fig. P14.48

14.3 STRESSES IN THIN-WALLED PRESSURE VESSELS

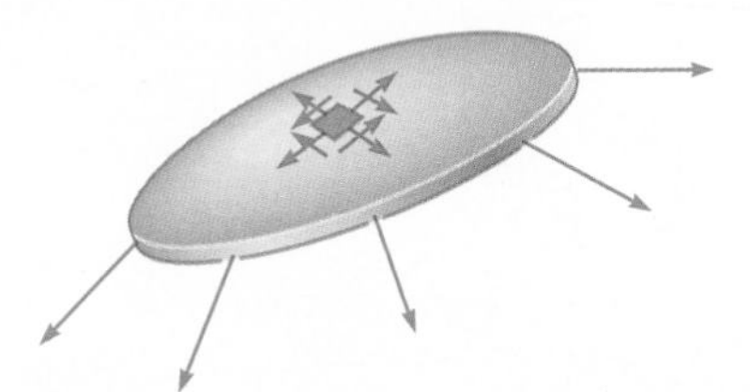

Fig. 14.19 Assumed stress distribution in thin-walled pressure vessels.

Thin-walled pressure vessels provide an important application of the analysis of plane stress. Since their walls offer little resistance to bending, it can be assumed that the internal forces exerted on a given portion of wall are tangent to the surface of the vessel (Fig. 14.19). The resulting stresses on an element of wall will be contained in a plane tangent to the surface of the vessel.

This analysis of stresses in thin-walled pressure vessels is limited to two types of vessels: cylindrical and spherical (Photos 14.3 and 14.4).

Photo 14.3 Cylindrical pressure vessels for liquid propane.
© *Clair Dunn/Alamy*

Photo 14.4 Spherical pressure vessels at a chemical plant.
© *Spencer C. Grant/PhotoEdit*

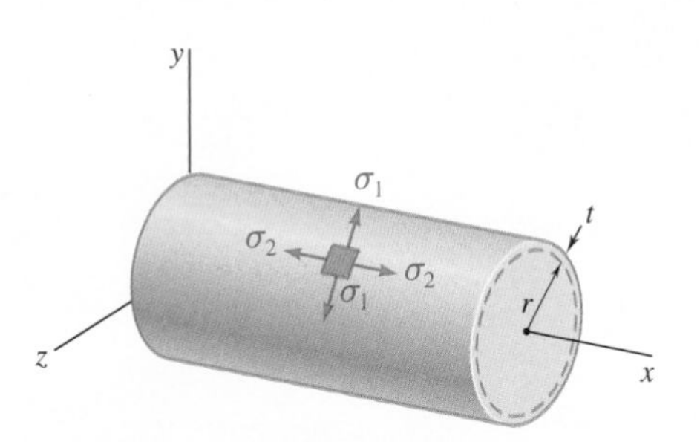

Fig. 14.20 Pressurized cylindrical vessel.

Cylindrical Pressure Vessels. Consider a cylindrical vessel with an inner radius r and a wall thickness t containing a fluid under pressure (Fig. 14.20). The stresses exerted on a small element of wall with sides respectively parallel and perpendicular to the axis of the cylinder will be determined. Because of the axisymmetry of the vessel and its contents, no shearing stress is exerted on the element. The normal stresses σ_1 and σ_2 shown in Fig. 14.20 are therefore principal stresses. The stress σ_1 is called the *hoop stress*, because it is the type of stress found in hoops used to hold together the various slats of a wooden barrel. Stress σ_2 is called the *longitudinal stress*.

To determine the hoop stress σ_1, detach a portion of the vessel and its contents bounded by the xy plane and by two planes parallel to the yz plane at a distance Δx from each other (Fig. 14.21). The forces parallel to the z axis acting on the free body consist of the elementary internal forces $\sigma_1\, dA$ on the wall sections and the elementary pressure forces $p\, dA$ exerted on the portion of fluid included in the free body. Note that the *gage pressure* of the fluid p is the excess of the inside pressure over the outside atmospheric pressure. The resultant of the internal forces $\sigma_1\, dA$ is equal to the product of σ_1 and the cross-sectional area $2t\, \Delta x$ of the wall, while the resultant of the pressure forces $p\, dA$ is equal to the product of p and the area $2r\, \Delta x$. The equilibrium equation $\Sigma F_z = 0$ gives

$$\Sigma F_z = 0: \qquad \sigma_1(2t\, \Delta x) - p(2r\, \Delta x) = 0$$

and solving for the hoop stress σ_1,

$$\sigma_1 = \frac{pr}{t} \tag{14.20}$$

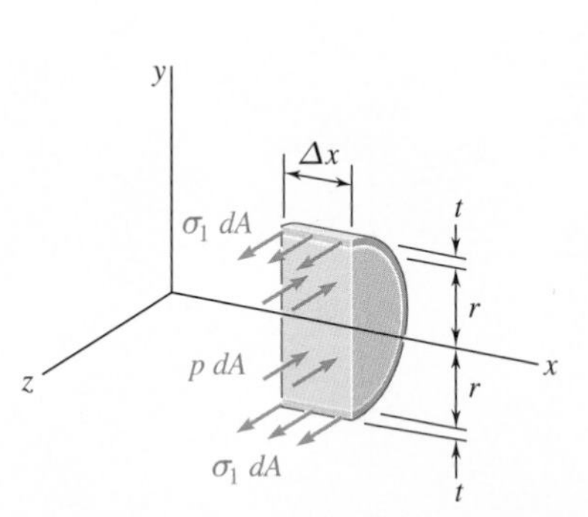

Fig. 14.21 Free-body diagram to determine hoop stress in a cylindrical pressure vessel.

To determine the longitudinal stress σ_2, pass a section perpendicular to the x axis and consider the free body consisting of the portion of the vessel and its contents located to the left of the section (Fig. 14.22). The forces acting on this free body are the elementary internal forces $\sigma_2 dA$ on the wall section and the elementary pressure forces $p\ dA$ exerted on the portion of fluid included in the free body. Noting that the area of the fluid section is πr^2 and that the area of the wall section can be obtained by multiplying the circumference $2\pi r$ of the cylinder by its wall thickness t, the equilibrium equation is:†

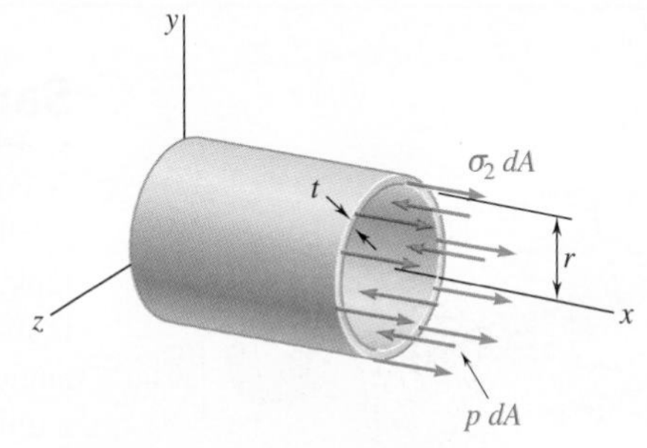

Fig. 14.22 Free-body diagram to determine longitudinal stress.

$$\Sigma F_x = 0: \qquad \sigma_2(2\pi r t) - p(\pi r^2) = 0$$

and solving for the longitudinal stress σ_2,

$$\sigma_2 = \frac{pr}{2t} \tag{14.21}$$

Note from Eqs. (14.20) and (14.21) that the hoop stress σ_1 is twice as large as the longitudinal stress σ_2:

$$\sigma_1 = 2\sigma_2 \tag{14.22}$$

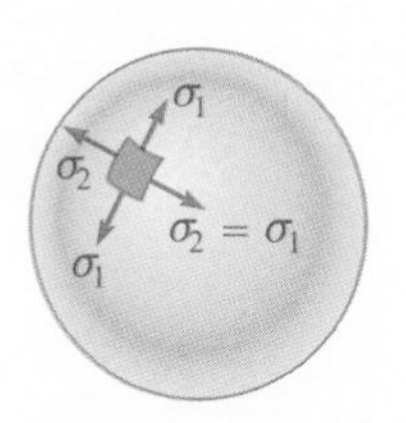

Fig. 14.23 Pressurized spherical vessel.

Spherical Pressure Vessels. Now consider a spherical vessel of inner radius r and wall thickness t, containing a fluid under a gage pressure p. For reasons of symmetry, the stresses exerted on the four faces of a small element of wall must be equal (Fig. 14.23).

$$\sigma_1 = \sigma_2 \tag{14.23}$$

To determine the stress, pass a section through the center C of the vessel and consider the free body consisting of the portion of the vessel and its contents located to the left of the section (Fig. 14.24). The equation of equilibrium for this free body is the same as for the free body of Fig. 14.22. So for a spherical vessel,

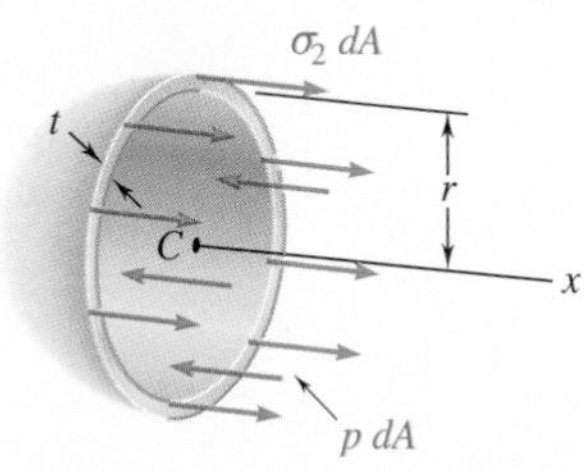

Fig. 14.24 Free-body diagram to determine spherical pressure vessel stress.

$$\sigma_1 = \sigma_2 = \frac{pr}{2t} \tag{14.24}$$

†Using the mean radius of the wall section, $r_m = r + \frac{1}{2}t$, to compute the resultant of the forces, a more accurate value of the longitudinal stress is

$$\sigma_2 = \frac{pr}{2t}\frac{1}{1 + \dfrac{t}{2r}}$$

However, for a thin-walled pressure vessel, the term $t/2r$ is sufficiently small to allow the use of Eq. (14.21) for engineering design and analysis. If a pressure vessel is not thin-walled (i.e., if $t/2r$ is not small), the stresses σ_1 and σ_2 vary across the wall and must be determined by the methods of the theory of elasticity.

Sample Problem 14.4

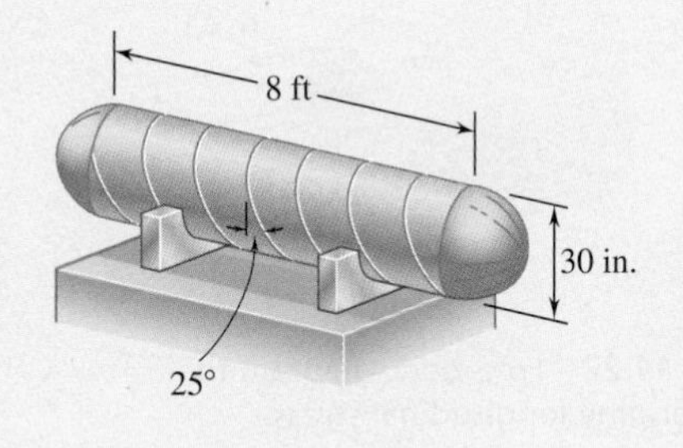

A compressed-air tank is supported by two cradles as shown. One of the cradles is designed so that it does not exert any longitudinal force on the tank. The cylindrical body of the tank has a 30-in. outer diameter and is made of a $\frac{3}{8}$-in. steel plate by butt welding along a helix that forms an angle of 25° with a transverse plane. The end caps are spherical and have a uniform wall thickness of $\frac{5}{16}$ in. For an internal gage pressure of 180 psi, determine (*a*) the normal stress in the spherical caps, (*b*) the stresses in directions perpendicular and parallel to the helical weld.

STRATEGY: Using the equations for thin-walled pressure vessels, you can determine the state of plane stress at any point within the spherical end cap and within the cylindrical body. You can then plot the corresponding Mohr's circles and use them to determine the stress components of interest.

MODELING and ANALYSIS:

a. Spherical Cap. The state of stress within any point in the spherical cap is shown in Fig. 1. Using Eq. (14.24), we write

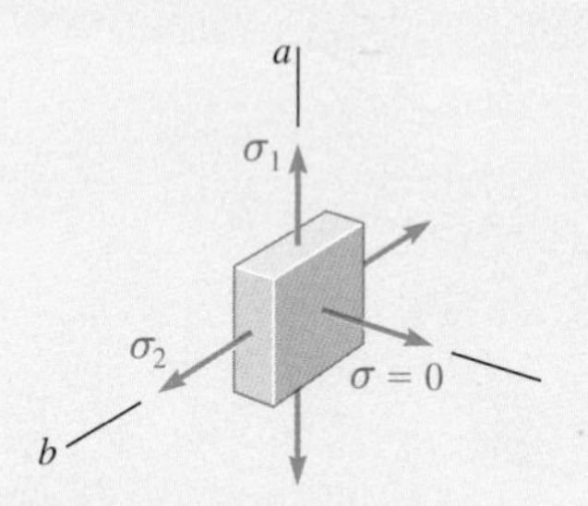

Fig. 1 State of stress at any point in spherical cap.

$$p = 180 \text{ psi}, t = \tfrac{5}{16} \text{ in.} = 0.3125 \text{ in.}, r = 15 - 0.3125 = 14.688 \text{ in.}$$

$$\sigma_1 = \sigma_2 = \frac{pr}{2t} = \frac{(180 \text{ psi})(14.688 \text{ in.})}{2(0.3125 \text{ in.})} \qquad \sigma = 4230 \text{ psi} \quad ◀$$

b. Cylindrical Body of the Tank. The state of stress within any point in the cylindrical body is as shown in Fig. 2. We determine the hoop stress σ_1 and the longitudinal stress σ_2 using Eqs. (14.20) and (14.22). We write

$$p = 180 \text{ psi}, t = \tfrac{3}{8} \text{ in.} = 0.375 \text{ in.}, r = 15 - 0.375 = 14.625 \text{ in.}$$

$$\sigma_1 = \frac{pr}{t} = \frac{(180 \text{ psi})(14.625 \text{ in.})}{0.375 \text{ in.}} = 7020 \text{ psi} \qquad \sigma_2 = \tfrac{1}{2}\sigma_1 = 3510 \text{ psi}$$

$$\sigma_{\text{ave}} = \tfrac{1}{2}(\sigma_1 + \sigma_2) = 5265 \text{ psi} \qquad R = \tfrac{1}{2}(\sigma_1 - \sigma_2) = 1755 \text{ psi}$$

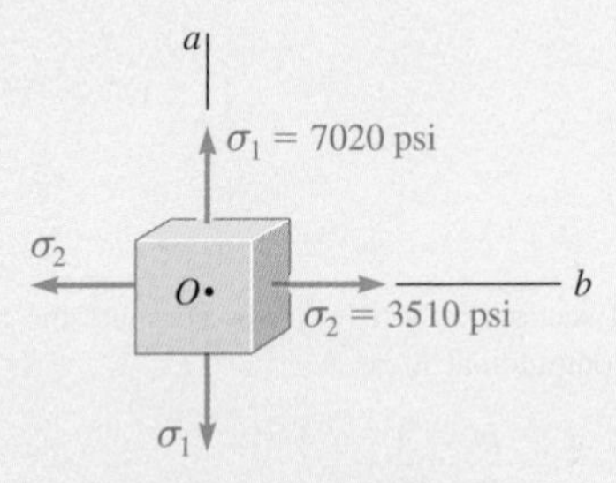

Fig. 2 State of stress at any point in cylindrical body.

(continued)

Stresses at the Weld. Noting that both the hoop stress and the longitudinal stress are principal stresses, we draw Mohr's circle as shown in Fig. 3.

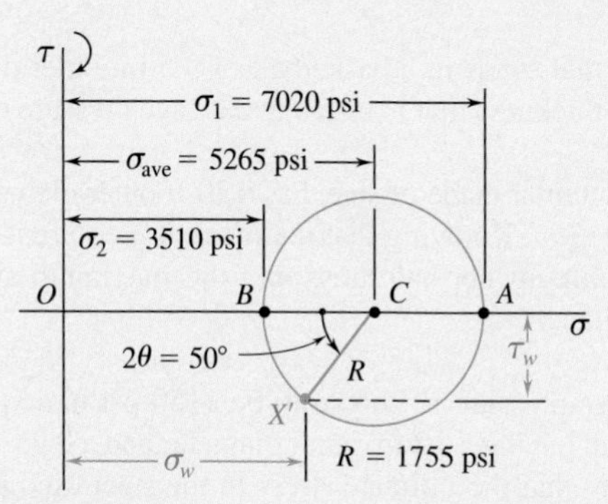

Fig. 3 Mohr's circle for stress element in cylindrical body.

An element having a face parallel to the weld is obtained by rotating the face perpendicular to the axis Ob (Fig. 2) counterclockwise through 25°. Therefore, on Mohr's circle (Fig. 3), point X' corresponds to the stress components on the weld by rotating radius CB counterclockwise through $2\theta = 50°$.

$$\sigma_w = \sigma_{ave} - R\cos 50° = 5265 - 1755\cos 50° \qquad \sigma_w = +4140 \text{ psi} \blacktriangleleft$$

$$\tau_w = R\sin 50° = 1755\sin 50° \qquad \tau_w = 1344 \text{ psi} \blacktriangleleft$$

Since X' is below the horizontal axis, τ_w tends to rotate the element counterclockwise. The stress components on the weld are shown in Fig. 4.

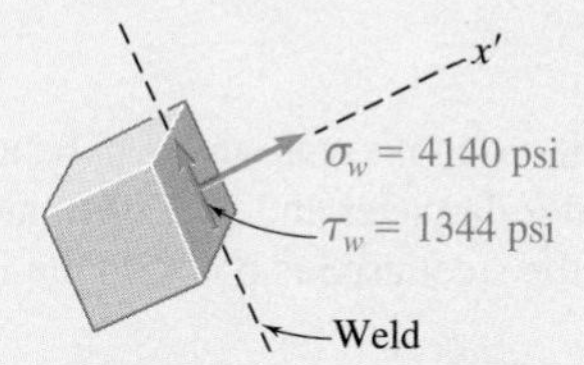

Fig. 4 Stress components on the weld.

Problems

14.49 Determine the normal stress in a basketball of 9.5-in. outer diameter and 0.125-in. wall thickness that is inflated to a gage pressure of 9 psi.

14.50 A spherical gas container made of steel has a 20-ft outer diameter and a wall thickness of $\frac{7}{16}$ in. Knowing that the internal pressure is 75 psi, determine the maximum normal stress and the maximum shearing stress in the container.

14.51 The maximum gage pressure is known to be 1150 psi in a spherical steel pressure vessel having a 10-in. outer diameter and a 0.25-in. wall thickness. Knowing that the ultimate stress in the steel used is $\sigma_U =$ 60 ksi, determine the factor of safety with respect to tensile failure.

14.52 A spherical gas container having an outer diameter of 5 m and a wall thickness of 22 mm is made of a steel for which $E = 200$ GPa and $\nu = 0.29$. Knowing that the gage pressure in the container is increased from zero to 1.7 MPa, determine (*a*) the maximum normal stress in the container, (*b*) the increase in the diameter of the container.

14.53 A spherical pressure vessel has an outer diameter of 3 m and a wall thickness of 12 mm. Knowing that for the steel used $\sigma_{all} = 80$ MPa, $E = 200$ GPa, and $\nu = 0.29$, determine (*a*) the allowable gage pressure, (*b*) the corresponding increase in the diameter of the vessel.

14.54 A spherical pressure vessel of 750-mm outer diameter is to be fabricated from a steel having an ultimate stress $\sigma_U = 400$ MPa. Knowing that a factor of safety of 4 is desired and that the gage pressure can reach 4.2 MPa, determine the smallest wall thickness that should be used.

14.55 Determine the largest internal pressure that can be applied to a cylindrical tank of 5.5-ft outer diameter and $\frac{5}{8}$-in. wall thickness if the ultimate normal stress of the steel used is 65 ksi and a factor of safety of 5.0 is desired.

14.56 The unpressurized cylindrical storage tank shown has a $\frac{3}{16}$-in. wall thickness and is made of steel having a 60-ksi ultimate strength in tension. Determine the maximum height h to which it can be filled with water if a factor of safety of 4.0 is desired. (Specific weight of water = 62.4 lb/ft^3.)

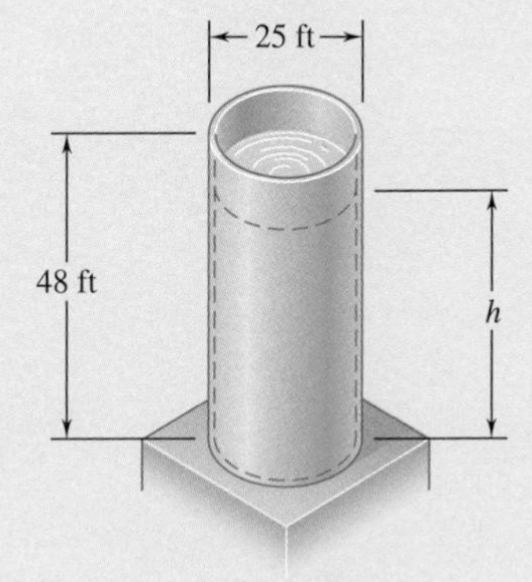

Fig. P14.56

14.57 For the storage tank of Prob. 14.56, determine the maximum normal stress and the maximum shearing stress in the cylindrical wall when the tank is filled to capacity ($h = 48$ ft).

14.58 The propane storage tank shown in Photo 14.3 has an outer diameter of 3.3 m and a wall thickness of 18 mm. At a time when the internal pressure of the tank is 1.5 MPa, determine the maximum normal stress in the tank.

14.59 A steel penstock has a 750-mm outer diameter, a 12-mm wall thickness, and connects a reservoir at A with a generating station at B. Knowing that the density of water is 1000 kg/m^3, determine the maximum normal stress and the maximum shearing stress in the penstock under static conditions.

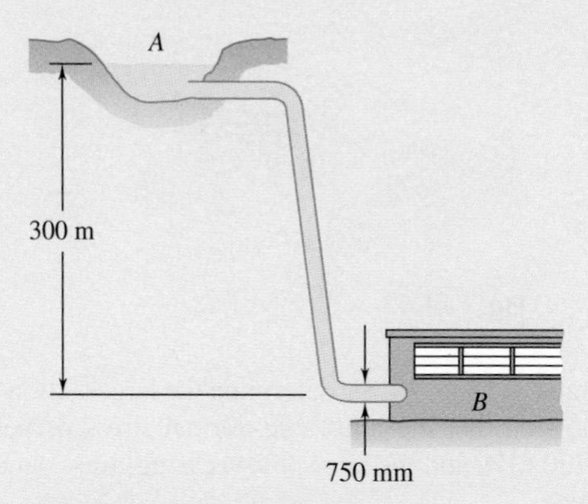

Fig. P14.59 and P14.60

14.60 A steel penstock has a 750-mm outer diameter and connects a reservoir at A with a generating station at B. Knowing that the density of water is 1000 kg/m^3 and that the allowable normal stress in the steel is 85 MPa, determine the smallest thickness that can be used for the penstock.

14.61 The cylindrical portion of the compressed air tank shown is fabricated of 0.25-in.-thick plate welded along a helix forming an angle $\beta = 30°$ with the horizontal. Knowing that the allowable stress normal to the weld is 10.5 ksi, determine the largest gage pressure that can be used in the tank.

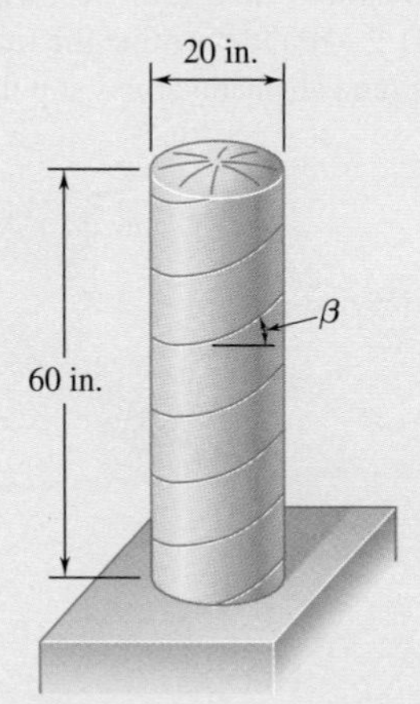

Fig. *P14.61*

14.62 For the compressed air tank of Prob. 14.61, determine the gage pressure that will cause a shearing stress parallel to the weld of 4 ksi.

14.63 The pressure tank shown has an 8-mm wall thickness and butt-welded seams forming an angle $\beta = 20°$ with a transverse plane. For a gage pressure of 600 kPa, determine (*a*) the normal stress perpendicular to the weld, (*b*) the shearing stress parallel to the weld.

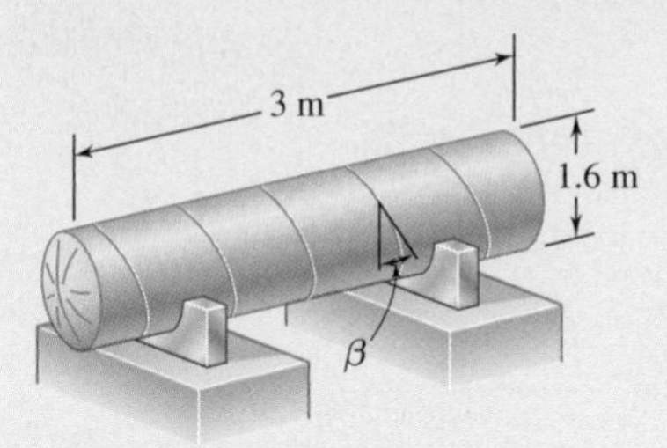

Fig. *P14.63*

14.64 For the tank of Prob. 14.63, determine the largest allowable gage pressure, knowing that the allowable normal stress perpendicular to the weld is 120 MPa and the allowable shearing stress parallel to the weld is 80 MPa.

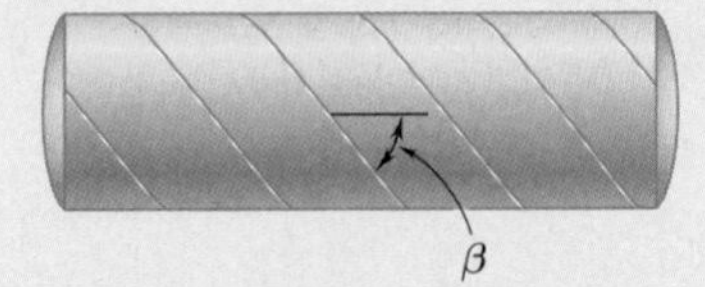

Fig. P14.65 and P14.66

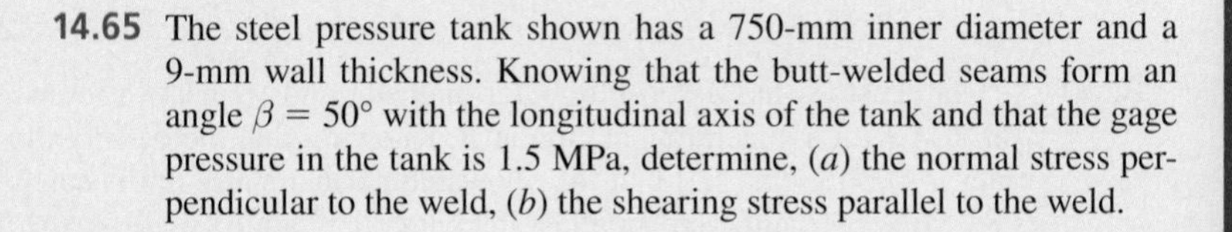

14.65 The steel pressure tank shown has a 750-mm inner diameter and a 9-mm wall thickness. Knowing that the butt-welded seams form an angle $\beta = 50°$ with the longitudinal axis of the tank and that the gage pressure in the tank is 1.5 MPa, determine, (*a*) the normal stress perpendicular to the weld, (*b*) the shearing stress parallel to the weld.

14.66 The pressurized tank shown was fabricated by welding strips of plate along a helix forming an angle β with a transverse plane. Determine the largest value of β that can be used if the normal stress perpendicular to the weld is not to be larger than 85 percent of the maximum stress in the tank.

14.67 The compressed-air tank *AB* has an inner diameter of 450 mm and a uniform wall thickness of 6 mm. Knowing that the gage pressure inside the tank is 1.2 MPa, determine the maximum normal stress and the maximum in-plane shearing stress at point *a* on the top of the tank.

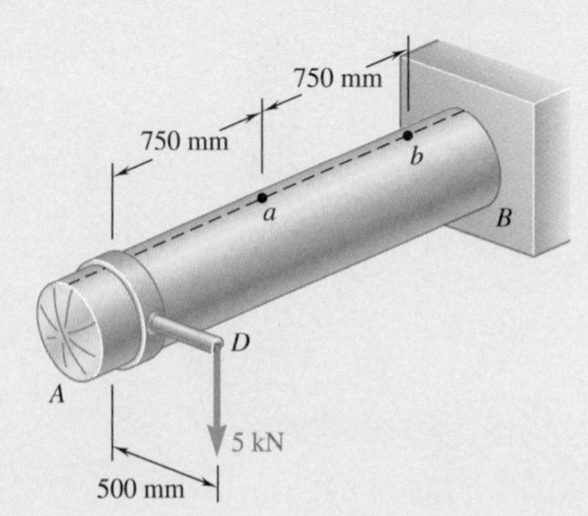

Fig. *P14.67*

14.68 For the compressed-air tank and loading of Prob. 14.67, determine the maximum normal stress and the maximum in-plane shearing stress at point *b* on the top of the tank.

14.69 A pressure vessel of 10-in. inner diameter and 0.25-in. wall thickness is fabricated from a 4-ft section of spirally welded pipe *AB* and is equipped with two rigid end plates. The gage pressure inside the vessel is 300 psi, and 10-kip centric axial forces **P** and **P′** are applied to the end plates. Determine (*a*) the normal stress perpendicular to the weld, (*b*) the shearing stress parallel to the weld.

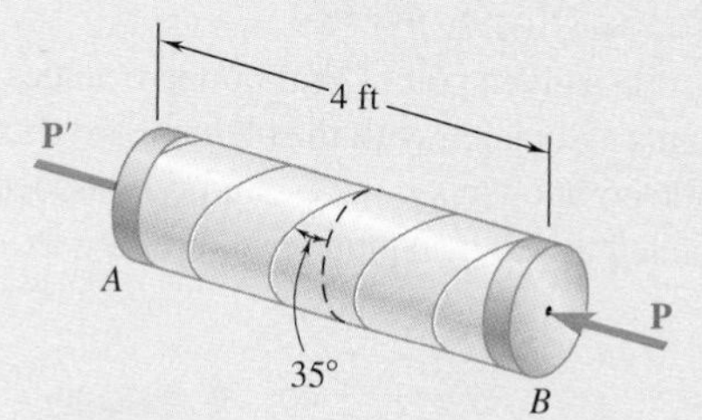

Fig. P14.69

14.70 Solve Prob. 14.69, assuming that the magnitude of *P* of the two forces is increased to 30 kips.

14.71 The cylindrical tank *AB* has an 8-in. inner diameter and a 0.32-in. wall thickness. Knowing that the pressure inside the tank is 600 psi, determine the maximum normal stress and the maximum in-plane shearing stress at point *K*.

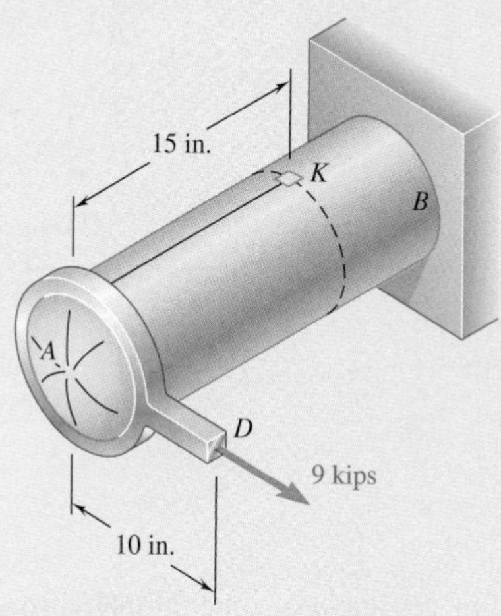

Fig. P14.71

14.72 Solve Prob. 14.71, assuming that the 9-kip force applied at point *D* is directed vertically downward.

Review and Summary

Transformation of Plane Stress

A state of *plane stress* at a given point Q has nonzero values for σ_x, σ_y, and τ_{xy}. The stress components associated with the element are shown in Fig. 14.25*a*. The equations for the components $\sigma_{x'}$, $\sigma_{y'}$, and $\tau_{x'y'}$ associated with that element after being rotated through an angle θ about the z axis (Fig. 14.25*b*) are

$$\sigma_{x'} = \frac{\sigma_x + \sigma_y}{2} + \frac{\sigma_x - \sigma_y}{2}\cos 2\theta + \tau_{xy}\sin 2\theta \tag{14.5}$$

$$\sigma_{y'} = \frac{\sigma_x + \sigma_y}{2} - \frac{\sigma_x - \sigma_y}{2}\cos 2\theta - \tau_{xy}\sin 2\theta \tag{14.7}$$

$$\tau_{x'y'} = -\frac{\sigma_x - \sigma_y}{2}\sin 2\theta + \tau_{xy}\cos 2\theta \tag{14.6}$$

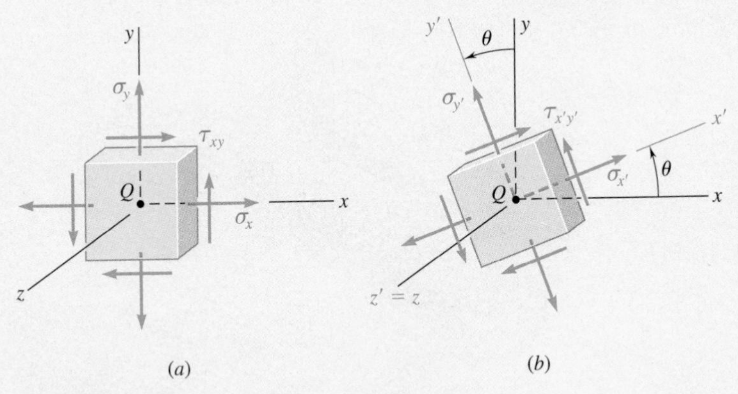

Fig. 14.25

The values θ_p of the angle of rotation that correspond to the maximum and minimum values of the normal stress at point Q are

$$\tan 2\theta_p = \frac{2\tau_{xy}}{\sigma_x - \sigma_y} \tag{14.12}$$

Principal Planes and Principal Stresses

The two values obtained for θ_p are 90° apart (Fig. 14.26) and define the *principal planes of stress* at point Q. The corresponding values of the normal stress are called the *principal stresses* at Q:

$$\sigma_{\max,\min} = \frac{\sigma_x + \sigma_y}{2} \pm \sqrt{\left(\frac{\sigma_x - \sigma_y}{2}\right)^2 + \tau_{xy}^2} \tag{14.14}$$

The corresponding shearing stress is zero.

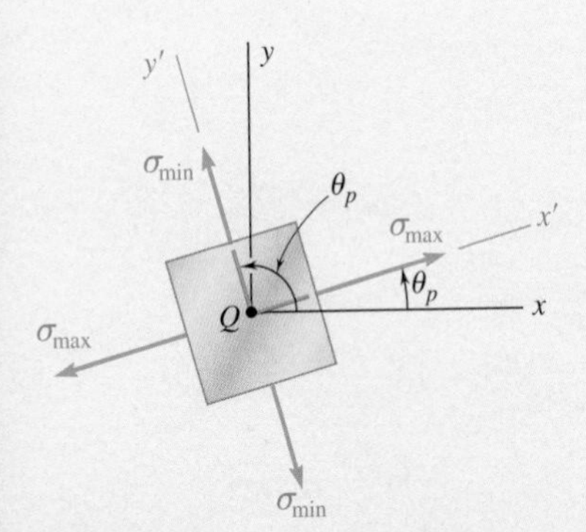

Fig. 14.26

Maximum In-Plane Shearing Stress

The angle θ for the largest value of the shearing stress θ_s is found using

$$\tan 2\theta_s = -\frac{\sigma_x - \sigma_y}{2\tau_{xy}} \tag{14.15}$$

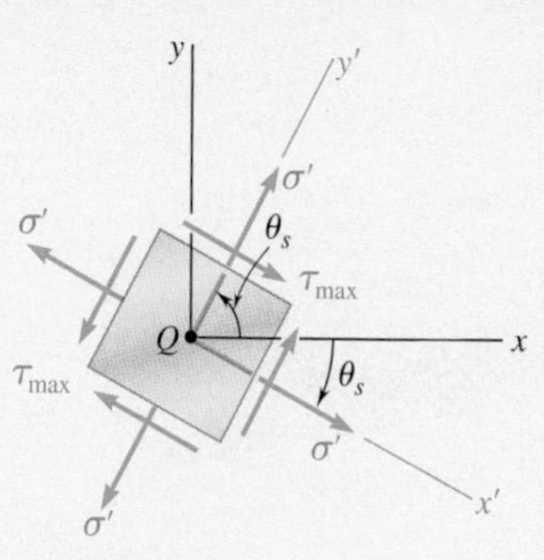

Fig. 14.27

The two values obtained for θ_s are 90° apart (Fig. 14.27). However, the planes of maximum shearing stress are at 45° to the principal planes. The maximum value of the shearing stress *in the plane of stress is*

$$\tau_{\max} = \sqrt{\left(\frac{\sigma_x - \sigma_y}{2}\right)^2 + \tau_{xy}^2} \tag{14.16}$$

and the corresponding value of the normal stresses is

$$\sigma' = \sigma_{\text{ave}} = \frac{\sigma_x + \sigma_y}{2} \tag{14.17}$$

Mohr's Circle for Stress

Mohr's circle provides an alternative method for the analysis of the transformation of plane stress based on simple geometric considerations. Given the state of stress shown in the left element in Fig. 14.28*a*, point

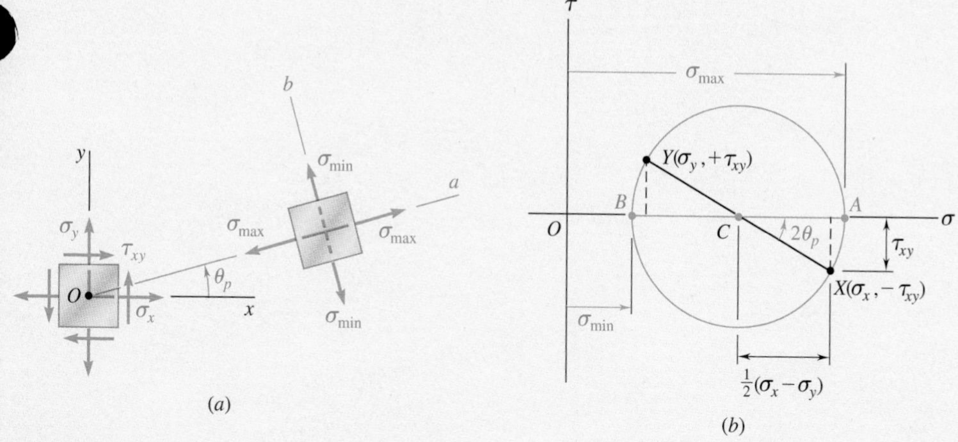

Fig. 14.28

X of coordinates σ_x, $-\tau_{xy}$ and point Y of coordinates σ_y, $+\tau_{xy}$ are plotted in Fig. 14.28*b*. Drawing the circle of diameter XY provides Mohr's circle. The abscissas of the points of intersection A and B of the circle with the horizontal axis represent the principal stresses, and the angle of rotation bringing the diameter XY into AB is twice the angle θ_p defining the principal planes, as shown in the right element of Fig. 14.28*a*. The diameter DE defines the maximum shearing stress and the orientation of the corresponding plane (Fig. 14.29).

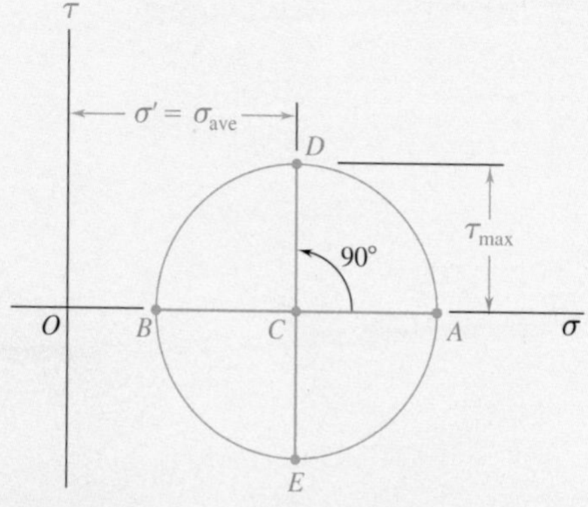

Fig. 14.29

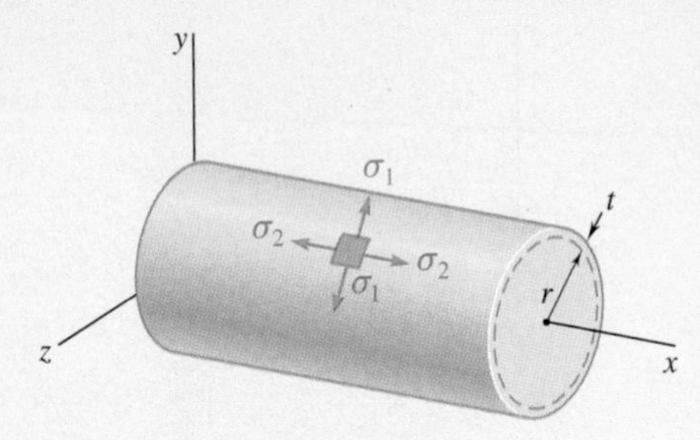

Fig. 14.30

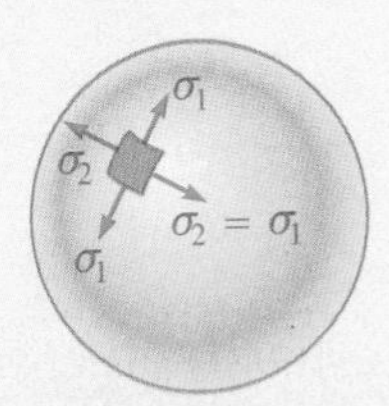

Fig. 14.31

Cylindrical Pressure Vessels

The stresses in *thin-walled pressure vessels* and equations relating to the stresses in the walls and the *gage pressure p* in the fluid were discussed. For a *cylindrical vessel* of inside radius r and thickness t (Fig. 14.30), the *hoop stress* σ_1 and the *longitudinal stress* σ_2 are

$$\sigma_1 = \frac{pr}{t} \qquad \sigma_2 = \frac{pr}{2t} \tag{14.20, 14.21}$$

Spherical Pressure Vessels

For a *spherical vessel* of inside radius r and thickness t (Fig. 14.31), the two principal stresses are equal:

$$\sigma_1 = \sigma_2 = \frac{pr}{2t} \tag{14.24}$$

Review Problems

14.73 Two members of uniform cross section 50 × 80 mm are glued together along plane *a–a* that forms an angle of 25° with the horizontal. Knowing that the allowable stresses for the glued joint are $\sigma = 800$ kPa and $\tau = 600$ kPa, determine the largest axial load **P** that can be applied.

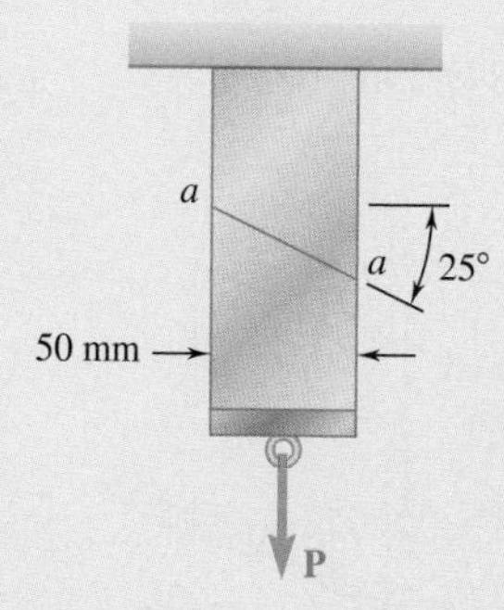

Fig. P14.73

14.74 For the state of stress shown, determine the range of values of θ for which the magnitude of the shearing stress $\tau_{x'y'}$ is equal to or less than 8 ksi.

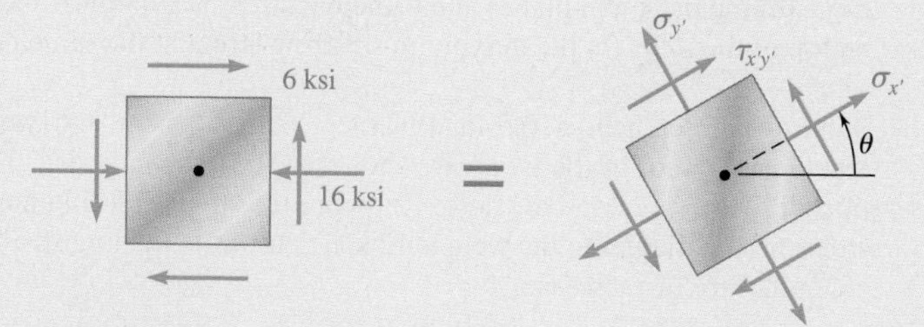

Fig. P14.74

14.75 Determine the range of values of σ_x for which the maximum in-plane shearing stress is equal to or less than 10 ksi.

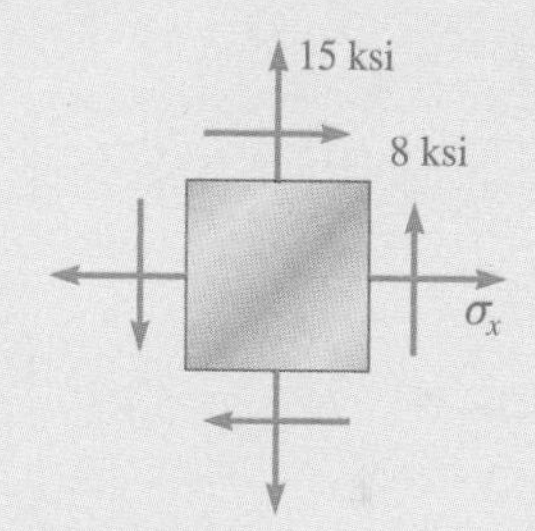

Fig. P14.75

14.76 For the state of stress shown, it is known that the normal and shearing stresses are directed as shown and that $\sigma_x = 14$ ksi, $\sigma_y = 9$ ksi, and $\sigma_{\min} = 5$ ksi. Determine (*a*) the orientation of the principal planes, (*b*) the principal stress $\sigma_{\max}$, (*c*) the maximum in-plane shearing stress.

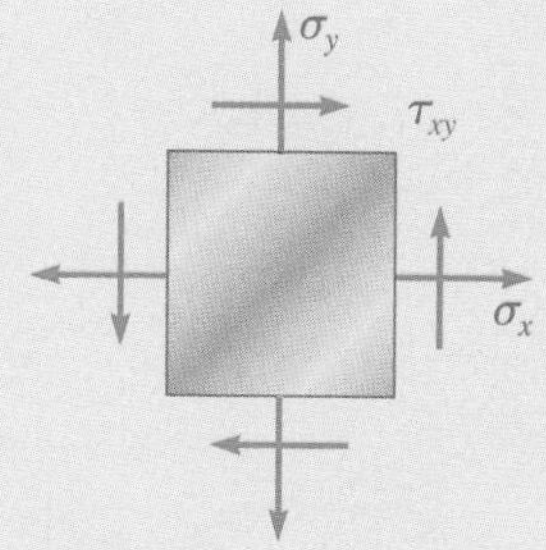

Fig. P14.76

14.77 Determine the principal planes and the principal stresses for the state of plane stress resulting from the superposition of the two states of stress shown.

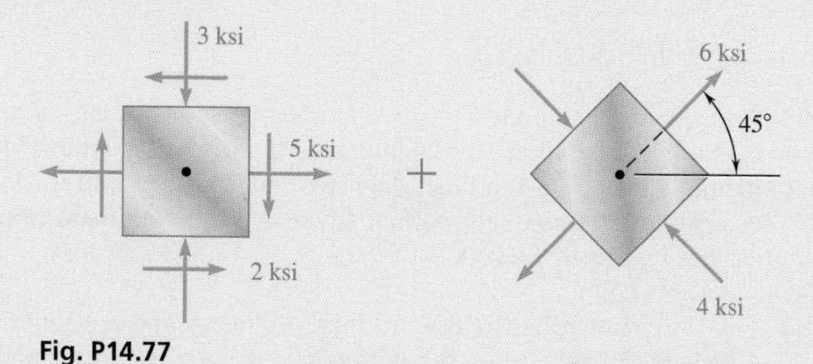

Fig. P14.77

14.78 A standard-weight steel pipe of 12-in. nominal diameter carries water under a pressure of 400 psi. (*a*) Knowing that the outside diameter is 12.75 in. and the wall thickness is 0.375 in., determine the maximum tensile stress in the pipe. (*b*) Solve part *a*, assuming that an extra-strong pipe is used, of 12.75-in. outside diameter and 0.5-in. wall thickness.

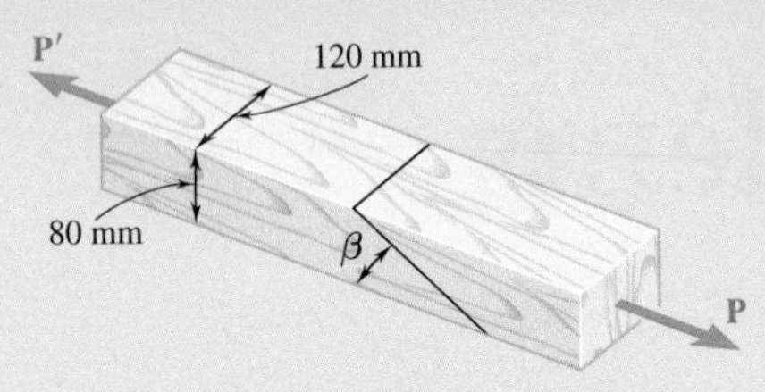

Fig. ***P14.79*** **and P14.80**

14.79 Two wooden members of 80 × 120-mm uniform rectangular cross section are joined by the simple glued scarf splice shown. Knowing that $\beta = 22°$ and that the maximum allowable stresses in the joint are, respectively, 400 kPa in tension (perpendicular to the splice) and 600 kPa in shear (parallel to the splice), determine the largest centric load **P** that can be applied.

14.80 Two wooden members of 80 × 120-mm uniform rectangular cross section are joined by the simple glued scarf splice shown. Knowing that $\beta = 25°$ and that the centric loads of magnitude $P = 10$ kN are applied to the member as shown, determine (*a*) the in-plane shearing stress parallel to the splice, (*b*) the normal stress perpendicular to the splice.

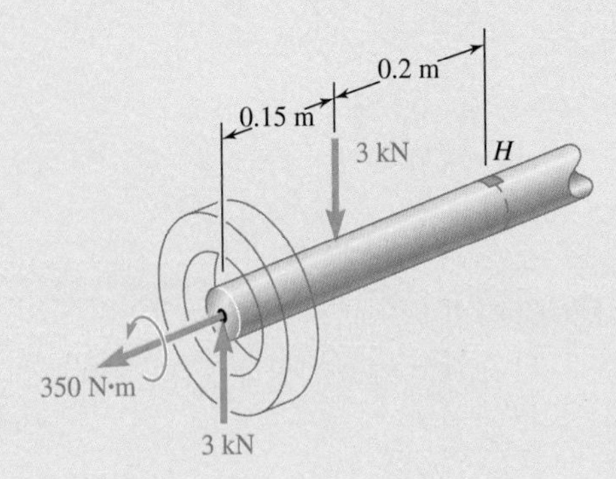

Fig. P14.81

14.81 The axle of an automobile is acted upon by the forces and couple shown. Knowing that the diameter of the solid axle is 32 mm, determine (*a*) the principal planes and principal stresses at point H located on top of the axle, (*b*) the maximum shearing stress at the same point.

14.82 Square plates, each of 0.5-in. thickness, can be bent and welded together in either of the two ways shown to form the cylindrical portion of the compressed air tank. Knowing that the allowable normal stress perpendicular to the weld is 12 ksi, determine the largest allowable pressure in each case.

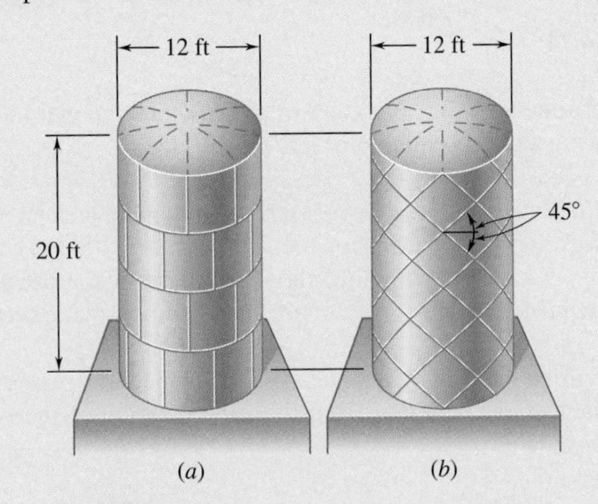

Fig. ***P14.82***

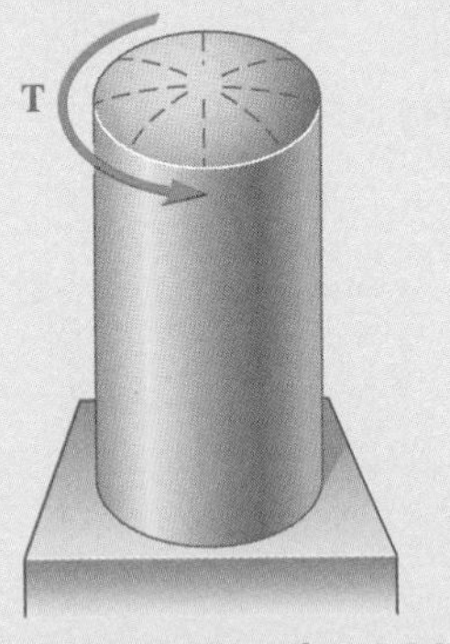

Fig. P14.83 and P14.84

14.83 A torque of magnitude $T = 12$ kN·m is applied to the end of a tank containing compressed air under a pressure of 8 MPa. Knowing that the tank has a 180-mm inner diameter and a 12-mm wall thickness, determine the maximum normal stress and the maximum in-plane shearing stress in the tank.

14.84 The tank shown has a 180-mm inner diameter and a 12-mm wall thickness. Knowing that the tank contains compressed air under a pressure of 8 MPa, determine the magnitude T of the applied torque for which the maximum normal stress is 75 MPa.

© Jetta Productions/Getty Images RF

15

Deflection of Beams

In addition to strength considerations, the design of this bridge is also based on deflection evaluations.

Objectives

In this chapter, you will:

- **Develop** the governing differential equation for the elastic curve, the basis for the techniques considered in this chapter for determining beam deflections.
- **Use** direct integration to obtain slope and deflection equations for beams of simple constraints and loadings.
- **Use** the method of superposition to determine slope and deflection in beams by combining tabulated formulae.
- **Apply** direct integration and superposition to analyze statically indeterminate beams.

Introduction

In the preceding chapters we learned to design beams for strength. This chapter discusses another aspect in the design of beams: the determination of the *deflection*. The *maximum deflection* of a beam under a given load is of particular interest, since the design specifications of a beam will generally include a maximum allowable value for its deflection. A knowledge of deflections is also required to analyze *indeterminate beams*, in which the number of reactions at the supports exceeds the number of equilibrium equations available to determine unknowns.

Recall from Sec. 11.2 that a prismatic beam subjected to pure bending is bent into a circular arc and, within the elastic range, the curvature of the neutral surface is

$$\frac{1}{\rho} = \frac{M}{EI} \tag{11.21}$$

where M is the bending moment, E is the modulus of elasticity, and I is the moment of inertia of the cross section about its neutral axis.

When a beam is subjected to a transverse loading, Eq. (11.21) remains valid for any transverse section, provided that Saint-Venant's principle applies. However, both the bending moment and the curvature of the neutral surface vary from section to section. Denoting by x the distance from the left end of the beam, we write

$$\frac{1}{\rho} = \frac{M(x)}{EI} \tag{15.1}$$

Knowing the curvature at various points of the beam will help us to draw some general conclusions about the deformation of the beam under loading (Sec. 15.1).

To determine the slope and deflection of the beam at any given point, the second-order linear differential equation, which governs the *elastic curve* characterizing the shape of the deformed beam (Sec. 15.1A), is given as

$$\frac{d^2y}{dx^2} = \frac{M(x)}{EI}$$

If the bending moment can be represented for all values of x by a single function $M(x)$, as shown in Fig. 15.1, the slope $\theta = dy/dx$ and the

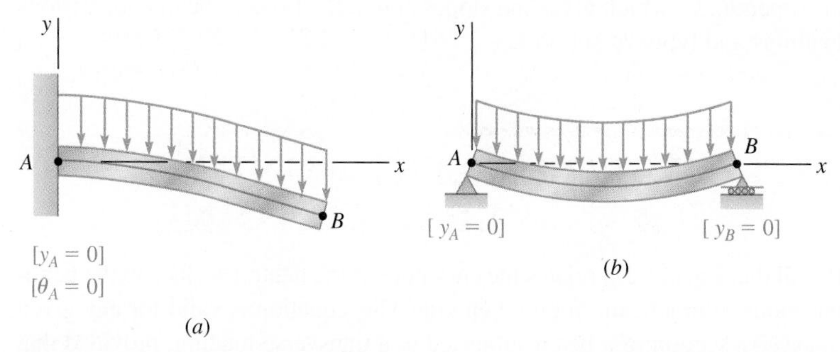

Fig. 15.1 Situations where bending moment can be expressed by a single function $M(x)$. (*a*) Uniformly-loaded cantilever beam. (*b*) Uniformly-loaded simply supported beam.

deflection y at any point of the beam can be obtained through two successive integrations. The two constants of integration introduced in the process are determined from the boundary conditions.

However, if different analytical functions are required to represent the bending moment in various portions of the beam, different differential equations are also required, leading to different functions defining the elastic curve in various portions of the beam. For the beam and loading of Fig. 15.2, for example, two differential equations are required: one for

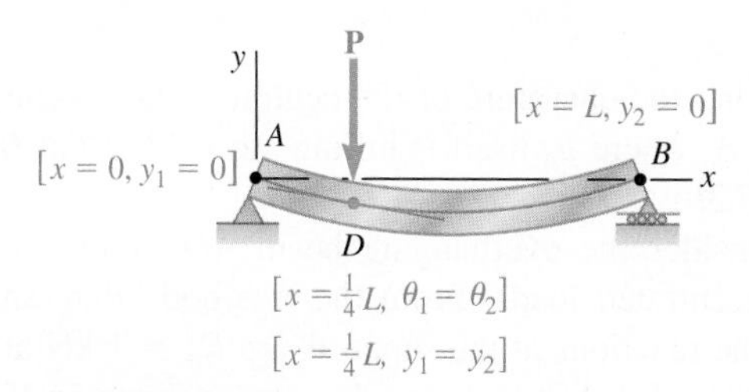

Fig. 15.2 Situation where two sets of equations are required.

the portion AD and the other for the portion DB. The first equation yields functions θ_1 and y_1, and the second functions θ_2 and y_2. Altogether, four constants of integration must be determined; two will be obtained with the deflection being zero at A and B, and the other two by expressing that the portions AD and DB have the same slope and the same deflection at D.

Sec. 15.1B shows that, in a beam supporting a distributed load $w(x)$, the elastic curve can be obtained directly from $w(x)$ through four successive integrations. The constants introduced in this process are determined from the boundary values of V, M, θ, and y.

Sec. 15.2 discusses *statically indeterminate beams* where the reactions at the supports involve four or more unknowns. The three equilibrium equations must be supplemented with equations obtained from the boundary conditions that are imposed by the supports.

The *method of superposition* consists of separately determining and then adding the slope and deflection caused by the various loads applied to a beam (Sec. 15.3). This procedure can be facilitated by the use of the table in Appendix C, which gives the slopes and deflections of beams for various loadings and types of support.

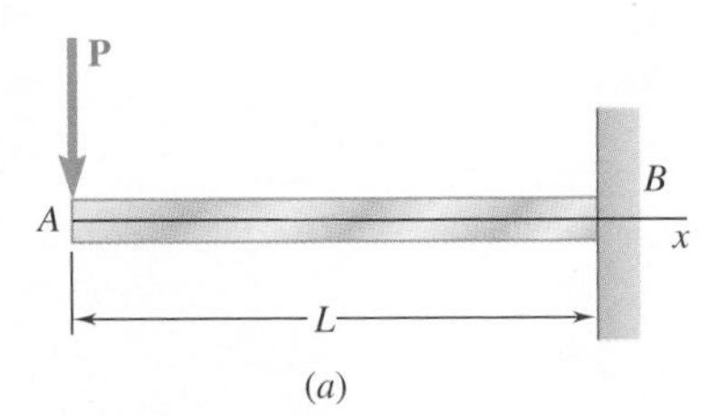

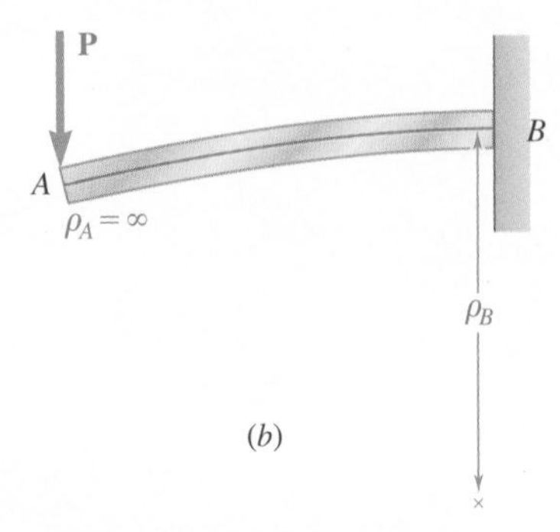

Fig. 15.3 (*a*) Cantilever beam with concentrated load. (*b*) Deformed beam showing curvature at ends.

15.1 DEFORMATION UNDER TRANSVERSE LOADING

Recall that Eq. (11.21) relates the curvature of the neutral surface to the bending moment in a beam in pure bending. This equation is valid for any given transverse section of a beam subjected to a transverse loading, provided that Saint-Venant's principle applies. However, both the bending moment and the curvature of the neutral surface vary from section to section. Denoting by x the distance of the section from the left end of the beam,

$$\frac{1}{\rho} = \frac{M(x)}{EI} \tag{15.1}$$

Consider, for example, a cantilever beam AB of length L subjected to a concentrated load **P** at its free end A (Fig. 15.3*a*). We have $M(x) = -Px$, and substituting into Eq. (15.1) gives

$$\frac{1}{\rho} = -\frac{Px}{EI}$$

which shows that the curvature of the neutral surface varies linearly with x from zero at A, where ρ_A itself is infinite, to $-PL/EI$ at B, where $|\rho_B| = EI/PL$ (Fig. 15.3*b*).

Now consider the overhanging beam AD of Fig. 15.4*a* that supports two concentrated loads. From the free-body diagram of the beam (Fig. 15.4*b*), the reactions at the supports are $R_A = 1$ kN and $R_C = 5$ kN. The corresponding bending-moment diagram is shown in Fig. 15.5*a*. Note from the diagram that M and the curvature of the beam are both zero at each end and at a point E located at $x = 4$ m. Between A and E, the bending

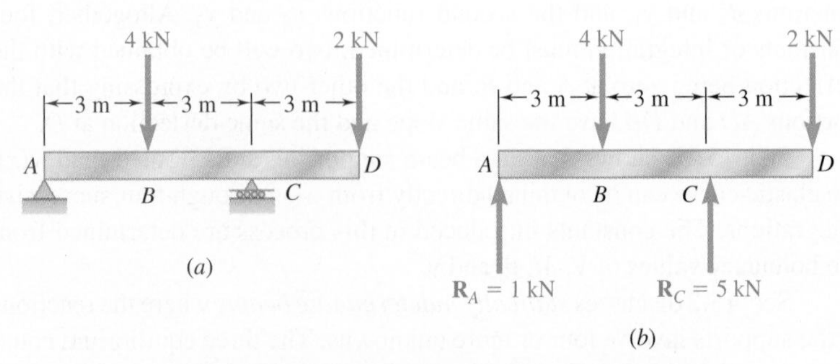

Fig. 15.4 (*a*) Overhanging beam with two concentrated loads. (*b*) Free-body diagram showing reaction forces.

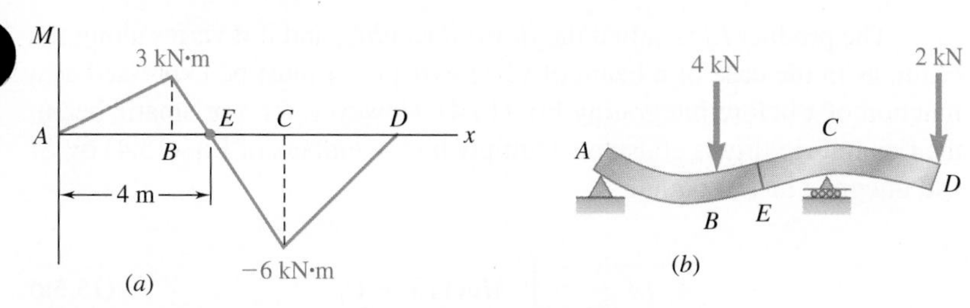

Fig. 15.5 Beam of Fig. 15.4. (*a*) Bending-moment diagram. (*b*) Deformed shape.

moment is positive, and the beam is concave upward. Between E and D, the bending moment is negative and the beam is concave downward (Fig. 15.5*b*). The largest value of the curvature (i.e., the smallest value of the radius of curvature) occurs at support C, where $|M|$ is maximum.

The shape of the deformed beam is obtained from the information about its curvature. However, the analysis and design of a beam usually requires more precise information on the *deflection* and the *slope* at various points. Of particular importance is the maximum deflection of the beam. Eq. (15.1) will be used in the next section to find the relationship between the deflection y measured at a given point Q on the axis of the beam and the distance x of that point from some fixed origin (Fig. 15.6). This relationship is the equation of the *elastic curve*, into which the axis of the beam is transformed under the given load (Fig. 15.6*b*).†

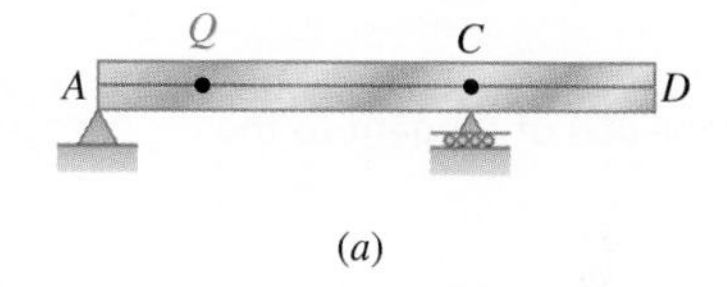

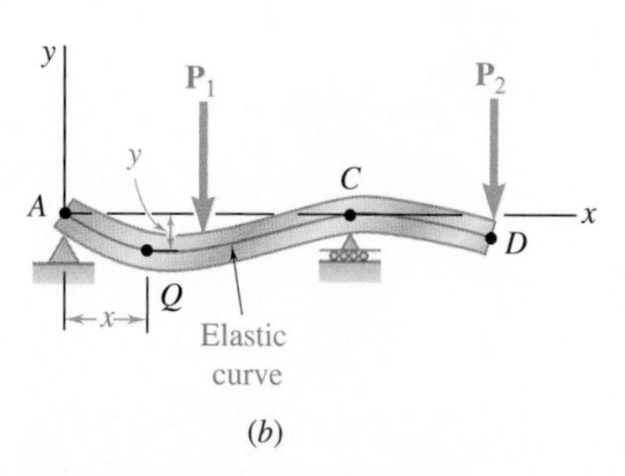

Fig. 15.6 Beam of Fig. 15.4. (*a*) Undeformed. (*b*) Deformed.

15.1A Equation of the Elastic Curve

Recall from elementary calculus that the curvature of a plane curve at a point $Q(x,y)$ is

$$\frac{1}{\rho} = \frac{\dfrac{d^2y}{dx^2}}{\left[1 + \left(\dfrac{dy}{dx}\right)^2\right]^{3/2}} \tag{15.2}$$

where dy/dx and d^2y/dx^2 are the first and second derivatives of the function $y(x)$ represented by that curve. For the elastic curve of a beam, however, the slope dy/dx is very small, and its square is negligible compared to unity. Therefore,

$$\frac{1}{\rho} = \frac{d^2y}{dx^2} \tag{15.3}$$

Substituting for $1/\rho$ from Eq. (15.3) into Eq. (15.1),

$$\frac{d^2y}{dx^2} = \frac{M(x)}{EI} \tag{15.4}$$

This equation is a second-order linear differential equation; it is the governing differential equation for the elastic curve.

†In this chapter, y represents a vertical displacement. It was used in previous chapters to represent the distance of a given point in a transverse section from the neutral axis of that section.

The product EI is called the *flexural rigidity*, and if it varies along the beam, as in the case of a beam of varying depth, it must be expressed as a function of x before integrating Eq. (15.4). However, for a prismatic beam, the flexural rigidity is constant. Multiply both members of Eq. (15.4) by EI and integrate in x to obtain

$$EI\frac{dy}{dx} = \int_0^x M(x)\,dx + C_1 \tag{15.5a}$$

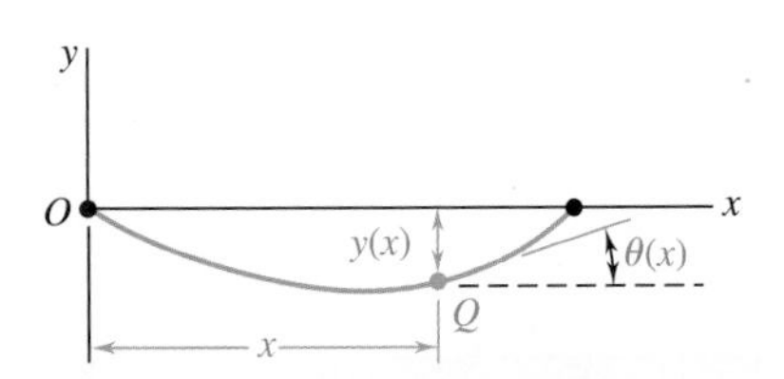

Fig. 15.7 Slope $\theta(x)$ of tangent to the elastic curve.

where C_1 is a constant of integration. Denoting by $\theta(x)$ the angle, measured in radians, that the tangent to the elastic curve at Q forms with the horizontal (Fig. 15.7), and recalling that this angle is very small,

$$\frac{dy}{dx} = \tan\theta \simeq \theta(x)$$

Thus, Eq. (15.5a) in the alternative form is

$$EI\,\theta(x) = \int_0^x M(x)\,dx + C_1 \tag{15.5b}$$

Integrating Eq. (15.5) in x,

$$EI\,y = \int_0^x \left[\int_0^x M(x)\,dx + C_1\right]dx + C_2$$

$$EI\,y = \int_0^x dx \int_0^x M(x)\,dx + C_1x + C_2 \tag{15.6}$$

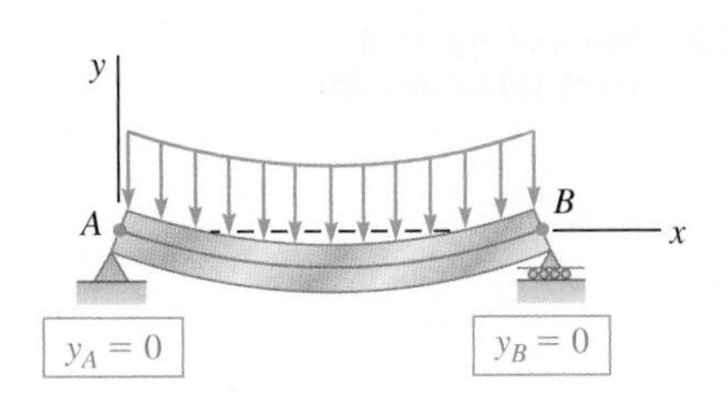

(*a*) Simply supported beam

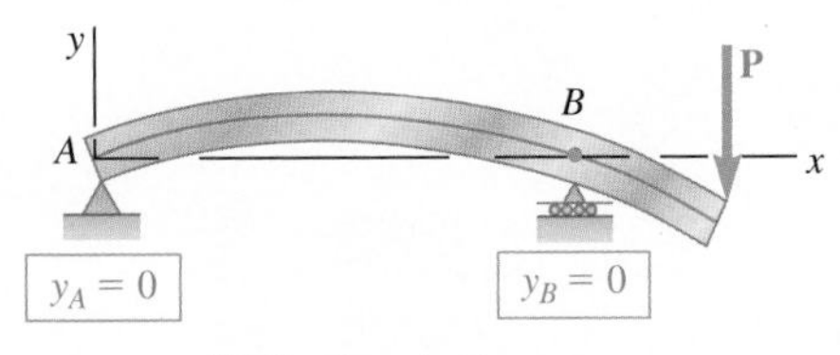

(*b*) Overhanging beam

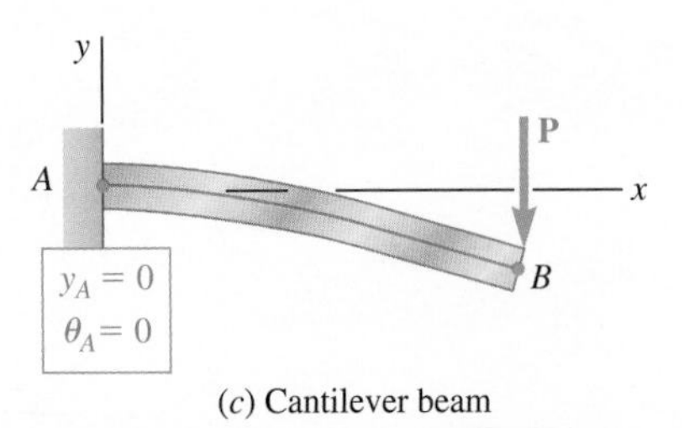

(*c*) Cantilever beam

Fig. 15.8 Known boundary conditions for statically determinate beams.

where C_2 is a second constant and where the first term in the right-hand member represents the function of x obtained by integrating the bending moment $M(x)$ twice in x. Although the constants C_1 and C_2 are as yet undetermined, Eq. (15.6) defines the deflection of the beam at any given point Q, and Eqs. (15.5a) or (15.5b) similarly define the slope of the beam at Q.

The constants C_1 and C_2 are determined from the *boundary conditions* or, more precisely, from the conditions imposed on the beam by its supports. Limiting this analysis to *statically determinate beams*, which are supported so that the reactions at the supports can be obtained by the methods of statics, only three types of beams need to be considered here (Fig. 15.8): (*a*) the *simply supported beam*, (*b*) the *overhanging beam*, and (*c*) the *cantilever beam*.

In Fig. 15.8*a* and *b*, the supports consist of a pin and bracket at A and a roller at B and require that the deflection be zero at each of these points. Letting $x = x_A$, $y = y_A = 0$ in Eq. (15.6) and then setting $x = x_B$, $y = y_B = 0$ in the same equation, two equations are obtained that can be solved for C_1 and C_2. For the cantilever beam (Fig. 15.8*c*), both the deflection and the slope at A must be zero. Letting $x = x_A$, $y = y_A = 0$ in Eq. (15.6) and $x = x_A$, $\theta = \theta_A = 0$ in Eq. (15.5b), two equations are again obtained that can be solved for C_1 and C_2.

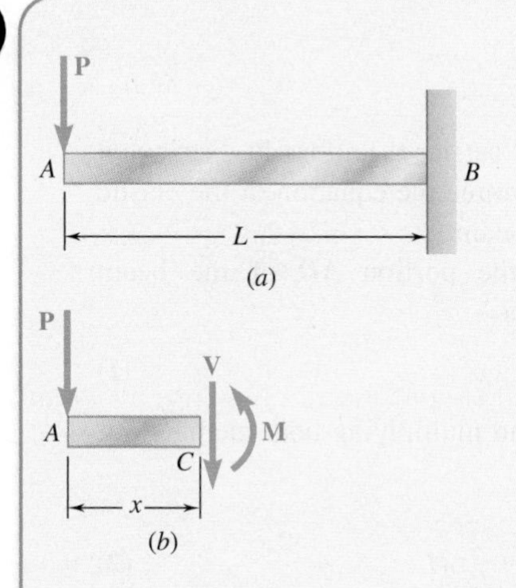

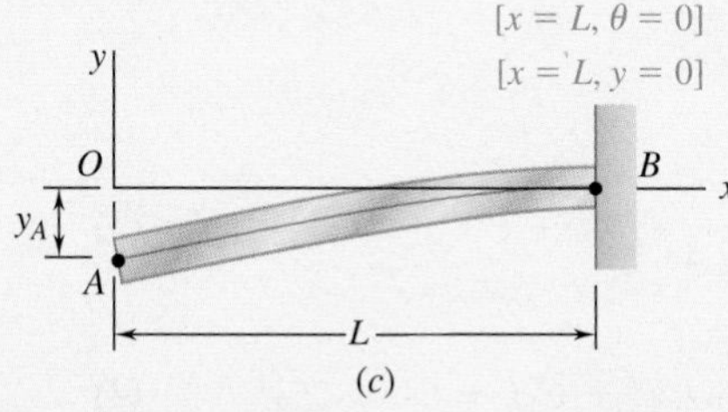

Fig. 15.9 (*a*) Cantilever beam with end load. (*b*) Free-body diagram of section *AC*. (*c*) Deformed shape and boundary conditions.

Concept Application 15.1

The cantilever beam *AB* is of uniform cross section and carries a load **P** at its free end *A* (Fig. 15.9*a*). Determine the equation of the elastic curve and the deflection and slope at *A*.

Using the free-body diagram of the portion *AC* of the beam (Fig. 15.9*b*), where *C* is located at a distance *x* from end *A*,

$$M = -Px \tag{1}$$

Substituting for *M* into Eq. (15.4) and multiplying both members by the constant *EI* gives

$$EI \frac{d^2y}{dx^2} = -Px$$

Integrating in *x*,

$$EI \frac{dy}{dx} = -\tfrac{1}{2}Px^2 + C_1 \tag{2}$$

Now observe the fixed end *B* where $x = L$ and $\theta = dy/dx = 0$ (Fig. 15.9*c*). Substituting these values into Eq. (2) and solving for C_1 gives

$$C_1 = \tfrac{1}{2}PL^2$$

which we carry back into Eq. (2):

$$EI \frac{dy}{dx} = -\tfrac{1}{2}Px^2 + \tfrac{1}{2}PL^2 \tag{3}$$

Integrating both members of Eq. (3),

$$EI\,y = -\tfrac{1}{6}Px^3 + \tfrac{1}{2}PL^2x + C_2 \tag{4}$$

But at B, $x = L$, $y = 0$. Substituting into Eq. (4),

$$0 = -\tfrac{1}{6}PL^3 + \tfrac{1}{2}PL^3 + C_2$$

$$C_2 = -\tfrac{1}{3}PL^3$$

Carrying the value of C_2 back into Eq. (4), the equation of the elastic curve is

$$EI\,y = -\tfrac{1}{6}Px^3 + \tfrac{1}{2}PL^2x - \tfrac{1}{3}PL^3$$

or

$$y = \frac{P}{6EI}(-x^3 + 3L^2x - 2L^3) \tag{5}$$

The deflection and slope at *A* are obtained by letting $x = 0$ in Eqs. (3) and (5).

$$y_A = -\frac{PL^3}{3EI} \qquad \text{and} \qquad \theta_A = \left(\frac{dy}{dx}\right)_A = \frac{PL^2}{2EI}$$

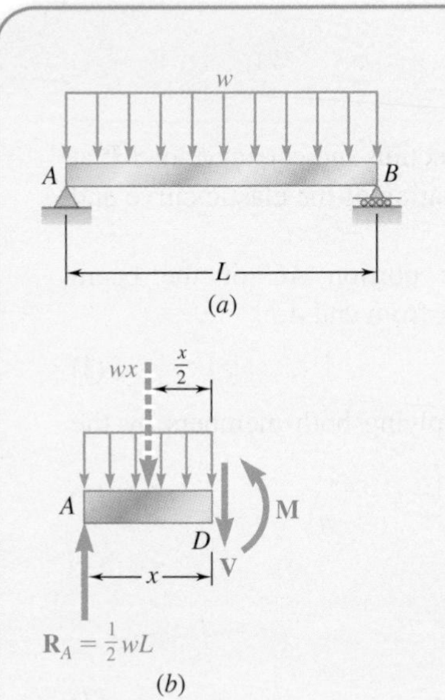

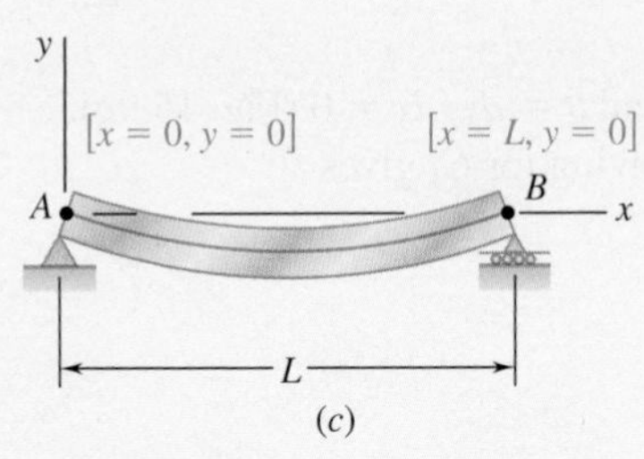

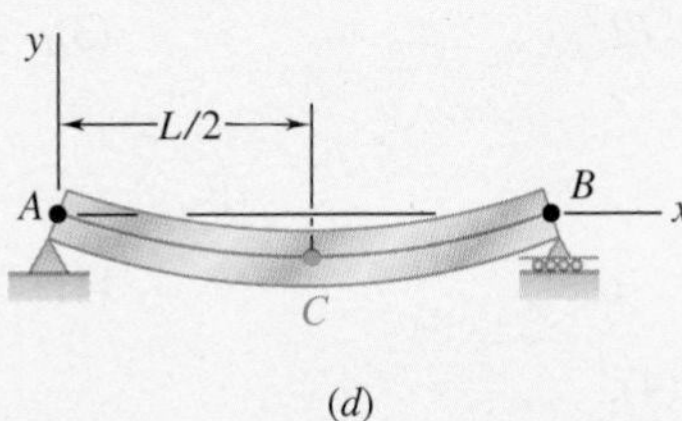

Fig. 15.10 (*a*) Simply supported beam with a uniformly distributed load. (*b*) Free-body diagram of segment *AD*. (*c*) Boundary conditions. (*d*) Point of maximum deflection.

Concept Application 15.2

The simply supported prismatic beam *AB* carries a uniformly distributed load *w* per unit length (Fig. 15.10*a*). Determine the equation of the elastic curve and the maximum deflection of the beam.

Draw the free-body diagram of the portion *AD* of the beam (Fig. 15.10*b*) and take moments about *D* for

$$M = \tfrac{1}{2}wLx - \tfrac{1}{2}wx^2 \quad (1)$$

Substituting for *M* into Eq. (15.4) and multiplying both members of this equation by the constant *EI* gives

$$EI\frac{d^2y}{dx^2} = -\frac{1}{2}wx^2 + \frac{1}{2}wLx \quad (2)$$

Integrating twice in *x*,

$$EI\frac{dy}{dx} = -\frac{1}{6}wx^3 + \frac{1}{4}wLx^2 + C_1 \quad (3)$$

$$EI\,y = -\frac{1}{24}wx^4 + \frac{1}{12}wLx^3 + C_1x + C_2 \quad (4)$$

Observing that $y = 0$ at both ends of the beam (Fig. 15.10*c*), let $x = 0$ and $y = 0$ in Eq. (4) and obtain $C_2 = 0$. Then make $x = L$ and $y = 0$ in the same equation, so

$$0 = -\tfrac{1}{24}wL^4 + \tfrac{1}{12}wL^4 + C_1L$$

$$C_1 = -\tfrac{1}{24}wL^3$$

Carrying the values of C_1 and C_2 back into Eq. (15.4), the elastic curve is

$$EI\,y = -\tfrac{1}{24}wx^4 + \tfrac{1}{12}wLx^3 - \tfrac{1}{24}wL^3x$$

or

$$y = \frac{w}{24EI}(-x^4 + 2Lx^3 - L^3x) \quad (5)$$

Substituting the value for C_1 into Eq. (3), we check that the slope of the beam is zero for $x = L/2$ and thus that the elastic curve has a minimum at the midpoint *C* (Fig. 15.10*d*). Letting $x = L/2$ in Eq. (5),

$$y_C = \frac{w}{24EI}\left(-\frac{L^4}{16} + 2L\frac{L^3}{8} - L^3\frac{L}{2}\right) = -\frac{5wL^4}{384EI}$$

The maximum deflection (the maximum absolute value) is

$$|y|_{max} = \frac{5wL^4}{384EI}$$

In both Concept Applications considered so far, only one free-body diagram was required to determine the bending moment in the beam. As a result, a single function of x was used to represent M throughout the beam. However, concentrated loads, reactions at supports, or discontinuities in a distributed load make it necessary to divide the beam into several portions and to represent the bending moment by a different function $M(x)$ in each. As an example, Photo 15.1 shows an elevated roadway supported by beams, which in turn will be subjected to concentrated loads from vehicles crossing the completed bridge. Each of the functions $M(x)$ leads to a different expression for the slope $\theta(x)$ and the deflection $y(x)$. Since each expression must contain two constants of integration, a large number of constants will have to be determined. As shown in the following Concept Application, the required additional boundary conditions can be obtained by observing that, while the shear and bending moment can be discontinuous at several points in a beam, the *deflection* and the *slope* of the beam *cannot be discontinuous* at any point.

Photo 15.1 A different function $M(x)$ would be required in each portion of the beams when a vehicle crosses the completed bridge.

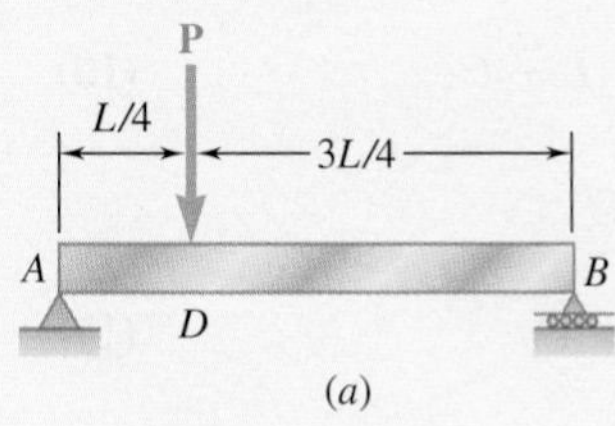

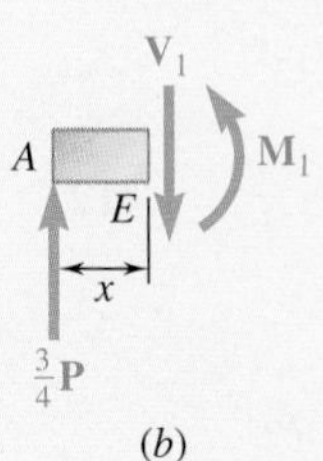

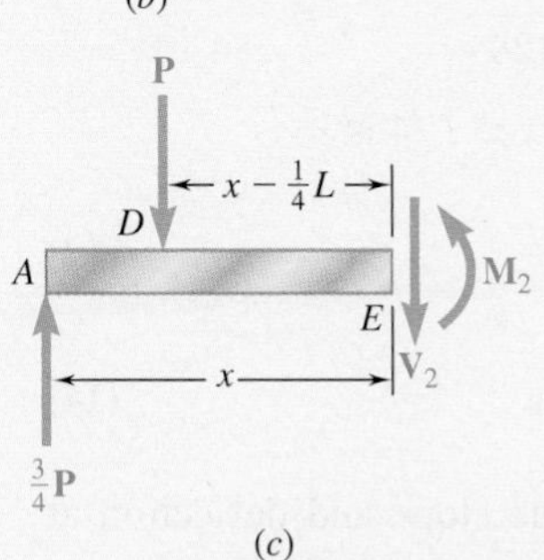

Fig. 15.11 (*a*) Simply supported beam with transverse load *P*. (*b*) Free-body diagram of portion *AE* to find moment left of load *P*. (*c*) Free-body diagram of portion *AE* to find moment right of load *P*.

Concept Application 15.3

For the prismatic beam and load shown (Fig. 15.11*a*), determine the slope and deflection at point *D*.

Divide the beam into two portions, *AD* and *DB*, and determine the function $y(x)$ that defines the elastic curve for each of these portions.

1. From *A* to *D* ($x < L/4$). Draw the free-body diagram of a portion of beam *AE* of length $x < L/4$ (Fig. 15.11*b*). Take moments about *E* to obtain

$$M_1 = \frac{3P}{4}x \quad (1)$$

and recalling Eq. (15.4), we write

$$EI\frac{d^2y_1}{dx^2} = \frac{3}{4}Px \quad (2)$$

where $y_1(x)$ is the function that defines the elastic curve *for portion AD of the beam*. Integrating in x,

$$EI\,\theta_1 = EI\frac{dy_1}{dx} = \frac{3}{8}Px^2 + C_1 \quad (3)$$

$$EI\,y_1 = \frac{1}{8}Px^3 + C_1x + C_2 \quad (4)$$

2. From *D* to *B* ($x > L/4$). Now draw the free-body diagram of a portion of beam *AE* of the length $x > L/4$ (Fig. 15.11*c*) and write

$$M_2 = \frac{3P}{4}x - P\left(x - \frac{L}{4}\right) \quad (5)$$

(continued)

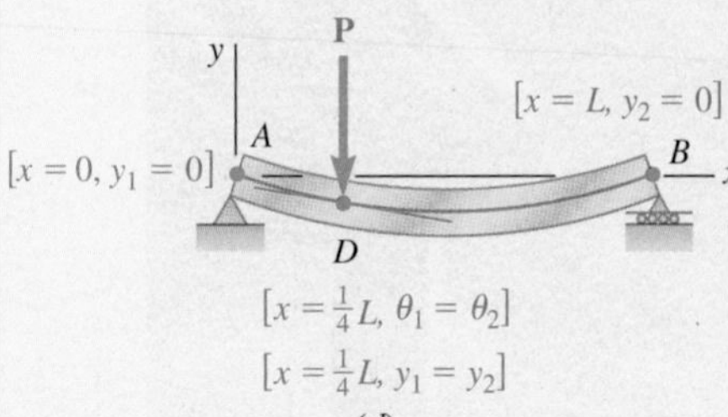

Fig. 15.11 *(cont.)* (*d*) Boundary conditions.

and recalling Eq. (15.4) and rearranging terms, we have

$$EI\frac{d^2y_2}{dx^2} = -\frac{1}{4}Px + \frac{1}{4}PL \quad \textbf{(6)}$$

where $y_2(x)$ is the function that defines the elastic curve *for portion DB of the beam*. Integrating in x,

$$EI\,\theta_2 = EI\frac{dy_2}{dx} = -\frac{1}{8}Px^2 + \frac{1}{4}PLx + C_3 \quad \textbf{(7)}$$

$$EI\,y_2 = -\frac{1}{24}Px^3 + \frac{1}{8}PLx^2 + C_3x + C_4 \quad \textbf{(8)}$$

Determination of the Constants of Integration. The conditions satisfied by the constants of integration are summarized in Fig. 15.11*d*. At the support *A*, where the deflection is defined by Eq. (4), $x = 0$ and $y_1 = 0$. At the support *B*, where the deflection is defined by Eq. (8), $x = L$ and $y_2 = 0$. Also, the fact that there can be no sudden change in deflection or in slope at point *D* requires that $y_1 = y_2$ and $\theta_1 = \theta_2$ when $x = L/4$. Therefore,

$[x = 0, y_1 = 0]$, Eq. (4): $\quad 0 = C_2 \quad$ **(9)**

$[x = L, y_2 = 0]$, Eq. (8): $\quad 0 = \frac{1}{12}PL^3 + C_3L + C_4 \quad$ **(10)**

$[x = L/4, \theta_1 = \theta_2]$, Eqs. (3) and (7):

$$\frac{3}{128}PL^2 + C_1 = \frac{7}{128}PL^2 + C_3 \quad \textbf{(11)}$$

$[x = L/4, y_1 = y_2]$, Eqs. (4) and (8):

$$\frac{PL^3}{512} + C_1\frac{L}{4} = \frac{11PL^3}{1536} + C_3\frac{L}{4} + C_4 \quad \textbf{(12)}$$

Solving these equations simultaneously,

$$C_1 = -\frac{7PL^2}{128}, \quad C_2 = 0, \quad C_3 = -\frac{11PL^2}{128}, \quad C_4 = \frac{PL^3}{384}$$

Substituting for C_1 and C_2 into Eqs. (3) and (4), $x \le L/4$ is

$$EI\,\theta_1 = \frac{3}{8}Px^2 - \frac{7PL^2}{128} \quad \textbf{(13)}$$

$$EI\,y_1 = \frac{1}{8}Px^3 - \frac{7PL^2}{128}x \quad \textbf{(14)}$$

Letting $x = L/4$ in each of these equations, the slope and deflection at point *D* are

$$\theta_D = -\frac{PL^2}{32EI} \quad \text{and} \quad y_D = -\frac{3PL^3}{256EI}$$

Note that since $\theta_D \neq 0$, the deflection at *D* is *not* the maximum deflection of the beam.

15.1B Determination of the Elastic Curve from the Load Distribution

Sec. 15.1A showed that the equation of the elastic curve can be obtained by integrating twice the differential equation

$$\frac{d^2y}{dx^2} = \frac{M(x)}{EI} \tag{15.4}$$

where $M(x)$ is the bending moment in the beam. Now recall from Sec. 12.2 that, when a beam supports a distributed load $w(x)$, we have $dM/dx = V$ and $dV/dx = -w$ at any point of the beam. Differentiating both members of Eq. (15.4) with respect to x and assuming EI to be constant,

$$\frac{d^3y}{dx^3} = \frac{1}{EI}\frac{dM}{dx} = \frac{V(x)}{EI} \tag{15.7}$$

and differentiating again,

$$\frac{d^4y}{dx^4} = \frac{1}{EI}\frac{dV}{dx} = -\frac{w(x)}{EI}$$

Thus, when a prismatic beam supports a distributed load $w(x)$, its elastic curve is governed by the fourth-order linear differential equation

$$\frac{d^4y}{dx^4} = -\frac{w(x)}{EI} \tag{15.8}$$

Multiply both members of Eq. (15.8) by the constant EI and integrate four times to obtain

$$\begin{aligned}
EI\frac{d^4y}{dx^4} &= -w(x) \\
EI\frac{d^3y}{dx^3} &= V(x) = -\int w(x)\,dx + C_1 \\
EI\frac{d^2y}{dx^2} &= M(x) = -\int dx \int w(x)\,dx + C_1x + C_2 \\
EI\frac{dy}{dx} &= EI\,\theta(x) = -\int dx \int dx \int w(x)\,dx + \frac{1}{2}C_1x^2 + C_2x + C_3 \\
EIy(x) &= -\int dx \int dx \int dx \int w(x)\,dx + \frac{1}{6}C_1x^3 + \frac{1}{2}C_2x^2 + C_3x + C_4
\end{aligned} \tag{15.9}$$

The four constants of integration are determined from the boundary conditions. These conditions include (*a*) the conditions imposed on the deflection or slope of the beam by its supports (see. Sec. 15.1A) and (*b*) the condition that V and M be zero at the free end of a cantilever beam or that M be zero at both ends of a simply supported beam (see. Sec. 12.2). This has been illustrated in Fig. 15.12.

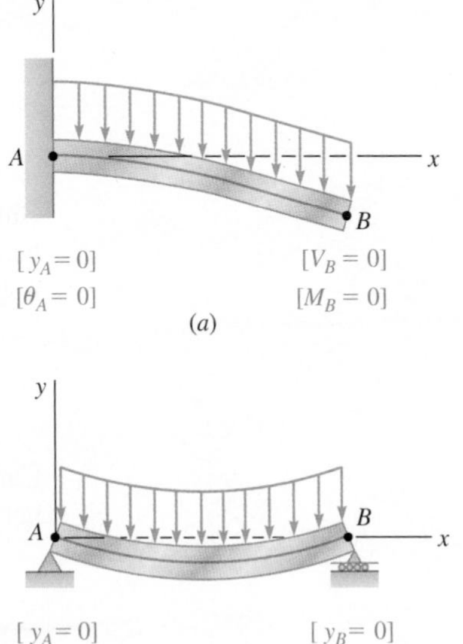

Fig. 15.12 Boundary conditions for (*a*) cantilever beam (*b*) simply supported beam.

This method can be used effectively with cantilever or simply supported beams carrying a distributed load. In the case of overhanging beams, the reactions at the supports cause discontinuities in the shear (i.e., in the third derivative of y), and different functions are required to define the elastic curve over the entire beam.

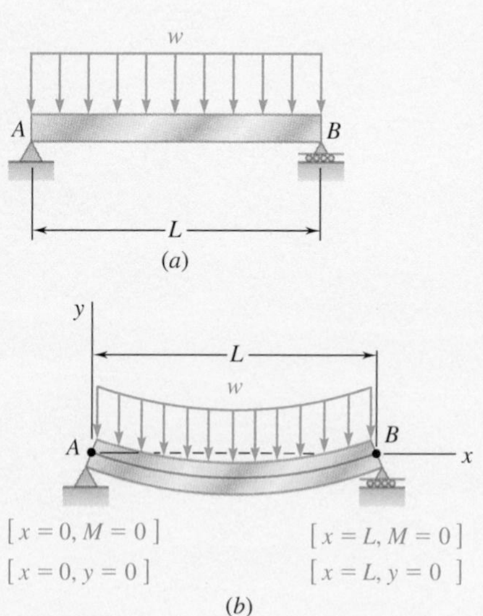

Fig. 15.13 (*a*) Simply supported beam with a uniformly distributed load. (*b*) Boundary conditions.

Concept Application 15.4

The simply supported prismatic beam AB carries a uniformly distributed load w per unit length (Fig. 15.13*a*). Determine the equation of the elastic curve and the maximum deflection of the beam. (This is the same beam and load as in Concept Application 15.2.)

Since w = constant, the first three of Eqs. (15.9) yield

$$EI\frac{d^4y}{dx^4} = -w$$

$$EI\frac{d^3y}{dx^3} = V(x) = -wx + C_1$$

$$EI\frac{d^2y}{dx^2} = M(x) = -\frac{1}{2}wx^2 + C_1x + C_2 \tag{1}$$

Noting that the boundary conditions require that $M = 0$ at both ends of the beam (Fig. 15.13*b*), let $x = 0$ and $M = 0$ in Eq. (1) and obtain $C_2 = 0$. Then make $x = L$ and $M = 0$ in the same equation and obtain $C_1 = \frac{1}{2}wL$.

Carry the values of C_1 and C_2 back into Eq. (1) and integrate twice to obtain

$$EI\frac{d^2y}{dx^2} = -\frac{1}{2}wx^2 + \frac{1}{2}wLx$$

$$EI\frac{dy}{dx} = -\frac{1}{6}wx^3 + \frac{1}{4}wLx^2 + C_3$$

$$EI\,y = -\frac{1}{24}wx^4 + \frac{1}{12}wLx^3 + C_3x + C_4 \tag{2}$$

But the boundary conditions also require that $y = 0$ at both ends of the beam. Letting $x = 0$ and $y = 0$ in Eq. (2), $C_4 = 0$. Letting $x = L$ and $y = 0$ in the same equation gives

$$0 = -\tfrac{1}{24}wL^4 + \tfrac{1}{12}wL^4 + C_3L$$

$$C_3 = -\tfrac{1}{24}wL^3$$

Carrying the values of C_3 and C_4 back into Eq. (2) and dividing both members by EI, the equation of the elastic curve is

$$y = \frac{w}{24EI}(-x^4 + 2Lx^3 - L^3x) \tag{3}$$

The maximum deflection is obtained by making $x = L/2$ in Eq. (3).

$$|y|_{max} = \frac{5wL^4}{384EI}$$

15.2 STATICALLY INDETERMINATE BEAMS

In the preceding sections, our analysis was limited to statically determinate beams. Now consider the prismatic beam *AB* (Fig. 15.14*a*), which has a fixed end at *A* and is supported by a roller at *B*. Drawing the free-body diagram of the beam (Fig. 15.14*b*), the reactions involve four unknowns, with only three equilibrium equations:

$$\Sigma F_x = 0 \quad \Sigma F_y = 0 \quad \Sigma M_A = 0 \tag{15.10}$$

Since only A_x can be determined from these equations, the beam is *statically indeterminate*.

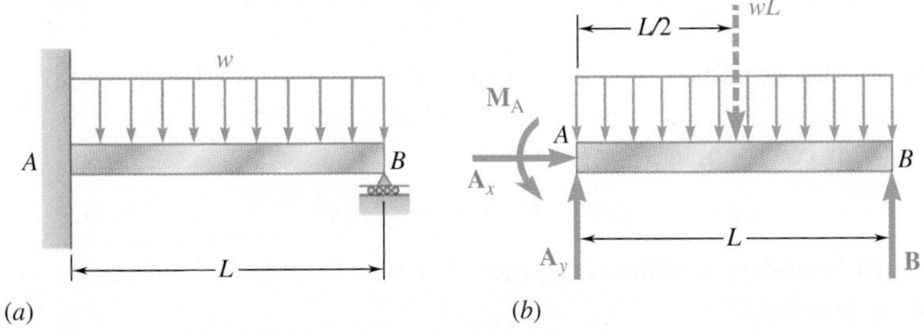

Fig. 15.14 (*a*) Statically indeterminate beam with a uniformly distributed load. (*b*) Free-body diagram with four unknown reactions.

Recall from Chaps. 9 and 10 that, in a statically indeterminate problem, the reactions can be obtained by considering the *deformations* of the structure. Therefore, we proceed with the computation of the slope and deformation along the beam. Following the method used in Sec. 15.1A, the bending moment $M(x)$ at any given point *AB* is expressed in terms of the distance *x* from *A*, the given load, and the unknown reactions. Integrating in *x*, expressions for θ and *y* are found. These contain two additional unknowns: the constants of integration C_1 and C_2. Altogether, six equations are available to determine the reactions and constants C_1 and C_2; they are the three equilibrium equations of Eq. (15.10) and the three equations expressing that the boundary conditions are satisfied (i.e., that the slope and deflection at *A* are zero and that the deflection at *B* is zero (Fig. 15.15)). Thus, the reactions at the supports can be determined, and the equation of the elastic curve can be obtained.

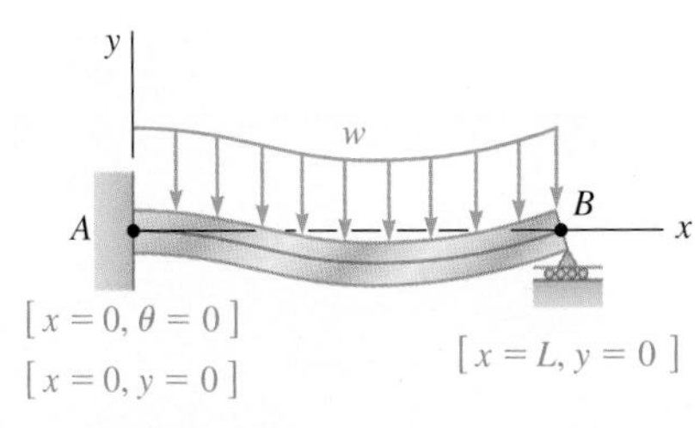

Fig. 15.15 Boundary conditions for beam of Fig. 15.14.

Concept Application 15.5

Determine the reactions at the supports for the prismatic beam of Fig. 15.14*a*.

Equilibrium Equations. From the free-body diagram of Fig. 15.14*b*,

$$\overset{+}{\rightarrow} \Sigma F_x = 0: \quad A_x = 0$$

$$+\uparrow \Sigma F_y = 0: \quad A_y + B - wL = 0 \tag{1}$$

$$+\circlearrowleft \Sigma M_A = 0: \quad M_A + BL - \tfrac{1}{2}wL^2 = 0$$

(continued)

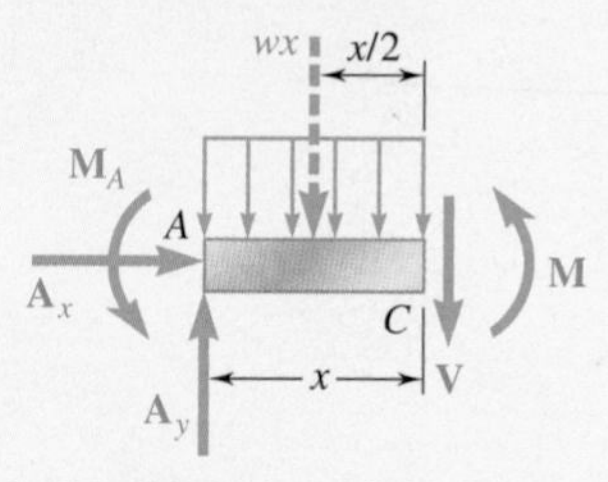

Fig. 15.16 Free-body diagram of beam portion AC.

Equation of Elastic Curve. Draw the free-body diagram of a portion of beam AC (Fig. 15.16) to obtain

$$+\circlearrowleft \Sigma M_C = 0: \qquad M + \tfrac{1}{2}wx^2 + M_A - A_y x = 0 \tag{2}$$

Solving Eq. (2) for M and carrying into Eq. (15.4),

$$EI\frac{d^2y}{dx^2} = -\frac{1}{2}wx^2 + A_y x - M_A$$

Integrating in x gives

$$EI\,\theta = EI\frac{dy}{dx} = -\frac{1}{6}wx^3 + \frac{1}{2}A_y x^2 - M_A x + C_1 \tag{3}$$

$$EI\,y = -\frac{1}{24}wx^4 + \frac{1}{6}A_y x^3 - \frac{1}{2}M_A x^2 + C_1 x + C_2 \tag{4}$$

Referring to the boundary conditions indicated in Fig. 15.15, $x = 0$, $\theta = 0$ in Eq. (3), $x = 0$, $y = 0$ in Eq. (4), and conclude that $C_1 = C_2 = 0$. Thus, Eq. (4) is rewritten as

$$EI\,y = -\tfrac{1}{24}wx^4 + \tfrac{1}{6}A_y x^3 - \tfrac{1}{2}M_A x^2 \tag{5}$$

But the third boundary condition requires that $y = 0$ for $x = L$. Carrying these values into Eq. (5),

$$0 = -\tfrac{1}{24}wL^4 + \tfrac{1}{6}A_y L^3 - \tfrac{1}{2}M_A L^2$$

or

$$3M_A - A_y L + \tfrac{1}{4}wL^2 = 0 \tag{6}$$

Solving this equation simultaneously with the three equilibrium equations of Eq. (1), the reactions at the supports are

$$A_x = 0 \qquad A_y = \tfrac{5}{8}wL \qquad M_A = \tfrac{1}{8}wL^2 \qquad B = \tfrac{3}{8}wL$$

In the previous Concept Application, there was one redundant reaction (i.e., one more than could be determined from the equilibrium equations alone). The corresponding beam is *statically indeterminate to the first degree*. Another example of a beam indeterminate to the first degree is provided in Sample Prob. 15.3. If the beam supports are such that two reactions are redundant (Fig. 15.17*a*), the beam is *indeterminate to the second degree*. While there are now five unknown reactions (Fig. 15.17*b*), four equations can be obtained from the boundary conditions (Fig. 15.17*c*). Thus, seven equations are available to determine the five reactions and the two constants of integration.

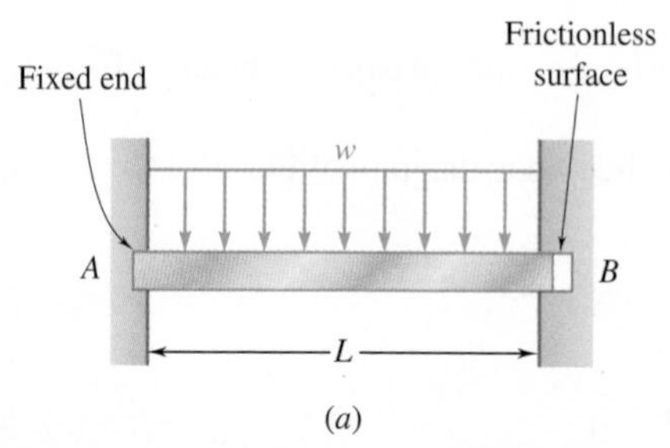

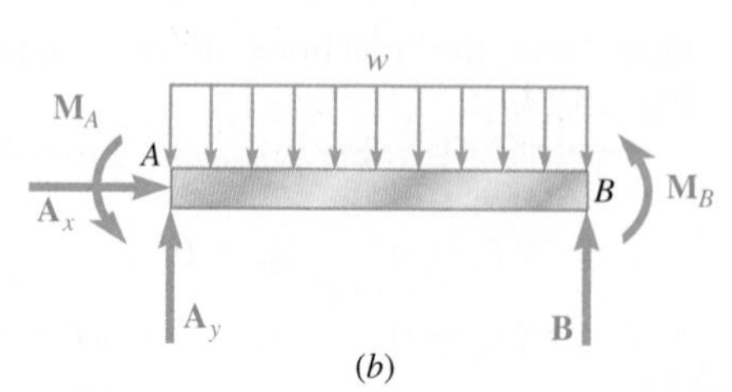

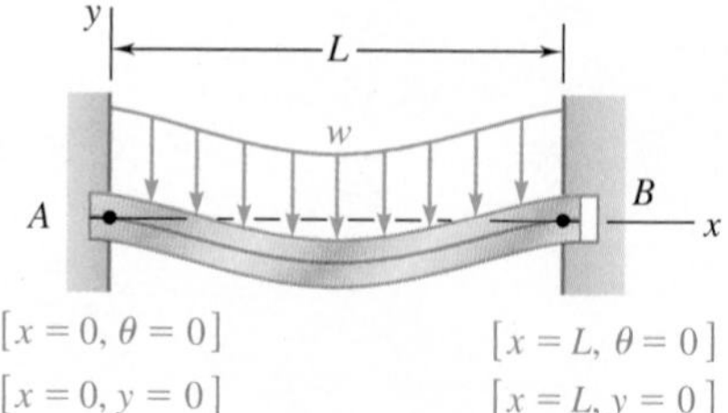

Fig. 15.17 (*a*) Beam statically indeterminate to the second degree. (*b*) Free-body diagram. (*c*) Boundary conditions.

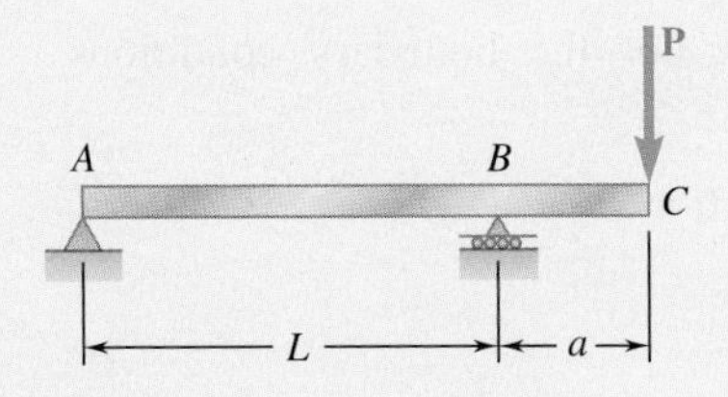

Sample Problem 15.1

The overhanging steel beam *ABC* carries a concentrated load **P** at end *C*. For portion *AB* of the beam, (*a*) derive the equation of the elastic curve, (*b*) determine the maximum deflection, (*c*) evaluate y_{max} for the following data:

W14 × 68	$I = 722 \text{ in}^4$	$E = 29 \times 10^6$ psi
$P = 50$ kips	$L = 15$ ft $= 180$ in.	$a = 4$ ft $= 48$ in.

STRATEGY: You should begin by determining the bending-moment equation for the portion of interest. Substituting this into the differential equation of the elastic curve, integrating twice, and applying the boundary conditions, you can then obtain the equation of the elastic curve. Use this equation to find the desired deflections.

MODELING: Using the free-body diagram of the entire beam (Fig. 1) gives the reactions: $\mathbf{R}_A = Pa/L \downarrow$ $\mathbf{R}_B = P(1 + a/L)\uparrow$. The free-body diagram of the portion of beam *AD* of length *x* (Fig. 1) gives

$$M = -P\frac{a}{L}x \qquad (0 < x < L)$$

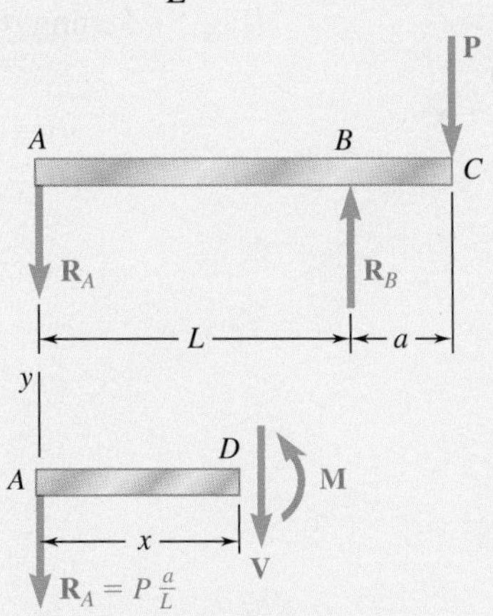

Fig. 1 Free-body diagrams of beam and portion *AD*.

ANALYSIS:

Differential Equation of the Elastic Curve. Using Eq. (15.4) gives

$$EI\frac{d^2y}{dx^2} = -P\frac{a}{L}x$$

Noting that the flexural rigidity *EI* is constant, integrate twice and find

$$EI\frac{dy}{dx} = -\frac{1}{2}P\frac{a}{L}x^2 + C_1 \qquad (1)$$

$$EI\,y = -\frac{1}{6}P\frac{a}{L}x^3 + C_1x + C_2 \qquad (2)$$

(continued)

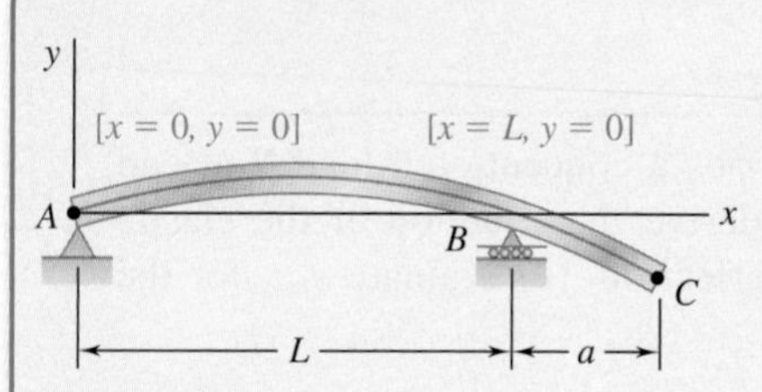

Fig. 2 Boundary conditions.

Determination of Constants. For the boundary conditions shown (Fig. 2),

$[x = 0, y = 0]$: From Eq. (2), $C_2 = 0$
$[x = L, y = 0]$: Again using Eq. (2),

$$EI(0) = -\frac{1}{6}P\frac{a}{L}L^3 + C_1L \qquad C_1 = +\frac{1}{6}PaL$$

a. Equation of the Elastic Curve. Substituting for C_1 and C_2 into Eqs. (1) and (2),

$$EI\frac{dy}{dx} = -\frac{1}{2}P\frac{a}{L}x^2 + \frac{1}{6}PaL \qquad \frac{dy}{dx} = \frac{PaL}{6EI}\left[1 - 3\left(\frac{x}{L}\right)^2\right] \quad \text{(3)}$$

$$EI\,y = -\frac{1}{6}P\frac{a}{L}x^3 + \frac{1}{6}PaLx \qquad y = \frac{PaL^2}{6EI}\left[\frac{x}{L} - \left(\frac{x}{L}\right)^3\right] \quad \text{(4)}$$

b. Maximum Deflection in Portion *AB*. The maximum deflection y_{max} occurs at point E where the slope of the elastic curve is zero (Fig. 3). Setting $dy/dx = 0$ in Eq. (3), the abscissa x_m of point E is

$$0 = \frac{PaL}{6EI}\left[1 - 3\left(\frac{x_m}{L}\right)^2\right] \qquad x_m = \frac{L}{\sqrt{3}} = 0.577L$$

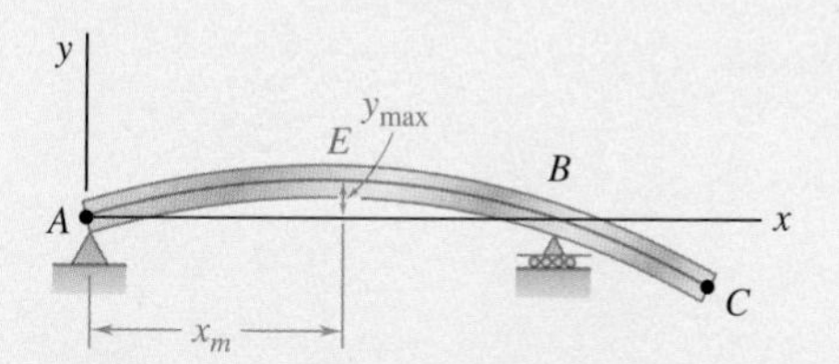

Fig. 3 Deformed elastic curve with location of maximum deflection.

Substitute $x_m/L = 0.577$ into Eq. (4):

$$y_{max} = \frac{PaL^2}{6EI}[(0.577) - (0.577)^3] \qquad y_{max} = 0.0642\frac{PaL^2}{EI}$$

c. Evaluation of y_{max}. For the data given, the value of y_{max} is

$$y_{max} = 0.0642\frac{(50\text{ kips})(48\text{ in.})(180\text{ in.})^2}{(29 \times 10^6\text{ psi})(722\text{ in}^4)} \qquad y_{max} = 0.238\text{ in.}$$

REFLECT and THINK: Because the maximum deflection is positive, it is upward. As a check, we see that this is consistent with the deflected shape anticipated for this loading (Fig. 3).

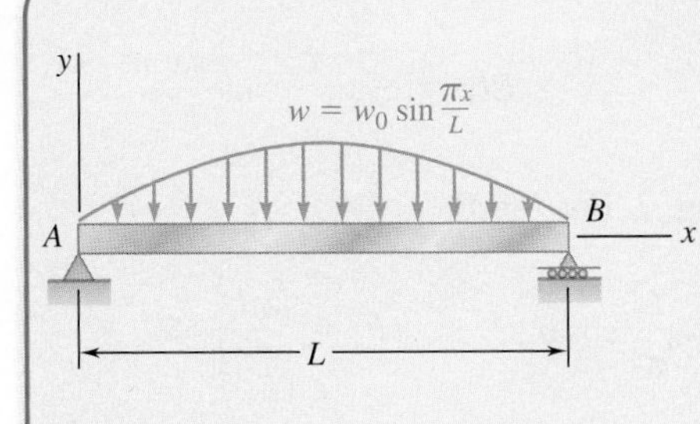

Sample Problem 15.2

For the beam and loading shown determine (*a*) the equation of the elastic curve, (*b*) the slope at end *A*, (*c*) the maximum deflection.

STRATEGY: Determine the elastic curve directly from the load distribution using Eq. (15.8), applying the appropriate boundary conditions. Use this equation to find the desired slope and deflection.

MODELING and ANALYSIS:

Differential Equation of the Elastic Curve. From Eq. (15.8),

$$EI \frac{d^4y}{dx^4} = -w(x) = -w_0 \sin \frac{\pi x}{L} \tag{1}$$

Integrate Eq. (1) twice:

$$EI \frac{d^3y}{dx^3} = V = +w_0 \frac{L}{\pi} \cos \frac{\pi x}{L} + C_1 \tag{2}$$

$$EI \frac{d^2y}{dx^2} = M = +w_0 \frac{L^2}{\pi^2} \sin \frac{\pi x}{L} + C_1 x + C_2 \tag{3}$$

Boundary Conditions: Refer to Fig. 1.

$[x = 0, M = 0]$: From Eq. (3), $C_2 = 0$

$[x = L, M = 0]$: Again using Eq. (3),

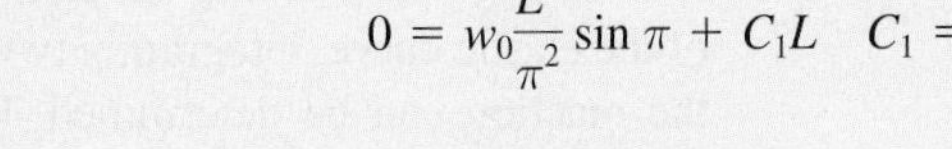

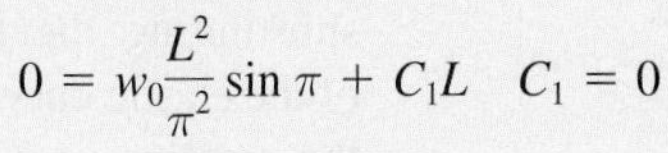

$$0 = w_0 \frac{L^2}{\pi^2} \sin \pi + C_1 L \qquad C_1 = 0$$

Thus,

$$EI \frac{d^2y}{dx^2} = +w_0 \frac{L^2}{\pi^2} \sin \frac{\pi x}{L} \tag{4}$$

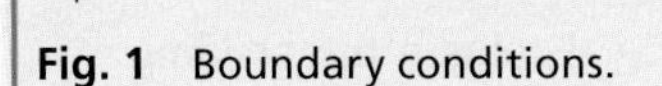

Fig. 1 Boundary conditions.

Integrate Eq. (4) twice:

$$EI \frac{dy}{dx} = EI\,\theta = -w_0 \frac{L^3}{\pi^3} \cos \frac{\pi x}{L} + C_3 \tag{5}$$

$$EI\, y = -w_0 \frac{L^4}{\pi^4} \sin \frac{\pi x}{L} + C_3 x + C_4 \tag{6}$$

Boundary Conditions: Refer to Fig. 1.

$[x = 0, y = 0]$: Using Eq. (6), $C_4 = 0$

$[x = L, y = 0]$: Again using Eq. (6), $C_3 = 0$

(continued)

a. Equation of Elastic Curve.

$$EIy = -w_0 \frac{L^4}{\pi^4} \sin \frac{\pi x}{L} \quad \blacktriangleleft$$

b. Slope at End *A*. Refer to Fig. 2. For $x = 0$,

$$EI\,\theta_A = -w_0 \frac{L^3}{\pi^3} \cos 0 \qquad \theta_A = \frac{w_0 L^3}{\pi^3 EI} \quad \blacktriangleleft$$

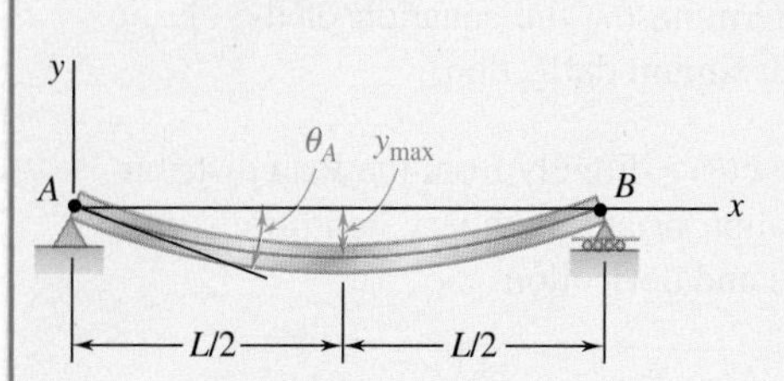

Fig. 2 Deformed elastic curve showing slope at *A* and maximum deflection.

c. Maximum Deflection. Referring to Fig. 2, for $x = \frac{1}{2}L$,

$$ELy_{max} = -w_0 \frac{L^4}{\pi^4} \sin \frac{\pi}{2} \qquad y_{max} = \frac{w_0 L^4}{\pi^4 EI} \downarrow \quad \blacktriangleleft$$

REFLECT and THINK: As a check, we observe that the directions of the slope at end *A* and the maximum deflection are consistent with the deflected shape anticipated for this loading (Fig. 1).

Sample Problem 15.3

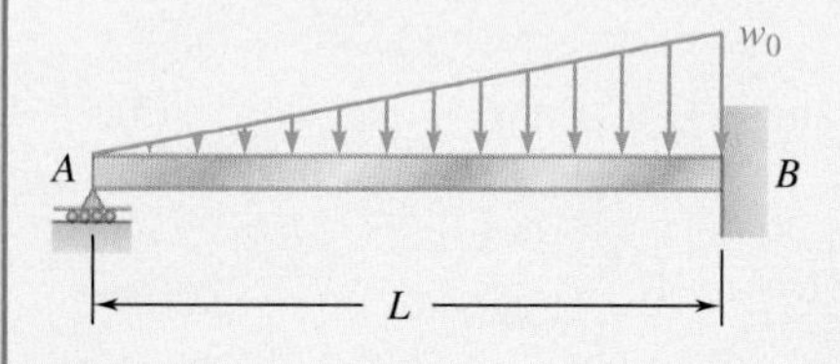

For the uniform beam *AB* (*a*) determine the reaction at *A*, (*b*) derive the equation of the elastic curve, (*c*) determine the slope at *A*. (Note that the beam is statically indeterminate to the first degree.)

STRATEGY: The beam is statically indeterminate to the first degree. Treating the reaction at *A* as the redundant, write the bending-moment equation as a function of this redundant reaction and the existing load. After substituting the bending-moment equation into the differential equation of the elastic curve, integrating twice, and applying the boundary conditions, the reaction can be determined. Use the equation for the elastic curve to find the desired slope.

MODELING: Using the free body shown in Fig. 1, obtain the bending moment diagram:

$$+\circlearrowright \Sigma M_D = 0: \qquad R_A x - \frac{1}{2}\left(\frac{w_0 x^2}{L}\right)\frac{x}{3} - M = 0 \qquad M = R_A x - \frac{w_0 x^3}{6L}$$

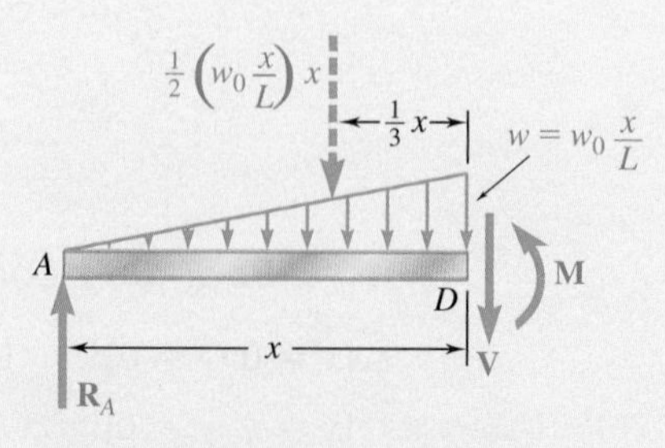

Fig. 1 Free-body diagram of portion *AD* of beam.

(continued)

ANALYSIS:

Differential Equation of the Elastic Curve. Use Eq. (15.4) for

$$EI\frac{d^2y}{dx^2} = R_A x - \frac{w_0 x^3}{6L}$$

Noting that the flexural rigidity EI is constant, integrate twice and find

$$EI\frac{dy}{dx} = EI\,\theta = \frac{1}{2}R_A x^2 - \frac{w_0 x^4}{24L} + C_1 \quad \text{(1)}$$

$$EI\,y = \frac{1}{6}R_A x^3 - \frac{w_0 x^5}{120L} + C_1 x + C_2 \quad \text{(2)}$$

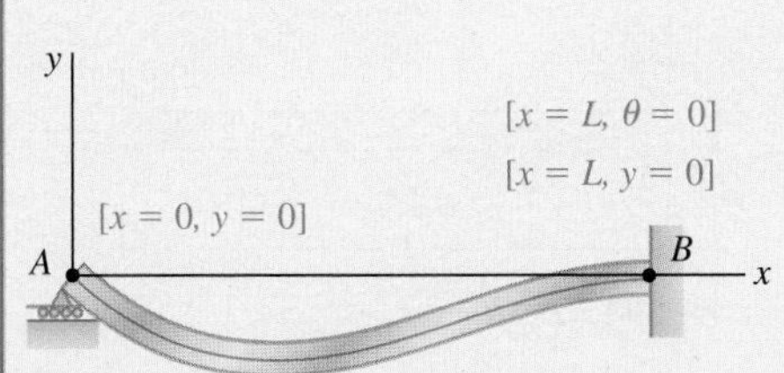

Fig. 2 Boundary conditions.

Boundary Conditions. The three boundary conditions that must be satisfied are shown in Fig. 2.

$$[x = 0, y = 0]: \quad C_2 = 0 \quad \text{(3)}$$

$$[x = L, \theta = 0]: \quad \frac{1}{2}R_A L^2 - \frac{w_0 L^3}{24} + C_1 = 0 \quad \text{(4)}$$

$$[x = L, y = 0]: \quad \frac{1}{6}R_A L^3 - \frac{w_0 L^4}{120} + C_1 L + C_2 = 0 \quad \text{(5)}$$

a. Reaction at *A*. Multiplying Eq. (4) by L, subtracting Eq. (5) member by member from the equation obtained, and noting that $C_2 = 0$, give

$$\tfrac{1}{3}R_A L^3 - \tfrac{1}{30}w_0 L^4 = 0 \qquad \mathbf{R}_A = \tfrac{1}{10}w_0 L\uparrow \quad ◀$$

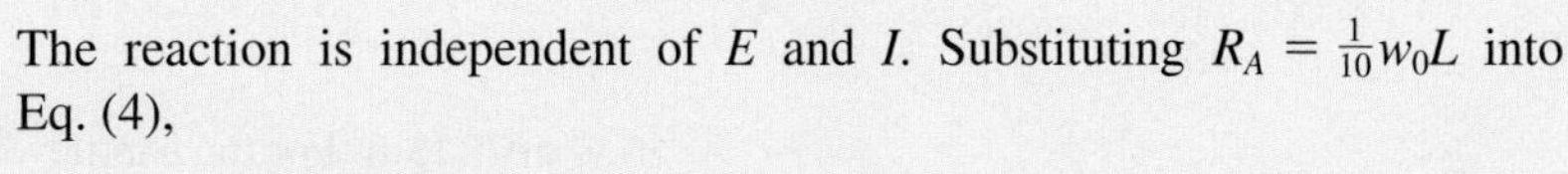

The reaction is independent of E and I. Substituting $R_A = \frac{1}{10}w_0 L$ into Eq. (4),

$$\tfrac{1}{2}(\tfrac{1}{10}w_0 L)L^2 - \tfrac{1}{24}w_0 L^3 + C_1 = 0 \qquad C_1 = -\tfrac{1}{120}w_0 L^3$$

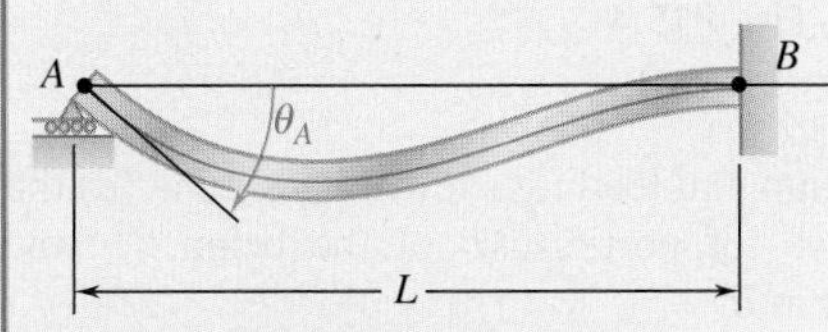

Fig. 3 Deformed elastic curve showing slope at *A*.

b. Equation of the Elastic Curve. Substituting for R_A, C_1, and C_2 into Eq. (2),

$$EI\,y = \frac{1}{6}\left(\frac{1}{10}w_0 L\right)x^3 - \frac{w_0 x^5}{120L} - \left(\frac{1}{120}w_0 L^3\right)x$$

$$y = \frac{w_0}{120EIL}(-x^5 + 2L^2x^3 - L^4x) \quad ◀$$

c. Slope at *A* (Fig. 3). Differentiate the equation of the elastic curve with respect to x:

$$\theta = \frac{dy}{dx} = \frac{w_0}{120EIL}(-5x^4 + 6L^2x^2 - L^4)$$

Making $x = 0$,

$$\theta_A = -\frac{w_0 L^3}{120EI} \qquad \theta_A = \frac{w_0 L^3}{120EI} ⦩ \quad ◀$$

Problems

In the following problems assume that the flexural rigidity *EI* of each beam is constant.

15.1 through 15.4 For the loading shown, determine (*a*) the equation of the elastic curve for the cantilever beam *AB*, (*b*) the deflection at the free end, (*c*) the slope at the free end.

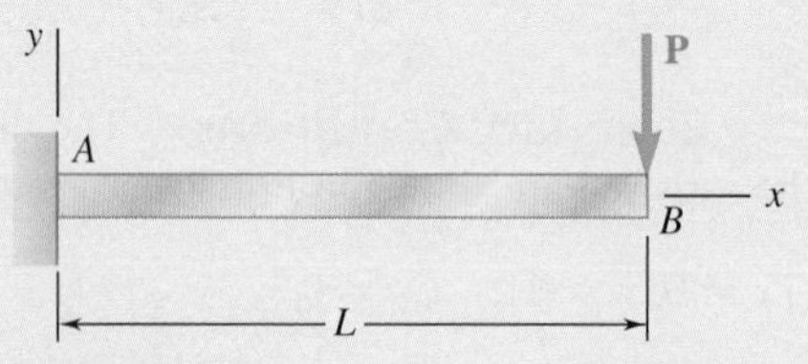

Fig. P15.1

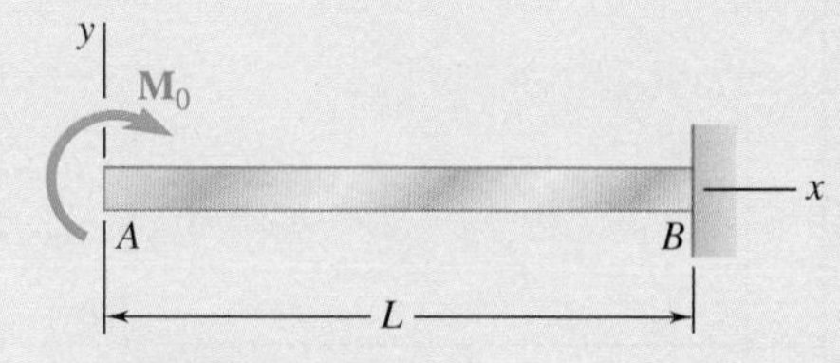

Fig. P15.2

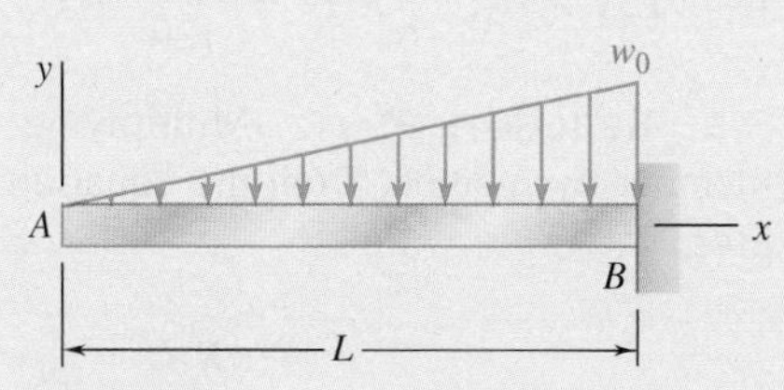

Fig. P15.3

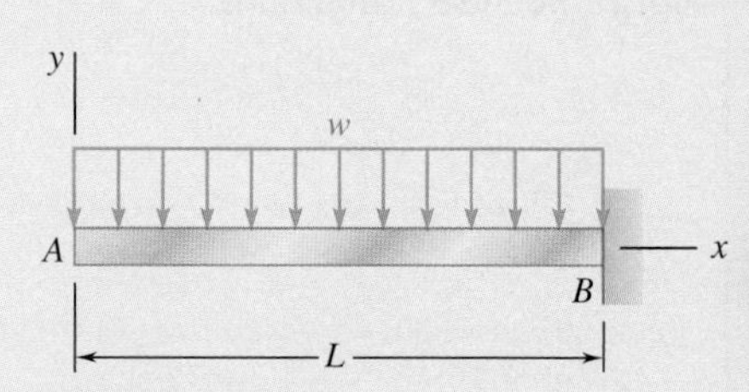

Fig. P15.4

15.5 and 15.6 For the cantilever beam and loading shown, determine (*a*) the equation of the elastic curve for portion *AB* of the beam, (*b*) the deflection at *B*, (*c*) the slope at *B*.

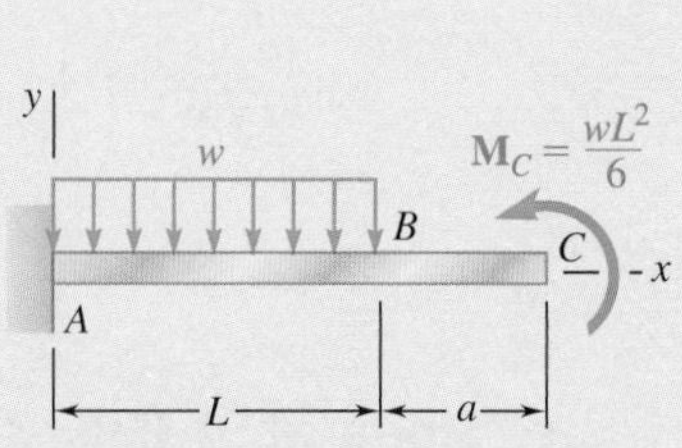

Fig. *P15.5*

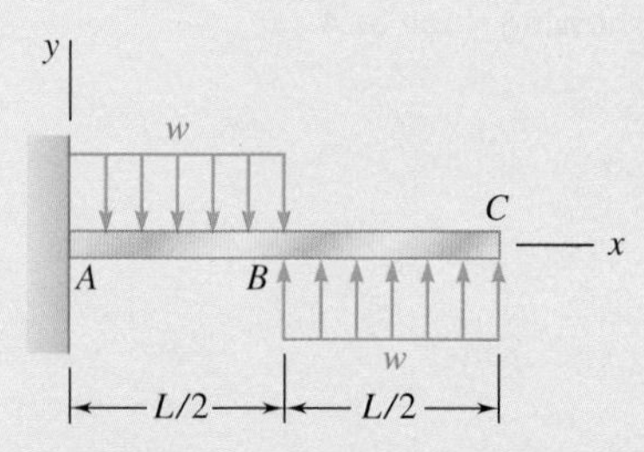

Fig. P15.6

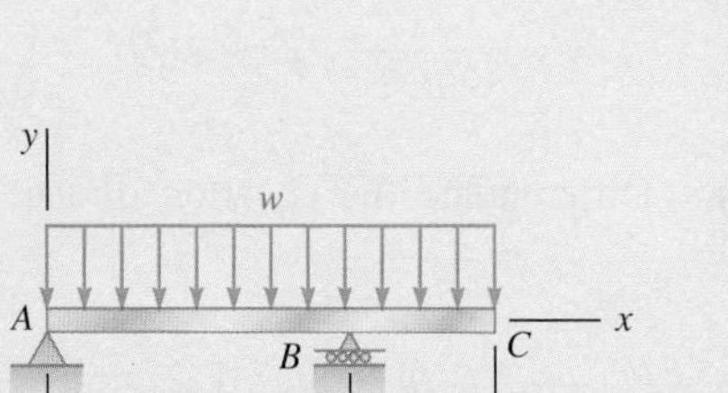

Fig. P15.7

15.7 For the beam and loading shown, determine (*a*) the equation of the elastic curve for portion *AB* of the beam, (*b*) the slope at *A*, (*c*) the slope at *B*.

15.8 For the beam and loading shown, determine (*a*) the equation of the elastic curve for portion *AB* of the beam, (*b*) the deflection at mid-span, (*c*) the slope at *B*.

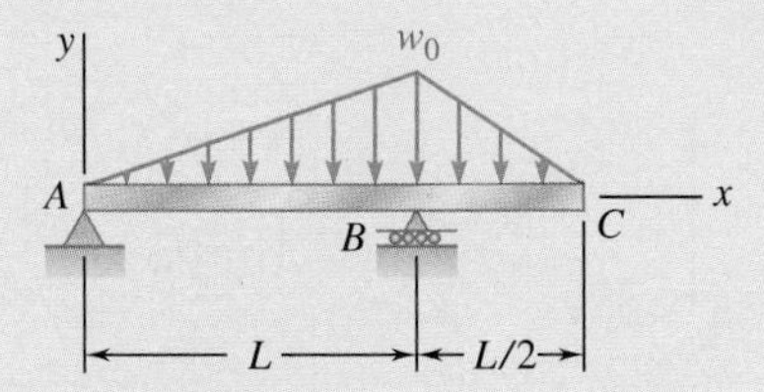

Fig. ***P15.8***

15.9 Knowing that beam *AB* is an S200 × 34 rolled shape and that $P = 60$ kN, $L = 2$ m, and $E = 200$ GPa, determine (*a*) the slope at *A*, (*b*) the deflection at *C*.

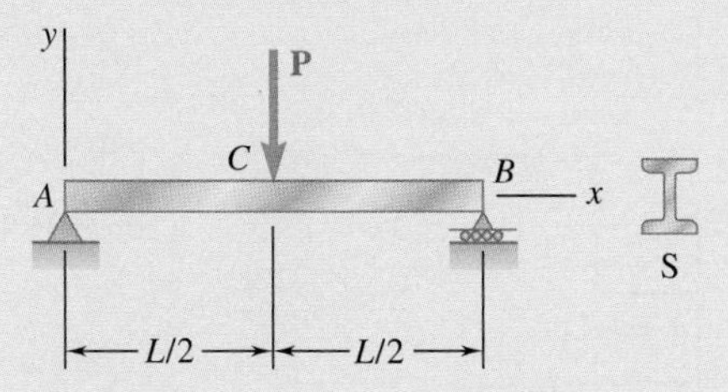

Fig. P15.9

15.10 Knowing that beam *AB* is a W10 × 33 rolled shape and that $w_0 = 3$ kips/ft, $L = 12$ ft, and $E = 29 \times 10^6$ psi, determine (*a*) the slope at *A*, (*b*) the deflection at *C*.

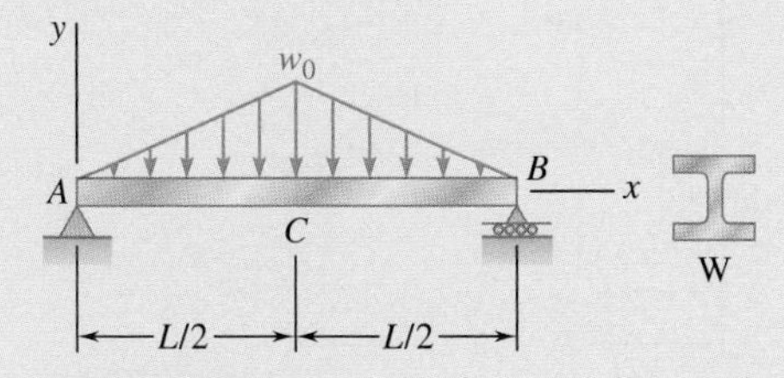

Fig. P15.10

15.11 (*a*) Determine the location and magnitude of the maximum deflection of beam *AB*. (*b*) Assuming that beam *AB* is a W360 × 64 rolled shape, $L = 3.5$ m, and $E = 200$ GPa, calculate the maximum allowable value of the applied moment $\mathbf{M}_0$ if the maximum deflection is not to exceed 1 mm.

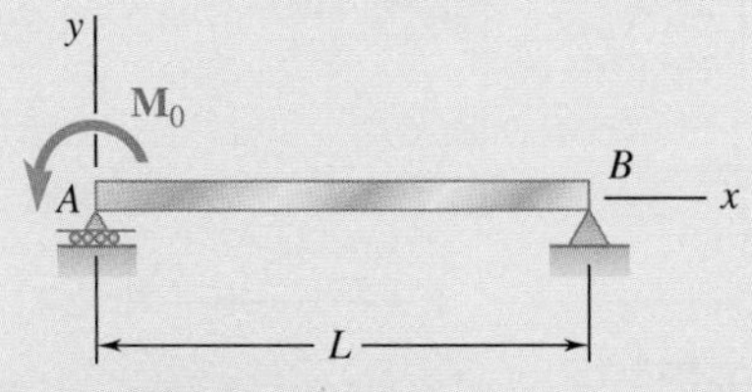

Fig. P15.11

15.12 For the beam and loading shown, (*a*) express the magnitude and location of the maximum deflection in terms of w_0, L, E, and I. (*b*) Calculate the value of the maximum deflection, assuming that beam AB is a W18 × 50 rolled shape and that $w_0 = 4.5$ kips/ft, $L = 18$ ft, and $E = 29 \times 10^6$ psi.

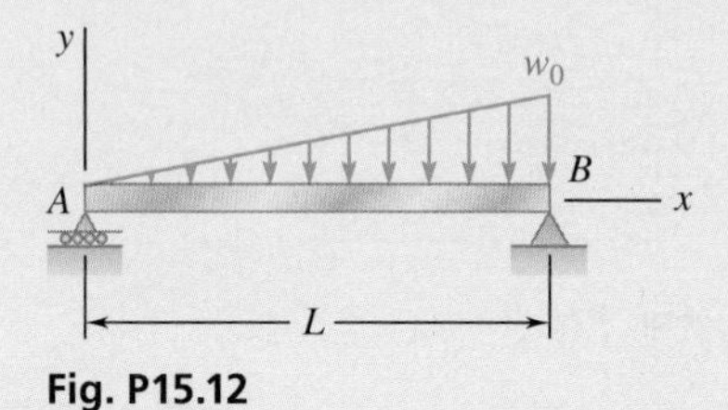

Fig. P15.12

15.13 For the beam and loading shown, determine the deflection at point C. Use $E = 200$ GPa.

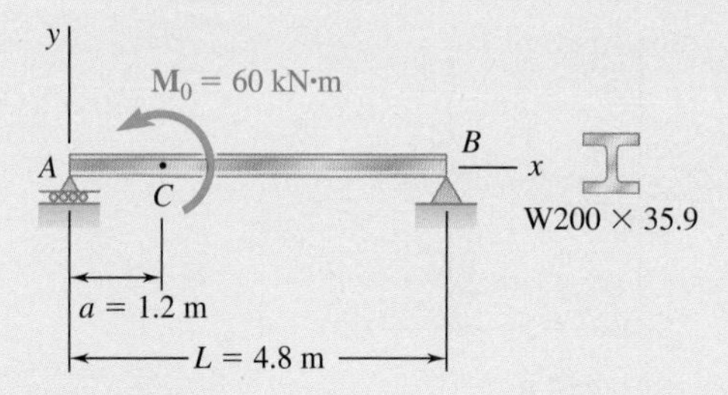

Fig. P15.13

15.14 For the beam and loading shown, determine the deflection at point C. Use $E = 29 \times 10^6$ psi.

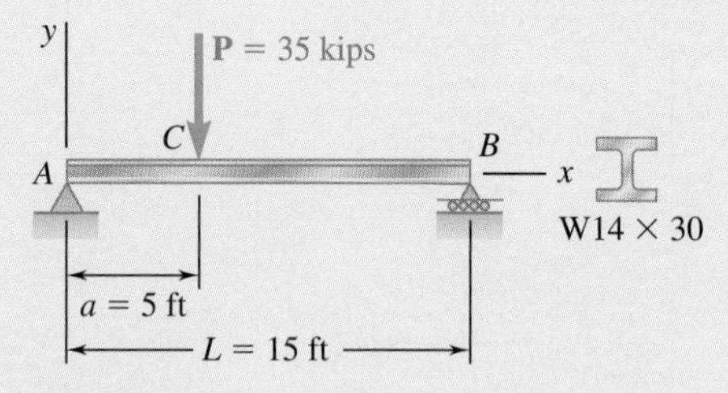

Fig. P15.14

15.15 For the beam and loading shown, determine (*a*) the equation of the elastic curve, (*b*) the slope at end A, (*c*) the deflection at the midpoint of the span.

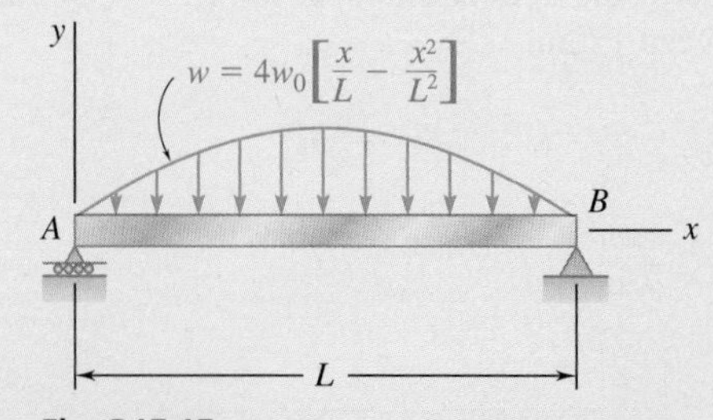

Fig. P15.15

15.16 For the beam and loading shown, determine (*a*) the equation of the elastic curve, (*b*) the deflection at the free end.

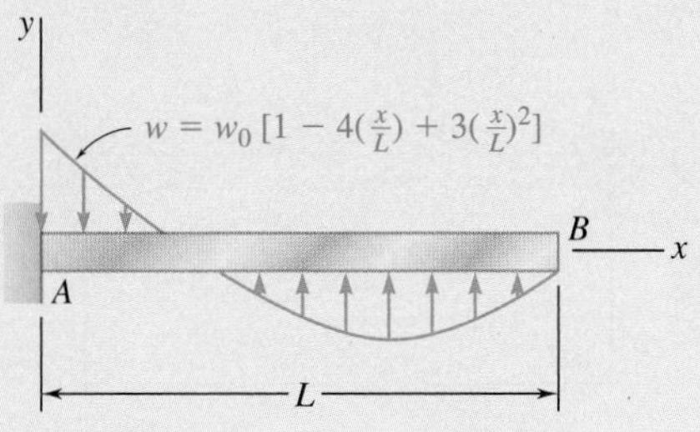

Fig. *P15.16*

15.17 through 15.20 For the beam and loading shown, determine the reaction at the roller support.

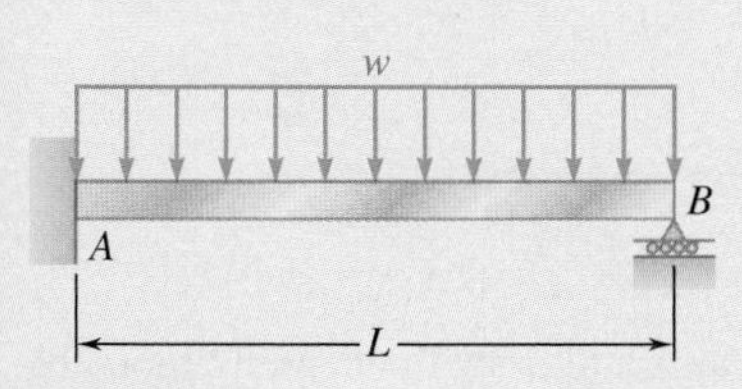

Fig. P15.17

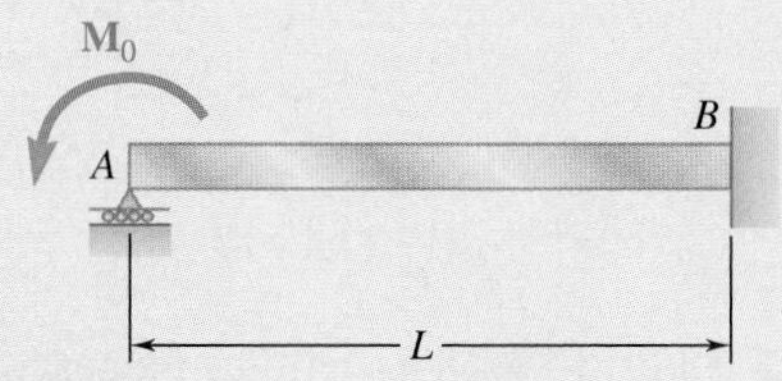

Fig. P15.18

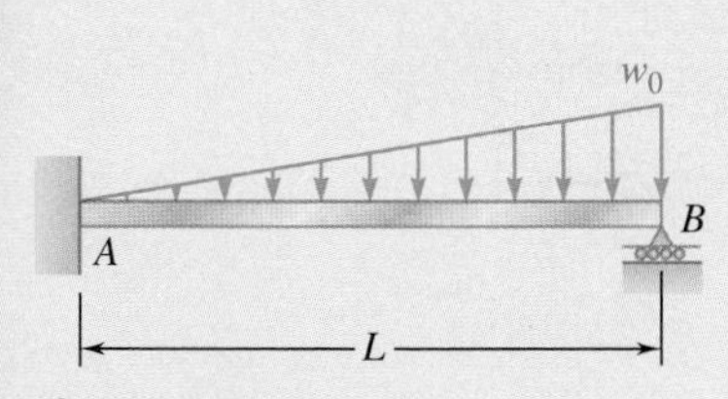

Fig. *P15.19*

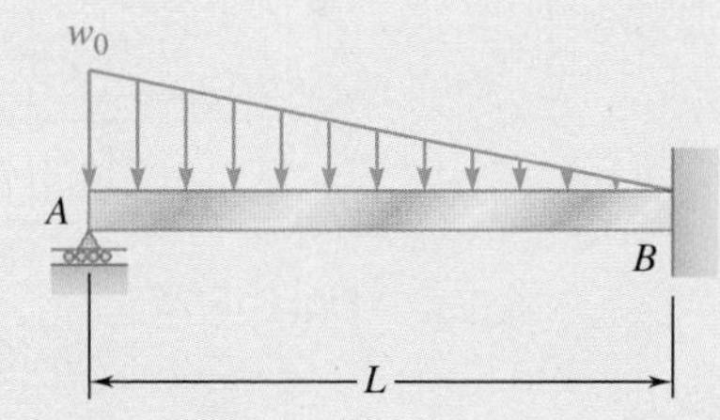

Fig. P15.20

15.21 and 15.22 Determine the reaction at the roller support, and draw the bending moment diagram for the beam and loading shown.

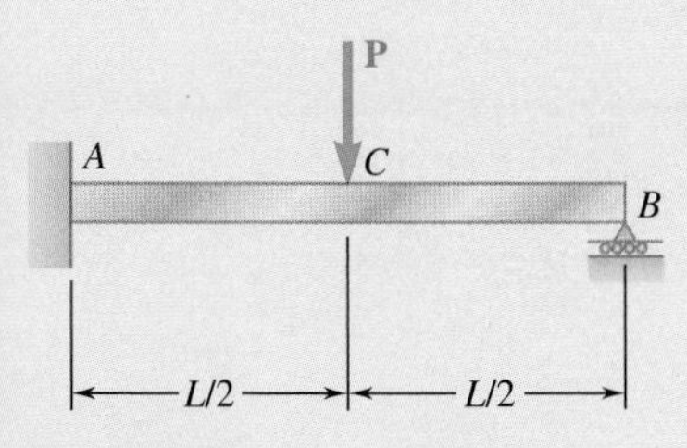

Fig. P15.21

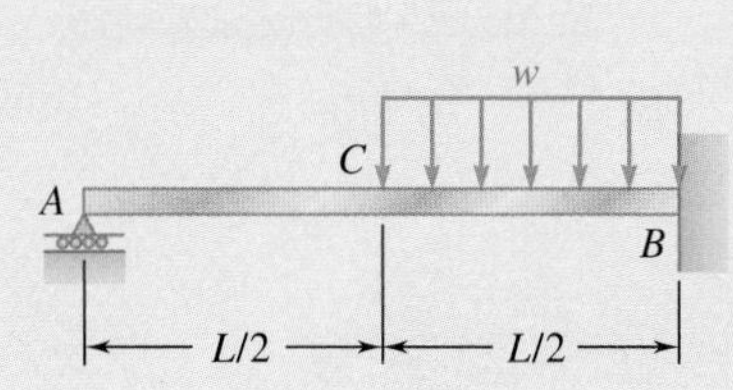

Fig. P15.22

15.23 and *15.24* Determine the reaction at the roller support and the deflection at point D if a is equal to $L/3$.

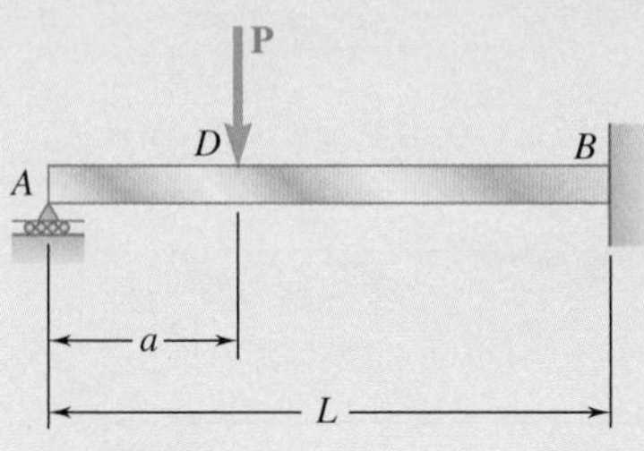

Fig. P15.23

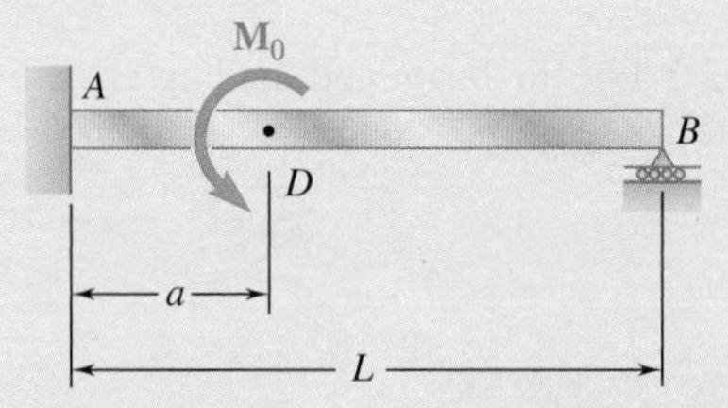

Fig. *P15.24*

15.25 and 15.26 Determine the reaction at A and draw the bending moment diagram for the beam and loading shown.

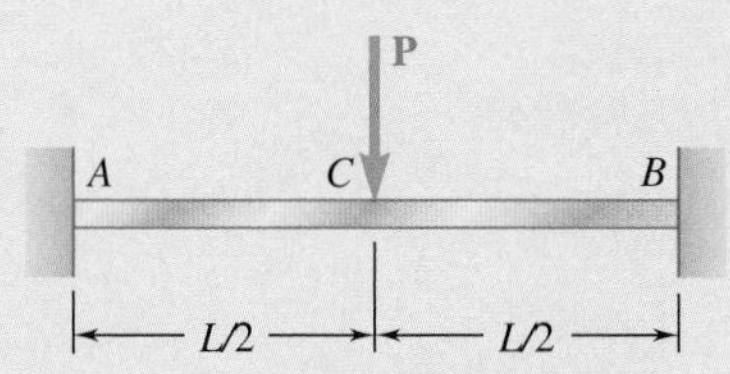

Fig. P15.25

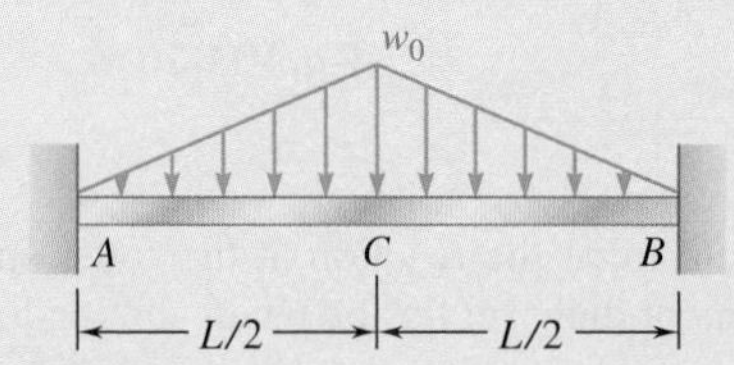

Fig. P15.26

15.3 METHOD OF SUPERPOSITION

15.3A Statically Determinate Beams

When a beam is subjected to several concentrated or distributed loads, it is convenient to compute separately the slope and deflection caused by each of the given loads. The slope and deflection due to the combined loads are obtained by applying the principle of superposition (Sec. 9.5) and adding the values of the slope or deflection corresponding to the various loads.

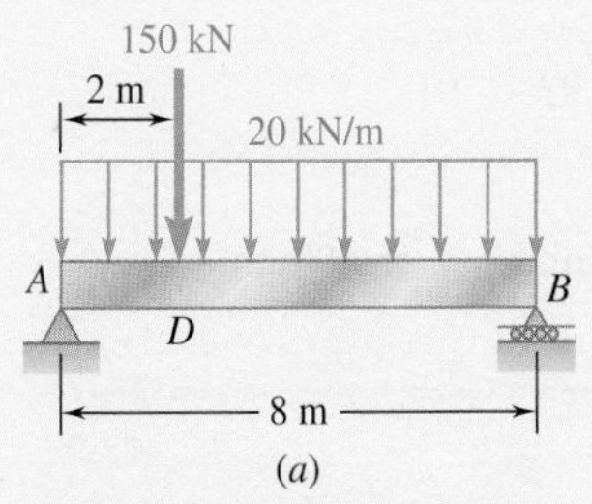

Fig. 15.18 (a) Simply supported beam having distributed and concentrated loads.

Concept Application 15.6

Determine the slope and deflection at D for the beam and loading shown (Fig. 15.18*a*), knowing that the flexural rigidity of the beam is $EI = 100\ \text{MN}\cdot\text{m}^2$.

The slope and deflection at any point of the beam can be obtained by superposing the slopes and deflections caused by the concentrated load and by the distributed load (Fig. 15.18*b*).

Since the concentrated load in Fig. 15.18*c* is applied at quarter span, the results for the beam and loading of Concept Application 15.3 can be used to write

$$(\theta_D)_P = -\frac{PL^2}{32EI} = -\frac{(150 \times 10^3)(8)^2}{32(100 \times 10^6)} = -3 \times 10^{-3}\ \text{rad}$$

$$(y_D)_P = -\frac{3PL^3}{256EI} = -\frac{3(150 \times 10^3)(8)^3}{256(100 \times 10^6)} = -9 \times 10^{-3}\ \text{m}$$

$$= -9\ \text{mm}$$

On the other hand, recalling the equation of the elastic curve obtained for a uniformly distributed load in Concept Application 15.2, the deflection in Fig. 15.18*d* is

$$y = \frac{w}{24EI}(-x^4 + 2Lx^3 - L^3x) \qquad (1)$$

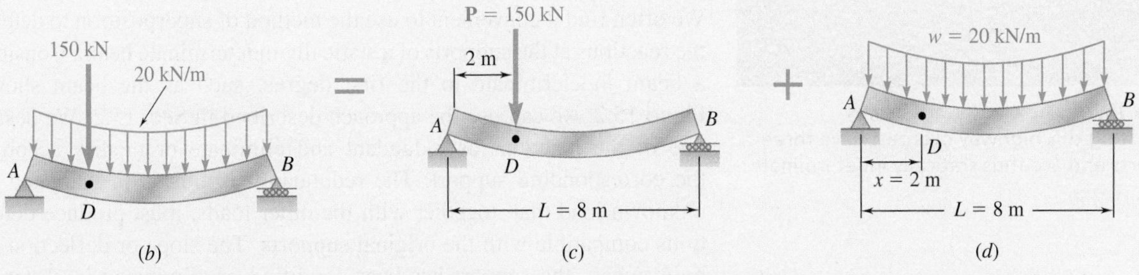

Fig. 15.18 (b) The beam's loading can be obtained by superposing deflections due to (c) the concentrated load and (d) the distributed load.

Differentiating with respect to x gives

$$\theta = \frac{dy}{dx} = \frac{w}{24EI}(-4x^3 + 6Lx^2 - L^3) \qquad (2)$$

(continued)

Making $w = 20$ kN/m, $x = 2$ m, and $L = 8$ m in Eqs. (1) and (2), we obtain

$$(\theta_D)_w = \frac{20 \times 10^3}{24(100 \times 10^6)}(-352) = -2.93 \times 10^{-3} \text{ rad}$$

$$(y_D)_w = \frac{20 \times 10^3}{24(100 \times 10^6)}(-912) = -7.60 \times 10^{-3} \text{ m}$$

$$= -7.60 \text{ mm}$$

Combining the slopes and deflections produced by the concentrated and the distributed loads,

$$\theta_D = (\theta_D)_P + (\theta_D)_w = -3 \times 10^{-3} - 2.93 \times 10^{-3}$$

$$= -5.93 \times 10^{-3} \text{ rad}$$

$$y_D = (y_D)_P + (y_D)_w = -9 \text{ mm} - 7.60 \text{ mm} = -16.60 \text{ mm}$$

To facilitate the work of practicing engineers, most structural and mechanical engineering handbooks include tables giving the deflections and slopes of beams for various loadings and types of support. Such a table is found in Appendix C. The slope and deflection of the beam of Fig. 15.18*a* could have been determined from that table. Indeed, using the information given under cases 5 and 6, we could have expressed the deflection of the beam for any value $x \leq L/4$. Taking the derivative of the expression obtained in this way would have yielded the slope of the beam over the same interval. We also note that the slope at both ends of the beam can be obtained by simply adding the corresponding values given in the table. However, the maximum deflection of the beam of Fig. 15.18*a* *cannot* be obtained by adding the maximum deflections of cases 5 and 6, since these deflections occur at different points of the beam.[†]

Photo 15.2 The continuous beams supporting this highway overpass have three supports and are thus statically indeterminate.

© John DeWolf

15.3B Statically Indeterminate Beams

We often find it convenient to use the method of superposition to determine the reactions at the supports of a statically indeterminate beam. Considering a beam indeterminate to the first degree, such as the beam shown in Photo 15.2, we can use the approach described in Sec. 15.2. We designate one of the reactions as redundant and eliminate or modify accordingly the corresponding support. The redundant reaction is then treated as an unknown load that, together with the other loads, must produce deformations compatible with the original supports. The slope or deflection at the point where the support has been modified or eliminated is obtained by computing the deformations caused by both the given loads and the redundant reaction and by superposing the results. Once the reactions at the supports are found, the slope and deflection can be determined.

[†]An approximate value of the maximum deflection of the beam can be obtained by plotting the values of y corresponding to various values of x. The determination of the exact location and magnitude of the maximum deflection would require setting equal to zero the expression obtained for the slope of the beam and solving this equation for x.

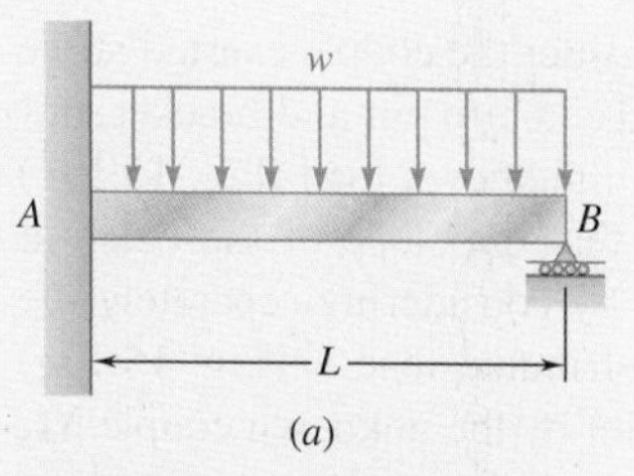

Fig. 15.19 (a) Statically indeterminate beam with a uniformly distributed load.

Concept Application 15.7

Determine the reactions at the supports for the prismatic beam and loading shown in Fig. 15.19*a*. (This is the same beam and loading as in Concept Application 15.5.)

We consider the reaction at B as redundant and release the beam from the support. The reaction $\mathbf{R}_B$ is now considered as an unknown load (Fig. 15.19*b*) and will be determined from the condition that the deflection of the beam at B must be zero. The solution is carried out by considering separately the deflection $(y_B)_w$ caused at B by the uniformly distributed load w (Fig. 15.19*c*) and the deflection $(y_B)_R$ produced at the same point by the redundant reaction $\mathbf{R}_B$ (Fig. 15.19*d*).

From the table of Appendix C (cases 2 and 1),

$$(y_B)_w = -\frac{wL^4}{8EI} \qquad (y_B)_R = +\frac{R_BL^3}{3EI}$$

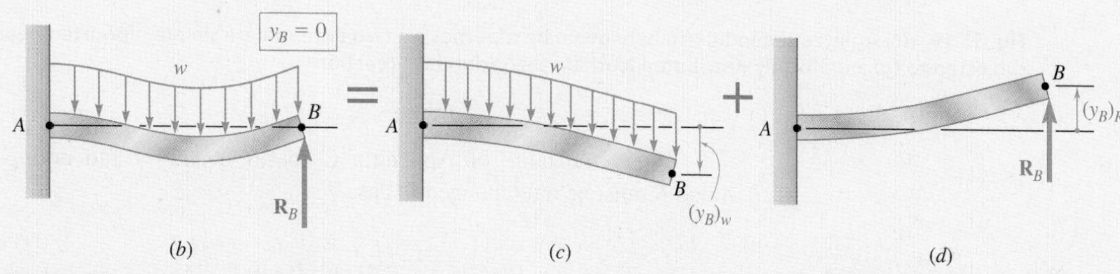

Fig. 15.19 (b) Analyze the indeterminate beam by superposing two determinate cantilever beams, subjected to (c) a uniformly distributed load, (d) the redundant reaction.

Writing that the deflection at B is the sum of these two quantities and that it must be zero,

$$y_B = (y_B)_w + (y_B)_R = 0$$

$$y_B = -\frac{wL^4}{8EI} + \frac{R_BL^3}{3EI} = 0$$

and, solving for R_B, $\quad R_B = \frac{3}{8}wL \quad \mathbf{R}_B = \frac{3}{8}wL\uparrow$

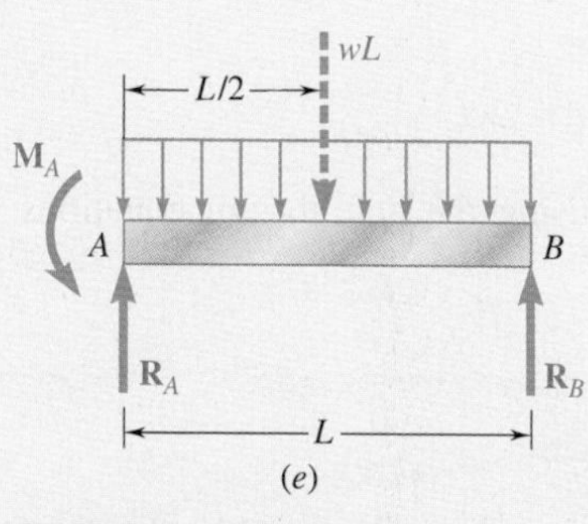

Fig. 15.19 (e) Free-body diagram of indeterminate beam.

Drawing the free-body diagram of the beam (Fig. 15.19*e*) and writing the corresponding equilibrium equations,

$+\uparrow\Sigma F_y = 0$: $$R_A + R_B - wL = 0 \quad (1)$$

$$R_A = wL - R_B = wL - \tfrac{3}{8}wL = \tfrac{5}{8}wL$$

$$\mathbf{R}_A = \tfrac{5}{8}wL\uparrow$$

$+\circlearrowleft\Sigma M_A = 0$: $$M_A + R_BL - (wL)(\tfrac{1}{2}L) = 0 \quad (2)$$

$$M_A = \tfrac{1}{2}wL^2 - R_BL = \tfrac{1}{2}wL^2 - \tfrac{3}{8}wL^2 = \tfrac{1}{8}wL^2$$

$$\mathbf{M}_A = \tfrac{1}{8}wL^2\circlearrowleft$$

(continued)

Alternative Solution. We may consider the couple exerted at the fixed end A as redundant and replace the fixed end by a pin-and-bracket support. The couple $\mathbf{M}_A$ is now considered as an unknown load (Fig. 15.19*f*) and will be determined from the condition that the slope of the beam at A must be zero. The solution is carried out by considering separately the slope $(\theta_A)_w$ caused at A by the uniformly distributed load w (Fig. 15.19*g*) and the slope $(\theta_A)_M$ produced at the same point by the unknown couple $\mathbf{M}_A$ (Fig 15.19*h*).

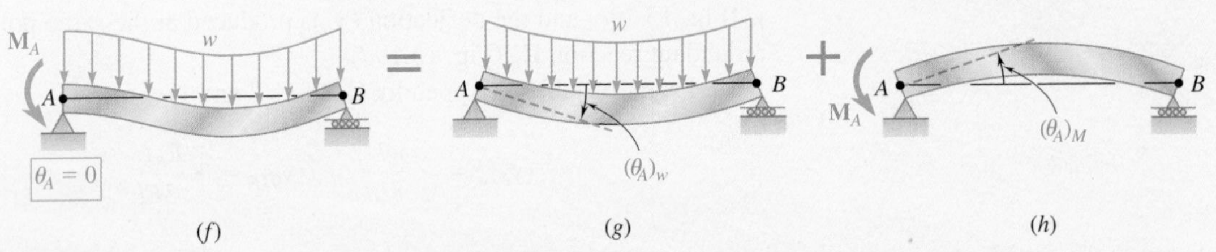

Fig. 15.19 (*f*) Analyze the indeterminate beam by superposing two determinate simply supported beams, subjected to (*g*) a uniformly distributed load, (*h*) the redundant reaction.

Using the table of Appendix C (cases 6 and 7) and noting that A and B must be interchanged in case 7,

$$(\theta_A)_w = -\frac{wL^3}{24EI} \qquad (\theta_A)_M = \frac{M_A L}{3EI}$$

Writing that the slope at A is the sum of these two quantities and that it must be zero gives

$$\theta_A = (\theta_A)_w + (\theta_A)_M = 0$$

$$\theta_A = -\frac{wL^3}{25EI} + \frac{M_A L}{3EI} = 0$$

where M_A is

$$M_A = \tfrac{1}{8}wL^2 \qquad \mathbf{M}_A = \tfrac{1}{8}wL^2 \circlearrowleft$$

The values of R_A and R_B are found by using the equilibrium equations (1) and (2).

The beam considered in the preceding Concept Application was indeterminate to the first degree. In the case of a beam indeterminate to the second degree (see Sec. 15.2), two reactions must be designated as redundant, and the corresponding supports must be eliminated or modified accordingly. The redundant reactions are then treated as unknown loads that, simultaneously and together with the other loads, must produce deformations that are compatible with the original supports. (See Sample Prob. 15.6.)

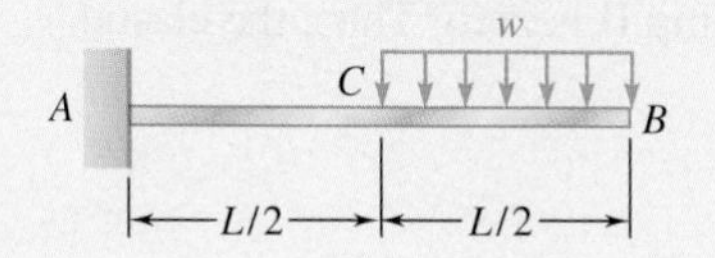

Sample Problem 15.4

For the beam and loading shown, determine the slope and deflection at point B.

STRATEGY: Using the method of superposition, you can model the given problem using a summation of beam load cases for which deflection formulae are readily available.

MODELING: Through the principle of superposition, the given loading can be obtained by superposing the loadings shown in the following picture equation of Fig. 1. The beam AB is the same in each part of the figure.

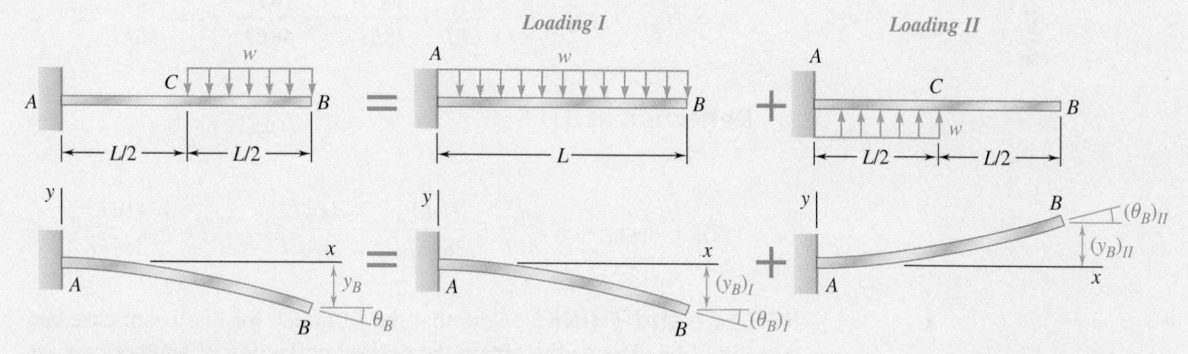

Fig. 1 Actual loading is equivalent to the superposition of two distributed loads.

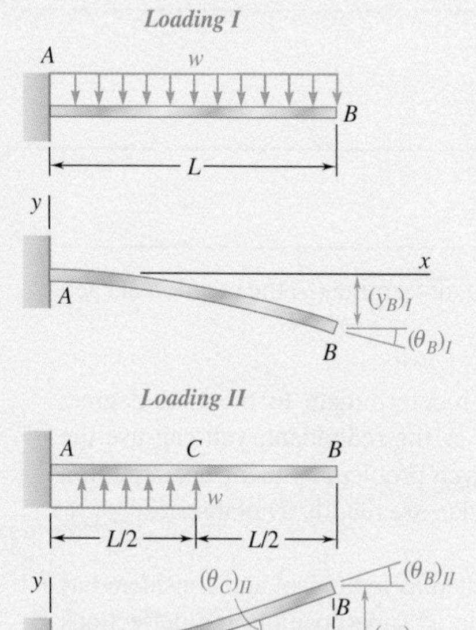

Fig. 2 Deformation details of the superposed loadings I and II.

ANALYSIS: For each of the loadings I and II (detailed further in Fig. 2), determine the slope and deflection at B by using the table of *Beam Deflections and Slopes* in Appendix C.

Loading I

$$(\theta_B)_{\text{I}} = -\frac{wL^3}{6EI} \qquad (y_B)_{\text{I}} = -\frac{wL^4}{8EI}$$

Loading II

$$(\theta_C)_{\text{II}} = +\frac{w(L/2)^3}{6EI} = +\frac{wL^3}{48EI} \qquad (y_C)_{\text{II}} = +\frac{w(L/2)^4}{8EI} = +\frac{wL^4}{128EI}$$

(continued)

In portion *CB*, the bending moment for loading II is zero. Thus, the elastic curve is a straight line.

$$(\theta_B)_{II} = (\theta_C)_{II} = +\frac{wL^3}{48EI} \qquad (y_B)_{II} = (y_C)_{II} + (\theta_C)_{II}\left(\frac{L}{2}\right)$$

$$= \frac{wL^4}{128EI} + \frac{wL^3}{48EI}\left(\frac{L}{2}\right) = +\frac{7wL^4}{384EI}$$

Slope at Point *B*

$$\theta_B = (\theta_B)_I + (\theta_B)_{II} = -\frac{wL^3}{6EI} + \frac{wL^3}{48EI} = -\frac{7wL^3}{48EI} \qquad \theta_B = \frac{7wL^3}{48EI} \ \measuredangle \ \blacktriangleleft$$

Deflection at *B*

$$y_B = (y_B)_I + (y_B)_{II} = -\frac{wL^4}{8EI} + \frac{7wL^4}{384EI} = -\frac{41wL^4}{384EI} \qquad y_B = \frac{41wL^4}{384EI} \downarrow \ \blacktriangleleft$$

REFLECT and THINK: Note that the formulae for one beam case can sometimes be extended to obtain the desired deflection of another case, as you saw here for loading II.

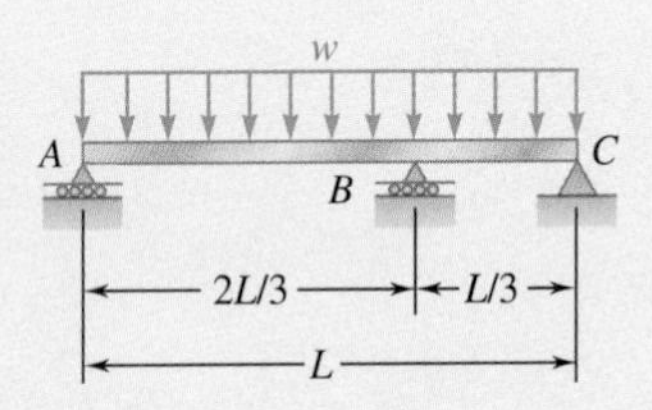

Sample Problem 15.5

For the uniform beam and loading shown, determine (*a*) the reaction at each support, (*b*) the slope at end *A*.

STRATEGY: The beam is statically indeterminate to the first degree. Strategically selecting the reaction at *B* as the redundant, you can use the method of superposition to model the given problem by using a summation of load cases for which deflection formulae are readily available.

MODELING: The reaction $\mathbf{R}_B$ is selected as redundant and considered as an unknown load. Applying the principle of superposition, the deflections due to the distributed load and to the reaction $\mathbf{R}_B$ are considered separately as shown in Fig. 1.

(continued)

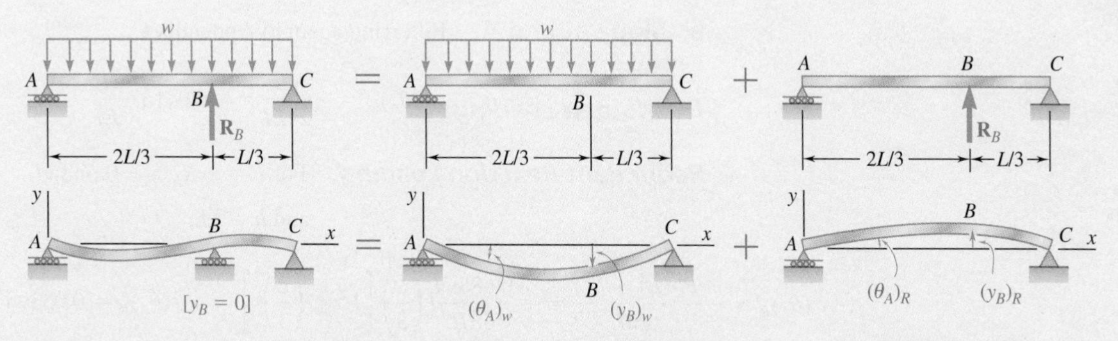

Fig. 1 Indeterminate beam modeled as superposition of two determinate simply supported beams with reaction at *B* chosen as redundant.

ANALYSIS: For each loading case, the deflection at point *B* is found by using the table of *Beam Deflections and Slopes* in Appendix C.

Distributed Loading. Use case 6, Appendix C:

$$y = -\frac{w}{24EI}(x^4 - 2Lx^3 + L^3x)$$

At point B, $x = \frac{2}{3}L$:

$$(y_B)_w = -\frac{w}{24EI}\left[\left(\frac{2}{3}L\right)^4 - 2L\left(\frac{2}{3}L\right)^3 + L^3\left(\frac{2}{3}L\right)\right] = -0.01132\frac{wL^4}{EI}$$

Redundant Reaction Loading. From case 5, Appendix C, with $a = \frac{2}{3}L$ and $b = \frac{1}{3}L$,

$$(y_B)_R = -\frac{Pa^2b^2}{3EIL} = +\frac{R_B}{3EIL}\left(\frac{2}{3}L\right)^2\left(\frac{L}{3}\right)^2 = 0.01646\frac{R_BL^3}{EI}$$

a. Reactions at Supports. Recalling that $y_B = 0$,

$$y_B = (y_B)_w + (y_B)_R$$

$$0 = -0.01132\frac{wL^4}{EI} + 0.01646\frac{R_BL^3}{EI} \qquad \mathbf{R}_B = 0.688wL\uparrow \blacktriangleleft$$

Since the reaction R_B is now known, use the methods of statics to determine the other reactions (Fig. 2):

$$\mathbf{R}_A = 0.271wL\uparrow \quad \mathbf{R}_C = 0.0413wL\uparrow \blacktriangleleft$$

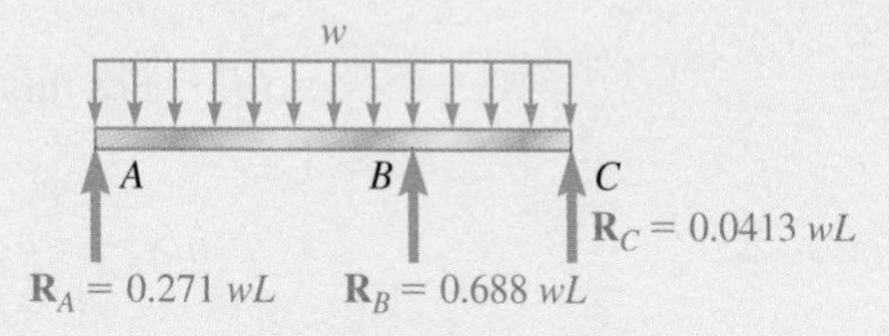

Fig. 2 Free-body diagram of beam with calculated reactions.

(continued)

b. Slope at End *A*. Referring again to Appendix C,

Distributed Loading. $(\theta_A)_w = -\frac{wL^3}{24EI} = -0.04167\frac{wL^3}{EI}$

Redundant Reaction Loading. For $P = -R_B = -0.688wL$ and $b = \frac{1}{3}L$,

$$(\theta_A)_R = -\frac{Pb(L^2 - b^2)}{6EIL} = +\frac{0.688wL}{6EIL}\left(\frac{L}{3}\right)\left[L^2 - \left(\frac{L}{3}\right)^2\right] \qquad (\theta_A)_R = 0.03398\frac{wL^3}{EI}$$

Finally, $\theta_A = (\theta_A)_w + (\theta_A)_R$

$$\theta_A = -0.04167\frac{wL^3}{EI} + 0.03398\frac{wL^3}{EI} = -0.00769\frac{wL^3}{EI}$$

$$\theta_A = 0.00769\frac{wL^3}{EI} ◄$$

Sample Problem 15.6

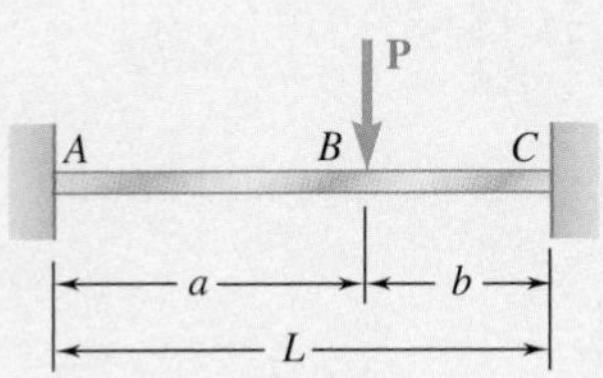

For the beam and loading shown, determine the reaction at the fixed support *C*.

STRATEGY: The beam is statically indeterminate to the second degree. Strategically selecting the reactions at *C* as redundants, you can use the method of superposition and model the given problem by using a summation of load cases for which deflection formulae are readily available.

MODELING: Assuming the axial force in the beam to be zero, the beam *ABC* is indeterminate to the second degree, and we choose two reaction components as redundants: the vertical force $\mathbf{R}_C$ and the couple $\mathbf{M}_C$. The deformations caused by the given load **P**, the force $\mathbf{R}_C$, and the couple $\mathbf{M}_C$ are considered separately, as shown in Fig. 1.

ANALYSIS: For each load, the slope and deflection at point *C* is found by using the table of *Beam Deflections and Slopes* in Appendix C.

Load P. For this load, portion *BC* of the beam is straight.

$$(\theta_C)_P = (\theta_B)_P = -\frac{Pa^2}{2EI} \qquad (y_C)_P = (y_B)_P + (\theta_B)_P b$$

$$= -\frac{Pa^3}{3EI} - \frac{Pa^2}{2EI}b = -\frac{Pa^2}{6EI}(2a + 3b)$$

(continued)

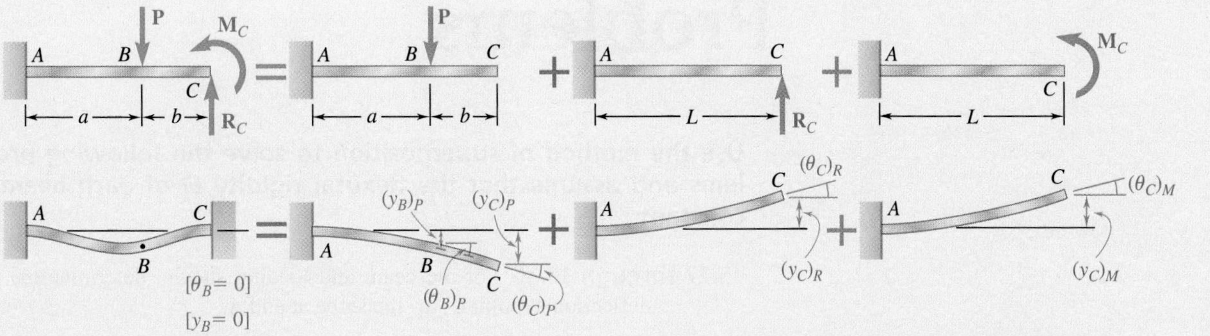

Fig. 1 Indeterminate beam modeled as the superposition of three determinate cases, including one for each of the two redundant reactions.

Force R_C $\quad (\theta_C)_R = +\dfrac{R_C L^2}{2EI} \qquad (y_C)_R = +\dfrac{R_C L^3}{3EI}$

Couple M_C $\quad (\theta_C)_M = +\dfrac{M_C L}{EI} \qquad (y_C)_M = +\dfrac{M_C L^2}{2EI}$

Boundary Conditions. At end C, the slope and deflection must be zero:

$$[x = L, \theta_C = 0]: \quad \theta_C = (\theta_C)_P + (\theta_C)_R + (\theta_C)_M$$

$$0 = -\frac{Pa^2}{2EI} + \frac{R_C L^2}{2EI} + \frac{M_C L}{EI} \tag{1}$$

$$[x = L, y_C = 0]: \quad y_C = (y_C)_P + (y_C)_R + (y_C)_M$$

$$0 = -\frac{Pa^2}{6EI}(2a + 3b) + \frac{R_C L^3}{3EI} + \frac{M_C L^2}{2EI} \tag{2}$$

Reaction Components at C. Solve Eqs. (1) and (2) simultaneously:

$$R_C = +\frac{Pa^2}{L^3}(a + 3b) \qquad \mathbf{R}_C = \frac{Pa^2}{L^3}(a + 3b) \uparrow ◀$$

$$M_C = -\frac{Pa^2 b}{L^2} \qquad \mathbf{M}_C = \frac{Pa^2 b}{L^2} \circlearrowright ◀$$

The methods of statics are used to determine the reaction at A, shown in Fig. 2.

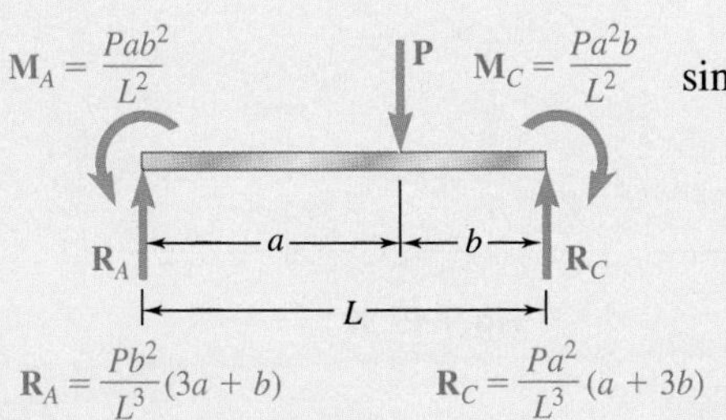

Fig. 2 Free-body diagram showing the reaction results.

REFLECT and THINK: Note that an alternate strategy that could have been used in this particular problem is to treat the couple reactions at the ends as redundant. The application of superposition would then have involved a simply-supported beam, for which deflection formulae are also readily available.

Problems

Use the method of superposition to solve the following problems and assume that the flexural rigidity *EI* of each beam is constant.

15.27 through 15.30 For the beam and loading shown, determine (*a*) the deflection at point *C*, (*b*) the slope at end *A*.

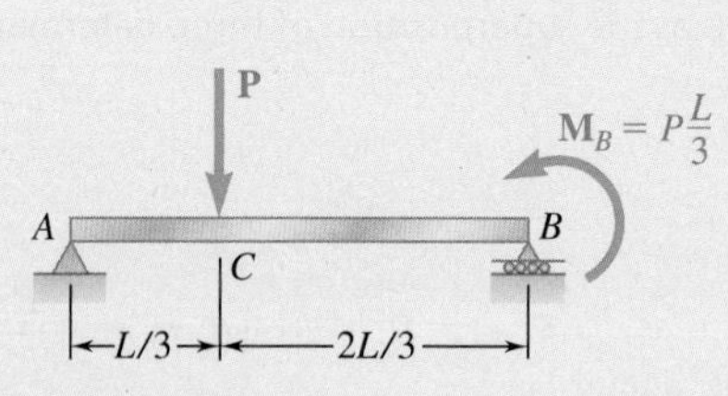

Fig. P15.27

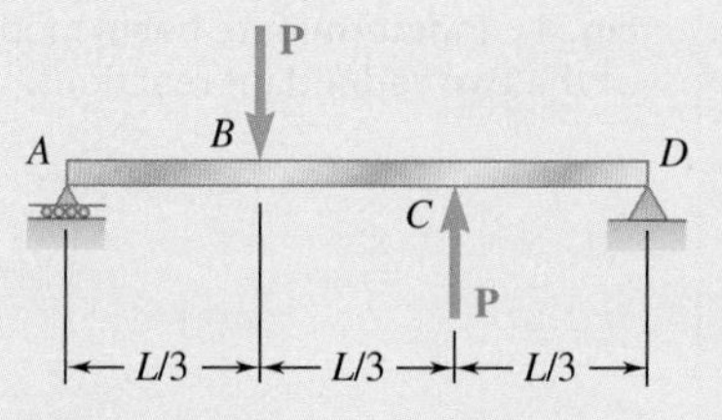

Fig. P15.28

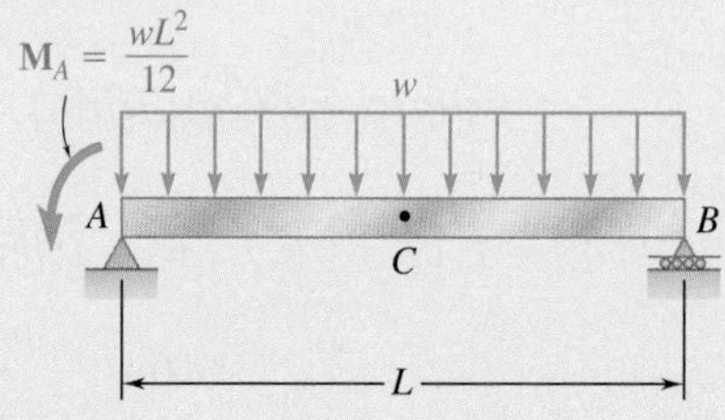

Fig. P15.29

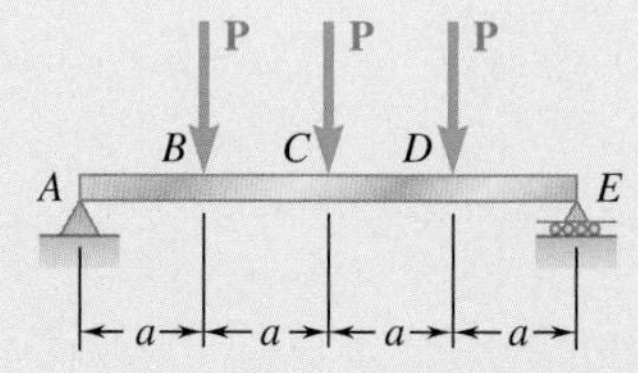

Fig. P15.30

15.31 and 15.32 For the cantilever beam and loading shown, determine the slope and deflection at the free end.

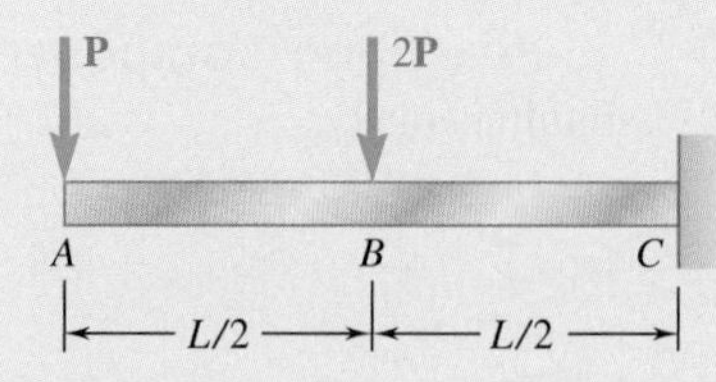

Fig. P15.31

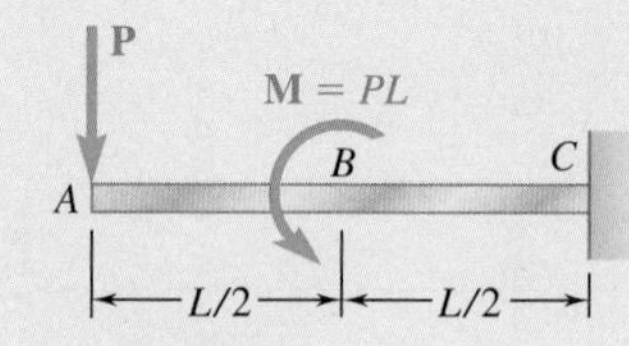

Fig. P15.32

15.33* and *15.34 For the cantilever beam and loading shown, determine the slope and deflection at point *C*.

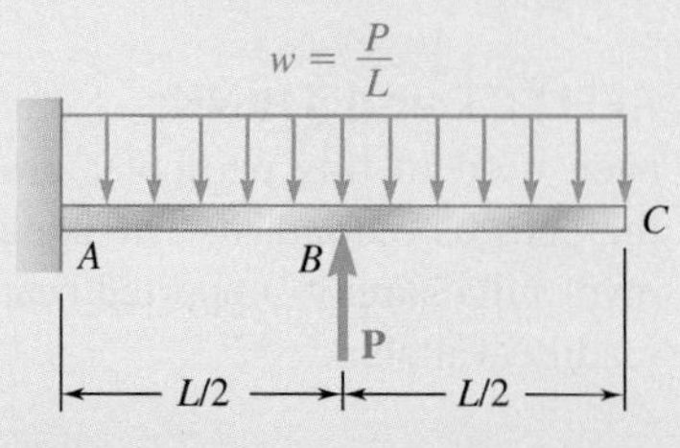

Fig. *P15.33*

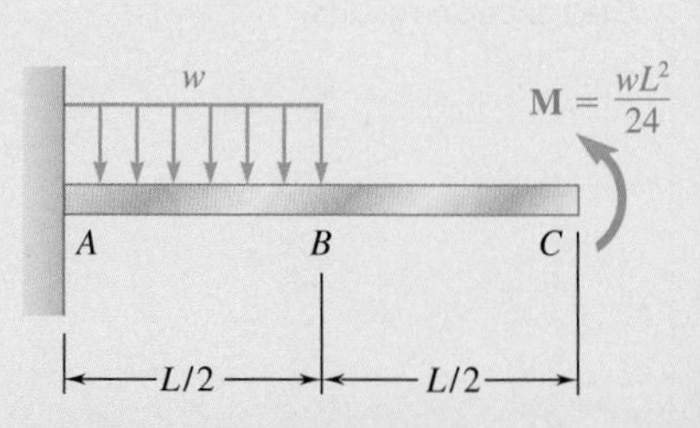

Fig. *P15.34*

15.35 For the cantilever beam and loading shown, determine the slope and deflection at end A. Use $E = 29 \times 10^6$ psi.

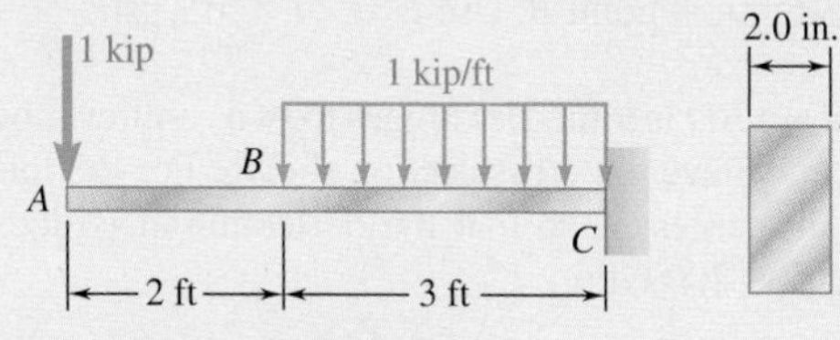

Fig. P15.35 and P15.36

15.36 For the cantilever beam and loading shown, determine the slope and deflection at point B. Use $E = 29 \times 10^6$ psi.

15.37 and *15.38* For the beam and loading shown, determine (*a*) the slope at end A, (*b*) the deflection at point C. Use $E = 200$ GPa.

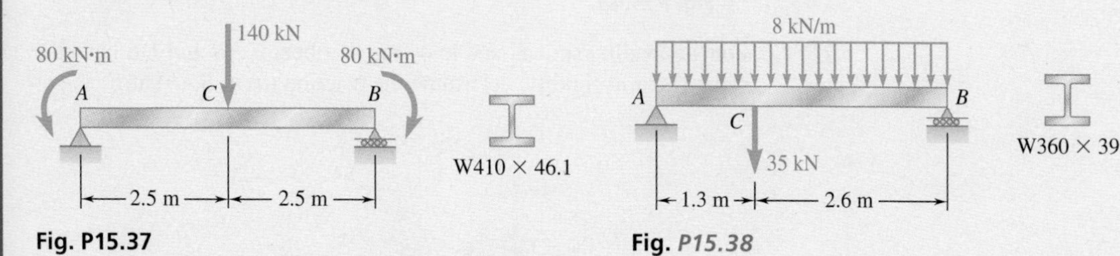

Fig. P15.37 **Fig. *P15.38***

15.39 and 15.40 For the uniform beam shown, determine (*a*) the reaction at A, (*b*) the reaction at B.

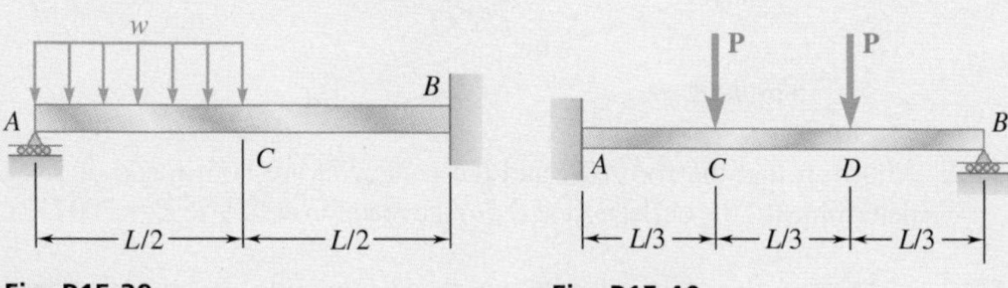

Fig. P15.39 **Fig. P15.40**

***15.41* and 15.42** For the uniform beam shown, determine the reaction at each of the three supports.

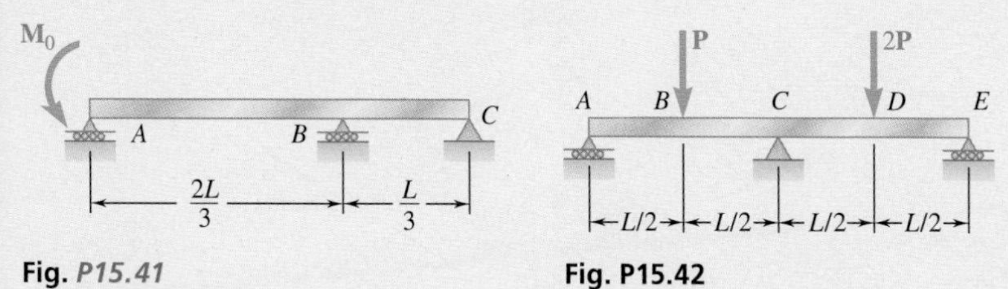

Fig. *P15.41* **Fig. P15.42**

15.43 and *15.44* For the beam shown, determine the reaction at B.

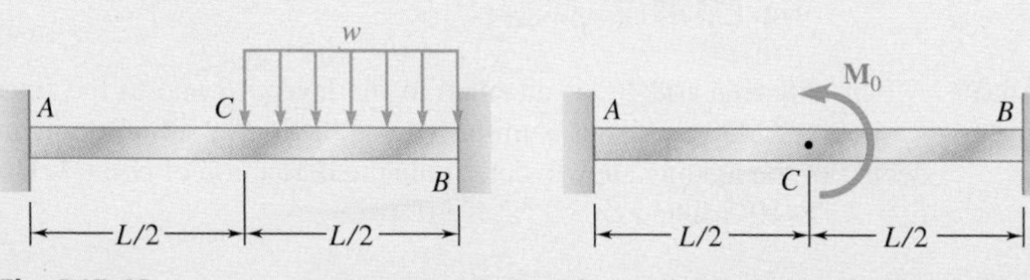

Fig. P15.43 **Fig. *P15.44***

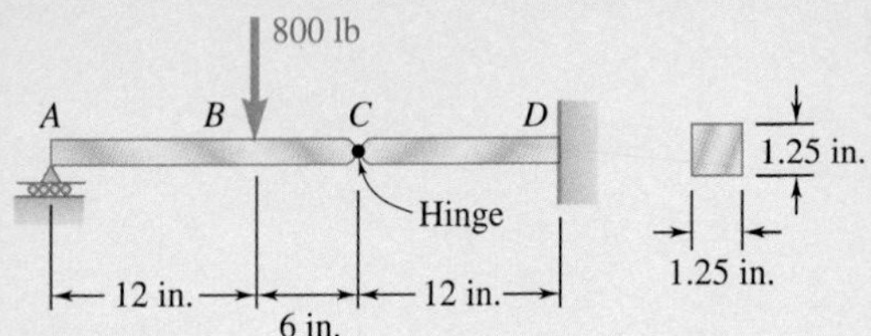

Fig. P15.45

15.45 The two beams shown have the same cross section and are joined by a hinge at C. For the loading shown, determine (*a*) the slope at point A, (*b*) the deflection at point B. Use $E = 29 \times 10^6$ psi.

15.46 A central beam BD is joined by hinges to two cantilever beams AB and DE. All beams have the cross section shown. For the loading shown, determine the largest w so that the deflection at C does not exceed 3 mm. Use $E = 200$ GPa.

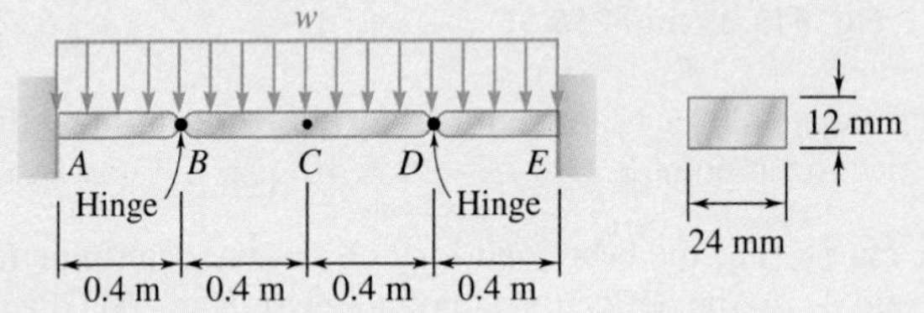

Fig. P15.46

15.47 For the loading shown, and knowing that beams AB and DE have the same flexural rigidity, determine the reaction (*a*) at B, (*b*) at E.

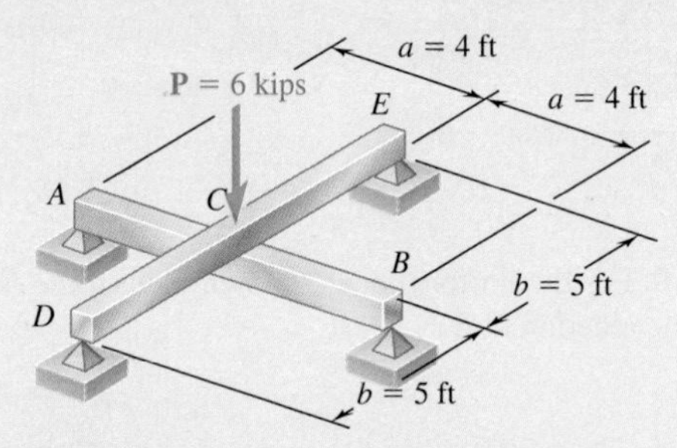

Fig. P15.47

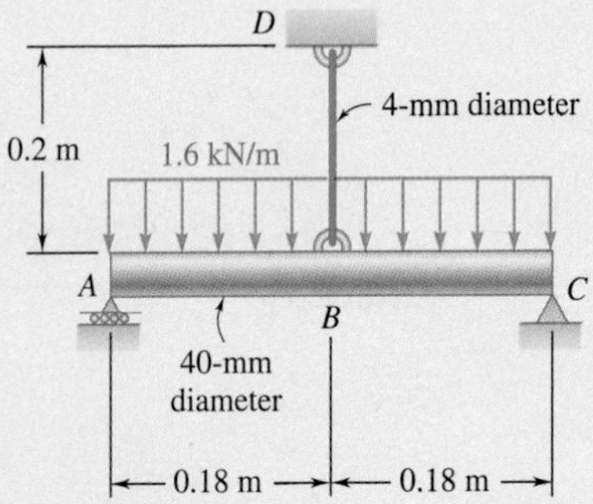

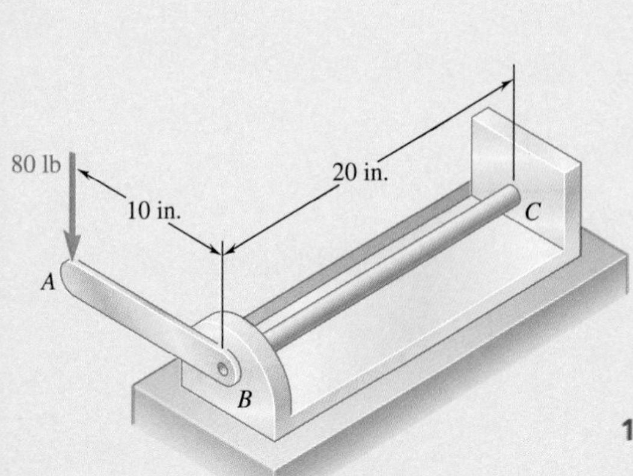

Fig. P15.48

15.48 Knowing that the rod ABC and the cable BD are both made of steel, determine (*a*) the deflection at B, (*b*) the reaction at A. Use $E = 200$ GPa.

15.49 A 16-mm-diameter rod ABC has been bent into the shape shown. Determine the deflection of end C after the 200-N force is applied. Use $E = 200$ GPa and $G = 80$ GPa.

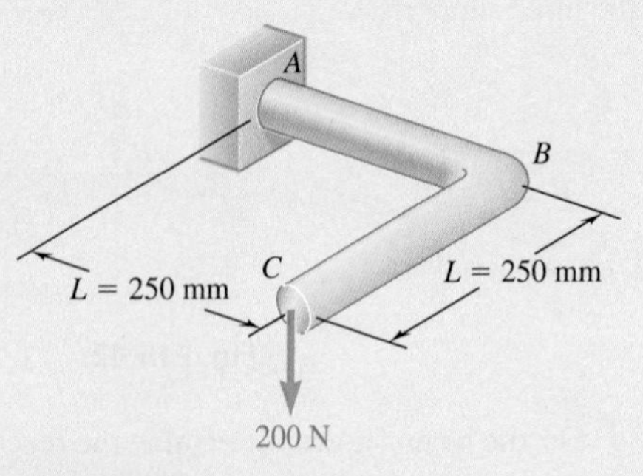

Fig. P15.49

Fig. P15.50

15.50 A $\frac{7}{8}$-in.-diameter rod BC is attached to the lever AB and to the fixed support at C. Lever AB has a uniform cross section $\frac{3}{8}$ in. thick and 1 in. deep. For the loading shown, determine the deflection of point A. Use $E = 29 \times 10^6$ psi and $G = 11.2 \times 10^6$ psi.

Review and Summary

This chapter was devoted to the determination of the slopes and deflections of beams under transverse loadings and applied moments. A mathematical method based on the method of integration of a differential equation was used to get the slopes and deflections at any point along the beam. Particular emphasis was placed on the computation of the maximum deflection of a beam under a given loading. This method also was used to determine support reactions and deflections of *indeterminate beams*, where the number of reactions at the supports exceeds the number of equilibrium equations available to determine these unknowns.

Deformation Under Transverse Loading

The relationship of the curvature $1/\rho$ of the neutral surface and the bending moment M in a prismatic beam in pure bending can be applied to a beam under a transverse loading, but in this case both M and $1/\rho$ vary from section to section. Using the distance x from the left end of the beam,

$$\frac{1}{\rho} = \frac{M(x)}{EI} \qquad \textbf{(15.1)}$$

This equation enables us to determine the radius of curvature of the neutral surface for any value of x and to draw some general conclusions regarding the shape of the deformed beam.

A relationship was found between the deflection y of a beam, measured at a given point Q, and the distance x of that point from some fixed origin (Fig. 15.20). The resulting equation defines the *elastic curve* of a beam. Expressing the curvature $1/\rho$ in terms of the derivatives of the function $y(x)$ and substituting into Eq. (15.1), we obtained the second-order linear differential equation

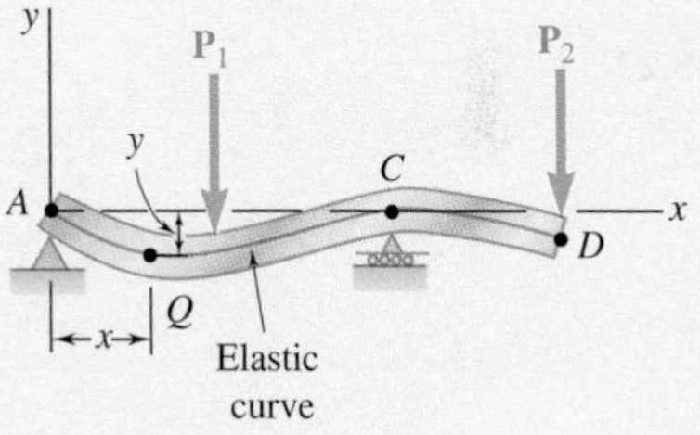

Fig. 15.20

$$\frac{d^2y}{dx^2} = \frac{M(x)}{EI} \qquad \textbf{(15.4)}$$

Integrating this equation twice, the expressions defining the slope $\theta(x) = dy/dx$ and the deflection $y(x)$ were obtained:

$$EI\frac{dy}{dx} = \int_0^x M(x)\,dx + C_1 \qquad \textbf{(15.5)}$$

$$EI\,y = \int_0^x dx \int_0^x M(x)\,dx + C_1 x + C_2 \qquad \textbf{(15.6)}$$

The product EI is known as the *flexural rigidity* of the beam. Two constants of integration C_1 and C_2 can be determined from the *boundary conditions* imposed on the beam by its supports (Fig. 15.21). The maximum deflection can be obtained by first determining the value of x for which the slope is zero and then computing the corresponding value of y.

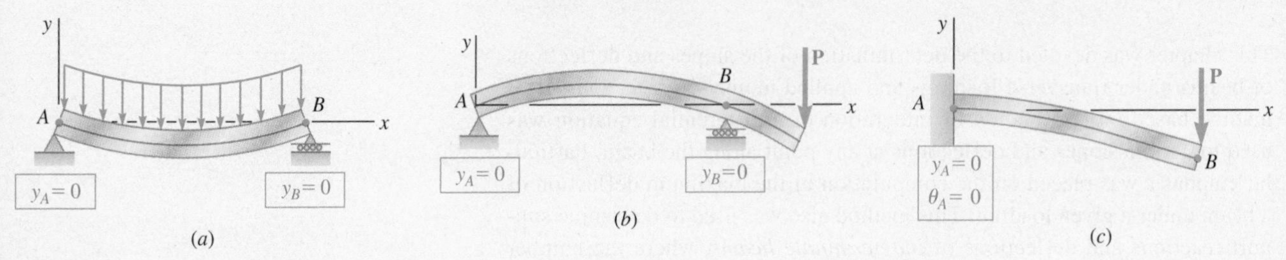

Fig. 15.21

Elastic Curve Defined by Different Functions

When the load requires different analytical functions to represent the bending moment in various portions of the beam, multiple differential equations are required to represent the slope $\theta(x)$ and the deflection $y(x)$. For the beam and load considered in Fig. 15.22, two differential equations are required: one for the portion of beam AD and the other for the portion DB. The first equation yields the functions θ_1 and y_1, and the second the functions θ_2 and y_2. Altogether, four constants of integration must be determined: two by writing that the deflections at A and B are zero and two by expressing that the portions of beam AD and DB have the same slope and the same deflection at D.

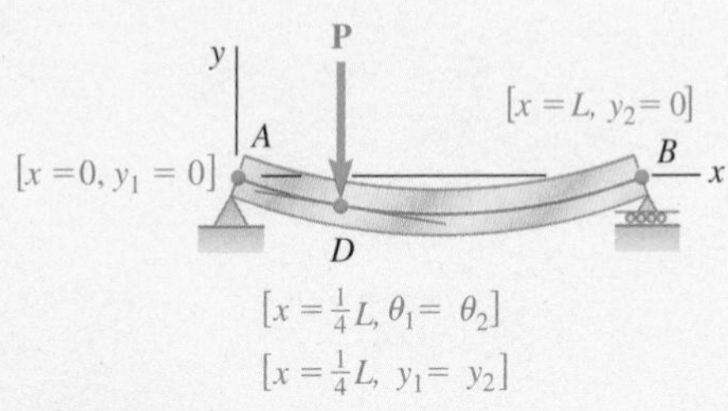

Fig. 15.22

For a beam supporting a distributed load $w(x)$, the elastic curve can be determined directly from $w(x)$ through four integrations yielding V, M, θ, and y (in that order). For the cantilever beam of Fig. 15.23a and the simply supported beam of Fig. 15.23b, four constants of integration can be determined from the four boundary conditions.

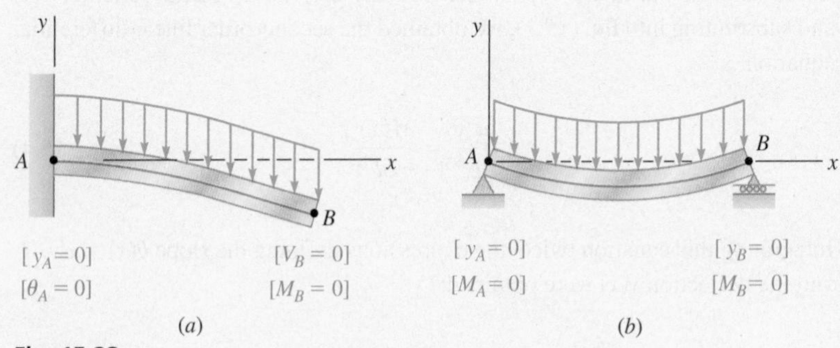

Fig. 15.23

Statically Indeterminate Beams

Statically indeterminate beams are supported such that the reactions at the supports involve four or more unknowns. Since only three equilibrium equations are available to determine these unknowns, they are supplemented with equations obtained from the boundary conditions imposed by the supports.

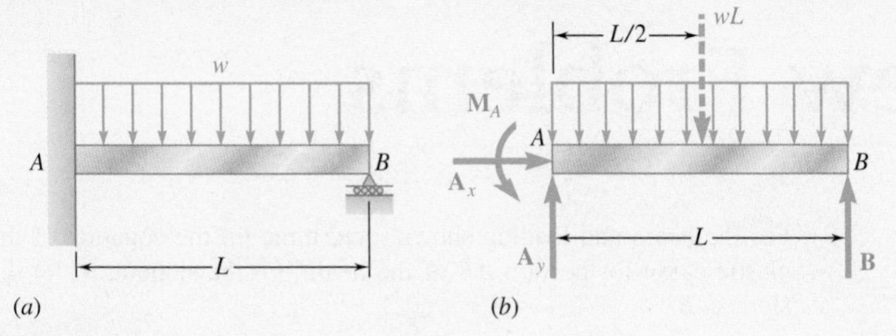

Fig. 15.24

For the beam of Fig 15.24, the reactions at the supports involve four unknowns: M_A, A_x, A_y, and B. This beam is *indeterminate to the first degree.* (If five unknowns are involved, the beam is indeterminate to the *second degree.*) Expressing the bending moment $M(x)$ in terms of the four unknowns and integrating twice, the slope $\theta(x)$ and the deflection $y(x)$ are determined in terms of the same unknowns and the constants of integration C_1 and C_2. The six unknowns are obtained by solving the three equilibrium equations for the free body of Fig. 15.24*b* and the three equations expressing that $\theta = 0$, $y = 0$ for $x = 0$, and that $y = 0$ for $x = L$ (Fig. 15.25) simultaneously.

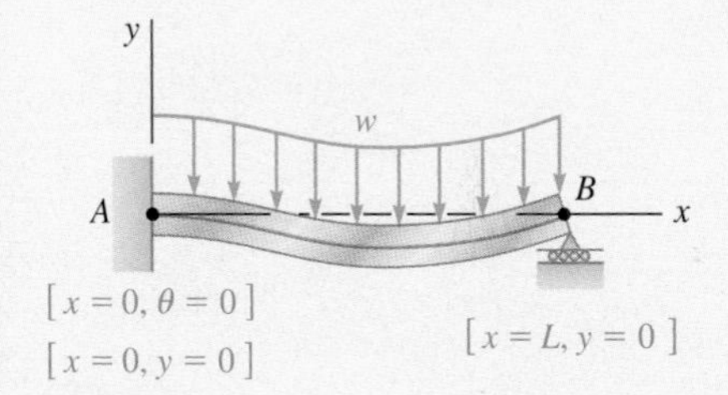

Fig. 15.25

Method of Superposition

The *method of superposition* separately determines and then adds the slope and deflection caused by the various loads applied to a beam. This procedure is made easier using the table of Appendix C, which gives the slopes and deflections of beams for various loadings and types of support.

Statically Indeterminate Beams by Superposition

The method of superposition can be effective for analyzing *statically indeterminate beams.* For example, the beam of Fig. 15.26 involves four unknown reactions and is indeterminate to the *first degree*; the reaction at B is chosen as *redundant*, and the beam is released from that support. Treating the reaction $\mathbf{R}_B$ as an unknown load and considering the deflections caused at B by the given distributed load and by $\mathbf{R}_B$ separately, the sum of these deflections is zero (Fig. 15.27). For a beam indeterminate to the *second degree* (i.e., with reactions at the supports involving five unknowns), two reactions are redundant, and the corresponding supports must be eliminated or modified accordingly.

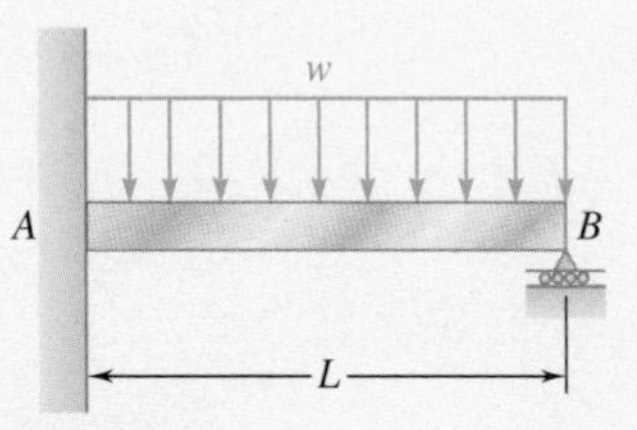

Fig. 15.26

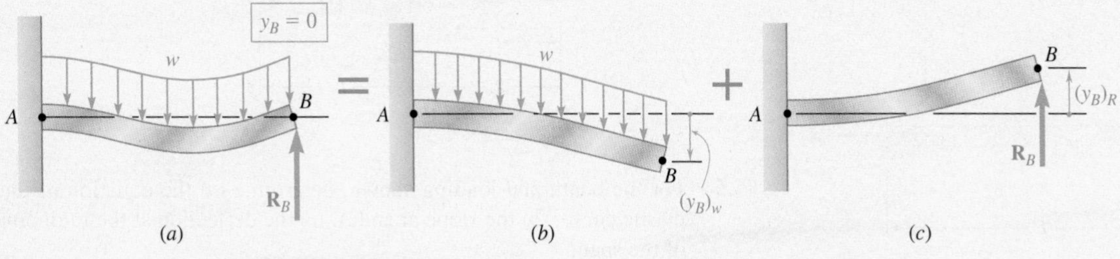

Fig. 15.27

Review Problems

15.51 For the beam and loading shown, determine (*a*) the equation of the elastic curve for portion AB of the beam, (*b*) the slope at A, (*c*) the slope at B.

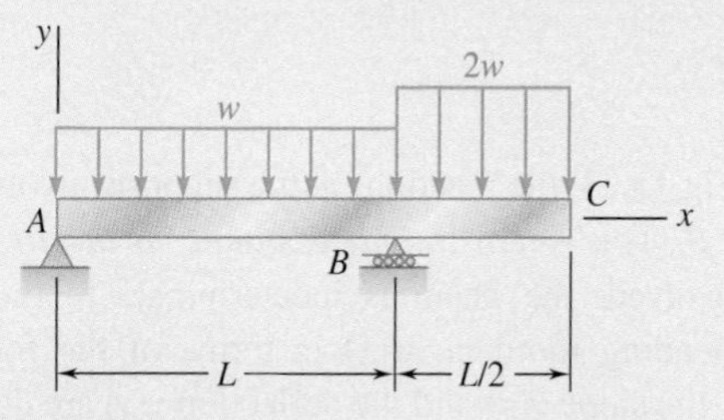

Fig. P15.51

15.52 (*a*) Determine the location and magnitude of the maximum absolute deflection in AB between A and the center of the beam. (*b*) Assuming that beam AB is a W460 × 113 rolled shape, $M_0 = 224$ kN·m, and $E = 200$ GPa, determine the maximum allowable length L so that the maximum deflection does not exceed 1.2 mm.

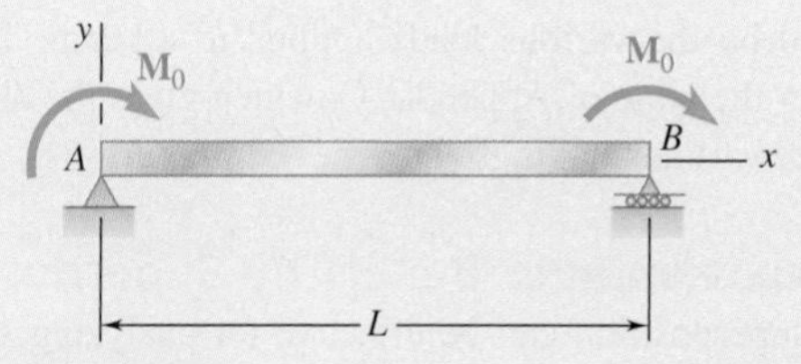

Fig. P15.52

15.53 Knowing that beam AE is an S200 × 27.4 rolled shape and that $P =$ 17.5 kN, $L = 2.5$ m, $a = 0.8$ m, and $E = 200$ GPa, determine (*a*) the equation of the elastic curve for portion BD, (*b*) the deflection at the center C of the beam.

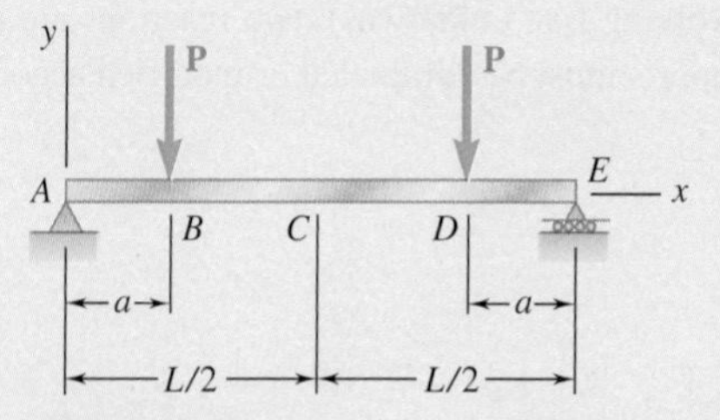

Fig. P15.53

15.54 For the beam and loading shown, determine (*a*) the equation of the elastic curve, (*b*) the slope at end A, (*c*) the deflection at the midpoint of the span.

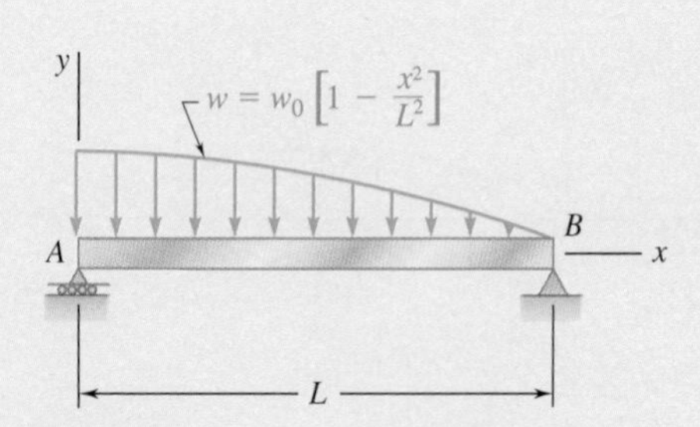

Fig. P15.54

15.55 For the beam shown, determine the reaction at the roller support when $w_0 = 6$ kips/ft.

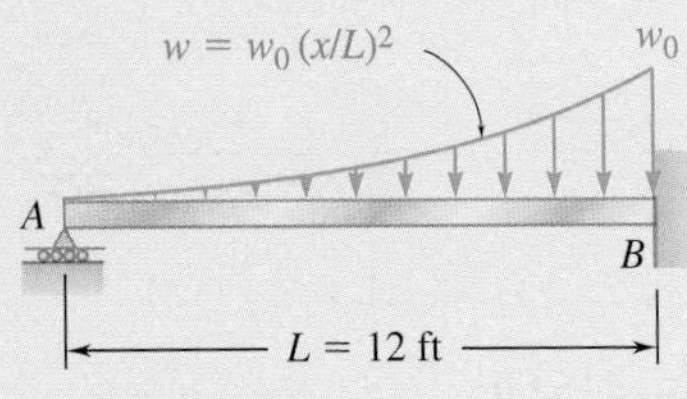

Fig. P15.55

15.56 Determine the reaction at the roller support and draw the bending-moment diagram for the beam and loading shown.

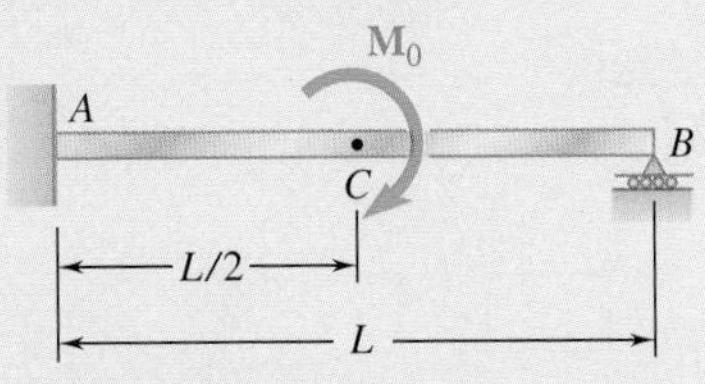

Fig. P15.56

15.57 For the cantilever beam and loading shown, determine the slope and deflection at the free end.

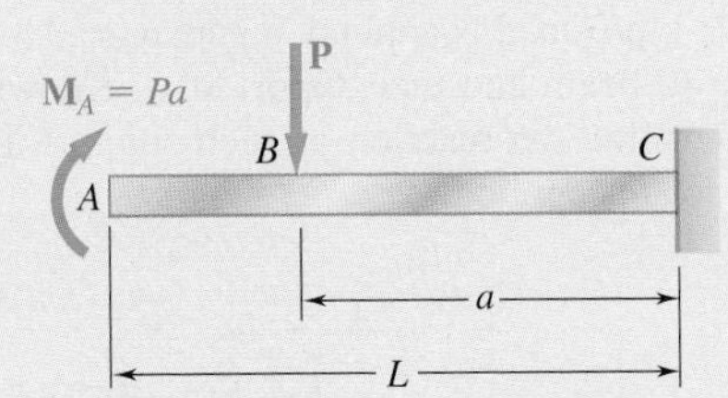

Fig. P15.57

15.58 For the beam and loading shown, determine (*a*) the deflection at point *C*, (*b*) the slope at end *A*.

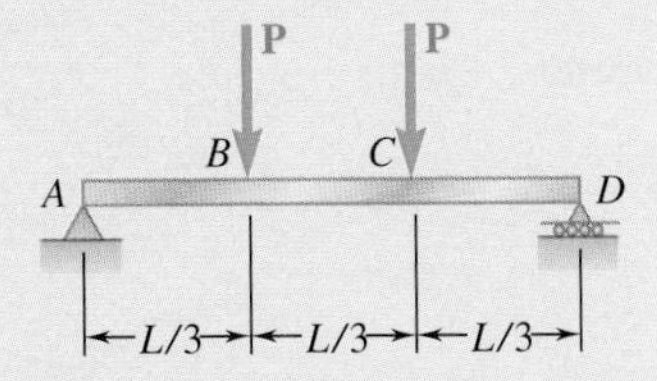

Fig. *P15.58*

15.59 For the cantilever beam and loading shown, determine the slope and deflection at point *B*. Use $E = 200$ GPa.

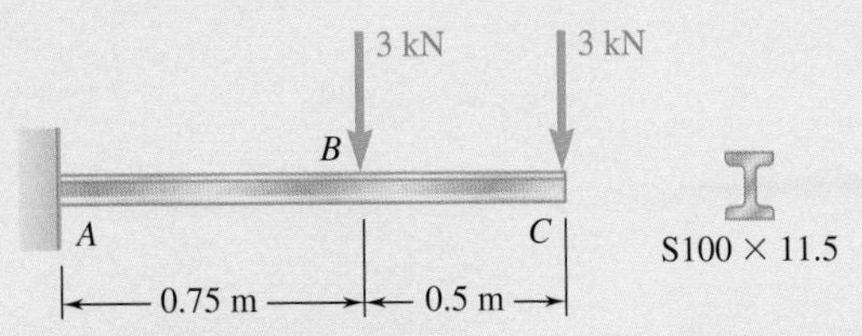

Fig. P15.59

15.60 For the uniform beam shown, determine the reaction at each of the three supports.

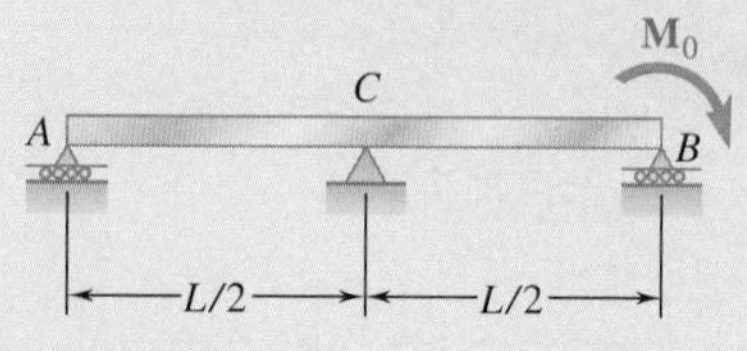

Fig. P15.60

15.61 The cantilever beam BC is attached to the steel cable AB as shown. Knowing that the cable is initially taut, determine the tension in the cable caused by the distributed load shown. Use $E = 200$ GPa.

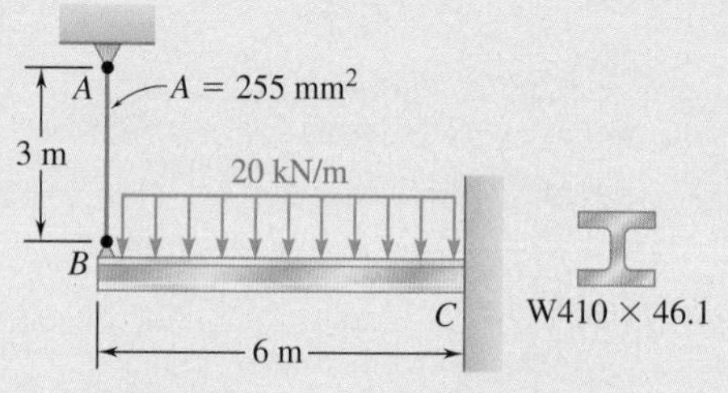

Fig. P15.61

15.62 Before the 2-kip/ft load is applied, a gap, $\delta_0 = 0.8$ in., exists between the W16 × 40 beam and the support at C. Knowing that $E = 29 \times 10^6$ psi, determine the reaction at each support after the uniformly distributed load is applied.

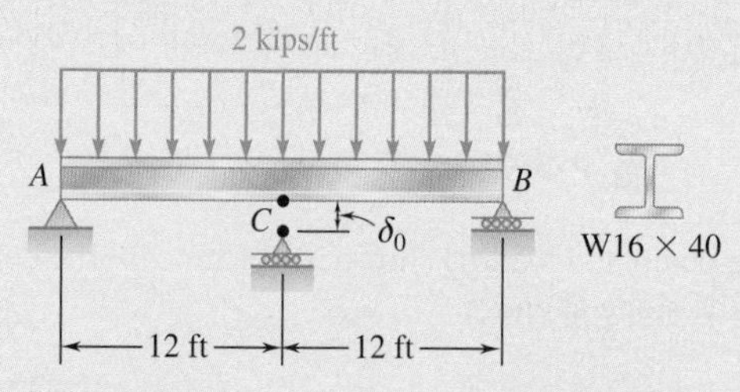

Fig. *P15.62*

© Jose Manuel/Getty Images RF

16 Columns

The curved pedestrian bridge is supported by a series of columns. The analysis and design of members supporting axial compressive loads will be discussed in this chapter.

Objectives

In this chapter, you will:

- **Describe** the behavior of columns in terms of stability.
- **Develop** Euler's formula for columns, using effective lengths to account for different end conditions.
- **Use** allowable-stress design for columns made of steel, aluminum, and wood.

Introduction

In the preceding chapters, we had two primary concerns: (1) the strength of the structure, i.e., its ability to support a specified load without experiencing excessive stress; (2) the ability of the structure to support a specified load without undergoing unacceptable deformations. This chapter is concerned with the stability of the structure (its ability to support a given load without experiencing a sudden change in configuration). This discussion is focused on columns, that is, the analysis and design of vertical prismatic members supporting axial loads.

In Sec. 16.1, the stability of a simplified model is discussed, where the column consists of two rigid rods connected by a pin and a spring and supports a load **P**. If its equilibrium is disturbed, this system will return to its original equilibrium position as long as P does not exceed a certain value P_{cr}, called the *critical load*. This is a *stable* system. However, if $P > P_{cr}$, the system moves away from its original position and settles in a new position of equilibrium. This system is said to be *unstable*.

In Sec. 16.1A, the *stability of elastic columns* considers a pin-ended column subjected to a centric axial load. *Euler's formula* for the critical load of the column is derived, and the corresponding critical normal stress in the column is determined. Applying a factor of safety to the critical load, we obtain the allowable load that can be safely applied to a pin-ended column.

In Sec. 16.1B, the analysis of the stability of columns with different end conditions is considered by learning how to determine the *effective length* of a column.

In the first sections of the chapter, each column is assumed to be a straight, homogeneous prism. In the last part of the chapter, real columns are designed and analyzed using empirical formulas set forth by professional organizations. In Sec. 16.2A, design equations are presented for the allowable stress in columns made of steel, aluminum, or wood that are subjected to a centric load.

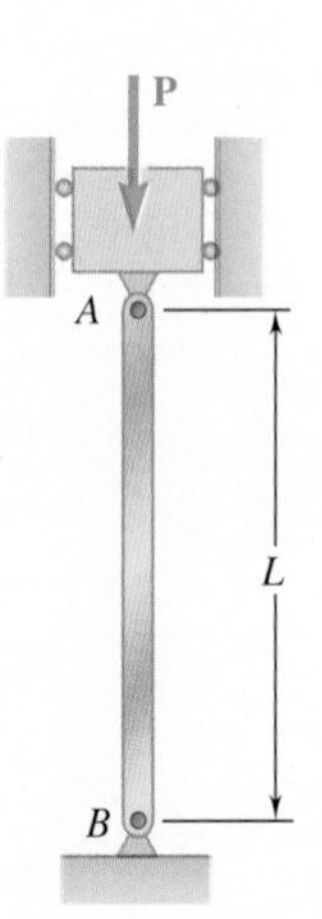

Fig. 16.1 Pin-ended axially loaded column.

16.1 STABILITY OF STRUCTURES

Consider the design of a column AB of length L to support a given load **P** (Fig. 16.1). The column is pin-connected at both ends, and **P** is a centric axial load. If the cross-sectional area A is selected so that the value $\sigma = P/A$ of the stress on a transverse section is less than the allowable stress σ_{all} for the material used and the deformation $\delta = PL/AE$ falls within

Photo 16.1 Laboratory test showing a buckled column.
Courtesy of Fritz Engineering Laboratory, Lehigh University

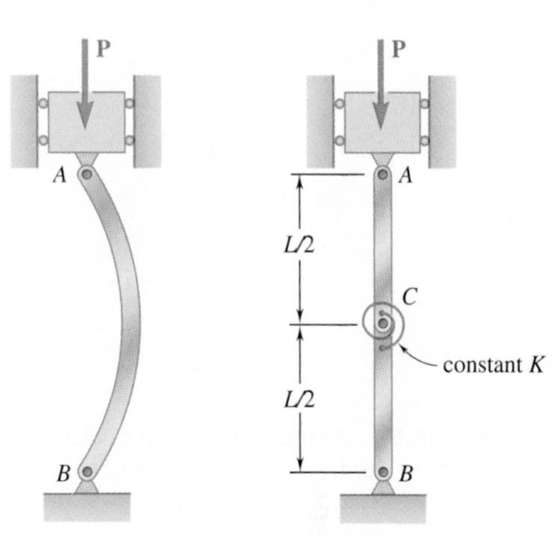

Fig. 16.2 Buckled pin-ended column.

Fig. 16.3 Model column made of two rigid rods joined by a torsional spring at *C*.

the given specifications, we might conclude that the column has been properly designed. However, it may happen that as the load is applied, the column *buckles* (Fig. 16.2). Instead of remaining straight, it suddenly becomes sharply curved such as shown in Photo 16.1. Clearly, a column that buckles under the load it is to support is not properly designed.

Before getting into the actual discussion of the stability of elastic columns, some insight will be gained on the problem by considering a simplified model consisting of two rigid rods *AC* and *BC* connected at *C* by a pin and a torsional spring of constant *K* (Fig. 16.3).

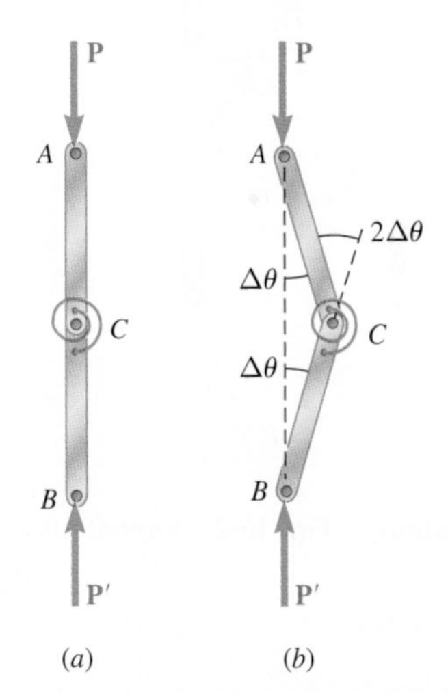

Fig. 16.4 Free-body diagram of model column (*a*) perfectly aligned (*b*) point *C* moved slightly out of alignment.

If the two rods and forces **P** and **P′** are perfectly aligned, the system will remain in the position of equilibrium shown in Fig.16.4*a* as long as it is not disturbed. But suppose we move *C* slightly to the right so that each rod forms a small angle $\Delta\theta$ with the vertical (Fig. 16.4*b*). Will the system return to its original equilibrium position, or will it move further away? In the first case, the system is *stable*; in the second, it is *unstable*.

To determine whether the two-rod system is stable or unstable, consider the forces acting on rod *AC* (Fig. 16.5). These forces consist of the couple formed by **P** and **P′** of moment $P(L/2)\sin\Delta\theta$, which tends to move the rod away from the vertical, and the couple **M** exerted by the spring, which tends to bring the rod back into its original vertical position. Since the angle of deflection of the spring is $2\,\Delta\theta$, the moment of couple **M** is

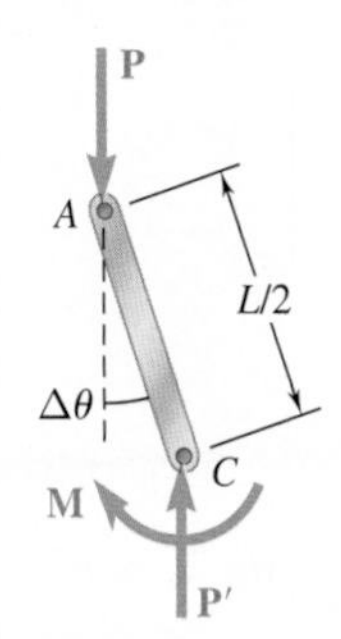

Fig. 16.5 Free-body diagram of rod *AC* in unaligned position.

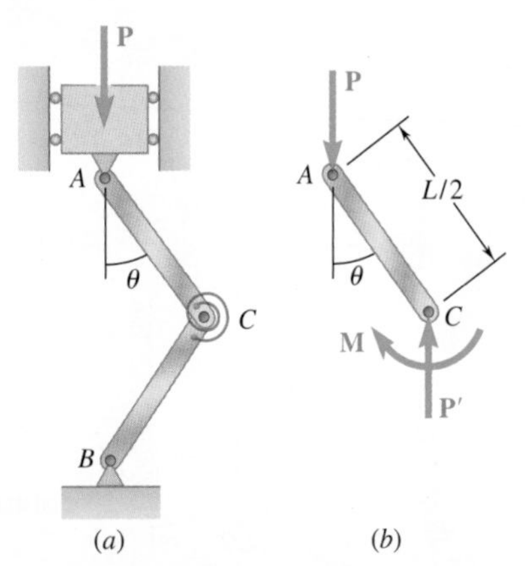

Fig. 16.6 (*a*) Model column in buckled position, (*b*) free-body diagram of rod *AC*.

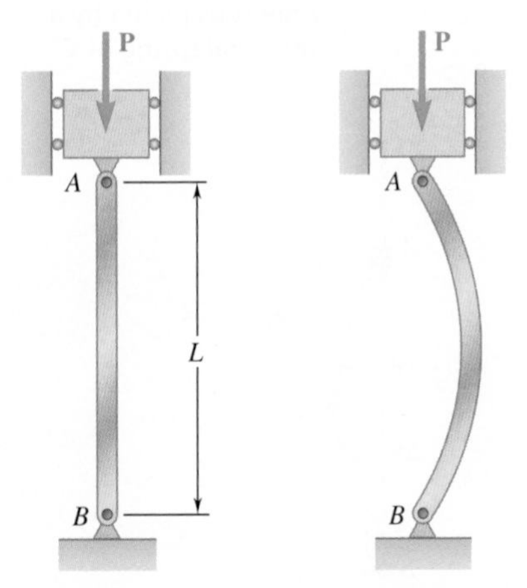

Fig. 16.1 (repeated). Fig. 16.2 (repeated).

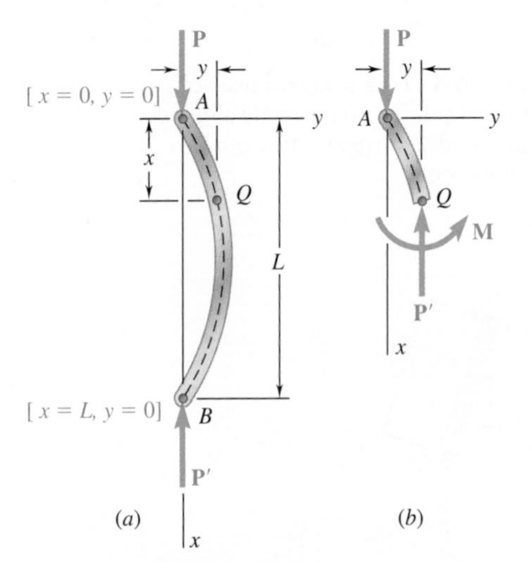

Fig. 16.7 Free-body diagrams of (*a*) buckled column and (*b*) portion *AQ*.

$M = K(2\,\Delta\theta)$. If the moment of the second couple is larger than the moment of the first couple, the system tends to return to its original equilibrium position; the system is stable. If the moment of the first couple is larger than the moment of the second couple, the system tends to move away from its original equilibrium position; the system is unstable. The load when the two couples balance each other is called the *critical load*, P_{cr}, which is given as

$$P_{cr}(L/2)\sin\Delta\theta = K(2\Delta\theta) \tag{16.1}$$

or since $\sin\Delta\theta \approx \Delta\theta$, when the displacement of *C* is very small (at the immediate onset of buckling),

$$P_{cr} = 4K/L \tag{16.2}$$

Clearly, the system is stable for $P < P_{cr}$ and unstable for $P > P_{cr}$.

Assume that a load $P > P_{cr}$ has been applied to the two rods of Fig. 16.3 and the system has been disturbed. Since $P > P_{cr}$, the system will move further away from the vertical and, after some oscillations, will settle into a new equilibrium position (Fig. 16.6*a*). Considering the equilibrium of the free body *AC* (Fig. 16.6*b*), an equation similar to Eq. (16.1) but involving the finite angle θ, is

$$P(L/2)\sin\theta = K(2\theta)$$

or

$$\frac{PL}{4K} = \frac{\theta}{\sin\theta} \tag{16.3}$$

The value of θ corresponding to the equilibrium position in Fig. 16.6 is obtained by solving Eq. (16.3) by trial and error. But for any positive value of θ, $\sin\theta < \theta$. Thus, Eq. (16.3) yields a value of θ different from zero only when the left-hand member of the equation is larger than one. Recalling Eq. (16.2), this is true only if $P > P_{cr}$. But, if $P < P_{cr}$, the second equilibrium position shown in Fig. 16.6 would not exist, and the only possible equilibrium position would be the one corresponding to $\theta = 0$. Thus, for $P < P_{cr}$, the position where $\theta = 0$ must be stable.

This observation applies to structures and mechanical systems in general and is used in the next section for the stability of elastic columns.

16.1A Euler's Formula for Pin-Ended Columns

Returning to the column *AB* considered in the preceding section (Fig. 16.1), we propose to determine the critical value of the load **P**, i.e., the value P_{cr} of the load for which the position shown in Fig. 16.1 ceases to be stable. If $P > P_{cr}$, the slightest misalignment or disturbance will cause the column to buckle into a curved shape, as shown in Fig. 16.2.

This approach determines the conditions under which the configuration of Fig. 16.2 is possible. Since a column is like a beam placed in a vertical position and subjected to an axial load, we proceed as in Chap. 15 and denote by *x* the distance from end *A* of the column to a point *Q* of its elastic curve and by *y* the deflection of that point (Fig. 16.7*a*). The *x* axis is vertical and directed downward, and the *y* axis is horizontal and directed to the right.

Considering the equilibrium of the free body AQ (Fig. 16.7b), the bending moment at Q is $M = -Py$. Substituting this value for M in Eq. (15.4) gives

$$\frac{d^2y}{dx^2} = \frac{M}{EI} = -\frac{P}{EI}y \tag{16.4}$$

or transposing the last term,

$$\frac{d^2y}{dx^2} + \frac{P}{EI}y = 0 \tag{16.5}$$

This equation is a linear, homogeneous differential equation of the second order with constant coefficients. Setting

$$p^2 = \frac{P}{EI} \tag{16.6}$$

Eq. (16.5) is rewritten as

$$\frac{d^2y}{dx^2} + p^2y = 0 \tag{16.7}$$

which is the same as the differential equation for simple harmonic motion, except the independent variable is now the distance x instead of the time t. The general solution of Eq. (16.7) is

$$y = A \sin px + B \cos px \tag{16.8}$$

and is easily checked by calculating d^2y/dx^2 and substituting for y and d^2y/dx^2 into Eq. (16.7).

Recalling the boundary conditions that must be satisfied at ends A and B of the column (Fig. 16.7a), make $x = 0$, $y = 0$ in Eq. (16.8), and find that $B = 0$. Substituting $x = L$, $y = 0$, obtain

$$A \sin pL = 0 \tag{16.9}$$

This equation is satisfied if either $A = 0$ or $\sin pL = 0$. If the first of these conditions is satisfied, Eq. (16.8) reduces to $y = 0$ and the column is straight (Fig. 16.1). For the second condition to be satisfied, $pL = n\pi$, or substituting for p from (16.6) and solving for P,

$$P = \frac{n^2\pi^2EI}{L^2} \tag{16.10}$$

The smallest value of P defined by Eq. (16.10) is that corresponding to $n = 1$. Thus,

$$P_{cr} = \frac{\pi^2EI}{L^2} \tag{16.11a}$$

This expression is known as *Euler's formula*, after the Swiss mathematician Leonhard Euler (1707–1783). Substituting this expression for P into Eq. (16.6), the value for p into Eq. (16.8), and recalling that $B = 0$,

$$y = A \sin \frac{\pi x}{L} \tag{16.12}$$

which is the equation of the elastic curve after the column has buckled (Fig. 16.2). Note that the maximum deflection $y_m = A$ is indeterminate.

This is because the differential Eq. (16.5) is a linearized approximation of the governing differential equation for the elastic curve.†

If $P < P_{cr}$, the condition $\sin pL = 0$ cannot be satisfied, and the solution of Eq. (16.12) does not exist. Then we must have $A = 0$, and the only possible configuration for the column is a straight one. Thus, for $P < P_{cr}$ the straight configuration of Fig. 16.1 is stable.

In a column with a circular or square cross section, the moment of inertia I is the same about any centroidal axis, and the column is as likely to buckle in one plane as another (except for the restraints that can be imposed by the end connections). For other cross-sectional shapes, the critical load should be found by making $I = I_{min}$ in Eq. (16.11a). If it occurs, buckling will take place in a plane perpendicular to the corresponding principal axis of inertia.

The stress corresponding to the critical load is the *critical stress* σ_{cr}. Recalling Eq. (16.11a) and setting $I = Ar^2$, where A is the cross-sectional area and r its radius of gyration gives

$$\sigma_{cr} = \frac{P_{cr}}{A} = \frac{\pi^2 EAr^2}{AL^2}$$

or

$$\sigma_{cr} = \frac{\pi^2 E}{(L/r)^2} \tag{16.13a}$$

The quantity L/r is the *slenderness ratio* of the column. The minimum value of the radius of gyration r should be used to obtain the slenderness ratio and the critical stress in a column.

Eq. (16.13) shows that the critical stress is proportional to the modulus of elasticity of the material and inversely proportional to the square of the slenderness ratio of the column. The plot of σ_{cr} versus L/r is shown in Fig. 16.8 for structural steel, assuming $E = 200$ GPa and $\sigma_Y = 250$ MPa. Keep in mind that no factor of safety has been used in plotting σ_{cr}. Also, if σ_{cr} obtained from Eq. (16.13a) or from the curve of Fig. 16.8 is larger than the yield strength σ_Y, this value is of no interest, since the column will yield in compression and cease to be elastic before it has a chance to buckle.

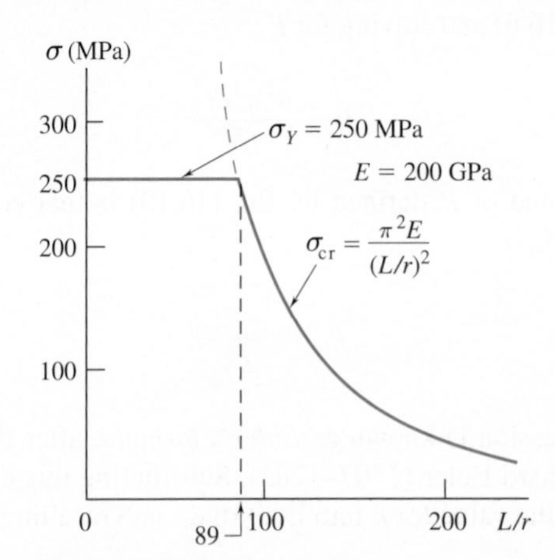

Fig. 16.8 Plot of critical stress.

†Recall that $d^2y/dx^2 = M/EI$ was obtained in Sec. 15.1A by assuming that the slope dy/dx of the beam could be neglected and that the exact expression in Eq. (15.3) for the curvature of the beam could be replaced by $1/\rho = d^2y/dx^2$.

Concept Application 16.1

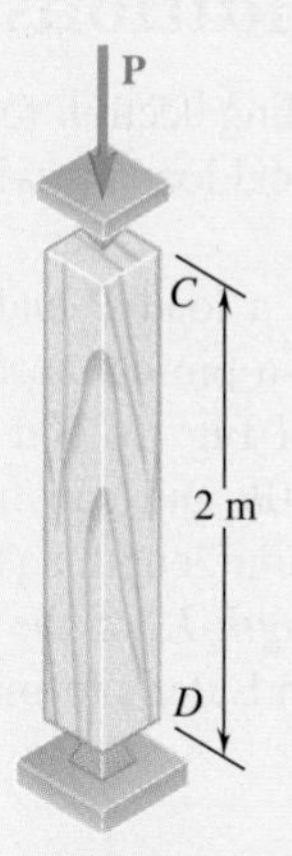

Fig. 16.9 Pin-ended wood column of square cross section.

A 2-m-long pin-ended column with a square cross section is to be made of wood (Fig 16.9). Assuming $E = 13$ GPa, $\sigma_{all} = 12$ MPa, and using a factor of safety of 2.5 to calculate Euler's critical load for buckling, determine the size of the cross section if the column is to safely support (*a*) a 100-kN load, (*b*) a 200-kN load.

a. For the 100-kN Load. Use the given factor of safety to obtain

$$P_{cr} = 2.5(100 \text{ kN}) = 250 \text{ kN} \qquad L = 2 \text{ m} \qquad E = 13 \text{ GPa}$$

Use Euler's formula, Eq. (16.11a), and solve for I:

$$I = \frac{P_{cr}L^2}{\pi^2 E} = \frac{(250 \times 10^3 \text{ N})(2 \text{ m})^2}{\pi^2(13 \times 10^9 \text{ Pa})} = 7.794 \times 10^{-6} \text{ m}^4$$

Recalling that, for a square of side a, $I = a^4/12$, write

$$\frac{a^4}{12} = 7.794 \times 10^{-6} \text{ m}^4 \qquad a = 98.3 \text{ mm} \approx 100 \text{ mm}$$

Check the value of the normal stress in the column:

$$\sigma = \frac{P}{A} = \frac{100 \text{ kN}}{(0.100 \text{ m})^2} = 10 \text{ MPa}$$

Since σ is smaller than the allowable stress, a 100 × 100-mm cross section is acceptable.

b. For the 200-kN Load. Solve Eq. (16.11a) again for I, but make $P_{cr} = 2.5(200) = 500$ kN to obtain

$$I = 15.588 \times 10^{-6} \text{ m}^4$$

$$\frac{a^4}{12} = 15.588 \times 10^{-6} \qquad a = 116.95 \text{ mm}$$

The value of the normal stress is

$$\sigma = \frac{P}{A} = \frac{200 \text{ kN}}{(0.11695 \text{ m})^2} = 14.62 \text{ MPa}$$

Since this is larger than the allowable stress, the dimension obtained is not acceptable, and the cross section must be selected on the basis of its resistance to compression.

$$A = \frac{P}{\sigma_{all}} = \frac{200 \text{ kN}}{12 \text{ MPa}} = 16.67 \times 10^{-3} \text{ m}^2$$

$$a^2 = 16.67 \times 10^{-3} \text{ m}^2 \qquad a = 129.1 \text{ mm}$$

A 130 × 130-mm cross section is acceptable.

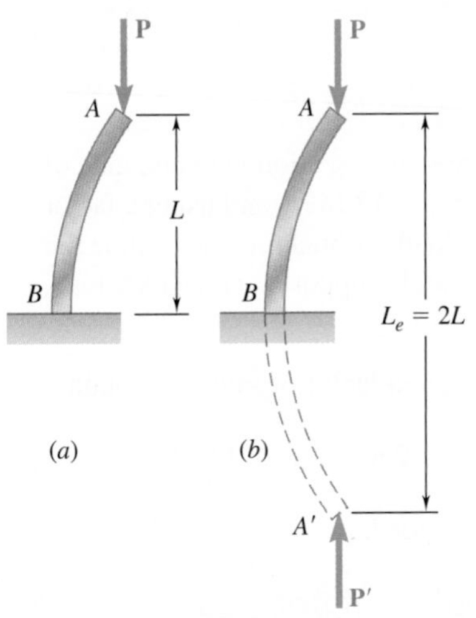

Fig. 16.10 Effective length of a fixed-free column of length L is equivalent to a pin-ended column of length $2L$.

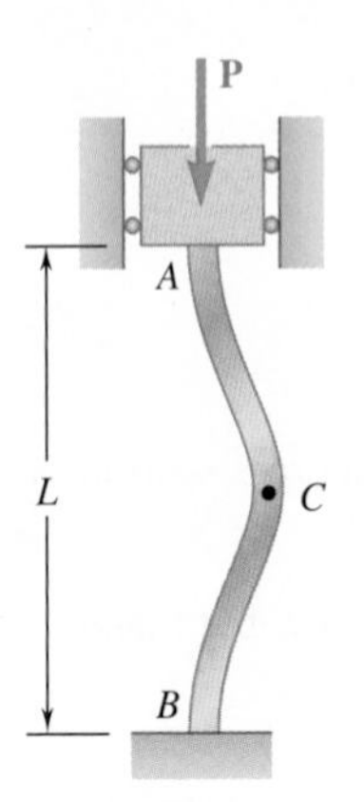

Fig. 16.11 Column with fixed ends.

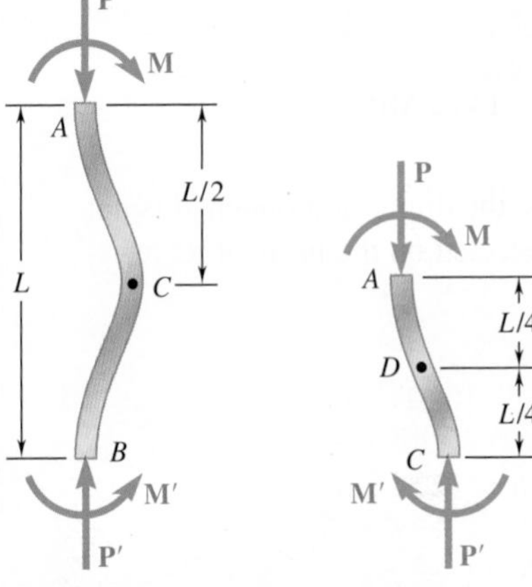

Fig. 16.12 Free-body diagram of buckled fixed-ended column.

Fig. 16.13 Free-body diagram of upper half of fixed-ended column.

16.1B Euler's Formula for Columns with Other End Conditions

Euler's formula (16.11) was derived in the preceding section for a column that was pin-connected at both ends. Now the critical load P_{cr} will be determined for columns with different end conditions.

A column with one free end A supporting a load **P** and one fixed end B (Fig. 16.10*a*) behaves as the upper half of a pin-connected column (Fig. 16.10*b*). The critical load for the column of Fig. 16.10*a* is thus the same as for the pin-ended column of Fig. 16.10*b* and can be obtained from Euler's formula Eq. (16.11a) by using a column length equal to twice the actual length L. We say that the *effective length* L_e of the column of Fig. 16.10 is equal to $2L$, and substitute $L_e = 2L$ in Euler's formula:

$$P_{cr} = \frac{\pi^2 EI}{L_e^2} \tag{16.11b}$$

The critical stress is

$$\sigma_{cr} = \frac{\pi^2 E}{(L_e/r)^2} \tag{16.13b}$$

The quantity L_e/r is called the *effective slenderness ratio* of the column and for Fig. 16.10*a* is equal to $2L/r$.

Now consider a column with two fixed ends A and B supporting a load **P** (Fig. 16.11). The symmetry of the supports and the load about a horizontal axis through the midpoint C requires that the shear at C and the horizontal components of the reactions at A and B be zero (Fig. 16.12a). Thus, the restraints imposed on the upper half AC of the column by the support at A and by the lower half CB are identical (Fig. 16.13). Portion AC must be symmetric about its midpoint D, and this point must be a point of inflection where the bending moment is zero. The bending moment at the midpoint E of the lower half of the column also must be zero (Fig. 16.14*a*). Since the bending moment at the ends of a pin-ended column is zero, portion DE of the column in Fig. 16.13*a* must behave like a pin-ended column (Fig. 16.14*b*). Thus, the effective length of a column with two fixed ends is $L_e = L/2$.

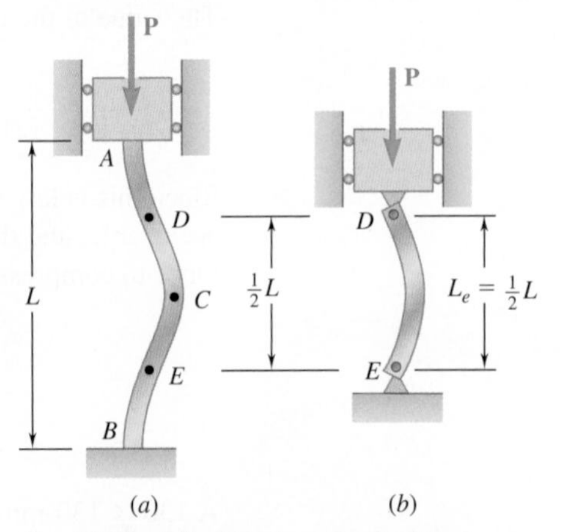

Fig. 16.14 Effective length of a fixed-ended column of length L is equivalent to a pin-ended column of length $L/2$.

In a column with one fixed end B and one pin-connected end A supporting a load **P** (Fig. 16.15), the differential equation of the elastic curve must be solved to determine the effective length. From the free-body diagram of the entire column (Fig. 16.16), a transverse force **V** is exerted at end A, in addition to the axial load **P**, and **V** is statically indeterminate. Considering the free-body diagram of a portion AQ of the column (Fig. 16.17), the bending moment at Q is

$$M = -Py - Vx$$

Substituting this value into Eq. (15.4) of Sec. 15.1A,

$$\frac{d^2y}{dx^2} = \frac{M}{EI} = -\frac{P}{EI}y - \frac{V}{EI}x$$

Transposing the term containing y and setting

$$p^2 = \frac{P}{EI} \tag{16.6}$$

as in Sec. 16.1A gives

$$\frac{d^2y}{dx^2} + p^2y = -\frac{V}{EI}x \tag{16.14}$$

This is a linear, nonhomogeneous differential equation of the second order with constant coefficients. Observing that the left-hand members of Eqs. (16.7) and (16.14) are identical, the general solution of Eq. (16.14) can be obtained by adding a particular solution of Eq. (16.14) to the solution of Eq. (16.8) obtained for Eq. (16.7). Such a particular solution is

$$y = -\frac{V}{p^2EI}x$$

or recalling Eq. (16.6),

$$y = -\frac{V}{P}x \tag{16.15}$$

Adding the solutions of Eq. (16.8) and (16.15), the general solution of Eq. (16.14) is

$$y = A \sin px + B \cos px - \frac{V}{P}x \tag{16.16}$$

The constants A and B and the magnitude V of the unknown transverse force **V** are obtained from the boundary conditions in Fig. (16.16). Making $x = 0$, $y = 0$ in Eq. (16.16), $B = 0$. Making $x = L$, $y = 0$, gives

$$A \sin pL = \frac{V}{P}L \tag{16.17}$$

Taking the derivative of Eq. (16.16), with $B = 0$,

$$\frac{dy}{dx} = Ap \cos px - \frac{V}{P}$$

and making $x = L$, $dy/dx = 0$,

$$Ap \cos pL = \frac{V}{P} \tag{16.18}$$

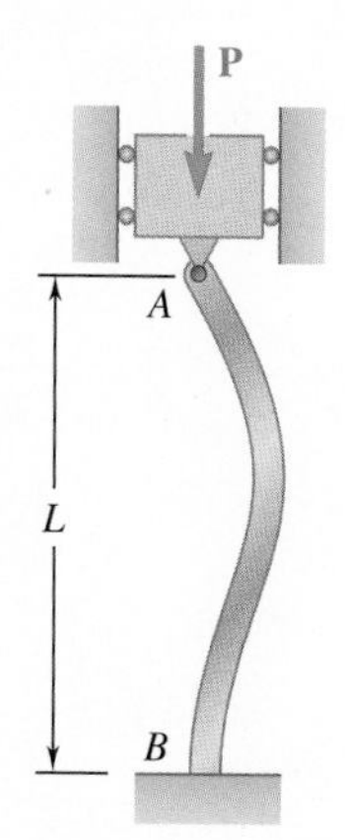

Fig. 16.15 Column with fixed-pinned end conditions.

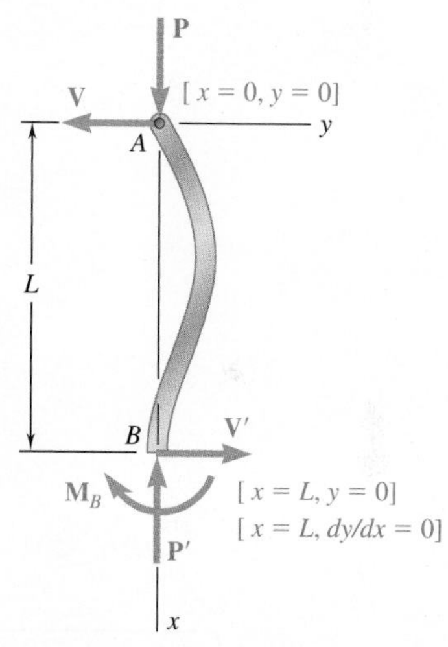

Fig. 16.16 Free-body diagram of buckled fixed-pinned column.

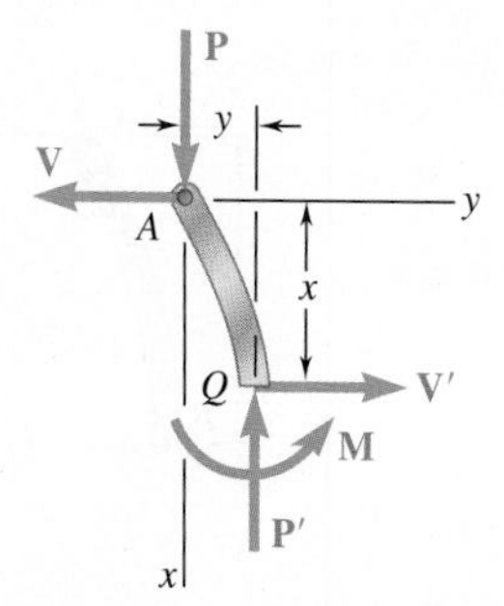

Fig. 16.17 Free-body diagram of portion AQ of buckled fixed-pinned column.

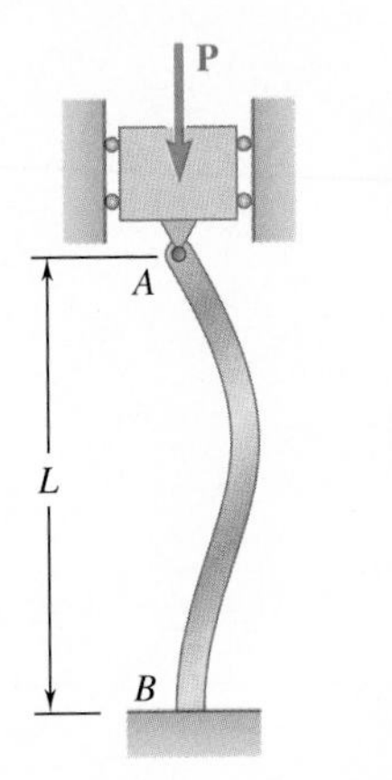

Fig. 16.15 (repeated).

Dividing Eq. (16.17) by Eq. (16.18) member by member, a solution like Eq. (16.16) can exist only if

$$\tan pL = pL \tag{16.19}$$

Solving this equation by trial and error, the smallest value of pL that satisfies Eq. (16.19) is

$$pL = 4.4934 \tag{16.20}$$

Carrying the value of p from Eq. (16.20) into Eq. (16.6) and solving for P, the critical load for the column of Fig. 16.15 is

$$P_{cr} = \frac{20.19EI}{L^2} \tag{16.21}$$

The effective length of the column is obtained by equating the right-hand members of Eqs. (16.11b) and (16.21):

$$\frac{\pi^2 EI}{L_e^2} = \frac{20.19EI}{L^2}$$

Solving for L_e, the effective length of a column with one fixed end and one pin-connected end is $L_e = 0.699L \approx 0.7L$.

The effective lengths corresponding to the various end conditions are shown in Fig. 16.18.

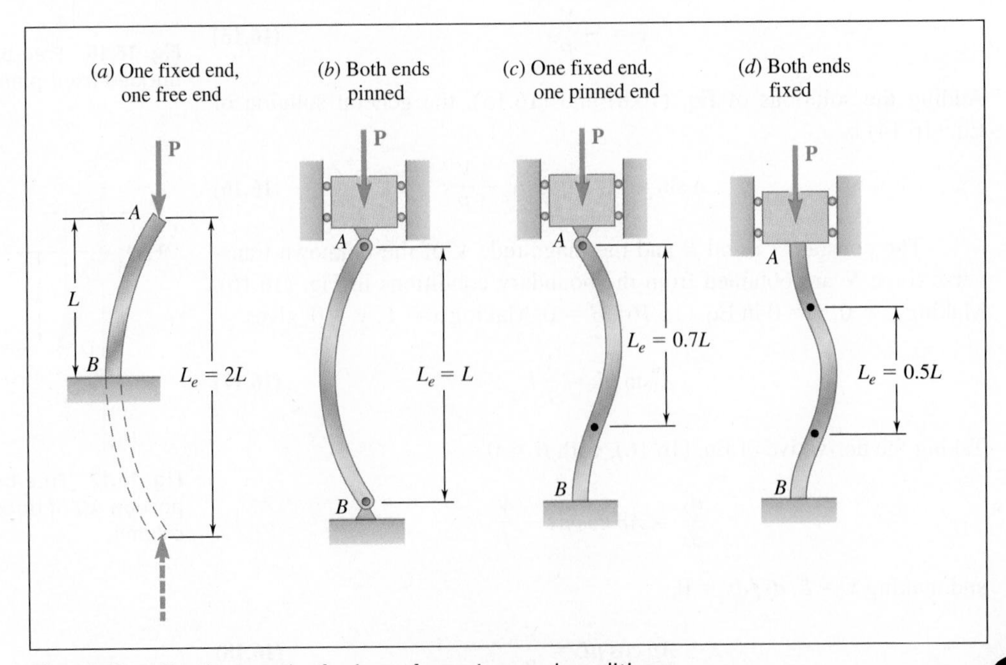

Fig. 16.18 Effective length of column for various end conditions.

Sample Problem 16.1

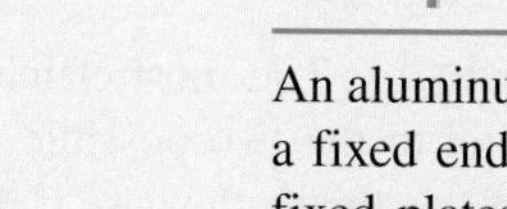
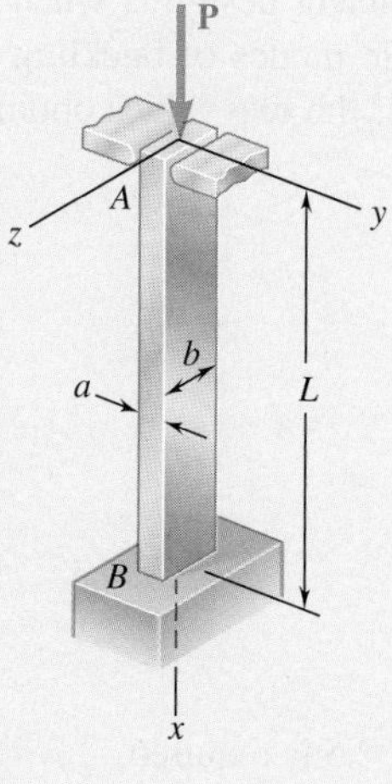

An aluminum column with a length of L and a rectangular cross section has a fixed end B and supports a centric load at A. Two smooth and rounded fixed plates restrain end A from moving in one of the vertical planes of symmetry of the column but allow it to move in the other plane. (*a*) Determine the ratio a/b of the two sides of the cross section corresponding to the most efficient design against buckling. (*b*) Design the most efficient cross section for the column, knowing that $L = 20$ in., $E = 10.1 \times 10^6$ psi, $P = 5$ kips, and a factor of safety of 2.5 is required.

STRATEGY: The most efficient design is that for which the critical stresses corresponding to the two possible buckling modes are equal. This occurs if the two critical stresses obtained from Eq. (16.13b) are the same. Thus for this problem, the two effective slenderness ratios in this equation must be equal to solve part *a*. Use Fig. 16.18 to determine the effective lengths. The design data can then be used with Eq. (16.13b) to size the cross section for part *b*.

MODELING:

Buckling in *xy* Plane. Referring to Fig. 16.18*c*, the effective length of the column with respect to buckling in this plane is $L_e = 0.7L$. The radius of gyration r_z of the cross section is obtained by

$$I_z = \tfrac{1}{12}ba^3 \quad A = ab$$

and since $I_z = Ar_z^2$, $$r_z^2 = \frac{I_z}{A} = \frac{\frac{1}{12}ba^3}{ab} = \frac{a^2}{12} \qquad r_z = a/\sqrt{12}$$

The effective slenderness ratio of the column with respect to buckling in the xy plane is

$$\frac{L_e}{r_z} = \frac{0.7L}{a/\sqrt{12}} \tag{1}$$

Buckling in *xz* Plane. Referring to Fig. 16.18*a*, the effective length of the column with respect to buckling in this plane is $L_e = 2L$, and the corresponding radius of gyration is $r_y = b/\sqrt{12}$. Thus,

$$\frac{L_e}{r_y} = \frac{2L}{b/\sqrt{12}} \tag{2}$$

(continued)

ANALYSIS:

a. Most Efficient Design. The most efficient design is when the critical stresses corresponding to the two possible modes of buckling are equal. Referring to Eq. (16.13b), this is the case if the two values obtained above for the effective slenderness ratio are equal.

$$\frac{0.7L}{a/\sqrt{12}} = \frac{2L}{b/\sqrt{12}}$$

and solving for the ratio a/b,

$$\frac{a}{b} = \frac{0.7}{2} \qquad \frac{a}{b} = 0.35 \blacktriangleleft$$

b. Design for Given Data. Since $F.S. = 2.5$ is required,

$$P_{cr} = (F.S.)P = (2.5)(5 \text{ kips}) = 12.5 \text{ kips}$$

Using $a = 0.35b$,

$$A = ab = 0.35b^2 \quad \text{and} \quad \sigma_{cr} = \frac{P_{cr}}{A} = \frac{12{,}500 \text{ lb}}{0.35b^2}$$

Making $L = 20$ in. in Eq. (2), $L_e/r_y = 138.6/b$. Substituting for E, L_e/r, and σ_{cr} into Eq. (16.13b) gives

$$\sigma_{cr} = \frac{\pi^2 E}{(L_e/r)^2} \qquad \frac{12{,}500 \text{ lb}}{0.35b^2} = \frac{\pi^2(10.1 \times 10^6 \text{ psi})}{(138.6/b)^2}$$

$$b = 1.620 \text{ in.} \qquad a = 0.35b = 0.567 \text{ in.} \blacktriangleleft$$

REFLECT and THINK: The calculated critical Euler buckling stress can never be taken to exceed the yield strength of the material. In this problem, you can readily determine that the critical stress $\sigma_{cr} = 13.6$ ksi; though the specific alloy was not given, this stress is less than the tensile yield strength σ_y values for all aluminum alloys listed in Appendix B.

Problems

16.1 Knowing that the spring at A is of constant k and that the bar AB is rigid, determine the critical load P_{cr}.

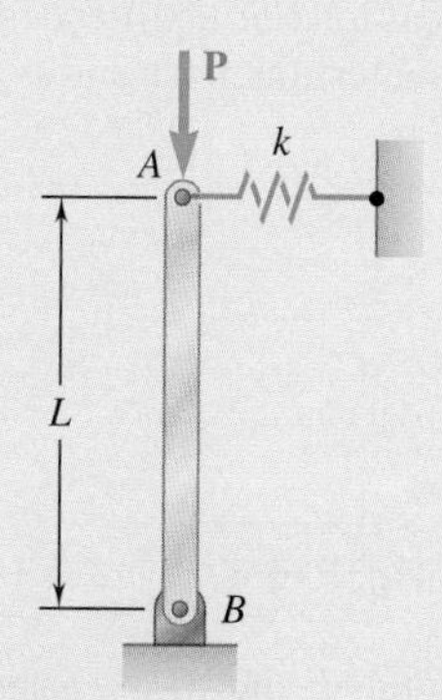

Fig. P16.1

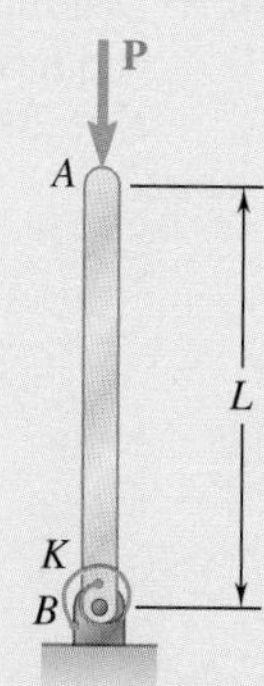

Fig. P16.2

16.2 Knowing that the torsional spring at B is of constant K and that the bar AB is rigid, determine the critical load P_{cr}.

16.3 Two rigid bars AC and BC are connected as shown to a spring of constant k. Knowing that the spring can act in either tension or compression, determine the critical load P_{cr} for the system.

16.4 Two rigid bars AC and BC are connected by a pin at C as shown. Knowing that the torsional spring at B is of constant K, determine the critical load P_{cr} for the system.

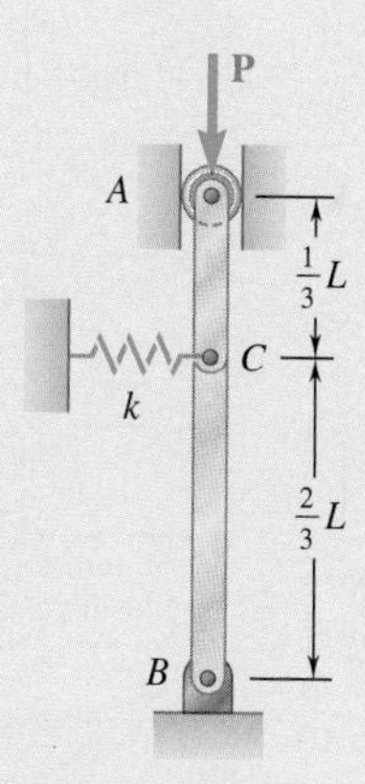

Fig. P16.3

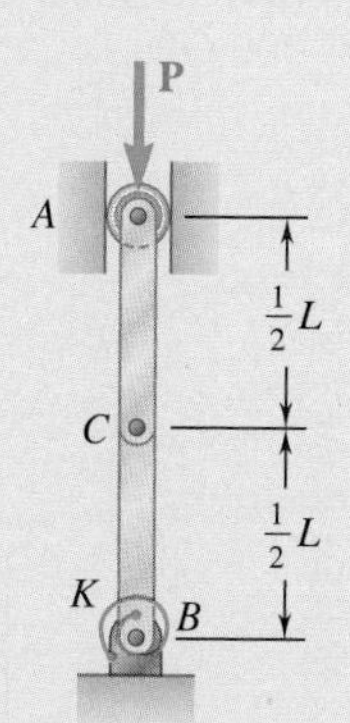

Fig. P16.4

16.5 The rigid bar AD is attached to two springs of constant k and is in equilibrium in the position shown. Knowing that the equal and opposite loads **P** and **P**′ *remain vertical*, determine the magnitude P_{cr} of the critical load for the system. Each spring can act in either tension or compression.

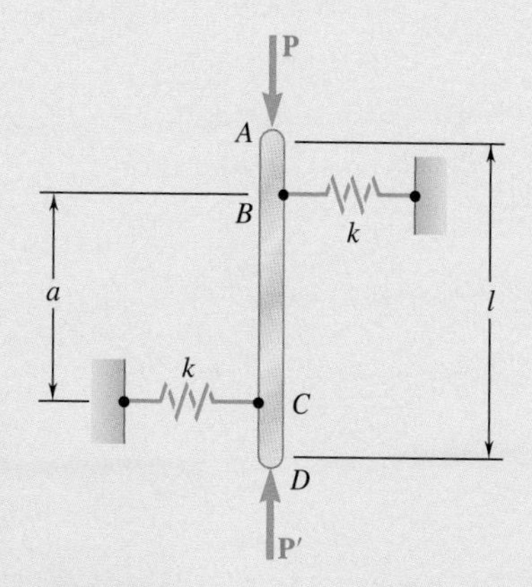

Fig. P16.5

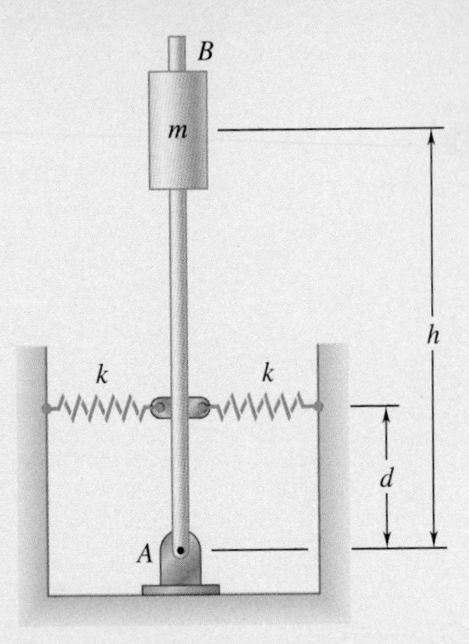

Fig. P16.6

16.6 The rigid rod AB is attached to a hinge at A and to two springs, each of constant k. If $h = 450$ mm, $d = 300$ mm, and $m = 200$ kg, determine the range of values of k for which the equilibrium of rod AB is stable in the position shown. Each spring can act in either tension or compression.

16.7 Determine the critical load of a round wooden dowel that is 48 in. long and has a diameter of (*a*) 0.375 in., (*b*) 0.5 in. Use $E = 1.6 \times 10^6$ psi.

16.8 Determine the critical load of a pin-ended steel tube that is 5 m long and has a 100-mm outer diameter and a 16-mm wall thickness. Use $E = 200$ GPa.

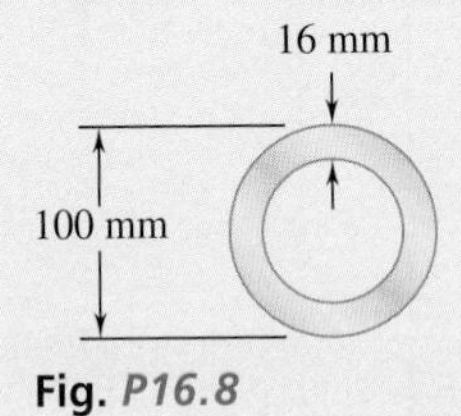

Fig. P16.8

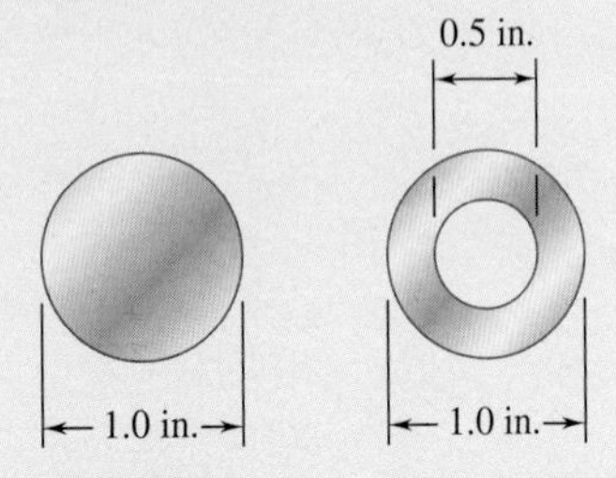

Fig. P16.9

16.9 A compression member of 20-in. effective length consists of a solid 1-in.-diameter aluminum rod. In order to reduce the weight of the member by 25%, the solid rod is replaced by a hollow rod of the cross section shown. Determine (*a*) the percent reduction in the critical load, (*b*) the value of the critical load for the hollow rod. Use $E = 10.6 \times 10^6$ psi.

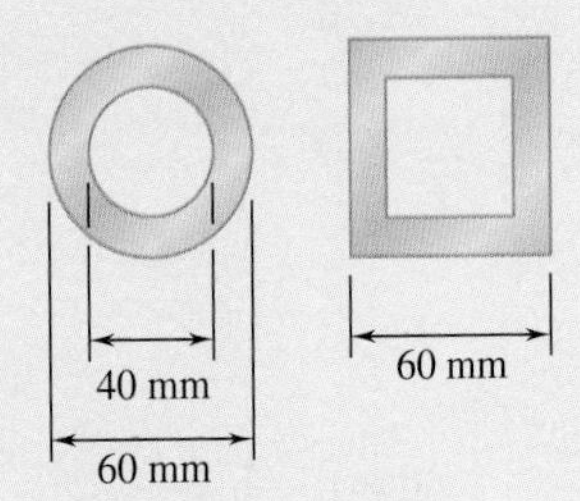

Fig. P16.10

16.10 Two brass rods used as compression members, each of 3-m effective length, have the cross sections shown. (*a*) Determine the wall thickness of the hollow square rod for which the rods have the same cross-sectional area. (*b*) Using $E = 105$ GPa, determine the critical load of each rod.

16.11 Determine the radius of the round strut so that the round and square struts have the same cross-sectional area, and compute the critical load for each. Use $E = 200$ GPa.

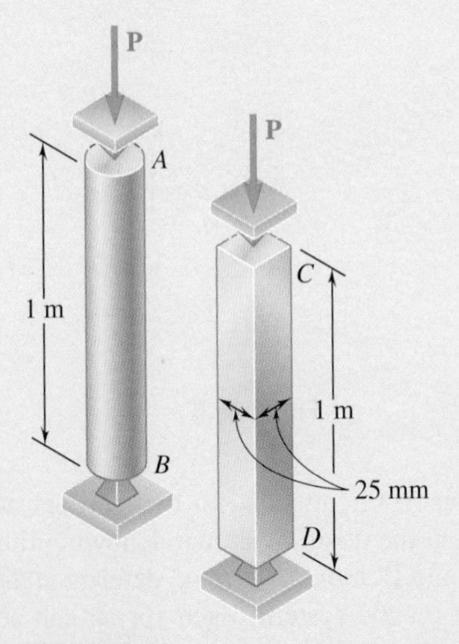

Fig. P16.11

16.12 A column of effective length L can be made by gluing together identical planks in either of the arrangements shown. Determine the ratio of the critical load using the arrangement a to the critical load using the arrangement b.

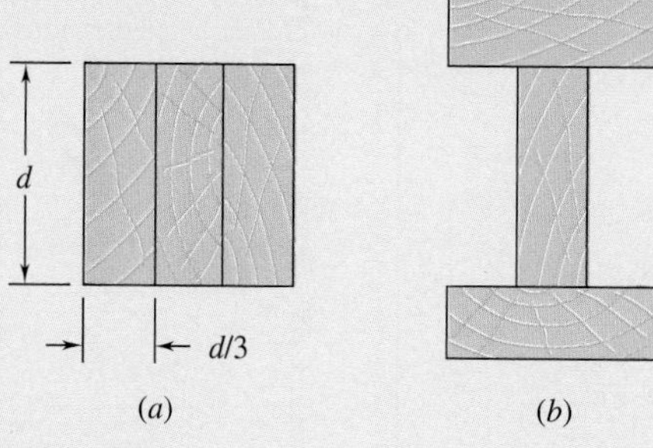

Fig. P16.12

16.13 A compression member of 7-m effective length is made by welding together two L152 × 102 × 12.7 angles as shown. Using $E = 200$ GPa, determine the allowable centric load for the member if a factor of safety of 2.2 is required.

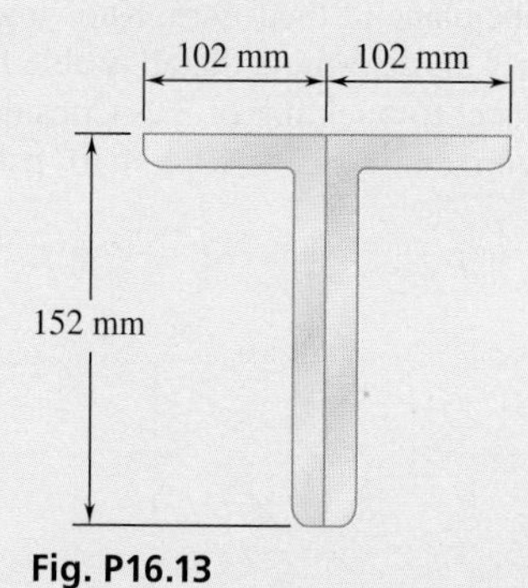

Fig. P16.13

16.14 A single compression member of 27-ft effective length is obtained by connecting two C8 × 11.5 steel channels with lacing bars as shown. Knowing that the factor of safety is 1.85, determine the allowable centric load for the member. Use $E = 29 \times 10^6$ psi and $d = 4.0$ in.

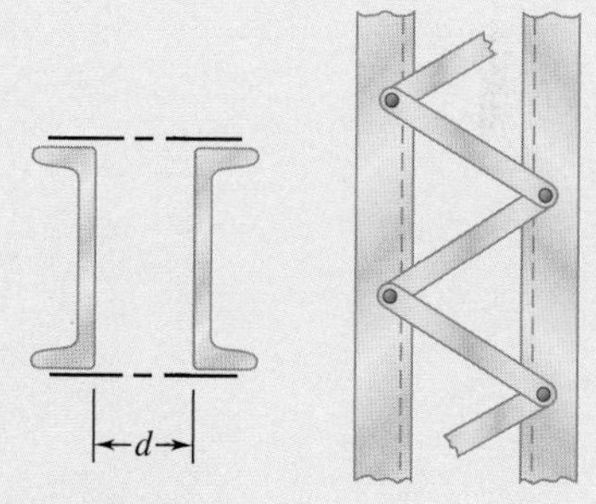

Fig. P16.14

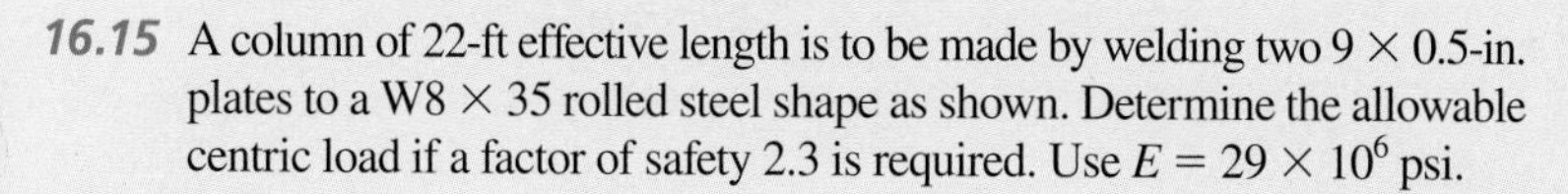

16.15 A column of 22-ft effective length is to be made by welding two 9 × 0.5-in. plates to a W8 × 35 rolled steel shape as shown. Determine the allowable centric load if a factor of safety 2.3 is required. Use $E = 29 \times 10^6$ psi.

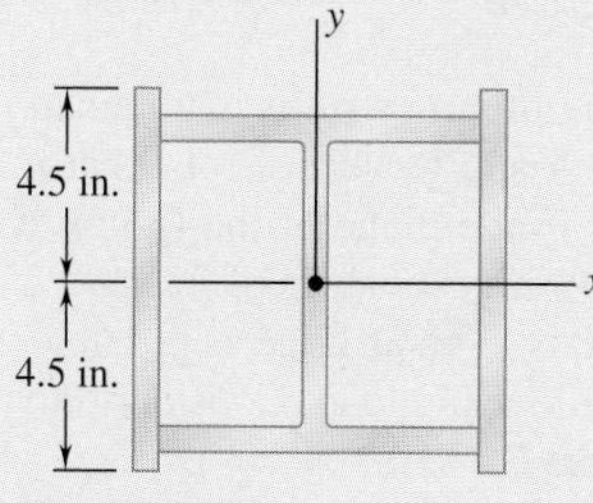

Fig. ***P16.15***

16.16 A column of 3-m effective length is to be made by welding together two C130 × 13 rolled-steel channels. Using $E = 200$ GPa, determine for each arrangement shown the allowable centric load if a factor of safety of 2.4 is required.

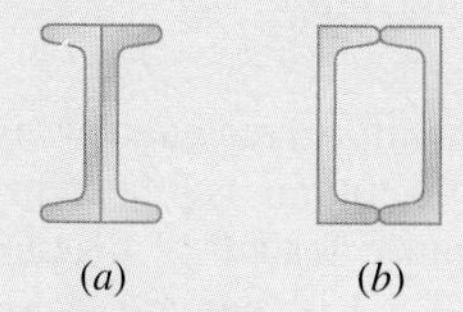

Fig. P16.16

16.17 Knowing that $P = 5.2$ kN, determine the factor of safety for the structure shown. Use $E = 200$ GPa and consider only buckling in the plane of the structure.

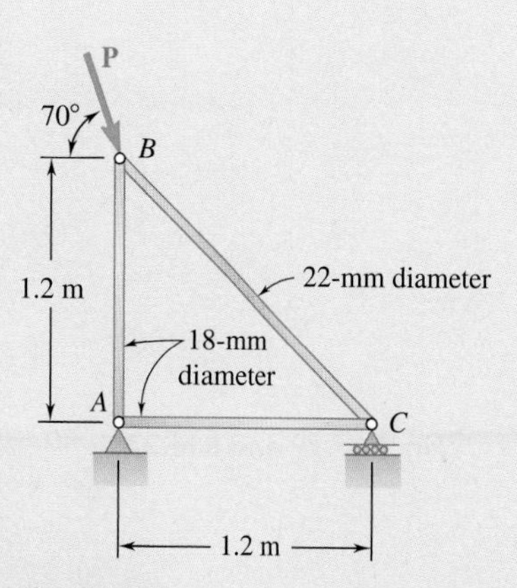

Fig. P16.17

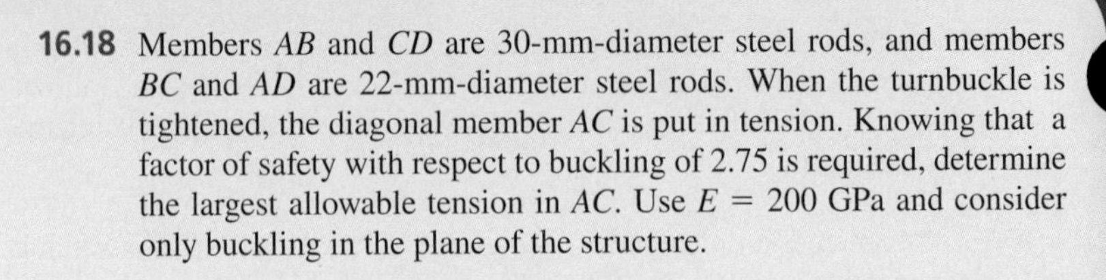
16.18 Members AB and CD are 30-mm-diameter steel rods, and members BC and AD are 22-mm-diameter steel rods. When the turnbuckle is tightened, the diagonal member AC is put in tension. Knowing that a factor of safety with respect to buckling of 2.75 is required, determine the largest allowable tension in AC. Use $E = 200$ GPa and consider only buckling in the plane of the structure.

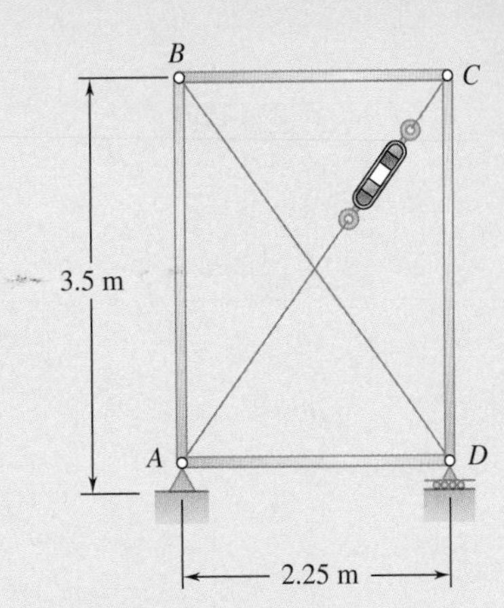

Fig. P16.18

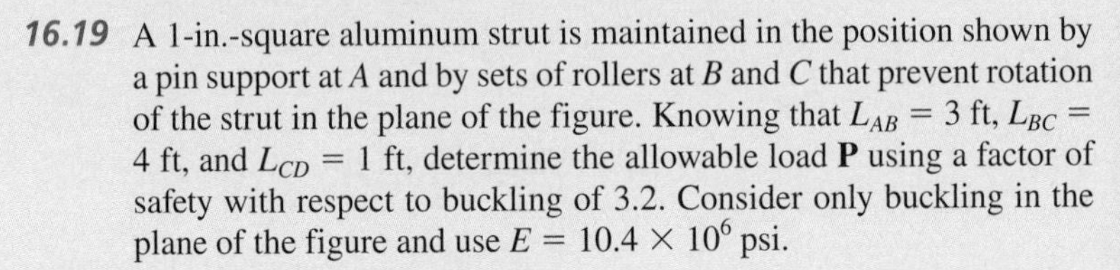
16.19 A 1-in.-square aluminum strut is maintained in the position shown by a pin support at A and by sets of rollers at B and C that prevent rotation of the strut in the plane of the figure. Knowing that $L_{AB} = 3$ ft, $L_{BC} = 4$ ft, and $L_{CD} = 1$ ft, determine the allowable load **P** using a factor of safety with respect to buckling of 3.2. Consider only buckling in the plane of the figure and use $E = 10.4 \times 10^6$ psi.

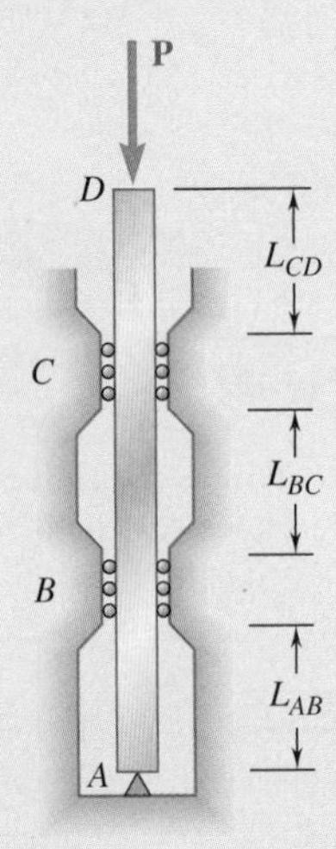

Fig. P16.19 and P16.20

16.20 A 1-in.-square aluminum strut is maintained in the position shown by a pin support at A and by sets of rollers at B and C that prevent rotation of the strut in the plane of the figure. Knowing that $L_{AB} = 3$ ft, determine (*a*) the largest values of L_{BC} and L_{CD} that can be used if the allowable load **P** is to be as large as possible, (*b*) the magnitude of the corresponding allowable load. Consider only buckling in the plane of the figure and use $E = 10.4 \times 10^6$ psi.

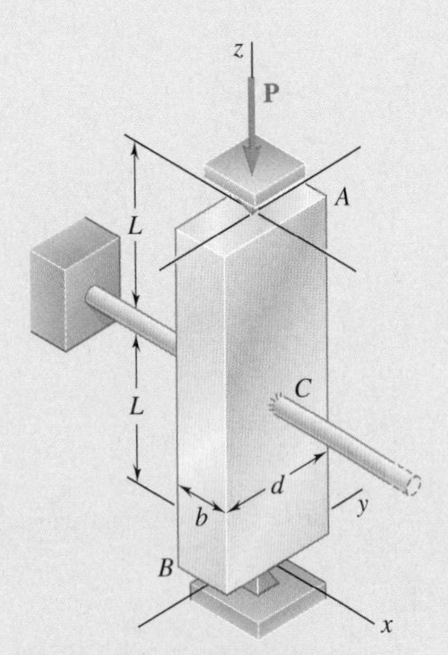

Fig. P16.21 and P16.22

16.21 Column ABC has a uniform rectangular cross section and is braced in the xz plane at its midpoint C. (*a*) Determine the ratio b/d for which the factor of safety is the same with respect to buckling in the xz and yz planes. (*b*) Using the ratio found in part *a*, design the cross section of the column so that the factor of safety will be 3.0 when $P = 4.4$ kN, $L = 1$ m, and $E = 200$ GPa.

16.22 Column ABC has a uniform rectangular cross section with $b = 12$ mm and $d = 22$ mm. The column is braced in the xz plane at its midpoint C and carries a centric load **P** of magnitude 3.8 kN. Knowing that a factor of safety of 3.2 is required, determine the largest allowable length L. Use $E = 200$ GPa.

16.23 A W8 × 21 rolled-steel shape is used with the support and cable arrangement shown. Cables *BC* and *BD* are taut and prevent motion of point *B* in the *xz* plane. Knowing that $L = 24$ ft, determine the allowable centric load **P** if a factor of safety of 2.2 is required. Use $E = 29 \times 10^6$ psi.

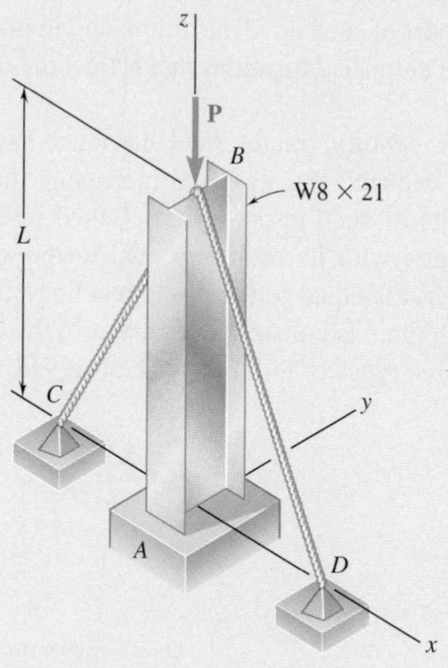

Fig. P16.23

16.24 Each of the five struts shown consists of a solid steel rod. (*a*) Knowing that strut (1) is of a 0.8-in. diameter, determine the factor of safety with respect to buckling for the loading shown. (*b*) Determine the diameter of each of the other struts for which the factor of safety is the same as the factor of safety obtained in part *a*. Use $E = 29 \times 10^6$ psi.

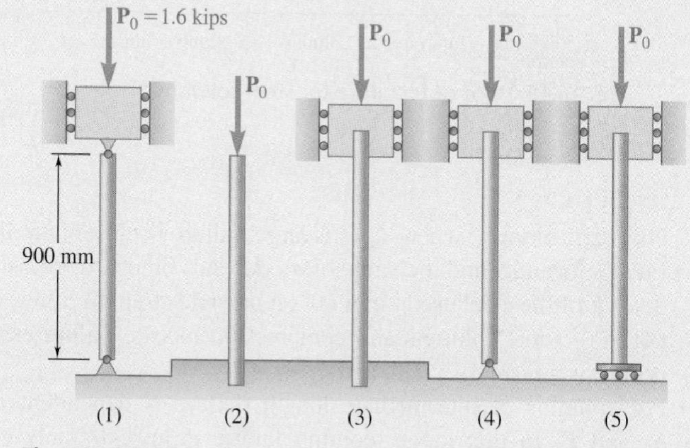

Fig. P16.24

16.2 CENTRIC LOAD DESIGN

The preceding sections determined the critical load of a column by using Euler's formula. We assumed that all stresses remained below the proportional limit, and the column was initially a straight, homogeneous prism. Real columns fall short of such an idealization, and in practice the design of columns is based on empirical formulas that reflect the results of numerous laboratory tests.

Over the last century, many steel columns have been tested by applying to them a centric axial load and increasing the load until failure occurred. The results of such tests are represented in Fig. 16.19 where a point has been plotted with its ordinate equal to the normal stress σ_{cr} at failure and its abscissa is equal to the corresponding effective slenderness ratio L_e/r. Although there is considerable scatter in the test results, regions corresponding to three types of failure can be observed.

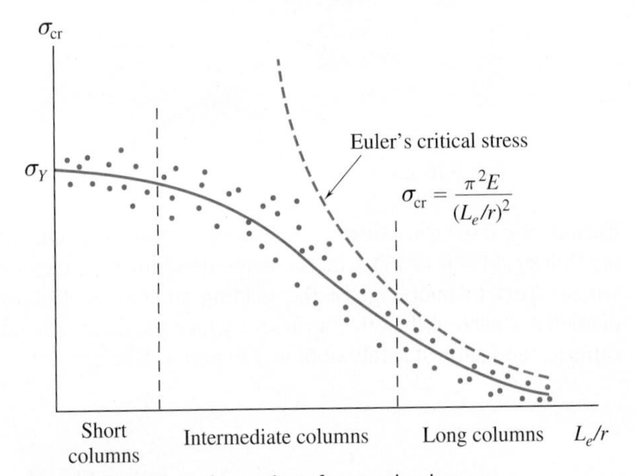

Fig. 16.19 Plot of test data for steel columns.

- For long columns, where L_e/r is large, failure is closely predicted by Euler's formula, and the value of σ_{cr} depends on the modulus of elasticity E of the steel used—but not on its yield strength σ_Y.
- For very short columns and compression blocks, failure essentially occurs as a result of yield, and $\sigma_{cr} \approx \sigma_Y$.
- For columns of intermediate length, failure is dependent on both σ_Y and E. In this range, column failure is an extremely complex phenomenon, and test data is used extensively to guide the development of specifications and design formulas.

Empirical formulas for an allowable or critical stress given in terms of the effective slenderness ratio were first introduced over a century ago. Since then, they have undergone a process of refinement and improvement. Typical empirical formulas used to approximate test data are shown in Fig. 16.20. It is not always possible to use a single formula for all values of L_e/r. Most design specifications use different formulas—each

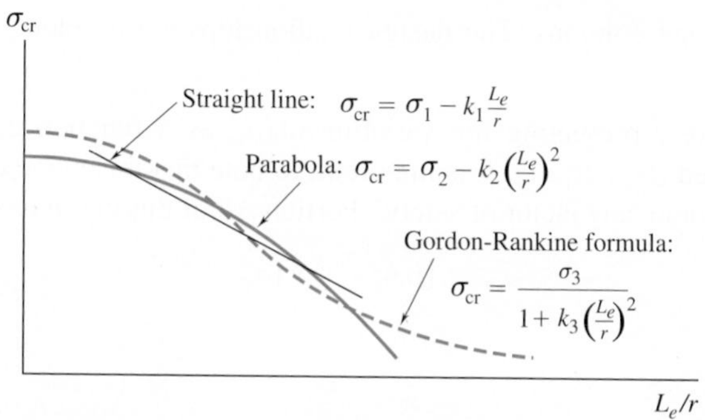

Fig. 16.20 Plots of empirical formulas for critical stresses.

with a definite range of applicability. In each case we must check that the equation used is applicable for the value of L_e/r for the column involved. Furthermore, it must determined whether the equation provides the critical stress for the column, to which the appropriate factor of safety must be applied, or if it provides an allowable stress.

Photo 16.2 shows examples of columns that are designed using such design specification formulas. The design formulas for three different materials using *Allowable Stress Design* are presented next, followed by formulas for the design of steel columns based on *Load and Resistance Factor Design.*†

(*a*) (*b*)

© *Steve Photo/Alamy RF* © *Peter Marlow/Magnum Photos*

Photo 16.2 (*a*) The water tank is supported by steel columns. (*b*) The house under construction is framed with wood columns.

16.2A Allowable Stress Design

Structural Steel. The most commonly used formulas for allowable stress design of steel columns under a centric load are found in the *Specification for Structural Steel Buildings* of the American Institute of Steel Construction.‡ An exponential expression is used to predict σ_{all} for columns of short and intermediate lengths, and an Euler-based relation

†In specific design formulas, the letter L always refers to the effective length of the column.

‡*Manual of Steel Construction,* 14th ed., American Institute of Steel Construction, Chicago, 2011.

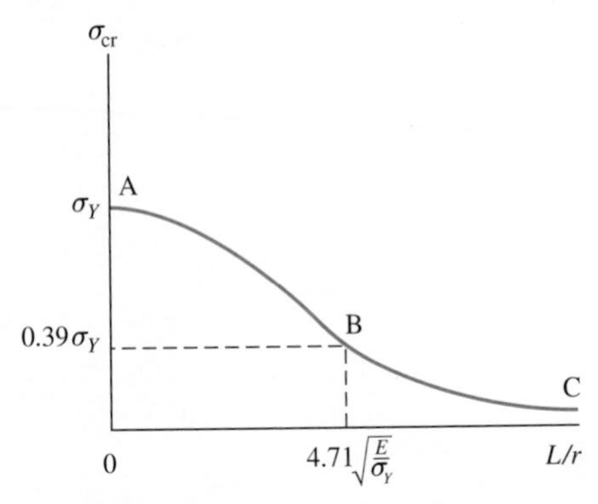

Fig. 16.21 Design curve for columns recommended by the American Institute of Steel Construction.

is used for long columns. The design relationships are developed in two steps.

1. A curve representing the variation of σ_{cr} as a function of L/r is obtained (Fig. 16.21). It is important to note that this curve does not incorporate any factor of safety.[§] Portion AB of this curve is

$$\sigma_{cr} = [0.658^{(\sigma_Y/\sigma_e)}]\sigma_Y \tag{16.22}$$

where

$$\sigma_e = \frac{\pi^2 E}{(L/r)^2} \tag{16.23}$$

Portion BC is

$$\sigma_{cr} = 0.877\sigma_e \tag{16.24}$$

When $L/r = 0$, $\sigma_{cr} = \sigma_Y$ in Eq. (16.22). At point B, Eq. (16.22) intersects Eq. (16.24). The slenderness L/r at the junction between the two equations is

$$\frac{L}{r} = 4.71\sqrt{\frac{E}{\sigma_Y}} \tag{16.25}$$

If L/r is smaller than the value from Eq. (16.25), σ_{cr} is determined from Eq. (16.22). If L/r is greater, σ_{cr} is determined from Eq. (16.24). At the slenderness L/r specified in Eq. (16.25), the stress $\sigma_e = 0.44\ \sigma_Y$. Using Eq. (16.24), $\sigma_{cr} = 0.877\ (0.44\ \sigma_Y) = 0.39\ \sigma_Y$.

2. A factor of safety must be used for the final design. The factor of safety given by the specification is 1.67. Thus,

$$\sigma_{all} = \frac{\sigma_{cr}}{1.67} \tag{16.26}$$

These equations can be used with SI or U.S. customary units.

By using Eqs. (16.22), (16.24), (16.25), and (16.26), the allowable axial stress can be determined for a given grade of steel and any given value of L/r. The procedure is to compute L/r at the intersection between the two equations from Eq. (16.25). For smaller given values of L/r, use Eqs. (16.22) and (16.26) to calculate σ_{all}, and if greater, use Eqs. (16.24) and (16.26). Figure 16.22 provides an example of how σ_e varies as a function of L/r for different grades of structural steel.

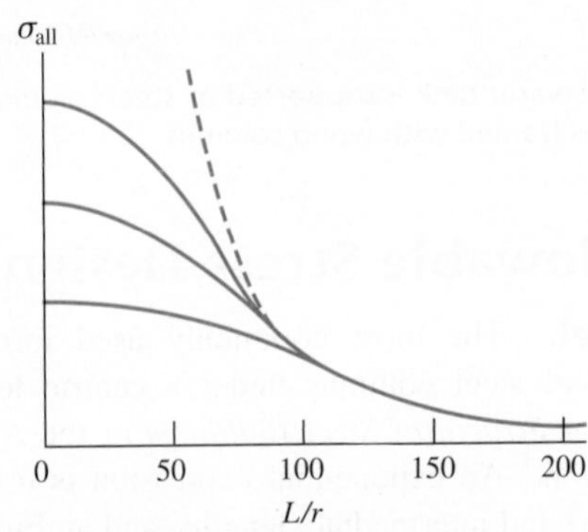

Fig. 16.22 Steel column design curves for different grades of steel.

[§]In the *Specification for Structural Steel Buildings*, the symbol F is used for stresses.

Concept Application 16.2

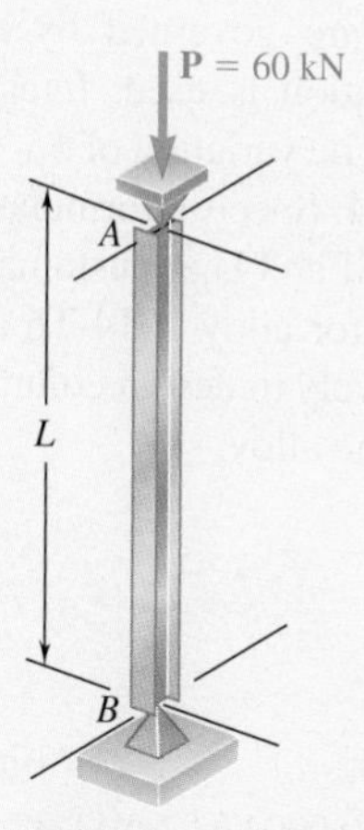

Fig. 16.23 Centrically loaded S100 × 11.5 rolled-steel member.

Determine the longest unsupported length L for which the S100 × 11.5 rolled-steel compression member AB can safely carry the centric load shown (Fig. 16.23). Assume $\sigma_Y = 250$ MPa and $E = 200$ GPa.

From Appendix C, for an S100 × 11.5 shape,

$$A = 1460 \text{ mm}^2 \qquad r_x = 41.7 \text{ mm} \qquad r_y = 14.6 \text{ mm}$$

If the 60-kN load is to be safely supported,

$$\sigma_{\text{all}} = \frac{P}{A} = \frac{60 \times 10^3 \text{ N}}{1460 \times 10^{-6} \text{ m}^2} = 41.1 \times 10^6 \text{ Pa}$$

To compute the critical stress σ_{cr}, we start by assuming that L/r is larger than the slenderness specified by Eq. (16.25). We then use Eq. (16.24) with Eq. (16.23) and write

$$\sigma_{\text{cr}} = 0.877\,\sigma_e = 0.877\frac{\pi^2 E}{(L/r)^2}$$

$$= 0.877\frac{\pi^2(200 \times 10^9 \text{ Pa})}{(L/r)^2} = \frac{1.731 \times 10^{12} \text{ Pa}}{(L/r)^2}$$

Using this expression in Eq. (16.26),

$$\sigma_{\text{all}} = \frac{\sigma_{\text{cr}}}{1.67} = \frac{1.037 \times 10^{12} \text{ Pa}}{(L/r)^2}$$

Equating this expression to the required value of σ_{all} gives

$$\frac{1.037 \times 10^{12} \text{ Pa}}{(L/r)^2} = 41.1 \times 10^6 \text{ Pa} \qquad L/r = 158.8$$

The slenderness ratio from Eq. (16.25) is

$$\frac{L}{r} = 4.71\sqrt{\frac{200 \times 10^9}{250 \times 10^6}} = 133.2$$

Our assumption that L/r is greater than this slenderness ratio is correct. Choosing the smaller of the two radii of gyration:

$$\frac{L}{r_y} = \frac{L}{14.6 \times 10^{-3} \text{ m}} = 158.8 \qquad L = 2.32 \text{ m}$$

Aluminum. Many aluminum alloys are used in structures and machines. The specifications of the Aluminum Association† provides formulas based on three slenderness ranges. Short columns are governed by material failure. For long columns, an Euler-type equation is used. Intermediate columns are governed by a quadratic equation. The variation of σ_{all} with L/r defined by these formulas is shown in Fig. 16.24. Specific formulas for the design of building structures are given in both SI and U.S. customary units for two commonly used alloys. The equations for alloy 2014-T6 apply to extrusions, but they can also be used conservatively to design columns with non-extruded cross sections made from this same alloy.

Alloy 6061-T6:

$$L/r \leq 17.8: \quad \sigma_{all} = [21.2] \text{ ksi} \quad \textbf{(16.27a)}$$

$$= [146.3] \text{ MPa} \quad \textbf{(16.27b)}$$

$$17.8 > L/r < 66.0: \sigma_{all} = [25.2 - 0.232(L/r) + 0.00047(L/r)^2] \text{ ksi} \quad \textbf{(16.28a)}$$

$$= [173.9 - 1.602(L/r) + 0.00323(L/r)^2] \text{ MPa} \quad \textbf{(16.28b)}$$

$$L/r \geq 66.0: \quad \sigma_{all} = \frac{51{,}400 \text{ ksi}}{(L/r)^2} \quad \sigma_{all} = \frac{356 \times 10^3 \text{ MPa}}{(L/r)^2} \quad \textbf{(16.29a,b)}$$

Alloy 2014-T6:

$$L/r \leq 17.0: \quad \sigma_{all} = [32.1] \text{ ksi} \quad \textbf{(16.30a)}$$

$$= [221.5] \text{ MPa} \quad \textbf{(16.30b)}$$

$$17.0 > L/r < 52.7: \sigma_{all} = [39.7 - 0.465(L/r) + 0.00121(L/r)^2] \text{ ksi} \quad \textbf{(16.31a)}$$

$$= [273.6 - 3.205(L/r) + 0.00836(L/r)^2] \text{ MPa} \quad \textbf{(16.31b)}$$

$$L/r \geq 52.7: \quad \sigma_{all} = \frac{51{,}400 \text{ ksi}}{(L/r)^2} \quad \sigma_{all} = \frac{356 \times 10^3 \text{ MPa}}{(L/r)^2} \quad \textbf{(16.32a,b)}$$

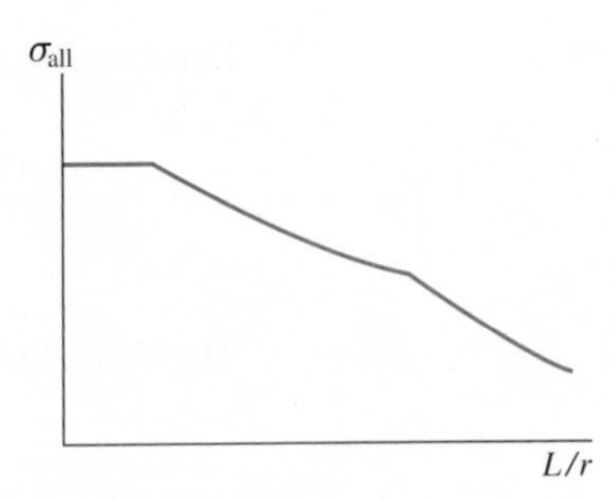

Fig. 16.24 Design curve for aluminum columns recommended by the Aluminum Association.

Wood. For the design of wood columns, the specifications of the American Forest & Paper Association‡ provide a single equation to obtain the allowable stress for short, intermediate, and long columns under centric

†*Specifications for Aluminum Structures,* Aluminum Association, Inc., Washington, D.C., 2015.

‡*National Design Specification for Wood Construction,* American Forest & Paper Association, American Wood Council, Washington, D.C., 2015.

loading. For a column with a *rectangular* cross section of sides b and d, where $d < b$, the variation of σ_{all} with L/d is shown in Fig. 16.25.

For solid columns made from a single piece of wood or by gluing laminations together, the allowable stress σ_{all} is

$$\sigma_{all} = \sigma_C C_P \tag{16.33}$$

where σ_C is the adjusted allowable stress for compression parallel to the grain.[§] Adjustments for σ_C are included in the specifications to account for different variations (such as in the load duration). The column stability factor C_P accounts for the column length and is defined by

$$C_P = \frac{1 + (\sigma_{CE}/\sigma_C)}{2c} - \sqrt{\left[\frac{1 + (\sigma_{CE}/\sigma_C)}{2c}\right]^2 - \frac{\sigma_{CE}/\sigma_C}{c}} \tag{16.34}$$

The parameter c accounts for the type of column, and it is equal to 0.8 for sawn lumber columns and 0.90 for glued laminated wood columns. The value of σ_{CE} is defined as

$$\sigma_{CE} = \frac{0.822E}{(L/d)^2} \tag{16.35}$$

where E is an adjusted modulus of elasticity for column buckling. Columns in which L/d exceeds 50 are not permitted by the *National Design Specification for Wood Construction.*

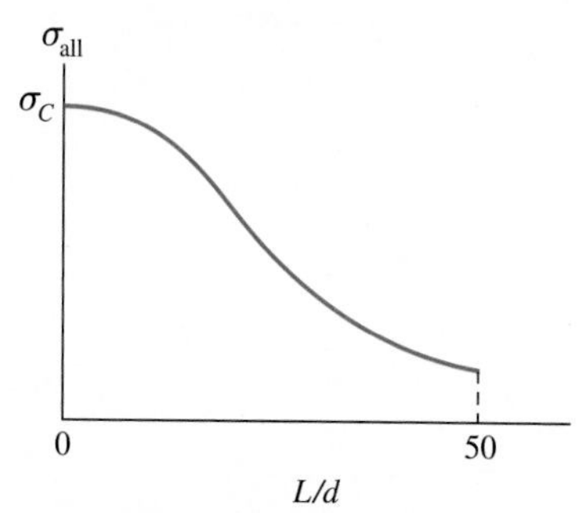

Fig. 16.25 Design curve for columns recommended by the American Forest & Paper Association.

Concept Application 16.3

Knowing that column AB (Fig. 16.26) has an effective length of 14 ft and must safely carry a 32-kip load, design the column using a square glued laminated cross section. The adjusted modulus of elasticity for the wood is $E = 800 \times 10^3$ psi, and the adjusted allowable stress for compression parallel to the grain is $\sigma_C = 1060$ psi.

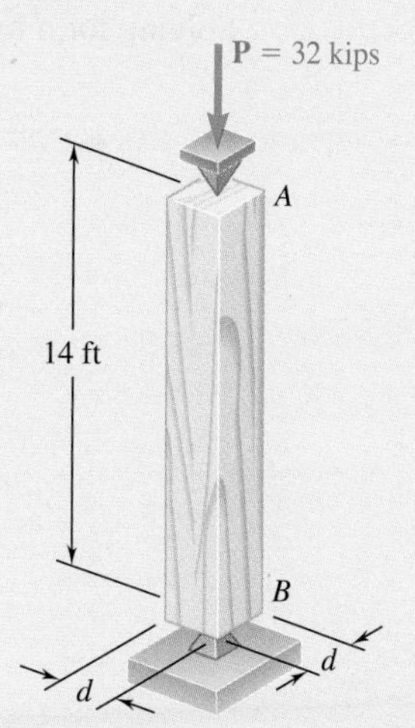

Fig. 16.26 Centrically loaded wood column.

[§]In the *National Design Specification for Wood Construction,* the symbol F is used for stresses.

Note that $c = 0.90$ for glued laminated wood columns. Computing the value of σ_{CE}, using Eq. (10.35), gives

$$\sigma_{CE} = \frac{0.822E}{(L/d)^2} = \frac{0.822(800 \times 10^3 \text{ psi})}{(168 \text{ in.}/d)^2} = 23.299d^2 \text{ psi}$$

Eq. (16.34) is used to express the column stability factor in terms of d, with $(\sigma_{CE}/\sigma_C) = (23.299d^2/1.060 \times 10^3) = 21.98 \times 10^{-3}\, d^2$,

$$C_P = \frac{1 + (\sigma_{CE}/\sigma_C)}{2c} - \sqrt{\left[\frac{1 + (\sigma_{CE}/\sigma_C)}{2c}\right]^2 - \frac{\sigma_{CE}/\sigma_C}{c}}$$

$$= \frac{1 + 21.98 \times 10^{-3}\, d^2}{2(0.90)} - \sqrt{\left[\frac{1 + 21.98 \times 10^{-3}\, d^2}{2(0.90)}\right]^2 - \frac{21.98 \times 10^{-3}\, d^2}{0.90}}$$

Since the column must carry 32 kips, Eq. (16.33) gives

$$\sigma_{\text{all}} = \frac{32 \text{ kips}}{d^2} = \sigma_C C_P = 1.060 C_P$$

Solving this equation for C_P and substituting the value into the previous equation, we obtain

$$\frac{30.19}{d^2} = \frac{1 + 21.98 \times 10^{-3}\, d^2}{2(0.90)} - \sqrt{\left[\frac{1 + 21.98 \times 10^{-3}\, d^2}{2(0.90)}\right]^2 - \frac{21.98 \times 10^{-3}\, d^2}{0.90}}$$

Solving for d by trial and error yields $d = 6.45$ in.

Note: The design formulas presented throughout Sec. 16.2 are examples of different design approaches. These equations do not provide all of the requirements needed for many designs, and the student should refer to the appropriate design specifications before attempting actual designs.

W10 × 39
$A = 11.5\ \text{in}^2$
$r_x = 4.27$ in.
$r_y = 1.98$ in.

Sample Problem 16.2

Column AB consists of a W10 × 39 rolled-steel shape made of a grade of steel for which $\sigma_Y = 36$ ksi and $E = 29 \times 10^6$ psi. Determine the allowable centric load **P** (*a*) if the effective length of the column is 24 ft in all directions, (*b*) if bracing is provided to prevent the movement of the midpoint C in the xz plane. (Assume that the movement of point C in the yz plane is not affected by the bracing.)

STRATEGY: The allowable centric load for part *a* is determined from the governing allowable stress design equation for steel, Eq. (16.22) or Eq. (16.24), based on buckling associated with the axis with a smaller radius of gyration since the effective lengths are the same. In part *b*, it is necessary to determine the effective slenderness ratios for both axes, including the reduced effective length due to the bracing. The larger slenderness ratio governs the design.

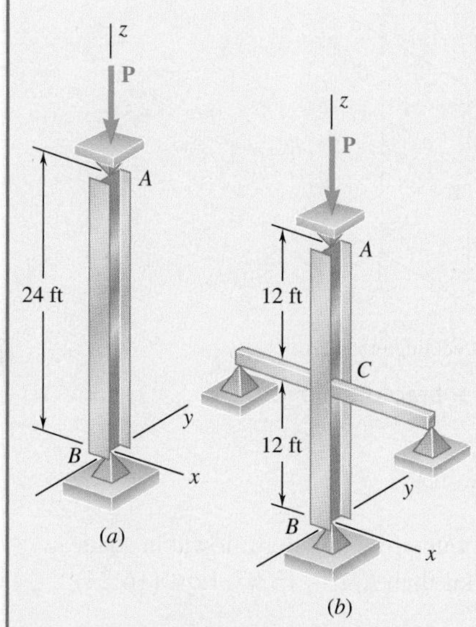

Fig. 1 Centrically loaded column (a) unbraced, (b) braced.

MODELING: First compute the slenderness ratio from Eq. (16.25) corresponding to the given yield strength $\sigma_Y = 36$ ksi.

$$\frac{L}{r} = 4.71\sqrt{\frac{29 \times 10^6}{36 \times 10^3}} = 133.7$$

ANALYSIS:

a. Effective Length = 24 ft. The column is shown in Fig. 1*a*. Knowing that $r_y < r_x$, buckling takes place in the xz plane (Fig. 2). For $L =$ 24 ft and $r = r_y = 1.98$ in., the slenderness ratio is

$$\frac{L}{r_y} = \frac{(24 \times 12)\ \text{in.}}{1.98\ \text{in.}} = \frac{288\ \text{in.}}{1.98\ \text{in.}} = 145.5$$

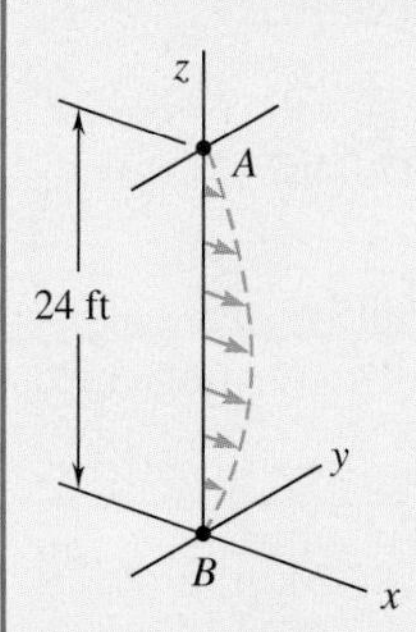

Fig. 2 Buckled shape for unbraced column.

Since $L/r > 133.7$, Eq. (16.23) in Eq. (16.24) is used to determine

$$\sigma_{cr} = 0.877\sigma_e = 0.877\frac{\pi^2 E}{(L/r)^2} = 0.877\frac{\pi^2(29 \times 10^3\ \text{ksi})}{(145.5)^2} = 11.86\ \text{ksi}$$

The allowable stress is determined using Eq. (16.26)

$$\sigma_{\text{all}} = \frac{\sigma_{cr}}{1.67} = \frac{11.86\ \text{ksi}}{1.67} = 7.10\ \text{ksi}$$

and

$$P_{\text{all}} = \sigma_{\text{all}}A = (7.10\ \text{ksi})(11.5\ \text{in}^2) = 81.7\ \text{kips}$$ ◀

(continued)

b. Bracing at Midpoint C. The column is shown in Fig. 1*b*. Since bracing prevents movement of point *C* in the *xz* plane but not in the *yz* plane, the slenderness ratio corresponding to buckling in each plane (Fig. 3) is computed to determine which is larger.

***xz* Plane:** Effective length = 12 ft = 144 in., $r = r_y = 1.98$ in.

$$L/r = (144 \text{ in.})/(1.98 \text{ in.}) = 72.7$$

***yz* Plane:** Effective length = 24 ft = 288 in., $r = r_x = 4.27$ in.

$$L/r = (288 \text{ in.})/(4.27 \text{ in.}) = 67.4$$

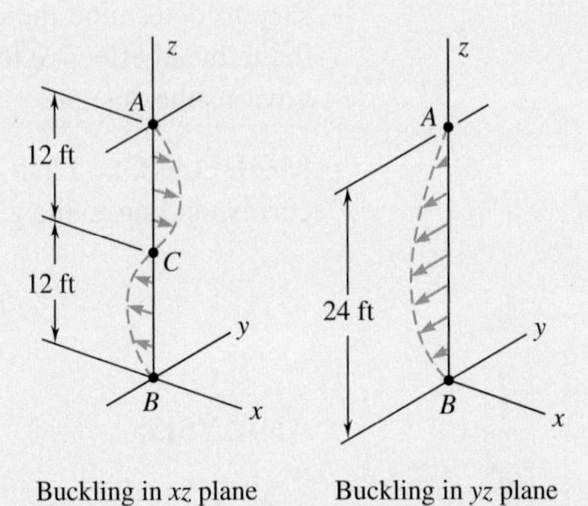

Fig. 3 Buckled shapes for braced column.

Since the larger slenderness ratio corresponds to a smaller allowable load, we choose $L/r = 72.7$. Since this is smaller than $L/r = 133.7$, Eqs. (16.23) and (16.22) are used to determine σ_{cr}:

$$\sigma_e = \frac{\pi^2 E}{(L/r)^2} = \frac{\pi^2(29 \times 10^3 \text{ ksi})}{(72.7)^2} = 54.1 \text{ ksi}$$

$$\sigma_{cr} = [0.658^{(\sigma_Y/\sigma_e)}]\, F_Y = [0.658^{(36 \text{ ksi}/54.1 \text{ ksi})}]\, 36 \text{ ksi} = 27.3 \text{ ksi}$$

The allowable stress using Eq. (16.26) and the allowable load are

$$\sigma_{\text{all}} = \frac{\sigma_{cr}}{1.67} = \frac{27.3 \text{ ksi}}{1.67} = 16.32 \text{ ksi}$$

$$P_{\text{all}} = \sigma_{\text{all}} A = (16.32 \text{ ksi})(11.5 \text{ in}^2) \qquad P_{\text{all}} = 187.7 \text{ kips} \blacktriangleleft$$

REFLECT and THINK: This sample problem shows the benefit of using bracing to reduce the effective length for buckling about the weak axis when a column has significantly different radii of gyration, which is typical for steel wide-flange columns.

Sample Problem 16.3

Using the aluminum alloy 2014-T6 for the circular rod shown, determine the smallest diameter that can be used to support the centric load $P = 60$ kN if (*a*) $L = 750$ mm, (*b*) $L = 300$ mm.

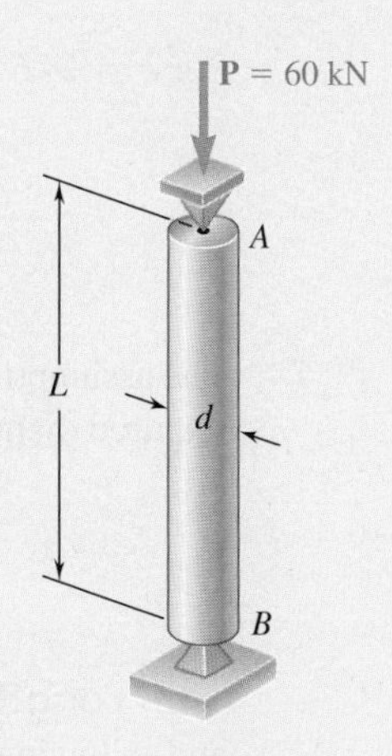

STRATEGY: Use the aluminum allowable stress equations to design the column, i.e., to determine the smallest diameter that can be used. Since there are two design equations based on L/r, it is first necessary to assume which governs. Then check the assumption.

MODELING: For the cross section of the solid circular rod shown in Fig. 1,

$$I = \frac{\pi}{4}c^4 \qquad A = \pi c^2 \qquad r = \sqrt{\frac{I}{A}} = \sqrt{\frac{\pi c^4/4}{\pi c^2}} = \frac{c}{2}$$

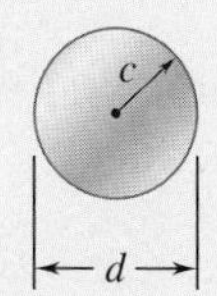

Fig. 1 Cross section of aluminum column.

ANALYSIS:

a. Length of 750 mm. Since the diameter of the rod is not known, L/r must be assumed. Assume that $L/r > 52.7$ and use Eq. (16.32). For the centric load **P**, $\sigma = P/A$ and write

$$\frac{P}{A} = \sigma_{\text{all}} = \frac{356 \times 10^3 \text{ MPa}}{(L/r)^2}$$

(continued)

$$\frac{60 \times 10^3\ \text{N}}{\pi c^2} = \frac{356 \times 10^9\ \text{Pa}}{\left(\dfrac{0.750\ \text{m}}{c/2}\right)^2}$$

$$c^4 = 120.7 \times 10^{-9}\ \text{m}^4 \qquad c = 18.64\ \text{mm}$$

For $c = 18.64$ mm, the slenderness ratio is

$$\frac{L}{r} = \frac{L}{c/2} = \frac{750\ \text{mm}}{(18.64\ \text{mm})/2} = 80.5 > 52.7$$

The assumption that L/r is greater than 52.7 is correct. For $L = 750$ mm, the required diameter is

$$d = 2c = 2(18.64\ \text{mm}) \qquad d = 37.3\ \text{mm} \blacktriangleleft$$

b. Length of 300 mm. Assume that $L/r > 52.7$. Using Eq. (16.32) and following the procedure used in part a, $c = 11.79$ mm and $L/r = 50.9$. Since L/r is less than 52.7, this assumption is wrong. Now assume that $L/r < 52.7$ and use Eq. (16.31b) for the design of this rod.

$$\frac{P}{A} = \sigma_{\text{all}} = \left[273.6 - 3.205\left(\frac{L}{r}\right) + 0.00836\left(\frac{L}{r}\right)^2\right]\ \text{MPa}$$

$$\frac{60 \times 10^3\ \text{N}}{\pi c^2} = \left[273.6 - 3.205\left(\frac{0.3m}{c/2}\right) + 0.00836\left(\frac{0.3m}{c/2}\right)^2\right] 10^6\ \text{Pa}$$

$$c = 11.95\ \text{mm}$$

For $c = 11.95$ mm, the slenderness ratio is

$$\frac{L}{r} = \frac{L}{c/2} = \frac{300\ \text{mm}}{(11.95\ \text{mm})/2} = 50.2$$

The second assumption that $L/r < 52.7$ is correct. For $L = 300$ mm, the required diameter is

$$d = 2c = 2(11.95\ \text{mm}) \qquad d = 23.9\ \text{mm} \blacktriangleleft$$

Problems

16.25 A steel pipe having the cross section shown is used as a column. Using allowable stress design, determine the allowable centric load if the effective length of the column is (*a*) 18 ft, (*b*) 26 ft. Use $\sigma_Y = 36$ ksi and $E = 29 \times 10^6$ psi.

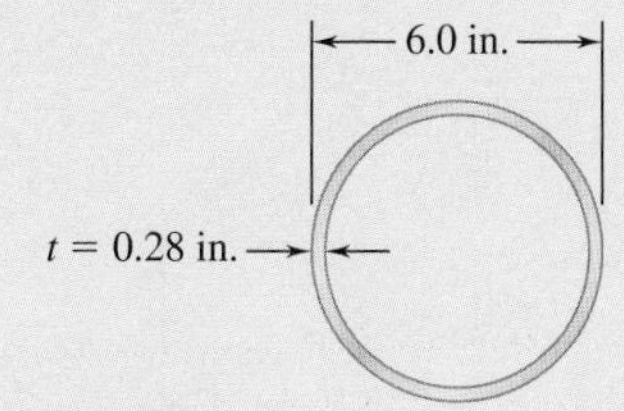

Fig. P16.25

16.26 A rectangular structural tube having the cross section shown is used as a column of 5-m effective length. Knowing that $\sigma_Y = 250$ MPa and $E = 200$ GPa, use allowable stress design to determine the largest centric load that can be applied to the steel column.

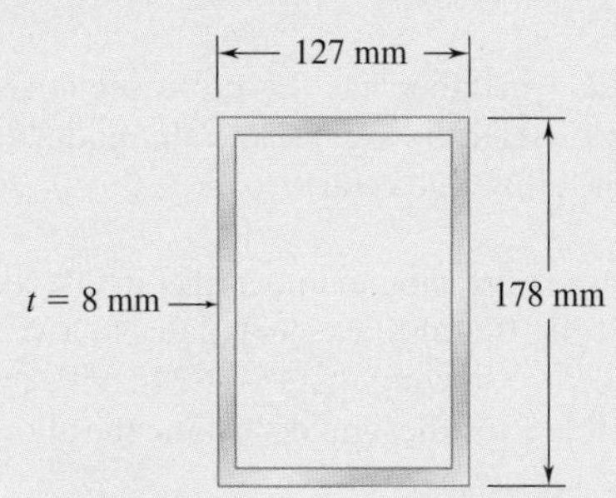

Fig. P16.26

16.27 Using allowable stress design, determine the allowable centric load for a column of 6-m effective length that is made from the following rolled-steel shape: (*a*) W200 × 35.9, (*b*) W200 × 86. Use $\sigma_Y = 250$ MPa and $E = 200$ GPa.

16.28 A W8 × 31 rolled-steel shape is used for a column of 21-ft effective length. Using allowable stress design, determine the allowable centric load if the yield strength of the grade of steel used is (*a*) $\sigma_Y = 36$ ksi, (*b*) $\sigma_Y = 50$ ksi. Use $E = 29 \times 10^6$ psi.

16.29 A column having a 3.5-m effective length is made of sawn lumber with a 114 × 140-mm cross section. Knowing that for the grade of wood used the adjusted allowable stress for compression parallel to the grain is $\sigma_C = 7.6$ MPa and the adjusted modulus $E = 2.8$ GPa, determine the maximum allowable centric load for the column.

16.30 A sawn lumber column with a 7.5 × 5.5-in. cross section has an 18-ft effective length. Knowing that for the grade of wood used the adjusted allowable stress for compression parallel to the grain is $\sigma_C = 1200$ psi and that the adjusted modulus $E = 470 \times 10^3$ psi, determine the maximum allowable centric load for the column.

16.31 Using the aluminum alloy 2014-T6, determine the largest allowable length of the aluminum bar AB for a centric load **P** of magnitude (*a*) 150 kN, (*b*) 90 kN, (*c*) 25 kN.

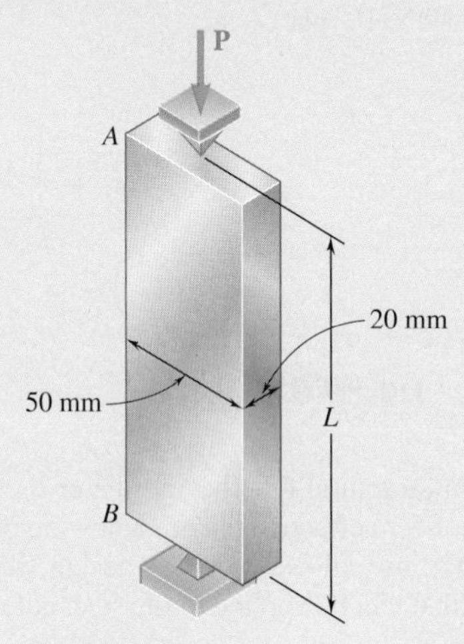

Fig. P16.31

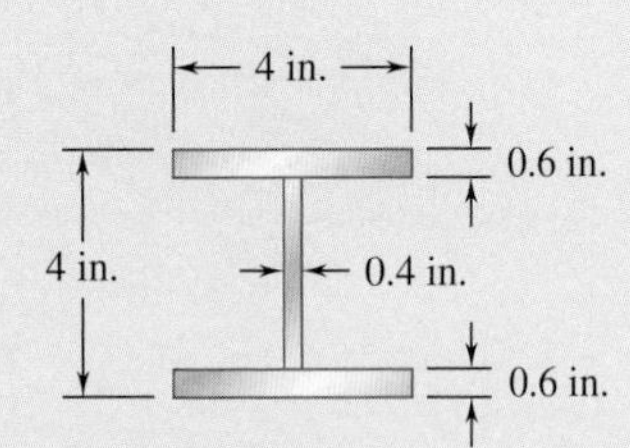

Fig. P16.32

16.32 A compression member has the cross section shown and an effective length of 5 ft. Knowing that the aluminum alloy used is 6061-T6, determine the allowable centric load.

16.33 and 16.34 A compression member of 9-m effective length is obtained by welding two 10-mm-thick steel plates to a W250 × 80 rolled-steel shape as shown. Knowing that $\sigma_Y = 345$ MPa and $E = 200$ GPa and using allowable stress design, determine the allowable centric load for the compression member.

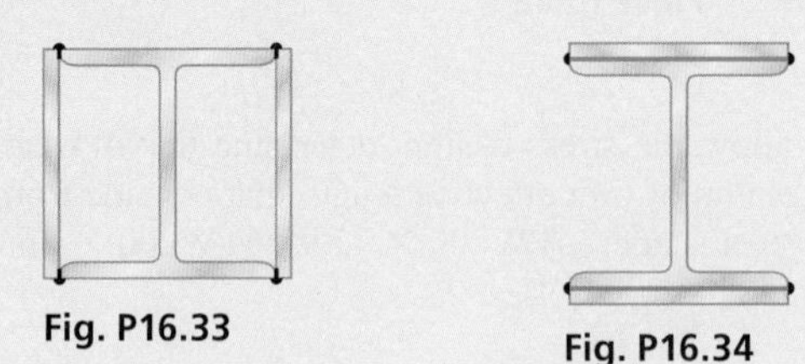

Fig. P16.33

Fig. P16.34

16.35 A compression member of 8.2-ft effective length is obtained by bolting together two L5 × 3 × $\frac{1}{2}$-in. steel angles as shown. Using allowable stress design, determine the allowable centric load for the column. Use $\sigma_Y = 36$ ksi and $E = 29 \times 10^6$ psi.

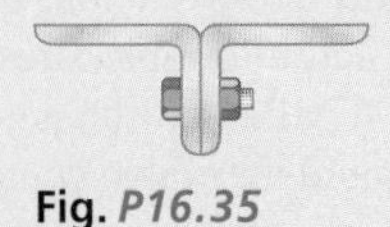

Fig. P16.35

16.36 A column of 18-ft effective length is obtained by connecting four L3 × 3 × $\frac{3}{8}$-in. steel angles with lacing bars as shown. Using allowable stress design, determine the allowable centric load for the column. Use $\sigma_Y = 36$ ksi and $E = 29 \times 10^6$ psi.

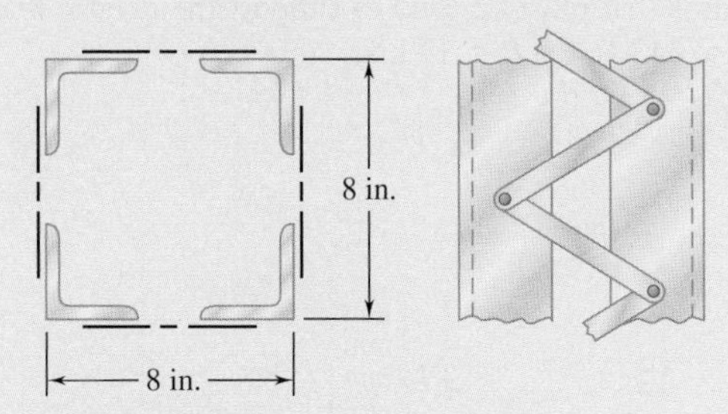

Fig. P16.36

16.37 A rectangular column with a 4.4-m effective length is made of glued laminated wood. Knowing that for the grade of wood used the adjusted allowable stress for compression parallel to the grain is $\sigma_C =$ 8.3 MPa and the adjusted modulus $E = 4.6$ GPa, determine the maximum allowable centric load for the column.

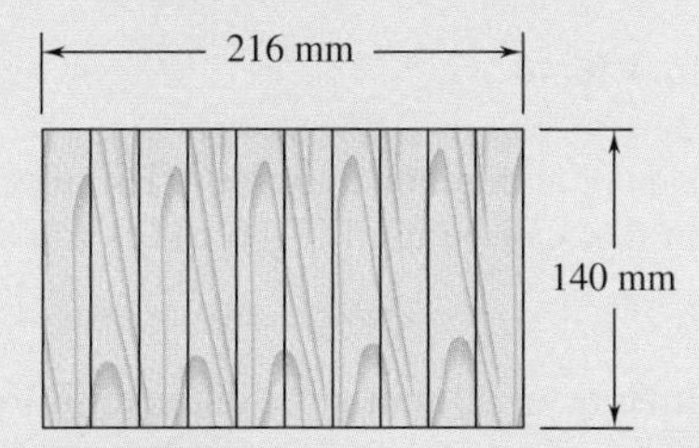

Fig. P16.37

16.38 An aluminum structural tube is reinforced by bolting two plates to it as shown for use as a column of 1.7-m effective length. Knowing that all material is aluminum alloy 2014-T6, determine the maximum allowable centric load.

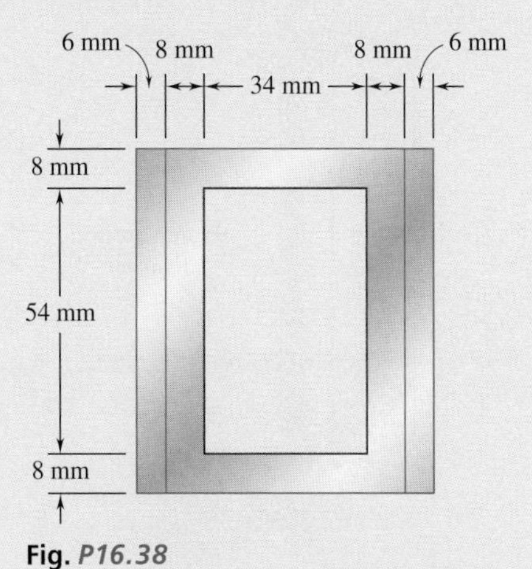

Fig. *P16.38*

16.39 An 18-kip centric load is applied to a rectangular sawn lumber column of 22-ft effective length. Using sawn lumber for which the adjusted allowable stress for compression parallel to the grain is $\sigma_C = 1050$ psi and the adjusted modulus is $E = 440 \times 10^3$ psi, determine the smallest cross section that may be used. Use $b = 2d$.

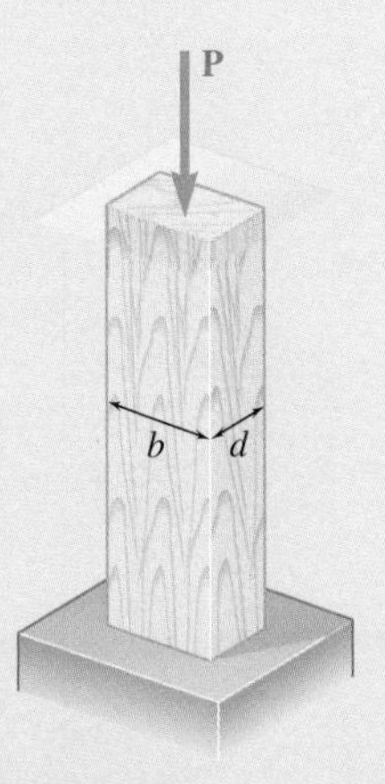

Fig. P16.39

16.40 A glue-laminated column of 3-m effective length is to be made from boards of 24 × 100-mm cross section. Knowing that for the grade of wood used, $E = 11$ GPa and the adjusted allowable stress for compression parallel to the grain is $\sigma_C = 9$ MPa, determine the number of boards that must be used to support the centric load shown when (*a*) $P = 34$ kN, (*b*) $P = 17$ kN.

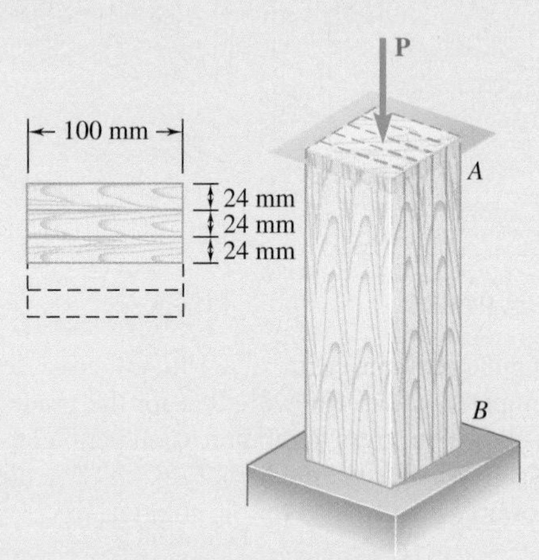

Fig. P16.40

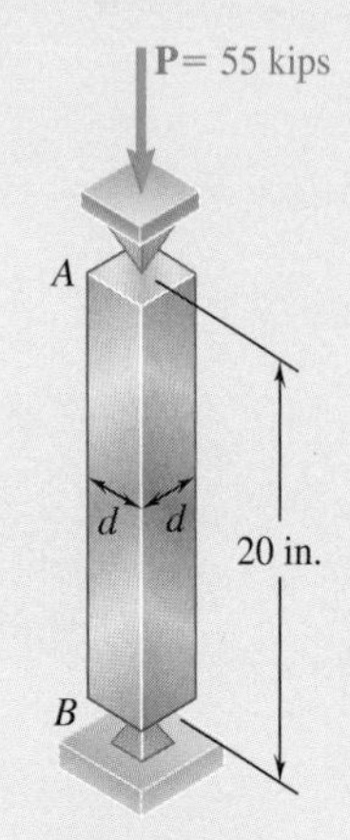

Fig. P16.41

16.41 For a rod made of aluminum alloy 2014-T6, select the smallest square cross section that can be used if the rod is to carry a 55-kip centric load.

16.42 An aluminum tube of 90-mm outer diameter is to carry a centric load of 120 kN. Knowing that the stock of tubes available for use are made of alloy 2014-T6 and with wall thicknesses in increments of 3 mm from 6 mm to 15 mm, determine the lightest tube that can be used.

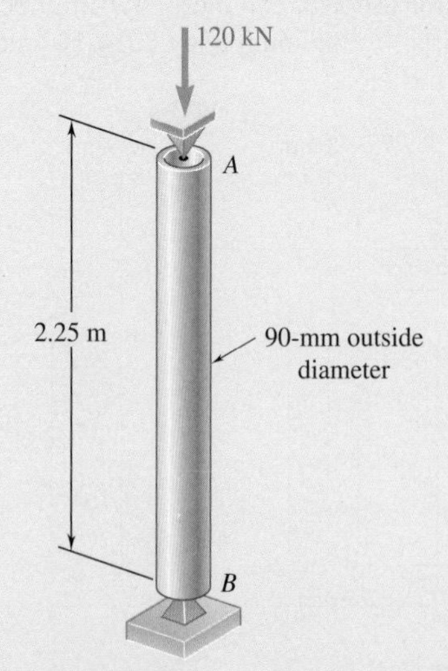

Fig. P16.42

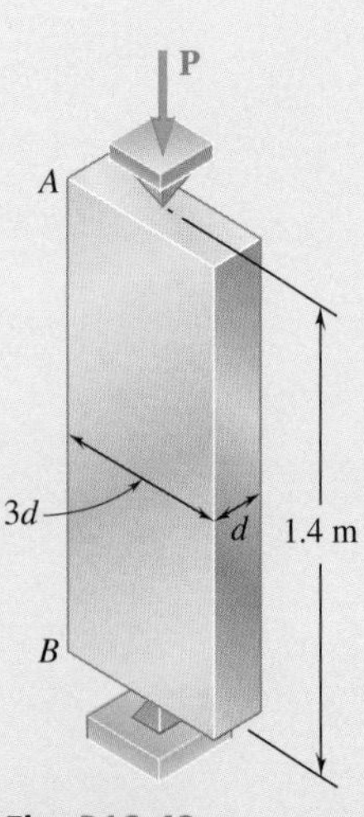

Fig. *P16.43*

16.43 A centric load **P** must be supported by the steel bar *AB*. Using allowable stress design, determine the smallest dimension *d* of the cross section that can be used when (*a*) $P = 108$ kN, (*b*) $P = 166$ kN. Use $\sigma_Y = 250$ MPa and $E = 200$ GPa.

16.44 A column of 4.5-m effective length must carry a centric load of 900 kN. Knowing that $\sigma_Y = 345$ MPa and $E = 200$ GPa, use allowable stress design to select the wide-flange shape of 250-mm nominal depth that should be used.

16.45 A column of 22.5-ft effective length must carry a centric load of 288 kips. Using allowable stress design, select the wide-flange shape of 14-in. nominal depth that should be used. Use $\sigma_Y = 50$ ksi and $E = 29 \times 10^6$ psi.

16.46 A square steel tube having the cross section shown is used as a column of 26-ft effective length to carry a centric load of 65 kips. Knowing that the tubes available for use are made with wall thicknesses ranging from $\frac{1}{4}$ in. to $\frac{3}{4}$ in. in increments of $\frac{1}{16}$ in., use allowable stress design to determine the lightest tube that can be used. Use $\sigma_Y = 36$ ksi and $E = 29 \times 10^6$ psi.

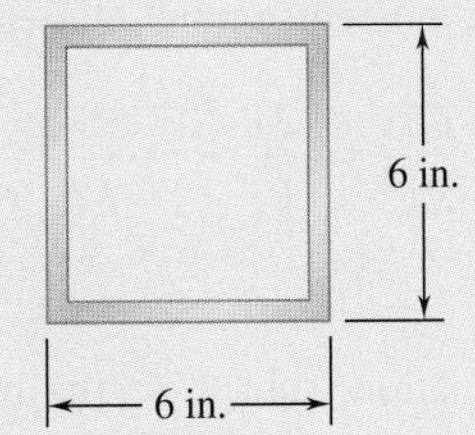

Fig. P16.46

16.47 Solve Prob. 16.46, assuming that the effective length of the column is decreased to 20 ft.

16.48 Two $3\frac{1}{2} \times 2\frac{1}{2}$-in. angles are bolted together as shown for use as a column of 6-ft effective length to carry a centric load of 54 kips. Knowing that the angles available have thicknesses of $\frac{1}{4}$, $\frac{3}{8}$, and $\frac{1}{2}$ in., use allowable stress design to determine the lightest angles that can be used. Use $\sigma_Y = 36$ ksi and $E = 29 \times 10^6$ psi.

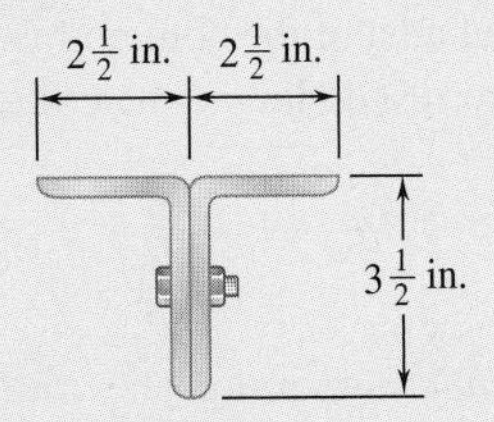

Fig. P16.48

Review and Summary

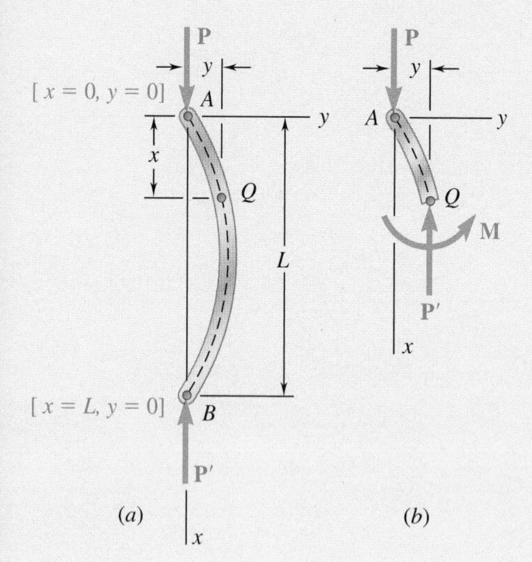

Fig. 16.27

Critical Load

The design and analysis of columns (i.e., prismatic members supporting axial loads), is based on the determination of the *critical load*. Two equilibrium positions of the column model are possible: the original position with zero transverse deflections and a second position involving deflections that could be quite large. The first equilibrium position is unstable for $P > P_{cr}$ and stable for $P < P_{cr}$, since in the latter case it was the only possible equilibrium position.

We considered a pin-ended column of length L and constant flexural rigidity EI subjected to an axial centric load P. Assuming that the column buckled (Fig. 16.27), the bending moment at point Q is equal to $-Py$. Thus,

$$\frac{d^2y}{dx^2} = \frac{M}{EI} = -\frac{P}{EI}y \qquad \textbf{(16.4)}$$

Euler's Formula

Solving this differential equation, subject to the boundary conditions corresponding to a pin-ended column, we determined the smallest load P for which buckling can take place. This load, known as the *critical load* and denoted by P_{cr}, is given by *Euler's formula*:

$$P_{cr} = \frac{\pi^2 EI}{L^2} \qquad \textbf{(16.11a)}$$

where L is the length of the column. For this or any larger load, the equilibrium of the column is unstable, and transverse deflections will occur.

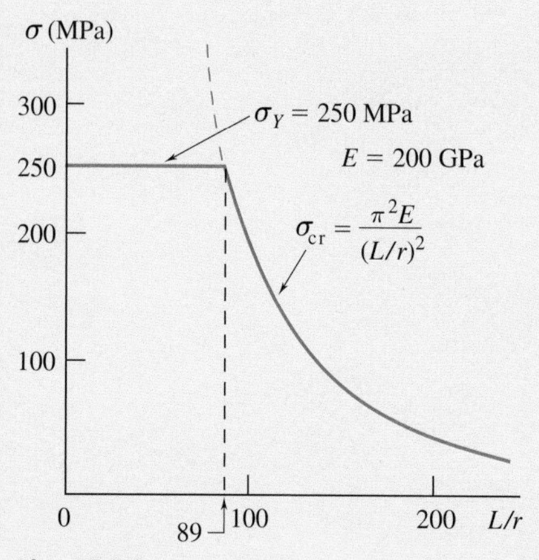

Fig. 16.28

Slenderness Ratio

Denoting the cross-sectional area of the column by A and its radius of gyration by r, the critical stress σ_{cr} corresponding to the critical load P_{cr} is

$$\sigma_{cr} = \frac{\pi^2 E}{(L/r)^2} \qquad \textbf{(16.13a)}$$

The quantity L/r is the *slenderness ratio*. The critical stress σ_{cr} is plotted as a function of L/r in Fig. 16.28. Since the analysis was based on stresses remaining below the yield strength of the material, the column will fail by yielding when $\sigma_{cr} > \sigma_Y$.

Effective Length

The critical load of columns with various end conditions is written as

$$P_{cr} = \frac{\pi^2 EI}{L_e^2} \tag{16.11b}$$

where L_e is the *effective length* of the column, i.e., the length of an equivalent pin-ended column. The effective lengths of several columns with various end conditions were calculated and shown in Fig. 16.18 on page 714.

Design of Real Columns

Since imperfections exist in all columns, the *design of real columns* is done with empirical formulas based on laboratory tests, set forth in specifications and codes issued by professional organizations. For *centrically loaded columns* made of steel, aluminum, or wood, design is based on equations for the allowable stress as a function of the slenderness ratio L/r.

Review Problems

16.49 A column with the cross section shown has a 13.5-ft effective length. Using a factor of safety equal to 2.8, determine the allowable centric load that can be applied to the column. Use $E = 29 \times 10^6$ psi.

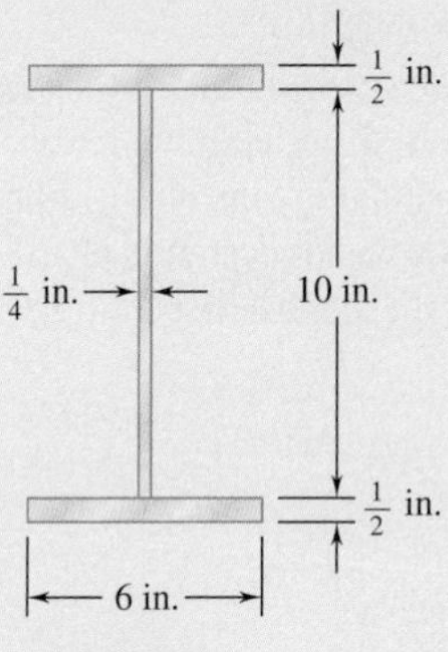

Fig. P16.49

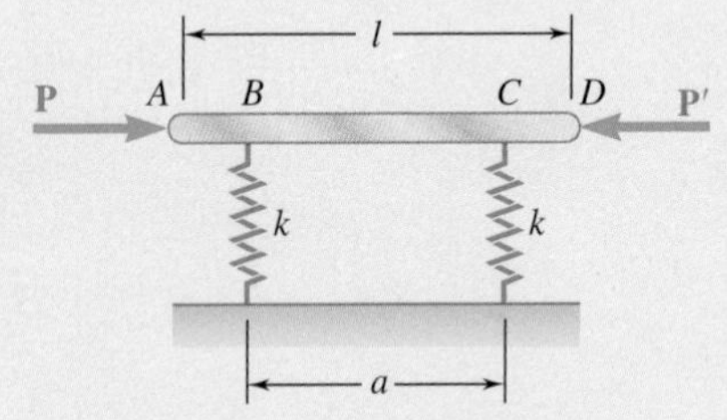

Fig. P16.50

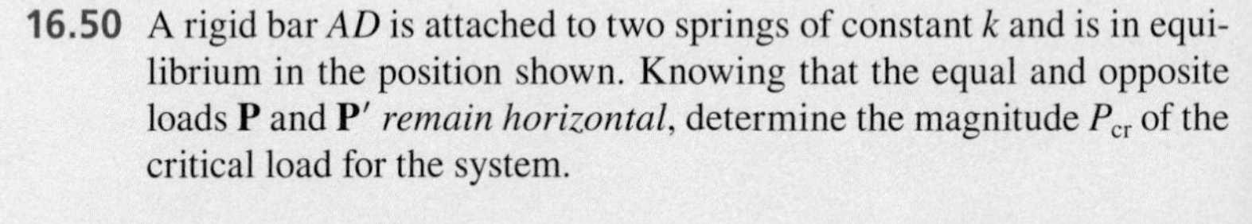

16.50 A rigid bar AD is attached to two springs of constant k and is in equilibrium in the position shown. Knowing that the equal and opposite loads **P** and **P′** *remain horizontal*, determine the magnitude P_{cr} of the critical load for the system.

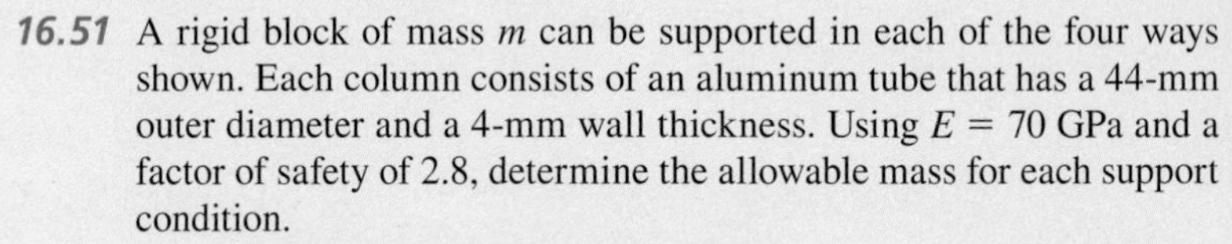

16.51 A rigid block of mass m can be supported in each of the four ways shown. Each column consists of an aluminum tube that has a 44-mm outer diameter and a 4-mm wall thickness. Using $E = 70$ GPa and a factor of safety of 2.8, determine the allowable mass for each support condition.

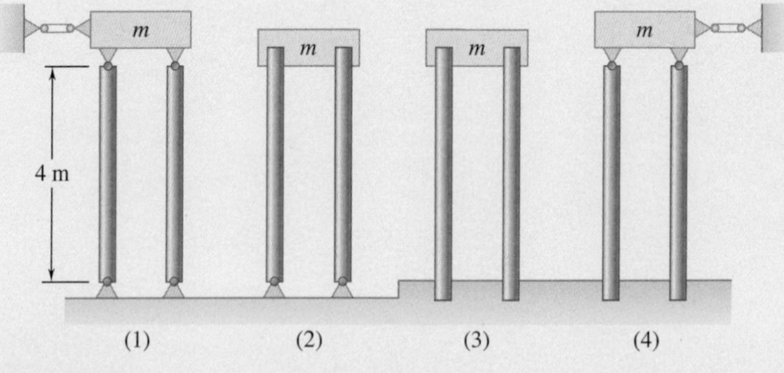

Fig. *P16.51*

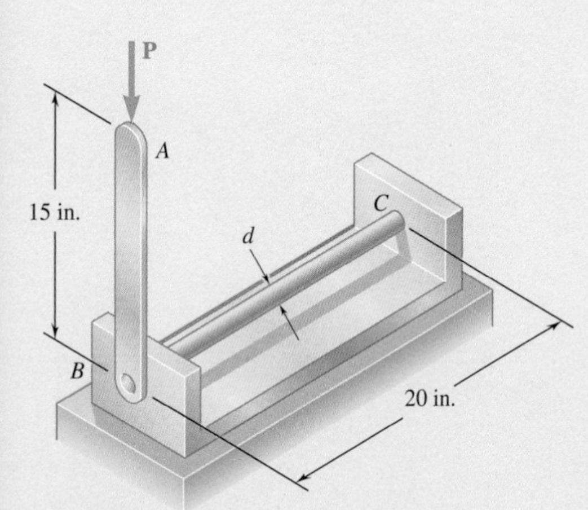

Fig. P16.52

16.52 The steel rod BC is attached to the rigid bar AB and to the fixed support at C. Knowing that $G = 11.2 \times 10^6$ psi, determine the diameter of the rod BC for which the critical load P_{cr} of the system is 80 lb.

16.53 Supports A and B of the pin-ended column shown are at a fixed distance L from each other. Knowing that at a temperature T_0 the force in the column is zero and that buckling occurs when the temperature is $T_1 = T_0 + \Delta T$, express ΔT in terms of b, L and the coefficient of thermal expansion α.

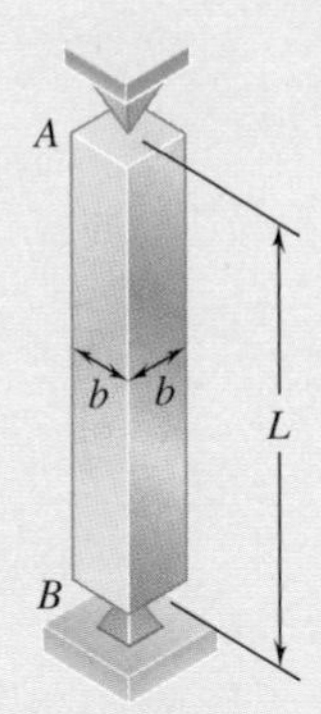

Fig. P16.53

16.54 Member AB consists of a single C130 × 10.4 steel channel of length 2.5 m. Knowing that the pins at A and B pass through the centroid of the cross section of the channel, determine the factor of safety for the load shown with respect to buckling in the plane of the figure when $\theta = 30°$. Use Euler's formula with $E = 200$ GPa.

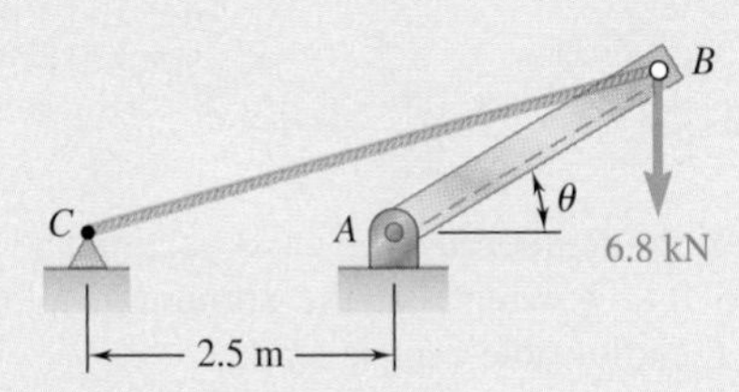

Fig. *P16.54*

16.55 (*a*) Considering only buckling in the plane of the structure shown and using Euler's formula, determine the value of θ between 0 and 90° for which the allowable magnitude of the load **P** is maximum. (*b*) Determine the corresponding maximum value of P knowing that a factor of safety of 3.2 is required. Use $E = 29 \times 10^6$ psi.

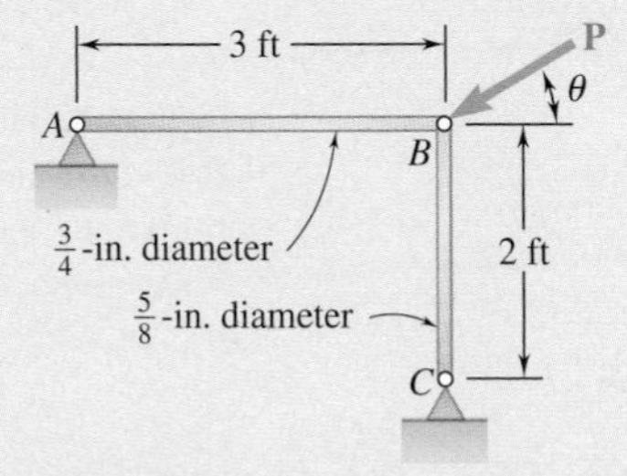

Fig. *P16.55*

16.56 The uniform aluminum bar AB has a 20 × 36-mm rectangular cross section and is supported by pins and brackets as shown. Each end of the bar may rotate freely about a horizontal axis through the pin, but rotation about a vertical axis is prevented by the brackets. Using $E = 70$ GPa, determine the allowable centric load **P** if a factor of safety of 2.5 is required.

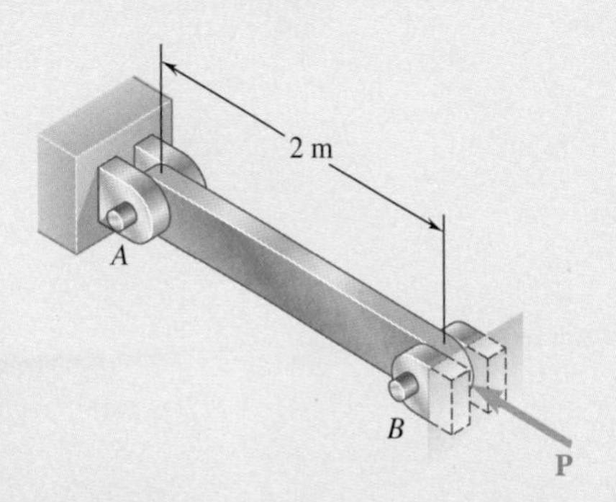

Fig. P16.56

16.57 Determine (*a*) the critical load for the steel strut, (*b*) the dimension d for which the aluminum strut will have the same critical load, (*c*) the weight of the aluminum strut as a percent of the weight of the steel strut.

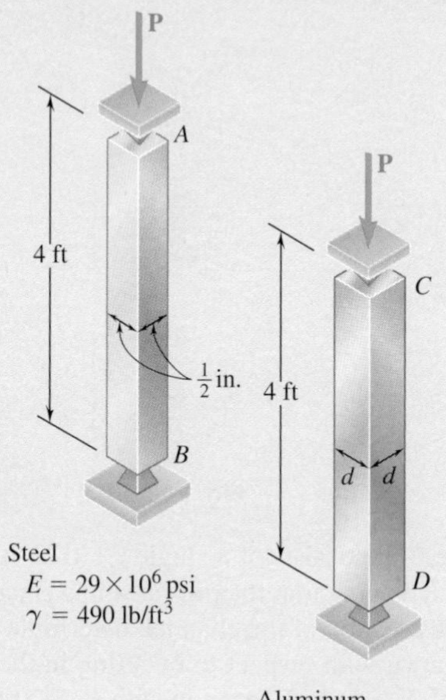

Fig. ***P16.57***

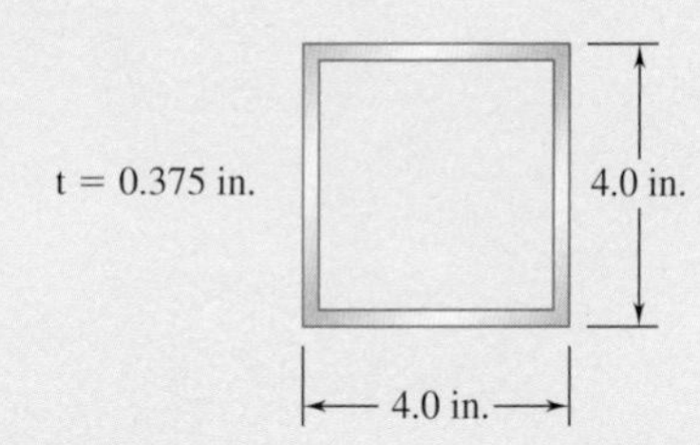

Fig. P16.58

16.58 A compression member has the cross section shown and an effective length of 5 ft. Knowing that the aluminum alloy used is 2014-T6, determine the allowable centric load.

16.59 A column is made from half of a W360 × 216 rolled-steel shape, with the geometric properties as shown. Using allowable stress design, determine the allowable centric load if the effective length of the column is (*a*) 4.0 m, (*b*) 6.5 m. Use $\sigma_Y = 345$ MPa and $E = 200$ GPa.

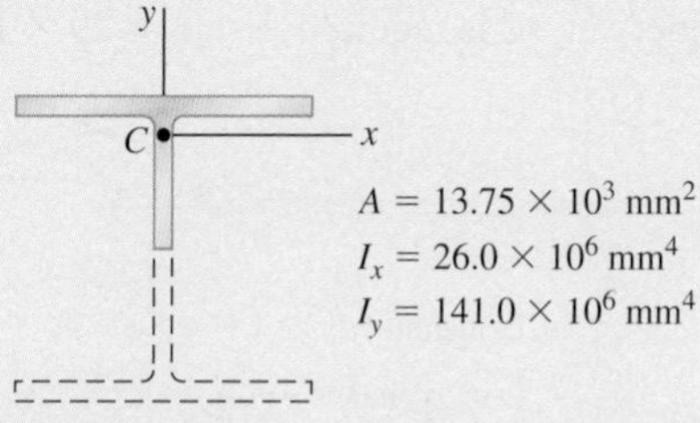

Fig. P16.59

16.60 A steel column of 4.6-m effective length must carry a centric load of 525 kN. Knowing that $\sigma_Y = 345$ MPa and $E = 200$ GPa, use allowable stress design to select the wide-flange shape of 200-mm nominal depth that should be used.

Appendices

[†]Courtesy of the American Institute of Steel Construction, Chicago, Illinois.

Appendix A Typical Properties of Selected Materials Used in Engineering[1,5]

(U.S. Customary Units)

		Ultimate Strength			Yield Strength[3]					
Material	Specific Weight, lb/in^3	Tension, ksi	Compression,[2] ksi	Shear, ksi	Tension, ksi	Shear, ksi	Modulus of Elasticity, 10^6 psi	Modulus of Rigidity, 10^6 psi	Coefficient of Thermal Expansion, 10^{-6}/°F	Ductility, Percent Elongation in 2 in.
Steel										
Structural (ASTM-A36)	0.284	58			36	21	29	11.2	6.5	21
High-strength-low-alloy										
ASTM-A709 Grade 50	0.284	65			50		29	11.2	6.5	21
ASTM-A913 Grade 65	0.284	80			65		29	11.2	6.5	17
ASTM-A992 Grade 50	0.284	65			50		29	11.2	6.5	21
Quenched & tempered										
ASTM-A709 Grade 100	0.284	110			100		29	11.2	6.5	18
Stainless, AISI 302										
Cold-rolled	0.286	125			75		28	10.8	9.6	12
Annealed	0.286	95			38	22	28	10.8	9.6	50
Reinforcing Steel										
Medium strength	0.283	70			40		29	11	6.5	
High strength	0.283	90			60		29	11	6.5	
Cast Iron										
Gray Cast Iron										
4.5% C, ASTM A-48	0.260	25	95	35			10	4.1	6.7	0.5
Malleable Cast Iron										
2% C, 1% Si, ASTM A-47	0.264	50	90	48	33		24	9.3	6.7	10
Aluminum										
Alloy 1100-H14 (99% Al)	0.098	16		10	14	8	10.1	3.7	13.1	9
Alloy 2014-T6	0.101	66		40	58	33	10.9	3.9	12.8	13
Alloy 2024-T4	0.101	68		41	47		10.6		12.9	19
Alloy 5456-H116	0.095	46		27	33	19	10.4		13.3	16
Alloy 6061-T6	0.098	38		24	35	20	10.1	3.7	13.1	17
Alloy 7075-T6	0.101	83		48	73		10.4	4	13.1	11
Copper										
Oxygen-free copper (99.9% Cu)										
Annealed	0.322	32		22	10		17	6.4	9.4	45
Hard-drawn	0.322	57		29	53		17	6.4	9.4	4
Yellow Brass (65% Cu, 35% Zn)										
Cold-rolled	0.306	74		43	60	36	15	5.6	11.6	8
Annealed	0.306	46		32	15	9	15	5.6	11.6	65
Red Brass (85% Cu, 15% Zn)										
Cold-rolled	0.316	85		46	63		17	6.4	10.4	3
Annealed	0.316	39		31	10		17	6.4	10.4	48
Tin bronze (88 Cu, 8Sn, 4Zn)	0.318	45			21		14		10	30
Manganese bronze (63 Cu, 25 Zn, 6 Al, 3 Mn, 3 Fe)	0.302	95			48		15		12	20
Aluminum bronze (81 Cu, 4 Ni, 4 Fe, 11 Al)	0.301	90	130		40		16	6.1	9	6

(Table continued on page A3)

Appendix A Typical Properties of Selected Materials Used in Engineering[1,5]

(SI Units)

Material	Density kg/m^3	Ultimate Strength			Yield Strength[3]		Modulus of Elasticity, GPa	Modulus of Rigidity, GPa	Coefficient of Thermal Expansion, $10^{-6}/°C$	Ductility, Percent Elongation in 50 mm
		Tension, MPa	Compression,[2] MPa	Shear, MPa	Tension, MPa	Shear, MPa				
Steel										
Structural (ASTM-A36)	7 860	400			250	145	200	77.2	11.7	21
High-strength-low-alloy										
ASTM-A709 Grade 345	7 860	450			345		200	77.2	11.7	21
ASTM-A913 Grade 450	7 860	550			450		200	77.2	11.7	17
ASTM-A992 Grade 345	7 860	450			345		200	77.2	11.7	21
Quenched & tempered										
ASTM-A709 Grade 690	7 860	760			690		200	77.2	11.7	18
Stainless, AISI 302										
Cold-rolled	7 920	860			520		190	75	17.3	12
Annealed	7 920	655			260	150	190	75	17.3	50
Reinforcing Steel										
Medium strength	7 860	480			275		200	77	11.7	
High strength	7 860	620			415		200	77	11.7	
Cast Iron										
Gray Cast Iron										
4.5% C, ASTM A-48	7 200	170	655	240			69	28	12.1	0.5
Malleable Cast Iron										
2% C, 1% Si,										
ASTM A-47	7 300	345	620	330	230		165	65	12.1	10
Aluminum										
Alloy 1100-H14										
(99% Al)	2 710	110		70	95	55	70	26	23.6	9
Alloy 2014-T6	2 800	455		275	400	230	75	27	23.0	13
Alloy-2024-T4	2 800	470		280	325		73		23.2	19
Alloy-5456-H116	2 630	315		185	230	130	72		23.9	16
Alloy 6061-T6	2 710	260		165	240	140	70	26	23.6	17
Alloy 7075-T6	2 800	570		330	500		72	28	23.6	11
Copper										
Oxygen-free copper										
(99.9% Cu)										
Annealed	8 910	220		150	70		120	44	16.9	45
Hard-drawn	8 910	390		200	265		120	44	16.9	4
Yellow-Brass										
(65% Cu, 35% Zn)										
Cold-rolled	8 470	510		300	410	250	105	39	20.9	8
Annealed	8 470	320		220	100	60	105	39	20.9	65
Red Brass										
(85% Cu, 15% Zn)										
Cold-rolled	8 740	585		320	435		120	44	18.7	3
Annealed	8 740	270		210	70		120	44	18.7	48
Tin bronze	8 800	310			145		95		18.0	30
(88 Cu, 8Sn, 4Zn)										
Manganese bronze	8 360	655			330		105		21.6	20
(63 Cu, 25 Zn, 6 Al, 3 Mn, 3 Fe)										
Aluminum bronze	8 330	620	900		275		110	42	16.2	6
(81 Cu, 4 Ni, 4 Fe, 11 Al)										

(Table continued on page A4)

Appendix A Typical Properties of Selected Materials Used in Engineering[1,5]

(U.S. Customary Units)

Continued from page A3

Material	Specific Weight, lb/in³	Ultimate Strength			Yield Strength[3]		Modulus of Elasticity, 10^6 psi	Modulus of Rigidity, 10^6 psi	Coefficient of Thermal Expansion, 10^{-6}/°F	Ductility, Percent Elongation in 2 in.
		Tension, ksi	Compression,[2] ksi	Shear, ksi	Tension, ksi	Shear, ksi				
Magnesium Alloys										
Alloy AZ80 (Forging)	0.065	50		23	36		6.5	2.4	14	6
Alloy AZ31 (Extrusion)	0.064	37		19	29		6.5	2.4	14	12
Titanium										
Alloy (6% Al, 4% V)	0.161	130			120		16.5		5.3	10
Monel Alloy 400(Ni-Cu)										
Cold-worked	0.319	98			85	50	26		7.7	22
Annealed	0.319	80			32	18	26		7.7	46
Cupronickel (90% Cu, 10% Ni)										
Annealed	0.323	53			16		20	7.5	9.5	35
Cold-worked	0.323	85			79		20	7.5	9.5	3
Timber, air dry[4]										
Douglas fir	0.017	15	7.2	1.1			1.9	.1	Varies	
Spruce, Sitka	0.015	8.6	5.6	1.1			1.5	.07	1.7 to 2.5	
Shortleaf pine	0.018		7.3	1.4			1.7			
Western white pine	0.014		5.0	1.0			1.5			
Ponderosa pine	0.015	8.4	5.3	1.1			1.3			
White oak	0.025		7.4	2.0			1.8			
Red oak	0.024		6.8	1.8			1.8			
Western hemlock	0.016	13	7.2	1.3			1.6			
Shagbark hickory	0.026		9.2	2.4			2.2			
Redwood	0.015	9.4	6.1	0.9			1.3			
Concrete										
Medium strength	0.084		4.0				3.6		5.5	
High strength	0.084		6.0				4.5		5.5	
Plastics										
Nylon, type 6/6, (molding compound)	0.0412	11	14		6.5		0.4		80	50
Polycarbonate	0.0433	9.5	12.5		9		0.35		68	110
Polyester, PBT (thermoplastic)	0.0484	8	11		8		0.35		75	150
Polyester elastomer	0.0433	6.5		5.5			0.03			500
Polystyrene	0.0374	8	13		8		0.45		70	2
Vinyl, rigid PVC	0.0520	6	10		6.5		0.45		75	40
Rubber	0.033	2							90	600
Granite (Avg. values)	0.100	3	35	5			10	4	4	
Marble (Avg. values)	0.100	2	18	4			8	3	6	
Sandstone (Avg. values)	0.083	1	12	2			6	2	5	
Glass, 98% silica	0.079		7				9.6	4.1	44	

[1]Properties of metals vary widely as a result of variations in composition, heat treatment, and mechanical working.

[2]For ductile metals the compression strength is generally assumed to be equal to the tension strength.

[3]Offset of 0.2 percent.

[4]Timber properties are for loading parallel to the grain.

[5]See also *Marks' Mechanical Engineering Handbook,* 11th ed., McGraw-Hill, New York, 2006; *Annual Book of ASTM,* American Society for Testing Materials, Philadelphia, Pa.; *Metals Handbook,* American Society for Metals, Metals Park, Ohio; and *Aluminum Design Manual,* The Aluminum Association, Washington, DC.

Appendix A Typical Properties of Selected Materials Used in Engineering[1,5]

(SI Units)

Continued from page A4

Material	Density kg/m³	Ultimate Strength: Tension, MPa	Ultimate Strength: Compression,[2] MPa	Ultimate Strength: Shear, MPa	Yield Strength[3]: Tension, MPa	Yield Strength[3]: Shear, MPa	Modulus of Elasticity, GPa	Modulus of Rigidity, GPa	Coefficient of Thermal Expansion, 10^{-6}/°C	Ductility, Percent Elongation in 50 mm
Magnesium Alloys										
Alloy AZ80 (Forging)	1 800	345		160	250		45	16	25.2	6
Alloy AZ31 (Extrusion)	1 770	255		130	200		45	16	25.2	12
Titanium										
Alloy (6% Al, 4% V)	4 730	900			830		115		9.5	10
Monel Alloy 400(Ni-Cu)										
Cold-worked	8 830	675			585	345	180		13.9	22
Annealed	8 830	550			220	125	180		13.9	46
Cupronickel (90% Cu, 10% Ni)										
Annealed	8 940	365			110		140	52	17.1	35
Cold-worked	8 940	585			545		140	52	17.1	3
Timber, air dry[4]										
Douglas fir	470	100	50	7.6			13	0.7	Varies	
Spruce, Sitka	415	60	39	7.6			10	0.5	3.0 to 4.5	
Shortleaf pine	500		50	9.7			12			
Western white pine	390		34	7.0			10			
Ponderosa pine	415	55	36	7.6			9			
White oak	690		51	13.8			12			
Red oak	660		47	12.4			12			
Western hemlock	440	90	50	10.0			11			
Shagbark hickory	720		63	16.5			15			
Redwood	415	65	42	6.2			9			
Concrete										
Medium strength	2 320		28				25		9.9	
High strength	2 320		40				30		9.9	
Plastics										
Nylon, type 6/6, (molding compound)	1 140	75	95		45		2.8		144	50
Polycarbonate	1 200	65	85		35		2.4		122	110
Polyester, PBT (thermoplastic)	1 340	55	75		55		2.4		135	150
Polyester elastomer	1 200	45		40			0.2			500
Polystyrene	1 030	55	90		55		3.1		125	2
Vinyl, rigid PVC	1 440	40	70		45		3.1		135	40
Rubber	910	15							162	600
Granite (Avg. values)	2 770	20	240	35			70	4	7.2	
Marble (Avg. values)	2 770	15	125	28			55	3	10.8	
Sandstone (Avg. values)	2 300	7	85	14			40	2	9.0	
Glass, 98% silica	2 190		50				65	4.1	80	

[1]Properties of metals very widely as a result of variations in composition, heat treatment, and mechanical working.

[2]For ductile metals the compression strength is generally assumed to be equal to the tension strength.

[3]Offset of 0.2 percent.

[4]Timber properties are for loading parallel to the grain.

[5]See also *Marks' Mechanical Engineering Handbook,* 11th ed., McGraw-Hill, New York, 2006; *Annual Book of ASTM,* American Society for Testing Materials, Philadelphia, Pa.; *Metals Handbook,* American Society of Metals, Metals Park, Ohio; and *Aluminum Design Manual,* The Aluminum Association, Washington, DC.

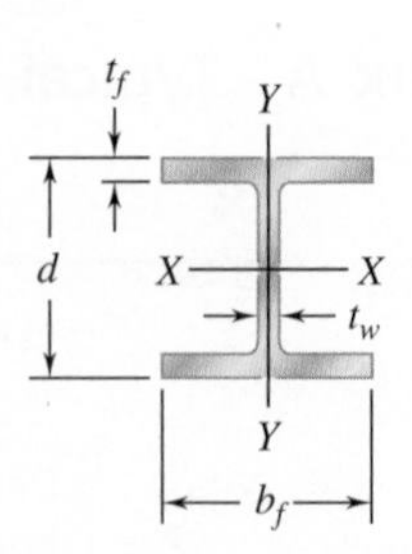

Appendix B Properties of Rolled-Steel Shapes (U.S. Customary Units)

W Shapes (Wide-Flange Shapes)

			Flange		Web	Axis X-X			Axis Y-Y		
Designation†	Area A, in^2	Depth d, in.	Width b_f, in.	Thickness t_f, in.	Thickness t_w, in.	I_x, in^4	S_x, in^3	r_x, in.	I_y, in^4	S_y, in^3	r_y, in.
W36 × 302	88.8	37.3	16.7	1.68	0.945	21100	1130	15.4	1300	156	3.82
135	39.7	35.6	12.0	0.790	0.600	7800	439	14.0	225	37.7	2.38
W33 × 201	59.2	33.7	15.7	1.15	0.715	11600	686	14.0	749	95.2	3.56
118	34.7	32.9	11.5	0.740	0.550	5900	359	13.0	187	32.6	2.32
W30 × 173	51.0	30.4	15.0	1.07	0.655	8230	541	12.7	598	79.8	3.42
99	29.1	29.7	10.50	0.670	0.520	3990	269	11.7	128	24.5	2.10
W27 × 146	43.1	27.4	14.0	0.975	0.605	5660	414	11.5	443	63.5	3.20
84	24.8	26.70	10.0	0.640	0.460	2850	213	10.7	106	21.2	2.07
W24 × 104	30.6	24.1	12.8	0.750	0.500	3100	258	10.1	259	40.7	2.91
68	20.1	23.7	8.97	0.585	0.415	1830	154	9.55	70.4	15.7	1.87
W21 × 101	29.8	21.4	12.3	0.800	0.500	2420	227	9.02	248	40.3	2.89
62	18.3	21.0	8.24	0.615	0.400	1330	127	8.54	57.5	14.0	1.77
44	13.0	20.7	6.50	0.450	0.350	843	81.6	8.06	20.7	6.37	1.26
W18 × 106	31.1	18.7	11.2	0.940	0.590	1910	204	7.84	220	39.4	2.66
76	22.3	18.2	11.0	0.680	0.425	1330	146	7.73	152	27.6	2.61
50	14.7	18.0	7.50	0.570	0.355	800	88.9	7.38	40.1	10.7	1.65
35	10.3	17.7	6.00	0.425	0.300	510	57.6	7.04	15.3	5.12	1.22
W16 × 77	22.6	16.5	10.3	0.760	0.455	1110	134	7.00	138	26.9	2.47
57	16.8	16.4	7.12	0.715	0.430	758	92.2	6.72	43.1	12.1	1.60
40	11.8	16.0	7.00	0.505	0.305	518	64.7	6.63	28.9	8.25	1.57
31	9.13	15.9	5.53	0.440	0.275	375	47.2	6.41	12.4	4.49	1.17
26	7.68	15.7	5.50	0.345	0.250	301	38.4	6.26	9.59	3.49	1.12
W14 × 370	109	17.9	16.5	2.66	1.66	5440	607	7.07	1990	241	4.27
145	42.7	14.8	15.5	1.09	0.680	1710	232	6.33	677	87.3	3.98
82	24.0	14.3	10.1	0.855	0.510	881	123	6.05	148	29.3	2.48
68	20.0	14.0	10.0	0.720	0.415	722	103	6.01	121	24.2	2.46
53	15.6	13.9	8.06	0.660	0.370	541	77.8	5.89	57.7	14.3	1.92
43	12.6	13.7	8.00	0.530	0.305	428	62.6	5.82	45.2	11.3	1.89
38	11.2	14.1	6.77	0.515	0.310	385	54.6	5.87	26.7	7.88	1.55
30	8.85	13.8	6.73	0.385	0.270	291	42.0	5.73	19.6	5.82	1.49
26	7.69	13.9	5.03	0.420	0.255	245	35.3	5.65	8.91	3.55	1.08
22	6.49	13.7	5.00	0.335	0.230	199	29.0	5.54	7.00	2.80	1.04

†A wide-flange shape is designated by the letter W followed by the nominal depth in inches and the weight in pounds per foot.

(Table continued on page A7)

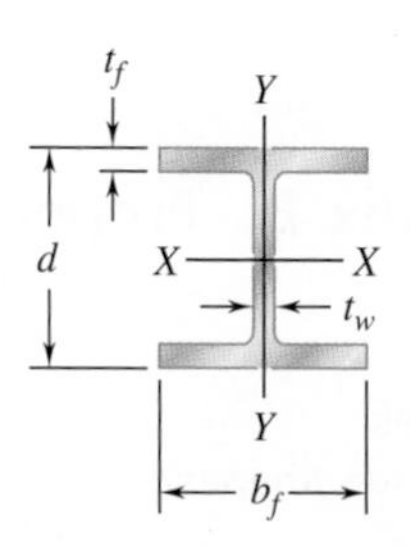

Appendix B Properties of Rolled-Steel Shapes

(SI Units)

W Shapes

(Wide-Flange Shapes)

			Flange		Web	Axis X-X			Axis Y-Y		
Designation†	Area A, mm²	Depth d, mm.	Width b_f, mm	Thickness t_f, mm	Thickness t_w mm	I_x 10⁶ mm⁴	S_x 10³ mm³	r_x mm	I_y 10⁶ mm⁴	S_y 10³ mm³	r_y mm
W920 × 449	57 300	947	424	42.7	24.0	8 780	18 500	391	541	2 560	97.0
201	25 600	904	305	20.1	15.2	3 250	7 190	356	93.7	618	60.5
W840 × 299	38 200	856	399	29.2	18.2	4 830	11 200	356	312	1 560	90.4
176	22 400	836	292	18.8	14.0	2 460	5 880	330	77.8	534	58.9
W760 × 257	32 900	772	381	27.2	16.6	3 430	8 870	323	249	1 310	86.9
147	18 800	754	267	17.0	13.2	1 660	4 410	297	53.3	401	53.3
W690 × 217	27 800	696	356	24.8	15.4	2 360	6 780	292	184	1 040	81.3
125	16 000	678	254	16.3	11.7	1 190	3 490	272	44.1	347	52.6
W610 × 155	19 700	612	325	19.1	12.7	1 290	4 230	257	108	667	73.9
101	13 000	602	228	14.9	10.5	762	2 520	243	29.3	257	47.5
W530 × 150	19 200	544	312	20.3	12.7	1 010	3 720	229	103	660	73.4
92	11 800	533	209	15.6	10.2	554	2 080	217	23.9	229	45.0
66	8 390	526	165	11.4	8.89	351	1 340	205	8.62	104	32.0
W460 × 158	20 100	475	284	23.9	15.0	795	3 340	199	91.6	646	67.6
113	14 400	462	279	17.3	10.8	554	2 390	196	63.3	452	66.3
74	9 480	457	191	14.5	9.02	333	1 460	187	16.7	175	41.9
52	6 650	450	152	10.8	7.62	212	944	179	6.37	83.9	31.0
W410 × 114	14 600	419	262	19.3	11.6	462	2 200	178	57.4	441	62.7
85	10 800	417	181	18.2	10.9	316	1 510	171	17.9	198	40.6
60	7 610	406	178	12.8	7.75	216	1 060	168	12.0	135	39.9
46.1	5 890	404	140	11.2	6.99	156	773	163	5.16	73.6	29.7
38.8	4 950	399	140	8.76	6.35	125	629	159	3.99	57.2	28.4
W360 × 551	70 300	455	419	67.6	42.2	2 260	9 950	180	828	3 950	108
216	27 500	376	394	27.7	17.3	712	3 800	161	282	1 430	101
122	15 500	363	257	21.7	13.0	367	2 020	154	61.6	480	63.0
101	12 900	356	254	18.3	10.5	301	1 690	153	50.4	397	62.5
79	10 100	353	205	16.8	9.40	225	1 270	150	24.0	234	48.8
64	8 130	348	203	13.5	7.75	178	1 030	148	18.8	185	48.0
57.8	7 230	358	172	13.1	7.87	160	895	149	11.1	129	39.4
44	5 710	351	171	9.78	6.86	121	688	146	8.16	95.4	37.8
39	4 960	353	128	10.7	6.48	102	578	144	3.71	58.2	27.4
32.9	4 190	348	127	8.51	5.84	82.8	475	141	2.91	45.9	26.4

†A wide-flange shape is designated by the letter W followed by the nominal depth in millimeters and the mass in kilograms permeter.

(Table continued on page A8)

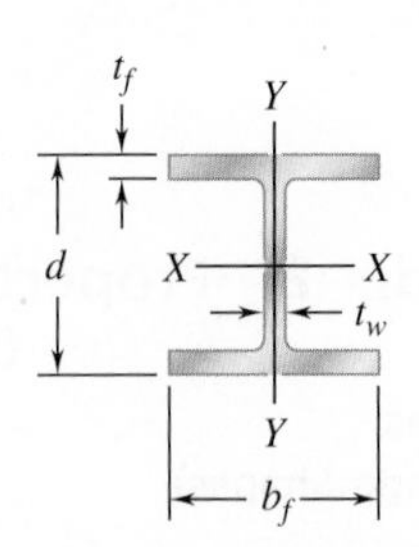

Appendix B Properties of Rolled-Steel Shapes

(U.S. Customary Units)

Continued from page A7

W Shapes
(Wide-Flange Shapes)

Designation†	Area A, in^2	Depth d, in.	Flange		Web Thickness t_w, in.	Axis X-X			Axis Y-Y		
			Width b_f, in.	Thickness t_f, in.		I_x, in^4	S_x, in^3	r_x, in.	I_y, in^4	S_y, in^3	r_y, in.
W12 × 96	28.2	12.7	12.2	0.900	0.550	833	131	5.44	270	44.4	3.09
72	21.1	12.3	12.0	0.670	0.430	597	97.4	5.31	195	32.4	3.04
50	14.6	12.2	8.08	0.640	0.370	391	64.2	5.18	56.3	13.9	1.96
40	11.7	11.9	8.01	0.515	0.295	307	51.5	5.13	44.1	11.0	1.94
35	10.3	12.5	6.56	0.520	0.300	285	45.6	5.25	24.5	7.47	1.54
30	8.79	12.3	6.52	0.440	0.260	238	38.6	5.21	20.3	6.24	1.52
26	7.65	12.2	6.49	0.380	0.230	204	33.4	5.17	17.3	5.34	1.51
22	6.48	12.3	4.03	0.425	0.260	156	25.4	4.91	4.66	2.31	0.848
16	4.71	12.0	3.99	0.265	0.220	103	17.1	4.67	2.82	1.41	0.773
W10 × 112	32.9	11.4	10.4	1.25	0.755	716	126	4.66	236	45.3	2.68
68	20.0	10.4	10.1	0.770	0.470	394	75.7	4.44	134	26.4	2.59
54	15.8	10.1	10.0	0.615	0.370	303	60.0	4.37	103	20.6	2.56
45	13.3	10.1	8.02	0.620	0.350	248	49.1	4.32	53.4	13.3	2.01
39	11.5	9.92	7.99	0.530	0.315	209	42.1	4.27	45.0	11.3	1.98
33	9.71	9.73	7.96	0.435	0.290	171	35.0	4.19	36.6	9.20	1.94
30	8.84	10.5	5.81	0.510	0.300	170	32.4	4.38	16.7	5.75	1.37
22	6.49	10.2	5.75	0.360	0.240	118	23.2	4.27	11.4	3.97	1.33
19	5.62	10.2	4.02	0.395	0.250	96.3	18.8	4.14	4.29	2.14	0.874
15	4.41	10.0	4.00	0.270	0.230	68.9	13.8	3.95	2.89	1.45	0.810
W8 × 58	17.1	8.75	8.22	0.810	0.510	228	52.0	3.65	75.1	18.3	2.10
48	14.1	8.50	8.11	0.685	0.400	184	43.2	3.61	60.9	15.0	2.08
40	11.7	8.25	8.07	0.560	0.360	146	35.5	3.53	49.1	12.2	2.04
35	10.3	8.12	8.02	0.495	0.310	127	31.2	3.51	42.6	10.6	2.03
31	9.12	8.00	8.00	0.435	0.285	110	27.5	3.47	37.1	9.27	2.02
28	8.24	8.06	6.54	0.465	0.285	98.0	24.3	3.45	21.7	6.63	1.62
24	7.08	7.93	6.50	0.400	0.245	82.7	20.9	3.42	18.3	5.63	1.61
21	6.16	8.28	5.27	0.400	0.250	75.3	18.2	3.49	9.77	3.71	1.26
18	5.26	8.14	5.25	0.330	0.230	61.9	15.2	3.43	7.97	3.04	1.23
15	4.44	8.11	4.01	0.315	0.245	48.0	11.8	3.29	3.41	1.70	0.876
13	3.84	7.99	4.00	0.255	0.230	39.6	9.91	3.21	2.73	1.37	0.843
W6 × 25	7.34	6.38	6.08	0.455	0.320	53.4	16.7	2.70	17.1	5.61	1.52
20	5.87	6.20	6.02	0.365	0.260	41.4	13.4	2.66	13.3	4.41	1.50
16	4.74	6.28	4.03	0.405	0.260	32.1	10.2	2.60	4.43	2.20	0.967
12	3.55	6.03	4.00	0.280	0.230	22.1	7.31	2.49	2.99	1.50	0.918
9	2.68	5.90	3.94	0.215	0.170	16.4	5.56	2.47	2.20	1.11	0.905
W5 × 19	5.56	5.15	5.03	0.430	0.270	26.3	10.2	2.17	9.13	3.63	1.28
16	4.71	5.01	5.00	0.360	0.240	21.4	8.55	2.13	7.51	3.00	1.26
W4 × 13	3.83	4.16	4.06	0.345	0.280	11.3	5.46	1.72	3.86	1.90	1.00

†A wide-flange shape is designated by the letter W followed by the nominal depth in inches and the weight in pounds per foot.

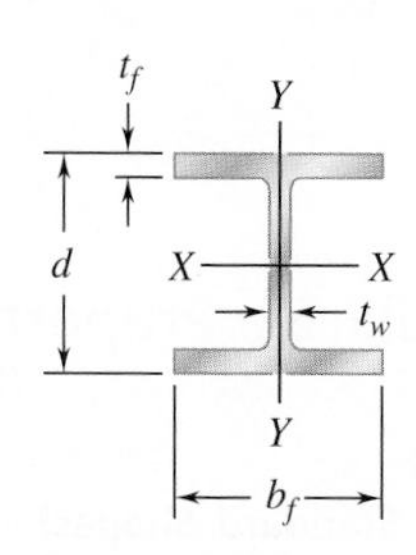

Appendix B Properties of Rolled-Steel Shapes
(SI Units)

Continued from page A8

W Shapes
(Wide-Flange Shapes)

			Flange		Web	Axis X-X			Axis Y-Y		
Designation[†]	Area A, mm^2	Depth d, mm	Width b_f, mm	Thickness t_f, mm	Thickness t_w, mm	I_x 10^6 mm^4	S_x 10^3 mm^3	r_x mm	I_y 10^6 mm^4	S_y 10^3 mm^3	r_y mm
W310 × 143	18 200	323	310	22.9	14.0	347	2 150	138	112	728	78.5
107	13 600	312	305	17.0	10.9	248	1 600	135	81.2	531	77.2
74	9 420	310	205	16.3	9.40	163	1 050	132	23.4	228	49.8
60	7 550	302	203	13.1	7.49	128	844	130	18.4	180	49.3
52	6 650	318	167	13.2	7.62	119	747	133	10.2	122	39.1
44.5	5 670	312	166	11.2	6.60	99.1	633	132	8.45	102	38.6
38.7	4 940	310	165	9.65	5.84	84.9	547	131	7.20	87.5	38.4
32.7	4 180	312	102	10.8	6.60	64.9	416	125	1.94	37.9	21.5
23.8	3 040	305	101	6.73	5.59	42.9	280	119	1.17	23.1	19.6
W250 × 167	21 200	290	264	31.8	19.2	298	2 060	118	98.2	742	68.1
101	12 900	264	257	19.6	11.9	164	1 240	113	55.8	433	65.8
80	10 200	257	254	15.6	9.4	126	983	111	42.9	338	65.0
67	8 580	257	204	15.7	8.89	103	805	110	22.2	218	51.1
58	7 420	252	203	13.5	8.00	87.0	690	108	18.7	185	50.3
49.1	6 260	247	202	11.0	7.37	71.2	574	106	15.2	151	49.3
44.8	5 700	267	148	13.0	7.62	70.8	531	111	6.95	94.2	34.8
32.7	4 190	259	146	9.14	6.10	49.1	380	108	4.75	65.1	33.8
28.4	3 630	259	102	10.0	6.35	40.1	308	105	1.79	35.1	22.2
22.3	2 850	254	102	6.86	5.84	28.7	226	100	1.20	23.8	20.6
W200 × 86	11 000	222	209	20.6	13.0	94.9	852	92.7	31.3	300	53.3
71	9 100	216	206	17.4	10.2	76.6	708	91.7	25.3	246	52.8
59	7 550	210	205	14.2	9.14	60.8	582	89.7	20.4	200	51.8
52	6 650	206	204	12.6	7.87	52.9	511	89.2	17.7	174	51.6
46.1	5 880	203	203	11.0	7.24	45.8	451	88.1	15.4	152	51.3
41.7	5 320	205	166	11.8	7.24	40.8	398	87.6	9.03	109	41.1
35.9	4 570	201	165	10.2	6.22	34.4	342	86.9	7.62	92.3	40.9
31.3	3 970	210	134	10.2	6.35	31.3	298	88.6	4.07	60.8	32.0
26.6	3 390	207	133	8.38	5.84	25.8	249	87.1	3.32	49.8	31.2
22.5	2 860	206	102	8.00	6.22	20.0	193	83.6	1.42	27.9	22.3
19.3	2 480	203	102	6.48	5.84	16.5	162	81.5	1.14	22.5	21.4
W150 × 37.1	4 740	162	154	11.6	8.13	22.2	274	68.6	7.12	91.9	38.6
29.8	3 790	157	153	9.27	6.60	17.2	220	67.6	5.54	72.3	38.1
24	3 060	160	102	10.3	6.60	13.4	167	66.0	1.84	36.1	24.6
18	2 290	153	102	7.11	5.84	9.20	120	63.2	1.24	24.6	23.3
13.5	1 730	150	100	5.46	4.32	6.83	91.1	62.7	0.916	18.2	23.0
W130 × 28.1	3 590	131	128	10.9	6.86	10.9	167	55.1	3.80	59.5	32.5
23.8	3 040	127	127	9.14	6.10	8.91	140	54.1	3.13	49.2	32.0
W100 × 19.3	2 470	106	103	8.76	7.11	4.70	89.5	43.7	1.61	31.1	25.4

[†]A wide-flange shape is designated by the letter W followed by the nominal depth in millimeters and the mass in kilograms per meter.

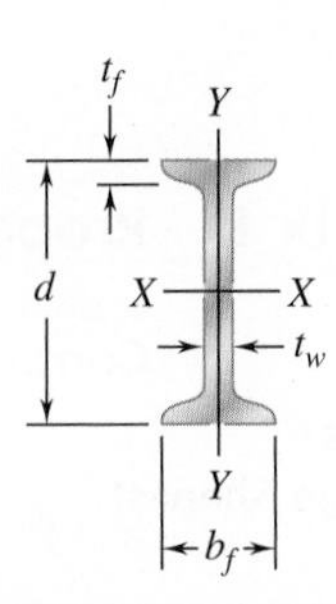

Appendix B Properties of Rolled-Steel Shapes
(U.S. Customary Units)

S Shapes
(American Standard Shapes)

			Flange		Web	Axis X-X			Axis Y-Y		
Designation†	Area A, in^2	Depth d, in.	Width b_f, in.	Thickness t_f, in.	Thickness t_w, in.	I_x, in^4	S_x, in^3	r_x, in.	I_y, in^4	S_y, in^3	r_y, in.
S24 × 121	35.5	24.5	8.05	1.09	0.800	3160	258	9.43	83.0	20.6	1.53
106	31.1	24.5	7.87	1.09	0.620	2940	240	9.71	76.8	19.5	1.57
100	29.3	24.0	7.25	0.870	0.745	2380	199	9.01	47.4	13.1	1.27
90	26.5	24.0	7.13	0.870	0.625	2250	187	9.21	44.7	12.5	1.30
80	23.5	24.0	7.00	0.870	0.500	2100	175	9.47	42.0	12.0	1.34
S20 × 96	28.2	20.3	7.20	0.920	0.800	1670	165	7.71	49.9	13.9	1.33
86	25.3	20.3	7.06	0.920	0.660	1570	155	7.89	46.6	13.2	1.36
75	22.0	20.0	6.39	0.795	0.635	1280	128	7.62	29.5	9.25	1.16
66	19.4	20.0	6.26	0.795	0.505	1190	119	7.83	27.5	8.78	1.19
S18 × 70	20.5	18.0	6.25	0.691	0.711	923	103	6.70	24.0	7.69	1.08
54.7	16.0	18.0	6.00	0.691	0.461	801	89.0	7.07	20.7	6.91	1.14
S15 × 50	14.7	15.0	5.64	0.622	0.550	485	64.7	5.75	15.6	5.53	1.03
42.9	12.6	15.0	5.50	0.622	0.411	446	59.4	5.95	14.3	5.19	1.06
S12 × 50	14.6	12.0	5.48	0.659	0.687	303	50.6	4.55	15.6	5.69	1.03
40.8	11.9	12.0	5.25	0.659	0.462	270	45.1	4.76	13.5	5.13	1.06
35	10.2	12.0	5.08	0.544	0.428	228	38.1	4.72	9.84	3.88	0.980
31.8	9.31	12.0	5.00	0.544	0.350	217	36.2	4.83	9.33	3.73	1.00
S10 × 35	10.3	10.0	4.94	0.491	0.594	147	29.4	3.78	8.30	3.36	0.899
25.4	7.45	10.0	4.66	0.491	0.311	123	24.6	4.07	6.73	2.89	0.950
S8 × 23	6.76	8.00	4.17	0.425	0.441	64.7	16.2	3.09	4.27	2.05	0.795
18.4	5.40	8.00	4.00	0.425	0.271	57.5	14.4	3.26	3.69	1.84	0.827
S6 × 17.2	5.06	6.00	3.57	0.359	0.465	26.2	8.74	2.28	2.29	1.28	0.673
12.5	3.66	6.00	3.33	0.359	0.232	22.0	7.34	2.45	1.80	1.08	0.702
S5 × 10	2.93	5.00	3.00	0.326	0.214	12.3	4.90	2.05	1.19	0.795	0.638
S4 × 9.5	2.79	4.00	2.80	0.293	0.326	6.76	3.38	1.56	0.887	0.635	0.564
7.7	2.26	4.00	2.66	0.293	0.193	6.05	3.03	1.64	0.748	0.562	0.576
S3 × 7.5	2.20	3.00	2.51	0.260	0.349	2.91	1.94	1.15	0.578	0.461	0.513
5.7	1.66	3.00	2.33	0.260	0.170	2.50	1.67	1.23	0.447	0.383	0.518

†An American Standard Beam is designated by the letter S followed by the nominal depth in inches and the weight in pounds per foot.

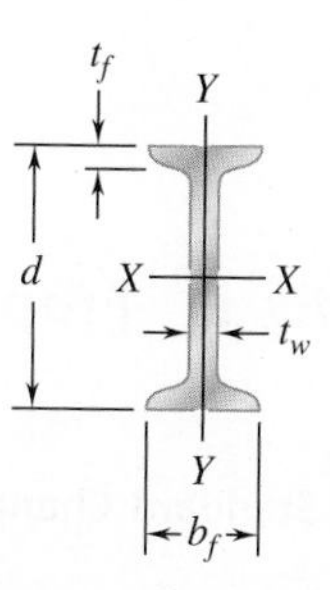

Appendix B Properties of Rolled-Steel Shapes

(SI Units)

S Shapes

(American Standard Shapes)

Designation[†]	Area A, mm^2	Depth d, mm	Flange Width b_f, mm	Flange Thickness t_f, mm	Web Thickness t_w, mm	Axis X-X I_x 10^6 mm^4	Axis X-X S_x 10^3 mm^3	Axis X-X r_x mm	Axis Y-Y I_y 10^6 mm^4	Axis Y-Y S_y 10^3 mm^3	Axis Y-Y r_y mm
S610 × 180	22 900	622	204	27.7	20.3	1 320	4 230	240	34.5	338	38.9
158	20 100	622	200	27.7	15.7	1 220	3 930	247	32.0	320	39.9
149	18 900	610	184	22.1	18.9	991	3 260	229	19.7	215	32.3
134	17 100	610	181	22.1	15.9	937	3 060	234	18.6	205	33.0
119	15 200	610	178	22.1	12.7	874	2 870	241	17.5	197	34.0
S510 × 143	18 200	516	183	23.4	20.3	695	2 700	196	20.8	228	33.8
128	16 300	516	179	23.4	16.8	653	2 540	200	19.4	216	34.5
112	14 200	508	162	20.2	16.1	533	2 100	194	12.3	152	29.5
98.2	12 500	508	159	20.2	12.8	495	1 950	199	11.4	144	30.2
S460 × 104	13 200	457	159	17.6	18.1	384	1 690	170	10.0	126	27.4
81.4	10 300	457	152	17.6	11.7	333	1 460	180	8.62	113	29.0
S380 × 74	9 480	381	143	15.8	14.0	202	1 060	146	6.49	90.6	26.2
64	8 130	381	140	15.8	10.4	186	973	151	5.95	85.0	26.9
S310 × 74	9 420	305	139	16.7	17.4	126	829	116	6.49	93.2	26.2
60.7	7 680	305	133	16.7	11.7	112	739	121	5.62	84.1	26.9
52	6 580	305	129	13.8	10.9	94.9	624	120	4.10	63.6	24.9
47.3	6 010	305	127	13.8	8.89	90.3	593	123	3.88	61.1	25.4
S250 × 52	6 650	254	125	12.5	15.1	61.2	482	96.0	3.45	55.1	22.8
37.8	4 810	254	118	12.5	7.90	51.2	403	103	2.80	47.4	24.1
S200 × 34	4 360	203	106	10.8	11.2	26.9	265	78.5	1.78	33.6	20.2
27.4	3 480	203	102	10.8	6.88	23.9	236	82.8	1.54	30.2	21.0
S150 × 25.7	3 260	152	90.7	9.12	11.8	10.9	143	57.9	0.953	21.0	17.1
18.6	2 360	152	84.6	9.12	5.89	9.16	120	62.2	0.749	17.7	17.8
S130 × 15	1 890	127	76.2	8.28	5.44	5.12	80.3	52.1	0.495	13.0	16.2
S100 × 14.1	1 800	102	71.1	7.44	8.28	2.81	55.4	39.6	0.369	10.4	14.3
11.5	1 460	102	67.6	7.44	4.90	2.52	49.7	41.7	0.311	9.21	14.6
S75 × 11.2	1 420	76.2	63.8	6.60	8.86	1.21	31.8	29.2	0.241	7.55	13.0
8.5	1 070	76.2	59.2	6.60	4.32	1.04	27.4	31.2	0.186	6.28	13.2

[†]An American Standard Beam is designated by the letter S followed by the nominal depth in millimeters and the mass in kilograms per meter.

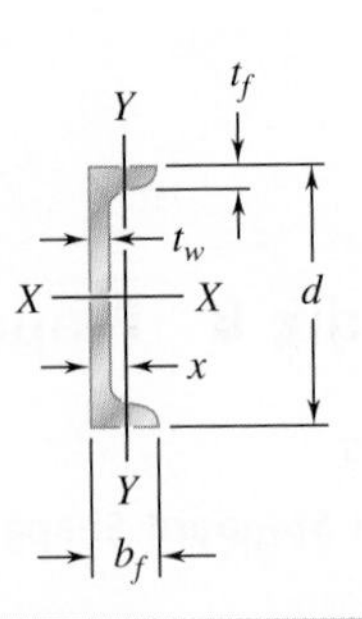

Appendix B Properties of Rolled-Steel Shapes (U.S. Customary Units)

C Shapes (American Standard Channels)

			Flange			Axis X-X			Axis Y-Y			
Designation†	Area A, in^2	Depth d, in	Width b_f, in	Thickness t_f, in	Web Thickness t_w, in	I_x, in^4	S_x, in^3	r_x, in	I_y, in^4	S_y, in^3	r_y, in	x, in
C15 × 50	14.7	15.0	3.72	0.650	0.716	404	53.8	5.24	11.0	3.77	0.865	0.799
40	11.8	15.0	3.52	0.650	0.520	348	46.5	5.45	9.17	3.34	0.883	0.778
33.9	10.0	15.0	3.40	0.650	0.400	315	42.0	5.62	8.07	3.09	0.901	0.788
C12 × 30	8.81	12.0	3.17	0.501	0.510	162	27.0	4.29	5.12	2.05	0.762	0.674
25	7.34	12.0	3.05	0.501	0.387	144	24.0	4.43	4.45	1.87	0.779	0.674
20.7	6.08	12.0	2.94	0.501	0.282	129	21.5	4.61	3.86	1.72	0.797	0.698
C10 × 30	8.81	10.0	3.03	0.436	0.673	103	20.7	3.42	3.93	1.65	0.668	0.649
25	7.34	10.0	2.89	0.436	0.526	91.1	18.2	3.52	3.34	1.47	0.675	0.617
20	5.87	10.0	2.74	0.436	0.379	78.9	15.8	3.66	2.80	1.31	0.690	0.606
15.3	4.48	10.0	2.60	0.436	0.240	67.3	13.5	3.87	2.27	1.15	0.711	0.634
C9 × 20	5.87	9.00	2.65	0.413	0.448	60.9	13.5	3.22	2.41	1.17	0.640	0.583
15	4.41	9.00	2.49	0.413	0.285	51.0	11.3	3.40	1.91	1.01	0.659	0.586
13.4	3.94	9.00	2.43	0.413	0.233	47.8	10.6	3.49	1.75	0.954	0.666	0.601
C8 × 18.7	5.51	8.00	2.53	0.390	0.487	43.9	11.0	2.82	1.97	1.01	0.598	0.565
13.7	4.04	8.00	2.34	0.390	0.303	36.1	9.02	2.99	1.52	0.848	0.613	0.554
11.5	3.37	8.00	2.26	0.390	0.220	32.5	8.14	3.11	1.31	0.775	0.623	0.572
C7 × 12.2	3.60	7.00	2.19	0.366	0.314	24.2	6.92	2.60	1.16	0.696	0.568	0.525
9.8	2.87	7.00	2.09	0.366	0.210	21.2	6.07	2.72	0.957	0.617	0.578	0.541
C6 × 13	3.81	6.00	2.16	0.343	0.437	17.3	5.78	2.13	1.05	0.638	0.524	0.514
10.5	3.08	6.00	2.03	0.343	0.314	15.1	5.04	2.22	0.860	0.561	0.529	0.500
8.2	2.39	6.00	1.92	0.343	0.200	13.1	4.35	2.34	0.687	0.488	0.536	0.512
C5 × 9	2.64	5.00	1.89	0.320	0.325	8.89	3.56	1.83	0.624	0.444	0.486	0.478
6.7	1.97	5.00	1.75	0.320	0.190	7.48	2.99	1.95	0.470	0.372	0.489	0.484
C4 × 7.2	2.13	4.00	1.72	0.296	0.321	4.58	2.29	1.47	0.425	0.337	0.447	0.459
5.4	1.58	4.00	1.58	0.296	0.184	3.85	1.92	1.56	0.312	0.277	0.444	0.457
C3 × 6	1.76	3.00	1.60	0.273	0.356	2.07	1.38	1.08	0.300	0.263	0.413	0.455
5	1.47	3.00	1.50	0.273	0.258	1.85	1.23	1.12	0.241	0.228	0.405	0.439
4.1	1.20	3.00	1.41	0.273	0.170	1.65	1.10	1.17	0.191	0.196	0.398	0.437

†An American Standard Channel is designated by the letter C followed by the nominal depth in inches and the weight in pounds per foot.

Appendix B Properties of Rolled-Steel Shapes
(SI Units)

C Shapes
(American Standard Channels)

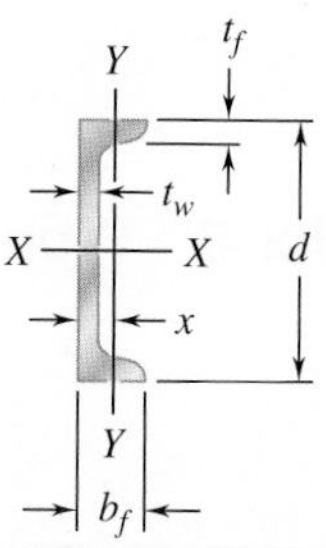

			Flange		Web	Axis X-X			Axis Y-Y			
Designation†	Area A, mm^2	Depth d, mm	Width b_f, mm	Thickness t_f, mm	Thickness t_w, mm	I_x 10^6 mm^4	S_x 10^3 mm^3	r_x mm	I_y 10^6 mm^4	S_y 10^3 mm^3	r_y mm	x mm
C380 × 74	9 480	381	94.5	16.5	18.2	168	882	133	4.58	61.8	22.0	20.3
60	7 610	381	89.4	16.5	13.2	145	762	138	3.82	54.7	22.4	19.8
50.4	6 450	381	86.4	16.5	10.2	131	688	143	3.36	50.6	22.9	20.0
C310 × 45	5 680	305	80.5	12.7	13.0	67.4	442	109	2.13	33.6	19.4	17.1
37	4 740	305	77.5	12.7	9.83	59.9	393	113	1.85	30.6	19.8	17.1
30.8	3 920	305	74.7	12.7	7.16	53.7	352	117	1.61	28.2	20.2	17.7
C250 × 45	5 680	254	77.0	11.1	17.1	42.9	339	86.9	1.64	27.0	17.0	16.5
37	4 740	254	73.4	11.1	13.4	37.9	298	89.4	1.39	24.1	17.1	15.7
30	3 790	254	69.6	11.1	9.63	32.8	259	93.0	1.17	21.5	17.5	15.4
22.8	2 890	254	66.0	11.1	6.10	28.0	221	98.3	0.945	18.8	18.1	16.1
C230 × 30	3 790	229	67.3	10.5	11.4	25.3	221	81.8	1.00	19.2	16.3	14.8
22	2 850	229	63.2	10.5	7.24	21.2	185	86.4	0.795	16.6	16.7	14.9
19.9	2 540	229	61.7	10.5	5.92	19.9	174	88.6	0.728	15.6	16.9	15.3
C200 × 27.9	3 550	203	64.3	9.91	12.4	18.3	180	71.6	0.820	16.6	15.2	14.4
20.5	2 610	203	59.4	9.91	7.70	15.0	148	75.9	0.633	13.9	15.6	14.1
17.1	2 170	203	57.4	9.91	5.59	13.5	133	79.0	0.545	12.7	15.8	14.5
C180 × 18.2	2 320	178	55.6	9.30	7.98	10.1	113	66.0	0.483	11.4	14.4	13.3
14.6	1 850	178	53.1	9.30	5.33	8.82	100	69.1	0.398	10.1	14.7	13.7
C150 × 19.3	2 460	152	54.9	8.71	11.1	7.20	94.7	54.1	0.437	10.5	13.3	13.1
15.6	1 990	152	51.6	8.71	7.98	6.29	82.6	56.4	0.358	9.19	13.4	12.7
12.2	1 540	152	48.8	8.71	5.08	5.45	71.3	59.4	0.286	8.00	13.6	13.0
C130 × 13	1 700	127	48.0	8.13	8.26	3.70	58.3	46.5	0.260	7.28	12.3	12.1
10.4	1 270	127	44.5	8.13	4.83	3.11	49.0	49.5	0.196	6.10	12.4	12.3
C100 × 10.8	1 370	102	43.7	7.52	8.15	1.91	37.5	37.3	0.177	5.52	11.4	11.7
8	1 020	102	40.1	7.52	4.67	1.60	31.5	39.6	0.130	4.54	11.3	11.6
C75 × 8.9	1 140	76.2	40.6	6.93	9.04	0.862	22.6	27.4	0.125	4.31	10.5	11.6
7.4	948	76.2	38.1	6.93	6.55	0.770	20.2	28.4	0.100	3.74	10.3	11.2
6.1	774	76.2	35.8	6.93	4.32	0.687	18.0	29.7	0.0795	3.21	10.1	11.1

†An American Standard Channel is designated by the letter C followed by the nominal depth in millimeters and the mass in kilograms per meter.

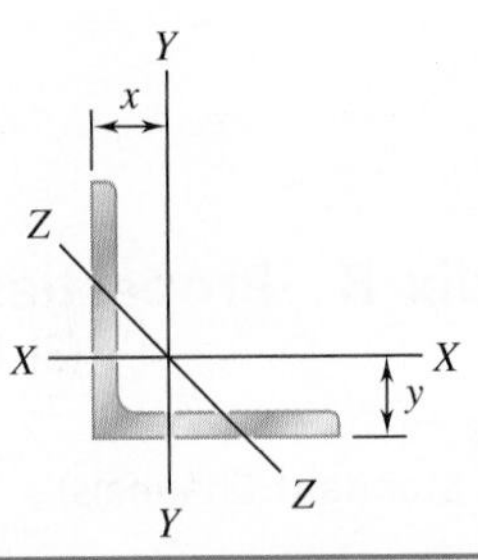

Appendix B Properties of Rolled-Steel Shapes (U.S. Customary Units)

Angles

Equal Legs

Size and Thickness, in.	Weight per Foot, lb/ft	Area, in²	Axis X-X and Axis Y-Y: I, in⁴	S, in³	r, in.	x or y, in.	Axis Z-Z: r_z, in.
L8 × 8 × 1	51.0	15.0	89.1	15.8	2.43	2.36	1.56
¾	38.9	11.4	69.9	12.2	2.46	2.26	1.57
½	26.4	7.75	48.8	8.36	2.49	2.17	1.59
L6 × 6 × 1	37.4	11.0	35.4	8.55	1.79	1.86	1.17
¾	28.7	8.46	28.1	6.64	1.82	1.77	1.17
⅝	24.2	7.13	24.1	5.64	1.84	1.72	1.17
½	19.6	5.77	19.9	4.59	1.86	1.67	1.18
⅜	14.9	4.38	15.4	3.51	1.87	1.62	1.19
L5 × 5 × ¾	23.6	6.94	15.7	4.52	1.50	1.52	0.972
⅝	20.0	5.86	13.6	3.85	1.52	1.47	0.975
½	16.2	4.75	11.3	3.15	1.53	1.42	0.980
⅜	12.3	3.61	8.76	2.41	1.55	1.37	0.986
L4 × 4 × ¾	18.5	5.44	7.62	2.79	1.18	1.27	0.774
⅝	15.7	4.61	6.62	2.38	1.20	1.22	0.774
½	12.8	3.75	5.52	1.96	1.21	1.18	0.776
⅜	9.80	2.86	4.32	1.50	1.23	1.13	0.779
¼	6.60	1.94	3.00	1.03	1.25	1.08	0.783
L3½ × 3½ × ½	11.1	3.25	3.63	1.48	1.05	1.05	0.679
⅜	8.50	2.48	2.86	1.15	1.07	1.00	0.683
¼	5.80	1.69	2.00	0.787	1.09	0.954	0.688
L3 × 3 × ½	9.40	2.75	2.20	1.06	0.895	0.929	0.580
⅜	7.20	2.11	1.75	0.825	0.910	0.884	0.581
¼	4.90	1.44	1.23	0.569	0.926	0.836	0.585
L2½ × 2½ × ½	7.70	2.25	1.22	0.716	0.735	0.803	0.481
⅜	5.90	1.73	0.972	0.558	0.749	0.758	0.481
¼	4.10	1.19	0.692	0.387	0.764	0.711	0.482
³⁄₁₆	3.07	0.900	0.535	0.295	0.771	0.687	0.482
L2 × 2 × ⅜	4.70	1.36	0.476	0.348	0.591	0.632	0.386
¼	3.19	0.938	0.346	0.244	0.605	0.586	0.387
⅛	1.65	0.484	0.189	0.129	0.620	0.534	0.391

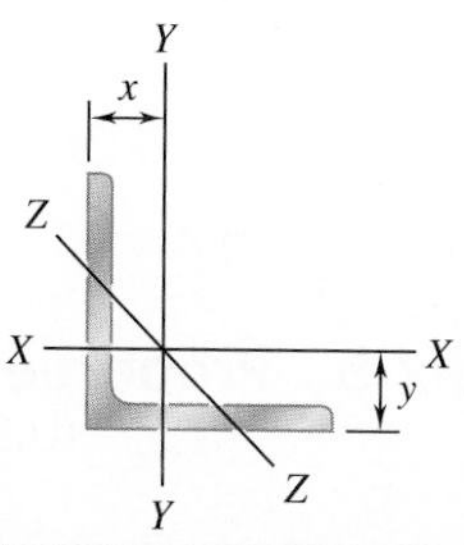

Appendix B Properties of Rolled-Steel Shapes
(SI Units)

Angles
Equal Legs

Size and Thickness, mm	Mass per Meter, kg/m	Area, mm^2	Axis X-X I 10^6 mm^4	S 10^3 mm^3	r mm	x or y mm	Axis Z-Z r_z mm
L203 × 203 × 25.4	75.9	9 680	37.1	259	61.7	59.9	39.6
19	57.9	7 350	29.1	200	62.5	57.4	39.9
12.7	39.3	5 000	20.3	137	63.2	55.1	40.4
L152 × 152 × 25.4	55.7	7 100	14.7	140	45.5	47.2	29.7
19	42.7	5 460	11.7	109	46.2	45.0	29.7
15.9	36.0	4 600	10.0	92.4	46.7	43.7	29.7
12.7	29.2	3 720	8.28	75.2	47.2	42.4	30.0
9.5	22.2	2 830	6.41	57.5	47.5	41.1	30.2
L127 × 127 × 19	35.1	4 480	6.53	74.1	38.1	38.6	24.7
15.9	29.8	3 780	5.66	63.1	38.6	37.3	24.8
12.7	24.1	3 060	4.70	51.6	38.9	36.1	24.9
9.5	18.3	2 330	3.65	39.5	39.4	34.8	25.0
L102 × 102 × 19	27.5	3 510	3.17	45.7	30.0	32.3	19.7
15.9	23.4	2 970	2.76	39.0	30.5	31.0	19.7
12.7	19.0	2 420	2.30	32.1	30.7	30.0	19.7
9.5	14.6	1 850	1.80	24.6	31.2	28.7	19.8
6.4	9.80	1 250	1.25	16.9	31.8	27.4	19.9
L89 × 89 × 12.7	16.5	2 100	1.51	24.3	26.7	26.7	17.2
9.5	12.6	1 600	1.19	18.8	27.2	25.4	17.3
6.4	8.60	1 090	0.832	12.9	27.7	24.2	17.5
L76 × 76 × 12.7	14.0	1 770	0.916	17.4	22.7	23.6	14.7
9.5	10.7	1 360	0.728	13.5	23.1	22.5	14.8
6.4	7.30	929	0.512	9.32	23.5	21.2	14.9
L64 × 64 × 12.7	11.4	1 450	0.508	11.7	18.7	20.4	12.2
9.5	8.70	1 120	0.405	9.14	19.0	19.3	12.2
6.4	6.10	768	0.288	6.34	19.4	18.1	12.2
4.8	4.60	581	0.223	4.83	19.6	17.4	12.2
L51 × 51 × 9.5	7.00	877	0.198	5.70	15.0	16.1	9.80
6.4	4.70	605	0.144	4.00	15.4	14.9	9.83
3.2	2.40	312	0.0787	2.11	15.7	13.6	9.93

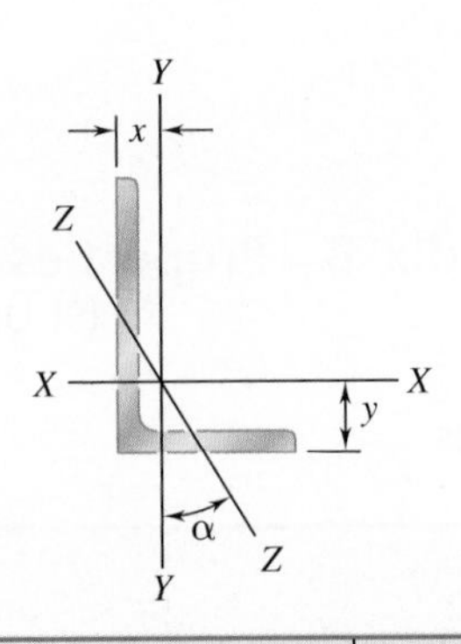

Appendix B Properties of Rolled-Steel Shapes
(U.S. Customary Units)

Angles
Unequal Legs

			Axis X-X				Axis Y-Y				Axis Z-Z	
Size and Thickness, in.	Weight per Foot, lb/ft	Area, in^2	I_x, in^4	S_x, in^3	r_x, in.	y, in.	I_y, in^4	S_y, in^3	r_y, in.	x, in.	r_z, in.	tan α
L8 × 6 × 1	44.2	13.0	80.9	15.1	2.49	2.65	38.8	8.92	1.72	1.65	1.28	0.542
¾	33.8	9.94	63.5	11.7	2.52	2.55	30.8	6.92	1.75	1.56	1.29	0.550
½	23.0	6.75	44.4	8.01	2.55	2.46	21.7	4.79	1.79	1.46	1.30	0.557
L6 × 4 × ¾	23.6	6.94	24.5	6.23	1.88	2.07	8.63	2.95	1.12	1.07	0.856	0.428
½	16.2	4.75	17.3	4.31	1.91	1.98	6.22	2.06	1.14	0.981	0.864	0.440
⅜	12.3	3.61	13.4	3.30	1.93	1.93	4.86	1.58	1.16	0.933	0.870	0.446
L5 × 3 × ½	12.8	3.75	9.43	2.89	1.58	1.74	2.55	1.13	0.824	0.746	0.642	0.357
⅜	9.80	2.86	7.35	2.22	1.60	1.69	2.01	0.874	0.838	0.698	0.646	0.364
¼	6.60	1.94	5.09	1.51	1.62	1.64	1.41	0.600	0.853	0.648	0.652	0.371
L4 × 3 × ½	11.1	3.25	5.02	1.87	1.24	1.32	2.40	1.10	0.858	0.822	0.633	0.542
⅜	8.50	2.48	3.94	1.44	1.26	1.27	1.89	0.851	0.873	0.775	0.636	0.551
¼	5.80	1.69	2.75	0.988	1.27	1.22	1.33	0.585	0.887	0.725	0.639	0.558
L3½ × 2½ × ½	9.40	2.75	3.24	1.41	1.08	1.20	1.36	0.756	0.701	0.701	0.532	0.485
⅜	7.20	2.11	2.56	1.09	1.10	1.15	1.09	0.589	0.716	0.655	0.535	0.495
¼	4.90	1.44	1.81	0.753	1.12	1.10	0.775	0.410	0.731	0.607	0.541	0.504
L3 × 2 × ½	7.70	2.25	1.92	1.00	0.922	1.08	0.667	0.470	0.543	0.580	0.425	0.413
⅜	5.90	1.73	1.54	0.779	0.937	1.03	0.539	0.368	0.555	0.535	0.426	0.426
¼	4.10	1.19	1.09	0.541	0.953	0.980	0.390	0.258	0.569	0.487	0.431	0.437
L2½ × 2 × ⅜	5.30	1.55	0.914	0.546	0.766	0.826	0.513	0.361	0.574	0.578	0.419	0.612
¼	3.62	1.06	0.656	0.381	0.782	0.779	0.372	0.253	0.589	0.532	0.423	0.624

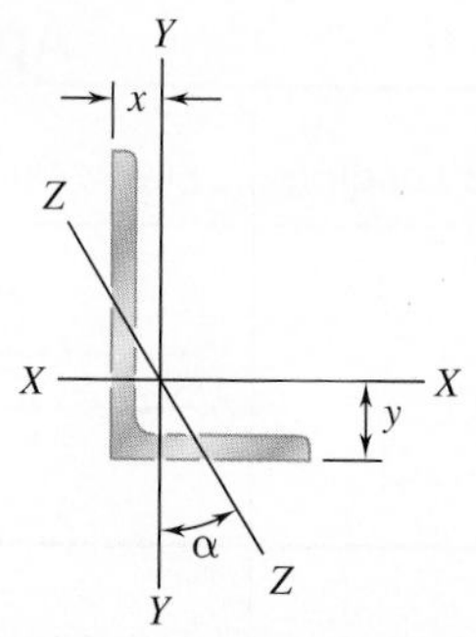

Appendix B Properties of Rolled-Steel Shapes
(SI Units)

Angles
Unequal Legs

			Axis X-X				Axis Y-Y				Axis Z-Z	
Size and Thickness, mm	Mass per Meter kg/m	Area mm^2	I_x 10^6 mm^4	S_x 10^3 mm^3	r_x mm	y mm	I_y 10^6 mm^4	S_y 10^3 mm^3	r_y mm	x mm	r_z mm	tan α
L203 × 152 × 25.4	65.5	8 390	33.7	247	63.2	67.3	16.1	146	43.7	41.9	32.5	0.542
19	50.1	6 410	26.4	192	64.0	64.8	12.8	113	44.5	39.6	32.8	0.550
12.7	34.1	4 350	18.5	131	64.8	62.5	9.03	78.5	45.5	37.1	33.0	0.557
L152 × 102 × 19	35.0	4 480	10.2	102	47.8	52.6	3.59	48.3	28.4	27.2	21.7	0.428
12.7	24.0	3 060	7.20	70.6	48.5	50.3	2.59	33.8	29.0	24.9	21.9	0.440
9.5	18.2	2 330	5.58	54.1	49.0	49.0	2.02	25.9	29.5	23.7	22.1	0.446
L127 × 76 × 12.7	19.0	2 420	3.93	47.4	40.1	44.2	1.06	18.5	20.9	18.9	16.3	0.357
9.5	14.5	1 850	3.06	36.4	40.6	42.9	0.837	14.3	21.3	17.7	16.4	0.364
6.4	9.80	1 250	2.12	24.7	41.1	41.7	0.587	9.83	21.7	16.5	16.6	0.371
L102 × 76 × 12.7	16.4	2 100	2.09	30.6	31.5	33.5	0.999	18.0	21.8	20.9	16.1	0.542
9.5	12.6	1 600	1.64	23.6	32.0	32.3	0.787	13.9	22.2	19.7	16.2	0.551
6.4	8.60	1 090	1.14	16.2	32.3	31.0	0.554	9.59	22.5	18.4	16.2	0.558
L89 × 64 × 12.7	13.9	1 770	1.35	23.1	27.4	30.5	0.566	12.4	17.8	17.8	13.5	0.485
9.5	10.7	1 360	1.07	17.9	27.9	29.2	0.454	9.65	18.2	16.6	13.6	0.495
6.4	7.30	929	0.753	12.3	28.4	27.9	0.323	6.72	18.6	15.4	13.7	0.504
L76 × 51 × 12.7	11.5	1 450	0.799	16.4	23.4	27.4	0.278	7.70	13.8	14.7	10.8	0.413
9.5	8.80	1 120	0.641	12.8	23.8	26.2	0.224	6.03	14.1	13.6	10.8	0.426
6.4	6.10	768	0.454	8.87	24.2	24.9	0.162	4.23	14.5	12.4	10.9	0.437
L64 × 51 × 9.5	7.90	1 000	0.380	8.95	19.5	21.0	0.214	5.92	14.6	14.7	10.6	0.612
6.4	5.40	684	0.273	6.24	19.9	19.8	0.155	4.15	15.0	13.5	10.7	0.624

Appendix C Beam Deflections and Slopes

Beam and Loading	Elastic Curve	Maximum Deflection	Slope at End	Equation of Elastic Curve
1 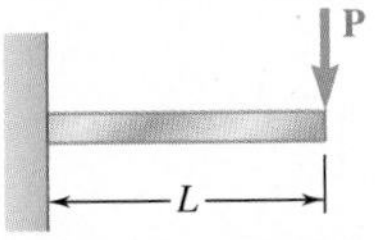	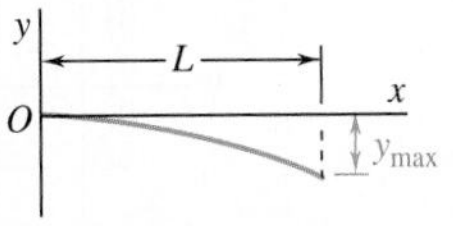	$-\dfrac{PL^3}{3EI}$	$-\dfrac{PL^2}{2EI}$	$y = \dfrac{P}{6EI}(x^3 - 3Lx^2)$
2 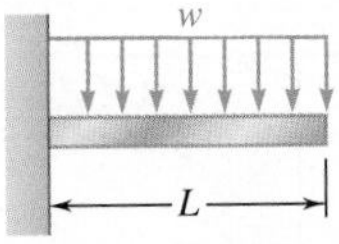	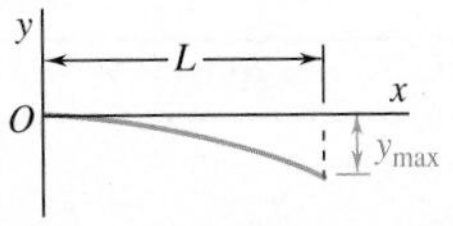	$-\dfrac{wL^4}{8EI}$	$-\dfrac{wL^3}{6EI}$	$y = -\dfrac{w}{24EI}(x^4 - 4Lx^3 + 6L^2x^2)$
3 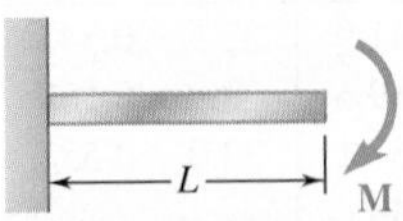	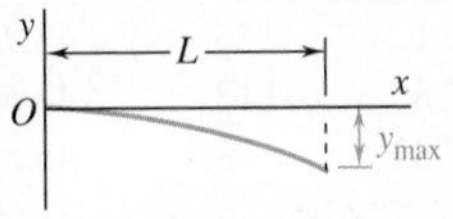	$-\dfrac{ML^2}{2EI}$	$-\dfrac{ML}{EI}$	$y = -\dfrac{M}{2EI}x^2$
4 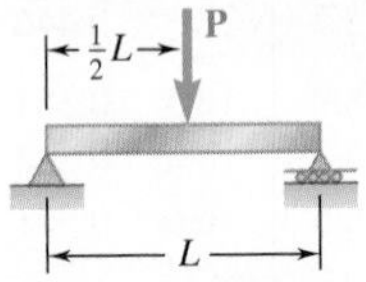	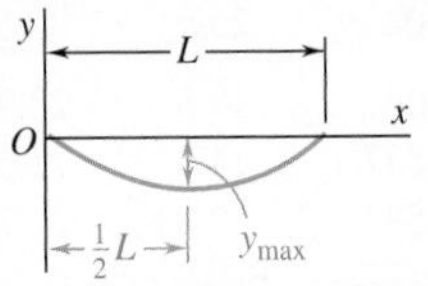	$-\dfrac{PL^3}{48EI}$	$\pm\dfrac{PL^2}{16EI}$	For $x \leq \frac{1}{2}L$: $y = \dfrac{P}{48EI}(4x^3 - 3L^2x)$
5 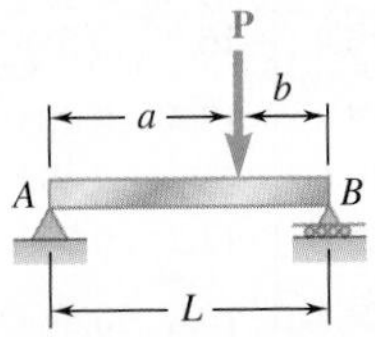	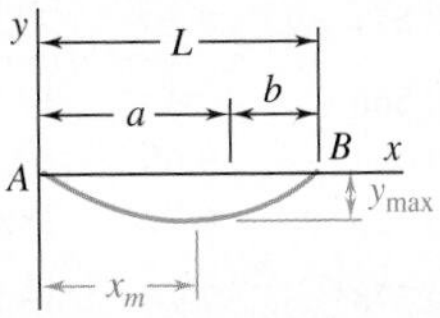	For $a > b$: $-\dfrac{Pb(L^2 - b^2)^{3/2}}{9\sqrt{3}EIL}$ at $x_m = \sqrt{\dfrac{L^2 - b^2}{3}}$	$\theta_A = -\dfrac{Pb(L^2 - b^2)}{6EIL}$ $\theta_B = +\dfrac{Pa(L^2 - a^2)}{6EIL}$	For $x < a$: $y = \dfrac{Pb}{6EIL}[x^3 - (L^2 - b^2)x]$ For $x = a$: $y = -\dfrac{Pa^2b^2}{3EIL}$
6 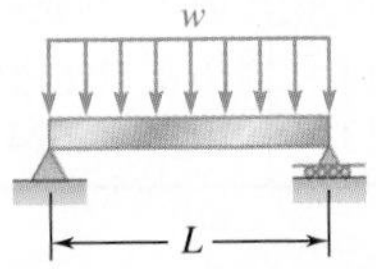	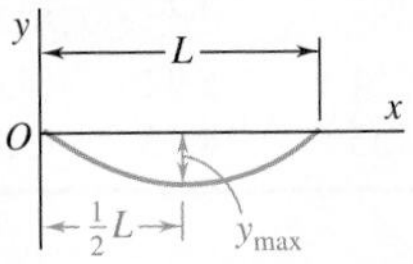	$-\dfrac{5wL^4}{384EI}$	$\pm\dfrac{wL^3}{24EI}$	$y = -\dfrac{w}{24EI}(x^4 - 2Lx^3 + L^3x)$
7 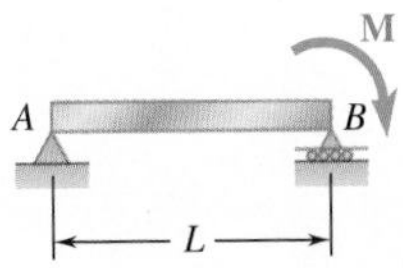	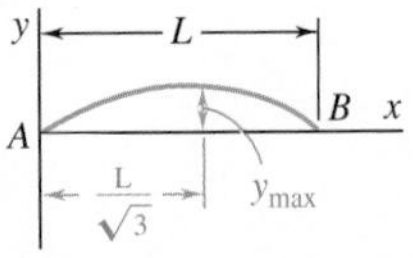	$\dfrac{ML^2}{9\sqrt{3}EI}$	$\theta_A = +\dfrac{ML}{6EI}$ $\theta_B = -\dfrac{ML}{3EI}$	$y = -\dfrac{M}{6EIL}(x^3 - L^2x)$

Index

D

E

F

G

H

I

J

K

L

M

N

O

P

Q

R

S

T

Answers to Problems

CHAPTER 2

2.1 1391 N ⦣ 47.8°.
2.2 906 lb ⦣ 26.6°.
2.3 20.1 kN ⦪ 21.2°.
2.4 8.03 kips ⦪ 3.8°.
2.5 (*a*) 76.1°. (*b*) 336 lb.
2.7 (*a*) 3660 N. (*b*) 3730 N.
2.9 (*a*) 37.1°. (*b*) 73.2 N.
2.10 2.66 kN ⦩ 34.3°.
2.11 2600 N ⦩ 53.5°.
2.12 414 lb ⦩ 72.0°.
2.13 139.1 lb ⦪ 67.0°.
2.14 8.03 kips ⦪ 3.8°.
2.16 (800 N) +640 N, +480 N; (424 N) −224 N, −360 N; (408 N) +192.0 N, −360 N.
2.17 (29 lb) +21.0 lb, +20.0 lb; (50 lb) −14.00 lb, +48.0 lb; (51 lb) +24.0 lb, −45.0 lb.
2.18 (40 lb) +20.0 lb, −34.6 lb; (50 lb) −38.3 lb, −32.1 lb; (60 lb) +54.4 lb, +25.4 lb.
2.19 (80 N) +61.3 N, +51.4 N; (120 N) +41.0 N, +112.8 N; (150 N) −122.9 N, 86.0 N.
2.20 (*a*) 523 lb. (*b*) 428 lb.
2.23 (*a*) 2190 N. (*b*) 2060 N.
2.24 654 N ⦩ 21.5°.
2.25 38.6 lb ⦣ 36.6°.
2.26 54.9 lb ⦩ 48.9°.
2.27 251 N ⦨ 85.3°.
2.29 (*a*) 177.9 lb. (*b*) 410 lb.
2.31 (*a*) 26.5 N. (*b*) 623 N.
2.32 (*a*) 5.22 kN. (*b*) 3.45 kN.
2.33 (*a*) 352 lb. (*b*) 261 lb.
2.34 (*a*) 586 N. (*b*) 2190 N.
2.35 (*a*) 500 lb. (*b*) 544 lb.
2.36 (*a*) 305 N. (*b*) 514 N.
2.38 (*a*) 16.73 kips. (*b*) 14.00 kips.
2.40 F_A = 1303 lb; F_B = 420 lb.
2.41 (*a*) 269 lb. (*b*) 37.0 lb.
2.43 (*a*) α = 35.0°; T_{AC} = 4.91 kN; T_{BC} = 3.44 kN. (*b*) α = 55.0°; $T_{AC} = T_{BC}$ = 3.66 kN.
2.44 (*a*) 784 N. (*b*) 71.0°.
2.45 (*a*) 1081 N. (*b*) 82.5°.
2.47 30.0 lb $\leq Q \leq$ 69.3 lb.
2.48 (*a*) 10.98 lb. (*b*) 30.0 lb.
2.49 68.6 in.
2.50 1.250 m.
2.51 (*a*) 300 lb. (*b*) 300 lb. (*c*) 200 lb. (*d*) 200 lb. (*e*) 150.0 lb.
2.54 (*a*) 1293 N. (*b*) 2220 N.
2.55 (*a*) 1048 N. (*b*) 608 N.
2.56 (*a*) −130.1 N; +816 N; +357 N. (*b*) 98.3°; 25.0°; 66.6°.
2.57 (*a*) +390 N; +614 N; +181.8 N. (*b*) 58.7°; 35.0°; 76.0°.
2.58 (*a*) +56.4 lb; −103.9 lb; −20.5 lb. (*b*) 62.0°; 150.0°; 99.8°.
2.59 (*a*) +37.1 lb; −68.8 lb; +33.4 lb. (*b*) 64.1°; 144.0°; 66.8°.
2.60 (*a*) −175.8 N; −257 N; +251 N. (*b*) 116.1°; 130.0°; 51.1°.
2.63 1050 N; 51.8°, 107.7°, 43.6°.
2.64 (*a*) 140.3°. (*b*) F_x = 79.9 lb, F_z = 120.1 lb; F = 226 lb.
2.65 (*a*) 118.2°. (*b*) F_x = 36.0 lb, F_y = −90.0 lb; F = 110.0 lb.
2.66 (*a*) 114.4°. (*b*) F_y = 294 lb, F_z = 855 lb; F = 1209 lb.
2.67 (*a*) F_x = 507 N, F_y = 919 N, F_z = 582 N. (*b*) 61.0°.
2.69 −165.0 N, 317 N, 238 N.
2.71 −0.820 kips, 0.978 kips, −0.789 kips.
2.72 515 N; θ_x = 70.2°, θ_y = 27.6°, θ_z = 71.5°.
2.73 515 N; θ_x = 79.8°, θ_y = 33.4°, θ_z = 58.6°.
2.75 913 lb; θ_x = 50.6°, θ_y = 117.6°, θ_z = 51.8°.
2.77 T_{AB} = 490 N; T_{AD} = 515 N.
2.78 130.0 lb.
2.79 137.0 lb.
2.80 13.98 kN.
2.81 9.71 kN.
2.82 T_{AB} = 201 N; T_{AC} = 372 N; T_{AD} = 416 N.
2.83 1868 lb.
2.85 T_{AB} = 571 lb; T_{AC} = 830 lb; T_{AD} = 528 lb.
2.86 T_{DA} = 119.7 lb; T_{DB} = 98.4 lb; T_{DC} = 98.4 lb.
2.87 T_{DA} = 14.42 lb; $T_{DB} = T_{DC}$ = 13.00 lb.
2.89 768 N.
2.90 T_{AB} = 30.8 lb; T_{AC} = 62.5 lb.
2.91 T_{AB} = 81.3 lb; T_{AC} = 22.2 lb.
2.92 960 N.
2.93 $0 \leq Q <$ 300 N.
2.95 W = 470 N; Q = 37.0 N.
2.96 T_{BAC} = 76.7 lb; T_{AD} = 26.9 lb; T_{AE} = 49.2 lb.
2.97 (*a*) 305 lb. (*b*) T_{BAC} = 117.0 1b; T_{AD} = 40.9 lb.
2.98 378 N.
2.99 T_{AB} = 65.6 lb; T_{AC} = 55.1 lb.
2.100 (*a*) 125.0 lb. (*b*) 45.0 lb.
2.102 (*a*) 1155 N. (*b*) 1012 N.
2.104 21.8 kN ⦩ 73.4°.
2.105 (102 lb) −48.0 lb, 90.0 lb; (106 lb) 56.0 lb, 90.0 lb; (200 lb) −160.0 lb, −120.0 lb.
2.107 203 lb ⦣ 8.46°.
2.108 (*a*) 1244 lb. (*b*) 115.4 lb.
2.110 27.4° $\leq \alpha \leq$ 222.6°.
2.112 1031 N↑.
2.113 956 N↑.
2.115 3090 lb.

CHAPTER 3

3.1 115.7 lb·in.
3.2 23.2°.
3.3 (*a*) 20.5 N·m↺. (*b*) 68.4 mm.
3.5 (*a*) 41.7 N·m↺. (*b*) 147.4 N ⦣ 45.0°.
3.6 (*a*) 41.7 N·m↺. (*b*) 334 N. (*c*) 176.8 N ⦣ 58.0°.

3.7 116.2 lb·ft.
3.9 1.120 kip·in.↺.
3.11 (*a*) 292 N·m↻. (*b*) 292 N·m↻.
3.12 2340 N.
3.14 (*a*) 9**i** + 22**j** + 21**k**. (*b*) 22**i** + 11**k**. (*c*) 0.
3.15 −(25.4 lb·ft)**i** − (12.60 lb·ft)**j** − (12.60 lb·ft)**k**.
3.16 (*a*) (28.8 N·m)**i** + (16.20 N·m)**j** − (28.8 N·m)**k**.
(*b*) −(28.8 N·m)**i** −(16.20 N·m)**j** + (28.8 N·m)**k**.
3.17 (2400 lb·ft)**j** + (1440 lb·ft)**k**.
3.18 −(153.0 lb·ft)**i** + (63.0 lb·ft)**j** + (215 lb·ft)**k**.
3.19 (7.50 N·m)**i** − (6.00 N·m)**j** − (10.39 N·m)**k**.
3.20 (492 lb·ft)**i** + (144.0 lb·ft)**j** − (372 lb·ft)**k**.
3.23 144.8 mm.
3.24 4.86 ft.
3.25 **P**·**Q** = +1; **P**·**S** = −11; **Q**·**S** = +10.
3.27 (*a*) 59.0°. (*b*) 648 N.
3.28 (*a*) 70.5°. (*b*) 135.0 N.
3.29 77.9°.
3.31 (*a*) 26.8°. (*b*) 26.8°.
3.33 (*a*) 67.0. (*b*) 111.0.
3.34 2.
3.35 M_x = 0; M_y = −162.0 N·m; M_z = 270 N·m.
3.37 283 lb.
3.39 1.252 m.
3.40 1.256 m.
3.41 61.5 lb.
3.42 6.23 ft.
3.43 1359 lb·in.
3.44 −2350 lb·in.
3.46 −111.0 N·m.
3.48 −176.6 lb·ft.
3.49 (*a*) 12.39 N·m↻. (*b*) 12.39 N·m↻. (*c*) 12.39 N·m↻.
3.50 (*a*) 336 lb·in.↺. (*b*) 28.0 in. (*c*) 54.0°.
3.51 (*a*) 7.33 N·m↺. (*b*) 91.6 mm.
3.52 (*a*) 26.7 N. (*b*) 50.0 N. (*c*) 23.5 N.
3.53 (*a*) 1170 lb·in.↺. (*b*) *A* and *D*, 53.1° ⊿, or *B* and *C* ◺ 53.1°.
(*c*) 70.9 lb.
3.54 1.125 in.
3.56 *M* = 15.30 lb·ft; θ_x = 78.7°; θ_y = 90.0°; θ_z = 11.30°.
3.57 *M* = 3.22 N·m; θ_x = 90.0°; θ_y = 53.1°; θ_z = 36.9°.
3.58 *M* = 2.72 N·m; θ_x = 134.9°; θ_y = 58.0°; θ_z = 61.9°.
3.59 *M* = 2150 lb·ft; θ_x = 113.0°; θ_y = 92.7°; θ_z = 23.2°.
3.61 (*a*) **F** = 30.0 lb↓; $\mathbf{M}_B$ = 150.0 lb·in.↺.
(*b*) $\mathbf{F}_B$ = 50.0 lb←; $\mathbf{F}_C$ = 50.0 lb→.
3.63 (*a*) $\mathbf{F}_B$ = 250 N ◺ 25.0°; $\mathbf{M}_B$ = 57.5 N·m↻.
(*b*) $\mathbf{F}_A$ = 375 N ◺ 25.0°; $\mathbf{F}_B$ = 625 N ◺ 25.0°.
3.65 $\mathbf{F}_A$ = 389 N ◺ 60.0°; $\mathbf{F}_C$ = 651 N ◺ 60.0°.
3.66 (*a*) **P** = 60.0 lb ⊿ 50.0°; 3.24 in. from *A*.
(*b*) **P** = 60.0 lb ⊿ 50.0°; 3.87 in. below *A*.
3.67 **F** = −(250 kN)**j**; **M** = (15.00 kN·m)**i** + (7.50 kN·m)**k**.
3.68 **F** = −(128.0 lb)**i** − (256 lb)**j** + (32.0 lb)**k**;
M = −(4.10 kip·ft)**i** + (16.38 kip·ft)**k**.
3.71 **F** = −(2.40 kips)**j** − (1.000 kips)**k**; **M** = −(12.00 kip·in.)**i** + (6.00 kip·in.)**j** − (14.40 kip·in.)**k**.
3.72 **F** = − (28.5 N)**j** + (106.3 N)**k**; $\mathbf{M}_O$ = (12.35 N·m)**i** − (19.16 N·m)**j** − (5.13 N·m)**k**.
3.73 (*a*) $\mathbf{R}_a$ = 600 N↓, $\mathbf{M}_a$ = 1000 N·m↺; $\mathbf{R}_b$ = 600 N↓,
$\mathbf{M}_b$ = 900 N·m↻; $\mathbf{R}_c$ = 600 N↓, $\mathbf{M}_c$ = 900 N·m↺;
$\mathbf{R}_d$ = 400 N↑, $\mathbf{M}_d$ = 900 N·m↺; $\mathbf{R}_e$ = 600 N↓,
$\mathbf{M}_e$ = 200 N·m↻; $\mathbf{R}_f$ = 600 N↓, $\mathbf{M}_f$ = 800 N·m↺;
$\mathbf{R}_g$ = 1000 N↓, $\mathbf{M}_g$ = 1000 N·m↺; $\mathbf{R}_h$ = 600 N↓,
$\mathbf{M}_h$ = 900 N·m↺. (*b*) (*c*) and (*h*).
3.74 Loading *f*.
3.75 (*a*) **R** = 600 N↓; 1.500 m. (*b*) **R** = 400 N↑; 2.25 m.
(*c*) **R** = 600 N↓; 0.333 m.
3.76 (*a*) 2.00 ft to the right of *C*. (*b*) 2.31 ft to the right of *C*.
3.78 Force-couple system at corner *D*.
3.80 **R** = 185.2 lb ⊿ 11.84°; 23.3 in. to the left of the vertical centerline (*y*-axis) of the motor.
3.81 (*a*) 34.0 lb ◺ 28.0°. (*b*) *AB*: 11.64 in. to the left of *B*; *BC*: 6.20 in. below *B*.
3.82 773 lb ⊿ 79.0°; 9.54 ft to the right of *A*.
3.83 (*a*) 0.365 m above *G*. (*b*) 0.227 m to the right of *G*.
3.84 (*a*) 0.299 m above *G*. (*b*) 0.259 m to the right of *G*.
3.85 $\mathbf{R}_A$ = (8.40 lb)**i** − (19.20 lb)**j** − (3.20 lb)**k**;
$\mathbf{M}_A$ = (71.6 lb·ft)**i** + (56.8 lb·ft)**j** − (65.2 lb·ft)**k**.
3.87 **R** = −(300 N)**i** − (240 N)**j** + (25.0 N)**k**;
M = −(3.00 N·m)**i** + (13.50 N·m)**j** + (9.00 N·m)**k**.
3.89 (*a*) 60.0°. (*b*) (20 lb)**i** − (34.6 lb)**j**; (520 lb·in.)**i**.
3.90 (*a*) Neither loosen nor tighten. (*b*) Tighten.
3.91 (*a*) **B** = −(80.0 N) **k**; **C** = −(30.0 N) **i** + (40.0 N) **k**.
(*b*) R_y = 0; R_z = −40.0 N. (*c*) when the slot head of the screw is vertical.
3.92 405 lb; 12.60 ft to the right of *AB* and 2.94 ft below *BC*.
3.94 **R** = 325 kN, *x* = −0.923 m; *z* = −0.615 m.
3.96 *x* = 2.32 m; *z* = 1.165 m.
3.97 8.97 lb ◺ 19.98°.
3.99 M_x = 78.9 N·m, M_y = 13.15 kN·m, M_z = −9.86 kN·m.
3.100 3.04 kN.
3.101 23.0 N·m.
3.102 (0.227 lb)**i** + (0.1057 lb)**k**; 63.6 in. to the right of *B*.
3.103 (*a*) **F** = 500 N ◺ 60.0°; **M** = 86.2 N·m↻.
(*b*) **A** = 689 N↑; **B** = 1150 N ◺ 77.4°.
3.105 (*a*) 71.1°. (*b*) 0.973 lb.
3.106 12.00 in.
3.108 $aP/\sqrt{2}$.

CHAPTER 4

4.1 (*a*) 245 lb↑. (*b*) 140.0 lb.
4.2 (*a*) 1761 lb↑. (*b*) 689 lb↑.
4.3 42.0 N↑.
4.5 1.250 kN ≤ *Q* ≤ 27.5 kN.
4.6 1.250 kN ≤ *Q* ≤ 10.25 kN.
4.7 2.00 in. ≤ *a* ≤ 10.00 in.
4.9 (*a*) 29.9 kips. (*b*) 33.0 kips ⊿ 31.5°.
4.10 (*a*) 150.0 lb. (*b*) 225 lb ⊿ 32.3°.
4.12 (*a*) 400 N. (*b*) 458 N ⊿ 49.1°.
4.13 (*a*) **A** = **B** = 37.5 lb↑. (*b*) **A** = 97.6 lb ⊿ 50.2°;
B = 62.5 lb←. (*c*) **A** = 49.8 lb ⊿ 71.2°; **B** = 32.2 lb ◺ 60.0°.

4.15 (*a*) 1.500 kN. (*b*) 1.906 kN ⦨ 61.8°.
4.16 (*a*) **A** = 150.0 N ⦨ 30.0°; **B** = 150.0 N ⦩ 30.0°.
(*b*) **A** = 433 N ⦪ 12.55°; **B** = 488 N ⦩ 30.0°.
4.17 T_{BE} = 50.0 lb; **A** = 18.75 lb →; **D** = 18.75 lb ←.
4.18 T = 80.0 N; **A** = 160.0 N ⦪ 30.0°; **C** = 160.0 N ⦩ 30.0°.
4.19 T = 69.3 N; **A** = 140.0 N ⦪ 30.0°; **C** = 180.0 N ⦩ 30.0°.
4.20 (*a*) 30.0 lb ⦩ 60.0°. (*b*) **A** = 20.2 lb ↑; **F** = 16.21 lb ↓.
4.23 (*a*) 11.20 kips. (*b*) $|M_E|$ = 28.8 kip·ft.
4.24 **C** = 28.3 kN ⦩ 45.0°; $\mathbf{M}_C$ = 4.30 N·m ↻.
4.25 (1) Completely constrained; determinate; equilibrium;
A = 120.2 lb ⦨ 56.3°; **B** = 66.7 lb ←.
(2) Improperly constrained; indeterminate; no equilibrium;
(3) Partially constrained; determinate; equilibrium;
A = 50.0 lb ↑; **C** = 50.0 lb ↑.
(4) Completely constrained; determinate; equilibrium;
A = 50.0 lb ↑; **B** = 83.3 lb ⦩ 36.9°; **C** = 66.7 lb →.
(5) Completely constrained; indeterminate; equilibrium;
$\mathbf{A}_y$ = 50.0 lb ↑.
(6) Completely constrained; indeterminate; equilibrium;
$\mathbf{A}_x$ = 66.7 lb →; $\mathbf{B}_x$ = 66.7 lb ←; ($\mathbf{A}_y$ + $\mathbf{B}_y$ = 100.0 lb ↑).
(7) Completely constrained; determinate; equilibrium;
A = 50.0 lb ↑; **C** = 50.0 lb ↑.
(8) Improperly constrained; indeterminate; no equilibrium.
4.26 (1) Completely constrained; determinate; equilibrium;
A = **C** = 196.2 N ↑.
(2) Completely constrained; determinate; equilibrium;
B = 0; **C** = **D** = 196.2 N ↑.
(3) Completely constrained; indeterminate; equilibrium;
$\mathbf{A}_x$ = 294 N →; $\mathbf{D}_x$ = 294 N ←.
(4) Improperly constrained; indeterminate; no equilibrium.
(5) Partially constrained; determinate; equilibrium;
C = **D** = 196.2 N ↑.
(6) Completely constrained; determinate; equilibrium;
B = 294 N →; **D** = 491 N ⦩ 53.1°.
(7) Partially constrained; no equilibrium.
(8) Completely constrained; indeterminate; equilibrium;
B = $\mathbf{D}_y$ = 196.2 N ↑; (**C** + $\mathbf{D}_x$ = 0).
4.27 **B** = 501 N ⦩ 56.3°; **C** = 324 N ⦪ 31.0°.
4.28 **A** = 446 lb ⦩ 7.73°; **B** = 442 lb →.
4.29 **A** = 82.5 lb ⦨ 14.04°; T = 100.0 lb.
4.31 **A** = 139.0 N ⦨ 62.4°; T = 69.6 N.
4.32 (*a*) 400 N. (*b*) 458 N ⦨ 49.1°.
4.33 (*a*) **A** = 150.0 N ⦨ 30.0°; **B** = 150.0 N ⦩ 30.0°.
(*b*) **A** = 433 N ⦪ 12.55°; **B** = 488 N ⦩ 30.0°.
4.34 (*a*) **P** = 24.9 lb ⦫ 30.0°. (*b*) **P** = 15.34 lb ⦨ 30.0°.
4.37 200 mm.
4.38 **C** = 270 lb ⦩ 56.3°; **D** = 167.7 lb ⦪ 26.6°.
4.40 (*a*) 59.2°. (*b*) T_{AB} = 0.596W; T_{CD} = 1.164W.
4.41 (*a*) 2P ⦩ 60.0°. (*b*) 1.239P ⦪ 36.2°.
4.42 $\tan \theta = 2 \tan \beta$.
4.43 (*a*) 49.1°. (*b*) 90.6 N ⦨ 60.0°.
4.44 **A** = 170.0 N ⦩ 33.9°; **C** = 160.0 N ⦨ 28.1°.
4.45 **A** = 170.0 N ⦫ 56.1°; **C** = 300 N ⦨ 28.1°.
4.47 **A** = 163.1 N ⦪ 55.9°; **B** = 258 N ⦩ 65.0°.
4.48 $\cos^2 \theta = 1/3\ [(R/L)^2 - 1]$.
4.50 32.5°.
4.51 **A** = (120.0 N)**j** + (133.3 N)**k**; **D** = (60.0 N)**j** + (166.7 N)**k**.
4.52 **A** = (125.3 N)**j** + (137.8 N)**k**; **D** = (62.7 N)**j** + (172.2 N)**k**.
4.53 **A** = (24.0 lb)**j** − (2.31 lb)**k**; **B** = (16.00 lb)**j** − (9.24 lb)**k**;
C = (11.55 lb)**k**
4.54 (*a*) 96.0 lb. (*b*) **A** = (2.40 lb)**j**; **B** = (214 lb)**j**.
4.56 (*a*) 1039 N. (*b*) **C** = (346 N)**i** + (1200 N)**j**;
D = −(1386 N)**i** − (480 N)**j**.
4.57 T_A = 30.0 lb; T_B = 10.00 lb; T_C = 40.0 lb.
4.59 (*a*) 121.9 N. (*b*) −46.2 N. (*c*) 100.9 N.
4.61 T_{DAE} = 520 lb; T_{BD} = 680 lb; **C** = (−120.0 lb)**i** + (120.0 lb)**j** + (1560 lb)**k**.
4.62 T_{DAE} = 832 lb; T_{BD} = 1088 lb; **C** = −(192.0 lb)**i** + (2496 lb)**k**.
4.63 T_{BD} = 780 N; T_{BE} = 390 N; **A** = −(195.0 N)**i** + (1170 N)**j** + (130.0 N) **k**.
4.65 **A** = −(56.3 lb)**i**; **B** = −(56.2 lb)**i** + (150.0 lb)**j** − (75.0 lb)**k**;
F_{CE} = 202 lb.
4.66 (*a*) 345 N. (*b*) **A** = (114.4 N)**i** + (377 N)**j** + (141.5 N)**k**;
B = (113.2 N)**j** + (185.5 N)**k**.
4.67 (*a*) 49.5 lb. (*b*) **A** = −(12.00 lb)**i** + (22.5 lb)**j** − (4.00 lb)**k**;
B = (15.00 lb)**j** + (34.0 lb)**k**.
4.70 (*a*) 462 N. (*b*) **C** = −(336 N)**j** + (467 N)**k**; **D** = (505 N)**j** − (66.7 N)**k**.
4.71 $\mathbf{F}_{CE}$ = 202 lb; M_A = (600 lb·ft)**i** + (225 lb·ft)**j**;
A = −(112.5 lb)**i** + (150.0 lb)**j** − (75.0 lb)**k**.
4.72 F_{CD} = 19.62 N; **B** = (−19.22 N)**i** + (94.2 N)**j**;
$\mathbf{M}_B$ = −(40.6 N·m)**i** − (17.30 N·m)**j**
4.73 T_{CF} = 200 N; T_{DE} = 450 N; **A** = (160.0 N)**i** + (270 N)**k**;
$\mathbf{M}_A$ = −(16.20 N·m)**i**.
4.74 **A** = (120.0 lb)**j** − (150.0 lb)**k**; **B** = (180.0 lb)**i** + (150.0 lb)**k**;
C = −(180.0 lb)**i** + (120.0 lb)**j**.
4.75 Equilibrium; **F** = 172.6 N ⦪ 25.0°.
4.76 Block moves down; **F** = 279 N ⦩ 30.0°.
4.77 Block moves up; **F** = 36.1 lb ⦪ 30.0°.
4.78 Block is in equilibrium; **F** = 36.3 lb ⦪ 30.0°.
4.80 (*a*) 18.09 lb →. (*b*) 14.34 lb ←.
4.81 31.0°.
4.82 46.4°.
4.83 (*a*) 403 N. (*b*) 229 N.
4.85 (*a*) 36.0 lb →. (*b*) 30.0 lb →. (*c*) 12.86 lb →.
4.87 (*a*) 58.1°. (*b*) 166.4 N
4.88 (*a*) 138.6 N. (*b*) Crate will slide.
4.90 (*a*) 275 N ←. (*b*) 196.2 N ←.
4.91 0.208.
4.93 (*a*) 43.6°. (*b*) 0.371W.
4.94 (*a*) 136.4°. (*b*) 0.928W.
4.95 1.225W.
4.97 $M_{max} = Wr\,\mu_s(1 + \mu_s)/(1 + \mu_s^2)$.
4.98 (*a*) 0.300Wr. (*b*) 0.349Wr.
4.99 (*a*) **A** = 20.0 lb ↓; **B** = 150.0 lb ↑.
(*b*) **A** = 10.00 lb ↓; **B** = 140.0 lb ↑.
4.101 T = 300 N; **B** = 375 N ⦪ 36.9°.
4.102 (*a*) 600 N. (*b*) **A** = 4.00 kN ←; **B** = 4.00 kN →.
4.104 (*a*) 225 mm. (*b*) 23.1 N. (*c*) **C** = 12.21 N.
4.105 1.300 ft.

4.107 (*a*) $\mathbf{A} = 0.745P \measuredangle 63.4°$; $\mathbf{C} = 0.471P \measuredangle 45.0°$.
(*b*) $\mathbf{A} = 0.812P \measuredangle 60.0°$; $\mathbf{C} = 0.503P \measuredangle 36.2°$.
(*c*) $\mathbf{A} = 0.448P \measuredangle 60.0°$; $\mathbf{C} = 0.652P \measuredangle 69.9°$.
(*d*) Rod is improperly constrained.
4.109 (*a*) 2.94 N. (*b*) 4.41 N.
4.110 (*b*) 2.69 lb.

CHAPTER 5

5.1 $\overline{X} = 42.2$ mm, $\overline{Y} = 24.2$ mm.
5.2 $\overline{X} = 3.27$ in., $\overline{Y} = 2.82$ in.
5.3 $\overline{X} = 1.045$ in., $\overline{Y} = 3.59$ in.
5.5 $\overline{X} = \overline{Y} = 5.06$ in.
5.6 $\overline{X} = \overline{Y} = 16.75$ mm.
5.7 $\overline{X} = -62.4$ mm, $\overline{Y} = 0$.
5.9 $\overline{X} = 3.20$ in., $\overline{Y} = 2.00$ in.
5.10 $\overline{X} = 10.11$ in., $\overline{Y} = 3.88$ in.
5.11 $\overline{X} = 0$, $\overline{Y} = 1.372$ m.
5.12 $\overline{X} = 386$ mm, $\overline{Y} = 66.4$ mm.
5.13 42.3×10^3 mm^3 for A_1, -42.3×10^3 mm^3 for A_2.
5.14 0.235 in^3 for A_1, -0.235 in^3 for A_2.
5.17 $\overline{X} = 40.9$ mm, $\overline{Y} = 25.3$ mm.
5.18 $\overline{X} = 3.38$ in., $\overline{Y} = 2.93$ in.
5.19 $\overline{X} = 172.5$ mm, $\overline{Y} = 97.5$ mm.
5.20 $\overline{X} = \overline{Y} = 4.90$ in.
5.21 120.0 mm.
5.23 (*a*) 5.09 lb. (*b*) 9.48 lb $\measuredangle$ 57.5°.
5.25 $\bar{x} = 2a/3$, $\bar{y} = 2h/3$.
5.26 $\bar{x} = 2a/5$, $\bar{y} = 3h/7$.
5.29 $\bar{x} = a(3 - 4 \sin \alpha)/6(1 - \alpha)$, $\bar{y} = 0$.
5.30 $\bar{x} = 0$, $\bar{y} = 4(r_2^3 - r_1^3)/3\pi(r_2^2 - r_1^2)$.
5.31 $\bar{x} = 2a/3(4 - \pi)$, $\bar{y} = 2b/3(4 - \pi)$.
5.32 $\bar{x} = \bar{y} = 9a/20$.
5.33 $\bar{x} = 17a/130$, $\bar{y} = 11b/26$.
5.34 $\bar{x} = a$, $\bar{y} = 17b/35$.
5.35 $\bar{x} = \bar{y} = 1.027$ in.
5.36 $\bar{x} = \bar{y} = (2a^2 - 1)/2a(1 + 2 \ln a)$.
5.37 (*a*) $V = 401 \times 10^3$ mm^3; $A = 34.1 \times 10^3$ mm^2.
(*b*) $V = 492 \times 10^3$ mm^3; $A = 41.9 \times 10^3$ mm^2.
5.39 (*a*) $V = 169.0 \times 10^3$ in^3; $A = 28.4 \times 10^3$ in^2.
(*b*) $V = 88.9 \times 10^3$ in^3; $A = 15.48 \times 10^3$ in^2.
5.41 31.9 liters.
5.42 0.0305 kg.
5.43 308 in^2.
5.44 (*a*) 8.10 in^2. (*b*) 6.85 in^2. (*c*) 7.01 in^2.
5.45 $V = 3.96$ in^2; $W = 1.211$ lb.
5.48 0.1916 kg.
5.49 (*a*) $\mathbf{R} = 7.60$ kN ↓, $\bar{x} = 2.57$ m.
(*b*) $\mathbf{A} = 4.35$ kN ↑; $\mathbf{B} = 3.25$ kN ↑.
5.51 $\mathbf{A} = 575$ lb ↑; $\mathbf{M}_A = 475$ lb·ft ↺.
5.53 $\mathbf{A} = 32.0$ kN ↑; $\mathbf{M}_A = 124.0$ kN·m ↺.
5.54 $\mathbf{B} = 1360$ lb ↑; $\mathbf{C} = 2360$ lb ↑.
5.55 $\mathbf{A} = 2860$ lb ↑; $\mathbf{B} = 740$ lb ↑.
5.56 $\mathbf{A} = 105.0$ N ↑; $\mathbf{B} = 270$ N ↑.
5.57 (*a*) $0.548L$. (*b*) $2\sqrt{3}$.
5.58 (*a*) $b/10$ to the left of base of cone.
(*b*) $0.01136b$ to the right of base of cone.
5.59 (*a*) $-0.402a$. (*b*) $h/a = 2/5$ or $2/3$.
5.60 27.8 mm above base of cone.
5.61 -0.0656 in.
5.63 40.3 mm.
5.65 $\overline{X} = \overline{Z} = 4.21$ in., $\overline{Y} = 7.03$ in.
5.66 $\overline{X} = 125.0$ mm, $\overline{Y} = 167.0$ mm, $\overline{Z} = 33.5$ mm.
5.69 $\overline{X} = 0.410$ m, $\overline{Y} = 0.510$ m, $\overline{Z} = 0.1500$ m.
5.70 $\overline{X} = 0$, $\overline{Y} = 10.05$ in., $\overline{Z} = 5.15$ in.
5.71 $\overline{X} = \overline{Z} = 0$, $\overline{Y} = 83.3$ mm above the base.
5.72 On center axis, 1.380 in. above base.
5.73 $\overline{X} = 19.27$ mm, $\overline{Y} = 26.6$ mm.
5.75 $\overline{X} = 20.6$ mm, $\overline{Y} = 23.4$ mm.
5.76 0.204 m or 0.943 m.
5.77 $\bar{x} = 1.607a$, $\bar{y} = 0.332$ h.
5.79 0.0900 in^3.
5.81 $\mathbf{B} = 3770$ lb ↑; $\mathbf{C} = 429$ lb ↑.
5.82 (*a*) 900 lb/ft. (*b*) 7200 lb ↑.
5.84 $\overline{X} = 61.1$ mm from the end of the handle.

CHAPTER 6

6.1 $F_{AB} = 4.00$ kN *C*; $F_{BC} = 2.40$ kN *C*; $F_{AC} = 2.72$ kN *T*.
6.2 $F_{AB} = 52.0$ kN *T*; $F_{BC} = 80.0$ kN *C*; $F_{AC} = 64.0$ kN *T*.
6.3 $F_{AB} = 720$ lb *T*; $F_{BC} = 780$ lb *C*; $F_{AC} = 1200$ lb *C*.
6.4 $F_{AB} = 900$ lb *T*; $F_{BC} = 720$ lb *T*; $F_{AC} = 780$ lb *C*.
6.6 $F_{AB} = 15.90$ kN *C*; $F_{AC} = 13.50$ kN *T*; $F_{CD} = 15.90$ kN *T*; $F_{BC} = 16.80$ kN *C*; $F_{BD} = 13.50$ kN *C*.
6.8 $F_{AB} = F_{BC} = 0$; $F_{AD} = F_{CF} = 7.00$ kN *C*; $F_{BD} = F_{BF} = 34.0$ kN *C*; $F_{DE} = F_{EF} = 30.0$ kN *T*; $F_{BE} = 8.00$ kN *T*.
6.9 $F_{AB} = F_{AE} = 671$ lb *T*; $F_{BC} = F_{DE} = 600$ lb *C*; $F_{AC} = F_{AD} = 1000$ lb *C*; $F_{CD} = 200$ lb *T*.
6.10 $F_{AB} = 4.00$ kN *T*; $F_{AD} = 15.00$ kN *T*; $F_{BD} = 9.00$ kN *C*; $F_{BE} = 5.00$ kN *T*; $F_{CD} = 16.00$ kN *C*; $F_{DE} = 4.00$ kN *C*.
6.11 $F_{AD} = 260$ lb *C*; $F_{BE} = 832$ lb *C*; $F_{CE} = 400$ lb *T*; $F_{DC} = 125.0$ lb *T*; $F_{AB} = 420$ lb *C*; $F_{AC} = 400$ lb *T*; $F_{BC} = 125.0$ lb *T*.
6.12 $F_{BA} = 0$; $F_{CA} = 22.4$ kips *T*; $F_{CB} = 60.0$ kips *C*; $F_{DA} = 41.2$ kips *T*; $F_{DC} = 40.0$ kips *C*.
6.13 $F_{AB} = F_{DE} = 8.00$ kN *C*; $F_{AF} = F_{FG} = F_{GH} = F_{EH} = 6.93$ kN *T*; $F_{BC} = F_{CD} = F_{BG} = F_{DG} = 4.00$ kN *C*; $F_{BF} = F_{DH} = F_{CG} = 4.00$ kN *T*.
6.15 $F_{AB} = F_{FH} = 1500$ lb *C*; $F_{AC} = F_{CE} = F_{EG} = F_{GH} = 1200$ lb *T*; $F_{BC} = F_{FG} = 0$; $F_{BD} = F_{DF} = 1000$ lb *C*; $F_{BE} = F_{EF} = 500$ lb *C*; $F_{DE} = 600$ lb *T*.
6.17 $F_{CD} = 30.0$ kN *T*; $F_{DG} = 32.5$ kN *C*; $F_{CG} = 0$; $F_{FG} = 32.5$ kN C; $F_{CF} = 19.53$ kN *C*; $F_{BC} = 45.0$ kN *T*; $F_{EF} = 48.8$ kN *C*; $F_{BF} = 6.25$ kN *T*; $F_{BE} = 24.0$ kN *C*; $F_{AB} = 60.0$ kN *T*; $F_{AE} = 37.5$ kN *T*.
6.18 $F_{AB} = F_{FH} = 7.50$ kips *C*; $F_{AC} = F_{GH} = 4.50$ kips *T*; $F_{BC} = F_{FG} = 4.00$ kips *T*; $F_{BD} = F_{DF} = 6.00$ kips *C*; $F_{BE} = F_{EF} = 2.50$ kips *T*; $F_{CE} = F_{EG} = 4.50$ kips *T*; $F_{DE} = 0$.
6.19 Truss of Prob. 6.24 is the only simple truss.
6.20 Neither truss is a simple truss.
6.21 *BC, CD, IJ, IL, LM, MN.*
6.24 *BF, BG, DH, EH, GJ, HJ.*
6.25 $F_{BD} = 36.0$ kips *C*; $F_{CD} = 45.0$ kips *C*.
6.26 $F_{FD} = 60.0$ kips *C*; $F_{GD} = 15.00$ kips *C*.
6.27 $F_{BD} = 216$ kN *T*; $F_{DE} = 270$ kN *T*.
6.29 $F_{DE} = 25.0$ kips *T*; $F_{DF} = 13.00$ kips *C*.
6.31 $F_{DE} = 38.6$ kN *C*; $F_{DF} = 91.4$ kN *T*.
6.32 $F_{CD} = 64.2$ kN *T*; $F_{CE} = 91.2$ kN *C*.
6.33 $F_{CE} = 7.20$ kN *T*; $F_{DE} = 1.047$ kN *C*; $F_{DF} = 6.39$ kN *C*.

6.34 $F_{EG} = 3.46$ kN T; $F_{GH} = 3.78$ kN C; $F_{HJ} = 3.55$ kN C.
6.35 $F_{AB} = 8.20$ kips T; $F_{AG} = 4.50$ kips T; $F_{FG} = 11.60$ kips C.
6.37 $F_{DF} = 40.0$ kN T; $F_{EF} = 12.00$ kN T; $F_{EG} = 60.0$ kN C.
6.39 $F_{AD} = 3.38$ kips C; $F_{CD} = 0$; $F_{CE} = 14.03$ kips T.
6.40 $F_{DG} = 18.75$ kips C; $F_{FG} = 14.03$ kips T; $F_{FH} = 17.43$ kips T.
6.41 $F_{DG} = 3.75$ kN T; $F_{FI} = 3.75$ kN C.
6.42 $F_{GJ} = 11.25$ kN T; $F_{IK} = 11.25$ kN C.
6.44 $F_{BE} = 10.00$ kips T; $F_{DE} = 0$; $F_{EF} = 5.00$ kips T.
6.45 $F_{BE} = 2.50$ kips T; $F_{DE} = 1.500$ kips C; $F_{DG} = 2.50$ kips T.
6.47 (*a*) Completely constrained, determinate.
(*b*) Completely constrained, indeterminate.
(*c*) Improperly constrained.
6.48 (*a*) Completely constrained, determinate.
(*b*) Partially constrained.
(*c*) Improperly constrained, indeterminate.
6.49 $F_{BD} = 255$ N C; $\mathbf{C}_x = 120.0$ N →, $\mathbf{C}_y = 625$ N ↑.
6.50 $F_{BD} = 780$ lb T; $\mathbf{C}_x = 720$ lb ←, $\mathbf{C}_y = 140.0$ lb ↓.
6.51 $F_{BD} = 375$ N C; $\mathbf{C}_x = 205$ N ←, $\mathbf{C}_y = 360$ N ↓.
6.52 $\mathbf{A}_x = 120.0$ lb →; $\mathbf{A}_y = 30.0$ lb ↑; $\mathbf{B}_x = 120.0$ lb ←, $\mathbf{B}_y = 80.0$ lb ↓; $\mathbf{C} = 30.0$ lb ↓; $\mathbf{D} = 80.0$ lb ↑.
6.53 $\mathbf{A} = 150.0$ lb →; $\mathbf{B}_x = 150.0$ lb ←, $\mathbf{B}_y = 60.0$ lb ↑; $\mathbf{C} = 20.0$ lb ↑; $\mathbf{D} = 80.0$ lb ↓.
6.55 (*a*) $\mathbf{A}_x = 300$ N ←, $\mathbf{A}_y = 660$ N ↑; $\mathbf{E}_x = 300$ N →; $\mathbf{E}_y = 90.0$ N ↑. (*b*) $\mathbf{A}_x = 300$ N ←, $\mathbf{A}_y = 150.0$ N ↑; $\mathbf{E}_x = 300$ N →, $\mathbf{E}_y = 600$ N ↑.
6.57 $\mathbf{B} = 152.0$ lb ↓; $\mathbf{C}_x = 60.0$ lb ←, $\mathbf{C}_y = 200$ lb ↑; $\mathbf{D}_x = 60.0$ lb →, $\mathbf{D}_y = 42.0$ lb ↑.
6.58 (*a*) $\mathbf{A}_x = 2700$ N →, $\mathbf{A}_y = 200$ N ↑; $\mathbf{E}_x = 2700$ N ←; $\mathbf{E}_y = 600$ N ↑. (*b*) $\mathbf{A}_x = 300$ N →, $\mathbf{A}_y = 200$ N ↑; $\mathbf{E}_x = 300$ N ←, $\mathbf{E}_y = 600$ N ↑.
6.59 (*a*) $D_x = 750$ N →, $D_y = 250$ N ↓; $\mathbf{E}_x = 750$ N ←, $\mathbf{E}_y = 250$ N ↑.
(*b*) $\mathbf{D}_x = 375$ N →, $\mathbf{D}_y = 250$ N ↓; $\mathbf{E}_x = 375$ N ←, $\mathbf{E}_y = 250$ N ↑.
6.61 (*a*) and (*c*) $\mathbf{B}_x = 24.0$ lb ←, $\mathbf{B}_y = 7.50$ lb ↓; $\mathbf{F}_x = 24.0$ lb →, $\mathbf{F}_y = 7.50$ lb ↑. (*b*) $\mathbf{B}_x = 24.0$ lb ←, $\mathbf{B}_y = 10.50$ lb ↑; $\mathbf{F}_x = 24.0$ lb →, $\mathbf{F}_y = 10.50$ lb ↓.
6.62 (*a*) $\mathbf{A} = 65.0$ lb ⦣ 22.6°; $\mathbf{C} = 120.0$ lb →; $\mathbf{G} = 60.0$ lb ←; $\mathbf{I} = 25.0$ lb ↑. (*b*) $\mathbf{A} = 65.0$ lb ⦣ 22.6°; $\mathbf{C} = 60.0$ lb →; $\mathbf{G} = 0$; $\mathbf{I} = 25.0$ lb ↑.
6.64 (*a*) 828 N T. (*b*) $\mathbf{C} = 1197$ N ⦣ 86.2°.
6.65 $\mathbf{A}_x = 176.3$ lb ←, $\mathbf{A}_y = 60.0$ lb ↓; $\mathbf{G}_x = 56.3$ lb →, $\mathbf{G}_y = 510$ lb ↑.
6.66 $\mathbf{A}_x = 56.3$ lb ←, $\mathbf{A}_y = 157.5$ lb ↓; $\mathbf{G}_x = 56.3$ lb →, $\mathbf{G}_y = 383$ lb ↑.
6.67 $\mathbf{D}_x = 13.60$ kN →, $\mathbf{D}_y = 7.50$ kN ↑; $\mathbf{E}_x = 13.60$ kN ←, $\mathbf{E}_y = 2.70$ kN ↓.
6.68 $\mathbf{A}_x = 45.0$ N ←, $\mathbf{A}_y = 30.0$ N ↓; $\mathbf{B}_x = 45.0$ N →, $\mathbf{B}_y = 270$ N ↑.
6.69 (*a*) $\mathbf{A} = 75.0$ kN ↑; $\mathbf{B} = 162.5$ kN ↑.
(*b*) $\mathbf{C} = 170.0$ kN ←; $\mathbf{D}_x = 170.0$ kN →, $\mathbf{D}_y = 25.0$ kN ↓.
6.70 (*a*) $\mathbf{A} = 12.50$ kN ↑; $\mathbf{B} = 187.5$ kN ↑.
(*b*) $\mathbf{C} = 30.0$ kN ←; $\mathbf{D}_x = 30.0$ kN →, $\mathbf{D}_y = 75.0$ kN ↓.
6.72 (*a*) 572 lb. (*b*) $\mathbf{A} = 1070$ lb ↑; $\mathbf{B} = 709$ lb ↑; $\mathbf{C} = 870$ lb ↑.
6.73 564 lb →.
6.74 275 lb →.
6.75 (*a*) 2860 N ↓. (*b*) 2700 N ⦣ 68.5°.
6.76 $T_{DE} = 81.0$ N; $\mathbf{B} = 216$ N ↓.
6.78 $\mathbf{D} = 30.0$ kN ←; $\mathbf{F} = 37.5$ kN ⦤ 36.9°.
6.80 $\mathbf{B} = 94.9$ lb ⦣ 18.43°; $\mathbf{D} = 94.9$ lb ⦤ 18.43°.
6.81 (*a*) 252 N·m ↻. (*b*) 108.0 N·m ↻.
6.82 (*a*) 3.00 kN ↓. (*b*) 7.00 kN ↓.
6.83 (*a*) 1261 lb·in. ↺. (*b*) $\mathbf{C}_x = 54.3$ lb ←, $\mathbf{C}_y = 21.7$ lb ↑.
6.85 (*a*) 475 lb. (*b*) 528 lb ⦤ 63.3°.
6.86 44.8 kN.
6.88 720 lb.
6.89 21.3 lb ↘.
6.91 140.0 N.
6.92 260 N.
6.94 (*a*) 10.00 kN ⦤ 2.58°. (*b*) 10.11 kN ⦤ 8.60°.
6.95 (*a*) 2.86 kips C. (*b*) 9.43 kips C.
6.96 (*a*) 4.91 kips C. (*b*) 10.69 kips C.
6.97 $F_{AC} = F_{BC} = 13.86$ kN T; $F_{AD} = F_{BE} = 6.93$ kN C; $F_{CD} = F_{CE} = 13.86$ kN T; $F_{AB} = 12.00$ kN C.
6.99 $F_{EH} = 22.5$ kips C; $F_{GI} = 22.5$ kips T.
6.100 $F_{HJ} = 33.8$ kips C; $F_{IL} = 33.8$ kips T.
6.101 7.36 kN C.
6.102 (*a*) $\mathbf{A}_x = 200$ kN →, $\mathbf{A}_y = 122.0$ kN ↑.
(*b*) $\mathbf{B}_x = 200$ kN ←, $\mathbf{B}_y = 10.00$ kN ↓.
6.103 $\mathbf{A}_x = 3.32$ kN ←, $\mathbf{A}_y = 14.26$ kN ↓; $\mathbf{C}_x = 3.72$ kN →, $\mathbf{C}_y = 14.26$ kN ↑.
6.104 31.3 lb.
6.106 (*a*) $\mathbf{E}_x = 2.00$ kips ←, $\mathbf{E}_y = 2.25$ kips ↑.
(*b*) $\mathbf{C}_x = 4.00$ kips ←, $\mathbf{C}_y = 5.75$ kips ↑.
6.108 Case (1) (*a*) $\mathbf{A}_x = 0$, $\mathbf{A}_y = 7.85$ kN ↑, $\mathbf{M}_A = 15.70$ kN·m ↺.
(*b*) $\mathbf{D} = 22.2$ kN ⦣ 45°.
Case (2) (*a*) $\mathbf{A}_x = 0$, $\mathbf{A}_y = 3.92$ kN ↑, $\mathbf{M}_A = 8.34$ kN·m ↺.
(*b*) $\mathbf{D} = 11.10$ kN ⦣ 45°.
Case (3) (*a*) $\mathbf{A}_x = 0$, $\mathbf{A}_y = 3.92$ kN ↑, $\mathbf{M}_A = 8.34$ kN·m ↺.
(*b*) $\mathbf{D} = 18.95$ kN ⦣ 45°.
Case (4) (*a*) $\mathbf{A}_x = 3.92$ kN →, $\mathbf{A}_y = 3.92$ kN ↑, $\mathbf{M}_A = 2.35$ kN·m ↻.
(*b*) $\mathbf{D} = 11.10$ kN ⦣ 45°.

CHAPTER 7

7.1 $a^3(h_1 + 3h_2)/12$.
7.2 $3a^3b/10$.
7.3 $ha^3/5$.
7.4 $a^3b/20$.
7.5 $a(h_1^2 + h_2^2)(h_1 + h_2)/12$.
7.6 $a^3b/6$.
7.9 $\pi ab^3/8$; $b/2$.
7.10 $0.525ah^3$; $1.202h$.
7.11 $1.638ab^3$; $1.108b$.
7.12 $3ab^3/35$; $0.507b$.
7.13 $\pi a^3b/8$; $a/2$.
7.14 $0.613a^3h$; $1.299a$.
7.17 (*a*) $J_O = 4a^4/3$; $r_O = 0.816a$. (*b*) $J_O = 17a^4/6$; $r_O = 1.190a$.
7.18 $J_O = 10a^4/3$; $r_O = 1.291a$.
7.20 $4ab(a^2 + 4b^2)/3$; $\sqrt{(a^2 + 4b^2)/3}$.
7.21 $(\pi/2)(R_2^4 - R_1^4)$; $(\pi/4)(R_2^4 - R_1^4)$.
7.23 $4a^3/9$.
7.24 $0.935a$.
7.25 390×10^3 mm^4; 21.9 mm.
7.26 46.0 in^4; 1.599 in.
7.27 501×10^6 mm^4; 149.4 mm.
7.30 46.5 in^4; 1.607 in.
7.31 150.3×10^6 mm^4; 81.9 mm.
7.32 185.4 in^4; 2.81 in.
7.33 3000 mm^2; 325×10^3 mm^4.
7.35 $\bar{I}_x = 191.3$ in^4; $\bar{I}_y = 75.2$ in^4.
7.36 $\bar{I}_x = 479 \times 10^3$ mm^4; $\bar{I}_y = 149.7 \times 10^3$ mm^4.

7.38 (*a*) 765 in^4. (*b*) 402 in^4.
7.39 (*a*) 11.57×10^6 mm^4. (*b*) 7.81×10^6 mm^4.
7.40 (*a*) 129.2 in^4. (*b*) 25.8 in^4.
7.41 $\bar{I}_x = 254$ in^4; $\bar{r}_x = 4.00$ in.; $\bar{I}_y = 102.1$ in^4; $\bar{r}_y = 2.54$ in.
7.43 $\bar{I}_x = 186.7 \times 10^6$ mm^4; $\bar{r}_x = 118.6$ mm;
$\bar{I}_y = 167.7 \times 10^6$ mm^4; $\bar{r}_y = 112.4$ mm.
7.44 $\bar{I}_x = 260 \times 10^6$ mm^4; $\bar{r}_x = 144.6$ mm;
$\bar{I}_y = 17.53 \times 10^6$ mm^4; $\bar{r}_y = 37.6$ mm.
7.45 $\bar{I}_x = 96.5$ in^4; $\bar{I}_y = 26.5$ in^4.
7.46 $\bar{I}_x = 9.54$ in^4; $\bar{I}_y = 104.5$ in^4.
7.47 $\bar{I}_x = 3.55 \times 10^6$ mm^4; $\bar{I}_y = 49.8 \times 10^6$ mm^4.
7.49 $b^3h/12$.
7.51 $0.0945ah^3$; $0.402h$.
7.53 $bh(12h^2 + b^2)/48$; $\sqrt{(12h^2 + b^2)/24}$.
7.54 $\bar{I}_x = 1.268 \times 10^6$ mm^4; $\bar{I}_y = 339 \times 10^3$ mm^4.
7.55 $\bar{I}_x = 1.874 \times 10^6$ mm^4; $\bar{I}_y = 5.82 \times 10^6$ mm^4.
7.56 $\bar{I}_x = 48.9 \times 10^3$ mm^4; $\bar{I}_y = 8.35 \times 10^3$ mm^4.
7.58 (*a*) 12.16×10^6 mm^4. (*b*) 9.73×10^6 mm^4.
7.60 $\bar{I}_x = 250$ in^4; $\bar{r}_x = 4.10$ in.; $\bar{I}_y = 141.9$ in^4; $\bar{r}_y = 3.09$ in.

CHAPTER 8

8.1 (*a*) 35.7 MPa. (*b*) 42.4 MPa.
8.2 $d_1 = 22.6$ mm.; $d_2 = 15.96$ mm.
8.3 (*a*) 12.73 ksi. (*b*) −2.83 ksi.
8.4 28.2 kips.
8.6 (*a*) 101.6 MPa. (*b*) −21.7 MPa.
8.7 1.084 ksi.
8.9 8.52 ksi.
8.10 4.29 in^2.
8.11 (*a*) 17.86 kN. (*b*) −41.4 MPa.
8.12 (*a*) 12.73 MPa. (*b*) −4.77 MPa.
8.13 308 mm.
8.14 43.4 mm.
8.16 12.57 kips.
8.17 10.82 in.
8.19 29.4 mm.
8.20 (*a*) 25.9 mm. (*b*) 271 MPa.
8.21 (*a*) 8.92 ksi. (*b*) 22.4 ksi. (*c*) 11.21 ksi.
8.22 (*a*) 10.84 ksi. (*b*) 5.11 ksi.
8.24 8.31 kN.
8.25 $\sigma = 489$ kPa; $\tau = 489$ kPa.
8.26 (*a*) 13.95 kN. (*b*) 620 kPa.
8.27 $\sigma = 70.0$ psi, $\tau = 40.4$ psi.
8.28 (*a*) 1.500 kips. (*b*) 43.3 psi.
8.30 (*a*) 180.0 kips. (*b*) 45.0°. (*c*) −2.50 ksi. (*d*) −5.00 ksi.
8.31 $\sigma = -21.6$ MPa, $\tau = 7.87$ MPa.
8.33 168.1 mm^2.
8.34 (*a*) 1.141 in. (*b*) 1.549 in.
8.35 (*a*) 3.35. (*b*) 1.358 in.
8.36 (*a*) 181.3 mm^2. (*b*) 213 mm^2.
8.38 1.800.
8.39 10.25 kN.
8.41 (*a*) 1.550 in. (*b*) 8.05 in.
8.42 3.47.
8.44 3.97 kN.
8.46 283 lb.
8.47 2.42.
8.48 2.05.
8.49 (*a*) 3.33 MPa. (*b*) 525 mm.
8.51 25.2 mm.
8.53 (*a*) −640 psi. (*b*) −320 psi.
8.54 3.09 kips.
8.55 (*a*) 94.1 MPa. (*b*) 44.3 MPa.
8.57 3.45.
8.59 $x_E = 24.7$ in.; $x_F = 55.2$ in.
8.60 1.683 kN.

CHAPTER 9

9.1 (*a*) 0.0303 in. (*b*) 15.28 ksi.
9.2 (*a*) 81.8 MPa. (*b*) 1.712.
9.3 (*a*) 0.546 mm. (*b*) 36.3 MPa.
9.4 (*a*) 9.82 kN. (*b*) 500 MPa.
9.5 (*a*) 0.381 in. (*b*) 17.58 ksi.
9.8 (*a*) 2.50 ksi. (*b*) 0.1077 in.
9.9 48.4 kips.
9.11 1.988 kN.
9.12 0.429 in.
9.13 (*a*) 9.53 kips. (*b*) 1.254×10^{-3} in.
9.14 (*a*) 32.8 kN. (*b*) 0.0728 mm.
9.15 (*a*) 0.01819 mm. (*b*) −0.0909 mm.
9.17 0.1812 in.
9.18 (*a*) −0.1549 mm. (*b*) 0.1019 mm ↓.
9.19 50.4 kN.
9.20 $S_{BD} = +79.4 \times 10^{-3}$ in.; $S_{DE} = +124.1 \times 10^{-3}$ in.
9.21 (*a*) 0.1767 in. (*b*) 0.1304 in.
9.24 14.74 kN.
9.25 (*a*) 65.1 MPa. (*b*) 0.279 mm.
9.26 (*a*) 287 kN. (*b*) 140.0 MPa.
9.27 Steel: −8.34 ksi; concrete: −1.208 ksi.
9.28 695 kips.
9.30 (*a*) 62.8 kN ← at *A*; 37.2 kN ← at *E*. (*b*) 46.3 μm →.
9.32 (*a*) $\mathbf{R}_A = 2.28$ kips ↑; $\mathbf{R}_C = 9.72$ kips ↑.
(*b*) $\sigma_{AB} = +1.857$ ksi; $\sigma_{BC} = -3.09$ ksi.
9.33 177.4 lb.
9.35 *A*: 0.525*P*; *B*: 0.200*P*; *C*: 0.275*P*.
9.36 *A*: 0.1*P*; *B*: 0.2*P*; *C*: 0.3*P*; *D*: 0.4*P*.
9.37 137.8°F.
9.39 $\sigma_S = -1.448$ ksi; $\sigma_C = 54.2$ psi.
9.40 (*a*) −98.3 MPa. (*b*) −38.3 MPa.
9.41 142.6 kN.
9.42 (*a*) $\sigma_{AB} = -5.25$ ksi; $\sigma_{BC} = -11.82$ ksi.
(*b*) 6.57×10^{-3} in. →.
9.44 (*a*) 52.3 kips. (*b*) 9.91×10^{-3} in.
9.45 (*a*) 201.6°F. (*b*) 18.0107 in.
9.46 (*a*) −116.2 MPa. (*b*) 0.363 mm.
9.48 (*a*) 21.4°C. (*b*) 3.67 MPa.
9.49 (*a*) 0.1973 mm. (*b*) −0.00651 mm.
9.52 94.9 kips.
9.53 1.99551 : 1.
9.54 (*a*) 0.0358 mm. (*b*) −0.00258 mm. (*c*) −0.000344 mm.
(*d*) −0.00825 mm^2.
9.55 (*a*) 5.13×10^{-3} in. (*b*) -0.570×10^{-3} in.
9.56 (*a*) 7630 lb compression. (*b*) 4580 lb compression.
9.58 (*a*) 0.0754 mm. (*b*) 0.1028 mm (*c*) 0.1220 mm.
9.62 16.67 MPa.
9.63 19.00×10^3 kN/m.
9.64 (*a*) 10.42 in. (*b*) 0.813 in.
9.65 (*a*) 13.31 ksi. (*b*) 18.72 ksi.
9.67 (*a*) 58.3 kN. (*b*) 64.3 kN.

9.68 (*a*) 87.0 MPa. (*b*) 75.2 MPa. (*c*) 73.9 MPa.
9.69 (*a*) 12.02 kips. (*b*) 108.0%.
9.70 23.9 kips.
9.72 36.7 mm.
9.73 1.219 in.
9.74 21.5 kN.
9.76 (*a*) 80.4 μm ↑. (*b*) 209 μm ↓. (*c*) 390 μm ↓.
9.77 0.536 mm ↓.
9.80 (*a*) 145.9°F. (*b*) 0.01053 in.
9.81 (*a*) −63.0 MPa. (*b*) −4.05 mm^2. (*c*) −162.0 mm^3.
9.82 a = 0.818 in., b = 2.42 in.
9.83 (*a*) 3/4 in. (*b*) 15.63 kips.

CHAPTER 10

10.1 641 N·m.
10.2 87.3 MPa.
10.3 (*a*) 7.55 ksi. (*b*) 7.64 ksi.
10.4 (*a*) 125.7 N·m (*b*) 181.4 N·m.
10.6 (*a*) 7.55 ksi. (*b*) 3.49 in.
10.8 (*a*) 1.292 in. (*b*) 1.597 in.
10.9 (*a*) 2.85 ksi. (*b*) 4.46 ksi. (*c*) 5.37 ksi.
10.10 (*a*) 3.19 ksi. (*b*) 4.75 ksi. (*c*) 5.58 ksi.
10.12 42.8 mm.
10.13 9.16 kip·in.
10.15 3.37 kN·m.
10.16 (*a*) 50.3 mm. (*b*) 63.4 mm.
10.17 *AB:* 42.0 mm; *BC:* 33.3 mm.
10.18 *AB:* 52.9 mm; *BC:* 33.3 mm.
10.20 1.189 kip·in.
10.21 (*a*) 55.0 MPa. (*b*) 45.3 MPa. (*c*) 47.7 MPa.
10.23 (*a*) 1.442 in. (*b*) 1.233 in.
10.24 4.30 kip·in.
10.25 (*a*) 2.83 kip·in. (*b*) 13.00°.
10.26 (*a*) 1.390°. (*b*) 1.482°.
10.28 9.38 ksi.
10.30 (*a*) 8.54° (*b*) 2.11°.
10.31 (*a*) 14.43°. (*b*) 46.9°.
10.32 6.02°.
10.33 12.22°.
10.34 3.78°.
10.36 $(T_A l/GJ)(1/n^4 + 1/n^2 + 1)$.
10.37 36.1 mm.
10.39 0.837 in.
10.40 1.089 in.
10.41 (*a*) 73.6 MPa. (*b*) 34.4 MPa. (*c*) 5.07°.
10.43 (*a*) 4.72 ksi. (*b*) 7.08 ksi. (*c*) 4.35°.
10.44 7.37°.
10.45 (*a*) *A:* 1105 N·m; *C:* 295 N·m.
(*b*) 45.0 MPa. (*c*) 27.4 MPa.
10.46 (*a*) T_A = 1090 N·m; T_C = 310 N·m.
(*b*) 47.4 MPa. (*c*) 28.8 MPa.
10.48 1.483 in.
10.49 12.44 ksi.
10.51 7.95 kip·in.
10.52 (*a*) 19.21 kip·in. (*b*) 2.01 in.
10.54 1.221.
10.56 127.8 lb·in.
10.58 τ_{AB} = 68.9 MPa; τ_{CD} = 14.70 MPa.
10.59 τ_{AB} = 10.27 MPa; τ_{CD} = 48.6 MPa.
10.60 12.24 MPa.

CHAPTER 11

11.1 (*a*) −61.6 MPa. (*b*) 91.7 MPa.
11.2 (*a*) −2.38 ksi. (*b*) −0.650 ksi.
11.3 80.2 kN·m.
11.4 24.8 kN·m.
11.6 (*a*) 1.405 kip·in. (*b*) 3.19 kip·in.
11.7 259 kip·in.
11.9 top: −14.71 ksi; bottom: 8.82 ksi.
11.10 top: −81.8 MPa; bottom: 67.8 MPa.
11.12 3.79 kN·m.
11.13 (*a*) 8.24 kips. (*b*) 1.332 kips.
11.15 42.6 kN.
11.16 4.11 kip·in.
11.17 7.67 kN·m.
11.18 20.4 kip·in.
11.19 106.1 N·m.
11.21 4.63 kip·in.
11.23 (*a*) σ = 75 MPa, ρ = 26.7 m. (*b*) σ = 125.0 MPa, ρ = 9.60 m.
11.24 (*a*) $\sigma_{max} = 6M/a^3$, $1/\rho = 12M/Ea^4$.
(*b*) $\sigma_{max} = 8.49M/a^3$, $1/\rho = 12M/Ea^4$.
11.25 1.240 kN·m.
11.26 887 N·m.
11.27 720 N·m.
11.29 335 kip·in.
11.30 689 kip·in
11.31 (*a*) 66.2 MPa. (*b*) −112.4 MPa.
11.32 (*a*) −56.9 MPa. (*b*) 111.9 MPa.
11.33 (*a*) −1.979 ksi. (*b*) 16.48 ksi.
11.36 43.7 m.
11.37 625 ft.
11.38 625 ft.
11.39 (*a*) 212 MPa. (*b*) −15.59 MPa.
11.40 (a) 210 MPa. (b) −14.08 MPa.
11.42 2.88 kip.ft.
11.43 (*a*) 24.1 ksi. (*b*) −1.256 ksi.
11.44 33.9 kip.ft.
11.46 (*a*) steel: 8.96 ksi; aluminum: 1.792 ksi; brass: 0.896 ksi.
(*b*) 349 ft.
11.48 (*a*) 54.1 MPa. (*b*) 130.2 MPa.
11.49 (*a*) $-2P/\pi r^2$ (*b*) $-5P/\pi r^2$.
11.50 (*a*) 4.87 ksi. (*b*) 5.17 ksi.
11.51 (*a*) −212 psi. (*b*) −637 psi. (*c*) −1061 psi.
11.52 (*a*) −102.8 MPa. (*b*) 80.6 MPa.
11.53 (*a*) 112.7 MPa. (*b*) −96.0 MPa.
11.54 (*a*) 130.2 MPa. (*b*) −110.0 MPa.
11.57 0.375d.
11.58 0.455 in.
11.59 (*a*) −0.750 ksi. (*b*) −2.00 ksi. (*c*) −1.500 ksi.
11.60 623 lb.
11.62 16.04 mm.
11.64 (*a*) 2.54 kN. (*b*) 17.01 mm to the right of loads.
11.65 (*a*) $-P/2at$. (*b*) $2P/at$. (*c*) $-P/2at$.
11.66 (*a*) 47.6 MPa. (*b*) −49.4 MPa. (*c*) 9.80 mm below top of section.
11.68 2.485 in. $< y <$ 4.561 in.
11.70 P = 44.2 kips, Q = 57.3 kips.
11.71 (*a*) 30.0 mm. (*b*) 94.5 kN.
11.72 (*a*) 5.00 mm. (*b*) 243 kN.
11.73 (*a*) 9.86 ksi. (*b*) −2.64 ksi. (*c*) −9.86 ksi.
11.74 (*a*) −3.37 MPa. (*b*) −18.60 MPa. (*c*) 3.37 MPa.
11.75 (*a*) −17.16 ksi. (*b*) 6.27 ksi. (*c*) 17.16 ksi.
11.76 (*a*) 7.20 ksi. (*b*) −18.39 ksi. (*c*) −7.20 ksi.

11.77 (*a*) 0.321 ksi. (*b*) −0.107 ksi. (*c*) 0.427 ksi.
11.78 (*a*) 57.8 MPa. (*b*) −56.8 MPa. (*c*) 25.9 MPa.
11.79 (*a*) 11.3° ∠. (*b*) 15.06 ksi.
11.80 (*a*) 57.4°. (*b*) 75.7 MPa.
11.81 (*a*) 9.59° ∠. (*b*) 76.5 MPa.
11.82 (*a*) 10.03°. (*b*) 54.2 MPa.
11.83 (*a*) 27.5° ∠. (*b*) 5.07 ksi.
11.84 (*a*) 32.9° ∠. (*b*) 61.4 MPa.
11.86 (*a*) $\sigma_A = -62.5$ psi; $\sigma_B = -271$ psi. (*b*) Does not intersect *AB*. (*c*) Intersects *BD* at 0.780 in. from *B*.
11.87 (*a*) $\sigma_A = 31.5$ MPa; $\sigma_B = -10.39$ MPa. (*b*) 94.0 mm above *A*.
11.89 0.1638 in.
11.91 91.3 kN.
11.93 (*a*) −39.3 MPa. (*b*) 26.2 MPa.
11.94 (*a*) 9.17 kN·m. (*b*) 10.24 kN·m.
11.96 (*a*) 152.25 kips. (*b*) $x = 0.595$ in., $z = 0.571$ in. (*c*) 300 μm.
11.97 73.2 MPa; −102.4 MPa.
11.99 (*a*) −1.526 ksi. (*b*) 17.67 ksi.
11.101 (*a*) 46.7°. (*b*) 80.2 MPa.
11.102 (*a*) 288 lb. (*b*) 209 lb.
11.104 (*a*) 1.414. (*b*) 1.732.

CHAPTER 12

12.1 (*a*) $V_{max} = Pb/L$, $V_{min} = -Pa/L$; $M_{max} = Pab/L$, $M_{min} = 0$. (*b*) $0 \le x < a$: $V = Pb/L$; $M = Pbx/L$; $a \le x < L$: $V = -Pa/L$; $M = Pa(L - x)/L$.
12.2 (*a*) $V_{max} = wL/2$, $V_{min} = -wL/2$; $M_{max} = wL^2/8$. (*b*) $V = w(L/2 - x)$; $M = wx(L - x)/2$.
12.3 (*a*) $|V|_{max} = w_0L/2$; $|M|_{max} = w_0L^2/6$. (*b*) $V = -w_0x^2/2L$; $M = -w_0x^3/6L$.
12.4 (*a*) $|V|_{max} = w(L - 2a)/2$; $|M|_{max} = w(L^2/8 - a^2/2)$. (*b*) $0 \le x \le a$: $V = w(L - 2a)/2$; $M = w(L - 2a)x/2$; $a \le x \le L - a$: $V = w(L/2 - x)$; $M = w[x(L - x) - a^2]/2$. $L - a \le x \le L$: $V = -w(L - 2a)/2$; $M = w(L - 2a)(L - x)/2$.
12.5 (*a*) 430 lb. (*b*) 1200 lb·in.
12.7 (*a*) 40.0 kN. (*b*) 40.0 kN·m.
12.9 (*a*) 3.45 kN. (*b*) 1125 N·m.
12.10 (*a*) 2000 lb. (*b*) 19200 lb·in.
12.11 (*a*) 12.00 kips. (*b*) 27.0 kip·ft.
12.12 (*a*) 900 N. (*b*) 112.5 N·m.
12.13 950 psi.
12.14 10.89 MPa.
12.15 129.2 MPa.
12.16 129.5 MPa.
12.17 9.90 ksi.
12.19 $|V|_{max} = 27.5$ kips; $|M|_{max} = 45.0$ kip·ft; $\sigma = 14.17$ ksi.
12.20 $|V|_{max} = 279$ kN; $|M|_{max} = 326$ kN·m; $\sigma = 136.6$ MPa.
12.23 $|V|_{max} = 28.8$ kips; $|M|_{max} = 56.0$ kip·ft; $\sigma = 13.05$ ksi.
12.24 $|V|_{max} = 1.500$ kips; $|M|_{max} = 3.00$ kip·ft; $\sigma = 2.11$ ksi.
12.25 (*a*) 866 mm. (*b*) 99.2 MPa.
12.26 (*a*) 819 mm. (*b*) 89.5 MPa.
12.27 (*a*) 1.260 ft. (*b*) 7.24 ksi.
12.29 See Prob. 12.1.
12.30 See Prob. 12.2.
12.31 See Prob. 12.3.
12.32 See Prob. 12.4.
12.33 See Prob. 12.5.
12.35 See Prob. 12.7.
12.37 (*a*) 1.800 kips. (*b*) 6.00 kip·ft.
12.38 (*a*) 880 lb. (*b*) 2000 lb·ft.
12.39 (*a*) 6.75 kN. (*b*) 6.51 kN·m.
12.40 (*a*) 600 N. (*b*) 180.0 N·m.
12.41 See Prob. 12.13.
12.42 10.89 MPa.
12.43 129.2 MPa.
12.44 See Prob. 12.17.
12.45 (*a*) $V = (w_0L/\pi)\cos(\pi x/L)$; $M = (w_0L^2/\pi^2)\sin(\pi x/L)$. (*b*) w_0L^2/π^2.
12.47 (*a*) $V = w_0(L^2 - 3x^2)/6L$; $M = w_0(Lx - x^3/L)/6$. (*b*) $0.0642w_0L^2$.
12.49 $|V|_{max} = 20.7$ kN; $|M|_{max} = 9.75$ kN·m; $\sigma_{max} = 60.2$ MPa.
12.50 $|V|_{max} = 16.80$ kN; $|M|_{max} = 8.82$ kN·m; $\sigma_{max} = 73.5$ MPa.
12.51 $|V|_{max} = 1670$ lb; $|M|_{max} = 2640$ lb·ft; $\sigma_{max} = 959$ psi.
12.52 $|V|_{max} = 8.00$ kips; $|M|_{max} = 16.00$ kip·ft; $\sigma_{max} = 6.98$ ksi.
12.54 $|V|_{max} = 9.28$ kips; $|M|_{max} = 28.2$ kip·in; $\sigma_{max} = 11.58$ ksi.
12.55 $|V|_{max} = 150$ kN; $|M|_{max} = 300$ kN·m; $\sigma_{max} = 136.4$ MPa.
12.57 $h = 173.2$ mm.
12.58 $h = 361$ mm.
12.60 $b = 6.20$ in.
12.62 $a = 6.67$ in.
12.63 W27 × 84.
12.64 W27 × 84.
12.65 W530 × 66.
12.66 W250 × 28.4.
12.67 S460 × 81.4.
12.69 S12 × 31.8.
12.71 C9 × 15.
12.72 C180 × 14.6.
12.73 3/8 in.
12.74 9 mm.
12.77 (*a*) 18.00 kips. (*b*) 72.0 kip·ft.
12.78 (*a*) 85.0 N. (*b*) 21.3 N·m.
12.80 $|V|_{max} = 342$ N; $|M|_{max} = 51.6$ N·m; $\sigma_{max} = 17.19$ MPa.
12.81 $|V|_{max} = 144.0$ kN; $|M|_{max} = 84.0$ kN·m; $\sigma_{max} = 99.5$ MPa.
12.84 $|V|_{max} = 30.0$ lb; $|M|_{max} = 24.0$ lb·ft; $\sigma_{max} = 6.95$ ksi.
12.85 11.74 in.
12.87 7.01 kips.

CHAPTER 13

13.1 92.6 lb.
13.2 326 lb.
13.3 738 N.
13.4 747 N.
13.5 193.2 kN.
13.7 12.01 ksi.
13.9 (*a*) 7.40 ksi. (*b*) 6.70 ksi.
13.10 (*a*) 3.17 ksi. (*b*) 2.40 ksi.
13.11 (*a*) 920 kPa. (*b*) 765 kPa.
13.12 (*a*) 8.97 MPa. (*b*) 8.15 MPa.
13.13 14.05 in.
13.14 88.9 mm.
13.17 (*a*) 31.0 MPa. (*b*) 23.2 MPa.
13.18 (*a*) 1.744 ksi. (*b*) 2.81 ksi.
13.19 32.7 MPa.
13.20 3.21 ksi.
13.22 2.00.
13.23 1.125.

13.24 1.500.
13.25 10.79 kN.
13.26 1.672 in.
13.27 (*a*) 59.9 psi. (*b*) 79.8 psi.
13.28 (*a*) 12.21 MPa. (*b*) 58.6 MPa.
13.29 (*a*) 95.2 MPa. (*b*) 112.8 MPa.
13.31 $\tau_a = 3.93$ ksi; $\tau_b = 2.67$ ksi; $\tau_c = 0.631$ ksi; $\tau_d = 1.022$ ksi; $\tau_e = 0$.
13.33 (*a*) 41.4 MPa. (*b*) 41.4 MPa.
13.34 (*a*) 18.23 MPa. (*b*) 14.59 MPa. (*c*) 46.2 MPa.
13.35 (*a*) 40.5 psi. (*b*) 55.2 psi.
13.36 (*a*) 2.67 in. (*b*) 41.6 psi.
13.37 9.05 mm.
13.39 7.19 ksi.
13.41 (*a*) 23.2 MPa. (*b*) 35.2 MPa.
13.42 10.76 MPa at *a*, 0 at *b*, 11.21 MPa at *c*, 22.0 MPa at *d*, 9.35 MPa at *e*.
13.43 (*a*) 2.025 ksi. (*b*) 1.800 ksi.
13.46 (*a*) 23.3 MPa. (*b*) 109.7 kPa.
13.48 (*a*) 1.323 ksi. (*b*) 1.329 ksi.
13.49 (*a*) 0.888 ksi. (*b*) 1.453 ksi.
13.50 (*a*) 155.8 N. (*b*) 329 kPa.
13.51 11.54 kips.
13.53 (*b*) $h = 225$ mm, $b = 61.7$ mm.
13.55 (*a*) 84.2 kips. (*b*) 60.2 kips.
13.56 (*a*) 379 kPa; (*b*) 0.
13.57 (*a*) 239 N; (*b*) 549 N.
13.58 1.167 ksi at *a*, 0.513 ksi at *b*, 4.03 ksi at *c*, 8.40 ksi at *d*.
13.60 53.9 kips.

CHAPTER 14

14.1 $\sigma = -0.521$ MPa, $\tau = 56.4$ MPa.
14.2 $\sigma = 32.9$ MPa, $\tau = 71.0$ MPa.
14.3 $\sigma = 9.46$ ksi, $\tau = 1.013$ ksi.
14.4 $\sigma = 10.93$ ksi, $\tau = 0.536$ ksi.
14.5 (*a*) −37.0°, 53.0°. (*b*) −13.60 MPa, −86.4 MPa.
14.7 (*a*) 14.0°, 104.0°. (*b*) 20.0 ksi, −14.00 ksi.
14.9 (*a*) 8.0°, 98.0°. (*b*) 36.4 MPa. (*c*) −50.0 MPa.
14.10 (*a*) 18.4°; 108.4°. (*b*) 100.0 MPa. (*c*) 90.0 MPa.
14.12 (*a*) −26.6°, 63.4°. (*b*) 5.00 ksi. (*c*) 6.00 ksi.
14.13 (*a*) $\sigma_{x'} = -2.40$ ksi; $\tau_{x'y'} = 0.1498$ ksi; $\sigma_{y'} = 10.40$ ksi. (*b*) $\sigma_{x'} = 1.951$ ksi; $\tau_{x'y'} = 6.07$ ksi; $\sigma_{y'} = 6.05$ ksi.
14.14 (*a*) $\sigma_{x'} = 9.02$ ksi; $\tau_{x'y'} = 3.80$ ksi; $\sigma_{y'} = -13.02$ ksi. (*b*) $\sigma_{x'} = 5.34$ ksi; $\tau_{x'y'} = -9.06$ ksi; $\sigma_{y'} = -9.34$ ksi.
14.16 (*a*) $\sigma_{x'} = -56.2$ MPa; $\tau_{x'y'} = -38.2$ MPa; $\sigma_{y'} = 86.2$ MPa. (*b*) $\sigma_{x'} = -45.2$ MPa; $\tau_{x'y'} = 53.8$ MPa; $\sigma_{y'} = 75.2$ MPa.
14.17 (*a*) −0.600 MPa. (*b*) −3.84 MPa.
14.18 (*a*) 217 psi. (*b*) −125.0 psi.
14.19 (*a*) 47.9 MPa. (*b*) 102.7 MPa.
14.20 (*a*) 18.4°. (*b*) 16.67 ksi.
14.22 $\sigma_a = 5.12$ ksi, $\sigma_b = -1.640$ ksi, $\tau_{max} = 3.38$ ksi.
14.24 205 MPa.
14.25 See 14.5 and 14.9.
14.26 $\theta_p = -26.6°$ and 63.4°; $\sigma_{max} = 190.0$ MPa; $\sigma_{min} = -10.00$ MPa, see 14.10.
14.27 (*a*) 59.0° and −31.0° (*b*) 17.00 ksi. (*c*) 3.00 ksi.
14.28 See 14.12.
14.29 See 14.13.
14.30 See 14.14.
14.32 See 14.16.
14.33 See 14.17.
14.34 See 14.18.
14.35 See 14.19.
14.36 See 14.20.
14.38 See 14.22.
14.40 205 MPa.
14.41 (*a*) 7.94 ksi. (*b*) 13.00 ksi, −11.00 ksi.
14.43 (*a*) −2.89 MPa. (*b*) 12.77 MPa, 1.226 MPa.
14.44 (*a*) −8.66 MPa. (*b*) 17.00 MPa, −3.00 MPa.
14.46 24.6°, 114.6°; 72.9 MPa, 27.1 MPa.
14.47 60°, −30°; $1.732\tau_0$, $-1.732\tau_0$.
14.48 $\frac{1}{2}\theta, \frac{1}{2}\theta + 90°$; $\sigma_0(1 + \cos\theta)$, $\sigma_0(1 - \cos\theta)$.
14.49 166.5 psi.
14.50 $\sigma = 10.25$ ksi; $\tau = 5.12$ ksi.
14.51 5.49.
14.52 (*a*) 95.7 MPa. (*b*) 1.699 mm.
14.53 (*a*) 1.290 MPa. (*b*) 0.0852 mm.
14.54 7.71 mm.
14.56 43.3 ft.
14.57 $\sigma_{max} = 16.62$ ksi; $\tau_{max} = 8.31$ ksi.
14.59 $\sigma_{max} = 89.0$ MPa; $\tau_{max} = 44.5$ MPa.
14.60 12.55 mm.
14.62 474 psi.
14.64 2.17 MPa.
16.65 (*a*) 44.2 MPa. (*b*) 15.39 MPa.
14.66 56.8°.
14.68 $\sigma_{max} = 45.1$ MPa, $\tau_{max(in\text{-}plane)} = 7.49$ MPa.
14.69 (*a*) 3.15 ksi. (*b*) 1.993 ksi.
14.71 $\sigma_{max} = 8.48$ ksi; $\tau_{max} = 2.85$ ksi.
14.72 $\sigma_{max} = 13.09$ ksi; $\tau_{max} = 3.44$ ksi.
14.74 $-45° \leq \theta \leq 8.13°$; $45° \leq \theta \leq 98.13°$.
14.75 3.00 ksi $\leq \sigma_x \leq$ 27.0 ksi.
14.77 $\theta_p = 18.40°$, 108.4°; $\sigma_{max} = 7.00$ ksi; $\sigma_{min} = -3.00$ ksi.
14.78 (*a*) 6.40 ksi. (*b*) 4.70 ksi.
14.80 (*a*) 399 kPa. (*b*) 186.0 kPa.
14.81 (*a*) $\theta_p = 18.90°$, 108.9°; $\sigma_{max} = 18.67$ MPa; $\sigma_{min} = -158.5$ MPa. (*b*) 88.6 MPa.
14.83 $\sigma_{max} = 68.6$ MPa, $\tau_{max(in\text{-}plane)} = 23.6$ MPa.
14.84 17.06 kN·m.

CHAPTER 15

15.1 (*a*) $y = -Px^2(3L - x)/6EI$. (*b*) $PL^3/3EI$ ↓. (*c*) $PL^2/2EI$ ⦩.
15.2 (*a*) $y = M_0(L - x)^2/2EI$. (*b*) $M_0L^2/2EI$ ↑. (*c*) M_0L/EI ⦩.
15.3 (*a*) $y = -w_0(x^5 - 5L^4x + 4L^5)/120EIL$. (*b*) $w_0L^4/30EI$ ↓. (*c*) $w_0L^3/24EI$ ⦨.
15.4 (*a*) $y = -w(x^4 - 4L^3x + 3L^4)/24EI$. (*b*) $wL^4/8EI$ ↓. (*c*) $wL^3/6EI$ ⦨.
15.6 (*a*) $y = w(L^2x^2/8 - x^4/24)/EI$. (*b*) $11wL^4/384EI$ ↑. (*c*) $5wL^3/48EI$ ⦨.
15.7 (*a*) $y = w(Lx^3/16 - x^4/24 - L^3x/48)/EI$. (*b*) $wL^3/48EI$ ⦨. (*c*) 0.
15.9 (*a*) 2.79×10^{-3} rad ⦩. (*b*) 1.859 mm ↓.
15.10 (*a*) 3.92×10^{-3} rad ⦩. (*b*) 0.1806 in. ↓.
15.11 (*a*) $x_m = 0.423L$, $y_m = 0.06415M_0L^2/EI$ ↑. (*b*) 45.3 kN·m.
15.12 (*a*) $x_m = 0.519L$, $y_m = 0.00652w_0L^4/EI$ ↓. (*b*) 0.229 in. ↓.
15.14 0.398 in. ↓.

15.15 (*a*) $y = w_0(x^6/90 - Lx^5/30 + L^3x^3/18 - L^5x/30)/EIL^2$. (*b*) $w_0L^3/30EI$ ↘. (*c*) $61w_0L^4/5760EI$ ↓.
15.17 $3wL/8$.
15.18 $3M_0/2L$ ↑.
15.20 $11w_0L/40$ ↑.
15.21 $\mathbf{R}_A = 11P/16$ ↑, $\mathbf{M}_A = 3PL/16$ ↺, $\mathbf{R}_B = 5P/16$ ↑, $\mathbf{M}_B = 0$; $M = -3PL/16$ at A, $M = 5PL/32$ at C, $M = 0$ at B.
15.22 $\mathbf{R}_A = 7wL/128$ ↑; $M = 0.0273wL^2$ at C, $M = -0.0703wL^2$ at B, $M = 0.0288wL^2$ at $x = 0.555L$.
15.23 $\mathbf{R}_A = 14P/27$ ↑; $y_D = 20PL^3/2187EI$ ↓.
15.25 $\mathbf{R}_A = \frac{1}{2}P$ ↑, $\mathbf{M}_A = PL/8$ ↺; $M = -PL/8$ at A, $M = PL/8$ at C, $M = -PL/8$ at B.
15.26 $\mathbf{R}_A = w_0L/4$ ↑, $\mathbf{M}_A = 5w_0L^2/96$ ↺; $M = -5w_0L^2/96$ at A, $M = w_0L^2/32$ at C, $M = -5w_0L^2/96$ at B.
15.27 (*a*) $8PL^3/243EI$ ↓. (*b*) $19PL^2/162EI$ ↘.
15.28 (*a*) $PL^3/486EI$ ↑. (*b*) $PL^2/81EI$ ↘.
15.29 (*a*) $wL^4/128EI$ ↓. (*b*) $wL^3/72EI$ ↘.
15.30 (*a*) $19Pa^3/6EI$ ↓. (*b*) $5Pa^2/2EI$ ↘.
15.31 $3PL^2/4EI$ ↗, $13PL^3/24EI$ ↓.
15.32 PL^2/EI ↗, $17PL^3/24EI$ ↓.
15.35 7.91×10^{-3} rad ↗; 0.340 in. ↓.
15.36 6.98×10^{-3} rad ↗; 0.1571 in. ↓.
15.37 (*a*) 0.601×10^{-3} rad ↘, (*b*) 3.67 mm ↓.
15.39 (*a*) $41wL/128$ ↑. (*b*) $23wL/128$ ↑; $7wL^2/128$ ↻.
15.40 (*a*) $4P/3$ ↑; $PL/3$ ↺. (*b*) $2P/3$ ↑.
15.42 $\mathbf{R}_A = 7P/32$ ↑; $\mathbf{R}_C = 33P/16$ ↑; $\mathbf{R}_E = 23P/32$ ↑.
15.43 $13wL/32$ ↑, $11wL^2/192$ ↻.
15.45 (*a*) 5.06×10^{-3} rad ↘. (*b*) 47.7×10^{-3} in. ↓.
15.46 121.5 N/m.
15.48 (*a*) 0.00937 mm ↓. (*b*) 229 N ↑.
15.49 9.31 mm ↓.
15.50 0.278 in. ↓.
15.52 (*a*) $0.211L$, $0.1604M_0L^2/EI$ ↓. (*b*) 6.08 m.
15.54 (*a*) $y = w_0(x^6 - 15L^2x^4 + 25L^3x^3 - 11L^5x)/360EIL^2$. (*b*) $11w_0L^3/360EI$ ↘. (*c*) $0.00916w_0L^4/EI$ ↓.
15.55 4.00 kips.
15.56 $9M_0/8L$ ↑; $M_0/8$ at A, $-7M_0/16$ just to the left of C, $9M_0/16$ just to the right of C, 0 at B.
15.57 $Pa(2L - a)/2EI$; ↘; $Pa(3L^2 - 3aL + a^2)/6EI$ ↑.
15.59 5.58×10^{-3} rad ↘, 2.51 mm ↓.
15.60 $\mathbf{R}_A = M_0/2L$ ↑; $\mathbf{R}_B = 5M_0/2L$ ↑; $\mathbf{R}_C = 3M_0/L$ ↓.
15.61 43.9 kN.

CHAPTER 16

16.1 kL.
16.2 K/L.
16.3 $2kL/9$.
16.4 K/L.
16.5 $ka^2(2l)$.
16.7 (*a*) 6.65 lb. (*b*) 21.0 lb.
16.9 (*a*) 6.25%. (*b*) 12.04 kips.
16.10 (*a*) 7.48 mm. (*b*) 58.8 kN for round, 84.8 kN for square.
16.12 1.421.
16.13 168.4 kN.
16.14 69.6 kips.
16.16 (*a*) 93.0 kN. (*b*) 448 kN.
16.17 2.27.
16.18 2.77 kN.
16.20 (*a*) $L_{BC} = 4.20$ ft; $L_{CD} = 1.050$ ft. (*b*) 4.21 kips.
16.22 657 mm.
16.23 29.5 kips
16.24 (*a*) 2.78. (*b*) $d_1 = 0.800$ in., $d_2 = 1.131$ in., $d_3 = 0.566$ in., $d_4 = 0.669$ in., $d_5 = 0.800$ in.
16.25 (*a*) 59.6 kips. (*b*) 31.9 kips.
16.26 414 kN.
16.27 (*a*) 220 kN. (*b*) 841 kN.
16.28 (*a*) 86.6 kips. (*b*) 88.1 kips.
16.31 (*a*) 251 mm. (*b*) 363 mm. (*c*) 689 mm.
16.32 79.3 kips.
16.33 1596 kN.
16.34 899 kN.
16.36 144.1 kips.
16.37 107.7 kN.
16.39 6.53 in.
16.40 (*a*) 4 boards. (*b*) 3 boards.
16.41 1.591 in.
16.42 9.00 mm.
16.44 W250 × 67.
16.46 3/8 in.
16.47 1/4 in.
16.48 L3-1/2 × 2-1/2 × 3/8.
16.49 70.2 kips.
16.50 $ka^2/2l$.
16.52 0.384 in.
16.53 $\pi^2b^2/12L^2\alpha$.
16.56 5.37 kN.
16.58 124.6 kips.
16.59 (*a*) 1529 kN. (*b*) 638 kN.
16.60 W200 × 46.1.

Centroids of Common Shapes of Areas and Lines

Shape		$\bar{x}$	$\bar{y}$	Area
Triangular area			$\frac{h}{3}$	$\frac{bh}{2}$
Quarter-circular area		$\frac{4r}{3\pi}$	$\frac{4r}{3\pi}$	$\frac{\pi r^2}{4}$
Semicircular area		0	$\frac{4r}{3\pi}$	$\frac{\pi r^2}{2}$
Semiparabolic area		$\frac{3a}{8}$	$\frac{3h}{5}$	$\frac{2ah}{3}$
Parabolic area		0	$\frac{3h}{5}$	$\frac{4ah}{3}$
Parabolic spandrel		$\frac{3a}{4}$	$\frac{3h}{10}$	$\frac{ah}{3}$
Circular sector		$\frac{2r \sin\alpha}{3\alpha}$	0	αr^2
Quarter-circular arc		$\frac{2r}{\pi}$	$\frac{2r}{\pi}$	$\frac{\pi r}{2}$
Semicircular arc		0	$\frac{2r}{\pi}$	πr
Arc of circle		$\frac{r \sin\alpha}{\alpha}$	0	$2\alpha r$

Moments of Inertia of Common Geometric Shapes

Shape		Moments of inertia
Rectangle	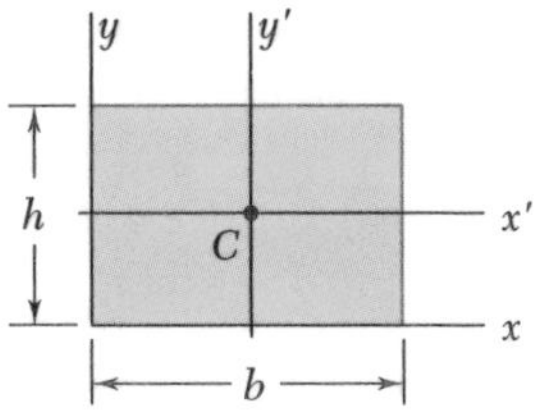	$\bar{I}_{x'} = \frac{1}{12}bh^3$ $\bar{I}_{y'} = \frac{1}{12}b^3h$ $I_x = \frac{1}{3}bh^3$ $I_y = \frac{1}{3}b^3h$ $J_C = \frac{1}{12}bh(b^2 + h^2)$
Triangle	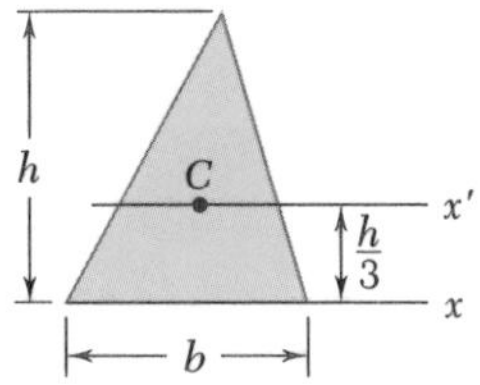	$\bar{I}_{x'} = \frac{1}{36}bh^3$ $I_x = \frac{1}{12}bh^3$
Circle	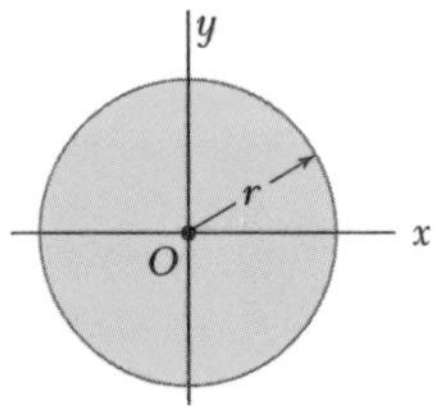	$\bar{I}_x = \bar{I}_y = \frac{1}{4}\pi r^4$ $J_O = \frac{1}{2}\pi r^4$
Semicircle	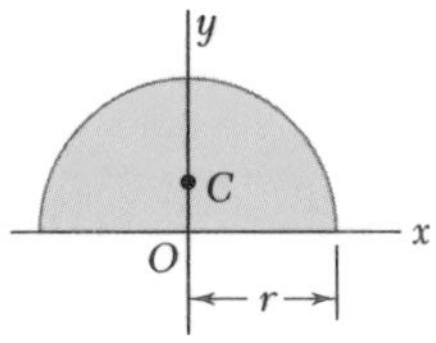	$I_x = I_y = \frac{1}{8}\pi r^4$ $J_O = \frac{1}{4}\pi r^4$
Quarter circle	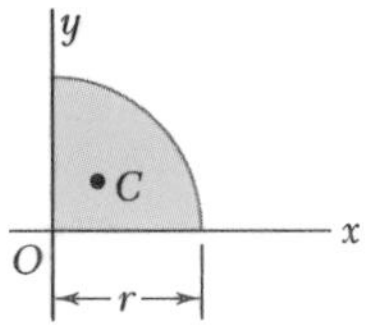	$I_x = I_y = \frac{1}{16}\pi r^4$ $J_O = \frac{1}{8}\pi r^4$
Ellipse	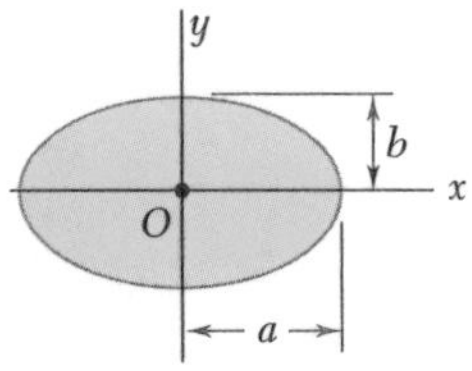	$\bar{I}_x = \frac{1}{4}\pi ab^3$ $\bar{I}_y = \frac{1}{4}\pi a^3b$ $J_O = \frac{1}{4}\pi ab(a^2 + b^2)$